AF815931

Semiconductor Lithography

Principles, Practices, and Materials

Wayne M. Moreau
General Technology Division
International Business Machines Corporation
Hopewell Junction, New York

Plenum Press • New York and London

Library of Congress Cataloging in Publication Data

Moreau, Wayne M.
Semiconductor lithography.

(Microdevices)
Includes bibliographical references and index.
1. Semiconductors. 2. Microlithography. 3. Photoresists. I. Title. II. Series.
TK7871.85.M583 1987 621.3815'2 87-29077

ISBN-13: 978-1-4612-8228-0 e-ISBN-13: 978-1-4613-0885-0
DOI: 10.1007/ 978-1-4613-0885-0

This limited facsimile edition has been issued for the purpose of keeping this title available to the scientific community.

10 9 8 7 6 5 4

Softcover reprint of the hardcover 1st edition 1988

A Division of Plenum Publishing Corporation
233 Spring Street, New York, N.Y. 10013

To my wife, Nancy, and
children, David and Deborah,
for their patience, cooperation, and understanding
and to Dr. Karl Weiss, graduate advisor,
for his inspiration that has lasted all these years

Preface

Semiconductor lithography is one of the key steps in the manufacturing of integrated silicon-based circuits. In fabricating a semiconductor device such as a transistor, a series of hot processes consisting of vacuum film deposition, oxidations, and dopant implantation are all patterned into microscopic circuits by the wet processes of lithography. Lithography, as adopted by the semiconductor industry, is the process of drawing or printing the pattern of an integrated circuit in a resist material. The pattern is formed and overlayed to a previous circuit layer as many as 30 times in the manufacture of logic and memory devices. With the resist pattern acting as a mask, a permanent device structure is formed by subtractive (removal) etching or by additive deposition of metals or insulators. Each process step in lithography uses inorganic or organic materials to physically transform semiconductors of silicon, insulators of oxides, nitrides, and organic polymers, and metals, into useful electronic devices. All forms of electromagnetic radiation are used in the processing. Lithography is a multidisciplinary science of materials, processes, and equipment, interacting to produce three-dimensional structures. Many aspects of chemistry, electrical engineering, materials science, and physics are involved. The purpose of this book is to bring together the work of many scientists and engineers over the last 10 years and focus upon the basic resist materials, the lithographic processes, and the fundamental principles behind each lithographic process. The first chapter describes the book's general organization into the two major categories of resist materials and lithographic processes. Every lithographic process step and resist composition is thoroughly covered. I have drawn upon the multitude of patents and published articles to show how resist material improvements have progressed and how lithographic problems have been solved.

The next generation of nanometer lithography and packaging will build upon and extrapolate from the experiences of today's micrometer-sized lithography. I hope this book on silicon micrometer lithography will serve as a basic reference and as a useful guide to semiconductor lithography for present and future tasks.

Wayne M. Moreau

Wappinger Falls, New York

Acknowledgments

The cooperation and courtesy of many technical societies, publishers, companies, and authors, throughout the world made this book possible, I would like to thank the following publishers and scientific societies for permission to redraw figures and to use published materials:

Academic Press
Allyn & Bacon
Butterworths
Cahners Publishing
Cannon Communications
Elsevier Science Press
Elsevier Sequoia
W. H. Freeman
Grampian–Morgan
Harcourt Brace Jovanovich
Lake Publishing
McGraw–Hill
Pergamon Press
Plenum Publishing
Science Forum
Technical Publishing
John Wiley & Sons

American Association for the Advancement of Science
American Chemical Society
American Institute of Metallurgical Engineers
American Institute of Physics
Electrochemical Society
Institute of Electronic and Electrical Engineers
Institute of Physics
Japanese Journal of Applied Physics
Optical Society of America
Society of Applied Spectroscopy
Society of Photographic Scientists and Engineers
Society of Photooptical Instrumentation and Engineering
Society of Polymer Engineering

The cooperation of the following companies in giving permission to use published material is appreciated:

American Telephone and Telegraph
DuPont
Eastman Kodak
International Business Machines
Japanese Electron Optics Lab
Perkin Elmer
Radio Corporation of America
VEECO

Many colleagues and individuals within the IBM Corporation itself provided support during this undertaking. I would like to thank the following:

Research Division	General Technology Division	
R. Feder	A. Agajanian	C. Lyons
E. Gipstein	E. Brown	H. Pfeiffer
M. Hatzakis	L. Bushnell	L. Plaushin
H. Hiroaka	M. Chow	H. Powers
S. MacDonald	F. DeAleo	A. Rapa
J. Shaw	N. Eib	A. Scaduto
N. Viswanathan	L. Gregor	M. Skinner
D. Webb	G. Hefferon	H. Sullivan
G. Willson	P. Hood	R. Tabone
	C. Johnson	E. Wajda
	S. Kirkup	E. Weber

I thank Dr. Ming Chow for his useful comments on the manuscript, and the terrific staff of the IBM East Fishkill General Technology Division Library for their untiring assistance. Finally, I am indebted to the IBM Corporation for providing the atmosphere, support, and incentive for the pursuit of technical excellence.

The cooperation and assistance of the technical staff of Plenum Publishing Company are appreciated. To all the named and unnamed companies, individuals, and societies, I am very grateful.

The following materials mentioned in the text are cited for technical information purposes only and are the registered trademarks of the companies listed:

Trademark	Company
AZ resists	American Hoechst, Somerville, New Jersey
FBM resists	Daikin Industries, Osaka, Japan
OMNR, OFPR, OEBR resists	Tokyo Ohko, Kawasaki, Japan
EK, KMER, KPR, KOR, KTFR, 820 resists	Eastman Kodak, Rochester, New York

Trademark	Company
HPR 206, Waycoat resists	Hunt Chemical, Palisades Park, New Jersey
SEL-N resists	Somar Corporation, Tokyo, Japan
CMS resist	Toyo Soda, Tokyo, Japan
EBR resist	Toray Industries, Tokyo, Japan
Riston resist Teflon film Mylar film Freon gas	E. I. DuPont and Company, Wilmington, Delaware
EB-46 resist	General Aniline and Film, New York, New York
COP, GMC resists	American Telephone and Telegraph, Murray Hill, New Jersey
MES resist	Japanese Synthetic Rubber, Tokyo, Japan
RE 400N, RE 5000P resists	Hitachi Chemical, Tokyo, Japan
FMR resist	Fuji Chemical, Tokyo, Japan
Varcum resin	Reichold Chemical, White Plains, New York
FC surfactant	Minnesota Mining and Manufacturing, Minneapolis, Minnesota
Cellosolve solvents	Union Carbide, Danbury, Connecticut
Monazoline surfactant	Mona Industries, Paterson, New Jersey

Contents

CHAPTER 3. Positive Radiation Resists

CHAPTER 4. Negative Photoresists

CHAPTER 5. Negative Radiation Resists

CHAPTER 8. Optical Exposure

CHAPTER 9. Radiation Exposure

CHAPTER 10. Developing Resist Images

CHAPTER 11. Postbake

CHAPTER 12. Additive Processes

CHAPTER 13. Subtractive Etching

CHAPTER 14. Stripping of Resists

CHAPTER 15. Process Controls

CHAPTER 16. Nonresist Processes

1

Introduction

1-1. SEMICONDUCTOR DEVICE MANUFACTURING

The production of a semiconductor device begins with a silicon wafer which after several thousand steps (Fig. 1-1-1) becomes a fully packaged and integrated electronic component. One-third of the cost of manufacturing involves lithography and hot processing. The other costs are associated with the packaging of the computer chip and the extensive electrical testing of all the components. For semiconductor manufacturing, the device fabrication steps involve (Fig. 1-1-2) the liquid processing operations of photolithography and the hot processing of wafers. The hot processes are divided into three major operations:

1. *Thin film deposition:* includes chemical vapor deposition of insulating films, plasma deposited and sputtered deposited films, and evaporated metal films
2. *Oxidations:* growth of diffusion barriers by thermal oxidation of silicon
3. *Diffusion/implantation:* tubular batch diffusion of dopants of P, B, and As and implantation with ion beams of dopants masked by resist or oxide windows

The rapid progress in the microminiaturization of electronic circuits is the result of a set of fabrication processes collectively known as silicon planar technology[1] (Fig. 1-1-3). In this technology, complex circuits are built into and on the surface of a wafer of silicon in a succession of process steps. Each step begins with coating the silicon wafer with a radiation-sensitive material called a resist. A pattern is delineated in this material by shining ultraviolet light or high-energy radiation onto it through a mask containing the desired pattern (Fig. 1-1-4). In positive-type resists, the exposed region is made soluble, whereas in negative resists, the exposed region remains insoluble and the background is dissolved away. The pattern of the polymer stencil is repeated in the metal film and insulator films by etching away the undesirable area. The metal can also be patterned by an additive lift-off process. After the resist has been stripped, the silicon is doped through the oxide mask to produce the desirable semiconductor transistor functions. In this manner, complex circuits are built into the silicon by a sequence of such steps.

The lithography steps used in silicon processing originated over 100 years ago in the graphic arts industry as a process for making printing plates. Lithography is the art of printing from a plane surface made of a flat stone or a metal plate onto which ink-receptive letters are formed by photoresists.[1, 16] The resists can be the

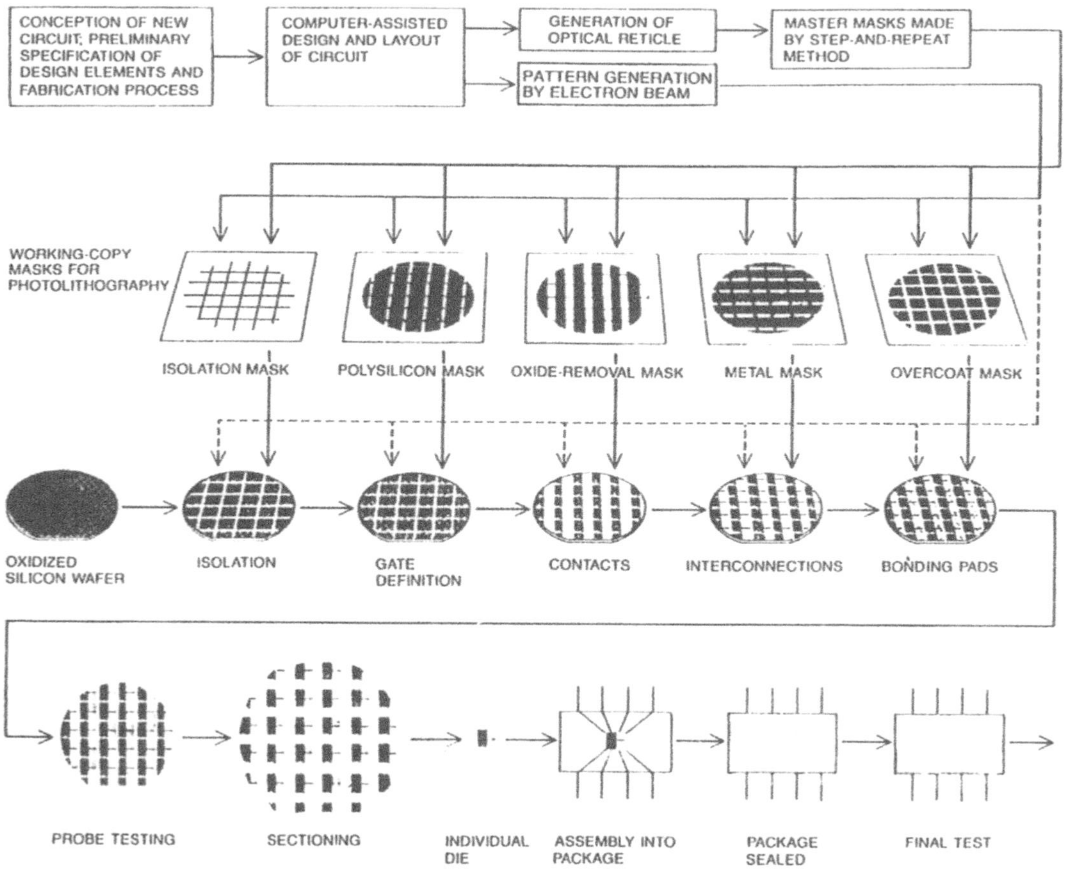

FIGURE 1-1-1. Outline of manufacture of large-scale integrated circuits. From W. Oldham, *Scientific American* Sept. 1977; copyright and courtesy of *Scientific American* Inc., all rights reserved.

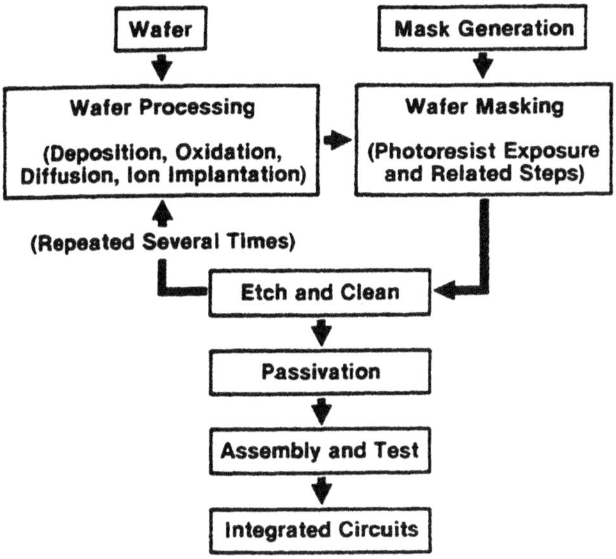

FIGURE 1-1-2. Schematic of general processes used in semiconductor device fabrication including hot oxidation, diffusion, and wet lithographic steps. Courtesy of H. Pogge, IBM.

ink-receptive material or they can be used as a mask for deeper etching. Historically, the electronics industry adopted the lithographic materials and etching processes used to fabricate printed circuit boards to produce the first silicon device—the transistor. Today, the term *microlithography* is used to describe the lithographic process of making micrometer-sized electronic devices.

Semiconductor lithography is concerned with the production of electronic components based on thin films ($\leqslant 1.0\ \mu m$ in thickness). The solid-state devices serve the purpose of electrical storage of information (memory) addressed by logic circuits. The basic functional electrical element of a modern electronic circuit is the transistor. The fabrication of a simple bipolar transistor involves the selective successive diffusion of *p*-type dopants and *n*-type dopants through openings in oxidized silicon.

Today, the *n–p–n* and *p–n–p* bipolar junction transistors are used in computers, calculators, communication and entertainment equipment, and process control systems. "Bipolar" refers to the fact that both types of charge carriers are used (Fig. 1-1-5). The *n–p–n* bipolar transistors are favored for circuits with a fast use time. They turn on quickly since electrons are more mobile than holes; the *p–n–p* transistors turn off quickly and are used in circuits in which a fast fall time is required.

Microminiaturization made it possible to employ thousands of transistors in a single circuit. This reduction in size of transistors has also brought about a dramatic increase in the speed of operation of electronic circuits and a lowering of the cost of computations. The ultimate speed of a switching device usually comes down to the time it takes to charge or discharge a capacitor, and the smaller the device the lower is its capacitance.[1, 2, 15] In addition, the time taken for a signal to propagate from one circuit to the next is now a significant portion of the total delay in the logic circuits of computers. Microminiaturization obviously reduces this source of delay.[1] Figure 1-1-6 shows the scaling factors in the shrinkage of a metal-on-silicon (MOS) device from the 1975 dimensions of $5.0\ \mu m$ to the limiting

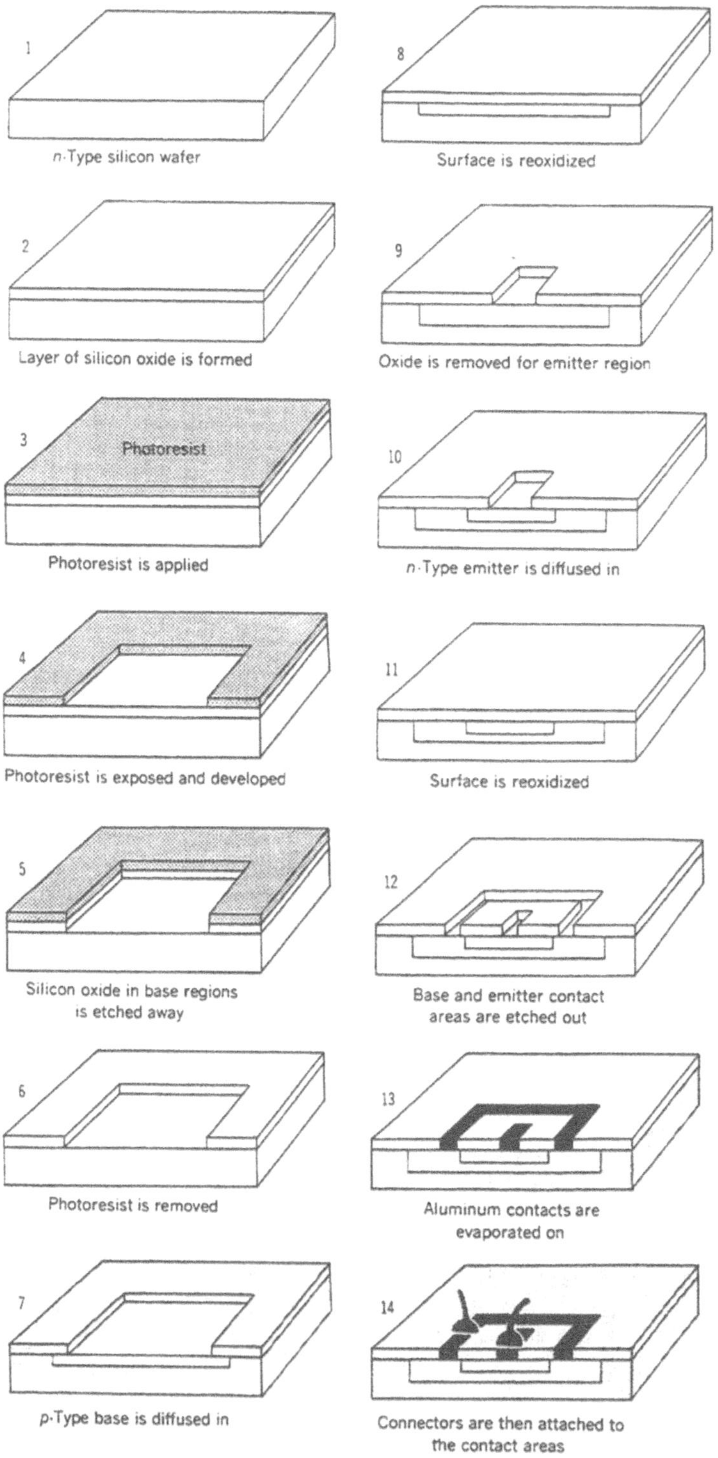

FIGURE 1-1-3. Planar transistor manufacturing process. From Ref. 8, L. Dunlap, *Chem. Eng. News* Nov. 30, 1970, p. 47; copyright and reproduced by permission of the American Chemical Society.

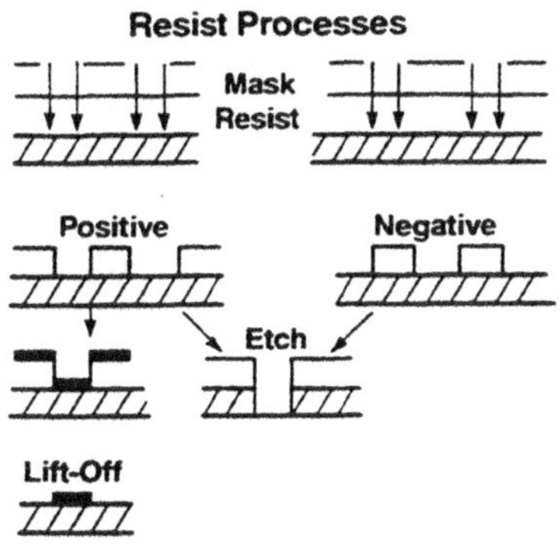

FIGURE 1-1-4. Negative and positive resist definition. Positive resists develop in the exposed region and usually remain soluble for lift-off. Negative resists remain in the exposed region but are insoluble and not suitable for lift-off. Both positive and negative are used for subtractive etching.

size of 0.25 μm projected for 1995. To accomplish the submicrominiaturization will require a 20-fold-better process control, a smaller linewidth bias, and a precise overlay (alignment of each level) in the lithography process sector. An appreciation of the reduction of the size of electronic components can be seen in the development of the pocket calculator,[2] from a large desktop machine to one that fits in the head of a pin (Fig. 1-1-7). In 1960, three or four logic circuits were contained on a circuit board 4 by 5 inches on a side. Today, hundreds of thousands of similar circuits may be contained on a chip a tenth of an inch square.[1] The benefit of all the size reduction by vertical large-scale integration (VLSI) has been the significant reduction in cost per circuit (computation) by several orders of magnitude (Fig. 1-1-8).[3] The explosive growth of the electronic industry has been called a modern industrial revolution. Since 1970, the growth in the number of components per chip has increased steadily by 10^6 and is projected (Fig. 1-1-9) to slowly reach one billion per chip by the year 2010.

In order to increase the production capacity of manufacturing and hopefully reduce costs, the silicon wafers have increased in size (Fig. 1-1-10)[4] and are expected by 1990 to be as large as 250 mm in diameter. The silicon wafers are the production vehicles for transistors. Each wafer contains several hundred chips of 5–10 mm on a side. In addition, each wafer contains electrical test sites, and alignment marks for the overlaying one circuit pattern level (Fig. 1-1-11) precisely

TABLE 1-1-1. Technology Challenges for VLSI

High-resolution precise registration lithography
Scaled tolerances
Improved etching processes
Reliable thin gate insulators
Shallow implanted doping profiles
Shorter time/lower temperature hot processes
Better interconnection technology in terms of:
Contact metallurgy
Electromigration immunity
Higher density, lower capacitance
Multiple wiring layers
Removal of process-induced radiation damage
Enhanced reliability

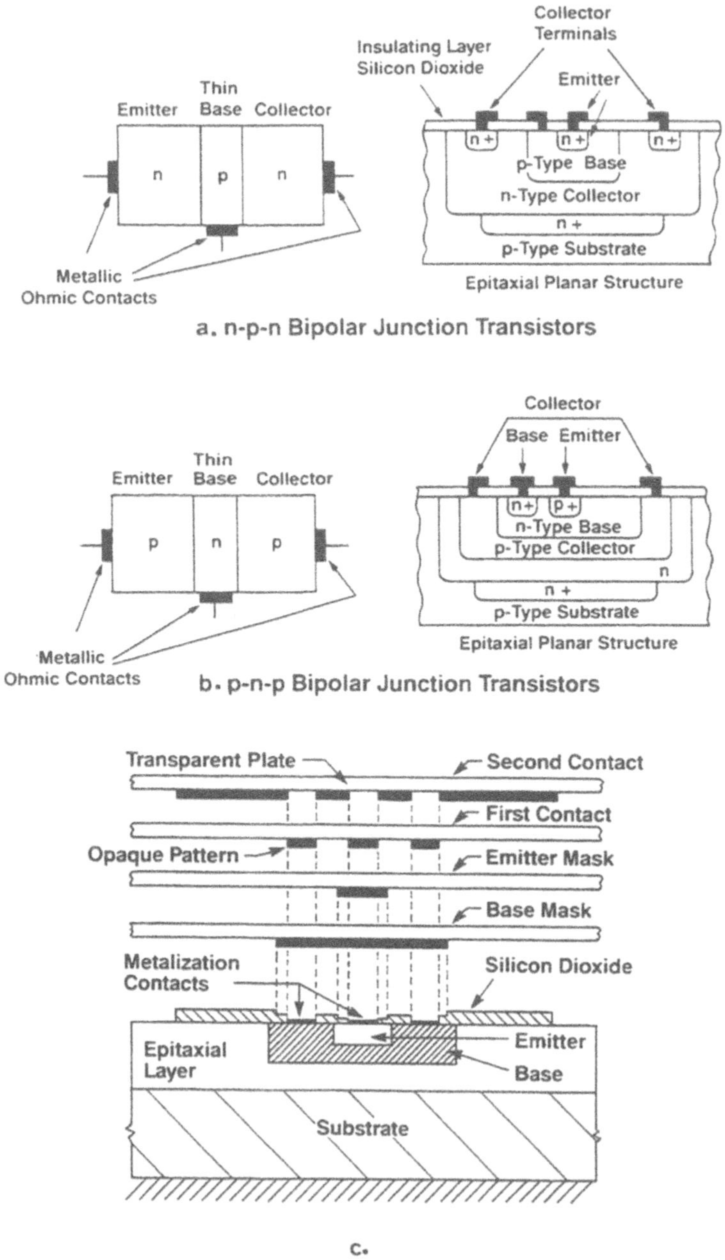

FIGURE 1-1-5. (a, b) Cross sections of elementary bipolar transistors. From J. Betts, *Physics for Technology*, copyright 1981 by Reston Publishing, p. 588, all rights reserved. (c) Photolithography masking steps to produce transistors on silicon wafers. From Ref. 13.

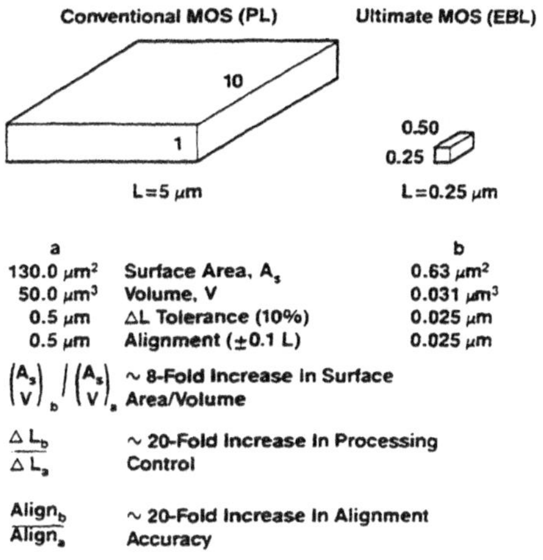

FIGURE 1-1-6. Shrinkage of MOS devices from 1975 dimensions of 5.0 μm to 0.25 μm by 1995 and the resultant need for 20-fold improvement in process control and alignment (overlay) accuracy. From Ref. 17, copyright 1980 and courtesy of Academic Press.

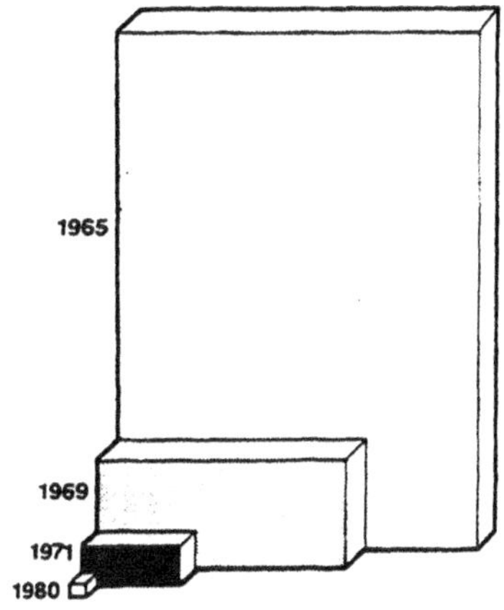

FIGURE 1-1-7. Size of circuitry required to produce calculator chip in 1965 and size reduction that occurred over the next 15 years. Adapted from D. Elliott, *Microelectronics* 6, 29 (1975), courtesy of Benn Electronics Publication, Ltd.

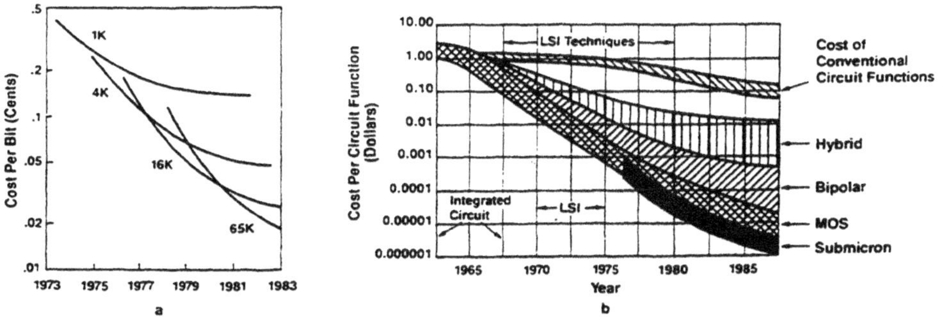

FIGURE 1-1-8. (a) Cost reduction of memory circuits by VLSI fabrication. From Ref. 3, copyright 1977 and courtesy of *Scientific American* Inc. (b) Cost reduction of many types of integrated circuits by microminiaturization. From Ref. 18; reproduced with permission of *Solid State Technology*, published by Technical Publishing, a company of Dun and Bradstreet.

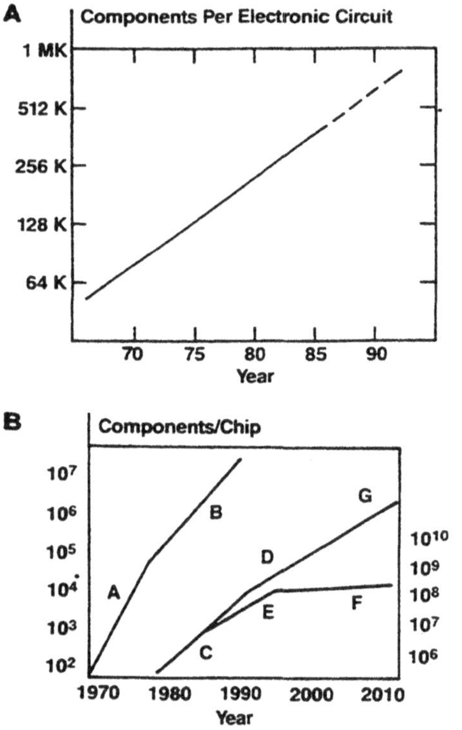

FIGURE 1-1-9. (A) Growth projection made in 1965 by Moore of the number of transistors per chip. From Ref. 19. (B) Transistors per chip doubled yearly over the period 1970–1980 (A) but the rate dropped in segment B to doubling every $1\frac{1}{2}$ years. Segments C, D, and G are projections assuming a size of 250 nm and projections E and F assume a size of 500 nm. From A. Robinson, *Science* 223, 268 (1984), copyright 1984 and reproduced by permission of AAAS.

on top of the previous layer. Reisman[5] has recently reviewed some of the major manufacturing challenges for the production of VLSI devices (Table 1-1-1). On top of the list of technology challenges is the lithography step–precise registration (the alignment of one layer to each other) and high-resolution patterns (the ability to make submicrometer dimensions within acceptable biases and tolerances). Moreover, the ability to precisely overlay one intermediate device level is now more important than fabricating a device to an absolute size. A registration accuracy of

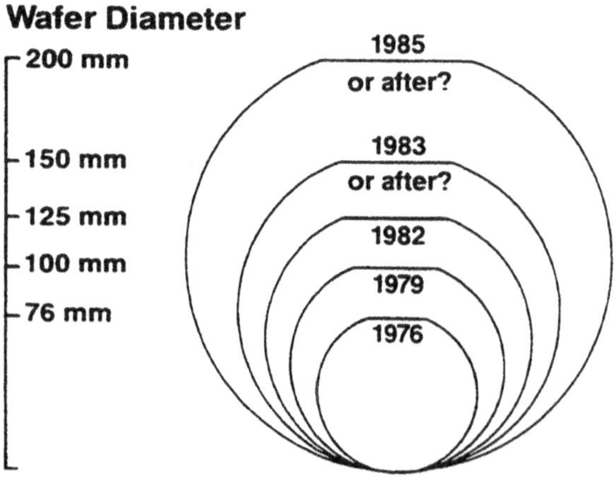

FIGURE 1-1-10. Growth in size of silicon wafers over the last 10 years. From Ref. 4, courtesy of Science Forum Inc.

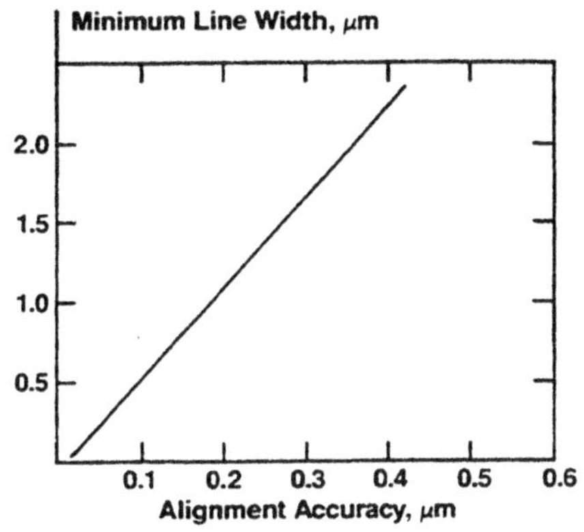

FIGURE 1-1-11. Alignment (overlay) required for VLSI devices. Accuracy is about one-fifth the critical minimum device size.

one-fourth the linewidth or better is required (Fig. 1-1-11). For optical exposure projection tools, registration is an optical mechanical problem of detecting a ray reflected from an engraved silicon wafer and coincidentally, aligning a pattern contained in a planar chrome photomask. Improved registration can be obtained with an electron beam exposure tool by an enhanced signal of backscattered electrons from an alignment mark. In an electron beam exposure system, the chip scan field is electronically adjusted to define the next chip pattern. For accurate metal circuitry, the precise alignment of the metal interconnection over the etched via holes of insulators is a critical step in silicon device fabrication. By a previous lithographic etching process, conventional types of tapered via hole structures are made in sputtered quartz films. Covering metal pads are extended in width to form pads, so that the etched via hole is smaller than the underlying pad (Fig. 1-1-12). By forming via holes and depositing metal patterns in the holes using a self-aligned single resist to do both the etching and lift-off process, it is possible to place the metal patterns closer and more precisely together in densified circuits. In silicon device fabrication,

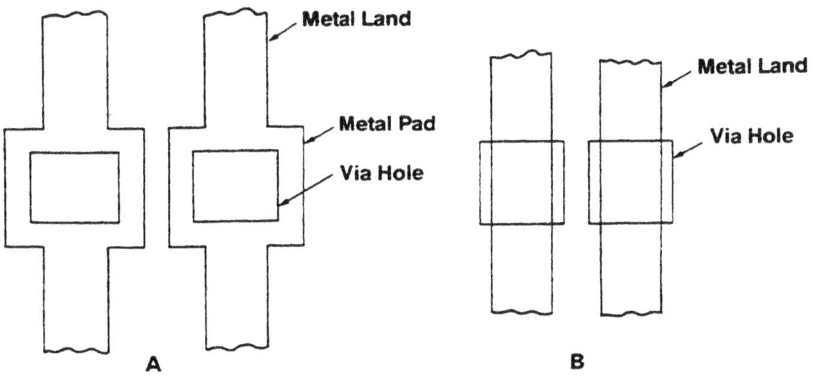

FIGURE 1-1-12. Concept of zero-overlap vias. (A) Conventional via hole design in which the first-level spacing is controlled by pad dimensions. (B) Zero-overlap vias which allow for closing spacing of adjacent lines. From Ref. 6, courtesy of IBM.

self-aligning layers and zero-overlap vias (Fig. 1-1-12) are examples of process and design innovations to enhance the superposition of layers and increase process yields and achieve denser high-performance circuits.

1-2. LITHOGRAPHY PROCESSING AND MATERIALS

For a lithographic operation, the interaction of materials, production tools, and processes are the necessary ingredients (Fig. 1-2-1). The resist materials perform the function of transforming a two-dimensional circuit design into a three-dimensional electrical circuit. The resists used in microlithography are generally formulated from organic film forming polymers photosensitized with small molecular additives (Table 1-2-1). Recently, other new types of resist have been made from inorganic silver compounds or polymers containing metallic elements for application in bilayer resists. Except for photosensitivity, all of the "resistance" properties of a resist are derived from the parent polymer component. Most of the photoresists are two-component formulations (Table 1-2-2) made of a polymer base and a photosensitizer. Practically all of the γ-radiation-exposed engineering plastics have been investigated as photon or electron beam resists (Table 1-2-3). Since high-energy sensitive electron, ion, or X-ray resists capture the radiation by mass absorption, a photosensitive additive is not required. In photoresists, the sensitizers are added at 2–20% loading by weight of base resin depending on their molar extinction coefficients. The casting solvent is usually chosen from an acetate family of organic sources. In resists, the images formed are due to a differential change in some property,

$$\text{Image} = \text{Property of exposed/property of unexposed} \rightarrow \text{Developer readout}$$

For positive resists, the increase in solubility in a liquid or plasma developer is a common image-forming step, while a decrease in solubility is noted for negative working resists. Other properties such as volatility, changes in adhesion and tackiness have been used as image differentials.

Exposure transforms the reactive sites in negative resist polymers into cross-

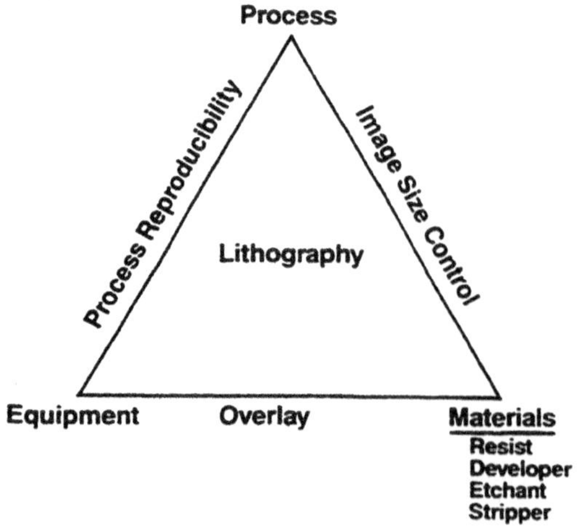

FIGURE 1-2-1. Interaction of materials, processes, and tools in lithography required to produce semiconductor devices.

TABLE 1-2-1. Composition of Typical Resists and Function of Components

Resist	Polymethylmethacrylate	Novolak	Cyclized rubber
Sensitizer		Diazonaphthoquinone	Bisazide
Casting solvent	Chlorobenzene	2-Ethoxyethylacetate	Xylene
Prebake	175°C	75–95°C	70–90°C
Developer	Methyl isobutyl ketone	0.25 N KOH	Xylene
Use	Positive *e*-beam, ion, X-ray resist	UV positive	UV negative
Function	*Polymer component* Film former Adhesive Etch resistance Plasma resistance Thermal resistance	*Sensitizer* Photosensitivity Spectral response	

linked insoluble materials. Photosensitive additives such as bisazides generate nitrene cross-linking agents that insert across polymer chains,

• PHOTOCROSSLINKING ADDITIVE

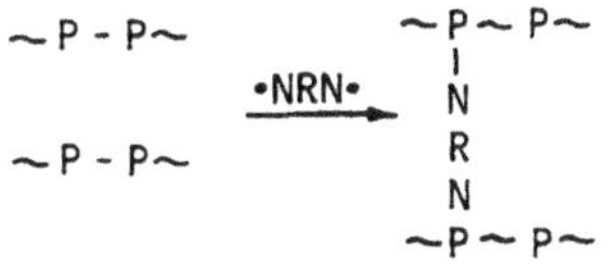

In a second type of cross-linking, the photosensitive additive can absorb energy and activate the pendant side chains of polymers, such as polyvinyl cinnamate, to undergo intermolecular cycloaddition and cross-link the chains,

•PHOTOCROSSLINKABLE SIDE CHAIN

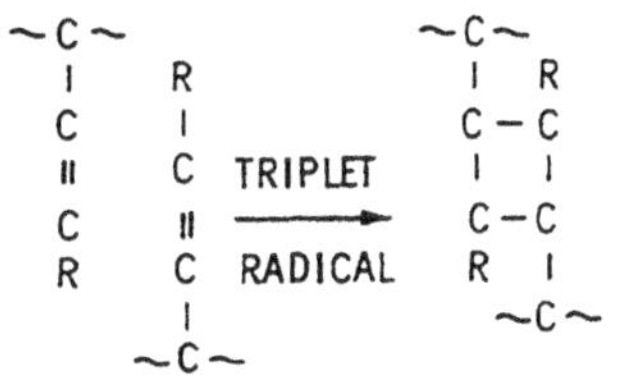

TABLE 1-2-2. Classification of Resists

Positive (P)			*Negative* (N)		
	Organic	Inorganic	Organometallic		
e-beam	Ion beam	X-ray	Deep-UV	Mid-UV	Near-UV
One-component			Two-component		
Polyacrylates (P)			Phenolic + diazoquinone (P)		
Polyolefinsulfones (P)			Rubber + azide (N)		
Polystyrenes (N)			Polyvinylcinnamate + sensitizer (N)		
Polysiloxanes (N)			Polycarbonates + onium (P, N)		
Single layer	Multilayer	Dry develop	Posttreatment		

TABLE 1-2-3. Classification of Engineering Materials Showing Polymer Classes[21]

Engineering materials							
Metals		*Nonmetals*					
		Polymers					
Basic	*Alloys*	*Thermosetting*	*Elastomers*	*Thermoplastics*	*Ceramics*	*Glasses*	*Others*
Aluminium	Steel	Alkyds	Acrylates	Acetates	Aluminas	Silicas	Fluids
Magnesium	Brass	Diallyl phthalates	Butyls	Acrylics	Beryllas	Soda limes	Greases
Beryllium	Bronze	Epoxies	Chlorosulfonated polyethylene (hypalon)	Cellulosics	Carbides	Leads	Lubricants
Titanium	All metals which are not basic metals	Melamines	Fluorocarbons	Chlorinated polyethers	Cordierites	Borosilicates	Oils
Copper		Phenolics	Fluorosilicons	Fluorocarbons	Forsterites	Aluminosilicates	Papers
Nickel		Polyesters	Polysulfides	Nylons (polyamides)	Lavas	Others	Others
Gold		Urethanes (rigid forms)	Neoprenes	Polycarbonates	Magnesias		
Iron		Silicones (rigid resins)	Nitriles	Polyethylenes and polypropylenes	Nitrides		
Others		Others	Silicones	Polyimides	Titanias		
			Styrene butadienes	Polyphenylene oxides	Steatites		
			Others	Polystyrenes	Zirconias		
				Polysulfones	Others		
				Vinyls			
				Others			

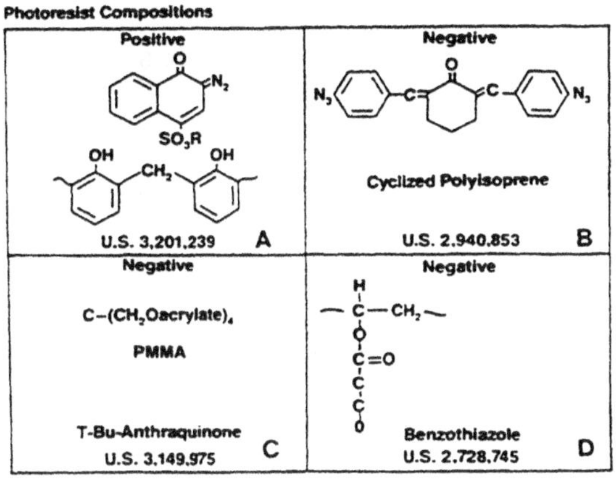

FIGURE 1-2-2. Examples of classic photoresist compositions. (A) Diazoquinone–novolak positive resist; (B) azide sensitized rubber negative resist; (C) dry film polymethylmethacrylate thick film negative photoresist; (D) polyvinylcinnamate sensitized negative photoresist.

In negative resists, common solvents remove the background unexposed regions. The most important negative photoresists are shown in Fig. 1-2-2. For microelectronic lithography, photosensitized polyvinyl cinnamate and azide cross-linked rubber were popular negative photoresists. Also shown is a very popular dry film resist consisting of a tetrafunctional monomeric acrylate dispersed in a polyacrylate binder. This resist is used exclusively for printed circuit board etching and electroplating. Gradually, the negative resists have been replaced by a positive photoresist composition of a diazoquinone-inhibited novolak resin (Fig. 1-2-2). The hydrophobic diazoquinone inhibitor (I) strongly absorbs the 365(i)-, 405(h)-, and 435(g)-nm line output of the mercury exposure lamps. The addition of the hydrophobic inhibitor (I) to the hydrophilic novolak (N) resin retards the dissolution of the mixture in alkaline developer. Upon photoexposure, the inhibitor is destroyed and rearranged into a hydrophilic acid (A) which together with the resin dissolves in the alkaline developer,

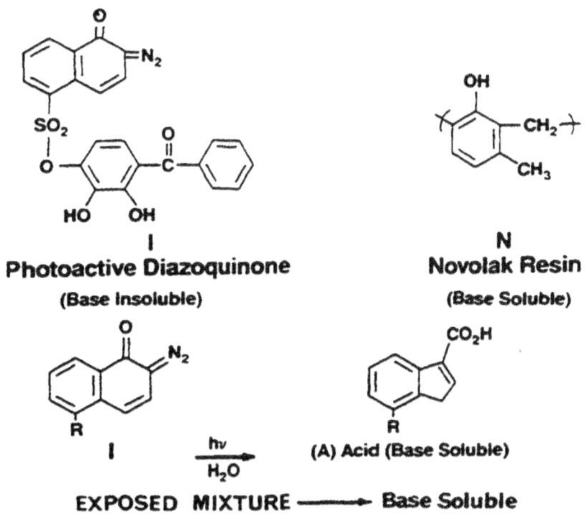

For electron beam exposure, simpler one-component electron beam resists are

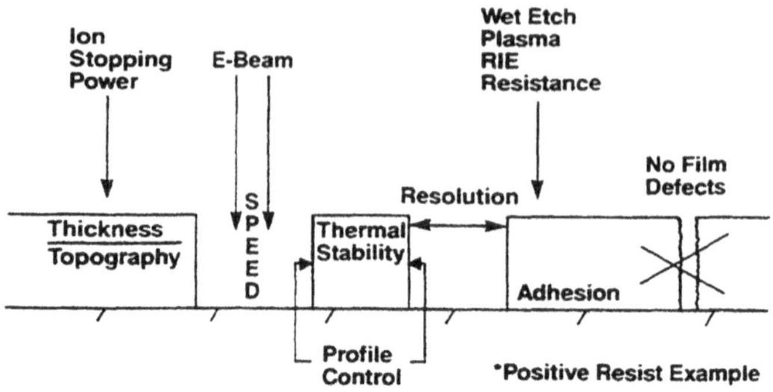

FIGURE 1-2-3. Resist requirements for submicrometer lithography using example of a positive electron beam resist.

formulated from chain scission polymers such as polymethylmethacrylate (PMMA),

Polymethylmethacrylate (PMMA)

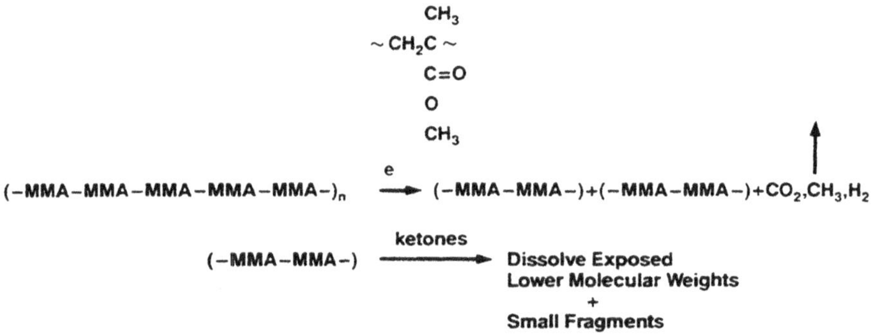

Exposure fragments the main chain into smaller molecular components that dissolve in a suitable developer.

As a family of recording films, the resists are very slow ASA speed materials of superb resolution (Table 1-2-4). Unfortunately, very intense sources are required for exposure. Such high-intensity sources limit the area of exposure of radiation to one chip ($1\ cm^2$) at a time with stepped exposure. Thus, the throughput of electron beams, ion beams, and X rays are highly dependent on the speed (dose requirement) of the resist. High speed is a basic requirement of a resist along with high resolution, etch resistance, and low defects (Fig. 1-2-3). Because of the distortion of the photomask pattern by the projection lens and the scattering of the irradiation beam inside the resist film, it is difficult for optical exposure tools to produce dimensions below $0.5\ \mu m$ in size. At this dimension, many forms of radiation (deep UV, X-ray, ion, and electron beams) become primary exposure contenders.[11]

For better resolution, the resist film thickness should be a fraction of the desired horizontal linewidth. Second, since high-speed resists are usually deficient in thermal and plasma resistance, new resist structures of bi- or trilayer form have been introduced to separate the imaging resist layer from the underlayer. The

TABLE 1-2-4. Photospeeds of Recording Processes

Process	ASA speed (ergs/cm²)⁻¹	Dose (ergs/cm²)	Resolution (lines/mm)
TV tube	10^4	10^{-4}	5
B & W film	10^2	10^{-2}	50
Xerox	1	1	100
HR emulsion	10^{-2}	10^2	1000
Photopolymers	10^{-3}	10^3	500
Photoresists	10^{-5}	10^5	5000
Printing inks	10^{-6}	10^6	50
UV curing	10^{-6}	10^6	Film
Diazo	10^{-7}	10^7	100
Thermography	10^{-7}	10^7	100
Kalvar	10^{-7}	10^7	100

underlayer can be chosen from any thermally stable polymer like a polyimide (Fig. 1-2-4). The multilayer resists offer better resolution, good step coverage, and excellent thermal and plasma resistance layers.

Over 50 chemical and photographic firms supply resist materials and associated etchants, developers, strippers, primers, and rinses. Processing with the resist chemicals is comparable to the techniques and materials used in the paint industry (Table 1-2-5). Paint formulations are similar in composition to negative photoresists. Many of the resist strippers are based on paint removers. Using process familiar to the paint industry, resist films are applied to a cleansed and primed surface by spin coating (Fig. 1-2-5). A puddle of liquid resist is applied to a silicon wafer and the excess liquid is leveled and spun off by centrifugal forces.

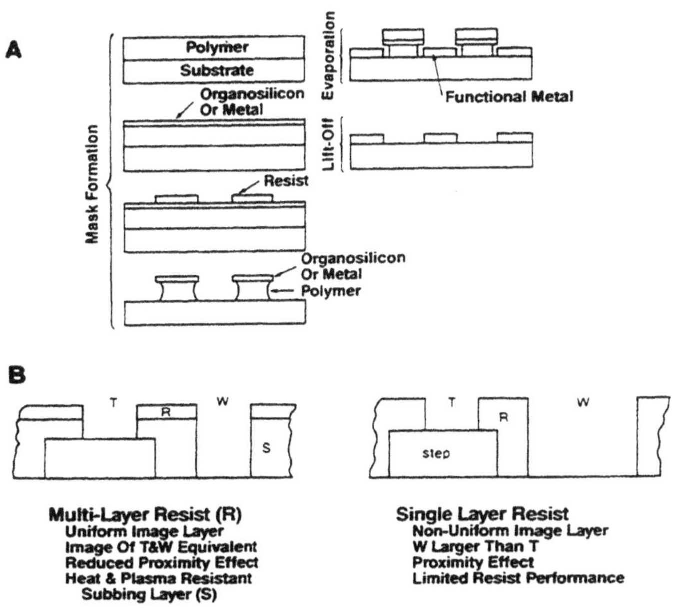

FIGURE 1-2-4. (A) Multilayer resist processing for metal lift-off. From Ref. 6, courtesy of IBM. (B) Advantages of multilayer resists over single-layer resists.

TABLE 1-2-5. Processing of Paint and Resist Films

	Paint	Resist
Primer	Surface clean	Preetch
	Primer paint	Silanes
Apply	Spray	Spin
	Brush	Spray
	Roll	Roll
		Dip
Dry	Air	Oven
	IR	IR
Expose	Degrade	Decompose
	Cross-link	Cross-link
Etch	Water resistent	Water resistent
Strip	Degrade	Oxidize
	Crack	Displace
	Paint stripper	
Composition	Polymer	Polymer
	Colorant	Sensitizer
	Solvent	Solvent

Spinning evaporates the casting solvent to partially dry the film. The film is mildly baked to further remove the solvent and then is exposed. After development, the film is further hard baked to improve its resistance and then the mask underlayer is etched by aqueous acids or gaseous reactive halocarbons that form volatile products. Alternatively, the metal patterned can be defined by an additive process involving metal deposition on top of the patterned resist followed by lifting of the

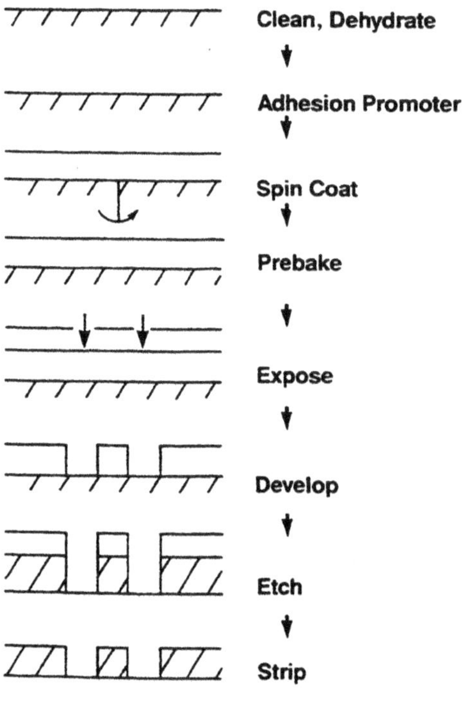

FIGURE 1-2-5. Typical process steps involved in lithographic processing with a resist mask definition, etching, and resist stripping.

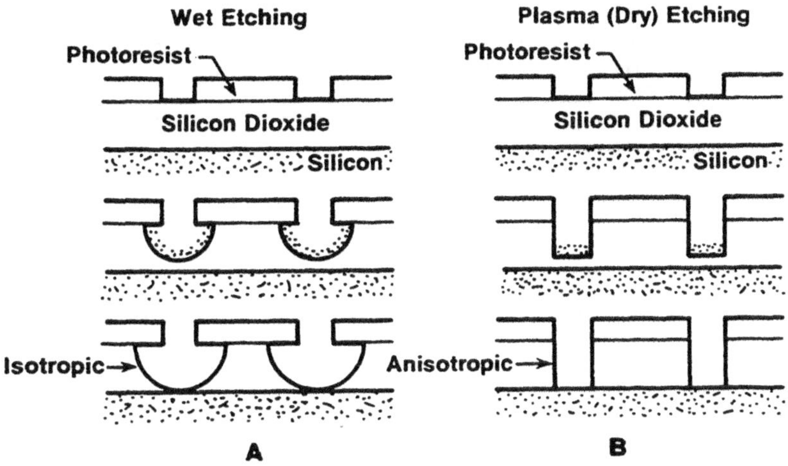

FIGURE 1-2-6. Schematic of (A) wet isotropic liquid-phase etching used since the early 1950s and (B) anisotropic gas ion beam plasma etching developed in the 1970s and dominating the present mode of etching. Courtesy of H. Pogge, IBM.

undesired metal by dissolution of the unexposed resist. Since normal wet etching is isotropic (proceeds equally in all directions), the images are distorted laterally to a width equivalent to the vertical etch depth (Fig. 1-2-6). Gas-phase etching with collimated beams of reactive ions achieves anisotropic etching in only the vertical direction. Wet etch biases can easily be of the order of the film thicknesses ($\pm$ 1 μm) whereas dry etch biases are an order of magnitude less (Fig. 1-2-7). With the increase in wafer sizes to 250 mm in diameter, the traditional batch processing of wafers is slowly converting to single-wafer etch tools with built-in process monitors. For lithographic processing, the important lithographic process controls are time, thickness, and temperature. Process controls improve yield by adjusting etch and development conditions to compensate for different substrate thickness and for variations in exposure or prebake. In the lithography process sector, automation (Fig. 1-2-8) is being used to shorten the turnaround time for processing custom circuits and for reducing contamination by elimination of operating personnel and excessive handling.

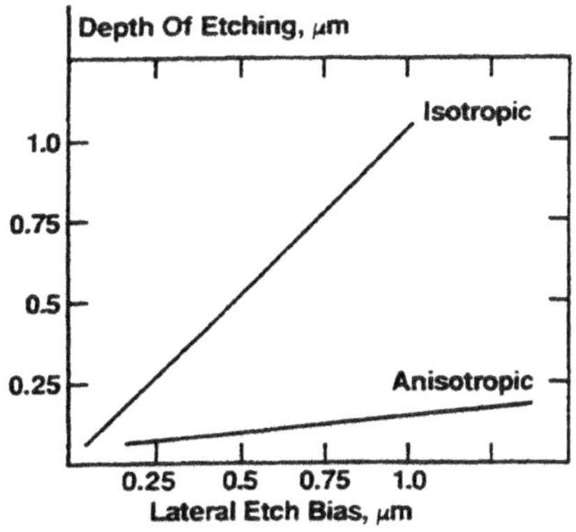

FIGURE 1-2-7. Anisotropic etch biases versus isotropic etch biases as a function of the depth of etching to lateral etch rates.

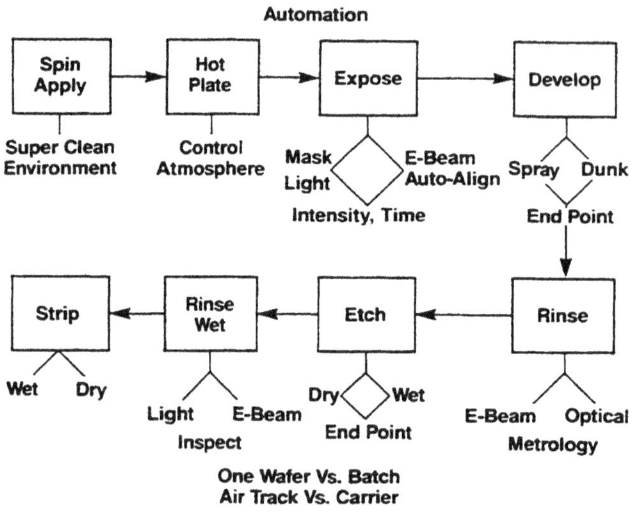

FIGURE 1-2-8. Layout of an automated lithography manufacturing sector.

1-3. LITHOGRAPHY COSTS AND PROCESS EQUIPMENT

The breakdown of the total cost of the processing of a MOS memory device is shown in Table 1-3-1. The lithography sector represents one-third of the process costs.[12] At present, a mix of exposure tools are being used to phase out scanners and phase in optical steppers. In the lithography costs (Fig. 1-3-1), the exposure tool represents over half of the capital investment. With exposure tools, optical units cost $1.0 M and $3.0 M for electron beams. Electron beams are being justified for the production of custom parts of low volume and high part number

TABLE 1-3-1. Production Cost per Square Inch of MOS Device[a]

Major equipment item	Cost (in dollars)	Percent of total: Step	Percent of total: Scan
Wafer scan expose, 2.5 μm	6.2	30	
Wafer step expose, 1.5 μm	10.4		43
Wafer step expose, 0.8 μm	19.4		(60)
LPCVD	0.7	3	2
High press oxidation	0.6	2	1
Ion implant (source/drain)	1.6	8	7
Ion implant (threshold)	3.9	20	17
Photoresist processing	0.9	5	4
Dry etch planar	2.4	12	11
Dry etch barrel	0.4	2	1
Nitride deposit	0.7	3	2
Sputter metal	3.1	15	13

[a] From Ref. 16.

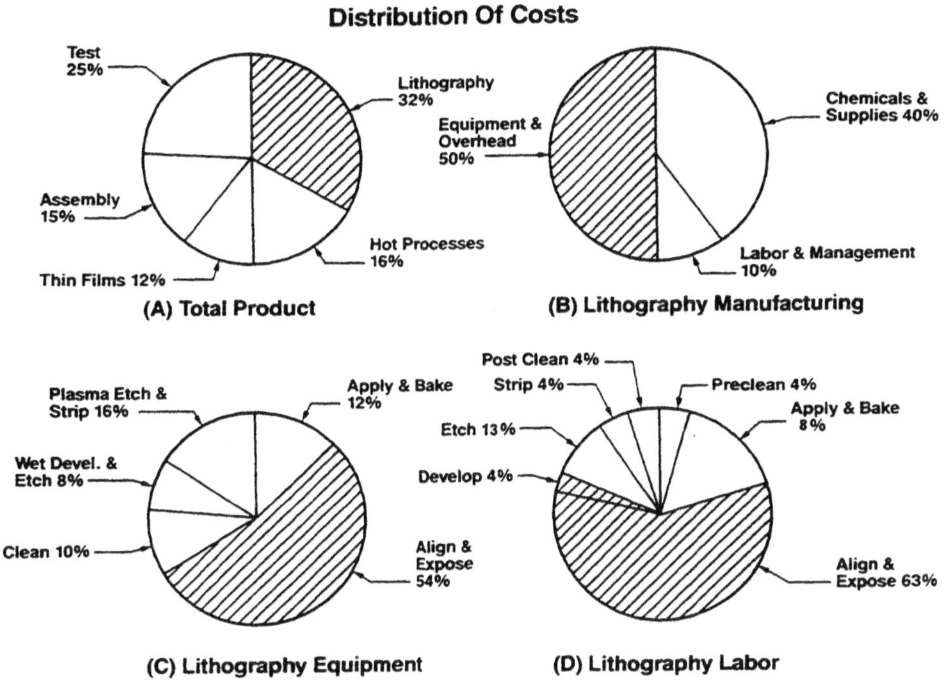

FIGURE 1-3-1. Breakdown of costs of semiconductor manufacturing and the costs of the lithography steps. From Ref. 12.

(Fig. 1-3-2). For custom silicon circuits, electron beams are used to personalize the metal interconnections of bipolar logic devices.[6] Electron beams are also used exclusively to fabricate chrome optical masks and masks for X-ray exposure (Fig. 1-3-3).[11] In the next 15 years, optical exposure tools are expected to handle the mass production of high-volume, low-part-number memory devices until the device dimensions reach 0.3 μm (Fig. 1-3-4). At that point, masked exposure using flood X-ray and electron beams or serialized electron beam exposure will take over.

Each exposure tool in turn dictates what type of resist and process are required

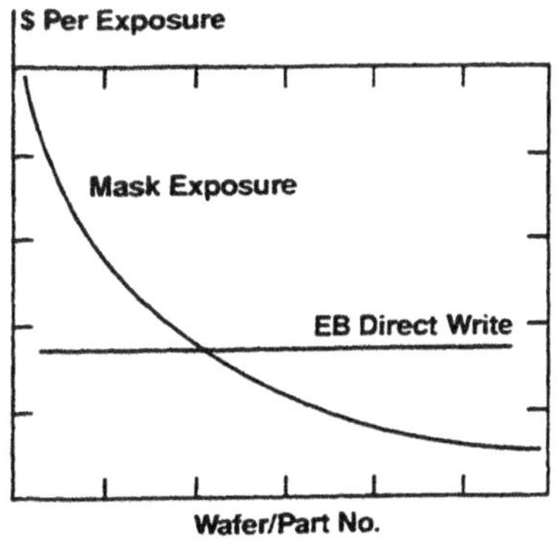

FIGURE 1-3-2. Low-volume high-part-number personalization of bipolar devices is cheaper by direct electron beam writing, while high-volume low-part-number is made at lower cost by optical mask printing. From Ref. 9, reproduced by permission of *Solid State Technology*, published by Technical Publishing, a company of Dun and Bradstreet.

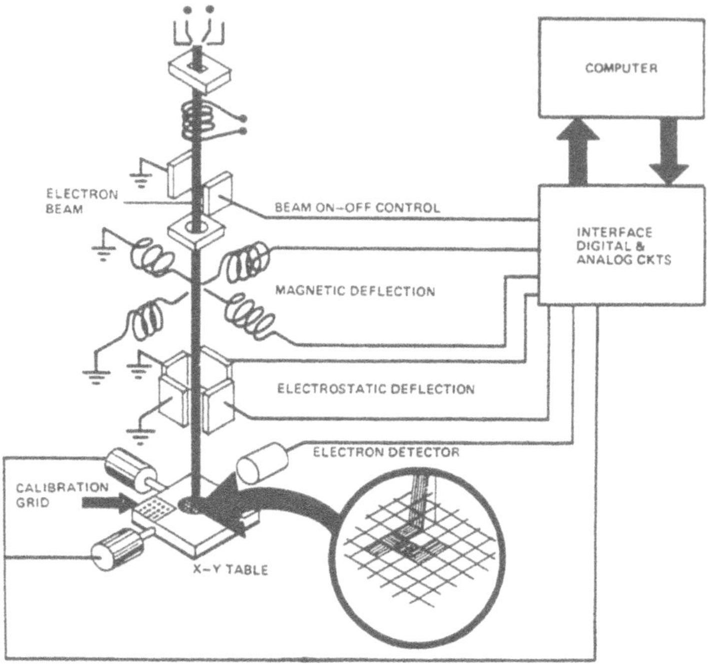

FIGURE 1-3-3. Schematic of round beam electron exposure tool. From Ref. 9, reproduced by permission of *Solid State Technology*, published by Technical Publishing, a company of Dun and Bradstreet.

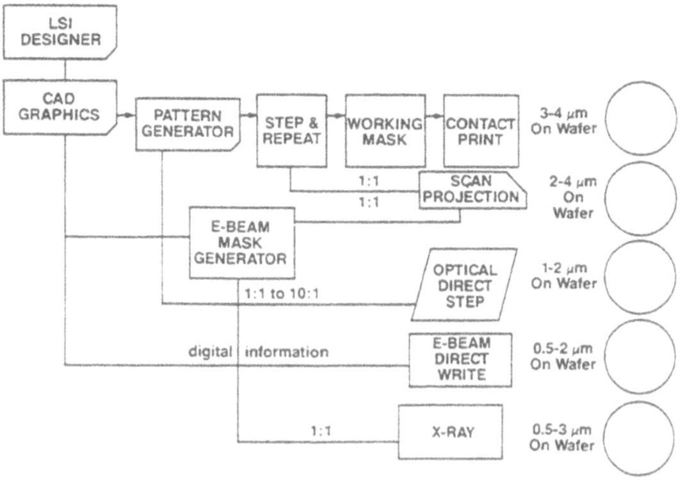

FIGURE 1-3-4. Linewidth capabilities of different exposure tools. From Ref. 10, reproduced by permission of *Solid State Technology*, published by Technical Publishing, a company of Dun and Bradstreet.

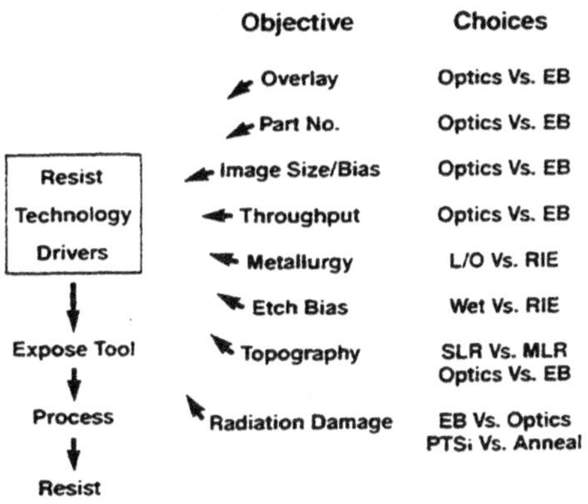

FIGURE 1-3-5. Process and equipment factors that drive the choice of resist materials and processes. The high cost of the exposure tool defines the choice of resist.

(Fig. 1-3-5). Although high throughput is the primary consideration, it may have to be compromised with lower throughput tools of higher resolution and better alignment. In order to achieve the lowest costs, lithography economics requires high yields (Fig. 1-3-6). In the end, the lithographic processing environment requires costly clean rooms and equipment to remove particulates causing defects from the wafer surfaces.

Lithography is a complex interactive arena, and indicated below are the six key elements of a lithography system and their influence on the device fabrication process:

1. Exposure tools: affect resolution, overlay, and costs
2. Resists: affect throughput and image quality
3. Resist develop process: affects image tolerance and image quality
4. Etching or lift-off: affects image tolerance and image quality
5. Metrology: affects process control
6. Contamination control: affects yield

High yields necessitate the systematic elimination and reduction of errors in linewidth control. Improved, precision alignment and the eradication of random fatal defects such as pinholes or surface particulates are also required. In general, more defects and lower yields can be expected (Fig. 1-3-7) with larger size chips and/or more device lithography levels.[7] To improve lithography yields, process controls such as develop–etch endpoint monitors and automated thickness controls are becoming increasingly important.

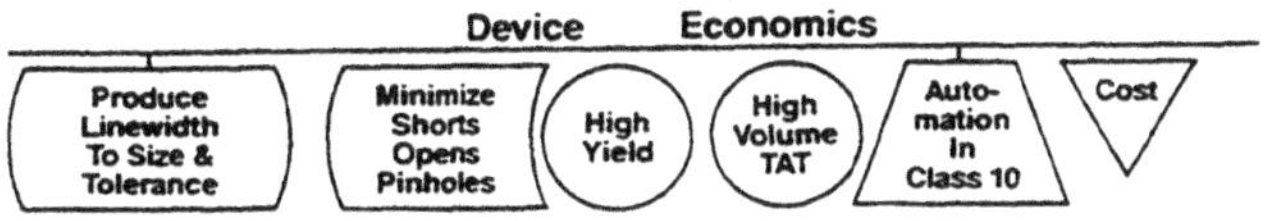

FIGURE 1-3-6. The major objective of semiconductor device manufacturing—high yield and high volume at low cost.

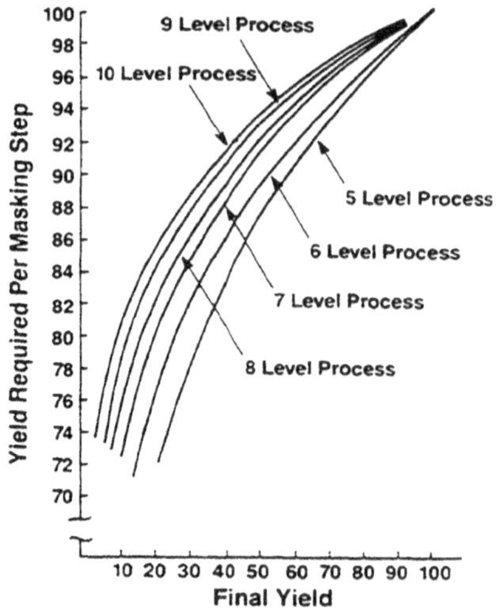

FIGURE 1-3-7. The greater the number of lithographic levels, the greater is the risk of more defects and lower yields. From Ref. 7, reproduced by permission of *Solid State Technology*, published by Technical Publishing, a company of Dun and Bradstreet.

1-4. ORGANIZATION OF THE BOOK

The major intent of this book is to explain and examine the principles, processes, and materials of semiconductor lithography. A global summary of the organization of this book can be found in Fig. 1-4-1 which outlines the total resist involvement in the lithography sector of semiconductor lithography. In this undertaking, we seek to organize the scattered literature and patents in lithography to see how technical problems are solved by a combination of material and process innovations. Initially, in Chapters 2–5, the structure-property of resist components of all categories of resists (positive photoresists, negative photoresists, positive radiation resists, and negative radiation resists) will be related to the sensitivity and performance of the resists. The next 11 chapters are devoted to the individual process steps and the controls of lithography.

In the first process chapter (Chapter 6), the preparation of the surface and coating techniques are discussed. As a practical concern, contamination control is included in great detail as for maintaining and for increasing the process yields. Once the film has been coated, the resist film is prebaked to remove the casting solvent and improve adhesion. Chapter 7 examines the consequences of prebake on subsequent processes and the thermal–chemical limits of prebake. The prebaked film is exposed by optical systems (Chapter 8) or high-energy radiation (Chapter 9) from electron beams, ion beams, or X rays. The physical optics of the exposure tools determine the edge slope of the incoming beam. Within the resist, the forward and reflective properties of the substrate scattering of the absorbed radiation influence the shape of the resist image. The developer reproduces the action of absorbed radiation. In Chapter 10, we return to the chemical aspects of lithography by using development models of positive and negative resists. For future processing, solventless plasma gas-phase development is also discussed. The net action of exposure and development is the formation of critical profiles (Fig. 1-4-2) prior to the permanent image transfer by additive or subtractive techniques. In resists, critical profile controls are necessary for subsequent dry etching, ion implantation, and lift-off processing.

The developed resist image is usually given a physical baking or chemical hardening step to improve its adhesion and etch resistance (Chapter 11). During postbake, the chemical changes and the different chemical treatments of diazonovolaks are examined. The postbaked image serves as a resist mask for additive lift-off (Chapter 12) and subtractive etching (Chapter 13). For device interconnections, metal patterns can be deposited by the additive process. In order to create profiles for lift-off and dry etching and overcome the image deficiencies of single film resists, multilayer structures are important for present and future submicrometer silicon device fabrication. Both bilayer and trilayer structures make use of stable composites of an image resist and stable underlayer. The image resist is not required to withstand subsequent etching conditions, since its image is transferred to the planarizing sublayer.

In lithography, wet and gas-phase subtractive etching are a very crucial step since the temporary resist image is permanently transferred into electrical patterns. Isotropic wet etching suffices to about 3-μm dimensions. For smaller dimensions in silicon and insulating films, anisotropic reactive ion etching is required. Over the last 10 years, anisotropic reactive ion etching has been the single most significant technical advance in lithography. Image etch biases can be controlled to less than ± 200 nm. Although dry gas-phase etching in barrel reactors mimics wet etching, the process incurs substantially lower costs and operates with better endpoint control. When chemically active halocarbons are ionized between parallel plate capacitors, an anisotropic electrical field is established which can etch vertical profiles in films of silicon, insulators, and A1 alloys. The total core of wet and dry etching is examined in detail in this chapter. After the image has been transferred to a permanent device structure or opening, the resist has served its role and is removed by liquid- or gas-phase processes (Chapter 14). Liquid stripping involves the chemical and physical breakdown of the resist film into an emulsifiable product. Plasma gas-phase stripping was the forerunner of dry etching and is a safer process than liquid stripping since the resists are converted into the oxidation products of carbon dioxide and water. Once the lithographic procees steps have been completed, in Chapter 15 we look back at the process controls which can be used to improve yield and process reproducibility. With the current trend to larger wafers and single wafer processing, resist develop and etch endpoint monitors and dynamic feedback process controls are necessary to ensure correct circuit dimensions, fast turnaround, and high yields. In Chapter 16, we examine alternative "resistless lithography" in which directed energy beams of lasers, ions, or electrons can directly etch semiconducting device layers or deposit metals without the need for a resist mask. These processes can also be used to repair faulty circuits on the device or correct defects in optical photomasks. Chapter 17 contains a conclusive summary of the key materials and processes used in silicon device lithography and the interdependencies of each process step. Future lithographic processes are also projected based on a current ranking of preferred resists and processes. Each chapter contains an individual summary and outlook.

This book is aimed at students and beginning technologists eager to apply chemical and physical principles in a research–development or in a practical process environment. For the experienced lithographer, the book serves as an invaluable guide to the scientific and patent literature from which improved lithography is

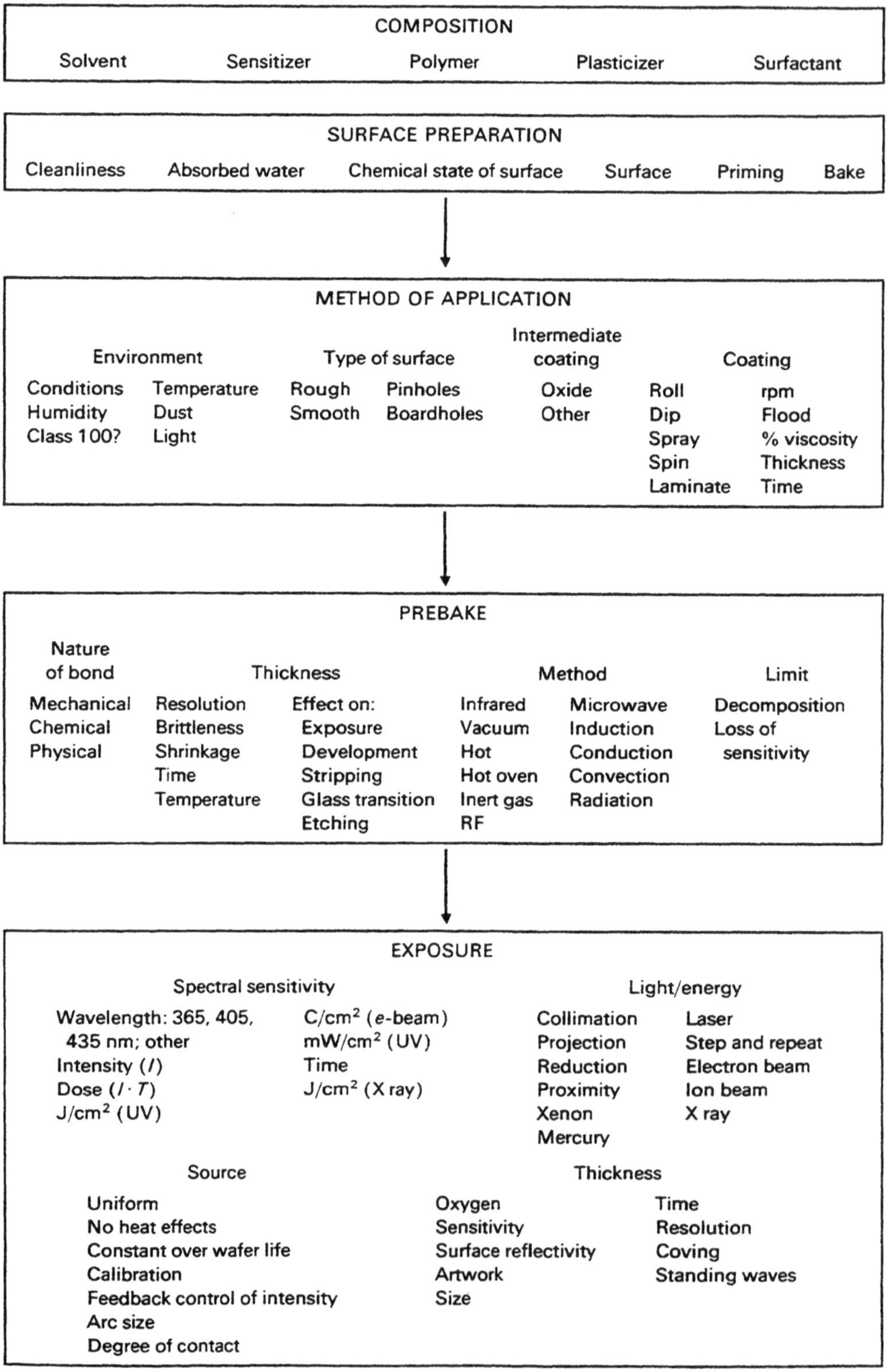
COMPOSITION
Solvent
Sensitizer
Polymer
Plasticizer
Surfactant
SURFACE PREPARATION
Cleanliness
Absorbed water
Chemical state of surface
Surface
Priming
Bake
METHOD OF APPLICATION
Environment
Conditions
Humidity
Class 100?
Temperature
Dust
Light
Type of surface
Rough
Smooth
Pinholes
Boardholes
Intermediate coating
Oxide
Other
Coating
Roll
Dip
Spray
Spin
Laminate
rpm
Flood
% viscosity
Thickness
Time
PREBAKE
Nature of bond
Mechanical
Chemical
Physical
Thickness
Resolution
Brittleness
Shrinkage
Time
Temperature
Effect on:
Exposure
Development
Stripping
Glass transition
Etching
Method
Infrared
Vacuum
Hot
Hot oven
Inert gas
RF
Microwave
Induction
Conduction
Convection
Radiation
Limit
Decomposition
Loss of sensitivity
EXPOSURE
Spectral sensitivity
Wavelength: 365, 405, 435 nm; other
Intensity (I)
Dose (I · T)
J/cm² (UV)
C/cm² (e-beam)
mW/cm² (UV)
Time
J/cm² (X ray)
Light/energy
Collimation
Projection
Reduction
Proximity
Xenon
Mercury
Laser
Step and repeat
Electron beam
Ion beam
X ray
Source
Uniform
No heat effects
Constant over wafer life
Calibration
Feedback control of intensity
Arc size
Degree of contact
Thickness
Oxygen
Sensitivity
Surface reflectivity
Artwork
Size
Time
Resolution
Coving
Standing waves

FIGURE 1-4-1

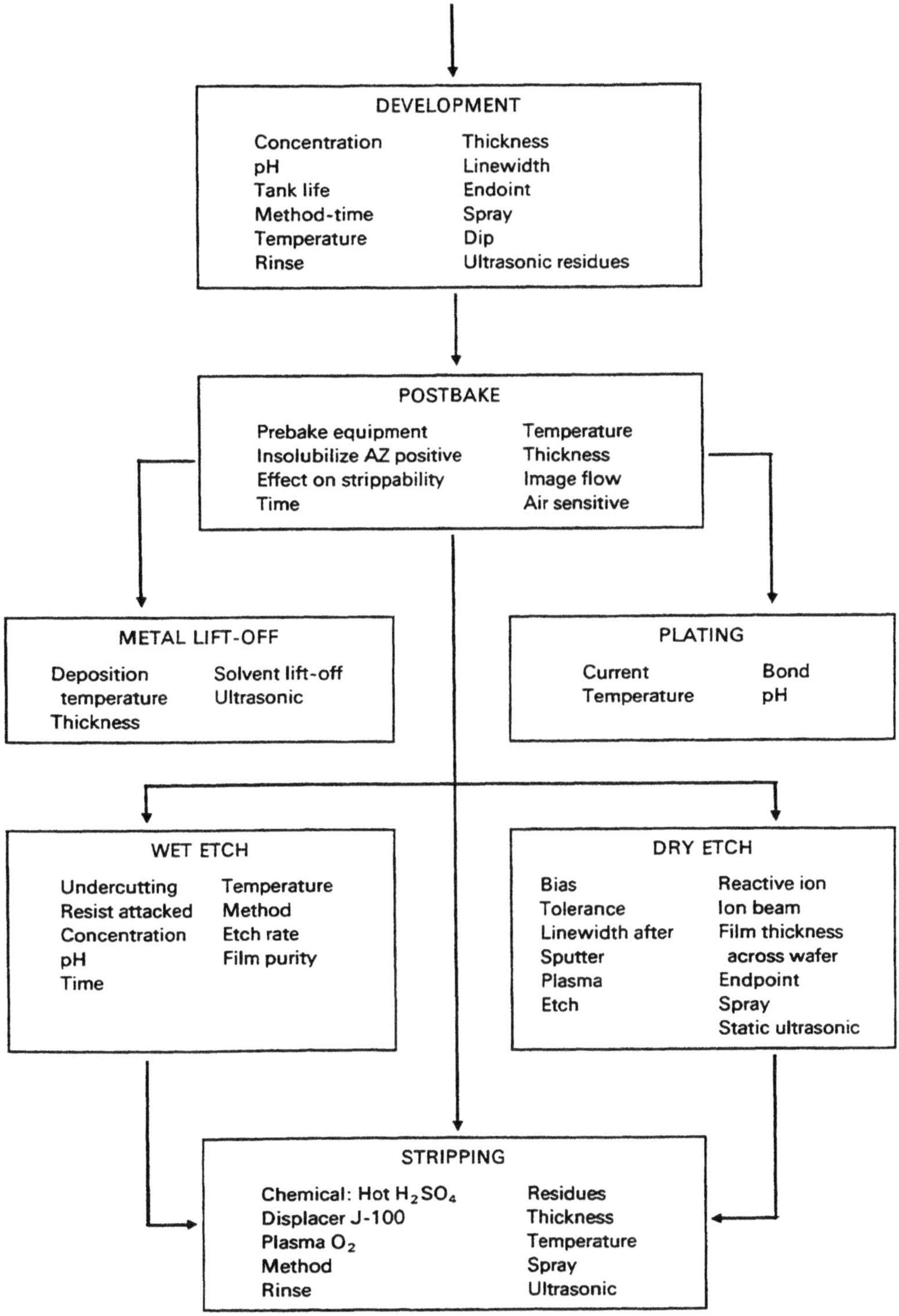

FIGURE 1-4-1 (*continued*)

TABLE 1-4-1. Semiconductor Lithography—Suggested Readings

Introduction to Microlithography, edited by L. Thompson, C. Willson, and M. Bowden, *ACS Symp. Ser.* **219**, 1983
Integrated Circuit Fabrication Technology, D. Elliott, McGraw–Hill, 1982
VLSI Electronics, Vols. 1–8, edited by N. Einspruch, Academic Press
Fine Line Lithography, edited by R. Newman, North-Holland, 1980
Electron Beam Technology and Microelectronics Fabrication, edited by G. Brewer, Academic Press, 1980
Handbook of Thin Film Technology, edited by L. Maissel and R. Glang, McGraw–Hill, 1973
Photoresist Materials and Processes, W. DeForest, McGraw–Hill, 1975
Light Sensitive Systems, J. Kosar, Wiley, 1964
Photosensitive Diazo Compounds, M. Dinaburg, Focal Press, 1970
Microphotography, G. Stevens, Wiley, 1966
Physics of Microfabrication, I. Brodie and J. Muray, Plenum Press, 1982

TABLE 1-4-2. Lithographic Journals and Societies

Society	Major journal
Electrochemical	*Journal of the Electrochemical Society* Sponsors Plasma Etching and International Electron, Ion Beam Conference every 2 years
American Vacuum Society	*Journal of Vacuum Science and Technology* Sponsors Electron, Ion Beam Conference every other year with IEEE
Institute of Electronic and Electrical Engineers	*IEEE Transactions on Electron Devices*
American Institute of Physics	*Journal of Applied Physics* *Applied Physics Letters*
American Chemical Society	Organic Coatings Symposiums *Journal of Applied Polymer Science*
Society of Plastics Engineers	RETEC Photopolymers *Polymer Engineering Science*
Society of Photographic Science and Engineering	*Photographic Science and Engineering*
Society of Photooptical Instrumentation and Engineering	SPIE Symposium and Proceedings
Microcircuit Engineering	Yearly Europe Symposium
Eastman Kodak	Yearly Microelectronics Symposium and Proceedings
Others	*Solid State Technology* *Semiconductor International* *Japanese Journal of Applied Physics* *Circuits Manufacturing* *Applied Optics* *Electronic Letters* *Japan Semiconductor News* *Japan Electronic Engineering* *Electronics*

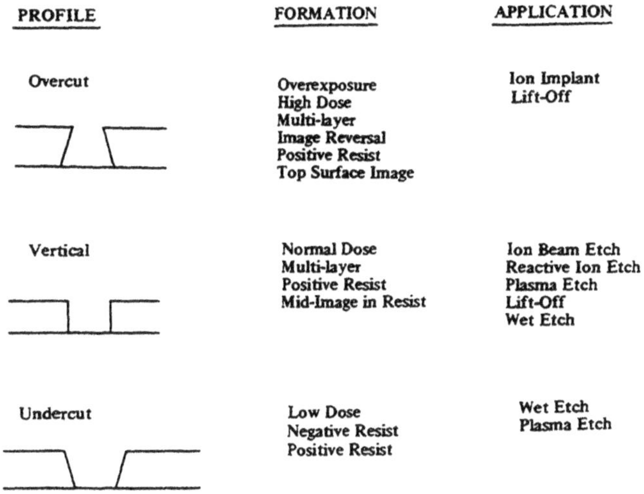

FIGURE 1-4-2. Preferred profiles in resist, their end use in additive or subtractive processes, and their principal methods of formation.

derived. For background materials, some suggested readings are listed in Table 1-4-1, and for current material, the technical societies and journals devoted to microlithography are given in Table 1-4-2.

REFERENCES

1. W. Hitinger, *Sci. Am.* August 1973, p. 48.
2. Y. Nishi, *Microelectron. J.* **12**, 5 (1981).
3. R. Noyce, *Sci. Am.* Sept. 1977, p. 68.
4. M. Furuichi and N. Kemura, *Jpn. Semicond. Technol. News* **2**(5), 25 (1983).
5. A. Reisman, *Proc. IEEE* **71**, 550 (1983).
6. L. Fried, J. Havas, J. Lechaton, J. Logan, P. Totta, and G. Paal, *IBM J. Res. Dev.* **26**, 364 (1982).
7. K. Levy and P. Sadland, *Solid State Technol.* May 1978, p. 50.
8. L. Funlap, *Chem. Eng. News* Nov. 30, 1970, p. 47.
9. E. Weber and R. Moore, *Solid State Technol.* May 1979, p. 61.
10. W. Bayleis, *Solid State Technol.* Febr. 1981, p. 132.
11. J. Trotel and B. Fay, in *Electron Beam Technology in Microelectronic Fabrication*, edited by G. Brewer, Academic Press, New York, 1980, p. 315.
12. N. Wu and B. Pickney, *Proceedings of the International Conference on Microlithography*, Paris, 1977, p. 139.
13. M. Cooke and S. Wiltstock, *Trans. S.Afr. Inst. Electr. Eng.* **64**, 194 (1973).
14. A. Robinson, *Science* **224**, 590 (1984).
15. R. Miller, *Solid State Technol.* July 1973, p. 25.
16. E. Ostroff, *J. Photogr. Sci.* **17**, 65 (1969).
17. R. Henderson, in *Electron Beam Technology in Microelectronic Fabrication*, edited by G. Brewer, Academic Press, New York, 1980, p. 220.
18. G. Madland, *Solid State Technol.* Aug. 1977, p. 93.
19. B. Deal and J. Early, *J. Electrochem, Soc.* **126**, 28C (1979).
20. C. Harper, *Handbook of Plastics and Elastomers*, McGraw-Hill, New York, 1975, p. 4.

2

Positive Photoresists

2-1. INTRODUCTION

After exposure to 150- to 800-nm radiation, positive photoresists develop more rapidly in the exposed region at a rate R which is about ten times greater than the unexposed rate, R_0. For various lithographic processing, the various preferred profiles (Table 2-1-1) are formed at different doses and R/R_0 development conditions.

At high doses (> 100 mJ/cm^2), the profile is dominated by the absorption of radiation, the reflected photons, and the quantum yield, Φ, of chemical transformation. The energy absorbed (E) per depth Z dominates the rate of dissolution (R) per depth,

$$dR/dZ = (dR/dE)(dE/dZ) \tag{2-1-1}$$

At moderate doses, both the developer elution term (dR/dE) and energy absorbed (dE/dZ) contribute to the formation of the resist profile. At low doses, the developer term dominates, profiles recede, and film thinning occurs (see Chapter 10). The photoresist edge slope (dR/dZ) of an image mask width, W,

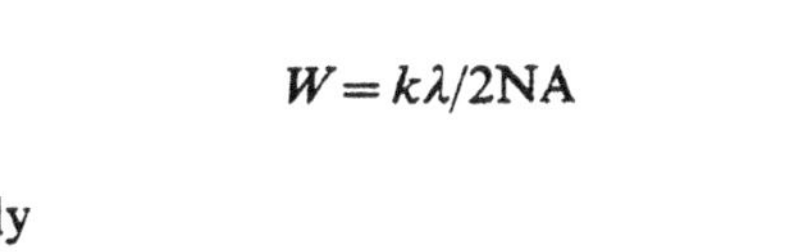

is influenced by the optics of the exposure tool.[1] For an optical system, the resolution (W) of an optical system is proportional to the exposure wavelength λ, the numerical aperture (NA), and the depth of focus x,

$$W = k\lambda/2\text{NA} \tag{2-1-2}$$

and Δx is approximately

$$\Delta x = \lambda/2(\text{NA})^2 \tag{2-1-3}$$

By using shorter exposure wavelengths (250 nm), smaller feature sizes can be resolved. Either the printed image bias (ΔW) or the ratio $\Delta W/W$ of the feature size

TABLE 2-1-1. Profiles, Expose–Develop Criteria, and Uses of Positive Resists

Profile	Dose	Developer influence	R/R_0	Uses	τ
	High	Low	>10	Lift-off Ion implant	3
	Medium	Moderate	5–10	Dry etch Lift-off Wet etch	2
	Low	Dominant	<5	Wet etch	1

must also be decreased if smaller features are to be printed. The ΔW is primarily a function of the edge slope, dR/dZ [Eq. (2-1-1)].

The exposure term dE/dZ is a function of the NA, λ, and ΔX, the depth of focus of the exposure tool,[1,69]

$$dE/dX = 2\text{NA}/\lambda[1 - \Delta X(\text{NA})^2/\lambda]^2 \tag{2-1-4}$$

The response of the resist to the energy deposited (E) is given by the contrast of the resist developer. For positive photoresists, the contrast is the slope of the percent film remaining versus log dose (Tables 2-2-5 and 2-2-6, Fig. 2-1-1).

The contrast ratio is $\tau = [\log_{10}(E_1/E_0)]^{-1}$, where E_1 is the exposure energy below which there is no resist removal in the developer and E_0 is the exposure energy at which no resist remains after development. In general, E_1 is independent of thickness, t, whereas E_0 will vary approximately with the reciprocal absorbance $(-E = 10^{\alpha t})$. Under these assumptions,

$$\tau = (\beta + \alpha t)^{-1} \tag{2-1-5}$$

where β is a constant and α is the resist absorption. To uniformly illuminate the

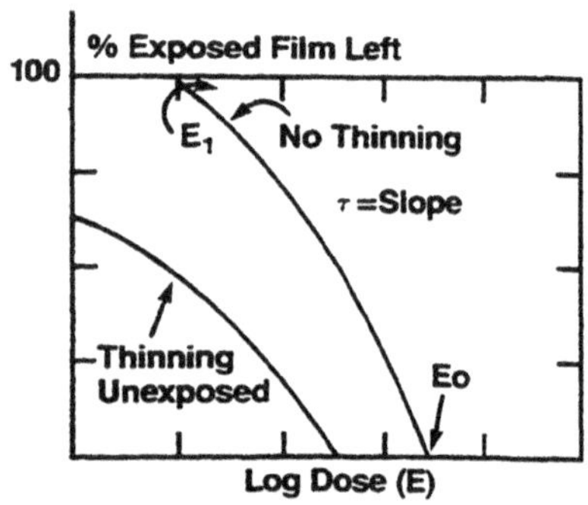

FIGURE 2-1-1. Contrast curve of positive resist.

TABLE 2-1-2. Characteristics of Optical Components for UV Exposure

Region	λ(nm)	Optics	Sources	Tools
Deep UV	150–300	Quartz Fluoride Vacuum	Laser Resonance Hg	Proximity Reflective
Mid UV	300–350	"Pyrex"	Hg Laser	Reflective Contact
Near UV	350–500	Glass	Hg	Steppers Reflective

depths of the resist, the absorbance should be about 0.4. As an upper limit, the resist contrast approaches 3–4 (see Tables 2-1-1, 2-2-5, and 2-2-6).

The developer-process term in Eq. (2-1-1) (dR/dE) is given by

$$dR/dE = -\tau/E_0 \tag{2-1-6}$$

thus, dR/dZ in Eq. (2-1-1) becomes

$$\frac{dR}{dZ} = \frac{\text{NA}}{\lambda(\beta + \alpha t)\, E_0}\left[(1-k)\left(\frac{\Delta X(\text{NA})^2}{\lambda}\right)\right]^2 \tag{2-1-7}$$

To minimize the edge slope, dR/dZ, and image bias, ΔW, the resist contrast (τ) should be as high as possible to counter the shorter exposure λ or the higher NA. At low doses, the developer term dominates and the contrast (τ) falls to < 1.5.

Positive photoresists can be exposed by light in three spectral regions dictated by the emission of mercury lamps, the transmission of mask blanks, and the availability of suitable exposure lens (Table 2-1-2). Since the mercury lamp intensity falls off in the deep UV, more expensive sources are needed to compensate the two- to fourfold loss of intensity (Fig. 2-1-2). The dominate emission lines are at 254, 313, 365, and 435 nm. Since resolution (smallest feature size) is directly proportional to λ [Eq. (2-1-2)], submicrometer lithography has been driven to the deep-UV region. The deep-UV region expands the ensemble of resists since enough photochemical energy is available (> 5 eV) to break any organic bonds.

Historically, the positive photoresists based on diazoquinones (DQ) and novolak (phenolic-formaldehyde resins) (N), were a spin-off of diazonium compounds used prior to 1940 for making blueprints and letterpresses. For the

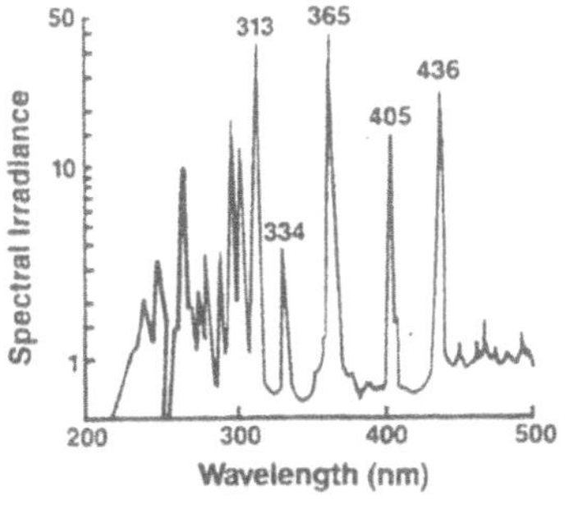

FIGURE 2-1-2. Mercury lamp output for resist exposure. From *Introduction to Microlithography*, edited by L. Thompson, M. Bowden, and C. Willson, American Chemical Society, 1983, p. 24, reproduced by permission of the American Chemical Society.

evolution of diazonovolak resists prior to 1965, the reader is referred to the treatises of Kosar,[1] Dinaburg,[2] and the review of Hanson.[3]

Since DQN absorbs all of the 365, 405, and 435-nm mercury lines and is not oxygen sensitive as the azide–rubber resists (Chapter 4), after 1975 they became the dominant photoresist. At least 16 companies now manufacture DQN resists comprised of a photoactive diazoquinone ester (DQ) and a phenolic novolak resin (N),

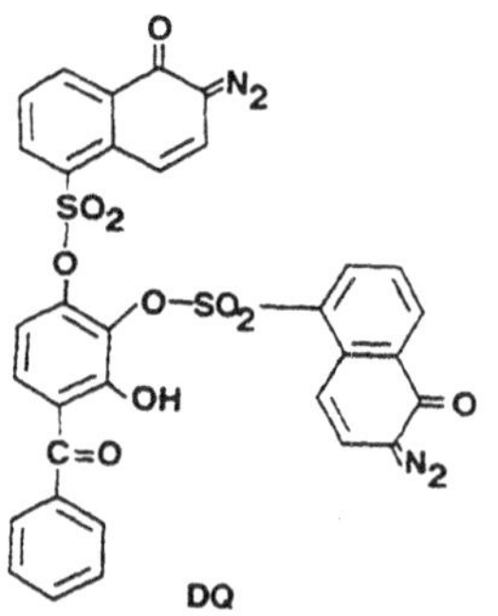

DQ

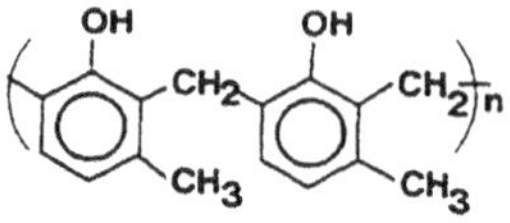

Novolak Resin

Positive photoresists are formulated from one or two major ingredients:

One component	Two components
PMMA	DQN
As_2S_3	Sensitizers + PMMA
Photoxidizable polymers	Nitroaldehydes + PMMA
Ablative exposure	Onium salts + scission polymers

In positive photoresists, the photochemically induced reactions for producing differential images in positive photoresists are based on the chemical transformation of small photoactive additives or polymeric materials:

$$\text{Nonpolar} \rightarrow \text{Polar} \tag{2-1-8}$$

$$\text{Polar} \rightarrow \text{Nonpolar} \tag{2-1-9}$$

$$\text{Polymer} \rightarrow \text{Monomer} \tag{2-1-10}$$

The DQN system is an outstanding example of a two-component resist which is photochemically transformed into a polar (base) soluble product:

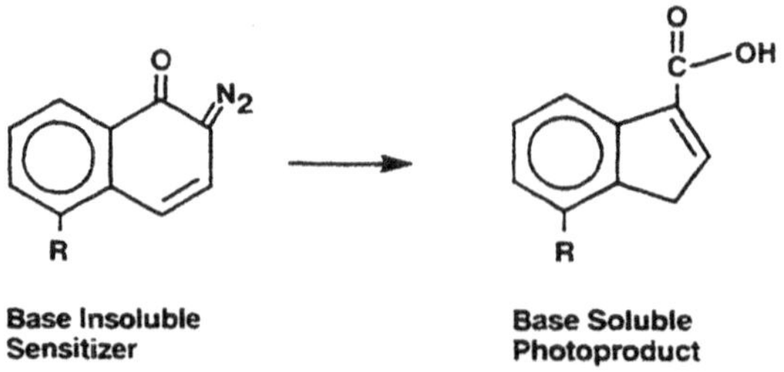

(2-1-11)

The hydrophilic resin is rendered alkali insoluble by the addition of a 15%

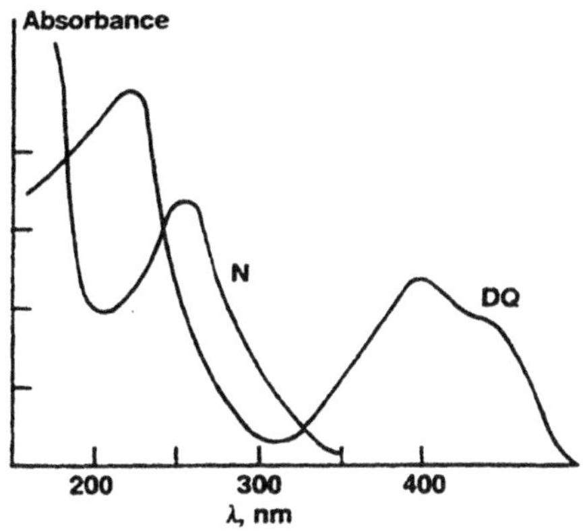

FIGURE 2-1-3. UV absorbance spectrum of novolak (N) and diazoquinone (DQ) sensitizer.

loading of DQ. In the DQN formulation, the photoresponsive part of the resist is imparted by the DQ photoactive component and process resistance furnished by the commercial resin (N). For exposure in the deep-UV (200–300 nm) region, other positive photoresists have evolved because of the high opacity of the novolak resin (Fig. 2-1-3) in this spectral region. These systems have been based on photoinduced chain scission of the well-known PMMA resist [Eq. (2-1-10)],

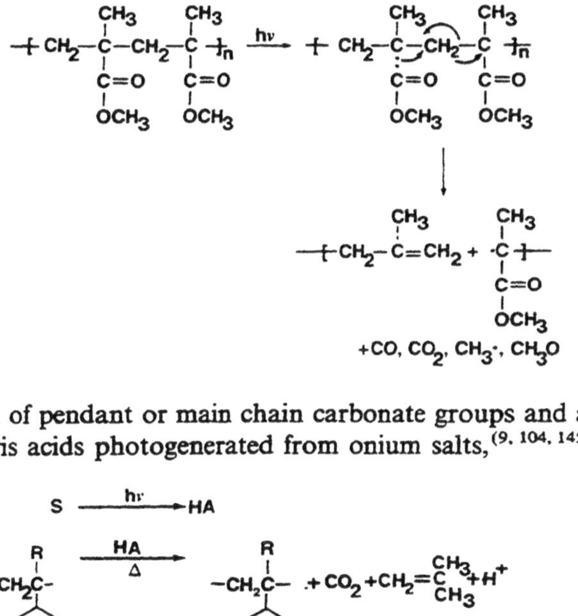

or on the scission of pendant or main chain carbonate groups and aldehyde groups catalyzed by Lewis acids photogenerated from onium salts,(9, 104, 145–150, 225, 231)

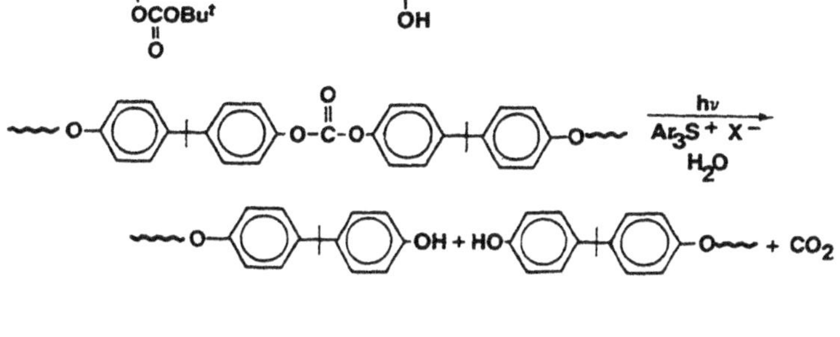

The onium-photosensitized resists exhibit high photosensitivity (5 mJ/cm^2) in the deep–mid UV region because of the image amplification factor that 1 mole (per mole of photons) of acid may cleave ten or more polymer chains. By contrast, 1 mole of DQN or PMMA resist requires up to 10 or more "moles" of photons, $> 75\ mJ/cm^2$ ($1\ J = 1$ mole of photons or one einstein, 10^{18} photons). The positive photoresists are dominated by DQN systems. We will discuss these resists along with PMMA and polymethyl isopropenyl ketones. Other acid-labile or photorearrangement systems will also be briefly discussed.

2-2. DIAZOQUINONE PHOTORESISTS

2-2-1. *General Performance*

The DQN resist consists of a smaller photoactive component of diazonaphthoquinone dispersed in a novolak resin. The photoactive dissolution inhibitor primarily absorbs the 365-, 405-, and 435-nm mercury lines (Fig. 2-1-3) but it is predominantly transparent in the mid-UV region unless blueshifted with alkyl sulfonyl esters (see below).

As resist systems, positive photoresists based on novolak resins have distinct advantages over negative cross-linking resists:

1. Aqueous alkaline (nonswelling) development which kinetically develops anisotropically from the heavily exposed interior of the image (Fig. 2-2-1) and subtends the edge-scattered radiation. On the other hand, negative resists develop from the outside scattered region revealing in the case of cross-linked resists the scattered radiation because swelling inhibits its

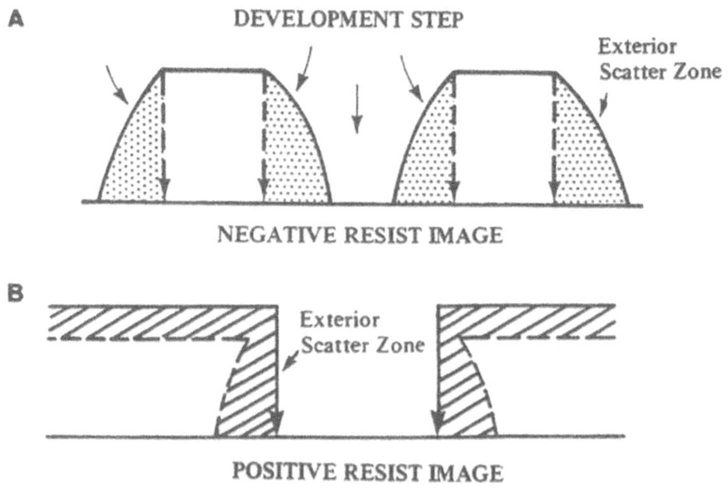

FIGURE 2-2-1. (A) Time-independent development of cross-linked negative resist fails to remove light scatter zone. (B) Development of positive resist rapidly removes exposed region and can be quenched to inhibit removal of lateral scattered exposed resist region.

removal. An image width to resist thickness of 1:1 can be achieved. Also, the developer is ecologically disposable.
2. Not sensitive to oxygen inhibition of the photochemical reactions.
3. Because of the phenolic resin component, the plasma etch resistance is the best of conventional resists.
4. Can be thermally stabilized to 200°C (see Chapter 13).
5. The DQN resists are reworkable, easily stripped, and suitable for lift-off.
6. Can be converted into an O_2 plasma-resistant mask for bilayer processing (Chapter 12).
7. Image reversal to a negative image is possible.

The main disadvantages of positive DQN resists are

1. Small process window in developer at moderate and low doses (see Chapter 10).
2. Moderate contrast $\tau < 2$ in many commercial DQ resists contributing to a large ΔW [Eq. (2-1-7)].[7, 196]
3. Average photospeed, 75 mJ/cm^2, because of low quantum yields of 0.2.
4. Poor adhesion on silicon surfaces. Primers are required.

2-2-2. *General Photochemistry and Solubility Properties*

When a hydrophobic DQ is added to an alkali-soluble resin (N), the resin dissolution is inhibited (Fig. 2-2-2). The photochemical destruction of the inhibitor and the release of uncomplexed resin induces upon development the dissolution of the exposed resist. Upon photolysis, the DQ generates in Wolff rearrangement a carbene (C) which rearranges to a ketene (K). The ketene reacts with adsorbed moisture present in the resin, to form an indene acid (A),

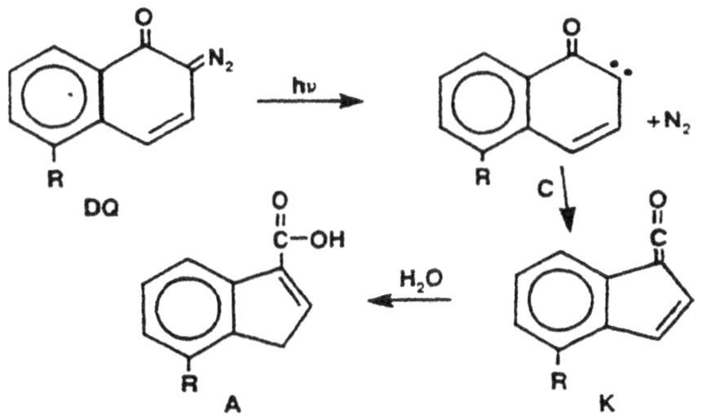

Upon formation of the indene acid products, the exposed resist becomes hydrophilic and is more easily wet by the aqueous developer. In addition, the evolved N_2 gas foams the exposed region to allow the developer to penetrate into the exposed regions.[60, 232] For DQ structures, the diester (B) and triester (T) sulfonic esters are more efficient inhibitors of dissolution than the monoester (M),[236, 238]

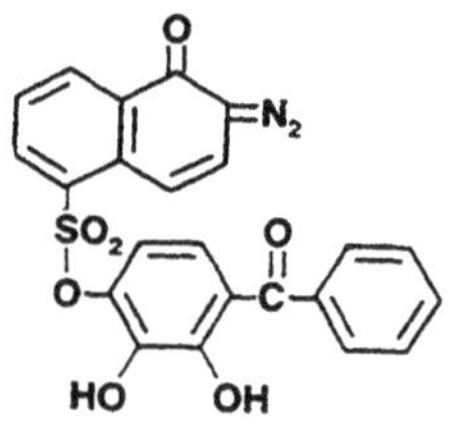

M

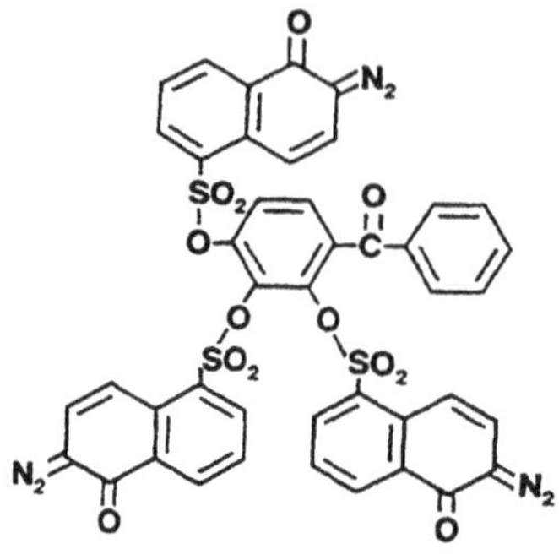

T

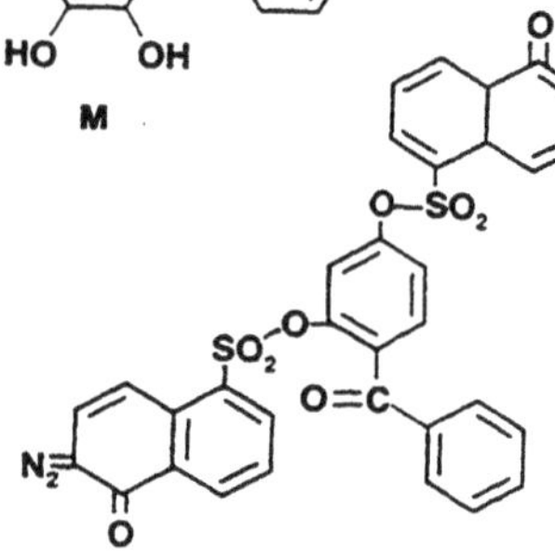

B

because of the better intermolecular (hydrogen bonding) complexation with the novolak resin chains,

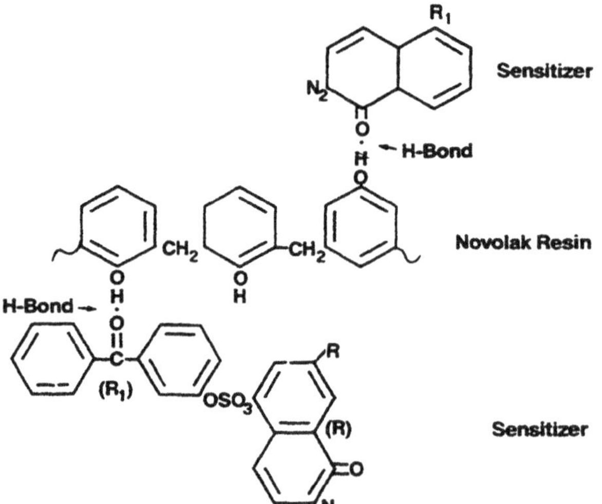

The DQs can also be attached to the novolak resin, to comprise a one-component resist, although they are sparingly soluble and require admixtures with novolak resins,

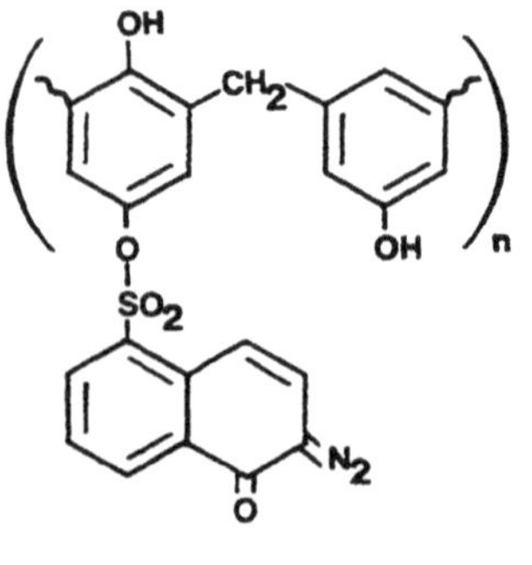

U.S. 3,046,120 (1962)

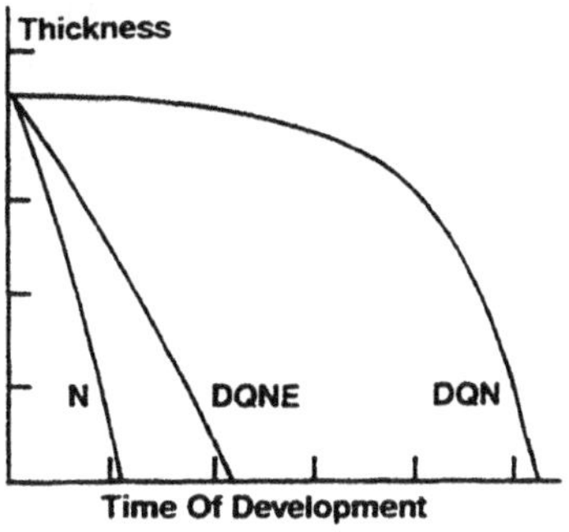

FIGURE 2-2-2. Thickness versus time of development of novolak (N) film, diazoquinone mixed with novolak (DQN), and exposed diazoquinone in novolak (DQNE).

For retarding the background dissolution, DQ's role in the novolak resin has been suggested to involve at the surface of the unexposed resist a diazonium azo-coupling reaction promoted by immersion in the alkaline developer,[4, 8]

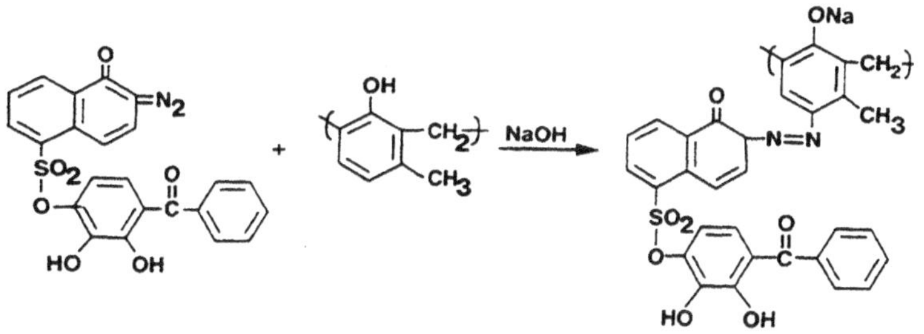

(2-2-1)

Historically, the coupling reaction [Eq. (2-2-1)] made DQs a substitute precursor for azo vat dyes and a direct replacement for diazonium salts as lithographic platemaking resists. Because the spectral absorbance of DQs was redshifted to 459 nm and the diazonium salts did not absorb beyond 365 nm, the DQN resists could be exposed with less expensive tungsten or fluorescent light sources. Accordingly, the DQN resists were formulated with high solid contents (50%) in order to cast mil-thick films for platemaking. Upon dilution to 25% total solids by weight, micrometer-thick films could be cast for semiconductor lithography. Table 2-2-1 list some of the key U.S. patents in the development of diazonaphthoquinone resists for printing plates, that later evolved into microelectronic-grade resist compositions. The 5-sulfonic acid ester of diazonaphthoquinone,

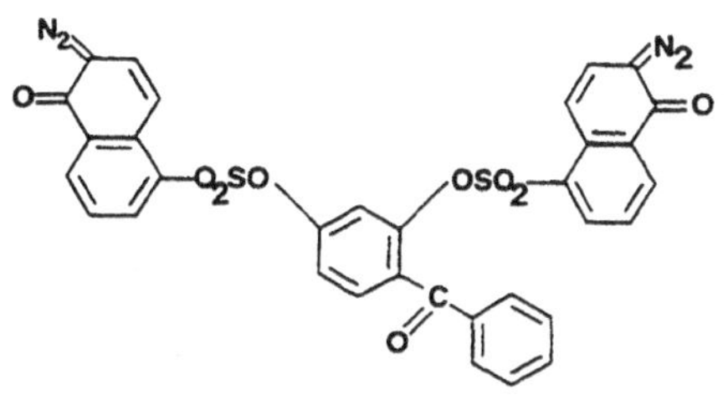

is the most prevalent derivative of the over 200 patented formulations. The 4-sul-

TABLE 2-2-1. Evolution of DQN Resists

Patent	Year	Assignee[a]
2,696,435	1954	Azoplate
2,702,243	1955	Azoplate
2,848,328	1958	Kodak
2,959,842	1960	Azoplate
2,971,842	1961	GAF
2,997,467	1961	GAF
3,028,239	1962	Azoplate (1)
3,102,809	1963	Azoplate
3,106,465	1963	Azoplate
3,130,047–9	1964	Azoplate
3,148,983	1964	Azoplate
3,661,582	1972	Western Electric (2)
3,666,473	1972	IBM (3)
3,950,173	1976	RCA (4)
4,093,461	1978	GAF (5)
4,284,706	1981	IBM (6)
4,377,631	1983	Hunt (7)
4,397,937	1983	IBM (8)
4,550,069	1986	Hoechst (9)

[a] *Notes*: (1) Benzophenone esters most widely used. (2) Additives to enhance *e*-beam sensitivity. (3) Optimizing novolak molecular weight. (4) Triester of (1) to enhance *e*-beam sensitivity. (5) Polyimide–high temperature composition. (6) Deep-UV Meldrum's acid. (7) *o*-, *m*-, *p*-cresol optimized novolaks. (8) Alkyl substituents for mid-UV. (9) PM acetate nontoxic-solvent.

fonic acid or quinoline rings have insufficient miscibility to inhibit the dissolution of novolak resins,

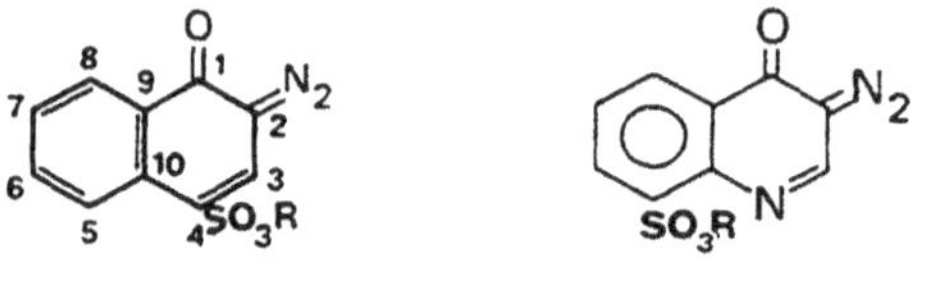

2-2-3. *Spectral Absorbance*

The spectral characteristics of DQs have been studied by Willson and coworkers.[9] The basic chromophore in the far UV is the *ortho*-naphthoquinone which strongly absorbs the 365- and 405-nm Hg lines (Fig. 2-2-3). The absorbance at 300–340 nm is mainly due to the aryl ring excitation while that at 340–450 nm is due to the quinone transitions. The low absorbance of aryl sufonic esters at 313 nm makes them unsuitable as mid-UV resist sensitizers. Alkyl sulfone diesters such as tricyclododecane derivatives blueshift (Fig. 2-2-4) the spectral absorbance for higher actinic activation at 313 nm.[9] In addition, the 5-aryl sulfoesters can be photosensitized in the mid-UV region to the 313-nm line, by the addition of pyrene[10]:

$$\text{Pyrene} \rightarrow \text{Pyrene (singlet)} \quad (2\text{-}2\text{-}2)$$

$$\text{Pyrene (singlet)} + \text{DQ} \rightarrow \text{DQ (singlet)} \quad (2\text{-}2\text{-}3)$$

$$\text{DQ (singlet)} \rightarrow \text{Photoproduct} \quad (2\text{-}2\text{-}4)$$

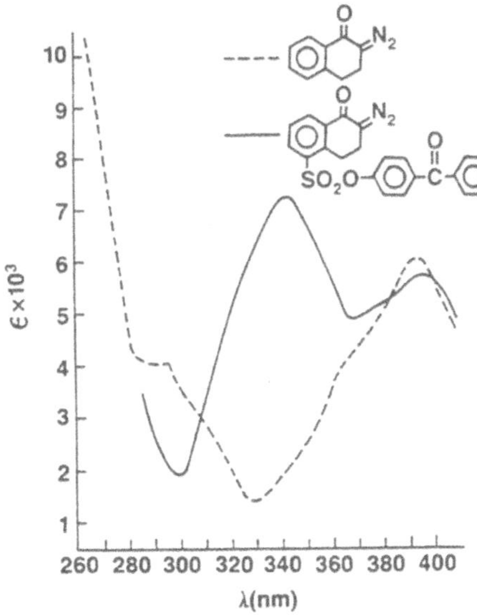

FIGURE 2-2-3. UV absorbance spectrum of unsubstituted diazoquinone and 5-aryl substituted. From Ref. 9, courtesy of the Society of Plastic Engineers.

For exposure with visible light, the redshift of the diazo band to the 500-nm visible region is claimed with phenanthrenequinone derivatives,[12]

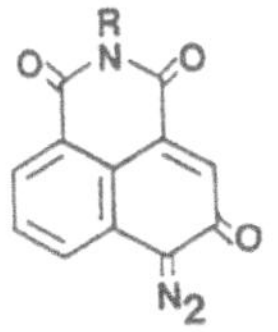

In mixtures with novolaks, the alkyl substituents of DQ are more soluble than the aryl substituents and the higher loading can improve the DQN contrast.[7, 179, 180]

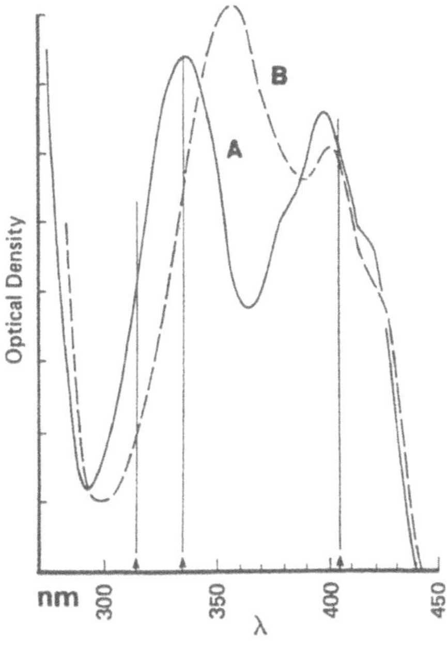

FIGURE 2-2-4. UV spectrum of (A) 5-alkyl and (B) 5-aryl sulfonate esters of diazonaphthoquinone. From Ref. 9, courtesy of the Society of Plastic Engineers.

High solubility is claimed (up to 40% by weight of novolak) for 1, 10-bis-4-hydroxyl phenyl decane 1,2-naphthoquinone 2-azido-5-sulfonate.[11, 14, 190]

2-2-4. *Resins for DQ Resists*

The DQs are generally admixed with *meta*-cresol resins derived from cresylic acid with molecular weight ranges from 500 to 20,000 (M_w) and wide polydispersity > 5.[242] Since the condensation polymerization of phenols and formaldehyde used in the synthesis of novolaks produces polydisperse resins of unpredictable solubility rates, high-and low-molecular-weight resins are blended to achieve reproducible dissolution rates.[2, 165, 172] Pure *meta*-cresol novolak,

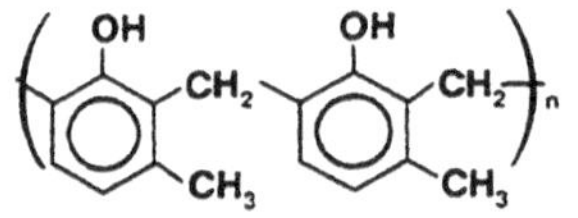

Meta-Cresol Novolak

is often used as a pure resin feedstock. The DQN resists formulated from the *meta*-cresol resin withstand flow temperatures of up to 150°C, while blends with *ortho* and *para* resins or other feedstocks flow at 125°C.[193, 213, 243] Novolak terpolymers with slight amounts of other alkyl groups such as *t*-butyl phenol have been used to increase the alkali resistance.[13-15, 113] Pure novolak resins are water-white polymers, but due to the oxidation step in driving off the excess formaldehyde, they become orange-brown due to quinone formation.[16, 227] For the thermal decomposition reaction, an activation energy of 30–40 kcal/mole has been measured.[65–67] Accordingly, a high-temperature postbake in air is used to opacify the film as a suitable underlayer that absorbs the near (300–400 nm)-UV light in a multilayer resist sandwich (Chapter 12),

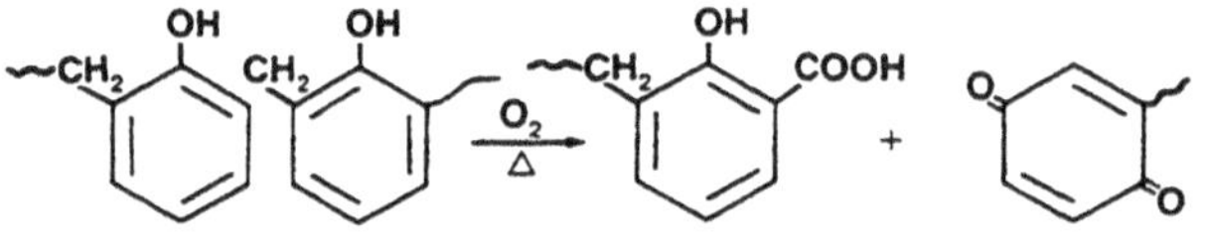

Because of the aryl constituents, the novolak resins absorb strongly in the deep-UV region. The UV spectra of 0.2-μm-thick films of pure *o*-, *m*-, and *p*-cresol-formaldehyde resin are shown in Fig. 2-2-5. A spectral window appears at 254 nm and the *para* novolak has the lowest absorbance (0.5 for a 1.0-μm-thick film). In order to take advantage of this window, an aliphatic DQ photosensitizer of Meldrum's acid and/or diazohomotetramic acid[212, 239] which absorbs strongly at 254 nm was used in a deep-UV formulation (Fig. 2-2-6[18]). Since the sensitizer does not absorb beyond 300 nm, the resolution loss due to, the absorbance of higher exposure wavelengths of polychromatic sources such as in scanning printers, is prevented. A dose of 50 mJ/cm^2 at 254 nm produced suitable images.[214] In this case the product of exposure is not necessarily an indene acid [Eq. (2-1-11)] but

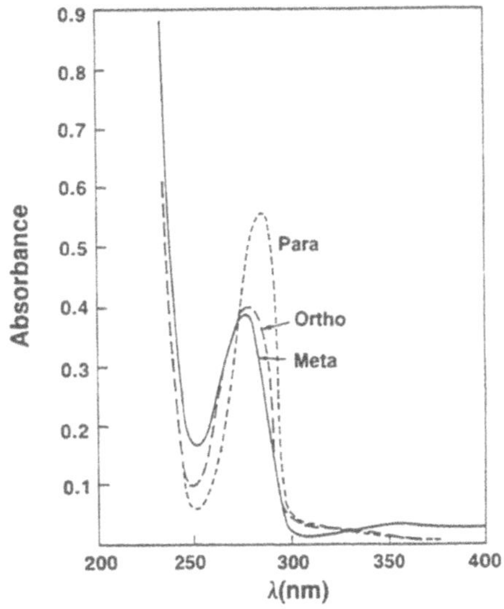

FIGURE 2-2-5. UV spectrum of 0.2-μm-thick films of *ortho*, *meta*, and *para* cresol formaldehyde resin. From Ref. 17, reproduced by permission of the Electrochemical Society, Inc.

solely a gaseous product of acetone and CO. The formation of a porous exposed region permits rapid infusion of the aqueous developer.[18, 61, 229]

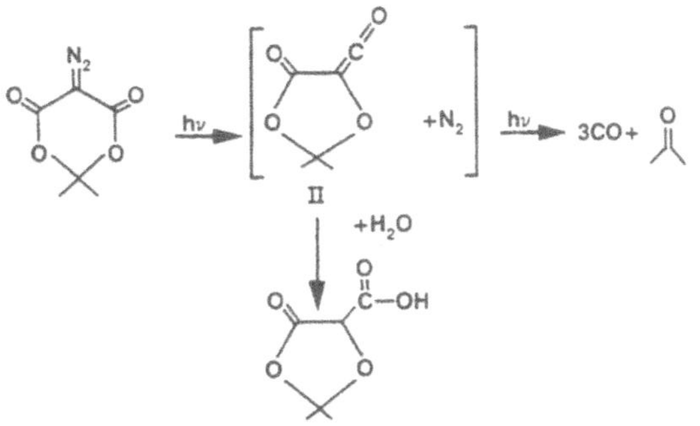

Another example of an inhibitor that does not form soluble products, can be found

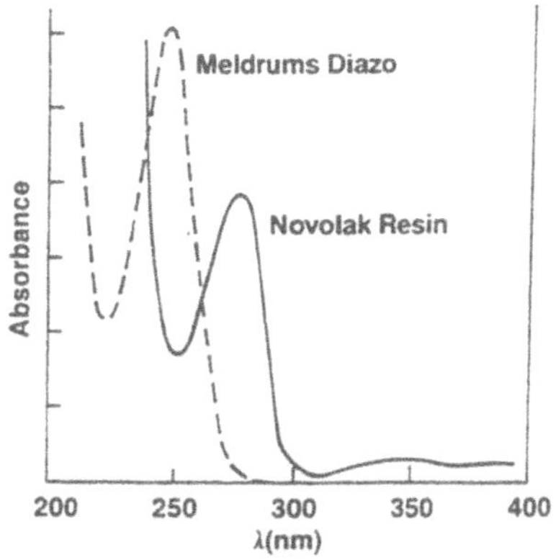

FIGURE 2-2-6. UV spectra of 5-diazo-Meldrum's acid in methanol and a film of novolak resin. From Ref. 18, copyright and courtesy of IEEE, 1981.

in the polymeric blend of polyolefin sulfones such as polymethyl pentene sulfone dispersed in novolak resins,

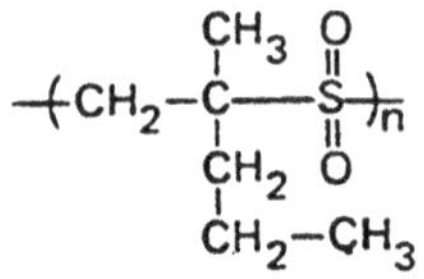

Exposure to an electron beam or photons forms a positive resist which developed in alkaline 0.25 N developer.[19–21] In the direct photochemical reaction at 254 nm, Hiraoka[21] found that the excited novolak resin (N^*) transferred the absorbed energy to decompose the transparent polysulfone,

$$N^* + RSO_2R \rightarrow SO_2 + N + R \tag{2-2-6}$$

Thus, these findings give evidence to the use of other novel photosensitive inhibitors besides the traditional diazonaphthoquinones.

Because of the high absorbance of phenolic resins in the deep UV, more transparent resins such as polyacrylate derivatives have been examined (Table 2-2-2). Other improvements of properties such as adhesion, thermal stability, and oxygen plasma resistance have also been investigated (Table 2-2-2). For example, silicon (silyl) novolaks have been synthesized as bilayer oxygen masks,[216, 233, 247]

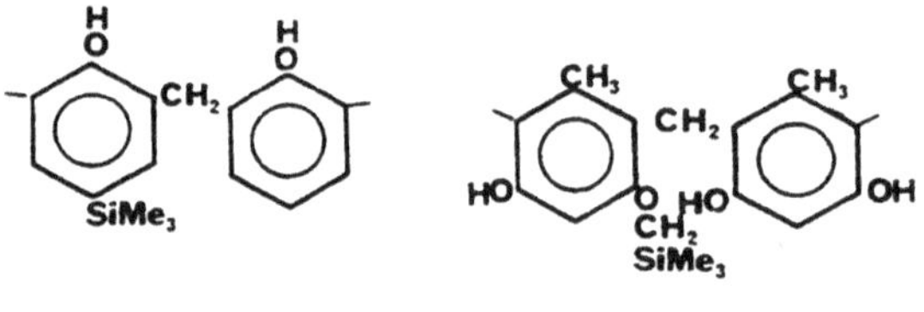

New high-temperature maleimide resins[230, 243] which do not melt at 200°C

TABLE 2-2-2. Substitute Resins for DQN

Polymer	References
Polyhydroxystyrenes	23, 185
Polyvinylphenol	24
Styrene-methacrylic acid	25, 186
PMMA-methacrylic acid	26
Styrene-acrylic acid	27
Esterified polyvinyl alcohol	28
Polyimide	29
Silylated novolak	30, 211

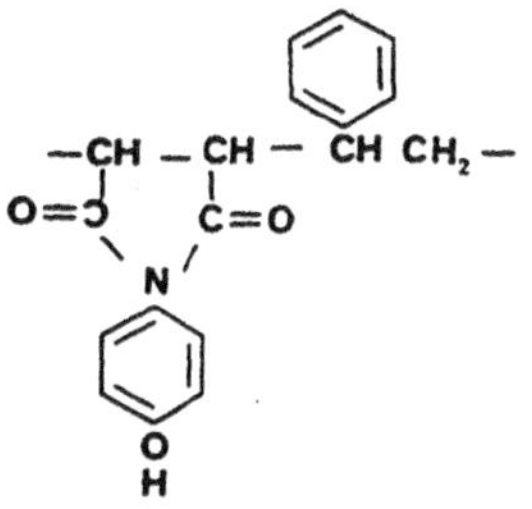

show promise at eliminating the posthardening steps required for DQN resists prior to ion implantation or for long reactive ion etching times.

With additives, others have sought to improve the performance of DQN resists. A typical DQN formulation consists of:

DQ	2 g
Novolak	8 g
Cellosolve acetate	20 g
Butyl acetate	2 g
Xylene	2 g
Water (trace)	0.05 g

Because of deficiencies in adhesion, speed, or other properties, addition of small or polymeric molecules has sought to improve or extend DQN resists (Table 2-2-3).

TABLE 2-2-3. Additives to DQN

Additive	Purpose	References
Imidazoles	Speed	31, 32
Phthalic anhydride	Speed	33–35, 178
Organic acids	Speed	36, 37, 53
Thiodipropionate	Speed	38
Organic acids	Adhesion	36, 37, 184
Silanes	Adhesion	39
Other resins	Adhesion	40–44
Imidazoles	Adhesion	45
SiO_2	Heat resistance	46
Peroxides	Heat resistance	47, 48
Siloxanes	Heat resistance	49
Aryl polymers	Plasma resistance	50
Castor oils	Flexibility	51
Dyes	Detection	52
SiO_2	Better developer	46
Surfactants	Reduce striations	54
Imidazoline	Negative resist	55, 56
Epoxy	Negative resist	57
Gold, resinate	Conductivity	58
Halotriazines	Visible image	181, 191, 199
Acetylated novolaks	Adhesion	182
Silazanes	Dry develop	209

Additives for improving adhesion are specifically discussed in Chapter 13, while additives to the developer are discussed in Chapter 10. Thus, the basic shortcomings of DQN of poor adhesion, thermal meting, and moderate photospeed have been addressed by traditional polymer technology, and for every major polymer problem there is a solution—an additive! Lastly, if new polymers or additives fail to improve the performance of novolak-based resists, posttreatments with heat, radiation, or chemicals (Chapter 11) can in most cases improve some basic properties such as thermal and plasma resistance. Currently, deep-UV exposure of the developed image is a popular method of stabilizing the resist image to 200°C.

2-2-5. *Chemical Mechanisms in DQN*

In exposure, unimolecular reactions are involved in the decomposition of DQ via the carbene (C) to ketene (K),

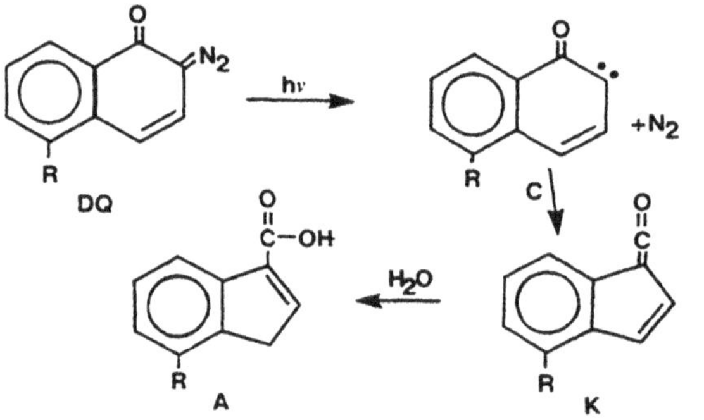

(2-2-7)

In flash photolysis studies of diazonaphthoquinones, the ketene has been identified by strong peaks at 330 nm,[59]

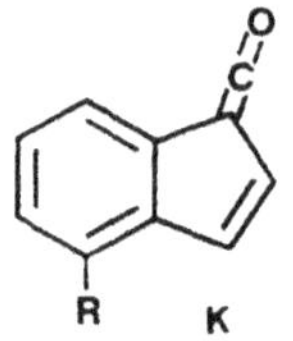

Bimolecular reactions can involve the simple addition of water to form the acid (A) or the coupling of any of the intermediates in Eq. (2-2-7) with any of the components of the formulation [DQ sensitizer and resin (N), Eq. (2-2-8)].[4, 164] For example, in the absence of moisture in vacuum exposure, cross-linking by esterification of the resin occurs,

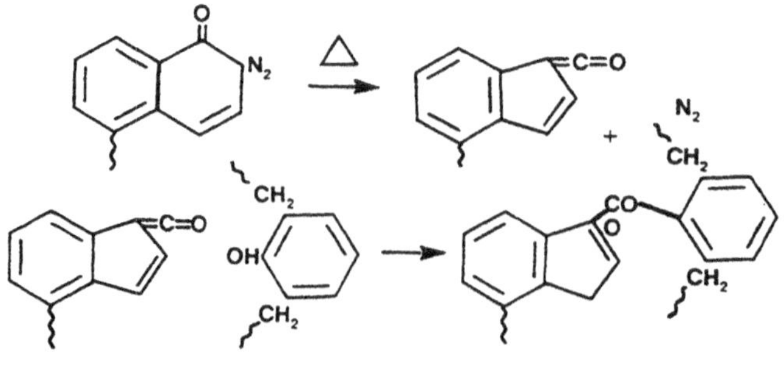

(2-2-8)

In the older literature, the indene acid product formed by the bimolecular reaction of ketene (K) with moisture is usually noted as the 1-indene carboxylic acid, but this has now been identified as the 3-indene carboxylic acid (B),[4, 167]

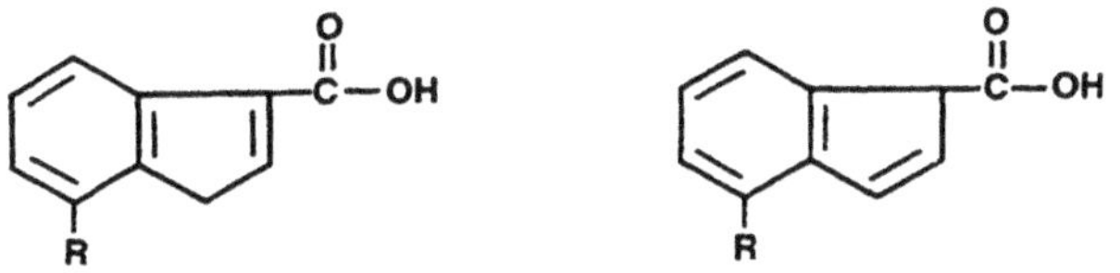

B A

In the solid state, the unimolecular reactions are favored because of the immobility (cage effect) of the intermediates or reactants (except with adsorbed water). Water is present at about 0.5% by weight of the N resin, and is presumably H-bonded,[164]

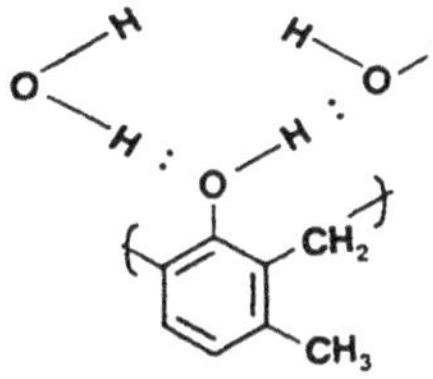

The bleaching of the DQ (Fig. 2-2-7) can be followed by the UV or IR loss of the diazo band,[197, 249] and the bleaching also can be used as an actinometer to determine the dose absorbed, or the spectral intensity of projection scanning tools.[194] The bleaching allows thick films with high absorbance (> 1.5) to be used, since light will eventually reach the substrate. Upon excitation of DQ, the excited

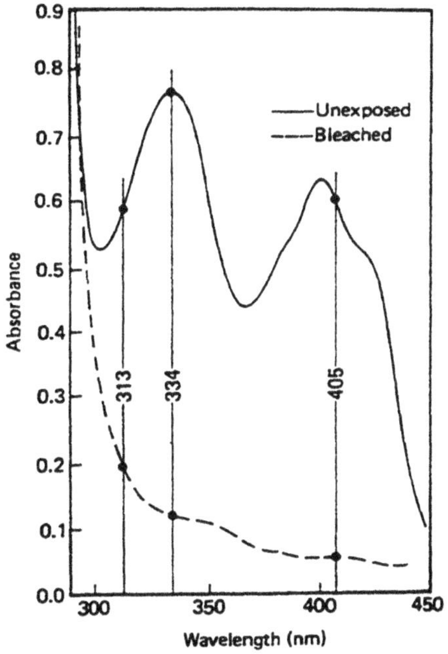

FIGURE 2-2-7. Bleaching of diazoquinone sensitizer after UV exposure. From Ref. 9, courtesy of the Society of Plastic Engineers.

singlet state photodecomposes through bending of the N_2 group out of the plane of the aromatic ring.[164, 168] Photochemical studies of the Wolff rearrangement of the ketocarbene into a ketene reveal that the reaction proceeds smoothly from the first excited singlet state of the diazoquinone[205, 206] and the triplet state does not generate the ketocarbene intermediate. Photosensitization proceeds more smoothly via the singlet state of pyrene than sensitization from the triplet state of benzophenone. The quantum yield for DQ loss gives an indication of the efficiency of photon utilization (Table 2-2-4). Only one of five photons converts the diazonaphthoquinone into product or about 10^{17} photons (or 100 mJ/cm^2) are required for every 10^{16} molecules of DQ. The inefficiency of the DQ may be due to an energy loss in the singlet state of a charge transfer transition which stabilizes the C–N bond against nitrogen expulsion[68] and out-of-plane bending,[160]

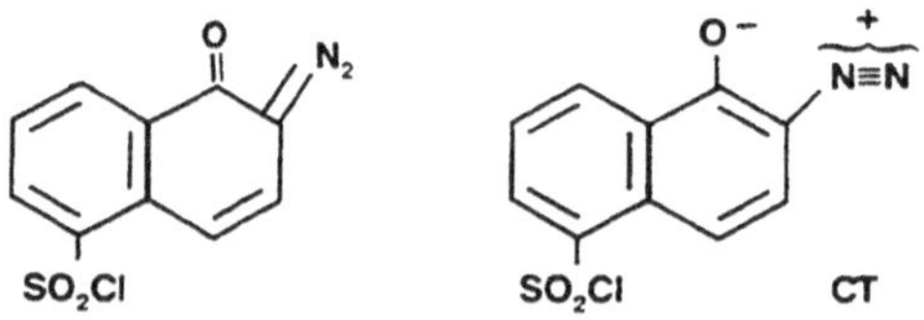

Fortunately, only about half of the added DQ has to be destroyed from the exposure to induce dissolution of the resist image.[6, 7, 249] The practical result of exposure and development is the contrast τ of the resist which depends on the energy deposited (dE) and the developer response (dR) terms in Eq. (2-1-1). The *dE* term depends on

1. Extinction coefficient, ε
2. Moles of sensitizer present
3. Quantum yield, Φ
4. λ and dose E_0 and E_1, numerical aperture (NA) and MTF of lens

The *dR* term is dependent on

1. Molecular weight and dispersity of resin[208, 222]
2. Prebake
3. Developer and temperature[229]

At a moderate dose of 100 mJ/cm^2, the contrasts τ shown in Table 2-2-5[69] were obtained. At these doses the developer term dominates the contrast. At a higher

TABLE 2-2-4. Quantum Yield of DQN

Quantum yield	Reference
0.15 (365 nm)	62
0.23 (405 nm)	62
0.16 (435 nm)	62
0.32 (405 nm)	6
0.28 (365 nm)	6
0.13–0.26 soln	5
0.32	63
0.56	64

TABLE 2-2-5. τ of DQN Resists[a]

Exposing bandwidth (Å)	Resist	τ
3400–4400	AZ-111	1.3
	AZ-1370	2.0
	AZ-2400	2.0
	HR-204	1.43
3000–3300	AZ-2400	1.67
	Kodak 809	1.61
	AZ-1370	1.2
	HR-204	1.0
2800–3300	AZ-2400	0.96
	AZ-1370	1.2
	HR-204	0.85

[a] Measured values of τ (as defined in text) for $f/3$ reflective projection printer. Substrate is silicon wafer, coated with 7000 Å of wet oxide. Resist thickness is 1.0 μm with standard processing.

dose of 200 mJ/cm^2, the contrast increases (Table 2-2-6[(70)]) and in this case the energy-deposited term dominates. For AZ-1350, a contrast has been reported as high as 5.0.[(191)]

If thin (< 0.5 μm) films are used, the DQN can even be exposed in the deep UV at 248 nm. Because of the high absorbance at 248 nm, for 1-μm or thicker films, high-power excimer lasers were used for exposure. No reciprocity was noted and 1-μm images could be contact printed.[(71)] The next generation of exposure tools will probably use flash exposure for deep-UV sources. By taking advantage of the high opacity of DQN in the region 200–300 nm, Lin has formed a portable conformable mask (PCM) of a bilayer resist with a sandwich of AZ-1350/PMMA. The image formed in the top layer of AZ-1350 by near-UV exposure is transferred into the PMMA sublayer with deep-UV exposure.[(72)] Submicrometer resolution is achievable by the PCM (see Chapter 12). In other applications, the darkening of AZ-1350 or DQN to ion beams has also been proposed to serve as a cheap organic mask to replace chrome on glass. In this case, the developed-on-glass images are exposed to heavy doses of ion beams[(73)] and become UV opaque.

Finally, the image reversal of diazonovolak resists has been used to improve the resolution of positive images since the scattered forward radiation is not reproduced. This can be accomplished in three basic ways: (1) by the addition of amines to catalyze the thermal decarboxylation of the indene acid salt photoproduct, followed by normal flood exposure; (2) by the simple predevelop

TABLE 2-2-6. τ of DQN Resists[a]

λ(nm)	AZ-1350	AZ-1450	Hunt 204
248	0.7	0.7	0.85
313	3.4	3.4	1.9
365	3.6	3.6	2
436	3.6	3.6	2.1

[a] From Ref. 70.

bake of the image prior to aqueous development[228]; and (3) by the addition of azides which absorb only in the deep UV.[200]

A coupling amine (imidazole) or NH_3[217] is added to react with the intermediate indene acid photoproduct[74, 75, 218, 220] formed by high doses of > 200 mJ/cm^2,

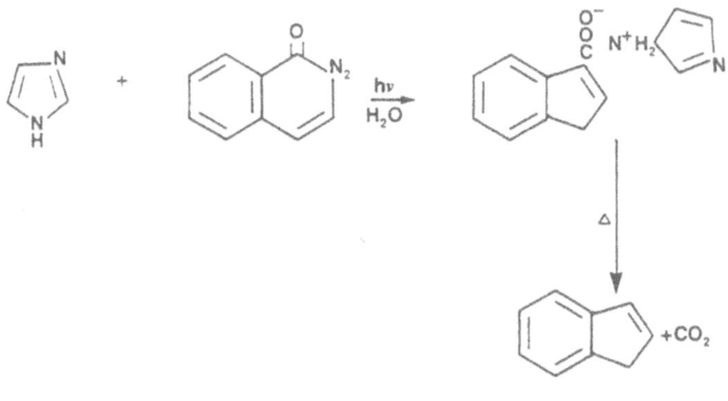

Before development, the indene acidic salt is decarboxylated by heating at 100°C. (219) The background (original unexposed) is then flood exposed to induce alkaline solubility.[74, 75] Direct exposure of AZ-1350 at 254 nm in a vacuum induces insolubilization of the exposed region due to the reaction of the ketene in the absence of water [Eq. (2-2-8)] with the novolak resin. Because of the image reversal, good lift-off profiles (Chapter 12) (see Fig. 2-2-8) can be formed by this process[166] as well as images free of standing waves.

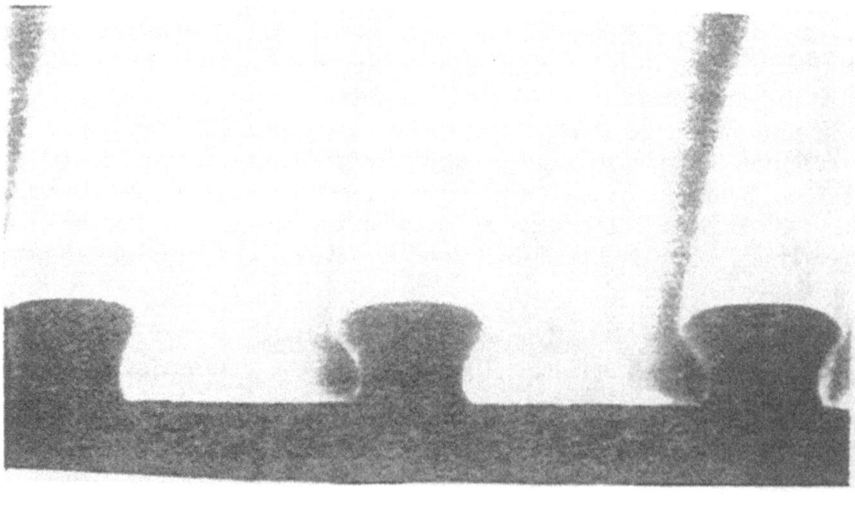

FIGURE 2-2-8. Image reversal of DQN resist into negative resist by flood exposure in vacuum at 254 nm of normally exposed but undeveloped image. From Ref. 166, courtesy of the Society of Plastics Engineers.

2-3. SHELF LIFE AND QUALITY CONTROL OF DQN

At room temperature, the shelf life of DQN resists is limited. If solutions are refrigerated, a shelf life of from 6 months up to 1 year can be expected. The user should follow the manufacturer's storage recommendations and be aware of the expiration date. Storage of prebaked films is affected by absorption of moisture which slows down the exposed and unexposed dissolution rate and leads to image residues, if the develop times are not compensated. An endpoint development monitor is an invaluable aid in protecting against such occurrences. The humidity effect is reversible. After standing in dry atmospheres, the development rate of the resist speeds up. Long-term storage of prebaked resist films is not recommended as a routine manufacturing process.

In bottled liquid resists, several thermolytic and hydrolytic reactions are envisioned to slowly occur:

1. Hydrolysis of the Cellosolve acetate to form an acid. Acids catalyze the nitrogen loss,(76)

$$CH_3COOC_2H_4OC_2H_5 + H_2O \rightarrow CH_3COOH + C_2H_5OC_2H_4OH$$

Ketone(207) or acid-free solvents promote a longer shelf life.

2. Thermolysis of the DQ followed by acid-induced azo dye formation,

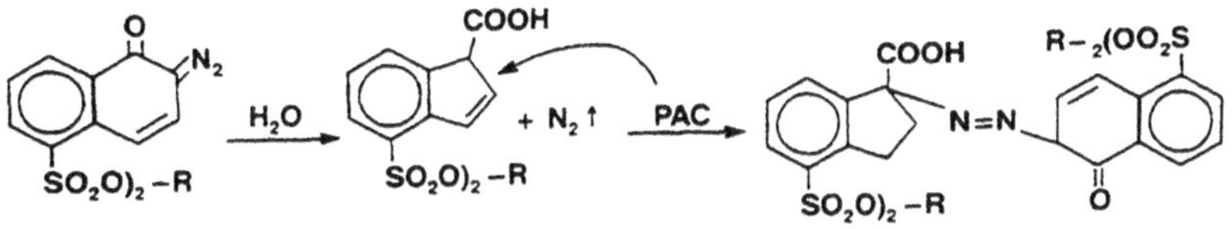

3. Azo dye coupling of the DQ and resin,

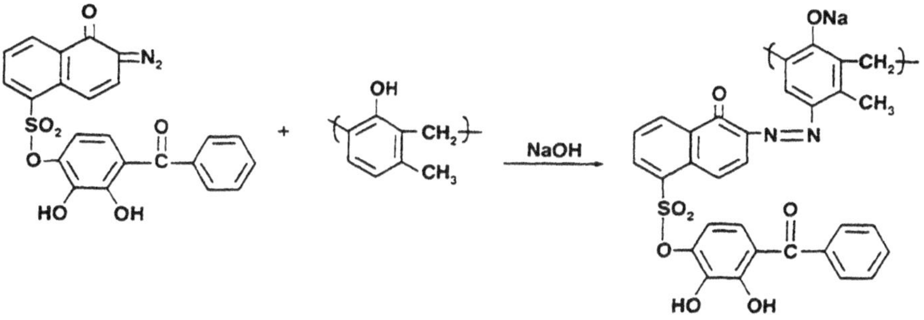

4. Hydrolysis of the sulfo-ester to form more soluble products,

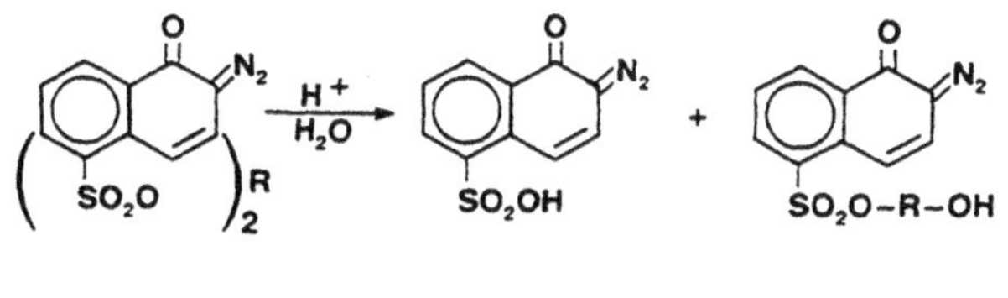

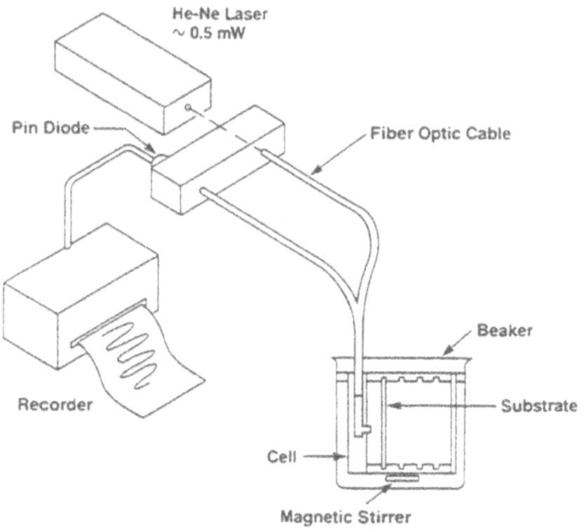

FIGURE 2-3-1. Laser endpoint of development of DQN monitor. The reflected laser light of the dissolving region produces interference patterns until the film is developed. Courtesy of Dr. C. G. Willson, IBM Research Division, San Jose, Calif.

In solution as opposed to films, the probability of bimolecular reactions is much higher. Water at 0.5% by weight of the resin or in the liquid resist corresponds to 0.3 M which is equal to the molarity of the DQ sensitizer in 1 liter of resist. Traces of acids appear to be one of the catalytic agents which leads to DQ decomposition and alteration of the photoresponse and dissolution rates. A quick test of resist aging involving no exposure is the measurement of film thinning rate in the endpoint apparatus shown in Fig. 2-3-1.[7, 77] In this arrangement, the reflected laser light monitors the film interference of both the exposed and unexposed resist. For the unexposed prebaked film, the rate of film thinning, R_0 (Fig. 2-3-1), in fresh developer can reveal loss of DQ and/or buildup of more soluble or insoluble species. If the resist is flood exposed on a silicon wafer with measured doses at a monochromatic λ (e.g., 365 nm), the exposed rate R can also be measured. Each node in Fig. 2-3-2 corresponds to a change in thickness of $\lambda/2n$ where n is the refractive index at 6238 Å (see Chapter 10). The superposition of old thinning curves with new curves indicates no changes in shelf life. Any changes in the UV spectra of DQN films cast to known thicknesses (e.g., 1 μm) on quartz can also reveal azo dye formation or the loss of DQ. In aged solutions, any change in resist performance or appearance can manifest itself in processing in many ways:

1. Alteration of R and R_0 rates or their ratio

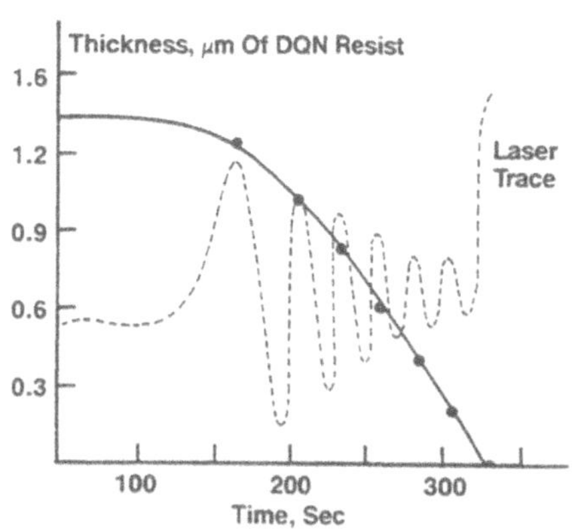

FIGURE 2-3-2. Endpoint trace of laser monitor (Fig. 2-3-1) of DQN development. Courtesy of Dr. C. G. Willson, IBM Research Division, San Jose, Calif.

2. Changes in actinic absorption (bleaching, shifts, new bands)
3. Turbidity—changes in flow, filter rate, gels or insolubles in filtrate
4. Increase in acid titer of resist
5. Residues in images after development
6. Different exposure or develop times
7. Changes in viscosity or thickness
8. Dewet of coated films
9. Nitrogen gas evolution upon opening of bottle

Since the resist formulations are proprietary, the quality control of resists rests with the manufacturer. Unfortunately, the manufacturers cannot simulate all of the exposure tools or anticipate all of the surfaces, etchants, and temperatures to which the user subjects the resist. Therefore, the user should perform functional tests to evaluate new lots of resists.

Because of variations in the rebake, exposure sources, and develop conditions, the photospeed (ASA) of a resist may vary widely. For better quality control, the film thinning (endpoint detect) apparatus should enhance the reproducibility of the develop cycle. In addition, calibrated monochromatic light sources coupled with hot plate bake at exact temperatures, should reduce the exposure and bake variance between manufacturers and user. Hopefully, the manufacturers will supply an ASA speed as a function of λ for a 1-μm-thick film coated on a known reflective Si wafer.

Some of the typical physical data supplied by manufacturers are given in Table 2-3-1. The resists are sold with different viscosity grades (centered at 3600 rpm for a specified range of spun thicknesses, e.g., 0.5–1.5 μm). The developer is generally supplied as a concentrate. Sometimes, confusion is created for the user since the chemical concentration (titer) of the developer is not given. In this case, the appropriate dilution should be stated as 1 to 4, which means one part of developer concentrate to four parts of water and not vice versa which occasionally happens. The developers can also be supplied as diluted materials but at this concentration (0.2–0.3 N) they have a shorter shelf life. Likewise, sometimes three or four different developers supplied by the manufacturer can also be used for the same resist. The best practical solution to the problem is for the user to measure the film thinning curves and ratio for the resist with a variety of developer conditions and determine the best contrast, τ [Eq. (2-1-6)], his or her required image profile and image bias (see Chapter 10). In the end, the quality control of liquid resists remains in the hands of the user, who must execute functional and analytical tests of the performance and shelf life of the product.

TABLE 2-3-1. Typical Physical Data on DQN

	Kodak 809	Hunt 206	AZ-2400
Solids, %	32	33	26
Viscosity, cS	23	41	18
Specific gravity	1.045	1.055	1.015
Percent water	<2	<0.5	<0.5
Metals, ppm	<10	<10	<10
Flash point, °C	58	46	46
Developer	809	LSI	2401

2-4. SUMMARY AND OUTLOOK ON DQN RESISTS

For microelectronic lithography, the DQN resists are the workhorse positive photoresists for the whole industry. Their major advantages are wide spectral sensitivity, excellent reactive ion etch resistance, aqueous development, and fair thermal stability. Certain deficiencies have been improved by additives, substitute resins, and posttreatments. By using a two-component mixture, the photosensitivity attributes of formulation are separated from the resin component. The DQN system firmly established the advantages of the two-component concept in a resist formulation. The photosensitivity is imparted by an organic functional group and the resist properties defined by the resin component. Spectral tuning by modification of the sensitizer structure allowed the resist to meet the needs of mid- and eventually deep-UV exposure. For the next decade, steppers operating at conventional wavelengths will continue to utilize DQN resists. Portable conformable masking can extend the resist into the deep-UV region if the image in DQN is opacified by deep-UV or chemical treatments.

As in any resist, there are specific areas of DQN resists requiring improvement:

1. *Photospeed and absorbance.* The quantum yield is fair and enhancement to 0.5–1.0 is potentially inherent. Substitute resins or narrow disperse resins are required to enhance the developer contribution to photosensitivity. For thick films ($> 3.0\ \mu m$), a tenfold lower extinction coefficient (ε) sensitizer is required. Image amplification perhaps by onium salts or acid precursors (see Section 2-8) may improve the photosensitivity to the 1–5 mJ/cm^2 range. More transparent resins are required for the deep UV. Contrasts of 4–5 are required for high resolution.
2. *Thermal stability.* The novolak resins melt at 80–120°C, and a competitive thermosetting reaction competes with thermoplastic flow to stabilize the images to 200°C. Several physical and chemical treatments are available for hardening. A more simplified approach would be to substitute high-temperature resins or resins which thermoset above the normal prebake conditions. For example, monodisperse novolaks are stable to 180°C.(222)
3. *Reactive ion etch resistance.* This cannot be expected to improve without resorting to metallic or inorganic substituents by direct incorporation in the novolak resin or by posttreatment of the developed resist via the hydroxyl group. Metal atoms incorporated into the resist will enhance the versatility and resolution of the resist in bilayer structures.(243, 247)
4. *Develop.* Dry develop will eventually be invented from the host of schemes now available. Modification of developer is another way to gain back the low quantum yield and contrast. Basic physical studies of the microfoaming contribution and physical dissolution mechanism are still lacking.
5. *Adhesion.* With the advent of dry etching, the DQN resist only has to withstand the action of its own developer. Adhesion continues to be addressed by priming the substrate and not by the modification of the resist. Because of the multifarious surfaces encountered in semiconductor technology, the adhesion problems will continue unless substitute resins can be formulated for each specific or the most difficult surfaces.

2-5. DEEP-UV EXPOSURE

2-5-1. *Optical Systems*

The major lithographic tooling costs for the mass production of VLSI circuits are associated with the optical exposure equipment. If circuits continue to shrink below 1.0 μm in size, substantial capital costs can be recovered by extending the established optical tools by:

1. Substitution of quartz or reflective optics for shorter wavelength exposure in the range 100–300 nm
2. Quartz or transparent masks and thicker chrome
3. Suitable resists with uniform 200- to 300-nm absorption (no resin absorbance as in DQN)
4. Compensation for standing waves and substrate reflectivity
5. A high-intensity source
6. Alignment better than ±0.3 μm
7. Contrast enhancing and multilayer resists

For the deep UV, new light sources are just being developed (Table 2-5-1). If excimer lasers can be retrofitted to scanning or stepper exposure systems, then the resist sensitivity will not be a factor as it is with the other sources. At present, in order to compensate for the loss of Hg intensity in the deep UV, a resist sensitivity of 1–10 mJ/cm^2 is required.

The simplest way to take advantage of the shorter exposure wavelengths in the deep UV and gain or extend resolution is to utilize proximity printing. In this

TABLE 2-5-1. Deep-UV Sources Including Excimer Lasers

Source	λ (nm)	Reference
Cd lamp	240	78
Microwave Hg	200–500	79
Cd-Hg	200–500	80
Excimer	see below	71, 81
D_2	200–300	82, 83
Xe-Hg	200–400	84
Xe flash	200–500	85
Br_2	180	85

Excimer lasers[a]

Laser gas	λ (nm)
F_2	157[a]
ArF	193
KrCl	222
KrF	249
ClF	284
XeCl	308
I_2	342
XeF	351, 353

[a] Ref. 158.

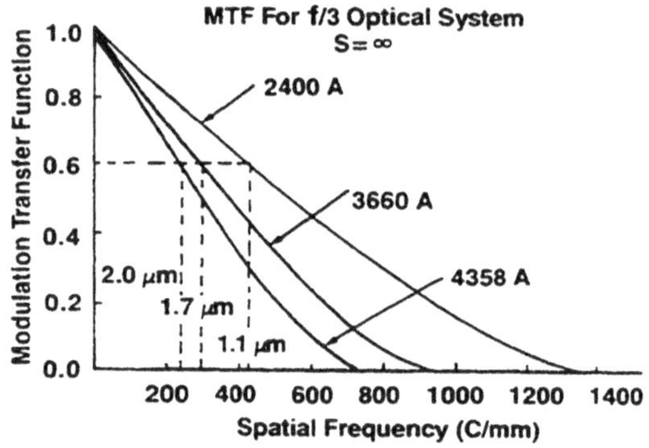

FIGURE 2-5-1. Incoherent light MTF as a function of exposure wavelength. From M. King, *IEEE Trans. Electron Devices* **ED-26**, 711 (1979), copyright and courtesy of IEEE.

exposure system, the mask is deliberately separated from the wafer by $> 10\ \mu m$. The diffraction model implies a basic limit in reproducing a square pattern of dimension X separated from an imaging surface by a distance r with an exposure wavelength of λ,[97]

$$r > BX^2/\lambda$$

For an image feature of $2\ \mu m$, and λ of $0.2\ \mu m$, a separation of $10\ \mu m$ can be tolerated. In projection printing, the resolving power of the lens is proportional to the numerical aperture by

$$\lambda/\mathrm{NA}$$

A shorter λ allows for a finer feature size at the same modulation transfer function (MTF) (Fig. 2-5-1).[88] However, the use of shorter exposure wavelength reduces the amount of tolerable defocus by

$$\lambda/2(\mathrm{NA})^2$$

leaving only a ± 4-μm focus for 1-μm features.[89, 90] In the range 200–300 nm, suitable glasses are not available for refractive optical exposure systems. Therefore, reflective mirror projection is effective in the deep UV because it minimizes chromatic aberration and because the optics does not alter the intensity of the pro-

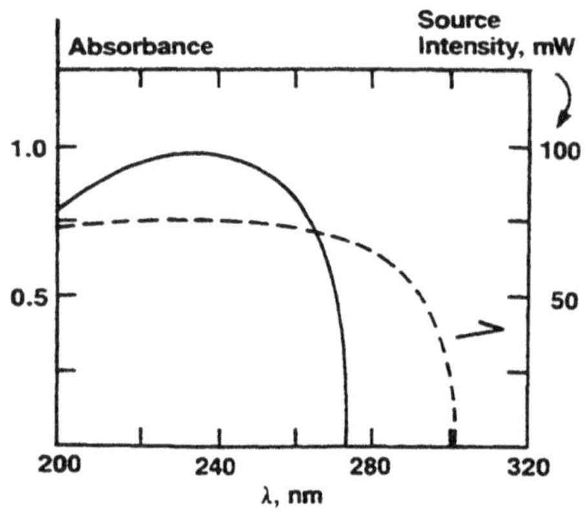

FIGURE 2-5-2. Idealized resist spectral absorbance in the deep UV and the spectral output of the source (no output above 280 nm).

TABLE 2-5-2. Absorbance and Percent Light Absorbed by a Resist Film (at a Specific λ)

Absorbance	% absorbed
3.0	99.9
2.0	99
1.0	90
0.5	70
0.1	20

jected deep-UV light.[90] Therefore, scanning projection can utilize deep-UV resists and extend resolution to the region of 0.5–1.0 μm, provided suitable flat wafers and planarized surfaces are used in device fabrication. If not, a combined deep-UV stepper and scanning tool can be expected to be engineered for this last remaining photochemical region.

2-5-2. *Resist Sensitometry*

In the deep-UV region, the extinction coefficient (ε) of most polymers and monomers increases due to more allowed transitions. For DQN (Fig. 2-1-3), the absorbance may exceed 2.0 for a 1-μm-thick film due to the combined resin and sensitizer absorbance. Both the spectral absorbance and the optical density of the resist regulate the penetration range of the exposing beam to capture the actinic output of the lamp. Ideally, both the source output and resist absorbance should be linear or constant over the desired wavelength (Fig. 2-5-2). In turn, the resist spectral absorbance (optical density) regulates the fraction of light absorbed (Table 2-5-2). The absorbance is dictated by Beer's law,

$$a = \varepsilon C l \tag{2-5-1}$$

where in a thickness l of 1.0 μm, an area of 1 cm^2 corresponds to 10^{-4} g in a volume of 10^{-4} ml or 10 M in absorber of 10^2 molecular weight. In Table 2-5-3, the effect of extinction coefficient on absorbance can be seen. Thus, the very strong $\pi \rightarrow \pi^*$ transitions of 10^4 (ε) cannot be used in a resist sensitizer without bleaching (except for ablation as in laser exposure which vaporizes its own path length.[81, 82] Although aromatic transitions at 254 nm are strong, weaker ketone $n \rightarrow \pi^*$ trans-

TABLE 2-5-3. ε and Absorbance

ε	Absorbance in 1μm film
20,000	20
2,000	2.0
200	0.2
2	0.002

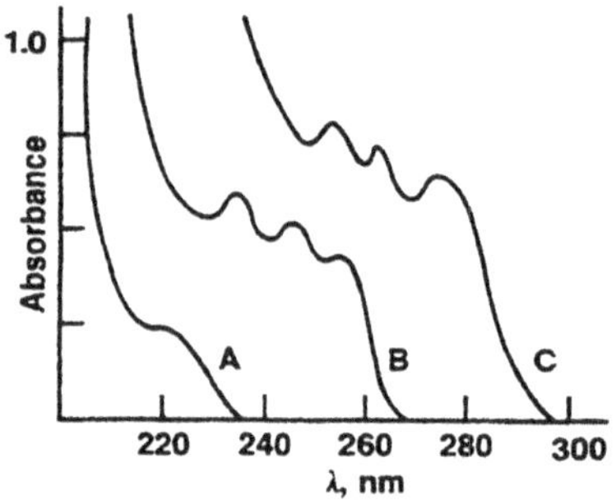

FIGURE 2-5-3. Absorption spectra of 1-μm-thick film of (A) PMMA, (B) polybenzylmethacrylate, and (C) polyphenylmethacrylate.

itions are more suitable ($\varepsilon < 1000$) but in general are of lower quantum yields (< 0.2). The effect of aryl substituents on PMMA,

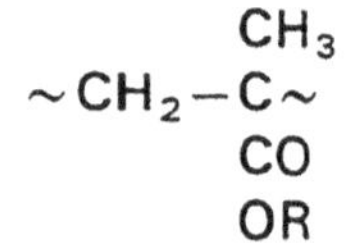

$R = CH_3, \phi, \phi CH_2$

can be seen in Fig. 2-5-3. Favorable absorbance is achieved at 254 nm for the benzylmethacrylate substituent. However, the quantum yield of methacrylates is low (< 0.03). (237) At the shorter λ of 200 nm, the number of photons per quantum is lower[(123)] than at 300 nm by about a factor of two. However, the chemical energy of each photon is higher. Thus, in the deep-UV region, the quantum yield of the resist must be greater than 0.2 to reduce (unless chemical or thermal amplification occurs) the resist doses to less than 150 mJ/cm^2. For scanning projection, a resist sensitivity of 10–50 mJ/cm^2 is desirable for sufficient throughput.

2-5-3. *Deep-UV Resists*

For the first deep-UV resist, the prime candidate was PMMA which was applied to proximity printing in 1970.[(87)] Later in 1975, PMMA was used as a deep-UV sensitive underlayer of a bilayer resist in combination with an AZ-1350 conformable mask.[(72)] Subsequently, PMMA and PMIPK were sensitized with aryl ketones to enhance their low absorbance. The speeds of these resists are low to moderate (100–500 mJ/cm^2).[(123)]

For order-of-magnitude speed enhancements, recent investigations have focused on acid-catalyzed polymer unzipping in order to overcome the cage recombination effects found in PMMA-type systems. We will examine the PMMA and PMIPK resists and the important new scission or catalyzed rearrangements.

2-6. PMMA AND PMIPK

2-6-1. *PMMA*

Polyesters and polyketones have been the most widely studied photodegradable polymers. Images are formed by the photoscission of the polymer backbone and the lowering of the molecular weight. For images with a steep profile,

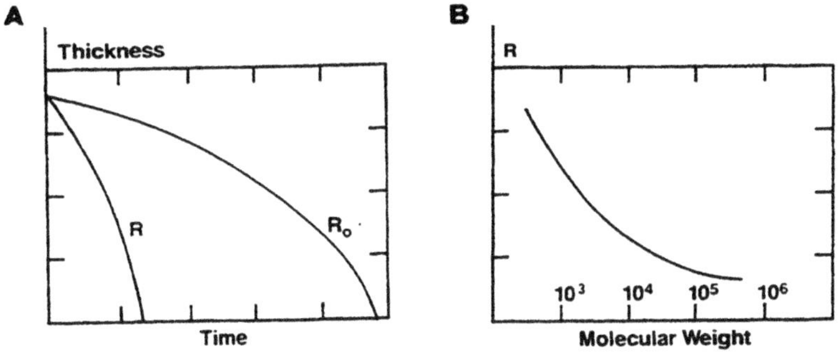

FIGURE 2-6-1. (A) Dissolution curve of exposed PMMA (R) and unexposed PMMA (R_0), (B) R as a function of molecular weight.

the exposed region should dissolve with an R at least tenfold faster than the unexposed dissolution rate R_0 (Fig. 2-6-1). The exposed and unexposed dissolution rate ratio (R/R_0) is inversely proportional to the initial (M_0) and to the final (exposed) molecular weight (M) of the polymer,

$$R/R_0 = (M_0/M)^{-a} \tag{2-6-1}$$

where a is the developer kinetic solubility parameter.

After a few chain breaks, random scission rapidly converges the dispersity of the polymer (M_w/M_n) to 2. The post-thermal-induced unzipping of the fragments is the reverse of polymerization,

$$\text{Polymer} \rightarrow \text{Monomer} \tag{2-6-2}$$

but is rarely achieved due to the chain immobility (cage effect). Unzipping is characterized by end group effects, monomer products, thermal activation, and a gradual change in molecular weight with dose.

In polyesters and polyketones, the direct absorption of energy results in Norrish type I and II cleavage of the main chain,[91, 92]

Type I

Type II

or side chain scission,

Type I

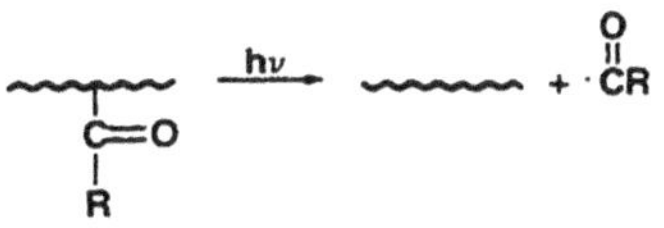

Type II

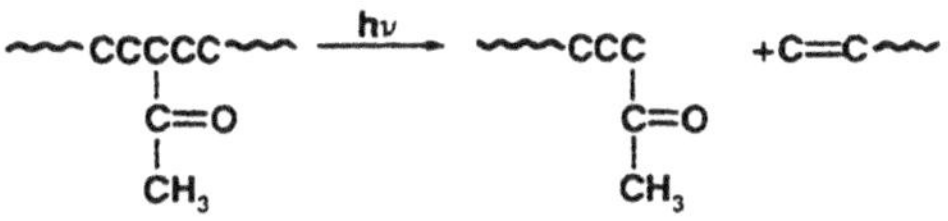

In solution, the quantum yields of chain scission can be 0.2–2.0, but in the solid state, the chain scission can be an order of magnitude lower due to the recombination of radicals in immobilized chains.[93] The propagating radical of photodegraded PMMA,

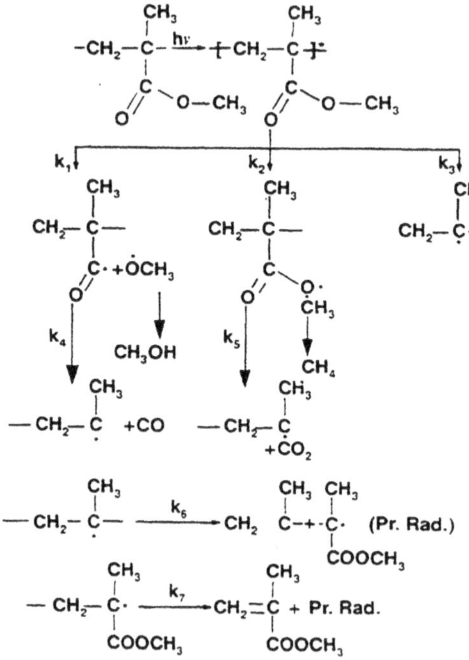

is formed by side chain rearrangement of a tertiary radical (k_6 and k_7) while side chain scission (k_1–k_5) form methanol, methane, and formate products.[94] The following radicals have been detected by electron spin resonance (ESR): CH_3, CHO, CH_3CO, $COOCH_3$, and $(CH_3)_2CCO_2CH_3$.[94, 95] Hiraoka[95] found the following products for PMMA photolysis at 254 nm:

Product	Quantum yield
H_2	0.06
CH_4	0.3
CHO	2.2
CO_2	2.2

No monomer was found although others report complete monomer formation with thick films and extensive exposure especially above the T_g (glass transition) of PMMA.[93] In PMMA, the four major reactions are:

1. Random main chain scission,

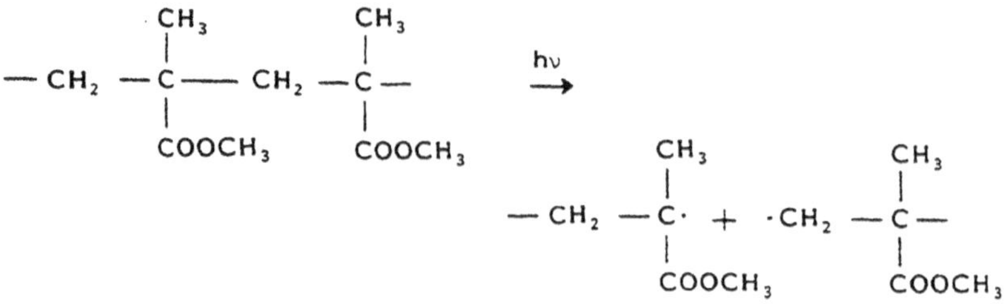

2. Photolysis of the ester side chain,

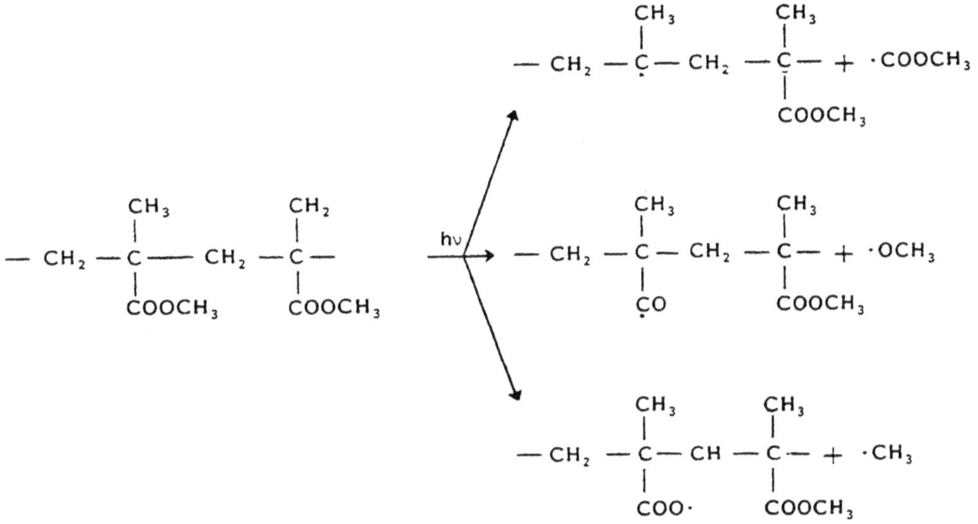

3. Photodissociation of α-methyl group,

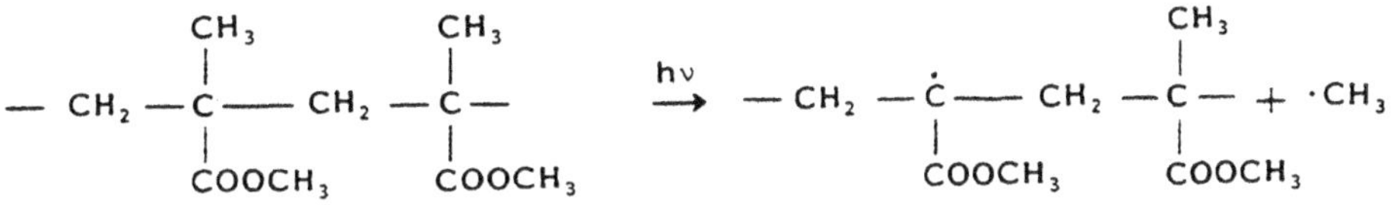

4. Unzipping of radical chain,

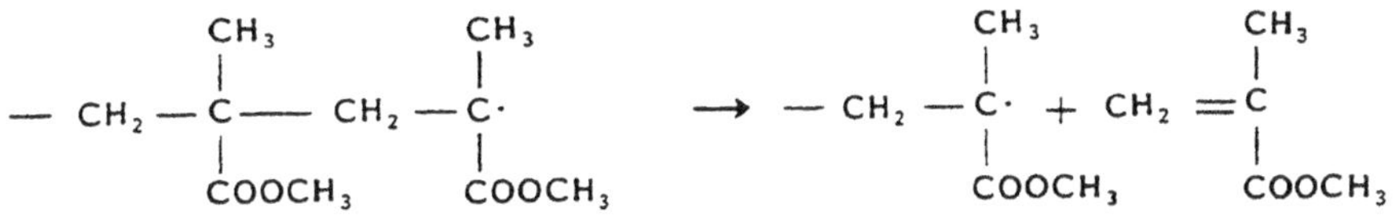

In deep-UV vacuum exposure, the products formed are methyl formate, methanol, and methyl methacrylate with quantum yields of 0.14, 0.48, and 0.20, respectively; and in air: methane, hydrogen, CO_2, and CO.[96] Under the same exposure conditions, the quantum yield for chain scission was 0.004 in air or vacuum.[96, 177] Intense deep-UV radiation from excimer lasers instantly volatizes PMMA into gaseous products.[175, 185–188, 204] The quantum yield of main chain scission is determined from the ratio of the initial number average molecular weight, M_i, to the final, M_f, molecular weight by the relation,

$$M_i/M_f - 1 = \phi_{cs} I_{abs} t C_0 \qquad (2\text{-}6\text{-}3)$$

where C_0 is the number of polymer chains initially present. The quantum yield (ϕ_{cs}) for main chain scission is only 0.04.[94, 224] while Fox also found a ϕ of 0.04, he determined a monomer quantum yield of 0.2, of approximately five unzippings per chain of radical intermediate. The total quantum yield of gaseous products,

$$\phi_{CH_3} + \phi d_{CH_3OH} + \phi_{HCOOCH_3} = 1.0$$

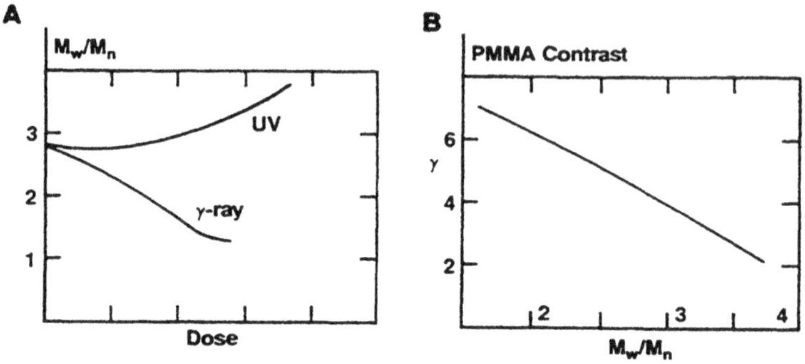

FIGURE 2-6-2. (A) Effect of γ (nonionizing) and deep-UV radiation on the dispersity (M_w/M_n) of PMMA with dose. (B) The contrast of PMMA as a function of initial polydispersity. From Ref. 97, copyright 1982 and reproduced by permission of John Wiley & Sons, Inc.

indicates that the PMMA excited state is dissociative and undergoes bond cleavage with unit quantum efficiency but recombination occurs upon immersion in the solvent developer and drastically reduces ϕ_{cs}. For imaging purposes, a tenfold decrease in molecular weight is required. Harada and co-worker(97) found that the dispersity after UV exposure did not decrease (Fig. 2-6-2). If the dispersity does not decrease, but increases, the resist contrast also increases (Fig. 2-6-2). They also found that *e*-beam exposure decreased the dispersity to 2. However, for random chain scission, the probability of scissions per chain increases with chain length. Viswanathan and Santini found that an initial higher molecular weight produced a higher solubility ratio at the same dose (Fig. 2-6-3).(98, 99) Likewise, Van Pelt(80) contended that while the molecular weight (Fig. 2-6-4) decreased by a factor of 10, the solubility rate increased by 10^2. The increase in solubility rate was postulated to be due to microfoaming by the trapped or volatilizing products.

Since PMMA is a slow deep-UV resist requiring doses > 250 mJ/cm^2, several means of enhancing the sensitivity have been investigated:

1. Increase the temperature of the resist above T_g to promote unzipping.(97, 99) The contrast also increases from 2 at 25°C to 7 at 140°C (Fig. 2-6-5), implying a convergence of dispersity.(97)
2. Increase the spectral absorbance by incorporation of photosensitizers such as *t*-butyl benzoic acid (Table 2-6-1 and Fig. 2-6-6). A side benefit of the 10% loading of aryl(192) photosensitizers is the reduction of the PMMA plasma erosion rate by a factor of 3.(80)

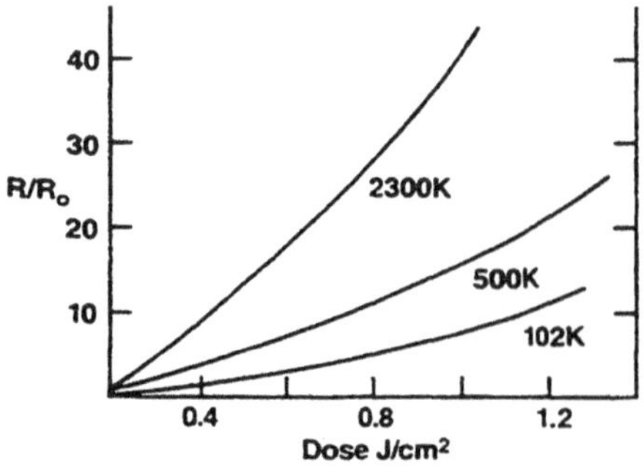

FIGURE 2-6-3. Solubility ratio of dissolution rate of exposed (R) to unexposed rate (R_0) for PMMA of various initial M_w after exposure to deep-UV (250 nm) light of increasing dose.(98,99)

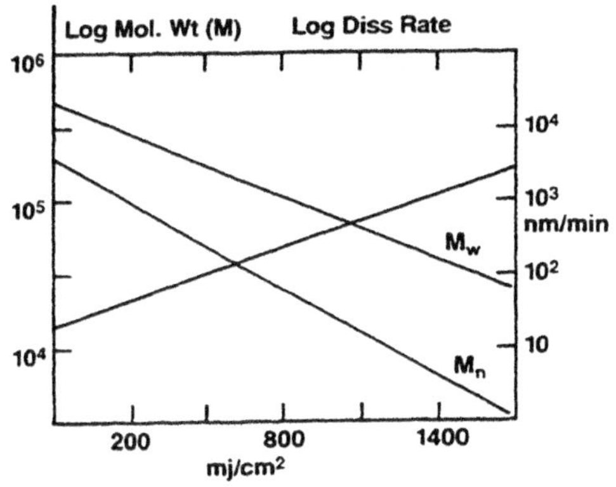

FIGURE 2-6-4. Molecular weights and solubility rates for deep-UV-exposed PMMA films. From Ref. 80, courtesy of SPIE.

3. Incorporation of comonomers with methyl methacrylate into the PMMA chain. Early work by Amerik and Guillet[104] demonstrated a tenfold enhancement of chain scissions by incorporation of methyl vinyl ketone into PMMA.[118, 246] A cyclic energy transfer was envisioned to occur for the ketone Norrish type II scission,[92]

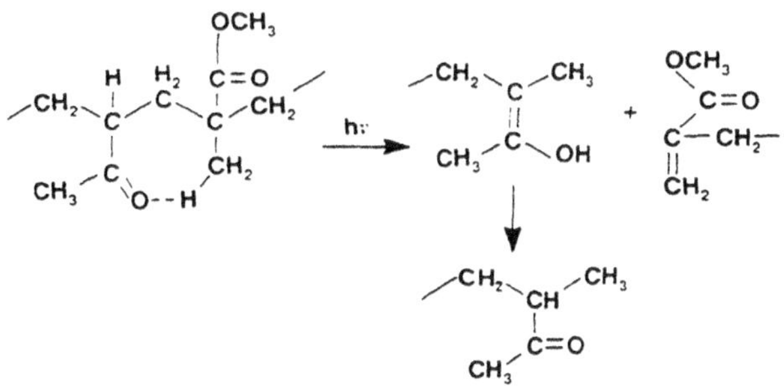

In this case, an intramolecular H-abstraction is favorable for atom transfer. Table 2-6-2 lists some of the speed enhancement achieved by PMMA copolymerization. The indenone-PMMA is a good example of increased

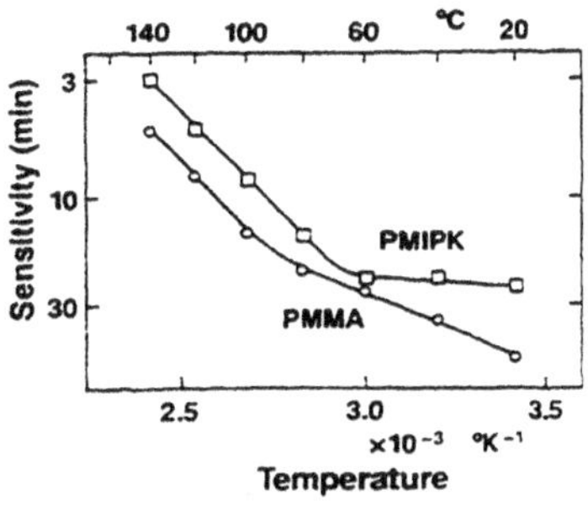

FIGURE 2-6-5. Temperature effect of exposure of heated films of PMMA and PMIPK exposed to deep-UV (254 nm) radiation on the sensitivity as measured by the exposure time required for imaging. From Ref. 97, copyright 1982 and reproduced by permission of John Wiley & Sons, Inc.

TABLE 2-6-1. Photosensitized PMMA

Additive	Dose (mJ/cm^2)	Reference
t-Butyl benzoic acid	150	100, 101
n-Acetyl carbazole	150	102
o-Nitrobenzoyl carboxylate	100	107
Aryl polymers	400	80

sensitivity obtained by attaching the ketone group to the main chain and not relying upon side chain scission,[103, 107]

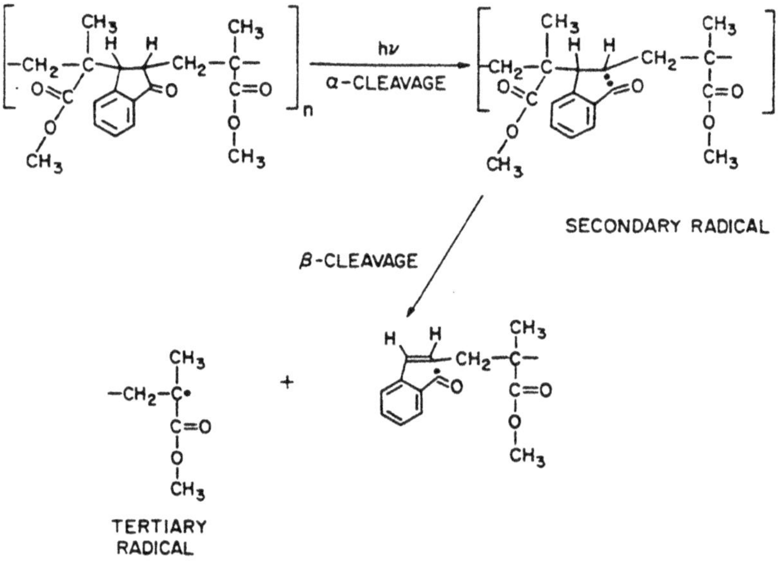

4. Use of pre-cross-linked PMMA. Instead of focusing upon increased scissions and increased R in the ratio of R/R_0, the sensitivity can also be increased if R_0 can be made smaller. Pre-cross-linking PMMA through

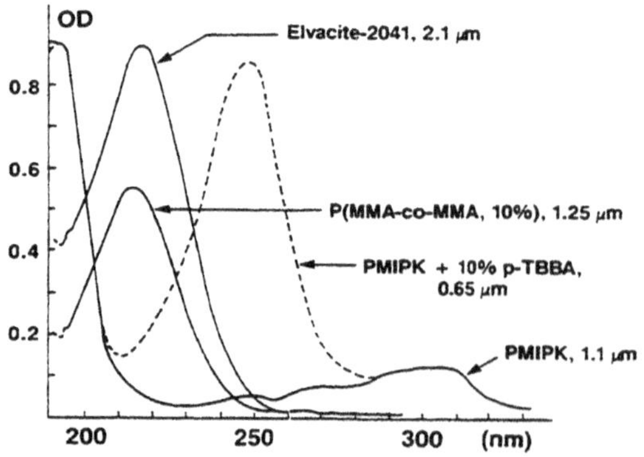

FIGURE 2-6-6. Absorption spectra of PMMA, PMMA-methacrylic acid (MMA-co-MMA), and PMIPK with and without *t*-butyl benzonic acid. From Ref. 80, courtesy of SPIE.

TABLE 2-6-2. Enhanced PMMA UV Sensitivity

Copolymer	Sensitivity increase	Reference
Methyl vinyl ketone	10 ×	104–106
Methyl isopropenyl ketone	7 ×	105
Indenone	20 ×	103, 113
α-Chloro-MA	7 ×	108
Phenyl isopropenyl ketone	10 ×	109
Acyloximo-MMA	10 ×	102, 111
Benzophenone-oxime	3 ×	110
Phenyl MA	5 ×	112

succinic anhydride links to insolubilize the unexposed region, has been used with methacrylic acid[202] and methacrylyl chloride,[114]

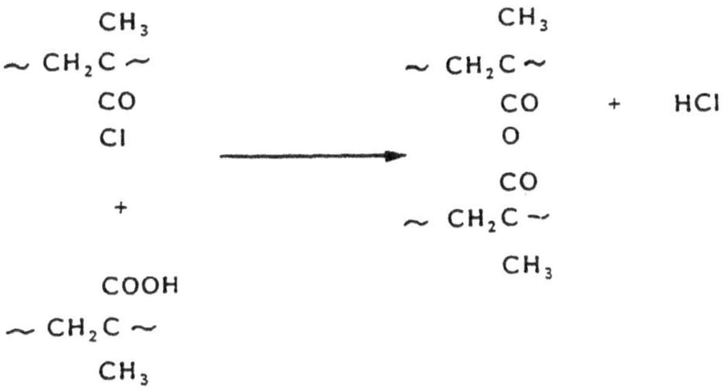

The unexposed resist is insoluble in the developer and behaves like a negative resist, i.e., it can be overdeveloped without thinning. The intermolecular cross-link can also be obtained from the precursors of glycidyl methacrylate[115, 116] or *t*-butyl methacrylate.[117] However, no major speed increase is noted, since main chain scission has to occur to a greater extent than rupture of the cross-link, unless higher absorbing chromophores (X) can be built into polymer cross-links,

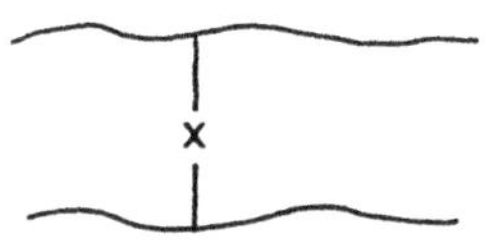

If only one intermolecular cross-link could be synthetically incorporated, a minimum of only two scissions would be required to release the original linear backbone (main chain).

2-6-2. *PMIPK*

Polymethyl isopropenyl and phenyl ketone are scissioned by a Norrish type I reaction,[110, 169, 246]

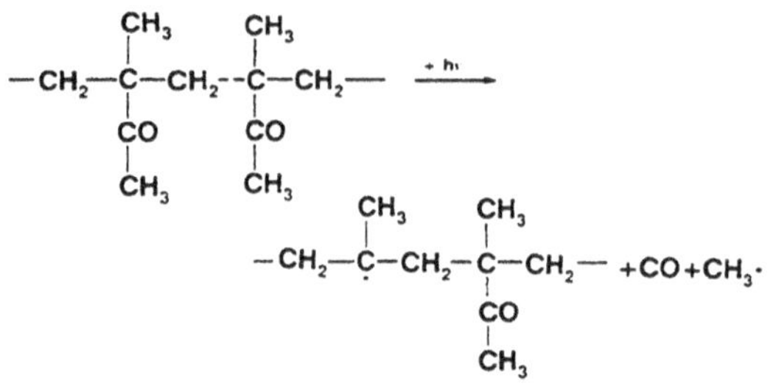

followed by a radical propagation,

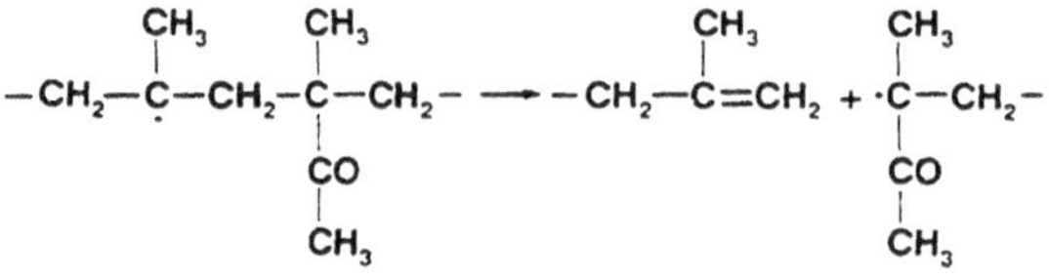

and further unzipping at elevated temperature,

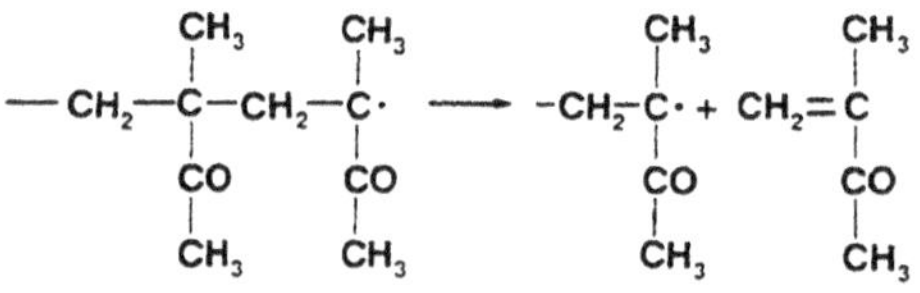

The extinction coefficient (ε) of PMIPK at 254 nm is tenfold higher than that of PMMA and quantum yields are an order of magnitude larger.[80] For exposure above the T_g, the quantum yield for chain scissions increases rapidly (Fig. 2-6-7).[93] However, in spite of the increased absorption in PMIPK, the net absorbance is still too low at 254 nm (Fig. 2-6-6), < 0.1, and photosensitizers are required to absorb a reasonable quantity of incident light. Increased absorbance and sensitivity (Fig. 2-6-6) are accomplished by the addition of PMMA-type sensitizers (Table 2-6-3). Copolymerization with phenylisopropenyl ketone also

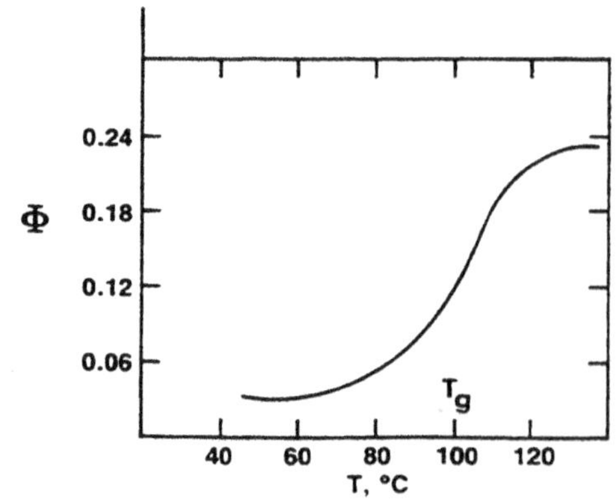

FIGURE 2-6-7. Quantum yield (Φ) for chain scissions of bipolymer of PMMA-methyl vinyl ketone as a function of the temperature of the irradiated film. From Ref. 92, reproduced by permission of the American Chemical Society.

TABLE 2-6-3. PMIPK Sensitization

Sensitization	Dose (mJ/cm²)	Reference
Copolymer *p*-MeO-PhIPK	80	109
t-Butyl benzoic acid	80	80, 100
Benzophenone	95	119
Benzoic acid	75	120

achieves good sensitivity (50–100 mJ/cm²).[105] Triplet energy transfer has been postulated as the major photosensitization pathway,[100]

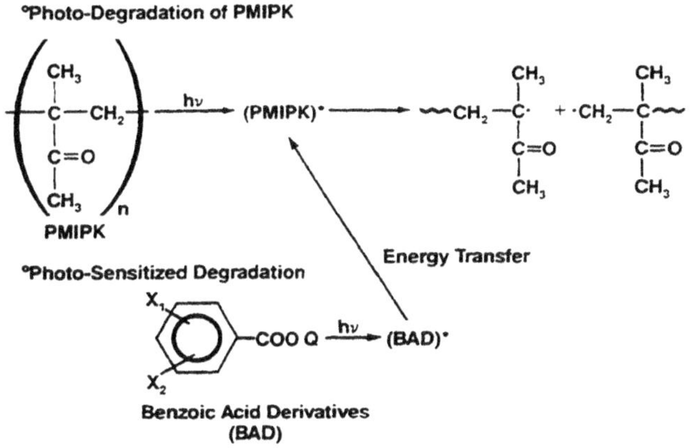

Since oxygen quenches the triplet state of many ketones, it should be excluded during PMIPK exposure.[121] The highest PMIPK sensitivity is obtained with the largest initial molecular weight.[119] In addition, the plasma resistance of PMIPK is about twofold better than that of PMMA due to the internal cyclization upon heating to stabilize the polymer chain against further thermal- and plasma-induced chain scission,

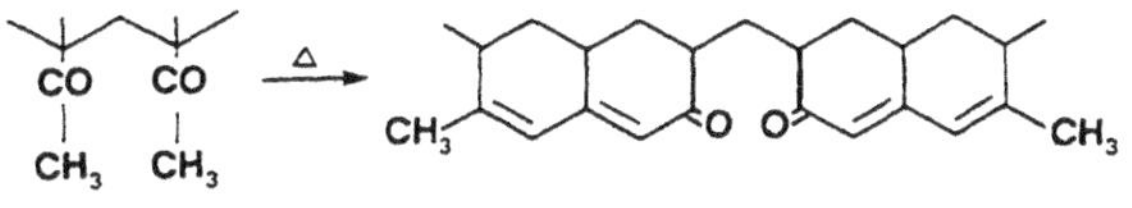

In addition, the internal cyclization prior to exposure has been used by Hiraoka and co-workers,[159] to not only heat-stabilize polymethacrylonitrile (PMAN) but to enhance the deep-UV absorbance by extending the conjugation. Heating a copolymer of PMAN and polymethacrylic acid produces intramolecular cyclization and insolubilizes the unexposed resist for deep-UV exposure,

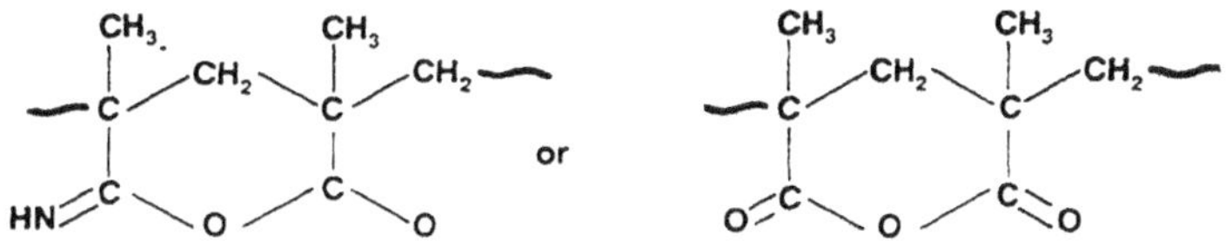

Heating a polymethacrylic acid film in the presence of NH_3[139] or a powder of PMMA and ammonia[221, 226] produces polyglutaramide,

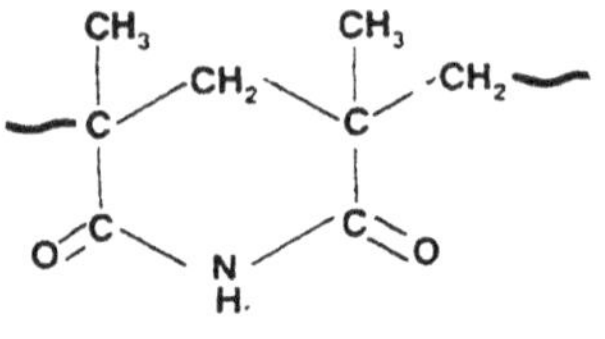

The polyglutaramide is more melt resistant (T_g of 160°C) than PMMA and is finding applications as a PCM underlayer[221] or as a replacement polymer for novolak resin mixed with a diazonaphthoquinone sensitizer. The cyclized polymers also exhibit about a sevenfold better plasma resistance than PMMA.

2-6-3. *Summary*

PMMA is a classic chain scission system with poor to fair sensitivity due to the low-extinction chromophores and low-quantum chain scissions. In PMMA, the quantum efficiency could be as high as 0.5 if the large recombination in the cage effect did not lower the amount of permanent scissions. To reduce the recombination upon development, the trapped radicals in PMMA require postpolymerization with imbided monomer grafting[122] or an *in situ* reaction with additives to prevent recombination. Further enhancement of sensitivity can be obtained by copolymerization with ketones or functional groups which scission in the main chain (not side chain).

Although PMIPK has an order of magnitude higher quantum yield than PMMA, because of its low extinction coefficient, it requires a photosensitizer compensation. In spite of the added photosensitizers, the sensitivity is still too low (100 mJ/cm^2) to be useful as a single film deep-UV resist. Improved thermal and plasma stability can be obtained in PMMA and polyketones by precyclization to glutaramide and methacrylonitrile derivatives. These systems will find applicability as sublayers in portable conformable bilayer resists.

2-7. NITROALDEHYDE RESISTS

The rearrangement reactions of *ortho* aryl nitro compounds are well-known reactions used for actinometry in the 250- to 350-nm region with a high quantum yield of 0.5,[94]

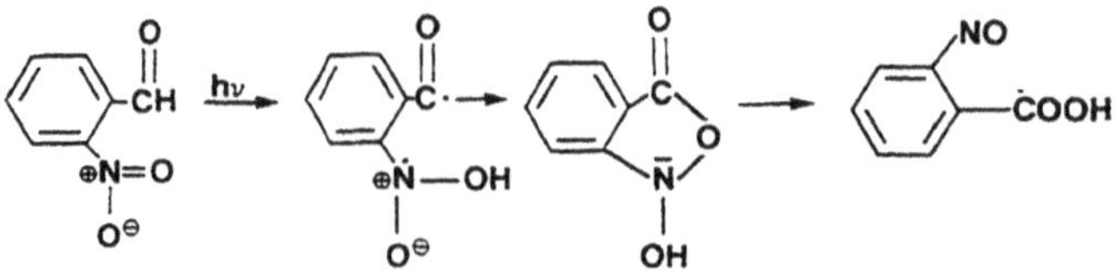

Incorporation of the *ortho*-nitrobenzyl ester onto a polymeric acrylate side chain produces a water-soluble acid, suitable for dry film resists,(124–127, 132, 189, 244)

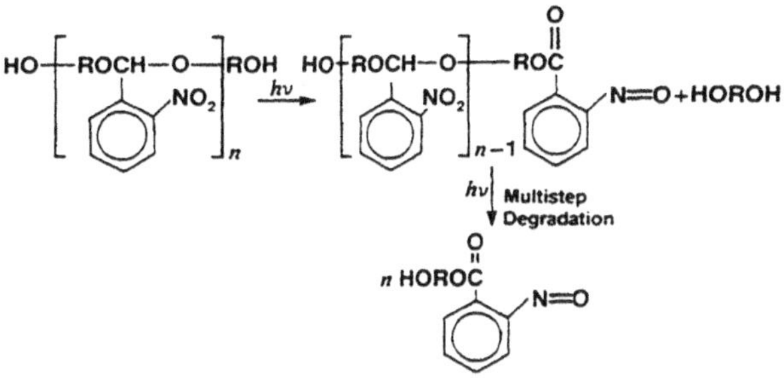

Unfortunately, the nitroso product also absorbs at the same wavelength to reduce the exposure efficiency. However, this inner filter effect has the side benefit of reducing the backscattered distortion by standing waves reflected from the substrate. If the *ortho* nitro group is incorporated at the end of an unzippable polymer (an unzippable polymer is one which is thermodynamically unstable at room temperature due to a low ceiling temperature) such as a polyaldehyde, end group cleavage produces a secondary image amplification (especially if heat is applied),(128, 129, 131)

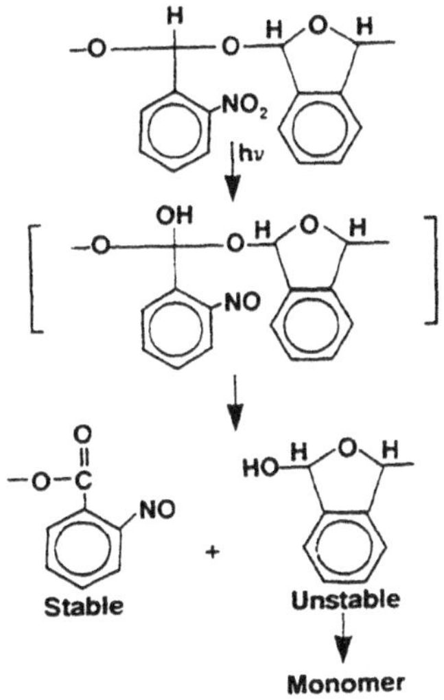

Due to the postunzipping reactions, only low doses of 25 mJ/cm^2 are required to initiate depolymerization.

2-8. PHOTOCATALYZED POSITIVE RESISTS

Polymers can be synthesized by three major reactions:

1. Free radical addition,[163]

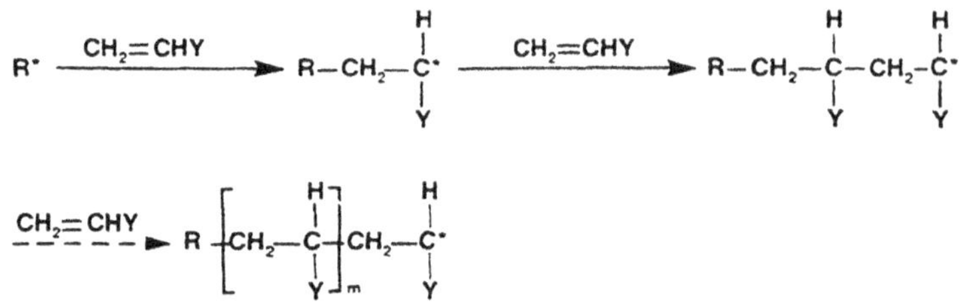

2. Anionic initiated addition,

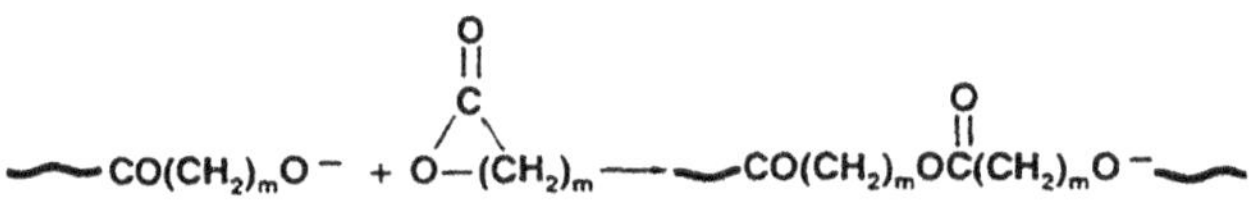

3. Cationic ring opening polymerization,

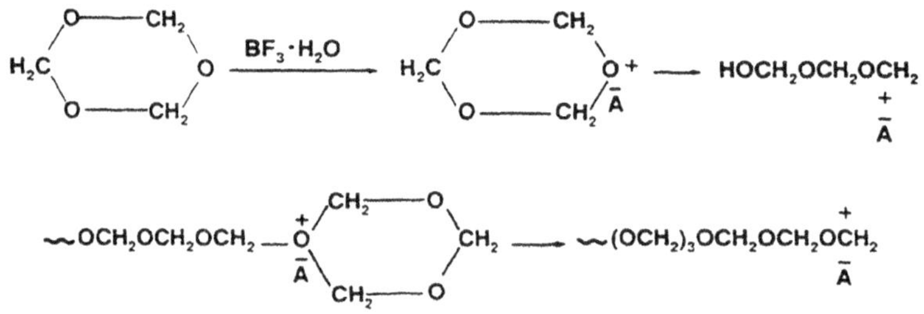

All of the above polymerization reactions can be photochemically reversed into useful imagery for positive photoresists,

$$\text{Polymer} \rightarrow \text{Monomer}$$

For example, PMMA and PMIPK are formed by free radical addition. Once the polymer chain is scissioned the 10^n or more links, it can be depolymerized (unzipped) by the action of heat, acid (cations or anions) if the decomposition is kinetically favorable. Thus, one photon could be amplified by 10^n if the scissioned fragment reverted to a monomer,

Dry development would also be possible as well as high-speed resists with doses less than 10 mJ/cm^2.

Polyaldehydes which are formed by free radical addition, cationic, or anionic initiation,

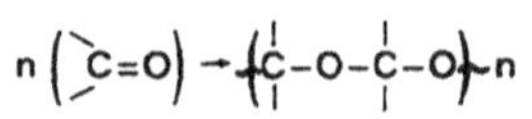

have low ceiling temperatures (−40°C) but are stable if end capped, e.g., Delrin (DuPont polymer). If the chain is scissioned by radiation, heat, or chemical agents (acid), they will revert back to monomer. This type of reversion has formed the basis of potential dry film, dry develop, and high-speed resists (Table 2-8-1). Ito and Willson[130, 131] describe a photosensitizer for unzippable polymers of triaryl onium salts (S),

$$\underset{\text{I}}{ArN_2^+X^-} \qquad \underset{\text{II}}{Ar_2I^+X^-} \qquad \underset{\text{III}}{Ar_3S^+X^-}$$

$$X^- = BF_4^-,\ PF_6^-,\ AsF_6^-,\ \text{and } SbF_6^-$$

which upon photolysis[162] (200–330 nm) and at high quantum yields (0.8)[195] release Lewis acids and hydrogen acids[143, 144]

$$Ar_3S^+X^- \xrightarrow{h\nu} [Ar_3S^+X^-]^{\cdot}$$

$$[Ar_3S^+X^-]^{\cdot} \longrightarrow Ar_2S^{+} + Ar\cdot$$

$$Ar_2S^+X^- + R\text{-}H \longrightarrow Ar_2\overset{+}{S}H\,X^- + R\cdot$$

$$Ar_2\overset{+}{S}H\,X^- \rightleftharpoons Ar_2S + H^+ + X^-$$

to further acid (A) catalyze the decomposition of polyphthalaldehyde (I),[203]

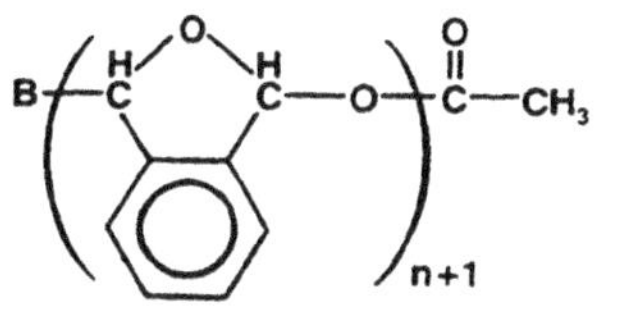

I

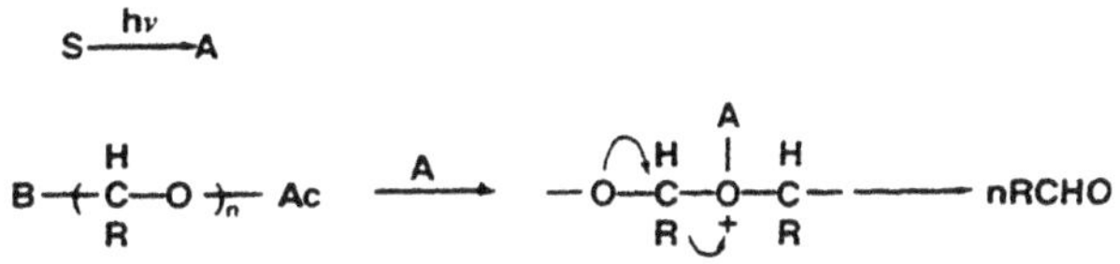

TABLE 2-8-1. Acid-Sensitized Positive Resists

Sensitive polymer	Acid source	Reference
Polyacetaldehyde	β-Naphthol	133
	Nitrobenzaldehyde	134
	Diazonium BF_4	135
	Polyvinylchloride	136
	Halomethanes	137
	Triarylarsenonium salts	130, 131
	Photooxidants	138
	Nitrobenzaldehyde	129
	Electron beam	139
Polyacetals	Diazonium salts	140
Polyorthonovolaks	Trichlorotriazine	141
Polyacyliminocarbonates	Trichlorotriazine	142

Doses as low as 2–5 mJ/cm^2 in deep UV could be used, representing a chemical amplification of 50 for a unique self-developing resist. In a further modification, main chain polycarbonates are scissioned into gaseous fragments by onium photocatalysis,[231,240]

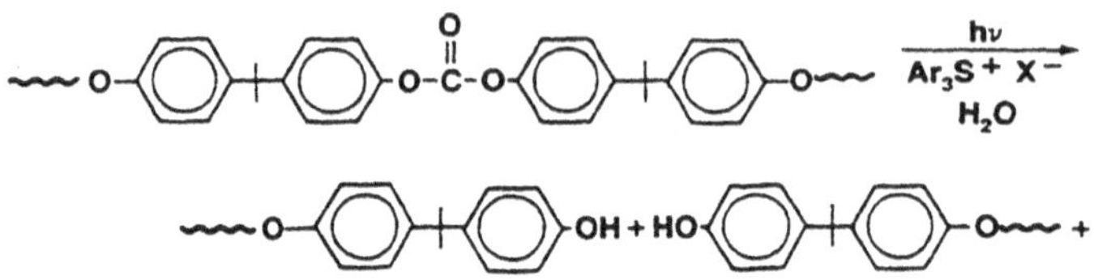

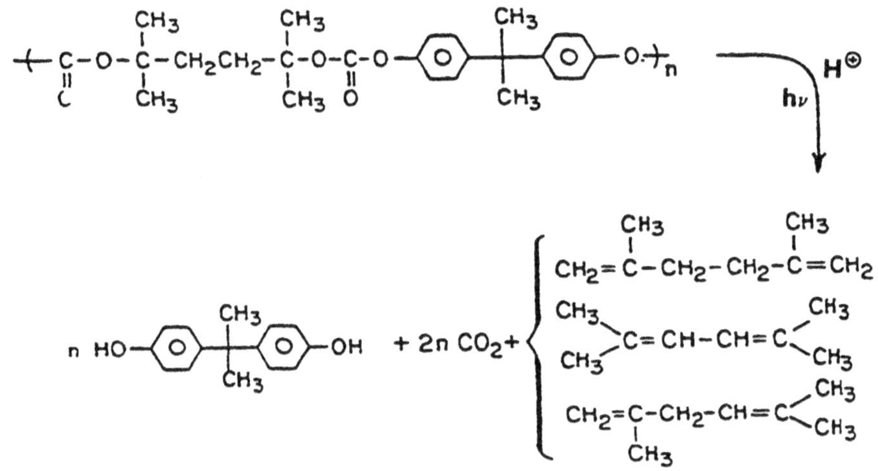

The image amplification demonstrated by acid catalysis represents an outstanding example of reverse synthetic engineering. Other catalyzed decompositions with new photoprecursors are likely to meet the challenge of formulating a 1 mJ/cm^2 deep-UV resist.

Instead of main chain scission, Ito and Willson also used the acid-catalyzed (onium salt) scission of *t*-butylcarbonato side chains of polyvinylphenol to form an alkali-soluble product[131] at doses of under 25 mJ/cm^2,

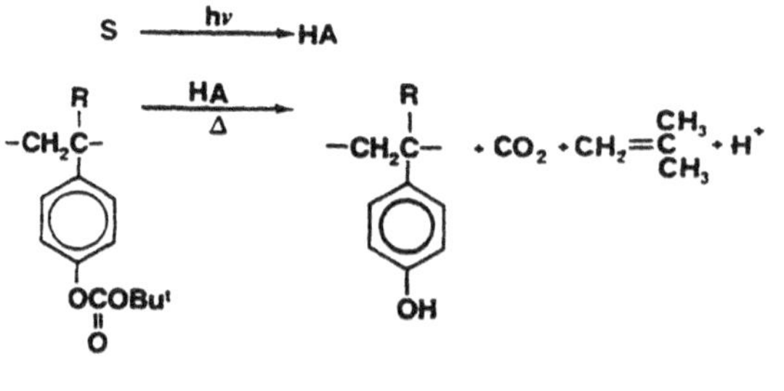

Later, main chain polycarbonates were used as resists.[235,240,248] Image amplification occurs since the acid is regenerated from the decomposing isobutene cation intermediate. Negative or positive images (nonswelling) can be formed with the appropriate nonpolar or polar developers.[245,246,250] Likewise, poly-*p*-formyloxyl-styrene decarbonylates to *p*-hydroxystyrene[205] and *p*-siloxystyrene to *p*-

hydroxystyrene.[216] Another potential resist which upon photolysis generates isobutene and carbon dioxide is poly-*t*-butyl vinylcarbamate,[152]

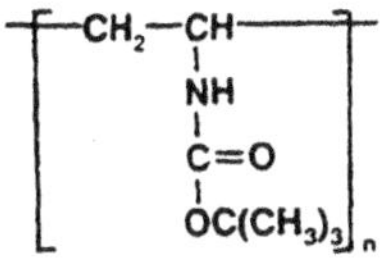

Crivello has pointed out[144] a host of polymers such as polyethers,[216] esters, acetals, and carbonates that could be scissioned (as well as cross-linked in the case of epoxies) with photogenerated acids,

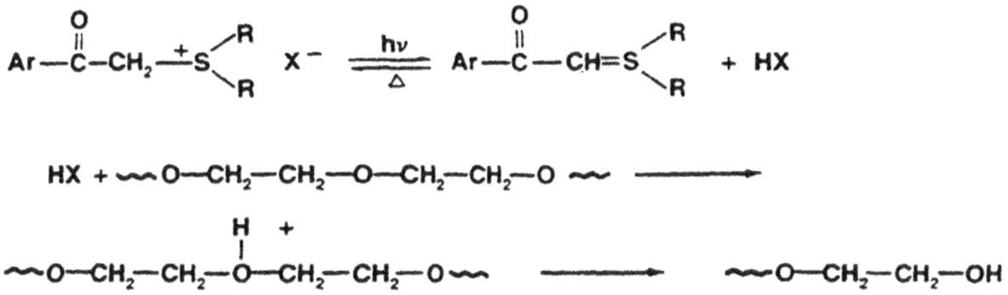

Historically, workers at GE had used photogenerated acids to bypass the resist step altogether and generated acids such as HF and HCl directly from photosensitive precursors dispersed in polymeric films (see page 881),[170, 223]

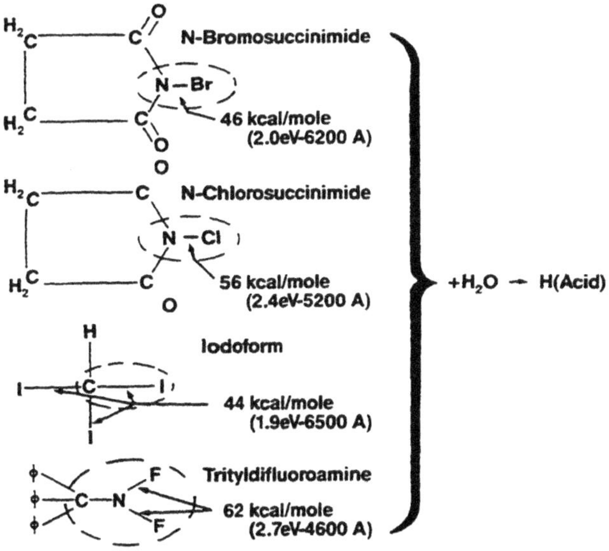

Orthoesters have been made water soluble by the photolysis of polychloro compounds which give rise to acid catalysis,[22, 132, 141, 142, 151]

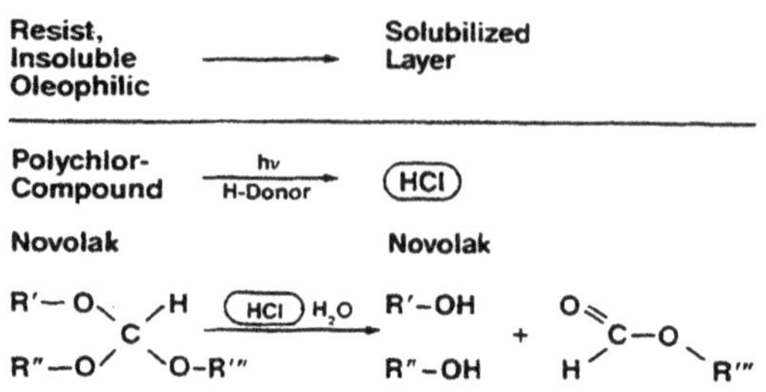

Other polymers formed by ring[163] opening polymerization are potential candidates for acid catalyzed unzipping (image-amplified positive photon or radiation resists). Some examples are as follows:

1. Polypropane sultam

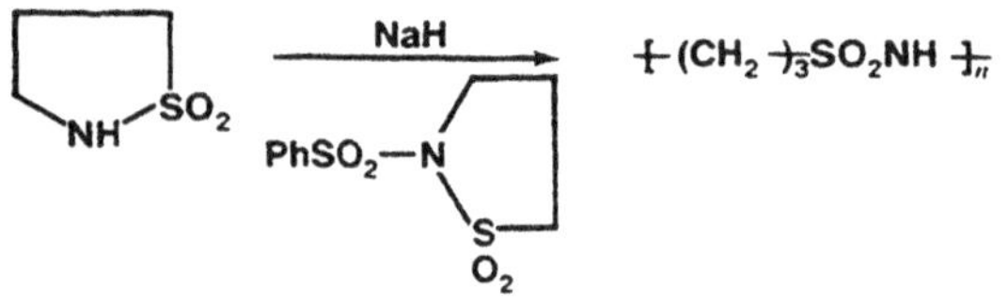

2. Polycarboanhydride

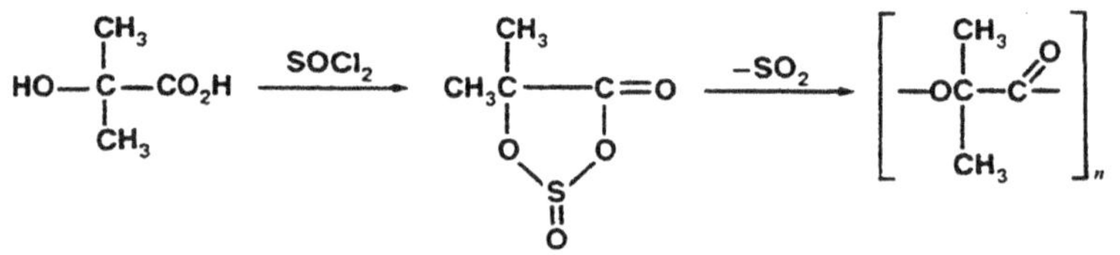

3. Polylactone

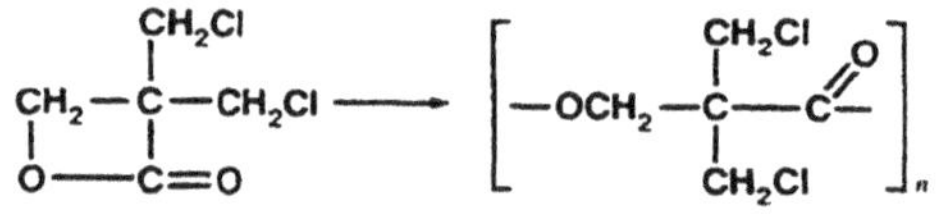

4. Polyhydroxyacetic acid

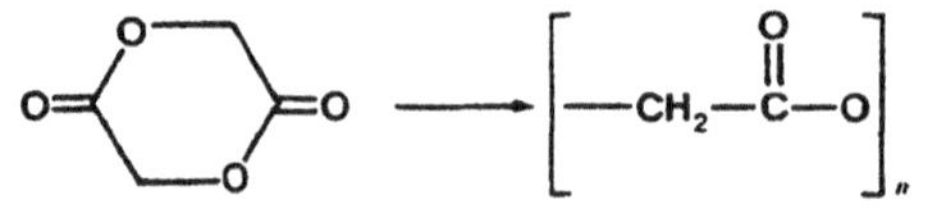

5. Polyoxetane

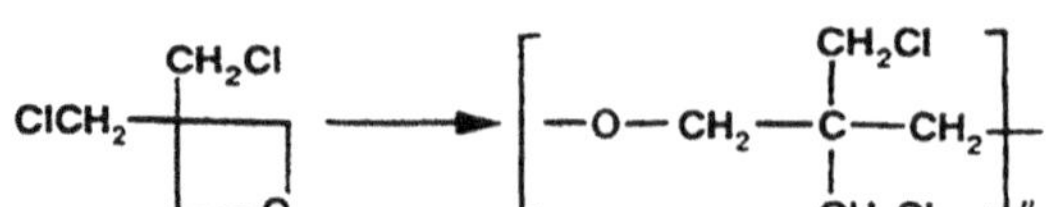

6. Polydimethylsilylene[171, 173, 215]

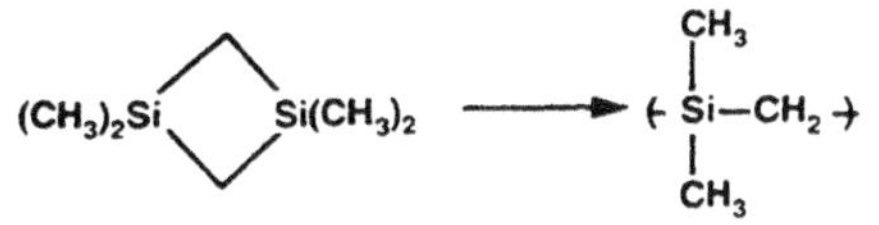

The polysilanes are useful as barrier layers or as top surface bilayer images.[173, 174, 201] Direct exposure with an excimer laser vaporizes the polysilanes.[210, 215] For other silicon-based positive photoresists, the utilization of the side chain elimination in preference to "cage"-inhibited main chain scission, was applied to a positive (oxygen dry develop) methacrylate ester of trimethylsilyl methylene acetophenoneoxime,[153]

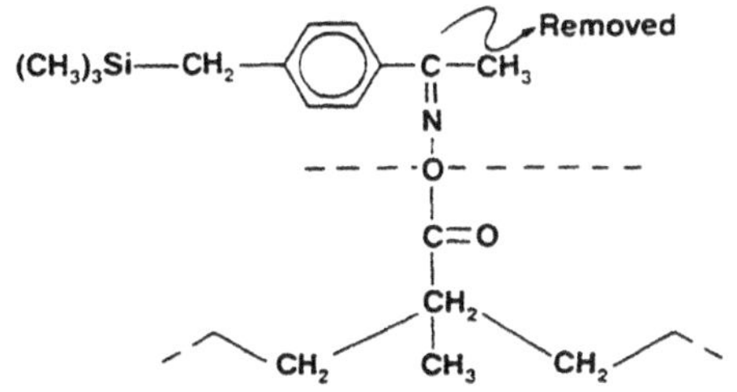

The release of silicon fragments after a 250 mJ/cm^2 deep-UV dose (followed by heat volatilization) provides an oxygen plasma developable resist. An oxygen plasma develop positive resist has also been derived from posttreatment of KTFR (azide-rubber resist) with tin and boron halides.[157]

2-9. MISCELLANEOUS POSITIVE DEEP-UV RESISTS

The degradation of polymer chains by photooxidation has been used for imaging polystyrene[192] and poly-*p*-xylylene,[154, 155]

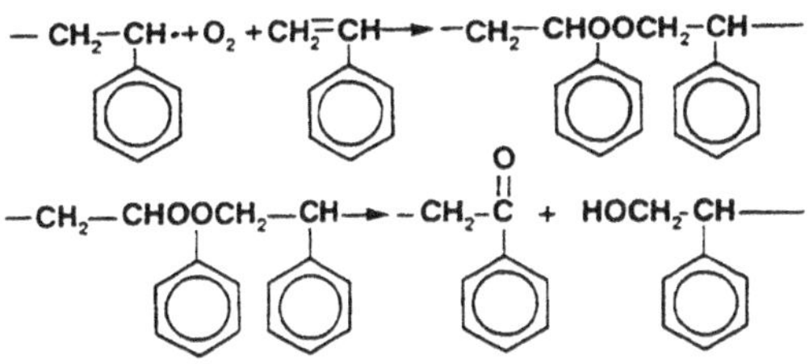

The interception of the scissioned chain by oxygen and peroxides can serve as the basis of image amplification if the peroxides are used for postgrafting reactions by treatment of the image with monomers,[122]

$$\text{Polymer} \rightarrow \text{Peroxides}$$

$$\text{Peroxides at surface} + \text{monomer} \rightarrow \text{Grafting}$$

Another system which uses oxygen is the photooxidation of organic sulfur compounds such as thioamides in a urea-formaldehyde resin base. The exposed area is removed by the action of alkali.[156] Photochromic spiropyran[157] resists and bisimidazoles[166] have been made into positive working resists. Inorganic photoresists are just being explored. Many photographic halides or salts decompose by self-oxidation reduction.[176] For example, arsenic trisulfide is photooxidized by exposure to the trioxide and is postheated to dry develop the image,[161]

Step 1—Light Enhanced Oxidation:

$$As_2S_3 + 3/2\ O_2 \xrightarrow{h\nu} As_2O_3 + 3S$$

Step 2—Vaporization:

$$As_2O_3 \xrightarrow{200°C} As_2O_3\uparrow$$

$$nS \xrightarrow{200°C} S_n\uparrow$$

2-10. OUTLOOK

Diazoquinones dispersed in novolak resins are the primary resist choice and will remain so as long as stepper tools can project submicrometer-size images. Their continued use in the mid UV or in multilayer resists will be predicated upon the available actinic output of the choice exposure steppers or scanners. In diazonaphthoquinone resists, many basic improvements such as mid-UV spectral response with new sensitizers, and greater thermal and plasma stability by additives, posttreatments, and substitute polymers extend the applicability of these systems. The incorporation of silicon as silicone novolaks or by posttreatments of DQN images with silicon compounds will provide a useful bilayer mask for submicrometer resolution.

In the deep-UV region, new excimer source could supply all the energy (intensity) ever needed. If these sources are not applicable, 1 mJ/cm^2 resists are needed to compensate for the lack of a suitable high-intensity source. In this regard, image amplification by acid catalysis is the primary means of achieving high sensitivity in the deep UV, provided proper etch and thermal resistance is maintained. If these types of resists are not plasma resistant, they may still be useful as image layers in multilayer resist configurations or as a photosensitive component in a blend of a three-component resist consisting of an etch- and thermal-resistant resin, a low-ceiling-temperature scissionable polymer, and an acid or photocatalytic agent.

REFERENCES

1. J. Kosar, *Light Sensitive Systems*, Wiley, New York, 1965.
2. M. Dinaburg, *Photosensitive Diazo Compounds*, Focal Press, New York, 1968.
3. P. Hanson, *Photogr. Sci. Eng.* **14**, 438 (1970),
4. J. Pacansky and J. Lyerla, *IBM J. Res. Dev.* **23**, 42 (1979); *J. Electrochem. Soc.* **124**, 862 (1977).
5. H. Meier and K. Zeller, *Angew. Chem. Int. Ed. Engl.* **14**, 32 (1975).
6. M. Kaplan and D. Meyerhofer, *RCA Rev.* **40**, 170 (1979); *Polym. Eng. Sci.* **20**, 1073 (1980).
7. D. Meyerhofer, *IEEE Trans. Electron Devices* **ED-22**, 921 (1980).
8. W. De Forest, *Photoresist Materials and Processes*, McGraw–Hill, New York, 1975, p. 54.
9. G. Willson, R. Miller, D. McKean, T. Tompkins, N. Clecak, and D. Hofer, SPE RETEC Photopolymers, Ellenville, N.Y., 1982, p. 111, Organic Coatings Proceedings, *Am. Chem. Soc.* **40**, 54 (1983), U. S. Patent 4,397,937 (1983), IBM; *Polym. Eng. Sci.* **23**, 1004 (1983).

10. W. Babie, M. Chow, and W. Moreau, *Am. Chem. Soc. Org. Coat. Prepr.* **48**, 53 (1983).
11. European Patent 68, 346, *Chem. Abstr.* **99**, 30749 (1983), Hoechst.
12. U. S. Patent 4,207,107 (1981), RCA.
13. U.S. Patent 4,173,470 (1979), Bell.
14. U.S. Patent 4,123,279 (1978), Fuji.
15. S. Farenholtz, *Am. Chem. Soc. Org. Coat.* **35**, 311 (1975).
16. A. Knop, *Applications of Phenolic Resins*, Springer Verlag, Berlin, 1979.
17. E. Gipstein, A. Duano, and T. Tompkins, *J. Electrochem. Soc.* **129**, 201 (1981).
18. B. Grant, N. Clecak, R. Tureg, and C. Willson, *IEEE Trans. Electron Devices* **ED-28**, 1300 (1981), U.S. Patent 4,284,706 (1980).
19. M. Bowden, L. Thompson, S. Farenholtz, and E. Doerries, *J. Electrochem. Soc.* **128**, 1304 (1981); U.S. Patent 4,289,845 (1982), Bell.
20. U.S. Patent 4,398,001 (1983), IBM.
21. H. Hiroaka and L. Welsh, *Am. Chem. Soc. Org. Coat. Prepr.* **48**, 48 (1983).
22. U.S. Patent 4,250,247 (1981), Hoechst.
23. Japanese Patent 75,127,619; *Chem. Abstr.* **85**, 12367 (1975).
24. Japanese Patent 76,120,712; *Chem. Abstr.* **86**, 13177 (1976).
25. French Patent 2,477,294; *Chem. Abstr.* **96**, 43856 (1982).
26. Japanese Patent 82,173,941; *Chem. Abstr.* **99**, 13965 (1983).
27. U.S. Patent 3,637,384 (1972), GAF; U.S. Patent 4,384,037 (1983), JSR.
28. U.S. Patent 3,890,153 (1975), Phillips.
29. U.S. Patent 4,339,521 (1981), Siemens; German Patent 2,631,535; *Chem. Abstr.* **86**, 198007 (1976).
30. Japanese Patent 83,48,045; *Chem. Abstr.* **99**, 80066h (1983).
31. U.S. Patent 3,661,582 (1972), Bell.
32. U.S. Patent 4,365,019 (1982), Kodak.
33. European Patent 70,201; *Chem. Abstr.* **98**, 189014 (1982).
34. U.S. Patent 4,115,128 (1981), Fuji.
35. U.S. Patent 4,307,173 (1979), Hoechst.
36. U.S. Patent 4,009,033 (1978), IBM.
37. U.S. Patent 4,036,644 (1979), IBM.
38. U.S. Patent 4,059,449 (1977), RCA.
39. Japanese Patent 77,126,221; *Chem. Abstr.* **89**, 120920 (1977).
40. Russian Patent 744,426.
41. U.S. Patent 4,148,654 (1982), Shipley.
42. Japanese Patent 80,129,341; *Chem. Abstr.* **80**, 129341 (1978).
43. U.S. Patent 3,130,047 (1964), Azoplate.
44. U.S. Patent 3,264,104 (1966), Azoplate.
45. U.S. Patent 3,827,908 (1974), IBM.
46. U.S. Patent 4,336,319 (1981), Fuji.
47. U.S. Patent 4,259,430 (1981), IBM.
48. U.S. Patent 4,089,766 (1979), IBM.
49. Japanese Patent 78,135,621; *Chem. Abstr.* **90**, 130686 (1979).
50. Russian Patent 731,413; *Chem. Abstr.* **93**, 177308 (1980).
51. U.S. Patent 3,634,082 (1972), Shipley.
52. German Patent 2,944,237; *Chem. Abstr.* **93**, 15922 (1980).
53. U.S. Patent 4,007,047 (1977), IBM.
54. J. Magerlin and D. Webb, *IBM J. Res. Dev.* **24**, 561 (1980).
55. U.S. Patent 4,104,070 (1980), IBM.
56. Japanese Patent 80,32,088; *Chem. Abstr.* **93**, 1995530 (1980).
57. Russian Patent 595,694; *Chem. Abstr.* **88**, 161485 (1978).
58. French Patent 2,017,782, RCA.
59. K. Nakumura, *Chem. Lett.* **1972**, 763 (1972).
60. A. Ouano, *Am. Chem. Soc. Org. Coat.* **48**, 42 (1983).
61. A. Ouano, *Polym. Eng. Sci.* **18**, 306 (1978).
62. D. Ilten and R. Sutton, *J. Electrochem. Soc.* **119**, 539 (1972).
63. B. Broyde, *J. Electrochem. Soc.* **117**, 1555 (1970).
64. A. Paramov, *Chem. Abstr.* **81**, 97720 (1974).
65. J. Shaw, M. Frisch, and F. Dill, *IBM J. Res. Dev.* **21**, 219 (1977).

66. *Chem. Abstr.* **84**, 31812 (1976).
67. *Chem. Abstr.* **84**, 4214 (1976).
68. *Chem. Abstr.* **82**, 111177 (1975).
69. T. Shankoff, J. Brunning, and R. Johnston, *Polym. Eng. Sci.* **20**, 1102 (1980).
70. D. Leers, *Solid State Technol.* March 1981, p. 91.
71. K. Jain, C. Willson, and B. Lin, *IEEE Electron Device Lett.* **EDL-3**, 53 (1982); *IBM J. Res. Dev.* **26**, 151 (1982).
72. B. Lin, *J. Vac. Sci. Technol.* **19**, 1313 (1981); U.S. Patent 4,360,585 (1983), GE.
73. U.S. Patent 4,239,787 (1981), Bell.
74. S. MacDonald, R. Miller, C. Willson, G. Feinberg, R. Gleason, R. Halverson, W. MacIntyre, and W. Motsiff, *Kodak Microelectronics*, 1982, p. 114.
75. U.S. Patent 4,104,070 (1978).
76. L. Katzisyma, *Russ. Chem. Rev.* **35**, 388 (1966).
77. C. Willson, in *Introduction to Microlithography*, edited by C. Willson and L. Thompson, *Am. Chem. Soc. Symp. Ser.* **219**, 88–148 (1983).
78. B. Griffing, *J. Vac. Sci. Technol.* **19**, 1423 (1981).
79. L. Mury, J. Matthews, and C. Wood, *SPIE Proc.* **334**, 241 (1982).
80. P. Van Pelt, *SPIE Proc.* **275**, 150 (1981).
81. Y. Kawamura, *J. Appl. Phys.* **53**, 6489 (1981), *Appl. Phys. Lett.* **40**, 374 (1982).
82. N. Veno, K. Sugita, S. Konishi, and K. Tanimoto, *Jpn. J. Appl. Phys.* **20**, L709 (1981).
83. K. Sugita, N. Ueno, S. Konishi, and Y. Suzuki, *J. Photogr. Sci. Eng.* **27**, 149 (1983).
84. J. Bachus, *Solid State Technol.* Feb. 1982, p. 124.
85. D. Doane, *Electrochem. Soc. Ext. Abstr.* No. 332 (1980).
86. T. McGrath, *Solid State Technol.* Dec. 1983, p. 165.
87. W. Moreau and P. Schmidt, *Electrochem. Soc. Ext. Abstr.* **138**, 459 (1970).
88. M. King, *IEEE Trans. Electron Devices* **ED-26**, 711 (1979).
89. A. Offner, *Photogr. Sci. Eng.* **23**, 374 (1979).
90. S. Iwamatsu and K. Asonami, *Solid State Technol.* May 1980, p. 81.
91. B. Ranby and J. Rabek, *Photodegradation, Photooxidation, and Photostabilization of Polymers*, Wiley, New York, 1975, pp. 143–164.
92. E. Dane and J. Gullet, *Macromolecules* **6**, 230 (1974).
93. J. MacCallum and C. Schoff, *Trans. Faraday Soc.* **67**, 2383 (1971), **67**, 2372 (1971).
94. A. Gupta, R. Liang, F. Tsay, and J. Moacanin, *Macromolecules* **13**, 1696 (1980).
95. H. Hiroaka, *IBM J. Res. Dev.* **21**, 121 (1977).
96. R. Fox, L. Issacs, and S. Stokes, *J. Polym. Sci.* **A1**, 1079 (1963).
97. K. Harada and S. Sugawara, *J. Appl. Polym. Sci.* **27**, 1441 (1982).
98. N. Viswanathan and H. Santini, *Kodak Microelectronics Seminar*, 1982, p. 47.
99. M. Lanagan, S. Lindsey, and N. Viswanathan, *Jpn. J. Appl. Phys.* **22**, 267 (1983).
100. M. Tsuda, Y. Nakamura, S. Oikawa, H. Nagata, Y. Yakota, H. Nakane, T. Tsunori, Y. Nakase, and T. Mifusi, *Photogr. Sci. Eng.* **23**, 290 (1979).
101. German Patent 2,911,286, Tokyo Ohko; *Chem. Abstr.* **92**, 86003 (1980).
102. C. Wilkins, E. Reichmanis, and E. Chandross, *J. Electrochem. Soc.* **127**, 2510 (1980); U.S. Patent 4,382,120 (1983), Bell.
103. E. Reichmanis, C. Wilkins, and E. Chandross, *J. Vac. Sci. Technol.* **19**, 1338 (1981).
104. Y. Amerik and J. Guillet, *Macromolecules* **4**, 375 (1971).
105. N. Grassie and A. Davidson, *Polym. Degr. Stabil.* **3**, 45 (1980).
106. U.S. Patent 3,853,814 (1974), J. Guillet.
107. C. Wilkins, E. Reichmanis, and E. Chandross, *J. Electrochem. Soc.* **129**, 2553 (1982).
108. European Patent 48,889, Siemens; *Chem. Abstr.* **97**, 31279 (1982).
109. K. Nate and T. Kobayashi, *J. Electrochem. Soc.* **128**, 1395 (1981).
110. Japanese Patent 83,93,047; *Chem. Abstr.* **99**, 96853 (1983).
111. T. Bowmer, C. Wilkins, E. Reichmanis, and M. Hellman, *Polymer* **20**, 2661 (1982).
112. Japanese Patent 83,49.942; *Chem. Abstr.* **99**, 61731 (1983), **99**, 30758 (1983).
113. British Patent 2,099,168; *Chem. Abstr.* **98**, 98822 (1983).
114. W. Moreau, *Opt. Eng.* **22**, 181 (1983).
115. Japanese Patent 83,48,048; *Chem. Abstr.* **99**, 30757 (1983).
116. U.S. Patent 4,278,754 (1981), Oki.

117. Y. Yamashita, K. Ogure, M. Kunishi, R. Kawazawa, S. Ohno, and Y. Mizokawi, *J. Vac. Sci. Technol.* **16**, 2026 (1979).
118. J. Guillet, *Macromolecules* **5**, 212 (1972).
119. U.S. Patent 4,276,369 (1981), Tokyo Ohko.
120. U.S. Patent 4,243,740 (1981), Tokyo Ohko.
121. J. Kwiwi and W. Schnabel, *Macromolecules* **11**, 468 (1978).
122. M. Gazard, A. Chapiro, J. Dubois, and M. Duchesne, *Polym. Eng. Sci.* **20**, 1069 (1980).
123. E. Chandross, C. Wilkins, E. Reichmanis, and M. Hartless, *Solid State Technol.* Aug. 1981, p. 81.
124. U. S. Patent 3,849,137 (1974).
125. H. Barzynski and D. Sanger, *Angew. Makromol. Chem.* **93**, 131 (1981).
126. U.S. Patent 4,086,210 (1978), Kodak.
127. C. Petropoulos, *J. Polym. Sci. Polym. Chem. Ed.* **15**, 1637 (1977).
128. U.S. Patent 3,991,033 (1976), DuPont.
129. U.S. Patent 4,150,989 (1979), DuPont.
130. H. Ito and C. Willson, in *Polymers in Electronics*, edited by T. Davidson, Am. Chem. Soc., 1984, p. 11; *Polym. Eng. Sci.* **23**, 1012 (1983).
131. H. Ito and C. Willson, SPE RETEC Photopolymers, Ellenville, N.Y., 1982, p. 331.
132. U.S. Patent 3,849,137 (1974), Badische.
133. U.S. Patent 3,917,483 (1975), Xerox.
134. U.S. Patent 4,108,839 (1978), DuPont.
135. U.S. Patent 3,963,491 (1976), Xerox.
136. U.S. Patent 3,915,706 (1975), Xerox.
137. U.S. Patent 3,984,253 (1976), Kodak.
138. U.S. Patent 3,923,514 (1975), Xerox.
139. U.S. Patent 3,940,507 (1976), RCA.
140. U.S. Patent 3,779,778 (1973), 3M.
141. U.S. Patent 4,248,957 (1981), Hoechst.
142. U.S. Patent 4,250,247 (1981), Hoechst.
143. J. Crivello and J. Lam, *J. Polym. Sci. Polym. Chem. Ed.* **17**, 2877 (1979).
144. J. Crivello, *Am. Chem. Soc. Org. Coat.* **48**, 65 (1983), *Polym. Eng. Sci.* **23**, 953 (1983).
145. U.S. Patent 4,193,799 (1979), GE.
146. U.S. Patent 4,245,029 (1981), GE.
147. U.S. Patent 4,256,828 (1981), 3M.
148. U.S. Patent 4,156,035 (1978), W. Grace.
149. S. Schlesinger, *Photogr. Sci. Eng.* **18**, 187 (1974), *Polym. Eng. Sci.* **14**, 513 (1974).
150. J. Burgess and D. Schaefer, *J. Vac. Sci. Technol.* **6**, 134 (1969).
151. H. Steppan, G. Buhr, and H. Vollmann, *Angew. Chem. Int. Ed. Engl.* **21**, 455 (1982), reference 50 therein.
152. A. Monahan, *Macromolecules* **1**, 408 (1968).
153. W. Meyer, B. Curtis, and H. Brunner, *Microelectron. Eng.* **1**, 29 (1983), U. S. Patent 4,443,044 (1984), RCA.
154. G. Sheldrick and O. Vogl. *Polym. Eng. Sci.* **16**, 68 (1976).
155. U.S. Patent 3,375,110 (1968), Union Carbide.
156. U.S. Patent 3,763,397 (1969), Kodak.
157. T. Wolf, G. Taylor, T. Venkatesan, and R. Kraetsch, Paper F-6, Electron, Ion, and Photon Beam Conference, Los Angeles, 1983.
158. H. Craighead, J. White, R. Howrad, L. Jackel, R. Behringer, J. Sweeney, and R. Epworth, Paper J-1, Electron, Ion, and Photon Beam Conference, Los Angeles, 1983; *J. Vac. Sci. Technol.* **B1**, 1186 (1983).
159. H. Hiroaka, L. Welsh, and J. Bargon, Paper F-1, Electron, Ion, and Photon Beam Conference, Los Angeles, 1983; *J. Vac. Sci. Technol.* **B1**, 1062 (1983).
160. M. Tsuida and S. Oikawa, *Photogr. Sci. Eng.* **23**, 177 (1979).
161. M. Janai and P. Rudman, *Photogr. Sci. Eng.* **20**, 234 (1976).
162. S. Pappas and J. Jilek, *Photogr. Sci. Eng.* **23**, 140 (1979).
163. W. Sorenson and T. Campbell, *Preparative Methods of Polymer Chemistry*, Wiley, New York, 1968, pp. 342–398.
164. J. Pacansky, *Polym. Eng. Sci.* **20**, 1049 (1980).

165. U.S. Patent 4,377,631 (1983), Hunt.
166. E. Vollenbroek, E. Spiertz, and H. Kroon, *Polym. Eng. Sci.* **23**, 925 (1983).
167. T. Gupta, *Eur. Polym. J.* **17**, 1127 (1981).
168. C. G. Willson, in *Introduction to Microlithography*, edited by L. Thompson, C. Willson, and M. Bowden, *Am. Chem. Soc. Symp. Ser.* **219**, 102–104 (1983).
169. B. Ranby and J. Rabek, *Photodegradation, Photooxidation, and Photostabilization of Polymers*, Wiley, New York, 1977, pp 323–325.
170. J. Burgess and D. Schafer, *J. Vac. Sci. Technol.* **6**, 135 (1969).
171. P. Trefonas, R. Miller, D. Hoefr, and R. West, *J. Polym. Sci. Polym. Lett. Ed.* **21**, 823 (1983).
172. T. Pampalone, *Solid State Technol.* June 1984, p. 115.
173. D. Hofer, A. Neureuther, C. Willson, and R. Miller, *SPIE Advances in Resist Technology* **469**, 108 (1984).
174. T. Deutsch and M. Geis, *J. Appl. Phys.* **54**, 7201 (1983).
175. J. Sheats, *Appl. Phys. Lett.* **44**, 1016 (1984).
176. U. S. Patent 4,454,221 (1984).
177. B. Dickens, *Polymer* **25**, 706 (1984).
178. U. S. Patent 4,444,195 (1984), Konishiroku Ind.
179. U. S. Patent 4,460,674 (1984), Konishiroku Ind.
180. M. Nakase, *Photogr. Sci. Eng.* **27**, 254 (1983).
181. Japanese Patent 58,190,946, *Chem. Abstr.* **100**, 165446 (1984).
182. Czechoslovakian Patent 213,016, *Chem. Abstr.* **100**, 219037 (1984).
183. German Patent 3,220,816, *Chem. Abstr.* **100**, 219047 (1984).
184. Russian Patent 1,068,879, *Chem. Abstr.* **100**, 219051 (1984).
185. European Patent Application EP 95,388, *Chem. Abstr.* **100**, 219051 (1984).
186. Japanese Patent 57,201,229, *Chem. Abstr.* **100**, 129885 (1984).
187. Japanese Patent 57,162,330, *Chem. Abstr.* **100**, 15329 (1984).
188. Anon, *Res. Discl.* **240**, 173 (1984), *Chem. Abstr.* **100**, 200822 (1984).
189. U. S. Patent 4,456,760 (1984), BASF.
190. G. Taylor, *Solid State Technol.* June 1984, p. 105.
191. D. Meyerhofer, in *Microcircuit Engineering*, edited by J. Cleaver, H. Ahmed, and G. Jones, Academic Press, New York, 1984, p. 305.
192. U.S. Patent 4,405,708 (1983), U. S. Phillips.
193. U. S. Patent 4,404,272 (1983), Hoechst.
194. U. S. Patent 4,474,864 (1984), IBM.
195. S. Pappas, B. Pappas, L. Gatechair, and W. Schnabel, *J. Polym. Sci.* **22**, 69 (1984).
196. T. Pampalone, M. Hanifan, S. Jain, and K. Krieger, *J. Electrochem. Soc.* **131**, 2670 (1984).
197. F. Branwell, S. Fahrenholtz, R. Jadujura, and C. Paley, *J. Chem. Educ.* **55**, 403 (1978).
198. U.S. Patent 4,460,674 (1984), Konishiroku Ind.
199. U.S. Patents 4,457,999 and 4,458,000 (1984), Am. Hoechst.
200. German Patent 3,337,315, *Chem. Abstr.* **101**, 181207 (1984).
201. M. Ishikawa, *J. Polym. Sci.* **22**, 669 (1984).
202. A. Vanier, *Mikroelektronika* **13**, 3119 (1984) (Izd. Sov. Radio), *Chem. Abstr.* **101**, 219624 (1984).
203. Japanese Patent 59,92,532, *Chem. Abstr.* **101**, 238180 (1984).
204. P. Dyer and J. Sidhu, *J. Appl. Phys.* **57**, 1420 (1985).
205. J. Frechet, T. Tessier, C. Willson, and H. Ito, *Macromolecules* **18**, 317 (1985).
206. T. Shankoff and A. Trozollo, *Photogr. Sci. Eng.* **19**, 173 (1975).
207. Japanese Patent 59,155,838, *Chem. Abstr.* **102**, 36775 (1985).
208. East German Patent 211,415, *Chem. Abstr.* **102**, 140893 (1985).
209. Japanese Patent 59,180,545, *Chem. Abstr.* **102**, 140893 (1985).
210. R. Miller, D. Hofer, D. McNean, C. Willson, P. West, and P. Trefonis, *Am. Chem. Soc. Symp. Ser.* **266**, 293 (1984).
211. Y. Saotome, H. Gokan, K. Saigo, M. Suzuki, and Y. Ohnishi, *J. Electrochem. Soc.* **132**, 909 (1985).
212. U.S. Patent 4,522,911 (1985), IBM.
213. U.S. Patent 4,529,682 (1985), Hunt.
214. U.S. Patent 4,522,911 (1985), IBM.
215. J. Ziegler, L. Harrah, and W. Johnson, *Proc. SPIE Int. Soc. Opt. Eng.* **539**, 166 (1985), U.S. Patent 4,587,205 (1986).
216. Japanese Patent 60,52,845 (1985), JSR; *Chem. Abstr.* **103**, 113359 (1985).

217. E. Alling and C. Stauffer, *SPIE Opt. Eng.* **539**, 194 (1985).
218. U.S. Patent 4,546,066 (1985), IBM.
219. U.S. Patent 4,544,627 (1985), Fuji.
220. H. Klose, R. Seguish, and W. Arder, *IEEE Trans. Electron Devices* **ED-32**, 1654 (1985).
221. U.S. Patent 4,524,121 (1985), Rohm & Haas.
222. German Patent 3,344,202 (1985), Merck; *Chem. Abstr.* **103**, 79517 (1985).
223. U. S. Patent 4,546,064 (1985), NAP.
224. A. Schultz, P. Frank, B. Griffing, and A. Young, *J. Polym. Sci.* **23**, 1749 (1985).
225. S. Pappas, *J. Imag. Technol.* **14**, 146 (1985).
226. M. de Grandpre, D. Vidusek, and M. Legenza, *SPIE Resist Technol.* **539**, 103 (1985).
227. Z. Katovic and M. Stefanic, *Ind. Eng. Chem. Prod. Res. Dev.* **24**, 179 (1985).
228. M. S. Pak, D. Mammato, S. Jarn, and D. Durham, SPE RETEC Photopolymers, Ellenville, N.Y., 1985, p. 247.
229. M. Watts and D. DeBreun, SPE RETEC Photopolymers, Ellenville, N.Y., 1985, p. 285.
230. S. Turner, R. Arcus, C. Houle, and W. Schleiger, SPE RETEC Photopolymers, Ellenville, N.Y., 1985, p. 35.
231. J. Frechet, F. Houlihan, F. Bouchard, E. Eichler, A. Hult, R. Allen, S. MacDonald, H. Ito, and C. Willson, SPE RETEC Photopolymers, Ellenville, N.Y., 1985, p. 1.
232. W. Hinsberg, C. Willson, and K. Kanazawa, *Proc. SPIE Resist Technol.* **539**, 6 (1985).
233. F. Buiguez, J. Guibert, C. Rosilic, A. Rosilio, F. Schue, R. Sagnes, B. Serres, L. Girial, W. Abou-Madi, and C. Montginoul, *Microcircuit Engineering*, edited by H. Beneking and A. Heuberger, Academic Press, New York, 1985, p. 471.
234. A. Furuta, M. Hanabata and Y. Uemiura, *J. Vac. Sci. Technol.* **B4**, 430 (1986)
235. F. Houlihan, F. Bouchard, J. Frechet, and C. Willson, *Macromolecules* **14**, 13 (1986).
236. U.S. Patent 4,596,763 (1986), Hoechst.
237. H. Phillips, A. Cole, Y. Liu, and T. Sitnik, *Appl. Phys. Lett.* **48**, 192, (1986); **48**, 212 (1986).
238. U.S. Patent 4,588,670 (1986), Hoechst.
239. U.S. Patent 4,601,969 (1986), IBM.
240. J. Frechet, F. Bouchard, F. Houlihan, B. Kryczka, E. Eichler, N. Clecak, and C. Willson, *J. Imag. Sci.* **30**, 59 (1986).
241. U. S. Patent 4,550,069 (1986), Hoechst (PM acetate).
242. P. Paniez and A. Weill, *Microelectron. Eng.* **4**, 57 (1985).
243. European Patent Appl. EP 140,273, *Chem. Abstr.* **104**, 197029 (1986).
244. U.S. Patent 4,551,416 (1985), AT & *T* Bell Labs.
245. C. Oscuh, E. Brahim, F. Jopf, M. McFarland, A. Mooring, and C. Wu, *SPIE Proc.* **631**, 68 (1986).
246. U. S. Patent 4,584,309 (1986), Allied.
247. E. Reichmanis, A. Novembre, R. Tarascon, and A. Shugard, *Polym. Mater. Sci. Eng.* **55**, 299 (1986); *SPIE Proc.* **631**, 40 (1986).
248. J. Frechet, T. Iizawa, F. Bouchard, M. Stanciulescu, C. Willson, and N. Clecak, *Polym. Mater. Sci. Eng.* **55**, 299 (1986).
249. T. Pampalone, A. Gilfillan and P. Zanzucchi, *J. Electrochem. Soc.* **133**, 1917 (1986).
250. R. Turner, K. Ahn, and C. Willson, *Polym. Mater. Sci. Eng.* **55**, 608 (1986).

3

Positive Radiation Resists

3-1. INTRODUCTION

3-1-1. Tool Contributions

Positive radiation resists are required for submicrometer lithography using high-energy (10–200 keV) X-ray, ion, and electron beams. Since the resolution of an exposure process is limited by the wavelength of forward and, moreover, backscattered radiation, high-energy rays of atom-size wavelengths are capable of resolution to 50 Å (Fig. 3-1-1). However, the practical resolution limit is about 500 Å. If the backscatter is reduced or eliminated using silicon membranes instead of thick wafers, nanometer-size devices can be fabricated.

Submicrometer lithography is driven by the lower bit costs of VLSI obtained with smaller and denser circuits.[1] Although the labor and material cost are about the same for processing a 2-μm or 0.5-μm circuit, the high costs of sophisticated systems for submicrometer exposure may override the other equivalent advantages. For example, optical steppers cost about $1000K but *e*-beam tools for 0.5-μm lithography can be as much as $3000K.[2,12] As the spot size of the *e*-beam tool decreases to 0.5 μm or less, more time is spent writing and the throughput decreases (Fig. 3-1-2) to 1–3 wafers/hr. The minimum exposure dose (E) is a product of beam intensity (A/cm^2, I) and the exposure time (t) per pixel,

$$E = (I)(t) \qquad (3\text{-}1\text{-}1)$$

The throughput of the tool is the product of the total of the number of chips per wafer, C, and the exposure time per chip, t_c, which is comprised of

$$t_c = \text{Exposure time} + \text{align time} + \text{overhead time} \qquad (3\text{-}1\text{-}2)$$

Typically, a throughput of 1–10 (125-mm wafers) per hour can be expected, depending on the resist dose (E), the beam current, the size of the beam, and whether vector or raster writing is used. Since the exposure time constitutes about 75% of t_c, high-speed resists are necessary to reduce the exposure time per chip and to lower the beam current for better edge slope (spot shape). For a reduction in resist exposure time, t (dose), of 5×, is about a 2× gain in wafer throughput. Because of the throughput disadvantage of electron beams compared to the optical

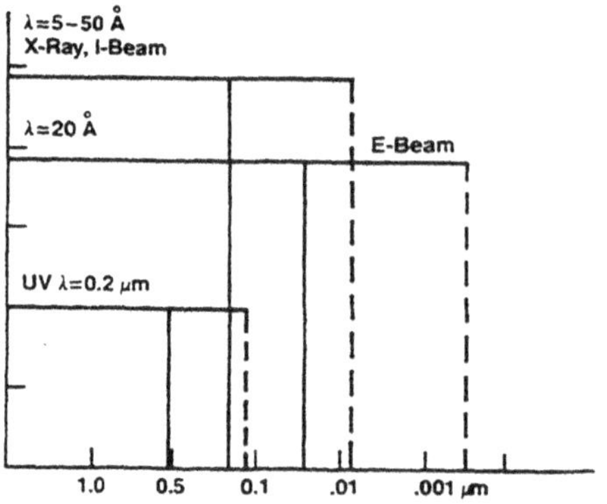

FIGURE 3-1-1. Theoretical (---) and practical (——) limits of exposure and lithography for a 0.1-μm-thick resist.

form of imaging (15–50 wafers/hr), the driving factors for high-energy lithography are improved alignment and the ease of handling the production of customized circuits involving high part number–low volume (Table 3-1-1).

High-energy lithography can be performed in two basic ways:

Projection	Scanning
X ray	*E*-beam
Ion beam	Ion beam
E-beam	

For exposure with masks, high-energy radiation projection is mainly confined to chip-size fields. Several chips can be exposed by X-ray and some photocathode *e*-beam tools, but only for large (> 2 μm) dimensions. The exposure masks are fabricated from heavy metals on semitransparent organic or inorganic membranes(3–5) (Fig. 3-1-3). In addition, the masks can also be a source of energy themselves such as the UV-excited TiO_2 photocathode which emits 10-keV electrons(6) (Fig. 3-1-4). As an *in situ* secondary mask, a portable conformable mask is formed in a heavy metal pattern(7) on top of a resist (Fig. 3-1-4). In proximity printing, a collimated beam of electrons,(8) ions,(19) or X rays passes through a shadow mask and is (diffractionless) printed into a resist. In X-ray printing (Fig. 3-1-5), the shape of the source, penumbra shadows, and the poor collimation limit the field of exposure.(8,360) Because of the high cost of mask fabrication and the instability of the mask to beam heating, the implementation of flood exposure by high-energy radiation has lagged a decade behind traditional optical exposure.(9,30)

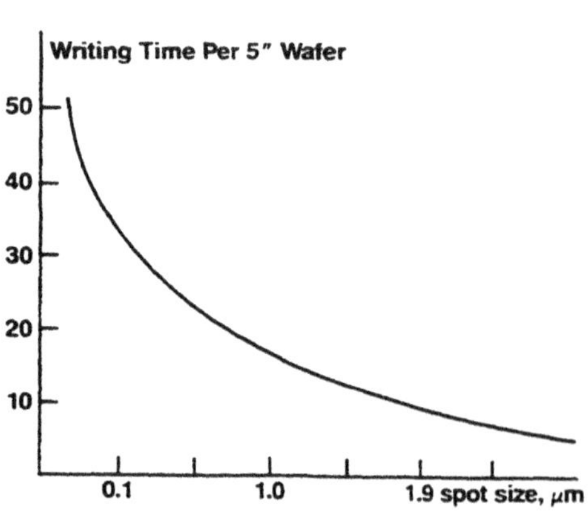

FIGURE 3-1-2. Writing time in an *e*-beam raster system. From Ref. 4, courtesy of SPIE.

TABLE 3-1-1. Comparison of Lithography Drivers for Various Exposure Tools

Optical lithography	High-energy lithography
Well-known principles	Part number capability
Simple mask	Nanometer resolution
Foundation in Ag photography	Automatic registration
Moderate cost	Nanometer alignment
Resistant resist	More resist candidates
No radiation damage	Necessary for photomask fabrication
Good throughput	

For the low volume (high part number), the personalization of master slice logic circuits[10] by direct *e*-beam wafer write is more economical than optical printing with a mask (Fig. 3-1-6). The high-resolution aspect of *e*-beam lithography will continue to be utilized to fabricate masks for optics, X-ray, and ion beam. Secondly, for *e*-beams, the ability to register each chip to $\pm 0.1\ \mu m$ (3σ) is a critical advantage over all other forms of exposure.

3-1-2. Resist Considerations

The major roles of the resist interaction with the exposure tool are

1. High speed for reasonable throughput
2. High resolution for submicrometer images

Besides these primary objectives, the resist must be a durable mask in the image transfer step (Fig. 3-1-7). High throughput is necessary to justify the tool costs. In X-ray exposure, attempts have been made to provide the highest-intensity sources that eliminate the need for a high-speed resist (Fig. 3-1-8). The X-ray exposure time per chip, T_c, is

$$T_c = \text{Expose time} + \text{align time} + \text{step time} \tag{3-1-3}$$

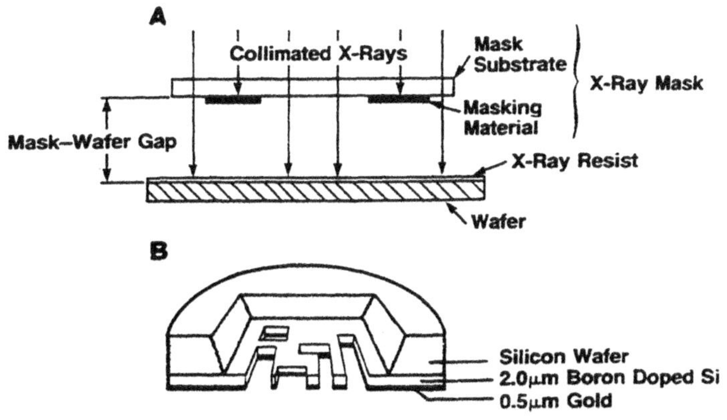

FIGURE 3-1-3. (A) X-ray mask of a thin membrane with a metal absorber. From Ref. 3. (B) Flood ion or *e*-beam mask of a perforated silicon membrane with metal absorber. From Ref. 5, courtesy of IBM.

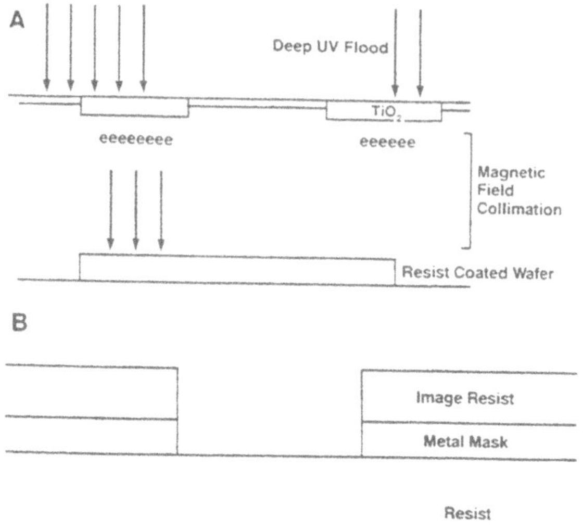

FIGURE 3-1-4. (A) Active photocathode mask of TiO_2 emitting electrons to expose resist. From Ref. 6. (B) Passive molecular mask of metal absorber prepatterned on top of resist to absorb electrons, protons, X rays, or ions. From Ref. 7.

If a 125-mm wafer contains 200 chips and the alignment time is 1 sec/chip, then 3 min/wafer is required for just the alignment step. If we choose a throughput of 30 wafers/hr and allow 0.2 sec/chip for the align and step process, the exposure should be 0.1 sec/chip or 20 sec/wafer. The exposure time (t) is

$$t = E_{\text{resist}}/\text{source fluence} \qquad (\text{particles/cm}^2) \tag{3-1-4}$$

In the example of X-ray exposure in Fig. 3-1-8, a short exposure can be accomplished by a very intense source of high fluence such as a synchrotron (Fig. 3-1-9). Conventional sources require a fast (1 mJ/cm^2, Table 3-1-2) resist to achieve even a throughput of 1 wafer/hr. However, high-intensity sources may be too costly and irreproducible.

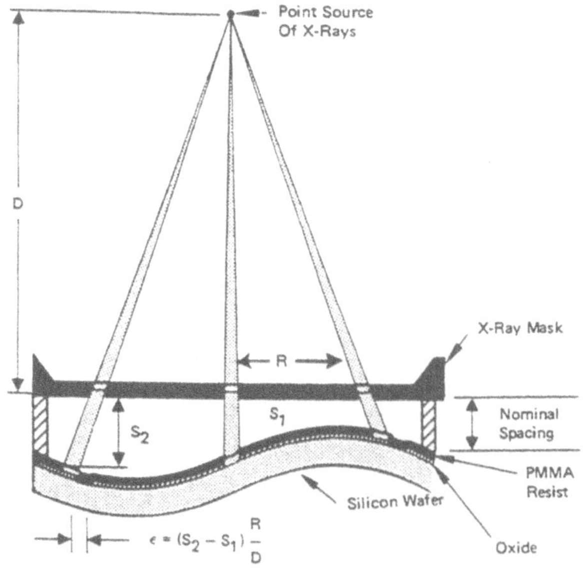

FIGURE 3-1-5. Runout (ε) due to penumbra effect, due to wafer distorted by thermal processing (S) and diffusion of source at edges. From J. McCoy, *Circuit Manuf.* Nov. 1977, p. 37, copyright and courtesy of Morgan–Grampian Publ. Co.

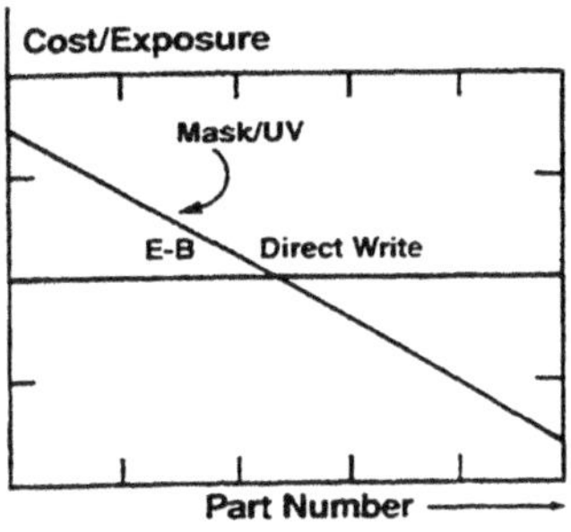

FIGURE 3-1-6. Cost of exposure versus part number for direct *e*-beam write or mask printing with UV. From Ref. 300, reprinted with permission of *Solid State Technology*, published by Technical Publishing, a company of Dun and Bradstreet.

For high-energy sources, lowering the dose required, and reducing the size of the image (pixel or beam size) leads to a statistical dilemma.[30] Since high-energy sources emit energetic particles off a filament in a random stream, potentially insufficient number of electrons (or particles) may strike a nanometer-diameter pixel. For a 1-cm^2 area, 6×10^{11} electrons strike the surface. For a 0.1×0.1-μm spot, only 0.6 electron strikes the surface within a shot noise (N) of uncertainty given by $\pm\sqrt{N}$. The resolution by Poisson statistics (Fig. 3-1-10) is a simple function of the dose,[11,12]

$$\text{Resolution limit} = (\text{Dose})^{-1/2} \qquad (3\text{-}1\text{-}5)$$

To achieve nanometer resolution (smaller features), the small pixels require a high dose equivalent to the present state-of-the-art resist sensitivities (Table 3-1-2), but high-energy exposure requires a faster resist to attain a minimum throughput. At low doses ($< 1\ \mu C/cm^2$), the number of pixels is so low that the polymer does not develop.[355] For isolated lines (Fig. 3-1-11), insufficient electrons are backscattered and an extra dose (intraproximity effect) is required to compensate for the inefficiency. Polymeric resists which only require top surface imaging and not bulk exposure such as monomer grafted PMMA[173] may overcome some of the shot noise limitations since the trapped radicals initiate surface isotropic polymerization.

Of the three exposure energies (electron, X-ray, and ion beam), the ion beam has the greatest efficiency since most of the energy (90%) can be deposited in a 1-μm-thick film without deterioration due to scattering (such as that encountered in *e*-beam exposure).[9] Electron beams and X rays deposit only 1–10% of the incident dose (Fig. 3-1-12). The high resolution of proton (H^+) exposure can be seen in negative bisazide KTFR images. For *e*-beam exposure, a negative cross-linking resist reveals all of the electron scatter. The backscattered electrons enlarge the base of KTFR about 50% of its height (Fig. 3-1-13A), while a proton exposure produces no image distortion due to the high absorbance and limited range of

TABLE 3-1-2. Resist Sensitivity and Exposure Sources

Present resist sensitivity	Exposure	Increased speed
200 mJ/cm^2	X-ray	1 mJ/cm^2
10 μC/cm^2	*e*-beam	1 μC/cm^2
1 μC/cm^2	Ion beam	0.01 μC/cm^2
75 mJ/cm^2	Photon	4 mJ/cm^2

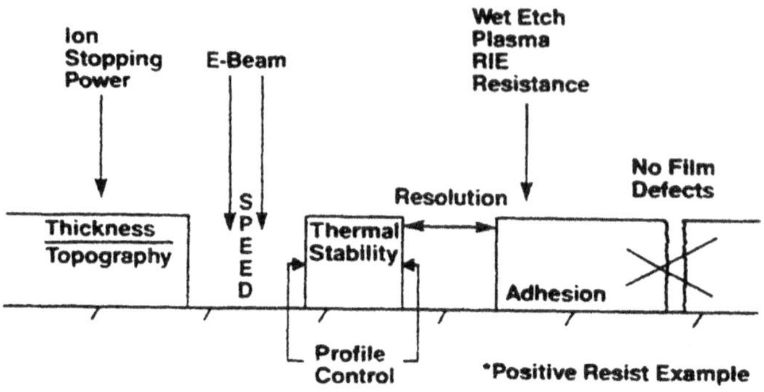

FIGURE 3-1-7. Resist requirements for processing.

secondary electrons (< 500 Å) (Fig. 3-1-13B). Because of the low energy and limited range of photoelectrons[54] (< 1000 Å), X-ray exposure also produces high-aspect-ratio images up to 20 to 1 of height to width (Fig. 3-1-14).[39,44] For isolated features, *e*-beam exposure can produce an aspect ratio of 5 to 1 in a positive PMMA resist (Fig. 3-1-15). However, in closely spaced lines, backscatter from adjacent lines (Fig. 3-1-16) obliterates the unexposed (interproximity effect) spaces.

The resolving power of all forms of exposure is a function of the modulation transfer function (MTF) of the lens.[22,24] Figure 3-1-17 shows the MTF of all forms of exposure. Because of backscatter effects, the electron beam (MTF) rapidly deteriorates below the 60% cutoff considered adequate for a resist.[24,25] Historically, multilayer resists were derived from the multilayers of the PMMA resist (Table 3-1-3), and they are required for submicrometer *e*-beam images.

Other *e*-beam resist process parameters such as the energy (keV) of exposure, the thickness of the resist, prebake, and developer can affect the resolution and the sensitivity of a positive resist. Depending on the dose, shallow or straight wall profiles can be produced in resist images (Fig. 3-1-18). At high doses the image profile is dominated by the absorbed radiation, while at low doses the developer activity on both the exposed and unexposed region dominates the shallower profile.

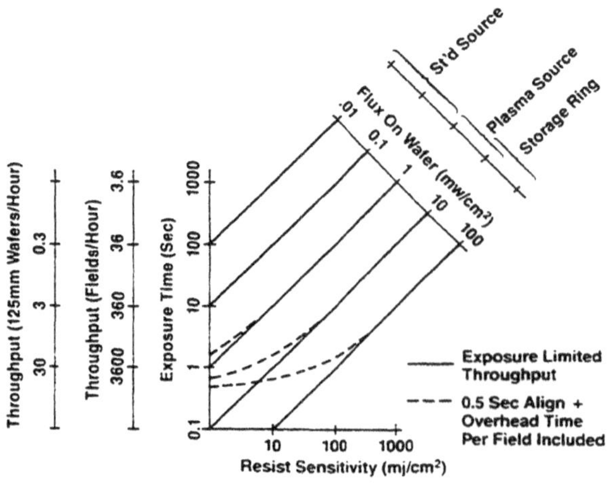

FIGURE 3-1-8. X-ray throughput as a function of beam intensity and resist sensitivity. Courtesy of W. Hicks, IBM.

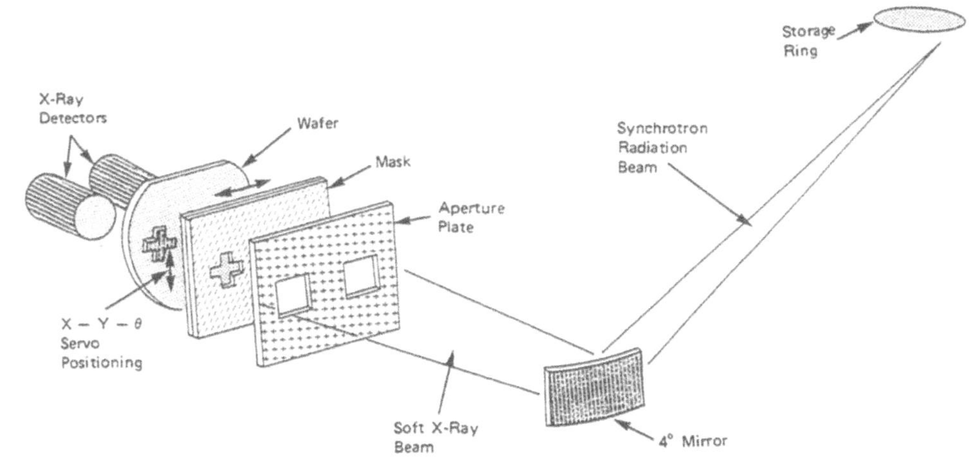

FIGURE 3-1-9. Synchrotron source for soft X-ray lithography. From J. McCoy, *Circuit Manuf.* Nov. 1977, p. 42, copyright and courtesy of Morgan–Grampian Publ. Co.

3-1-3. High-Energy Radiation Absorption

The absorption of high-energy radiation occurs by collisions of the incoming rays with the atomic electrons of the resist solid and by nuclear collisions in the case of ion-beam exposure (Fig. 3-1-19). Electrons and X-rays lose their energy by the photoelectric effect (Fig. 3-1-20) through the excitation of orbital electrons. The electrons of the resist molecules are set in motion with a kinetic energy equal to

$$E = hv - Q \tag{3-1-6}$$

where Q is the binding energy of the electron in its orbital. The ionizing radiation hv is usually larger than Q. The electron is expelled from the orbital and a positive ion is left behind,

$$AB \rightarrow AB^{+} + e^{-} \tag{3-1-7}$$

as in PMMA degradation,

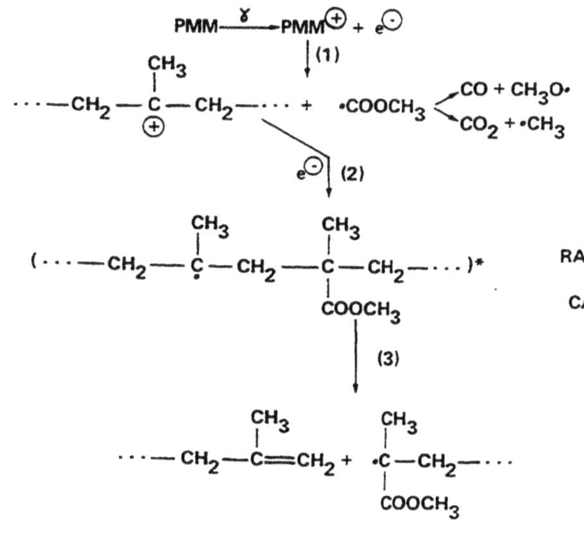

RADIATION MECHANISM

CATION FORMATION

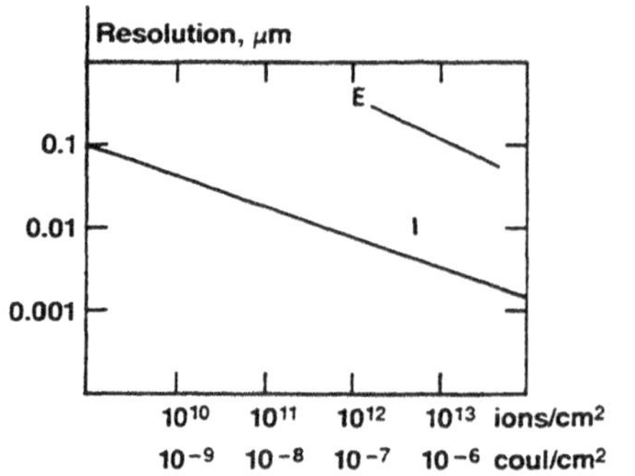

FIGURE 3-1-10. Resolution–statistical limit of resist versus incident (I) ion[12] and electron (E) dose.[13a]

As shown by many mass spectra studies, the positive ion can dissociate into free radicals and ejected electrons,

$$AB^+ \rightarrow A^+ + B \tag{3-1-8}$$

$$AB^+ \rightarrow A^+ + B^+ + e^- \tag{3-1-9}$$

The liberated electrons lose their kinetic energy (25–50 eV) by exciting other molecules along its path. For a 10-keV Al X-ray source, the range of photoelectrons in PMMA is 1000 Å and for a 20-keV (4.6 Å, Rh) source, about 2800 Å.[26] For resists containing heavier atoms (Cs, Th), the range is of the order of 100–500 Å. Heavier inorganic resists have a better chance of absorbing more X rays without scatter than organic carbon-based resists.[14] The absorption of X rays is very wavelength dependent (Fig. 3-1-21) and sources have to be tuned to match the dominant absorbers in the resist.

For *e*-beam absorption, the stopping power for electrons in a solid is given by the Bethe relation,[27]

$$-dE/dS = 7.9 \times 10^4\ Z/EA \ln 2E/J \qquad \text{keV/cm} \tag{3-1-10}$$

where Z is the mean atomic number, A is the atomic weight, J is the mean

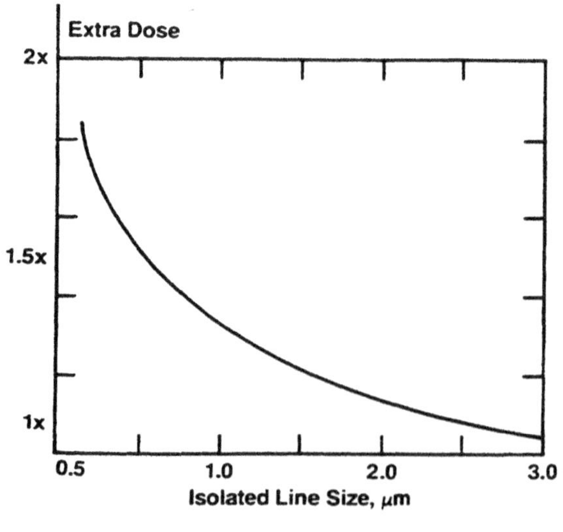

FIGURE 3-1-11. Intraproximity effect–extra dose required for line size.

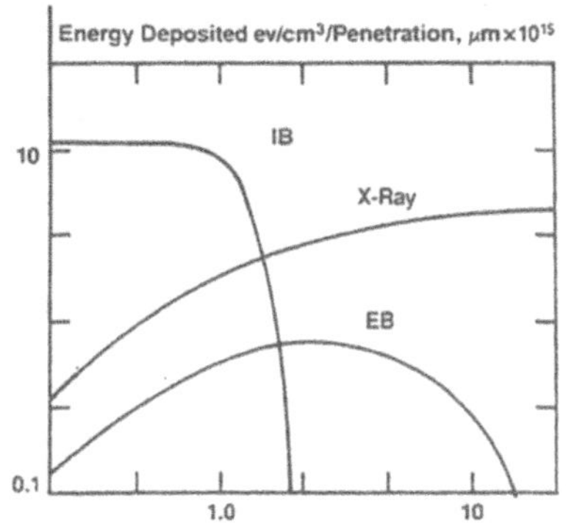

FIGURE 3-1-12. Energy deposited by 100-keV ion beam, 10-Å X rays and 20-keV *e*-beam into organic resist.

ionization potential, and S is the path length traveled by electrons which for backscattered electrons >1 keV can be of the order of several micrometers. Since the majority of incident electrons are deposited in the substrate, the atomic nature of the substrate can influence the sensitivity and profile of the resist. On a gold substrate (Fig. 3-1-22), more backscatter distorts the profile, while on silicon with lower backscatter, the profile shows steep walls. In electron absorption, although the energy loss is dominated by excitation and ionization of orbital electrons, the direction of electrons is subtended by small angular nearly elastic scatter from the nuclei. As the high energy cascades to lower energy levels and emits the secondary low-energy electrons (<50 eV), the Bethe approximation does not hold, since a small number of collisions brings the electrons to rest.[347] The liberated photoelectrons lose their kinetic energy by exciting other molecules along its path. Crewe has calculated that the radius of damage of photoelectrons is approximately 10× the radius of the polymer and the number of polymers damaged, N, depends exponentially on the number of incident electrons,[32]

$$N = N_0[1 - \exp(-tn/AN_0)] \qquad (3\text{-}1\text{-}11)$$

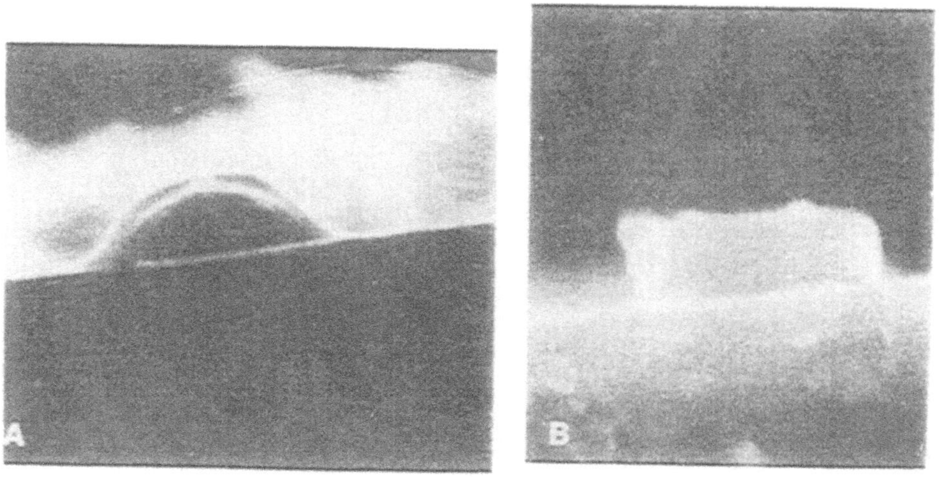

FIGURE 3-1-13. (A) Image in 1.0-μm-thick KTFR (Kodak) resist exposed to 25-keV electron beam. (B) Image in 1.0-μm-thick KTFR exposed to 125-keV proton ion beam.

TABLE 3-1-3. Historical Contributions of PMMA to High-Energy Lithography

Contribution	Reference
First *e*-beam positive resist	15
First ion-beam exposure	16
First X-ray exposure	17
First lift-off	18, 21
First multilayer resist	19
Small fabricated metal lines (17 nm)	20

where N_0 is the original number of possible grams per centimeter, t is the thickness of the film resist, and Λ is the mean free path of the scattered electron. The maximum value of N_0 would be

$$N_0 = 3t/4\pi R^3 \tag{3-1-12}$$

The exponential dependence can be characterized by a "$1/e$" dose given by

$$N_e = N_0 \Lambda/t = 3\Lambda/4\pi R^3 \tag{3-1-13}$$

From this equation we can see that the speed of a photoresist depends inversely on the cube of the grain size with the only free parameter being the mean free path, Λ. The average energy loss deposited between inelastic collisions is characterized by a mean free path between collisions.

For organic materials (e.g., polyethylene) the mean free path is about 400 Å at 20 keV so that

$$N_e = 10^{-6}/R^3 \qquad (e^-/\text{cm}^2) \tag{3-1-14}$$

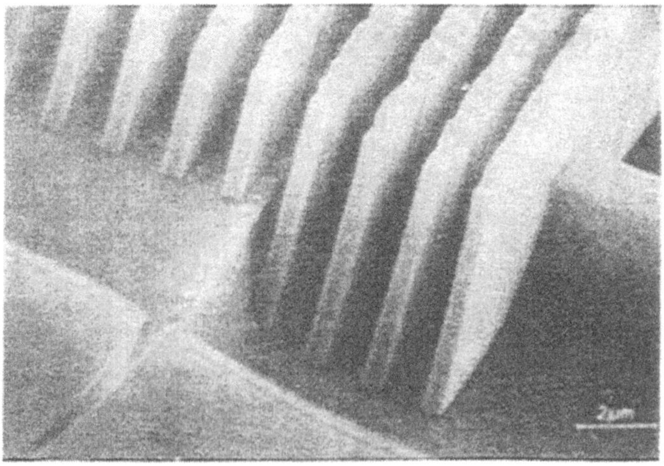

FIGURE 3-1-14. Steep-aspect-ratio X ray printed in PMMA. From H. Luthje, "X-Ray Lithography for VLSI," courtesy of *Philips Tech. Rev.* **41**, 150 (1983).

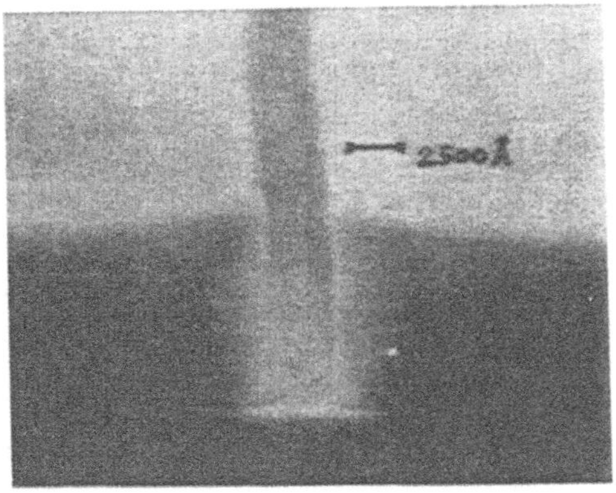

FIGURE 3-1-15. Image in 1.2-μm-thick PMMA (0.22 μm wide) exposed at 25 keV in e-beam. From M. Hatzakis, *J. Vac. Sci. Technol.* **12**, 1276 (1975), copyright and courtesy of AIP.

or

$$\text{Dose} = 1.5 \times 10^{-25}/R^3 \qquad (\text{C/cm}^2) \qquad (3\text{-}1\text{-}15)$$

The mean free path for inelastic collisions depends on the values of Z, the average atomic number, and the beam voltage, V,

$$\Lambda \propto Z^{1/2} J/V \qquad (3\text{-}1\text{-}16)$$

and one can therefore gain quantum efficiency by increasing Z and decreasing V. From these relations, we see that the speed of the resist is proportional to the inverse cube of the grain size. The largest grain size is set by the required edge resolution. To illustrate the role of the maximum grain-size effect for a given exposure, we estimate the electron flux (F) for a 1-cm^2 chip with 0.1-μm^2 pixel

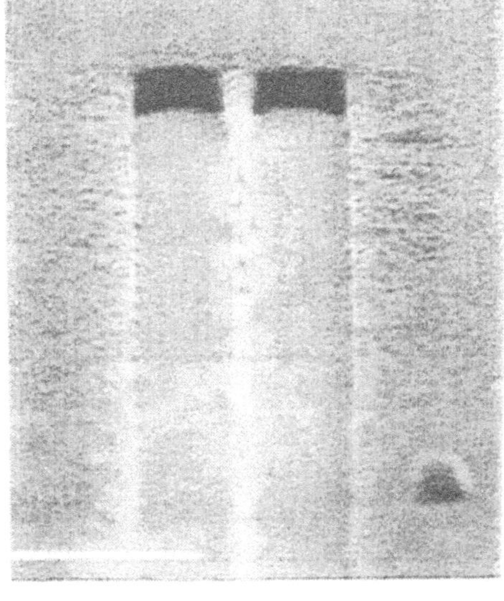

FIGURE 3-1-16. Interproximity effect—the 1-μm land between 2-μm spaces has shrunk to 0.5 μm.

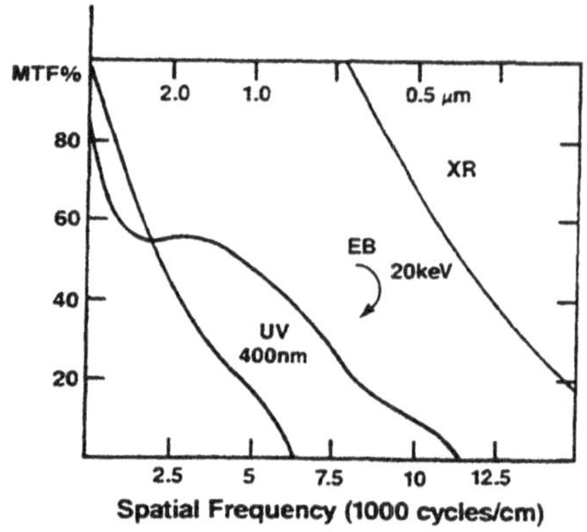

FIGURE 3-1-17. Comparison of MTF of X-ray, *e*-beam, and UV exposure of resist.(24,25)

elements(108) and a 0.1-μm edge resolution. Here the maximum grain size is ⩽ 1 μm. Thus,

$$F = 10^{-6}/10^{-15} = 10^{9}\ \text{electrons/cm}^2 \text{ or } 10^{-10}\ \text{C/cm}^2 \tag{3-1-17}$$

For ion-beam exposure, the nuclear and electronic stopping (Fig. 3-1-23) is treated by LSS (Linhard–Scharff–Schott) theory.(23) At higher ion-beam energies (> 100 keV), the energy loss is mainly by nuclear collisions.(31) The wavelength for X-ray exposure is $12.4/E$ in keV, $0.39/E^{1/2}$ for *e*-beam, and $(m/M^{1/2})\,(0.39/E^{1/2})$ for ion-beam exposure, where m is the mass of the electron and M is the mass of the ion. Since ions lose their energy through inelastic collisions with the atomic electrons of the solid and by nuclear collisions with the atomic nuclei,(29,334,335) the heavier ions like O^+ are more effective than can be accounted for on the basis of relative energy loss (LSS). This nonlinear result has been interpreted as self-induced ionization of adjacent molecules aside the linear energy transfer (LET) path of an ion.(340) Along the path, the proximity of charged ions increases the sensitivity (charge deflection) of absorption resulting in an ion pair formation. Therefore, resists exposed with high-energy ions require less particles per square centimeter than can be accounted for by inelastic collisions.(31)

3-1-4. Radiation Chemical Reactions

Electron impact processes which produce electronic excitation as well as ionization, involve electrons at all energies below the maximum for the particular incident radiation. For electrons of 200 eV or less energy, experiments on electron

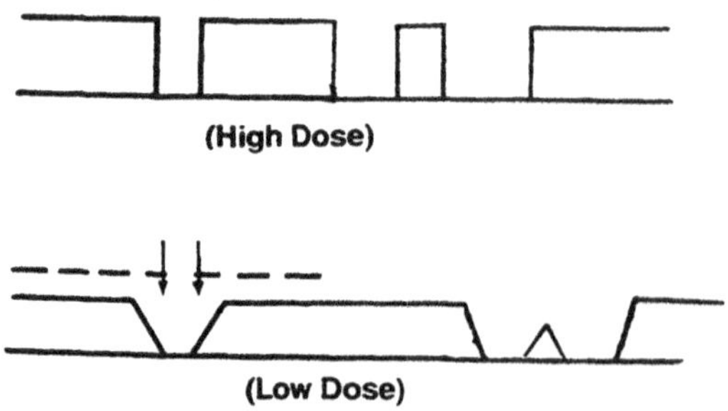

FIGURE 3-1-18. Positive resist images produced at high dose dominated conditions and low dose developer dominated conditions.

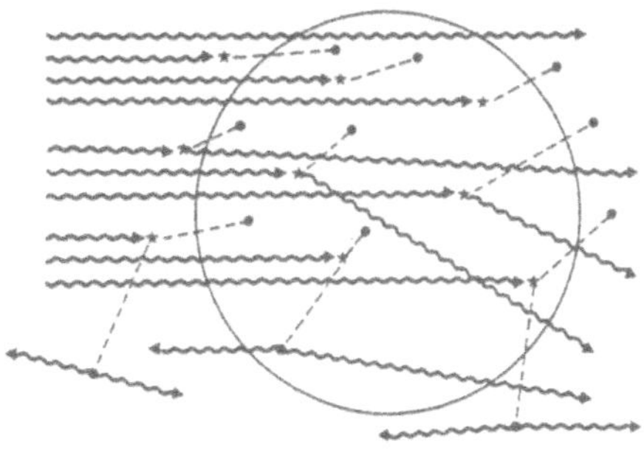

FIGURE 3-1-19. Incoming high-energy electromagnetic waves (E) (electron-beam, X rays, ions) strike resist polymer molecule and lose energy by secondary (e) electron emission. From Ref. 301.

energy loss spectroscopy have demonstrated that the excitation of molecules by electron impact is governed by the same selection rules that apply to photochemical reactions. Totally allowed optical transitions may be expected under electron impact. Even optically forbidden transitions (singlet to triplet, Chapter 4) may be produced by electrons with energy in the range of 100 eV through the process of electronic energy excitation.[33]

Thus, ionizing radiation produces a variety of intermediates:

1. Ionization to form cationic species, A^+, AB^+
2. Electronic excitations in allowed and forbidden transitions
3. Electronic attachment to form anionic species

There are two main processes by which an electron loses its freedom; Combination with a positive ion,

$$AB^+ + e^- \rightarrow AB^*$$

to form the "photochemical" excited state or by dissociative electron capture

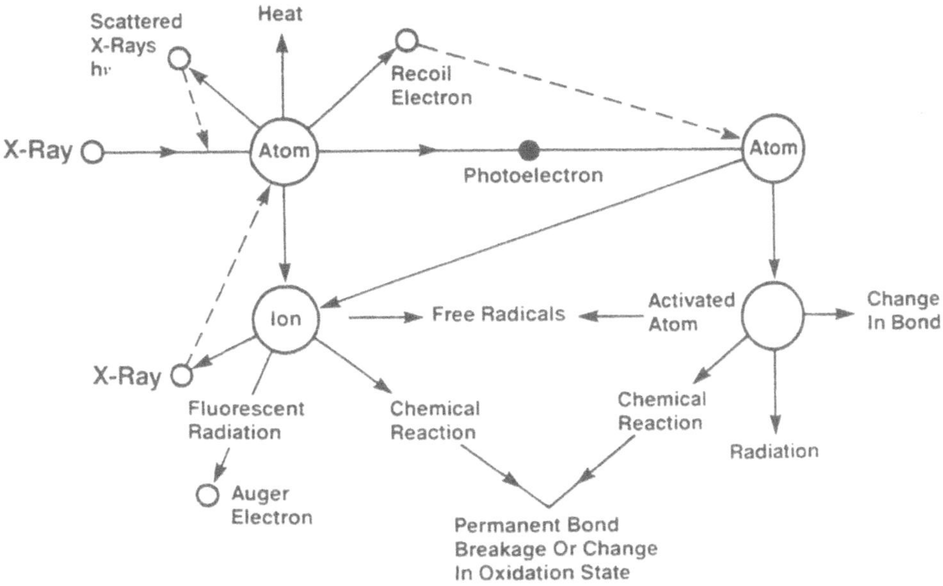

FIGURE 3-1-20. Schematic of the interaction of X rays with the atoms of a resist molecule. From Ref. 301.

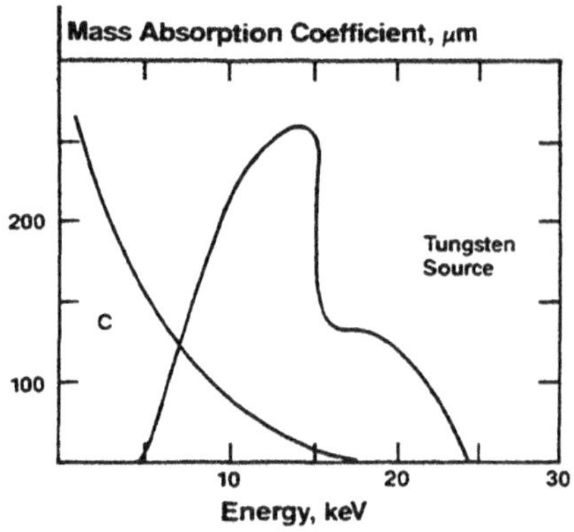

FIGURE 3-1-21. X-ray source output of a conventional X-ray anode and the X-ray absorption spectrum of a chlorinated polyether (C). From Ref. 22, by permission of John Wiley & Sons, Inc.

especially with atoms (Cl) or groups (CN) (Table 3-1-4) that have high electron affinity,

$$AB + e^- \rightarrow A + B^-$$

For example, in PMMA,

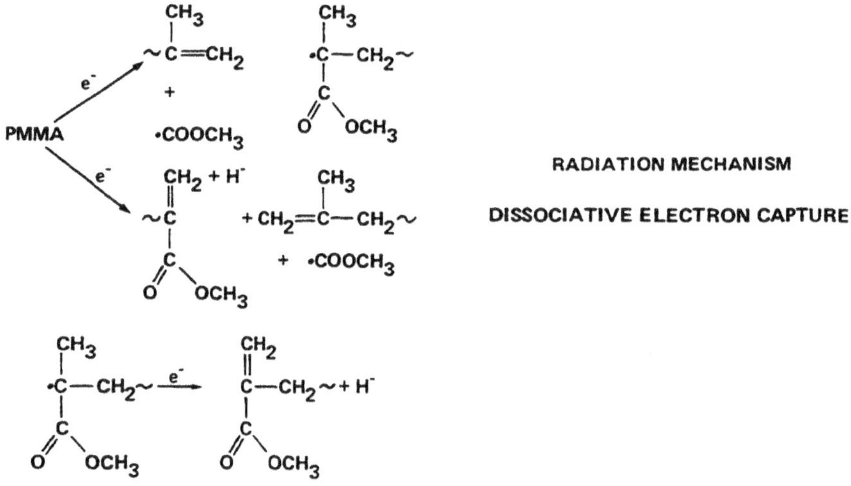

TABLE 3-1-4. Electron Affinities of Atoms[34] and Radicals

Atom	Electron affinity (eV)	Radical	Electron affinity (eV)
H	0.75	OH	2.1
C	1.7	NH_2	1.2
O	2.2	CH_3	1.2
S	2.4	CN	3.6
F	3.6		
Cl	3.8		
Br	3.6		
I	3.2		

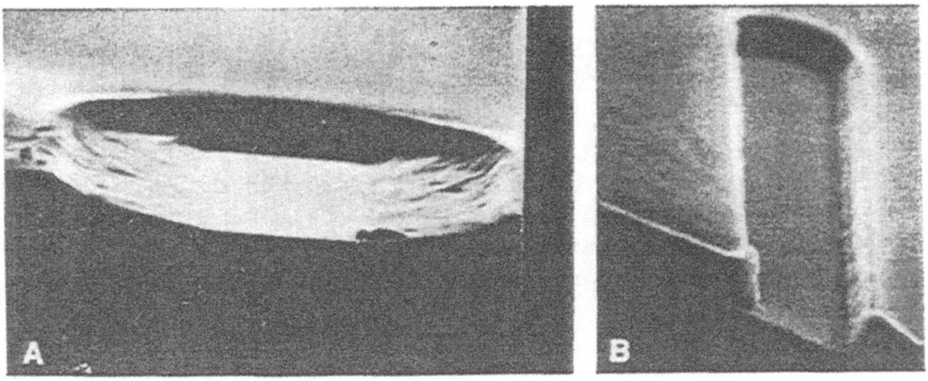

FIGURE 3-1-22. Positive *e*-beam resist exposed at same dose of 20 keV (A) on gold substrate and (B) on silicon substrate.

the introduction of strong electron-attracting groups (Cl, CN) in the ∝ position of PMMA,

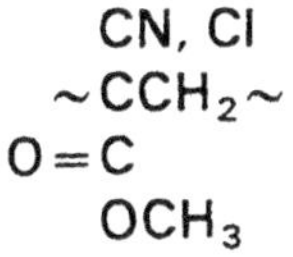

lowers the ionization potential and enhances the secondary capture to cause threefold more chain scissions in PMMA,

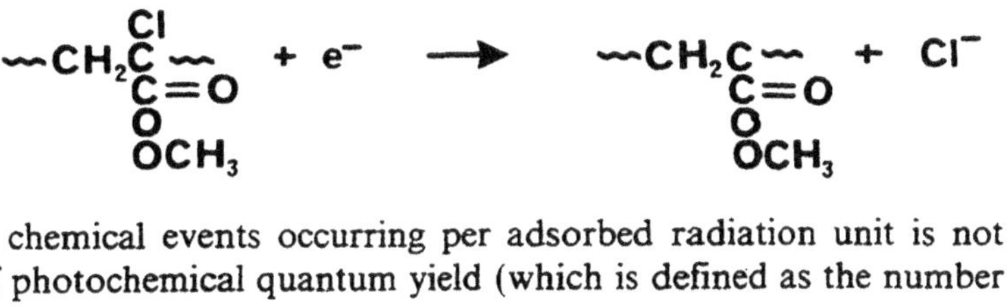

The number of chemical events occurring per adsorbed radiation unit is not defined on a basis of photochemical quantum yield (which is defined as the number

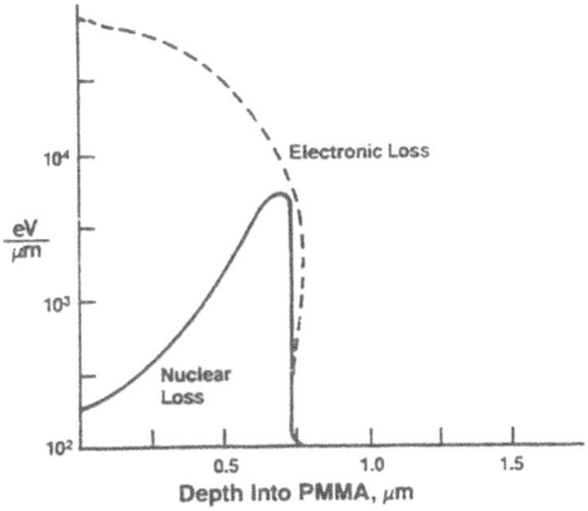

FIGURE 3-1-23. Nuclear and electronic energy loss of 60-keV H^+ ions into PMMA. From Ref. 23, courtesy of AIP.

of products formed or reactants consumed per mole of photon adsorbed) but on the basis of the radiation G value,

$$G = \text{Number of chemical events/100 eV absorbed} \qquad (3\text{-}1\text{-}18)$$

The chemical events can be scissions (G_s) or cross-links (G_x). For PMMA, a photon quantum yield (Φ) of 0.02 corresponds to a G value of 1. For a diazoquinone exposed to photons, a Φ of 0.2 corresponds to a G value of 7. For example, an approximate equivalency of e-beam exposure (10–25 keV) for other forms of exposure is as follows:

$$5\ \mu\text{C/cm}^2 = 125\ \text{mJ/cm}^2\ (\text{X-ray})$$
$$5\ \mu\text{C/cm}^2 = 6 \times 10^3\ e^-/\text{cm}^2$$
$$5\ \mu\text{C/cm}^2 = 10\ \text{Mrad}\ \gamma\ (\text{Co 60})$$
$$5\ \mu\text{C/cm}^2 = 1 \times 10^4\ \text{J/kg of absorber}$$

G_s values range from about 0.1 to 20 with most polymeric scissions about 1–3. For e-beam exposure on wafers, corrections must be made for backscattered electrons(25–28,355) as well as the forward absorption, Grun range. The energy deposition at penetration depth z is

$$dE/dt = (J_0/e)(\delta v/\delta z)_z \qquad \text{eV/cm}^3\ \text{sec} \qquad (3\text{-}1\text{-}19)$$

with $\delta v/\delta z$ the specific energy dissipation in eV/cm, J_0 the incident current density, and e the electron charge. The specific dissipation can be written as(26)

$$dv/dz = AV_a/R_G f(x) \qquad \text{eV/cm} \qquad (3\text{-}1\text{-}20)$$

for a beam voltage V_a (keV) and the Grun range, R_G, given by

$$R_G = (0.046/p)\ V_a^{1.75} \qquad \mu\text{m} \qquad (3\text{-}1\text{-}21)$$

and $f(x)$ the depth–dose function in terms of the normalized penetration ($f = z/R_G$). For a polymeric resist, the density of the target is p and nearly unity.

Since the ionizing radiation transfers its energy to molecules via charged par-

TABLE 3-1-5. Reaction Products of PMMAN-Anhydride Exposure[a]

Product	Quantum efficiency of UV photolysis at 254 nm (percent)	G of ^{60}Co γ-ray radiolysis (molecules/100 eV)	G of e-beam exposure at 2.5 keV (molecules/100 eV)
H_2	0.06	0.005	4.7
CH_4	0.3	0.003	0.9
CO	2.2	4.2	9
CO_2	2.2	6.5	7

[a] From Ref. 35.

ticles liberated in the medium, the probability of dissipating energy in a particular part of the molecule is proportional to the electron density in that region.

While primary events are considered to occur at random in a given polymer, specific chemical factors do prevail at weak links or at highly absorbing end groups. Thus, the site at which a chemical bond is broken is not necessarily the site at which energy absorption occurred. Consequently, different forms of exposure may produce different products due to the difference in absorption coefficients or excited state population. For example, Hiroaka[35] found different reaction products for irradiation of PMMAN (polymethylmethacrylic acid anhydride) with various exposure sources (Table 3-1-5).

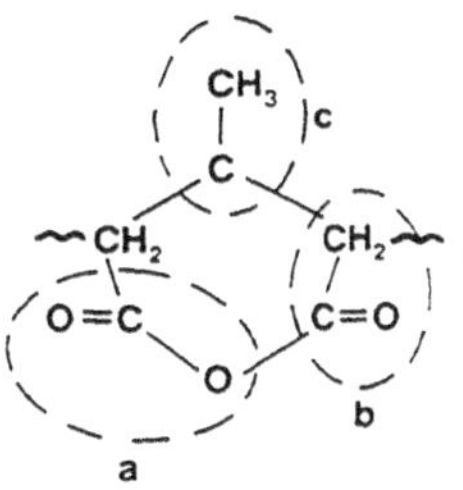

PMMAN breakdown

The complete analysis of the breakdown of macromolecules after high-energy irradiation is a series of complex analysis involving mass spectroscopy, photoelectron spectroscopy, vacuum UV photochemistry, and gel permeation chromatography. For example, the direct analysis of molecular weight fragments of polymers of up to 12,000 atomic mass units has been achieved by direct mass spectroscopy.[36]

In linear positive degrading resists, the scission of many different backbones consisting of C–C, C–S, C–O, and C–N bonds occurs respectively in such resists as PMMA, polyolefin sulfones, polyaldehydes, and polyimines (pre-cross-linked polymethacrylamide). From studies of the mass spectra of alkanes and small molecules, definite patterns have been established in the breakdown of primary, secondary, and tertiary substituted carbons. As early as 1954, Miller [*J. Polym. Sci.* **14**, 503 (1954)] pointed out (Table 3-1-6) that tertiary substituted polymers are the most likely to degrade. Thus, a tabulation can be made of commercial plastics and their propensity to degrade (positive resists) or cross-link (negative resists).

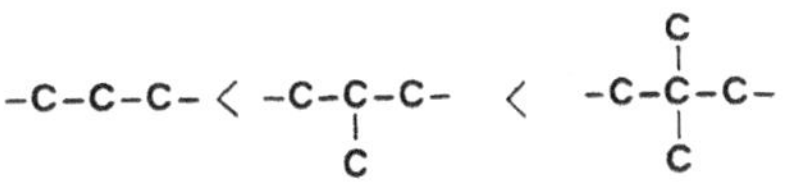

It can be expected that polymers which are degraded by γ radiation can be easily formulated into positive or negative working *e*-beam, ion-beam, or X-ray resists. Furthermore, the breakdown of neopentanes supports Miller's early observations,

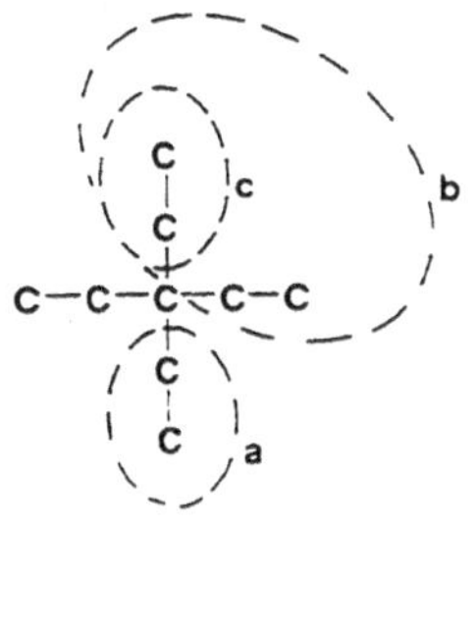

TABLE 3-1-6. Behavior of Solid Polymers under Radiation (in the Absence of Oxygen)[a]

Polymers which cross-link	Polymers which degrade
Polymethylene	Polyisobutylene
Polyethylene	Polytetrafluorethylene (PTFE)
Polypropylene	Polymonochlorotrifluorethylene (Kel F)
Polystyrene	Poly-α-methyl styrene
Polyvinyl chloride (?)	Polyvinylidene chloride
Polyvinyl alkyl ether	
Polyvinyl methyl ketone	
Chlorinated polyethylene	
Chlorosulfonated polyethylene	
Polyvinyl acetate	
Polyacrylonitrile	Polymethacrylonitrile
Polyacrylic acid and esters (polymethyl acrylate)	Polymethylacrylic acid and esters (polymethyl methacrylate)
Polyacrylamide	Polymethacrylamide
Rubber	
Polybutadiene	
Polychloroprene (Neoprene)	
Polyamides	
Nylon	
Polycaprolactam	
Copolymers	Cellulose
Styrene-butadiene	Cellulose derivatives:
Styrene-acrylonitrile	acetate, nitrate, etc.
Vinyl chloride-vinylidene chloride	Copolymers
Polydimethyl siloxane	Butyl rubber
Polyphenyl siloxane	
Polyethylene oxide	

[a] From Ref. 292.

which occurs by multiple bond scissions at bonds—a, b, c—and probably accounts for the variety of products observed by Hiroaka (Table 3-1-5) from the degradation of PMMAN. In the primary breakdown by ionization, the radicals formed from alkanes such as polypropylene, polyisobutylene, and polyisoprene are ionized,[34]

$$R + e^- \rightarrow R^+ + e$$

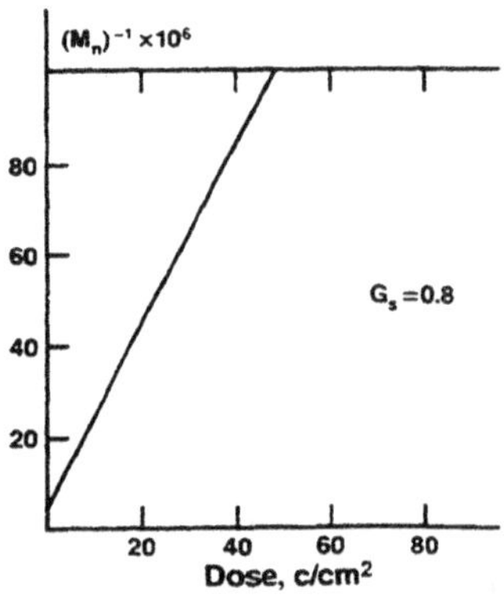

FIGURE 3-1-24. Determination of G value of PMMA by plot of inverse scissioned number average molecular weight (M_n) versus dose.

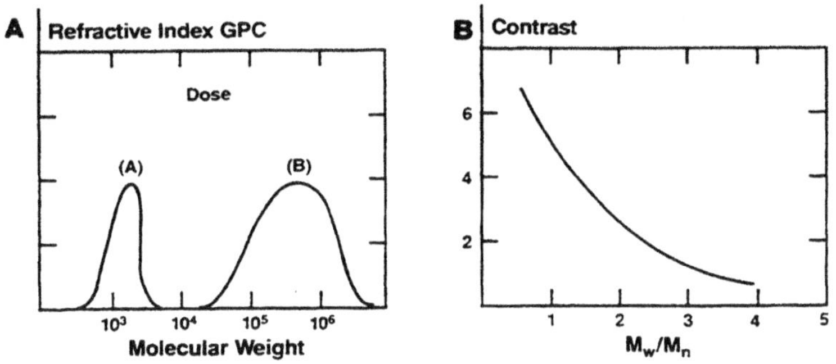

FIGURE 3-1-25. (A) Molecular weight of positive PMMA resist by GPC before (B) and after (A) exposure. (B) Contrast of PMMA as a function of dispersity.

in order of their higher substitution:

t-C_4H_9	6.9 eV (ionization potential)
sec-C_3H_7	7.9 eV (ionization potential)
C_2H_5	8.7 eV (ionization potential)
CH_3	10.0 eV (ionization potential)

The amount of permanent scissions (G_s) is measured by the ratio of initial molecular weight M_0 (number average) to final molecular weight M_n,[37,355,359]

$$1/M_n = 1/M_0 + (G_s - 4G_x)/9.65 \times 10^5 Q \qquad (3\text{-}1\text{-}22)$$

Higher doses (Q) can cause cross-linking G_x in polymers like PMMA and poly-α-fluoromethacrylates.[154] A plot of reciprocal molecular weight versus dose determines the G_s (scission) value (Fig. 3-1-24). Typical values range from 1 for PMMA to 10 for polyolefin sulfones. To achieve an R/R_0 of 5, it is necessary to expose PMMA with sufficient dose to cause ten chain scissions. For PMMA, the amount of those Q and the number of chain scissions combined with developer of solubility dissolution parameter (a) determines the film thinning ratio,

$$R/R_0 = k(M_0/M)^{-a} = (1 + kQG_s/M_0)^{-a} \qquad (3\text{-}1\text{-}23)$$

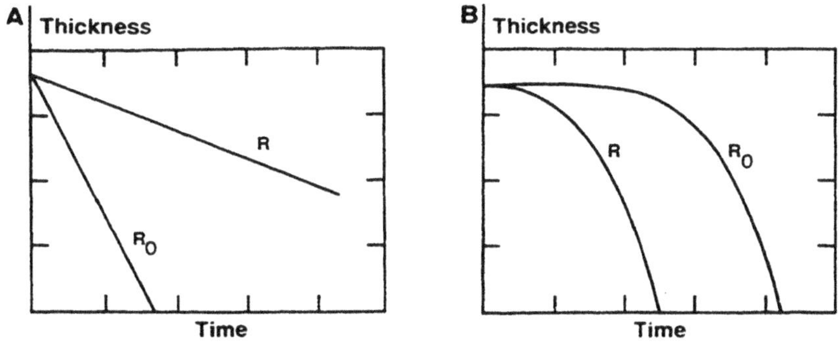

FIGURE 3-1-26. (A) Linear dissolution resist; (B) nonlinear with induction before dissolution. R are exposed and R_0 are unexposed dissolution rates.

High initial molecular weights of low dispersity (Fig. 3-1-25) improve the contrast and increase the sensitivity. After exposure, the dispersity rapidly converges to 2, while the resist contrast (γ) improves. For a 1 $\mu C/cm^2$ resist, the sensitivity requires the G_s to be > 10, unless a swelling developer ($a > 3$) can be found such as in polyhexafluorobutyl methacrylate.(51,58) However, swelling severely limits the resolution of features to dimensions of 2–3 μm. The choice of developer can also influence the resist contrast (Fig. 3-1-27 and Chapter 10). For example, amyl acetate induces a nonlinear development rate (induction effect) in PMMA (Fig. 3-1-26B) while methyl isobutyl ketone (MIBK) (Fig. 3-1-26A) dissolves both the exposed and unexposed regions in a linear fashion. The nonlinear resist system does not reproduce the topside scatter as aggressively as the MIBK-developed images.

There are two families of positive radiation resists—one-component and two-component resists derived from the compositions of positive photoresists:

One-component	Two-component
Polyacrylates	Novolak + diazoquinone
Polyolefin sulfones	Novolak + polyolefin
Polyaldehydes	Sulfones
	Onium salts + polycarbonates

We will concentrate on these resists and examine their response to high-energy radiation. It is expected that they all respond in a predictable manner to all forms of high-energy radiation. However, there are some exposure conditions or environments which can change the resist speed or alter the course of a radiation reaction. For example,

1. Electron-beam exposure is performed under high vacuum (10^{-8} mm) while X-ray and ion-beam exposure are done at slight vacuum or in an inert gas atmosphere. If traces of oxygen are present, the resist sensitivity or linewidth may be altered by reaction with oxygen.
2. Electron beams of > 10 A/cm^2 may heat the resist on chrome/glass substrates to > 100°C during exposure and enhance the resist sensitivity.(118,119)

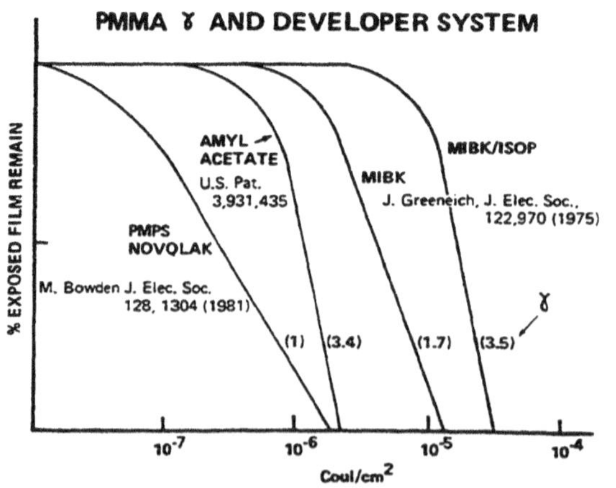

FIGURE 3-1-27. PMMA contrast plot with different developers.

3. Electronic energy transfer in two-component novolak–diazoquinone resists has been suggested to occur[38] which is sensitive to the ratio of each component. A low loading of diazoquinone (DQ) will favor a lower dose since the novolak resin transfers its electronic energy more efficiently to the DQ component.
4. Secondary factors such as the energy (keV) of the ray, backscatter from the substrate, and the beam intensity (reciprocity failure) must be kept in mind.
5. Finally, from the user's requirement of the resist, whether thinning can be or is tolerated for a particular resist image can introduce uncertainties of 2–5× in the true resist speed (minimum dose quoted).

3-2. X-RAY POSITIVE RESISTS

3-2-1. Basic Exposure Tool

In X-ray printing, the mask after alignment is deliberately offset above the wafer (10 μm) to prolong its life. Because the X-ray wavelength is small, diffraction effects can be ignored and simple geometrical considerations are used in relating the image to the pattern on the mask (Fig. 3-2-1). The opaque parts of the mask cast shadows onto the wafer below, but the edge of the shadow is not absolutely sharp because of the finite extent (S) of the X-ray source (diameter of the focal spot of the electrons on the anode) at distance D from the mask. If the gap between mask and wafer is set at g, the blur σ is given by

$$\sigma = g(S/D) \qquad (3\text{-}2\text{-}1)$$

Typical values are $g = 20\ \mu$m, $S = 3$ mm, and $D = 40$ cm, with the resolution determined (1) by the blur (σ), (2) by the minimum linewidth achievable in mask fabrication, and (3) by the contrast of the resist. In a high-contrast resist, the edges can be considerably sharper than indicated by σ.[40] As the gap increases, the prin-

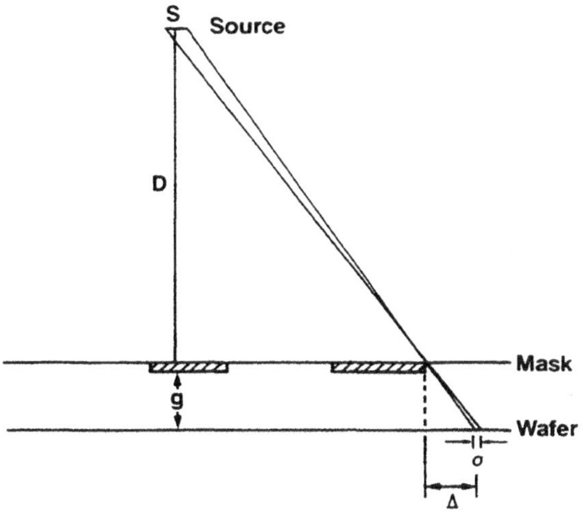

FIGURE 3-2-1. Shadow effects at edges of X-ray field of exposure which distort images (runout). Reprinted with permission of *Solid State Technology*, published by Technical Publishing, a company of Dun and Bradstreet.

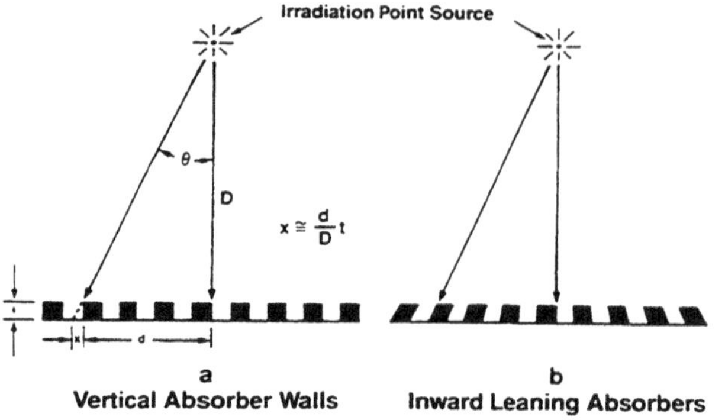

FIGURE 3-2-2. Beveled outer edges of absorber mask reduce shadow effects in X-ray printing. From Ref. 46, by permission of the Electrochemical Society, Inc.

ted image bias deteriorates and the exposure time becomes longer. In order to reduce the lateral distortion from a point source, inward leaning mask absorbers have been used (Fig. 3-2-2).[46] In the source (S) configuration, the electron gun used to excite the X-ray source is usually at right angles to the target (Fig. 3-2-3),[41] although parallel arrangements have also been used in which a flood e-beam excites a target in a parallel fashion.[42] Comparatively, the output of conventional sources is low:

Conventional	0.1–1.0 mW/cm^2
Pulsed[143]	10–100 mW/cm^2
Synchrotron[44]	100–1000 mW/cm^2

Since a continuous semitransparent membrane is used for a mask, at least 50% of the radiation is absorbed by the mask (Figs. 3-2-4 and 3-2-5). Pulsed X-ray sources have been shown to cause sufficient thermal stresses to permanently warp the mask and laterally distort submicrometer geometries.[85,353] In order to balance the contrast of the transmitted beam with the exposure (intensity) time, some of the X-ray radiation passes through the opaque portion of the mask. The minimum contrast between the transparent and the opaque absorbing portion of the mask to

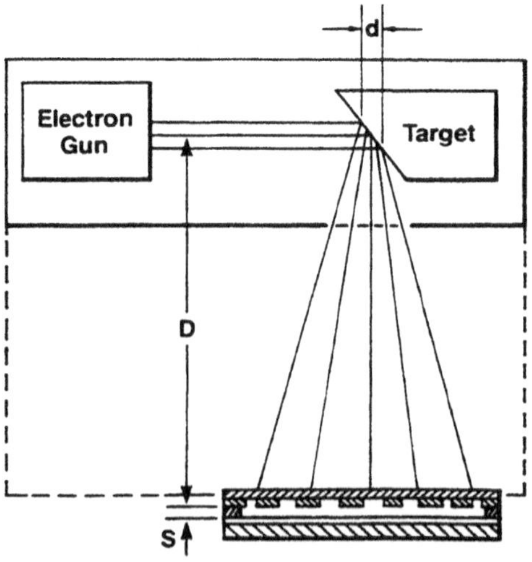

FIGURE 3-2-3. Right-angle X-ray source. From Ref. 41.

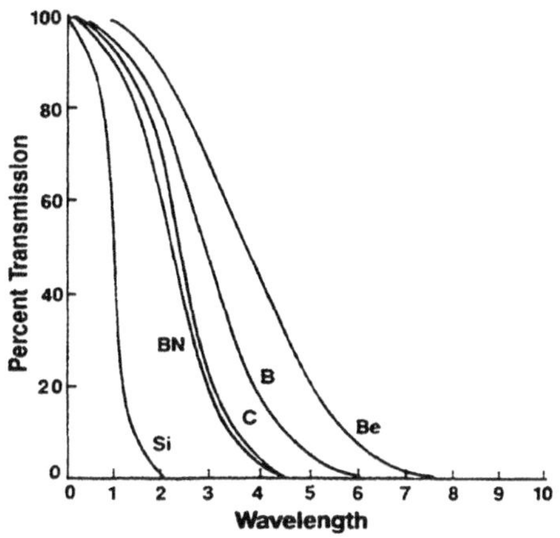

FIGURE 3-2-4. Transmission of several X-ray mask substrate materials, 0.02 cm thick. Courtesy of R. Feder, IBM.

expose the resist is about 4 to 1 (60% modulation) (Fig. 3-2-6).[46] Secondary electrons emanating from the Au absorber of the mask can cause edge blurring.[45] To reduce the emission of secondary electrons, topcoats of polyimide are used over the gold patterns to absorb the ejected photoelectrons which can cause an image distortion of 0.2–0.4 μm.[45,326,351] By using longer-wavelength (> 10 Å) sources (Fig. 3-2-7), only nondistorting photoelectrons are ejected from the mask. Once the X rays strike the substrate, secondary photoelectrons of short range are backscattered which can cause coving at the foot of an image in a positive resist (Fig. 3-2-8).[334]

Many critical problems remain to be solved before X-ray printing becomes a viable process (Table 3-2-1).[47] For a minimum throughput of 2–5 (125-mm) wafers/hr, a resist with 1 mJ/cm^2 sensitivity (Fig. 3-1-8) is definitely required for conventional (*e*-beam excited) X-ray sources.[65]

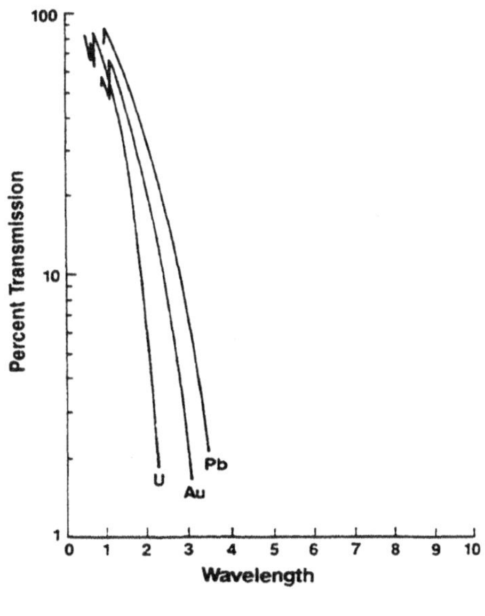

FIGURE 3-2-5. Transmission of several X-ray absorbers on mask 2 μm thick. Courtesy of R. Feder, IBM.

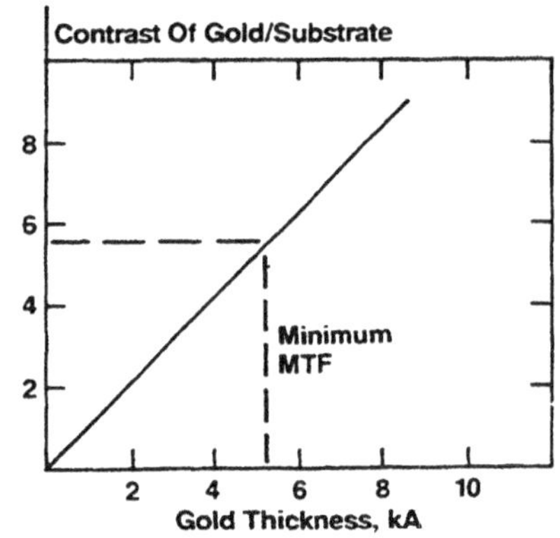

FIGURE 3-2-6. Minimum MTF and contrast from gold thickness. From Ref. 46, reproduced by permission of the Electrochemical Society, Inc.

3-2-2. X-Ray Positive Resists

The speed (minimum dose, Q) of an X-ray positive resist is a direct function of the efficiency of X-ray absorption by the resist and the number of bonds broken (G_s),

$$\text{Speed} \propto \text{Quanta absorbed } (I_a)/G_s \tag{3-2-2}$$

The radiation absorbed (I_a) is a function of the atomic masses, A, of the resist and the X-ray exposure wavelength (Fig. 3-2-9)[73]:

$$I_a = I_0 \exp(-\mu_m t) \tag{3-2-3}$$

where μ_m are the sum of the atomic (A_i) absorption coefficient of a resist of density p and thickness t,

$$\frac{\mu}{p} = \frac{\sum n_i \left(\frac{\mu}{p}\right)_i A_i}{\sum n_i A_i} \tag{3-2-4}$$

TABLE 3-2-1. X-Ray Lithography[a]

Issues	Solutions
Mask technology	
Membrane stability (0.01 μm)	Titanium (opaque and transparent viewing) mask
Defect density (< 0.5/cm²)	Silicon carbide (transparent) mask
Patterning distortion (0.03 μm)	Boron nitride (transparent) mask
Overlay error	
Overlay requirement (0.1–0.2 μm)	Geometrical optical approach
Alignment reqirement (0.02 μm)	Physical optics alignment technology
X-ray resist	
Sensitivity (2–4 mJ/cm²)	Commercial developments (EK88 and COP)
Dry etch compatibility	Advanced resist developments
X-ray source	
High brightness (10 and 20 kW)	Conventional X-ray source (tungsten rotating anode)
Small spot size (1.5–3.0 mm)	Plasma source

[a] From Ref. 47.

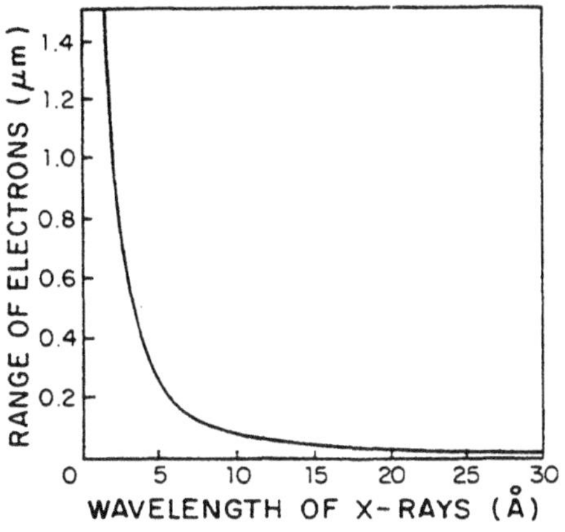

FIGURE 3-2-7. Range of photoelectrons emitted from absorber mask as a function of backside X-ray wavelength. Adapted from Refs. 3 and 343.

An empirical approximation for the X-ray absorption coefficient of polymers containing C, N, and O polymers is

$$\mu = \sum 0.089\rho^{3.05} Z^{4.1} d/A \tag{3-2-5}$$

where Z is the atomic number, A is the atomic weight, ρ is the wavelength, and d is the partial density of each element. In Table 3-2-2, an example of the matching of ρ of source and the atomic orbitals of the resist atoms is shown for poly-1-butene sulfone, $C_4H_8SO_2$,

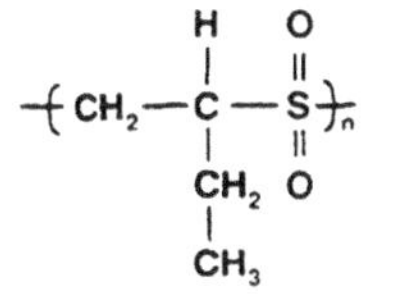

and poly-1-hexene sulfone, $C_6H_{12}SO_2$. For the Al source, the larger S atoms effectively absorb the longer-wavelength X rays. For chlorobutene sulfone polymer, $C_4H_7SO_2Cl$, Kaplan[48] calculated the X-ray cross sections to show the large contribution of heavier atoms in the absorption of photoelectrons.[302] Because of closed shells (*K*, *L*, *M*,...), a lower absorption is noted and the highest unfilled atomic subshells contribute to the greater absorbance (Table 3-2-3). Each element in the resist has a favorable absorption edge (Fig. 3-2-9 and Table 3-2-4). For the Al source, the most likely event is the ionization of the C and O orbitals, and for

TABLE 3-2-2. Mass Absortion (μ/p) for Polyolefin Sulfones[a]

	μ/p (cm^2/g) 4.37 Å (Pd, *L*)	F^b	μ/p (cm^2/g) 8.34 Å (Al, *K*)	F
$C_4H_8SO_2$	522	6.6	836	11.5
$C_6H_{12}SO_2$	437	5.2	768	8.9

[a] From Ref. 48.
[b] Fraction (%) of absorption in a 1-μm-thick film.

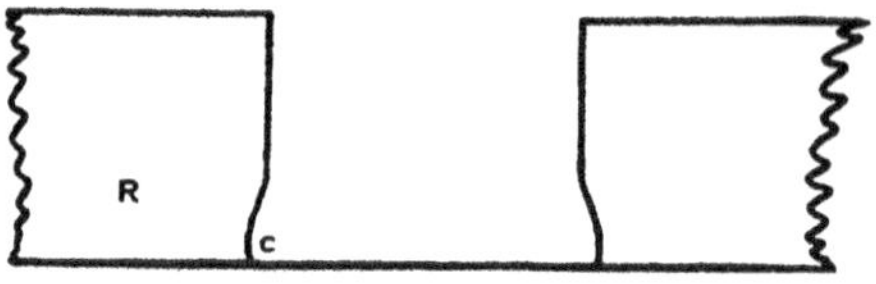

FIGURE 3-2-8. Slight coving (c) produced at resist (R) substrate interface in positive X-ray resist due to backscattered photoelectrons.

the Pd source, the ionization of the S and Cl atoms in the polymer, hopefully, occur with expulsion of SO_2 and the concomitant breakdown of the polyolefin sulfone,

$$R\ SO_2\ R' \longrightarrow [RSO_2R']^{+} + e^{-}$$

$$[RSO_2R']^{\dot{+}} \longrightarrow RSO_2^{+} + \cdot R',\ R^{+} + SO_2$$

An extensive tabulation of the X-ray absorption coefficients of organic resist atoms and all elements can be found in Table 3-2-5 and Fig. 3-2-10, respectively. In order to increase the X-ray absorption of carbonaceous polymers, heavy metal salts of U and thallium (Tl) have been complexed with polymethacrylic (MA) acid,[49,50,341]

CH_3
$\sim CCH_2 \sim$
$C{=}O$
O
Tl

and the X-ray sensitivity of PMMA-MA was reduced by an order of magnitude to 25–50 mJ/cm^2. In this case, the absorption of X rays by the metal caused side chain scission and destroyed the metal salt bridge releasing a polar species.[49] Except for resists containing heavy metals,[52] the absorption of high-energy[51] electrons (25 keV) or X rays leads to similar values of resist sensitivity (Table 3-2-6 and Fig. 3-2-11[51]). In polyvinylferrocene, the heavy metal (Fe) absorbance of X rays probably accounts for the increased X-ray sensitivity over *e*-beam exposure.[52]

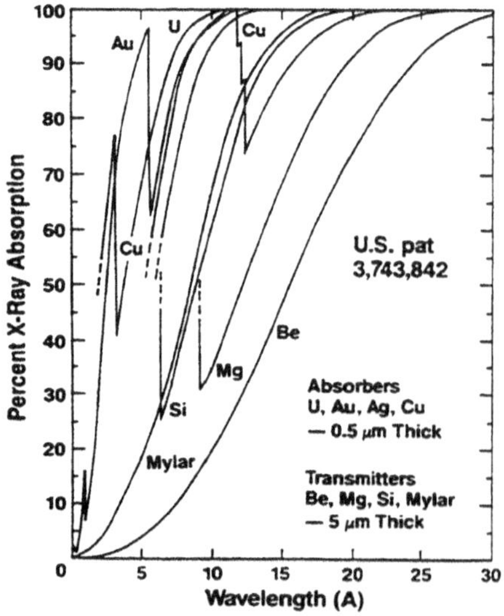

FIGURE 3-2-9. Transmission of X-ray substrates and absorption of X-ray mask material.

TABLE 3-2-3. X-Ray Photoelectron Photoionization for Chlorobutene Sulfone[a]

	Photoionization	
Orbitals	Al (1487 eV)	Pd (2837 eV)
C 1*s*–2*p*	0.243	0.042
O 1*s*–2*p*	0.345	0.065
S 1*s*–3*p*	0.183	0.407
Cl 1*s*–3*p*	0.237	0.486

[a] From Ref. 48.

The broad categories of X-ray positive resists are as follows:

One-component	Two-component
Polyacrylates	Novolak + DQ
Polyaldehydes	Novolak + polyolefin sulfones
Polyolefin sulfones	Onium salts + polycarbonates

In the popular acrylate family there are three general categories: (1) substituted (Table 3-2-7), (2) metal acrylates (Table 3-2-8), and (3) pre-cross-linked (Table 3-2-9).

By incorporation of higher-absorbing halogen atoms into PMMA, the resist sensitivity has been increased by an order of magnitude. The fast *e*-beam resists of polyolefin sulfones, polycarbonates, and polyaldehydes are also the fastest X-ray resists (Table 3-2-10). However, the positive acrylate and sulfone resists are poor reactive etch masks for single-layer masking. In two-component resists such as the DQN resins, the resins absorbed more radiation than the DQ. If X-ray exposure of DQN is compared with optical exposure to the same degree of inhibitor (DQ) bleaching, the optical DQN resists are twice as fast in dissolution than the X-ray exposed DQN.[38,66] Apparently the resin undergoes a cross-linking or an esterification reaction with decomposed DQN under X-ray exposure to inhibit dissolution.

TABLE 3-2-4. X-Ray Absorption Edge of Elements[a]

Element	Atomic Number	Highest X-ray, λ (Å)
C	6	44
N	7	31
O	8	23
F	9	18
Si	14	7
P	15	6
S	16	5
Cl	17	4
Br	35	7

[a] Data from Refs. 75 and 76.

TABLE 3-2-5. μ_m of Selected Elements at Seven Target λ[a]

		μ_m (cm^2/g)						
		Pd	Rh	Mo	Si	Al	Cu	C
Element	Z	4.36 Å	4.60 Å	5.41 Å	7.13 Å	8.34 Å	13.36 Å	44.70 Å
C	6	100	116	184	402	627	2,714	2,373
N	7	155	180	285	622	970	4,022	3,903
O	8	227	264	416	908	1415	5,601	6,044
F	9	323	376	594	1298	2022	6,941	8,730
Si	14	1149	1337	2115	279	423	1,959	36,980
P	15	1400	1630	2579	371	564	2,405	41,280
S	16	1697	1975	232	483	733	3,079	47,940
Cl	17	2013	197	301	628	953	3,596	50,760
Br	35	1500	1730	2649	3680	1456	3,101	32,550
Fe	26	630	726	1118	2330	3536	10,690	13,300
Sn	50	675	772	1160	2318	3437	9,623	6,332
Tl	81	1083	1213	1032	1904	2697	6,276	13,030
Most absorbing elements		Cl S Br P Si Heavy atoms	S Br P Si Heavy atoms	Br P Si Heavy atoms F	Br Heavy atoms F O Cl	Heavy atoms F Br O N	Heavy atoms F O N Cl	Cl S P Si Heavy atoms
Least absorbing elements		C N O	C N Cl	C S N	Si P C	Si P C	Si P C	C N O

[a] After Ref. 53.

The fastest X-ray positive resists have been reported by Ito and Willson based on the image amplification by Lewis acids released from triphenyl onium (SbF_6) salts to catalyze the acidolysis of *p*-*t*-butyl carbonate of styrene,[69]

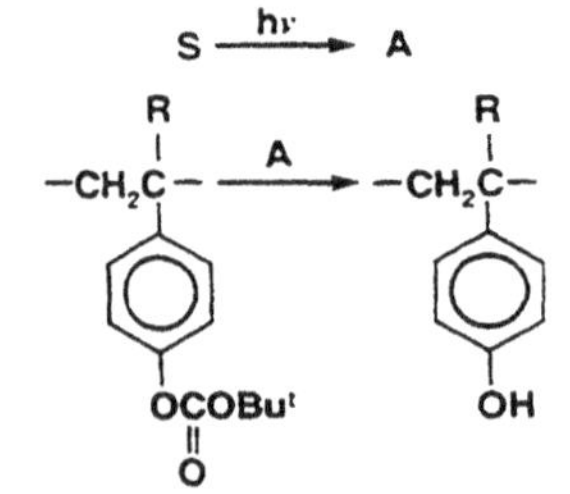

Upon postexposure baking, the acid produced by the onium salts catalyzed the decarbonylation of the side chain group to form polyvinylphenol. A quantum amplification of 50 enabled speeds of 5 mJ/cm^2 to be attained. Both positive and negative images obtained by developing with different polarity solvents. Using a similar catalysis that involves main chain unzipping in polyphthaldehyde,

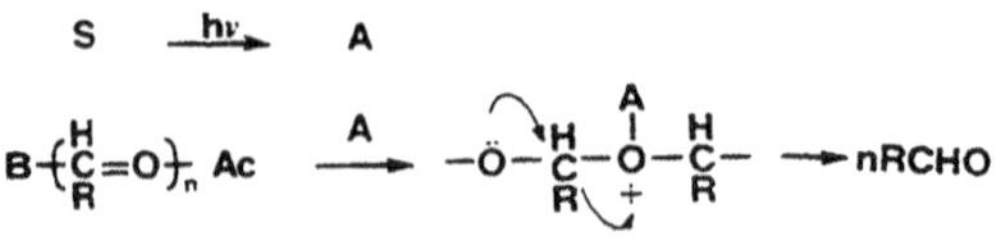

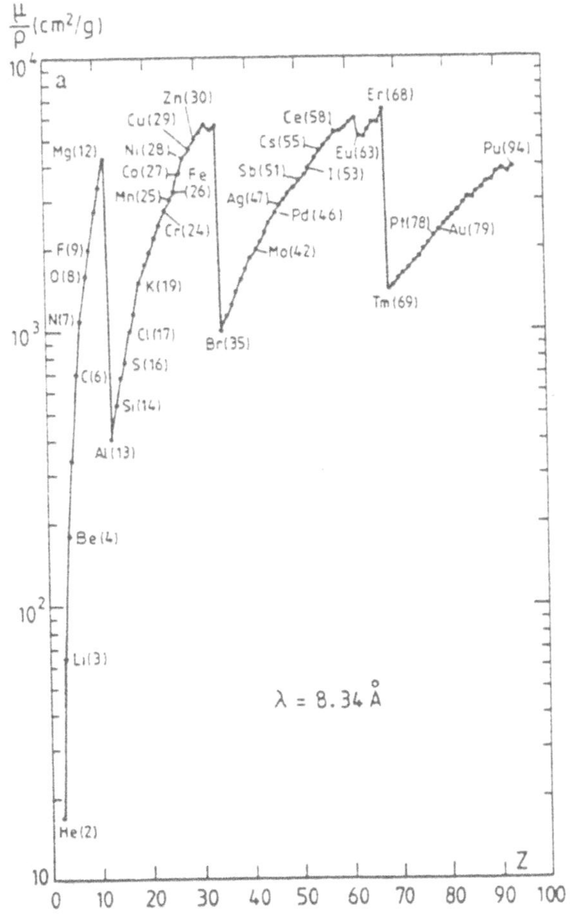

FIGURE 3-2-10. X-ray mass absorption coefficients versus atomic number Z for the Al 8.34 Å line. From Ref. 55, reproduced by permission of the American Chemical Society.

TABLE 3-2-6. Electron-Beam and X-ray Dose of Various Resists (J/cm^3)[a]

Resist	*e*-beam	X ray	*e*-beam/X ray
PMMA	400	500	1
Polybutene sulfone	80	14	5
Polyvinylferrocene	4400	150	30
Epoxy	26	25	1

[a] From Ref. 52.

TABLE 3-2-7. Halogenated PMMA and Other Substituents

Resist	Speed (mJ/cm^2)	Reference
PMMA	300–1000	45, 49, 53, 58
Polyhexafluorobutyl methacrylate	60	51, 55
Poly dimethyltetrafluoro propyl methacrylate	100	51, 55
Poly-2-fluoro ethyl methacrylate	75	55
Poly-2-chloroethylmethacrylate	75	55
Poly-2,2,2-trifluoro ethyl methacrylate	50	55
Poly-2-bromo ethyl methacrylate	75	57
PMMA-MA	200	49, 57
PMMA-MA-anhydride	100	56
Polymethacrylonitrile	100	59
Polythiomethacrylate	125	103
Polybutylcyanoacrylate	20	338

TABLE 3-2-8. Metal-Containing Acrylates

Metal	Speed (mJ/cm^2)	Reference
Cs, Tl (PMMA-MA)	25	49
V, Cr, Au (PMMA-MA)	100	50
Au on top of PMMA	50	60
PMMA + ZnI_2	18	61

TABLE 3-2-9. Pre-cross-linked Acrylate Resists

Resist	Speed (mJ/cm^2)	Reference
PMMA-methacrylyl chloride	200	62
Poly-α-cyanoacrylate-amide ethyl acrylate	50	63

TABLE 3-2-10. High-Speed X-Ray Resists

Resist	Speed (mJ/cm^2)	Reference
Poly-1-butene SO_2	14	52, 64
Poly-hexyl-butyraldehyde	12	51
Polyisobutyl aldehyde	25	51
Polymethylisopropenyl ketone	15	63

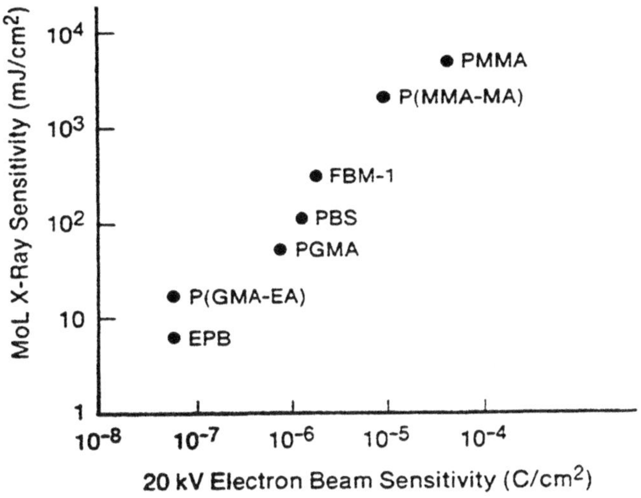

FIGURE 3-2-11. X-ray sensitivity versus *e*-beam sensitivity. From Ref. 51.

images were dry developed after doses of 5 mJ/cm^2.(69) Unfortunately, most of the fast X-ray resists are not stable to reactive ion etching, and processing with costlier multilayer resists will have to be utilized. Another potential fast resist is the image amplification of PMMA by treatment of the latent image (after a dose of only 1 mJ/cm^2) with acrylic monomers to graft interpenetrating networks into the exposed region.(77)

In summary, positive X-ray resists have attempted to enhance the sensitivity of resists. Almost one order of magnitude of sensitivity has been achieved with acrylates and two orders (1–5 mJ/cm^2) with low-ceiling-temperature aldehyde-based resists. However, no practical single-layer resist with adequate sensitivity, contrast, and reactive ion resistance has been developed. To meet these objectives, multilayer resists will be required. Increased resist absorbance has been achieved by incorporation of high-atomic-number atoms. If no resist advances are imminent, then X rays will have to rely upon high-intensity sources such as the synchrotron to be economical in meeting the throughput requirements of >25 200-mm wafers/hr. In addition, advanced alignment capabilities to better than ± 0.1 μm and thermally stable masks are required to complement the exposure process.

3-3. POSITIVE ION-BEAM RESISTS

Ion-beam lithography consists of flood exposure with ions of H^+, He^{2+}, or Ar^+ through a mask of Au on a silicon membrane or point-by-point exposure with a scanning source of liquid Ga metal(70,71) (Fig. 3-3-1). The mask is proximity gapped(79) about 20 μm from the wafer but for submicrometer work, contact is necessary,(70) because isolated images are unsupported by the etched-back mask and complementary masks are partitioned into two separate components.

Since ions are absorbed by a factor of 10–100 more effectively than electrons, 10–100× fewer ions than electrons per square centimeter are required to expose resists (10^{10}–10^{12} ions/cm^2 or 0.01–1 μC/cm^2).(72) A good correlation between

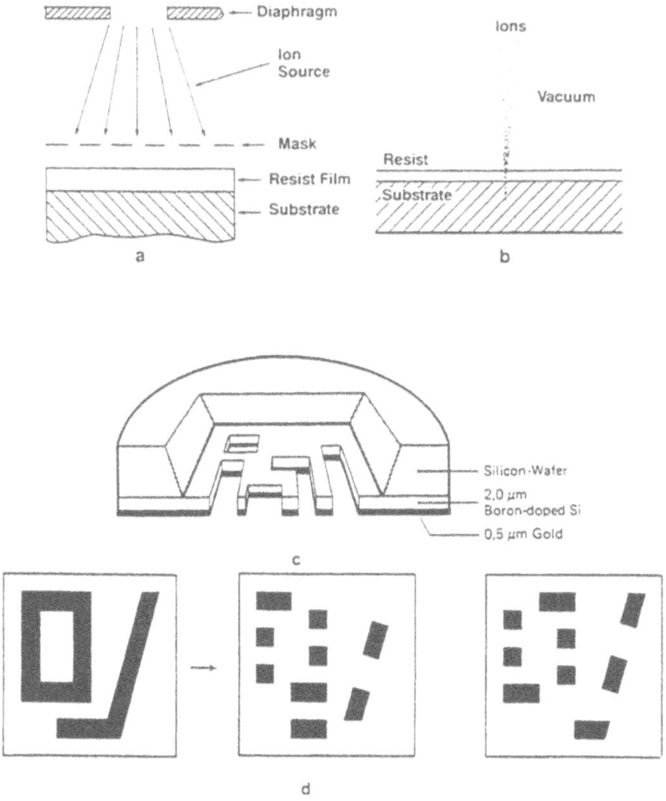

FIGURE 3-3-1. (a) Flood ion-beam exposure of resist through the stencil foil mask shown in (c). (b) Focused ion-beam source for scanning exposure of resist. (c) Stencil mask made by backetching of silicon wafer. (d) Principle of partitioning mask into complementary patterns to handle the donut hole problem. From Ref. 5, courtesy of IBM.

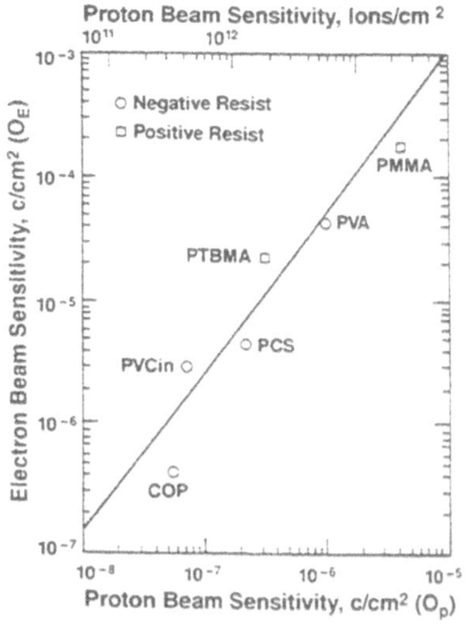

FIGURE 3-3-2. Electron-beam sensitivity versus ion-beam sensitivity. After Brault and Miller,[80] courtesy of SPE.

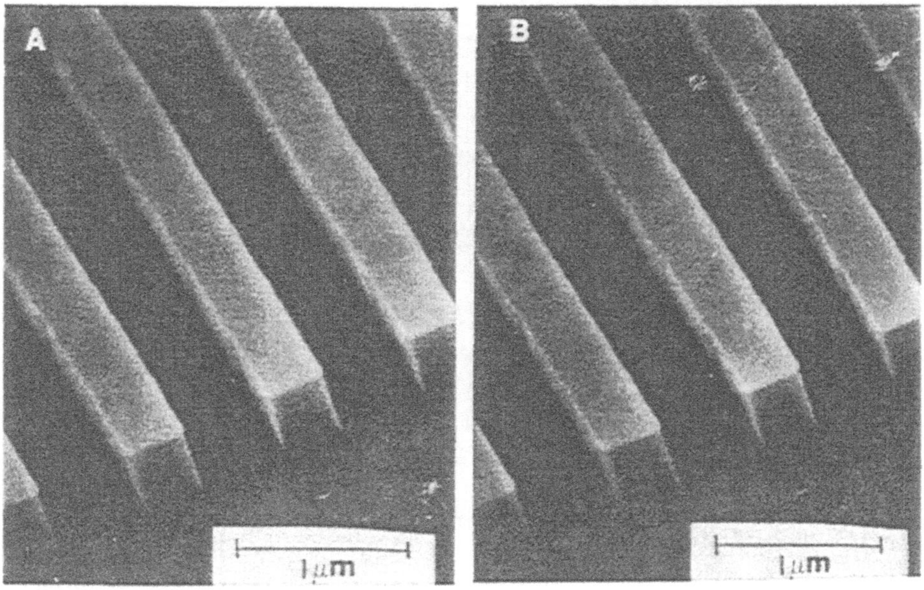

FIGURE 3-3-3. 100-keV H^+ exposure of AZ-1350 at (A) 10^{12} ions/cm^2 and (B) 10^{13} ions/cm^2.

proton exposure and *e*-beam exposure was demonstrated by Brault and Miller (Fig. 3-3-2).[80] Since a proton source can emit greater than 100 mW/cm^2 (> 1 A/cm^2), an exposure time per field of microseconds per chip or seconds per wafer could be achieved provided a stable mask and submicrometer alignment exist. Absolute straight-walled images (< 0.1-μm edge distortion due to mask fabrication) can be fabricated in negative or positive resists (Fig. 3-3-3).[74] Even a 10× overexposure of the AZ-1350 resist did not change the linewidth. The inter-

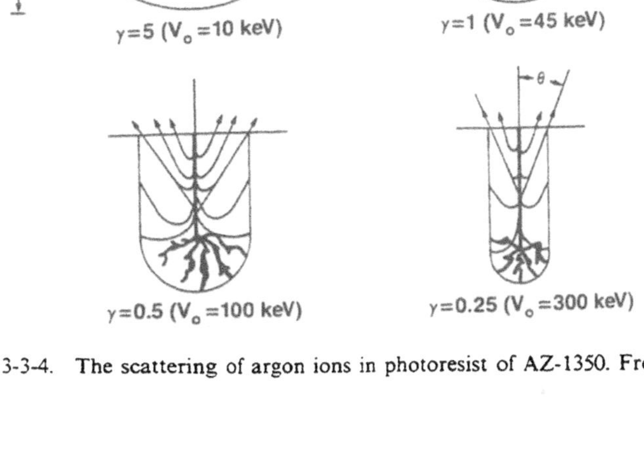

FIGURE 3-3-4. The scattering of argon ions in photoresist of AZ-1350. From Ref. 78.

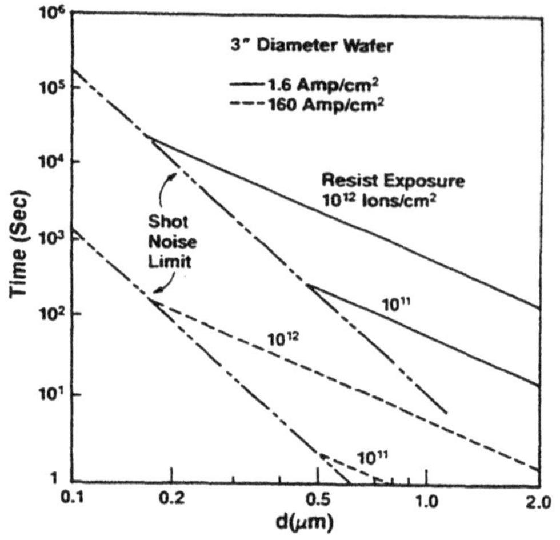

FIGURE 3-3-5. Time for exposing a complete 3-inch-diameter wafer at two different ion-beam current densities as a function of beam diameter d and resist exposure sensitivity. From Ref. 31.

proximity effect[23,78] prevalent in *e*-beam exposure is not evident because of the lack of backscattered electrons. The absorption of light ions is similar to the absorption of 5-keV electrons with a penetration of only 1 μm/100-keV protons into PMMA (Fig. 3-1-10). In ion-beam exposure, the absorbed radiation is partitioned into separate electronic and nuclear collisions.[111,112] The molecular weight (M_n) after absorption of dose Q, is given by

$$M_n = M_0/1 + kQ[G_{ei}(s)\, S_{ei}(z) + G_n(s)\, S_n(z)] \qquad (3\text{-}3\text{-}1)$$

where $G_{ei}(s)$ and $G_n(s)$ are the radiation yields for chain scission caused by electronic collisions and nuclear collisions, respectively. At very low energy (Fig. 3-3-4), the ion backscatter is evident for 5-keV Ar^+ exposure of AZ-1350[78] but none occurs at higher voltages.

For exposing resists directly, focused ion beams have limited applicability because the field of exposure is less than 1 mm². It has been estimated to require several hours (Fig. 3-3-5) to expose a moderately fast (10^{12} ions/cm² or 0.1 μC/cm²) resist. Exposing with a faster resist at 10^{11} ions/cm² runs into shot noise limitations.[31] With scanning ion beams, the beam deflection is slow compared to electron beams and the lens resolution in MTF is only good to 1 μm in a 5-mm square chip.[329] At the present time, the main use of ion beams is in the repair of optical photomasks. A secondary lithographic application of metal ion sources such as Si[81] or Ga is to top surface implant only 1000 Å deep into PMMA. Since the ion-implanted layer resists O_2 plasma, the image is reversed and transferred into PMMA via reactive ion etching (Fig. 3-3-6).[82,83] Positive images have been formed in Ge–Se inorganic resists by ion bombardment.[113] *In situ* masks,[84,95] consisting of an intermediate metallic layer in a trilayer resist,

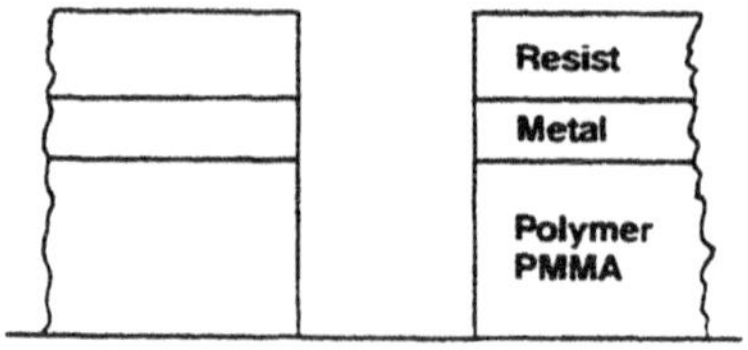

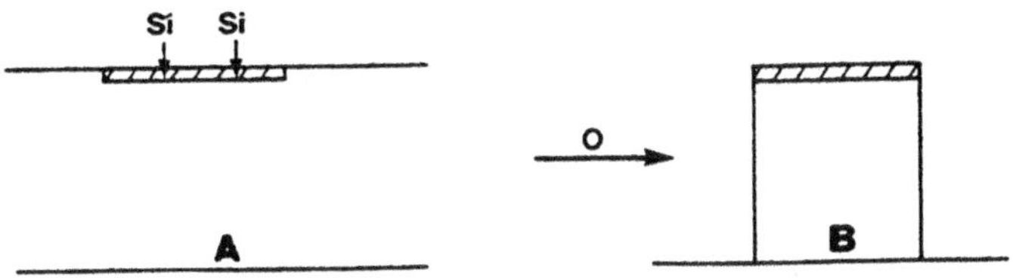

FIGURE 3-3-6. (A) Image reversal in PMMA resist by Si or Ga ion-beam exposure implant into top surface and (B) anisotropic dry development of image with an oxygen plasma.

have also been used to expose the PMMA sublayer using ion beams as the image transfer step (Chapter 12).

In resolution studies of positive resists exposed to ions, the contrast (γ) (Fig. 3-3-7) is used to predict the response of the latent image in the resist,

$$dR/dZ = (dR/dE)(dE/dZ) \tag{3-3-2}$$

The first term on the right is the developer dependence on the rate of development into the film and the second term is the energy (E) deposited (dose) per depth (Z). Even though there is little lateral scatter, the contrast of PMMA does not improve over electron exposure. Brault and Miller found the same γ of 2.2 for protons deposited in PMMA as for 20-keV electrons. The range of secondaries is only 100 Å from 100-keV H^+, but 5000 Å from 20-keV electrons. However, the contrast does not reflect the scatter, but is controlled by the polydispersity of the polymer and the choice of developer.[317,318,325,339] With narrow-dispersity negative resists, they found improved contrast. However, other studies of PMMA have shown that the contrast is dependent on the ion (H^+, Ga^+, or Ar^+) source and is also controlled by the polydispersity of PMMA, PMIPK, or FPM (Table 3-3-1).[89] A contrast of 10 for *e*-beam exposure has been reported at high doses for PMMA[95] in which the developer contribution is negligible [dR/dE, Eq. (3-3-2)]. Ryssell *et al.*[89] and Moriwaki *et al.*[90] found that ion-beam and *e*-beam exposure of PMMA and FPM produced the same contrast, but the use of larger ions (Ar^+, He^+) reduces the contrast due to a larger backscatter contribution[78] (Fig. 3-3-4). With H^+ beams of 10^{12} ions/cm^2, lines 0.25 μm by 1.3 μm wide lines were fabricated in high-molecular-

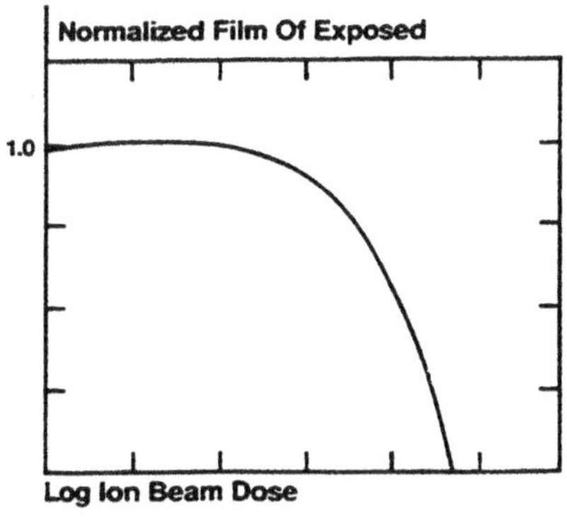

FIGURE 3-3-7. Contrast curve of positive resist exposed to H^+ ions.

TABLE 3-3-1. Ion-Beam Contrast of Positive Resists

Resist	Exposure ion	γ	Reference
PMMA[a] 10^6 M_w	H^+ 120 keV	6	89
PMMA 10^6 M_w	He 120 keV	2	89
PMMA 10^6 M_w	Ar 120 keV	2	89
PMMA 2041	H^+ 120 keV	3.4	89
PMMA 2041	He 120 keV	2.0	89
PMMA 2041	Ar 120 keV	2.0	89
PMMA 2041	H^+ 180 keV	3.0	96
PMMA 2041	H^+ 100 keV	2.2	80
FPM	H^+ 120 keV	2.9	89
FPM	H^+ 50 keV	3.3	90

[a] Fractional PMMA.

weight PMMA.(89) Reciprocity failure(88) has been reported in PMMA at very high incident intensities (10^{17} ions/cm^2), probably due to the trapping of reaction products or cross-linking of the resist. For pulsed X-ray exposure, Pearlman and Riordan(43) found no reciprocity failure in PMMA resists exposed to pulsed X rays.(106,107)

Of all the reported resists, the most sensitive positive resist is poly-1-butene sulfone requiring 4×10^{11} O^+ ions/cm^2.(86,87) Again, these high-speed resists are not favorable for submicrometer work. Thus, high-intensity sources will be required for favorable throughput.

Ancillary uses of ion-beam exposure have been to harden DQN(99,101) or PMMA resists(303) for reactive ion etch or other mask applications (Table 3-3-2). In conjunction with ion implantation of B, P, and As, the imaged resist also serves as a stopping power mask.(100,342)

Ion-beam exposure represents the ideal source due to the insignificant forward and backscattering, and little radiation damage of silicon is evident with the beams barely penetrating the resists. Since the ions are highly absorbed, the resist sensitivity is not a throughput factor provided a suitable high-intensity source, a thermally stable mask, and an accurate (± 0.1 μm) alignment scheme can be established.

TABLE 3-3-2. Uses of Ion Beam Applied to Resists

Use	Reference
Convert PMMA with Si^+ into deep-UV mask	91
Harden PMMA for RIE	98
Convert AZ-1350 into UV mask	92–94, 97
Harden AZ-1350 for RIE	98
Image reversal of PBS	102
Induce ionic conductivity	104
Deposit Si from polysilane, metals(319)	108
Produce antireflective coatings	105
Study diffusion of gases in resist	109

3-4. ELECTRON-BEAM POSITIVE RESISTS

PMMA, first reported as an *e*-beam resist by Haller *et al.* in 1968,[110] inspired the development of all high-energy radiation resists (Table 3-1-3). No (1–2 μm) single-layer film has ever matched the resolution of PMMA, and contrasts greater than 10 have been reported.[95] Over the last 15 years, the significant milestones in the improvements of PMMA resists (Fig. 3-4-1) have been a 100-fold enhanced sensitivity, the improved thermal and reactive ion resistance, and the feasibility of an all-dry deposit and develop resist lithography. PMMA resists for *e*-beam lithography have been critically reviewed by Moreau[59] and the radiation chemistry of PMMA has been discussed by Chapiro,[291] Charlesby,[292] and Dole.[122] In the last 20 years, about 100 papers have appeared on the radiation and thermal chemistry of PMMA. However, PMMA is not the only positive *e*-beam resist (Table 3-4-1).

Because of the wide availability of acrylate monomers, many substituted PMMA derivatives have been scrutinized as potential *e*-beam resists. All disubstituted vinyl monomers,

$$\sim\underset{\displaystyle Y}{\overset{\displaystyle X}{C}}CH_2\sim$$

are subject to chain scission. Polymers with low ceiling temperatures ($< 25°C$) can unzip to smaller fragments after random scission of the main chain. In theory, any polymer can revert back to a monomer,

$$\text{Polymer} \rightarrow \text{Monomer} \qquad (3\text{-}4\text{-}1)$$

by cationic, anionic, or radical depropagation to the gain of 10–10^4 depending on radical diffusion out of the cage to overcome recombination or termination reactions:

$$\text{Exposure of polymer} \rightarrow \text{Polymer fragments} \rightarrow \text{Polymer} \qquad (3\text{-}4\text{-}2)$$

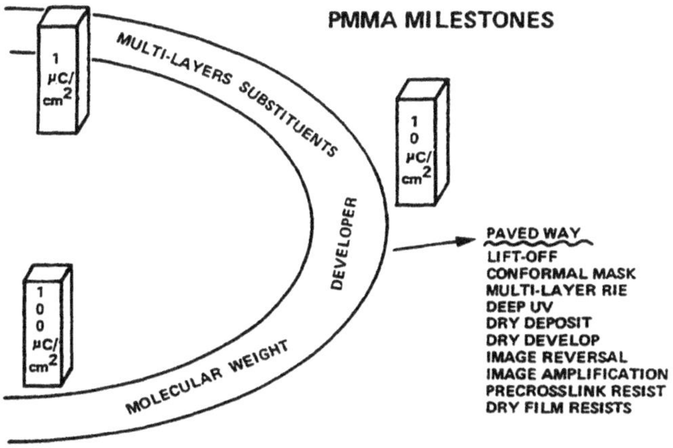

FIGURE 3-4-1. Significant historical role of PMMA resist on all lithographic processes evolved out of *e*-beam exposure. The sensitivity enhancement factors of molecular weight effects, developer, and substituent groups which paved the way for PMMA.

TABLE 3-4-1. Family Tree of Positive Electron-Beam Resists

One-component	Two-component
PMMA and derivatives	DQN
Polyolefin sulfones	Novolaks + polyolefin sulfone
Polyaldehydes	PMMA + additives
Polyisocyanates	Polycarbonates + onium salts
Polyketones	
Cellulose	

3-4-1. Electron-Beam Tool and Process Considerations

There are two major objectives of *e*-beam lithography:

1. *High resolution–precise alignment* (Fig. 3-4-2). The resolution (r) should exceed the thickness (t) of the resist and alignment should be better than one-fifth of the resolution. With electron beams, an alignment to $\pm 0.1\ \mu m$ has been achieved.[130] In the 10- to 30-keV range of exposure, the major detriment to high resolution is the backscatter of electrons from the substrate (Fig. 3-4-3). Gaussian-shaped beams or poor spot overlap can lead to edge scallop and underlapped patterns (Fig. 3-4-4). On the other hand, the lack of backscatter from nearest neighboring patterns leads to insufficient exposure of submicrometer isolated lines (Figs. 3-4-5 and 3-4-6). Narrow ($< 1\ \mu m$) lines do not absorb the same backscatter as larger lines and therefore require a two- to threefold higher dose or a longer develop time.

Closely pitched features suffer from an overdose of backscatter from adjacent patterns, causing the interproximity effect (Fig. 3-4-7) and an obliteration of unexposed spaces. If the silicon substrate is deliberately thinned to reduce the backscatter (Fig. 3-4-8), a negligible interproximity effect has been noted. To reduce the interproximity effect, the major methods are as follows:

a. Apply proximity corrections of dose and spot size or use etched-back silicon membranes.[114]
b. Use multilayer resist with a metal barrier layer and a 0.1- to 0.2-μm-thick image layer[115] (Chapter 12).
c. Use beams of 50–100 keV.[116]

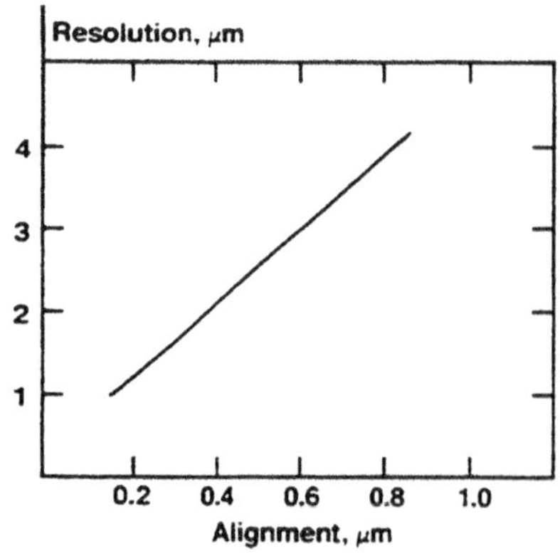

FIGURE 3-4-2. Resolution and alignment requirements for device fabrication.

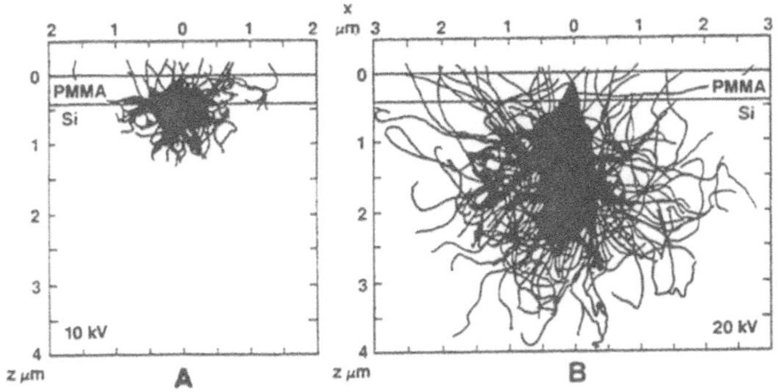

FIGURE 3-4-3. Backscatter from low-energy (10 and 20 keV) electrons. Monte Carlo simulated trajectories of 100 electrons in PMMA resist on silicon. From D. Kyser and N. S. Viswanathan, *J. Vac. Sci. Technol.* **12**, 1305 (1975), courtesy of AIP.

d. Treat top surface image of resist (first 1000 Å deep) with silicon or metal monomers and dry develop.[117]
e. Use high-contrast (γ) resist.

2. *Reasonable throughput.* A throughput of at least > 10 (125-mm) wafers/hr is needed to amortize the multimillion dollar cost of equipment. For a positive *e*-beam resist, the G_s value of the resist is the dominant factor and a G_s of at least 5 or more is required for a 10^{-6} C/cm^2 dose. For all types of *e*-beams, the throughput have been recently reviewed by Pfeiffer[128] and Varnell *et al.*[1] The throughput depends on the type of exposure (raster or vector), beam size, pattern density, align and overhead time, and resist sensitivity. A figure of merit was given considering several commercial and in-house scanning tools. Shaped spot, vector *e*-beam can

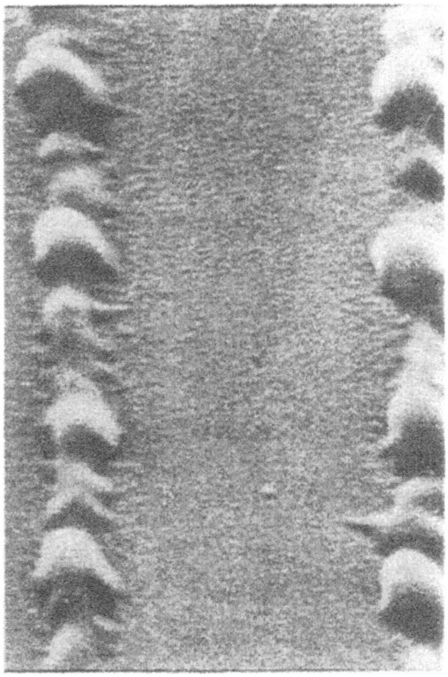

FIGURE 3-4-4. Poor spot overlap in Gaussian-shaped beam as revealed by a negative resist.

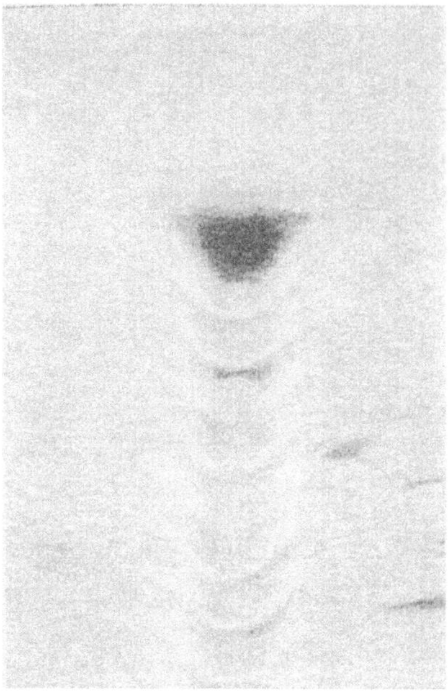

FIGURE 3-4-5. Isolated line intraproximity effect —insufficient backscattered electrons from one spot to expose adjacent spot.

provide faster writing and has better image control than raster–Gaussian spot systems.(129,130)

Other ancillary problems which can arise from *e*-beam exposure (especially on quartz masks) are beam heating of the resist causing image bloom,(118,119) outgassing from the resist, poisoned LaB_6 cathodes,(120) and bubbling of the resist over registration marks which distorts the alignment signal.(129–131) The electron beam heating of silicon and glass substrates depends on the instantaneous power of the beam and the thermal diffusivity of the resist/substrate (Table 3-4-2).(119) Faster resists (1 μC/cm^2) are required to lower the incident dose and reduce the temperature rise. For unzipping resists such as polysulfones or polyaldehydes, the

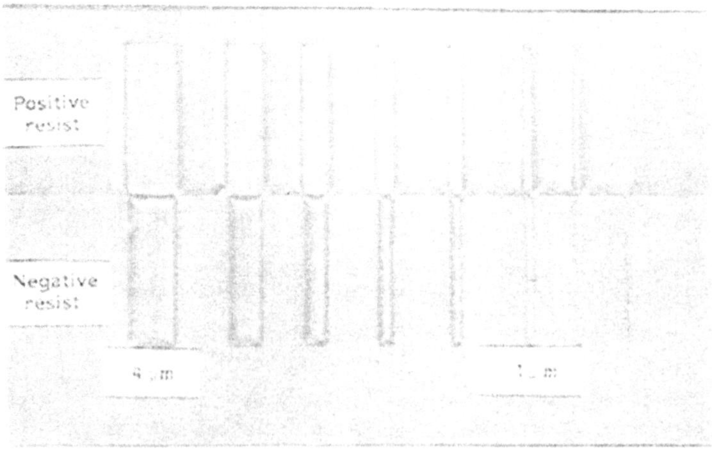

FIGURE 3-4-6. Intraproximity effect—lack of exposure in isolated lines. 1-μm line requires 1.5-fold more dose than 3.0-μm line to develop in same time for positive or same thickness for negative.

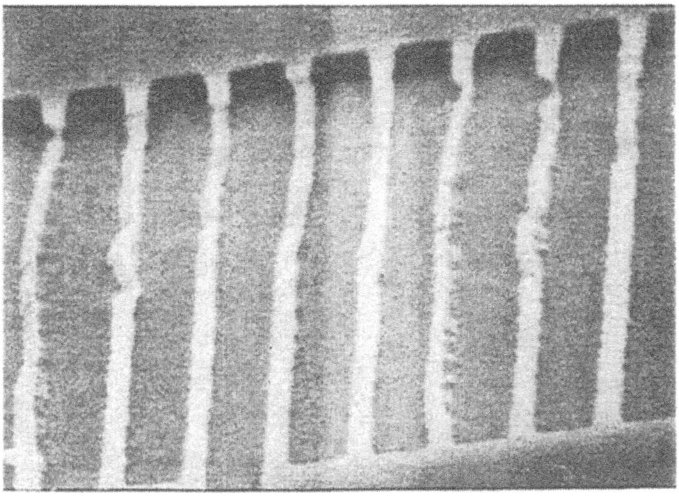

FIGURE 3-4-7. Interproximity effect of *e*-beam positive resist image with 2-μm exposed space and 1-μm unexposed line. Severe loss of unexposed land results in adhesion failure.

internal heating can lead to enhanced resist sensitivity, but always at the risk of blooming at the corners of patterns or in lines and spaces with pitches of less than 2 μm.

For a positive *e*-beam resist process, a single-layer resist costs the least, but proximity effects force one to use multilayer resists for submicrometer features. The plasma-thermal durability of aromatic, inorganic, or metal-organic resists to reactive ion etching is at least a factor of 5 better than alkyl acrylate, sulfone, or aldehyde resists. In mask making, ultrathin resists (100–300 nm) and 10 keV are used to reduce the proximity effect (which is not dose programmable on some tools). The proximity effect appears to be best handled by a materials process solution (multilayer resists).

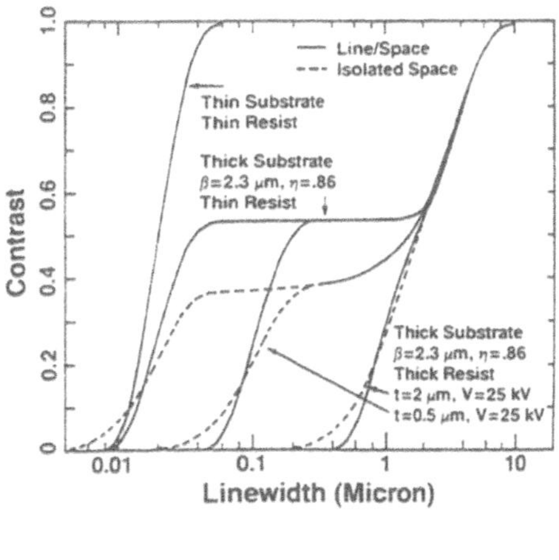

FIGURE 3-4-8. Contrast for line and space patterns for isolated lines on Si and Si membrane for thick and thin resist. From Ref. 132, copyright 1981 by IEEE and reproduced by permission.

TABLE 3-4-2. Temperature Rise in Si and Quartz Wafers

	V (kV)	I (A/cm^2)	Dose (μC/cm^2)	Size of square 5 × 5 (μm)	2 × 2 (μm)	1 × 1 (μm)
Silicon	20	50	10	96[a]	34	13
	20	25	10	56	18	7
	40	140	20	188	63	22
	40	70	20	129	35	10
Glass	20	50	10	179[b]	178	159
	20	25	10	190	184	142
	40	140	20	169	169	159
	40	70	20	178	175	168

[a] Temperature rise on silicon under a shaped beam (in °C).
[b] Temperature rise on glass under a shaped beam (in °C).

3-4-2. One-Component Positive Radiation Resists

The breakdown of linear polymers can occur as in PMMA by:

1. Random homolytic scission of main chain,

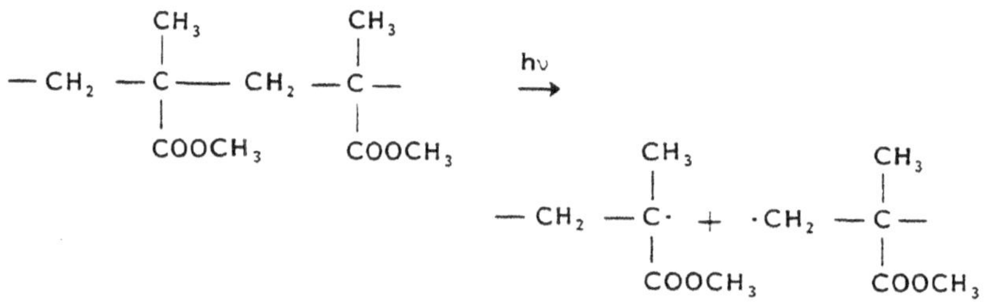

2. Breakdown of ester side chain,

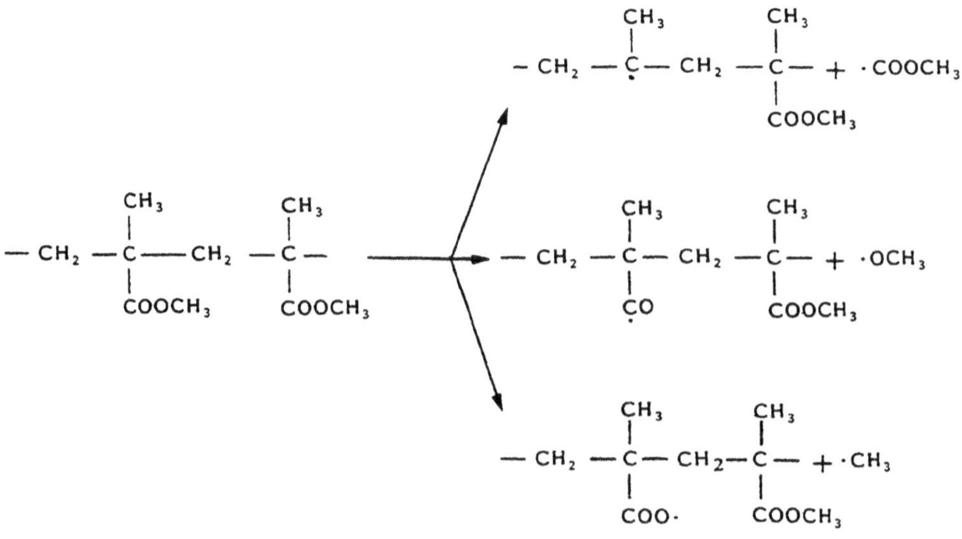

3. Elimination of α-methyl group,

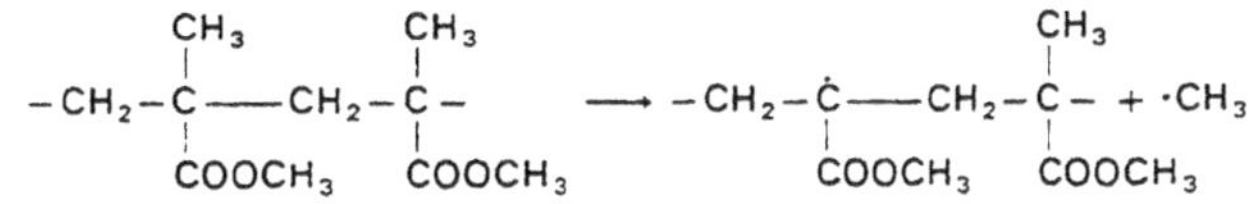

4. End chain depolymerization (unzipping),

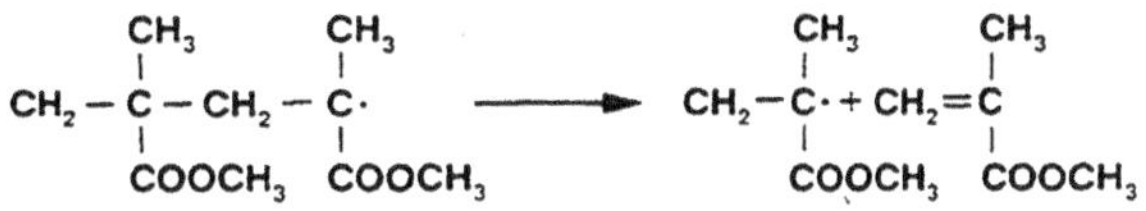

The depropagation process of unzipping ends by disproportionation,

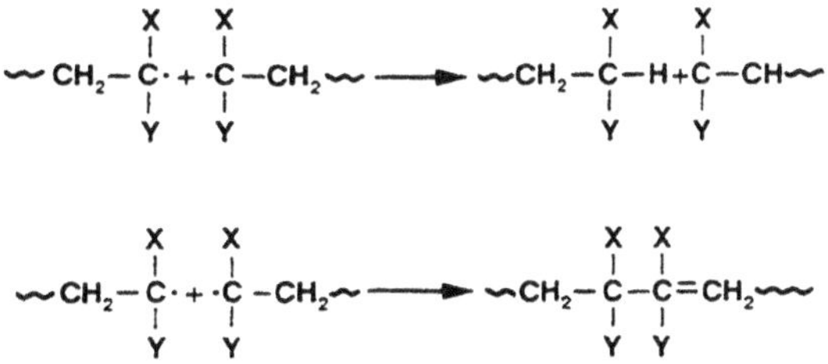

or recombination upon liquid development,

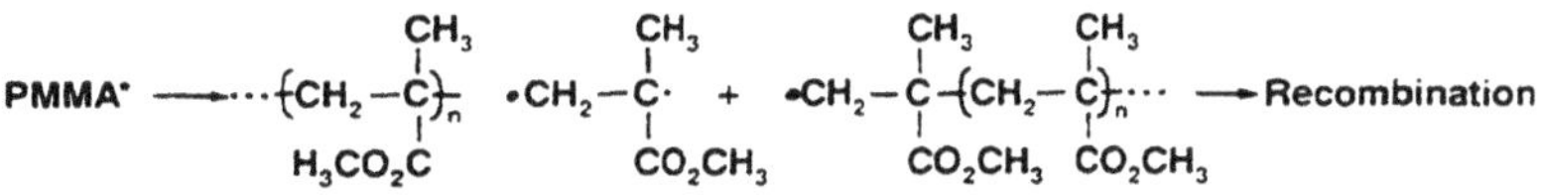

In PMMA, the G_s value of scission has been estimated from studies of stereoregular polymers to be as high as 6–10, but because of recombination and racemization, the G_s is reduced to unity (cage effect),[188]

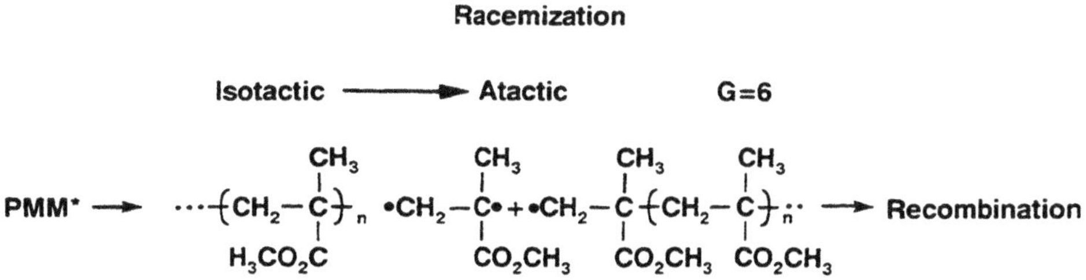

Initiation of random chain scission is a necessary condition of depolymerization because tertiary carbons are produced which are capable of further unzipping. Initiation at end groups is also possible if the weak links are present as in the photochemical example of capped polyaldehydes,[69]

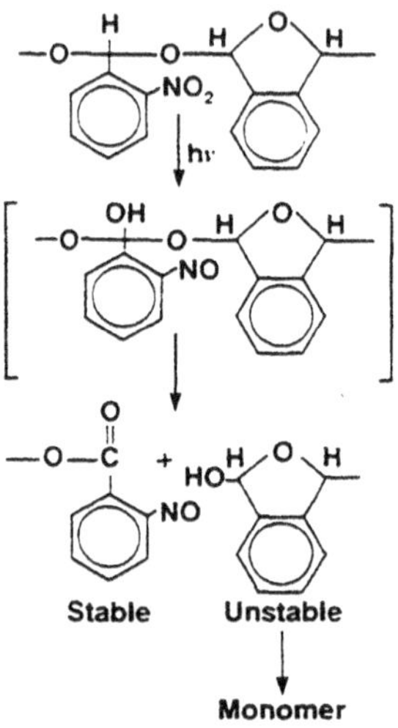

The basic theories of chain scission start with a distribution function for a polymer chain and proceed to apply a weighted average for the convergence of dispersity. to the most probable value of 2 (Fig. 3-4-9). The reader is referred to the treatises of Dole[122] and Kelen[121] for the current theories of random and unzipping reactions. In random scissions, the probability increases with chain length and high-molecular-weight polymers are selected for better resist sensitivity.[234,236] While free radical depropagation is common in vinyl degradation[123] because of the high probability of ionization, cationic chain scission mechanisms are equally as probable for the dissociation of PMMA,[124-126]

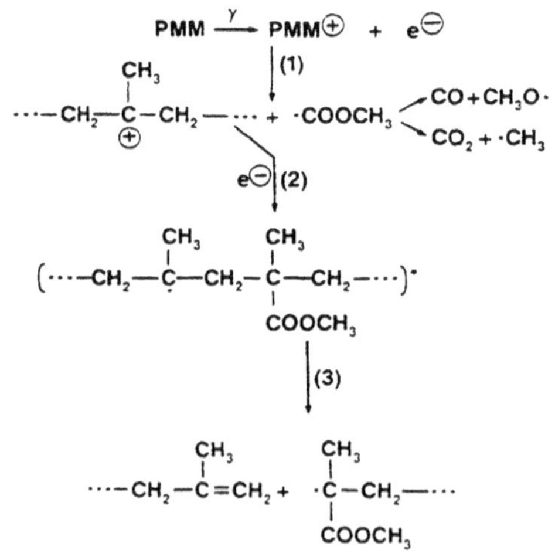

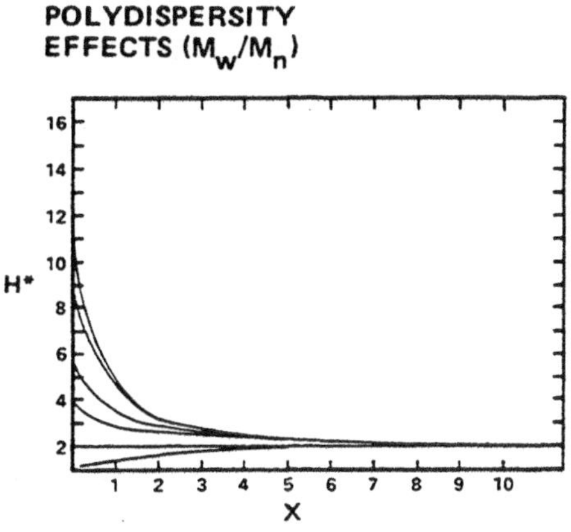

FIGURE 3-4-9. Heterogeneity index of exposed resist, H^*, plotted against X for various initial (unexposed) heterogeneity index, H. From A. Ouano, *IBM J. Res. Dev.* March 1977, p. 150, copyright and courtesy of IBM.

and with polyolefin sulfones,[127]

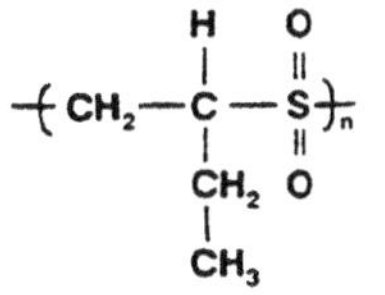

$$R\ SO_2\ R' \longrightarrow [RSO_2R']^{\dot{+}} + e^-$$

$$[RSO_2R']^{\dot{+}} \longrightarrow RSO_2^+ + \cdot R' \quad R^+ + SO_2$$

Emanating from PMMA a host of derivatives have arisen to increase the sensitivity of PMMA or to improve its processibility (reactive ion resistance and dry develop capability). Both electron-donating and -attracting groups have been used,[137–143,147,148,327]

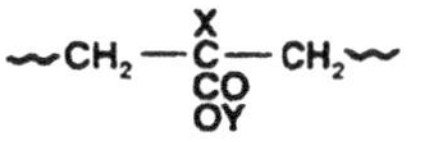

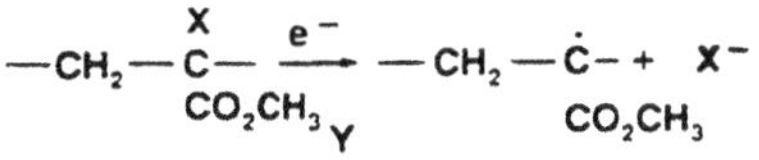

where

X = CN, Cl, Br, F, I, NO_2, CF_3, COOOH (acceptors)
X = CH_3, alkyl, O, NH_2, CH_2COOH, SiH_3 (donors)
Y = Alkyl, aryl, silyl, stannyl, halolalkyl, epoxy

Because of their high electron affinity (Table 3-1-4), three electron-attracting substituents—α-Cl, α-Cr, and α-CF_3—have enhanced the G_s to the 3–6 range presumably by electron dissociative capture.[309] More secondary electrons are absorbed along the impact track than with other substituents. After exposure, the scissioned products are removed by carefully selecting developers which maximize the differential dissolution rates.[235] When PMMA was first used in the late 1960s, a dose of $100\ \mu C/cm^2$ was necessary for methyl isobutyl ketone (MIBK)/isopropanol developer. At a lower dose ($10\ \mu C/cm^2$), the development time was too long (> 1 hr) to be practical. In the original scission model of Ku and Scala,[133] it was assumed that Q is independent of the initial molecular weight, M_0, i.e.,

$$M_0/M_n - 1 = M_0/M_n \tag{3-4-3}$$

for a nominal ($100\ \mu C/cm^2$) dose (Q). At lower doses, Ting[14] pointed out that the full relation [Eq. (3-4-4)] must be used. Since the dissolution of PMMA, which has a high T_g (glass transition temperature), proceeds without swelling (except for $M_n > 10^6$ K), the ratio of the exposed (R) to the unexposed (R_0) rate can provide a basic resist exposure model,[10,134]

$$R/R_0 = k(M_0/M_n)^{-a} = (1 + kQG_sM_0)^a \tag{3-4-4}$$

where a is the solubility dissolution parameter and G is the scission efficiency. If a molecular weight decrease of 10× in M_0/M_n can be achieved ($R/R_0 > 5$), a dose of $2.5\ \mu C/cm^2$ could be used for PMMA[135] starting with a high molecular weight ($M_n > 650{,}000$) and using $G = 1.4$ and $a = 1.5$.

Since the early 1970s, the sensitivity of PMMA has been enhanced 100-fold to the 1–$10\ \mu C/cm^2$ range by three major means:

1. Increase G_s in main chain or side chain scission by introduction of electron acceptors (Table 3-4-3). Pittman *et al.*[136] have done extensive γ radiolysis studies of haloacrylates and cyanoacrylates. The highest G_s value of 6 has been measured for α-Cl acrylates and values of 2–5 for α-CF_3 substituents.[151,152,159,160,171,229]

TABLE 3-4-3. Acrylate Substituents and Sensitivity

X[a]	Y	$\mu C/cm^2$	G_s	Reference
CH_3	COOH	20	3	49, 148
CH_3	CN	15	5	138
CH_3	Anhydride	10	5	144, 294
CH_3	COO Pb	20	—	145
CH_3	Itaconic acid	10	6	146
CH_3	$COOCF_3$	3	1	51, (148–150), 160
CH_3	COOF propyl	1	2	151
CH_3	COOF Bu	2	3	159
Cl	$COOCH_3$	10	5	140, 157
CN	$COOCH_3$	10	3	141
NO_2	$COOCH_3$	10	3	142
CF_3	$COOCH_3$	15	3	136, 143, 156
Cl	CN	10	4	154
CN	COOEt	5	2	147
CH_3	COOSn	50	—	155

[a] Alpha substituent.

Because of their high electron affinity (Table 3-1-4), the substituents enhance the capture of secondary electrons, which is followed by resonance dissociation,[169]

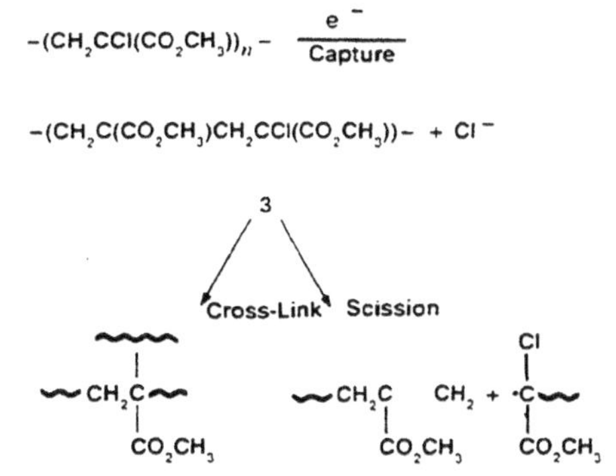

4

In addition, some mild cross-linking with α-Cl acrylates has been noted.[136] The introduction of α-CF_3 or α-Cl groups in the backbone or fluoro groups in the side chain raises the G_s to the 2–5 range.[336,344] Pittman *et al.*[169,348] studied a whole series of fluorine and chlorine methacrylate homo- and co-polymers,

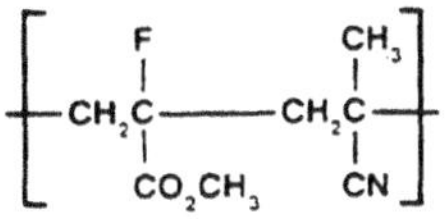

MFA-co-MCN

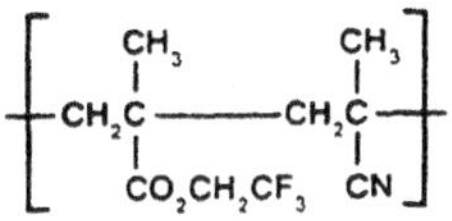

TFEM-co-MCN

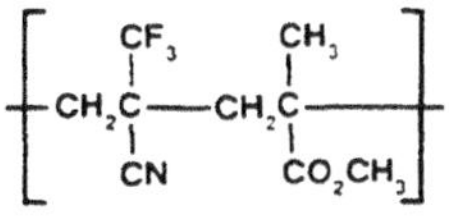

TFMAN-co-MMA

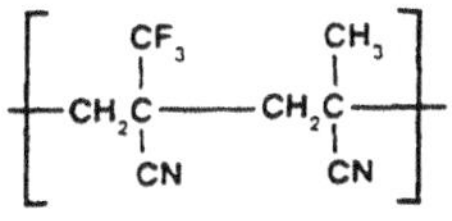

TFMAN-co-MCN

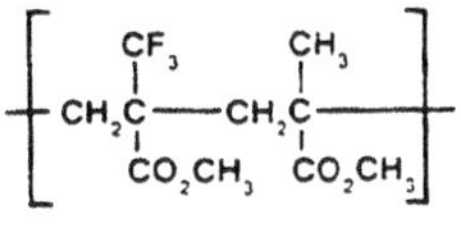

TFMMA-co-MMA

and found cross-linking as well as scissions in the homopolymers. The high sensitivity of fluorobutylmethacrylates[149,151,152,159,232] cannot be accounted for on the basis of a G_s of 1 or molecular weight effects[154] but is more likely due to an enhancing developer which swells the unexposed region. The other aspect of the work of Pittman *et al.* was the detailed study of the effect of the percentage of fluoromonomer (α-TFMMA) on the copolymer G_s value (Fig. 3-4-10).[154,156,348] The G_s reaches a limiting value dependent on the following sequences of monomers, (A) and (B), in the bipolymer:

- Alternating,

-ABABABABAB-

- Random,

-ABAAABBABAA-

- Blocky,

-AABBBBABBAAAA-

For random scission, the blocky is the most inefficient whereas an alternating polymer has the highest efficiency[320–322] (which makes polyolefin sulfone copolymers very efficient):

Alternating > Random > Blocky

If weak links (W) such as CO_2, CO, NO_2, or O_2 are present, then a minimum scission would be required if the weak link W were positioned in a chain such as

-AAAAWAAAAWAAAAAAW-

Weak links such as peroxide,[175] O–O, could convert any chain into a scissionable polymer. A dose of only 10^{-8} C/cm^2 has been reported for a PMMA-peroxide resist.[175]

Following the work of Pittman, the additive principle of the summation of the G_s values of individual monomers in copolymers was proposed by Willson *et al.*[171] For example, the α-CF_3 group which has a G_s of 2.5 is copolymerized with PMMA-MA-anhydride[144] (G_s of 4.5) to enhance the terpolymer sensitivity to a G_s of 9 (hopefully),

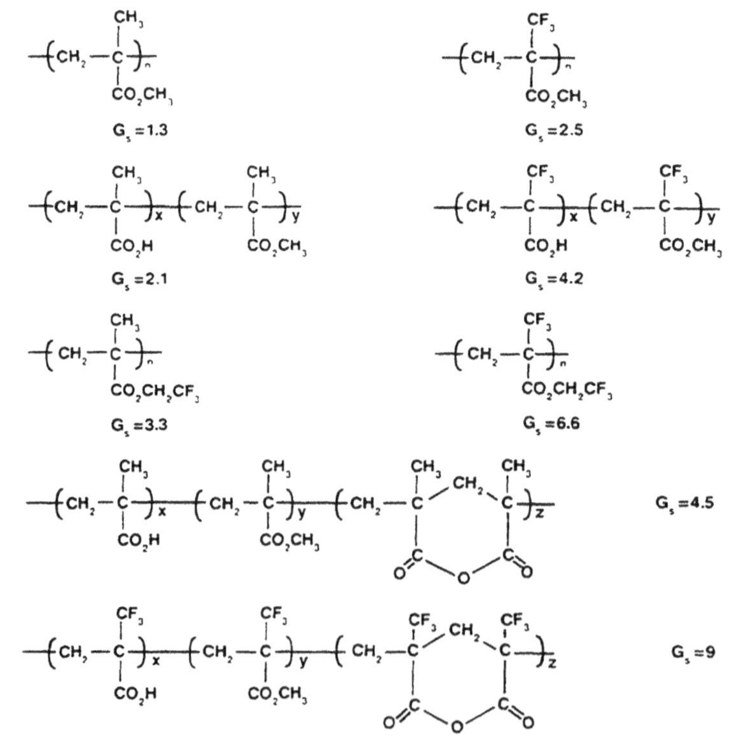

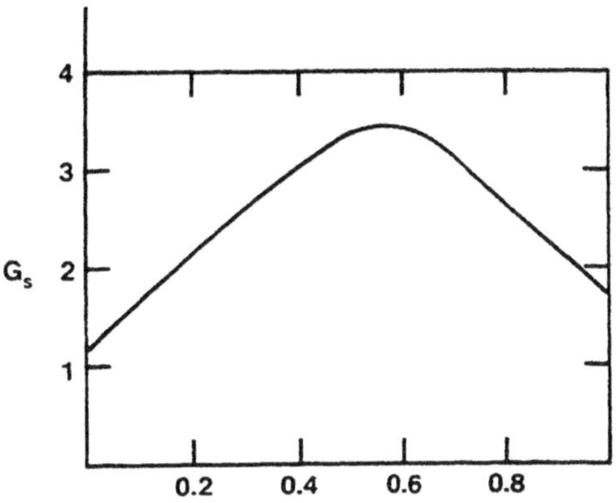

FIGURE 3-4-10. Mole fraction (M_F) of TFMA in bipolymer of P(TFMMA-MMA) versus *G*(scission). From Ref. 154, reproduced by permission of John Wiley & Sons, Inc.

Since no substituents or copolymers with PMMA have achieved a $G_s > 6$, other steps have to be taken:

2. Increase the initial molecular weight[161–164,177] [Eq. (3-4-4)] and reduce the polydispersity (< 1.5) since the weight average molecular weight rapidly converges to 2.[165–167,170,172]

3. Choose developers with a high *a* value such as amyl acetate (Fig. 3-4-11) which also improves the contrast, in PMMA and in fluorobutylmethacrylates[168,169] (see Chapter 10).

The primary factors of *G* value, molecular weight, and dispersity, and choice of developer have allowed sensitivities of PMMA as low as 1.0 $\mu C/cm^2$ for 1×10^6 molecular weight[157,173–176] which is predicted by Eq. (3-4-4).[164] In summary, the primary means for enhanced PMMA speed have been:

High G_s and substituents	Increase R
High molecular weight	Decrease R_0
High-*a* developer	Increase R/R_0

These primary effects have paved the way for enhanced PMMA sensitivity (Fig. 3-4-12) as well as the side benefits of multilayer resists. In addition, many

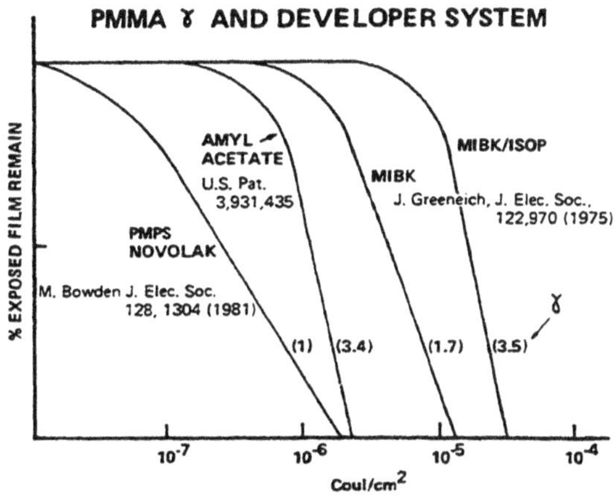

FIGURE 3-4-11. Developer enhancement of PMMA sensitivity and contrast.

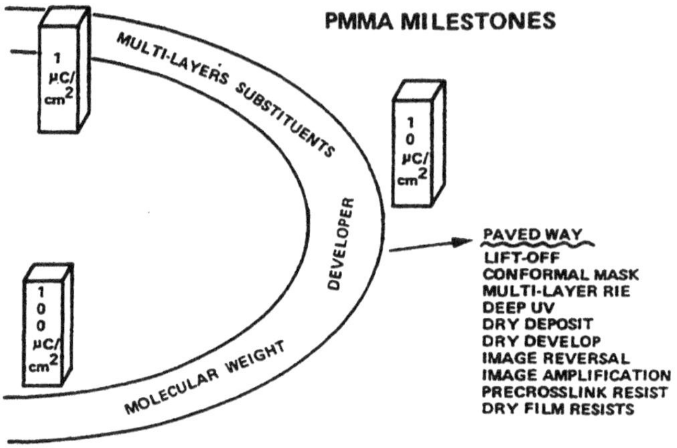

FIGURE 3-4-12. Major enhancement of the sensitivity of PMMA by use of initial highest molecular weight, improved developers, and electron-withdrawing substituents.

secondary effects (Table 3-4-4) have contributed to enhanced PMMA *e*-beam resist sensitivity and potentially in other resists. Some examples of these secondary factors are as follows:

1. Polydisperse mixtures high in M_w converge rapidly at low doses to $M_w/M_n = 2$.[166] Mixtures including syndiotactic–isotactic complex,[178] and mixtures of PMMA with other polymers have been used.[179]
2. Copolymers which fragment to release more soluble components in the exposed region—if PMMA copolymers fragment to release portions with a low T_g, a higher R can be achieved since the low-T_g fragments dissolve faster. The prime examples are PMMA-fluoro and butylmethacrylates[180] and PMMA-isobutylene[138] (Fig. 3-4-13). Copolymers of PMMA-MA methacrylic acid,[145,158] methacrylic acid anhydride,[144] and itaconic anhydride[146] can fragment to release more polar soluble fragments which dissolve rapidly in polar developers.
3. Polymers that release copious gas products (Fig. 3-4-14) foam the resist and accelerate infusion of developer. In polymethacrylic acid anhydride copolymers with PMMA, the side chain fragmentation produces gaseous products 2–3× higher than main chain (no recombination of cage effect due to outgassing of fragments) scission.[35] The trapped products or out-

TABLE 3-4-4. Secondary Contributions to Enhanced PMMA Sensitivity

Factor	Effect
Polydisperse	$M_w/M_n \rightarrow 2$, higher R
Microfoam	$G_{gas} > 10$, higher R
T_g of product	Higher R
T_c (ΔH)	Further scission, higher R
Electron absorbers	Higher R
Pre-cross-link	Lower R_0
Multilayer resist	Lower R_0, higher R/R_0

FIGURE 3-4-13. Sensitivity of alkyl and fluoroalkyl methacrylates. Relation between electron sensitivity and thermal stability of linear polymers and CMR, PBzMA: Polybenzyl methacrylate; PEMA: Polyethyl methacrylate; PCyHMA: Polycyclohexyl methacrylate; PiBMA: Poly-isobutyl methacrylate; PFBMA: Poly-hexafluorobutyl methacrylate; PnPMA: Poly-*n*-butyl methacrylate. From Ref. 150, by permission of the Electrochemical Society Inc.

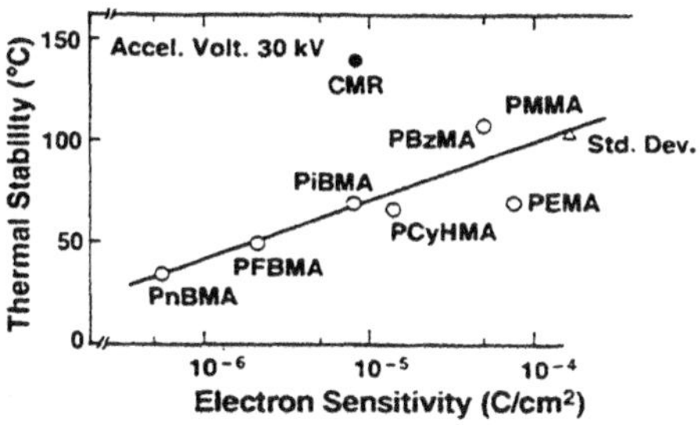

gassing can foam the exposed region increasing the porosity to developer infusion.[181]

4. The steric strain factor of disubstituted carbon on the PMMA backbone. PMMA exhibits a higher (120°) than normal (109°) dihedral angle from steric repulsion by α-substituents. Polymethylmethacrylic acid anhydride (PMMAN) is severely strained in the six-membered ring,[183,294]

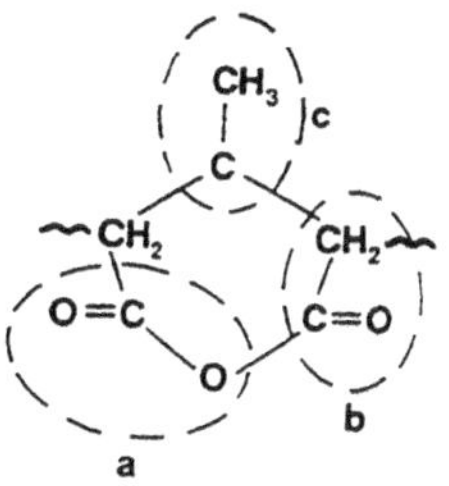

and radiation fracture relieves the ring strain.

5. Exposing scissionable polymers at elevated temperatures above T_c and above T_g can induce mobility out of the cage and induce unzipping rather than recombination.[182,184,186,188] Charlesby reported a G_s value of 10 for PMMA[185,186] exposed 10°C above its T_g of 104°C.
6. Use of fast acrylates as an image layer in bilayer schemes such as a PMMA-MA/PMMA sandwich can enhance sensitivity to the 2–5 $\mu C/cm^2$ range.[158,187]
7. Use of a pre-cross-linked resist to inhibit R_0 (see next section).
8. Adding metals such as Sn with higher electron absorption than carbon.[155,233,313]

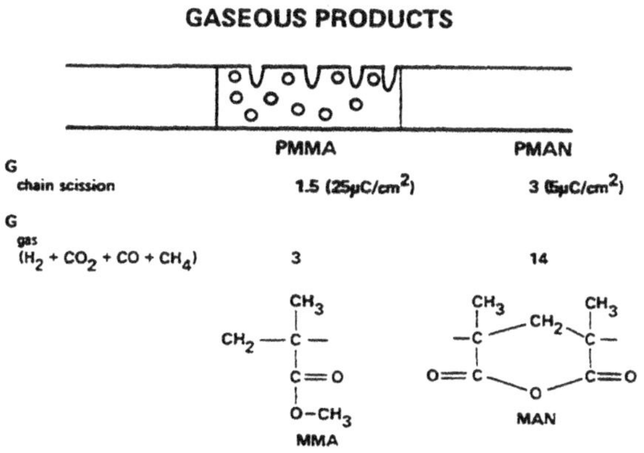

FIGURE 3-4-14. Higher amount of gaseous products from polymethacrylic anhydride (PMAN) versus PMMA as measured by G_{gas}.

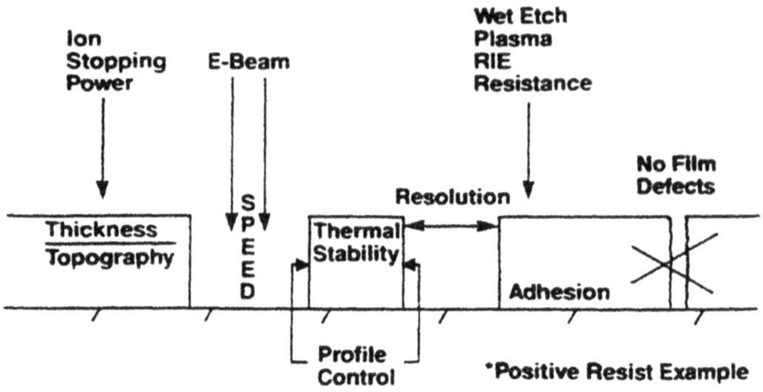

FIGURE 3-4-15. Resist requirements for submicrometer lithography.

The above factors have contributed to the enhancement of PMMA and positive radiation resists. Besides the basic speed factor, other processing factors (Fig. 3-4-15) such as thermal and plasma stability, good adhesion, and low film defects must be achieved for submicrometer lithography. I will address these factors for PMMA resists in the next section.

3-4-3. Pre-cross-linked PMMA Resist

PMMA has a reported T_g of 106°C and begins to melt at 150°C. Because of electron exposure at the edges (lower M_w), image profiles begin to recede at 130°C. Functional groups appear unable to raise the melt stability to 200°C. High melt stability is obtained by the introduction of cyclic anhydride[144,294] or glutarimide[190] groups which raise the melt stability to 170°C. Since the inherent melt stability of linear acrylates is limited to about 10°C, permanent intermolecular cross-links are the next structural order to achieve melt stability.

After Roberts first cross-linked PMMA with succinic anhydride links[191] by the intermolecular reaction of polymethacrylic acid (MA) and polymethacrylyl chloride (MACl),

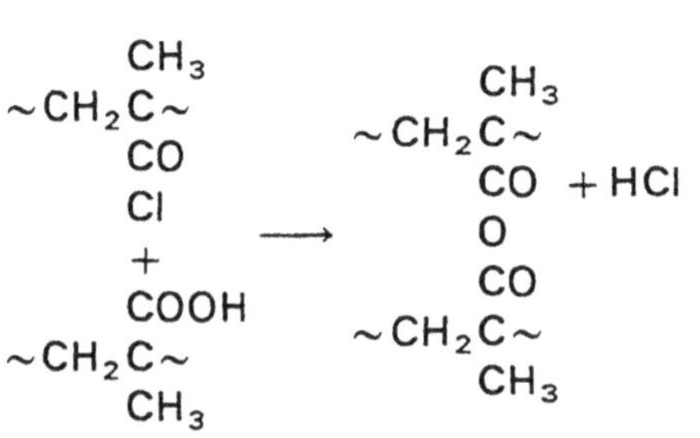

a host of pre-cross-linked PMMA have appeared including additional aryl[193] groups to enhance the reactive ion stability. The cross-linking can be performed by heat, UV, or by chemical reactions.[303] Once cross-linked, the unexposed resists are virtually insoluble and stable to 150–250°C. However, the resist prebake step is critical since the sensitivity of the resist (10–50 $\mu C/cm^2$) is proportional to the amount of cross-links. Prior to exposure, the exposed region represents a polymer of infinite molecular weight.[194] After exposure, the imaged region dissolves in a developer, and the unexposed resist remains untouched by the developer. Unfortunately, moderate to high doses (20–50 μC/cm^2) are necessary to achieve dissolution. Other reactive groups forming cyclic and cross-linked anhydrides, imides, or epoxies are:

Methacrylamide (MAM)[190,197]
Methacrylonitrile (MAN)[198,315,354,356]
α-Cyano ethyl acrylate (PCA)[63,350]
Amido ethyl acrylate[198]
t-Butyl methacrylate (TBMA)[200–202,311]
Glycidyl methacrylate (GM)[199]
α-Ethyl chloride (EC)[204]
Itaconic acid (IA)[203,316,337]

For additional reactive ion etch resistance, copolymers of PMMA or polystyrene have been used.[308,310,323,328] A summary of pre-cross-linked systems is given in Table 3-4-5.

The major advantages of pre-cross-linked systems are[323] (1) enhanced melt resistance to 150 to 250°C; (2) a low R_0 and wide development latitude; (3) enhanced adhesion; (4) increased plasma resistance with aromatic groups; and (5) the elevation of T_g and ceiling temperature to increase the thermal stability to 200°C. A resist sensitivity of 5 to 50 μC/cm^2 has been maintained. In principle, high sensitivity could be achieved if only two cross-links per chain could be synthesized and broken by the irradiation to release soluble chains.

TABLE 3-4-5. Pre-cross-linked Acrylates

Resist	Dose (μC/cm^2)	Reference
PMMA-MA + PMMA-MACl	20	191
PMMA-MA + PMMA-MA-MACl	19	205
PMMA-MA-MACl	0.5	206
PMMA-MA-MACl	10	207
P(trifluoro-α-ClA)	7	153
P(benzyl IMA-MACl)	20	208, 354
PMMA-hydroxyethyl MA	*a*	209
PMMA-MA	*b*	199
PMA-Cr acetonate	5	210
PMMA-MA-Tl	—	211
PMMA-MACM-MACl	8	192, 195, 196, 212
P(trichloroethyl MA)	7	153
PMMA-MACl	*c*	213
PMMA-GM	UV, 10	214
PTBMA	10	215
PMAM	1	197, 216
PMAM	10, *d*	190
PMMA	*e*	217, 332
PECMA	10	218
PMMA-polyimide	UV	60

[a] Cured with hexanedioic acid.
[b] Cured with epoxy monomers.
[c] Cured with neopentyl glycol.
[d] Treat PM acid with NH_3.
[e] Cross-link with UV of bisazides.

The major disadvantages of pre-cross-linked resists are the difficulty of prebake replication of reproducing cross-links (hence, sensitivity), film thinning and the formation of pinholes during prebake, the difficulty of stripping for lift-off, and the low to moderate sensitivity.

3-4-3-1. Plasma Stability of PMMA. While single-component resists with high sensitivity (low T_c) such as polyolefin sulfones exhibit poor plasma resistance, PMMA and derivatives, being of moderate sensitivity, have fair plasma resistance. When compared to commercial diazonovolaks, the etch rates of acrylates are about two- to fivefold faster. PMMA will be further stabilized if the wafer temperatures are kept below a T_g of 100°C or a T_c of 150°C and the lowest-molecular-weight polymer is used. In a plasma, active species, such as chlorine, fluorine, or atomic oxygen, readily attack the backbone, perhaps by abstraction reactions. Sputtering by argon also contributes to nucleus displacement and volatization.

Recent studies have shown that PMMA can be used to etch insulators if oxygen- and fluorine-deficient halocarbons are used as etchants. With these etchants, fluorocarbon polymers deposit on the resist sidewall to inhibit lateral erosion. In plasma erosion, similar evidence as the photon and *e*-beam decomposition[189] (chain scission and radical intermediates) has been obtained from molecular weight and radical trapping experiments on PMMA.[237] If random scission predominates, lower-molecular-weight acrylates would be more stable.[220–222] Increased plasma resistance in acrylate series has been achieved by introduction of groups of cyano,[228] cyclic imides,[219] anhydrides, aryl, and metal groups into, onto, or mixed in PMMA and derivatives (Table 3-4-6).

For acrylate resists, the incorporation of > 5% by weight of an element which forms nonvolatile oxides (e.g., Si, Ti, Sn, Mg) inhibits O_2 plasma etching and finds applicability in bilayer resist processing (Chapter 12).[233] The most significant heat and thermal stabilization has been achieved by pre-cross-linking the resist and soaking it in metal salts (10× over PMMA control) which form nonvolatile halides, e.g., Na^+, K^+, Mg^{2+}.[224]

3-4-3-2. Processing. Enhanced adhesion has been noted by treatment of PMMA with oxygen or fluorine plasma[229,230] and improved adhesion has also been achieved by copolymerization with isobutylene to provide H-bonding groups and stronger van der Waals forces (Fig. 3-4-16) to attach to the silanol and silica

TABLE 3-4-6. Plasma Resistance Improvements in Acrylates

Resist	Remarks	Reference
PMMA	*a*	221
PMMA	*b*	223
PMMA-MA	*c*	224
Phenyl IMA	—	222
PMMA-benzyl IMA-MACl	*d*	225–227
PMMA-α-F MAN	—	154

[a] DPPH and antioxidant additives.
[b] Soak in epoxy monomer.
[c] Cross-link with silanes; soak in NaOH.
[d] Pre-cross-link in prebake.

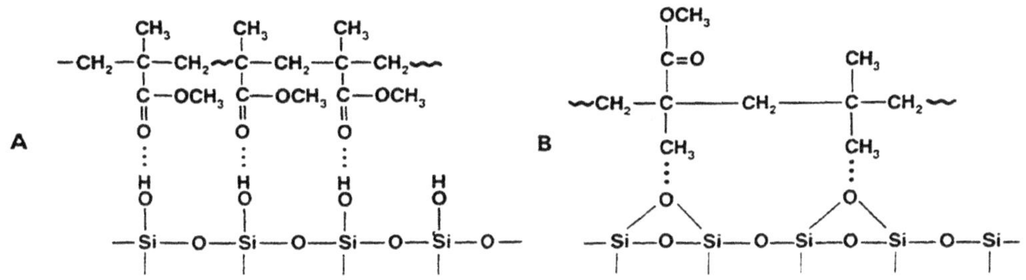

FIGURE 3-4-16. (A) H-bonding of PMMA to silanol surface. (B) van der Waals bonding of MMA-isobutylene bipolymer to siloxane surfaces.

surfaces.[231] For an aqueous processible acrylate, PMMA side chains have been modified with diazoquinone groups,[238]

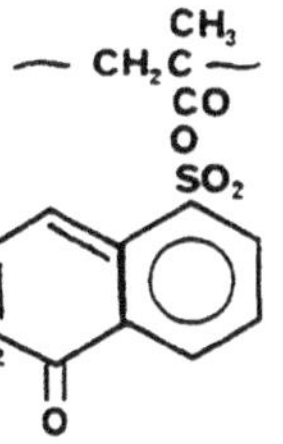

The resolution of PMMA was enhanced by mixing 5 wt% of SiO_2 powder into the resist to prevent lateral scattering.[293] Increased resolution is also claimed by dispersing Fe_2O_3 into PMMA and, prior to exposure, prealigning the wafer in a magnetic field.[295]

3-4-3-3. Outlook for PMMA. PMMA and the acrylate family of resists have been pioneering and versatile materials for significant advances in high-resolution lithography. Many advances in sensitivity have been achieved. Multilayer resists can take advantage of speed enhancement. Synthetic introduction of functional groups (e.g., CN, Cl) will continue to improve the lithographic importance of acrylate resists and their outstanding contrast will continue to pave the way for customized submicrometer and nanometer lithography.

3-5. POLYOLEFIN SULFONES AND OTHER ELECTRON-BEAM POSITIVE RESISTS

Polyolefin sulfones, polyisocyanates, and polyaldehydes are classic examples of polymers that are thermodynamically unstable at room temperature.[240] Because of the need for high-speed exposure for direct silicon write, mask making, and video disk recording with low-beam-current guns, polyolefin sulfones (Table 3-5-1) were examined as potential positive *e*-beam resists.[296,307] The tertiary or disubstituted olefins copolymerized with SO_2 are very unstable due to steric crowding and a strained dihedral angle in the following order:

Tertiary > Secondary > Primary

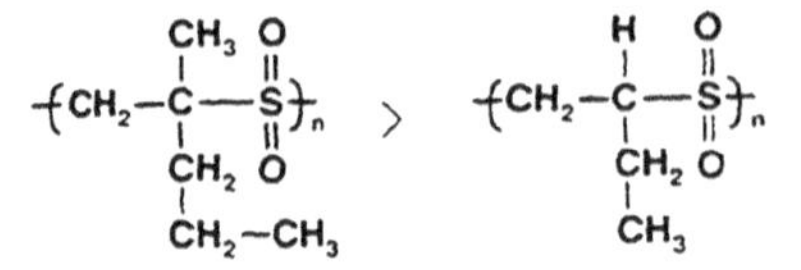

Upon exposure to γ rays, polyolefin sulfones were initially reported by Brown and O'Donnell[239] to revert to monomer with a G_s of 5–15 and a G_{SO_2} of 20. In the synthesis, the SO_2 group enters the bipolymer as an alternating[240] group due to charge transfer formation,

$$\text{Olefin} + SO_2 \rightarrow \text{Olefin-}SO_2 \rightarrow \text{Polymer}$$

The tertiary substituted olefins such as polymethylpentene sulfone have the lowest T_c and the highest G_s values (10–20). However, if all the secondary and tertiary olefin sulfones are exposed at elevated temperatures above their T_c and T_g, equal G_s are found.[251] At the elevated temperatures, the radical intermediates diffuse away from each other and unzip to olefins and SO_2.

TABLE 3-5-1. Representative Patents[a] Issued on Polyolefin Sulfone Resists[b]

Patent	Assignee	Use
1,421,805 (Br. Pat.)	IBM	Olefin sulfones
2,321,684 (Ger. Pat.)	Bell	Olefin sulfones
3,884,696	Bell	UV, styrene sulfone
3,893,127	RCA	Dry develop video disk
3,898,350	IBM	Terpolymer
3,916,036	IBM	Sensitized decomposition
3,935,331	RCA	Dry to prevent cracking
3,935,332	RCA	Developers
3,964,909	RCA	UV ketoolefins
4,007,295	RCA	Improved adhesion
4,045,318	RCA	UV harden
4,054,454	RCA	UV-sensitive ketoolefin
4,153,741	RCA	Cycloolefin sulfone
4,252,886	RCA	Bicyclooctane SO_2
4,262,073	RCA	Acetylene SO_2
4,263,386	RCA	Dust with phosphors
4,289,845	Bell	PMPS-novolak
4,330,671	RCA	Aziridene SO_2
4,396,702	RCA	P-vinylsilane-SO_2
4,397,939	RCA	P-vinylacetate-methylcyclopentene-SO_2
4,398,001	IBM	Terpolymer-novolak
4,409,317	Hitachi	PMPS-novolak
4,476,216	Amdahl	Poly-1-butene sulfone plasma harden
4,497,891	IBM	Dry develop polyolefin SO_2-novolak
4,513,077	Hitachi	*m*-Cresol/*p*-cresol (50–50)-PMPS

[a] U.S. patents unless otherwise indicated
[b] Source: *Chemical Abstracts*.

For poly-1-butene sulfone the G_s values in air are higher than radiolysis in vacuum:

Temp. (°C)	G (fracture)
Vacuum	
−196	9.2
0	10.7
30	12.2
Air	
−78	18.1
0	22.2
30	23.7

In air, the O_2 gas probably oxidizes the scissioned radicals, preventing their recombination. The thermolysis, in air, also promotes more permanent chain scissions than in vacuum baking.[250] In the degradation reaction, free radicals have been detected by electron spin resonance (ESR)[251]:

Mechanistically, the degradation, suggested by Bowmer and O'Donnell,[290] includes the depropagation via free radical and cationic intermediates, followed by the recombination of the free olefin back into polymer, and the isomerization of the olefin.[251,252] However, Bowmer and Bowden reported no isomerization of 2-methyl-1-pentene sulfone, and if any does occur, then it proceeds through a charge transfer complex formed between the olefin donor (D) and the SO_2 acceptor where D* is the isomerized olefin (*cis* to *trans*),[252]

$$D + A \rightarrow DA^* \rightarrow D^* + A$$

With *e*-beam exposure, Pacansky and co-workers reported that very little olefin is formed and SO_2 is expunged from PMPS, leaving the polyolefin fragments to be recombined and cross-linked.[253,254] But later, Bowden and Bowmer[256] found very little polyolefin remaining after exposure of PMPS.

The *e*-beam radiolysis of PMPS is intensity dependent.[246] This dependence is consistent with classical depropagation kinetics, which follows an $I^{1/2}$ dependence,[255]

$$\frac{d(\mathrm{PMPS})}{dt} = \frac{(IG_s)^{1/2}\, PSO_2\cdot}{kt} \qquad (3\text{-}5\text{-}1)$$

Heating (post-expose-predevelop bake) the polyolefin sulfones after an exposure of $> 1\ \mu C/cm^2$, or during exposure, produces image amplification, due to the thermal induced unzipping of scissioned fragments. Unfortunately, recombination of olefin radicals occurs due to the surface immobility of the fragments.

Some of the practical problems of processing polyolefin sulfones are:

1. Swelling of developed images has been noted with the result that resolution in thick films is not as good as in thinner (0.5-μm) films.[243]
2. Film cracking can also occur in some cases of thicker films, > 0.5 μm thick. Terpolymers of cyclopentene and hexane with SO_2 are self-plasticized, and can be used at thicker films without cracking.[249,250,259] Film cracking can be reduced by using a high-molecular-weight polybutene sulfone in conjunction with spray development.[241–249,297] Another technique is to drive the moisture out of the film in a vacuum overnight bake (10^{-6} mm), and if storage is required, the films should be kept in a dry box.[258,260]
3. When moisture is present on the substrate, poor adhesion to chrome surfaces has been noted.[296] For bonding polyolefin sulfones to chrome, a nanometer layer of gold has been used as an adhesion promoter on video disks.[262]
4. The reactive ion etch resistance is poor (five- to tenfold worse than AZ-1350). Films of polyolefin sulfones can be used to plasma etch chrome, if the wafer temperature is kept below 50°C. Polystyrene sulfone is more plasma resistant than the alkyl derivatives, but it is a very slow *e*-beam resist[257,261] (> 50 μC/cm^2). For bilayer processing in an oxygen plasma, oxygen plasma-resistant polysulfones are formed by polymerization of vinylsilane or allyltrimethyl silane and SO_2.[306,358,362]
5. A low contrast (γ) of 0.3–1.0 has been reported[263] and doses higher than 5 μC/cm^2 are required to increase the contrast.

The poor reactive ion etch resistance of polyolefin sulfones is attributed (Chapter 14) to their low empirical formula (C_6) and low T_c when compared to C_{12} aryl polymers. By mixing the polymethyl pentene sulfones (PMPS) with an excess of novolak resins, their plasma resistance was reported to be equivalent to AZ-1350,[264,268]

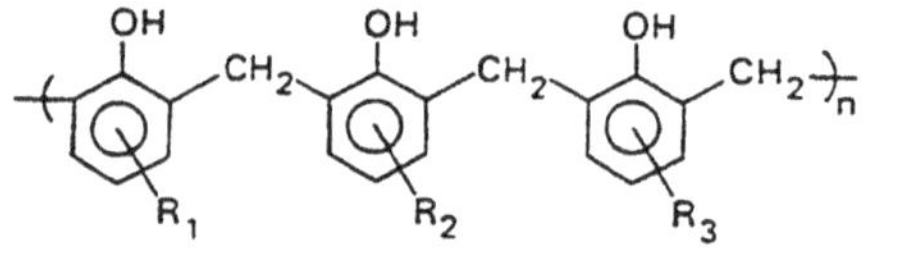

Matrix Resin: Novolak Copolymer

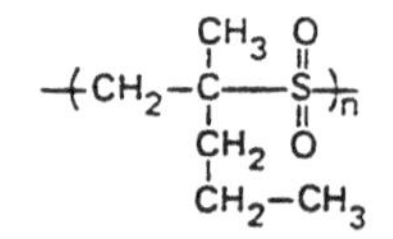

Sensitizer: 2-Methylpentene Sulfone

In the mixture, the PMPS acts as a diazoquinone, inhibiting novolak dissolution. Upon exposure, the inhibitor is destroyed and the exposed region dissolves in an alkaline developer. If the exposure is followed by a predevelop bake, the sensitivity increases from 16 μC/cm^2 (without bake) to 4 μC/cm^2.[264] However, the contrast is still low ($\gamma = 1$) when compared to PMMA (Fig. 3-5-1).

In the absence of intermolecular forces, most polymers are immiscible. The

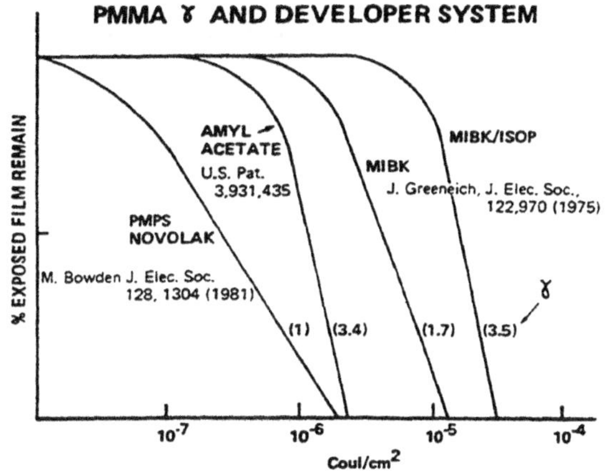

FIGURE 3-5-1. PMPS–novolak contrast versus contrast of PMMA (with different developers).

compatibility of mixing two polymers (PMPS + novolak) is critically dependent on the following factors:

1. Molecular structure of olefin
2. Molecular structure of novolak
3. Casting solvent

PMPS does not phase separate when mixed with a *t*-butyl novolak terpolymer, but is immiscible with *m*-cresol novolaks (Chapter 6). For mixtures of PMPS with novolaks, the acetate family of solvents is preferred. "Cellosolve" acetate (2-ethoxyethyl acetate) or isoamyl acetate are the preferred casting solvents. For mixing PMPS with commercial cresylic acid novolaks, a modification of PMPS by copolymerization with methylallylether sulfone,

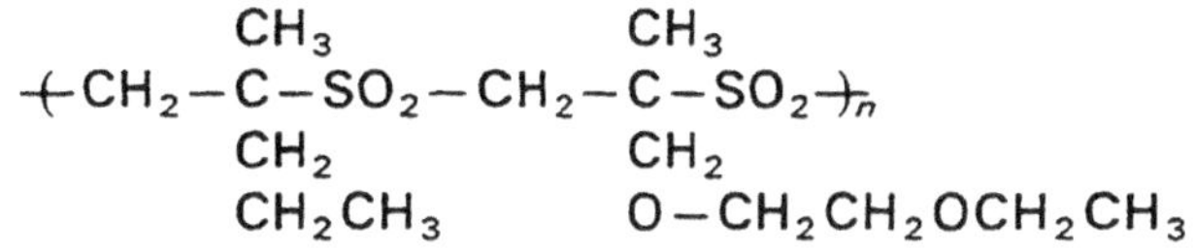

increases the miscibility.[263] A composition of 50/50 (*p*-cresol, *m*-cresol novolak) resin mixed with PMPS in isoamyl acetate[314] gives an *e*-beam sensitivity of 1 $\mu C/cm^2$ while a mixture of PMPS with a 70/30 resin exhibits a sensitivity of only 5 $\mu C/cm^2$.[349]

In the miscible polyolefin sulfone–novolak resist, since polymer–polymer miscibility is not very common (negative entropic factors), a molecular complex may be formed between the polyolefin sulfone and the novolak resin.[267] If diglyme, a bisdiether, is used as a casting solvent for a mixture of a polyolefin sulfone and an *m*-cresol novolak, phase separation occurs (see Fig. 6-5-7). If Cellosolve acetate is used, no phase separation occurs. Within the (D) donor (phenol group on resin) and (A) acceptor sulfone and a solvent (S), three competitive complexes can be formed:

$$D + A \rightarrow D{-}A \quad k_1$$
$$D + S \rightarrow D{-}S \quad k_2$$
$$A + S \rightarrow A{-}S \quad k_3$$

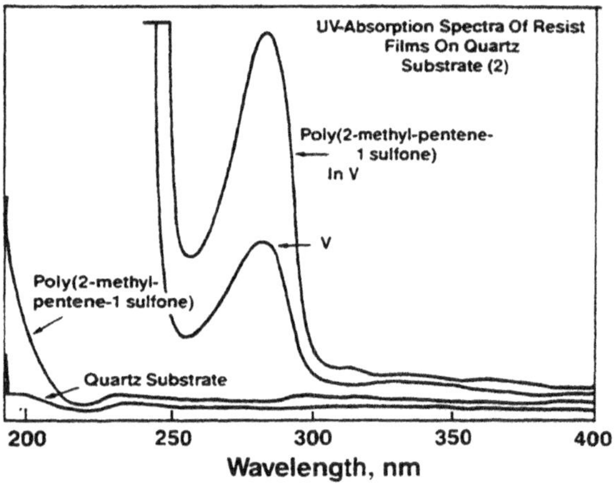

FIGURE 3-5-2. Charge transfer in UV spectrum formed by mixing polyolefin sulfone and Varcum (V) novolak resin. From Ref. 269a, by permission of the American Chemical Society.

If the solvent (S) bonds more strongly, in the case of diglyme to the polysulfone (A), than a casting solvent of Cellosolve acetate, then the donor novolak cannot effectively complex sufficient polysulfone chains and phase separation occurs. With diglyme as a casting solvent, the polysulfones aggregate, and phase separation occurs in cast films. A weak solvent nonpolar acetate such as isoamyl acetate is needed to allow k_1 to predominate. When a polyolefin sulfone was mixed with a novolak resin, Hiroaka and Welsh[269a] noticed an increase in absorbance at 265 nm (Fig. 3-5-2). In charge transfer complexes, the existence of new UV bands in mixtures of alkyl sulfones as acceptors, is well known.[269b] Furthermore, complexation of the novolak resin (D) with the hydrophobic polyolefin sulfone (A), inhibits novolak resin dissolution in alkaline developer. Upon exposure to radiation, the SO_2 group of the polyolefin sulfone is volatized, destroying the charge transfer complex and releasing the hydrophilic phenolic resin to dissolve. The release of free SO_2 and olefin may also foam the exposed region, enhancing the infusion of developer into the exposed region since the dissolution cannot be accounted on the basis of only chain scission.

The mixing of polyolefin sulfone with novolak resin is an outstanding example of two-component resists. The resin component offers resistance to plasma and heat, while the radiation-sensitive inhibitor determines the speed.

Another fast single-component resist of low ceiling temperature has been formulated from polyaldehydes.[305,333] Previously, polyaldehydes have been examined as deep-UV resists, and dry-developed resists.[271] As deep-UV resists, the polyaldehyde sensitivity is enhanced by the presence of onium salts (S), which form Lewis acids (A),[69]

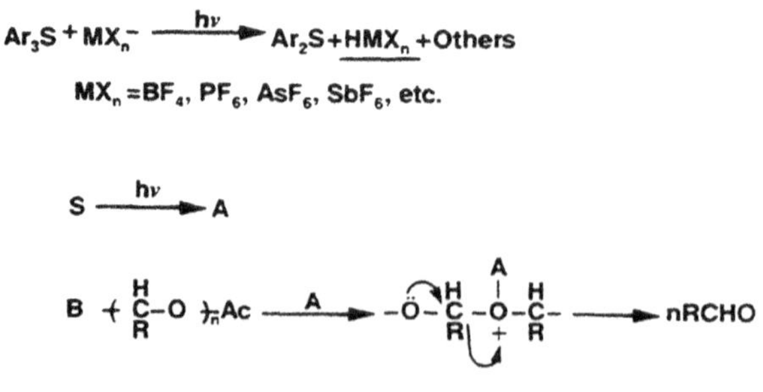

to catalyze the decomposition of polyaldehydes. For *e*-beam exposure, polymers such as polyphthalaldehyde, develop images in dry film, instantly after doses[(269c)] of 3 $\mu C/cm^2$. Polymers such as polyoxymethylene have G_s values of 10–20,[(270)]

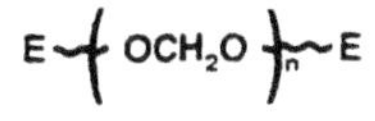

The onium salts catalyze the side chain decomposition of poly-*t*-butoxycarbonylstyrenes,[(272,330,361)]

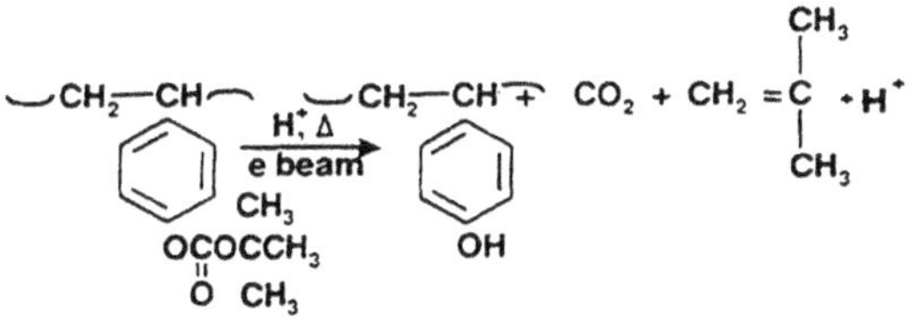

In addition, main chain polycarbonates[(357)] were also cationically decomposed, after a mixture with onium salts was *e*-beam exposed at 2 $\mu C/cm^2$. For the carbonate resists, polar solvents develop the polyvinylphenol (positive), while a nonpolar solvent removes the unexposed (negative) resist. An image amplification of at least 50 G_s occurs since the acid is regenerated from the isobutene intermediate to decarboxylate all of the carbonate groups into phenols.

Some of the miscellaneous positive *e*-beam resists are listed in Table 3-5-2. The major commercial positive *e*-beam resists have been diazoquinone–novolak (DQN) and the polyolefin sulfones (discussed above). Although a G_s of 7 was calculated for commercial DQN resists, the *e*-beam sensitivity is two orders of magnitude lower than the equivalent optical dose (Fig. 3-5-3).[(38,281,282,285,286,288)] Apparently, a side product of sensitizer–resin coupling occurs, to inhibit dissolution,[(284)]

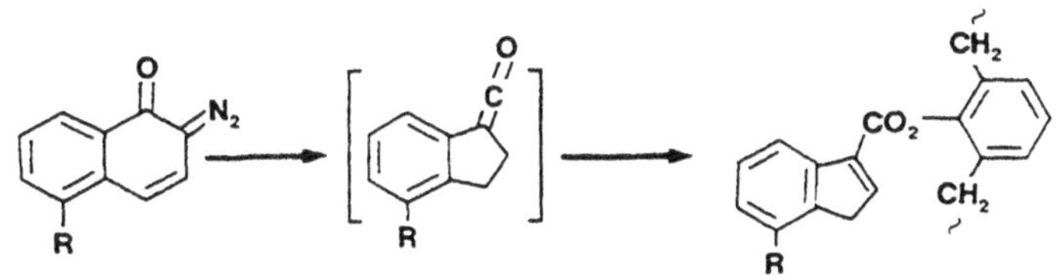

TABLE 3-5-2. Miscellaneous Positive Electron-Beam Resists

Resist	Dose ($\mu C/cm^2$)	Reference
Allyl diglycol carbonate	1	287
Polymethyl isopropenyl ketone	5	273
Nitrocellulose	3	274
Cellulose acetate	5	274
Polycarbonate	10	275
Polyisocyanates	10	276
Polyisopropenyl acetate	2	277
Polyisobutylene	2	278
Polyoxetanes	2	279

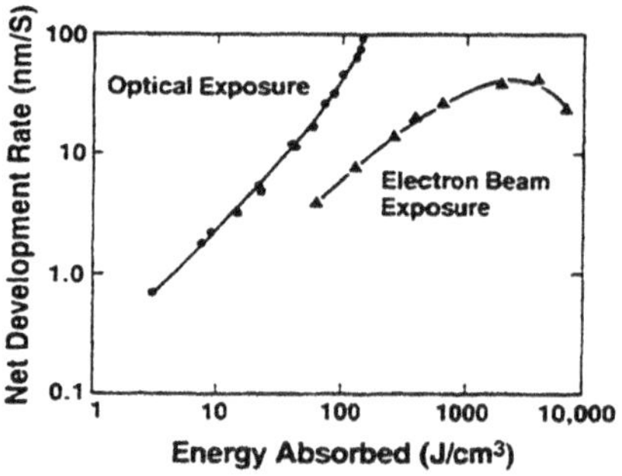

FIGURE 3-5-3. Faster dissolution rate of diazoquinone resist after absorption of equivalent UV and *e*-beams. From M. Kaplan and D. Meyerhofer, *Polym. Eng. Sci.* **20**, 1073 (1980), copyright and courtesy of SPE.

The use of chlorocresol resins is claimed[282] to enhance the DQN sensitivity to 5–10 $\mu C/cm^2$. The chlorocresol resin may form HCl, which adds to the ketene intermediate, preventing resin coupling. Normally, the commercial DQN resists require doses of $> 30\ \mu C/cm^2$ but if a low-temperature (14°C) developer is used, doses as low as 10 $\mu C/cm^2$ can be used.[282–284,352] Furthermore, a DQN formulation made with purified DQ[345] or the addition of imidazole has reduced the *e*-beam doses to the range of 10–20 $\mu C/cm^2$.[346] Finally, image reversal of DQN by *e*-beam exposure,[289] followed by flood UV has been suggested as a means of cutting down on the *e*-beam writing area (pattern factor).[246] Another recent positive *e*-beam resist system is a mixture of *p*-vinylphenol with a phenylester ($PhCO_2Ph$) sensitizer.[331]

3-6. PRACTICAL POSITIVE RESISTS

Ever since commercial *e*-beam mask-making tools have been available, the use of resists has grown in conjunction with the need for 10× and 1× photomasks. The direct *e*-beam fabrication of masks is easier than silicon write, since no alignment or thick resists, have to be used. The photomask requirements for memory circuits DRAM are shown in Table 3-6-1. The speed of the resist is not too important, since a 10× mask requires several hours to expose; but this is a great gain over the several-day turnaround for reticle photorepeaters. The options for direct *e*-beam mask making are shown in Fig. 3-6-1. In direct silicon write, the *e*-beam is preferred for low-volume and high-part-number custom circuits and for fabricating all types of radiation printing masks. Future X-ray, ion, and flood *e*-beam, as well as optical mask, because of negative windage, will require the submicrometer capability of an *e*-beam mask maker.

TABLE 3-6-1. Photomasks for VLSI DRAMs[a]

DRAM Integration	64K bit	256K bit	1M bit
Design rule (μm)	3	2	1–1.5
Tolerance ($\pm \mu m$)	0.20	0.15	0.10
Photomask process	Optical EB master/copy	EB master	Next generation
Dry etching ratio (%)	Optical/EB reticle	EB reticle	EB reticle
Wafer stepper ratio (%) in transfer	30–50	70–80	90
	20–30	50	70

[a] From Ref. 299.

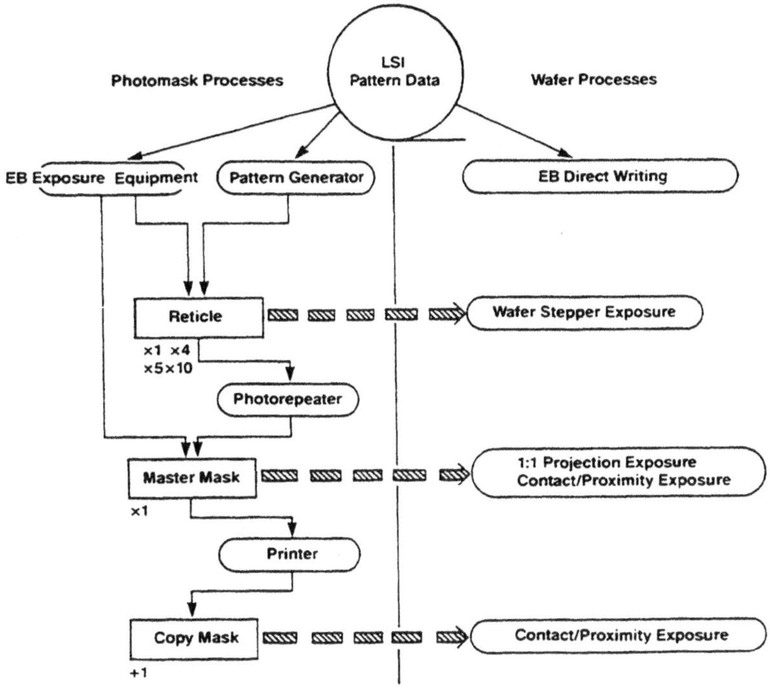

FIGURE 3-6-1. Electron-beam mask-making options. From Y. Hasimoto, *Jpn. Semicond. Technol. News* **2**, 18 (Oct. 1983), copyright and courtesy of Science Forum Inc.

For mask making, the resists that have evolved have focused on only one factor—speed. Contrast, adhesion, etch resistance, ease of development, shelf life, and reproducibility have been neglected since the wet etching of chrome is routinely practiced for 10× single segments. The available positive *e*-beam resists are listed in Table 3-6-2. Spray development without an endpoint is common practice.

A summary of the performance of several commercial resists on a MEBES

TABLE 3-6-2. Commercial Positive Electron-Beam Resists[a]

Resist	Name	Manufacturer	10 keV ($\mu C/cm^2$)	γ
Fluoroallyl methacrylates	FBM-120 FBM-110 FPM	Daikin Kogyo	1–2	2–5
PMMA-methacrylonitrile	OEBR-1013	Tokyo Ohka	5	2
Poly-2,2,2-trichloro ethyl methacrylate	EBR-1	Toray	3	2
Poly-2,2,2-ethyl α-acrylate	EBR-9	Toray	1	2
Polycyano ethyl acrylate-amido ethyl acrylate	FMRE102	Fuji	5	2
PMMA-*t*BMA	CP3	Tokyo Ohka	4	2
Polybutene SO_2	PBS	Mead	1	1
Diazoquinone–novolak	Many	Many	20	2
PMPS-novolak	RE-5000P	Hitachi	2	1
PMMA	950K	KTI	50	5

[a] Data from Refs. 297 and 298.

TABLE 3-6-3. Sensitivity, Contrast, and Resolution of Resists[a]

Resist		Sensitivity ($\times 10^{-7}$ C/cm^2)	Contrast (γ)	Exposure[b] volume to obtain designed linewidth	Resolution (μm)				
					Line and space	Single space	Single line	Window	Island
OEBR-100		2.1	0.99	3.0	1.0	1.5	0.5	1.5	1.0
COP		2.4	0.82	2.1	1.0	1.5	1.0	1.5	1.0
SEL-N MH		3.7	0.93	3.8	0.75	0.75	0.75	1.0	0.75
CMS-EX(S)		8.0	1.51	7.0	0.75	1.0	0.75	1.5	0.75
PBS		6.5	1.17	14.0	0.5	0.5	0.5	0.5	0.75
FBM-120	Developing for 60 sec	5.0	7.01	18.0	1.5	0.75	1.5	1.5	2.0
	Developing for 120 sec	2.0	—	5.0	1.5	0.75	1.5	1.5	2.0
EBR-9	Developing for 6 min	19.0	3.48	28.0	0.5	0.5	0.5	1.0	0.75
	Developing for 7 min	13.0	1.95	20.0	0.5	0.75	0.75	1.0	1.0

[a] From Ref. 297.
[b] $\times 10^{-7}$ C/cm^2.

(trademark Bell Labs) tool, found by Inoue[297] and co-workers, is given (Table 3-6-3) for negative COP, CMS-EX, and positive FBM-120, EBR-9, and PBS. With the introduction of scanning 1× optical projection, the demand for higher-resolution *e*-beam positive resists with Cr dry etching began. Even though the contrast is high ($\gamma = 2$–5) for FBM (fluorobutyl methacrylate), the resolution is poor, due to swelling of the images. Since the contrast is measured in a large pad exposure,[80] it does not assess the interproximity effects, and micrometer resolution. Because of swelling, the images are restricted to micrometer-size geometries. Thus, contrast cannot be truly measured without factoring in the MTF, the beam edge slope, and the resulting resist wall profiles, at images close to film thickness.

For chrome mask making, improved positive and negative resists are needed. If a resist cannot perform well at the mask making level, for silicon processing, the problems will be worse. Multilayer resists are still required to correct for the proximity effect and eventually the mask-making tool will also require extensive proximity tool corrections and higher beam currents.

After examining all of the positive *e*-beam resists, an assessment can be made against the performance characteristics of an "ideal" resist (Table 3-6-4). In practical lithographic processes, there are trade-offs in performance,

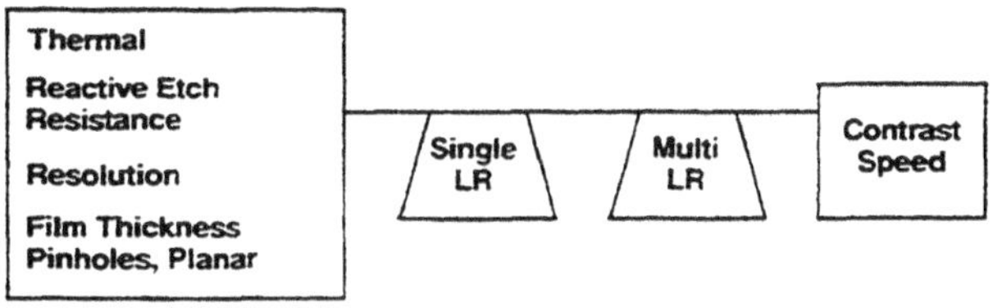

If we examine the positive *e*-beam or radiation resists, listed in Table 3-6-5, we conclude that: (1) the PMMA resist has the highest resolution at the slowest speed and (2) the two-component resists of DQN and polysulfone–novolak mixtures offer excellent plasma resistance and thermal stability (once deep UV hardened, Chap-

TABLE 3-6-4. Characteristics of a High-Performance Resist

Process step	Performance
Dry deposit or liquid conformal planarizing	Rapid deposition, pinhole < $1/cm^2$, adhesive, uniform ± 75 Å, (0.1–3.0 μm)
Prebake	None
Expose	1 $\mu C/cm^2$, 25 keV, *e*-beam 5 mJ/cm^2, 200–300 μm UV, X ray
Resolution	Contrast $\gamma \geqslant 3.0$ 1.0-μm pitch, through width ⩽ thickness
Develop	> 95% film retention ±10% process window Dry or develop
Thermal stability	250°C and soluble for lift-off
Etch rate, reactive ion etch	Equal or less than substrate (ideal no loss, anisotropic)
Strippability	O_2 or liquid
Shelf life	> 6 months at 25°C

TABLE 3-6-5. Assessment of Positive Radiation Resists

Resist type	Speed ($\mu C/cm^2$)	Resolution in 1.0 μm	RIE relative to SiO_2	WH (flow) (°C)
PMMA	10–100	0.25	4×	130
PMMA-halo derivative	3–10	1.0	4×	100
PMMA-cross-linked	10–100	0.5	4×	180
Polysulfone	1–10	1.0	10×	75
Polyaldehydes	1–10	1.0	10×	75
Polysulfone–novolak	5–15	1.0	2×	150–250[a]
DQN	10–30	1.0	1×	150–250[a]

[a] Posttreatments.

ter 13). However, the polycarbonate or other image-amplified resists will eventually replace the novolak mixtures because of their inherent image-amplified speed in an aryl-containing system. Thus, the upgrading of a resist performance can be accomplished in two basic ways:

Multilayer resist ⟷ Additives to resist or Posttreat resist

In multilayer resists, speed can be traded for better resist thermal and plasma performance.

Resist Dosimetry

Exposure to high-energy radiation requires accurate measurements ($\pm 2\%$) of dose (intensity × time), if submicrometer linewidths are to be controlled to ± 0.1-μm bias or less (3σ). Fortunately, the edge slope and intensity of the *e*-beam, for example, can be accurately measured, since the radiation is very monochromatic. Flood X-ray and ion-beam measurements are less accurate, and usually require a secondary assessment to ascertain the physical flux.

Chemical actinometry is based on the quantitative measurement of the quantum yield or *G* value of a chemical reaction, which is generally independent of

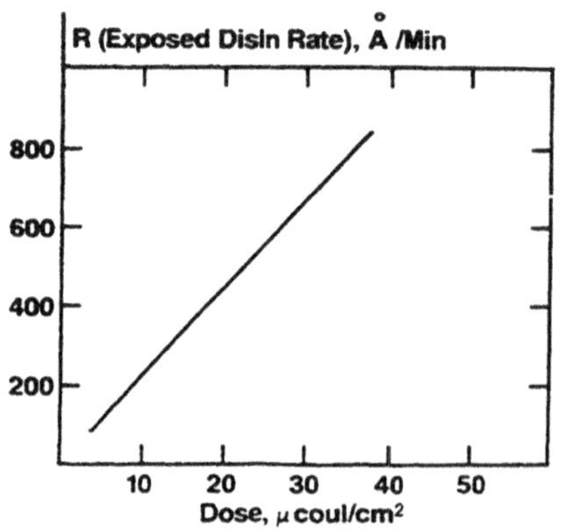

FIGURE 3-6-2. PMMA dissolution rate *R* as a function of 20-kV incident *e*-beam dose.

wavelength. By itself, the resist dissolution response is a form of physical actinometry which is a good indicator of the incident dose. For example, PMMA prebaked at 160°C, exposed, and developed in methyl isobutyl ketone at constant temperature (Fig. 3-6-2) is fairly linear in dissolution, and dose response. If a film-thinning apparatus can be set up (Fig. 2-3-1), accurate dosimetry, ±2%, can be established for X-ray, *e*-beam, and ion-beam exposure.

For chemical actinometry, the direct UV measurement of the bleaching of dyes is another convenient method. For example, X-ray badge-dosimeters bleach or change color, when exposed to megarad levels. Thin dyed films (50 μm thick), cast from nylon in organic solvents, have been used as a universal lithographic dosimeter.[280] The bleaching or darkening of AZ-1350 has also been used to calibrate the flux of an ion implanter,[304] and even the etch pits in plastic films, usually associated with neutron bombardment, have also been used[287] for ion-beam dosimetry. A convenient X-ray dosimeter is a dispersion of a diazoquinone,

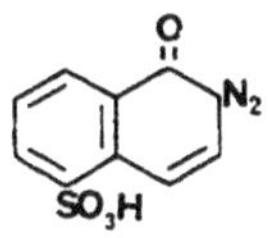

DQ

in polyvinylpyrollidone (DQ-P). The organic, insoluble layer can be cast on top of or under resists,

The bleaching of the diazoquinone can be monitored by UV analysis at 365 nm, or the loss of diazo band in the infrared can be used. As an underlayer, for example, for X-ray dosimetry,

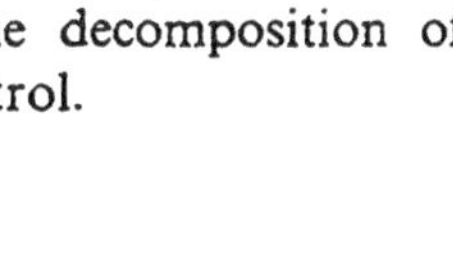

the absorbance of X rays by the resist can be measured by absorbing the transmitted radiation in the DQ-P layer, stripping the resist layer in organic solvents, and measuring spectroscopically the decomposition of DQ in comparison with the uncoated but exposed DQ control.

3-7. OUTLOOK

Positive radiation resists are pioneering materials. They will continue to improve with many kinetic and thermodynamic studies of chain scission, molecular rearrangement, and the physical interaction of the radiation products with the developer.

Since free radical chain scissioning polymers have a limited G_s value of about 15, increased speed by chemical amplification is a new and exciting area of investigation for future positive resists. From the exposure tool viewpoint, the ability of software and hardware to handle the interproximity effect problems for *e*-beam exposure will determine the fate of single-film positive radiation resists. If the proximity effect can be solved by beam dose corrections, then a single-layer resist will be the potential choice for low-density patterns such as via holes. In this case, a moderate-speed (5–10 $\mu C/cm^2$) resist with the performance of DQNs will be needed. For high resolution, PMMA is still the best material. If proximity cannot be handled properly or economically by exposure tricks, then multilayer resists, with more costly process layers and steps, will be required.

REFERENCES

1. G. Varnell, P. Shah, and R. Havemann, *Proc. IEEE* **71**, 612 (1983).
2. D. Herriott, *J. Vac. Sci. Technol.* **20**, 781 (1982).
3. D. Spears and H. Smith, *Solid State Technol.* July 1972, p. 21.
4. E. Weber, *SPIE Proc.* **333**, 95 (1982).
5. H. Bohlen, J. Greschner, J. Keyser, W. Kulcke, and P. Nehmiz, *IBM J. Res. Dev.* **26**, 568 (1982).
6. R. Speidel and M. Mayr, *Optik* (Stuttgart) **48**, 247 (1977).
7. M. Hatzakis, D. Hofer, and T. Chang, *J. Vac. Sci. Technol.* **16**, 1631 (1979); U. S. Patent 4,267,259 (1981), IBM.
8. M. Heritage, *J. Vac. Sci. Technol.* **12**, 1135 (1975).
9. W. Brown, T. Venkatesan, and A. Wagner, *Solid State Technol.* Aug. 1981, p. 60.
10. L. Fried, J. Havas, P. Totta, J. Logan, G. Paal, and J. Lechaton, *IBM J. Res. Dev.* **26**, 362 (1982).
11. B. El Karek and A. Pattletz, *J. Vac. Sci. Technol.* **15**, 1047 (1978).
12. H. Ryssel, *Appl. Phys. A* **27**, 239 (1982); G. Bird, *Photogr. Sci. Eng.* **25**, 246 (1981).
13. (a) J. Greeneich, in *Electron Beam Technology in Microelectronics Fabrication*, edited by G. Brewer, Academic Press, New York, 1980, p. 209; (b) J. Trotel and B. Fay, *ibid.* p. 309.
14. C. Ting, *Record of the 11th Symposium on Electron, Ion, and Laser Beam Technology*, edited by R. Thomley, San Francisco Press, San Francisco, 1971, p. 345.
15. U. S. Patent 3,535,137 (1971), IBM.
16. R. Selliger and W. Fleming, *J. Appl. Phys.* **45**, 1416 (1974).
17. D. Spears and H. Smith, *Electron. Lett.* **8**, 102 (1972).
18. U. S. Patents 3,934,057; 4,024,293 (IBM).
19. S. Beaumont, P. Bower, T. Tamamura, and C. Wilkinson, *Appl. Phys. Lett.* **38**, 436 (1981).
20. U. S. Patent 4,004,044 (1976), IBM.
21. M. Hatzakis, *J. Electrochem. Soc.* **116**, 1033 (1969).
22. B. Mallon, *J. Polym. Sci. Part A* **6**, 2637 (1968).
23. L. Karapiperis, I. Adesida, C. Lee, and E. Wolfe, *J. Vac. Sci. Technol.* **19**, 1259 (1981).
24. W. Buckley and G. Hughes, *J. Electrochem. Soc.* **128**, 1106 (1981).
25. M. Natase and Y. Matsumoto, *Photogr. Sci. Eng.* **23**, 215 (1979).

25a. R. Heidenreich, *J. Appl. Phys.* **48**, 1418 (1977).

26. S. Bernacki and H. Smith, *IEEE Trans. Electron Devices* **ED-22**, 421 (1975); R. Heidenreich, J. Ballantyne, and L. Thompson, *J. Vac. Sci. Technol.* **12**, 1284 (1975).
27. T. Matsukawa, *J. Appl. Phys.* **45**, 733 (1974); R. Hawryluk, H. Smith, A. Soares, and A. Hawryluk, *J. Appl. Phys.* **46**, 2528 (1975).
28. T. Kato, *J. Vac. Sci. Technol.* **19**, 1279 (1981).
29. J. Ziegler, *Handbook of Stopping Cross Sections for All Elements*, Pergamon Press, Elmsford, N.Y., 1980.
30. M. Levi, *IEEE Electron Device Lett.* **EDL-1**, 194 (1980).
31. W. Brown, T. Venkatesan, and A. Wagner, *Solid State Technol.* Aug. 1981, p. 60.
32. A. Crewe, *J. Vac. Sci. Technol.* **16**, 258 (1979); M. Tsuda, *Polym. Eng. Sci.* **17**, 390 (1977).

33. L. Dorfman, *J. Chem. Educ.* **58**, 84 (1981).
34. F. Field and F. Franklin, *Electron Impact Phenomena*, Academic Press, New York, 1957.
35. H. Hiroaka, *IBM J. Res. Dev.* **21**, 121 (1977).
36. C. McNeal, *J. Am. Chem. Soc.* **103**, 1609 (1981).
37. R. Kieb, *J. Phys. Chem.* **63**, 1838 (1959).
38. M. Kaplan and D. Meyerhofer, *RCA Rev.* **40**, 166 (1979); *Polym. Eng. Sci.* **20**, 1073 (1980).
39. K. Kawabuchi, S. Sakurai, and M. Yoshimi, *Electron. Lett.* **19**, 287 (1983).
40. R. Watts, *Solid State Technol.* May 1979, p. 69.
41. U. S. Patents 3,743,842 (1973), MIT; 4,037,111 (1979), Bell; 4,035,522 (1978), IBM.
42. K. Okada, *J. Vac. Sci. Technol.* **17**, 1233 (1980).
43. J. Pearlman and J. Riordan, *J. Vac. Sci. Technol.* **19**, 1190 (1981).
44. B. Fay, I. Trotel, Y. Petroff, R. Pinchaux, and P. Thiry, *Appl. Phys. Lett.* **29**, 370 (1976).
45. (a) K. Heinrich, H. Betz, A. Heuberger, and S. Pongratz, *J. Vac. Sci. Technol.* **19**, 1254 (1981). (b) T. Ono and A. Ozawa, *J. Vac. Sci. Technol.* **B2**, 68 (1984). (c) Y. Saitoh, H. Yoshihara, I. Watametre, and S. Nakayama, *J. Vac. Sci. Technol.* **B2**, 63 (1984).
46. W. Buckley, *J. Electrochem. Soc.* **128**, 1116 (1981).
47. C. Fencil and G. Hughes, *Circuits Manuf.* Dec. 1982, p. 22.
48. M. Kaplan, SPE RETEC Photopolymers, Ellenville, N.Y., 1982, p. 249; *Polym. Eng. Sci.* **23**, 957 (1983).
49. I. Haller, R. Feder, M. Hatzakis, and E. Speller, *J. Electrochem. Soc.* **126**, 154 (1979).
50. R. Feder, *IBM Tech. Discl. Bull.* **18**, 2343 (1975).
51. K. Murcase, M. Kakuchi, and S. Sugawara, Microcircuit Eng. Conf., Paris, 1977, p. 261.
52. L. Thompson and M. Bowden, *J. Electrochem. Soc.* **121**, 1500 (1974); *J. Appl. Phys.* **44**, 4039 (1973).
53. G. Taylor, *Solid State Technol.* May 1980, p. 69.
54. Y. Saithoh, V. H. Yoshihara, and I. Watanabe, *Jpn. J. Appl. Phys.* **21**, L52 (1982).
55. A. Eranian, A. Coutlet, E. Datamanti, and J. Dubois, *Am. Chem. Soc. Symp. Ser.* **151**, 275 (1981); U. S. Patent 4,268,590 (1981), Thomson-CSF.
56. W. Moreau, W. Moyer, D. Merritt, M. Hatzakis, L. Pederson, and D. Johnson, *J. Vac. Sci. Technol.* **16**, 1989 (1979).
57. Y. Bocko, *Sov. Phys. Tech. Phys.* **27**, 1416 (1982).
58. M. Kakuchi, V. S. Sagawara, K. Murase, and K. Matsuyama, *J. Electrochem. Soc.* **124**, 1648 (1977).
59. W. Moreau, *Opt. Eng.* **22**, 181 (1983).
60. E. Spiller, J. Topalian, E. Castellani, L. Romankiw, R. Feder, and M. Heritage, *Solid State Technol.* April 1976, p. 65.
61. Japanese Patent 80,17,112; *Chem. Abstr.* **93**, 16986 (1980).
62. F. Asmussen, H. Betz, J. Chen, A. Heuberger, S. Pongratz, H. Sotobayashi, and W. Schindel, *J. Electrochem. Soc.* **130**, 180 (1983).
63. P. Blais, *Electrochem. Soc. Ext. Abstr.* **80**, 835 (1980).
64. P. Lenzo and E. Spencer, *Appl. Phys. Lett.* **24**, 289 (1974).
65. P. Burggraaf, *Semicond. Int.* Sept. 1983, p. 60.
66. J. Lane, Proc. 1983 Int. Conf. Electron, Ion, Photon Beams; *J. Vac. Sci. Technol.* **B1**, 1072 (1983); S. Farenholtz, *J. Vac. Sci. Technol.* **19**, 1111 (1981).
67. M. Bowden, L. Thompson, S. Farenholtz, and E. Doerries, *J. Electrochem. Soc.* **128**, 1304 (1981).
68. U. S. Patent 4,409,319 (1983), IBM.
69. H. Ito and G. Willson, *Org. Coat. Appl. Polym. Sci. Proc.* **48**, 60 (1983).
70. N. Economou, D. Flanders, and J. Donnelly, *J. Vac. Sci. Technol.* **19**, 1172 (1981).
71. R. Seliger, J. Ward, V. Wang, and R. Kiebena, *Appl. Phys. Lett.* **34**, 310 (1979).
72. P. Grigaitis and L. Pranevicus, *Appl. Phys. Lett.* **34**, 310 (1979); *Jpn. J. Appl. Phys.* **20**, L261 (1981).
73. P. Bohn, J. Taylor, and H. Guckel, *Anal. Chem.* **53**, 1082 (1981); *J. Vac. Sci. Technol.* **17**, 735 (1980).
74. D. Rensch, R. Seliger, G. Csanky, R. Olney, and H. Stover, *J. Vac. Sci. Technol.* **16**, 1897 (1979); L. Karapiperis and C. Lee, *Appl. Phys. Lett.* **35**, 395 (1978).
75. P. Sullivan and J. McCoy, *IEEE Trans. Electron Devices* **ED-23**, 412 (1976).
76. J. Woodhouse, A. Fields, and A. Bucklow, *J. Phys. D* 7, 483 (1974).
77. M. Gazard, J. Dubois, L. Duchesne, and A. Chapiro, *Polym. Eng. Sci.* **20**, 1069 (1980).

78. K. Kanaya, K. Shimizu, and Y. Ishikawa, *J. Phys. D* **1**, 1657 (1968).
79. K. Csepregi, F. Iberl, and P. Eichinger, *Appl. Phys. Lett.* **37**, 630 (1980).
80. R. Brault and L. Miller, *Polym. Eng. Sci.* **20**, 1064 (1980), **23**, 941 (1983); *J. Electrochem. Soc.* **128**, 1159 (1981).
81. I. Adesida, L. Rathburn, J. Chinn, and E. Wolf, *J. Vac. Sci. Technol.* **21**, 666 (1982).
82. G. Taylor, T. Venkatesan, A. Wagner, B. Wilkens, and D. Barr, *J. Vac. Sci. Technol.* **19**, 1379 (1981); U. S. Patent 4,377,437 (1983), Bell.
83. H. Kurvano, K. Yoshida, and S. Yamazaki, *Jpn. J. Appl. Phys.* **19**, L615 (1980).
84. D. Tenant, *J. Vac. Sci. Technol.* **B1**, 493 (1983).
85. B. Hunt and R. Buhrman, *J. Vac. Sci. Technol.* **19**, 1310 (1981).
86. A. Wagner, *Solid State Technol.* May 1983, p. 97.
87. T. Hall, A. Wagner, and L. Thompson, *J. Appl. Phys.* **53**, 3997 (1982).
88. A. Macrander, *SPIE Proc.* **333**, 142 (1982).
89. H. Ryssel, K. Haberger, and H. Kranz, *J. Vac. Sci. Technol.* **19**, 1358 (1981).
90. K. Moriwaki, H. Aritome, S. Namba, and L. Karapiperis, *Jpn. J. Appl. Phys.* **20**, L881 (1981).
91. B. MacIver, *J. Electrochem. Soc.* **129**, 827, 2384 (1982).
92. Japanese Patent 80,73,047, *Chem. Abstr.* **93**, 248282 (1980).
93. Japanese Patent 80,22,864, *Chem. Abstr.* **93**, 58286 (1980).
94. Y. Okuyama, T. Hashimoto, and T. Koguchi, *J. Electrochem. Soc.* **125**, 1293 (1978).
95. K. Harada, T. Tamamura, and O. Kogure, *J. Electrochem. Soc.* **129**, 2576 (1982).
96. Y. Wada, M. Migitaka, K. Mochiji, and H. Obayashi, *J. Electrochem. Soc.* **130**, 1127 (1983).
97. U. S. Patent 4,239,787 (1980), Bell.
98. K. Mochiji, Y. Wada, and H. Obayashi, *J. Electrochem. Soc.* **129**, 2556 (1982).
99. Y. Iida, V. H. Okabayashi, and K. Suzuki, *Jpn. J. Appl. Phys.* **16**, 1313 (1977).
100. U. S. Patent 3,920,483 (1975), IBM.
101. German Patent 2,930,416, *Chem. Abstr.* **93**, 228638 (1980).
102. Japanese Patent 80 28078, *Chem. Abstr.* **93**, 58291 (1980).
103. Japanese Patent 79,115,128, *Chem. Abstr.* **92**, 67739 (1979).
104. S. Forrest, V. M. Kaplin, P. Schmidt, T. Venkatesan, and A. Lovinger, *Appl. Phys. Lett.* **41**, 708 (1982), **43**, 30 (1983); *J. Appl. Phys.* **54**, 3150 (1983).
105. European Patent Application 193,022 (1980), IBM; E. Spiller, *Appl. Opt.* **19**, 3022 (1980).
106. K. Suzuki, *Electron. Lett.* **16**, 178 (1980).
107. K. Galloway, *J. Electrochem. Soc.* **127**, 1862 (1980).
108. T. Venkatesan, T. Wolf, D. Allara, B. Wilkens, and G. Taylor, *Appl. Phys. Lett.* **43**, 934 (1983).
109. T. Venkatesan, W. Brown, C. Murry, K. Marcantonio, and B. Wilkens, *Polym. Eng. Sci.* **23**, 931 (1983).
110. I. Haller, M. Hatzakis, and R. Srinivassan, *IBM J. Res. Dev.* May 1968, p. 251; U. S. Patent 3,535,137 (1973), IBM.
111. M. Komuro, N. Atoda, and H. Kawakatsu, *J. Electrochem. Soc.* **126**, 483 (1979).
112. G. Mladenov and B. Emmonoth, *Appl. Phys. Lett.* **38**, 1000 (1981).
113. K. Balasubramanyan, L. Karapiperis, C. Lee, and A. Rouff, *J. Vac. Sci. Technol.* **19**, 18 (1981).
114. (a) J. Phang and H. Ahmed, *J. Vac. Sci. Technol.* **16**, 1756 (1979); (b) M. Parikh, *J. Vac. Sci. Technol.* **15**, 935 (1978); (c) C. Shaw, *J. Vac. Sci. Technol.* **19**, 1286 (1981); (d) T. Chang, *J. Vac. Sci. Technol.* **12**, 1271 (1975); (e) I. Adesida and T. Everhart, *J. Appl. Phys.* **51**, 5994 (1980); (f) M. Rosenfield, A. Neureuther, and C. Ting, *J. Vac. Sci. Technol.* **19**, 1242 (1981); (g) R. Hawryluk, *J. Vac. Sci. Technol.* **19**, 1 (1981); (h) G. Owen and P. Rissman, *J. Appl. Phys.* **54**, 3573 (1983); (i) P. Vettiger, *J. Vac. Sci. Technol.* **B1**, 1378 (1983); (j) M. Nakase, *IEEE Trans. Electron Devices* **ED-27**, 1460 (1980).
115. B. Lin, *Solid State Technol.* May 1983, p. 105.
116. A. Broers, *J. Electrochem. Soc.* **128**, 166 (1981); D. Kyser and C. Ting, *J. Vac. Sci. Technol.* **16**, 1759 (1979).
117. M. Yoshimi, K. Kawabuchi, T. Takigawa, M. Takahashi, and Y. Kato, *Electron. Lett.* **18**, 880 (1982), **16**, 621 (1980).
118. A. Iranmanesh and R. Pease, *Electrochem. Soc. Ext. Abstr.* **82-1**, 507 (1982).
119. H. Ralph, G. Duggan, and R. Elliott, *Electrrochem. Soc. Ext. Abstr.* **82-1**, 514 (1982); T. Lin, *IBM J. Res. Dev.* Sept. 1967, p. 257.
120. A. Avidenko and M. Malev, *Vacuum* **27**, 583 (1977).
121. T. Kelen, *Polymer Degradation*, Van Nostrand, Princeton, N.J., 1983.

122. M. Dole, editor, *Radiation Chemistry of Macromolecules*, Vols. I and II, Academic Press, New York, 1973.
123. T. Tada, *J. Electrochem. Soc.* **129**, 1070 (1982); S. Schlick and L. Kevan, *J. Appl. Polym. Sci.* **27**, 319 (1982).
124. A. Torikai, H. Kato, and Z. Kuri, *J. Polym. Sci.* **14**, 1065 (1976); M. Tabata, *J. Polym. Sci.* **21**, 3257 (1983).
125. Y. Yamamoto, *Macromolecules* **10**, 1316 (1977).
126. M. Tabata, *J. Polym. Sci. Polym. Chem. Ed.* **21**, 3257 (1983).
127. M. Bowden, L. Stillwagon, T. Thompson, and T. Bowmer, *Org. Coat. Appl. Polym. Sci. Proc.* **48**, 161 (1983).
128. H. Pfeiffer, *Microcircuit Eng.*, 1983, Paper A-1.
129. E. Weber, *SPIE Proc.* **333**, 94 (1982).
130. R. Moore, G. Colloma, H. Pfeiffer, E. Weber, and O. Woodward, *J. Vac. Sci. Technol.* **19**, 950 (1981); *Solid State Technol.* Sept. 1983, p. 127.
131. M. Miyazaki, N. Saitou, and C. Munakata, *J. Phys. E* **14**, 194 (1981).
132. A. Broers, *IEEE Trans. Electron Devices* **ED-28**, 1268 (1981).
133. H. Ku and L. Scala, *J. Electrochem. Soc.* **116**, 980 (1969).
134. J. Greeneich, *J. Electrochem. Soc.* **122**, 970 (1975).
135. U. S. Patent 3,996,393 (1975), IBM.
136. C. Pittman, M. Uedo, C. Chen, J. Kwiatkowski, C. Cook, and J. Helbert, *J. Electrochem. Soc.* **128**, 1758 (1981), and early references therein.
137. J. Helbert, C. Cook, J. Pittman, and C. Chen, *J. Electrochem. Soc.* **126**, 694 (1979); U. S. Patent 3,914,462 (1974), Hitachi.
138. E. Gipstein, W. Moreau, and O. Need, *J. Electrochem. Soc.* **123**, 1105 (1973).
139. U. S. Patent 3,898,350 (1974), IBM.
140. U. S. Patent 4,011,351 (1978), IBM.
141. Y. Hatano, Proc. 8th Int. Conf. Electron, Ion Beams, 1978, p. 332.
142. U. S. Patent 4,004,043 (1978), IBM.
143. U. S. Patent 3,931,435 (1976), IBM.
144. W. Moreau, W. Moyer, D. Merritt, M. Hatzakis, L. Pederson, and D. Johnson, *J. Vac. Sci. Technol.* **16**, 1989 (1979); U. S. Patent 4,087,569 (1979), IBM.
145. M. Hatzakis and D. Webb, *J. Vac. Sci. Technol.* **16**, 2008 (1979); U. S. Patent 4,156,745 (1981), IBM.
146. C. Anderson, Y. Namaste, F. Rodriguez, and K. Obendorf, *Am. Chem. Soc. Org. Coat. Prepr.* **48**, 156 (1983).
147. U. S. Patent 4,279,984 (1982), Matsushita.
148. U. S. Patent 3,914,462 (1973), Hitachi.
149. M. Kakuchi, S. Sugawara, K. Murase, and K. Matsuyama, *J. Electrochem. Soc.* **124**, 1648 (1977); *Rev. Electr. Commun. Lab.* **27**, 1113 (1979).
150. T. Kitakohji, *J. Electrochem. Soc.* **126**, 1881 (1979).
151. T. Kitayama, *Jpn. Semicond. Technol. News* **2**(4), 2 (1983).
152. Japanese Patent 80,58,211, *Chem. Abstr.* **93**, 141027 (1980).
153. T. Tada, *J. Electrochem. Soc.* **126**, 1829 (1979), **130**, 912 (1983); U. S. Patent 4,268,607 (1981), VLSI.
154. C. Chen, C. Pittmann, and J. Helbert, *J. Polym. Sci. Polym. Chem. Ed.* **18**, 169 (1980); *J. Vac. Sci. Technol.* **16**, 1992 (1979); U. S. Patents 4,304,840 and 4,302,259 (1981), Honeywell; J. Lai, *J. Electrochem. Soc.* **129**, 1596 (1982).
155. M. Yamada and S. Hattori, *Jpn. J. Appl. Phys.* **20**, 1969 (1981).
156. C. Pittmann, J. Helbert, and C. Chen, *J. Polym. Sci.* **18**, 3413 (1980).
157. J. Helbert, *Polym. Eng. Sci.* **17**, 402 (1977); French Patent 2,240,200, TI.
158. M. Hatzakis, *J. Vac. Sci. Technol.* **16**, 1984 (1979).
159. U. S. Patent 4,125,672 (1978), Nippon.
160. H. Ito, D. Miller, and C. Willson, *Macromolecules* **15**, 915 (1982).
161. U. S. Patent 3,996,393 (1975), IBM.
162. R. Harris, *J. Electrochem. Soc.* **120**, 272 (1973).
163. N. Atoda, M. Komeiro, and H. Kawakatsu, *J. Appl. Phys.* **50**, 3707 (1979).
164. L. Gavens, B. Wu, D. Hess, A. Bell, and D. Soong, *J. Vac. Sci. Technol.* **B1**, 481 (1983).
165. J. Lai and L. Sheperd, *J. Appl. Polym. Sci.* **20**, 2367 (1976).

166. E. Gipstein, O. Need, A. Ouano, and D. Johnson, *IBM J. Res. Dev.* **21**, 143 (1977).
167. M. Ballauf and B. Wolf, *Macromolecules* **14**, 654 (1981).
168. K. Harada, T. Tamamura, and O. Kogure, *J. Electrochem. Soc.* **129**, 2576 (1982).
169. C. Pittman, M. Ueda, C. Chen, C. Cook, J. Helbert, and J. Kwatkowski, *J. Electrochem. Soc.* **128**, 1759 (1981); *Macromolecules* **11**, 1104 (1978).
170. N. Grassie, *Polym. Degr.* **3**, 209 (1981).
171. G. Willson, H. Ito, D. Miller, and T. Tessier, SPE RETEC Photopolymers, 1982; *Polym. Eng. Sci.* **23**, 1000 (1983).
172. N. Viswanathan, *J. Polym. Sci. Polym. Chem. Ed.* **14**, 1553 (1976).
173. J. Dubois, M. Gazard, and C. Duchesne, Int. Conf. Microlithogr., Paris, 1977, p. 289.
174. Japanese Patent 78,21,286, *Chem. Abstr.* **89**, 138401 (1979).
175. Japanese Patent 78,51,284, Tokyo Shibaura; *Chem. Abstr.* **90**, 31974h (1978).
176. French Patent 2,304,933, *Chem. Abstr.* **91**, 220371 (1979), Thompson CSF.
177. Japanese Patent 79,41,719, *Chem. Abstr.* **91**, 999536 (1979), Cho LSI.
178. U. S. Patent 4,051,271 (1978), AGIST.
179. U. S. Patent 3,931,435 (1976), IBM.
180. K. Murase, Int. Conf. Microlithogr., Paris, 1977, p. 265.
181. A. Ouano, *Polym. Eng. Sci.* **18**, 306 (1978), **20**, 160 (1980).
182. K. Harada, *J. Macromol. Sci.* **C3**, 347 (1969).
183. D. Yoon, IBM Internal Report RJ1684 (1975).
184. W. Moreau and P. Schmidt, *Electrochem. Soc. Ext. Abstr.* **70**, 459 (1970).
185. A. Charlesby and R. Blackburn, *Nature* **210**, 1036 (1966).
186. H. Harada and S. Sugawara, *J. Appl. Polym. Sci.* **27**, 1441 (1982).
187. U. S. Patents 3,934,057 and 4,024,293, IBM.
188. G. Geuskeins, *Macromol. Chem.* **160**, 343 (1973), **165**, 273 (1973).
189. J. Gardella and D. Hercules, *Anal. Chem.* **52**, 2273 (1980).
190. J. Bargon, E. Gipstein, H. Hiroaka, and L. Welsh, *J. Appl. Polym. Sci.* **22**, 3397 (1978); U. S. Patent 4,084,569 (1979).
191. E. Roberts, *Am. Chem. Soc. Org. Coat. Prepr.* **33**, 359 (1973); U. S. Patents 4,061,832 (1978) and 3,981,985 (1976), U.S. Phillips; *Polym. Eng. Sci.* **23**, 968 (1983).
192. U. S. Patent 4,276,365 (1981), Fujitsu.
193. U. S. Patent 4,273,857 (1981), Fujitsu.
194. M. Suzuki and Y. Ohnishi, *J. Electrochem. Soc.* **129**, 402 (1982).
195. T. Tada, *J. Electrochem. Soc.* **126**, 1635 (1979).
196. F. Asmussen, H. Betz, J. Chen, A. A. Heuberger, S. Pongrantz, H. Sotobayashi, and W. Schnabel, *J. Electrochem. Soc.* **130**, 180 (1983).
197. U. S. Patent 4,121,936 (1978), Fuji.
198. U. S. Patent 4,279,984 (1981), Matsushita.
199. Japanese Patent 80,52,051, *Chem. Abstr.* **93**, 177306 (1980).
200. German Patent 2,946,206, *Chem. Abstr.* **93**, 248264 (1980).
201. Japanese Patent 80,151,637, *Chem. Abstr.* **94**, 165703 (1981).
202. U. S. Patent 4,264,715 (1981), VLSI.
203. U. S. Patent 4,276,365 (1981), Fujitsu.
204. U. S. Patent 4,268,607 (1981), VLSI.
205. T. Kitakohji, Y. Yoneda, K. Kitamura, H. Okuyama, and K. Muiakawa, *J. Electrochem. Soc.* **126**, 1881 (1979).
206. German Patent 2,537,932, *Chem. Abstr.* **91**, 132155 (1979).
207. Japanese Patent 79,66,836, *Chem. Abstr.* **92**, 119729 (1980).
208. Japanese Patent 80,159,436, *Chem. Abstr.* **94**, 112562 (1982).
209. Japanese Patent 80,108,656, *Chem. Abstr.* **94**, 112563 (1982).
210. Japanese Patent 78,125,023, *Chem. Abstr.* **89**, 144309 (1978).
211. U. S. Patent 3,984,582, IBM; I. Haller, *J. Electrochem. Soc.* **126**, 154 (1979).
212. Japanese Patent 79,66,122, *Chem. Abstr.* **91**, 166416 (1979).
213. Japanese Patent 78,32,718, *Chem. Abstr.* **89**, 188990 (1978).
214. Y. Yamashita, *J. Vac. Sci. Technol.* **16**, 2026 (1979); *Jpn. J. Appl. Phys.* **20**, L22 (1981); German Patent 2,849,996, *Chem. Abstr.* **92**, 102322 (1980); Japanese Patent 79,29,627, *Chem. Abstr.* **91**, 66338 (1979).
215. A. Mura, *J. Vac. Sci. Technol.* **16**, 1996 (1979); Japanese Patent 80,151,637, *Chem. Abstr.* **94**, 165703 (1980).

216. S. Matsuda, *Polym. Eng. Sci.* **17**, 410 (1977).
217. Japanese Patent 80,90,950, *Chem. Abstr.* **94**, 55936 (1980).
218. U. S. Patent 4,096,290 (1979), IBM.
219. German Patent 2,437,422, *Chem. Abstr.* **86**, 24468 (1977).
220. J. Chinn, I. Adesida, E. Wolfe, and R. Tiberio, *J. Vac. Sci. Technol.* **19**, 1418 (1981).
221. K. Harada, *J. Appl. Polym. Sci.* **26**, 3395 (1981); *J. Electrochem. Soc.* **128**, 491 (1981).
222. L. Pederson, *J. Electrochem. Soc.* **129**, 206 (1982).
223. Japanese Patent 80,142,337, *Chem. Abstr.* **94**, 112566 (1980); U. S. Patent 4,289,573 (1981), IBM.
224. U. S. Patent 4,289,573 (1981), IBM.
225. European Patent 8,787, *Chem. Abstr.* **93**, 228635 (1980).
226. Japanese Patent 80,159,436, *Chem. Abstr.* **94**, 217649 (1980).
227. K. Harada, O. Kogure, and K. Murase, *IEEE Trans. Electron Devices* **ED-29**, 518 (1982).
228. J. Lai, *J. Electrochem. Soc.* **129**, 1596 (1982).
229. Japanese Patent 81,16,416, *Chem. Abstr.* **95**, 159909 (1981).
230. Japanese Patent 81,39,539, *Chem. Abstr.* **95**, 12409 (1981).
231. E. Gipstein, W. Moreau, and O. Need, *J. Electrochem. Soc.* **123**, 1105 (1976).
232. J. Aronte, *Fujitsu Sci. Tech. J.* June 1978, p. 143.
233. M. Yamada and S. Hattori, *Jpn. J. Appl. Phys.* **20**, 1969 (1981).
234. Japanese Patent 81,95,903, *Chem. Abstr.* **96**, 60868 (1981).
235. Japanese Patent 77,93,493, *Chem. Abstr.* **88**, 56981 (1977), **86**, 198006 (1976).
236. J. Tamano, *Plasma Chem. Plasma Process.* **1**, 261 (1981).
237. M. Gazard, S. Haltori, M. Iida, and S. Morita, *Polym. Eng. Sci.* **20**, 1069 (1980).
238. U. S. Patent 4,193,797 (1981), DuPont.
239. J. Brown and J. O'Donnell, *Macromolecules* **3**, 265 (1970), **5**, 109 (1972).
240. N. Tokura, in *Encyclopedia of Polymer Science and Technology*, Wiley, New York, 1969, pp. 460–485.
241. L. Thompson, M. Bowden, J. Ballantyne, R. Pease, L. Yau, S. Skinner, and J. Frackoviak, in *Microcircuit Engineering*, edited by H. Ahmed and W. Nixon, Cambridge University Press, London, 1981, p. 239.
242. M. Bowden and L. Thompson, *J. Appl. Polym. Sci.* **17**, 3211 (1973).
243. M. Bowden and L. Thompson, *Appl. Polym. Symp.* **23**, 99 (1974).
244. M. Bowden, *J. Polym. Sci. Polym. Symp.* **49**, 221 (1975).
245. M. Bowden, L. Thompson, and J. Ballantyne, *J. Vac. Sci. Technol.* **12**, 1295 (1975).
246. M. Bowden and L. Thompson, *Polym. Eng. Sci.* **14**, 5251 (1974), **17**, 269 (1977).
247. M. Bowden and L. Thompson, *J. Electrochem. Soc.* **121**, 1620 (1974).
248. R. Himics and M. Kaplan, *Am. Chem. Soc. Org. Coat. Prepr.* **35**, 266 (1975); *Polym. Eng. Sci.* **17**, 406 (1977).
249. E. Gipstein, W. Moreau, G. Chiu, and O. Need, *J. Appl. Polym. Sci.* **21**, 677 (1977).
250. M. Bowden, L. Thompson, W. Robinson, and M. Biolsi, *Macromolecules* **15**, 1417 (1982).
251. T. Bowmer and J. O'Donnell, *Polymer* **22**, 71 (1981) and references therein; *Radiat. Phys. Chem.* **17**, 177 (1981); *J. Polym. Sci.* **19**, 45 (1981).
252. T. Bowmer and M. Bowden, *Am. Chem. Soc. Org. Coat. Appl. Polym. Sci.* **48**, 171 (1983).
253. A. Gutierrez, J. Pacansky, and R. Kroeker, ACS *Org. Coat. Appl. Polym. Sci.* **46**, 520 (1982).
254. J. Pacansky, R. Krocker, E. Gipstein, and A. Gutierrez, *Electrochem. Soc. Ext. Abstr.* **82-1**, 447 (1982).
255. J. Calvert and J. Pitts, *Photochemistry*, Wiley, New York, 1966, p. 653, and references to Norrish rotating sector method for radical lifetime.
256. M. Bowden and T. Bowmer, *Am. Chem. Soc. Org. Coat. Appl. Polym. Sci.* **48**, 161 (1983).
257. U. S. Patent 3,884,696 (1975), Bell.
258. U. S. Patent 3,935,331 (1976), RCA.
259. U. S. Patent 3,898,350 (1975), IBM.
260. U. S. Patent 4,007,295 (1977), RCA.
261. M. Bowden and L. Thompson, *J. Electrochem. Soc.* **121**, 1620 (1974).
262. U. S. Patent 4,007,295 (1977).
263. M. Bowden, L. Thompson, J. Ballantyne, R. Pease, L. Yau, J. Frackoviak, and J. Skinners, in *Microcircuit Engineering*, edited by W. Nixon, Pergamon Press, New York, 1981.
264. M. Bowden, L. Thompson, S. Farenholtz, and F. Doerries, *J. Electrochem. Soc.* **128**, 1304 (1981), U. S. Patent 4,289,845 (1982), Bell.
265. U. S. Patent 4,398,001 (1983), IBM.

266. H. Shiraishi, *Am. Chem. Soc. Org. Coat. Appl. Polym. Sci.* **48**, 178 (1983); *Am. Chem. Soc. Symp. Ser.* **242**, 170 (1983); U. S. Patent 4,409,317 (1981), Hitachi.
267. S. Farenholtz and T. Kwei, *Macromolecules* **14**, 1076 (1981), **15**, 937 (1982); *J. Vac. Sci. Technol.* **19**, 1111 (1981).
268. M. Bowden, *J. Polym. Sci.* **26**, 1424 (1981).
269. (a) H. Hiroaka and L. Welsh, *Am. Chem. Soc. Org. Coat. Appl. Polym. Sci.* **48**, 48 (1983); (b) R. Andrews and R. Kufer, *Molecular Complexes in Organic Chemistry*, Holden–Day, San Francisco, 1964; (c) U. S. Patent 3,940,507 (1976), RCA.
270. S. Torikai, *J. Polym. Sci. Part A* **2**, 239 (1964); *J. Polym. Sci.* **7**, 186 (1969).
271. U. S. Patents 3,915,704, 3,915,706, 3,917,483, Xerox.
272. H. Ito and C. Willson, *Am. Chem. Soc. Org. Coat Appl. Pol. Sci.* **48**, 60 (1983).
273. A. Levine, M. Kaplan, and E. Polinak, *Polym. Eng. Sci.* **14**, 118 (1974); U. S. Patent 4,012,536 (1978), RCA.
274. U. S. Patent 3,985,914 (1976), IBM.
275. U. S. Patent 3,961,099 (1975), IBM.
276. A. Levine, M. Kaplan, and J. Fech, *J. Polym. Sci.* **11**, 311 (1973).
277. Japanese Patent 78,78,835, *Chem. Abstr.* **90**, 64496 (1978).
278. Japanese Patent 78,115,832, *Chem. Abstr.* **90**, 130685 (1978).
279. Japanese Patent 78,81,115, *Chem. Abstr.* **90**, 64498 (1978).
280. J. Pacansky and J. Lyerla, *IBM J. Res. Dev.* **23**, 42 (1979); *J. Electrochem. Soc.* **126**, 860 (1979).
281. J. Shaw, M. Fritsch and F. Dill, *IBM J. Res. Dev.* **21**, 219 (1977).
282. J. Shaw and M. Hatzakis, *J. Electrochem. Soc.* **126**, 2026 (1979).
283. J. Shaw and M. Hatzakis, *IEEE Trans. Electron Devices* **ED-25**, 426 (1978).
284. U. S. Patent 4,409,319 (1983), IBM.
285. J. Pacansky and H. Cocufal, *J. Am. Chem. Soc.* **102**, 402 (1980).
286. J. Pacansky and A. Guetterrez, IBM Internal Report RJ 3627 (1982).
287. D. O'Sullivan and C. Willson, *J. Electrochem. Soc.* **129**, 811 (1982).
288. J. Shaw and M. Hatzakis, *J. Vac. Sci. Technol.* **19**, 1343 (1981).
289. S. Farenholtz, *J. Vac. Sci. Technol.* **19**, 1111 (1981); U. S. Patent 4,173,470 (1979), Bell.
290. T. Bowmer and J. O'Donnell, *J. Macromol. Sci.* **A17**, 243 (1982).
291. A. Chapiro, *Radiation Chemistry of Polymers*, Interscience, New York, 1962.
292. A. Charlesby, *Atomic Radiation and Polymers*, Pergamon Press, New York, 1960.
293. Japanese Patent 76,148,417, *Chem. Abstr.* **86**, 131180 (1976).
294. U. S. Patent 4,074,031 (1978), IBM.
295. Japanese Patent 76,148,418, *Chem. Abstr.* **86**, 131179 (1976).
296. U. S. Patent 3,900,737 (1975), Bell.
297. M. Inoue, *Jpn. Electron. Eng.* May 1982, p. 60.
298. S. Fok and G. Hong, Proceedings of Kodak Microelectronics Seminar, 1983.
299. Y. Hashimoto, *Jpn. Semicond.Technol. News* **2**, 22 (1983).
300. E. Weber and R. Moore, *Solid State Technol.* May 1979, p. 61.
301. G. Clark, *Applied X-rays*, McGraw–Hill, New York, 1955, p. 216.
302. K. Mochiji, *Appl. Phys. Lett.* **45**, 251 (1984).
303. H. Gecim, *Electron. Lett.* **20**, 598 (1984).
304. J. Golin and J. Glaze, *Solid State Technol.* Aug. 1984, p. 137.
305. European Patent Application 96,895, *Chem. Abstr.* **100**, 94542 (1983).
306. U. S. Patent 4,396,702 (1983), RCA.
307. U. S. Patent 4,397,939 (1983), RCA.
308. E. Roberts, *Polym. Eng. Sci.* **23**, 968 (1983).
309. U. S. Patent 4,414,313 (1983), Honeywell.
310. Japanese Patent 58,184,944, *Chem. Abstr.* **100**, 165447 (1983).
311. Japanese Patent 57,139,738, *Chem. Abstr.* **100**, 15327 (1983).
312. U. S. Patent 4,405,710 (1983), Cornell.
313. Japanese Patent 57,205,739, *Chem. Abstr.* **101**, 15036 (1984).
314. German Patent 3,322,886, *Chem. Abstr.* **100**, 129921 (1982).
315. U. S. Patent 4,415,653 (1983), Honeywell.
316. Y. Namatse, F. Rodriguez, C. Anderson, and K. Obendorf, *SPIE Proc.* **469**, 53 (1984).
317. N. Tung, *J. Electrochem. Soc.* **131**, 2152 (1984).
318. J. Jensen and C. Slayman, *SPIE Proc.* **471**, 53 (1984).

319. H. Hiroaka, *SPIE Proc.* **469**, 127 (1984).
320. V. Sharma, S. Affrossmann, and R. Pethrick, *Polymer* **25**, 1087 (1984).
321. V. Sharma, S. Affrossmann, and R. Pethrick, *Polymer* **25**, 1090 (1984).
322. V. Sharma, S. Affrossmann and R. Pethrick, *Polymer* **25**, 1087 (1984).
323. E. Roberts, *Solid State Technol.* June 1984, p. 135.
324. G. Taylor, *Solid State Technol.* June 1984, p. 124.
325. J. Jensen, *Solid State Technol.* June 1984, p. 145.
326. V. Aristov, in *Microcircuit Engineering* edited by J. Cleaver, H. Ahmed, and G. Jones, Academic Press, New York, 1983, p. 343.
327. W. Schnabel and H. Sotobayashi, *Prog. Polym. Sci.* **9**, 297 (1983).
328. E. Roberts and C. Fuller, in *Microcircuit Engineering*, edited by J. Cleaver, H. Ahmed, and G. Jones, Academic Press, New York, 1983, p. 297.
329. J. Cleaver, P. Heard, and H. Ahmed, in *Microcircuit Engineering*, edited by J. Cleaver, H. Ahmed, and G. Jones, Academic Press, New York, 1983, p. 133.
330. U.S. Patent 4,491,628 (1985), IBM.
331. Japanese Patent 58,114,030, *Chem. Abstr.* **101**, 81659 (1984).
332. Japanese Patent 58,116,532, *Chem. Abstr.* **101**, 81660 (1984).
333. European Patent Application 109,617, *Chem. Abstr.* **101**, 81666 (1984).
334. W. Schnabel, S. Klamunzer, H. Sotobayashi, F. Asmussen, and Y. Tabata, *Macromolecules* **17**, 2108 (1984).
335. S. Equsa, K. Ishigure, and Y. Tabata, *Macromolecules* **12**, 939 (1979).
336. G. Babu, A. Narula, J. Chien, and S. Hsu, *J. Polym. Sci.* **22**, 195 (1984).
337. Y. Namatase, S. Obendorf, C. Anderson, P. Krasicky, F. Rodriguez, and R. Tiberio, *J. Vac. Sci. Technol.* **B1**, 1160 (1983).
338. A. Eranian, A. Cordlet, F. Damatani, and J. Dubois, in *Microcircuit Engineering*, Academic Press, New York, 1984, p. 404.
339. I. Aesida, M. Zhang, and E. Wolfe, *J. Electron. Mater.* **13**, 689 (1984).
340. G. Mladenov, B. Emmoth, and M. Braun, *Vacuum* **34**, 551 (1984).
341. K. Valiev, *Sov. Microelectron.* **12**, 1 (1983).
342. K. Valiev, *Sov. Microelectron.* **12**, 101 (1983).
343. K. Murata, *J. Appl. Phys.* **57**, 575 (1985).
344. G. Babu, A. Narula, S. Hsu, and J. Chien, *Macromolecules* **17**, 2749 (1984).
345. U. S. Patents 3,852,771 (1974); 4,005,437 (1977), RCA.
346. U. S. Patent 3,661,582 (1972), Western Electric.
347. P. Clifford, N. Green, and M. Pilling, *J. Phys. Chem.* **89**, 925 (1985).
348. E. Wallace, C. Chen, C. Pittmann, J. Kwiatkowski, C. Cook, and J. Helbert, *Polym. Eng. Sci.* **25**, 83 (1985).
349. U. S. Patent 4,513,077 (1985), Hitachi.
350. Japanese Patent 60,26,337 (1985), Fujitsu; *Chem. Abstr.* **103**, 62586 (1985).
351. Japanese Patent 60,55,337 (1985), NEC; *Chem. Abstr.* **103**, 62594 (1985).
352. H. Yamashita and Y. Todokoro, *J. Vac. Sci. Technol.* **B3**, 1004 (1985).
353. H. Yamashita and Y. Todokoro, *Electron. Lett.* **21**, 645 (1985).
354. K. Mochyi, T. Kimura, and H. Oboyashi, *Appl. Phys. Lett.* **46**, 387 (1985).
355. N. Samoto, R. Shizumu, and H. Hashimoto, *Jpn. J. Appl. Phys.* **24**, 482 (1985).
356. M. Suzuki, Y. Ohnishi, and A. Furuta, *J. Electrochem. Soc.* **132**, 1390 (1985).
357. U. S. Patent 4,476,217 (1985), Honeywell.
358. S. MacDonald, H. Ito, and C. Willson, *Microelectron. Eng.* **1**, 269 (1983).
359. A. Schiltz, A. Weill, and P. Paneez, in *Microcircuit Engineering*, edited by H. Beneking and H. Beureuther, Academic Press, New York, 1985, p. 545.
360. T. Hayasaka, S. Ishihara, H. Kinoshita, and N. Takeuchi, *J. Vac. Sci. Technol.* **B3**, 1581 (1985).
361. J. Frechet, F. Houlihan, and C. Willson, *Polym. Mater. Sci. Eng.* **53**, 268 (1985).
362. A. Gowdz, H. Craighead, and M. Bowden, SPE RETEC Photopolymers, Ellenville, N.Y., 1985, p. 157.

4

Negative Photoresists

4-1. INTRODUCTION

4-1-1. *Chemical Reactions*

Negative photoresists are polymers or compositions that retain the exposed region after development. Upon exposure, the exposed region becomes more insoluble (slower dissolving) than the unexposed region. The insolubilization can be brought about by:

1. Increase in molecular weight due to polymerization or cross-linking
2. Change in polarity due to formation of more polar or more nonpolar functional groups
3. Change in oxidation state of ion or the permanent ionization of a charge transfer complex

The most prevalent negative photoresists are based on random cross-linking (Fig. 4-1-1) of main chains or pendant side chains. Each linkage requires one or more photochemical events to activate cross-linking. In the liquid state, direct monomer polymerization can achieve low doses (high sensitivity) by quantum amplification. The primary photochemical event is the formation of an initiator (I) which commences polymerization by addition to a monomer. In turn, the monomer continues to propagate 10^4 monomer units to form a linear vinyl polymer [Eq. (4-1-2)]. A theoretical maximum photoresist speed of 1 mJ/cm^2 has been calculated from a random polymerized liquid monomer system initiated by one photochemical event.[1, 163] Unfortunately, liquid monomers do not form tack-free films at room temperature and therefore are not practical resist formulations.

$$II \rightarrow 2I'' \qquad (4\text{-}1\text{-}1)$$

$$I'' + \text{monomers} \rightarrow \text{Polymers} \qquad (4\text{-}1\text{-}2)$$

However, when the monomers are dispersed in a solid film, the cross propagation reaction greatly subsides, since the propagating radicals are immobilized by the high viscosity of the medium and gains of only 10 × in degrees of polymerization are realized. A recent modification of classical free radical polymerization for negative resists is the innovation of top surface postpolymerization. Free radicals or cations (from decomposition of onium salts)

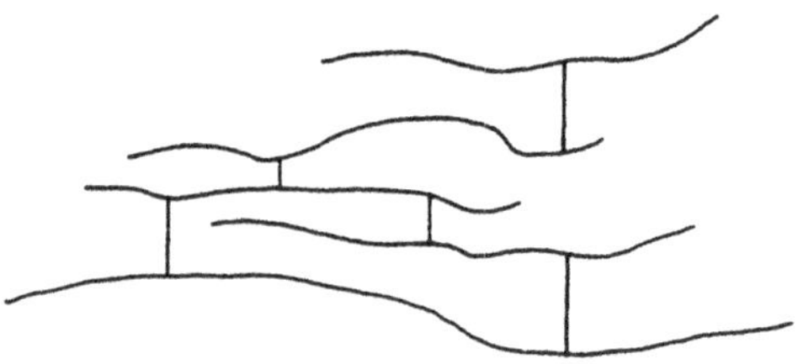

FIGURE 4-1-1. Conventional cross-linking of randomly orientated polymer chains. Each cross-link requires one or more separate radiation events. For the five polymer chains shown, five to ten radiation events are required. From Ref. 1.

are created at the surface of a polymer by radiation. The film is brought in contact with monomers which polymerize on top of the image. Very low doses ($< 5\ \mathrm{mJ/cm^2}$) can be converted into useful bilayer resists if silicon monomers are used.

Instead of polymerizing linear vinyl monomers into soluble polymers, branched monomers, such as polyfunctional acrylates,[13b]

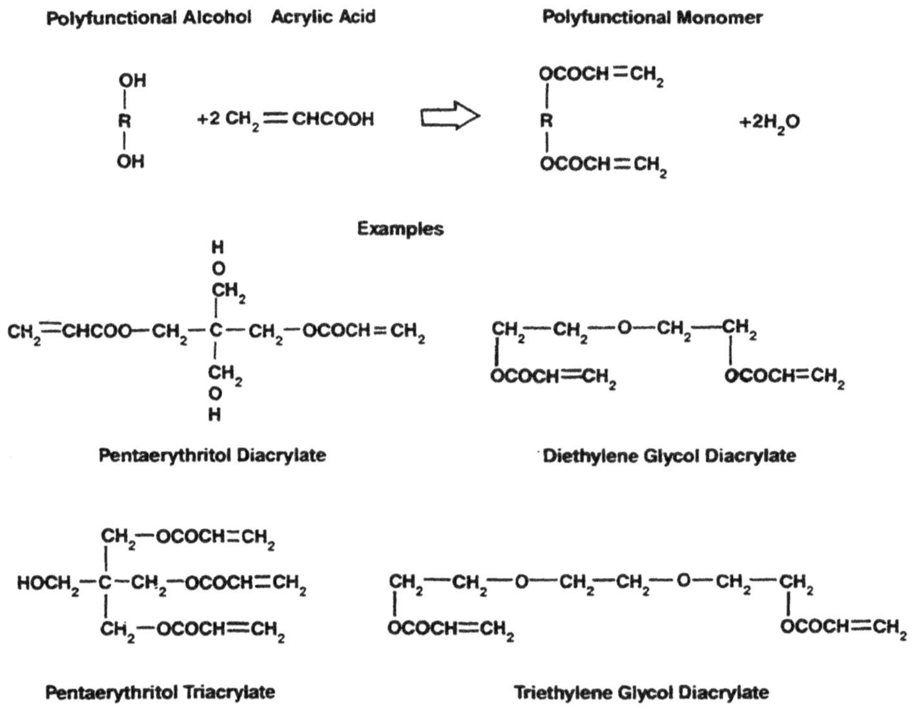

can undergo more rapid changes in molecular weight due to directional branching and cross-linking. Nevertheless, in an amorphous medium, the vinyl sites may not be in sufficient proximity to achieve high quantum gain. The ideal state[1] of ordered polymerization is one in which the vinyl groups or cross-linking sites are arranged as in a crystal lattice in close ordered proximity (Fig. 4-1-2). Only in a few cases such as solid monomer crystals or Langmuir–Blodgett monomolecular films,

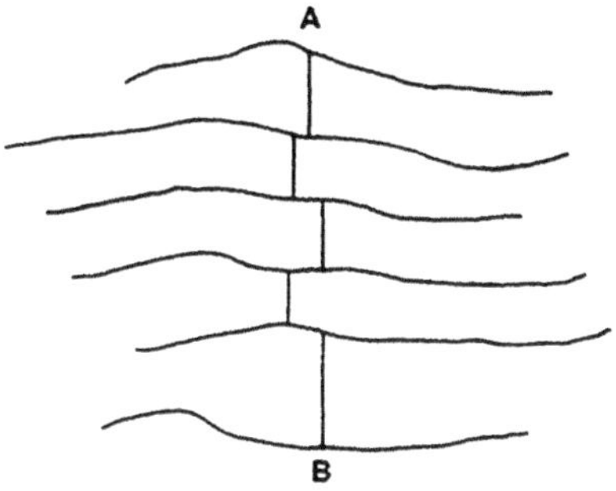

FIGURE 4-1-2. A single radiation event at A in orientated polymer chains induces five polymer chains to cross-link until inhibited at B. Chain reaction results in network formation and a quantum gain (yield) of 5. One photon produces five chemical reactions. From Ref. 1.

do such states exist. Of all the types of polymerization (Fig. 4-1-3), ordered polymerization is the most effective means of rapidly converting monomers or monomer appendages into cross-linked or high-molecular-weight polymers. However, practical resists rely mainly on random interchain cross-linking to achieve insolubilization.

Upon exposure, the stages of insolubilization pass through agglomeration and then gelation (Fig. 4-1-4). A sol (soluble state) may exist in the gel state owing to the polydispersity of polymers, i.e., the occurrence of low-molecular-weight components. The polydispersity is the statistical ratio of the weight average (M_w) to the number average (M_n) molecular weight and ranges from 1.0 (monodisperse) to about 15. Most free radical polymers exhibit polydispersity in the range of 1.5–3. Ideally, the monodisperse polymers and highest-molecular-weight polymers maximize the gain, sensitivity, and resolution of cross-link networks, but lower-molecular-weight polymers are less prone to swelling and thus are of higher resolution than their higher-molecular-weight components. In practice, more polydisperse polymers are used because of the high synthetic cost of obtaining monodisperse polymers.

4-1-2. *Exposure and Contrast*

The nonuniformity and scattering (forward and reflective) of exposure limit resolution in a cross-linking negative resist since the topside of the film is overexposed to achieve insolubilization and adhesion to the substrate (Fig. 4-1-5). Typically, the forward[2] and backscattered radiation enlarge the base of an image of a typical negative resist and the linewidths continue to enlarge with exposure.

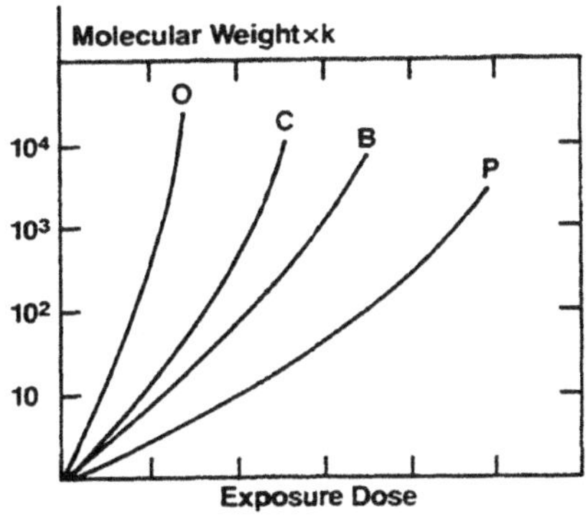

FIGURE 4-1-3. Changes in polymer molecular weight as a function of dose for: O, ordered cross-links; C, random cross-links; B, branching; and P, linear polymerization.

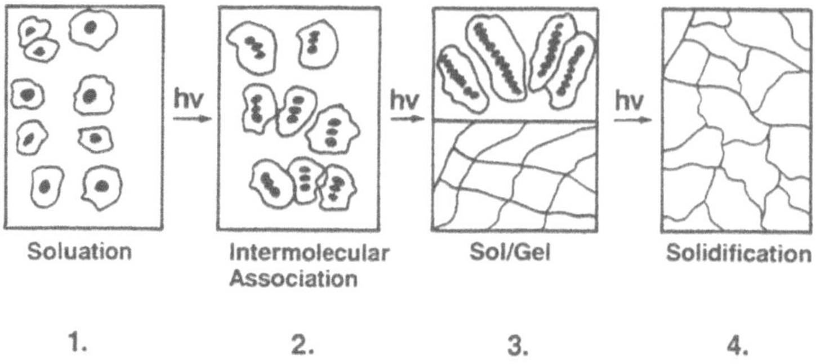

FIGURE 4-1-4. Stages of photoinsolubilization of cross-linking resist.

Ideally, a negative resist could be severely overexposed with radiation scattering and be developed without linewidth enlargement, since after several cross-links are formed, additional cross-linking does not lead to further insolubilization.

Line broadening or resolution loss in negative photoresists would be less of a problem if transparent substrates could be used and the resist was exposed from the inside out (Fig. 4-1-6). Underexposed resists should develop off the wafer (cases 2 and 3), but in practice only isolated islands are developed. In most cases, the isolated lines wiggle from the topside stresses and large pads remain because the swollen top layer redeposits on the substrate. If the resists are exposed on reflective

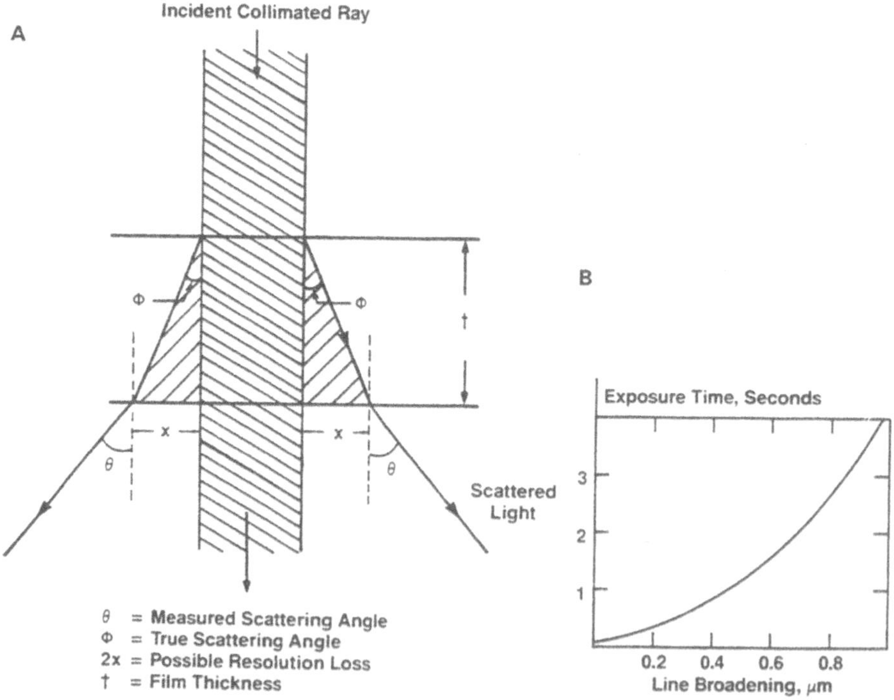

FIGURE 4-1-5. (A) Resolution loss in resist due to forward and backscattered radiation. (B) Line broadening in negative KMER with exposure. From Ref. 2, reproduced by permission of the Electrochemical Society Inc.

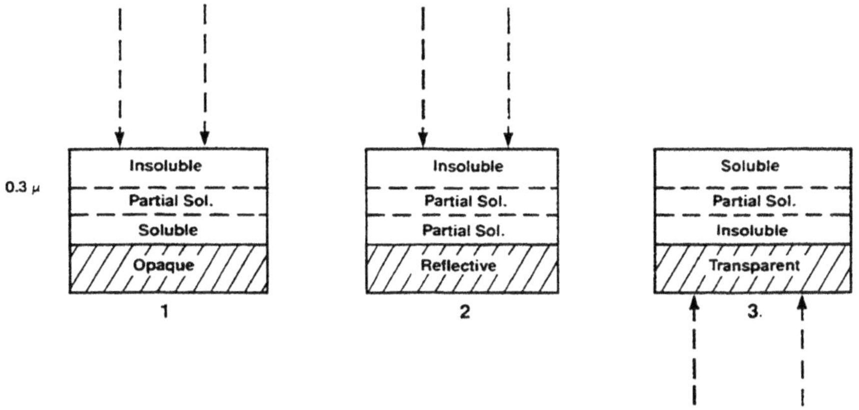

FIGURE 4-1-6. Stages of insolubilization of negative resist. (1) Top exposure on opaque exposing substrate. (2) Top exposure on reflective substrate. (3) Backside exposure on transparent substrate.

substrates, the backscattered radiation also widens the base and reduces the pitch (i.e., the resolution between closely spaced lines).

The formation of a gel state at the interface of a resist exposed on a glass slide (Fig. 4-1-6) represents the minimum dose for the resist. Charlesby[1] calculated that for a vinyl side chain reacting by random cross-linking, that at the gel point fraction (g),[3,4]

$$g = 1 - \exp[\sigma g f(g, i)] \tag{4-1-3}$$

where σ is the cross-linking coefficient (about 2 for typical cross-linking) and i is the activity of an inhibitor which terminates polymer radicals. Reiser and Pitts[5] derived a relation to connect the gel fraction with the cross-linking density (p),

$$g = 1 - \sum n f(n)(1 - pg)^n \tag{4-1-4}$$

where n is the degree of polymerization and $f(n)$ is the molecular weight distribution given by

$$f(n) = \exp - (\ln n/2B^2) \tag{4-1-5}$$

where B is the polydispersity, given by the ratio half-width (Fig. 4-1-7).[5]

The gel dose E_g measures the gel fraction. Experimentally, the gel fraction is equivalent to the normalized film thickness, d/d_0, where d is the developed film thickness and d_0 is the undeveloped cast film thickness.[3] For backside exposure (Fig. 4-1-8), the gel point dose[5] E_g is

$$E_g = dS/AM_w\Phi \tag{4-1-6}$$

where S is the specific gravity of the film, A is the absorption coefficient of energy (photons), and Φ is the quantum yield of cross-linking. Photoresists of the highest initial molecular weight (M_w)[203] are the most sensitive. For the topside exposure

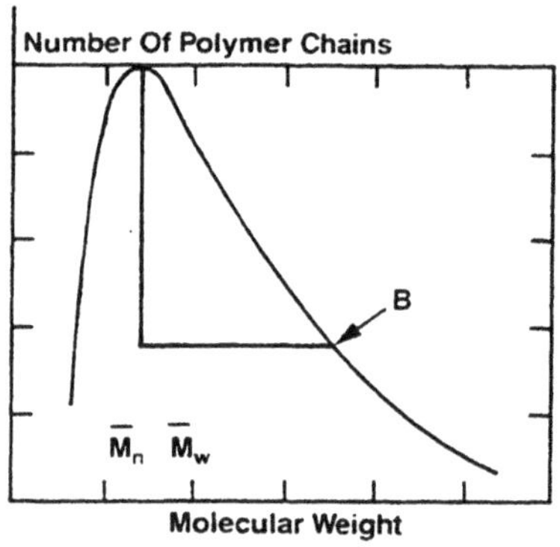

FIGURE 4-1-7. Typical molecular weight distribtion curve for polymers of dispersity *B*, half-width. From Ref. 5.

(Fig. 4-1-8) the last 30% of the film nearest the substrate is never fully exposed (unless overexposed) and a work point must be used to extrapolate to the retention at the silicon interface. In topside exposure, the upper layers absorb more radiation and adhesion to the wafer only occurs in the overexposed state. Underexposure produces more sol than gel at the silicon interface. Small features with lower surface-to-volume ratio (1×1 μm versus 10×5 μm) may develop off and only large pads will remain.

Higher doses (2–5 ×) are generally required for topside exposure (work point) in order to compensate for the absorbance (especially with nonbleaching sensitizers) by a factor of 10^{-D} for an exposure work dose, E_s,

$$E_s = E_g(1/TD) \tag{4-1-7}$$

where T is the transmittance of the film and D is the optical density given by Beer's law,

$$D = \Delta x(C - C_0)\, d \tag{4-1-8}$$

where Δx is the difference in photosensitizer absorbance before and after exposure.[10] The optimum photospeed and most uniform exposure occur at an optical density of 0.43[6] and the resist optical contrast is the simple inverse of D or a maximum of 2.3.[5, 205] Films with a high optical density (> 1.0) have lower contrast and require heavy overexposure whereas resists with lower optical density

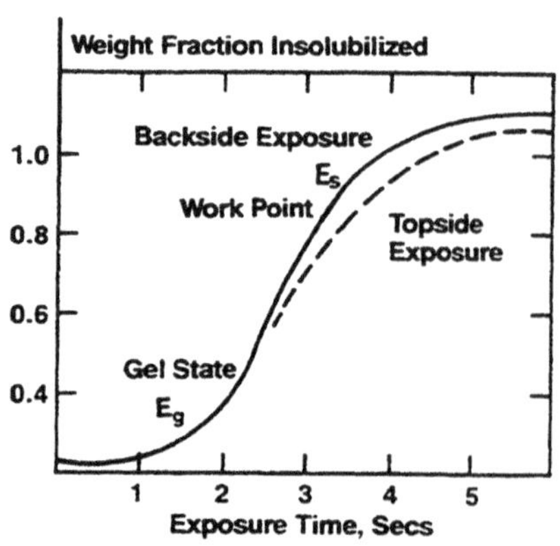

FIGURE 4-1-8. Photoinsolubilization of Kodak KTFR resist by exposure on a glass slide by frontside and backside exposure of a 1.0-μm-thick film.

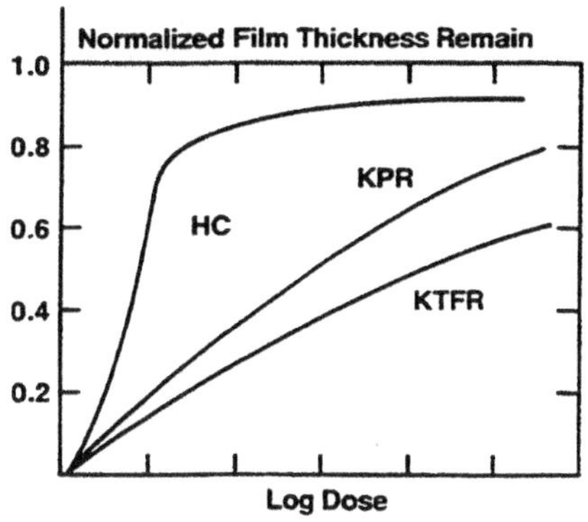

FIGURE 4-1-9. Contrast curves of Kodak KPR resist (contrast of 1.2), Kodak KTFR resist (contrast of 0.8), and HC, a high-contrast (4.0) resist.

require more exposure but have higher contrast (resolution). Low absorbance and high resolution can be artificially obtained by using the thinnest resist film and by thinning the same resist upon development. Film thinning curves determine the resist response function to over- and underexposure. The slope of the plot in Fig. 4-1-9 (or tangent at the abscissa) is the important parameter γ, or contrast,

$$[d(d/d_0)/dE] = \gamma = 2\exp(-B^2)/E_g \tag{4-1-9}$$

which is related to the polydispersity (B) of the resist (Fig. 4-1-7).[5-7] Polymers with low dispersity (1.1–1.3) have a tangent of 75° or a contrast of 3.5. Since most commercial resists are polydisperse ($B = 2$–5) (Table 4-1-1), they have low contrast ($\gamma < 2$). Resists with a high contrast respond up to a theoretical limit[3] (Fig. 4-1-10) of 3.5, to form more steeper profile images than low-contrast (0.5–1.5) negative photoresists.[8]

In optical exposure, the incoming beam is distorted by diffraction into a sinusoidal beam (Fig. 4-1-11). Mathematically, the modulated beam is described by a Fourier series[11] of modulation transfer function (MTF) defined by the ratio of the spatial distribution of exposure and the energy deposited in the resist (Fig. 4-1-12).[12] In order to print 1-μm lines and spaces, at least 60% MTF of the lens must be achieved. The MTF of the resist is about 60% (Fig. 4-1-13) and is given by the contrast ratio,[12]

$$(\text{CMTF})_{\text{resist}} = (I_{100} - I_0)/(I_{100} + I_0) \tag{4-1-10}$$

where I_{100} is E_s and I_0 is E_g (Fig. 4-1-8) and the resist thickness contrast is defined as

$$\gamma = \log(E_g/E_s)^{-1} \tag{4-1-11}$$

TABLE 4-1-1. Contrast and Minimum Dose of Commercial Photoresists

Resist	Contrast	E_g (mJ/cm^2)	Reference
KPR	0.9	70	6, 7
KTFR	0.5	6	5
KOR	0.5	4	5

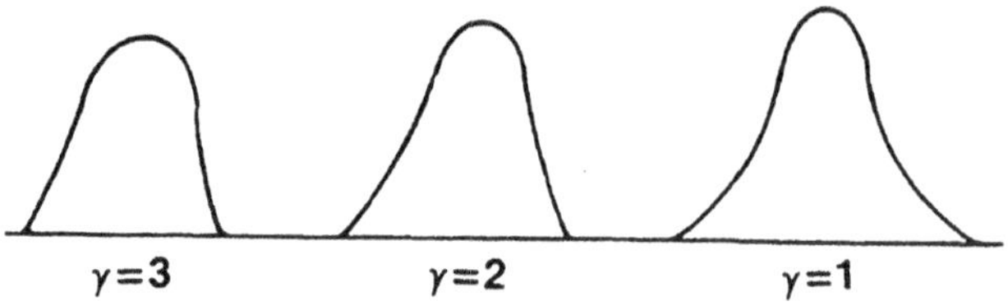

FIGURE 4-1-10. Line cross-section profiles in negative photoresists for contrast of 3, 2, and 1 in 1.0-μm-thick resist film.

Since the photoresist contrasts are less than 2, the minimum $(CMTF)_{resist}$ is 0.8–1.0 or at least a 1.0 $(MTF)_{lens}$ which restricts resolution to about 2.5-μm lines and spaces for 0.5-μm-thick films. Even though low contrast means low resolution, the resolution requirements for negative resists in the years 1960–1980 did not exceed 3 μm, so no basic improvements were motivated by higher resolution. The solution to higher resolution was to use thinner resists (< 0.5 μm thick) at the risk of high pinholes and poor step coverage. When reactive ion (plasma) etching came into dominant use (1980), negative photoresist use significantly diminished in favor of positive DQN photoresists.

4-1-3. *Photosensitizers*

Since most polymers are transparent to wavelengths above 330 nm or are photochemically inert, photosensitizers are added to provide an actinic response in the region of 365–435 nm. The photospeed of a cross-linking resist is determined by the quantum yield (Φ) of the sensitizer[5] (ABC in Table 4-1-2). Most of the normal UV resists (360–460 nm) contain a polymer and a photoactive component which can undergo homolysis or rearrangement reactions within itself and then react with the polymer base resin; the sensitizer can also transfer electronic energy to activate the polymer (P). In the primary photochemical processes (Table 4-1-2), ions, neutrals, and radicals are formed.

For primary processes, quantum yields are about 0.1–1.0. A yield of 0.5 is considered to be excellent. Once the hot product is formed (singlet or triplet state), the secondary quantum yields of bimolecular reactions (energy transfer) become impor-

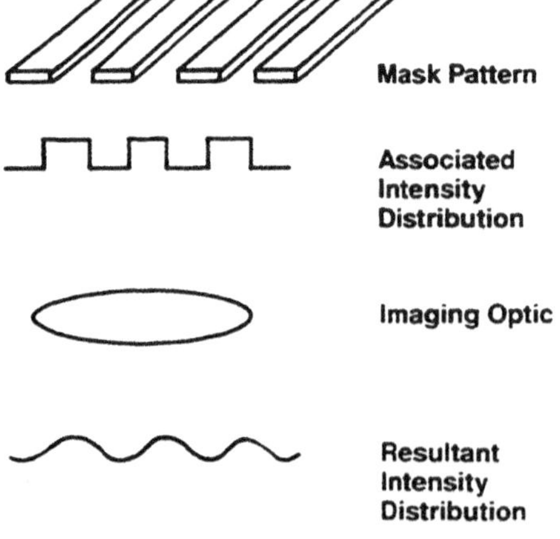

FIGURE 4-1-11. Profiles of intensity of light exposing negative photoresist after passing through mask and lens. From J. Cuthbert, *Solid State Technol.* Aug. 1977, p. 59, reproduced by permission of *Solid State Technology*, published by Technical Publishing, a company of Dun and Bradstreet.

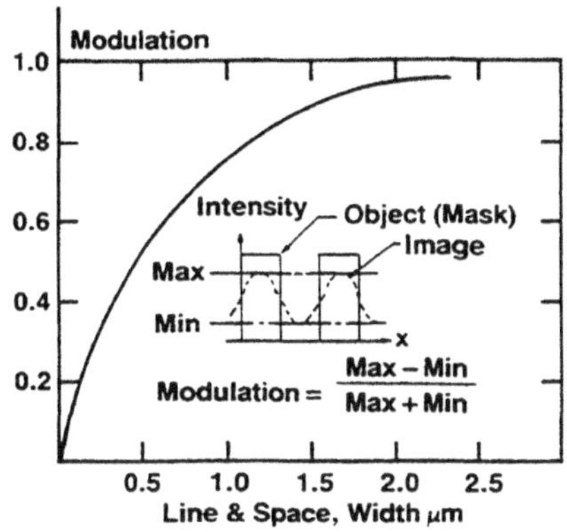

FIGURE 4-1-12. Modulation transfer function for lens of 0.35 numerical aperture at exposure of 405 nm. From Ref. 12, copyright and courtesy of AIP.

tant in transfer reactions from the photosensitizer (S) to activate the intermolecular dimerization of polyvinylcinnamate (P) polymers,

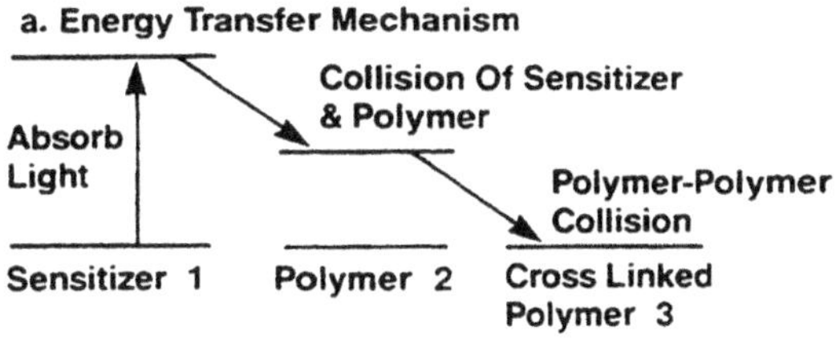

b.

$$S + h\nu \rightarrow S^{*}$$
$$S^{*} + P \rightarrow P^{*} + S$$
$$\text{or} \quad \left[\begin{array}{l} P^{*} + P \rightarrow P{-}P \\ 2P^{*} \rightarrow P{-}P \end{array}\right.$$

In photoactivated systems induced by free radical sensitizers such as azides in rubber (Fig. 4-1-14), the sensitizer (S) becomes incorporated into the cross-linking between the polymer chains (P) of KTFR (trademark of Eastman Kodak),

$$S + h\nu \rightarrow \cdot S \cdot$$
$$\cdot S \cdot + 2P \rightarrow P{-}S{-}P$$

$$\overset{*}{P} + P \rightarrow P{-}P$$

The maximum quantum yield value ($\Phi = 2$) is obtained by the bimolecular reaction,

$$P^{*} + P \rightarrow P{-}P \tag{4-1-12}$$

but because of misorientation and the high viscosity of the film (cage recombination), the quantum yield is rarely achieved.

The rate of disappearance of the photoactive component in photoresists or vinyl groups in photopolymers can be treated as a first-order polymerization[5] kinetics or an nth-order kinetics[9] by the generalized equation,[169]

$$-dm/dt = km^{n} I \qquad (n = 1, 2, \ldots) \tag{4-1-13}$$

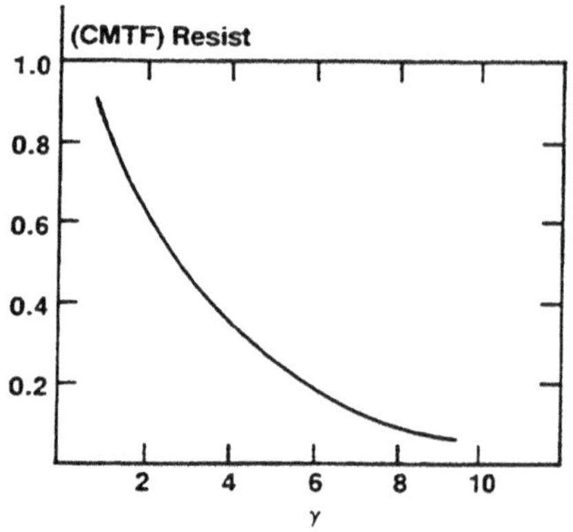

FIGURE 4-1-13. Critical modulation transfer function of resist as a function of resist contrast. From Ref. 12, courtesy of AIP.

where m is the concentration of polymerizable or cross-linking sites at time t, k is the rate constant, and I is the intensity of absorbed light [Eq. (4-1-7)] given by Beer's law. Multiple wavelength exposure with high absorbance ($>$ 1.0) leads to top surface cross-linking and greater induced stresses (wrinkling of the resist image upon development).(169) If the quantum yields of the photosensitizer are equivalent at the principal exposure wavelengths (365, 405, and 435 nm), then the treatment of the reaction kinetics leads to the conclusion that the highest contrast is obtained by exposure at the lowest absorbing wavelengths (405 nm in KTFR).(169)

4-1-4. *Commercial Resist Compositions*

The driving force for the commercial development of photoresists was the need to replace natural colloids with synthetically reproducible and stable polymeric materials. Polyvinyl alcohol was the first colloid replacement for albumin (dichromate sensitized) which was followed by polyvinylcinnamate (Table 4-1-3), azide-cross-linked rubber, and at present azide-insolubilized polyvinylphenols which are nonswelling in aqueous developers. In Fig. 4-1-14 are shown basic patents for polyvinylcinnamate (KPR, trademark of Eastman Kodak) and azide

TABLE 4-1-2. Primary Photochemical Processes[a,b]

(S_1) ABC → (T_1) ABC	→	$AB\cdot + C\cdot$	Dissociation into radicals
	→	E + F	Decomposition into molecules
	→	ACB	Intermolecular rearrangement
	→	ABC^* (S_1) ABC^* (T_1) or ABC^* (S_0)	Photoisomerization
	RH →	$(ABCH)\cdot + R\cdot$	Hydrogen ion abstraction
	ABC →	$(ABC)_2$	Photodimerization
	D →	ABC + prod	Photosensitization
	→	$ABC^+ + e^-$	Photoionization
	→	$ABC^{+ (or -)} + D^{- (or +)}$	External electron transfer
	→	$AB^+ + C^-$	Internal electron transfer

[a] From J. Calvert and J. Pitts, *Photochemistry*, Wiley, 1966, p. 367.
[b] S_1, singlet; T_1, triplet.

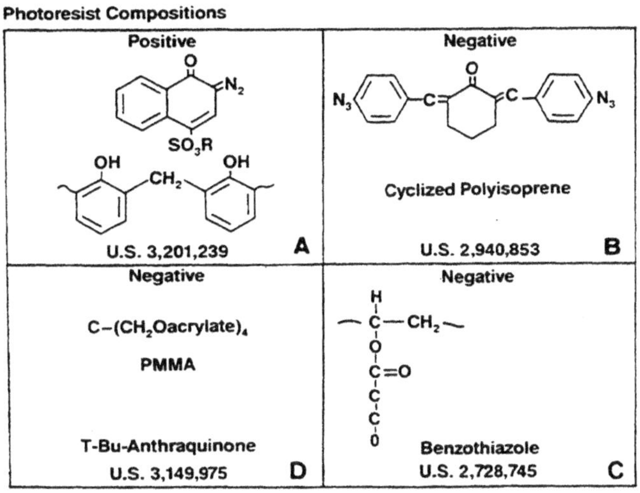

FIGURE 4-1-14. Major photoresist compositions. (A) Diazoquinone-sensitized novolak resin; (B) azide-sensitized rubber; (C) polyvinylcinnamate with benzothiazole photosensitizer; (D) photopolymer dry film resist consisting of a polyfunctional acrylate monomer dispersed in PMMA and photosensitized by anthraquinones.

rubber (KTFR) along with a dry film packaging resist (Riston, trademark of DuPont Co.) based on a tetrafunctional acrylate dispersed in PMMA resist.

In the beginning, semiconductor technology adopted the KPR polyvinylcinnamate (printing plate) lithographic resist as the primary etch mask. Later in the early 1970s, azide-rubber (ANR) with its superior adhesion and faster speed became popular. The dry film resist (e.g., Riston also was introduced at that time to solve tenting problems in copper circuit boards.

The main advantages of negative photoresists are:

1. Wide developer process latitude, insensitive to overdevelopment.
2. Excellent adhesion and wet etch resistance.
3. Compensating image feature size (Fig. 4-1-15). The unexposed space ($-R$) is restored to the correct dimensions (M) by the undercutting isotropic wet etch (E).
4. Vast amount of compositions. Since the paint industry, UV curing inks and adhesives use intrinsic compositions which harden, many polymeric

TABLE 4-1-3. Evolution of Negative Photoresists

Resist	Year	Key U.S. patent
Polyvinyl alcohol	1928	1,672,156
Polyvinylcinnamate	1951	1,728,745
Polyvinylcinnamylidene acetate	1960	2,716,102
Azide-cyclized rubber	1960	2,940,853
PMMA (tetraacrylate) dry film	1965	3,149,975
Polyimide-MMA	1980	4,180,404
Polyvinyl phenol-azide	1979	4,148,655
Inorganic Ag resist	1981	4,269,935

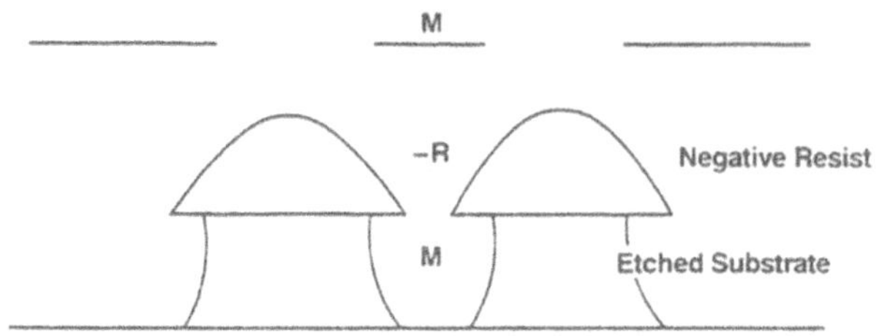

FIGURE 4-1-15. Self-compensation of image size in negative resist. The narrower image (−R) in the negative resist image is smaller than the mask opening (M). After isotopic etching, the transferred image is etched back to the dimensions of the mask (M).

compositions could be formulated into negative photoresists. Second, the recent introduction of deep-UV exposure tools (200–300 nm) eliminates the need for a photosensitizer.[5]

The disadvantages of negative photoresists are:

1. Resolution limited by film thickness, topside exposure, forward and backscattered radiation, and swelling in developer. The thicker the resist, the greater the scattered radiation and overdose needed to insolubilize the resist at the substrate interface (Fig. 4-1-16), especially at high optical densities. Thinner films restrict backscattering, but are more prone to pinholes. Typical resolution is about 3 × the film thickness (3-μm lines and spaces in 1-μm-thick negative photoresist).
2. Compositions quenched by oxygen infusion.
3. Very few radical sensitizers sensitive to 435 nm for stepper exposure.
4. Insoluble for lift-off metal process. All types of engineering plastics have been made into negative photoresists by direct irradiation or by the use of photosensitizers (Table 4-1-4). Some of the polymers due to their low melting points form tacky films and could not be used for contact printing.

With the advent of dry etching and deep-UV contactless printing, new requirements have been imposed upon negative photoresists. Submicrometer resolution is required which is out of the grasp of conventional polymeric cross-linking negative photoresists.

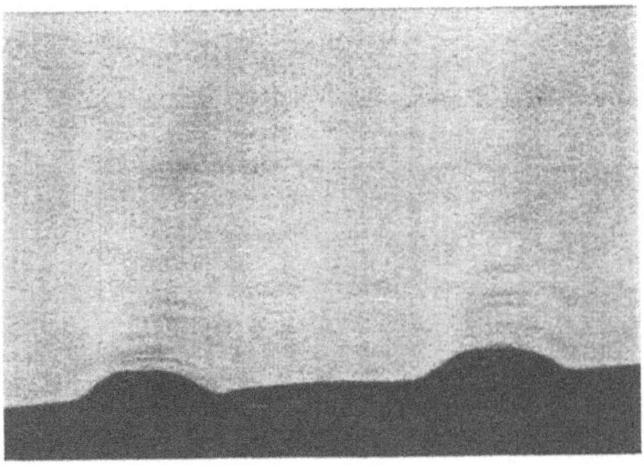

FIGURE 4-1-16. Cross section of photoimage in Kodak KTFR negative photoresist. (Sample broken at liquid nitrogen temperatures to facilitate breaking.)

TABLE 4-1-4. Types and Classification of Negative Photoresist Polymers

Thermoset	Elastomers	Thermoplastics
Diallylphthalates	Acrylates	Acetates
Epoxies	Polysulfides	Acrylates
Phenolics	Neoprenes	Cellulosics
Urethanes		Polycarbonates
Silicones		Polystyrenes
Polyimides		

Dry etching requires the polymer to withstand halogen radicals and increasing temperatures up to 200°C. Only aromatic and silicone-based polymers (or metal organic) are able to offer good plasma etch resistance. Thus, bilayer or multilayer resists may be required for continued use of negative photoresists.

We will examine in the next sections the polyvinylcinnamate systems, azide-sensitized resists, high-temperature resists, and dry film resists as classic examples of negative photoresists. We will also discuss recent innovations such as dry develop, inorganic resists, and deep-UV photoresists. For the classical literature prior to 1970 on negative photoresists, the reader is referred to a treatise by Kosar(13a) and a review article by Delzenne.(14)

4-2. POLYVINYLCINNAMATE

Polyvinylcinnamate was the first all-organic photoresist, being comprised of a cinnamate group attached to polyvinyl alcohol. The net photochemical reaction is energy transfer to the cinnamate group followed by dimerization,

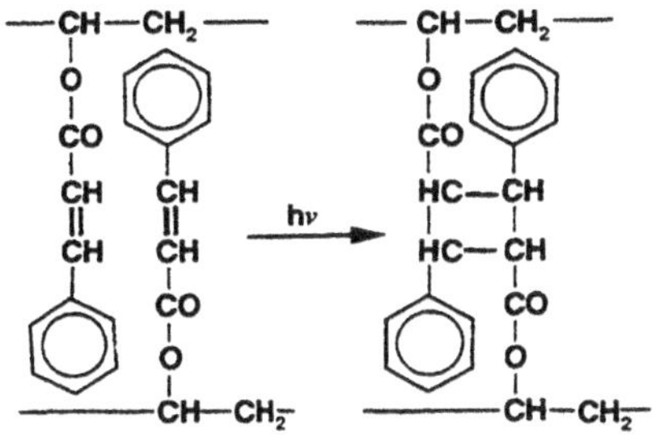

The ability of the cinnamate group to react in the bimolecular dimerization is controlled by the rigidity and geometry of the polymer. In crystalline cinnamic acid, the quantum yield (Φ) is 1.4.(15) In a polymer film or in solution, the quantum yields are low (≤ 0.25).

As the cross-linking proceeds, the reactive cinnamoyl groups become depleted, more quanta are captured in the top layers by nonreactive sites, and the quantum yield decreases in the reaction (Fig. 4-2-1).(10)

The trapped solvent left after prebake is often overlooked. Most prebakes for KPR are 80°C for less than 20 min. Murski and Simoliunas(17) reported a new triplet sensitizer which does not sublime after prebake at 120°C for 1 hr. By baking

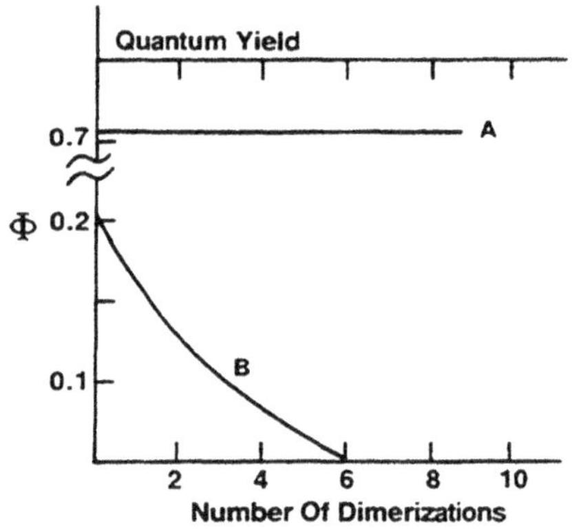

FIGURE 4-2-1. Quantum yield of cycloaddition between cinnamyl groups in (A) crystalline cinnamic acid and in (B) solid polymer film of polyvinylcinnamate. From Ref. 10, reproduced by permission of the American Chemical Society.

the KPR-type resist with the new sensitizer at 120°C, closer chain packing occurs and sensitivity for dimerization is enhanced.

In the solid polymer film, the influence of the resin glass transition temperature, T_g, was studied by grafting cinnamate groups onto rubber. The resist sensitivity increased with a decrease in T_g.[18] The influence of the T_g was also noted in a study of a one-phase system consisting of a nitrosensitizer copolymerized with a cinnamate group onto a methacrylate backbone,[19]

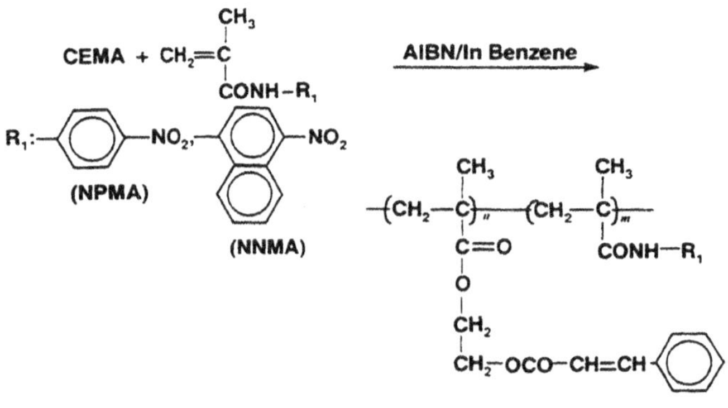

The sensitized side chain resist was tenfold more photosensitive than the control which consisted of a free nitrosensitizer and the homocinnamate polymer. The lower T_g permitted efficient energy transfer along the chain and allowed the cinnamate group to overlap and dimerize. Thus, in the cinnamate resist the stereochemistry of the interproximity of cinnamate groups, the excited state mobility, and the side chain motion can significantly influence the quantum efficiency of dimerization in the cinnamate resist. For example, polyvinylcinnamate (PVC) is much less efficient in a solid film than in an ordered crystal. Other side reactions such as *cis–trans* isomerization and radical formation also decrease the quantum efficiency by partitioning the absorbed energy into non-cross-linking reactions.

By itself, unsensitized PVC absorbs strongly in the deep-UV region due to the unsaturated ketone chromophore (Fig. 4-2-2). Sensitization into the visible region has been empirically attempted with over 300 candidates. One of the most efficient is the naphthothiazoline (NS) derivative of cinnamate itself (Fig. 4-2-2) which absorbs strongly at 405 nm,

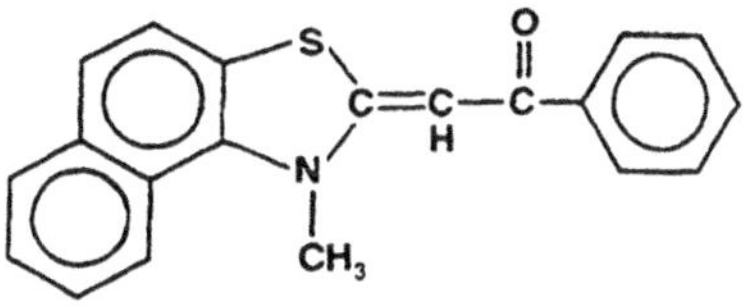

NS

In an attempt to further extend[22] the spectral sensitivity of PVC, cinnamylidene groups[175] were introduced,

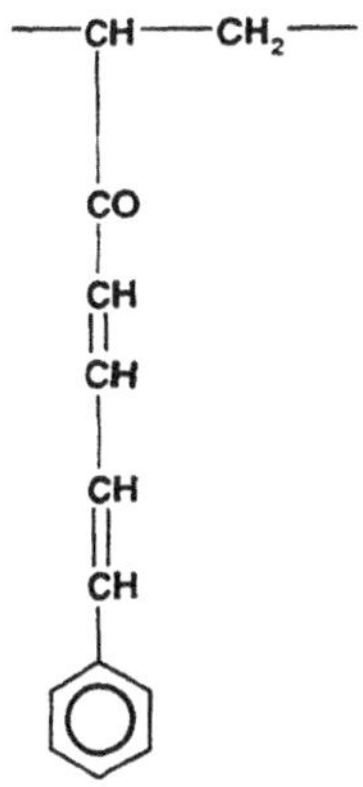

which lowered the PVC triplet level to 36 kcal/mole and allowed for sensitization to 540 nm and for exposure with visible light using an argon laser.[20] In this case, the quantum yield for dimerization was at 0.5 which is better than in PVC.[21]

The transfer of electronic energy from the absorbing sensitizer (S*) excited at 405 nm involves singlet or triplet states,

$$S^* + PVC \rightarrow PVC^* + S \tag{4-2-1}$$

Direct insight into the energy levels of the cinnamate groups can be gained by observation of the fluorescence (excited singlet state) emission and phosphorescence excited triplet state emission of the sensitizer (NS). In the presence of PVC (Fig. 4-2-3), the sensitizer phosphorescence is diminished and the cinnamate phosphorescence is enhanced. The vibrational structure ($1600\ cm^{-1}$) in the cinnamate phosphorescence at 510 nm is due to the conjugated ketone chromophore. The good overlap between the sensitizer triplet level at 54 kcal/mole and the PVC triplet level at 48 kcal/mole allows for efficient energy transfer (Fig. 4-2-4). Many sensitizers have been examined for PVC sensitization. An extensive study of PVC sensitization was reported by Moreau (Table 4-2-1).[23] The NS system has the right energy levels and a long lifetime to be the most efficient photosensitizer.

One of the main factors in the photosensitization mechanism is the triplet level of the sensitizer. The choice for a speed-enhancing sensitizer was in the past a brute force technique. Of the 15 sensitizers listed in Table 4-2-1, a striking increase is

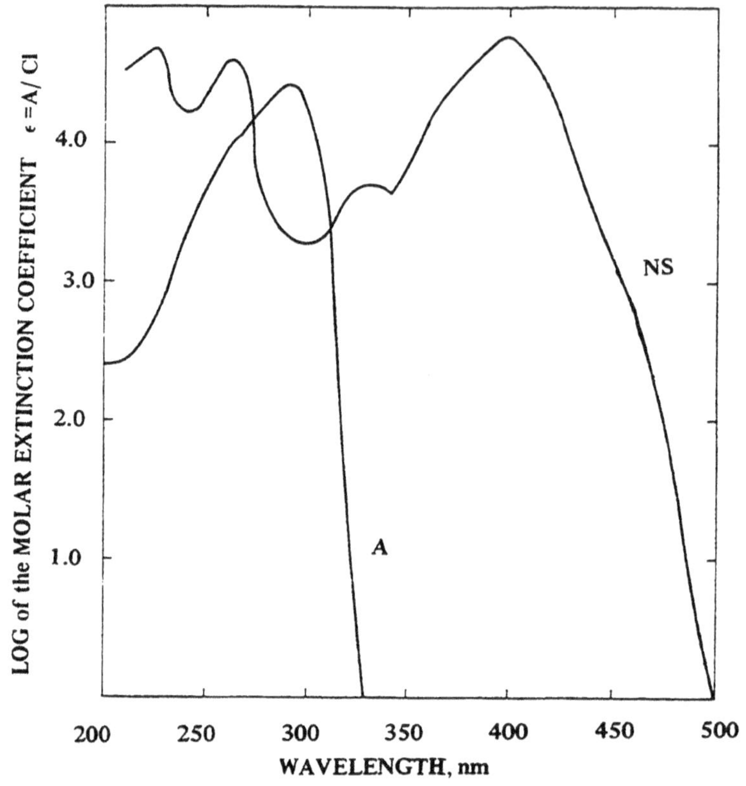

FIGURE 4-2-2. Molar absorption spectra of (A) polyvinylcinnamate and (NS) naphthothiazoline photosensitizer.

noted for several materials. The necessary factors for efficient triplet sensitization are:

1. A triplet donor level as close as possible to the cinnamate level (50–55 kcal/mole)
2. A high quantum yield of triplet sensitizer population, $\phi_p/\phi_f > 1$
3. A finite triplet lifetime, $T_t > 0.01$ sec

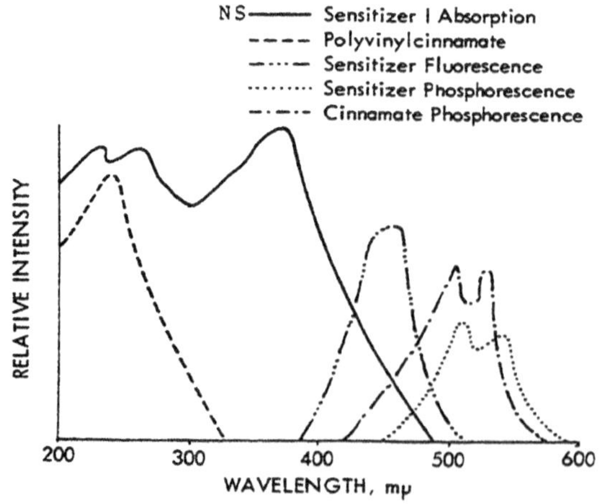

FIGURE 4-2-3. Phosphorescence emission spectra of polyvinylcinnamate and naphthothiazoline sensitizer (NS).

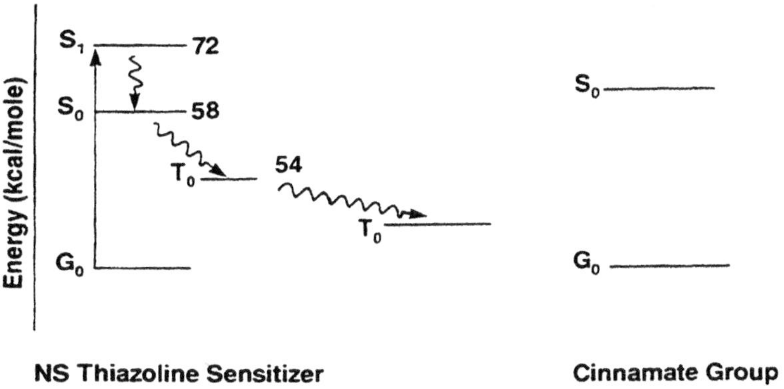

FIGURE 4-2-4. Triplet–triplet energy transfer from photosensitizer to polyvinylcinnamate.

All three factors should be present in the sensitizer system. The aromatic sensitizers ($\pi \rightarrow \pi^*$), for example, have competing singlet states that fluoresce (ϕ_f).

The speed ratio in Table 4-2-1 is relative to unsensitized PVC. Speed factors, such as the optical density of the sensitizer, the concentration of the sensitizer, and the use of a line or a continuous light source, are not explicitly controlled in the testing of patent formulations. These variations can introduce a factor of five in the speed ratio.

The speed of the sensitized resist in photographic terms is the inverse of the energy absorbed by the sensitizer,

$$S = k(I_{\text{abs}}\, t)^{-1} \tag{4-2-2}$$

The energy absorbed (E) is related to the quantum yield (Φ) of dimer formation,

$$\Phi = \text{P–P(dimer moles)}/I_a t \tag{4-2-3}$$

TABLE 4-2-1. Polyvinylcinnamate (PVC) Photosensitizers for the Region 300–500 nm

Sensitizer	Speed[a]	E_t (kcal)	E_s (kcal)	$E_s - E_t$ (kcal)	τ_t (sec)	ϕ_p/ϕ_f
Unsensitized PVC	1					
Naphthalene	3	61	92	31	2.3	0.11
Anthracene	8	42	67	25	10^{-2}	1.5
Phenanthrene	14	62	83	21	3.3	1.1
Chrysene	18	56	67	11	2.5	0.51
Nitrobenzene	8	60	—	—	10^{-3}	—
Anthrone	31	72	78	6	10^{-2}	>1000
Pyranthrone	7	72	78	6	10^{-3}	>1000
Benzil	50	53	—	—	—	—
Benzophenone	20	70	74	4	10^{-3}	>1000
p-Nitroaniline	100	55	—	—	0.1	>1000
Acetophenone	3	74	69	-5	10^{-3}	>1000
NS sensitive	750	54	58	4	0.5	100
Michler's ketone	640	61	—	—	6.0	>1000
p-Nitrobiphenyl	200	58	—	—	0.5	>1000

[a] Data from J. Kosar, *Light Sensitive Systems*, Wiley, New York, 1965, p. 143.

In this respect, the speed (Q) of a photoresist reaction is proporoional to the quantum yield Φ of sensitized product formation,

$$Q = k\Phi(\text{P–P})^{-1} \tag{4-2-4}$$

A high-speed material would exhibit a high quantum yield and a low formation of cross-linking necessary for insolubilization.

If the triplet mechanism is operative in PVC, it establishes a limit to the spectral domain of sensitization in the visible region (<450 nm). The sensitizer must have at least 50 kcal/mole of triplet energy. Since the triplet levels of most molecules are at least 24 kcal/mole lower than the absorption maximum, sensitizers that absorb beyond 480 nm, i.e., crystal violet, cannot be used. However, the efficiency of triplet energy transfer is about one-half of the direct dimerization of PVC (Table 4-2-2) and doses of $> 100\ \text{mJ/cm}^2$ are required to compensate for the inefficiency of triplet energy transfer. In the literature a wide variation in Φ exists since the film thicknesses are not corrected for substrate reflectivity and film absorbance[10] [Eqs. (4-1-7) and (4-1-10)]. In the polymer film, the quantum yield for dimerization is an order of magnitude lower in the polymer. Reiser and Egerton[16] have pointed put the unfavorable geometries in the excited state for spin inversion,

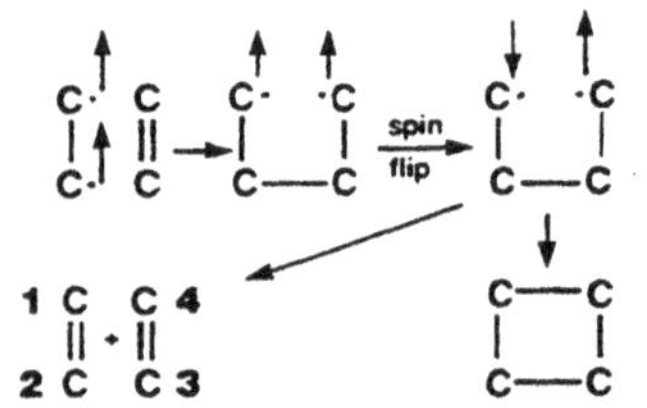

The first step produces a biradical intermediate in which the free electrons retain their spin correlation for some time. The second C–C bond, which completes the cycle, can be formed only if at the moment of spin inversion atoms 1 and 4 are still within reach of each other. If that is not the case, the systems reverts to its ground state of two isolated double bonds. However, during the time interval prior to spin inversion, the repulsion between the parallel spins at atoms 1 and 4 will tend to deform the biradical into a nonreactive configuration, and thermal energy is needed to counteract this tendency. Thus, the repulsive spin speed interaction is equivalent to an energy barrier separating the excited triplet state from the cyclic products.

TABLE 4-2-2. Quantum Yields of PVC Dimerization

Quantum yield	Reference
0.26	16
0.42	24
0.09 (sensitized)	24
0.5	25
0.5	23
0.25 (sensitized)	23
0.15 (cinnamylidene)	24

Other attempts to enhance PVC sensitivity from 500 mJ/cm^2 to < 100 mJ/cm^2 have included the replacement of the phenyl end group with sulfur and nitrogen derivatives,[14] but no major improvements were noted.

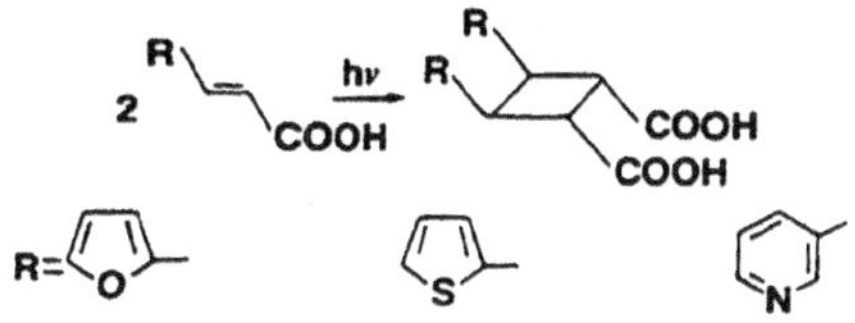

The trade-off of sensitivity for thermal stability and alkaline etch resistance has been synthetically manipulated by condensation polymerization,

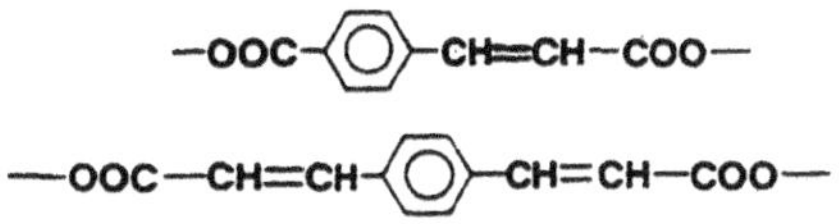

in which the cinnamate group of chalcones is incorporated into the main chain.[26] High-temperature phosphates, sulfonates,

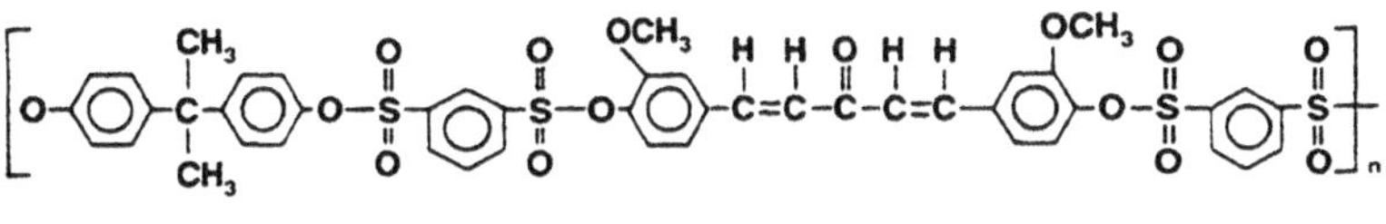

and carbonates have been used as the condensation linkages. Delzenne[14] has pointed out many patent examples. These resists can now be used directly in the deep UV because of the strong chalcone chromophores.

Since the cinnamate group is usually grafted onto polydisperse polymers, narrow-molecular-weight PVC ($B < 2$) is difficult to prepare.[27] Vinylcinnamate forms cyclopolymers instead of linear polymers. Cationic polymerization of substituted ethoxy cinnamates has been used to produce narrower-dispersity polymers with 100 mJ/cm^2 sensitivity compared to 400 mJ/cm^2 for PVC.[28,37]

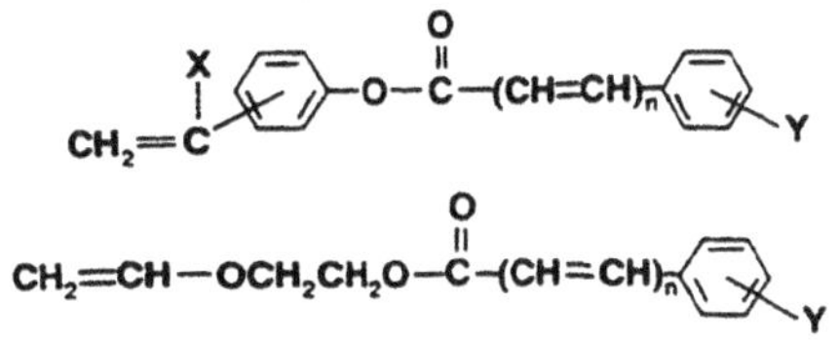

X=H, CH_3
Y=H, CH_3, NO_2, OCH_3, Cl
Z=H, CH_3
n=1 or 2

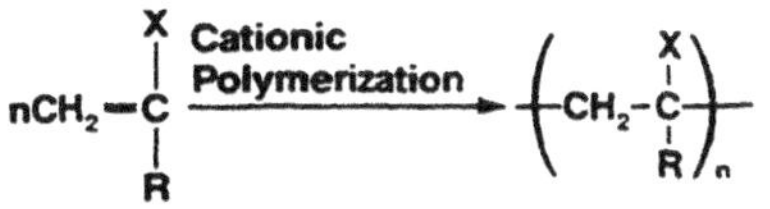

In the deep-UV cross-linking of PVC directly, free radicals have been detected,[14]

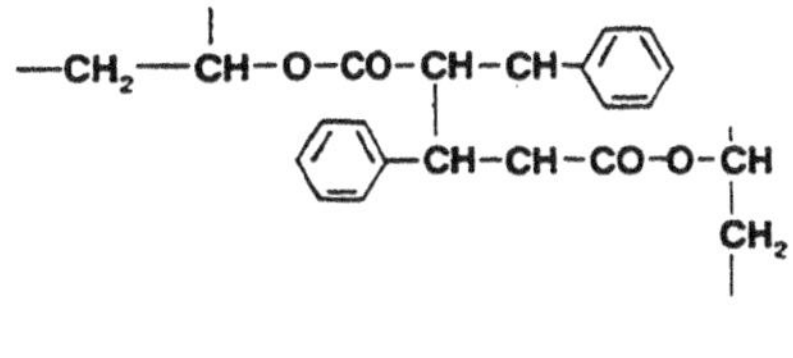

The free radicals may be quenched by the oxygen present. The postdevelop thermal-induced hardening of PVC is thought to occur by the free radical mechanism.[29] Indeed, the exposure of PVC in the presence of O_2 reduces its photosensitivity. Ozone has also been shown to insolubilize stored PVC films.[30] Overcoating KPR films with polyvinyl butyral eliminates scumming and improves the exposure in air.[31] The scumming may be due to peroxide formation in the ozone attack on the double bond followed by thermally induced dimerization.

Free radical sensitizers (bisazides) enhance the sensitivity of PVC to 100 mJ/cm^2. The cross-linking nitrenes add intermolecularly across pendant cinnamate.[32] However, the shelf life of PVC with free radical sensitizers may be very limited. Storage of KPR (PVC) films on Cu boards is also susceptible to desensitization to copper ion reaction with the NS sensitizer. The boards must be pretreated with chelates to getter the free copper ions.[33]

For further improvements of PVC, many homologues have been investigated. Since the quantum yield of PVC films falls a factor of 10 below the expected values of cinnamic acid (Table 4-2-), several attempts have been made to find new or more efficient photodimerizing systems. Phenyl acrylic acid on a vinyl backbone[34] was found to be about equivalent to PVC, but when incorporated into a condensation polyester backbone a speed of 5 × sensitized PVC or about 100 mJ/cm^2 was achieved.[203] To date, the most sensitive system is based on the 1967 observations of DeBroer and Breslow of the efficient triplet state dimerization of diphenycyclopropenes,[35]

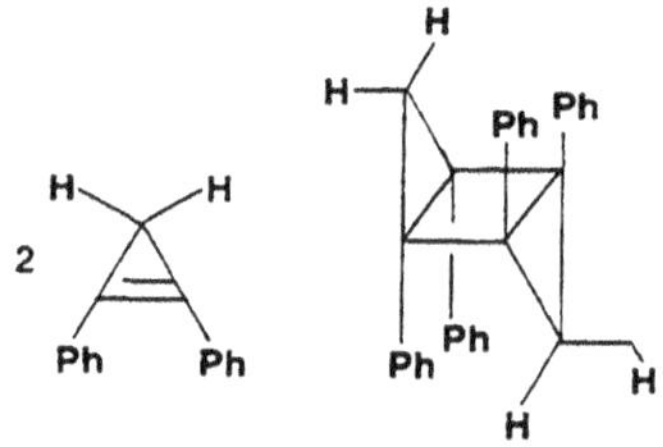

The cycloproprene group was introduced into a polymer backbone by attachment to polyvinyl alcohol in the traditional manner of PVC,[36]

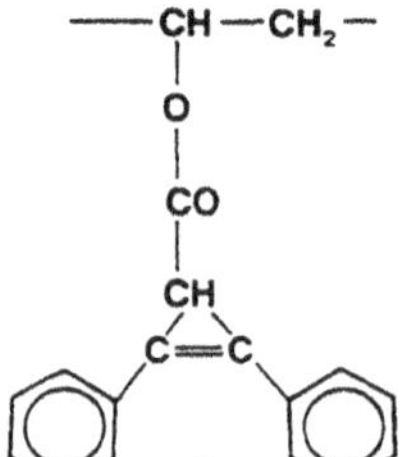

A

The triplet sensitized quantum yield of dimerization of A is about 1.8. The higher efficiency appears to be due to the severe ring strain of the three-membered ring and the longer triplet lifetime than found in PVC.[36] However, the photospeeds did not achieve high sensitivity (75 mJ/cm^2).

The PVC resist represents the first example of incorporating a well-known molecular dimerization reaction into a polymeric medium. In many cases, the efficiency of dimerization was low, due to the stiffness of the side chains and the unfavorable overlap of intermolecular cinnamate groups in the presence of small photosensitizer additives. To bring pendant groups into more favorable geometric positions, the direct grafting of both the cinnamate group and its own triplet sensitizer into a terpolymer chain of a low-T_g polymer was a new and novel synthetic attempt to bring molecular groups into favorable reaction cages. The incorporation of both groups is not without its problems. Since the triplet sensitizers do not bleach upon absorbance, insufficient light may reach the silicon wafer surface. A balance has to be struck between the absorbance of the sensitizer and a deficiency of groups present for energy transfer. In this respect, new dimerizing systems have been explored which undergo bleaching. These are based on Schenk adducts,[38]

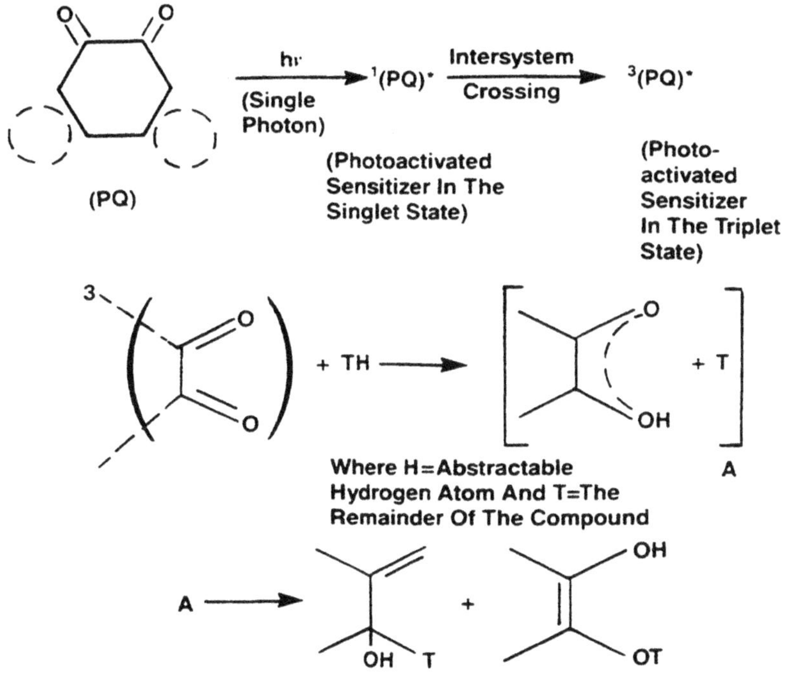

If the quinone is a bischromophore,

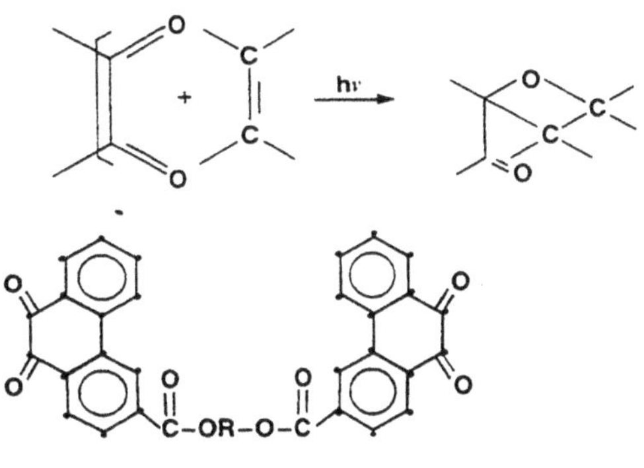

polymers with abstractable hydrogens such as polyvinylbenzoate can be insolubilized and transformed into deep- and mid-UV resists.[39] The search for dimerizing deep UV continues and is a formidable task because the molar extinction coefficient at 254 nm of conjugated carbonyl groups approaches 10^4 or an optical density of > 2.0 for a film of 0.5-μm thickness.

As a pioneering resist, PVC resists replaced natural polymers with a synthetically reproducible and durable wet etch resist which was insensitive to oxygen during exposure. Unfortunately, because of its low photospeed of > 100 mJ/cm^2 (although sufficient for contact printing), the PVC resist was too slow for modern projection printers and, hence, the resist has been discontinued as a commercial product. In spite of the many interesting photochemical aspects of its bimolecular reactivity and the many attempts to enhance its photosensitivity, nevertheless, PVC-based resists await further investigations, discoveries, and speed enhancement before they will be restored to their former prominence.

4-3. AZIDE-SENSITIZED RESISTS

4-3-1. *Azide-Activated Formulations*

Azides are nitrogen trimers which are relatives of classical diazonium compounds used for blueprints. Under the action of light or heat, azides are capable of losing nitrogen to give **1**,

$$RN_3 \xrightarrow{h\nu} \underset{1}{[RN]} + N_2$$

R may be alkyl, aryl acyl, or sulfonyl.

The nitrene intermediate state may exist either in the triplet biradical state (**2**) or in the singlet state (**3**),

$$\underset{2}{R-\dot{\underset{..}{N}}\cdot} \qquad \underset{3}{R-\ddot{\underset{..}{N}}}$$

The nitrenes are very reactive and undergo many insertion, addition, or reduction reactions:

1. H-abstraction

$$RN\!: + 2\dot{R}H \longrightarrow RNH_2$$

2. Insertion

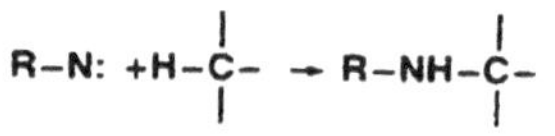

3. Addition

4. Dimerization

$$R-N\!:+R-N\!:\longrightarrow R-N{=}N-R$$

5. Isomerization

$$RCH_2N\!:\longrightarrow RC{=}NH$$

6. Reduction

$$RN\!: + XH_2 \longrightarrow RNH_2 + X_2$$

7. Oxidation (oxygen effect)

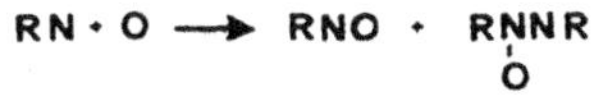

The side oxidation reaction with oxygen occurs at the surfaces of films and quenches the cross-linking reactions.[43] The addition of bisazides,

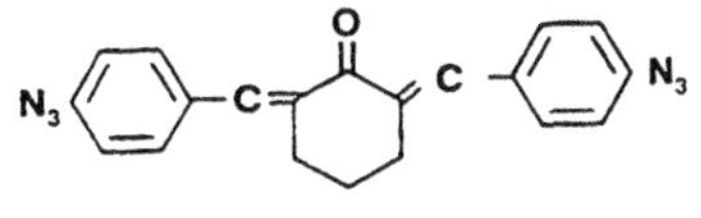

to olefin polymers, will form dinitrene cross-links,

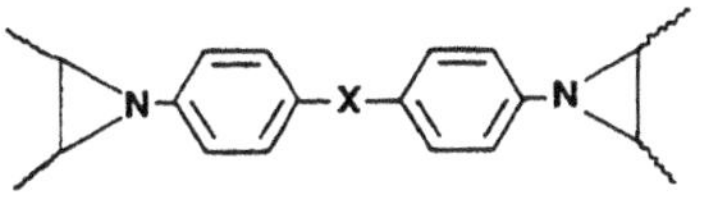

Both water-soluble azides (WSA),

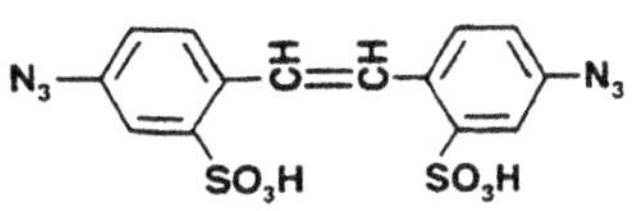

and oil-soluble azides (OSA),[40]

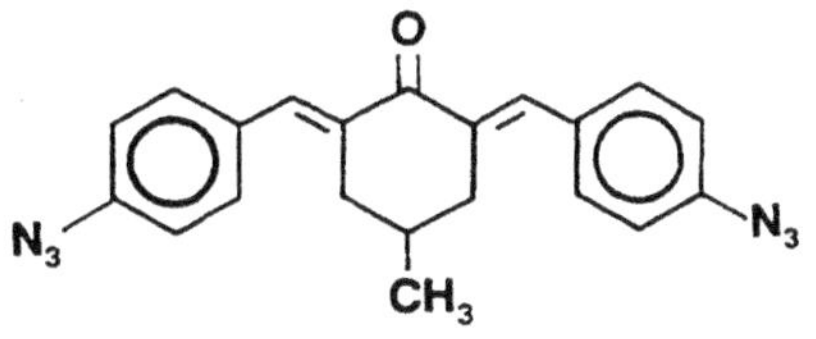

have been used as cross-linking agents.

The azide group is a weak chromophore (ε of 10^3) but when incorporated as part of a conjugated ketone (OSA) with an ε of 10^4 at 365 nm, the azide (group) spectral sensitivity can be extended as far as 450 nm (Fig. 4-3-1).

The azide group can also be incorporated as a polymer side chain,

Azidopolymers **Bis Phenol A**

Polystyrene

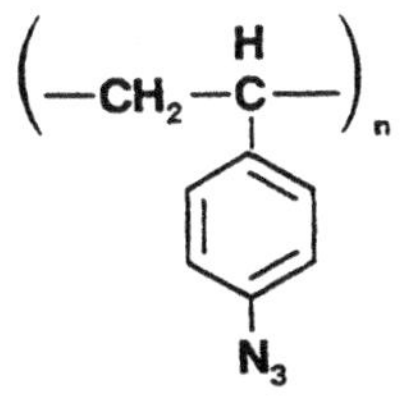

U.S. 2,870,011 (1959)

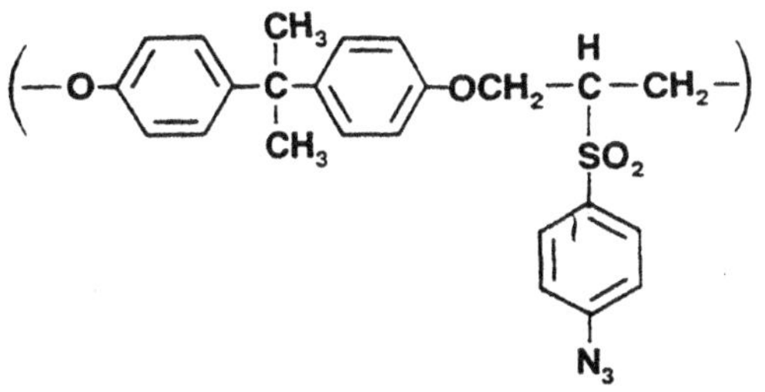

British 1,108,612 (1968)

By themselves, the bisazide crystals (OSA) form negative resist images (Fig. 4-3-2) from the polymerization of bisnitrenes into polyazos,

$$N_3RN_3 \rightarrow n(NRN) \rightarrow (NRN)_n$$

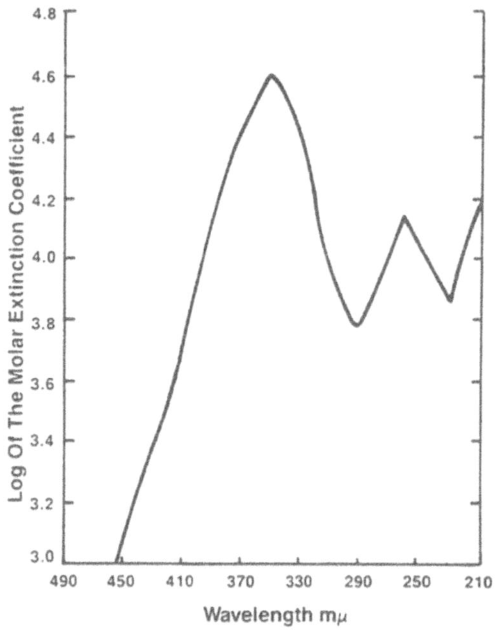

FIGURE 4-3-1. Absorption spectrum of OSA (2,6-azido-benzylidene-4-methyl cyclohexanone).

However, the images are full of defects due to the crystallization of the cast sensitizer on the surface.

When OSA was mixed with cyclized natural rubber (KMER) or cyclized polyisoprene (KTFR), the formulations became popular replacements in 1970 for PVC. The azide rubber resists exhibit excellent adhesion, good photospeed (50 mJ/cm^2), and withstand acid or alkaline etchants. The versatile azide sensitizer (OSA) can insolubilize a variety of polymers such as high-temperature polyimides,

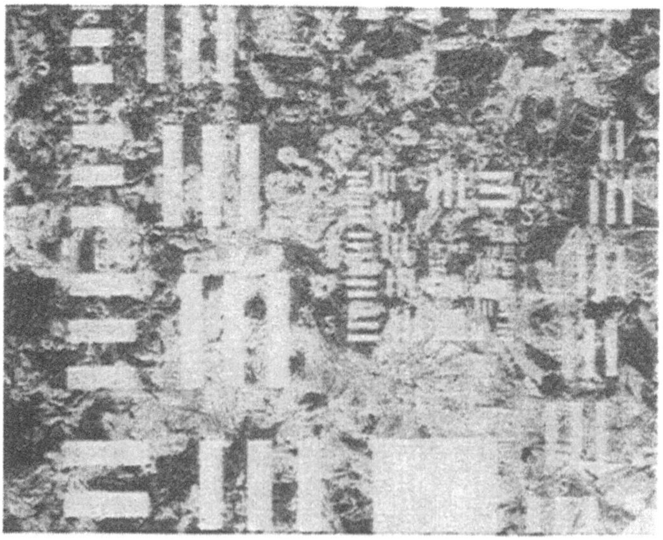

FIGURE 4-3-2. Film of azide sensitizer alone as negative photoresist. Due to crystallization, the image are discontinuous in the exposed region. Sufficient polymerization occurs to form a polyazo compound.

deep-UV resists, and form the basis of O_2 dry develop resists. Thus, formulation chemists can concentrate on the polymeric aspects of the resist knowing that cross-linking could be achieved by the bisazide addition.

4-3-2. *Photochemistry of Azides*

At 254 nm, the bisazide chromophore in the absorption spectrum of Fig. 4-2-5 occurs as an $n \rightarrow \pi^*$ transition. Absorption by the conjugated ketone of OSA at 365 nm is followed by intersystem crossing to a dissociative triplet level. However, the singlet state is also dissociative and contributes equally to nitrene formation. The singlet level for phenyl azide is 100 kcal/mole while the triplet level lies at 75 kcal/mole (3 eV).[42] The azide dissociates in consecutive steps to form the hot bisnitrene radical. The quantum yield for the first nitrene (k_1),

$$N_3RN_3 \xrightarrow{k_1} N_3RN\!: \xrightarrow{k_2} :\!NRN\!: + 2N_2$$

is about 0.4 and about 0.5–1.0 for k_2 for a total of 1.0 in the consecutive but not biphotonic reaction.[43, 44, 46] The intermediate nitrene absorbs about half the radiation of the parent azide.[43]

In the presence of cyclized rubber with polyunsaturated groups,

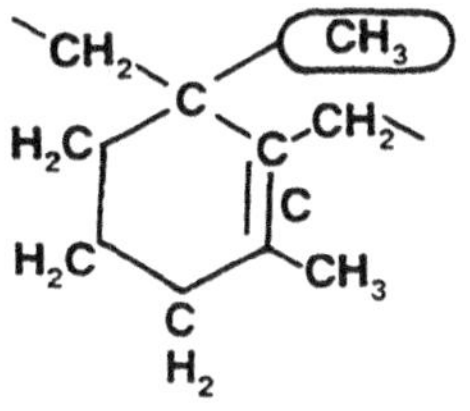

the nitrene addition or insertion reactions lead to insolubilization[161] of the matrix. Some of the nitrenes are wasted by formation of amines[45] which also absorb at the

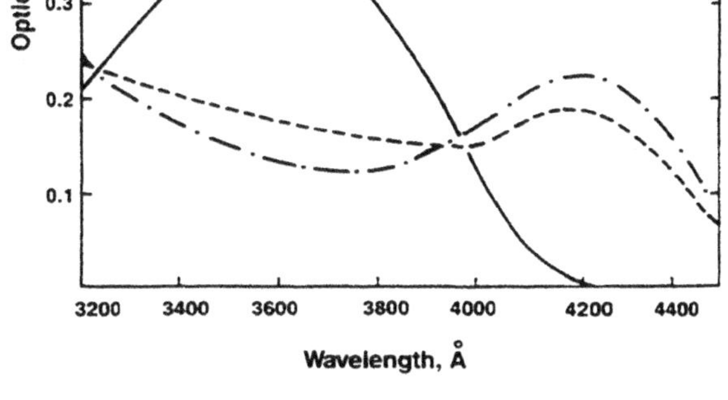

FIGURE 4-3-3. Absorption spectra of Kodak KTFR after exposure and after postbake.

TABLE 4-3-1. Effect of Unsaturation of Base Polymer on the Photosensitivity of Azide-Sensitized Resist

Resist polymer	Percent unsaturation	mJ/cm^2
Cyclized polyisoprene	14	120
Polyvinylcinnamate	6	170
Polyvinylacetate	0	2000

same wavelengths (Fig. 4-3-3). The following example shows the intermolecular insertion addition of bisnitrenes to 1,2-polybutadiene,[161]

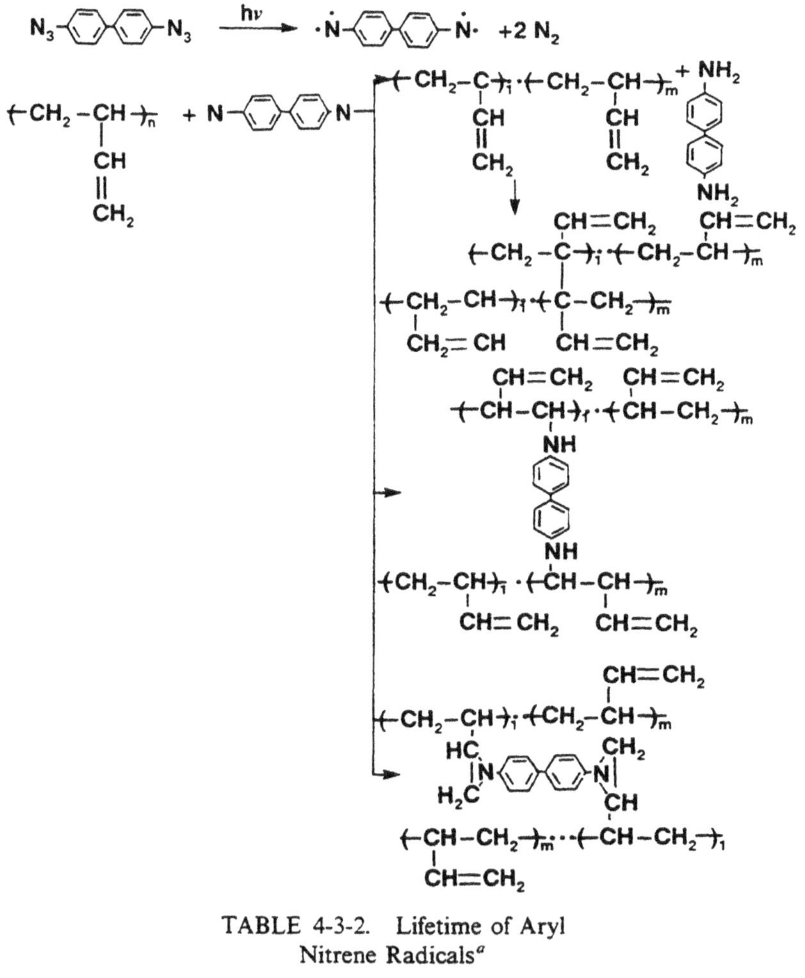

TABLE 4-3-2. Lifetime of Aryl Nitrene Radicals[a]

Medium	Lifetime (sec)
Alcohol	10^{-6}
Paraffin oil	10^{-3}
PMMA	2
PMMA, O_2 present	10^{-3}

[a] From Ref. 50.

Both natural and synthetic rubber have served as the base resin for commercial formulations. Synthetically, the *cis*-polyisoprene[47–49] is partially cyclized,

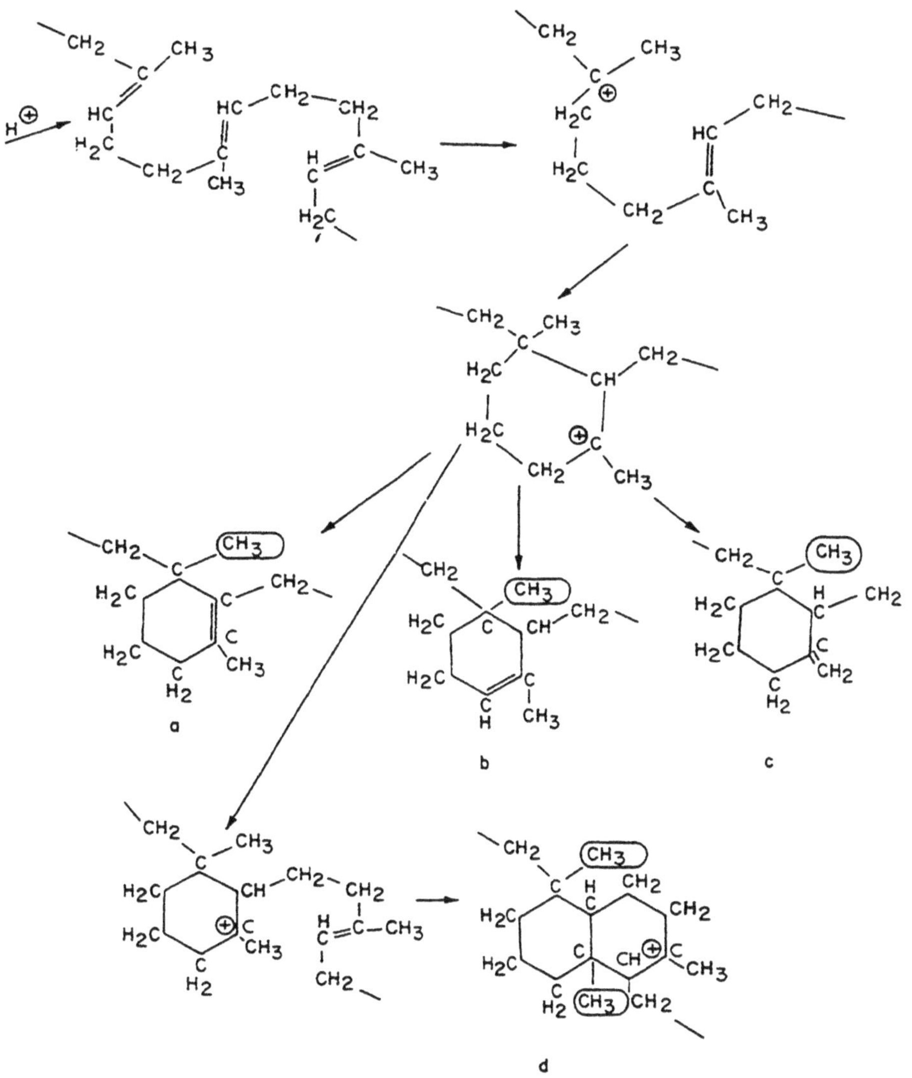

to remove the film tackiness by raising the T_g to about 50°C. The preferred ratio of external to internal double bonds in bicyclic cyclized rubber contains about 10% of a and b.[47–49] The ring strain in the bicyclic cyclohexene ring is relieved by nitrene addition. Polymers without double bonds require an order of magnitude greater dose to insolubilize the resists with the OSA sensitizer (Table 4-3-1).[32]

About 2 crosslinks of azide per chain are required for complete insolubilization which is equivalent to a quantum yield of 0.5 for an incident dose of about 10^{15} photons, 50 mJ/cm^2.[46] The fate of the nitrene radical and its reactivity depends on the mobility in the matrix. In Table 4-3-2, the lifetime of aryl nitrene radicals in various media is given.[50] The preservation of the nitrene in the absence of O_2 is controlled by the diffusion (viscosity of the media). In a rubber system with a T_g of

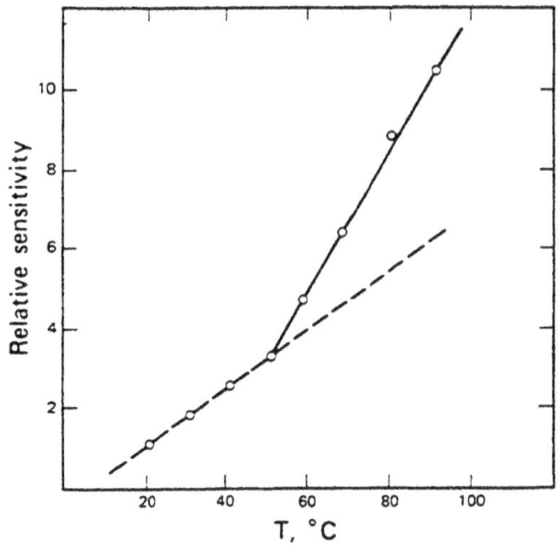

FIGURE 4-3-4. Increased photosensitivity of negative azide resist by exposure of films at elevated temperatures.

50°C, the radicals are more mobile and reactive than in PMMA (T_g 100°C). In rubber, the nitrene mobility is determined by residual solvent and the polymer chain packing density.[51] Exposure of KTFR at temperatures above its T_g of 50°C leads to a large gain in photosensitivity (Fig. 4-3-4).[23] The abrupt increase in sensitivity is apparent at the T_g of cyclized rubber at 50°C. The immobility of the nitrene radical in the crystalline cage also leads to the insolubility of the polymer by the self-polymerization of the nitrene radical into an azo polymer,

$$n(\mathrm{NRN}) \rightarrow (\mathrm{NRNNRN})_n$$

The formation of the azo polymer can be followed by the disappearance (X-ray) of the crystallinity of single crystals of OSA in the cast rubber film. The nitrene also attacks the ground-state azide to amplify the nitrene:

$$\mathrm{N_3RN_3} + \cdot\mathrm{NRN_3} \rightarrow \mathrm{N_3RNNRN_3} + \mathrm{N_2}$$

Thermally, the nitrene radical can also be generated to cross-link the rubber (at high prebake) and generate similar bisamine by-products (Fig. 4-3-5).[45] In fact, Shimizu and Bird demonstrated that thermal cross-linking is more efficient than the photochemical means (Fig. 4-3-5).[45] The simultaneous heating of the rubber resist

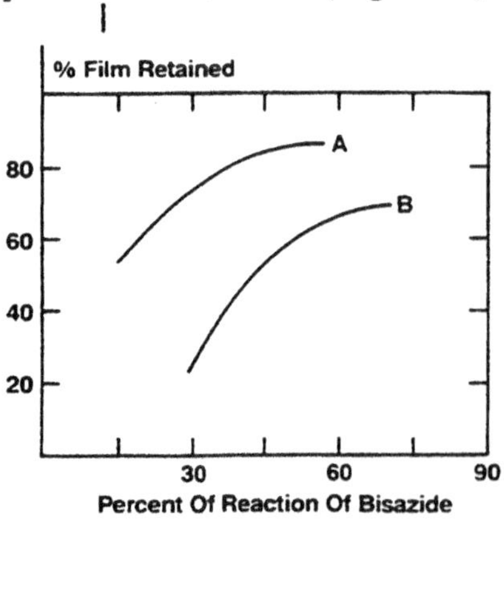

FIGURE 4-3-5. Cross-linking versus reacted bisazide in photo- and thermoreactions of photoresist. Photoresist: 3% bisazide, 1.2-μm coating on Si. (A) Thermal reaction at 100°C in air. (B) Photoreaction by 365-nm dosing at 25°C at 3.6 mJ/cm^2 min under N_2. From Ref. 45, copyright 1978 and reprduced by permission of the Electrochemical Society Inc.

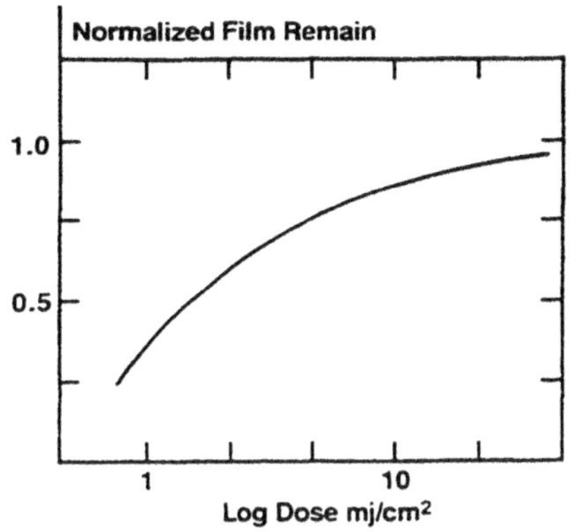

FIGURE 4-3-6. Contrast curve for KTFR. From Ref. 9.

above the T_g and the incipient higher nitrene mobility leads to more efficient insolubilization.

4-3-3. *Rubber and Phenolic Based Systems*

Although the photospeed of KTFR is fair (about 75–125 mJ/cm^2), the contrast is low[9] (0.5) (Fig. 4-3-6) due to the high polydispersity.[47, 48] The wide dispersity also leads to pinholes and swelling in the development step.[52, 53] Fractionation of the rubber reduces pinholes by a factor of 3–4. For a monodisperse rubber resin, cyclized polybutadiene appears to be easier to synthesize than cyclized polyisoprene.[54, 55] Polymers, such as polyvinylcyclohexene, would be a more sensitive and controlled polymer system of known olefin content.[201, 209]

Since negative KTFR resist films swell, thin submicrometer films are used to resolve 2-μm lines.[28] To improve the contrast due to the bleaching of the bisazide, organic dyes have been added to the formulations.[197] The addition of novolaks to KTFR and KPR is claimed to reduce swelling.[62] Chlorinated developer solvents are also claimed to reduce swelling.[61] In addition, the rinsing of the swollen image in butyl acetate as well as the addition of surfactant to the developer reduces swelling.[63–65] Insoluble gels which lead to residues and pinholes are removed by electrophoresis and centrifugation.[66] Water-soluble azides mixed with water-soluble polymers as polyvinyl alcohol[67] and polyvinylpyrollidone[68, 69] do not swell, but have poor sensitivity. To reduce swelling and improve the resolution of 1.0-μm-thick rubber-based resists, polymers of polycyclic norborene of high T_g(150°C) are used.[209] The best nonswelling system has been the mixture of bisazides and novolak resins exposed in the 250 to 450-nm region (depending on the absorption of a particular bisazide) and developed in alkaline solutions.[71–73]

The reaction of aryl azides (I) with phenols has led to a practical water-developed negative resist. Both *p*-vinylphenol (II)[74, 191–196] and novolak resins[71, 73, 76–78] have been used as resins,

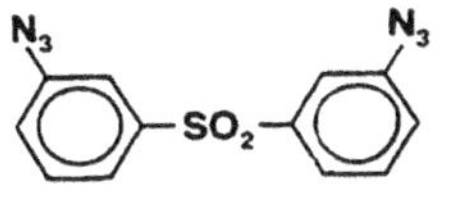

(I)

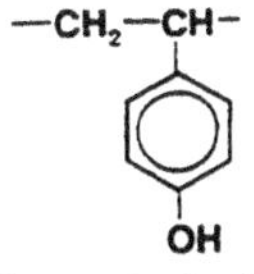

(II)

Structural Formulas For Resist Components: (I), 3,3′-diazidodiphenyl sulfone; (II), poly(*p*-vinylphenol).

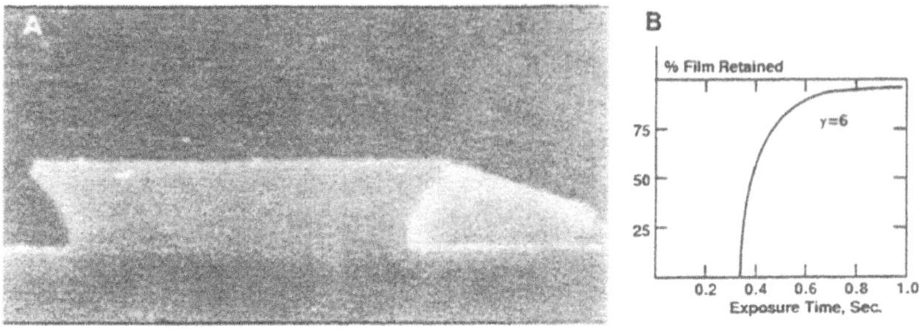

FIGURE 4-3-7. (A) Exposure profile of 1-μm lines and spaces in azide-*p*-(MRS) vinylphenol resist. (B) Contrast curve of MRS resist. Both from Ref. 71, copyright and courtesy of IEEE.

Good contrast and 1.0-μm resolution have been obtained (Fig. 4-3-7) due to the fact that the optical density is high (no backscatter effects) and the developer etches away the interior of the profile at the silicon substrate.[71, 119] Although lift-off profiles can be obtained by undercutting the profile, tight development process controls are necessary to prevent excessive image undercutting.[71–75, 191] Other nonswelling resists based on polyvinylphenol use a pseudoazide of diaziridine to cross-link the polymer chains,[174]

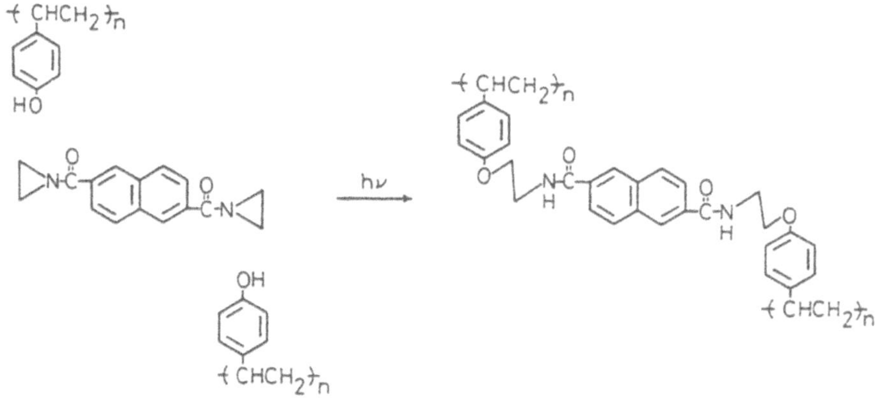

The reaction of the nitrene with the phenolic resin involves a new compound thought to be a quinoimine,[75]

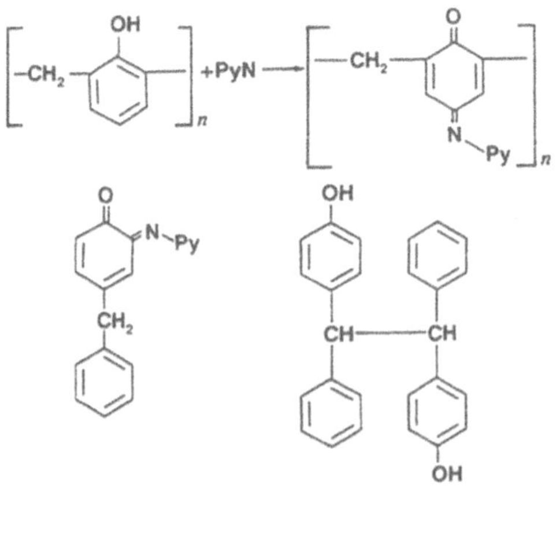

A dose of about 200 mJ/cm^2 at 313 nm is used. If oxygen is present, reactions occur with the nitrenes to form nitroso, azo, and azo-oxy derivatives.

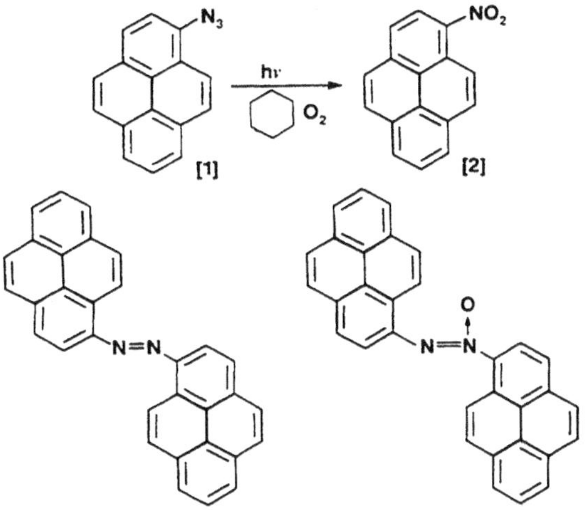

The versatility of the bisazide as a sensitizer has been demonstrated for a variety of polymers (Table 4-3-3).

The silicone-containing triallylsilane system,[182, 208]

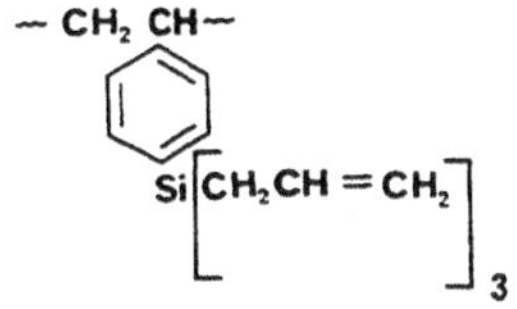

has a pendant vinyl group for facile cross-linking with bisazides. The developed image serves as an oxygen-resistant mask for bilayer processing (Chapter 12).[206]

For a dry developable resist (Chapter 10), a high loading of 30% bisazide is used in conjunction with a polymethylisopropenyl ketone base. The bisamine photoproduct and the cross-linked amine are envisioned as quenchers of singlet oxygen. Thus, the exposed region is slowly eroded in oxygen plasma (see also

TABLE 4-3-3. Azide-Based Photoresists

Use	Polymer base	Reference
Mid-UV	Phenolic	71–73
Deep-UV	Phenolic	74–78
High-temperature resist	Polyimide	79, 80
Aqueous develop	Polyvinyl alcohol	81, 82
UV	Polyvinyl cinnamate	83, 84
Deep-UV	Polyvinyl ether	85
Dry develop	Polymethyl ketone	86, 87
Dry develop	Polystyrene	88
Bilayer resist	Polytriallylsilane	206

Chapter 10).[86, 87] The oxygen plasma quencher is formed by irradiation and baking,[86]

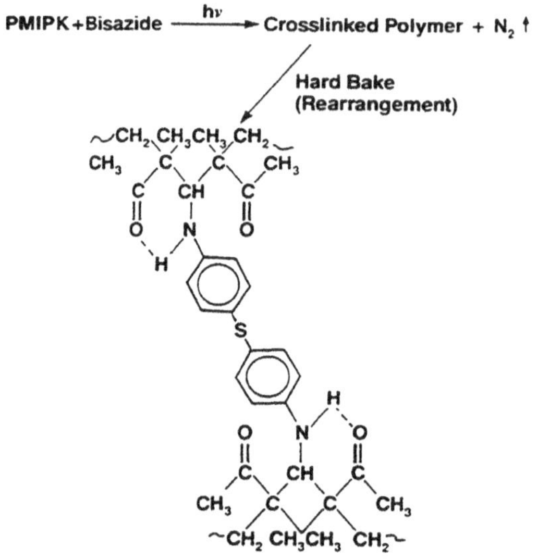

The irradiated and unirradiated regions are thought to produce etch rate differences due to the formation of a quenching product (Table 4-3-4).[86] The amine by-product inhibits the removal of the exposed region and permits the image to be reversed upon dry development (Fig. 4-3-8).

4-3-4. *Photospeed of Rubber Azide*

The photosensitivity of azides is hampered by an inner filter effect (Fig. 4-3-3) for speed and also for control of linewidth. The partial bleaching of the azide on reflective substrates permits the excess radiation to scatter and shrink the unexposed spaces. The addition of nonbleaching azo-aniline and fluorescent dyes to absorb the reflected 365 nm is claimed to reduce halation.[56]

The actinic sensitivity of KTFR was originally thought[57] to be sufficiently sensitive in the 405–435 nm region to enable one to switch from polychromatic contact printing to monochromatic exposure at 405 and 435 nm. However, the peak

TABLE 4-3-4. An Example of a Dry Develop Process[a]

Process	Irradiated part	Unirradiated part
Prebake	PMIPK + azide	PMIPK + azide
Exposure	Cross-linked polymer	PMIPK + azide
Postbake	Effectively desensitized cross-link polymer	PMIPK
Development (O_2 plasma)	Effective quenching of the excited states formed by photon, electron, or ion impact oxidative degradation	Main chain cleavage in the excited state formed by photon, electron, or ion impact oxidative degradation
Result	Low etch rate	High etch rate

[a] From Ref. 86.

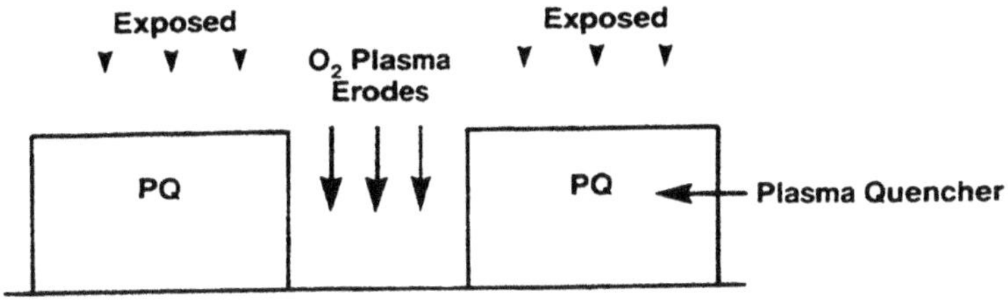

FIGURE 4-3-8. Formation of a quencher of oxygen plasma from photoexposure to permit plasma development of unexposed resist.

sensitivity lies mainly at 365 nm and exposure times at the higher wavelengths require 4–10× more dose (Fig. 4-3-9).

Several attempts have been made to shift the spectral sensitivity of KTFR:

1. Increase the conjugation of the ketone group in the bisazide sensitizer with a cinnamylidene group. Sensitivity to 405 nm was achieved (Fig. 4-3-10).[32, 58]
2. Triplet sensitization by the addition of triplets with 40–50 kcal/mole has very limited gain. Tsuda *et al.*[86] calculate a 25 kcal/mole (500 nm) triplet level in phenyl azide if the nitrogen groups are isoplanar with the aryl ring. Fluorenone (55 kcal/mole, triplet) has been noted to impart some sensitivity at 475 nm.[59]
3. More effective triplet sensitization has been claimed with azidostyrene polymers and thiapyrylium perchlorate sensitizers.[60, 70] The azide group was incorporated as part of a pendant polymer chain (Table 4-3-5). For these compositions, the azide polymers have used carbonyl azides and sulfonyl azides as functional groups.[13, 14] Although many patents were issued over the 1960–1970 period, in general, the polymers are not too soluble, require additional sensitizers, or resins, and are prone to cross-link in storage.

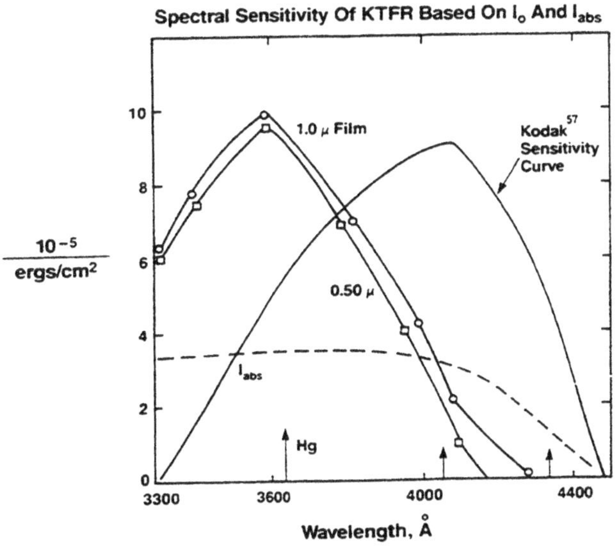

FIGURE 4-3-9. Spectral sensitivity of KTFR based on I_0 and I_{abs}.

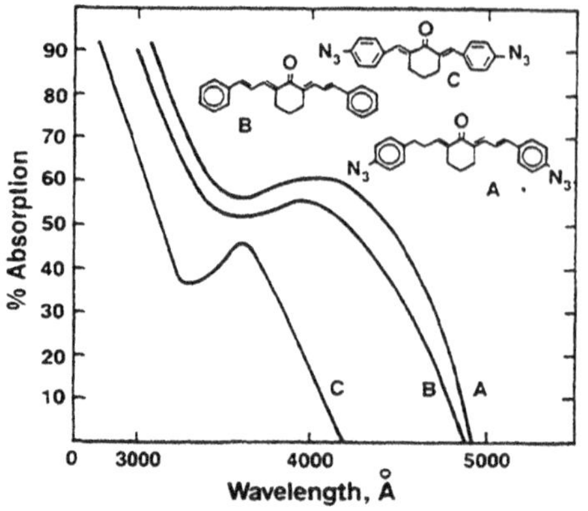

FIGURE 4-3-10. Spectral sensitivity of redshifted azide resists. Data from Refs. 32 and 58.

4-3-5. *Oxygen Effect in Azide Resists*

With the introduction of proximity printing and projection printing in early 1970, KTFR became the prime resist candidate. Exposure in air was noted to require longer exposure times and produced film thinning. The film thinning was constant independent of the initial film thickness (Fig. 4-3-11). While a 1.0-μm-thick film lost 55% of its original thickness, the loss of the uppermost portion of 0.45 μm of the film from a 0.5-μm initial film results in total film obliteration. In the UV spectrum a new photoproduct appears after exposure in air (Fig. 4-3-12).[46] Exposure in air produces nitroso, azoxybenzene, and nitro products from nitrene oxidation.[93] Exposure of KTFR in air at various intensities (Fig. 4-3-13) indicates that reciprocity failure occurs, i.e., the product of intensity (I) and time (t) of exposure is not an equivalent dose (E) at all levels of intensity,[41]

$$E \neq It \qquad (4\text{-}3\text{-}1)$$

Absorbed oxygen is present in the film and it also can be supplied by diffusion from the atmosphere into the film during exposure. Froschule and Backhus[94] found that the top layer of KTFR of thickness W is sensitive to the intensity of exposure and the concentration of oxygen given by the partial pressure, P_0,

$$W = (P_0/I)^{1/2} \qquad (4\text{-}3\text{-}2)$$

Oxygen is supplied by diffusion and infusion from the exposed region (Fig. 4-3-14).

TABLE 4-3-5. Examples of Azide Polymers

Azide polymer	Reference
Azidosulfonyl	88
Azidocarbonyl	89
Azidonitrobenzoyl	90
Azidostyrene	91
Azidocinnamylidene	92

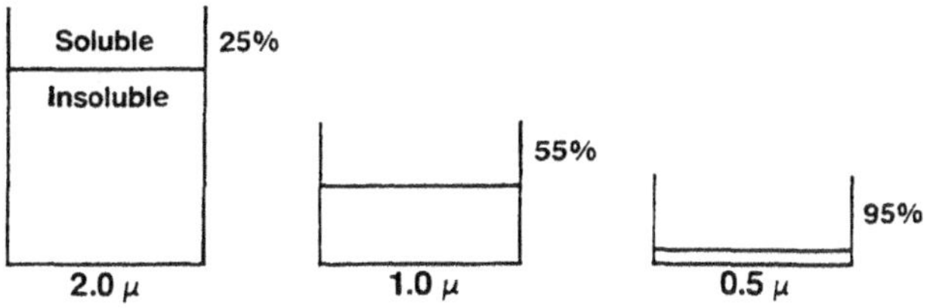

FIGURE 4-3-11. Constant film loss in the oxygen effect for azide resist exposed in air. The oxygen diffuses into the film and reacts with the nitrene intermediate before it can cross-link the rubber polymer.

As nitrogen is released from the photochemical decomposition of the sensitizer, the oxygen diffuses in to intercept the nitrene,

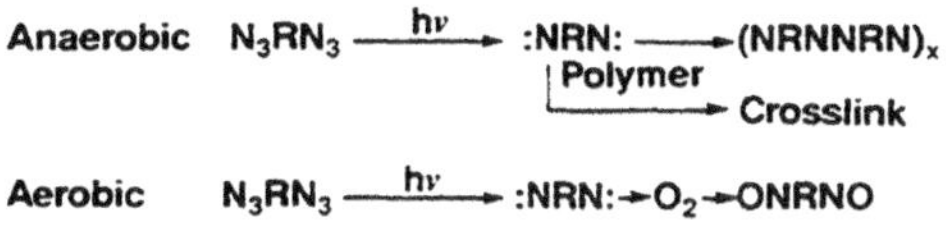

If oxygen is excluded from the top surface by flushing with an inert gas or blocked by a polymer topcoat, then the infusion of dissolved oxygen in the film from the unexposed region is beneficial for inhibiting forward scattered radiation and enhancing the resolution of the resist. The lightly scattered zones do not insolubilize due to the infusion of oxygen from the dark regions. This principle is also used in dry film photopolymerizing systems such as Riston (Fig. 4-3-15).[95] The oxygen-sensitive reaction is protected from atmospheric oxygen by a polymer film. Preconditioning of the prelaminated film in CO_2 to remove oxygen from within the polymer matrix resulted in a loss of resolution upon processing.

The diffusion of oxygen into the film has been modeled by Chong[96] using the example given in Fig. 4-3-16,

$$dO_2/dt = DO_2 d^2C/dx^2 - \Phi I_0 e^{-\varphi z} - D_{N_2}\, d^2C_0/dx^2$$

where

D = Diffusion rate of oxygen, 10^{-7} cm^2/sec
C_0 = Dissolved oxygen in resist film, 4×10^{-4} M
C = Diffused oxygen into resist film
x = Diffusion depth of oxygen
z = Thickness of resist
I_0 = Intensity of exposure, 100 W/cm^2
φ = Extinction coefficient of sensitizer, 10^3
ϕ = Quantum yield of sensitizer

The results of the steady-state equations of oxygen balance and radical concentration are shown in Fig. 4-3-17. The consumption of oxygen radicals depends on the photochemical rate of nitrene formation and the location of radicals generated in the film. To a depth of about 40% of the film, the oxygen effect dominates. Deeper in the film, diffusion cannot supply enough oxygen to quench the reaction. If the optical density is low throughout the film, the oxygen effect is severe since more oxygen is present than the rate of photogenerated radicals. At high optical

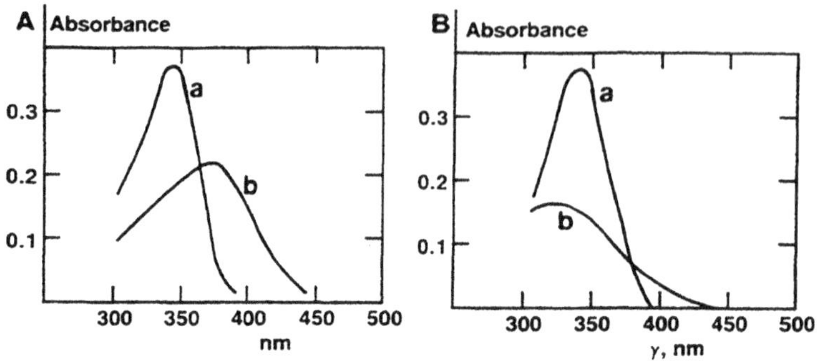

FIGURE 4-3-12. (A) Azide UV absorption spectrum before (a) and after (b) irradiation in nitrogen, and (B), before (a) and after (b) irradiation in oxygen. From Ref. 46, reproduced by permission of the Electrochemical Society Inc.

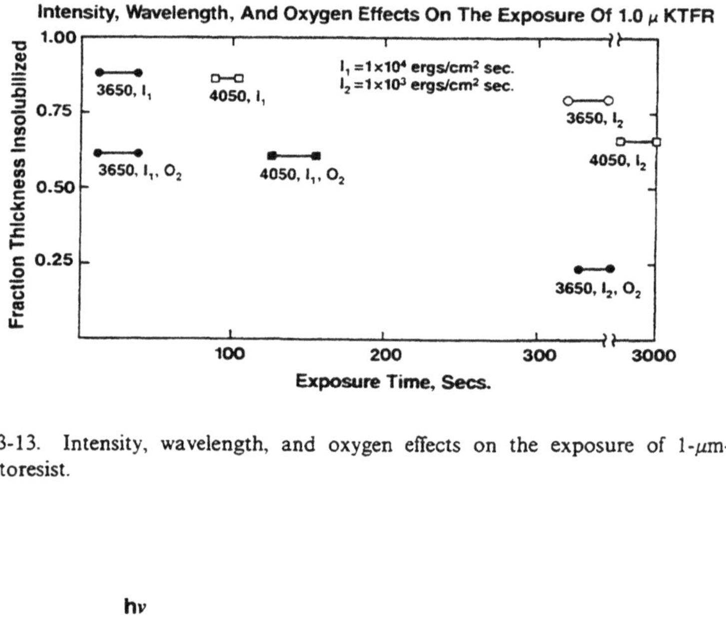

FIGURE 4-3-13. Intensity, wavelength, and oxygen effects on the exposure of 1-μm-thick KTFR negative photoresist.

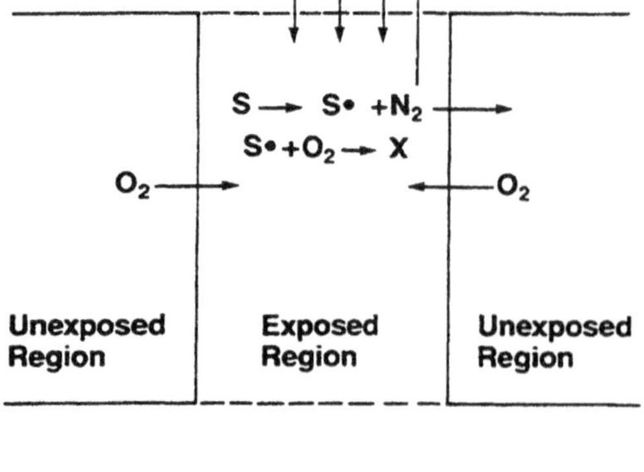

FIGURE 4-3-14. Diffusion of oxygen into ambient during exposure and internally from unexposed region into the exposed region to subtended stray light photoreactions and improved resolution of oxygen-sensitive dry film resists.

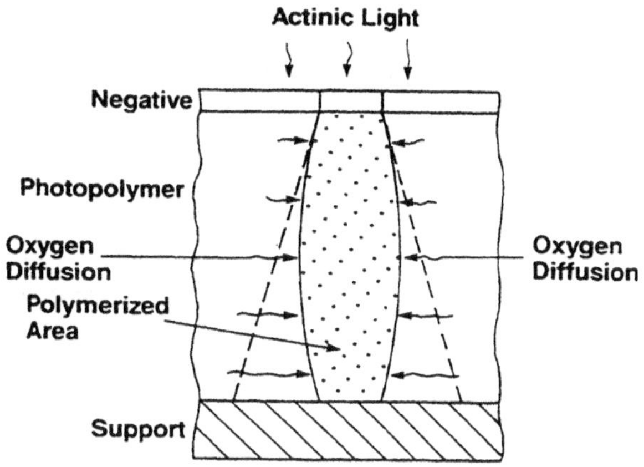

FIGURE 4-3-15. Enhanced resolution in PMMA photopolymer dry film resist due to O_2 interdiffusion into scattered region to prevent insolubilization. From Ref. 95, courtesy of SPE.

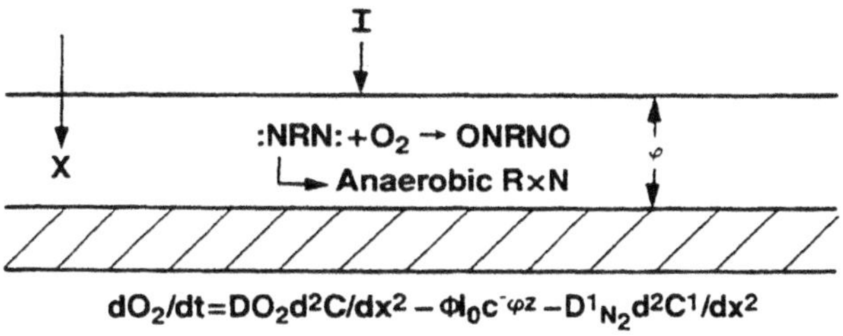

FIGURE 4-3-16. Steady-state calculations of oxygen consumption during exposure of free radical monomer polymerizing dry film resist during exposure of film in air. From Ref. 96, by permission of John Wiley & Sons, Inc.

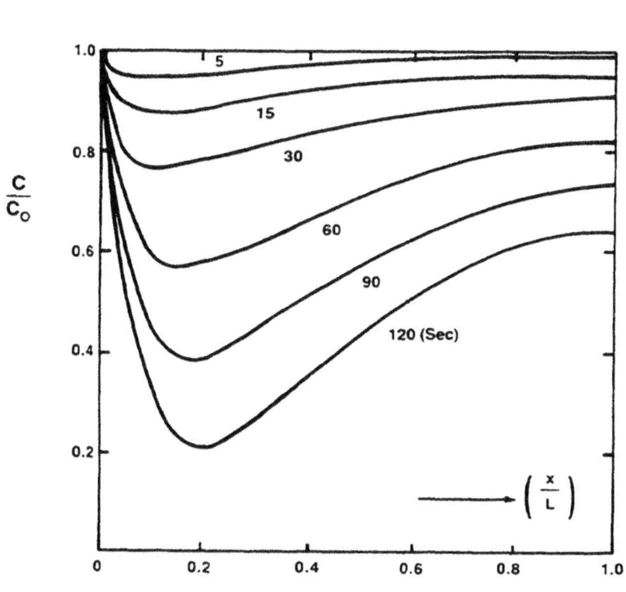

FIGURE 4-3-17. Depth of penetration of oxygen into resist film to consume free radicals which are generated at a rate proportional to the absorbed radiation and the intensity of the exposure source. From Ref. 96, reproduced by permission of the Electrochemical Society Inc.

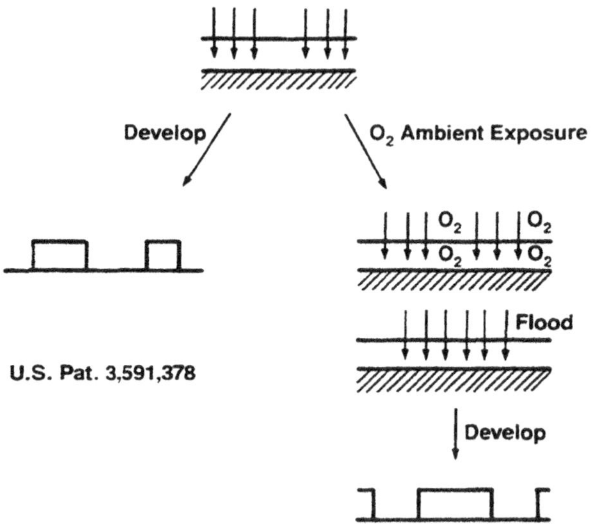

FIGURE 4-3-18. Image reversal of KTFR by low-intensity exposure in air followed by flood exposure in nitrogen.

density (> 1.0), excess radicals are present (assuming good quantum yields). By exposure with high-intensity radiation, there rate of radical generation,

$$\phi I_0 \exp(-\varphi x) \ggg DO_2$$

is much greater than the rate of oxygen quenching of the radical intermediates. This can be accomplished by flash lamp exposure[97] or in scanning sources such as the Perkin–Elmer 200 which concentrate the exposure intensity in small (mm^2) areas at $> 500\ mW/cm^2$ and scan the beam across the wafer. Other means of suppressing the oxygen effect are topcoatings of water-soluble and easily removed polymers[162] and preconditioning the resist in inert atmospheres (Table 4-3-6).

The oxygen effect is most severe at low light intensity ($< 10\ mW/cm^2$). Image reversal of KTFR takes advantage of the oxygen effect by exposing with low intensities to react all of the azide with oxygen followed by flood exposure (Fig. 4-3-18) at higher intensity to cross-link the previously unexposed region. If a contact printing relief mask is used which purposely traps oxygen at the top surface of the resist after exposure at low intensity, image reversal can also be demonstrated (Fig. 4-3-19). Thus, the oxygen effect can be put to advantageous use once the full implications of the mechanism are well understood.

TABLE 4-3-6. Control of Oxygen Effects

Topcoat	Diffusion
Thick film	Diffusion
Blanket exposure	Getter/intensity
Flash exposure	Intensity/diffusion
Precondition	Displacement
Anaerobic	Displacement
Polymer	Impermeability
Temperature	Diffusion/solvability/kinetics

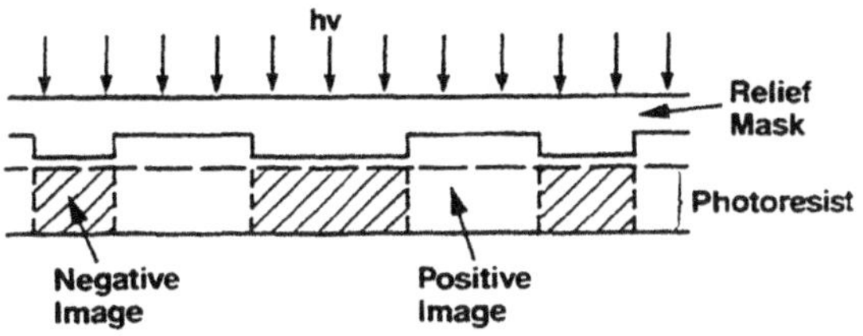

FIGURE 4-3-19. Relief mask for trapping oxygen at surface of photoresist and reversing image in irradiated areas which have sufficient oxygen trapped to diffuse in and kill reaction. The regions in direct contact with transparent upside-down mask form normal negative resist image in the absence of oxygen (as in normal contact printing).

4-4. MISCELLANEOUS RADICAL—BASED PHOTORESISTS

With the acceptance of deep-UV exposure, first demonstrated by Moreau and Schmidt in 1970,[98] many polymers such as novolaks, polystyrenes, and polyesters could be cross-linked with deep-UV radiation. Polymers with unsaturated main chains such as diallyl esters of dicarboxylic acids,[99]

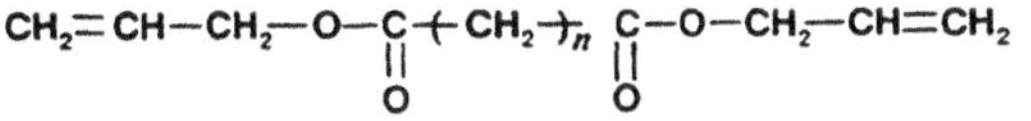

are highly absorbing in the deep-UV beam. Since deep-UV exposure is a good facsimile of *e*-beam exposure, an *e*-beam resist of polychlorostyrene has been converted into a deep-UV resist,[166]

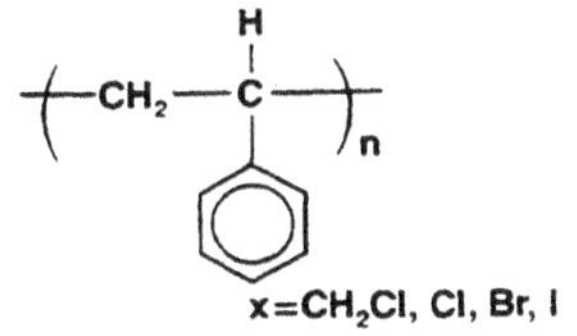

Upon excitation at 254 nm, the chlorine group is cleaved and the chlorine radical abstracts hydrogen to induce backbone cross-linking.[99–103] Other functional groups which have been attached to the phenyl group include cinnamate,[202] and phenylsiloxymethacrylate[206] which forms a useful bilayer resist mask. Since chlorinated polystyrenes can be synthesized from monodisperse polystyrene, improved resolution with a minimum of swelling can be obtained in polychlorostyrene-based resists.[104] However, the oxygen effect is still prominent in these systems. In some polymers such as polysiloxanes, the oxygen promotes cross-linking in the deep UV.[105] Most of the deep-UV resists will find applicability as image layers in multilayer resists.

Image reversal of normal positive photoresists of diazoquinone–novolaks (DQN)[200, 207] or polyvinylphenol carbonates is another means of overcoming the oxygen sensitivity of some of the cross-linking resists as well as reducing the swelling problem with organic solvent developable resists.[198] The image reversal can be accomplished in the following ways:

1. Direct UV exposure of DQN in the absence of water induces ketene addition to the novolak resin.[106] Polymeric diazoquinones in the presence of hydroxyl resins efficiently cross-link by esterification.[107]
2. Addition of imidazole or amines followed by heating and flood exposure prevents ketene hydrolysis and promotes decarboxylation to insolubilize the exposed region[108–110] as shown in the following steps studied by MacDonald *et al.*[164] The exposed product (A) is insolubilized via reaction (B) followed by normal flood solubilization of the background (C),

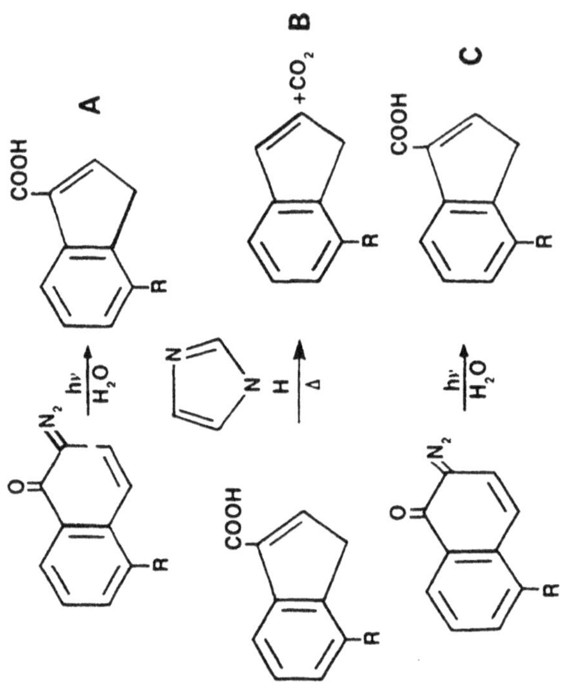
A
B
C
COOH
R
$+CO_2$
hν
H_2O
H
Δ

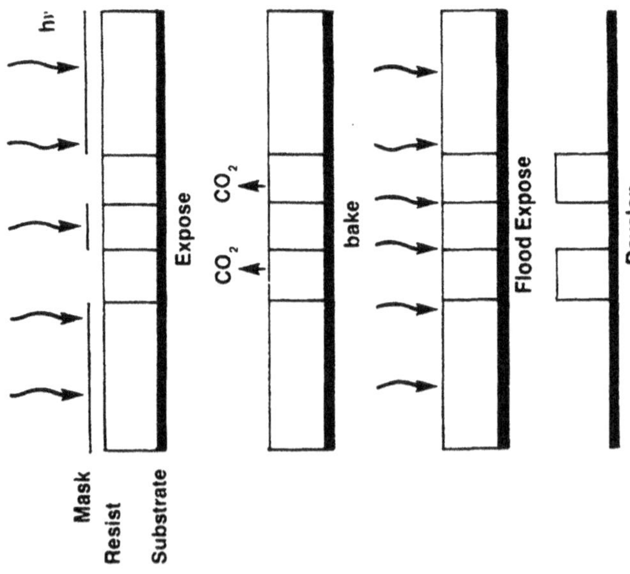
Mask
Resist
Substrate
hν
Expose
CO_2
bake
Flood Expose
Develop

3. Development of DQN in organic solvent removes the unexposed resist,[198–200]

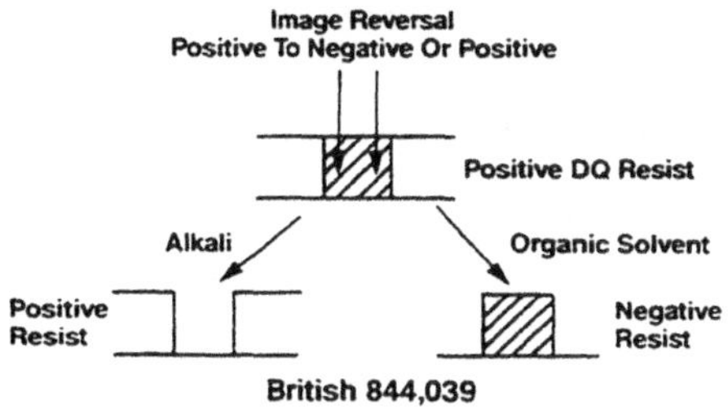

4. Polyvinylphenol backbone polymers with pendant ester and carbonate groups are decarboxylated back into polyvinylphenol by the action of Lewis acids generated from the photodecomposition of onium salts at low doses of 10 mJ/cm^2.[204] Both positive and negative images can be produced with the appropriate developer. High resolution is evident since the images are not cross-linked and unlikely to swell.

The advantages of all these systems are as follows: (1) lack of swelling, (2) no oxygen effects, (3) no reproduction of standing or reflected rays, and (4) the option of a positive or negative resist with one system. The DQN also exhibit wide spectral sensitivity. The main disadvantage is the higher exposure dose (> 200 mJ/cm^2) required to convert all of the DQ into product, although recent studies have shown that the addition of imidazole is not necessary and images can be reversed at normal DQN doses with a careful predevelop baking at 115°C[207] followed by a stronger developer.

4-5. HIGH-TEMPERATURE NEGATIVE PHOTORESISTS

The formulation of high-temperature[138] negative photoresists is driven by packaging needs for stable insulators and solder dam masks. Polyimides have been

TABLE 4-5-1. High-Temperature Negative Photoresist Compositions

Resist	Reference
Polyimide siloxane	167, 181
Polydiallylphthalate	114
Polyimide-MMA	115, 140
Polyimide + PMMA	116
Polyimide-MMA + azides	117, 177
Polyamino acid + diazoquinone	118
Polyoxazoles + diazoquinone	119
Polyimide-cinnamate	120
Polyimidazole-MMA	121
Polyimide + benzophenone	122
Polyoxazoles-MMA	123
Polycarbonates	124
Polysulfones-MMA	125
Polyacetylenes	126, 127

a popular choice of polymer films for insulators and lift-off processing.[111] In the usual processing, the polyimide is cured, overcoated with a resist, and then patterned by plasma etching or liquid etchants such as hydrazine or ethylenediamine. Recently, formulations can be directly imaged as resists[139, 183–190] and find applications as permanent dielectrics.

The first high-temperature negative photoresists were dichromated polyvinyl alcohol or dichromated polyimide which can be thermally converted into high-temperature polymers. Unfortunately, high exposure doses were required for these classic compositions (> 1 J/cm^2). More successfully, polyimides are converted into resists by grafting photoactive acrylate (MMA) or cinnamate groups onto the imide precursors and reacting through the pendant side chain,[112, 113] followed by thermal cyclization.[139] In one case, preimidized polyimide has been formulated into a high-temperature resist.[190] Rubner *et al.*[112] converted polyimides into photosensitive groups by attachment of acrylate side chains,

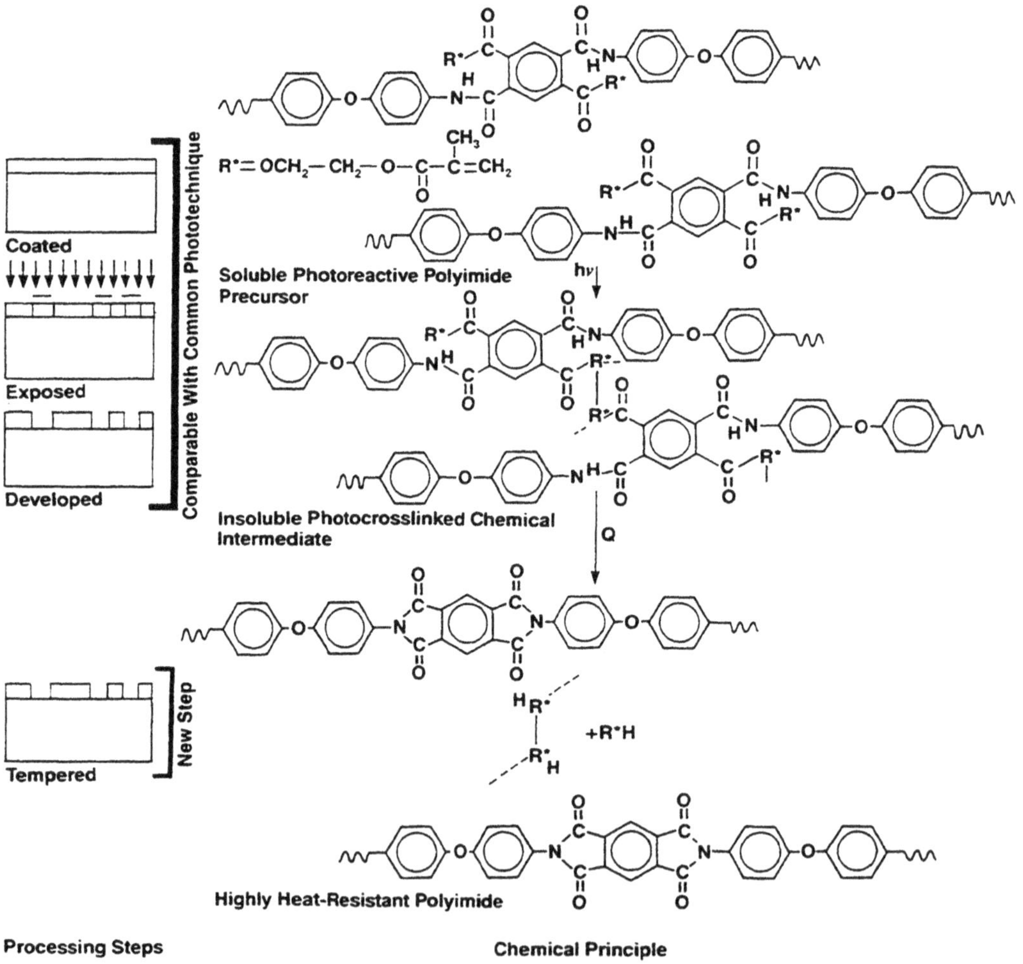

Representative patents of high-temperature resists are given in Table 4-5-1. The resolution of most of the resists is about three times their thickness which is adequate for images > 5 μm wide. In most cases, an oxygen effect in exposure is noted and the actinic sensitivity is usually confined to 365 nm. About 30% of the film is lost by imidization.[112] By themselves, the polyimides are good reactive ion masks and continued development in this area should be focused on obtaining submicrometer resolution at higher contrast with more monodisperse resists. In summary, the incorporation of reactive groups onto photoinsensitive but high-temperature backbones has been successfully demonstrated in the many examples of high-temperature resists.

4-6. PHOTOPOLYMERIZATION

Photoinitiated polymerization has never been applied to semiconductor lithography because of the oxygen effect and the tackiness of the films loaded with 50% tetrafunctional monomers. Recent interest in post-image amplification by grafting of monomers into irradiated resists has revived interest because of the potential for high-speed resist formulations.[128] For semiconductor device lithography, photopolymerization has not been uniquely applied because of the oxygen exposure effect and the inherent tackiness of the films loaded with 50% tetrafunctional monomers. Recent investigations of the grafting of monomers into irradiated resists have renewed interest in image amplification, because of the potential for high-sensitivity resists.[128] In addition, onium salts have been applied to the Lewis acid cationic cross-linking of acrylamide monomers, epoxides, and polyethers.[199, 200]

For packaging applications, photopolymerization has been uniquely applied in a dry film resist sandwich.[176] The tacky photopolymerizable mixture consists of a cross-linkable tetrafunctional monomer such as pentaerythritol tetraacrylate,

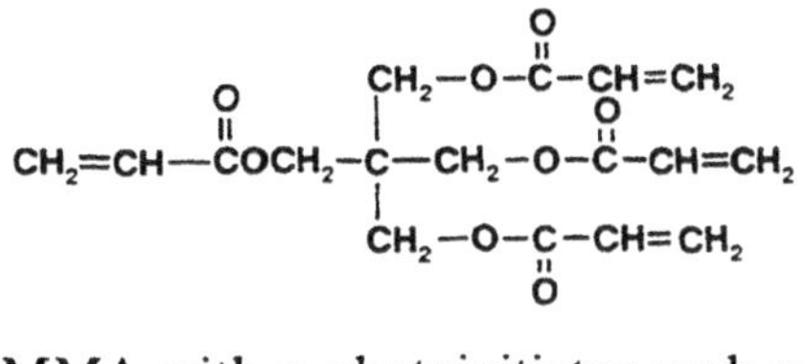

in a polymer base of PMMA with a photoinitiator such as *t*-butyl anthraquinone. The sandwich is covered by a laminating polyolefin and Mylar topcoat[130] (Figs. 4-6-1 and 4-6-2). The film is exposed with the Mylar topcoat in place to prevent external oxygen from interacting during exposure (Fig. 4-3-14). For films

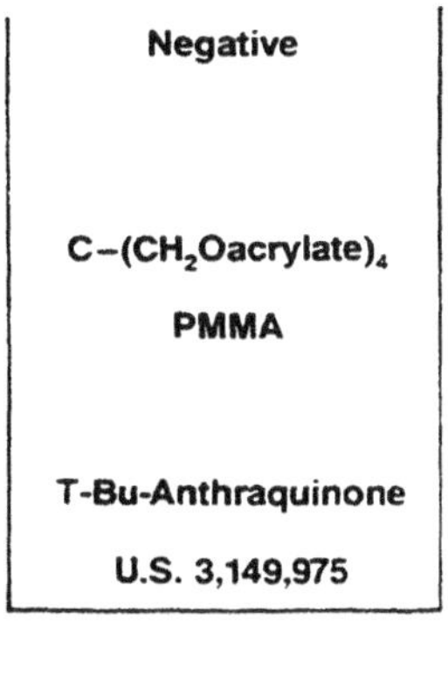

FIGURE 4-6-1. Composition of negative dry film resist consisting of PMMA polymer loaded at 50% level with pentaerythritol acrylate monomer and 360-nm sensitizer of *t*-butyl anthraquinone, U.S. Patent 3,149,975.

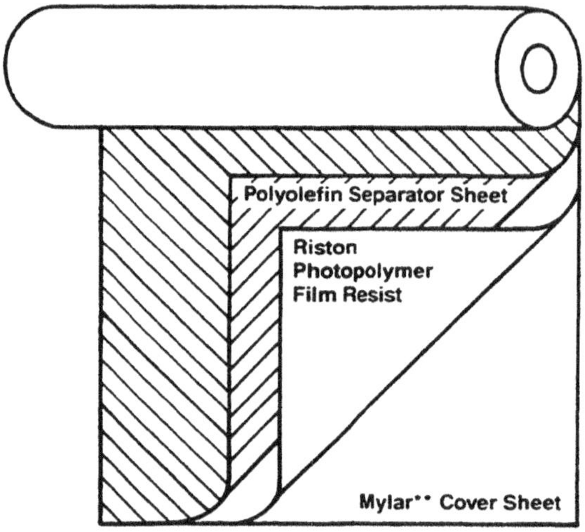

FIGURE 4-6-2. Laminated sheets in sandwich of Dupont Dry Film Riston Resist. (Dupont technical literature.)

free of pinholes and defects, the quality control rests with the manufacturer and not the traditional photoresist user. After exposure, the topcoat is removed and the film is developed in a variety of solvents including alkaline aqueous solutions for the acrylic acid-based formulations. Over 100 compositions(131) have been patented with various prepolymers and photoinitiators (Table 4-6-1) including recent water-developable composition(132) and a dry develop(133) peel-apart resist in which the

TABLE 4-6-1A. Common Sensitizers

Class	Typical examples
Carbonyl compounds	Benzoin, quinones, benzophenone
Azo compounds	Diazonium compounds
Organic sulfur compounds	Mercaptans, disulfides
Redox systems	Per acids, Fe^{2+}, Fe^{3+}
Halogen compounds	Silver halides, CBr_4, $HgBr_2$
Dyes	Eosin/amine, cyanine dyes, methylene blue
Organometallics	Metal alkyls
Metal carbonyls	$M(CO)_x$
Photoconductors	ZnO
Triplet	Benzophenone, benzothiazoles
Singlet	Eosin dyes

TABLE 4-6-1B. Commonly Used Prepolymers in Photolithography

Acrylates
Acrylonitriles
Allyl esters
Cellulose esters
Epoxies
Ethylenic (unsaturated) hydrocarbons
Polyhydroxy esters (e.g., pentaerythritol tetramethacrylate, ethylene glycol diacrylate)
Polyolefins
Vinyl compounds (esters, acids)
Vinylidene compounds

unexposed resist preferentially adheres to the oxygen barrier topcoat. After exposure, the image is simply developed by pulling off the top layer. The general order of photospeed cure is[129]

$$\text{Acrylic} > \text{Methacrylic} > \text{Vinyl} > \text{Allylic}$$

The dry film resists solved a basic problem of the failure of liquid resists to "tent" over and cover holes in copper board for selective electroplating. Dry film resists adequately bridged and protected the holes (Fig. 4-6-3).[134] Although the dry film costs are slightly[137] higher than those of liquid resists, this is offset by higher yields and fewer process steps. Lower defects are possible since the manufacturer can coat and control the level of defects. In addition, the user does not have to worry about coating and prebaking. The sensitivity of the dry film supplied in various mil thicknesses is about 200 mJ/cm^2 at 365 nm.

Photografting is another example of photopolymerization with a potential for reducing exposure requirements below 1 mJ/cm^2.[168] Even in the presence of oxygen, photosensitive dyes can form initiator sites for grafting,

$$\text{Dye(T)} + O_2 \rightarrow \text{Peroxy-dye}$$

$$\text{Peroxy-dye} + \text{monomer} \rightarrow \text{Polymer graft}$$

Trapped radicals or peroxides serve as centers to amplify the surface image by contacting the exposed resist with monomer vapors (Fig. 4-6-4). Onium salts[170–173, 175] can also be used for initiating cationic polymerization especially with silicone monomers for bilayer resists. Radicals can amplify photopolymerization by a factor of 10^4. Alternatively, trapped acids or ions could initiate anionic cationic or condensation reactions with silicone monomers (M) which could subsequently be dry developed in an oxygen plasma (Fig. 4-6-4). Contact of KTFR nitrenes with silicone monomers could surface graft images for dry develop. Since a top surface image is formed, high resolution is possible. One of the earliest examples of top surface imaging was the formation of hydroperoxides on the surface of polyesters followed by immersion in acrylic acid.[135] Finally, in the last example, the photopolymerization of adsorbed monomers has good potential

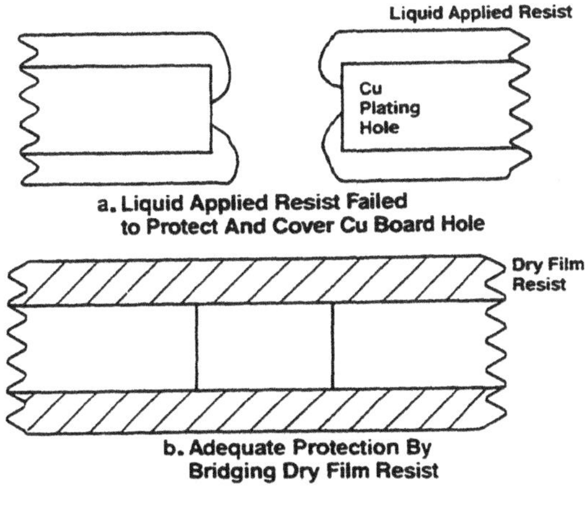

FIGURE 4-6-3. (a) Failure of liquid resist to bridge and protect holes in copper circuit boards against subsequent electroplating. (b) Dry film resist adequately bridges and protect holes.

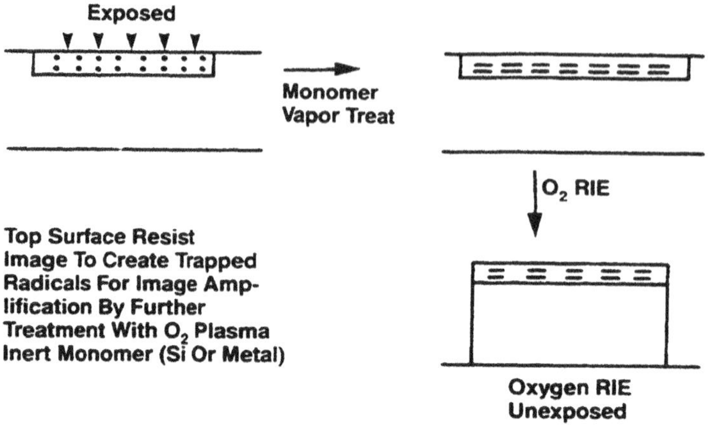

FIGURE 4-6-4. High-resolution submicrometer lithography with image-treated negative photoresist. The radiation is absorbed to a depth of only 100 nm of an etch-resistant mask or in the top layer of a bilayer mask. The image can be amplified in quantum yield if treated with a reactive monomer. If the monomer also contains Si or oxygen-resistant metal, then the image can be anisotropically dry developed in plasma.

for image amplification. Absorbed monomers such as butadiene have been exposed at 254 nm to grow resists on the surface of a wafer.(136) Silicone monomers could also be polymerized to grow diffusion masks in place (see Chapter 16).

4-7. CHARGE TRANSFER RESISTS

Cross-linking of negative photoresists has been the dominant mechanism in the last decade. However, cross-linking induces swelling which limits resolution to about 3× the cast film thickness (Fig. 4-7-1).(143) Instead of cross-linking, insolubilization can be brought about by a change in the polarity of a polymer or small molecule. If a hydrophobic polyolefin could be oxidized to an acid, a hydrophilic water-soluble polymer would remain upon development in organic solvents,

$$O_2 + RC{=}CR \rightarrow RCOOH$$

Polycarbonates or polyesters could be converted to polyacids, in the same manner as DQs are converted into indene acids (Chapter 2).

A very drastic change in polarity occurs if a nonpolar species is converted into an ionic species. Organic donors such as amines readily transfer electrons to organic halogen acceptors to form a whole class of organic molecular complexes called charge-transfer or donor–acceptor complexes,(142)

$$\text{Donor} + \text{acceptor} \rightarrow D^+ + A^-$$

If the molecular complex is exposed in the charge transfer UV band, complete

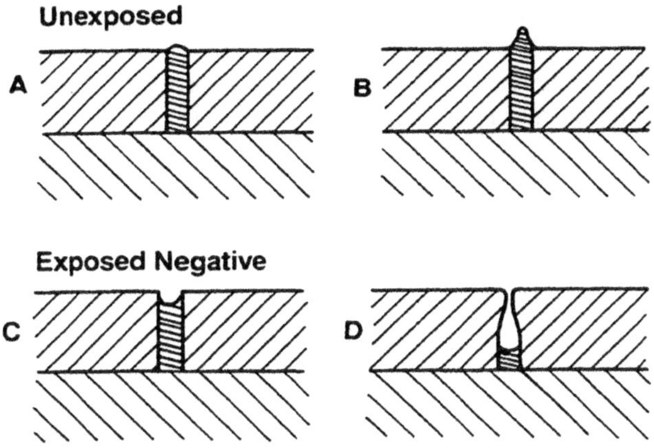

FIGURE 4-7-1. Development of narrow space in negative resist. (A) Slight swelling of unexposed; (B) increased swelling; (C) partial swelling; (D) internal swelling of exposed constricts opening. Adapted from G. Stevens, *Microphotography*, copyright John Wiley & Sons, Inc., 1967, by permission of John Wiley & Sons, Inc.

charge transfer (CT) can be brought about.[141] Kaufman and co-workers made negative resists using polymeric donor and carbon tetrabromide acceptor,[144–146]

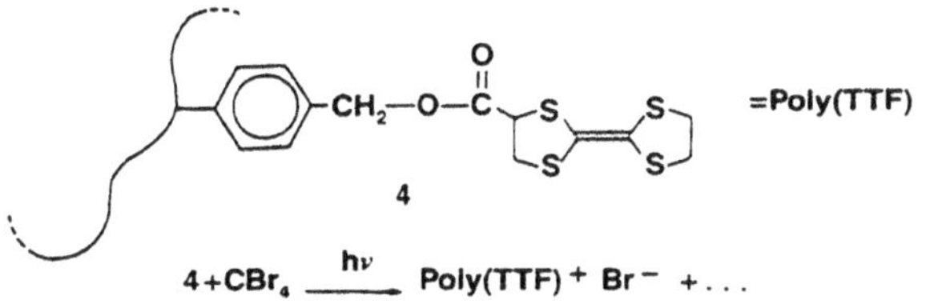

Since the resists do not swell, the resolution is good. One other aspect of enhanced resolution is the self-absorption of the exposure wavelength (Fig. 4-7-2).[146] Light of 405 nm, which would expose the base of a normal negative resist and shrink the unexposed spaces, is filtered out by the growth of the CT band.

Earlier composition based on CT molecular complexes consisted of amines and carbon tetrabromide.[147] The shelf life of the liquid CT resists is poor due to dark reactions. Other examples are polystyrenes as donors and halocarbons as acceptors.[148, 149] The CT polymers are a unique example of a new polarity mechanism for future resists especially if the species (charges) can initiate topside cationic or anionic polymerization to further amplify the images and reduce the incident dose requirement.

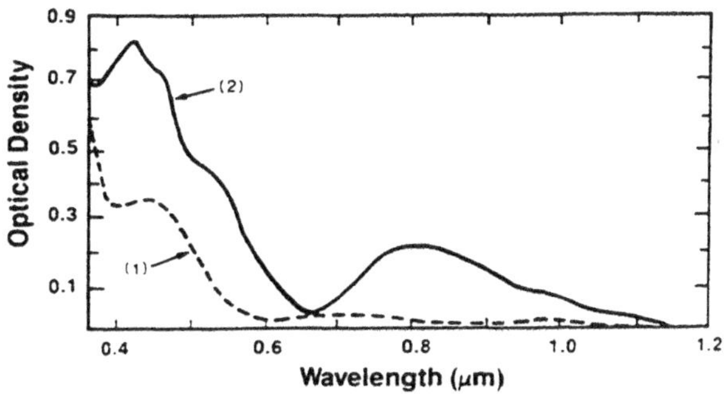

FIGURE 4-7-2. Absorption spectrum of charge transfer negative resist of PSTTF-CBr_4 complex before (1) and after (2) irradiation. From Ref. 146; figure originally presented at the 1982 Spring ECS meeting in Montreal, Canada; reproduced by permission of the Electrochemical Society Inc.

4-8. INORGANIC RESISTS

Inorganic resists represent a form of CT resist, but in this case, usually involving complete oxidation reduction (CT),

$$AgBr \rightarrow Ag + Br_2$$

Many inorganic salts darken on exposure and are useful for information storage such as chalcogenide glass. In most cases, the image formation used the principle[(149)] of energy-enhanced or -initiated doping by silver. Once the free silver atom is formed, it diffuses and alloys with As_2S_3 or the free Ag-Se chalcogenide to become etches resistant ("photodoping effect") (Fig. 4-8-1).[(150)] Since the absorption of light, because of the very high extinction coefficient of metallic components, is confined to the top 500-nm depth, very high resolution is possible if atomic-size molecules can be deposited in continuous films. Hopefully, pinhole-free films can be deposited by vacuum techniques. The sensitivity for the liquid-developed resist is low, but recent advances in dry plasma developing have reduced the dose required to less than 50 mJ/cm^2 (Table 4-8-1). A recent symposium has focused on inorganic resists.[(151, 165)]

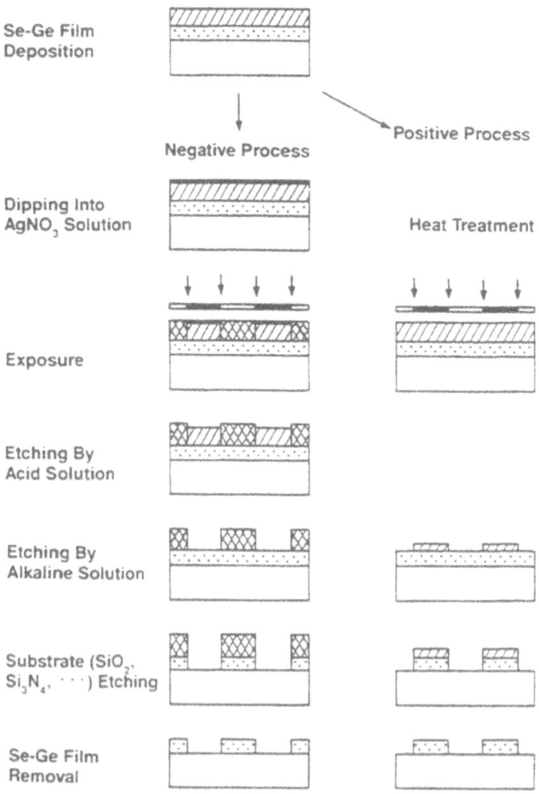

FIGURE 4-8-1. Process steps for negative photoresist of Se-Ge. From Ref. 150, courtesy of AIP.

TABLE 4-8-1. Inorganic Photoresists

Resist	Dose (mJ/cm^2)	Reference
Ag/As_2S_3	1000	152
Ag_2Ge/Ge	380	153, 154
Ge-Se-Ag	10	155
Ag-GeSe	200[a]	156
Ag-GeSe	80[a]	157
AgBr	0.005	158
Metal acrylates	1000	159
$Zn(CH_3)_2$	1000	160

[a] Plasma developed.

The Ag-doped Ge-Se responds to light of 200–460 nm.(209) Upon exposure, photodoping forms Ag_2Se which is removed by immersion in KI,

$$Ag_2Se + 4KI_3 + H_2O \rightarrow 2AgI + 4KI + 6HI + H_2SeO_4$$

The AgI is dissolved as a complexed polyiodide salt,

$$AgI + xKI \rightarrow K_xAgI_{x+1}$$

After the excess Ag has been removed, the latent image can be dry developed in a plasma of SF_6 or CF_4.(178) Inorganic photoresists possess a number of exceptional properties which make them attractive for submicrometer lithography: (1) high contrast due to the photoresist edge-sharpening effect, (2) high absorbance in thin 100-nm-thick films for top surface or bilayer masking (portable conformable deep-UV masks), and (3) excellent plasma resistance. The edge-scarpening effect is caused by the lateral diffusion of Ag into the exposed region due to the lowering of the coulombic barrier in the illuminated zone.(180) Ag diffusion into the exposed zone from the unexposed zone compensates for the diffraction effects of photobleaching. Because of the edge sharpening by diffusion, resolution to the atomic-size level has been achieved by contact printing with photons of 200–460 nm. Particles of Ag or bismuth (10 Å in size) were evaporated on top of Ag-Ge-Se resists and images were successfully contact printed.(179) The main drawbacks of inorganic resists are their poor photospeed (100–500 mJ/cm^2) and the difficulty of evaporating submicrometer particle-sized films with consistent stoichiometry. Although the dry film of silver bromide has outstanding sensitivity ($< 10\ mJ/cm^2$) due to the high chemical amplification of the silver photographic materials, the large grain size ($> 1\ \mu m$), however, limits the resolution to about 2 μm.(159) Chalcogenides of selenium are the most popular negative inorganic resists, but they require a source of exact (alloy) stoichiometric composition before the vacuum deposition of the resist. The future use of inorganic resists lies in multilayer structures since they are both nonerodable plasma masks(153, 154) and deep-UV photomasks. For inorganic resists, more sensitivity and better spectral absorbance are required.

4-9. SUMMARY AND OUTLOOK

Negative photoresists based on cross-linking reactions were the pioneering resists that fostered the chemists' contribution to microelectronics. In the period of 1960–1975, the cinnamate and azide resists made all semiconductor circuits of 3- to 10-μm size. They were good adhesive materials with wide development latitude. Further attempts to improve their resolution with more monodisperse polymers are hampered by the fact that the cinnamate and rubber-based polymers are synthesized by multistep reactions from polydisperse precursors.

The bisazide cross-linking agent has proven to be a versatile sensitizer paving the way for high-temperature, deep-UV, and dry develop resists. Dry film resists based on photopolymerization reaction solved tenting problems in circuit board fabrication and also pointed out the important fact that the manufacturer of resists could provide the user with defect-free coatings and complete processing line equipment.

New and novel imaging mechanisms such as charge transfer, acidolysis of esters and carbonates, and inorganic resists are a grand departure from the past 20 years of classical cross-linking negative photoresists. Fast resists based on imaging amplified grafting look promising for high-resolution top surface resists. The continued coupling of negative resists with plasma processing both for superior resolution and as plasma/thermal-resistant etch masks will be necessary for their future competition with positive resist systems for submicrometer lithography using deep-UV optics. Lastly, continued conversion of high-temperature polymers into negative photoresists will provide important new materials for the next hurdle—submil lithography for packaging.

REFERENCES

1. A. Charlesby, *Atomic Radiation and polymers*, Pergamon Press, Elmsford, N.Y., 1960.
2. F. Laming and R. Straile, *J. Electrochem. Soc.* **119**, 1745 (1972).
3. N. Atoda and H. Kawakatsu, *J. Electrochem. Soc.* **123**, 1519 (1976).
4. T. Kobayashi and E. Arai, *J. Appl. Phys.* **52**, 4785 (1981).
5. A. Reiser and E. Pitts, *J. Photogr. Sci.* **29**, 87 (1981).
6. A. Reiser and E. Pitts, *Photogr. Sci. Eng.* **20**, 225 (1976).
7. N. Parham, E. Pitts, and A. Reiser, *Photogr. Sci. Eng.* **21**, 145 (1977).
8. J. Kirk and G. Filmore, *Appl. Opt.* **11**, 2347 (1972).
9. J. Freijilich and J. Clair, *J. Opt. Soc. Am.* **67**, 92 (1977).
10. P. Egerton, J. Trigg, E. Hyde, and A. Reiser, *Macromolecules* **14**, 100 (1981).
11. M. Nakase and Y. Matsumoto, *Photogr. Sci. Eng.* **23**, 215 (1979).
12. K. Tai, R. Vadimsky, C. Kemmer, J. Wagner, A. Timko, and V. Lambertini, *J. Vac. Sci. Technol.* **17**, 1169 (1980).
13. (a) J. Kosar, *Light Sensitive Systems*, Wiley, New York, 1965. (b) W. DeForest, *Photoresist Materials and Processes*, McGraw–Hill, New York, 1975, p. 168.
14. G. A. Delzenne, in *Encyclopedia of Polymer Science and Technology*, Suppl. Vol. 1, edited by H. Mark and N. Bikales, Interscience, New York, 1976, p. 401.
15. J. Rennert, *Photochem. Photobiol.* **19**, 173 (1975).
16. A. Reiser and P. Egerton, *Macromolecules* **12**, 670 (1979).
17. K. Murski and S. Simoliunas, *Electrochem. Soc. Ext. Abstr.* **80-2**, 838 (1980).
18. C. Azuma, K. Sanui, and N. Ogata, *J. Appl. Polym. Sci.* **27**, 2065 (1982).
19. T. Nishibuko, *J. Polym. Sci.* **21**, 2025 (1983).
20. U.S. Patent 2,716,102 (1958), Eastman Kodak.
21. H. Tanaka and Y. Sato, *J. Polym. Sci.* **10**, 3279 (1972).

22. S. Merril and C. Unruh, *J. Appl. Polym. Sci.* **7**, 273 (1963).
23. W. Moreau, *Am. Chem. Soc. Polym. Prepr.* **10**, 362 (1969).
24. T. Shankoff and A. Trozzolo, *Photogr. Sci. Eng.* **19**, 173 (1975).
25. Y. Kirsh, *Russ. J. Phys. Chem.* **39**, 1002 (1965).
26. J. Arcesi, F. Rauner, and J. Williams, *J. Appl. Polym. Sci.* **15**, 513 (1971).
27. R. Agnihotri, F. Hood, L. Lesoine, and J. Offenbach, *Photogr. Sci. Eng.* **15**, 141 (1971).
28. M. Kato and H. Nakane, *Photogr. Sci. Eng.* **23**, 207 (1979).
29. K. Nakamura, T. Sakata, and S. Kikuchi, *Bull. Chem. Soc. Jpn.* **41**, 1765 (1968).
30. E. Engvall, Kodak Photoresist Seminar, May 1969 and Oct. 1976.
31. U.S. Patent 2,652,273 (1972), IBM.
32. W. Moreau, R. Cox, and N. Clecak, *Polym. Eng. Sci.* **14**, 95 (1981).
33. U.S. Patent 3,716,390 (1971), DuPont.
34. P. Egerton, J. Trigg, E. Hyde, and A. Reiser, *Macromolecules* **14**, 95 (1981).
35. E. De Broer and R. Breslow, *Tetrahedron Lett.* **1967**, 1033 (1967).
36. M. Mijovic, P. Beymor, T. Shawe, K. Petrak, and A. Reiser, *Macromolecules* **15**, 1464 (1982).
37. B. Gong, Y. Te, M. Gu, and Q. Zhang, *J. Vac. Sci. Techol.* **16**, 1980 (1979).
38. A. Schonberg, *Preparative Organic Photochemistry*, Springer Verlag, Berlin, 1968, p. 114.
39. U.S. Patent 4,302,527 (1981), Kodak.
40. U.S. Patent 2,940,853 (1962).
41. W. Moreau, SPE RETEC Photopolymers, Ellenville, N.Y., 1970, p. 137.
42. J. Swenton, T. Ikeler, and B. Williams, *J. Am. Chem. Soc.* **92**, 3103 (1970).
43. A. Reiser, G. Bowes, and R. Horne, *Trans. Faraday Soc.* **63**, 3162 (1970).
44. T. Iwayanagi, T. Kohashi, and S. Nonogaki, *J. Electrochem. Soc.* **127**, 2759 (1980).
45. S. Shimizu and G. Bird, *J. Electrochem. Soc.* **124**, 1394 (1977).
46. S. Shimizu and G. Bird, *J. Electrochem. Soc.* **126**, 273 (1979).
47. R. Agnihotri, D. Falcon, F. Hood, L. Lesoine, C. Needham, and J. Offenbach, *Photogr. Sci. Eng.* **16**, 443 (1972).
48. W. Hunter and P. Crabtree, *Photogr. Sci. Eng.* **13**, 271 (1969).
49. U.S. Patents 3,669,662 and 3,669,669 (1972), IBM.
50. A. Reiser, R. Egerton, and E. Pitts, *Trans. Faraday Soc.* **65**, 3265 (1969), **64**, 1806 (1968).
51. T. Zueva, *Chem. Abstr.* **98**, 188958 (1983).
52. R. Walters and R. Brechner, *J. Electrochem. Soc.* **119**, 1703 (1972).
53. U.S. Patent 3,615,952 (1971), RCA.
54. Japanese Patent 83,40,547, *Chem. Abstr.* **99**, 13971 (1983).
55. U.S. Patent 4,294,908 (1981), Japanese Synthetic Rubber Co.
56. U.S. Patent 4,268,603 (1981).
57. R. Lussow, *J. Vac. Sci. Technol.* **6**, 18 (1969).
58. U.S. Patent 3,695,886 (1972), IBM.
59. J. Frejilich and R. Knoesel, *Appl. Opti.* **18**, 1135 (1979).
60. U.S. Patent 3,493,380 (1970), Kodak.
61. German Patent 2,817,256, Allied Chem.; *Chem. Abstr.* **90**, 195637b (1976).
62. U.S. Patent 3,493,380 (1970), Kodak.
63. Japanese Patent 80, 155, 353, *Chem. Abstr.* **94**, 200872 (1980); Japanese Patent 80,87,143, *Chem. Abstr.* **94**, 10032 (1980).
64. Russian Patent 708,287, *Chem. Abstr.* **91**, 207140 (1978).
65. Japanese Patent 79,140,535, *Chem. Abstr.* **92**, 119732 (1979).
66. F. Laming and R. Straile, *J. Electrochem. Soc.* **120**, 292 (1973). L. Hirobe and S. Iwamatsu, *J. Electrochem. Soc.* **126**, 1426 (1979).
67. U.S. Patent 3,884,703 (1974), Hitachi.
68. U.S. Patent 4,299,910 (1981), RCA.
69. U.S. Patent 4,254,197 (1981), Hitachi.
70. U.S. Patent 3,072,485 (1963); 3,475,176 (1969), Kodak.
71. T. Matsuzuwa and H. Tomioka, *IEEE Electron Device Lett.* **EDL-2**, 90 (1981).
72. I. Iwayangi, T. Kohashi, S. Nonogaki, T. Matsuzawa, K. Douta, and H. Yanazawa, *IEEE Trans. Electron Devices* **ED-28**, 1306 (1981), *J. Electrochem. Soc.* **127**, 2759 (1980).
73. T. Matsuzawa and H. Tomioka, *IEEE Trans. Electron Devices* **ED-28**, 1284 (1981), **ED-30**, 1780 (1983).
74. U.S. Patent 4,148,655 (1979), Oji Paper.
75. T. Tsunoda, T. Yamoka, Y. Osabe, and Y. Hata, *Photogr. Sci. Eng.* **20**, 188 (1977).

76. U.S. Patent 4,268,603 (1981), Tokyo Ohko.
77. Japanese Patent 82,84,452, Sanyo; *Chem. Abstr.* **98**, 25528 (1982).
78. Japanese Patent 82,70,530, *Chem. Abstr.* **87**, 44226f (1982).
79. Japanese Patent 78,127,723, *Chem. Abstr.* **90**, 144308 (1978).
80. U.S. Patent 4,287,294 (1981), Siemens.
81. U.S. Patent 4,399,910 (1981), RCA.
82. U.S. Patent 4,254,197 (1981), Hitachi.
83. Japanese Patent 80,77,741, *Chem. Abstr.* **94**, 100300 (1980).
84. Japanese Patent 82,105,736, Daceil Ind.; *Chem. Abstr.* **99**, 30759r (1982).
85. German Patent 3,239,613, *Chem. Absdtr.* **99**, 13922 (1982).
86. M. Tsuda, S. Oikawa, W. Kanai, K. Hashimoto, A. Yokata, K. Nuino, I. Hijikata, A. Uehara, and H. Nakane, *J. Vac. Sci. Technol.* **19**, 1351 (1981), **19**, 259 (1981).
87. M. Tsuda, M. Yabuta, S. Oikawa, A. Yokota, and H. Nakane, in *Microcircuit Engineering*, edited by J. Cleaver, H. Ahmed, and G. Jones, Academic Press, New York, 1984, p. 371.
88. European Patent Application EP 70,198, Hitachi; *Chem. Abstr.* **98**, 152811b (1981).
89. U. S. Patents 3,467,523, 3,462,268, 3,455,689, 3,453,108 (1969), GAF.
90. U.S. Patent 3,728,305 (1968).
91. T. Nishibuko, T. Kizawa, I. Imagawa, and K. Kobayashi, *J. Polym. Sci. Polym. Chem. Ed.* **19**, 2705 (1981).
92. U.S. Patent 3,002,003 (1961), Kodak.
93. W. Likowsky, in *Nitrene*, edited by R. Abramovitch and B. Davis, Wiley, New York, 1979; *Chem. Rev.* **64**, 149 (1964). S. Patai, *Chemistry of the Azido Group*, Interscience, New York, 1969. G. Abbe, *Chem. Rev.* **68**, 345 (1968).
94. E. Froschule and R. Backhus, *Solid State Electron.* **14**, 95 (1971).
95. D. Smith, *Polym. Eng. Sci.* **12**, 264 (1968).
96. J. Chong, *J. Appl. Polym. Sci.* **13**, 241 (1969).
97. U.S. Patent 3,705,055 (1971), Western Electric.
98. W. Moreau and P. Schmidt, *Electrochem. Soc. Ext. Abstr.* **70-2**, 459 (1970).
99. U.S. Patent 4,267,258 (1981), Fujitsu.
100. U.S. Patent 4,286,049 (1981), Nippon.
101. O. Kogure, *Jpn. J. Appl. Phys.* **21**, 206 (1982).
102. S. Iwamura and S. Sugaware, *Jpn. J. Appl. Phys.* **21**, 776 (1982).
103. U.S. Patent 4,267,258 (1982), Fujitsu.
104. S. Iwamura, *J. Electrochem. Soc.* **126**, 1628 (1979).
105. D. Turner, *Macromolecules* **14**, 216 (1981).
106. J. Pacansky and J. Lyerla, *IBM J. Res. Dev.* **23**, 42 (1979).
107. U.S. Patent 3,533,796 (1970), GAF.
108. U.S. Patent 4,104,070 (1978).
109. Japanese Patent 80,32,088, *Chem. Abstr.* **93**, 195530y (1980).
110. G. Willson, in *Introduction to Microlithography*, edited by L. Thompson, C. Willson, and M. Bowden, *Am. Chem. Soc. Symp. Ser.* **219**, 121 (1983).
111. L. White, *J. Electrochem. Soc.* **130**, 1543 (1983).
112. R. Rubner, H. Ahne, E. Kuhn, and G. Kolodzeij, *Photogr. Sci. Eng.* **23**, 303 (1979).
113. U.S. Patent 4,218,283 (1980), Hitachi.
114. U.S. Patent 3,883,352 (1975), W. Grace.
115. U.S. Patents 4,208,477 (1980), 4,316,974 (1981), Ashahi.
116. U.S. Patent 3,753,270 (1974), DuPont.
117. U.S. Patents 3,957,512, 4,045,223, 4,088,489, 4,287,294, Re 30,186, Siemens.
118. U.S. Patent 4,093,461 (1979), Siemens.
119. U.S. Patent 4,399,521 (1982), Siemens.
120. U.S. Patent 4,310,641 (1982), Siemens.
121. U. S. Patent 4,332,882 (1982), Siemens.
122. U.S. Patent 4,242,437 (1980), Bayer.
123, U.S. Patent 4, 332,883 (1982), Siemens.
124. U.S. Patent 3,043,802 (1962), Bayer.
125. Japanese Patent 81,50,928, *Chem. Abstr.* **96**, 13661 (1981).
126 U.S. Patent 3,725,064 (1973), GAF.
127. U.S. Patent 3,594,175 (1972), GE.
128. M. Gazard, A. Chapiro, M. Duchesne, and J. Dubois, *Polym. Eng. Sci.* **20**, 1069 (1980).

129. C. Roffey, *Photopolymerization of Surface Coatings*, Wiley, New York, 1982, p. 137.
130. DuPont Co. Technical Literature on Riston, Wilmington, Del.
131. U.S. Patents 3,469,982, 3,547,651, 3,784,378, DuPont.
132. U.S. Patents 4,415,651, 4,415,652, (1983), DuPont.
133. U.S. Patent 4,289,841 (1981), DuPont.
134. D. Keleman, *Solid State Technol.* **19**, 37 (1976).
135. U.S. Patent 3,847,609 (1974), DuPont.
136. U.S. Patent 3,650,744 (1972), GE.
137. J. Bauer, *Electron. Packag. Prod.* May 1978, p. 87.
138. E. Eisenbach, *Angew. Makromol. Chem.* **109**, 101 (1982).
139. H. Steppan, G. Buhr, and H. Vollmann, *Angew. Chem. Int. Ed. Engl.* **21**, 455 (1982).
140. U. S. Patent 4,410,612 (1983), DuPont.
141. G. Briegleb, *Donor–Acceptor Complexes*, Springer Verlag, Berlin, 1966.
142. L. Andrews and R. Keefer, *Molecular Complexes in Organic Chemistry*, Holden–Day, San Francisco, 1964.
143. G. Stevens, *Microphotography*, Wiley, New York, 1967, p. 22.
144. F. Kaufman, *IBM J. Res. Dev.* **25**, 303 (1981); U.S. Patent 4,338,392 (1982), IBM.
145. D. Hofer, F. Kaufman, S. Kramer, and A. Aviram, *Appl. Phys. Lett.* **37**, 314 (1980).
146. F. Kaufman and S. Kramer, *Electrochem. Soc. Ext. Abstr.* **82-1**, 443 (1982).
147. U.S. Patent 3,147,117 (1964), Horizons.
148. U.S. Patents 3,522,044, 3,525,616 (1970), 3.529,966 (Kalle), 3,592,646 (1971), Kodak.
149. G. Chern and I. Lauks, *J. Appl. Phys.* **53**, 6979 (1982) and the many references therein.
150. Y. Yoshikawa, O. Ochi, H. Nagai, and Y. Mizushima, *Appl. Phys. Lett.* **29**, 677 (1976).
151. *Electrochem. Soc. Ext. Abstr.* **82-1**, 152–180 (1982).
152. M. Chang, T. Hou, J. Chen, K. Kolowicz, and J. Zemel, *J. Vac. Sci. Technol.* **16**, 1973 (1979).
153. K. Tai, R. Vadimsky, C. Kemmer, A. Timko, and V. Lambertini, *J. Vac. Sci. Technol.* **17**, 1169 (1980).
154. K. Tai, R. Sinclair, R. Vadimsky, J. Moran, and R. Rand, *J. Vac. Sci. Technol.* **16**, 1977 (1979).
155. A. Ruoff and K. Balasubramayan, *J. Vac. Sci. Technol.* **19**, 1374 (1981).
156. A. Yoshikawa, O. Ochi, and Y. Mizushima, *Appl. Phys. Lett.* **36**, 107 (1980).
157. P. Huggett, K. Frick, and H. Lehmann, *Appl. Phys. Lett.* **42**, 592 (1983).
158. J. Lavine, G. Goldberg, and A. Das. *IEEE Trans. Electron Devices* **ED-28**, 1311 (1981); U.S. Patent 4,269,935 (1981), Ionomet.
159. A. Pastor, G. Toniginami, R. Pastor, S. Wong, and R. Chew, *Thin Solid Films* **67**, 9 (1980).
160. D. Ehrlich, R. Osgood, and T. Deutsch, *J. Vac. Sci. Technol.* **B2**, 23 (1982).
161. T. Yamaoka, Y. Osobe, and T. Tsunoda, in *Modification of Polymers*, *Am. Chem. Soc. Symp. Ser.* **121,** 185 (1980).
162. W. Moreau, *IBM Tech. Discl. Bull.* **12**, 1418 (1970).
163. R. Jacobson, *J. Photogr. Sci.* **31**, 1 (1983).
164. S. MacDonald, C. Willson, R. Millert, W. MacIntyre, W. Motsiff, and R. Gleason, Proceedings of Kodak Microelectronics, Interface 82, p. 114.
165. *Proceedings of the Symposium on Inorganic Resist Systems*, edited by D. Doane and A. Heller, Electrochemical Society, 1982.
166. S. Inamura, T. Tamamura, and O. Kogure, *Polym. J.* **16**, 391 (1984).
167. G. Davis, *Am. Chem. Soc. Symp. Ser.* **242, 259** (1984).
168. Japanese Patent 58,55,926, *Chem. Abstr.* **101**, 46303 (1983).
169. V. Srinivasan and S. Babu, *Photogr. Sci. Eng.* **28**, 175 (1984).
170. M. Tsunooka, T. Ueda, and M. Tanaka, *J, Polym. Sci.* **22**, 2217 (1984).
171. S. Pappas, B. Pappas, L. Gatechair, and W. Schnabel, *J. Polym. Sci.* **22**, 69 (1984).
172. W. Watt. H. Hoffmann, H. Pobiner, I. Schkolnick, and L. Yang, *J. Polym. Sci.* **22**, 1789 (1984).
173. S. Pappas, L. Gatechair, and J. Jilek, *J. Polym. Sci.* **22**, 77 (1984).
174. S. Nishimoto, *J. Polym. Sci.* **22**, 323 (1984).
175. K. Iwata, T. Hagiwara, and H. Matsuzawa, *J. Polym. Sci.* **22**, 215L (1984).
176. S. Heilmann, *J. Polym. Sci.* **22**, 1179 (1984).
177. N. Yoda and H. Hiramoto, *J. Macromol. Sci. Chem.* **21**, 1641 (1984).
178. P. Huggett and H. Lehman, *J. Electron. Mater.*, **14**, 205 (1985).
179. A. Yoshikawa, O. Ochi, T. Mizushima, and T. Hisaki, *Jpn. J. Appl. Phys.* **20**, L81 (1981); U.S. Patent 4,350,541 (1982), Nippon Tel.
180. W. Leung, A. Neureuther, and W. Oldham, *J. Vac. Sci. Technol.* **B3**, 310 (1985).

181. U.S. Patent 4,515,887 (1985), GE.
182. A. Tanaka, M. Morita, and K. Onose, *Jpn. J. Appl. Phys.* **24**, L112 (1985).
183. European Patent 131,992 (1985), Phillips; *Chem. Abstr.* **103**, 176539 (1985).
184. Japanese Patent 59,222,833 (1984), Hitachi; *Chem. Abstr.* **102**, 176538 (1984).
185. German Patent 3,342,851 (1985), Merck; *Chem. Abstr.* **103**, 150965 (1985).
186. Japanese Patent 59,220,730 Ube (1985), *Chem. Abstr.* **102** 195201 (1985).
187. U.S. Patent 4,481,340 (1984).
188. Japanese Patent 59,212,382, *Chem. Abstr.* **103**, 45827 (1985).
189. N. Yoda and H. Hiramoto, *J. Macromol. Chem.* **121**, 1641 (1984).
190. G. Rohde, M. Riediker, A. Schaffner, and J. Bateman, *Proc. SPIE Opt. Eng.* **539**, 175 (1985).
191. Japanese Patent 60,177,740 (1985), Maruzen Oil; *Chem. Abstr.* **103**, 62584 (1985).
192. S. Kobuchi, A. Isobe, D. Makino, T. Iwayanagi, H. Hasimoto, and S. Nonogaki, *Proc. SPIE Opt. Eng.* **539**, 182 (1985).
193. European Patent 135,900 (1985), Hunt; *Chem. Abstr.* **103**, 62599 (1985).
194. Japanese Patent 60,70,439 (1985), Hitachi; *Chem. Abstr.* **103**, 62598 (1985).
195. Japanese Patent 60,86,543 (1985), Hitachi; *Chem. Abstr.* **103**, 113320 (1985).
196. German Patent 3,325,022 (1985), Hoechst; *Chem. Abstr.* **103**, 79521 (1985).
197. Japanese Patent 59,214,847 (1984), Nagase; *Chem. Abstr.* **102**, 195202 (1984).
198. U.S. Patent 4,356,254 (1984), Hitachi.
199. Japanese Patent 59,222,833 (1984), Hitachi; *Chem. Abstr.* **102**, 176538 (1984).
200. U.S. Patent 4,196,003 (1982), Fuji.
201. G. Bendkit, *SMIE Resist Technol.* **539**, 242 (1985).
202. Japanese Patent 59,216,140 (1985), AIST; *Chem. Abstr.* **103**, 62581 (1985).
203. K. Iwata, T. Hagiwara, and H. Matsuzawo, *J. Polym. Sci.* **23**, 2361 (1985).
204. C. Willson, H. Ito, J. Frechet, T. Tessier, and F. Houlihan, *J. Electrochem. Soc.* **133**, 181 (1986).
205. S. Babu and R. Srinivasan, *SPIE Resist Technol.* **539**, 36 (1985).
206. Y. Ohnishi, M. Suzuki, K. Saigo, Y. Satoire, and H. Gokan, *SPIE Resist Technol.* **539**, 62 (1985).
207. M. Spak, D. Mammato, S. Jain, and D. Durham, SPE RETEC Photopolymers, Ellenville, N.Y., 1985, p. 247.
208. U.S. Patent 4,564,576 (1986), NRC.
209. U.S. Patent 4,571,375 (1986), G. Benedikt.

5

Negative Radiation Resists

5-1. INTRODUCTION

In Chapter 4, we discussed negative photoresist compositions and polymeric factors which affect resolution; in Chapter 3, we considered exposure tools and radiation chemistry principles; and in Chapter 10, we discuss the swelling of negative resists in a developer. These chapters form the principal backgrounds for the processing aspects of negative radiation resists.

Negative radiation resists are abundantly more available than positive radiation resists, because of two main factors:

1. Many monosubstituted vinyl polymers are available with a wider variety of substituents,

$$\sim CH_2{=}\overset{H}{C}X\sim$$

   than for disubstituted vinyl monomers,

$$\sim CH_2{=}\overset{Y}{C}X\sim$$

   Many vinyl formulations are already available as negative photoresists, inks, dry film resists, and *e*-beam cured protective coatings.
2. Monosubstituted vinyl polymers are easier to polymerize than are disubstituted vinyl polymers, because of less steric crowding.

Negative resists have primarily been based on the insolubilization induced by random cross-linking of polydisperse (a) or monodisperse (b) polymers,

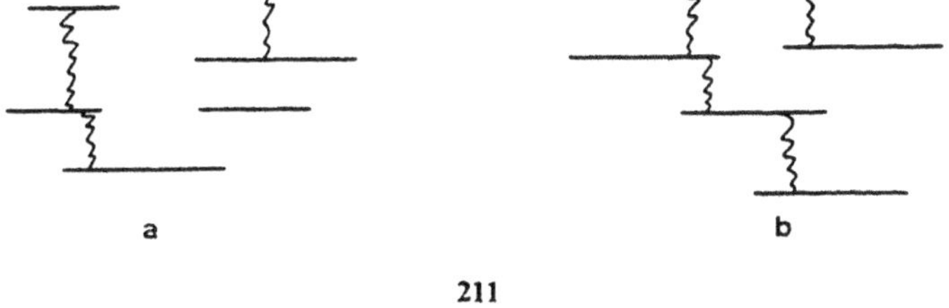

a b

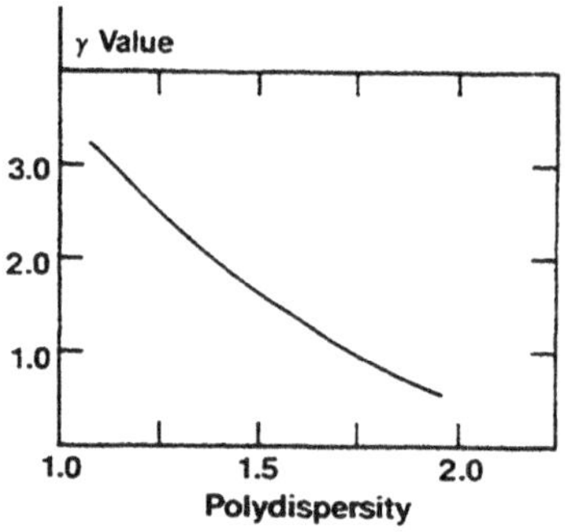

FIGURE 5-1-1. Effect of polydispersity on contrast of polyglycidylmethacrylate negative *e*-beam resists. Fom Ref. 2, courtesy of the *Japanese Journal of Applied Physics*.

Although the random cross-linking is more effective with high-molecular-weight polymers ($M_w > 200{,}000$), the images are more prone to swelling. While lower-molecular-weight resists do not swell as much, they can be an order of magnitude lower in sensitivity. Monodisperse polymers have higher contrast (Fig. 5-1-1) than polydisperse ($M_w/M_n > 2$) polymers.[1,2]

For several types of initial molecular weight distributions, Flory[3] and Charlesby[4] derive formulas that relate the insoluble fraction (g_0) as a function of the cross-linking index, x, defined to be the average number of cross-links per molecule, g,

$$g = 1 - \exp(-xg_0) \tag{5-1-1}$$

The average number of cross-links per molecule is related to the dose,

$$g = 1.0 \times 10^{-6} Q G_x M_w \tag{5-1-2}$$

where Q is the absorbed dose in megarads (10 ergs/g), and G_x is the number of cross-links per 100 eV. For cross-linking, the dose Q (a minimum) decreases with increasing molecular weight. A typical response curve for a negative resist is shown in Fig. 5-1-2.

The minimum gel dose is E_g and the maximum dose is E_1. The contrast (γ) is the slope of the line from E_1 to E_g. At the work dose, E_0, a 50% thinning of the original prebaked film thickness is used to reduce the backscatter contribution at the base of the image. The contrast (slope of the plot in Fig. 5-1-2) is exponentially related to the polydispersity (M_w/M_n) of the polymer,

$$\gamma = 2/E_g \exp(-M_w/M_n) \tag{5-1-3}$$

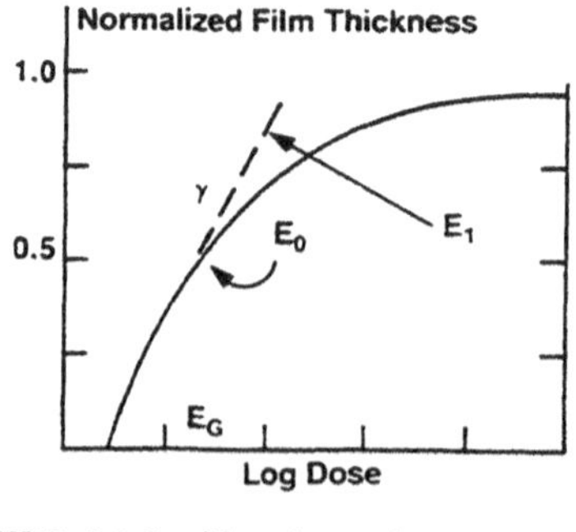

FIGURE 5-1-2. Negative resist contrast curve.

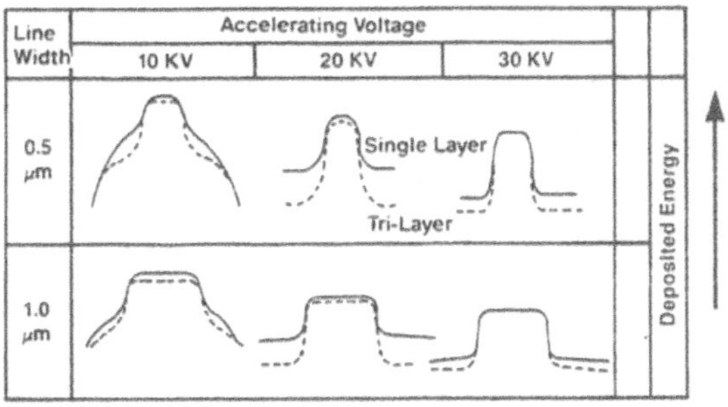

FIGURE 5-1-3. Energy deposited in a negative *e*-beam resist as a single film and in a trilayer stack for various linewidths and exposure voltages. From Ref. 4.

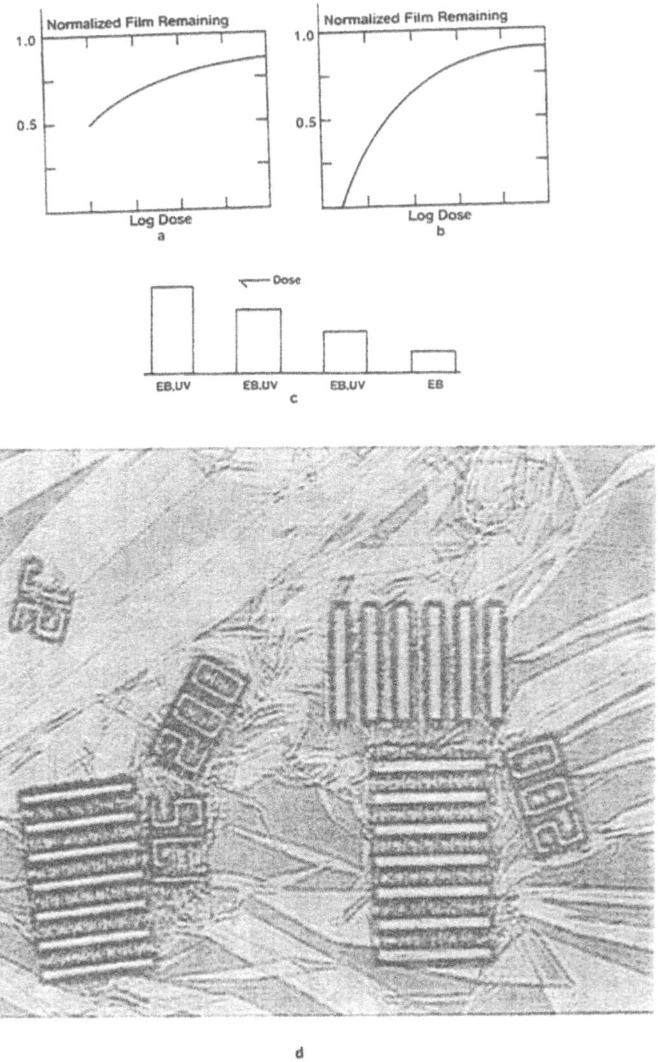

FIGURE 5-1-4. (a) Film remaining in negative resist after topside UV exposure; (b) after *e*-beam exposure; (c) comparison of cross section of negative resists after UV and *e*-beam exposure; (d) optically underexposed negative azide-rubber resist developed to show images float off in developer without sufficient exposure to cross-link at the interface and adhere to the substrate.

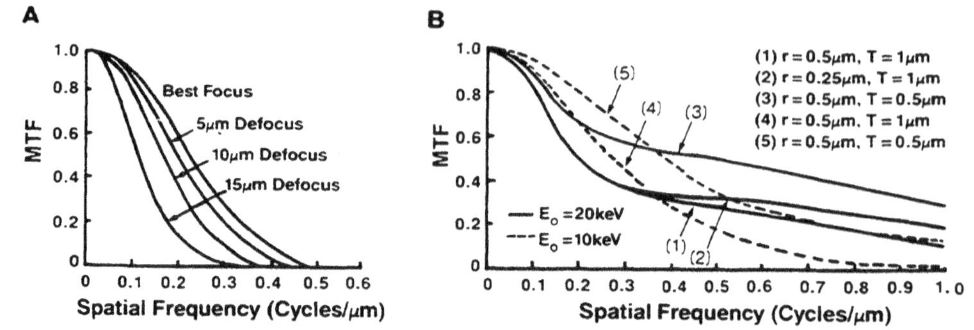

FIGURE 5-1-5. MTF for (A) optical lens exposure and (B) *e*-beam exposure. (0.2 cycle/μm = 2.5 μm linewidth.) From Ref. 5, courtesy of SPSE.

The contrast is also a function of resist thickness and energy (keV) of the beam. The narrower the dispersity (1.0–1.5), the higher is the contrast which reaches a theoretical maximum of 3.5 with a monodisperse polymer.[1]

For films of 1.0-μm thickness, the base width of an image will widen to a trapezoid shape (about 50%) depending on the energy of exposure (Fig. 5-1-3). Thinner films are less prone to backscatter, especially when used as an image layer in a trilayer resist (Fig. 5-1-3). In *e*-beam exposure, the backscatter effect is more significant than in optical exposure. For optical exposure, the last 30% of the film (depending on the optical density) near the substrate may not receive sufficient radiation to become insolubilized (Fig. 5-1-4). Thus, optical films are inherently overexposed to achieve interfacial adhesion (insolubilization). Electron-beam exposure, as well as X-ray, involves backscattered radiation as the dominant form of exposure, since a staircase of resist is left after developing the underexposed resist (Fig. 5-1-4).

In *e*-beam exposure, the backscatter problem is very severe, because the modulation transfer function (MTF) of the negative resist falls off severely between patterns closer than 2.5 μm (Fig. 5-1-5). The backscatter causes the interproximity effect and scumming between exposed lands. In the resist, the lower-energy electrons are more effectively absorbed and less dose is required (Fig. 5-1-6).

Because of the isotropic nature of backscattering, the developer (Fig. 5-1-7) for a negative cross-linking resist encounters the edge of scattered radiation as it dissolves the unexposed resist. Initially, upon development of a positive resist, the highly exposed region of anisotropic radiation is removed inside the image, and the lightly scattered radiation is not developed if the developer is quenched (Fig. 5-1-7).

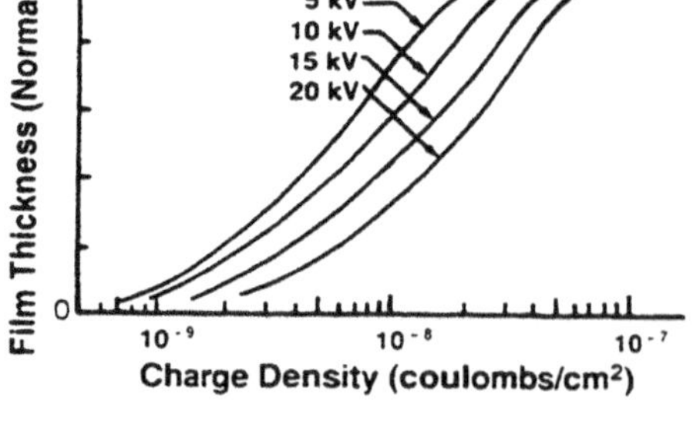

FIGURE 5-1-6. Effect of energy of exposure on contrast and sensitivity of epoxidized butadiene negative *e*-beam resist. From Ref. 76, courtesy of the Electrochemical Society Inc.

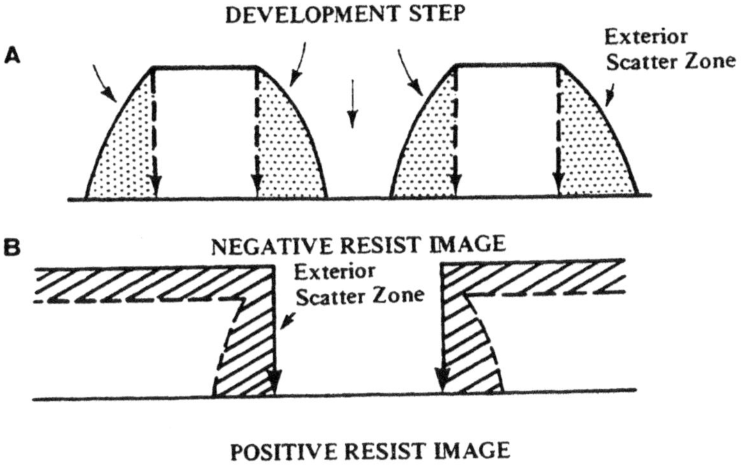

FIGURE 5-1-7. (A) Development of negative cross-linking resist is subtended by the exterior swollen scatter zone. (B) Positive-type resist can be subtended on a time-of-development basis to not reveal exterior scatter zone.

In thick (> 1 μm) negative resists, small unexposed spaces (< 1 μm) may fail to develop (Fig. 5-1-8). The isotropic swelling of the exposed polymer constricts the contiguous unexposed spaces. However, thick films are required for adequate step coverage and for dry etch resistance in silicon device lithography. In thick (> 1.0 μm) negative resists, the resolution is limited in resist images to about 2.0-μm spaces.

If thick (> 0.3 μm) low-T_g polymers (< 50°C) such as polyisobutylenes, polysiloxanes, and polyglycidylmethacrylates are properly exposed, the images readily swell upon development (Fig. 5-1-9). The cross-linked rubbery polymers are still elastic and deform upon slight absorption of the developer. In *e*-beam exposure, closely spaced lines (unexposed spaces) narrow with increasing exposure (Fig. 5-1-10). Gaps also suffer from severe interproximity effects (Fig. 5-1-10). The residues may be removed, if not too severe, by oxygen plasma descum. For most cases, the reduction of proximity effects is done by tool corrections of dose and beam size, or by using thinner negative resist films (< 0.2 μm), such as in multilayer

FIGURE 5-1-8. Swelling of fine exposed lines in negative resist adjacent to exposed lines. (A) Unexposed negative resist swells slightly more than background exposed region. (B) The differential swelling increases. (C) Unexposed resist washes from the surface of line. (D) Internal pressure in the exposed but swollen line causes it to bulge inwards and hinder removel. From Ref. 6, copyright 1967 and reproduced by permission of John Wiley & Sons, Inc.

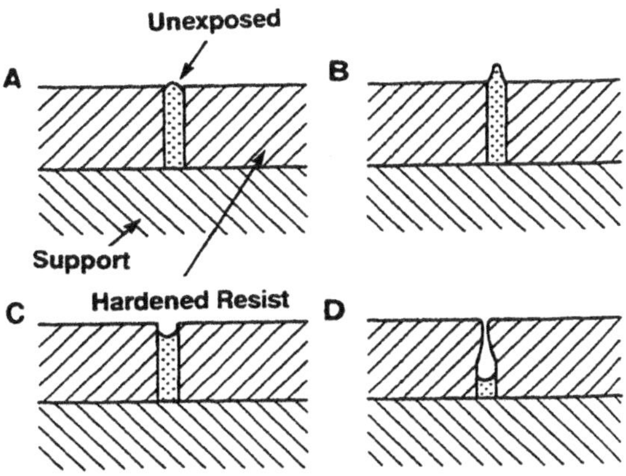

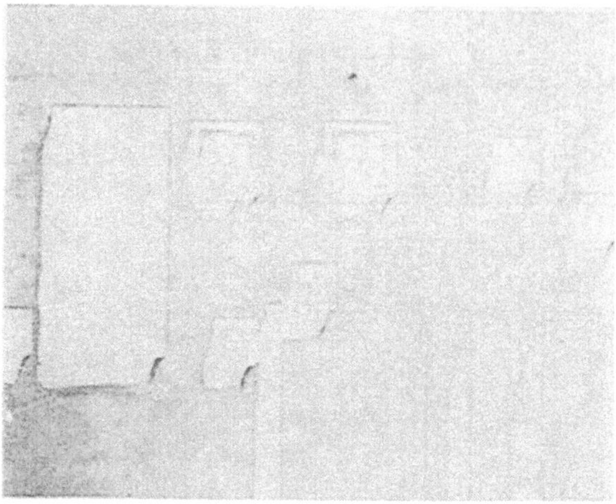

FIGURE 5-1-9. Very swollen images in low-T_g polyisobutene resist (T_g of −35°C).

resists of polysilicone polymers on top of a planarizing layer. With negative resists, the proximity correction is simplified since the developer time dependency as found in positive resist corrections is not a factor.

Directly, a quick assessment of the interproximity effect can be readily made by measuring the image enlargement of an isolated line with increasing dose. The enlargement per dose estimates the growth of a line exposed next to another line (interproximity effect).[6,110] An opposite problem, the lack of backscatter, occurs because of the falloff of the MTF below 2.0 μm (Fig. 5-1-5). Because of the subtended backscatter and the lack thereof of backscatter contributions from neighbors, narrow lines are thinner and require more dose to compensate the loss of vertical dimensions (Fig. 5-1-11). The ineffectiveness of the dose at submicrometer features

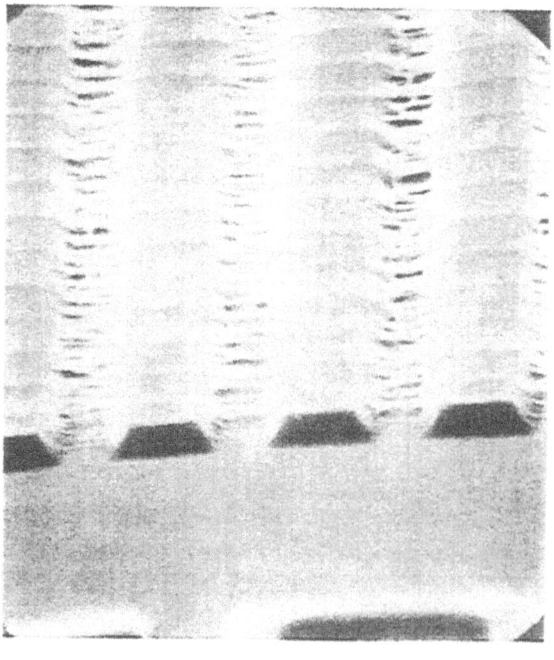

FIGURE 5-1-10. Interproximity effects in polyglycidylmethacrylate *e*-beam resist due to scattering from adjacent exposed lines.

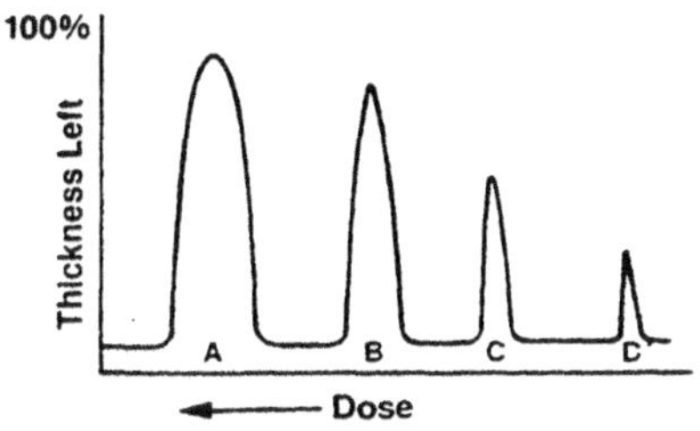

FIGURE 5-1-11. Falloff of thickness remaining and linewidth due to introproximity effect in a negative *e*-beam resist exposed with beam decreasing in width from (A) 2 μm, (B) 1 μm, (C) 0.5 μm, and (D) 0.2 μm.

is known as the intraproximity effect. More dose is needed in the features smaller than 1 μm, and thus the resist sensitivity is overestimated if flood exposure data are used.

In negative resists, the developer process latitude for micrometer geometries is very wide, since the linewidth is independent of develop time (Fig. 5-1-12). No further film loss occurs of the exposed resist. However, linewidths can grow because of postdiffusional cross-linking polymerization of the latent image on the exposed wafer inside the *e*-beam chamber (Fig. 5-1-13) unless quenched by opening the system to air (oxygen). Exposed wafers or masks have to be stored in a vacuum after exposure to be sure that all features have grown to the same size. Epoxy-based (COP) resists postpolymerize but chlorostyrene-based resists (Fig. 5-1-13) do not exhibit the severe linewidth distortion problem.

For negative photoresists, several modes of insolubilization are available. Cross-linking vinyl, epoxy, allyl, and siloxane are the fastest resist systems,

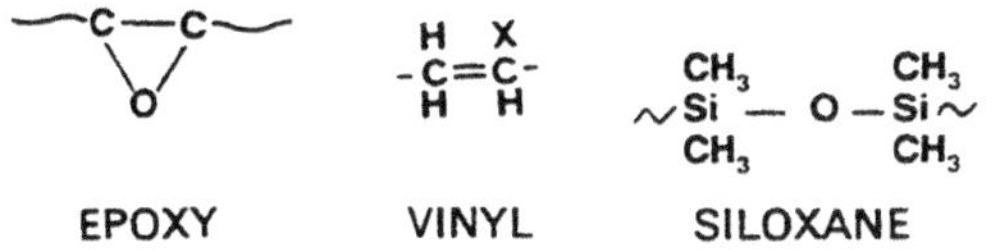

The principal negative resist systems are as follows:

1. Side chain cross-linking of pendant groups, such as epoxy in polyglycidylmethacrylate (PGMA), allyl in polydiallylisophthalate (PDIP), and vinyl ($-C=C-$) in polyvinylcinnamate,

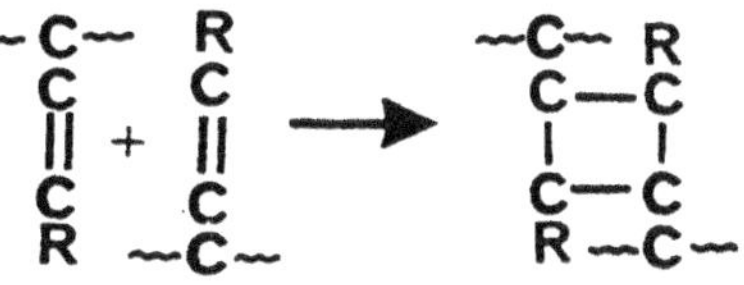

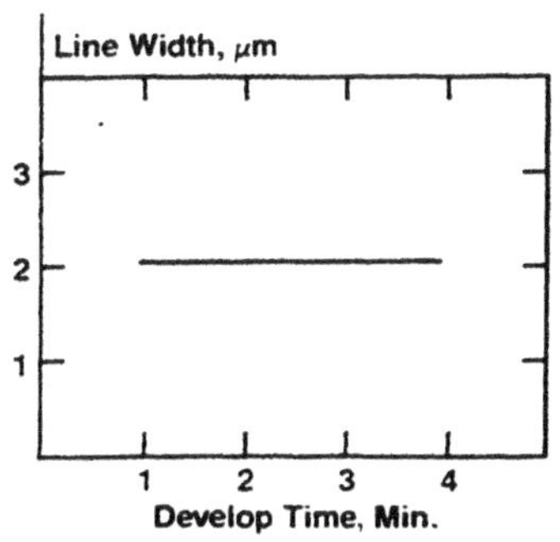

FIGURE 5-1-12. Typical linewidth dependency of development of a negative cross-linking *e*-beam resist. The linewidth is independent of development time.

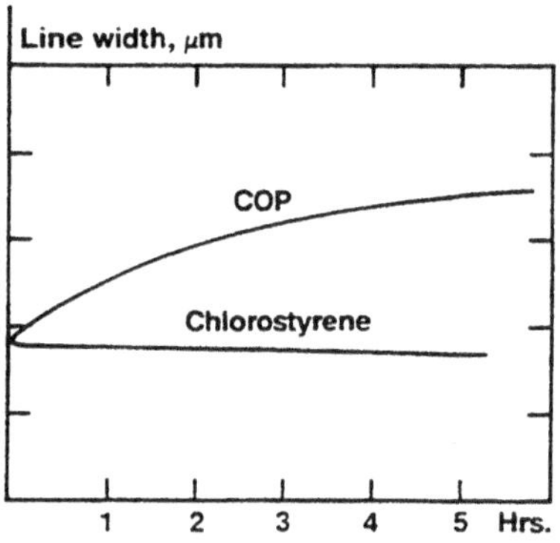

FIGURE 5-1-13. Postcure of negative *e*-beam resists of COP (trademark of AT&T) and chlorostyrene after exposure. Lines can grow with time with some resists after exposure due to post-cross-linking reactions especially in low-T_g polymers with long-lived intermediates. Adapted from Ref. 87, courtesy of AIP.

2. Main chain cross-linking by side chain scission,

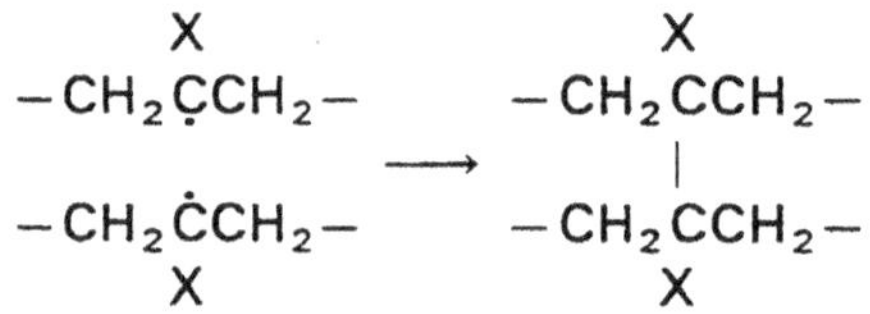

In Table 5-1-1, examples[9] are given of vinyl polymers and their radiation G_x cross-linking yields (number of cross-links per 100 eV absorbed radiation). Because silicon- and metal-containing polymers resist oxygen plasma, they are preferred for use as image layers in bilayer resists. If we assume a certain percentage of cross-links are derived from side chain rupture, the monitoring of side chain products such as H_2 or HCl from polyvinylchloride can give an indication of cross-linking efficiency. Directly, the measure of cross-linking density is difficult because the polymers are insoluble and indirect inferences are derived from swelling measurements and by solid-state NMR (nuclear magnetic resonance). The fastest negative cross-linking resists are the epoxies, with speeds (doses, Q) of 10^{-7} C/cm^2 followed by allyl and siloxane polymers with sensitivities of 1 μC/cm^2.

3. Cross-linking additives such as bisepoxies and bisazides, and divinyl benzene, can insolubilize linear polymers such as styrenes, novolaks, vinyl, allyl, and epoxies. The azide-induced cross-linking of phenolic resins is an important negative

TABLE 5-1-1. Vinyl Negative Resists

Resist	G_x
Polyethylene	1.0
Polyvinyl chloride	0.4
Polystyrene	0.05
Polymethacrylate	0.6
Polyvinyl alcohol	0.4
Polyvinyl acetate	0.5
Polyvinyl methyl ether	0.7
Polydimethyl siloxane	2–6
Polyazidostyrene	2.5
Epoxy	50–200
Polybutadiene	1–2

resist system because the resist can be developed in an aqueous alkaline medium without swelling,

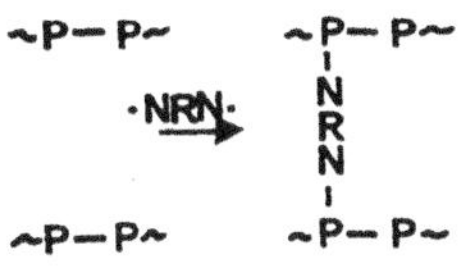

4. Polymerization of monomers in a polymeric matrix such as vinylsilane, vinylcarbazole, and reactive monomers can be achieved with X rays in inert gases[33] or with topcoats to exclude oxygen interference. Electron-beam exposure cannot be used because of outgassing at high vacuum, unless a temporary polymer overcoat is used. Tetrafunctional monomers cross-link more efficiently than monofunctional systems,

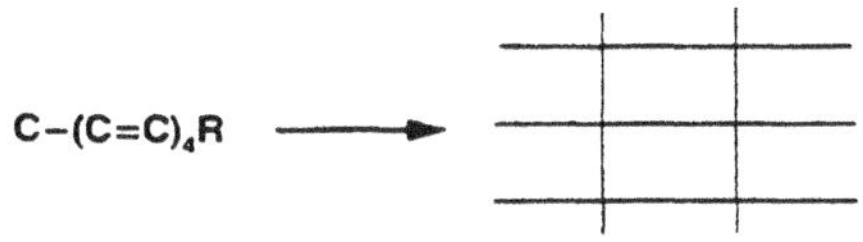

5. Molecular rearrangement to form a less soluble species. Triarylsulfonium salts catalyze the decarbonylation of carbonate, ester, and acid side chains to form systems which are not cross-linked and can be developed in organic or, in some cases, in aqueous media,

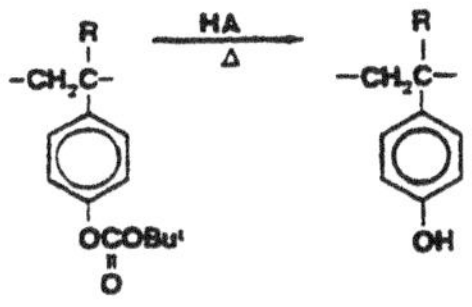

6. Formation or decomposition of charge transfer complexes of polymeric components which develop without swelling,

$$\text{Donor} + \text{acceptor} \rightarrow D^{+} + A^{-}$$

7. Postgrafting or the formation of interpenetrating networks by the reactions of monomers with trapped radicals.

8. Ordered cross-linking in monolayers deposited by the Langmuir method (Chapter 6). There is a potential for exposure amplification (doses $< 1\ \mu C/cm^2$), if the pendant triple bonds are properly orientated.

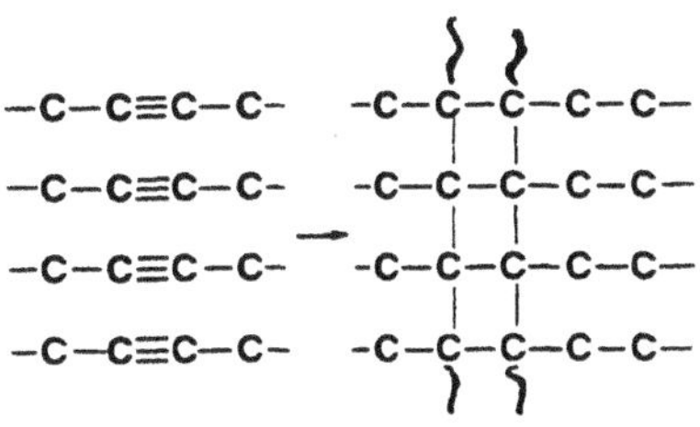

In normal amorphous polymers, random cross-linking with a low level of main

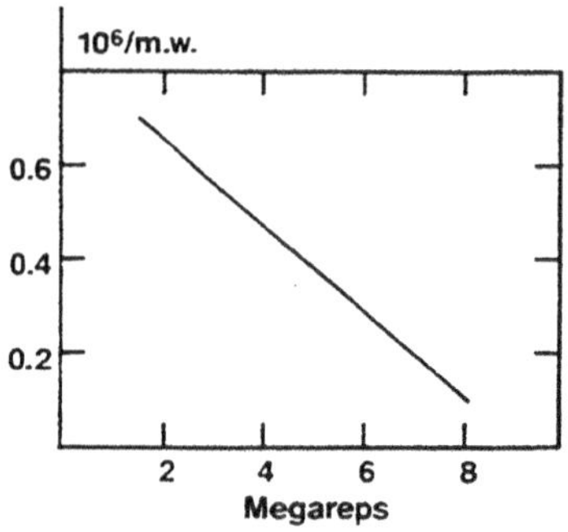

FIGURE 5-1-14. *G*-value determination of cross-linking of polystyrene. From Ref. 8.

chain scission, results in a gradual increase in M_w from the initial molecular weight, M_0,

$$1/M_w = 1/M_0 + G_x kQ/W \qquad (M_w \gg M_0) \qquad Q = k/M_0 G_x \tag{5-1-4}$$

where W is the cross-link density. A plot of inverse final molecular weight versus dose yields the G_x (Fig. 5-1-14). A high M_0 and high G_x favor a minimum dose (Q). The G_x values range from 0.1 for polyethylene to > 10 for polyepoxides. Depending on the lifetime of the intermediate radicals, the presence of O_2 during or after exposure can quench intermediates by radical interception and reduce the amount of cross-linking, by generally forming peroxides and ketones,

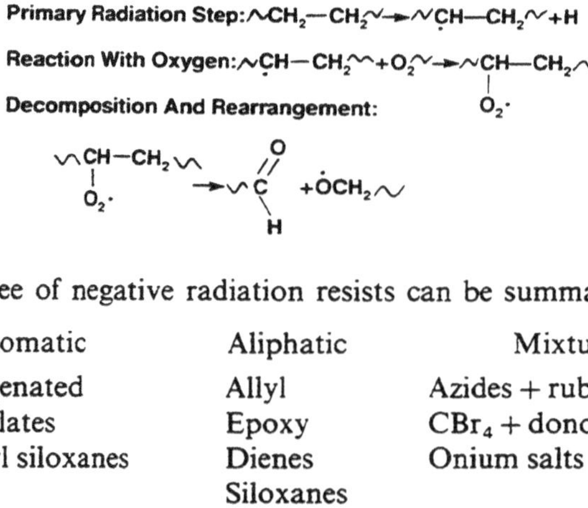

The family tree of negative radiation resists can be summarized as follows:

Aromatic	Aliphatic	Mixtures
Halogenated	Allyl	Azides + rubber
Phthalates	Epoxy	CBr_4 + donor
Phenyl siloxanes	Dienes	Onium salts + epoxies
	Siloxanes	
	Vinyl	

Negative radiation resists are required to have some resistance to reactive ion beams. Since the erosion rate is proportional to molecular empirical size (C_nH_n), aryl polymers are more resistant than aliphatic polymers. Thus, for single-film resists for silicon processing, the current trend is to synthesize negative resists from styrene or even vinylnaphthalene to provide etch resistance in submicrometer-thick film. Silicon-containing resists are popular as bilayer masks. Epoxy-onium-based resists show little swelling and have high sensitivity.

In the next section, we will examine the negative resists, exposable by X rays, ion and electron beams.

5-2. X-RAY RESISTS

X rays offer the promise of higher resolution than optical or *e*-beam exposure, since back-reflected photoelectrons have a range of only 1000 Å.[10,35] Figure 5-2-1 shows the contrast of KTFR (azide rubber) in response to exposure by X rays, *e*-beams, ion beams, and photons. Ion beams gave the highest contrast, followed by X rays. After all the commercial photoresists were examined and found to have low sensitivity ($>150\ \mathrm{mJ/cm^2}$), the first fast X-ray negative resist (25–$75\ \mathrm{mJ/cm^2}$) was reported by Taylor and co-workers based on chloropropylacrylate cross-linking side chains.[11,12] Later, commercial *e*-beam epoxy and allyl resists were explored. Since resolution was not of initial concern (3- to 10-μm linewidths), low-T_g polymers were used. The resists, e.g., polyglycidylmethacrylate-ethylacrylate (COP), swelled and the resolution was limited to about 3 μm. For any resolution, very thin films ($<0.5\ \mu$m) had to be used in multilayer resist configurations. A later development was the incorporation of monomers for dry plasma development (Chapter 10).

The early and the continuing goal for X-ray resists was high sensitivity, because conventional sources are weak and require several minutes to deliver a $100\ \mathrm{mJ/cm^2}$ dose. Two basic means of increasing the sensitivity are to increase the absorption of X rays with high-atomic-number species, and to tune the X-ray emission source to the absorbing elements in the resist. Foremost is to increase X-ray absorption, which follows a Beer's law-type dependence on density (ρ), thickness (l), and mass absorption coefficient (μ_m), according to

$$I = I_0 \exp(-\mu_m \rho l) \tag{5-2-1}$$

where I_0 and I are the intensities before and after passage through the absorbing layer. To optimize the absorption, and minimize I, one must increase the density and μ_m of the film. Since μ_m for a polymer (μ_{mp}) is given by

$$\mu_m = \sum A_i \mu_{mi} \Big/ \sum A_i \tag{5-2-2}$$

where A_i and μ_{mi} are the atomic weights and mass absorption coefficients, respectively, for the i elements in the polymer. The mass absorption of polymers, μ_{mp}, is

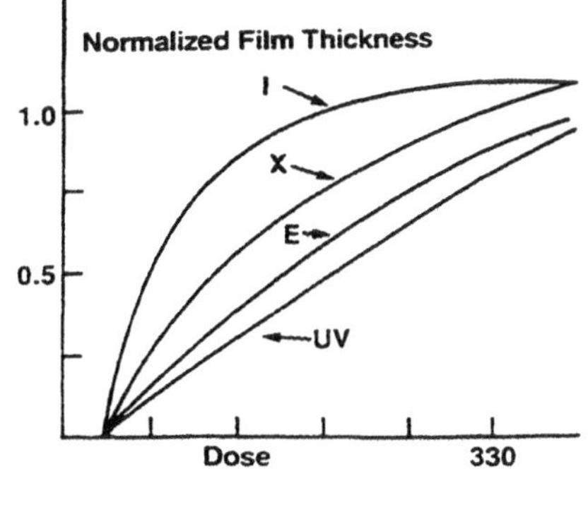

FIGURE 5-2-1. Contrast curves for Kodak KTFR negative resist exposed to I, ion beam of H; X, X rays; E, electron beam; and UV, ultraviolet, 435 nm.

maximized by the incorporation of high weight percentages of absorbing elements ($Z > 25$). The elemental absorption μ_{mi} is given by

$$\mu_{mi} = C\lambda^n$$

where C and n are constants, dependent on the element and the type of absorption event ($K, L, M,\ldots$). The introduction of heavy elements has been one basic approach. Workers at Bell used the Fe[14] in polyvinylferrocene at 155 mJ/cm^2 and the Cl[11,12] in 2,3-dichloro-1-propyl acrylate to achieve a sensitivity of 15 mJ/cm^2. Following these discoveries, many workers studied (Table 5-2-1) the introduction of chlorine and fluorine groups. Exposures of < 60 sec/wafer could be achieved at doses < 20 mJ/cm^2. The most practical resists that evolved are polychloromethylstyrene and a family of polychlorinated aromatic polymers. If post-UV cured, the resist images of polychlorostyrenes are stable at 300°C and offer excellent reactive ion etch resistance. However, the sensitivity is traded for resolution (Table 5-2-2). Even though sensitivities to X rays have been enhanced by an order of magnitude versus the nonhalogenated control, the contrasts are low and most resists are used at 50% thinning.

For some other radical-based resists, in some cases, the sensitivity of the resist to O_2 during exposure went unnoticed. The true X-ray speed of resists has to be corrected for the oxygen effect (quenching) and the presence of secondary emission from the mask.[35] Polyvinylalcohol (water removable) topcoats are now used to suppress the oxygen effect.[187,191]

While the introduction of heavier atoms and the tuning of the spectral source have been basic approaches, the introduction of highly sensitive ($G_x > 10$) epoxy or

TABLE 5-2-1. Halogenated X-Ray Negative Resists

	mJ/cm^2	γ	Reference
Acrylate			
Poly-2,3-dichloro-1-propyl acrylate (DCP)	5	1.1	11
DCP + polyglycidyl methacrylate-ethyl acrylate (COP)	2	0.6	15
DCP + vinyl carbazole	4.5	0.5	16
DCP + allyl acrylate	1.5	0.6	13
DCP + acrylsiloxane	3.0	0.6	17
Poly-2,3-dibromopropyl acrylate (PBrPA)	500	1.0	18
PBrPA-GMA	400	1.0	18
Chloroethers			
Poly-2-chloroethyl vinyl ether (CEVE)	180	1.5	19
CEVE-vinyloxyethyl acrylate	18	1.1	19
CEVE-isobutyl vinyl ether	72	1.1	19
Teflon-propyl vinyl ether	20	0.9	2
Styrenes			
Polychloromethylated	8, 19	1.1	21, 48
	15	1.5	22, 23
Iodinated	10	2.3	24
Tetrathiafulvalene CBr_4	50	2.3	25
Heavy atoms			
Polychloroprene	11	0.5	11
Poly-2,3-epithiopropyl MMA	80	1.3	26
Barium acrylate	16	1.0	27

TABLE 5-2-2. PCMS X-Ray Performance Summary[a]

	Low M_w (30k)	Medium M_w (200k)	High M_w (440k)
Sensitivity, $D_g^{1/2}$ (mJ/cm^2)	468	50	20
Contrast, γ	1.72	1.36	1.15
Resolution (μm)	0.5	0.75	1.0
Dry etch resistance	$\geqslant$AZ-1350	$\geqslant$AZ-1350	$\geqslant$AZ-1350
Resist profile thermal stability (°C)	$\geqslant$300	$\geqslant$300	$\geqslant$300

[a] From Ref. 48.

allyl groups is another classic means of sensitivity enhancement (Table 5-2-3). However, the contrast of these resists is low and they are not very suitable as single-layer resists for silicon processing.

For the best resolution using an image layer in a bilayer stack, the successful exposure of chloracrylate resists, impregnated with oxygen plasma-resistant silicone or aryl monomers, led to a whole series of dry-developed X-ray resists. The unexposed region (Fig. 5-2-2) is heated to volatilize the monomer, prior to oxygen dry develop. During exposure, a critical balance of O_2 level,[68] monomer loading, and the adjustment of O_2 plasma erodability are necessary to subdue the 30% exposed film. The oxygen susceptibility of the reaction is balanced by the inclusion of a trace of O_2 in the N_2 ambient exposure atmosphere to balance the internal O_2 dissolved in the film diffusing into the exposed region (as in DuPont Riston, Fig. 4-3-10). A critical level of oxygen is necessary to quench the postpolymerization just after X-ray exposure.

A second reason for the oxygen effect has been proposed by Okada and co-

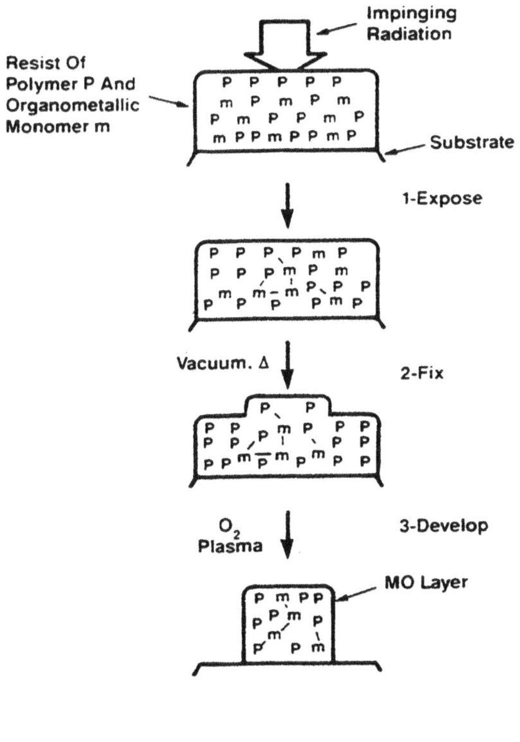

FIGURE 5-2-2. Basic concept of dry developing a polymer by polymerizing a volatile oxygen plasma-resistant monomer in a host polymer followed by heating to sublime the additive from the unexposed region. Subsequently, the image is formed by oxygen plasma etching of the unexposed region. From Ref. 15, reproduced by permission of *Solid State Technology*, published by Technical Publishing, a company of Dun and Bradstreet.

TABLE 5-2-3. Conventional X-Ray Negative Resists

Resist	mJ/cm^2	γ	Reference
PGMA	50	1	7, 28, 29
COP	5, 116, 185	0.8	14, 30, 31
PGMA-EA-MMA	25	0.8	30, 32
PMMA-allyl MA	20	1.3	31
SEL-N-PGMA-ester	10	0.8	23
Polybutadiene	30	1.3	14
Epoxidized polybutadiene	1, 2, 5	0.6	14, 19
Polydiallylisophthalate	14	1.2	14

workers.[7,185] The electrons ejected from the gold mask cause secondary ionization of air (O_2) above the resist (Fig. 5-2-3). The excited oxygen ions diffuse, recombine, and further dissociate by secondary electron capture to emit secondaries which diffuse and quench the resist reaction (to the depth of interdiffusion). Increasing the flow of air between the mask to wafer gap, increased the background exposure, as predicted by a lateral diffusion model. Exposure in nitrogen,[184] applying a topcoat of polymer on the resist or overcoating the masks with polyimide, lowered the gold emission and reduced the "oxygen effect."

On X-ray steppers, exposure of resists should be done with awareness of the postreactivity of some materials and accordingly compensate with a postcure period, to ensure that all chips "enlarge" to the same dimensions.[162] The radicals formed by exposure of epoxy (PGMA) have been shown by electron spin resonance to be long lived (several hours).[34] To inhibit postpolymerization, free radical quenchers (DPPH) were added, but at a fivefold loss of sensitivity. Among the

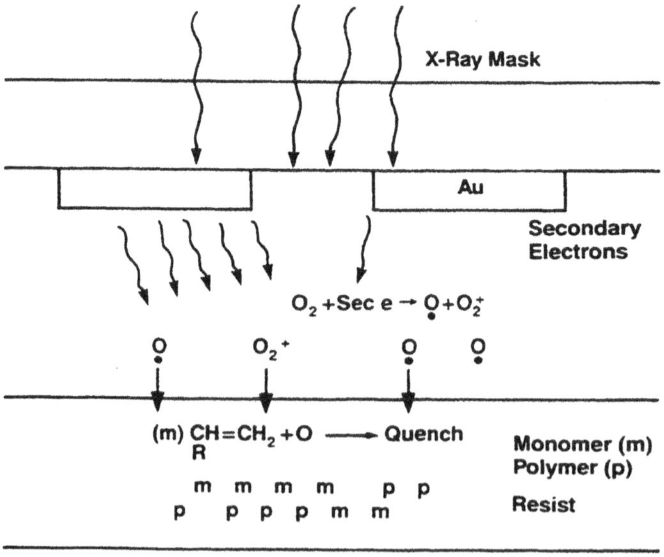

FIGURE 5-2-3. Oxygen effect for X-ray exposure caused by secondary ionization of oxygen gas in the proximity gap between the mask and the resist-coated wafer. The excited gas diffuses to the resist surface and reacts with (quenches) the monomer polymerization inside the resist film. From Ref. 7, reproduced by permission of the Electrochemical Society Inc.

susceptible resists are epoxidized butadiene, SEL-N, COP, CER, and cyclized butadiene.

A final new type of resist for X-ray exposure is the recent extension of the monomer image amplification of PMMA. By postpolymerization treatment of the X-ray-exposed resist with (negative tone) an acrylic acid monomer, fast speeds of 1 mJ/cm^2[150,155] have been obtained. However, the monomer grows isotropically on the image surface and the image reversal occurs upon development in nonpolar solvents.

5-3. ION-BEAM RESISTS

With exposure energies of 100–200 keV, ion beams deposit most of their energy in a 1- to 2-μm-thick organic resist film (Fig. 5-3-1). The deposited energy is transferred by collisions into secondary electrons. This translates to very little secondary electron emission with a range of only 100 Å and no serious backscatter (no proximity) effects. In fact, excellent straight walls at high contrast (Fig. 5-2-1) can be obtained, even with the conventional KTFR bisazide rubber resist which swells in the developer. Figure 5-3-2 shows the backscatter effect in *e*-beam-exposed KTFR and the lack thereof in proton-beam exposure.

Ion beam studies of negative resists have been performed at Bell[36,42,43] and Hughes[37] laboratories. Working with conventional H^+beams, Brault and Miller[37] found that the sensitivities of negative *e*-beam resists were proportionally a factor of ten lower in dose (Fig. 5-3-3) than required for *e*-beam exposure. For monodisperse polystyrene, the minimum dose is proportional to the increasing molecular weight. The contrast increased to 2.3 for monodisperse polystyrene; however, this is lower than the contrast of 3.5 later found[38,39,159] for *e*-beam-exposed monodisperse polystyrene.

For ion-beam resists, high contrast is apparently not necessary because of the short range of secondaries. Workers at Bell utilized mega-ions (1.5 MeV) H^+, He^+, and O^+ to study linear energy transfer (LET) in resists (Fig. 5-3-4). The contrast increases with the lower energy H^+ and He^+ ions, and further increases upon O^+ exposure. The sensitivity of plain novolak resin to electrons[40] requires a dose of about 10^{-1} C/cm^2 but only 10^{-4} C/cm^2 of O_2^+ ions. Inorganic resists of Ge-Se were found to require about 10^{13} $ions/cm^2$ or 10^{-6} C/cm^2 for 100-Å films.[41] Impregnating the first 100-Å depth of PMMA with Ga ions or Si ions allows the exposed region to withstand oxygen plasma and be developed as a top surface

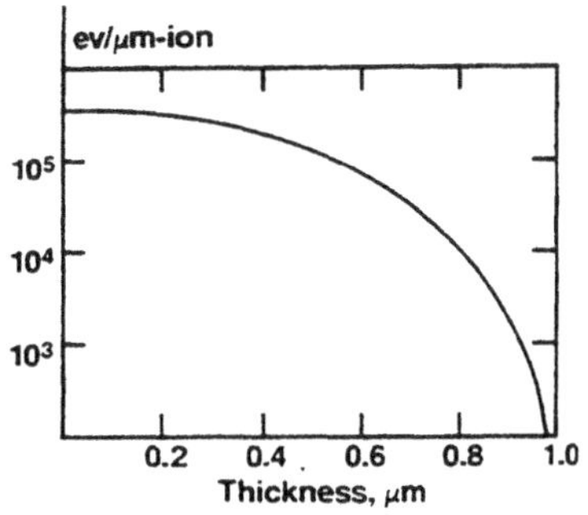

FIGURE 5-3-1. Energy deposited in PMMA by a 100-keV proton beam as a function of resist depth.

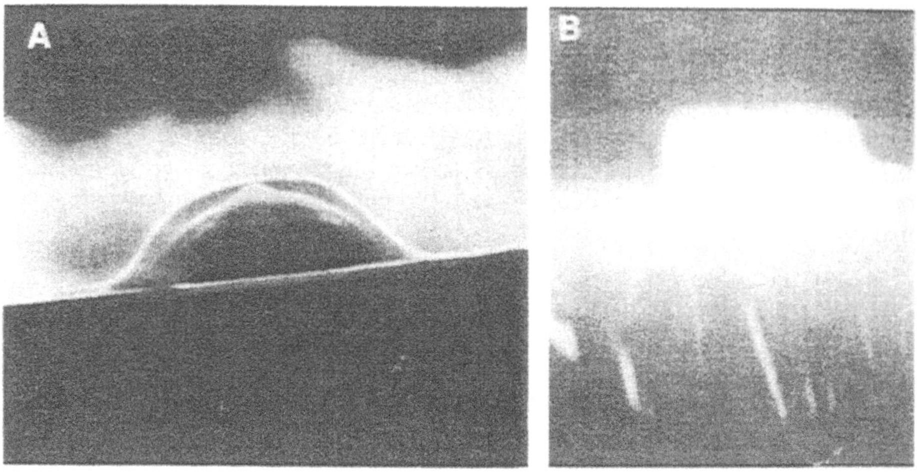

FIGURE 5-3-2. KTFR images produced by exposure with (A) 25-keV electrons and (B) 125-keV protons.

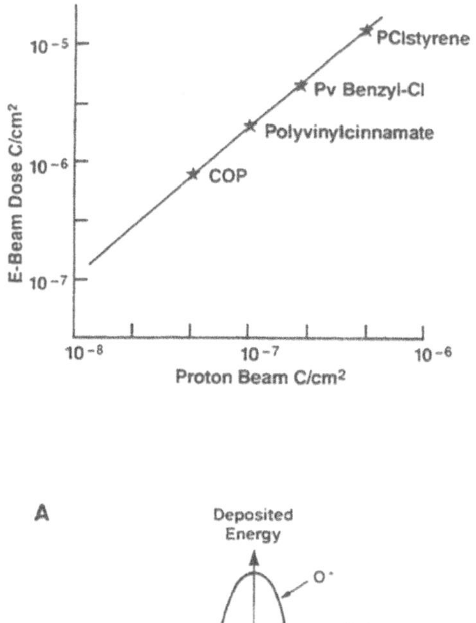

FIGURE 5-3-3. Sensitivity correlation between *e*-beam and ion-beam dose for negative resists. From Ref. 37, courtesy of SPE.

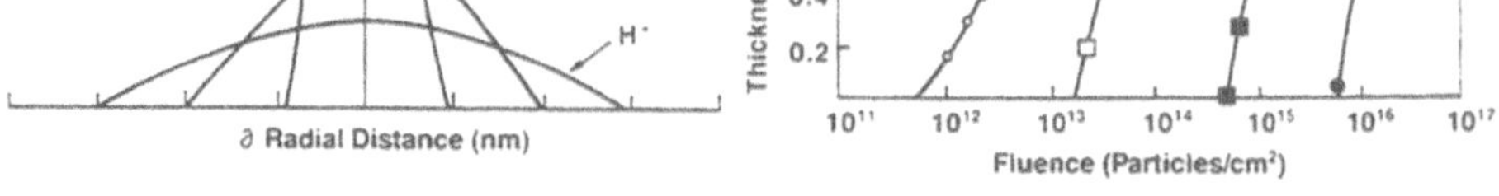

FIGURE 5-3-4. (A) Profile of energy deposition by ions in resist. From Ref. 36, courtesy of AIP. (B) Comparison of exposure of novolak resin by ions of 1.5-MeV and 20-keV electrons. From Ref. 44, reprinted by permission of *Solid State Technology*, published by Technical Publishing, a company of Dun and Bradstreet.

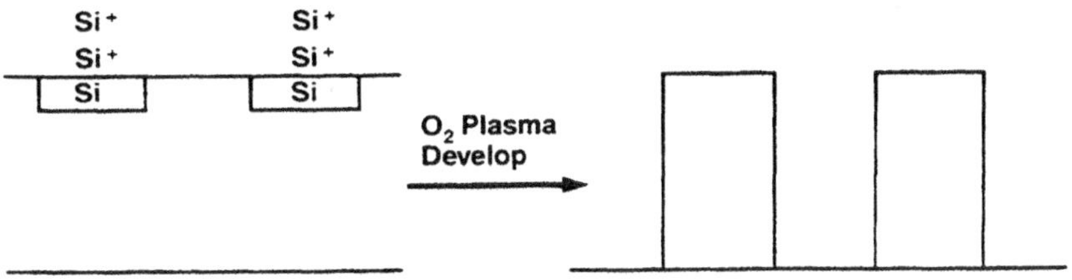

FIGURE 5-3-5. Dry-develop ion-beam-exposed (Si^+) resist. The top layer of the resist is exposed to ions followed by oxygen plasma develop. From Ref. 45, courtesy of AIP.

bilayer resist (Fig. 5-3-5).[42] In addition, impregnating a silicon wafer directly (without a resist cover) with Ga ions permits selective KOH etching.

Ion-beam lithography offers proximity-free printing of negative resists. Further (Chapter 9) technical developments in stabilized masks, better alignment, and fast steppers will be required for submicrometer production lithography.

5-4. ELECTRON-BEAM RESISTS

Initially, because of their wide abundance from commercial sources, more polymers were evaluated as negative resists than as positive systems. When it was found that negative cross-linking resists exhibited all of the proximity effects, many hopes for direct silicon wafer write in 1-μm-thick films were abandoned. Since the backscattered electrons exit at an average angle of 45°, the base of the image is wider than at the top surface (Fig. 5-4-1). The range of backscattered electrons is comparable to the feature size.

The major families of negative *e*-beam resists are primarily the vinyl carbon backbone systems with reactive side chains (X),

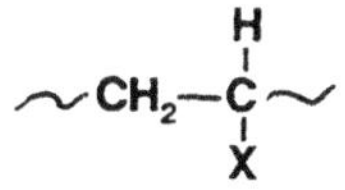

where X is chlorostyrene, epoxy, and allyl.

In the α position, H is a preferred substituent, but it can also be a methyl group, if the other side chain cross-linking dominates over main chain scission. Some of the simple early negative resists are shown in Table 5-4-1.

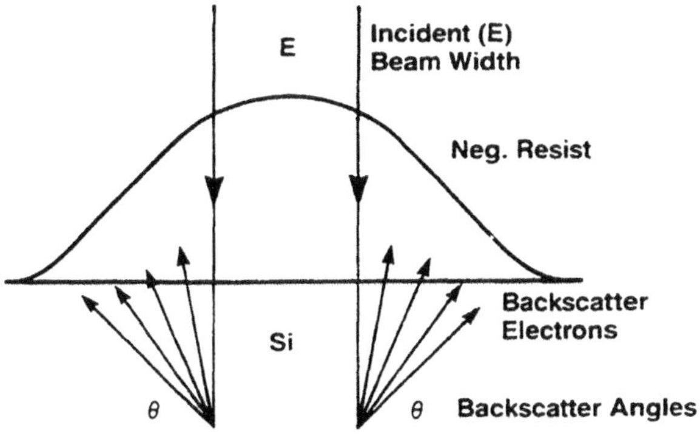

FIGURE 5-4-1. Backscattered electrons broaden negative resist images.

TABLE 5-4-1. Early Negative Electron-Beam Resists

	Structure
Methyl vinyl ketone (PMVK)	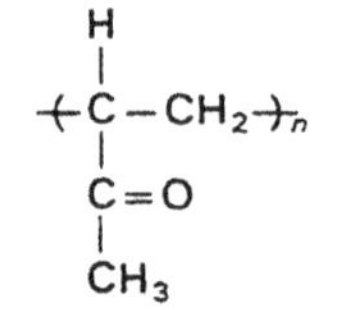
Vinyl methyl ether (PVME)	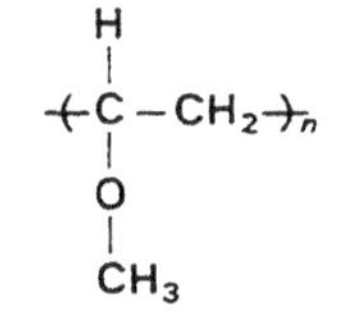
Vinyl isopropyl ether (PVIE)	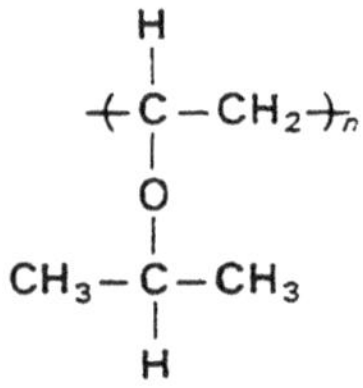
Vinyl-*t*-butyl ether (PVTBE)	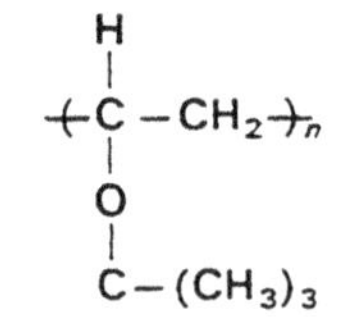
Vinyl acetate (PVA)	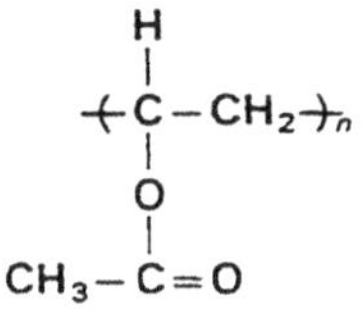
Vinyl chloride-acetate-azide (PVCAAz)	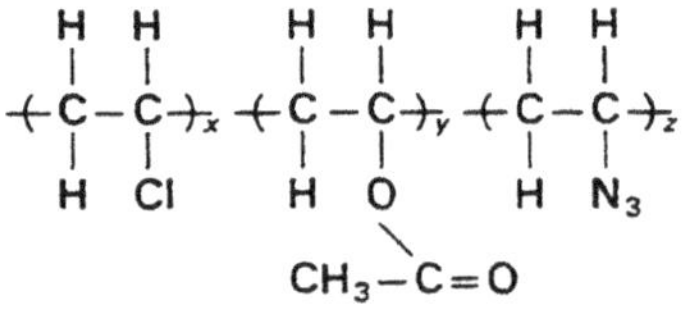

In the absence of O_2 during exposure, the epoxy and allyl side chains form long-lived radicals, which are subject to postpolymerization (Fig. 5-1-13). For the vinyl-based resists, the longer the side chain, the lower is the T_g and the greater is the susceptibility to swelling. Most of the alkyl-based negative resists are poor reactive ion etch masks, but they are acceptable for wet etching of chrome mask of 10× single segments of 20- to 30-μm dimensions. The recent trend is to use plasma-assisted dry etching and, therefore, more resistant materials are sought. For direct silicon etching, an etch resistance equivalent to phenolic positive based resists is required. Polystyrene and its chlorinated derivatives form the basis of a good reactive ion-resistant *e*-beam resist.

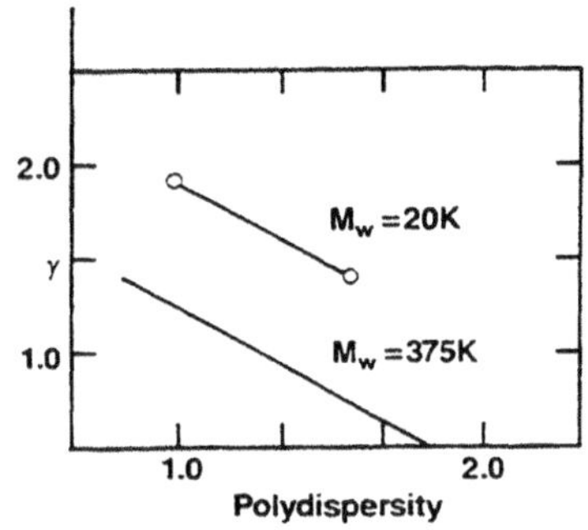

FIGURE 5-4-2. Contrast versus polydispersity for polychlorostyrene resist. From Ref. 49, courtesy of AIP.

5-4-1. Styrene-Based Resists

The most resistant layers for reactive ion etching have been derived from chlorinated polystyrenes (A) and poly-α-methylstyrenes (B),[67,158,166,183]

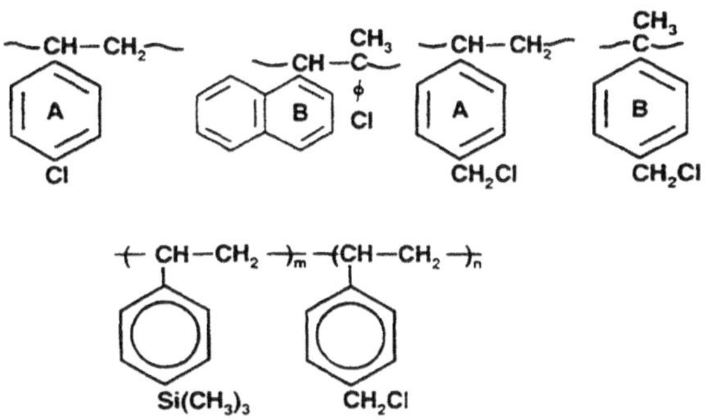

The chlorine group is introduced, by postchlorination or by chloromethylation of narrow-dispersity polystyrene, poly-α-methylstyrene, and poly-*p*-methylstyrene.[166,169] Chlorostyrene or chloromethylstyrene can also be directly polymerized or copolymerized with other aryl polymers, such as siloxanes,[181,190] and polyvinylnaphthalene. Likewise, brominated,[189] iodinated,[176] and fluorinated[174] polystyrenes have also been investigated (see Table 5-4-3). Iodostyrene is also a sufficiently oxygen plasma-resistant (Chapter 12) image layer in bilayer resist processing since a nonvolatile I_2O_5 forms in the oxygen plasma.

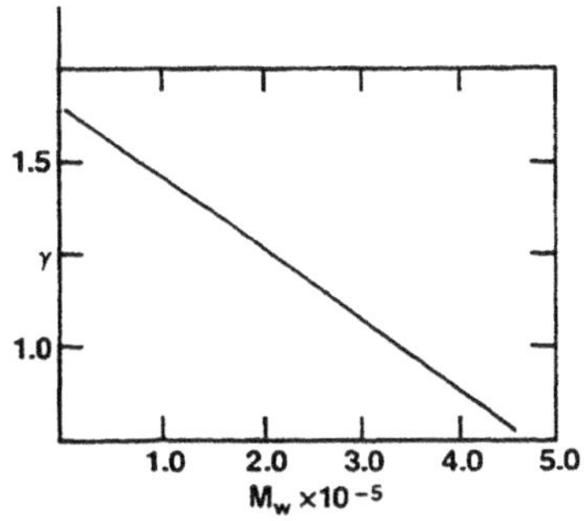

FIGURE 5-4-3. Contrast versus M_w for PCMS. From Ref. 49, courtesy of AIP.

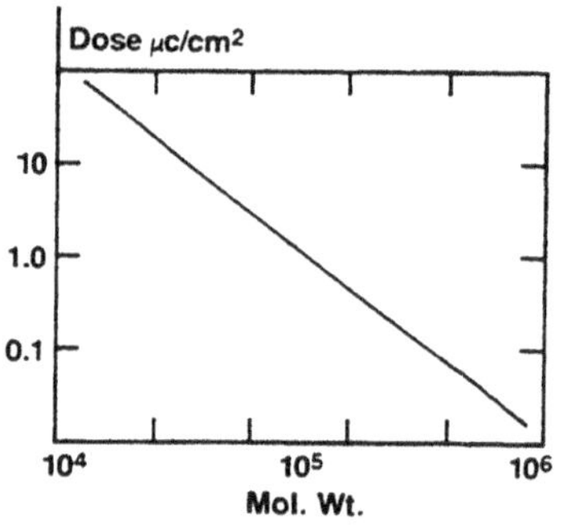

FIGURE 5-4-4. Sensitivity of PCMS as a function of molecular weight. From Ref. 49, courtesy of AIP.

In polystyrene-based resists, the contrast increases with monodispersity [Fig. 5-4-2; Eq. (5-1-3)]. As expected [Eq. (5-1-3)], the contrast also increases with decreasing molecular weight (Fig. 5-4-3). However, the lower sensitivity of the lower-molecular-weight chlorostyrenes is often traded for higher resolution (contrast) (Fig. 5-4-4).

Even though the base styrene polymer has a T_g of 100°C, which is 50°C higher than the epoxy-based resists, swelling of images upon development is still evident. Swelling may be suppressed by a good choice of developer.[189] As such, developers which are poor thermodynamic solvents do not appreciably swell the resist polymer. Kamosida and co-workers[50] examined the swelling of PCMS (chloropolymethylated styrene) from the viewpoint of the H-bonding solubility-interaction parameter (σ) of Flory. Swelling limits pitches to about 3.0 μm. Interrelated to swelling (Chapter 10) is the isotropic volume parameter, ϕ_2, which is given by the swelling ratio ($1/\phi_2$). For thermodynamically good solvents, the swelling ratio is the highest, and ranges from 1.0 to 2.0 (total swelling). Good solvents for styrenes, with a solubility parameter of 9.0, such as xylene, swell the most (Table 5-4-2). Poor solvents minimize swelling.[50,186,188] While the swelling ratio agrees quite favorably with the solubility parameter of the developer for poorly

TABLE 5-4-2. Swelling Solvents for PCMS and Resolution[a]

Solvent	σ (cal/cm^3)$^{1/2}$	$(1/\phi_2)^{1/3}$ (swelling ratio)
Xylene	8.6	1.8
Cellosolve acetate	8.7	1.4
Cyclohexanone	9.9	1.7
Octane	7.6	1.1[b]

Resolution of PCMS and developer

Developer	σ	H-bond	Observed swelling ratio	Calculated swelling ratio	Resolution (μm)[c]		
					1L 1S	2L 1S	3L 1S
m-Xylene	8.7	Poor	1.75	1.69	1.2	3.4	5.0
2-Ethoxy ethyl acetate	8.7	Moderate	1.47	1.67	1.0	2.6	3.0

[a] Modified from Ref. 50.
[b] Did not develop images.
[c] L 1S, equal line and space; 2L 1S, exposed lines 2× the space (unexposed).

TABLE 5-4-3. Polystyrene (PS) Negative Electron-Beam Resists

Resist	$M_w \times 10^{-4}$	D^a	$\mu C/cm^2$	γ	Reference
PS	10	1.0	100	2.5	54
PS	100	1.1	9	2.2	54
PS	10	1.0	1000	2.3	39
PS	43	1.0	300	2.2	54
PS	63	1.0	1000	3.5	38
PS + azides	102	1.2	1.3	3.0	55
Polyiodostyrene	24	2.5	10	1.6	56
Polydiallylisophthalate	11	1.8	5	1.4	57
Poly-Cl-PS	25	2.0	1.9	1.2	58
Poly-Cl-PS	70	2.0	3	3.5	59
Poly-4Cl-PS	15	1.9	3	1.5	60
Poly-4Cl-PS	65	1.5	3	1.5	61
Polyvinylbenzyl Cl	9.6	1.1	2.5	2.2	54
Poly-CMS	270	3.0	2	1.0	62, 167
Poly-CMS	0.7	1.1	39	3.0	62, 63
Poly-CMS	20	1.3	7	2.1	49
Poly-α-MCMS	19	1.2	33	3.0	64
Poly-α-NCMS	8.3	1.9	42	4.1	64
Poly-α-MCMS	188	1.2	3	2.5	64
PS-GMA	22	1.9	3	0.9	56, 65, 151
Polyvinylnaphthalene (PVN)	19	1.6	200	1.5	66
Poly-α-MCMS-PVN	80	1.3	10	2.1	67

[a] *D*, polydispersity.

H-bonded solvents, it fails to simulate the behavior of moderately H-bonded solvents with PCMS. Better resolution is obtained for PCMS with 2-ethoxyethyl acetate as a developer than xylene (Table 5-4-2) which has the same solubility parameter.

Since fewer cross-links are present, higher-molecular-weight polymers exhibit more swelling than lower-molecular-weight polymers. The trade-off of molecular weight and sensitivity for resolution is evident in polychlorostyrenes. Consequently, the choice of developer depends on the swelling ratio, the molecular weight and polydispersity, and the T_g of the resist. For the chlorinated styrenes, the best contrast (3–4) and submicrometer resolution were exhibited by narrower-dispersity polychlorinated-*p*-methylstyrene (Fig. 5-4-5).[169]

The polychlorostyrenes do not exhibit postimage (Fig. 5-1-13) growth due to the short lifetime of the radicals and the immobility of the intermediates in the matrix of the high-T_g (115°C) polymer.[34,156,192] In order to suppress swelling, higher-T_g (170°C) polymethylstyrenes have been chlorinated and used as resists. Higher resolution was obtained but at lower sensitivity.[51,156] Otherwise, the linewidth enlargement of chloromethylated styrene with high molecular weight, forces a trade-off of sensitivity for resolution. A very-low-molecular-weight (30,000) PCMS at a dose of 200 $\mu C/cm^2$ grows in width for an isolated 1.0-μm line, from 0.6 to 1.2 μm. Because of the proximity effect, a 1.0-μm line next to a large exposed pad (Fig. 5-4-6) grows at a faster slope, at a factor of 2 lower dose. The highest contrast of 3–4 is reported for *para*-chlorostyrene[38] and poly-α-MCS.[64] Nanometer structures can be fabricated in poly-α-MeClMS, if a thin membrane of silicon is used as the substrate to eliminate the silicon backscatter.[74]

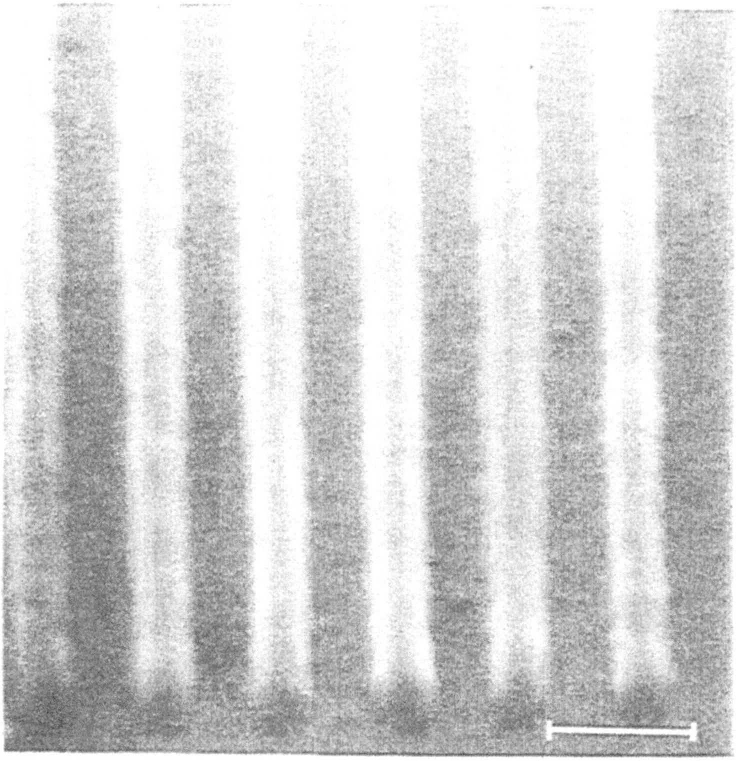

FIGURE 5-4-5. High-resolution images of 0.5 μm by 0.5 μm (2 $\mu C/cm^2$ dose) in chlorinated poly-*p*-methylstyrene negative resist. From M. Hartney, R. Tarascon, and A. Novembre, *J. Vac. Sci. Technol.* **B3**, 30 (1985), courtesy of AIP and Dr. M. Hartney, AT&T. Scale bar: 1 μm.

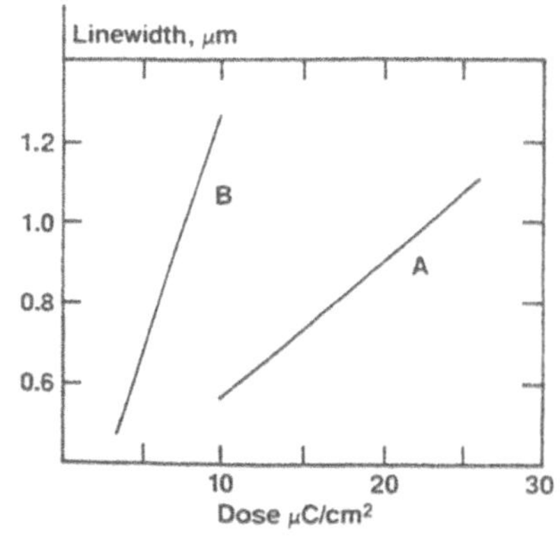

FIGURE 5-4-6. Growth of negative resist line in PCMS of (B) 1-μm width (next to an exposed line, interproximity effect) and (A) isolated line (interproximity effect). From Ref. 52, courtesy of AIP.

The enhanced sensitivity of chlorostyrenes (10 $\mu C/cm^2$ versus 100 $\mu C/cm^2$ for unsubstituted styrenes) is due to the lower ionization potential of halogenated styrenes, and the efficient electron capture of secondary electrons.(53,158) In addition, exposure of a heated resist increases sensitivity.(157) Mechanistically, the dissociation of chlorine from the aryl ring, as well as α-H elimination from the backbone, or from chlorines on the main chain, leads to a bi- or monoradical cross-linking,(54,192)

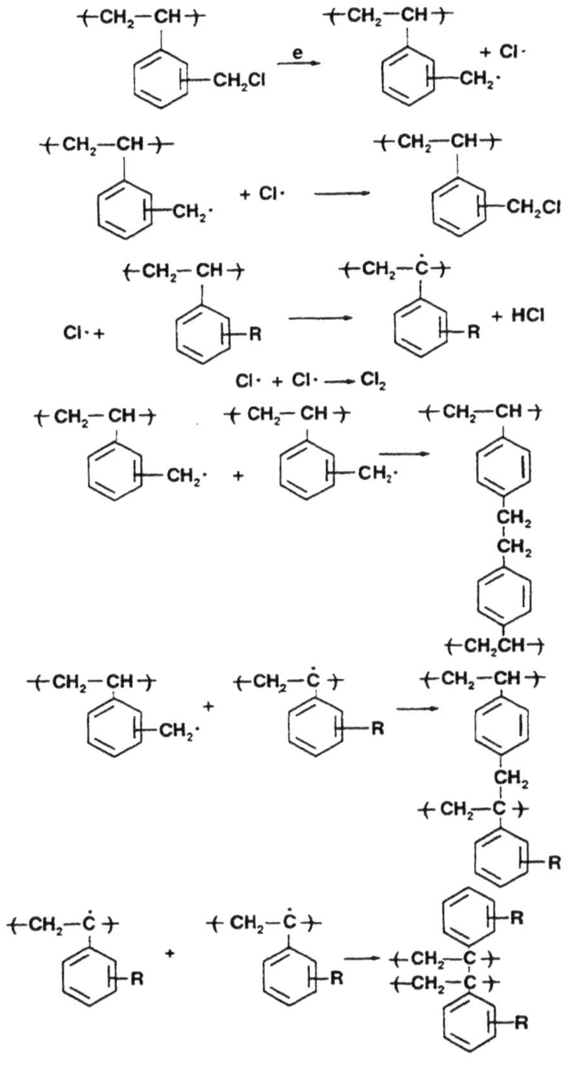

In other polystyrene-based negative resists, the simple addition of bisazides to monodisperse polystyrene, produced both high sensitivity (1 $\mu C/cm^2$) and a high contrast of 3.(55) In the films, bisazide compounds distribute to a great extent in the vicinity of the substrate during the prebake process. An excess of azide was found to decrease the sensitivity of the resist at the interface. Therefore, the backscattered electrons do not cause the isotropic cross-linking and near-vertical walls at 0.5-μm lines and spaces were obtained.(193)

Besides the chlorostyrenes, early workers sought to increase polystyrene sensitivity by copolymerization with butadiene,

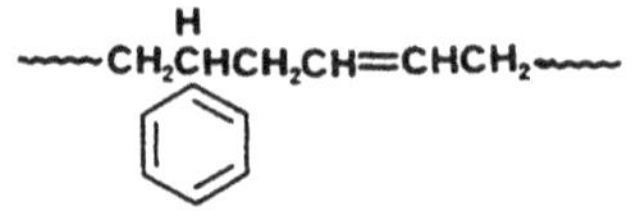

The base sensitivity was enhanced from 100 $\mu C/cm^2$ to 10 $\mu C/cm^2$.[69,71] With the high molecular weights, the further addition of epoxy monomers increased the speed to 5 $\mu C/cm^2$.[70] Copolymerization of polystyrene with glycidylmethacrylate led to 5 $\mu C/cm^2$ sensitivity,[77]

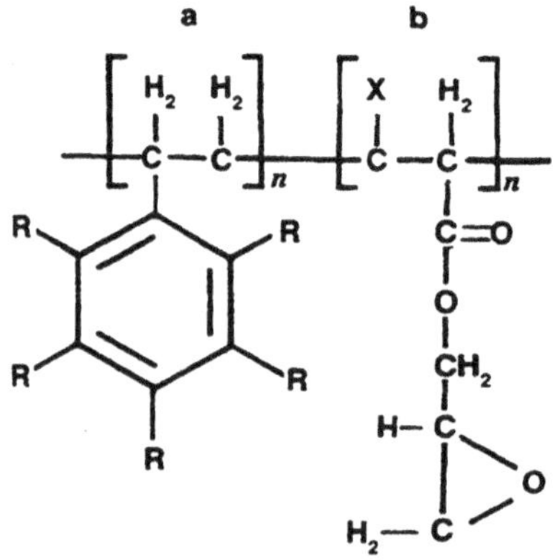

Later, copolymerization with allylmethacrylate gave 1 $\mu C/cm^2$ sensitivity,[73,179]

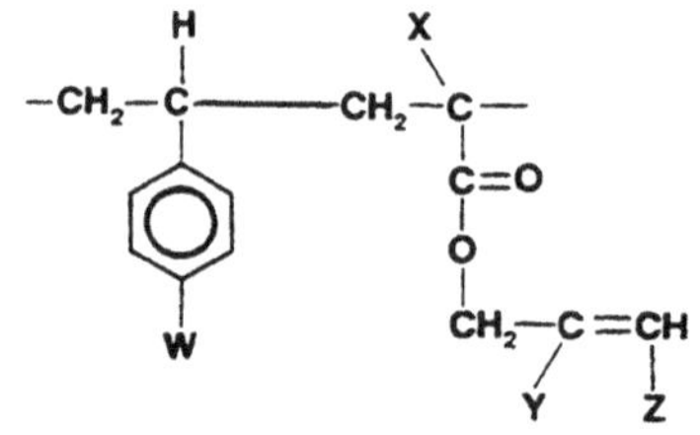

in which structures, the first monomeric unit is styrene (where W = H) or its substituted derivatives, in which W can be F, Cl, Br, I, NO_2, OCH_3, and the like. The second monomeric unit is a methacrylic, or an acrylic-type structure, in which X can be H, CH_3, or Cl; Y can be either H or CH_3; and Z can be H or CH_3.

The chlorostyrenes can be made more thermally stable, up to 300°C, by post-deep-UV hardening.[48] Incorporation of Sn, Si,[154,160] or metal chlorostyrenes provides a resist suitable for layer resist masking, or for O_2 plasma etchable systems.[75,175,177]

5-4-2. Epoxy-Based Resists

Epoxy-based resists are very fast resists owing to the ring strain of the epoxy group and the ease of ring opening to addition cross-linking,

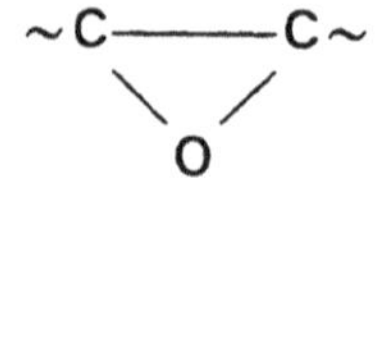

From gamma radiolysis studies, G_x values of 200 have been estimated.[76] The epoxy group can be introduced in the main chain as the epoxidized rubber of polyisoprene,

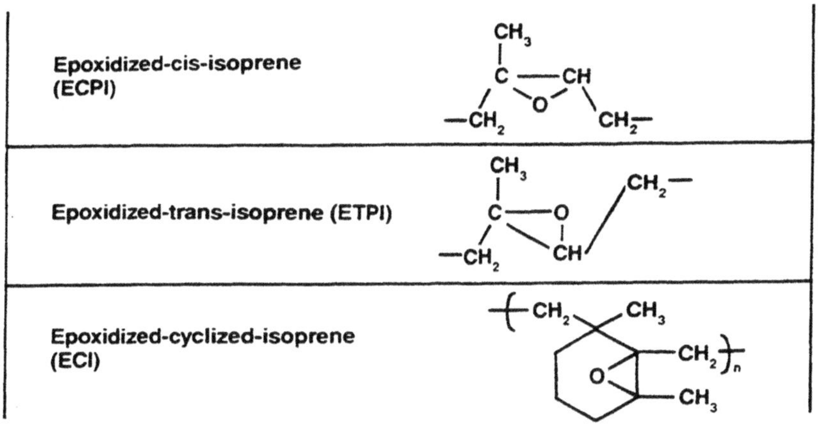

or into butadiene,

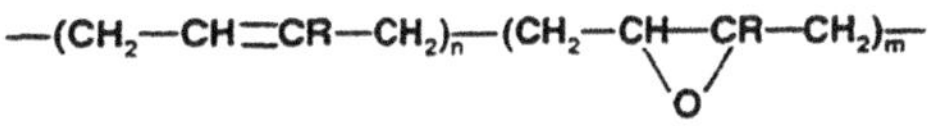

or as a side chain in glycidylmethacrylate copolymer (COP),

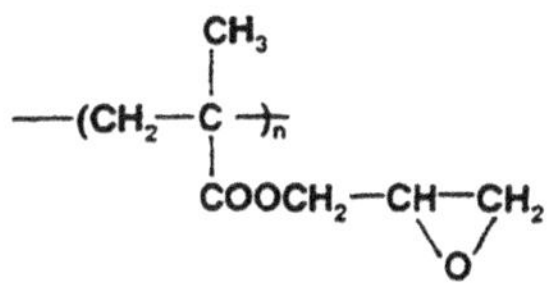

The glycidyl acrylate and methacrylate (GMA) side chain group has been conveniently introduced as a vinyl monomer,

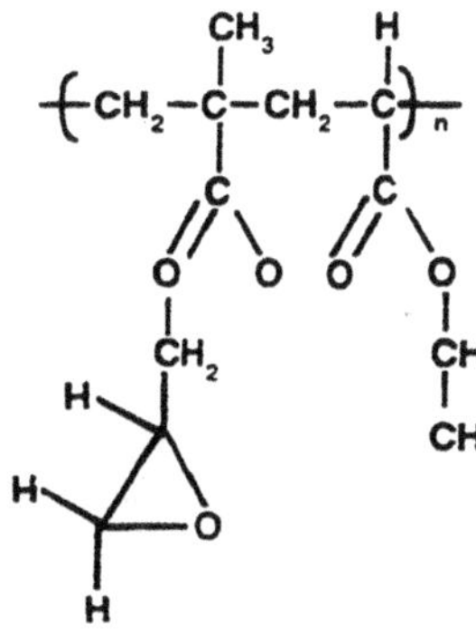

attached to an epoxy (glycidyl) side chain, generally as a copolymer, GMC,[72]

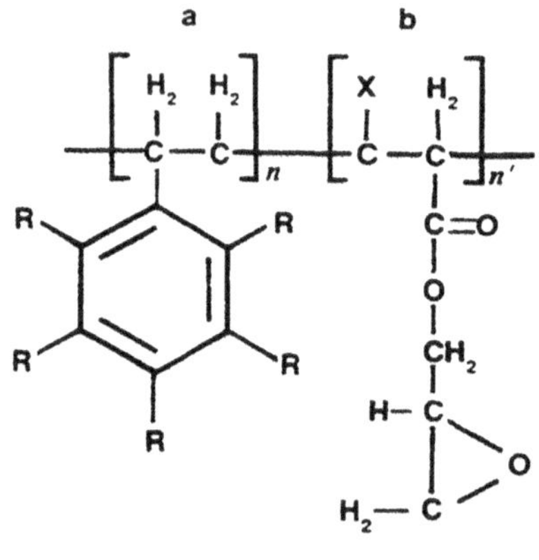

in which X = H, CH_3, C_2H_5, or other alkyl groups containing from 1 to 4 carbons; at least one of the R substituents is a grouping represented as $-R'Z$ in which the $R'Z$ grouping need not be identical and in which Z = halogen atom (F, Cl, Br, I); R' = any grouping containing from 1 to 4 atoms bridging the Z halogen to the aromatic ring. In general, any other R substituents are hydrogen, methyl or higher alkyl homologues. Copolymers with ethyl acrylate have formed the basis of a popular 10× photomask-making resist called COP.[77]

Since pure GMA has a poor shelf life, attempts were made to stabilize the resist by copolymerization to form a tetrapolymer,[78]

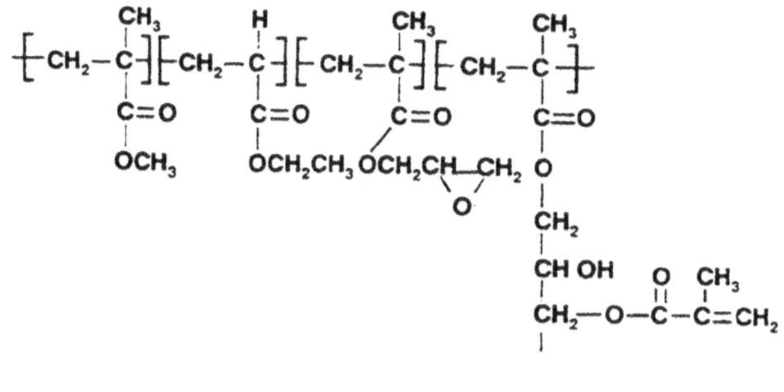

If stabilized with antioxidants, pure GMA can be used as a resist[74] and the contrast increases in the more monodisperse polymers (Fig. 5-4-7).[79] For improved reactive dry etch resistance, grafting of glycidyl groups onto a novolak resin overcomes the poor plasma resistance of PGMA but at the expense of lower sensitivity, 5–10 $\mu C/cm^2$ (Fig. 5-4-7).[170] For image reversal into a positive resist, thermally cross-linked GMA can be exposed by deep UV or higher *e*-beam (10–20 $\mu C/cm^2$) and developed to form a positive image.[80]

Typical exposure doses for epoxy-based resists are 1 $\mu C/cm^2$ with a 50% film loss. Oxygen descum is necessary to remove the protruding feet (Fig. 5-4-8).

Requiring doses of only 0.1 $\mu C/cm^2$, epoxidized butadienes are the fastest *e*-beam resists ever reported (Fig. 5-4-9).[79] The *G* values based on the infrared loss

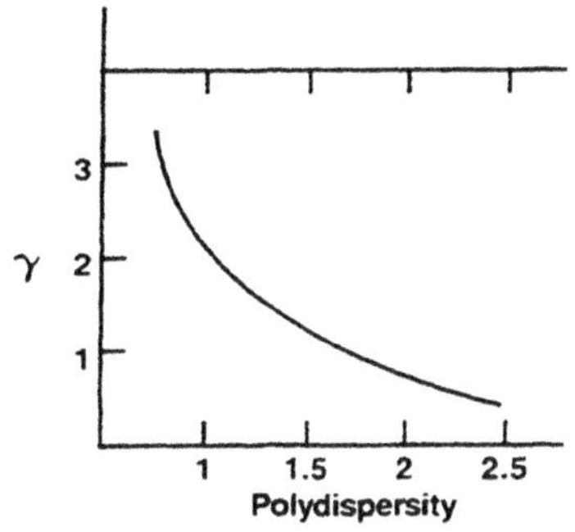

FIGURE 5-4-7. Contrast versus polydispersity for polyglycidylmethacrylate. From Ref. 79, courtesy of the *Japanese Journal of Applied Physics*.

of epoxy and the degree of swelling have been estimated to be as high as 200.[79] Cross-linking involves ring opening polymerization,[18]

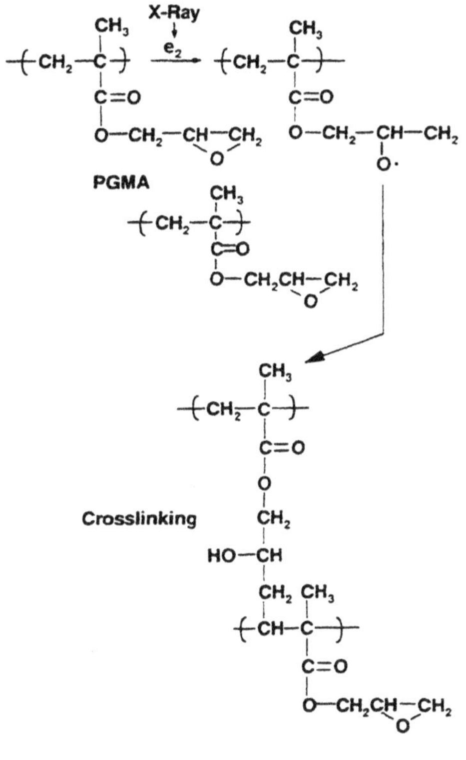

If an interchain reaction is considered (Fig. 5-4-10), at least ten epoxy groups cross-polymerize per scission of one epoxy chain. If the epoxy groups are arranged in an ordered overlap, large amplification may occur. In the opening of the epoxy ring, the high heat of polymerization of 33 kcal/mole contributes to the thermodynamic sensitivity factor. Even higher speeds (10^{-9}–10^{-10} C/cm^2) could potentially be obtained, by using onium salts to open the epoxy groups.[82–86] The onium salts form Lewis acids under the action of light or electrons.[159] The Lewis acids catalyze epoxy ring opening and induce image amplification (lower doses).[194] Thus, two forms of image amplification are possible with epoxy or thiaurane (episulfide) rings[81,165].

FIGURE 5-4-8. Residual feet (scumming) in epoxy negative resist.

1. Epoxy ring opening and self additions
2. Lewis acid cationic ring opening and addition as depicted by Dubois *et al.*,[81]

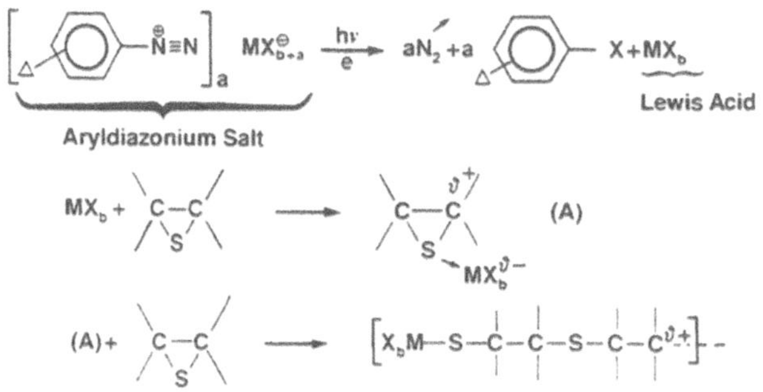

Inside a vacuum, images of exposed resists based on epoxy, allyl, and vinyl polymers grow isotropically unless exposure to air quenches the reaction. For long exposures, such as 1–2 hr/mask, wafers have to be postconditioned in vacuum in order for the linewidths to enlarge to a consistent size. Even different chips on a mask may have different size, due to different sequences of beam writing. Free radical trapping agents such as diphenylpicrylhydrazil (DPPH) slow down the rate of postpolymerization, but with a five- to tenfold loss in sensitivity (Fig. 5-4-11).[87,88]

5-4-3. Allyl-Based Resists

Vinyl allyl groups,

$$-(C{=}C)-C-$$

are synthesized by the polymerization of groups, such as allylmethacrylate, or by

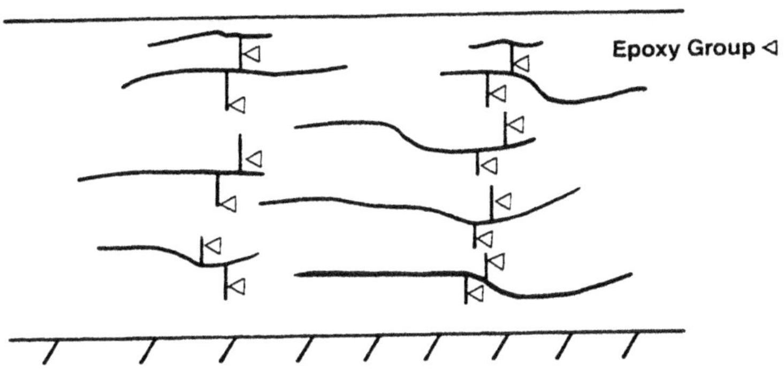

FIGURE 5-4-9. Schematic of ordered cross-linking in epoxy resists promoted by strategic overlap of side chains.

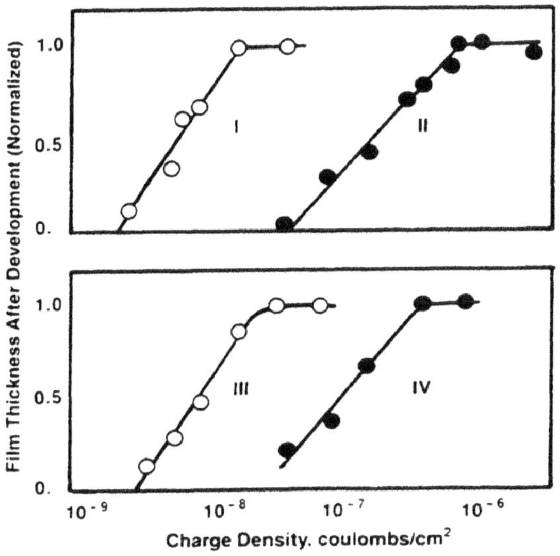

FIGURE 5-4-10. Exposure contrast curves of epoxidized negative resists. I, epoxidized polybutadiene; II, polybutadiene; III, epoxidized polyisoprene; IV, polyisoprene. From Ref. 76, reproduced by permission of the Electrochemical Society Inc.

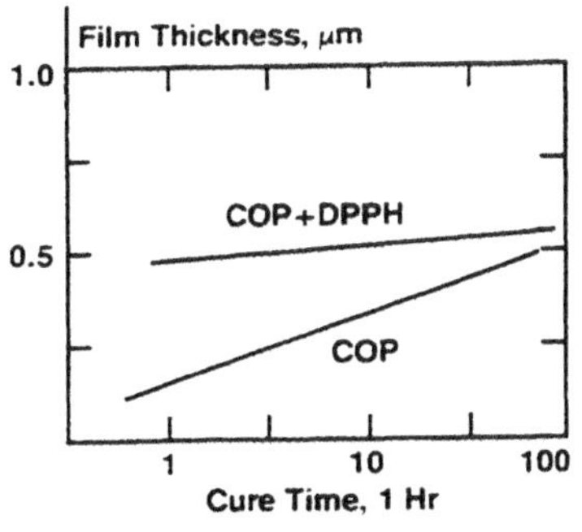

FIGURE 5-4-11. Growth of linewidth of *e*-beam-exposed COP negative *e*-beam resist with and without addition of DPPH. From Ref. 87, courtesy of AIP.

Table 5-4-4. Allyl Negative Electron-Beam Resists

Resist	$\mu C/cm^2$	Reference
Diallylmelamine	10	89
Allylisocyanate	5	90
Allylmethacrylate	1	91, 92, 172
Allylmaleate	0.1	93, 94
Diallylisophthalate	10	95
Styrene-allyl MA	5	96, 97

the grafting of allyl alcohol onto polymer chains.[178] The net result is a polymer with dangling double bonds (Table 5-4-4) which can undergo intermolecular free radical addition cross-linking at low doses. Since the resists are subject to postpolymerization, have a low T_g, and require O_2 plasma descumming, their practical use remains in the fabrication of 10× photomasks by *e*-beam exposure.

5-4-4. Charge Transfer Resists

These resists offer a new mechanism of insolubilization with a radical departure from the traditional cross-linked swollen negative resists. Upon exposure of molecules with low ionization, polar salts are formed which do not swell upon development. For example, polyvinylpyridine salts have been used at 20 $\mu C/cm^2$[92,98,185,186,194] wherein the *e*-beam exposure induces a transfer from pyridinium cation to the anion, to form iodine radicals,

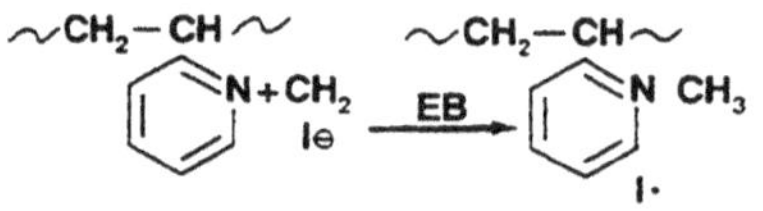

Water develops the unexposed regions. Earlier, workers at Horizons[149] demonstrated negative photoresists could be formed from amine-CBr_4 charge-transfer resists. Recent findings have shown that organic sulfur compounds of tetrathiafulvalene (donor) form salts with bromides, which become insoluble upon exposure,[99]

$$\pi(\text{Donor})_{1-n}\,(\pi\text{Donor}^+)_n\,(X^-)_n \xrightarrow{\text{e-beam}} \pi\text{-Donor} + X_2\uparrow$$

For a polymeric homologue, a side chain of an electron donor of

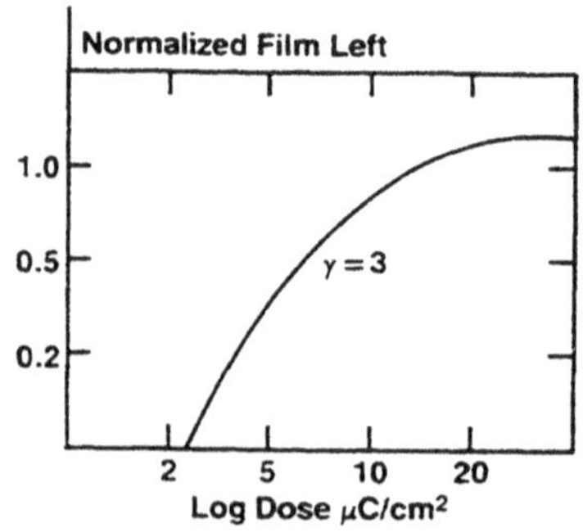

FIGURE 5-4-12. PSTTF contrast curve. From Ref. 100, courtesy of AIP.

tetrathiafulvalene is attached onto polyvinylbenzyl chloride to complex with CBr_4, which results in a moderately sensitive resist, with a contrast of 2.3 at a dose of 2.3 $\mu C/cm^2$ (Fig. 5-4-12),[100]

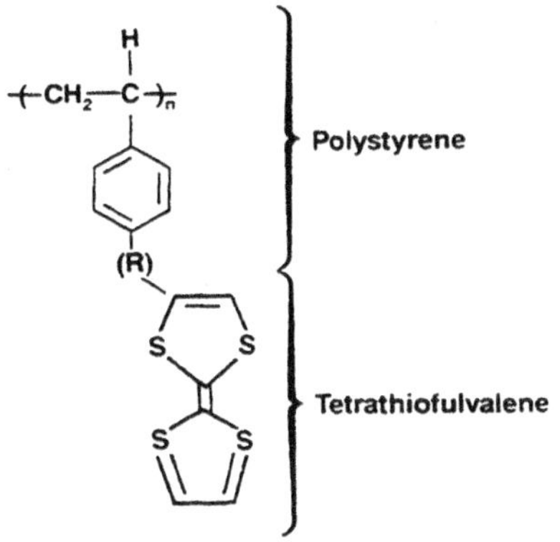

$$\text{PSTTF} + CBr_4 \xrightarrow{e} \text{PS[TTF]}^+ \ Br^{\ominus}$$

While the contrast is good, it is not outstanding for these new types of negative resist. However, although swelling is not present, as in polychloromethylstyrenes, proximity corrections are required for *e*-beam exposure. Thus, even with a new image formation mechanism and the absence of swelling, the backscattered effects of proximity cannot be overcome in a 1.0-μm-thick single-film resist. Multilayer resists with thinner imaging layers are required.

5-4-5. Silicone Negative Resists

The incorporation of the silicon atom into a carbon backbone via polydimethylsiloxane (PDMS) produced a fast resist of 1 $\mu C/cm^2$ (G_x of 6) and the first good plasma-resistant O_2 bilayer resist. Presumably, the high sensitivity of PDMS,

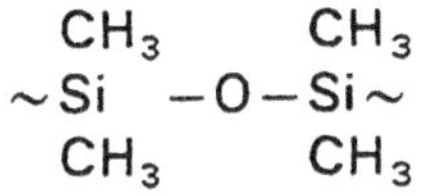

is due to the weak Si-C bond (35–65 kcal/mole) and the formation of diradicals for efficient cross-linking,[4]

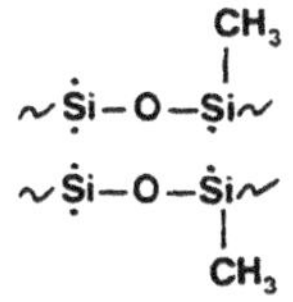

In fact, silicone pump oil was the first monomer resist and it was called the contamination resist, due to the deposition of backstreaming diffusion pump oil inside a SEM chamber.[109] As early as 1970, the siloxanes were investigated as a potential replacement for oxidized silicon diffusion masks.[104–107] The developed pattern was cremated in oxygen to form a silica diffusion barrier. Later, the reported use of polydimethylsiloxane by Hatzakis and co-workers[119] as a bilayer resist image (1000 Å thick) to reduce the proximity effect, prompted the investigation of the incorporation of Si and other metals (B, Ti, Sn) in a host of polymers as plasma O_2 resists.[181]

Roberts studied polymethylcyclosiloxane and found a sensitivity of 100 $\mu C/cm^2$.[105] Incorporation of vinyl groups increased the sensitivity by 100-fold.[108,110] The methylsiloxanes are ultralow-T_g polymers ($-145°C$), which readily flow at 100°C and are not suitable as the top layer in a bilayer stack for lift-off (Chapter 12). More stable polymers are obtained by the replacement of the methyl groups in PDMS with chloromethyl phenyl siloxane, which raises the T_g to 150°C with slight loss of *e*-beam speed (4 $\mu C/cm^2$),[75,161,190]

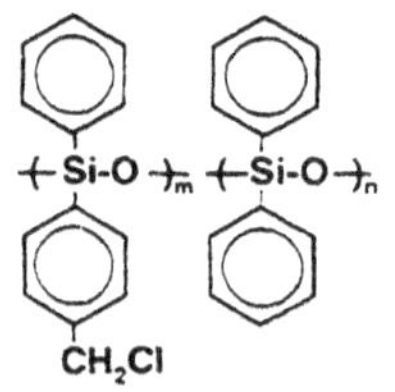

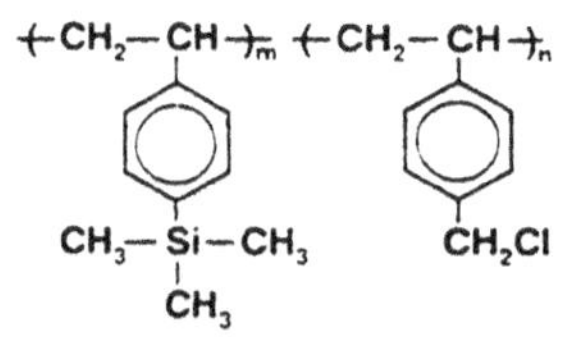

Because of internal cross-links, cyclic siloxanes (**I**),

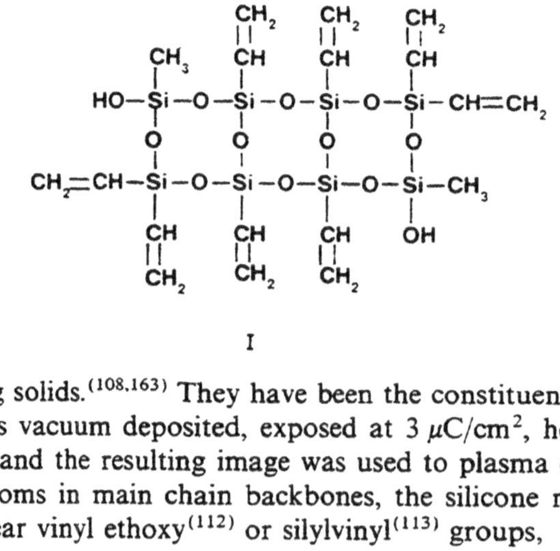

I

are higher melting solids.[108,163] They have been the constituent of a unique all-dry lithography. **I** was vacuum deposited, exposed at 3 $\mu C/cm^2$, heated to remove the unexposed resist, and the resulting image was used to plasma etch Al.[111]

Besides Si atoms in main chain backbones, the silicone resists can also have side chains of linear vinyl ethoxy[112] or silylvinyl[113] groups,

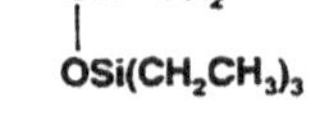

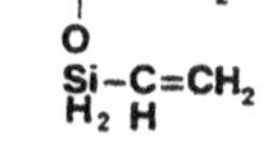

The silicone systems are summarized in Table 5-4-5.

Silicon and other transition metals,[181] such as Sn and Ti, can be incorporated into resists for plasma O_2 dry developing, or more importantly, as an image/barrier layer for bilayer resists.[152] About 7 wt% or more metal is needed to form in the oxygen plasma, a metal oxide as such SiO_2. Since Si can replace carbon in most organic monomers, the metal can be incorporated as part of a polymer chain (siloxanes and silylstyrenes) or as a small sensitive molecule (silyldiazoquinone, silylazide, or silylacrylate).[154] Metals may also be incorporated by posttreatment

TABLE 5-4-5. Silicone Negative Resists

Resist	$\mu C/cm^2$	Reference
Dimethyl siloxane	1–3	107, 114, 119
Pump oil	100	115
Vinyl methyl siloxane	1	114, 119
Phenyl methyl siloxane	4	114
Chlorodiphenyl siloxane	10	75
Vinyl silyl	1	113
Vinyl triacetoxyisilane	5	112
Silicone monomers	X ray	17
Cinnamate siloxane	UV	116
Epoxy-siloxane	UV	117
Acrylate-siloxane	UV	118

of the exposed or developed image with silanes, tin halides, silicone monomers, or by soaking in metal salts.(153) Direct introduction of metals can also be done by ion implantation or by decomposition of metal (alkyls) with a laser (Chapter 16). If the silicon treatment or layer of resist is confined to a thin layer or a top surface image, resolution to 0.1 μm should be achieved. This can be done by:

1. Use of a metallic resist in bilayer resist
2. Use of a low-penetrating ion or *e*-beam
3. Use of a highly absorbing resist, which confines the absorbed radiation to the top of the resist
4. Expose and diffuse in a reactive metal reagent, which is diffusion limited and confined to the top of the resist

Silicone-based resists are the best candidates for future submicrometer lithography, because of their superior O_2 plasma resistance and the inherent high resolution of bilayer resists. Their melt resistance has to be increased to 200°C for application as a lift-off mask. The contrast of most of the siloxanes is low and the sensitivity increases with molecular weight.(119,178) In addition, the postpolymerization effect or O_2 sensitivity has not been reported. The most practical use of siloxanes has been in bilayer resists, which use a 2000-Å film image layer to limit the proximity effect.(119,120) In this case, the siloxane acts as both an image layer and an O_2 plasma etch barrier layer,

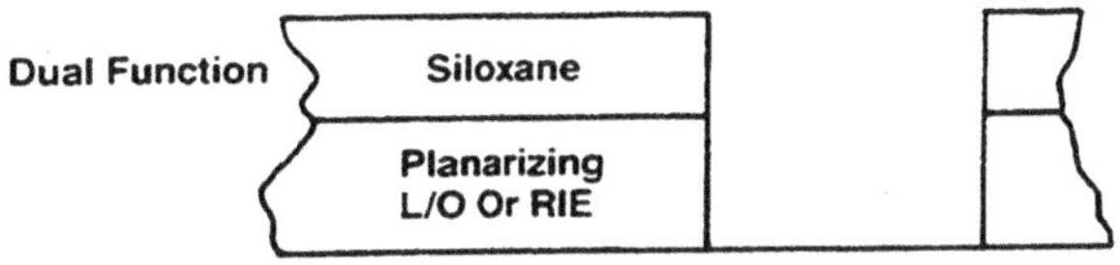

For lift-off, the chloromethyl phenyl siloxane polymers and silyl novolaks are much more flow-resistant stable masks (T_g of 150°C) than polydimethylsiloxane (T_g −145°C).(75) Both lift-off and reactive ion etching can be performed with a suitable underlayer.

As a form of image amplification, direct pattern formation has been demonstrated by exposing PMMA or polystyrene in the presence of silicone monomer(121,160) (similar to the early work of Gazard *et al.*(150) on acrylic acid graft)

but high doses (100 $\mu C/cm^2$) are required. The image is transferred by O_2 reactive ion etching. If a sensitivity of 1 $\mu C/cm^2$ can be achieved, a viable system of top surface imaging can be obtained for superior resolution.

5-4-6. *Miscellaneous Resists*

Polyarylazides have been estimated to have a G_x of 16.(101) With this high G_x, the azido resists or bisazide sensitized polymers should be 1 $\mu C/cm^2$ resists, although little work has been reported. In an attempt to raise the T_g and enhance the thermal stability to 200°C, polyamide-epoxies were made, but a sensitivity of only 10 $\mu C/cm^2$ was achieved.(102) Polybutadiene was also examined as a resist and found to have low sensitivity (100 $\mu C/cm^2$).(103)

Monomolecular soaplike vinyl stearates have been examined, in an attempt to arrange vinyl groups, for Langmuir–Blodgett ordered polymerization. The ultimate high-resolution resist, 10 Å thick, may result (Chapter 6). At present, doses of 3–5 $\mu C/cm^2$ have been demonstrated.(122)

Chlorinated polyvinylether is a fast *e*-beam resist requiring only 2 $\mu C/cm^2$, but is subject to postexpose polymerization.(19) Chlorinated polyethylene proved to be a very slow (100 $\mu C/cm^2$) resist.(123)

Inorganic resists of Ag,(125) WO_3,(126) and Ag-Ge-Se(41) have been examined as bilayer resists. Only 100 Å is required to serve as an O_2 mask. Doses of > 10 $\mu C/cm^2$ are needed.(124) Evaporated AgBr has the potential for becoming the fastest resist at doses of only 10^{-9} C/cm^2.(125) Unfortunately, the large grain size of AgBr in vacuum-deposited films limits the resolution to about 1–2 μm.

5-5. TWO-COMPONENT RESISTS

Two-component *e*-beam resists have been the conventional photoresists, such as KTFR and KPR photoresists exposed at doses of 10 $\mu C/cm^2$.(127) At high doses (> 100 $\mu C/cm^2$), followed by flood UV exposure, positive DQNs have been reversed into negative resists.(148,171) Similarly, azides have been added to polystyrene to achieve 2 $\mu C/cm^2$ sensitivity and high resolution.(193) Azides have also been used to

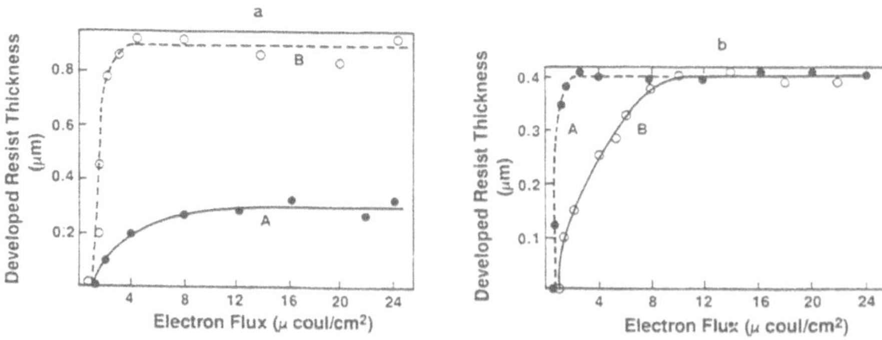

FIGURE 5-5-1. (a) Exposure of KTFR films to 15-keV electrons; thickness curves: A, 360 nm; B, 1200 nm. From Ref. 128, reproduced by permission of the Electrochemical Society Inc. (b) Exposure of KPR films of 600-nm initial thickness to 5 keV (curve A) and 15 keV (curve B).

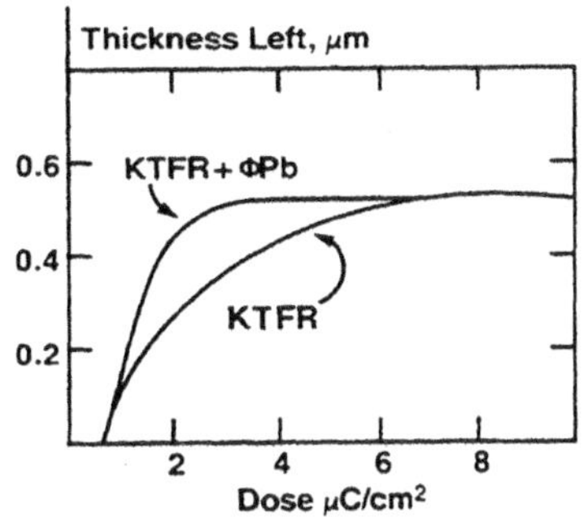

FIGURE 5-5-2. Improved contrast of KTFR by addition of hexaphenyl lead, U.S. Patent 3,594,120.

cross-link novolaks and upon development in an aqueous alkaline solvent, no swelling of the images is observed.[180]

For conventional cross-linking resists, Broyde[128] was one of the first to point out the optimum resolving thickness for negative resists, based on the energy of exposure, and penetration range into the polymer. If the film is thin, backscatter dominates (Fig. 5-5-1). At a certain thickness (B, 1200 nm), the forward and backscatter are equal and the highest contrast is found (Fig. 5-5-1). Exposure at 10 keV, for images as small as 1.0 μm, requires no proximity corrections. Since there is no concern for registration in Cr photomask making, at present, *e*-beam mask-making tools use 10-keV exposure without the need for proximity effect corrections.

The diffusion range of backscattered electrons can be artificially adjusted by adding heavy metal salts, such as hexaphenyl lead to KTFR (Fig. 5-5-2). Thus, the proximity effect can be reduced in any resist exposed to electrons by adding high-absorbing additives, as Au or lead metal or salts. Heavy metals have been directly incorporated into DQN by attachment through the phenolic group.[173] Sn maleate can be added to KPR (Fig. 5-5-3) to absorb electrons more uniformly throughout the film, by creating secondary electron emitters. Lowering of the energy to 5 keV also improves contrast (Fig. 5-5-4) by limiting the range of backscattered electrons.[128]

The depth–dose relationship has been modeled by Heidenreich and co-workers[129] as a function of energy and resist thickness. The optimum resist thickness is equal to the diffusion range, so that forward absorption (equivalent to maximum of standing waves) is equal to the diffusion range.

For mask making using a submicrometer-thick resist, 10 keV is the best choice, because the backscatter range is limited. For thicker resists on silicon, 25 keV is the better choice for registered writing.

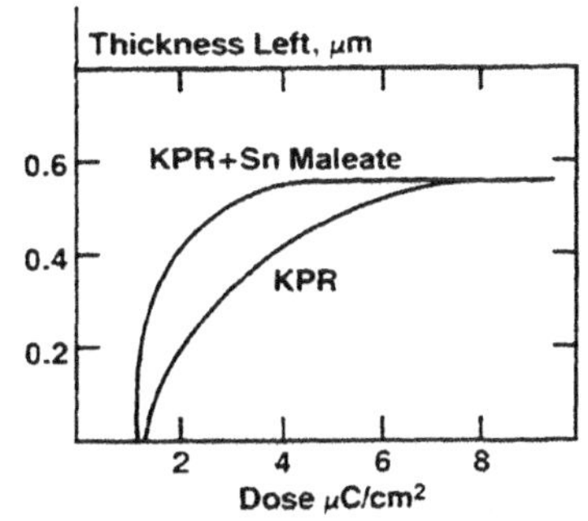

FIGURE 5-5-3. Improved constrast of KPR to *e*-beam by addition of metal salt, U.S. Patent 3,594,120.

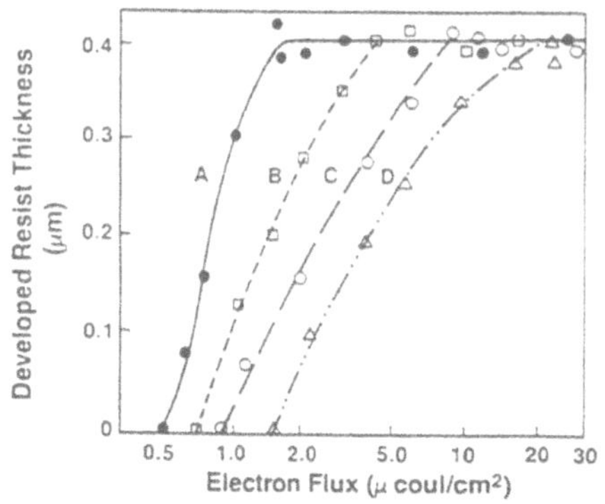

FIGURE 5-5-4. Effect of energy of electron beam on contrast of KTFR A: 5 keV, B: 10 keV, C: 15 keV, D: 20 keV. From Ref. 128, reproduced by permission of the Electrochemical Society Inc.

The effect of thickness in the forward and backscatter is given by the electron-beam modulation transfer function (EBTF). According to the double-Gaussian model, the EBTF is given by[131]

$$\mathrm{EBTF}(f) = \frac{1}{1+\eta}\left[\exp(-\pi^2\beta_{f\,\mathrm{eff}}^2 f^2) + \eta\exp(-\pi^2\beta_b^2 f^2)\right]$$

where β_f is the forward scatter width, β_b is backscatter width, and η is the backscatter–forward exposure efficiency ratio of the target material. The finite resist thickness is taken into account, by averaging the MTFs of the system in focus on top of the resist ($t = 0$) and out of focus on the bottom of the resist (t). Defocusing is treated as an aberration: The defocused MTF is the product of the focused MTF and the Fourier transform of the Airy disk,

$$F(f) = (1/\pi R_f)\, J_1(2\pi R_f) \tag{5-5-1}$$

where R is the disk radius and J_1 is the Bessel function of first order. Therefore, for finite resist thickness,[130]

$$\mathrm{MTF}_t = (1 + F/2)\,\mathrm{MTF}_0 \tag{5-5-2}$$

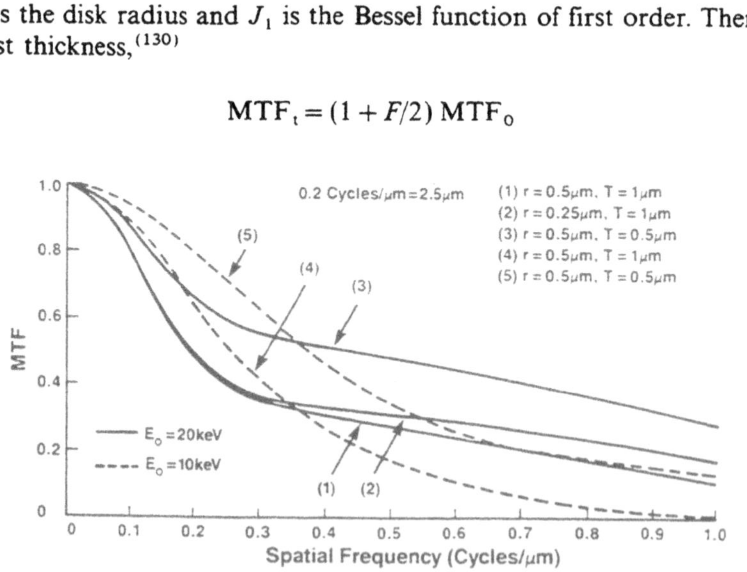

FIGURE 5-5-5. MTF of *e*-beam exposure as a function of resist film thickness (T). From Ref. 132, courtesy of SPSE.

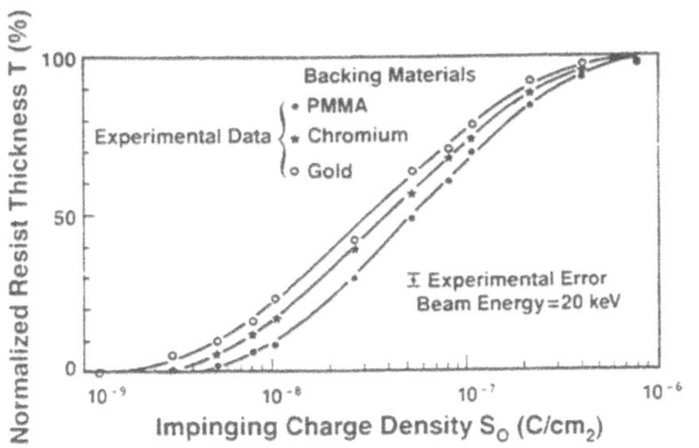

FIGURE 5-5-6. Effect of substrate backscatter of electrons on sensitivity of epoxybutadiene. From N. Saitou, *Jpn. J. Appl. Phys.* **11**, 1061 (1972), courtesy of the *Japanese Journal of Applied Physics*.

The effect of film thickness on MTF is shown in Fig. 5-5-5. In this case, thinner films are required to enhance the MTF. In a bilayer resist, the backscatter coefficient of an organic layer on top of another organic polymer is 0.04. For a resist on Cr, the backscatter is 0.28 and on top of Au is 0.51. The change in sensitivity for these substrates with a negative epoxy resist is shown in Fig. 5-5-6. Because of the higher backscatter from gold, the resist sensitivity is enhanced. For various substrates, increasing the atomic number generates more effective backscatter (Table 5-5-1). Backscatter from Cu produces a more effective energy deposition than for KTFR on Al (Fig. 5-5-7), while backscatter from Au produces a smaller spot, in the sense that the range of the backscattered electron is shorter, due to the lower energy of the electron reflected back into the resist. To study backscatter effects, Heidenreich and Thompson exposed various substrates to an *e*-beam, in the presence of a vinyl ferrocene monomer.[135] The size of the monomer deposit is used to assess the relative contribution of the backscatter electron range (Fig. 5-5-8). With the longer range of secondaries backscattered from Si, the largest radius of monomer deposit was formed. Thus, poorer resolution in cross-linking negative resists is obtained on low-atomic-mass substrates.

For exposing any resist over various substrates, backscatter measurements are necessary for the proximity dose corrections. For complicated structures of several metal substrates in Josephson devices, Vettiger *et al.*[136] describe an actual on-the-fly feedback loop, which measures the backscatter intensity and corrects the beam dose. This system was useful for negative resist traversing several different metals, on one exposure level.

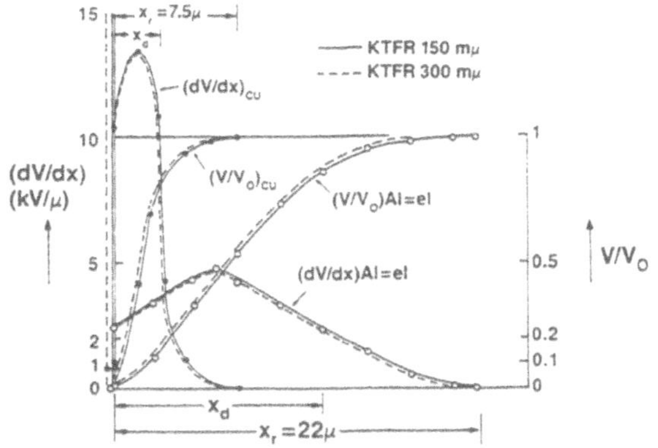

FIGURE 5-5-7. Calculated energy dissipation and energy loss in photoresist films of different thickness on Al substrate and Cu substrate. From Ref. 133.

TABLE 5-5-1. Effect of Atomic Number (Z) on Dose for a Negative Resist

Z	Dose reduction
14	1.0
30	1.31
45	1.42
60	1.45
75	1.55

Heavy metal films have also been used, not only to reduce the backscatter proximity effect, but to prevent radiation damage to device underlevels. For such purposes, a heavy metal barrier layer is used in a trilayer resist (Fig. 5-5-9).[137,164] The heavy metal or opaque barrier layer also prevents X rays and photons, which reflect from the Si or device layers, from exposing and widening the image of the image layer in a trilayer resist.[138]

Negative *e*-beam resists are more difficult to expose uniformly (especially with different pitches and variable isolated lines) than positive resists, since they reveal all of the backscattered electrons. One method[139] of proximity correction adjusts the dose (spot time) to handle the intraproximity (isolated line) effect falloff (Fig. 5-5-10).[136] By lowering the spot time, the base width of the image shrinks to achieve better resolution at equivalent final remaining thickness. The other correction to the proximity effect is to adjust the size of the beam to counter the interproximity effect (Fig. 5-5-11). In practice, both corrections are found necessary and a combination for negative resists is to reshape the distorted edge slope, by providing extra dose at the edge of the pattern (Fig. 5-5-12).[141]

For most negative cross-linking resists, oxygen plasma (or CF_4 for siloxanes) descum appears to be necessary, as well as using resists which do not postpolymerize after exposure.[146] Use of negative resists with films thicker than 0.5 μm limits the resolution in direct silicon write to about 3-μm pitches. Multilayer resists are a better materials solution to handle the proximity corrections. For a negative resist in a single film, the resolution is about 5× the thickness.[134]

Several key attempts to use nonorganic developable resists (to reduce swelling) have been based on aqueous-based developable phenolic resins (Chapter 4), and the

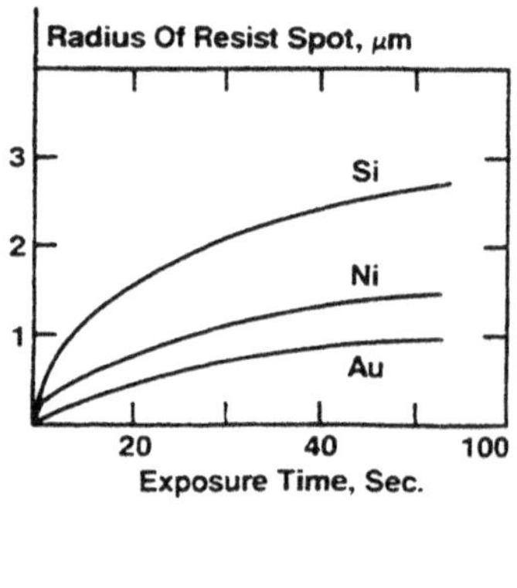

FIGURE 5-5-8. Effect of substrate on exposure of monomer of vinyl ferrocene by *e*-beams. The backscatter from Si (low atomic number) produces electrons of higher energy and longer range which is recorded by polymerizing monomer. From Ref. 135, courtesy of AIP.

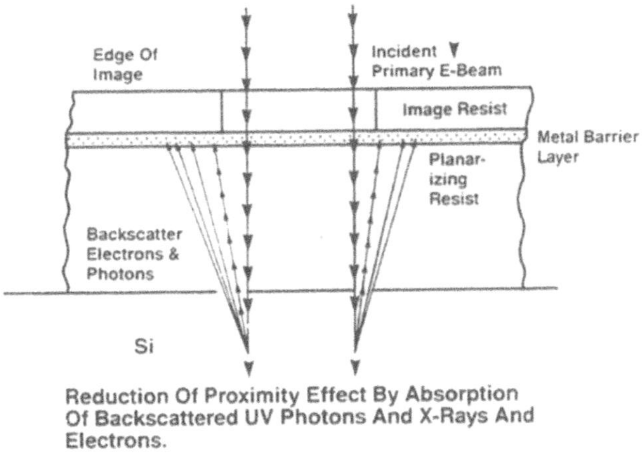

FIGURE 5-5-9. Superposition of a metal barrier layer in a trilayer resist to absorb backscattered electrons and reduce the interproximity effect in the top resist image layer.

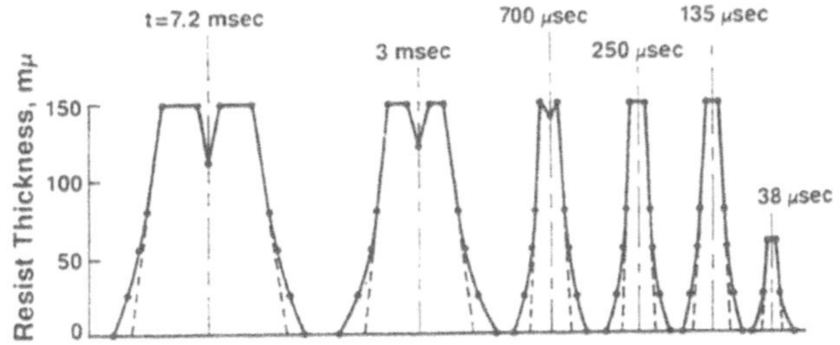

FIGURE 5-5-10. Thickness of KTFR remaining after exposure by a 1.0-μm beam at different exposure times (t). From Ref. 133.

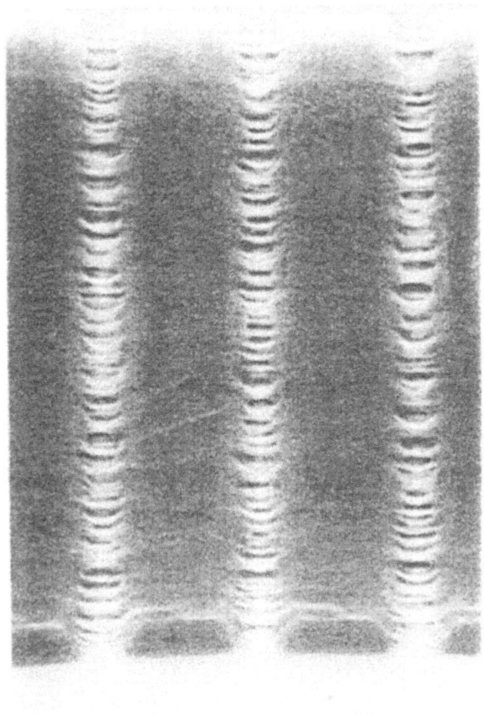

FIGURE 5-5-11. Interproximity effect in negative e-beam resists. Unexposed regions are filled in with resist exposed by backscattered electrons.

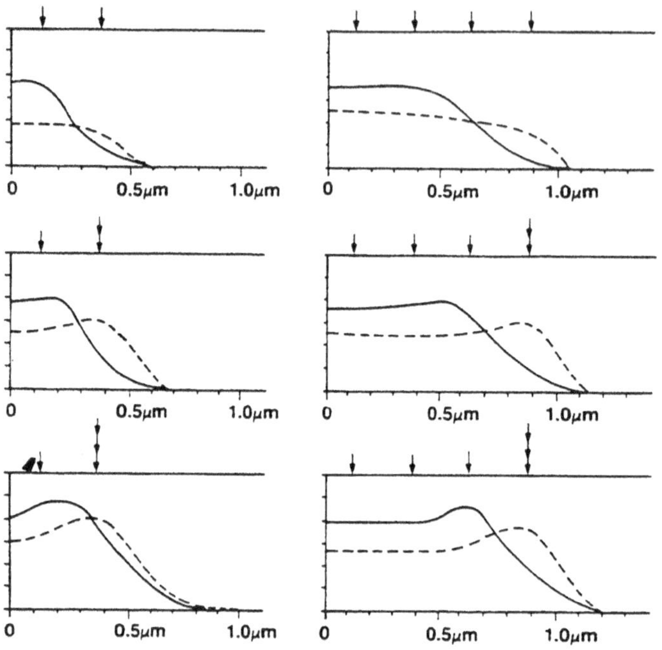

FIGURE 5-5-12. Exposure compensation of negative resist at edges of line to steepen profiles. ---, calculated; ——, experimental. From Ref. 141, courtesy of AIP.

onium salt decarbonylation of *t*-butylcarbanatostyrene (Chapters 2 and 4). These are the basic approaches for future high-resolution negative resists not based on classical cross-linking reactions.

5-6. PRACTICAL USE OF NEGATIVE ELECTRON-BEAM RESISTS

The practical use of negative *e*-beam resists has been confined to submicrometer films for chrome mask making. At present, many problems exist, such as pinholes, postpolymerization, oxygen descum, poor thermal and plasma resistance, resolution limited by swelling, low contrast, and proximity effects. A few

TABLE 5-6-1. Standard Deviation of the Feature Size[a]

	Standard deviation (3σ)	
	5-inch mask	6-inch mask
Resist pattern (μm)	0.11	0.11
After etching pattern (μm)		
Wet	0.20	0.24
Dry		
Planar	0.15	0.18
Barrel	0.50	—

[a] Modified from Ref. 144.

TABLE 5-6-2. Commercial Negative Electron-Beam Resists[a]

Trade name	Polymer	Supplier	T_g (°C)
OEBR-100	Polyglycidal MMA	Tokyo Ohka	78
COP	GMA-ethyl acrylate	ATT/Mead	20
CMS	Chloromethyl polystyrene	Toyo Soda	120
SEL-N	PGMA-maleic methylate	Somar	40
RE 4000N	Iodinated polystyrene	Hitachi	156
GMC-I	GMA-3-chlorostyrene	ATT/Mead	80
MES	Chlorinated polymethylstyrene	JSR	120
EK-88	Allylmethacrylate PMMA	Kodak	72
GAF EB-46	Unknown	GAF	−12

[a] From Ref. 145.

TABLE 5-6-3. Main Parameters and Process Conditions for Negative Electron-Beam Resists[a]

Resist	COP	OEBR-100	SEL-N type B
Molecular weight distribution	2.0	1.4	2.6
Sensitivity at 10 kV (C/cm^2)	1.8×10^{-7}	1.6×10^{-7}	2.7×10^{-7}
Contrast	0.8	1.1	0.8
Resolution			
1-μm gap	Not good	Not good	Good
1-μm L/S	Fairly good	Not good	Good
2-μm gap	Good	Good	Good
2-μm L/S	Good	Good	Good
Adhesion to chromium	Fairly good	Very good	Poor
Prebaking	90 °C, 30 min	80 °C, 10 min	80 °C, 30 min
Developing (spray)			
Developer	MEK:Ethanol = 5:2	MEK:Ethanol = 7:1	MEK:Ethanol = 1:1
Rinse 1	MIBK:IPA = 1:3	MIBK	IPA
Rinse 2	IPA		
Postbaking	180 °C, 30 min	140 °C, 30 min	140 °C, 30 min
Descumming (O_2 1.0 Torr, 100 W)	15 min	15 min	10 min
Stripping (H_2SO_4/H_2O_2)	30 min	30 min	10 min

[a] From Ref. 147.

TABLE 5-6-4. Sensitivity, Contrast, and Resolution of Electron-Beam Resists[a]

Resist	Dose (μC/cm^2)	γ	Line/space	Line
OEBR-100	0.2	1.0	1.5 μm	0.5 μm
COP	0.2	0.8	1.5 μm	1.0 μm
SEL-N	0.4	0.9	1.0 μm	0.75 μm
CMS-EXS	0.8	1.5	1.0 μm	0.75 μm

[a] From Ref. 147.

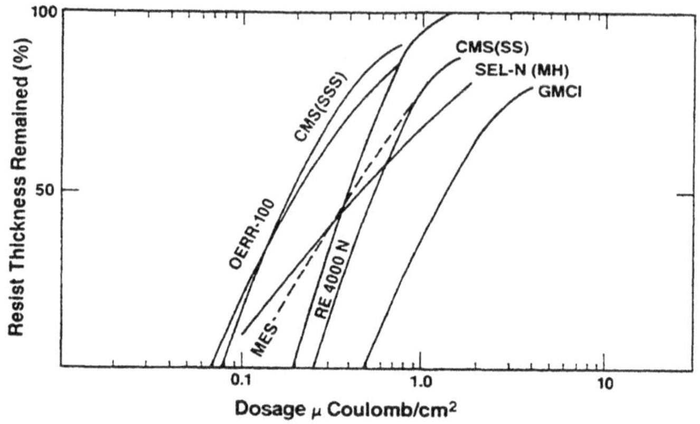

FIGURE 5-6-1. Contrast curves for negative *e*-beam resists exposed at 10 keV. From Ref. 145, reproduced by permission.

silicon circuits of 2- to 3-μm size have been made in laboratory experiments[142,143] with chlorostyrenes but no single-film resist appears to be directly involved in manufacturing production. Most of the resists continue to be used in chrome photomask fabrication.

For a 2-μm image generation on chrome, made by an epoxy resist, the standard image deviation in a 0.3-μm-thick film is about 0.1 μm (Table 5-6-1). In Table 5-6-2, the commercially available mask-making resists are given. Since dry etching has taken over the process of Cr etching, aromatic groups have been introduced into resist polymers for resistance improved plasma over aliphatic resist systems.

The contrast curves for some of the negative resists are shown in Fig. 5-6-1. Resolution is limited to about 1 μm with contrasts less than 2.[146] Most of the negative resists are used to support stepper photomasks made at the 10× level. If in the future 1× masks are required, single-film resists will likely be replaced by multilayer resist structures.

Typical process conditions for negative mask-making resists are given in Table 5-6-3. Oxygen descumming is necessary to reduce line interproximity effects. To control radical postpolymerization, the plates are left for 30 min in vacuum before exiting to development.

The performance of several *e*-beam negative resists on MEBES is given in Table 5-6-4. Because no proximity corrections are available, patterns of 1- to 2-μm linewidth were inferior and difficult to reproduce without 0.3-μm or more distortions. Resist systems with no proximity correction and no postpolymerization effects are required for producing 1- to 2-μm geometries.

5-7. OUTLOOK

Negative radiation resists have struggled to be effective as 1-μm-thick films for direct silicon writing, and thus they have been primarily used for chrome mask making with thinner (0.4 μm) films. Even exposure by X rays has failed to produce a suitable high-resolution reactive ion resist. The only practical use of single-film negative resists has been in photomask fabrication at large (10×) dimensions.

Basic resolution limitations remain in conventional negative resists. The cross-linking of organic systems is accompanied by swelling in organic developers. Secondly, after the resist has been exposed, postpolymerization is a major curing problem with many fast resists. Other means of insolubilization, such as charge-transfer or molecular rearrangement, are required for formulating nonswelling resists.

Aqueous developable systems have the potential to reduce swelling. Two-component negative resists have a better chance of achieving the desirable resist properties in the base polymer while the sensitivity is incorporated in the sensitizer component.

X-ray lithography needs a good 1 mJ/cm^2 resist for meeting the throughput goals of 10 (200-mm) wafers/hr. Electron-beam exposure requires negative resists, less sensitive to proximity effects and higher contrast. Top surface imaging, with postmonomer grafting, offers high speed and high resolution.

Since negative resists have a wider development process latitude than positive resists, nonswelling negative resists should be developed to exploit this advantage.[175] The second role of negative resists will be as an image layer in bi- or multilayer resists. Image transfer by deep-UV exposure or plasma etching completes the final image development. For bilayer structures, the silicon or transition metal-containing resists are the best candidates for submicrometer imaging, since the image can be developed and anisotropically transferred to a planarizing resist. With thin silicone image layers (0.2 μm), the proximity effect is reduced to less than 0.15 μm in bilayer resists. We can expect more and more organometallic resists of the negative type for use in future *e*-beam exposure tools.

REFERENCES

1. A. Reiser and E. Pitts, *J. Photogr. Sci.* **29**, 187 (1981).
2. Y. Taniguchi, Y. Hatano, H. Shiraishi, S. Horigome, S. Nonogaki, and K. Navoaka, *Jpn. J. Appl. Phys.* **18**, 1143 (1979).
3. P. Flory, *Mechanics of Chain Polymers*, Wiley, New York, 1969.
4. A. Charlesby, *Atomic Radiation and Polymers*, Pergamon Press, Elmsford, N.Y., 1960.
5. M. Nakase and Y. Matsumoto, *Photogr. Sci. Eng.* **23**, 215 (1979).
6. G. Stevens, *Microphotography*, Wiley, New York, 1967, p. 24.
7. K. Okada, K. Suzuki, and J. Matsui, *J. Electrochem. Soc.* **130**, 1947 (1983).
8. A. Shultz, *J. Polym. Sci.* **21**, 325 (1956).
9. A. Chapiro, *Radiation Chemistry of Polymeric Systems*, Wiley, New York, 1962.
10. S. Bernachi and H. Smith, *IEEE Trans. Electron Devices* **ED-22**, 42 (1975).
11. G. Taylor, G. Coquin, and S. Somekh, *Polym. Eng. Sci.* **17**, 420 (1977).
12. U. S. Patent 4,061,829 (1978), Bell.
13. G. Taylor, *Solid State Technol.* May 1980, p. 73.
14. L. Thompson, E. Feit, M. Bowden, P. Llengza, and E. Spencer, *J. Electrochem. Soc.* **121**, 1500 (1974).
15. G. Taylor and J. Moran, *J. Vac. Sci. Technol.* **16**, 2014 (1979).
16. G. Taylor, T. Wolf, and M. Goldrick, *J. Electrochem. Soc.* **128**, 361 (1981).
17. G. Taylor, T. Wolf, and J. Moran, *J. Vac. Sci. Technol.* **19**, 872 (1981).
18. T. Yamaoka, T. Tsunoda, and Y. Goto, *Photogr. Sci. Eng.* **23**, 196 (1979).
19. S. Imamura, S. Sugawara, and K. Murase, *J. Electrochem. Soc.* **124**, 1139 (1977).
20. K. Murase, K. Kakuchi, and S. Sugawara, Proceedings of the International Conference on Microlithography, Paris, 1977, p. 261.
21. S. Imamura, *J. Electrochem. Soc.* **126**, 1619 (1979).

22. S. Imamura and S. Sugawara, *Jpn. J. Appl. Phys.* **21**, 776 (1982), *J. Electrochem. Soc.* **131**, 1122 (1984).
23. Y. Saitoh, H. Hoshihara, and I. Watanabe, *Jpn. J. Appl. Phys.* **21**, L52 (1982).
24. H. Shiraishi, Y. Tanaguchi, S. Horigome, and S. Nonogaki, *Polym. Eng. Sci.* **20**, 1054 (1980).
25. D. Hofer, F. Kaufman, S. Kramer, and A. Aviram, *Appl. Phys. Lett.* **37**, 314 (1980).
26. J. Dubois and M. Gazard, Proc. 8th Electron, Ion Beam Sci. Technol., Electrochemical Society, 1978, p. 303.
27. R. Brault, Proc. 6th Int. Conf. Electron Ion Beams, Electrochemical Society, 1976, p. 63.
28. M. Okabe, Y. Kitamura, and Y. Furukawa, Proc. 10th Electron Ion Beam Sci., Electrochemical Society, 1982, p. 302.
29. K. Okada, *Jpn. J. Appl. Phys.* **20**, L22 (1981).
30. P. Blais, C. Dixon, T. O'Keefe, J. Ostroski, T. Meek, and M. Picher, *Electrochem. Soc. Ext. Abstr.* **80**, 835 (1980).
31. Z. Tan, C. Petropolous, and F. Rauner, *J. Vac. Sci. Technol.* **19**, 1348 (1981).
32. U. S. Patent 3,770,433 (1973), Bell.
33. J. Moran and G. Taylor, *J. Vac. Sci. Technol.* **16**, 2014–2020 (1979).
34. T. Tada, *J. Electrochem. Soc.* **129**, 1070 (1982).
35. Y. Ohnishi, M. Itoh, K. Mizano, H. Gokan, and S. Fujiwara, *J. Vac. Sci. Technol.* **19**, 1141 (1981).
36. T. Hall, A. Wagner, and L. Thompson, *J. Vac. Sci. Technol.* **16**, 1889 (1979).
37. R. Brault and L. Miller, *Polym. Eng. Sci.* **20**, 1064 (1980), **23**, 941 (1983).
38. J. Lai and L. Shepard, *J. Electrochem. Soc.* **126**, 696 (1979).
39. K. Itaya, K. Shibayama, and T. Fujimoto, *J. Electrochem. Soc.* **129**, 663 (1982).
40. U. S. Patent 4,383,026 (1982), Bell.
41. K. Balasulbramanyam, L. Kapapiperis, C. Lee, and A. Ruoff, *J. Vac. Sci. Technol.* **19**, 21 (1981), *Microelectron. J.* **14**, 35 (1983).
42. U. S. Patent 4,377,437 (1983), Bell.
43. T. Venkatesan, G. Taylor, A. Wagner, B. Wilkens, and D. Barr, *J. Vac. Sci. Technol.* **19**, 1379 (1981).
44. W. Brown, T. Venkatesan, and A. Wagner, *Solid State Technol.* Aug. 1981, p. 60.
45. E. Wolf, I. Aesida, L. Ratburn, and J. Chinn, *J. Vac. Sci. Technol.* **21**, 661 (1982).
46. P. LaMarche, R. Levi-Setti, and Y. Wang, *J. Vac. Sci. Technol.* **B1**, 1056 (1983); H. Kuwano, *J. Appl. Phys.* **551**, 1149 (1984).
47. I. Berry and A. Caviglia, *J. Vac. Sci. Technol.* **B1**, 1059 (1983).
48. H. Choong and F. Kahn, *J. Vac. Sci. Technol.* **B1**, 1066 (1983).
49. H. Choong and F. Kahn, *J. Vac. Sci. Technol.* **19**, 1121 (1981).
50. Y. Kamoshida, M. Koshiba, H. Yoshimoto, Y. Harita, and K. Harada, *J. Vac. Sci. Technol.* **B1**, 1156 (1983).
51. T. Tamamura, S. Imamura, O. Kogure, S. Sugawara, and K. Sukegawa, *Am. Chem. Soc. Org. Coat. Appl. Polym. Sci.* **48**, 150 (1983).
52. P. Rissman, E. Liui, and G. Owen, *J. Vac. Sci. Technol.* **B1**, 1014 (1983).
53. J. Steelhammer and W. Wentworth, *J. Chem. Phys.* **51**, 1802 (1969).
54. S. Imamura, T. Tamamura, K. Harada, and S. Sugawara, *J. Appl. Polym. Sci.* **27**, 937 (1982).
55. J. Jagt and A. Seviriens, *Polym. Eng. Sci.* **20**, 1082 (1980).
56. H. Shiraishi, Y. Tanaguchi, S. Horigome, and S. Nonogaki, *Polym. Eng. Sci.* **20**, 1054 (1980).
57. Y. Yoneda, K. Kitamura, J. Naito, T. Kitakohji, H. Okuyama, and K. Murakawa, *Polym. Eng. Sci.* **20**, 1110 (1980).
58. L. Thompson and E. Doerries, *J. Electrochem. Soc.* **126**, 1699 (1978).
59. J. Liutkus, J. Shaw, M. Hatzakis, and J. Paraszczak, *Polym. Eng. Sci.* **23**, 1047 (1983).
60. E. Feit and L. Stillwagon, *Polym. Eng. Sci.* **16**, 1997 (1979).
61. E. Feit and L. Stillwagon, *Polym. Eng. Sci.* **20**, 1058 (1980).
62. S. Inamura, *J. Electrochem. Soc.* **126**, 1628 (1979).
63. U. S. Patent 4,286,049 (1982), Nippon.
64. K. Sukegawa, T. Tamamura, and S. Sugawara, *J. Electrochem. Soc.* **129**, 1835 (1982); 10th Electron Ion Beam Conf., Electrochemical Society, 1982, p. 183.
65. L. Thompson, L. Stillwagon, and E. Doerries, *J. Vac. Sci. Technol.* **15**, 940 (1978).
66. Y. Ohnishi, K. Tanigaki, and S. Fugawara, *J. Vac. Sci. Technol.* **19**, 1136 (1981).
67. Y. Ohnishi, K. Tanigaki, and S. Fugawara, *Am. Chem. Soc. Org. Coat. Appl. Polym. Sci.* **48**, 152 (1983).

68. U. S. Patent 4,426,439 (1984), Fujitsu.
69. H. Ku and L. Scala, *J. Electrochem. Soc.* **116**, 980 (1969).
70. U. S. Patent 4,061,799 (1977), TI.
71. U. S. Patent 3,916,035 (1975), TI.
72. U. S. Patents 4,130,424 (1978) and 4,262,081 (1981), Bell.
73. U. S. Patent 4,318,976 (1982), TI; J. Shy, *Polym. Eng. Sci.* **23**, 980 (1983).
74. T. Tamamura, S. Imamura, K. Harada, and S. Sugawara, *J. Electrochem. Soc.* **129**, 1831 (1982).
75. M. Morita, A. Tanaka, S. Imamura, T. Tamamura, and O. Kogure, *Jpn. J. Appl. Phys.* **22**, L659 (1983).
76. T. Hirai, Y. Hatano, and S. Nonogaki, *J. Electrochem. Soc.* **118**, 669 (1971).
77. L. Thompson, J. Ballantyne, and E. Feit, *J. Vac. Sci. Technol.* **12**, 1280 (1975).
78. J. Bartelt and E. Feit, *J. Electrochem. Soc.* **122**, 541 (1975).
79. Y. Taniguchi, Y. Hatano, H. Shiraishi, S. Horigome, S. Nonogagki, and K. Naroaka, *Jpn. J. Appl. Phys.* **18**, 1143 (1979).
80. E. Feit, L. Thompson, C. Wilkins, M. Wurtz, E. Doerries, and L. Stillwagon, *J. Vac. Sci. Technol.* **16**, 1997 (1979).
81. J. Dubois, A. Eranian, and E. Datamanic, Proc. 8th Electron, Ion Beam Sci. Technol., Electrochemical Society, 1978, p. 303.
82. J. Crivello and J. Lam, *Polym. Eng. Sci.* **23**, 953 (1983), *J. Polym. Sci.* **17**, 2877 (1979), U. S. Patents 4,256,828 (1981) and 4,102,687 (1978), GE.
83. U. S. Patents 4,090,936 (1978) and 4,256,828 (1981), 3M.
84. U. S. Patent 4,156,035 (1979), W. Grace.
85. S. Pappas and J. Jilek, *Photogr. Sci. Eng.* **23**, 141 (1979).
86. S. Schlesinger, *Polym. Eng. Sci.* **14**, 513 (1974).
87. Y. Ohnishi, M. Itoh, K. Mizuno, H. Gokan, and S. Fugawara, *J. Vac. Sci. Technol.* **19**, 1141 (1981).
88. U. S. Patent 4,299,911 (1981), SOMAR.
89. U. S. Patent 3,627,599 (1971), RCA.
90. G. Paal, U. Strohle, and G. Keelhorn, *J. Electrochem. Soc.* **120**, 1717 (1973).
91. U. S. Patent 4,289,824 (1981), Kodak.
92. Z. Tan, S. Georgia, C. Petropoulous, and F. Rauner, *J. Vac. Sci. Technol.* **19**, 1348 (1981), *Polym. Eng. Sci.* **23**, 931 (1983).
93. U. S. Patent 3,703,402 (1972), GE.
94. M. Cole, D. Skelley, and B. Wagner, *IEEE Trans. Electron Devices* **ED-22**, 417 (1975).
95. J. Bartelt, *Am. Chem. Soc. Org. Coat.* **33**, 390 (1973).
96. U. S. Patent 4,267,258 (1981).
97. U. S. Patent 4,318,976 (1982), TI.
98. K. Lee, W. Jensen, and P. Cukor, *Am. Chem. Soc. Org. Coat. Prepr.* **45**, 143 (1981), *SPIE Proc.* **333**, 15 (1982), U. S. Patent 4,456,678 (1984), GTE.
99. U. S. Patents 4,338,392 and 4,247,622, IBM.
100. D. Hofer, S. Kramer, F. Kaufman, and A. Avriam, *Appl. Phys. Lett.* **37**, 314 (1980), **36**, 423 (1980), **40**, 90 (1982).
101. T. Yaniaka, T. Tsunodo, and Y. Goto, *Photogr. Sci. Eng.* **23**, 196 (1979).
102. N. O. Gato, K. Saniu, C. Azuma, H. Tanaka, K. Oguchi, T. Nakida, and Y. Takakashi, *J. Appl. Polym. Sci.* **28**, 699 (1983).
103. L. Thompson, E. Feit, and R. Heidenreich, *Polym. Eng. Sci.* **14**, 529 (1974).
104. E. Roberts, Third Electron Ion Beam Conf., Electrochemical Society, 1968, p. 571.
105. E. Roberts, *J. Electrochem. Soc.* **120**, 716 (1973).
106. Y. Yatsui, T. Nakato, and K. Umehara, *J. Electrochem. Soc.* **116**, 94 (1969).
107. M. Gazard, *Am. Chem. Soc. Org. Coat.* **33**, 372 (1973).
108. J. Dubois and M. Gazard, *Electrochem. Soc. Ext. Abstr.* **72-1**, 127 (1972).
109. U. S. Patents 3,236,707 and 3,222,670 (1966).
110. J. Paraszczak, D. Kern, J. Bucchignano, E. Arthur, and M. Rosenfield, *J. Vac. Sci. Technol.* **B1**, 1372 (1983).
111. V. Korchov, T. Martynova, and V. Danilovich, *Thin Solid Films* **101**, 369 (1983).
112. U. S. Patent 4,301,231 (1981), Toray.
113. U. S. Patent 4,237,208 (1980), RCA.
114. N. Atoda and H. Kawakatsu, *J. Electrochem. Soc.* **123**, 1519 (1976).
115. U. S. Patent 4,197,332 (1980), IBM.

116. U. S. Patent 3,296,126 (1967), GE.
117. U. S. Patent 3,516,764 (1970), U.S. Army.
118. U. S. Patent 3,872,940 (1972), Dai Kokyo.
119. J. Shaw, M. Hatzakis, and J. Liutkus, RETEC SPE Photopolymers, Ellenville, N.Y., 1982, p. 285; *Polym. Eng. Sci.* **23**, 1054 (1983).
120. J. Paraszczak, J. Shaw, M. Hatzakis, E. Babich, and F. Arthur, *SPIE Proc.* **393**, 57 (1983).
121. M. Morita, S. Imamura, T. Tamamura, O. Kogure, and K. Murase, *J. Vac. Sci. Technol.* **B1**, 1171 (1983).
122. A. Barraud, C. Rossilo, and A. Ruaudel-Teixier, *J. Vac. Sci. Technol.* **16**, 2003 (1979), *Thin Solid Films* **68**, 99 (1980), U. S. Patent 4,235,958 (1980).
123. J. Krasewski and R. Gabrys, *J. Appl. Polym. Sci.* **25**, 1131 (1980).
124. B. Singh, S. Beaumont, A. Webb, P. Bower, and C. Wilkinson, *J. Vac. Sci. Technol.* **B1**, 1174 (1983).
125. A. Shepp, G. Goldberg, and J. Lavine, *IEEE Electron Devices Lett.* **EDL-1**, 61 (1980), *SPIE Proc.* **393**, 34 (1983).
126. M. Babu and T. Ikeda, *Jpn. J. Appl. Phys.* **20**, L149 (1981).
127. R. Thornley and T. Sun, *J. Electrochem. Soc.* **112**, 1151 (1965).
128. B. Broyde, *J. Electrochem. Soc.* **16**, 1241 (1969).
129. R. Heidenreich, L. Thompson, E. Feit, and C. Melliar-Smith, *J. Appl. Phys.* **44**, 4039 (1973).
130. W. Stickel and G. Langer, *J. Vac. Sci. Technol.* **B1**, 1008 (1983).
131. T. Chang, *J. Vac. Sci. Technol.* **12**, 1271 (1975).
132. M. Nakase and Y. Natsumoto, *Photogr. Sci. Eng.* **23**, 215 (1979).
133. K. Kanaya and H. Yamazaki, *Optik* (Stuttgart) **25**, 474 (1967).
134. S. Schiller and M. Newman, Third Int. Conf. Electron Ion Beam, Electrochemical Society, 1968, p. 481.
135. R. Heidenreich and L. Thompson, *Appl. Phys. Lett.* **22**, 279 (1973).
136. P. Vettiger, T. Forester, and D. Kern, *J. Vac. Sci. Technol.* **B1**, 1378 (1983).
137. T. Kitayama, *Jpn. Semicond. Technol. News* **2**, 6 (1983).
138. S. Neustrolov and E. Sokolov, *Russ. Chem. Rev.* **41**, 795 (1972).
139. N. Nakayama, Y. Machida, S. Furiya, S. Yamamoto, and T. Hitsatsugu, Eighth Int. Conf. Electron, Ion, and Photon Beams, Electrochemical Society, 1978, p. 252; *Electrochem. Soc. Ext. Abstr.* **82-1**, 509 (1982).
140. B. Lin, *J. Vac. Sci. Technol.* **12**, 1289 (1975), **12**, 1284 (1975).
141. H. Nakata, T. Kato, A. Murata, Y. Hirai, and K. Nagami, *J. Vac. Sci. Technol.* **19**, 1248 (1981).
142. P. Shah, G. Pollack, R. Miller, G. Varnell, U. Lee, R. Love, S. Wood, and R. Robbins, *J. Vac. Sci. Technol.* **19**, 905 (1981).
143. G. Varnell and P. Shah, *Proc. IEEE* **71**, 612 (1983).
144. Y. Watakale, T. Kato, A. Shigetomi, and H. Morimoto, *J. Vac. Sci. Technol.* **21**, 1005 (1982).
145. S. Fok and G. Hong, Proc. 1983 Kodak Microelectronic Seminar, San Diego, p. 159.
146. G. Martel and W. Thompson, *Semicond. Int.* Jan. 1979, p. 78.
147. M. Inoue, *Jpn. Electron. Eng.* May 1982, p. 61.
148. T. Berker and D. Casey, *IEEE Trans. Electron Devices* **ED-29**, 524 (1982).
149. U. S. Patents 3,046,125, 3,042,516, 3,042,518, Horizons.
150. M. Gazard, A. Chapiro, J. Dubois, and C. Duchesne, *Polym. Eng. Sci.* **20**, 1069 (1980).
151. A. Novembre and M. Bowden, *Polym. Eng. Sci.* **23**, 975 (1983).
152. J. Shaw, M. Hatzakis, E. Babich, and J. Liutkus, *Polym. Eng. Sci.* **23**, 1047 (1983).
153. G. Taylor, T. Wolf, and L. Stillwagon, *Solid State Technol.* Feb. 1984, p. 145.
154. M. Suzuki, K. Sargo, H. Gokan, and Y. Ohnishi, *J. Electrochem. Soc.* **130**, 1962 (1983).
155. W. Liu, J. Corelli, A. Steckl, J. Moore, and J. Silverman, *Appl. Phys. Lett.* **44**, 973 (1984).
156. S. Imamura, T. Tamamura, K. Sukegawa, O. Kogure, and S. Sugawara, *J. Electrochem. Soc.* **131**, 1122 (1984).
157. Japanese Patent 58,16,234, *Chem. Abstr.* **101**, 15307 (1984).
158. K. Tanigaki, Y. Ohnishi, and S. Fugawara, *Am. Chem. Soc. Symp. Ser.* **242**, 177 (1983).
159. P. Whipps, in *Microcircuit Engineering*, edited by J. Cleaver, H. Ahmed, and G. Jones, Academic Press, New York, 1984, p. 279.
160. M. Morita, S. Imamura, T. Tamamura, O. Kogure, and R. Murase, *J. Electrochem. Soc.* **131**, 653 (1984).
161. M. Morita, S. Imamura, T. Tamamura, O. Kogure, and R. Murase, *J. Electrochem. Soc.* **131**, 2402 (1984).

162. J. Lane, *SPIE Proc.* **448**. 119 (1984).
163. C. Rosilla, A. Rosilla, and F. Buiguez, *Microelectron. Eng.* **1**, 197 (1983).
164. Japanese Patent 58,105,140 (1983), *Chem. Abstr.* **101**, 120489 (1983).
165. M. Tsunooka, T. Ueda, and M. Tanaka, *J. Polym. Sci.* **22**, 2217 (1984).
166. Y. Tabata, S. Tagawa, and M. Washio, in *Materials for Microlithography*, edited by L. Thompson, C. Willson, and J. Frechet, *Am. Chem. Soc. Symp. Ser.* **266**, 151 (1984).
167. R. Taracson, M. Hartney, and M. Bowden, *Am. Chem. Soc. Symp. Ser.* **266**, 361 (1984).
168. A. Ledwith, M. Mills, P. Hendy, A. Brown, S. Clements, and R. Moody, *J. Vac. Sci. Technol.* **B3**, 373 (1985).
169. M. Hartney, R. Tarascon, and A. Novembre, *J. Vac. Sci. Technol.* **B3**, 360 (1985).
170. F. Lin, B. Gong, Y. Ye, and C. Ku, *J. Vac. Sci. Technol.* **B3**, 373 (1985).
171. U. S. Patent 4,508,813 (1985), Fujitsu.
172. U. S. Patent 4,513,076 (1985), GAF.
173. C. Lewis, *J. Polym. Sci.* **21**, 1543 (1983), U. S. Patent 4,517,276 (1985), Varian.
174. German Patent 3,414,104, *Chem. Abstr.* **102**, 87568 (1984).
175. Japanese Patent 59,146,047, *Chem. Abstr.* **102**, 70253 (1984).
176. T. Ueno, H. Shiraishi, T. Twayangi, and S. Nonogaki, *J. Electrochem. Soc.* **132**, 1168 (1985).
177. M. Hartney and A. Novembre, *SPIE Advances in Resist Technology and Processing* **539**, 90 (1985); SPE RETEC Photopolymer, Ellenville, N.Y., 1985, p. 211.
178. R. Brault, R. Kubena, and R. Metzger, *SPIE Advances in Resist Technology and Processing* **539**, 70 (1985).
179. R. Daly, M. Hanrahan, and R. Blevins, *SPIE Advances in Resist Technology and Processing* **539**, 138 (1985).
180. M. Toukhy, *SPIE Advances in Resist Technology and Processing* **539**, 131 (1985).
181. Y. Ohnishi, M. Suzuki, K. Saigo, Y. Satome, and H. Gokan, *SPIE Advances in Resist Technology and Processing* **539**, 62 (1985), *J. Electrochem. Soc.* **133**, 977 (1986).
182. H. Gokan, Y. Saotome, K. Saigo, M. Suzuki, and Y. Ohnishi, *J. Electrochem. Soc.* **132**, 909 (1985).
183. S. Imamura, T. Tamamura, and O. Kogure, *Polym. J.* **16**, 441 (1984).
184. U. S. Patent 4,508,813 (1985), Fujitsu.
185. K. Okada and J. Matsui, *Jpn. J. Appl. Phys.* **24**, 772 (1985).
186. U. S. Patent 4,474,869 (1984), Hughes; 4,587,203 (1986).
187. Y. Suzuki, *J. Vac. Sci. Technol.* **B3**, 1009 (1985).
188. U. S. Patent 4,535,054 (1985).
189. Japanese Patent 60,15,636, Maruzen; *Chem. Abstr.* **102**, 212703 (1985).
190. Japanese Patent 60,80,852, NTT; *Chem. Abstr.* **103**, 79513 (1985).
191. Y. Suzuki, *Solid State Technol.* **28**(5), 197 (1985).
192. K. Tanigaki, M. Suzuki, Y. Saotome, Y. Ohnishi, and K. Tateishi, *J. Electrochem. Soc.* **132**, 1678 (1985).
193. K. Tanigaki, M. Suzuki, and Y. Ohnishi, *J. Vac. Sci. Technol.* **B3**, 1594 (1985).
194. U. S. Patent 4,456,678 (1985), GTE.

6

Surface Preparation and Coating

6-1. INTRODUCTION

The coating of a device surface with resist is similar to the common practice of applying paint films. First, the correct paint must be selected for the intended purpose of resisting heat, radiation, physical forces, or chemical action. Next, the surface is cleaned to remove contaminants and a primer is applied to improve the adhesion of the paint film. Then, a variety of coating techniques such as spray, roll, dip, and spin can be used to apply the paint film. After drying, the paint may be hardened by the action of heat, electron beams, or by chemical cross-linking reactions. While these steps comprise the coating of paint films, similar processes are used to coat resist films.

The films used in fabricating semiconductor devices are formed by deposition (metals, silicon nitride, polysilicon, quartz) or oxidation (oxidized silicon). Besides these standard surfaces, other surfaces such as films of polymers in multilayer resists may be encountered in the lithographic process sector.

Since the surfaces of the films are in a pristine state immediately after deposition, an immediate application of a resist coating would avoid all the costly wafer cleaning steps. Unfortunately, there is a long time lag between hot processes (deposition and oxidations) and wet lithography operations, before a wafer is coated with a resist. With the many storage and handling steps a wafer undergoes prior to the lithography operation, contamination inevitably occurs. Eventually, the contamination can lead to the loss of active device area due to pinholes or opaque defects (extraneous material) on the wafer. As such, contamination can exist on the wafer surface as particulates or films composed of metallic, inorganic, or organic constituents (Table 6-1-1) and mixtures thereof.

Particulates can be defined as discrete pieces of matter ranging in size from 100-nm diameter up to granules visible to the human eye. The film-forming contaminants may be atomic, ionic, or polymeric in character. Even surface dirt as small as one-fifth the size of the finest circuit can be fatal to the device's performance. Moreover, invisible film contamination causes even larger yield losses such as gross resist adhesion failure during development or etching operations. Secondly, the prevention, detection, monitoring, and identification of invisible contaminants require sophisticated and costly equipment. In the first step, it is preferable to eliminate or prevent contamination by providing a total system to detect, monitor, and remove or reduce the sources of contaminants to a tolerable level.

TABLE 6-1-1. Semiconductor Device Films

Metallic	Organic	Inorganic
Cu	Polyimide	Oxidized Si
Al	Baked DQN	Silicon nitride
Cr	Silicones	Polysilicon
Si	Resists	Quartz
Au		Doped oxides
Permalloy		GaAs
Mo		Metal silicides
Alloys		
W		

After cleaning the surface to restore the film to its pristine condition, the resists are generally applied by spin coating to a thickness of 0.5–5.0 μm. For optical alignment considerations, submicrometer images exposed by optical steppers require resist films to be as uniform as ± 100 Å in order to avoid overlay errors in a stepper which constantly senses the alignment marks beneath the resist film.[1] Figure 6-1-1 shows that a nonuniformity of > ±200 Å in a resist thickness deposited over oxide, can vary the reflectivity by a factor of 2 and significantly alter the throughput in wafers per hour (WPH).

In the resist process, many physical properties of an optical, mechanical, electrical, and chemical nature are dependent on the static and dynamic resist thickness during the course of lithographic operations (Table 6-1-2). Secondly, many process steps are dependent on the diffusion of small molecules *into* or *out* of the film (apply, bake, expose, develop, and strip). Several of the key resolution and resist protection factors are illustrated in Fig. 6-1-2. Both single-film and multilayered resists are used to produce integrated circuits. A thick resist film provides better adhesion, reduced pinholes, and greater protection for reactive ion erosion (RIE). However, thinner films are favored for higher[2,3] resolution (Fig. 6-1-3) because of the reduced forward and backscattered radiation. For example, for a line and space patch[4] of 1-μm by 1-μm widths exposed by a 25-keV *e*-beam, a bias of 0.5 μm is evident for a 1.5-μm-thick film (Fig. 6-1-4). For a 0.5-μm-thick film, the bias is only ±0.21 μm.

Negative resists,[5] such as azide-rubber based (KMER, Waycoat, and OMR-83), because of swelling and exposure by backreflected radiation, tend

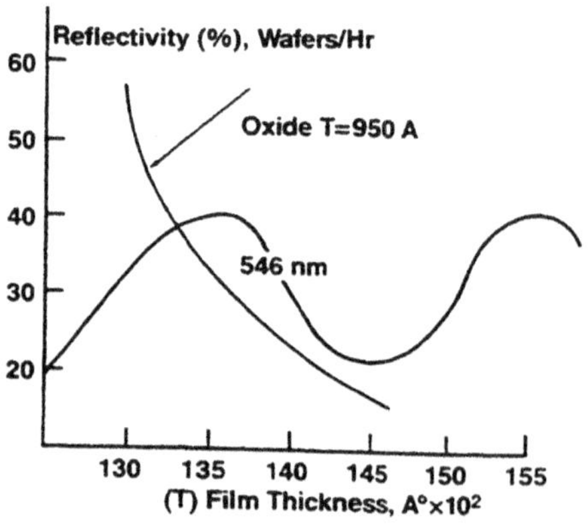

FIGURE 6-1-1. Reflectivity and throughput of an optical stepper versus resist thickness for alignment at 546 nm. From Ref. 1, courtesy of Cahners Publishing Co.

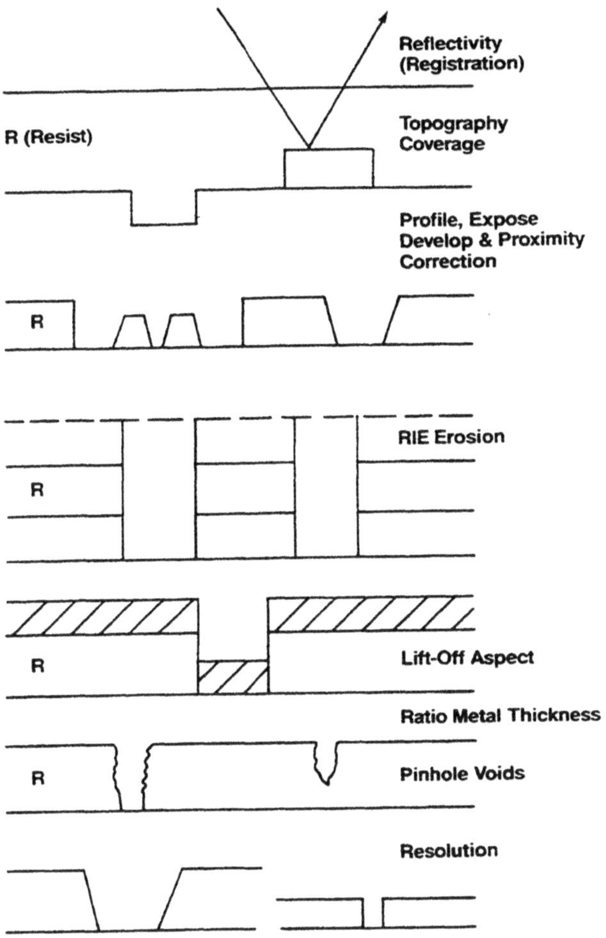

FIGURE 6-1-2. The importance of resist thickness in various lithographic processes.

to be of poorer resolution than positive diazoquinone–novolak (DQN) resists (Fig. 6-1-5). Therefore, thinner negative resists are used at the risk of pinholes and insufficient plasma erosion protection.

Thus, trade-offs are likely to occur in resist operations. Some of the basic resist thickness trade-offs and the operating ranges of thicknesses for processing resists are given in Table 6-1-3. Because of the thick film requirements of plasma etching, or the thick film requirements for lift-off (Chapter 12) and sufficient step coverage, submicro-meter-thick resist films are not likely to be used.[6] In order to provide sufficient protection from plasma and ion erosion, the resist thickness should be about twice the thickness of the semiconductor films to be patterned.

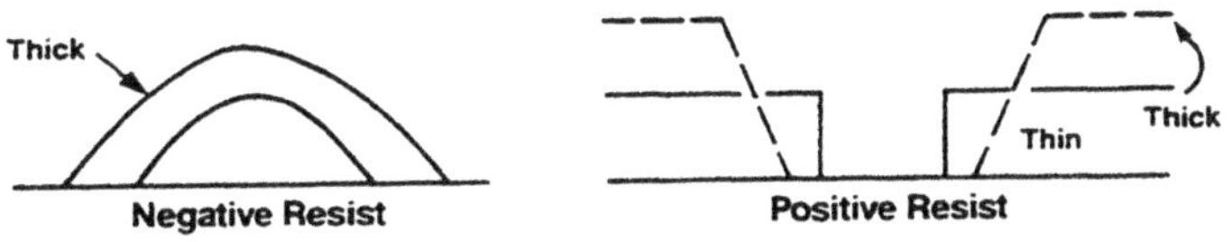

FIGURE 6-1-3. Effect of resist film thickness on resist resolution. Thicker films, especially negative cross-linking types, have poor resolution.

TABLE 6-1-2. The Thickness Factors in Resist Process Steps

Process step	Thickness factor
Apply	Step
	Planarization
	Pinholes
	Spin coating
Prebake	Time
	Temperature
Expose	Dose
	Contrast
	Resolution
	Alignment
Develop	Time
	Film thinning ratio
Etch	Selectivity in plasma
Lift-off	Aspect ratio
Strip	Time
Ion implant	Stopping power

It is instructive to compare the influence of two resist film thicknesses (0.5 and 2.0 μm) on the performance of a resist (Table 6-1-4). The better resolution and smaller image biases found in the thinner (0.5 μm) film are exchanged for the poorer resist performance in etching, defect level, and in lift-off.

Thinner and more uniform films with zero defects will be required for denser submicrometer circuitry. For better resist performance, multilayer resists trade off thinner recording layers for better resist performance in the underlayer material. Ultraclean classrooms and process equipment with low particle generation will be needed to deposit defect-free resist films.

Many process steps are time dependent on the absolute resist thickness and good uniformity from wafer to wafer. In Table 6-1-5, the dependency of resist thickness (d) versus time (t) can take several forms: linear, exponential, or a power function thickness. The exposure and development of resists have the most critical thickness dependency.

In the static or dynamic mode, the accurate measurement of resist thickness in developing a latent image is vital to reproducible processing. While the vertical resist thickness provides the "resistance" in processing, the horizontal components of an image are time-dependent processes (exposure, develop, and etch time).

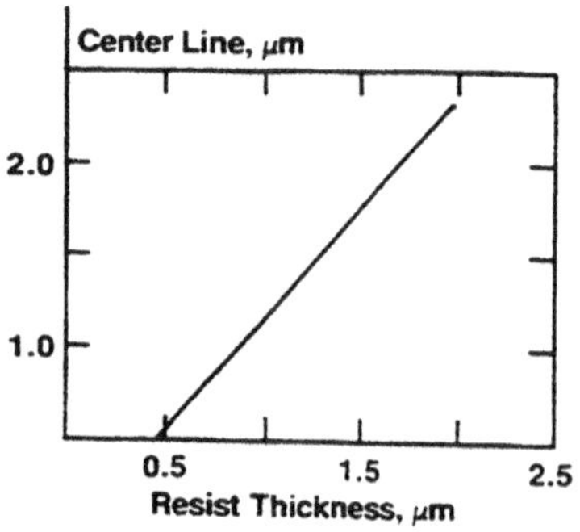

FIGURE 6-1-4. Center linewidth of a 1.0-μm by 1.0-μm array versus resist thickness. Courtesy of N. Eib, IBM.

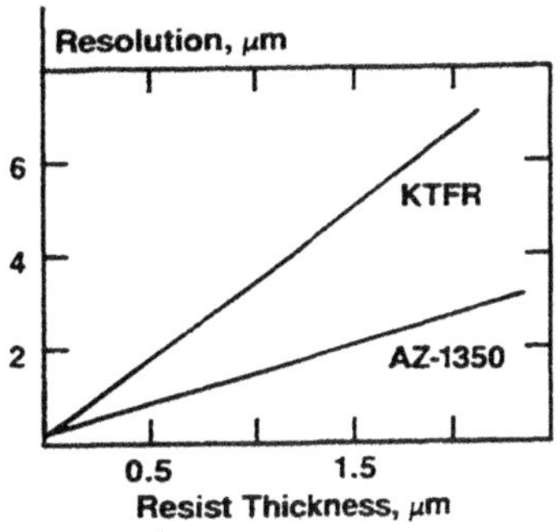

FIGURE 6-1-5. Resolution versus thickness of resist for commercial Kodak KTFR negative photoresist and Azoplate AZ-1350. From Ref. 5, courtesy of SPSE.

Process engineers seek to minimize isotropic biases by limiting the horizontal components.

The resist coating must cover and planarize steep topography. As a device is fabricated, layers of insulation and metals deposit in a conformable fashion (Fig. 6-1-6) and create an uneven surface prior to resist lithography. The ability of a resist coating to be conformal or planarizing (Fig. 6-1-7) is dependent on the step height, the radial orientation of the step (parallel or perpendicular to the center line of the wafer), the distance between steps, and the viscoelastic properties of the dried resist film. Deposition of resists from the gas phase tend to be conformal because each monomolecular layer is put down consecutively. In the liquid-phase processes, excess resist is applied to the wafer surface and planarized by spinning to level the coating. For optical submicrometer printing, the planarized coating is not ideal, because the depth of focus of printing is close to the film thickness (2–3 μm). Because of the variation of resist thickness in the vicinity of a step (Fig. 6-1-8), images tend to be constricted.

Conformal resist coatings are rare. In order to overcome the imaging limitations imposed upon a single resist layer[7] (standing waves), multilayer resists (Chapter 12) which separate the image layer from the "resist layer" are becoming

TABLE 6-1-3. Uniformity and Range of Thickness of Semiconductor Films

Consideration	Range (μm)	Uniformity (Å)
Thickness		
Present	0.3–4.0	±100
Future	0.1–1.5	±30
Step coverage	0.5–1.5	
Planarization	0.5–1.5	
Pinholes	0.1–0.5	$<1/cm^2$
Striations		±25
Aspect ratio[a]	1.0–2.0	
Etch selectivity[b]	2:1 to 0.5:1	
Film thinning ratio[c]	6 to 20	
Resolution	1/5 to film thickness	

[a] Lift-off process.
[b] Plasma etch rate of resist to underlayer.
[c] Exposure dissolution/unexposed dissolution (positive resist example).

TABLE 6-1-4. Effect of Film Thickness on Resist Performance or Processing of Positive Resist[a]

Performance	Thickness	
	0.5 μm	2.0 μm
Resolution	0.5 μm	2.5 μm
Pinholes	5/cm^2	1/cm^2
Adhesion	Worse	Better (less stress)
Etch resistance		Better
Uniformity of film	±50 Å	±150 Å
Prebake	Shorter	
Exposure time	2 sec	3.5 sec
Develop time	60 sec	175 sec
Strip time	15 min	20 min
Lift-off		Easier
Melt resistance		Higher

[a] Hypothetical positive resist example.

more popular; for use in lift-off[8] and dry etching.[9] Dyes are added to the underlayers to absorb the reflected UV of the exposing beam. Planarizing coatings, such as polyimides,[10] are useful as dielectric isolation coatings, or as underlayers for multilayer resists.[11] In multilayer resists, the image layer (Fig. 6-1-9) is separated from the planarizing layer to provide maximum resolution from the resist.

TABLE 6-1-5. Thickness (t)–Time Dependencies (d) in Processing Resists

Process step	Time factor
Apply	Spin time $t \propto d^{2/3}$
Prebake	$t \propto d^2$
Expose	Dose (t) $\propto$ exp(d)
Develop	$t \propto d$ or exp(d)
Postbake	$t \propto d^2$
Etch rate	$t \propto d$
Strip rate	$t \propto d^2$

TABLE 6-1-6. Resist Apply Techniques

Technique	Range (μm)	Uniformity (%)
Spin	0.001–10	±1
Spray	1–20	±10
Roll	1–20	±5
Dip	1–15	±10
Laminate	5–100	±10
Plasma deposit	0.01–2	±5
Monomer polymerize	0.001–0.1	±2
Vacuum deposit	0.001–1.0	±2
Langmuir films	0.00001–0.3	±2

Besides resist films, other polymeric coatings are used in semiconductor manufacture. These include planarizing high-temperature underlayers, barrier layers for multilayer masks, and insulating dielectrics such as polyimides to replace sputtered quartz. Specialty polymers have been applied in four basic areas: (1) photomasks, (2) silicon wafers, (3) multilayer resists, and (4) dielectric layers. Some examples are given below. The polymers may contain additives such as dyes for antireflection purposes.

Masks

Photomask protective coating
Mask cleaning coating
Photomask release layer
Photomask pellicle membrane
X-ray membrane
X-ray secondary emission absorber

Wafer

Backside etching protector
Dicing protective coating
Dielectric isolation
Spin on dopant
Contrast-enhancing layer
Oxygen effect prevention coating

Multilayer resists

O_2 plasma barrier layer
Antireflective underlayer
Planarizing underlayer
Antireflective O_2 barrier layer
Planarizing release layer

In many cases, the techniques for depositing resists have been adopted from the paint[81] industry. A liquid resist can be coated by a variety of solvent evaporative techniques. Vacuum deposition is limited to stable monomers or inorganic resists. In Table 6-1-6, the various coating techniques are ranked in order of preference. Their useful range of thickness and uniformity are also included.

Spin coating of resist liquids of 20–200 centistokes (viscosity) is the predominant and most uniform process. For coating a resist, the process steps which are given below use surface cleaning, priming, coating, and baking, followed by some form of measurement or assessment.

Step	Measure
Clean wafer	Contamination
Rinse and dry	Wettability
Prime surface	Wettability
Apply coat	Viscosity
Prebake	Percent residual solvent
Measure	Thickness, pinholes, contamination

The cleanliness of the wafer, the environment, and the process are reflected in the defect levels, such as pinholes in the film. Figure 6-1-10 illustrates the pinhole

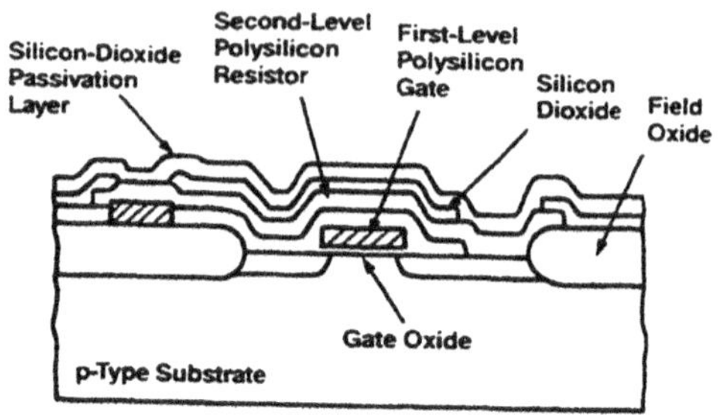

FIGURE 6-1-6. Rough terrain of a 16K memory cell requiring resist step coverage. From *Electronics*, August 18, 1977, p. 10; copyright 1977 by McGraw-Hill, all rights reserved, reproduced by permission.

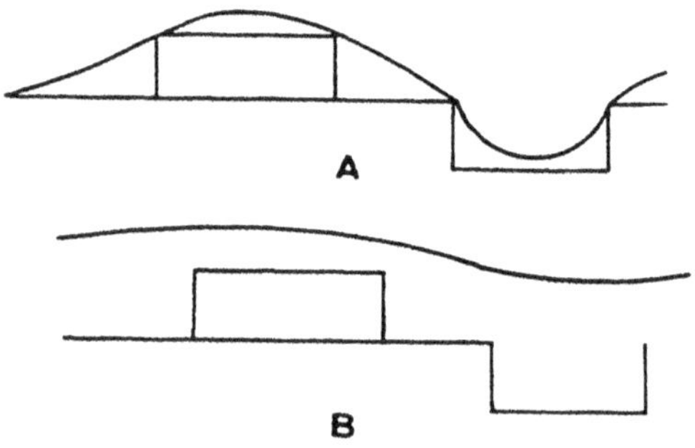

FIGURE 6-1-7. (A) Failure of conformable resist coating of insufficient thickness to cover steep or deep topography features on semiconductor devices. (B) Traditional coverage of steep or deep topography by planarizing resist coating applied by spin casting.

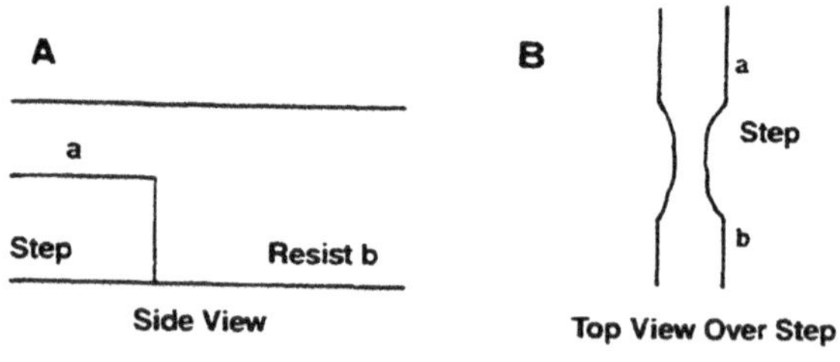

FIGURE 6-1-8. (A) Cross section of planarizing resist coating traversing a step. (B) Traditional problems of printing single film resist with optical exposure over a step. Image distortion occurs due to standing (reflective) light and variation in thickness.

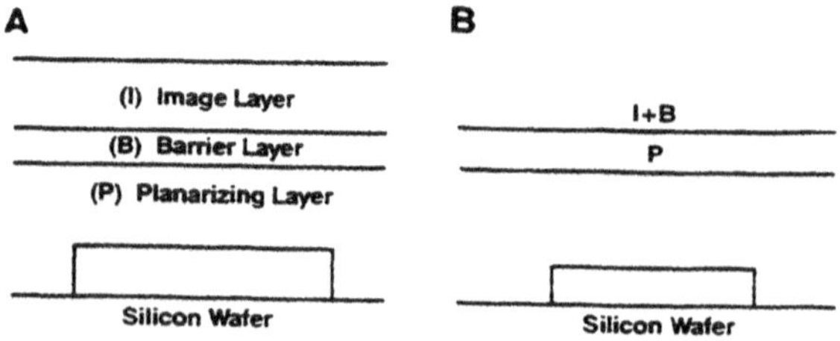

FIGURE 6-1-9. Multilayer resist configurations to overcome dogbone effect in Fig. 6-1-8. (A) Trilayer resist of image layer, oxygen-resistant plasma barrier layer, and planarizing stable underlayer. The image layer is transferred to the underlayer by a series of RIE processes (see Chapter 12). (B) A bilayer configuration which combines the image and barrier layer into one layer.

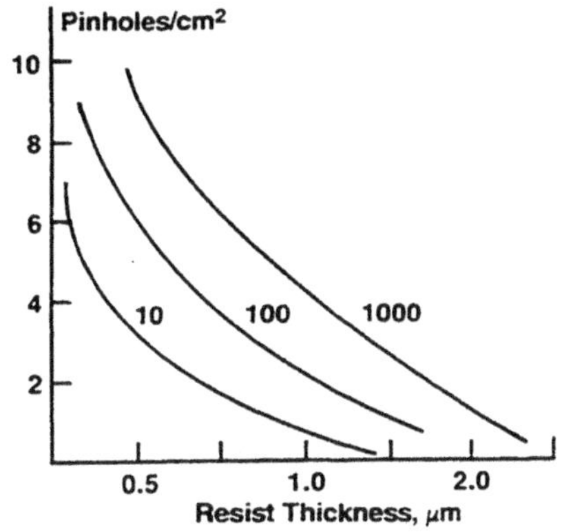

FIGURE 6-1-10. Pinholes in resist films cast under different classes of clean room conditions versus resists thickness.

level in a resist film spun cast in clean rooms of classes 10–1000. Class 10 rooms have only 10 particles/ft^3 in the air of less than 0.5-μm diameter. In general, the cleaner the work area, the lower is the defect level. However, the thinner the film, the greater is the likelihood of defects reaching the underlying substrate film. The control of the liquid- and gas-borne contamination by filtration and cleaning is vital to most phases of semiconductor processing. Furthermore, the defect analysis of device failure has established a direct relationship between the surface-deposited contaminants and the probability of device failure. In turn, the yield of fabricated devices is an exponential function of the cumulative defects.

Since spin coating is the dominant process for applying resists in the 0.5- to 4.0-μm thickness range, we will concentrate on the physical and chemical aspects of the process. First, we will discuss contamination and surface cleaning, spin coating, and then thickness and defect measurements. Afterwards, alternate coating techniques will be discussed. The chemistry of primers such as silanes will be deferred to Chapter 13. Some of the plasma and organic-based stripping processes to rework wafers can be found in Chapter 14.

6-2. CLEANING THE WAFER SURFACE

6-2-1. *Contamination and General Consequences*

Every surface is covered with foreign substances. Some are contaminants because they form film defects or eventually cause device failure. Human hair is a very large contaminant (100-μm diameter) compared to the size of a silicon chip (Fig. 6-2-1). On a silicon wafer, random deposits can originate from the wafer itself (edge defects), process equipment, storage boxes, personnel, and the room air. Most particles are undetectable by the human eye and require sophisticated light scattering and diffusion counters to detect their presence in the diameter range of 0.3–10 μm. In turn, particles give rise to random defects such as etched pinholes, unetched opaque spots, and interlevel shorts or opens.

Invisible surface films are a costly form of contamination. An absolutely clean surface of silicon can only be obtained by cleavage in a high vacuum. With ordinary silicon wafer processing, the surface is covered with one or more molecular layers of oxygen, water, and adsorbents. The surface is said to be clean (1) uniform, (2) volatile, soluble, or convertible into volatiles, (3) noninterfering with resist adhesion, development, lift-off, or etching, and (4) unable to affect the electrical or

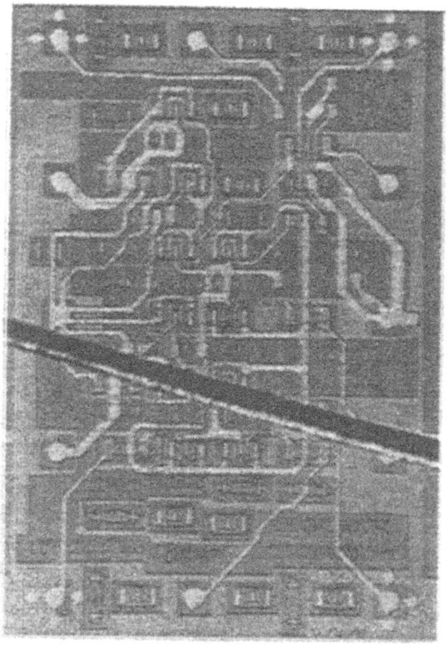

FIGURE 6-2-1. Human hair (100 μm) as a fatal defect relative to the size of devices on a semiconductor chip. Courtesy of R. Tabone, IBM.

reliability properties of the finished device. The essence of the concept of cleanliness, then, lies in its systematic control. Some examples (see Table 6-1-1) of invisible contaminant films are plasticizers from any plastic storage or processing equipment, deposits of pump oil, lubricants, plasma etch deposits, and adsorbed metal salts. Particulate contaminants as small as a fraction of the smallest circuit can be the cause of fatal defects. Some examples are shown in Fig. 6-2-2. At present, the tolerable defects for 1985 circuits (0.5 μm) become, in 1990, the new circuit dimensions.(13,24,228)

In the fabrication of chrome masks, laser light scan detectors are used to monitor surface contamination. A dirty working environment is the major contributor of a photo-limited yield (Fig. 6-2-3A). Particle contamination is estimated to be the cause of 20% integrated circuit loss (Fig. 6-2-3B).

At the packaging level, a clean assembly area is vital to maintaining the highest yield for placing chips on a circuit board (Fig. 6-2-3C).

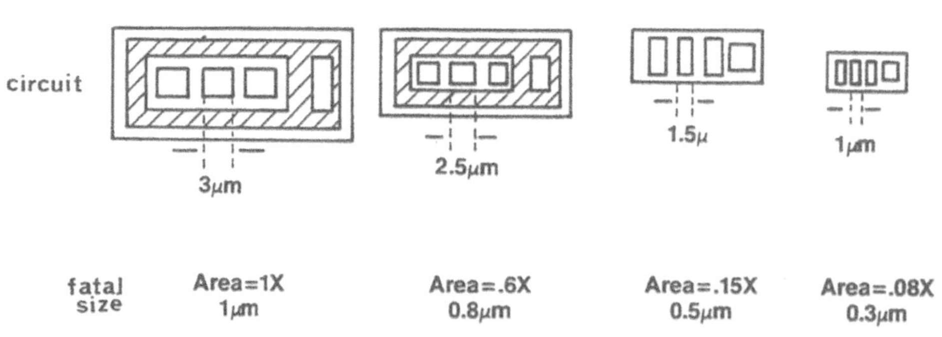

FIGURE 6-2-2. Fatal defects relative to the size of circuit features in logic devices. Courtesy of A. Rapa, IBM.

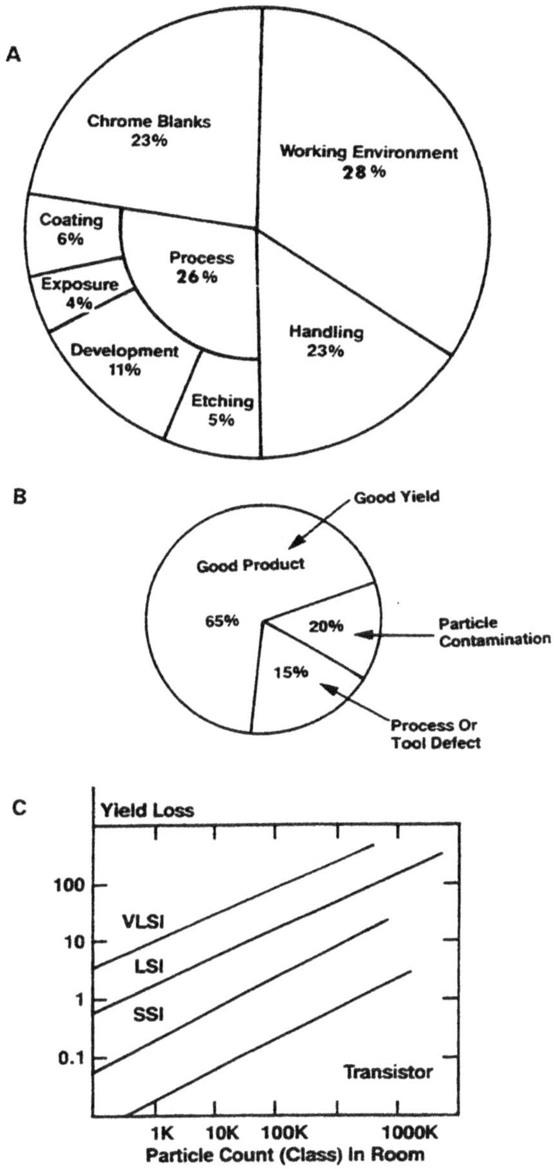

FIGURE 6-2-3. (A) Causes of defects in the resist fabrication of chrome masks. From Ref. 12. (B) Yield losses in semiconductor device fabrication due in part to particle contamination from the different classes of clean rooms. From Ref. 13, courtesy of *Semiconductor International*, published by Cahners Publishing Co. (C) Yield losses in packaging due to environmental defects introduced in classes of clean room conditions. From Ref. 13.

As chips become larger (Fig. 6-2-4A) and metal lines come closer together (Fig. 6-2-4B), and for good yields, the number of contaminants (defects) on a surface of a wafer and the size of the defects must be reduced. As shown in Fig. 6-2-4, wafers with smaller chips yield 75% defect-free chips, while wafers with large chips yield only 33% good chips. Typical in today's integrated circuit industry is a chip of 0.5 cm/side and a process defect density of 5 defects/cm^2. A chip yield of 30% can be expected.[13] Since defects have limited the growth of chip sizes, the only way to increase functional integration has been to pack more circuitry onto one chip with closely spaced features. A dust particle causing a pinhole in a resist line used for lift-off would thus leave extra metal and cause a short (Fig. 6-2-4B).

The battles against contamination are waged on three fronts: (1) prevention by

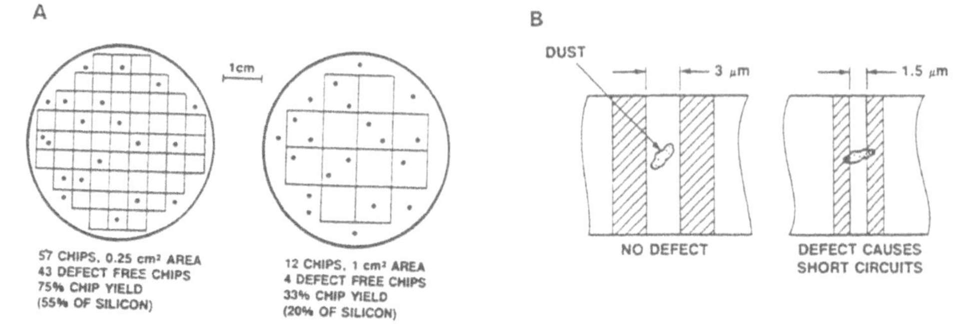

FIGURE 6-2-4. (A) Defect density is lower for smaller chips. (B) Fatal metallic dust particle causing short in smallest pitch circuit lines. From Ref. 227.

filtration and design, (2) monitoring and identification of sources, and (3) removal by physical and/or chemical means. Entropic forces drive the transport of contaminants to a wafer surface where they accumulate as a hierarchy of deposits (nonpolar on top of polar materials; see Fig. 6-2-7). Contaminant control is a system and all of the materials, chemicals, and equipment working inside a clean room are dependent variables. The most difficult task that persists is to identify the chemical structure and source of contamination. From the analytical data, the source of contamination can be identified and its reoccurrence prevented.

6-2-2. *Causes and Mechanisms of Surface Contamination*

Defect-causing particles that reach the wafer surface are generated by four major sources (Fig. 6-2-5). These sources in turn require parameters that operate as control systems:

Major particle sources	Control parameters
Resist process equipment	Design, materials, operation
Process chemicals	Purity, filtration, packaging
Clean room environment	Design, filtration, monitors
Operating personnel	Garments, number, training

The interaction of the sources as seen in Fig. 6-2-6 involves the control of major factors of personnel, equipment, and utilities to achieve a superclean room. From these sources, the relative number of particles generated varies from one process

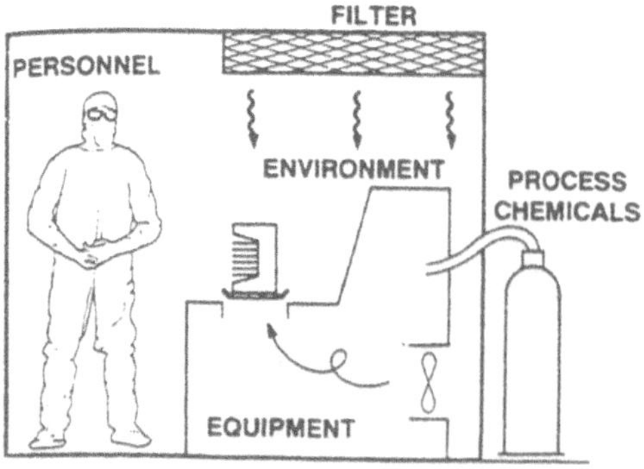

FIGURE 6-2-5. The four major sources of particles in a wafer fab line. From Ref. 227.

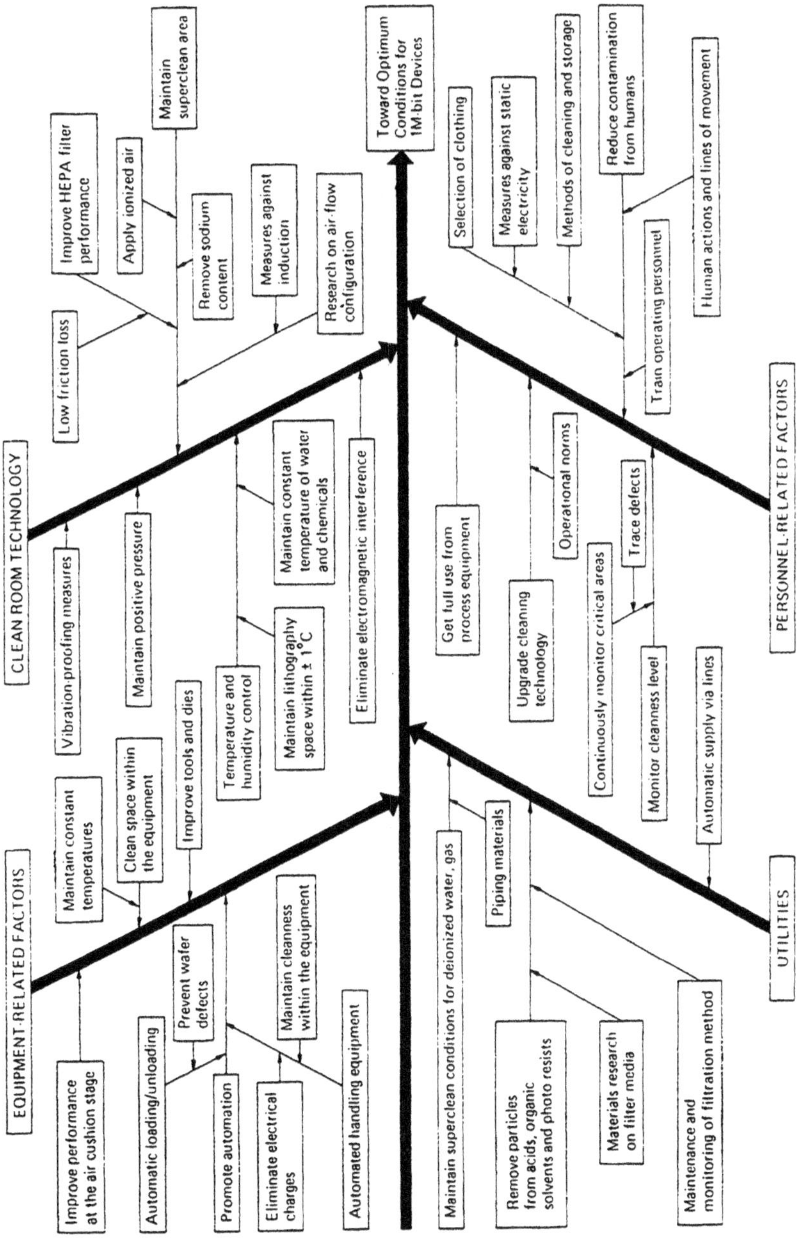

FIGURE 6-2-6. Cause-and-effect diagram for achieving superclean room. From T. Hashimoto, Challenging Class 0 Problems in Superclean Environments for 1M-bit Devices, *Jpn. Semicond. Technol. News* **2**, No. 2, April 1983, p. 20, copyright and courtesy of Science Forum Inc.

area to another. The relationships between cleanliness of the environment and the fallout of particulates are complex and depend on many factors like airflow and direction, turbulences caused by equipment, human traffic, and the duration of exposure of wafers to the fallout. In order to keep the entire facility at extremely high levels of cleanliness, a compromise of cost is to use small isolated clean spaces to protect the wafers from defect-causing particles.

The transport of aerosol particles and their deposition rate on a wafer can be estimated from gas-phase kinetics.[13] If we assume a class 100 clean room (100 particles/ft^3), the number of particles colliding with a unit area of surface per unit of time is given by Boltzmann statistics,

$$N = N'(kt/2\pi M) \tag{6-2-1}$$

where N' is the number of particles per unit volume of mass M. If we assume a particle density of 2, this gives a deposition rate of 1 particle/cm^2-hr. A device with 20% active area at a fatal limit of 1 defect/cm^2, will be ruined by a 5-hr exposure in a class 100 room.[13] In air, particles with less than a 0.5-μm diameter exhibit more violent Brownian motion and collide to form an agglomerate.

A clean surface is one that contains no significant amounts of undesirable material; thus, what constitutes a clean surface depends on the device requirements.[14] Since cleaning is often associated with good adhesion, contamination is associated with poor adhesion and pinholes or voids in the resist film. Contamination in resist processing can arise in several areas: in the coating process, printed in the resist, or redeposited during developing or etching. Every piece of physical equipment and every person emits, adsorbs, and reemits particles or contaminants. Eventually, personnel will have to be removed from lithography operations. On the wafer surface, contaminants can be present as solids, liquids, gases and as mixtures (Table 6-2-1). Contaminants, especially solids, tend to aggregate on top of other contaminants. The hierarchy of spontaneously absorbed layers on a wafer surface is shown in Fig. 6-2-7.[20]

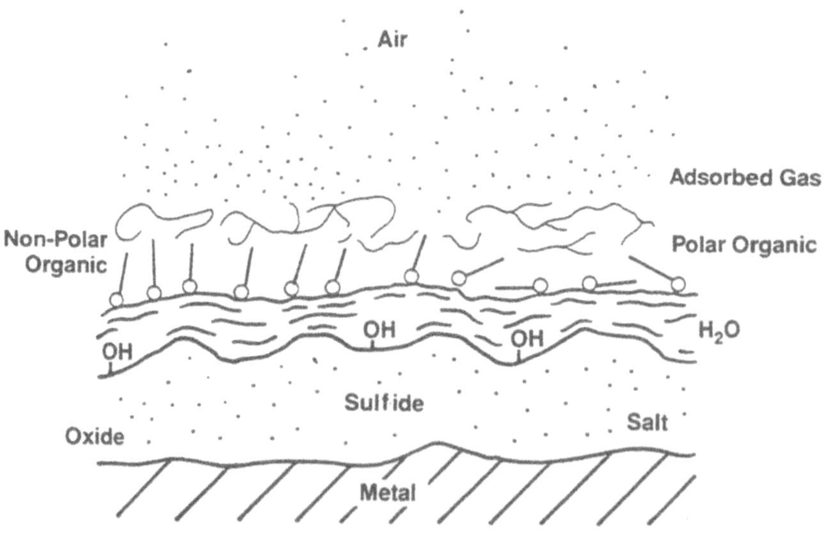

FIGURE 6-2-7. Hierarchy of different types of contaminants adsorbed on a metal surface. From Ref. 20, courtesy of Plenum Press.

TABLE 6-2-1. Contaminants on a Wafer Surface

Solid	Liquid	Gas
Hair, lint[a]	Oil	Fumes
Flakes[b]	Water	Water
Resist	Plasticizer	Ozone[16]
Salts		
Redeposit		

[a] Human origin.[15]
[b] Glass, silicon, metal.

The worst type of contaminants are metal flakes and other conductive particles. Metals are opaque to UV light. If they were to adhere to the top of a resist or near the surface, they would act as an opaque mask to exposure. Upon development, pinholes could form in a region of the exposed negative resist. A positive-acting resist would form underdeveloped areas. In plasma-etched films, extra material would remain. Unetched aluminum could cause electrical shorts. On a wafer surface, the adsorbed metals are resistant to oxidizing plasma, oxidizing liquid strippers, and organic-based strippers.

Surface contaminants can be broadly classified as molecular, ionic, and atomic[26]:

1. *Molecular*. Waxes, resins, oil, silicone release agents, plastic wafer carriers and boats emitting plasticizers and moulding materials, organic soils, and water. Organic contaminants on silicon devices may cause polarization and ionic drift due to transport of protons. Water-insoluble organic compounds tend to make the surface hydrophobic and thus make removal of trapped ions difficult. As the primary step, removal of the molecular contaminants should be considered.[27 28] Hydrocarbons are removed by oxidation[44] but silicone derivatives are resistant and require HF or CF_4 plasma.
2. *Ionic*. Ions such as Na^+ are readily absorbed. The ions damage devices by raising the surface conductivity. The removal of chemisorbed ions requires special displacement reactions with reagents such as hydrogen peroxide.[28]
3. *Atomic*. These are mainly metallic Cu, Ag, and Au. It is necessary to remove the underlayer (slightly) by etching or sputter cleaning.[14]

Solids adhere by a variety of forces[165]:

Bonding	Energy
Depression	1 eV
Capillary	3 eV
Electrostatic	10 eV
Coulombic	5 eV
van der Waals	0.5 eV

The interactions between solids which bring about adhesion can be classified into three stages[229]:

1. *Stage* A—Long-range interaction of van der Waals forces, electrostatic and magnetic attractions, which act to bring the particle to the surface and establish the adhesive contact area.

2. *Stage* B—Forces, including Stage A type, which establish the adhesive area. These are interfacial reactions such as diffusion, condensation, diffusive mixing, and mutual dissolution at the interface. Some forces such as capillary action establish a liquid-to-solid bridge between the surface and the solid particulate.
3. *Stage* C—Very-short-range forces which add to the adhesion. These include all types of chemical bonds, i.e., hydrogen bonds, charge-transfer, ionic, covalent, and metallic.

Electrostatic forces predominate for large particles and van der Walls forces prevail for the smaller particles. Charged particles have an additional coulombic force to attract them to the surface. A triboelectric series of silicon process materials has established a preferred charge for specific silicon processing materials (Table 6-2-2). The double electric layer on a wafer surface is a strong electrostatic force for bonding organics.[22]

The polymeric resist surface can literally act as "fly paper," because of the presence of polar groups (OH, COOH) which increase electrostatic forces.[21] In particular, low-melting resists such as polydimethylsiloxane readily embed par-

TABLE 6-2-2. Triboelectric Series of Si Process Materials[a]

Material	Charge
Air	Positive ↑
Human hands	
Asbestos	
Rabbit fur	
Glass	
Mica	
Human hair	
Nylon	
Fur	
Lead	
Silk	
Aluminum	
Paper	
Cotton	
Steel	
Wood	
Amber	
Sealing wax	
Hard rubber	
Nickel, copper	
Brass, silver	
Sulfur	
Acetate rayon	
Polyester	
Polyethylene	↓
Polypropylene	Negative
Polyvinylchloride	
Silicon	
Teflon	

[a] Modified from Ref. 205.

TABLE 6-2-3. Adhesion of Dust Particles[a]

Particle size (μm)	Relative adhesive force
500	1
50	100
5	10,000
0.5	1,000,000

[a] From Ref. 23.

ticulates. With decreasing particulate size and increasing room humidity ($>$ 50%), the adhesion of particles to the wafer increases.[21] Intermolecular forces depend mainly on $(\text{separation})^{-1}$ and the surface-to-volume ratio of the particle (Table 6-2-3).

In a clean room, electrostatic precipitators, static neutralizers,[164] and humidity controls[21, 158] are necessary to reduce adhesion of particles to the resist or silicon surface.

Adhesive forces are also the greatest when the two contiguous surfaces have the same polarity, e.g., hydrophobic dust on a hydrophilic surface. If, after cleaning, the oxidized silicon wafers are immediately primed with silanes, the surface tension of the wafer is lowered, and the surface becomes hydrophobic. The attraction for hydrophilic (polar) dust is lowered. In an ideal process, the resist is applied (such as in plasma system) in a vacuum chamber attached to a vacuum system which deposits semiconductor films.

On a wet surface or in the liquid state, solids can have an additional capillary force to act as a bridge between the two solids. An immersed system may experience a shielding of the capillary and the van der Waals forces, so that the total forces are less than in the dry state. In the liquid state, the solvent replaces air, and the attractive force is diminished by competition between the depositing solid and the wafer surface. Figure 6-2-8 shows the poorer adhesion of diamond particles to glass, when cast from nonpolar cyclohexane (hydrocarbon) than when cast from the more polar cyclohexanol.[29] A greater number of the smaller diamond particles adhere to the surface of the glass. Polar solvents are good surface-active agents to displace particles. Addition of traces of detergents lowers the adhesion of particles (surface-active agents) and assists in the displacement of films and particles.[228] In liquid deposition, the adhesion is governed mainly by molecular forces. In solid deposition

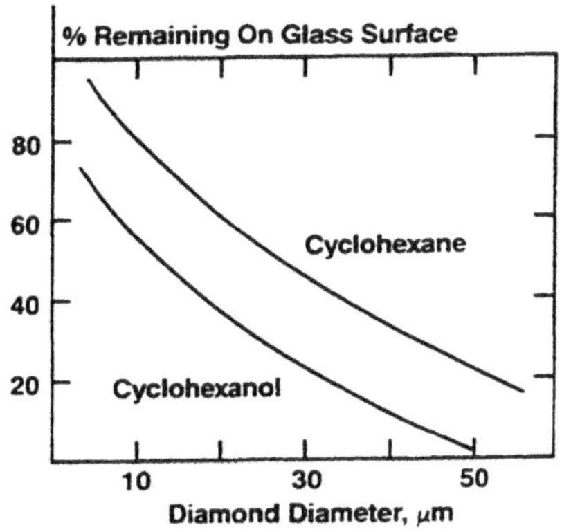

FIGURE 6-2-8. Adhesion of diamond particles to the surface of glass. The particles are cast from a solution of cyclohexane and cyclohexanol. From Ref. 29, courtesy of Plenum Press.

from air, the adhesive force increases with smaller size particles, but in a liquid the force is proportional to the size of the particle. Thus, from liquid contact we are likely to find larger micrometer particles.[23] In liquids, we are likely to find the following contaminants[30] suspended or dissolved:

Suspended		
Organic	Inorganic	Dissolved
Bacteria	Salt	Salt
Algae	Diatoms	Water
Resist	Glass	Oils
Resins	Colloidal metal, Si flakes	Plasticizers
Pollen		

Resist developers frequently contain surfactants to aid in wetting polymeric surfaces and they also prevent the redeposition of developing resist. Solid deposition from liquids occurs by:

1. Evaporation—Solvent removal after rinsing and drying, spin drying and spin coating resist, oil from air guns.
2. Sedimentation—Setting on the bottom of the container.
3. Precipitation—Loss of one solvent component or salting out due to change in the dielectric constant (polymer gels, coagulation).
4. Centrifugation—Spin coating and spin drying.
5. Langmuir adsorption—Solids suspended on liquid surface adhere to withdrawing wafer, oil films as well.
6. Condensation—Gas such as steam condenses on surface. Solid-to-solid deposition (also gas to solid) occurs by a migration mechanism or in the case of gas to solid by sublimation or condensation.

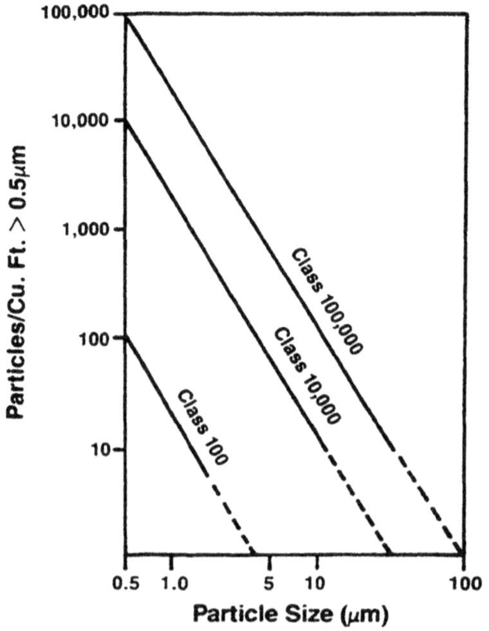

FIGURE 6-2-9. Clean room classification standards (Federal Standard 209B).

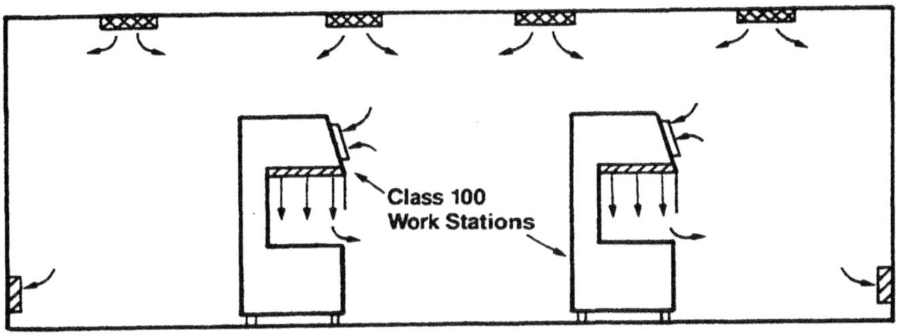

FIGURE 6-2-10. Schematic of class 100 clean room. Courtesy of A. Rapa, IBM.

Removal of contamination involves rupture of the adhesive bond by displacement or the total destruction of the contaminant by physical or chemical means. In general, submicrometer particles are greater in density and bond more strongly than large particles. Thus, the task of removing killer-size particles which are fatal to 1-μm circuits becomes a more difficult task. The problems are also compounded by the difficulty of detection and counting of submicrometer particles with light scattering equipment.

Furthermore, other factors also contribute as sources of contamination. Personnel[17] including their garments and gloves are a significant source of contamination in a clean room. Up to 6000 particles/hr[17] emanate from each wafer. Secondary contributing factors are the excessive heat and vibration, stray light, and noisy equipment. Tweezer or handling damage has been reduced by air track or contactless handling.[18] Single wafer processing,[19] in more rapid steps, is evident for better process control and lower defects.

In the next section we will examine the operation of clean rooms and the removal of contaminants from the air and from process liquids. Subsequently, in order to achieve cleanliness, the methods of wafer cleaning, and the detection of contaminants will be discussed.

6-2-3. *Removal of Contaminants*

6-2-3-1. *Clean Rooms.* Clean rooms or clean stations are compartments into which the airborne particulates are filtered to remove the aerosol content to an acceptable low level. Federal Standard 209B defines the particle population[31] (Fig. 6-2-9).

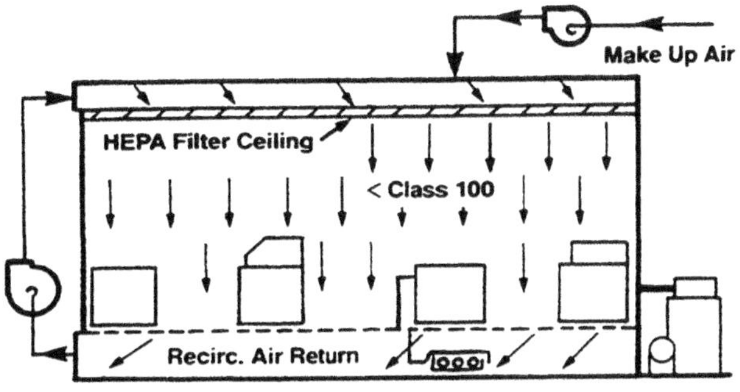

FIGURE 6-2-11. Vertical laminar flow clean room. Courtesy of A. Rapa, IBM.

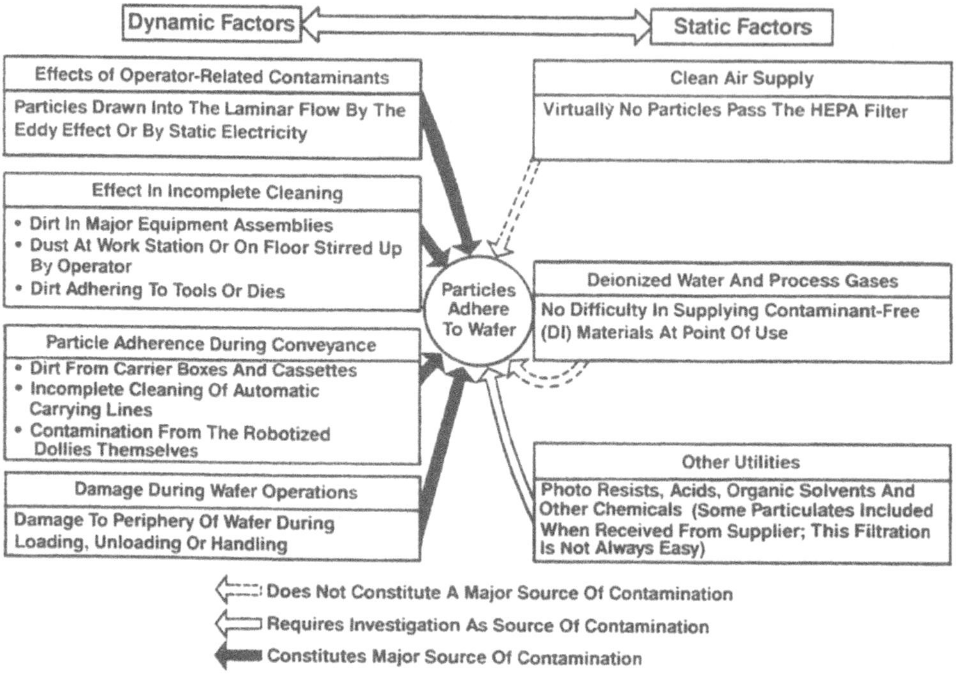

FIGURE 6-2-12. Routes by which contaminants reach wafer surface. From T. Hashimoto, Challenging Class 0 Problems in Superclean Environments for 1M-bit Devices, *Jpn. Semicond. Technol. News* **2**, No. 2, April 1983, p. 20, copyright and courtesy of Science Forum Inc.

In the inexpensive clean room, only the work station is filtered (Fig. 6-2-10). The service area and contaminating pieces of equipment are located outside the filtered work area. If the air flow is not directed, isotropic mixing limits the effectiveness of the work station. A more effective system is the laminar flow (Fig. 6-2-11) which subtends the lateral motion and confines the contaminant to the area being generated.[25]

To achieve a class 10 clean room, the whole room is cleaned by laminar flow. In order to achieve a class 0 clean room, strict control of other sources of particles such as personnel, process equipment, and services is necessary (Fig. 6-2-12). For the reduction of particulates from personnel constantly entering a clean room, the replacement of operators with robots and automated wafer handling are the next

TABLE 6-2-4. Clean Room Performance[a]

Type	Cost	Range	Control of temperature and humidity[158]	Air changes/hr
Random flow	Low	10k–100k	Fair	20/hr
Horizontal laminar flow	Low	100	Fair	100 fpm
Recirculated laminar flow	Medium	100	Good	100 fpm
Perforated ceiling laminar flow	Medium	100–500	Good	10–35 fpm
HEPA ceiling and perforated floor	High	100	Excellent	100 fpm

[a] Modified from Ref. 125.

TABLE 6-2-5. Requirements of Clean Rooms for Semiconductor Operations[a,b]

	Class				
Application	10	100	1000	10,000	100,000
Crystal growth	– – –	– – ———			
Diffusion		———			
Etching	– – –	———			
Alignment	– – –	———			
Surface treat	– – –	———			
Metallization	– – –	———			
Assembly and test		———	———	———	———
Materials			———		
Polishing		———			
Packaging					———
Storage of product				———	

[a] Modified from I. Hayakawa, Air Purification for Cleanrooms, *Jpn. Semicond. Technol. News* 3(4), 15 (1984).
[b] Particle sizes are 0.5 μm or larger.

step to total cleanliness, since many contaminants reach a wafer surface via handling and operator-borne debris (Fig. 6-2-12).[13, 15, 17, 24, 25] A summary of clean rooms and their performance is given in Table 6-2-4.[125]

In the complete fabrication of a packaged semiconductor device, the lithography operation (Table 6-2-5) should have the highest degree of cleanliness. As the density of memory circuits approaches 4M bits, the removal of 40-nm-diameter particles (as well as their detection!) will be a basic necessity. For every significant increase (Fig. 6-2-13) in memory capacity by 2^n the critical particle sizes are an order of magnitude smaller (50 nm) than the optical wavelength (500 nm) used in a conventional 1000 × microscope. Slower, costly, and more sophisticated particle surface analyzers are required to keep abreast of the submicroscopic defects.

In a clean room operation, air filtration is achieved as a result of three basic mechanisms[228]: (1) interception and sieving effects in which the particle is larger than the HEPA filter pore sizes, (2) impaction effects where the particles actually hit the filter fibers as they are thrown out of the airstream by centrifugal force, and (3) diffusion of particles to the filter fibers where they are attracted and held by a variety of surface-substrate interactions. In liquid filtration, only the first stage of interception is achieved. HEPA filters are effective for particles of about 1-μm diameter. For submicrometer particles, the most effective collection mechanism involves electrostatic-enhanced filtration. A corona system is used to charge and agglomerate the dust to increase the efficiency. As device manufacturing moves toward smaller geometries, the particles of concern will decrease in size from the present 0.1-μm diameter to 0.01-μm. Electrostatic filtration is expected to be used in

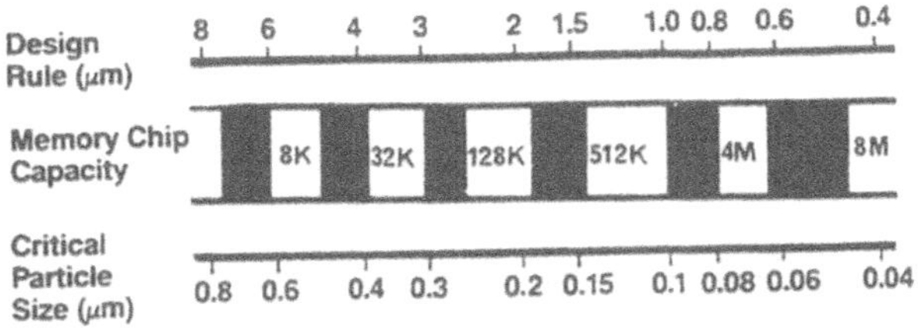

FIGURE 6-2-13. Critical defect (particle size) level for memory devices. From *Microelectronics and Manufacturing*, Dec. 1984, p. 9, copyright and courtesy of Lake Publishing Co.

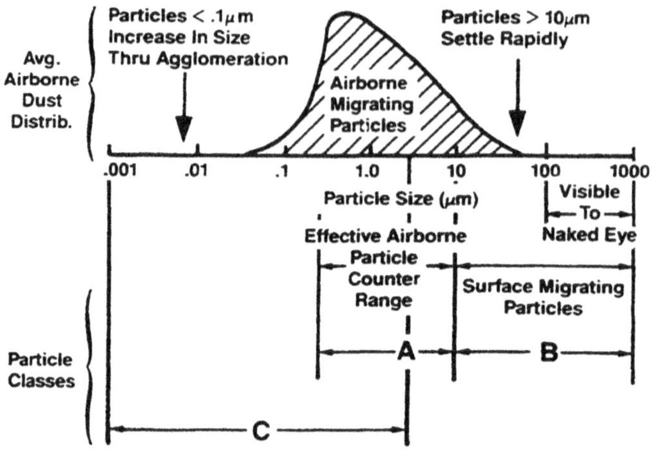

FIGURE 6-2-14. Particle size and particle classes. (A) Airborne, (B) surface migrating, (C) surface energy directed. Courtesy of A. Rapa, IBM.

the removal of atmospheric dust. If the particles escape filtration, particulate contamination can migrate by three major machanisms:

1. Type A—Particles remain suspended in air for relatively long periods and move primarily along airflow lines (airborne migration).
2. Type B—Particles are too heavy to remain airborne and migrate by surface transfer.
3. Type C—Particles which are attracted to clean surfaces by surface energy conditions (e.g., electrostatic, surface energy directed).

Type A particles are counted by light scatter, airborne[31] detectors. Types B and C are detected by sedimentation (deposition on clean wafers and analysis by microscopy and chemical surface-analytical techniques) (see Table 6-2-9). The particle size distribution of the contaminating migration of A, B, and C types, their stimulating forces, and the magnitude of airborne distribution are summarized in Fig. 6-2-14 using a data base of ASHRE survey. It is to be noted that peak sizes in

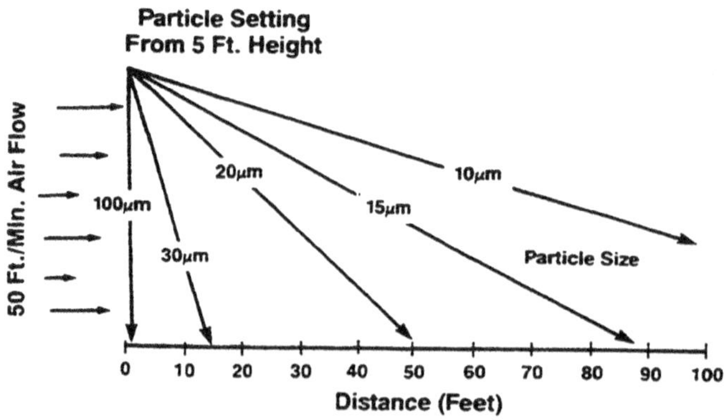

FIGURE 6-2-15. The distance traveled by airborne particles settling from a height of 5 ft. Courtesy of A. Rapa, IBM.

TABLE 6-2-6. Typical Sources of Particles[a]

Activity	Particle size (μm)
Rubbing ordinary painted surface	90
Sliding metal surfaces (nonlubricated)	75
Crumping or folding paper	65
Rubbing an epoxy painted surface	40
Seating screws	30
Belt drive	30
Writing with ballpoint pen on ordinary paper	20
Handling passivated metals such as fastening materials	10
Vinyl fitting abraded by a wrench	8
Rubbing the skin	4

[a] From Ref. 13.

aerosol A are near 1 μm or less in diameter. Type C which transfer by surface energetics are more of a problem for submicrometer devices because of their high susceptibility to submicrometer contaminants. In a clean room, the large heavy particles [Eq. (6-2-1)] travel the shortest distance (Fig. 6-2-15).

Inside a clean room, human activity and the mechanical abrasion or contact between solids generate particles. Some examples are given in Table 6-2-6. The movement of personnel, wafer, and equipment generates activity that creates particulates and movement of matter by:

1. *Vibration*—causes particles to move along a surface or break loose from a surface.
2. *Abrasion* (friction)—heat and shearing can embed or interdiffuse materials for interfacial transfer. Extensive abrasion pulverizes a surface into a fine powder.
3. *Proximity transfer*—electrical charges on surfaces close to each other act to transfer particles between them.

For future device processing, the maintenance, care, and design of improved clean rooms of class 1 or lower, especially in the coating areas, will be concerns for VLSI semiconductor lithography.

6-2-3-2. *Liquid Contamination.* Many process liquids come in contact with the bare wafer or resist-coated surface. These include organic liquids and aqueous solutions which contain acids, bases, oxidants, and other proprietary ingredients. At least 15–50 fluids of various composition are used to clean wafers, prime surfaces, deposit resists or other polymers, develop, rinse, etch, and strip the resist. Thus, the reduction or removal of insoluble and soluble impurities in the processing fluids before or during use is basic insurance for the prevention of particulate or film-forming contamination.

The photoresist liquid itself may form solid gels or insolubles due to dark reactions. Processing baths become increasingly polluted with time as photoresist fragments, oxide and silicon flakes, and other process-generated particles accumulate. Soluble impurities such as moisture, silicone oils, plasticizers, and metal ions may be absorbed from the packaging containers or dispensing tanks. Even trapped bubbles, from point-of-use filtration, or the shaking of the resist

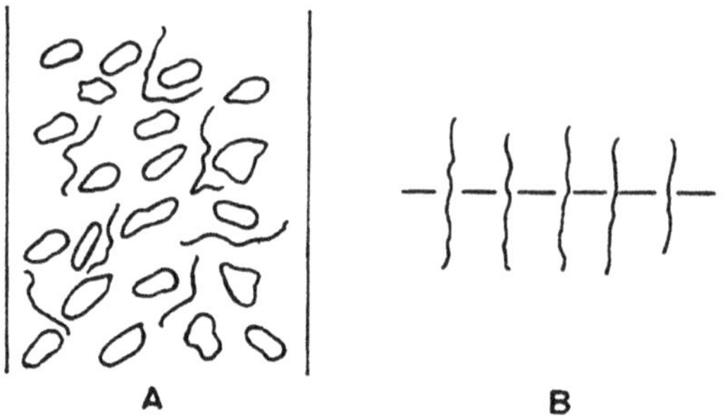

FIGURE 6-2-16. (A) Gelatinous polymers are trapped by tortuous path in a depth filter. (B) Gelatinous (gel) polymers elongate and pass through absolute sieve filters.

bottle prior to dispensing, can lead to pinholes in resist coatings. While in class 100 clean rooms[32] airborne particulate counts amount to 3 particles/liter of density of 2, liquids contain about 100,000 particles/liter. Liquids also contain dissolved ions and dissolved organics which can be more serious contaminants since they are undetected by particle counters and form adhesive layers. The particle count of 100,000/liter seems very high but if translated into a solid of 0.6 μm in size (density of 2), this is equivalent to 10 parts per million (ppm). A level of 10 ppm amounts to a deposition of 1 mg/liter. Since liquids are heavily contaminated compared to clean room air, effective filtration is the key to control.

Ultrafiltration of resist liquids has progressed. Manufacturers now supply resists filtered through 0.2-μm-diameter absolute filters. As soon as the bottles are opened, they are subject to contamination. Sealed, disposable packages are now available to dispense resist liquids. Nevertheless, the storage or poor manufacture of resists can lead to insoluble solids including gelatinous particles which have to be removed. In the cases of high-molecular-weight polymeric resists ($M_w > 250{,}000$), the solutions are very viscous (> 100 cS) and filtration is a slow process. While absolute sintered silver or fluorocarbon submicrometer filters remove solids, the gels readily deform and pass through. Gels[39] are more effectively removed by ultracentrifugation,[195] electrostatic treatment of the photoresist,[38] and depth filtration. Insoluble gels present in the resist usually lead to pinholes (negative resists) or residues in developed images[32] (positive resists). Depth filters of diatomaceous earth[34] or folded fluorocarbon membranes[36] provide a tortuous path to trap gels (Fig. 6-2-16).

Filter capacity is not dependent on the amount of liquid filtered but on the number of particles removed, according to the relation[217]

$$p = \exp(-n)$$

where p is the fraction of the original particulate level and n is the number of liquid volume turnovers. By passage of the same volume of a liquid through a series of smaller filters equivalent in surface area to a larger filter, a more effective technique of filtration is obtained. The results of photoresist filtration through a 0.2-μm-diameter fluorocarbon depth filter are shown in Table 6-2-7.[36]

The effective area of depth filters is huge (> 1 ft^2) and they have a much larger capacity than a single membrane. By using reverse osmosis, the ultrafiltration of process liquids to 0.001 μm is possible, at least with water.[33] Unfortunately, filtration is never complete since resist liquids tend to re-form insolubles and point-of-use[204] filtration is usually required. The detection of gels by light scattering is a

TABLE 6-2-7. Depth Filtration of Resists

Particle size (μm)	Counts: Unfiltered	Counts: Filtered
⩾0.5	2800	<1
⩾1.0	1800	<1
⩾2.5	780	<1
⩾5.0	<1	<1

practical means of qualitative monitoring. As a word of caution, since pressurized vessels are used to assist in resist filtration, gas bubbles remain trapped in fresh filtered resist and liquid degassing or settling is necessary to prevent bubbles from forming pinholes in spin-coated resists of high viscosity.

Filtration of other process liquids such as acids, alcohols, developer, and etchant is now a standard process for electronic-grade chemicals.(35) Without 0.2-μm filtration, most of the process liquids contain high counts of particles.(42) Table 6-2-8 gives the results of a study of the filtration of 50 ml of electronic liquids before and after 0.2-μm filtration.(40–42)

Effectively, the filtration reduced the number of particles in water by several orders of magnitude. However, the accurate counting and sizing of particles smaller than 1 μm in diameter remains a difficult task.

For routine process liquids, filtration is a preventative step to reduce particulates before and during lithographic processing. Dissolved contaminants such as mobile sodium ions require sophisticated detection methods (ion chromatography) and distillation of the process liquid to avoid contamination. Accordingly, in the first place, the prevention and avoidance of the contamination by foreign dissolved materials is at present a better processing philosophy.

TABLE 6-2-8. Particles in Semiconductor Process Liquids[a]

	Before filtration Size (μm/50 ml)				
	2	5	15	25	50
Water	31k	4k	150	50	10
Xylene	9k	3k	240	80	20
Nitric acid	63k	23k	5k	3k	900
	After filtration (0.2-μm filter) Size (μm/50 ml)				
	2	5	15	25	50
Water	80	30	5	5	3
Xylene	60	20	10	5	3
Nitric acid	5k	1.5k	300	200	80

[a] Modified from Ref. 42.

6-2-3-3. *Removal of Contaminants on the Wafer Surface.* Surface contaminants appear as films or as particles of different shapes, diameters, thicknesses, origins, and with random frequency in the amount of material. On the wafer surface, the contaminants are generally irreproducible in composition or in their location. We can expect a mixture of layers of materials composed of organic, inorganic, or metallic elements and compounds (see Fig. 6-2-7). Some of the likely contaminants and their sources are listed in Table 6-2-9.

While the filtration of gases and liquids is a preventive step to reduce surface contaminants,(32) dissolved solids remain a potential problem. Therefore, cleaning to remove molecular, ionic, polymeric, and atomic contaminants, prior to resist coating, is standard practice in semiconductor lithography. In some cases, reworking is also necessary to recover a defective processed resist or for reclaiming wafers. Cleaning involves a combination of physical and chemical steps(24):

Mechanical	Physical	Chemical
Displace	Dissolve	Oxidize
Dislodge	Vaporize	Etching
Spin, scrub	Sputter	Reduce
Brush	Displace	
Vacuum		

The sequence of cleaning steps for bare silicon consists of the following procedures(24):

1. Removal of organics that interfere with aqueous reagent contact with the surface
2. Removal of the oxide layers that interfere with the cleaning action
3. Removal of any metallic or ionic impurities

These cleaning procedures do not serve to remove particulates. When particles are

TABLE 6-2-9. Examples of Surface Contaminants Encountered in Semiconductor Lithography

Contamination	Examples	Sources
Organic	Silicone oil	Hand cream, pump oil
	Silicone grease	Release agents, lubricants
	Polymer films	Reworked wafers
		Bad primers
		Solvent deposits
		Dry etch deposits
	Humans	Skin, hair, lint
	Water	Bacteria
Inorganic	Oxide films	O_2, H_2O, in deposition
		O_2 plasma ash
		Excess heat
	Salts	Aerosols
		Outside air
		Developers, etchants, process liquids
	Water	Humidity
Metallic	Si	Wafer
	Fe, Al	Plasma equipment

present on the wafer surface, they should be first removed by brushing,[49] ultrasonics,[48] waterwheels, or high-pressure sprays. This is followed by the appropriate chemical sequences, rinsing, vacuum bake, and dry storage.

Physical methods for cleaning also employ dissolution, volatilization, plasmolysis, and chemical polishing. To minimize redeposition, vapor-phase degreasing is preferred over solvent sprays or dipping. Ultrasonic cavitation assists in microscopic dislodgement and the addition of detergents assists in particle removal by endropic action.[228]

Chemical removal involves powerful oxidants such as oxygen plasma, hydrogen peroxide,[28, 43] and persulfates (Caro's acid). The results of a study of a multitude of resist stripping solutions and processes using the breath condensation test are given in Table 6-2-10.[45] Most of the resist was removed by Caro's[43] or oxidizing acids.[44] Since about 20 Å of residues remains as detected by ellipsometry, aqueous hydrogen peroxide–ammonia solution was recommended as the final cleaning procedure.[45] Plasma or ozone stripping was just as effective.[14]

The battle of contamination control and removal is a contest between the prevention and the elimination of defects. However, the greatest risk involves the cleaning operation itself. Does the cleaning process remove some contaminants but introduce new types? In this case, the prevention of contaminantion is better than the cure.

6-2-4. *Detection of Contaminants on the Wafer Surface*

The detection, analysis, and monitoring of contaminants are necessary procedures in the prevention, diagnosis, and elimination of recurrent processing defects. Since contaminants are not usually the result of chemical reactions, their direct identification can lead directly to the physical sources of the material, whether they are generated by a process, tool, or material, or are introduced from outside the clean room environment. Table 6-2-11 lists some of the most common physical (surface) identification and chemical analytical techniques available to the line process engineer.[188, 189]

Most of the analytical techniques are not very good for organic structures with complex stoichiometry. Since contaminants may be mixtures, analytical techniques are limited, unless some surface separation can be effected. The most common analytical techniques are characterized in Table 6-2-12.

TABLE 6-2-10. Effectiveness of Resist Strippers and Cleaners[a]

Treatment	Rank
Dichromate	1
Caro's acid	1
Sulfonic acid	1
Plasma O_2	1
Ashing (air)	2
HNO_3	3

[a] From Ref. 45.

TABLE 6-2-11. Abbreviations for Surface Analysis[a]

AES	Auger electron spectroscopy[50]
APS	Appearance potential spectroscopy
$\Delta\phi$	Change in work function
ELL	Ellipsometry
EELS	Electron energy loss spectroscopy
EMP	Electron microprobe analysis
EPR	Electron paramagnetic resonance
ESD	Electron simplified desorption
ESR	Electron spin resonance
FEM	Field emission microscopy
FIM	Field ion microscopy
IMMA	Ion microprobe mass analysis
ISS	Ion scattering spectroscopy
LEED	Low-energy electron diffraction
MS	Mass spectroscopy
NMR	Nuclear magnetic resonance
RBS	Rutherford backscattering
RHEED	Reflection high-energy electron diffraction
SAM	Scanning Auger microscopy
SEM	Scanning electron microscopy
SIMS	Secondary ion mass spectroscopy
TD	Thermal desorption
TEM	Transmission electron microscopy
UHV	Ultrahigh vacuum
UPS	UV photoelectron spectroscopy
XPS	X-ray photoelectron spectroscopy (ESCA)
XRD	X-ray diffraction
XRF	X-ray fluorescence

[a] From Ref. 46.

A localized composition can be quickly analyzed as to whether it is organic, inorganic, or metallic in nature. The techniques are generally not fast, or of sufficient resolution to analyze submicrometer defects. On a production line, the faster a contaminant can be identified, the less impact it will have if steps are taken to filter or remove the source.

Very localized defects such as pinholes, cracks, pits, nicks, or random defects require some highlighting to be discernible. A survey by Kern[26] of the useful

TABLE 6-2-12. Surface Analysis Techniques[a]

Method	Probe (μm)	Depth (μm)	Limits (ppm)
XRF	10^4	>3	1
EMP	1	0.03–10	100
MS	10–100	1–10	0.01
ISS	10^2–10^3	A[b]	100
SIMS	10^3	A	0.1
IMMA	1–100	A	0.01
AES	25–100	A	0.01
SAM	0.$\frac{1}{4}$–0.3	A	0.1
ESCA	10^4	A	1000

[a] From Ref. 47.
[b] A, atomic layers.

methods for detecting local defects in dielectric films is applicable to resist films as well (Table 6-2-13). To detect contaminants, qualitative procedures use light scattering at low angles and the water wetting test.[45] Pinholes (as discussed later) require careful testing.

On-line monitoring of surface contamination on wafers, uses bright reflected monochromatic light, and laser scanning systems.[214] Scanning electron microscopy (SEM) allows the detection of submicrometer particles but is a slow process and sometimes requires the sample's destruction if low energies (< 5 keV) cannot be used. Laser scanning systems operate to a 0.1- to 0.3-μm limit but complex interpretations[228] are necessary to unravel the data so as to determine the shape of the particles. Because of different particle shapes, surface roughness, transparency, and index of refraction, scattered light is significantly different between a transparent resist particle and an opaque particle.[228] Light scattering is also used to monitor aerosol particles in clean rooms.

Sampling and counting techniques for airborne particles[31] are not absolute, since any shape will scatter light. At present, particles smaller than 0.5 μm are difficult to detect but future clean room standards require a detection accuracy of 0.1 μm (Federal Standard 209B) by 1986.[31] Light scatter detectors use broadband or monochromatic sources to illuminate moving particles which scatter light and are detected by photodiodes.[31] Many measurements are necessary to establish the

TABLE 6-2-13. Survey of Useful Methods for Detecting and Characterizing Localized Defects in Dielectric Films[a]

Optical methods	Brightfield, darkfield, phase-contrast microscopy
	Nomarski differential interference-contrast microscopy
	Multiple-beam interferometry
	Light scattering
Electrical methods	Standard dielectric breakdown
	Self-healing dielectric breakdown
	Electrochemical autography
	Electrolytic copper decoration
	Electrolytic gas bubble formation
	Electrophoretic decoration
	Corona-charging decoration
	Nematic liquid-crystal light scattering in electric field
	Current noise of metal strips over dielectric
Electronic methods	Scanning electron microscopy
	Transmission electron microscopy
	Replica electron microscopy
	Electron diffraction
Mechanical methods	Surface profilometry by stylus instruments
	Sectioning techniques
Chemical methods	Selective solution etching of films or substrates
	Preferential high-temperature gas-phase etching
Localized composition analytical microvolume methods	Electron-probe microanalysis
	Ion microprobe analysis
	Electron scanning chemical analysis
	Scanning Auger microprobe analysis
	Laser scanning photoemission
	Neutron activation combined with autoradiography
	X-ray and electron diffraction

[a] From Ref. 26.

precision of a Gaussian distribution of particles. On-line automated equipment is essential to detect the peak pollution periods in a clean room. Aerosol counting methods for clean rooms have recently been reviewed by Donovan *et al.*[232, 233] The four basic instruments are:

- OPC (optical particle counter)—light scatter
- APS (aerodynamic particle sizer)—inertia
- EMA (electric mobility analyzer)—electric mobility
- DB (diffusion battery)—Brownian motion

The OPC and APS are useful in the micrometer particle size range. The EMA and DB are used in conjunction with condensation nuclei counter and are able to detect particles in the range 0.01–0.5 μm.

In liquids, there are four general ways to count and measure particles of 0.5- to 20-μm diameter[30]:

1. Image replication—particles are collected on a filter, sized under a microscope or an automated scanner—difficult, tedious, and hard to see transparent particles.
2. Light scatter—laser light reflected, qualitative, smaller particles hidden by larger, ultrapure standards difficult to prepare particle-free solvent.
3. Light absorption—turbidity, gross levels.
4. Ultrasonic devices—acoustic sound[37] waves reflected. Unaffected by refractive index such as resist gels.

In an application of dynamic acoustic spectroscopy, acids, solvents, and resists were assayed for particles. While the counting defines ranges of particles from about 0.5 μm to 5 μm, the various process liquids were monitored with in-line detectors. For commercial negative and novolak positive resists, the counts per liter were taken[37] after 0.2-μm filtration:

	Size		
	0.6 μm	2.0 μm	4.0 μm
Positive resist	60k	50k	25k
Negative resist	70	20	0

These counts are equivalent to 10,000 per liter or 1 ppm. In novolak resists, resin synthesis produces gels during the synthetic step to drive off the formaldehyde reactant.[64] Residual acids also catalyze further cross-linking during resin storage.

Various amounts of particle and surface contaminants are transported by air-borne, surface migration, and liquid deposition. Particle prevention by filtration and counting is necessary for wafer cleaning and the maintenance of clean surfaces. In the submicrometer range, the improved detection of particle ranges is necessary for lowering the defects of lithographic processing for future submicrometer device fabrication.

6-3. STORAGE OF CLEANED WAFERS

In single-wafer processing, wafers proceed immediately after cleaning and drying into the primer coating operation. Most lithographic wet processing is done in the batch mode. After cleaning, the wafers are transported in boats. Because of queing and rework, a supply of cleaned wafers is usually necessary for production. An often-neglected aspect of cleaning is the storage of wafers. Good on-line process storage means keeping contaminants away from the surface of the wafer.[14] The problem is to identify the undesirable contaminants and eliminate them from the storage environment.[20]

Since clean surfaces can be easily recontaminated, it is not unusual to find that the storage container itself outgasses or degrades to emit solids, gases, ions, and plasticizers. Plastics are not pure materials and usually contain mobile additives. Hydrocarbon vapors are the most common contaminants. In general, plastics are the source. Quite commonly, the silicone mold release agents used in the plastic manufacture of wafer storage boxes may seriously contaminate a wafer surface and cause massive dewetting or pinholes. Since the silicones are not removed by oxidation, chlorinated solvents and fluorine plasma are more effective cleaning agents. Thus, the storage of clean wafers in metal or glass containers with hermetic seals is often employed.[14]

Contamination Control System

Contamination control is a total system commitment.[13, 228] The clean room is a sophisticated process chamber which is designed to reduce particles and contaminants to a tolerable level. The prevention of contamination is the basic processing goal of the equipment, personnel, and chemicals operating in that environment. As a last resort, cleaning is used to prepare the silicon surface prior to resist application.

Eventually, the entire lithography operation will use total automation in class 0–1 areas. No personnel will be needed for operation inside the room and robotics and automatic wafer transport will be controlled by external computer systems.[230] Thus, the resist will be processed as if the material were radioactive and required processing in a vacuum or dry box. Automation is expected to increase yields by lowering process defects and by tightening process control.

In the following sections we will examine the physical and practical aspects of techniques available for coating resists.

6-4. ADHESION PROMOTERS

The chemistry of adhesion promoters is extensively discussed in Chapter 13. Briefly, the resists are low-energy (< 36 dynes/cm^2) polymers which adhere weakly to hydrophobic semiconductor surfaces covered by adsorbed water.[51] In most cases, water is removed by heating the wafer at elevated temperatures.[244] After removal of surface water, the silanol groups on the oxidized silicon wafer are esterified with reactive silicone primers to lower the surface tension of the oxide surface,[15]

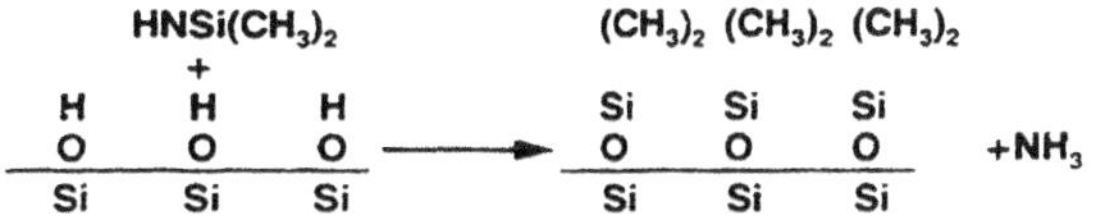

In Table 6-4-1, a comprehensive list of adhesion promoters of priming films is

TABLE 6-4-1. Adhesion Treatments or Additives to Promote Adhesion of Diazoquinone–Novolak Resists

	Reference
1. Treatments	
a. Treat by oxidizing in $KMnO_4$ (add to developer)	53
b. Use rubber resists OMNR (Tokyo OHKO) under AZ-1350	54
c. Baker and exposer AZ-1350 sublayer under AZ-1350 (plasma ash after)	55
d. Treat surface with cyanoacrylate	56
e. Polymerize silanes with *e*-beam	57
2. Additives to AZ-1350	
a. Alkylimidazolines	58
b. Butyric acid	59
c. Alkyl acids	60
d. Polyhydroxystyrene	61
e. Polyvinylmethylether	62
f. *t*-Butyl novolaks	61a
g. Epoxy silanes	61b
h. Epoxy polymers	61c
i. Silane	61d

given for preparation of the surface for the weaker adhering diazoquinone–novolak resists. For surfaces other than oxides (Table 6-4-2), a polymeric adhesion promoter,[247] conversion to an oxide state as with Si_3N_4, or additives to the DQN resist are employed. Surface silanol groups can also become hydrophobic by reacting in the vapor phase with Grignard agents, alcohols, diazomethane and other organic compounds[63, 113, 114]:

$$-SiOH + ROH \rightarrow -SiOR + H_2O$$

Primers can be applied by dipping, spraying, spin coating, or vapor phase[244] priming. Since the reaction of silanol with HMDS requires several minutes at elevated temperatures,

$$2SiOH + [(CH_3)_3Si]_2NH \rightarrow 2SiOSi(CH_3)_3 + NH_3$$

room-temperature treatments only serve to clean particulates from the surface of the wafer. Since HMDS boils at 125°C, it is best applied and reacted by vacuum

TABLE 6-4-2. Adhesion Treatments for Other Dielectric ($Non-SiO_2$) Surfaces

Layer	Treatment	Reference
Si_3N_4	Plasma oxidize	61e
Si_3N_4	AZ-1350 sublayer	55
Polysilicon	Plasma oxidize	61e
Al_2O_3	Cr complexes	61f
PSG	AZ-1350 sublayer	55
PSG	Polyvinylmethylether added to resist	62

distillation or by wafer treatments at reduced pressure and elevated temperature. A test of contact angle measurements should be run to determine the effectiveness of the priming method.[51, 52]

6-5. PHYSICAL CHEMISTRY OF SPIN COATING

Spin coating is the dominant and most reproducible resist coating process (Table 6-1-6). Spin coating encompasses the dynamics of centrifugal forces, the rheology of viscous polymers (Newtonian and non-Newtonian), and the dynamics of solvent transport.[200] Empirically, the spin speed versus thickness curve is parametric (Fig. 6-5-1) and linear on a log–log scale, with the slope (a) independent of concentration (or viscosity) (Fig. 6-5-2).[65] The slope of the log plot found in various resist investigations is given in Table 6-5-1 for the spin speed velocity, ω^a. The a parameter appears to be close to the Mark–Houwink constant for the viscosity (η) dependency on molecular weight (M),

$$\eta = kM^a \tag{6-5-1}$$

For Newtonian fluids, the thickness[234] varies with the inverse half power of the spinning speed ($a = 0.5$). The amount of fluid dispensed or the speed of rotation (ω) has no effect on the final film thickness, as long as sufficient fluid is present to cover about two-thirds of the wafer surface area.[67, 76] As long as the rpm is less than 6000, the thickness of the spun film is independent of the diameter of the wafer and the nature of the surface.[67, 77, 234] Most of the resist liquids are Newtonian,[73, 78] exhibiting the same viscosity, independent of the shear rate.[68]

The other dependent variable unnoticed by most workers, until laboratories began producing their own resists, was the nature (evaporation rate, θ solvent) of the casting solvent.[66, 68, 74] In spin coating, low-boiling solvents, such as acetone, produce thick and cloudy films, due to the absorption of water and the precipitation of the polymer from the solvent.[68] Laminar air flow, during spinning, evaporates more solvent and produces thicker films.[71] As the spinning speed and/or wafer diameter increases, turbulent flow, and concomitant nonuniform mass transfer of solvent, can develop on the wafer, leading to a nonuniform coating thickness.[234] At the higher solid content ($> 10\%$), non-Newtonian models better fit the spin speed dependence than the assumed Newtonian model.[200] Since

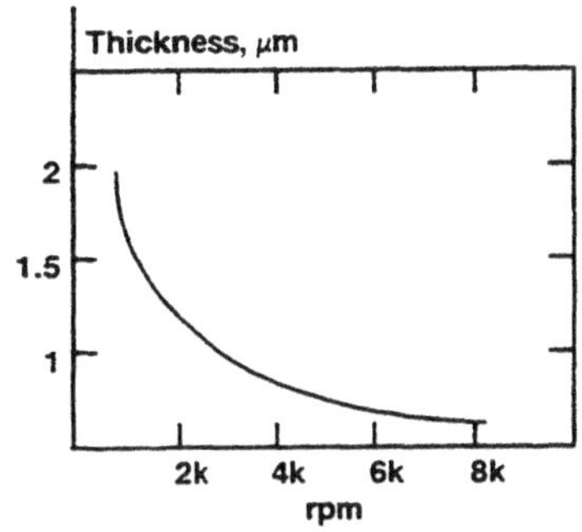

FIGURE 6-5-1. Typical spin speed versus resist thickness curve.

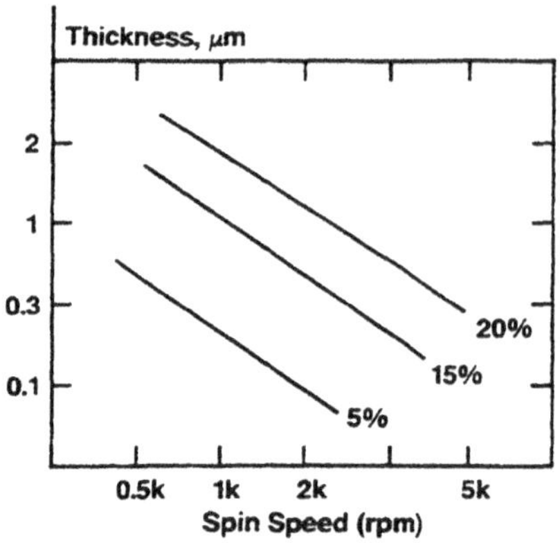

FIGURE 6-5-2. Log–log plot of spin speed versus thickness for percent solids in resist solution. From Ref. 65, copyright 1978 and courtesy of AIP.

saturating the vapor with solvent produces thinner films,[66] the evaporation rate[200, 245] of the solvent controls the final film thickness.[65–68, 74, 79] If there were no evaporation, then after the centrifugal forces have leveled the film, the film thickness would decrease with continued spinning to zero thickness.[65] Figure 6-5-3 shows the separation of the drying forces operating during spin coating. Evaporation does not become significant until 10 sec has elapsed and the evaporation of the solvent has a negligible effect on cooling the wafer.[243]

The exponent a in Eq. (6-5-1) appears to be widely[234] representative of most resist polymers, except for a polyimide of 200-cP viscosity, where an exponent of 0.8 was noted (Table 6-5-1).[77] The empirical formula for the spun film thickness (t) includes a solid or viscosity, η, term,

$$t = \eta^b \omega^{-1/2} / t^* \tag{6-5-2}$$

The spreading time, t^*, is usually of the order of seconds, and can be estimated from the interference colors exhibited by a (drying) spun wafer. The time (t^*) is inversely proportional to the spin speed, ω (Fig. 6-5-4).[65] During spinning, the film levels in a few seconds and then dries by evaporation. The value of b appears to be about 2.[68, 69, 73]

TABLE 6-5-1. Velocity Coefficient a in Spin Coating

a	Resist	Reference
0.5	Polyvinyl butyral	66
0.5	Hunt HR100	67
0.8	Polyimide	67
0.5	Styrene-epoxy	68
0.5	KTFR	69
0.8	KTFR	70
0.7	KTFR	71
0.6	KMER	72
0.7	Nylon	73
0.5	AZ-1350	65
0.5	PMMA	74
0.5	Polyimide	74
0.5	PMMA	75

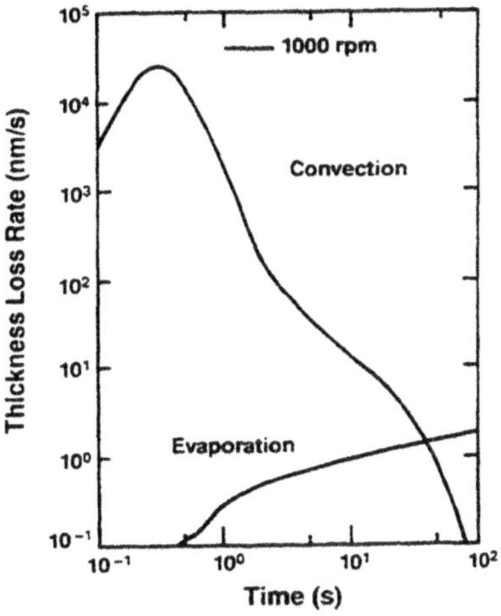

FIGURE 6-5-3. Forces operating during spin coating to level the film. From Ref. 200, copyright and courtesy of AIP.

Spin coating of a film (Fig. 6-5-5), encompasses the dynamics of centrifugal force, and the diffusion of solvent through the leveled film.[200] If we assume a flat disk not subject to gravitational or Corelios effects, and with smooth acceleration (no vibration during spinning which gives rise to radial striations), the centrifugal force, F, acting on the resist of wafer of radius r is

$$F = m\omega^2 r \tag{6-5-3}$$

Opposing the centrifugal force is the viscous resistance. The solution[80] to equations of the opposing forces for thickness (h) as a function of the liquid height (h_0) for a resist of density ρ yields

$$h = \frac{h_0}{[1 + (4\rho\omega^2/3\eta)\, h_0^2\, t]^{1/2}} \tag{6-5-4}$$

Non-Newtonian liquids behave similarly[199] up to $100\ \text{sec}^{-1}$ (7000 rpm).[78] For the fluid layer, a decrease in h_0 by a (square root)$^{-1}$ is noted. In a fluid of h_0 with a Gaussian cross section, a thickness uniformity of $\pm 3\%$ can be readily obtained in time t,

$$t = \frac{3\eta}{4\rho\omega^2 h_0^2} \tag{6-5-5}$$

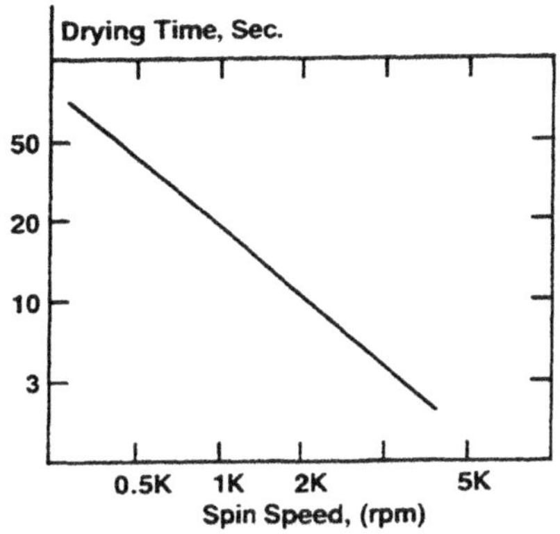

FIGURE 6-5-4. Log–log plot of drying time for a resist versus spin speed. From Ref. 65, copyright 1978 and courtesy of AIP.

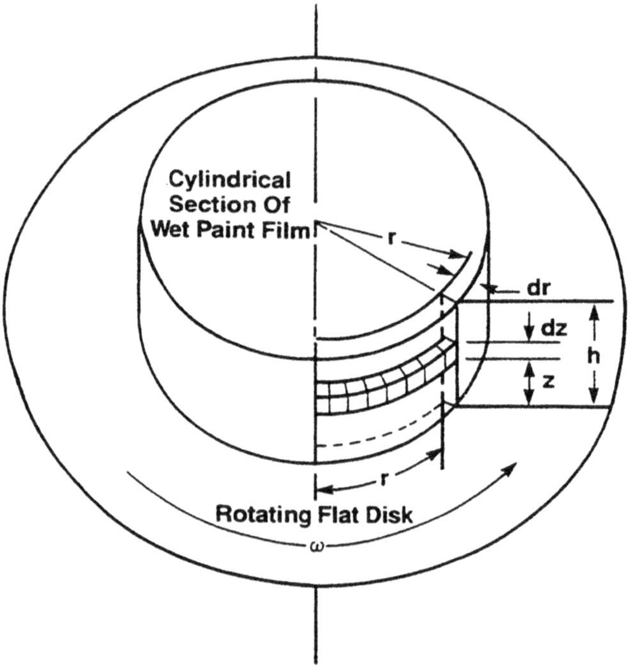

FIGURE 6-5-5. Schematic diagram of a liquid applied to a spinning disk. After Ref. 81, copyright 1979 and courtesy of John Wiley & Sons, Inc.

According to this equation, a more viscous solution will produce a thicker film and a larger volume (h_0) of applied liquid will thin out faster, than a thinner one. Since the early models of spin coating[73, 78, 80] neglected solvent evaporation, Meyerhofer empirically applied an evaporation rate factor of $e^{1/2}$ from studies of the air flow, $f^{1/2}$, dependency on thickness,

$$\frac{dL}{dt} = -(1-c)\frac{2\omega^2 h^3}{3v} - e \tag{6-5-6}$$

$$\frac{dS}{dt} = -c\frac{2\omega^2 h^3}{3v} \tag{6-5-7}$$

where L and S are the solid and liquid components, v is the volume, and c is the concentration (wt%) of solids. Equation (6-5-7) tells us that the rate at which a film thins out is in no way influenced by how far it is away from the center of a wafer.

Another way to include evaporation is to use the volume fraction of initial solids of ϕ_0, and ϕ at time t. The rate of film thinning, during spinning, is proportional to the first-order diffusion of solvent out of the film,[79]

$$\phi(t) = \phi_0 \exp(-\alpha t) \tag{6-5-8}$$

The formation of a film h, with α close to 0.5, is

$$\frac{dh}{dt} = \frac{\alpha h \phi_0 e^{-\alpha t}}{1-\phi_0 e^{-\alpha t}} - \frac{2\rho\omega^2 h^3}{3\eta} \tag{6-5-9}$$

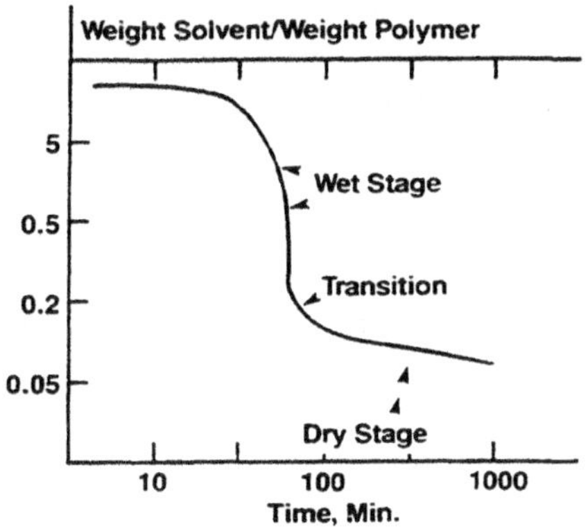

FIGURE 6-5-6. Solvent retained versus drying time for a typical liquid cast polymer film. From Ref. 81, copyright 1979 and courtesy of John Wiley & Sons, Inc.

which states that the film thickness is proportional to the rate of solvent evaporation, and directly proportional to the viscosity or molecular weight of the film. In thick films (Fig. 6-5-6), the evaporation of a solvent from a dip-coated film (50 μm thick) proceeds in two stages.[81] The first stage is the loss of most of the solvent followed by a slow drying period. In spin coating, similar stages are found but on a second time scale. In a matter of seconds, the solvent evaporates from a very thin film. In spin coating, when h_0 has dropped about one-third, evaporation takes over, and the solid concentration reaches its final value where flow has ceased completely, due to the high viscosity.[65] At this stage, about 30% of the solvent is left, and evaporation during spinning reaches the transition stage, and then slowly dries (Fig. 6-5-6). An equation of fit for spin coating (which is also useful for prebake in Chapter 7) is[66]

$$\log C = A \log h^2/t + B \tag{6-5-10}$$

where C is the concentration of retained solvent, proportional to film thickness, h^2. If the thickness of the spun cast film is doubled, the film will retain more solvent after spinning, and will require four times as long to dry, to the same residual solvent content as a thinner cast film.

Since solvents are denser than air, they tend to remain in the vicinity of the film, unless flushed by air currents. A marked cooling accompanies evaporation, and heat must be extracted from the film, to satisfy the latent heat of vaporization (10–12 kcal/mole). Solvents with a higher heat of vaporization are slower evaporating (relative to butyl acetate) (Table 6-5-2). However, silicon is a good

TABLE 6-5-2. Heat of Vaporization (ΔH) and Evaporation Rates of Resist Casting Solvents

Solvent	ΔH (kcal/mole)	Evaporation rate[83]
Diglyme	11.1	16
Cellosolve acetate	10.4	21
Methyl cellosolve	9.9	32
Xylene	10.1	63
Butyl acetate	9.8	100
Toluene	9.0	240

thermal conductor, and a negligible cooling of the surface occurs.[234] In the evaporative drying, the controlling kinetic factors are the size of solvent molecules (molar volume), and the diffusion rate of the solvent in the polymer. Since the rate of diffusion is inversely proportional to the molar volume of the solvent, low-boiling solvents dry most rapidly by evaporation and produce the thickest uneven films. Some mixtures do not obey Raoult's law and show a slower evaporation rate than individual components.[82] Many solvents are available,[83] but because of health concerns, the choices are narrowed to one or two materials. Solvents with ether, ester, and acetate groups are widely used in resist formulations.

6-5-1. *Solvents for Spin Coating*

Besides high solubility for all the resist solids, the choice of solvents for spin coating may also be required to satisfy, or meet other performance requirements such as:

- Low metallic content and residue-free deposits
- Long shelf life and low water content
- Low outgassing in electron beam[109]
- Liquid surface tension < silicon surface tension
- Nonquencher of excited sensitizer (not photochemically active)
- High θ temperature
- Moderate cost
- Low toxicity
- Striation-free coatings

Solvents with low surface tension ($<$ 30 dynes/cm) will allow the resist to wet the primed surfaces. For a particular polymer–solvent combination, a theta (θ) temperature exists, below which phase separation can occur, because polymer cohesive forces are stronger than the polymer–solvent intermolecular forces. The so-called poor solvents are theta solvents, in which polymers are highly coiled[84] and the solvent parameter [a, Eq. (6-5-1)] is close to 0.5.[85] In a good solvent, the polymer chains are unfolded.[86]

Polymers adsorb on the wafer surface more strongly when cast from poor solvents than good solvents.[220] In a good solvent such as cyclohexane, polystyrene adsorbs more strongly on mica than when cast from toluene, since the good solvent is also likely to be adsorbed on the surface. The amount of water in the liquid resist controls the DQN adhesion[222] and should be kept below 0.1% by weight. In the literature, there is some conjecture on the nature of the free volume or morphology of films cast from good or poor solvents. Poor-solvent-cast film may pack polymers closer together, and be more resistant to solvent developer. However, more residual solvent may be trapped and increase the permeability to developer solvents.

In general, theta solvents should be avoided, especially in high-molecular-weight materials such as PMMA. If very high (10^6) molecular weights[87] are used, a thixotropic phenomenon may occur, in which the spun film thickness is independent of spin speed or concentration. The molecular dispersity, at high-molecular-weight ends, may also be subject to non-Newtonian shearing[219] and the different molecular weights may deposit as laminar layers.[96]

Casting PMMA from different solvents, even after baking above the T_g, can affect subsequent dissolution rates.[88] Theta solvents make good developers

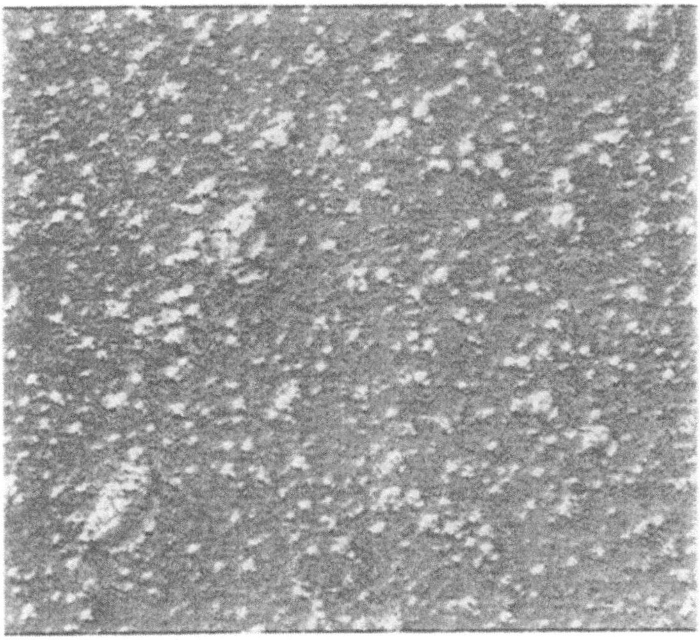

FIGURE 6-5-7. Polymethylpentene sulfone mixed at 15 wt% with a novolak resin and spun cast from diglyme. Severe phase separation occurs.

because the high-molecular-weight components are sparingly soluble.[95] For dip coating, theta solvents are better[187] than good solvents because of low adhesive interference.[187]

Polymer–solvent interaction is very evident in mixtures of polymeric resists. Polymer–polymer miscibility is rare in polymer systems. In a study of polyolefin sulfone mixed as a polymeric sensitizer with phenolic resins,[112] certain solvents compete for polymer–polymer intermolecular attraction and cause phase separation. Figure 6-5-7 is an example of a polyolefin sulfone–phenolic novolak resist cast from diglyme ether. Casting the same mixture from Cellosolve acetate produces homogeneous films.[89, 90] In the small molecule–polymer system of DQN, aggregation of sensitizer can occur at 20–40% loading, if sufficient solvent is present to allow migration, during spin drying.[75]

If it is desirable[10, 246] to planarize a resist over steps or protrusions (depending on their pitch), then a low-viscosity liquid (less than 8% solid) or a fast evaporating solvent should be avoided, since too much film shrinkage occurs upon drying. Since film leveling is limited to about 20% film shrinkage, a high-solid-content resist should be used. For multilayer resist underlayers or other wafer coatings, the best planarization is achieved at high rpm and high viscosity.[10] Polymers with low T_g are also good planarizers since disruptions can be melted in prebaking.

Residual stress from solvent casting of any type (spin, roll, spray, dip) is evident in polymer films.[91] Centrifugal forces can radially orientate the polymer and produce radial striations (see Chapter 8). Upon prebaking, the film compacts and the residual internal stress can give rise to adhesion failure, swelling, and cracking in the developer (Chapters 10 and 12). However, spin coating, as opposed to dip coating, is not subject to *adsorption*[187] of the polymer on the surface. The dynamics of the process freeze-dry the film. By contrast, if a wafer is dipped in a liquid resist and withdrawn, the resist does not adhere, in many cases, to the surface. If the resist does adhere, excellent adhesion is obtained. A combination of dip coating followed by spin coating minimizes the wasteful quantities applied in spin coating a wafer in a horizontal position.[221]

The choice of casting solvent, besides satisfying the spin coating aspects, must also meet several other expectations in the functional performance of a resist such as:

- High contrast
- Good adhesion
- Good film thinning ratio
- Minimum dose–high sensitivity

Casting DQN resin from Cellosolve acetate produces no induction and lower contrast (Fig. 6-5-8B) while casting the same composition from diglyme produces an induction effect and higher contrast (Fig. 6-5-8A).[95] Since the unexposed film takes a longer time to dissolve in a developer, a higher-contrast resist is obtained.

Spin coating is not restricted to organic resists. Inorganic films of binary chalcogenide glasses of As_2S_3 have been coated from amines.[133]

In the classic DQ formulations, the synthesis of very soluble DQ sensitizers with sulfonic acid esters of cyclohexyl, gallic acid, or other side chains, has achieved 50% by weight solubility in novolaks[92,93] and higher-contrast resists.[94] In the synthesis, sensitizers are intentionally formulated to be noncrystalline and film formers. Even mixtures of DQs are used to prevent crystallization. The incompatibility of DQ sensitizer, resin, and casting solvent can give rise to radial striations during spin coating. Protrusions of sine waves of 200–500 Å peak at the surface, but are wider in spacing inside the film (Fig. 6-5-9); The striations are due to the uneven loss of solvent, and mechanical vibrations. As the solvent evaporates, the sensitizer molecules aggregate. Secondly, if a strong polymer–solvent interaction exists, insufficient solvent is left to interact and disperse the DQ sensitizers. High spin speed ($>$ 600 rpm) causes uneven air turbulence and induces striations from the fluc-

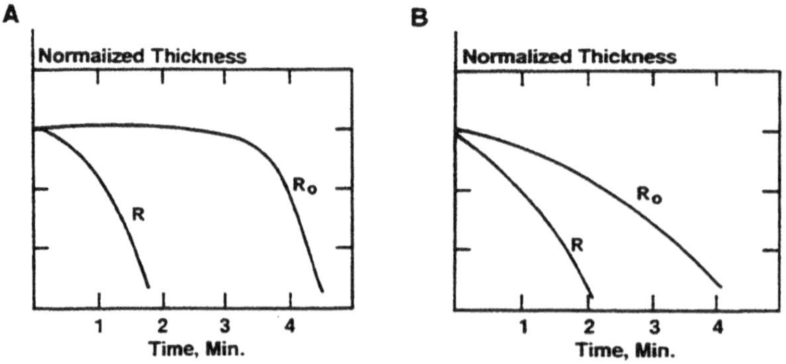

FIGURE 6-5-8. Dissolution of exposed (R) and unexposed (R_0) DQN of resist cast from (A) diglyme and (B) cyclohexanone at same dose.

FIGURE 6-5-9. Striations and pinholes in spin coated, exposed, and developed DQN resist.

tuation of the evaporating liquid. Single-component resists such as PMMA do not exhibit striations.

Casting DQN from Cellosolve acetate produces striations, while films cast from diglyme form smooth coatings. If a solvent atmosphere is present during spin coating, radial striations are reduced. Placing a prebaked film in a solvent atmosphere also tends to level striations. For DQN resists, the addition of small amounts of surfactants[202] or silicone oils, will reduce striations below 50 Å.[97] Also, additives of high-boiling benzyl acetate[236] and octanol[237] are used to reduce striations in novolak resists. Striations over alignment marks can raise havoc for stepper alignment (Fig. 6-1-1). During spin coating, the amplitude of striations of commercial DQN resists, such as AZ-1300, AZ-1400, and Hunt-200 series, increases between waves with thicker films (Fig. 6-5-10)[98] and with lower accelerations (Fig. 6-5-11).[99]

Of all the coating techniques (dip, spin, spray, roll, or vacuum deposit), spin coating prevails as the most dynamic, most uniform process, capable of ±100-Å or less variation in absolute cast film thickness. For photoresists, the choice of casting solvents and/or additives to reduce striations and internal phase separation, is critical for high resist performance in optical exposure. For multilayer resists, the natural planarization of the undercoat is a major benefit of spin coating. Failure to planarize occurs at high rpm and low solid contents, at critical orientated steps on the wafer, and with high-T_g polymers.[223-226]

During spin coating, the alignment of molecules, such as in monomolecular[117-122] films, is not possible because the solvents are not retained long enough[96] for chain ordering. If possible, vinyl polymerization, for example, could be obtained, with high photospeed, if groups could be arranged in a lamellar

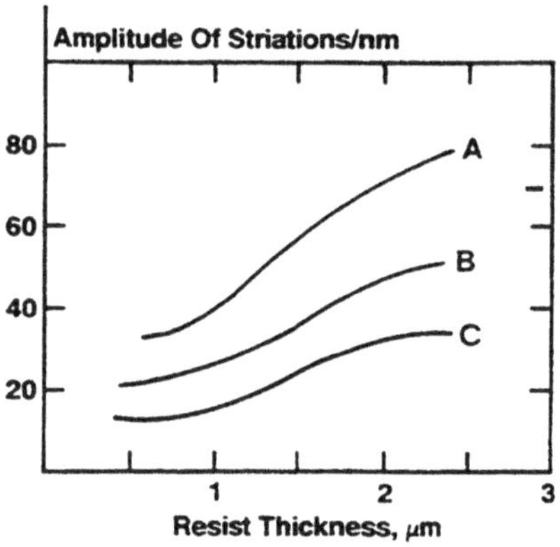

FIGURE 6-5-10. Amplitude of striations versus resist thickness for (A) AZ-1350H, (B) HPR 206, and (C) AZ-1470. From Ref. 99, courtesy of IBM.

layer. The solvent plays an important role in adhesion and resist performance. Residual solvents left after prebake can subsequently after the rate of dissolution of positive resists (and hence contrast).

6-5-2. *Practical Aspects of Spin Coating*

An empirical equation with various parametric factors has been develop to include solvent evaporation rates:

$$t = K\eta^{0.4}\omega^{-1/2}(E\lambda/C_p)^{0.6}$$

where E is the evaporation rate, λ is the latent heat of evaporation, and C_p is the heat capacity of the solvent.[66]

In actual operation, the practical aspects of spin coating deal with mechanical vibrations, torque, the shape and aerodynamics of the spin bowl, dispense tubes and cycles, and troubleshooting defects in the spin-coated film. Some of the problems encountered in spin coating and their possible causes are summarized in Table 6-5-3.

During spin coating, an air turbulence is created, which leads to rapid drying of the liquid resist at the edges of the wafer. The lips of the resist break off during wafer handling. Air vanes have been patented to eliminate lip buildup.[100] Blowing

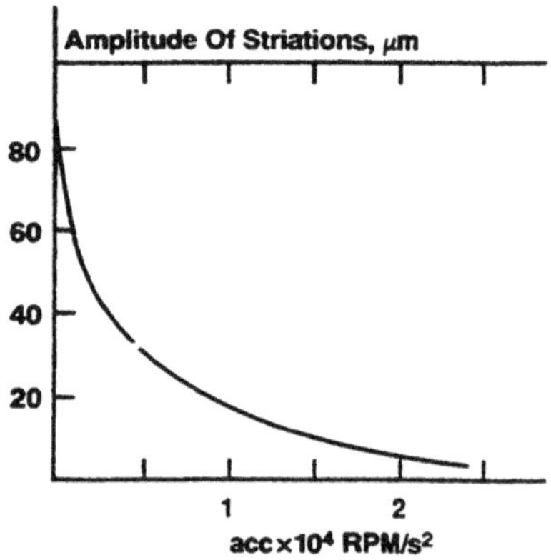

FIGURE 6-5-11. Amplitude of striations in AZ-1350 DQN resist versus acceleration of spin coater. From Ref. 99, courtesy of IBM.

TABLE 6-5-3. Spin Coating Problems and Their Causes

Problems	Possible causes
Comets	Spun-off dirt
Craters	Silicones in resist, splashback
Cobwebs	High molecular weight, air flow
Cloudy films	Moisture in air
Backside coatings	Excess liquid, noncentering
Edge dewet	Poor priming
Pinholes	Contaminants
Lip buildup	Air flow

air over the wafer during spinning improves film uniformity.[211] If that fails, the lip may be removed by a capillary spray gun pulsating a resist stripper on the periphery of the wafer.[216, 235]

In spin coating, particulate or invisible contamination can exist in the liquid resist, on the wafer surface, or be transported during spin coating, which leads to pinholes, craters, and microdewets (Fig. 6-5-12). Invisible species, such as silicone oils or organic contaminants, can cause defects. If no precautions are taken during resist dispensing, air bubbles can be trapped in the resist, and form pockets or pinholes, during spin coating. Very clean, class 10 rooms (or less) will be necessary to maintain a low number of defects ($< 1\ cm^2$) (Fig. 6-1-10).

If HMDS is spin applied in the same bowl used to apply the DQN resist, NH_3 can form from the silylation of the novolak resin or reactants, which induces diazocoupling within the subsequent spun-cast DQN resists. A separate station or vapor priming should be used. Since the DQN resist coating is sensitive to moisture, a constant 20–40% relative humidity[158] should be maintained. For high-volume uniformity of spun-cast films, the control of air flow, acceleration, speed, and feedback calibration is required. Nonspherical wafers give rise to elliptical patterns of nonuniform thickness. If a circular chuck is built with a recess to

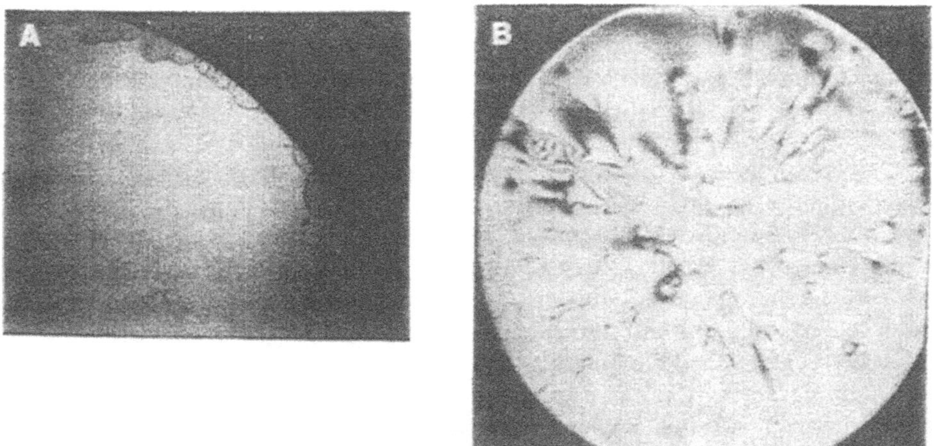

FIGURE 6-5-12. (A) Edge dewet and (B) comets and pinholes in spin-coated resist.

accommodate the plate, a uniform coating can be produced.[108] Prespinning the casting solvent prior to resist application improves adhesion of the resist.[210]

During spin coating, cratering can occur owing to liquid flow away from areas of low interfacial tension. Excess silane adhesion promoter may form soluble[114] polysiloxanes, which are redissolved and redeposited during spin coating. This can cause dewetting and cratering.

From an economic viewpoint, the spin coating is a very costly process. From a dispensed volume of 10 ml (2.5 g of solids), 1 mg or μliter of film is deposited on a wafer. This is a waste of over 99.9% of the material. At an annual resist expenditure of $240 million by 1986,[102] and growing at a 16% rate per year, over $200 million is wasted. Recovery and reconstitution or recycling (as underlayer resists) may be possible. If one resist (except for viscosity grades) was used on all device levels, then the resist could be more readily collected and recycled. A uniform resist layer with less waste can be applied in a dynamic mode,[77] by slowly rotating the wafer, and ramping to the final rpm.[103, 104] In this regard, a noteworthy innovation is the use of ultrasonic energy to create atomic aerosols, to deposit a minimum volume of a commercial resist on a wafer.[105] A combination of dip followed by spin in a vertical position is claimed to reduce the excess volume applied in spin coating.[231, 238] Accordingly, more basic improvements in spray, dip, roll, and laminate, which conserve volume, will be necessary to cut costs.

A reduction in radial striations in thick (2–15 μm) coatings is claimed by allowing the resist prior to spinning to stand on a wafer for 15 sec for 2-μm-thick, and 45 sec for 15-μm-thick[106] coating. With regard to pinholes in resist films, a reduction is noted when the resist can be double-coated (see next sections). The mutual insolubility of different resists is important in buildingup layers for multilayer resists.

Instead of coating clear liquids, dispersions of PMMA[107] and spin on have been successfully spin coated. Finally, free-standing films for studies of electron backscattering can be formed by spin coating wire screens.[108]

6-6. OTHER RESIST COATING TECHNIQUES

Besides spin coating (Table 6-1-6), other liquid- and gas-phase techniques have been used to cast micrometer films of resists or other polymers. Although spray, roll, and dip coating are regularly used in printed circuit technology, they are poor in uniformity (Table 6-1-6) when compared to spin coating. However, they offer a huge savings in using a minimum volume of resist.

Dip coating was a traditional process for printed circuits, until the bridge problem (failure to cover plating holes) became severe. A laminated,[110] dry film resist of PMMA solved the bridge problem and offered a material that the manufacturer could supply as a clean pinhole-free coating.[111] Dip coating also suffers from the "wedge effect," due to gravitational flow[239] and drying. This effect can be reduced, if a solvent atmosphere is maintained above the liquid.

Roller coating[115, 215] requires precise, flat rollers to avoid mechanical damage and to allow transfer of a minimum volume of a preformed resist film. Some of the commercial resist solvents may evaporate, and not be suitable for spray, roll, or dip coating. They are designed for spin coating and not for other forms of coating.

Spray coating depends on the size of the aerosol droplet, which may coalesce on the surface into irregular, larger droplets. The dynamic surface tension may cause further aggregation and the "orange peel" effect in coatings.[101] More fundamental work is necessary to perfect spraying technique, to become a replacement for spin coating.

6-6-1. *Langmuir–Blodgett Films*

The Langmuir–Blodgett technique[126] for formation of monomolecular layers of amphiphilic molecules on inorganic substrates, has been applied to ultraresolution nanolithography. Since the range of backscattered electrons limits the resolution of *e*-beam resists (proximity effect), nanometer-thick films would minimize the effect. Because of unique surface adsorption, the molecules are stacked in layers. The reactive sites are so sufficiently close that runaway polymerization of vinyl or acetylenic monomers may produce high-speed resists (10^{-9} C/cm^2). Since Cemal *et al.*[117] introduced the idea of employing molecules which could be polymerized by UV light, it has been found that polydiacetylenes[201] can be stacked and combined into polymers with metal-like conductivity[118]:

$$R-C\equiv C-C\equiv C-C\equiv C-X$$

$$R = \text{Alkyl} \qquad X = \text{Acid}$$

The Langmuir–Blodgett technique is based on the physical chemistry of soap molecules, which have a hydrophilic-acid end and a hydrophobic-hydrocarbon end:

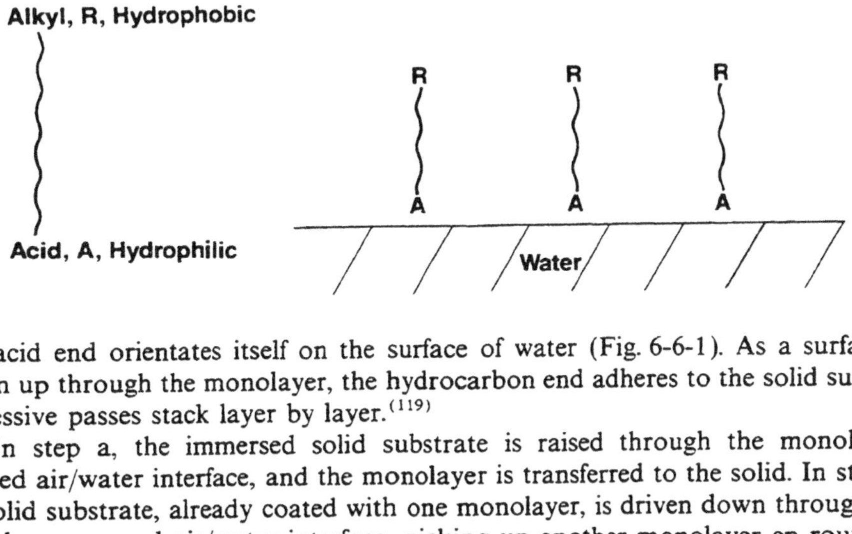

The acid end orientates itself on the surface of water (Fig. 6-6-1). As a surface is drawn up through the monolayer, the hydrocarbon end adheres to the solid surface. Successive passes stack layer by layer.[119]

In step a, the immersed solid substrate is raised through the monolayer-covered air/water interface, and the monolayer is transferred to the solid. In step b, the solid substrate, already coated with one monolayer, is driven down through the monolayer-covered air/water interface, picking up another monolayer en route. In step c, the solid substrate, now coated with two monolayers, is once more raised through the monolayer-covered air/water interface, with a third monolayer being transferred. The arrangement allows chemists to study two- and three-dimensional polymerization on an atomic scale.[121] A recent symposium covers all aspects of Langmuir–Blodgett films.[119–125]

The high-resolution aspect of these films continues to arouse interest as resists. Presently, all of the resists are negative working, relying on cross-linking or chain

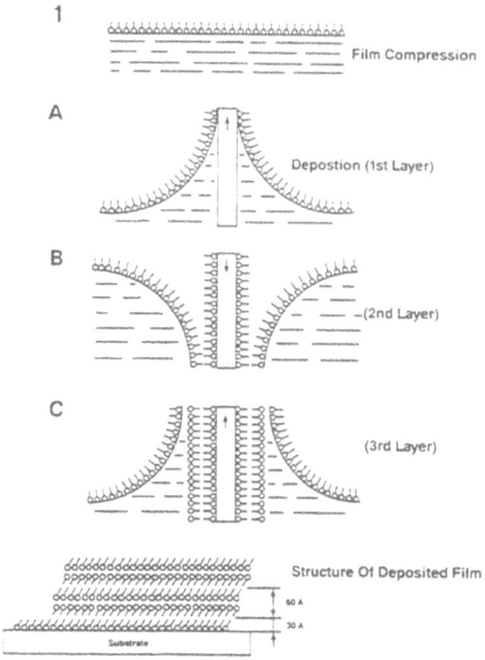

FIGURE 6-6-1. Schematic of monolayer Langmuir–Blodgett deposition of resist. (1) Spreading of film by compression; (A) deposition of first layer; (B) deposition of second layer; (C) deposition of third layer. From Ref. 123, courtesy of AIP.

polymerization. If pinhole-free films could be formed in monomolecular layers, then infinite resolution is possible. However, the subsequent image transfer by etching is limited by pinholes or erosion during plasma etching.

To deposit thick films (0.5 μm), a continuous deposition apparatus requires practical engineering considerations for an automated process.(119) The films conform to the surface and may be useful for etching metal films that traverse steps.

The goal for Langmuir films would be a high-speed resist since, in theory, all the vinyl groups could be prearranged in a chain. One initiation would be amplified by 10^4. At present, most of the speed achieved (10^{-6} C/cm^2 for *e*-beam) is due to the high absorption of the secondary photoelectrons and backscattered electrons from the substrate. In Langmuir–deposited films, pinholes are claimed to be reduced, at 450-Å thickness, by an order of magnitude, than in similar films formed by spin coating.(124) Until the need arises for nanometer devices, this technique has high potential for top surface imaging. The current major application will likely be in the synthesis of superconducting polymer films,(118) or in the fabrication of thin-film capacitors.(127)

6-6-2. *Gas-Phase Deposition*

Ideally, one would like to be able to coat a resist inside a vacuum, on a sputtered-clean wafer which is uncontaminated by the laboratory or factory air and environment. In the absence of solvent, excellent adhesion may be obtained, because no solvent competes for the bonding sites. If heated in a vacuum, unfortunately, most positive resists thermally degrade and negative resists cross-link. Specialized monomers or "stable" resists are required, before vacuum deposition and its conformal character will be able to compete with liquid resists.

For vacuum-deposited resists, active work progresses in inorganic Ag resists and a few organic monomer resists.[212] For example, molten polymers[116] adhere more strongly than solvent-cast film, and are the basis for the latest surge of hot-melt adhesives. Gas-phase deposition, either of neutral or excited molecules, may produce excellent adhesives, since no competing solvent is involved. Contamination-free substrates, precleaned by sputtering,[14] can be maintained in an atomic clean state prior to deposition. Adhesion promoters could be initially applied, followed by a resist coating.[126, 128] With different types of polymers, gas-phase deposition involves the following methods:

- Vacuum deposition
- Plasma deposition
- Adsorption–*in situ* radiation deposition
- Image amplification by monomer contact

Direct resist formation by adsorption and *in situ* exposure of patterns are discussed in Chapter 16 and image amplification is discussed in Chapter 5. Vacuum deposition or plasma deposition are standard methods of producing semiconductor or metal films. Ideally, a resist should be able to be deposited and developed by vacuum (all-dry) lithography. Since there is little exposure to particles, or contaminants in the air, defect-free films may be formed. Both organic monomers (RC = CR) and polymers, small molecules, and inorganic resists have been vacuum deposited. Prebake is eliminated and pinholes should be nonexistent, if class 1 conditions are present. The conformal nature of the coating also covers surface particles and provides a useful alternative (such as in step coverage) to planarizing films deposited by spin coating.

In order for a film of resist to be vacuum deposited, it should meet the following requirements[129]:

1. It should have a sufficiently high vapor pressure at moderate (100–200°C) temperatures.
2. No decomposition at vapor temperature.
3. Little interaction with vacuum walls.
4. Good adhesion at deposition temperatures (< 100°C).

Table 6-6-1 lists representative vacuum-deposited resists.

Inorganic resists are primarily deposited by classical vacuum methods. For organic resists, the octavinylsiloxane is an interesting resist, which can also be dry developed after exposure, by simply heating in vacuum.[129] The prepolymer is deposited at 150 Å/sec. Films of 5000-Å thickness with a uniformity of 2% were coated on A1, exposed at 3 $\mu C/cm^2$, and developed by heating. Al etching was per-

TABLE 6-6-1. Vacuum-Deposited Resists

Resists	Reference
Octavinylsilsesquioxane	129
Rare earth diketonates	130
Dry Ag	131
Diazoquinone	132
As_2S_3	135
Ag-Se-Ge	136, 137

formed with chlorine plasma. These resists will find significant use as image layers in bilayer resists or as the O_2 barrier level in trilayer resists.[11] Secondly, the process is simple and without the complicated side reactions of plasma-deposited films. For its simplicity, the siloxane system is ideal for all dry processing.

Plasma-deposited organic films have been around for several decades. In semiconductor fabrication, they are finding continued applications in multilayer resists, as adhesion promoters, encapsulants, and dielectrics.[138, 139] Siloxane polymers are deposited for the passivation of semiconductor devices, and to make a new polymeric, insulated, gate-field effect transistor.[140] If the polymers are fired in O_2, a diffusion mask similar to oxidized silicon can be created.

In plasma deposition, the adsorption of monomer occurs on the surface, followed by activation by the plasma. In other downstream reactors, radicals are created, which adsorb and polymerize on a surface.[141] However, the polymerization is not clean and is usually accompanied by branching and cross-linking. Even in PMMA, the plasma-deposited films are lower in sensitivity than solvent-cast films.[142, 143] All-dry (vacuum) lithography has been achieved by deposition of PMMA[142, 143] or PMMA-tin methacrylates, followed by plasma development. Thus, plasma chemistry has performed the total lithographic process. For vacuum deposition, more resist systems have to be synthesized to take advantage of the dry process capability.

Polymers can also be deposited by RF sputtering.[141] In plasma etching (Chapter 13), the deposition of fluorocarbon polymers from the etchant polymerization is a desirable mechanism of anisotropic etching. Although the plasma-deposited films are still subject to internal stresses since no liquid developer is used, the plasma dry development step is insensitive to adhesion failure by lifting.[144] New plasma resists are needed to exploit this new coating technology.

Instead of using a plasma to excite the monomer surface (see Chapters 10 and 16), images of polymers can be grown on a surface with directed radiation. The main factors are the deposition rates, and the contamination of the exposure tool by the vapors. The deposition rate is usually proportional to an $(\text{intensity})^{1/2}$ power, following classical polymerization kinetics,[145] If a flash dose is used, the monomer converts to a polymer:

$$\text{Monomer} \rightarrow \text{Polymer} \tag{6-6-1}$$

If continuous radiation is used, some equilibrium mixture of monomer and polymer is usually reached,

$$\text{Polymer} \rightarrow \text{Monomer} \tag{6-6-2}$$

For patterned exposure (Fig. 6-6-2), the unexposed monomer is removed by heating.[146] Unfortunately, the long exposure time (hours) required to build up a film, limits the utility of the technique. High resolution is possible, if the adsorbed radiation can be confined to the top image layer ($<$ 100-nm thickness) of the deposited resist possibly in a bilayer stack (see Chapter 12).

In principle, high-speed resists, based on a gain of 10^4, can be achieved by initiation of reaction (6-6-2). Flash exposure with a high-power laser in the deep UV would confine the energy to a two-dimensional layer with infinite resolution (similar to Langmuir–Blodgett resists). A step in this direction has been reported by

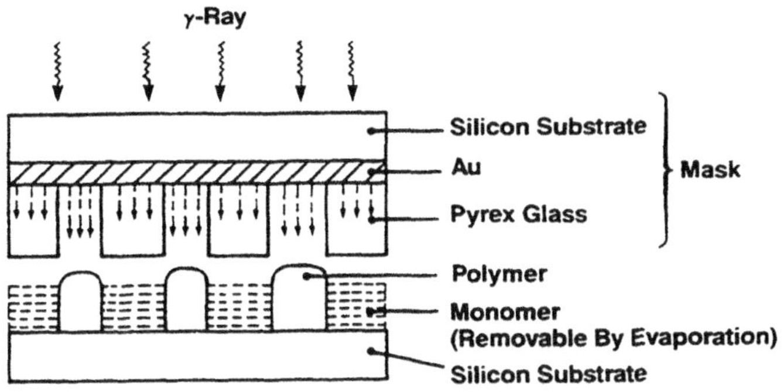

FIGURE 6-6-2. Schematic diagram of *in situ* polymer deposition by irradiation of adsorbed monomer. From H. Hiroaka, *Appl. Phys. Lett.* **31**, 503 (1977), copyright and courtesy of AIP.

Tsao and Ehrlich,[147] who polymerized methyl methacrylate with a focused deep-UV laser. Deposition rates of 10 Å/sec were achieved for a mercury lamp and 400 Å/sec for the laser source. No development is necessary.

Dry deposit techniques have the potential for defect-free, conformable masking and high resolution is possible with nanometer-thick films. All-dry resist processing has been demonstrated. For process simplicity, the simple, vacuum-deposited, siloxane resist, as a single film or as a bilayer mask, looms as a significant system for future device fabrication.

6-7. RESIST FILM THICKNESS MEASUREMENTS

After the resist film has been coated and baked, the film thickness is measured by four basic techniques as done for dielectric thin films (Table 6-7-1).[148, 240] The simplest technique is to measure the weight of the resist (1 mg/cm^2/μm of unit density) on a coated surface. The average value (excluding planarization over steps) is then computed. Using a piezoelectric transducer setup, the weight loss of resist with time can be used to monitor development kinetics.

The second simplest technique is the mechanical stylus, which is used to sense, versus a reference plane, the step height of a scratch or pattern in a resist. Surface striations to ± 50 Å have been measured by this technique.[98, 99] For accurate measurements, the careful selection of stylus size, a clean scratch mark, flat wafers, and a vibration-free table are necessary. In actual operation, the stylus may become embedded in soft resists ($T_g < 35°C$) and care should be taken to avoid the imper-

TABLE 6-7-1. Film Thickness Techniques

Technique	Accuracy
Volume–density	±500 Å
Optical interference	±50 Å
Optical absorption	±200 Å
Mechanical stylus	±200 Å
SEM cross section	±200 Å

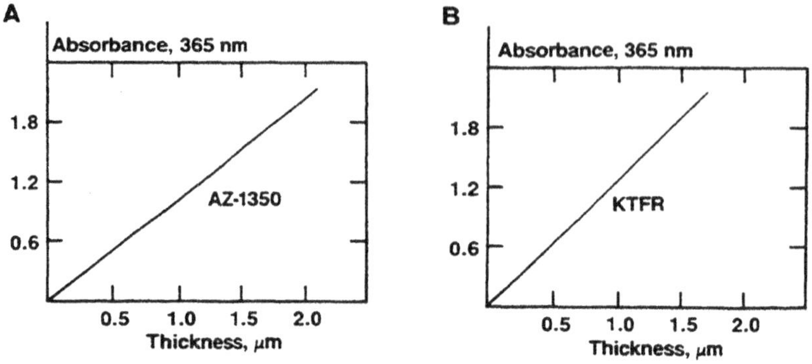

FIGURE 6-7-1. Beer's law plot for (A) AZ-1350 DQN positive photoresist and (B) KTFR negative photoresist. Thickness increases linearly with absorbance.

fection. Kinetic measurements can be made, if the stylus is placed on top of a resist and a capillary layer of developer is flowed over the exposed resist. Since the stylus measurement does not depend on the refractive index or the optical nature of the substrate, the technique is an excellent laboratory tool for resist development work. It is also used to confirm optical measurement techniques. Optical absorption measurements rely on Beer's law:

$$A = \log(I_0/I) = (\text{Concn})\,(\varepsilon)\,(t) \tag{6-7-1}$$

where A is the absorbance, ε is the molar extinction coefficient (liters/mole-cm), and t is the thickness. Films on "transparent" substrates are monitored by infrared (IR) or UV spectroscopy of appropriate bands such as the diazoquinone or azide sensitizer band at 365 nm as shown in Fig. 6-7-1.

On silicon wafers, IR radiation can be used, but the accuracy of IR is only about ±5%. For UV analysis, transparent windown on oxidized silicon wafers can be made by backetching the silicon.[149] Absorption measurements are very useful for quality control, since the sensitizer level can be ascertained (if a calibrated spinner is used to cast the film). By exposing the same wafer to radiation, followed by bleached absorption measurement, the photospeed of the resist[150, 151] can be ascertained.

The last method of film thickness measurement utilizes the optical reflection of light. The reflected rays produce interference patterns, characteristic of the film refractive index and thickness (Fig. 6-7-2). Interferometry, ellipsometry, and reflectance are all based on the disturbances of optical rays in a transparent medium.[241] The basic equation for the thickness (d) measured by reflectance is[152]

$$d = \frac{N\lambda}{(2\mu^2 - \sin^2 i)^{1/2}} \tag{6-7-2}$$

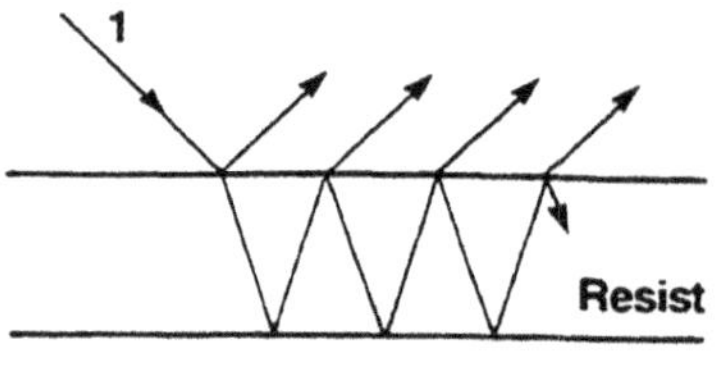

FIGURE 6-7-2. Optical interference from silicon wafer surface from reflections of light at the air–resist and resist–silicon surfaces.

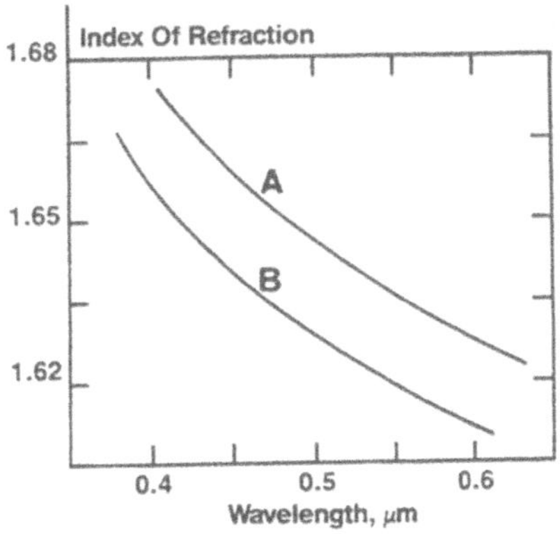

FIGURE 6-7-3. Change in the refractive index of a DQN resist before (A) and after (B) UV exposure as measured on VAMFO. From Ref. 153.

where N is the order of reflected light, λ is the refractive index of the film, and i is the angle of beam incidence.

The wavelength of light can be varied by a prism[154] or a filter wheel.[151, 155] By measuring the wavelengths at which maximum intensity are noted, the thickness can be calculated, if the refractive index of the film is known. The refractive index can be measured, using Eq. (6-7-2), if the angle of incidence is varied,[153] or by prism couplers.[154, 156] The refractive index changes with wavelength[153] (Fig. 6-7-3) and with photoexposure[157] (Fig. 6-7-3).

An extensive collection of refractive indexes of resists, polymers, and semiconductor films can be found in the work of Warnecke and Lopresti.[153] The refractive index of common resists is given in Table 6-7-2. For endpoint[155] determination of resist development, the dynamic mode of film thinning is monitored.[194] Red laser light is used at 6238 Å.

In all optical treatments, the refractive index is assumed to be isotropic and independent of residual solvent. Since exposure causes chemical changes, the

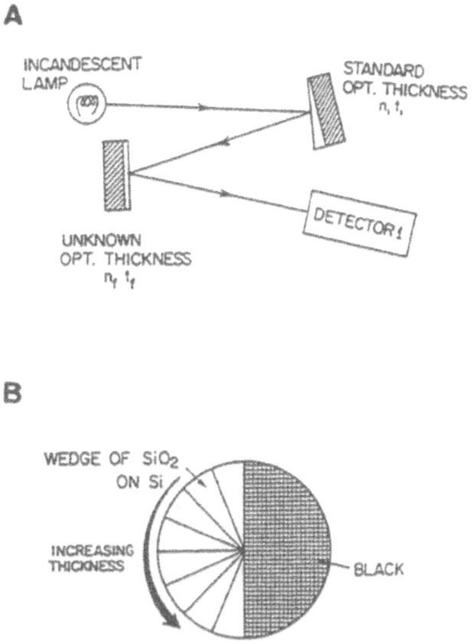

FIGURE 6-7-4. (A) Principle of comparative reflectance thickness measurement. The reflected intensity is a maximum when the standard and the unknown film thicknesses are equal. (B) Reference thickness wedge of oxide etched to different thicknesses. From Ref. 159, courtesy of Institute of Physics and Dr. J. Sandercock of RCA.

TABLE 6-7-2. Refractive Index of Unexposed Resist at 6238 Å[a]

Resist	Refractive index
KTFR rubber	1.591
KPR cinnamate	1.619
PMMA	1.495
AZ-1350 (novolak)	1.631
SiO_2	1.4602

[a] From Ref. 153.

accurate measurement of refractive index (to 6 places) might be used as a dose sensitometer. Optical treatment of the interference from multilayers such as

Resist	(1)
Oxidized silicon	(0)
Silicon wafer	

requires the solution of the dual refractive layer relation,

$$T_1 = \frac{N\lambda - 2t_0(\mu_0^2 - \sin^2 i)^{1/2}}{2(\mu_1^2 - \sin^2 i)^{1/2}}$$

where 1 refers to the resist layer and 0 to the oxide layer. Standing waves are the result of interference between oxide and resist.[7] The interference[65] colors exhibited by spin-coated resists as they dry, or by poorly coated nonuniform films, are the result of multiple interference.

A simple arrangement using white light to measure resist or film thickness is shown in Fig. 6-7-4. When the angle of beam incidence is less than 10°, Eq. (6-72) reduces to

$$t = \frac{\lambda_N N}{2\mu}$$

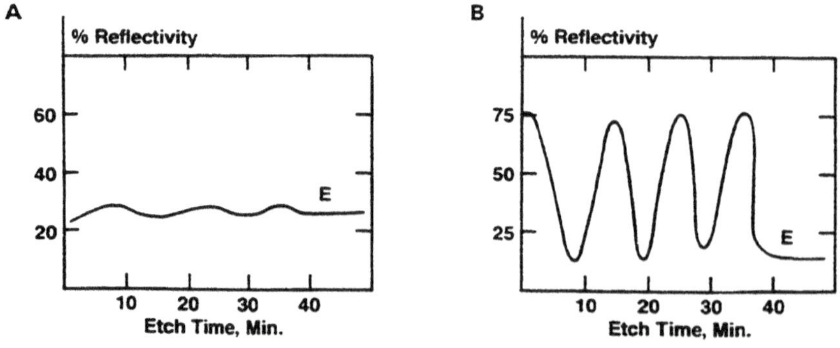

FIGURE 6-7-5. (A) Time to endpoint of etching a polysilicon/oxide/silicon sandwich. (B) Determination of endpoint by laser reflectivity interference pattern. From Ref. 162, reproduced by permission of the Electrochemical Society Inc.; figure originally presented at the Spring 1983 ECS meeting.

TABLE 6-7-3. Thickness Monitoring Resist Processing[a]

Process step	Thickness (Å) Positive	Negative
PR spin	14,955	14,890 ± 14
Bake, 65 °C, 15 min	—	14,463
Bake, 80 °C, 20 min	12,290	14,446
Expose		8,969 (11,500)
Develop	10,153	8,336 ± 692

[a] From Ref. 163.

By measuring the wavelength (λ_N) at which orders (N) of interference are observed, the thickness can be found. An etched, oxide silicon wafer (etched to different wedge thicknesses) is used as a reference grid. A partially developed resist could be used as well (Fig. 6-7-4).(159)

Ellipsometric measurements of resists use similar equations as (6-7-2), but are capable of higher accuracy to ±5–10 Å. For resist processes, residues, as thin as 20 Å left after resist stripping, have been investigated to determine the effectiveness of strippers.(43) Thicker films up to 1 μm, such as PMMA on GaAs, can also be measured.(160) Opaque films are monitored *in situ* by etching diffraction gratings.(208)

Measurements of not only the initial resist thickness, but of the rate of development (R) and of the unexposed film loss (R_0), are very important process controls to monitor resist dose–development conditions, and resist erosion during reactive ion etching.(161, 194, 208) For these applications, computer programs have been written to handle the calculation of the thickness from interference measurements of up to four optical layers, such as resist/oxide/polysilicon/silicon.(162) A typical etch rate pattern in the multilayers is shown in Fig. 6-7-5.

As stated before, in the introduction section and highlighted in Tables 6-1-2 to 6-1-4, resist thickness measurements in the static and dynamic modes are safeguards against large image biases, and to avert catastrophic process failures. As an example of process characterization, Flowers and Hughes(163) used a commercial IBM 7840

TABLE 6-7-4. Effect of Ambient on Film Thickness Loss for Azide Resist

Ambient	Percent thickness decrease Contact	Proximity
Air	45	38
Nitrogen	13	15

The resist thickness changes with different etchants were also examined:

HF, SiO_2	No change
Al etch	+776 Å
Si etch	7000 Å

film thickness analyzer to measure the thicknesses of wafer films at every step in device processing. In Table 6-7-3 are summarized the results of their film thickness measurements at each resist processing step.

The positive diazonovolak photoresist loses more solvent after baking than the negative azide-rubber resist. In one case, the negative resist loss of 4000 Å was discovered to be an "oxygen effect" during exposure. Oxygen combines (Chapter 4) with the photogenerated nitrene intermediate in azide-rubber resists, to prevent topside polymerization. In a separate study, the oxygen effect was examined (Table 6-7-4) for different exposure tools. Exposure under a blanket of nitrogen suppresses the oxygen effect. The resist thickness after etching was also examined:

HF etch of SiO_2	No change
H_3PO_4 etch of Al	+ 776 Å

Since the resist became swollen in Al etch, it may be an undercut factor, if the swelling perturbs the resist–Al interface (loss of adhesion). These examples point out the useful diagnostic, troubleshooting, and process optimization information that can be gleaned from simple thickness measurements.

6-8. PINHOLES IN RESIST FILMS

After the resist has been coated, baked, and its thickness measured, a pinhole count should be taken to ascertain potential defects. Further counts after exposure, develop, and etch steps are also required. However, pinhole counting is a difficult task, since pinholes are smaller than the circuit size. The counting usually requires a decoration or highlighting, to even ascertain their presence, let alone their density, location, and causes. For example, if we cast a 1.0-μm-thick resist film, we can

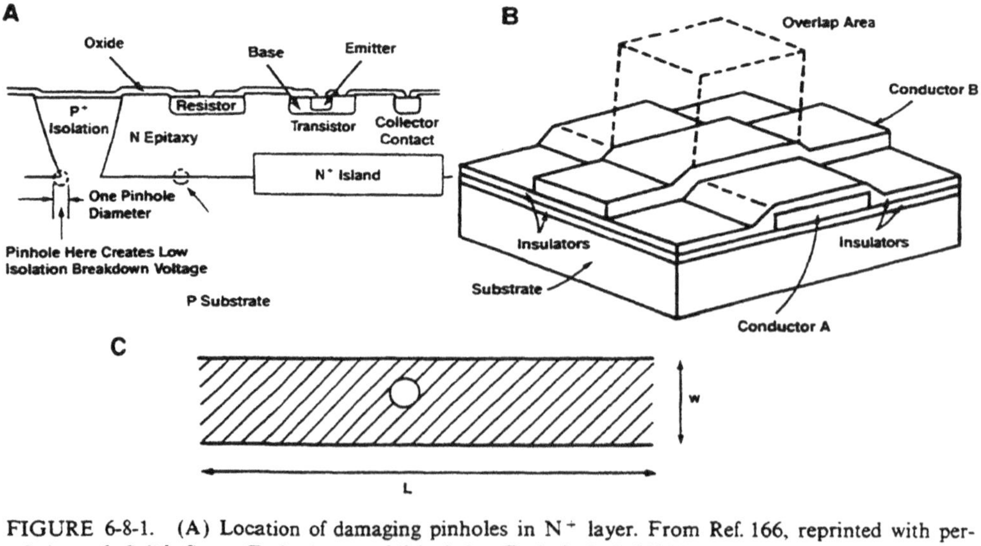

FIGURE 6-8-1. (A) Location of damaging pinholes in N^+ layer. From Ref. 166, reprinted with permission of *Solid State Technology*, published by Technical Publishing, a company of Dun and Bradstreet. (B) Location of pinholes in overlap area of metal lines. (C) Negligible and fatal size of pinholes in metal conductors of (B). (B, C) from Ref. 197, courtesy of IBM.

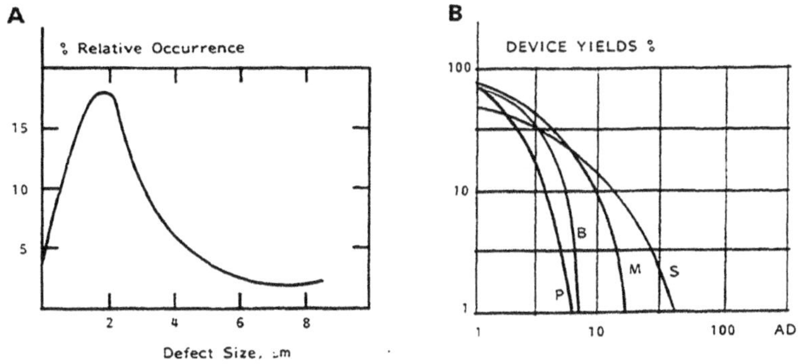

FIGURE 6-8-2. (A) Relative occurrence of resist photodefects and their size. (B) Device yields versus the average number of defects per device AD, for the Poisson (P), Bose–Einstein (B), Murphy (M), and Seeds (S) models. From S. Harrell, Photolithography—the Next Generation, Paper 1-14, Semiconductor Industry Conference, Oct. 1983, Palm Springs, Calif., Dataquest, Inc. (by permission).

expect to find 1–2 pinholes of 0.5-μm diameter per cm^2. If we cast a 0.5-μm film, we may find 5–10 pinholes/cm^2. Film thinning uncovers more subsurface defects not present as initially cast (but covered defects). Pinholes are indicative of process contamination and film thinning during processing. In this section, we will discuss the impact of pinholes, their detection, some of the causes, and the means of reducing their numbers.

Pinholes are a permanent or potential defect in a resist film, which can be transferred by etching or lift-off. In addition, pinholes can also be created by exposure, or upon physical contact of the resist surface with another surface. Particles dislodged during development are a third source of pinholes. Figure 6-8-1 shows the location of damaging pinholes (poly-to-poly short) in an isolation mask. A pinhole in an etched metal line may eventually lead to shorting or cause metal-to-metal shorts.[170]

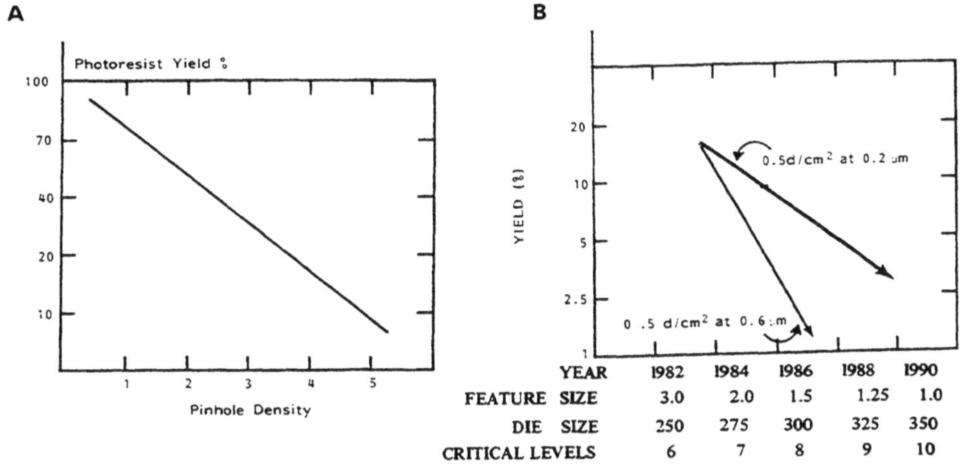

FIGURE 6-8-3. (A) Photoresist yield versus pinhole density. From Ref. 166. (B) Expected defect limit for 1M-bit DRAM assuming a tenfold increase to class 1 clean room conditions. Reprinted from *Microcontamination* 3, No. 5, May 1985, used with permission of the Institute for Microcontamination Control, Santa Clara, Calif., and Canon Communcations, Inc.

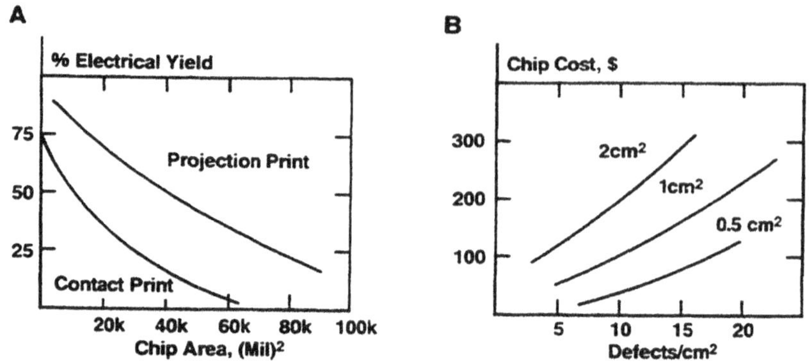

FIGURE 6-8-4. (A) Device yield in contact and projection print versus chip area. (B) Defect density versus chip cost versus chip area. From Ref. 172.

The critical pinhole level[(197)] to cause opens in two closely spaced Al lines (proximity effect) is less than the sum of the critical areas of two single conductors. This implies that manufacturers may cram photopatterns closer together to lower the defect susceptibility to opens for the same size pinholes. Experimentally, the critical area is monitored by measuring opens in Al conductors of varying lines and spaces.

A pinhole is defined as a small circular void in an oxide or resist caused by a tiny bubble, contaminants on the surface, or dust particles in the resist. On the wafer surface, the area affected is quite small. If a pinhole is one-half of the smallest circuit width,[(243)] the pinhole is considered to be fatal. Lawson[(166)] assumed that if the pinholes were randomly distributed, that the probability of occurrence is small, and that the probability of one defect does not depend on the occurrence of another. The probability of defects depends on a Poisson distribution:

$$p(k, d) = \exp(-d)\, d^k/k! \qquad (6\text{-}8\text{-}1)$$

where $p(k, d)$ is the probability of finding a circuit with k pinholes, when the average pinhole density is d. A solution to Eq. (6-8-1) yields a gamma distribution,

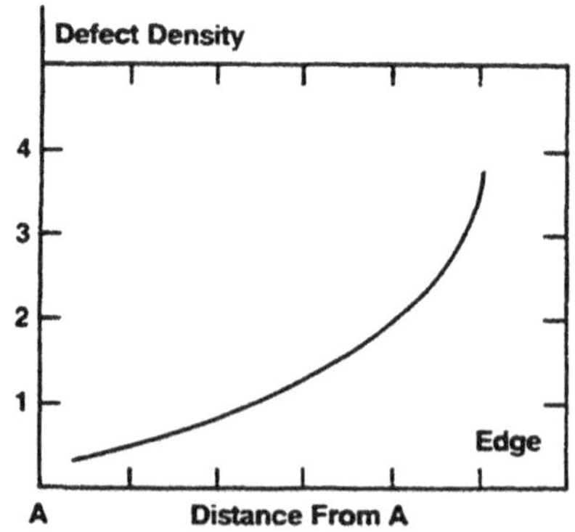

FIGURE 6-8-5. Radial dependence of defects in chips on a wafer. More defects are present at the circumference.

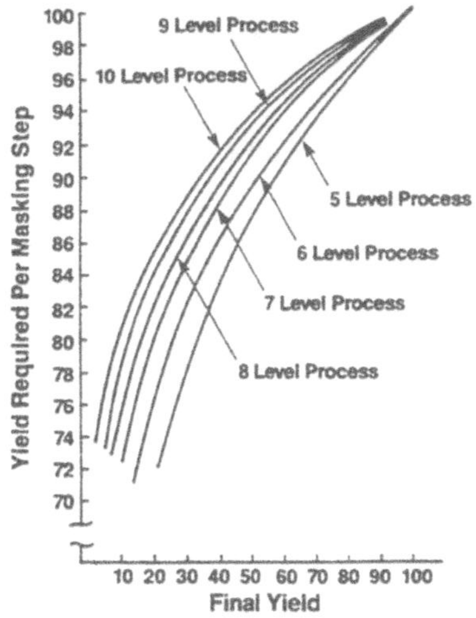

FIGURE 6-8-6. Device yield versus the number of photolithographic masking steps (levels). From Ref. 171.

which states that yield is an exponential function of defect density, D,

$$\text{Yield} = \exp(-D) \tag{6-8-2}$$

Various statistical models[212, 234] can be used to relate defects such as pinholes to device yields (Fig. 6-8-2). The greater the number of pinholes, the lower is the photo-yield (Fig. 6-8-3) for the more susceptible devices. Hemmert[167] modified Eq. (6-8-2) to explain a contradictory fact proposed by Lawson's result: that the log of yield linearly decreases with increasing area (Fig. 6-8-4).[172] As the chip area increases, the yield decreases.[192, 203] The underlying problem with Eq. (6-8-2) is that defects are averaged over a whole wafer, rather than at chip locations on the wafer.[243] On a wafer, good chips are located[168] in the middle of a wafer. The outer edge produces the lowest yield (Fig. 6-8-5). Modification by Hemmert of Eq. (6-8-2) gives[167]

$$\text{Yield} = 1/[\text{area(defect density)} + 1] \tag{6-8-3}$$

which approximates the yield in Fig. 6-8-3. Inclusion of the number of masking levels (n) leads to Bose–Einstein statistics in which the yield decreases with increasing n (Figs. 6-8-2B and 6-8-6):

$$\text{Yield} = 1/[\text{area(defect density)} + 1]\, n \tag{6-8-4}$$

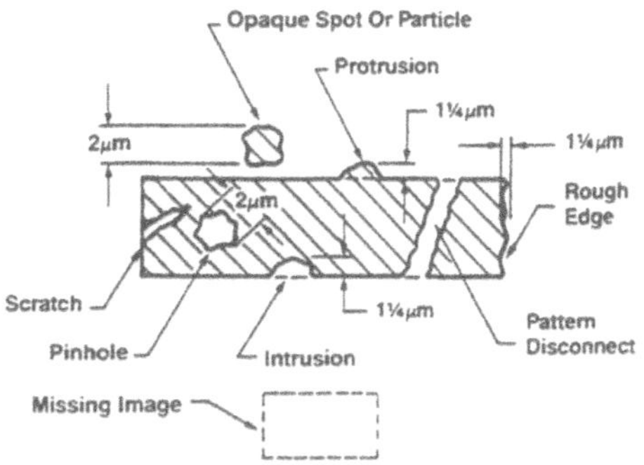

FIGURE 6-8-7. Schematic of photomask defects. From Ref. 173, copyright Bell Labs, 1974, by permission.

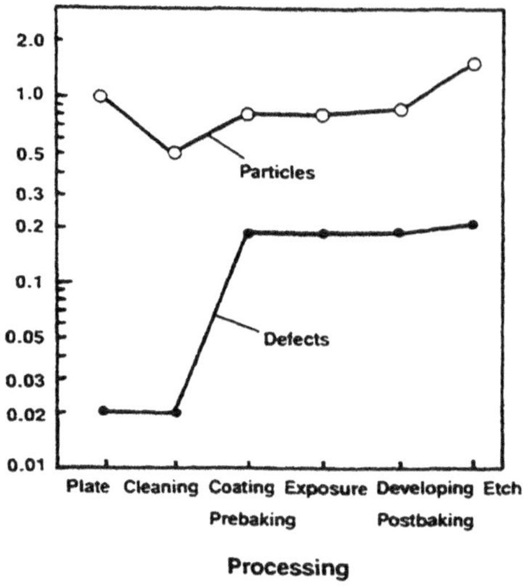

FIGURE 6-8-8. Particles on a resist surface versus the number of defects at each process step. From Ref. 190.

In the next 10 years, chips (dies) are expected to grow in size from 25 cm^2 to 100 cm^2 and the devices shrink to the submicrometer size. Thus, defects will have to be reduced to 0.2 μm in size to maintain a 10% yield (Fig. 6-8-3B).[(218)] Special circuits to monitor the shorts between conductive paths,[(168)] and a pinhole array capacitor,[(169)] can be built to characterize the frequency of pinhole defects. Due to an accumulation of defects, the greater the number of lithographic steps, the lower is the yield (Fig. 6-8-6) from the defects in coated films, resist processing, and inherent to the photomasks.[(171)] However, some processes such as spin coating are self-cleansing, by centrifuging particles from the center of the wafer, outwardly to the edge where they fly off and leave a defect.

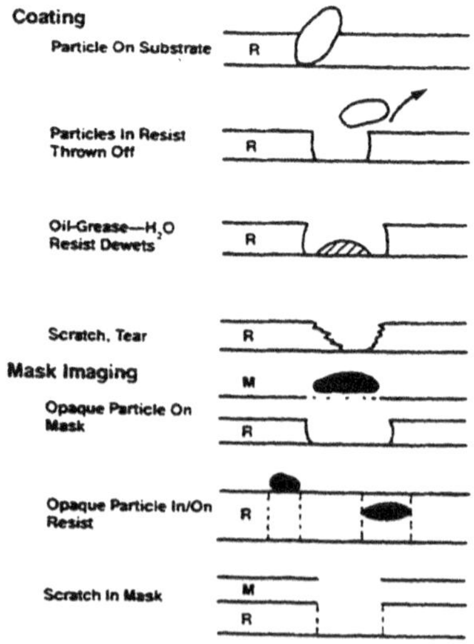

FIGURE 6-8-9. Potential causes of pinholes in resist films.

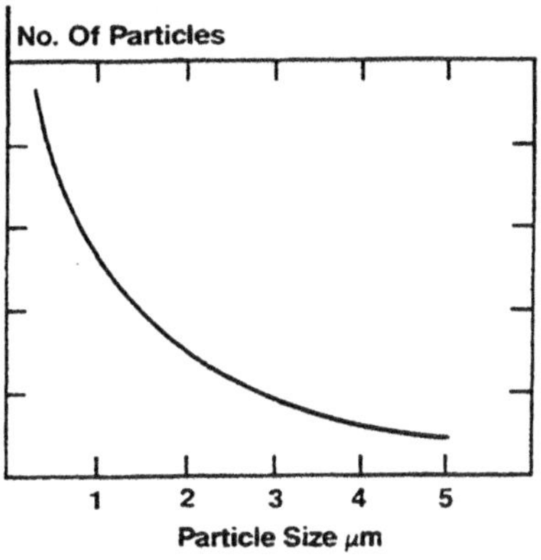

FIGURE 6-8-10. Typical particle size distribution in clean rooms. From Ref. 25, reprinted with permission of *Solid State Technology*, published by Technical Publishing, a company of Dun and Bradstreet.

6-8-1. *Causes of Pinholes*

Some of the particular causes of pinholes are recurrent with surface contaminants and particles on the surface or in the resist. Trapped air or particles in the interior of the resist film may become dislodged after developing the resist. In addition, defects[193] in the photomask (Fig. 6-8-7) may print in contact printing or may not image if at the diffraction size limit of a projection printer.[170]

Some of the potential causes of pinholes are shown in Figs. 6-8-8 and 6-8-9. As can be seen in Fig. 6-8-8, the higher the particle count on the surface, the higher is the number of defects in resist processing.[190] As the particle size of contaminants reaches submicrometer dimensions (Fig. 6-8-10), the number of particles landing on a surface in a clean room increases in density. Pinholes smaller than 1 μm may not print, unless high-numerical-aperture steppers are used.[170] Mask defects (Fig. 6-8-7) are of two types: opaque particles on the mask, or pinholes in chrome. The defects on a wafer can be classified into four types[170, 196, 197]:

1. A defect anomaly that is not transferred from the mask to the wafer (present in coating)
2. A detected anomaly that is transferred from the mask to the wafer, but has no effect on electrical yield or circuit reliability
3. A detected anomaly that has a significant effect on reliability
4. A detected anomaly that has a significant effect on yield

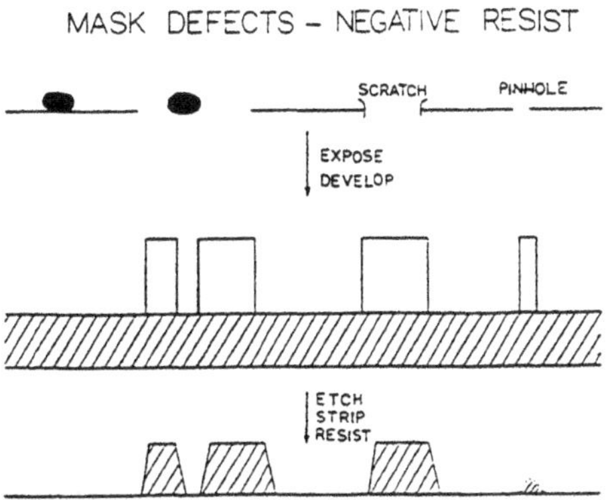

FIGURE 6-8-11. Potential printable defects in printing a negative photoresist.

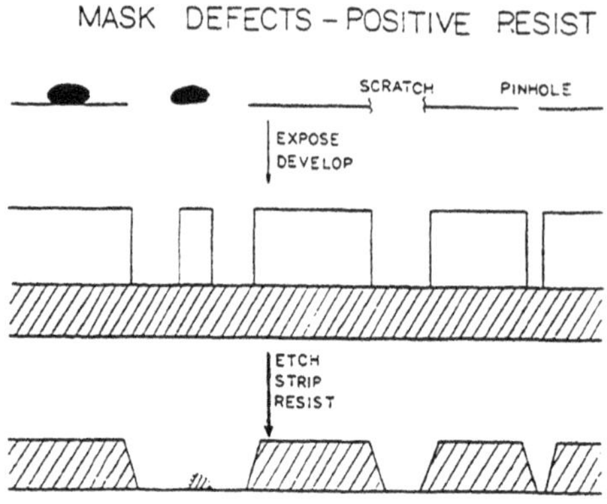

FIGURE 6-8-12. Potential printable defects in imaging a positive photoresist.

In a negative resist (Fig. 6-8-11), opaque particles on the mask or on top of the resist cause pinholes, while a positive resist prints mask pinholes, and is not as susceptible to opaque dirt on the clear portions of the mask (Fig. 6-8-12). In contact printing, photoresist pickup on the mask is also a common problem and masks are specially coated with low-surface-energy polymers to reduce the adhesion of the resist.

Particles present on the wafer as surface contaminants, and particles present in the resist, are likely to be spun off, or eventually (after resist developing) give rise to pinholes. Figure 6-8-13 shows the pinhole count in resists coated on wafers stored in three different classes of clean rooms. The more contaminated the surface, the greater is the occurrence of pinholes.[190] Since submicrometer particles adhere more strongly than larger particles,[23] a higher density of submicrometer defects can be expected in processing future devices. Thin resists also give rise to more pinholes, independent of the type of resist (Fig. 6-8-14). The drastic increase in pinholes below 0.5-μm thickness, reflects a practical limit for resist fabrication. In chrome mask fabrication, resists as thin as 0.25 μm are used. Pinholes as large as 1 μm may soon be the upper tolerable mask defect.[170] Although a single film of a high-contrast resist suppresses printable defects,[209] top surface imaging may be necessary to overcome the pinhole limit of single-layer submicrometer-thick resist films.

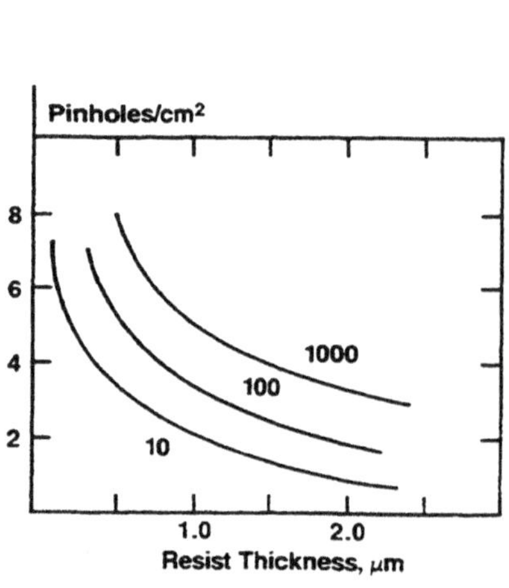

FIGURE 6-8-13. Pinholes (0.3-μm diameter) in resist coated on a silicon wafer processed in three different classes of clean rooms.

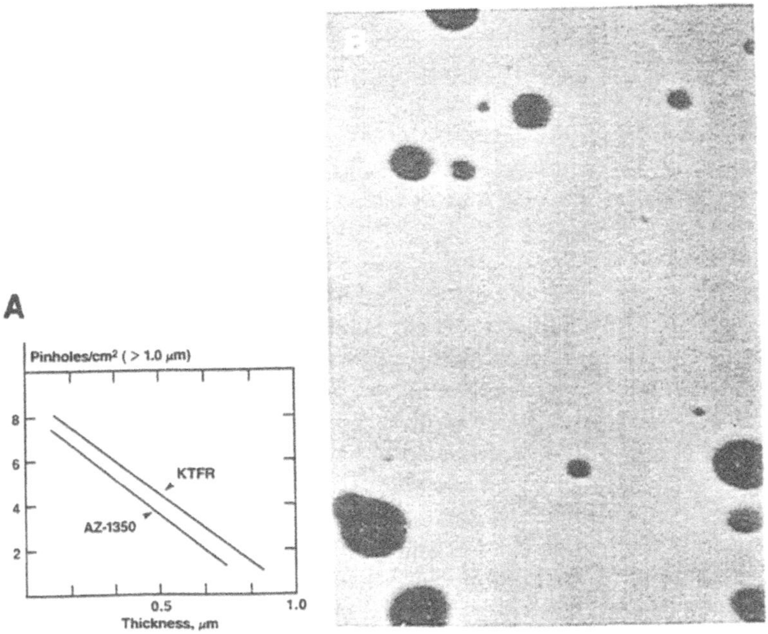

FIGURE 6-8-14. (A) Pinholes in negative KTFR and positive AZ-1350 resist cast under same conditions. Pinholes are not dependent on resist but on cleanliness of surfaces and environment (assuming no contaminants in resist. From Ref. 174. (B) Pinholes in thinned resist (thinned by develop and expose conditions for DQN resist to lose one-half of original 450-nm thickness).

6-8-2. *Measurement of Pinholes in Resist Films*

Accurate pinhole measurement is difficult, since even the oxide control may have a similar or larger pinhole count. The detection of the pinhole size, location, and frequency is a tedious process, generally involving destructive techniques. The main question to be answered concerns the statistical legitimacy of the test. Table 6-8-1 lists some of the chemical, physical, and electrical tests for pinholes. In the tests, the resist is generally coated on a thermal oxide of silicon with a known background level of pinholes. The resist is processed and pinholes are etched into the silicon. After the resist is stripped, the pinholes are measured. For reliable measurements, pinhole-free controls (oxidized blank wafers) are necessary.

In the pinhole determination, most of the pinhole tests are destrictive in highlighting the pinholes. A classic silicon etch is ethylene diamine-pyrocatechol for

TABLE 6-8-1. Pinhole Detection Techniques

Test[26]	Nature	Time	Resolution
MOS Al dot[177]	Capacitance[191]	Days	Submicrometer
H_2 formation	Electrolysis	Day	Micrometer
Si photoconductance	Electrolysis[194]	Hour	Submicrometer
Ionic conductance	Electrolyte	Hours	Submicrometer
Liquid crystal	Optical	Hours	Micrometer
Si etch[206,207]	Etching	Day	Micrometer
Cu decoration	Electroplate	Hours	Micrometer
Microscope	Optical	Day	Micrometer

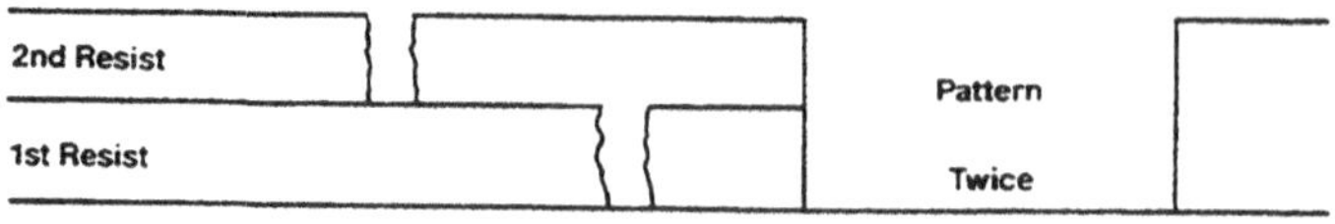

FIGURE 6-8-15. Statistical mismatch of pinholes by double resist processing of negative resists. The first layer is processed and overcoated with second resist layer. The mask is realigned, resist exposed and processed to mismatch pinholes. [U.S. Patent 3,506,441 (1970), RCA.]

etching pits in the 111 plane of silicon.(178) Chlorine, CF_4(198) gas, or NH_4F-acetic acid(176) is also used. In another method that detects only the presence of pinholes, capacitance techniques are used which involve evaporating Al(191) on top of SiO_2, to form a MOS capacitor, and detecting the pinholes as interlevel shorts.

A novel and nondestructive electrolytic conductance test can also be done with ionic liquids on top of photoresist-coated Si wafers. The ionic liquid drop is held by capillary action at the end of an electrode.(241) While the silicon is activated by exposure to light, the electrode is scanned across the surface of the resist. Pinholes are detected by conductance to the silicon surface and the frequency of current peaks is proportional to the pinhole defect density. The current peaks provide data relative to defect size and location. For resist material production and evaluation, the effectiveness of resist filtration and pinholes in developed resist films are routinely tested by this technique as part of the quality control of resist materials. Spun-cast resist films tested by this technique show a low pinhole count (1–2/cm^2). Upon resist development and loss of 200 nm of unexposed positive resist, the pinhole level increases to 6–8/cm^2. This type of information provided by the non-destructive photoelectrolytic conductance test is very valuable in selection of a pinhole-free resist and monitoring its pinhole count under process conditions such as after reactive ion etching or before metal lift-off deposition.

6-8-3. *Reduction and Control of Pinholes in Resist Films*

The task of reduction of pinholes is more one of prevention than of actual elimination, since the reliable accuracy of rapid testing is limited to about 1 μm. Nevertheless, there are some observations and procedures which should minimize their occurrence:

1. Use thickest resist as possible (Figs. 6-8-13 and 6-8-14).
2. Thoroughly clean surface, chemically and physically.
3. Reduce particles and contaminants in coat area (Fig. 6-8-13).

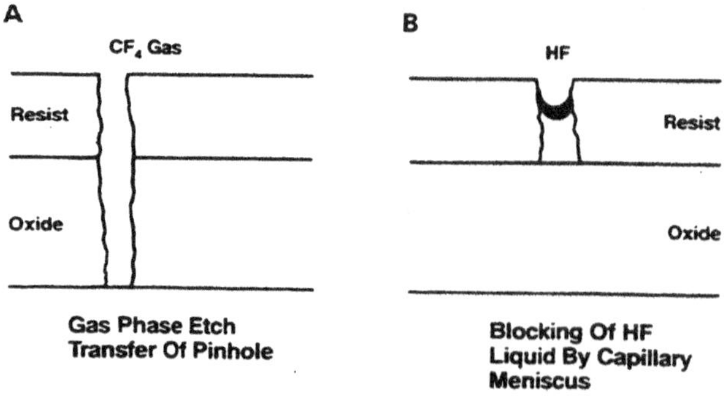

FIGURE 6-8-16. Transfer of pinholes in plasma etching but not in liquid etching processing.

FIGURE 6-8-17. Trilayer resist. The sandwich consists of a bottom planarizing layer (PR) coated with a silicon-containing layer (C) on top of which is coated a resist layer (R). The resist layer is imaged (1) and the image is transferred through the barrier layer by CF_4 plasma etching (2). The final image is formed by O_2 plasma etching of the planarizing layer.

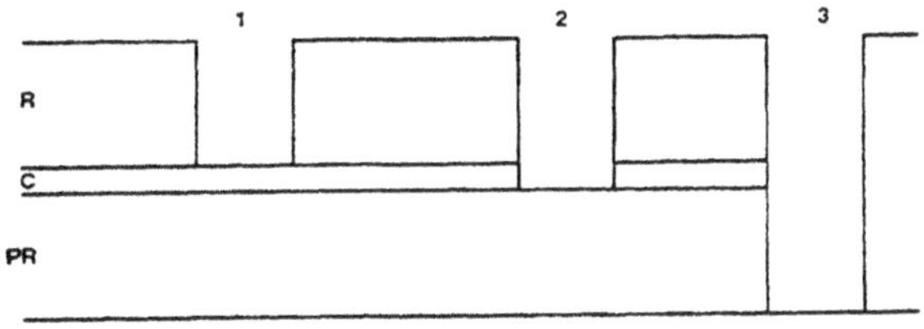

4. Ultrafiltration of resist to remove gels and insolubles.
5. Lubricate or coat contact print masks with adhesive layer.[175, 176]
6. Use highest prebake to melt resist.[176, 179]
7. Use resist with low melting point, such as polysiloxanes, and polyfluoroacrylates, to flow and close pinhole or surround defect.[180]
8. Fractionate polymer, such as PMMA, to remove developer-leachable low molecular weights.[179]
9. Do not thin resist in developer.
10. Use double coat resist process.[183]
11. Vacuum or plasma deposit resist to overcoat particles.[129]
12. Posttreat DQN image with surface treatment (e.g., silylation) to fill pinholes with an oxygen-resistant film prior to bilayer processing (see Chapter 12).
13. Use top surface imaging of thick resists.

The double coat resist process (Fig. 6-8-15) relies on the statistical mismatch of pinholes[182] in separate resist layers.

Spin coating is a self-cleansing process. The first layer cleanses itself. With a negative resist, the process is facilitated by insolubilization of the primary layer. The resist can be recoated (after expose and development) without dissolution. Positive PMMA resists can be recoated without processing, since the polymers are baked 100°C above the T_g, and all the solvent is driven out.[181] Otherwise, positive[185] DQ resists require processing of the primary coating, followed by baking, to insolubilize the DQN underlayer. Another layer is applied, realigned, exposed, and redeveloped.[187] A baked AZ-1350 underlayer has also been used as an adhesion promoter,[184, 185] followed by normal AZ-1350 processing after which the baked underlayer is opened by the action of the developer or if necessary by brief O_2 plasma etching. Multilayer resists also serve to reduce pinholes, by misalignment of the pinholes in each cast layer. However, there is some concern in multilayer resist processing that the plasma etch transfer steps are more susceptible to pinhole transfer than are liquid developers or etchants, which may not etch pinholes due to capillary constriction (Fig. 6-8-16).

Double layers of DQN can be coated, however, if a plasma polymerized CF_4 layer is deposited on top of the first layer, followed by a second DQN layer, exposure, develop, O_2 develop, and continued development (Fig. 6-8-17).[186]

6-9. SUMMARY AND OUTLOOK OF RESIST COATINGS

Spin coating is the dominant resist coating process with a uniformity of ±100 Å. In spin-coated single-film resists, planarized layers are the product, which

TABLE 6-9-1. Adhesion Factor Considerations

Condition of Surface	Contamination Cleansing method Storage condition
Surface states	Native oxide Native hydroxide Mixed states Adsorbed water
Primer	Remove wt water Esterify (silylate) surface Polymerize on surface Polymeric adhesive
Resist type	Polar (H-bonds) Nonpolar (dispersion bonds) Mixed adhesion
Process factors	Resist thickness Humidity Prebake, postbake Developer, etchant

creates depth-of-focus and reflectivity problems for exposing patterns running over steps. However, the planarizing properties are a payback for multilayer resists and planarization etch-back techniques. The failure to planarize with dye layers or contrast-enhancing layers can cause problems of notching and image distortion.

Conformal coatings deposited by vacuum, plasma, or Langmuir techniques are just being investigated with new inorganic or organic resists. They may prove to be useful in obtaining pinhole-free films and proper step coverage. Dry-processed resists will be a secondary process to complement vacuum deposition.

The importance of thickness as part of the general process "t" variables (temperature, time, and thickness) cannot be overemphasized. Both static and dynamic endpoint measurements are invaluable for setting up a resist process, as well as its proper daily monitoring.

For obtaining good process yields with the present VLSI devices and the future ULSI circuits, the proper cleaning of wafer surfaces, the maintenance of clean solid surfaces, and ultrafiltered process liquids (including resist) and gases, are technical necessities that justify their costs. Better, rapid, and more accurate submicrometer detectors for particulates on or in solids, liquids, and gases, and better pinhole detectors, are required to make contamination control a quantitative engineering process. If pinholes can be reduced ($< 1/\text{cm}^2$) in 0.5-μm-thick films, then higher resolution and smaller print biases may be obtained for etch-resistant resists. At the development or wet etching steps, good adhesion must be obtained, especially with the DQN family of resists. For review, the critical adhesion factors are summarized in Table 6-9-1.

Alternate coating techniques have not achieved the uniformity of spin coating, but they offer the economic gain of using lesser amounts of liquid resist per wafer. The rising cost of specialized resists will foster research into new ways to reduce the volume of liquid resist necessary to coat a wafer.

The casting solvent plays an important role in two-component resists in preventing phase separation during spin coating. Residual solvents, even after prebake, affect the dissolution properties and contrast of the resist. The controlling factor in reducing resist thicknesses is the nonplanarity of devices, and the failure of a resist to planarize steps better than 75% of its height. Since thinner resist films enhance resolution, basic improvements in planarizing the silicon device itself, and better etch conditions to avoid resist erosion, will serve to reduce the thickness of resist films. To avoid further pinholes, clean surfaces and controlled environments will be necessary to limit defects, improve yield, and produce reliable circuits. In the future, pinholes may be larger than the device size. Their detection and control will be an interesting challenge for equipment manufacturers and for operational use by process engineers. Finally, statistical defect analysis and yield modeling will aid in the design of more reliable circuits that tolerate some level of submicrometer defects which can be expected in the next generation of semiconductor lithography manufacturing processes.

REFERENCES

1. D. Angel, *Semicond. Int.* April 1983, p. 72.
2. M. Bowden, *J. Electrochem. Soc.* **128**, 195C (1981).
3. A. Broers, *IEEE Trans. Electron Devices* **ED-28**, 1268 (1981).
4. T. Chang, C. Codella, and R. Lange, *IEEE Trans. Electron Devices* **ED-28**, 1428 (1981).
5. M. Kato and N. Nakane, *Photogr. Sci. Eng.* **23**, 207 (1979).
6. D. Widmann and H. Binder, *IEEE Trans. Electron Devices* **ED-22**, 467 (1975).
7. S. Middlehoek, *IBM J. Res. Dev.* **14**, 117 (1970).
8. B. Lin, *Solid State Technol.* May 1983, p. 105.
9. J. Moran and D. Maydan, *J. Vac. Sci. Technol.* **16**, 1620 (1979).
10. L. White, *J. Electrochem. Soc.* **130**, 1543 (1983); *J. Vac. Sci. Technol.* **B1**, 1235 (1983).
11. A. Miligram, *J. Vac. Sci. Technol.* **B1**, 490 (1983).
12. Kodak Photoresist Seminar Proc., 1973, p. 69.
13. T. O'Neill, *Semicond. Int.* Dec. 1980, p. 49.
14. D. Mattox, *Thin Solid Films* **53**, 81 (1978).
15. M. Dahlstrom, *Semicond. Int.* April 1983, p. 110.
16. W. Jones, *Solid State Technol.* Jan. 1974, p. 56.
17. J. Lange, *Semicond. Int.* April 1983, p. 125.
18. J. Paivanas and J. Hassan, *Solid State Technol.* April 1980, p. 149.
19. D. Bauman, *Solid State Technol.* March 1973, p. 45.
20. K. Mittal, in *Surface Contamination*, edited by K. Mittal, Plenum Press, New York, 1979, p. 1.
21. M. Whitfield, in *Surface Contamination*, edited by K. Mittal, Plenum Press, New York, 1979, p. 73.
22. B. Derajguin, *J. Adhes.* **4**, 65 (1977).
23. A. Zimon, *Adhesion of Dust and Powders*, Consultants Bureau, Plenum Press, New York, 1982.
24. J. Amick, *Solid State Technol.* Nov. 1976, p. 47.
25. M. Brubach, *Solid State Technol.* Feb. 1972, p. 46.
26. W. Kern, *Solid State Technol.* March 1974, p. 35; *Solid State Technol.* April 1974, p. 35; *RCA Rev.* **34**, 655 (1973).
27. D. Burkman, *Semicond. Int.* July 1981, p. 103.
28. W. Kern and D. Puotinen, *RCA Rev.* June 1970, p. 187.
29. D. Brandeth and R. Johnson, in *Surface Contamination*, edited by K. Mittal, Plenum Press, New York, 1979, p. 83.
30. M. Balzaks and S. Walker, *Semicond. Int.* April 1982, p. 101.
31. P. Burggraaf, *Semicond. Int.* July 1982, p. 35.
32. A. Weiss, *Semicond. Int.* July 1982, p. 55.
33. R. deFilippi, in *Filtration*, edited by C. Orr, Dekker, New York, 1977, p. 476.
34. R. Akers and A. Ward, in *Filtration*, edited by C. Orr, Dekker, New York, 1977, p. 169.
35. R. Iscoff, *Semicond. Int.* April 1983, p. 88.

36. Millipore Co., Bedford, Mass., technical literature.
37. D. Tolliver, N. Davenport, and L. Abts, *Solid State Technol.*, Sept. 1982, p. 116.
38. K. Hirobe and S. Iwamatsu, *J. Electrochem. Soc.* **126**, 1426 (1979).
39. U.S. Patent 4,132,559 (1979), Motorola.
40. E. Julef, *Microelectronics* **8**, 35 (1977).
41. E. Julef, *Microelectronics* **7**, 17 (1976).
42. E. Julef, W. McCloud, E. Hulse, and W. Fawcett, *Solid State Technol.* Sept. 1982, p. 82.
43. L. Kaplan and B. Bergin, *J. Electrochem. Soc.* **127**, 386 (1980).
44. F. Pintchovski, J. Price, R. Tobin, J. Peavey, and K. Kobold, *J. Electrochem. Soc.* **126**, 1428 (1979).
45. D. Peters and C. Deckert, *J. Electrochem. Soc.* **126**, 883 (1979).
46. G. Harrabee and T. Shaffner, *Anal. Chem.* **53**, 163R (1981).
47. R. Honig, *Thin Solid Films* **31**, 89 (1976).
48. R. Seltzer, *Circuits Manuf.* Nov. 1975, p. 54.
49. B. Whitsel, *Microelectron. Manuf. Test.* Dec. 1983, p. 13.
50. C. Stapper, *IBM J. Res. Dev.* **27**, 549 (1983).
51. T. Meguro, H. Kurita, and T. Ito, *J. Electrochem. Soc.* **128**, 1379 (1981).
52. S. Jenekhe, *Polym. Eng. Sci.* **23**, 830 (1983); R. Piacente, *J,. Electrochem. Soc.* **133**, 979 (1986).
53. German Patent 2,447,225, IBM; *Chem. Abstr.* **85**, 102423 (1975).
54. U.S. Patent 4,142,636 (1981), Mitsubishi.
55. U.S. Patent 4,332,881 (1982), Bell.
56. Eastman Kodak Technical Literature on Applications of Eastman 910 Adhesive.
57. J. Helbert, *SPIE Proc.* **333**, 24 (1982).
58. U.S. Patent 4,587,196 (1986), Hunt.
59. *Kodak Res. Discl.* **149**, 27 (1976); *Chem. Abstr.* **86**, 362590 (1977).
60. U.S. Pat., 4,036,644 (1979), IBM.
61. Japanese Patent 75,127,619, *Chem. Abstr.* **85**, 12367 (1975); (a) S. Farenholtz, *Am. Chem. Soc. Org. Coat.* **35**, 306 (1975); (b) Japanese Patent 77,126,211, *Chem. Abstr.* **89**, 120920 (1978); (c) Japanese Patent 80, 129,341 (Hitachi), *Chem. Abstr.* **94**, 148396 (1982); (d) Japanese Patent 78, 135621 (Hitachi), *Chem. Abstr.* **90**, 130686 (1979); (e) U.S. Patent 4,176,003 (1979), NCR; (f) U.S. Patent 3,176,390 (1973), Bell.
62. U.S. Patents 3,634,082 and 4,148,654 (1982), Shipley Co.
63. H. Yanazawa, H. Utsugi, N. Hashimoto, and M. Ashikawa, *Jpn. J. Appl. Phys. Suppl.* **2**, Part I, 753 (1974); Proc. 6th Int. Vac. Congr., 1974.
64. A. Knop, *The Chemistry and Applications of Phenolic Resins*, Springer-Verlag, Berlin, 1979.
65. D. Meyerhofer, *J. Appl. Phys.* **49**, 3993 (1978).
66. B. Chin, *Polym. Eng. Sci.* **23**, 399 (1983).
67. W. Daughton and F. Givens, *J. Electrochem. Soc.* **129**, 177 (1982).
68. J. Lai, *Polym. Eng. Sci.* **19**, 1117 (1979).
69. G. Damon, Proceedings of the Second Kodak Seminar on Microminiaturization, 1967.
70. G. Schwartz, Proceedings of Kodak Seminar on Microminiaturization, 1965.
71. R. Kelly, Proceedings of Kodak Seminar on Microminiaturization, 1965.
72. C. Taylor, NASA Circ. CR 597, U.S. Government Publ. 6-2003, 1966.
73. B. Washo, *IBM J. Res. Dev.* **21**, 190 (1977).
74. R. Malangone and C. Needham, *J. Electrochem. Soc.* **129**, 2881 (1982).
75. K. Law, *Polymer* **23**, 162 (1982).
76. H. Denk, Kodak Interface Microelectronics Proceedings, 1976, p. 28.
77. F. Givens and W. Daughton, *J. Electrochem. Soc.* **126**, 269 (1979).
78. A. Acrivos, M. Shah, and E. Petersen, *J. Appl. Phys.* **31**, 963 (1960).
79. R. Eaton and F. Willeboordse, *J. Coat. Technol.* **52**, 63 (1980).
80. A. Emslie, F. Bonner, and L. Peck, *J. Appl. Phys.* **29**, 858 (1958).
81. T. Patton, *Paint Flow and Pigment Dispersion*, Wiley, New York, 1979, p. 585.
82. E. Hansen, *Chem. Technol.* Sept. 1972, p. 547.
83. L. Hoy, *J. Paint Technol.* **42**, 76 (1970).
84. F. Billmeyer, *Textbook of Polymer Science*, Wiley, New York, 1971, p. 41.
85. M. Kurata, in *Polymer Handbook*, edited by J. Brandup and E. Immergut, Wiley, New York, 1975, Chapter 4–1.
86. D. Patterson, *Polym. Eng. Sci.* **22**, 64 (1982).
87. L. Gavens, B. Wu, D. Hess, A. Bell, and D. Soong, *J. Vac. Sci. Technol.* **B1**, 481 (1983).

88. J. Greeneich, *J. Electrochem. Soc.* **122**, 970 (1975).
89. M. Bowden, *J. Polym. Sci.* **26**, 1424 (1981).
90. M. Bowden, L. Thompson, S. Farenholtz, and E. Doerries, *J. Electrochem. Soc.* **128**, 1304 (1981).
91. S. Croll, *J. Appl. Polym. Sci.* **23**, 847 (1979).
92. U.S. Patent 4,266,001 (1981), Hoechst.
93. U.S. Patent 4,174,222 (1981), Tokyo Ohko.
94. D. Meyerhofer, *IEEE Trans. Electron Devices* **ED-27**, 921 (1980).
95. W. Moreau and W. Moyer, *IBM Tech. Discl. Bull.* **23**, 991 (1980).
96. Y. Cohen and S. Reich, *J. Polym. Sci.* **19**, 599 (1981).
97. J. Magerlein and D. Webb, *IBM J. Res. Dev.* **24**, 554 (1980).
98. D. Leers, *Solid State Technol.* March 1981, p. 90.
99. P. Frasch and K. Saremski, *IBM J. Res. Dev.* **26**, 561 (1982).
100. U.S. Patent 4,086,870 (1978), IBM.
101. R. Smith, *Ind. Eng. Chem. Prod. Res. Dev.* **22**, 67 (1983).
102. S. Stinson, *Chem. Eng. News* Aug. 8, 1983, p. 7.
103. German Patent 2,637,105, *Chem. Abstr.* **88**, 129873h (1978).
104. U.S. Patent 3,695,911 (1967), Aleo.
105. U. S. Patent 4,290,384 (1981), Perkin Elmer.
106. U.S. Patent 4,281,057 (1981), IBM.
107. D. Cowley, C. Price, and C. Hardy, *Polymer* **21**, 1356 (1980).
108. J. Little, T. Callicott, and E. Arakawa, *Rev. Sci. Instrum.* **51**, 1581 (1980).
109. A. Avidienko and M. Malev, *Vacuum* **276**, 583 (1977).
110. W. De Forest, *Photoresist Materials and Processes*, McGraw–Hill, New York, 1975.
111. U.S. Patent 4,069,076 (1976), DuPont.
112. S. Farenholtz and T. Kwei, *Macromolecules* **14**, 1076 (1981), **15**, 937 (1982).
113. I. Haller, *J. Amer. Chem. Soc.* **100**, 8050 (1978).
114. W. Bascom, *Macromolecules* **5**, 792 (1972).
115. U. S. Patent 3,535,157 (1967), Shipley.
116. D. Owens and R. Wendt, *J. Appl. Polym. Sci.* **13**, 1741 (1969).
117. A. Cemal, T. Fort, and J. Lando, *J. Polym. Sci. Part A* **10**, 2061 (1972).
118. G. Wegner, *Angew. Chem. Int. Ed. Engl.* **20**, 361 (1981).
119. A. Barraud and M. Vandeveyer, *Thin Solid Films* **99**, 221 (1983); I. Peterson, *Thin Solid Films* **109**, 371 (1983).
120. A. Broers and M. Pmerantz, *Thin Solid Films* **99**, 323 (1983).
121. K. Fukuda, Y. Shibasaki, and H. Nakahara, *Thin Solid Films* **99**, 87 (1983).
122. G. Fariss, J. Lando, and S. Rickert, *Thin Solid Films* **99**, 305 (1983).
123. A. Barraud, *Thin Solid Films* **99**, 317 (1983).
124. A. Barraud, C. Rosilio, and A. Raudel-Texier, *Thin Solid Films* **68**, 91, 99 (1980), *Solid State Technol.* Aug. 1979, p. 124.
125. A. Barraud, *Thin Solid Films* **85**, 77 (1981).
126. G. Roberts, P. Vincett, and W. Barlow, *Phys. Technol.* **12**, 69 (1981).
127. A. Barraud and A. Rosilio, *Thin Solid Films* **31**, 243 (1976).
128. K. Mittal and D. O'Kane, *J. Adhes.* **8**, 93 (1976).
129. A. Korchov, T. Martynova, and V. Danilovich, *Thin Solid Films* **101**, 369 (1983).
130. USSR Patent 983,835, *Chem. Abstr.* **98**, 98824t (1981).
131. U.S. Patent 4,371,608 (1981), Ionomet.
132. U.S. Patent 3,751,285 (1973), Kalle.
133. G. Chern and I. Lauks, *J. Appl. Phys.* **53**, 6979 (1982).
134. G. Ray, S. Peny, D. Burriesi, M. O'Toole, and E. Liu, *J. Electrochem. Soc.* **129**, 2152 (1982).
135. M. Chang and J. Chem, *Appl. Phys. Lett.* **33**, 892 (1978).
136. S. Yoshikawa, O. Ochi, and Y. Mizushima, *Appl. Phys. Lett.* **36**, 107 (1980).
137. P. Huggett, K. Frick, and H. Lehmann, *Appl. Phys. Lett.* **42**, 592 (1983).
138. A. Bell and J. Hollahan, *Techniques and Applications of Plasma Processing*, Wiley, New York, 1974.
139. M. Shen and A. Bell, *Am. Chem. Soc. Symp. Ser.* **108**, 197 (1979).
140. Y. Segui, A. Bui, and M. Maisonnuere, *Thin Solid Films* **33**, 35 (1976), *J. Appl. Phys.* **51**, 5055 (1980), **50**, 6567 (1979).
141. L. Martinui and H. Brederman, *Vacuum* **33**, 253 (1983), **31**, 285 (1981).
142. S. Morita, J. Tamano, S. Hattori, and M. Ieda, *J. Appl. Phys.* **51**, 3938 (1980).

143. J. Tamano, K. Yoneda, S. Morita, and S. Hattori, *Plasma Chem. Plasma Process.* **1**, 261 (1981).
144. A. Morinka and Y. Asano, *J. Appl. Polym. Sci.* **27**, 2139 (1982).
145. J. Calvert and J. Pitts, *Photochemistry*, Wiley, New York, 1966, p. 653.
146. H. Hiroaka, *Appl. Phys. Lett.* **31**, 503 (1977).
147. J. Tsao and D. Ehrlich, *Appl. Phys. Lett.* **42**, 997 (1983).
148. W. Pliskin and S. Zanin, in *Handbook of Thin Film Technology*, edited by L. Maissel and R. Glang, McGraw–Hill, New York, 1970.
149. R. Powell and M. Morad, *J. Appl. Phys.* **49**, 2499 (1978).
150. F. Loprest and E. Fitzgerald, *Photogr. Sci. Eng.* **15**, 261 (1971).
151. J. Shaw and M. Hatzakis, *IEEE Trans. Electron Devices* **ED-25**, 425 (1978).
152. O. Heavens, *Optical Properties of Thin Solid Films*, Butterworths, London, 1955.
153. A. Warnecke and P. Lopresti, *IBM J. Res. Dev.* **13**, 256 (1973).
154. A. Adams, D. Schineke, and C. Capio, *J. Electrochem. Soc.* **126**, 539 (1979).
155. F. Dill and K. Konnerth, *IEEE Trans. Electron Devices* **ED-22**, 452 (1975).
156. D. Wood and J. Fleming, *Rev. Sci. Instrum.* **53**, 43 (1982).
157. M. Kato and H. Nakane, *Photogr. Sci. Eng.* **23**, 209 (1979).
158. L. Harriman, *Solid State Technol.* June 1983, p. 155.
159. J. Sandercock, *J. Phys. E* **16**, 866 (1983); U.S. Patent 4,355,905 (1982).
160. R. Scheps, *J. Electrochem. Soc.* **130**, 655 (1983).
161. M. Sternheim, W. Van Gelder, and A. Hartman, *J. Electrochem. Soc.* **130**, 655 (1983).
162. P. Herman and R. Shutz, *J. Electrochem. Soc.* **131**, 881 (1984).
163. P. Flowers and H. Hughes, *Semicond. Int.* Jan. 1981, p. 79.
164. P. Burggraf, *Semicond. Int.* Sept. 1981, p. 39.
165. J. Pudvin, in *Environmental Control in Electronic Manufacturing*, edited by P. Morrison, Van Nostrand, Princeton, N.J., 1973, Chapter 6.
166. T. Lawson, *SCP Solid State Technol.* July 1966, p. 22.
167. H. Hemmert, *Solid State Electron.* **24**, 511 (1981).
168. C. Mallory, *Solid State Technol.* Nov. 1983, p. 121; A. Marsh, *SPIE Proc.* **470**, 55 (1984).
169. M. Buehler, *Solid State Technol.* Nov. 1983, p. 131.
170. D. Angel and P. Johnson, *Semicond. Int.* March 1980, p. 60.
171. K. Levy and P. Sandland, Proceedings of Kodak Seminar on Microelectronics, 1977.
172. *Electron. Packag. Prod.* Dec. 1974, p. 62.
173. R. Wasson, Proceedings of Kodak Seminar on Microelectronics, 1973, p. 66.
174. Y. Mimemura, *Rev. Electron. Commun.* **20**, 641 (1972).
175. T. Matsuzawa, *J. Electrochem. Soc.* **128**, 185 (1981).
176. D. Flowers and R. Hughes, *J. Electrochem. Soc.* **124**, 1599 (1977).
177. R. Bowman, Proceeding of Kodak Seminar on Microelectronics, 1974, p. 1.
178. R. Finne and D. Klein, *J. Electrochem. Soc.* **114**, 965 (1967).
179. K. Morita, *Fujitsu Sci. Tech. J.* June 1978, p. 43.
180. M. Kakuchi, *Rev. Electron. Commun.* **27**, 1113 (1979).
181. S. Morita and K. Hattori, *J. Appl. Phys.* **51**, 3938 (1980).
182. G. Thomas, *Solid State Electron.* **11**, 378 (1968).
183. U.S. Patent 4,394,437 (1983).
184. U.S. Patent 3,873,313 (1977).
185. U.S. Patent 4,332,881 (1982), Bell.
186. D. Bobkin and B. Cantos, *IEEE Electron Device lett.* **EDL-2**, 222 (1981).
187. M. Kawaguchi and A. Tokahashi, *Macromolecules* **16**, 1465 (1983).
188. C. Evans and R. Blattner, *Semicond. Int. Nov.* 1980, p. 109.
189. W. Andrus, J. Buono, and A. Wisnewski, *Semicond. Int.* Nov. 1980, p. 71.
190. M. Dan, I. Tanabe, M. Hoga, and S. Torisawa, Kodak Microelectronics Seminar, 1979, p. 103.
191. S. Iwamatsu, *J. Electrochem. Soc.* **129**, 224 (1982).
192. M. Gulett, *Semicond. Int.* March 1981, p. 87.
193. J. Hollenstein, *Electron. Packag. Prod.* April 1978, p. 90.
194. A. Dalisa, *Appl. Phys. Lett.* **17**, 208 (1970).
195. F. Laming and P. Straile, *J. Electrochem. Soc.* **120**, 292 (1973).
196. B. Whitsel, *Microelectron. Manuf. Test. Dec.* 1983, p. 13.
197. C. Stapper, *IBM J. Res. Dev.* **27**, 549 (1983).
198. T. Meguro, H. Kurita, and T. Itoh, *J. Electrochem. Soc.* **128**, 1379 (1981).

199. S. Jenekhe, *Polym. Eng. Sci.* **23**, 830 (1983).
200. W. Flack, D. Hess, D. Soong, and A. Bell, *J. Appl. Phys.* **56**, 1199 (1983).
201. U.S. Patent 4,439,514 (1984), University Patents.
202. Japanese Patent 58,105,143, *Chem. Abstr.* **101**, 101231 (1983).
203. W. Morgan and J. Burnett, *Semicond. Int.* June 1984, p. 137.
204. J. Lee, *Semicond. Int.* June 1984, p. 101.
205. D. Tavers, *Microelectron. Manuf. Test.* Nov. 1984, p. 13.
206. M. Maeda and H. Nakamura, *Thin Solid Films* **112**, 279 (1984).
207. W. Shannon, *RCA Rev.* **31**, 431 (1970).
208. G. Mendes, L. Cescato, J. Freijlich, E. Braga, and A. Mannamma, *Thin Solid Films* **117**, 1079 (1984).
209. C. Fu and D. Dameron, *IEEE Electron Device Lett.* **EDL-5**, 398 (1984).
210. Japanese Patent 58,205,148, *Chem. Abstr.* **101**, 201532 (1984).
211. Japanese Patent 59,11,895, *Chem. Abstr.* **101**, 201544 (1984).
212. U.S. Patent 3,627,599 (1971), RCA.
213. W. Ham, *RCA Rev.* **39**, 231 (1978).
214. P. Giese, *Solid State Technol.* Nov. 1983, p. 163.
215. R. Bond and H. Naguib, *SPIE Proc.* **221**, 74 (1980).
216. N. Durrant and P. Jenkins, *Microcontamination* April 1985, p. 44.
217. D. Johnson, *Semicond. Int.* April 1985, p. 168.
218. J. Burnett, *Microcontamination* May 1985, p. 32.
219. J. Yu, J. Zakin, and G. Patterson, *J. Polym. Sci.* **23**, 2493 (1979).
220. P. Luckham and J. Klein, *Macromolecules* **18**, 721 (1985).
221. Japanese Patent 60,02,947, *Chem. Abstr.* **102**, 14905 (1985).
222. Japanese Patent 59,182,439, *Chem. Abstr.* **102**, 53988 (1985).
223. L. White, *Proc. SPIE* **539**, 29 (1985).
224. L. White, *J. Electrochem. Soc.* **132**, 3051 (1985).
225. R. Wilson and P. Piacente, *Electrochem. Soc. Proc.* **85-6**, 124 (1985), *J. Electrochem. Soc.* **133**, 981 (1986).
226. Y. Lin, A. Purdes, S. Saller, and W. Hunter, *J. Appl. Phys.* **55**, 1110 (1984).
227. S. Gunarwardena, R. Haven, U. Kaempf, M. Parikh, B. Tullis, and J. Victor, *J. Environ. Sci.* **27**(3), 23 (1984).
228. S. Hoenig, *Solid State Technol.* Sept. 1985, p. 129.
229. R. Bowling, *J. Electrochem. Soc.* **132**, 2208 (1985).
230. J. Burnett, *Solid State Technol.* Sept. 1985, p. 121.
231. E. Baker, *Microelectron. Manuf. Test.* July 1985, p. 20.
232. R. Donovan, B. Locke, and D. Ensor, *Solid State Technol.* Sept. 1985, p. 139.
233. R. Donovan, B. Locke, C. Osburn, and A. Caviness, *J. Electrochem. Soc.* **132**, 2730 (1985).
234. P. Sukanek, *J. Imag. Technol.* **11**, 184 (1985).
235. U.S. Patent 4,510,176 (1985), AT&T.
236. Japanese Patent 60,24,545, *Chem. Abstr.* **102**, 22940 (1985), JSR.
237. European Patent 126,266, *Chem. Abstr.* **103**, 14589 (1985), Allied.
238. Japanese Patent 60,02,947, *Chem. Abstr.* **102**, 14095 (1985), Fujitsu.
239. M. Gibson and J. Freijlich, *Thin Solid Films* **128**, 161 (1985).
240. A. Piegari and E. Masetti, *Thin Solid Films* **124**, 249 (1985).
241. U.S. Patent 4,473,795 (1984), IBM.
242. U.S. Patent 4,514,440 (1985), Allied.
243. A. Ferris-Prabhu, *IEEE Trans. Electron Devices* **ED-32**, 1727 (1985).
244. L. White, *RCA Rev.* **47**, 345 (1986).
245. I. Matsuba and K. Matsumoto, *IEEE Trans. Electron Devices* **ED-33**, 1236 (1986).
246. R. Wilson and P. Piacente, *J. Electrochem. Soc.* **133**, 981 (1986).
247. M. Levinson and C. Wilkins, *J. Electrochem. Soc.* **133**, 619 (1986).

7

Prebake (Softbake)

7-1. INTRODUCTION

Prebake, also known as softbake or preexpose bake, is the physical process of conversion of a liquid-cast resist into a solid film. The prebake step involves the physical removal of the casting solvent without the degradation of the resist components. By removing the casting solvent from the film, a solid state is formed which prevents mixing of the exposure products with the unexposed zone (reactants). In addition, during prebake, temporary adhesion of the resist is also established. In some specialized resists, prebake is also used to chemically convert resist polymers, such as PMMA anhydride and polymethacrylonitrile, into pre-cross-linked resists,[1,58,59,62]

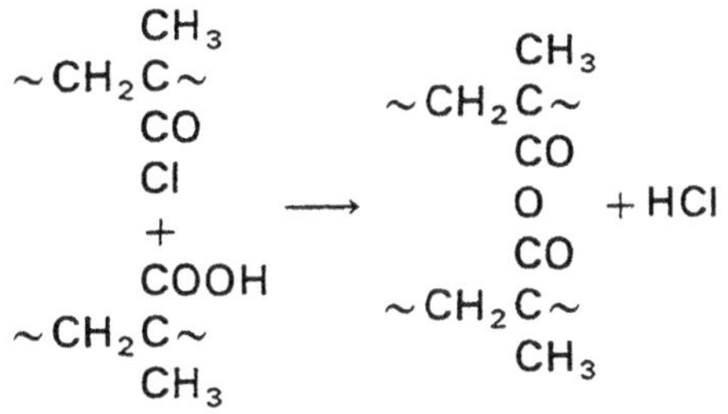

On the other hand, solventless resists, such as plasma-deposited PMMA[2] or evaporated resists of inorganic selenides,[3] require no prebake. Prebake is not to be confused with postexposure bake (predevelop bake), which is used to remove standing waves or amplify resist images.[4,8]

Since the prebake process can have an effect on subsequent processes such as exposure and development of the image, the prebake process should be precise, uniform across the wafer, and reproducible. For example, in the subsequent exposure step, if the prebake is too high, the radiation-sensitive component may be partially degraded, and higher doses will be required. If the prebake is too low, the casting solvent may interface with the radiation chemistry and different exposure–developer combinations may be required. In the development step, the dissolution of both the exposed and the unexposed areas are influenced by the following prebake factors:

TABLE 7-1-1. T_g and T_d of Common Resists

Resist	T_g (°C)	T_d (°C)
Rubber	54	200
Azide		100
Novolak	70–120	150–300
Diazoquinone		100
PMMA	105	200
Polybutene SO_2	65	125

1. The amount of residual solvent left after prebake and its distribution throughout the film.[5,6]
2. Surface (of resist) induction effects due to resist polymer oxidation or cross-linking reactions.[7,8]
3. The formation of a scum layer in the exposed (positive) or unexposed region (negative), due to cross-linking or sensitizer degradation.
4. Resist undercutting or interfacial adhesion failure in liquid development, due to residual solvent and the thermal stress of prebake.
5. Phase separation, crystallization, or precipitation of components or additives.
6. Stress cracking in films.[9,24]
7. Trapped solvents also change the rate of reactive ion etching[10] or sputter etching.[15]
8. Trapped solvents can also lower the T_g and resist flow point.
9. Poor planarization, film shrinkage, and pinholes in resist film.[11,12]
10. Swelling of the negative resist upon development due to residual solvent.[27]

In drying a film consisting of a polymer, sensitizer, and solvents, the efficiency of solvent removal and the upper prebake temperature are determined by several physical and chemical factors. The T_g (glass transition temperature) and the T_d (decomposition temperature) or any components determine a practical upper limit. In Table 7-1-1, the T_g and T_d for common resists are given. For efficient solvent removal, the prebake temperature should be above the T_g of the polymer, but should not be high enough to decompose the photo- or radiation-sensitive components. Secondly, the prebake temperature if possible should be above the boiling

TABLE 7-1-2. Evaporation Rates and Solvents for Resists

Resist	Casting solvent	Evaporation rate[a]	Boiling point (°C)
Polybutene SO_2	Methyl 2-ethoxyethyl acetate	31	145
DQN	Cellosolve acetate	20	156
	Butyl acetate	100	126
	Diglyme	18	162
PMMA	Chlorobenzene	102	126
Rubber azide	Xylene	62	140

[a] Relative to butyl acetate.

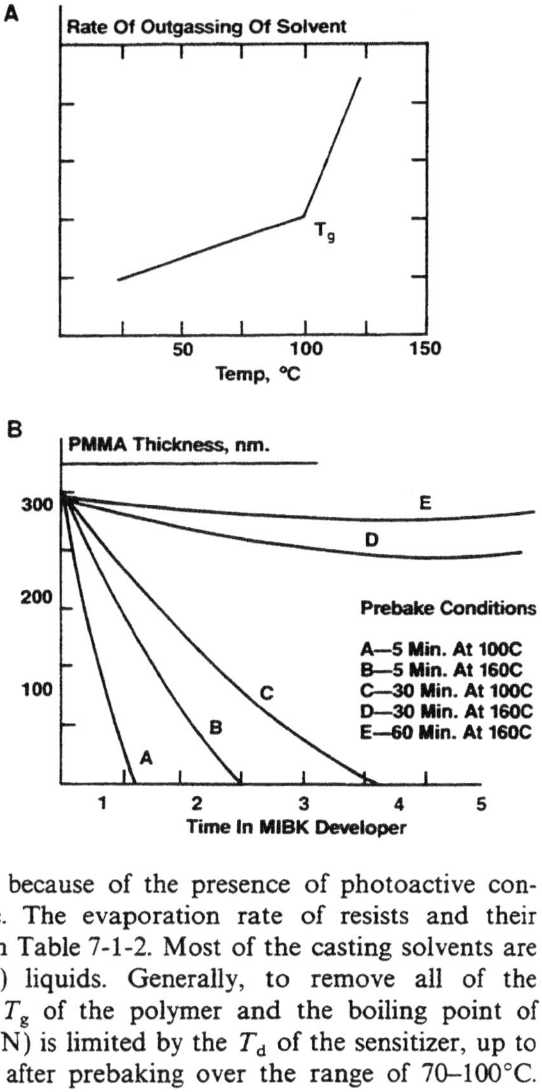

FIGURE 7-1-1. (A) Drying of PMMA (rate of outgassing) film heated below and above T_g. (B) Dissolution rate of PMMA film in MIBK after prebaking at and above T_g. From J. Greeneich, *J. Electrochem. Soc.* **122**, 970 (1975), reproduced by permission of the Electrochemical Society.

point of the solvent. In some cases because of the presence of photoactive constituents, this may not be possible. The evaporation rate of resists and their common casting solvents are given in Table 7-1-2. Most of the casting solvents are combustible high-boiling (> 125°C) liquids. Generally, to remove all of the solvent requires heating above the T_g of the polymer and the boiling point of the solvent. Since diazonovolak (DQN) is limited by the T_d of the sensitizer, up to 20% solvent by weight can remain after prebaking over the range of 70–100°C. Thus, prebaking AZ-1350 for 30 min at 100°C destroys 50% of the sensitizer and requires a 2× photodose to induce dissolution in the developer.

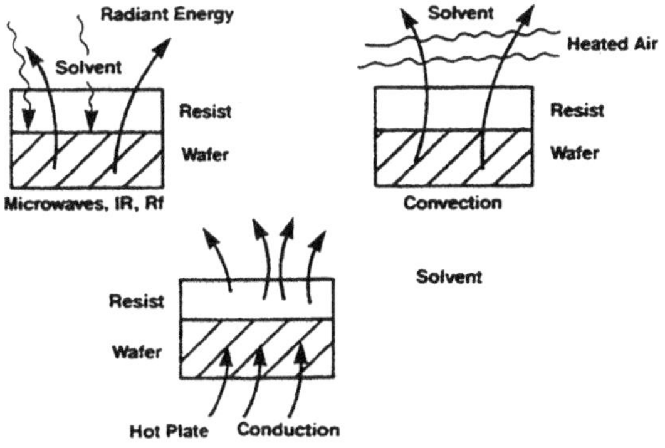

FIGURE 7-1-2. Different methods of prebaking. Adapted from Ref. 39.

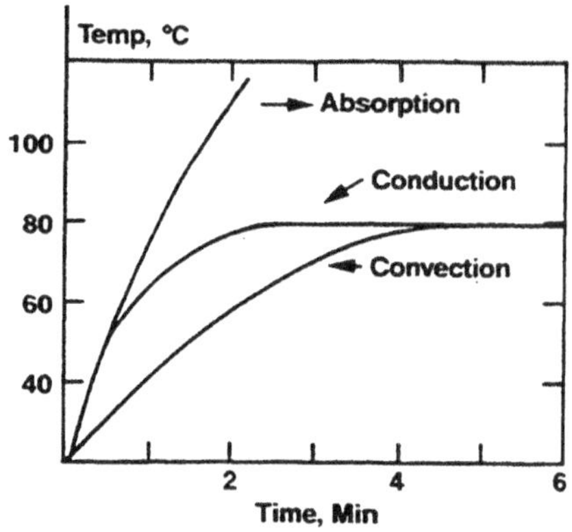

FIGURE 7-1-3. Rate of heating of resist film on silicon wafer versus mode of heat input.

The glass transition (T_g) is a second-order transition, marking the disruption of intermolecular bonds, and is characterized by a change from a brittle glass to a flexible plastic. Melting and boiling are examples of first-order transitions. Baking a polymer such as PMMA above its T_g causes a shrinkage of its free volume, excluding solvent from its intermolecular voids. Above the T_g, the solvent diffuses rapidly out of the film (Fig. 7-1-1).

In the prebake of DQN, the major problems are to measure and reproduce the level of trapped solvent in the process. Hopefully, the solvent is trapped and distributed uniformly throughout the film. For PMMA, which lacks a photoactive heat-sensitive component, in order to obtain high resist contrast, a prebake as high as 160°C is required to remove all of the solvent and induce a retarded dissolution effect at the top surface of the unexposed film.[14]

In terms of equipment, the basic unit of baking a boatful of wafers is the traditional convection oven. For single wafers, the new concepts of baking in an automated line employ rapid heating by external infrared radiation. Three basic types of heating can be used: convection, conduction, and absorption or radiant energy (Fig. 7-1-2). Kinetically, two basic types of heating can be employed. Steady-state temperatures can be reached using hot plate conduction or (slower) air convection ovens. Infrared and microwave heating are rapid and ramp to a peak temperature. Ramp heating methods have a tendency to overshoot the desired temperature (Fig. 7-1-3). In addition, the absorption of infrared radiation is also critically dependent on the chemical structure and emissivity of the films deposited on the wafer.

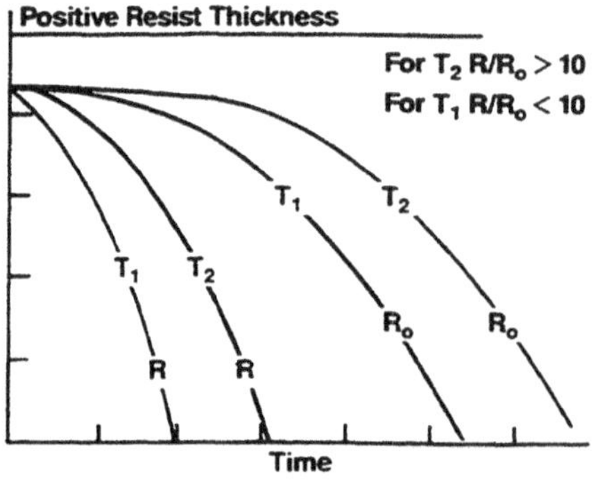

FIGURE 7-1-4. Effect of prebake temperature (T) of constant time on the dissolution rate of exposed (R) and unexposed (R_0) positive resist.

For a successful prebake, the critical factor is the reproducibility and uniformity of the rate of dissolution of the exposed (R) and unexposed (R_0) positive resist (Fig. 7-1-4). If the optimum prebake[12] conditions are chosen such that R/R_0 is greater than 10, then a larger exposure and development window can be accommodated. In conjunction with the dose develop conditions, *in situ* measurements (see Fig. 7-2-15) of the resist dissolution rates monitor the process contribution of prebake, exposure, and development. For example, the development time can be adjusted to compensate for process (prebake fluctuations). Thus, prebake plays the vital role of preparing the resist in an isotropic fashion to produce a three-dimensional image. Prebake also removes the solvent at the interface and improves adhesion. Throughout the prebake step, the chemical and physical integrity of the sensitizer and polymer should remain unaffected by the process. In the subsequent sections we will examine the kinetic, thermodynamic, and chemical reactions which control and limit prebake, as well as the practical equipment and process control that are involved.

7-2. KINETICS OF PREBAKING

In a sense, the drying of polymer films is the reverse of the stripping or development (permeation) process. In glassy polymers, the process of solvent removal occurs by assuming a Fick's law dependency. In Fig. 7-2-1, a model of the three stages of solvent loss is shown. The flux of solvent arriving at the resist surface is given by Fick's first law,[16]

$$\text{Flux} = \text{g/cm}^2\text{-sec} = D_i C/X \qquad (7\text{-}2\text{-}1)$$

where D_i is the diffusion coefficient of the casting solvent(s) in the resist and C/X is the concentration gradient. The diffusion coefficient for the solvent depends on the volume fraction (a free volume contribution),[5,13]

$$\ln(1/D/D_0) = K + M/V_1 \qquad (7\text{-}2\text{-}2)$$

where K and M are material constants related to the solvent–polymer pair, D and D_0 are the diffusion coefficients in the swollen and unswollen polymer, respectively, and V_1 is the solvent volume fraction in the polymer. Equation (7-2-2) assumes constant diffusivity and diffusion path length. Numerical solutions must be used to take into account the evaporation of solvent at the top surface, the reduction of polymer film thickness as the solvent evaporates, and the change in K as a function of solvent concentration and temperature.[17–20]

As the solvent is removed, the free volume decreases and the solvent must diffuse through smaller holes. The size of the free volume in polymers can vary with the T_g and intermolecular forces. If the same solvent of butyl acetate is used to cast four resins:

- Polyvinyl chloride copolymer (PVC)
- Nitrocellulose
- Polystyrene
- Polymethylmethacrylate

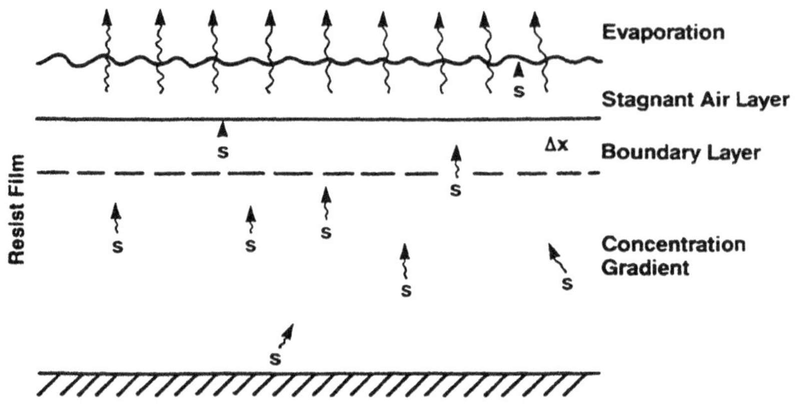

FIGURE 7-2-1. Different stages of diffusion of a solvent (s) exiting a resist film at the surface and in the bulk.

the retention of the solvent is the highest in PVC (Fig. 7-2-2). In PVC, the high T_g (smaller free volume) and the hydrogen bonding of the ester solvent account for the greater retention and longer dry time.

Inside the drying film, the changes in the diffusion coefficient are considered to occur in two stages. In the first stage of the wet spun-cast film, the increased ease (lower T_g by solvent plasticization) of polymer segmental motion leads to an immediate increase in the diffusion coefficient, D. But, in the second stage, at the surface of the resist, the surface concentration, C, is close to zero. Thus, the diffusion coefficient at any point is an exponential function of C,

$$D_i = D_0 \exp(KC) \tag{7-2-3}$$

where D_0 is the diffusion coefficient in the dry polymer (absence of solvent). The rate of diffusion of a solvent diffusing to the surface of a resist, increases with

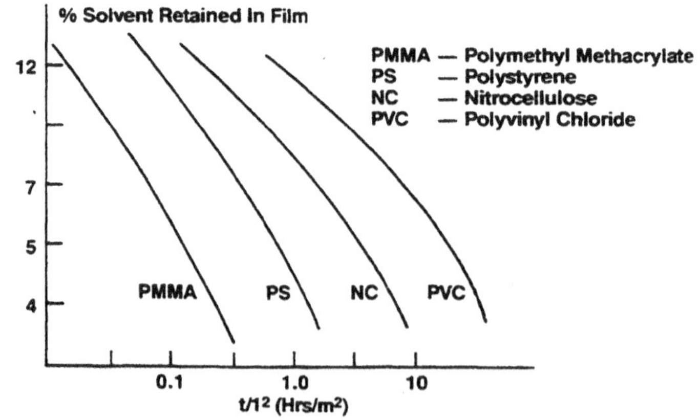

FIGURE 7-2-2. Retention of butyl acetate in resins dried at 23°C. From Ref. 18, copyright 1975 and courtesy of the Federation for Coating Technology.

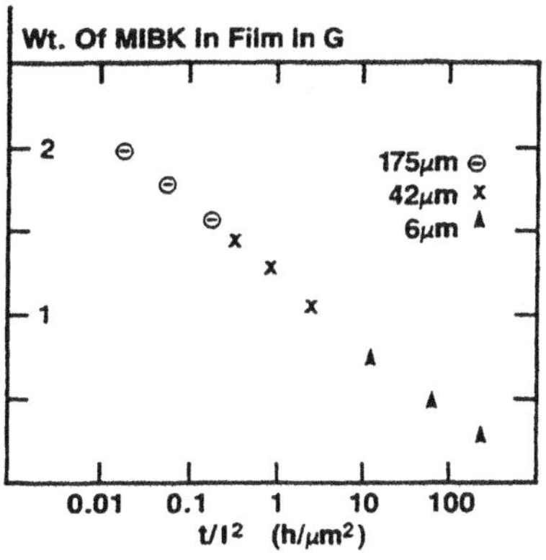

FIGURE 7-2-3. Log weight versus log t/l^2 for MIBK in PVC of various thickness. From Ref. 18.

increased concentration of retained solvent.[17] The time-dependent solution of Eq. (7-2-1) for the percent solvent retained, S/S_∞, at thickness l is

$$S/S_\infty = (Dt/l^2)^{1/2}(\text{error function}) \tag{7-2-4}$$

A value of D can be obtained from a graph of S/S_∞ versus $(t/l^2)^{1/2}$ (Fig. 7-2-3). Another form of Eq. (7-2-4) is

$$S/S_\infty = \exp(-D\pi^2 t/l^2)(8/\pi^2) \tag{7-2-5}$$

which predicts an exponential dependency of the two-stage rate of drying (Fig. 7-2-4). In the wet stage, solvent diffuses through the swollen holes in the polymer chain. The efficiency of removal is still low. For a factor of ten in time, the solvent content drops by 50% (Fig. 7-2-3). In theory (constricted state), the final stage (10% of retained solvent) of diffusion is very slow and several orders of magnitude in time are required to reduce the solvent level below 5% by weight remaining (Fig. 7-2-6). Figure 7-2-5 shows data obtained on the rate of outgassing of Cellosolve acetate (as measured in a cell connected to a gas chromatograph) versus the percent retained solvent in a 1.5-μm-thick novolak resin film (weight measurement), at a constant temperature of 85°C. As the film dries, the rate of removal decreases exponentially.[64] Similar results have been obtained with many

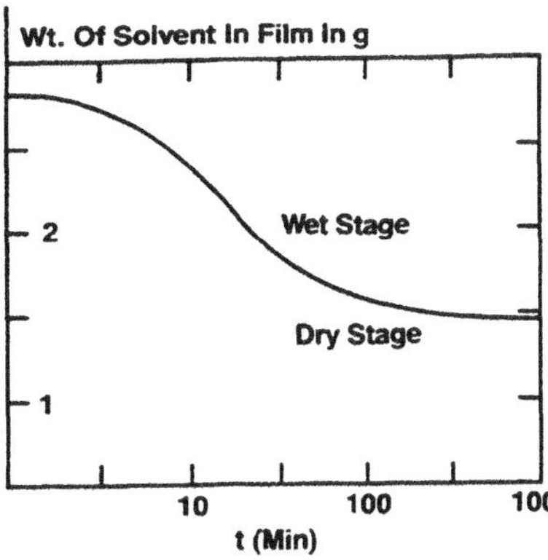

FIGURE 7-2-4. Release of MIBK from PVC film. From Ref. 18, copyright 1975 and courtesy of the Federation for Coating Technology.

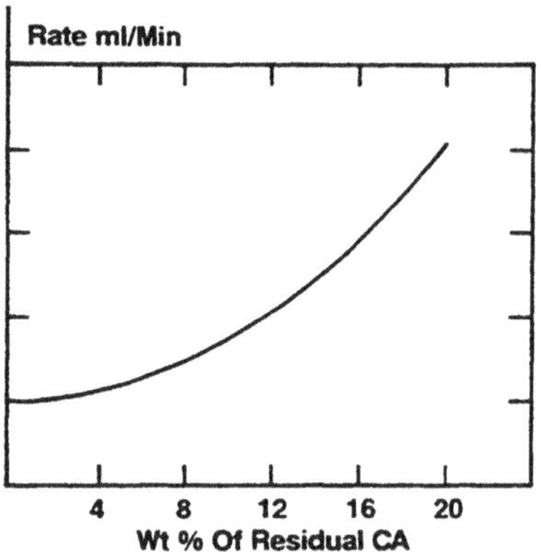

FIGURE 7-2-5. Outgassing rate of Cellosolve acetate from a drying DQN resist film. The rate is faster in the wet film.

polymer solvent systems. Thus, the rate of removal of solvent is rapid with wet films and decreases as the solvent concentration is depleted within the film.

In a resist-polymer solvent system, the movement of solvent during drying is by diffusion. Diffusion becomes the rate-controlling reaction and the drying rate is independent of what is happening at the surface of the resist. The rate is independent of the flow rate of air in a convection oven or the vacuum pressure in a vacuum bakeout. Temperature is a major driving force, increasing the diffusion constants by the Arrhenius activation energy,

$$D_i = D_0 \exp(-\Delta H/kT) \tag{7-2-6}$$

For liquid diffusion in a polymer matrix, the activation energies are about 12–24 kcal/mole, roughly increasing 2–3× for every 10°C rise.[13] Furthermore, the retention of solvent is not controlled by the evaporation rate or boiling point of the solvent, but by the size of the molecule (molar volume), and the bonding to the resist polymer. The activation energy is independent of the size of the diffusant molecule, if it is comparable in size to the monomer unit of the polymer.[13] In Fig. 7-2-6, the retention of individual solvents of different molecular sizes is shown for ethyl, butyl, isobutyl acetate and cyclohexanone in a vinyl chloride copolymer.

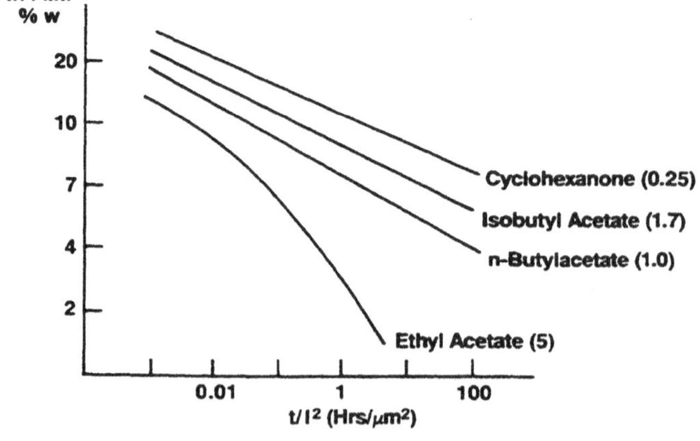

FIGURE 7-2-6. Release of solvent from PVC film. Evaporation rates relative to butyl acetate. From Ref. 18, copyright 1975 and courtesy of the Federation for Coating Technology.

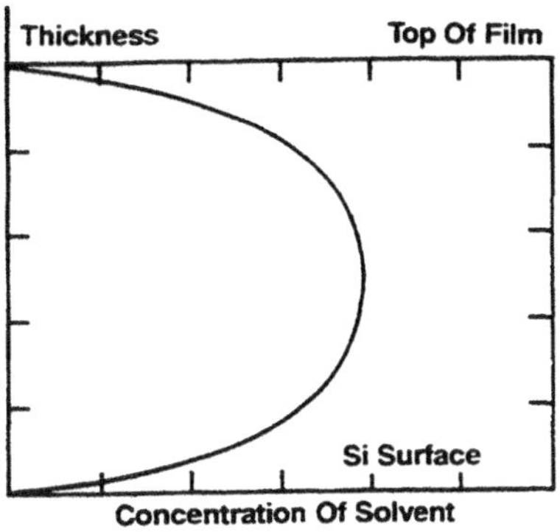

FIGURE 7-2-7. Concentration profile of solvent in resist after prebake. From Ref. 6, reprinted with permission of *Solid State Technology*, published by Technical Publishing, a company of Dun and Bradstreet.

Even though isobutyl acetate is a twofold faster evaporating liquid than butyl acetate, the solvent due to its chain branching and larger cross section does not hop from consecutive free volume holes, as readily as the smaller molecules of butyl or ethyl acetate. Cyclohexanone, a nonplanar cyclic molecule, has even a larger cross section and is strongly retained.[18,43]

For the actual resist system of DQN, Batchelder and Piatt measured the dissolution of diazonovolak resist at different hot-plate prebake temperatures and implied from the unexposed dissolution rates that the retained solvent exhibited a maximum concentration cross section as shown in Fig. 7-2-7.[6] The complementary dissolution rates are shown in Fig. 7-2-8.

At the resist surface, the resist has lost more internal solvent, due to evaporation during spin coating. The convection air flow across the wafer also increases surface drying, forming a dried crust. At the substrate, where the heat is initially the highest, the solvent rapidly diffuses toward the surface. However, the heat flow into the interior of the resist drops off because of the lower thermal conductivity of the resist. At the mid-depth of the film, the solvent diffusion piles up due to the drier outer surface crust. Reflected in the corresponding rates of unexposed film dissolution (Fig. 7-2-8), the R_0 decreases with increasing loss of solvent at higher prebake temperatures.

For most of the resists of the DQN type, the effect of prebake or unexposed loss is not noticed, since the dose develop conditions are chosen to retain 95% of prebake film. With the loss of solvent from the film, the average dissolution rate of DQN would be expected to decrease with loss of solvent. Figure 7-2-9 compares the

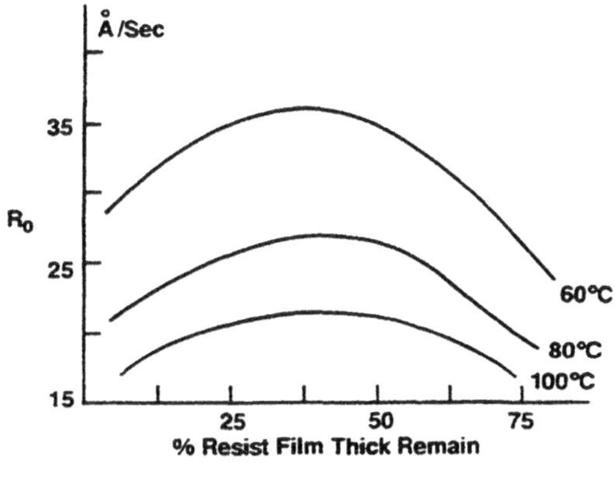

FIGURE 7-2-8. Dissolution rates of unexposed (R_0) of DQN resist after prebaking at different temperatures. From Ref. 6, reprinted with permission of *Solid State Technology*, published by Technical Publishing, a company of Dun and Bradstreet.

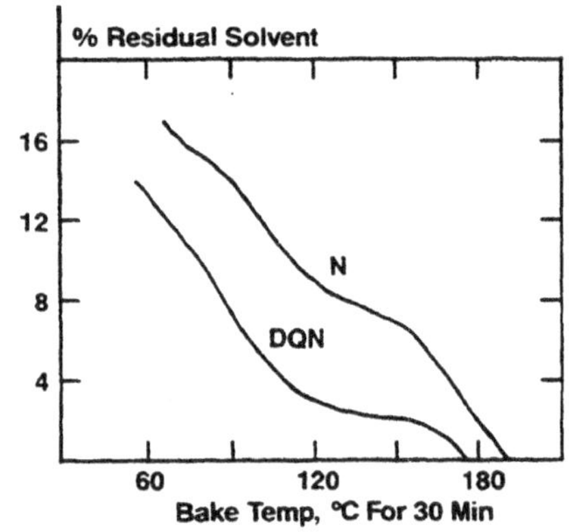

FIGURE 7-2-9. Solvent content of a film of DQN resist and of the novolak (N) resin itself after various prebakes. Both films cast to the same initial thickness.

dissolution rates of bare novolak resin and with 20% added DQ. In this case, the addition of DQ influences the rate of diffusion of casting solvent out of the film. The sensitizer bonds more strongly to the resin,

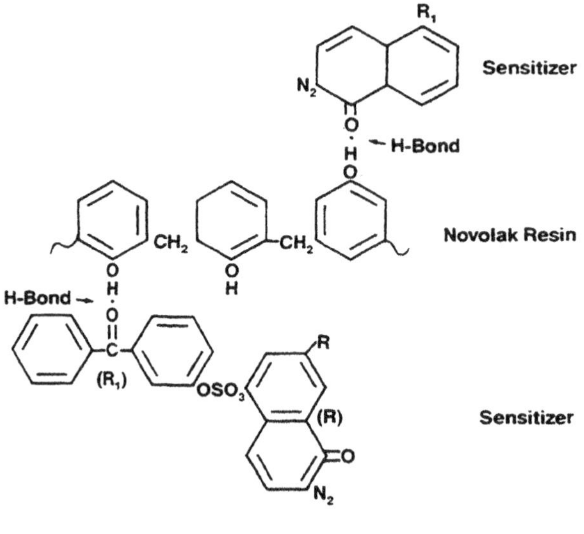

(7-2-7)

than the casting of Cellosolve acetate releasing the displaced solvent to diffuse out of the film. The addition or retention of small hydrophilic molecules can have a significant effect on the dissolution rate of solvent out of the film, or conversely the

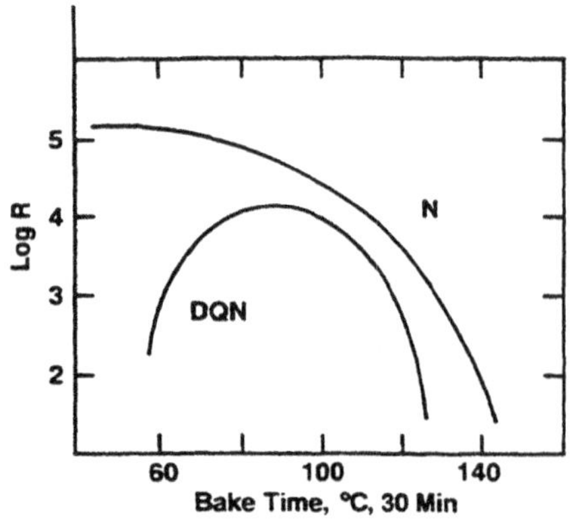

FIGURE 7-2-10. Dissolution rate (R) of prebaked DQN and of the bare novolak (N) resin alone.

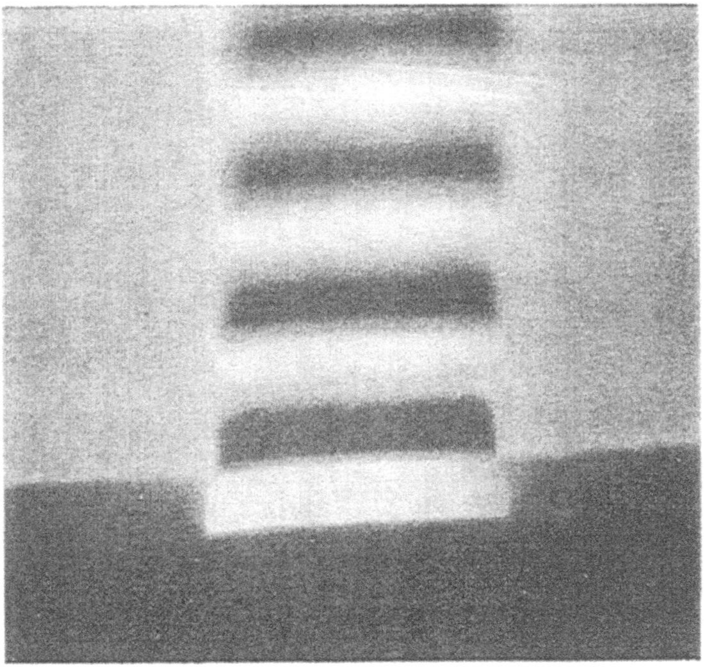

FIGURE 7-2-11. Lift-off profile formed in a DQN resist by top surface inhibition of the unexposed film. The film surface was rapidly baked to retard dissolution.

infusion of the developer into the film.[19] In the bare novolak resin, the Cellosolve acetate strongly hydrogen bonds to the resin and requires a longer heating cycle for release. The retention of different casting solvents for DQN resist adversely affects the adhesion of the resist. For example, diglyme, a bisether, bonds more strongly than Cellosolve acetate. Consequently, the adhesion is poor and resist films may lift from high-energy surfaces. Another anomaly occurs on prebaking of DQN, due to thermal decomposition of DQ and the formation of soluble acid. At 100°C, 50% of DQ is converted into a soluble acid,[22]

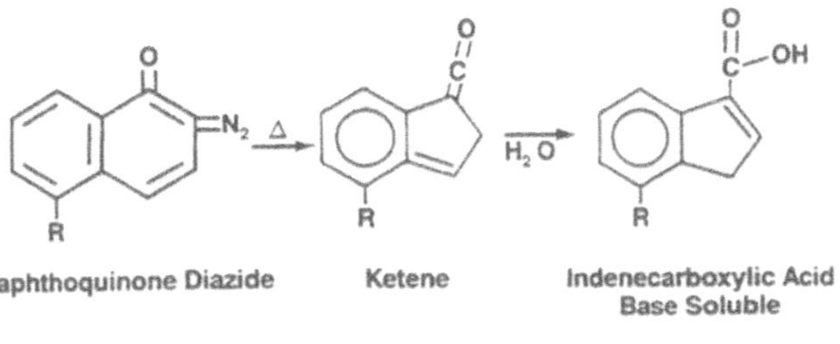

(7-2-8)

With higher prebake, one would expect a general decrease in dissolution (R_0) (Fig. 7-2-10), but if considerable indene acid is formed, the rate increases until at 110°C, the acid esterifies the resin and the resin oxidation commences.[23] Figure 7-2-10 shows the dissolution rate of prebaked DQN in aqueous developer versus films of the bare resin. Gradually, the rate of dissolution of the resin

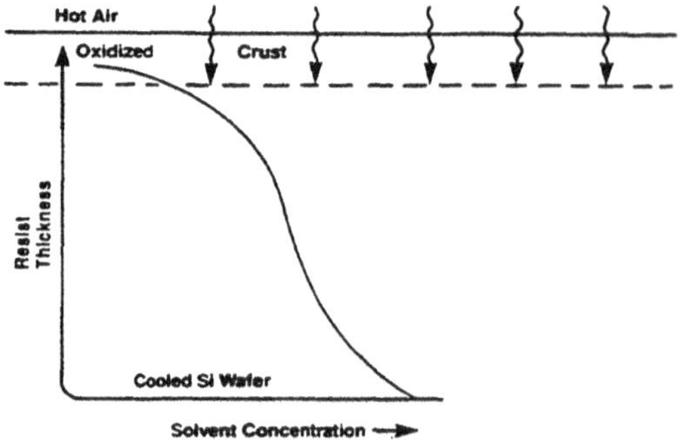

FIGURE 7-2-12. Trapped solvent in depth of DQN resist that was top surface baked.

decreases because insoluble oxidation products are formed. At the surface, the oxidation of the resin occurs,

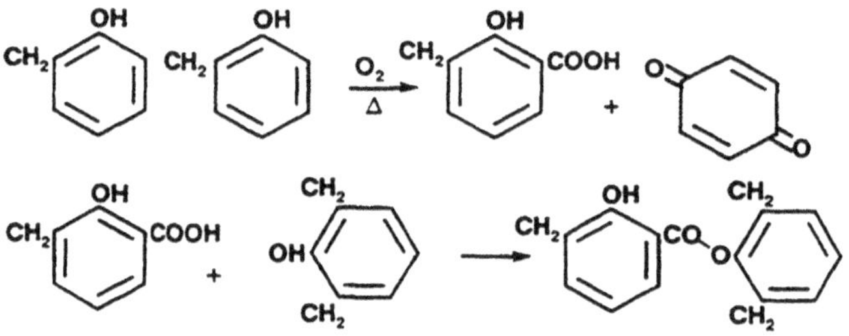 (7-2-9)

producing an inhibiting effect on unexposed dissolution (a surface crust). In fact, profiles suitable for lift-off can be created at high prebake by taking advantage of the surface crust (Fig. 7-2-11). Prebaking DQN at 120°C produces the highest contrast. Prebaking AZ-1350J with hot air at the surface, while simultaneously cooling the Si substrate, produced a dried surface crust which retarded loss of solvent from the interior of the film (Fig. 7-2-12). Subsequently, upon exposure and development, the top surface image is undercut by the isotropic development of the "wet interior." Profiles suitable for lift-off can thus be produced.

In the DQN resist, the dissolution of the exposed, R, is composed of the unexposed rate, R_0, and indene acid formed,

$$R = R_0 + R_{\text{exposed}} \tag{7-2-10}$$

Although the trapped solvent contributes a small amount to the dissolution

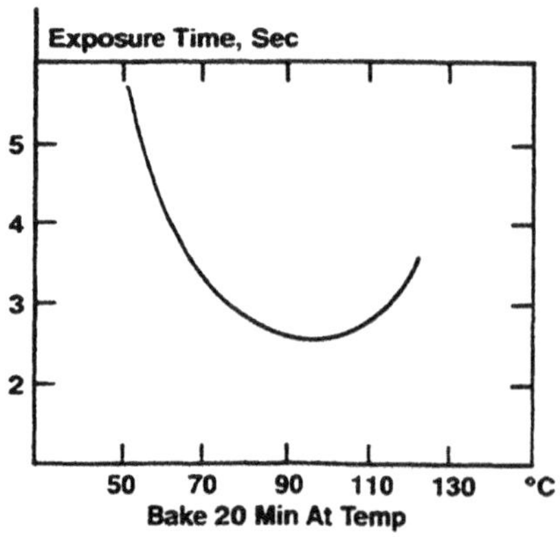

FIGURE 7-2-13. Exposure time required to maintain the same dissolution rate of the exposed DQN resist baked at different temperatures.

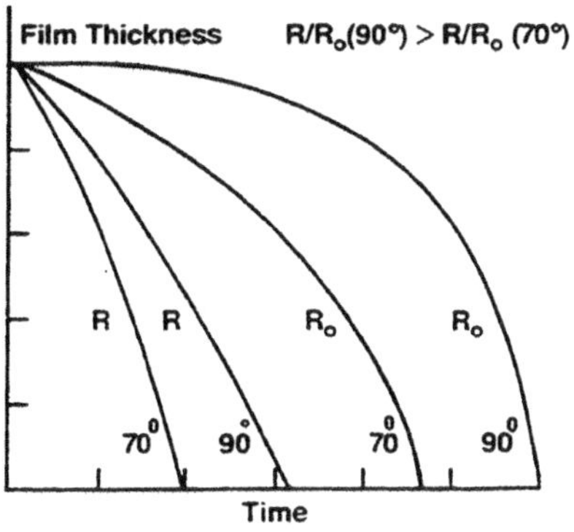

FIGURE 7-2-14. Dissolution rate of exposed (R) DQN resist and of unexposed (R_0) resist baked at various prebake temperatures. The higher prebake creates an induction at the surface of the resist and the developing rate is slower.

rate, the amount of solvent left after prebake can still influence the R_0 especially at low dose conditions of $R/R_0 < 10$. If the develop time is fixed by changing the dose (exposure time), the minimum dose is obtained by prebaking the DQN resist between 70 and 85°C (Fig. 7-2-13). Below 60°C prebake, the remaining solvent interferes with indene acid formation. Above 100°C, the complicated reactions of the sensitizer decomposition and the resin reactions compete to inhibit the dissolution. In actual practice, a range of 75–90°C is used. Since the prebake can cause variations in dose and development rate, it is important that dissolution monitors establish both the R and R_0 film thinning curves (Fig. 7-2-14). With the endpoint apparatus shown in Fig. 7-2-15, the induction effect of inhibited DQN unexposed dissolution (Fig. 7-2-14) can be optimized by examining various prebake tools and/or process conditions (time–temperature–loading effects). In actual practice, every DQN resist formulation should have a specified prebake to maximize its contrast and if possible for enhanced tool exposure throughput (minimize the exposure time). Both extensive sensitizer decomposition and the resin degradation are to be avoided. The monitoring of the dissolution rate of the unexposed and exposed resist, adjusts for fluctuations in prebake, thickness, and dose–developer combinations.

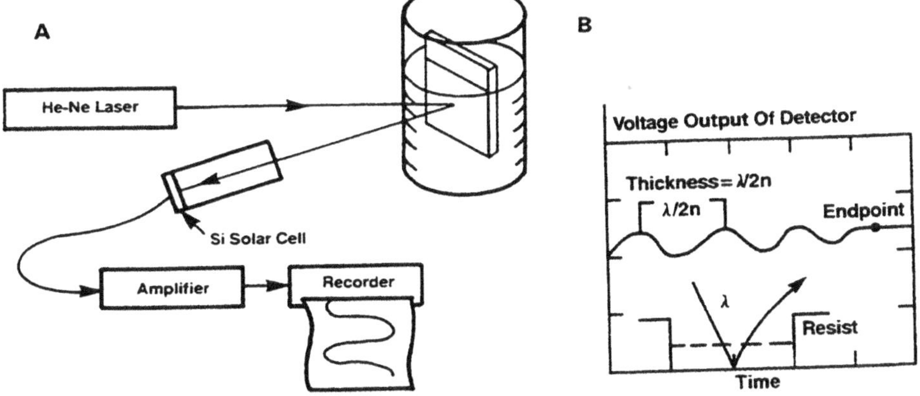

FIGURE 7-2-15. (A) Apparatus for monitoring the dissolution rate and endpoint of developing a DQN resist. The laser beam is reflected from the exposing region into a detector. An interference pattern (B) is produced which corresponds to thickness versus time of development. From Ref. 7, courtesy of and copyright by IEEE.

7-3. PHYSICAL CHANGES IN PREBAKING

Due to solvent loss and polymer melting, cast films undergo physical changes, such as shrinkage, planarization,[12] stress, and reorientation. The shrinkage of resist films on the order of 5–10% is common (Fig. 7-3-1).

7-3-1. Residual Stress Strain

Within the coating, the volume of solvent lost, ΔV, is responsible for internal bulk strain equivalent to an isotropic linear strain,[26]

$$\Delta V/V = \phi_s - \phi_r(1 - \Delta V/V) = 3\varepsilon \tag{7-3-1}$$

where ϕ_s and ϕ_r are the volume fraction of the solvent in spin-coated and in the prebaked film, respectively. The strain, ε, is

$$\varepsilon = \phi_s - \phi_r/3(1 - \phi_r) \tag{7-3-2}$$

and σ, the residual internal stress, is

$$\sigma = E/1 - v \tag{7-3-3}$$

where v is Poisson's ratio and E is Young's modulus of the coating. The contraction of the film during prebake is responsible for the nonplanarization of the resist over steps. Resists with high (40%) solid content and less solvent show minimal effects. Prior to and during spin coating, the coating is mobile enough to flow and allow the free volume change required by solvent evaporation. The surface area is constrained to its original size by adhesion to the substrate. Because of the resist bonding to the substrate, only vertical, one-dimensional contraction is possible. A tensile stress develops, which is proportional to the amount of residual solvent. The volume change appears as a change in thickness. If during prebake the polymer can no longer flow, further solvent loss produces internal stress in the plane of the coating. The thickness is still unconstrained and can contract in response to the component of stress-reducing adhesion.

Since the glass transition temperature is rarely exceeded during prebake, the

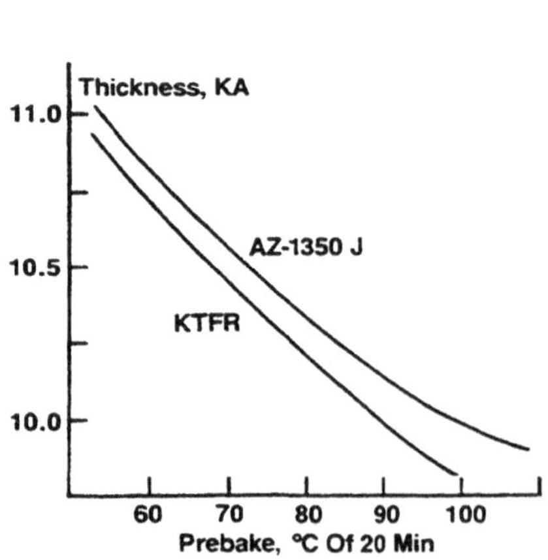

FIGURE 7-3-1. Shrinkage of resist films after prebaking at various temperatures.

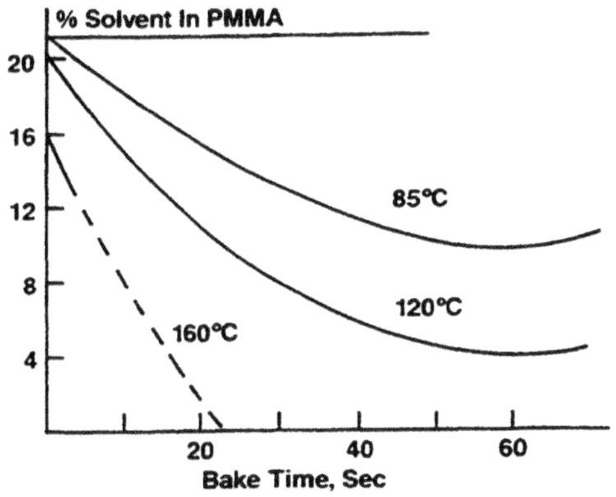

FIGURE 7-3-2. Drying (prebake effects) curve of a 1.5-μm-thick film of PMMA and the solvent retained in the film. From Ref. 5.

DQN resist cannot readily be flowed so as to relieve internal stress. However, the retention of 10–15% of Cellosolve acetate reduces some internal stress due to a plasticizing effect. A residual stress of 10 dynes/cm^2 was measured[28] and is of the same order as polymer stresses measured for polyisobutylmethacrylate (T_g of 45°C) as 4.5×10^9 and 1×10 dynes/cm^2 for polystyrene of T_g 105°C.[26] The lower-molecular-weight forms of the same polymer, cast better stress-free films. For ultrahigh-molecular-weight PMMA ($M_n > 10^6$), Hess and co-workers were unable to relieve the stresses by prebaking at 160°C. Although the PMMA stress was relieved by swelling of the film in the developer, the resolution of the images was limited by image distortion.[29] By contrast, Atoda and co-workers overcame the problems of Hess, by prebaking PMMA at 200°C for 1 hr.[30] In this case, no swelling was encountered, since sufficient flow occurred during prebake to relieve the internal stress. During the drying of the film, movement of the solvent can also orientate polymer chains to produce optical birefringence and in azide polystyrene *e*-beam resist, the drying can also partition the sensitizer to accumulate near the wafer interface.[31,32]

An added benefit of prebaking in some resists is the formation of an insoluble surface layer. Heating a resist, such as PMMA, about 70°C above T_g of 105°C, removes the solvent. However, the time of baking is also critical. A prebake at 160°C for 5 min is not sufficient to make the development rate (R_0) independent of the casting solvent or prebake time, as noted by Greeneich (Fig. 7-3-2).[33] At least 30 min is required at 160°C. Even though most of the surface solvent is removed in a matter of seconds (Fig. 7-3-2), the densification of the polymer film, to constrict

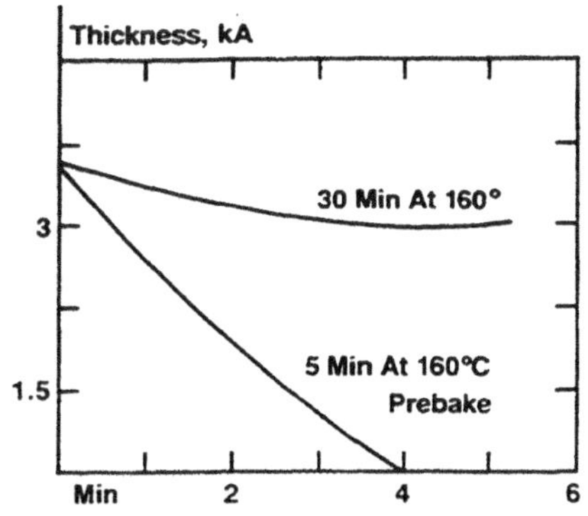

FIGURE 7-3-3. Dissolution rate of films of PMMA in methyl isobutyl ketone developer after prebaking at various times and temperatures. High bake above the T_g is necessary to obtain constant dissolution rate of unexposed film. From Ref. 33, reproduced by permission of the Electrochemical Society Inc.

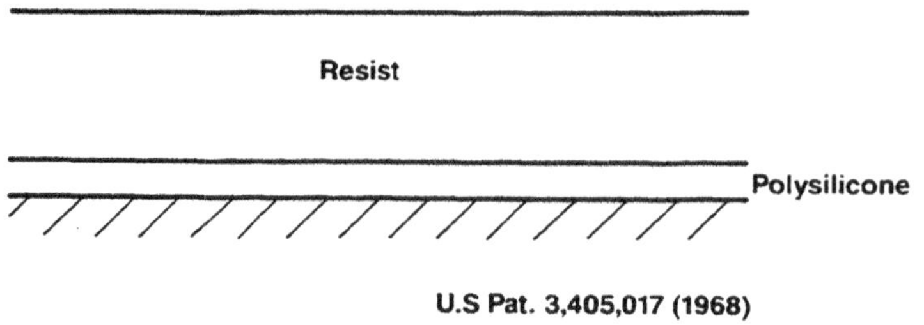

FIGURE 7-3-4. A resist primer for adhesion improvement consisting of a layer of polysilicone.

the solvent holes (vacancies), requires a long time for polymer relaxation and flow. Figure 7-3-3 of Greeneich shows that a prebake of 160°C for 5 min, according to Fig. 7-3-2 of Ouano, removes all the solvent, but does not retard dissolution, until a prebake of at least 30 min is used. Rapid prebake under several minutes, such as with microwave and infrared sources, may not allow sufficient time for polymer chain motion and densification. In this case, a slower prebake using a hot plate or convection oven (even on overnight) is not out of the question for prebake. With excessively high prebake, the dangers are the chain scission (lowering of the molecular weight) of PMMA or the oxidation (if done in O_2) of the fragments to lower contrast (R/R_0). A long prebake time also allows for polymer flow, the closure of pinholes, and the planarization of the film. For a stepper exposure and alignment, films as uniform as ±100 Å are required for precise processing without interference interruptions.[11,12]

Since DQN resists cannot be baked 50°C above their T_g of 70–100°C, or for long times, the internal stress weakens the adhesion upon development or in wet etching. Stress cracking has also been noted.[27] A polysilicone primer serves to relieve the internal stress (Fig. 7-3-4). The polysilicone layer may be formed by the hydrolysis of silanes with surface adsorbed water or by the esterification of silanol groups (Chapter 13). Since the resultant layer polysiloxanes are low-T_g (<0°C) polymers with high elastic modulus, the strain is relieved from the bonded DQN top layer. Other low-melting polymeric resists, such as negative azide-rubber with a T_g of 54°C, or polysiloxanes and polyepoxyacrylates, can be annealed of most of the internal stress by heating to 80–90°C. Finally, solventless coatings avoid the stress from solvent loss, but they may still be highly stressed such as plasma-deposited PMMA or polysiloxanes. Prebake may still be necessary to remove internal stresses.[34]

7-3-2. Measurement of Residual Solvent in Prebaked Films

Since less than a milligram of solvent remains, the measurement of residual solvent is somewhat of a difficult task. A 1-μm-thick film of resist coated on a 75-mm-diameter wafer weights about 3 mg. For a solvent residue of 1–5% by weight, an accuracy of micrograms is required.[63] Secondly, the assessment of the distribution of the solvent in the cross section of the film is the most difficult task, since the solvent and polymers are similar in electromagnetic absorption. As a last resort, kinetic analysis by the physical thinning of the film to estimate trapped solvent can be used.[6] The average solvent retained can be measured by several techniques (Table 7-3-1). Accurate techniques are required so that the

TABLE 7-3-1. Measurement of Residual Solvents in Resist Films

Technique	Resist	Reference
Mass spectometry	DQN	37
Thickness shrinkage	DQN	38–40
Gravimetric	DQN	5, 40
	PMMA	5
Gas chromatograph		38
Development rate, R_0	SQN	5, 22, 39
	PMMA	5, 33
Linewidth	DQN	39, 41
Exposure dose	DQN	42
ERA	Polymers	35, 36

reproducibility and uniformity of the prebake across the wafer can be measured, independent of dose and development.

Techniques such as evaporation rate analyses (ERA) employ the counting of radioactivity of C_{14} solvents. Previously, the technique was used for surface adhesion studies on metals(35) and for solvent evaporation from paint films.(36)

For better accuracy, the complete dissolution of the prebaked film in an appropriate solvent and subsequent analysis by gas chromatography can achieve $\pm 0.1\%$ accuracy. The gravimetric and ERA methods are accurate to about $\pm 1\%$ residual solvent while kinetic measurements of the rates of dissolution are good to about $\pm 2\%$ residues. Linewidth or dose–development methods are less accurate and suffer from the variations in the other process steps. If highly accurate refractive index (6 significant figures) measurements could be made, a rapid physical method may be used to monitor the prebake(31,32) conditions. In the absence of analytical equipment, a functional test of prebake is the measurement of the dissolution rates at fixed thickness, dose, and develop times, using endpoint dissolution (Fig. 7-2-15). The second type of assessment involves the contrast curve measurements with variable dose and development time to measure the γ of a resist as a function of prebake conditions. Prebake conditions which maximize the contrast will offer maximum process latitude and minimum linewidth bias.(44,45) Furthermore, the evaluation of different prebake methods can simultaneously concern many variables(63):

1. Different sublayers and their reflectivity of heat
2. Prebake temperature
3. Prebake time
4. Fixed or variable exposure
5. Fixed development time or variable
6. Different thicknesses
7. Batch-to-batch and wafer uniformity
8. Resist thinning rates and ratios
9. Adhesion during develop

Minimizing the experimentation and deciding on both chemical-physical and functional resist tests are formidable tasks. With automated processing, wafer track process, or boat processing, the throughput factors of time and cost are strong driving forces. The baseline (most reproducible) technique is the isothermal hot plate. Other techniques have to be evaluated relative to the benefits derived in cost or performance. In addition, the accurate temperature measurements of the resist surface, such as with fluoro-optical techniques,[46,47] ensure that a steady-state resist temperature has been reached. A typical contribution of $\pm 5°C$ variation in prebake temperature results in a ± 0.15-μm bias to the developed linewidth, at a fixed dose and develop time.[41]

The final application of residual solvent measurements is involved in the cases of resist reformulation, due to one or more of the following applications:

1. Reduce cost while maintaining performance
2. Improve performance such as elimination of radial striations
3. Solubilize a new sensitizer or resin
4. Improve resist adhesion
5. Substitute solvents because of new safety, health, and air pollution restrictions

The evaporation rate (ER) of solvents is proportional to the vapor pressure (P), and the molecular weight (M) of the solvent,

$$\mathrm{ER} = KP(M)^{1/2}/2\pi RT \tag{7-3-4}$$

It is quite common to express the evaporation rate of solvents relative to butyl acetate,[21]

$$(\mathrm{ER})_{\mathrm{rel}} = 0.823(MP)^{1/2} \tag{7-3-5}$$

In some cases, the evaporation of blends is not a linear sum of the rates of each component. Only in the case of ideal liquids or gases do deviations not occur from Raoult's law. For positive deviation, the rate of evaporation of the blend is slower than a single, faster-evaporating component.[44] Blends of fast- and slow-evaporating solvents are important in allowing sufficient trapped solvent to remain in order for the film to level during spin drying.

7-4. PREBAKING POSITIVE RESISTS

Prebaking PMMA is invariant to the casting solvent and time, as long as prebake temperatures of 180–200°C are used for at least 30 min. During prebake, polymer flow occurs to relieve stress, close pinholes, planarize the film, and produce a desirable induction effect upon dissolution in a developer.

The prebake of DQN resists raises special problems of balancing the residual solvent against the thermal instability of the sensitizer and resin components. In Section 7.3, we discussed the implications of residual solvent and resin sensitizer reactions. In the next section, we will discuss the implications of sensitizer decomposition.

The thermal decomposition of the sensitizer of DQ leads to formation of ketene [Eq. (7-2-8)] and acid, followed by the addition of the ketene to esterify the phenolic portion of the resin,[48,49]

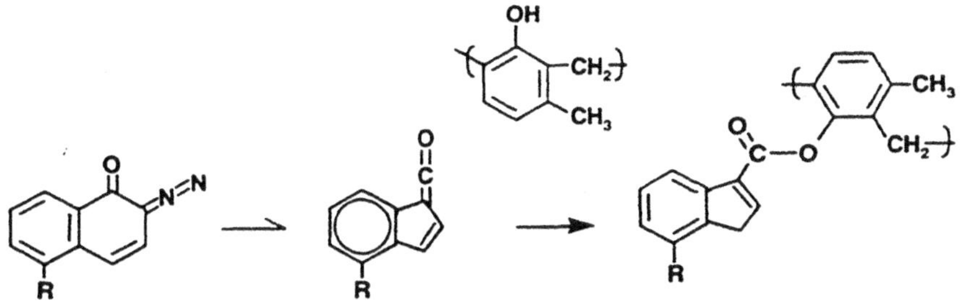

The ketene can also form dioxole from the dimerization of the ketene,[50,52]

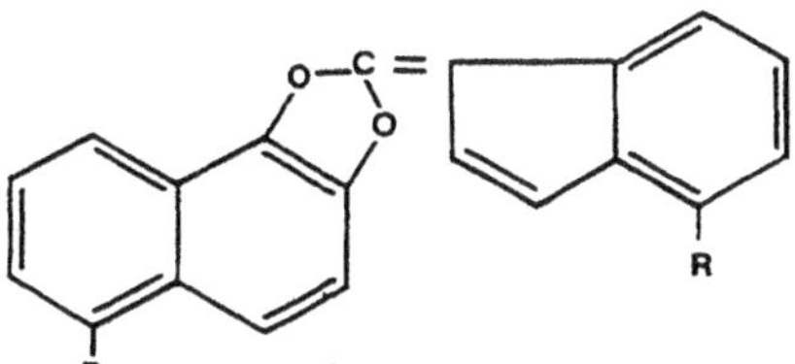

Depending on the extent of degradation, some of the products retard dissolution and require a compensatory longer develop time or stronger developer. For overbake conditions, the photosensitivity of the resist may be obliterated or residues may remain at the surface of the developed image. A convenient monitor of the extent of photoactive DQ (bleaching) decomposition is ultraviolet spectroscopy at 365–435 nm (Chapter 2). Several workers have tried to follow the thermal decomposition on the assumption that the thermolysis of DQ is a first-order bleaching, similar to photobleaching. The thermally produced by-products are optically insensitive in the 320- to 450-nm region.[22,25] Thermally, the problem of decomposition could be avoided if DQs stable to 130°C were synthesized.

The DQ photoactive components are thermally unstable above 70°C and decompose with an activation energy of 33 kcal/mole.[37] Thus, a prebake at 70°C for 2 hr, destroys 4% of the DQ, while at 90°C for 2 hr, all of the DQ is destroyed. Therefore, a 10°C change can have a significant effect on the extent of decomposition and the subsequent rate of resist dissolution.[51] For prebakes above

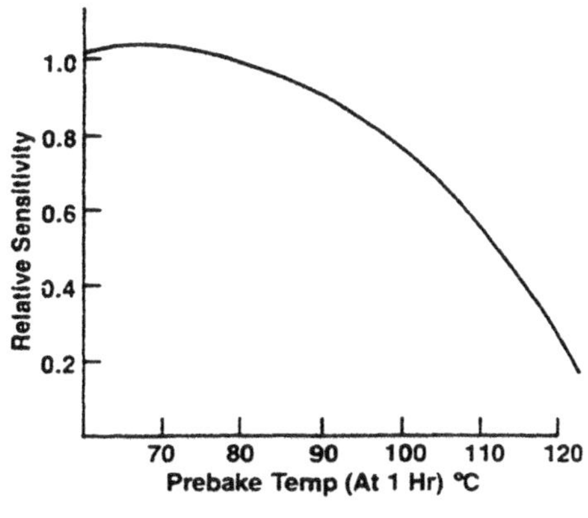

FIGURE 7-4-1. The relative sensitivity (ratio of exposed time required to expose time of prebaked film) for films of AZ-1350 prebaked at various temperatures. From Ref. 52.

105°C, the thermal decomposition products absorb more strongly at 404 nm than do the photodecomposition products. Baking AZ-1350J for 1-hr intervals from 70 to 120°C lowers the resist sensitivity by a factor of 5 (Fig. 7-4-1). Prebake at 70°C avoids sensitizer decomposition and this can be supplemented by a postexpose (predevelop bake) at 100°C to improve adhesion and reduce standing waves.(4) Drying of DQN films in a vacuum at room temperature leads to a threefold enhancement of resist sensitivity when compared to normal thermal prebake at 85°C.(58) However, the effect of vacuum bake was not noted for azide rubber or polyimide-type negative resists. In the case of DQN, the same overall level of solvent remained after vacuum or thermal prebake. Presumably, the vacuum bake may dry the film at the top of the resist and leave a skin induction layer. More studies of the concentration of casting solvent in the depth of the film are required to understand the differences between the vacuum and thermal prebake on the exposure speed of DQN resists.

7-5. PREBAKE EQUIPMENT AND PROCESSES

The throughput of various types of prebake equipment can be rapid or slow, depending on the time required to achieve steady-state conditions. Some systems such as microwave or infrared may be more difficult to regulate (Fig. 7-5-1) and often exceed the desired temperature. However, the time of heating is so short that little sensitizer decomposition may occur. Basically, there are four major modes of prebake (Table 7-5-1). Microwave and infrared are sensitive to metallic films on the silicon wafer. Conduction and convection are relatively insensitive to the nature of the semiconductor substrate. To some degree, all of the techniques heat the substrate. Infrared heating depends on the absorptivity of the substrate and must be adjusted to reach some even temperatures (Fig. 7-5-2).(54) The air inside an infrared oven is also a thermal contribution for convection drying.

Rapid baking can be achieved by microwave energy(55,56) in the form of electromagnetic energy in the frequency range from 2×10^8 to 3×10^{11} cps. By themselves, microwaves do not contain heat; their energy is converted to heat, only when they pass through certain types of materials. In single-crystal silicon, the charge carriers (electron + holes) may absorb microwave energy and, thereby, move to energies well above their thermal equilibrium levels. The carriers at higher levels will lose the energies to the lattice and excite the thermal vibrations of the lattice. The normal modes of lattice vibration in a quantum unit of energy are called phonons. In the long-wavelength optical modes, the phonons can be directly created by the lattice absorption of radiant energy. It is due to the excitation of normal modes of vibration, that microwave energy is converted to heat silicon. The

TABLE 7-5-1. Methods of Prebake

Method	Heat transfer	Reference
Microwave	Rotation	39, 55, 56
Infrared	Vibration, conduction	38, 54
Convection	Hot air	38
Conduction	Conductivity	38

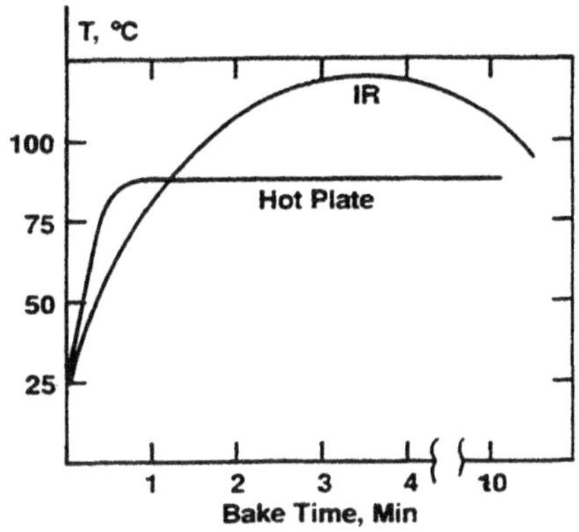

FIGURE 7-5-1. The temperature profile of a silicon wafer coated with resist in an infrared bake setup and on top of a hot plate.

nonabsorbing photoresist is insensitive to microwave energy. It is dried and cured in a microwave oven by conductive heating. In other words, each silicon slice is converted into a miniature hot plate. Semiconductor silicon begins to heat the instant the oven is turned on and heats continuously, until the oven is turned off.

Infrared heating is faster than air convection heating, because the silicon wafer absorbs energy, and the resist dries from the inside out.[57] IR lamps emit strongly in the 2- to 10-μm range, where resists absorb strongly. Unfortunately, in thick resists of DQN ($>2\ \mu$m) stagnant solvent-laden air, heated by the IR panels, causes skinning of the resist film and defeats the main advantage of drying from the inside out. Since inside the IR oven the (positive pressure) filtered air eliminates dust contamination, the IR system needs fresh air to replace the saturated solvent air. Thus, ramp heating methods, such as IR or microwave, require the careful control of heating time to adjust if possible for the differential absorption of electromagnetic energy by different substrates.

In prebaking energy considerations, the heat balance equation and film temperature change, ΔT, for time interval, $\Delta\phi$, is given by[17]

$$(M_s C_s + M_P C_P + M_{Si} C_{Si})\,\Delta T/\Delta\phi = h_{Si}(T_g - T)\,\Gamma k_g(P_s - P_g) \qquad (7\text{-}5\text{-}1)$$

where T_g is the temperature of the gas above the resist surface, P is the vapor pressure of the solvent (s) and gas (g) above wafer, Γ is the solvent heat of

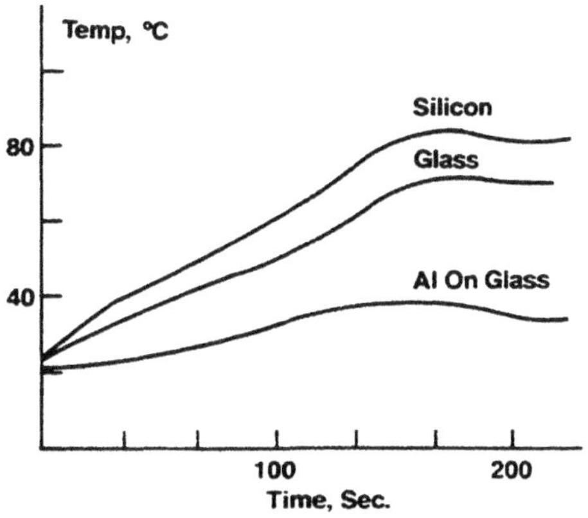

FIGURE 7-5-2. Temperature profiles of wafer made of silicon, glass, or aluminum-coated glass as the film is baked by infrared. From Ref. 54, reproduced by permission of the Electrochemical Society Inc.

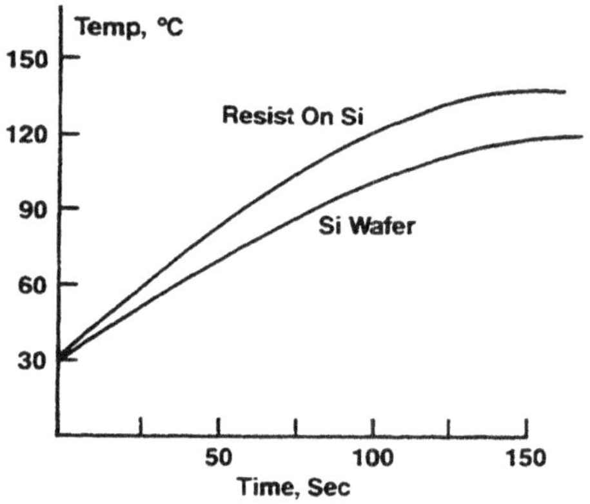

FIGURE 7-5-3. Temperature profile of infrared-baked Si wafer with and without photoresist coating. From Ref. 54, copyright 1983 and reproduced by permission of the Electrochemical Society Inc.

vaporization, and h_{Si} is the coefficient of heat transfer. The time, λ, for achieving a steady-state temperature is approximately

$$\lambda = PC_P d^2/K \tag{7-5-2}$$

where P, C_P, d, and K are the density, specific heat, thickness, and thermal conductivity of silicon. For baking, the time and temperature profiles for silicon, with and without resist, are in the 100-sec range (Fig. 7-5-3). Because of its larger mass and density, the silicon wafer dominates the heat rise.

For microwave baking, a magnetron supplies the excitation frequency (similar to the RF power source used in plasma etching; Chapter 13). Because of standing microwaves, the samples are rotated during use. When compared to the other conventional methods, microwave is a factor of 10 to 1000× faster, but the processed linewidth control (Fig. 7-5-4) (as a function of exposure time for DQN) appears to be the worst for microwave drying.[39]

Baking in a convection oven produces inconsistent results, due to loading effects, poor recovery, air flow, and poor circulation. For contact printing of DQN with an $R/R_0 > 25$, the prebake was invariant to 10–15°C variations. With exposures by scanning or steppers, the exposure times are near a minimum, and R/R_0 is close to 10. In this case, the process window for prebake decreases and more consistent and uniform processes are sought.

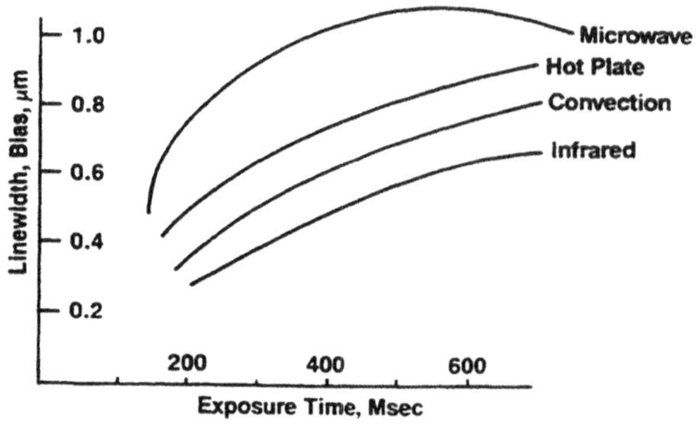

FIGURE 7-5-4. The linewidth enlargement (bias of image from mask dimensions) of DQN resist processed by different methods of prebaking. From Ref. 39, courtesy of SPIE.

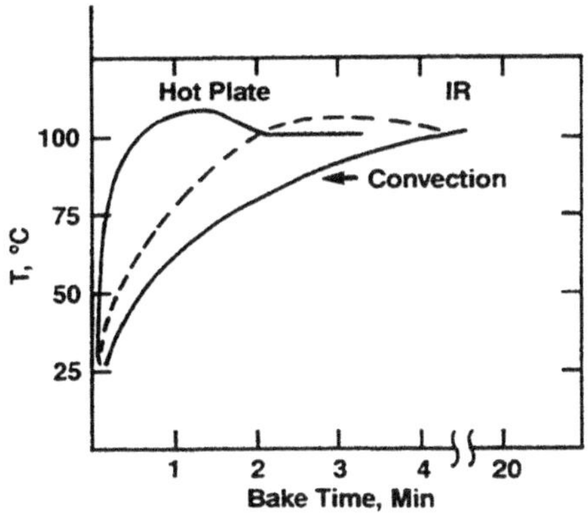

FIGURE 7-5-5. The time–temperature profiles of silicon wafer baked by different methods of heating. From Ref. 38, reproduced by permission.

If IR is tuned to achieve a steady-state temperature, then heat must first be absorbed by the substrate to a short depth (6–20 μm). In a comparative study of the prebake of the Kodak 820 DQN resist at 110–130°C for less than 60 sec, Macbeth[38] concluded that all three (IR, conduction, and convection) were essentially equivalent (Fig. 7-5-5). If the amount of residual solvents or the percentage of decomposed DQ (by IR or UV analysis) is used to tune the prebake time, similar results can be expected. The real question is the wafer-to-wafer reproducibility of heat input and transfer. In prebaking, the time should be chosen to achieve steady-state conditions. Overall, the hot plate turns out to be the fastest and most consistent method of prebaking.[60]

For process implications, a prebake difference of ± 1°C introduces a dimensional change of 0.008 μm[50] in a DQN resist. In a typical process controlled to ± 5°C, a dimensional change in image size of ± 0.1 μm was found.[50] For DQN resists, the prebake time has much less of an effect on the linewidth control than temperature (Fig. 7-5-6). With isothermal heating conditions, a control of ± 2°C for DQN will reduce the bias to a value of 0.05 μm. Other high-bake resists such as PMMA are generally insensitive to precise prebake conditions, as long as the temperature (160–190°C) is high and the time sufficient (>45 min). After prebaking, the wafers should be slowly cooled[61] and stored in dry, superclean environments to await the subsequent exposure step.

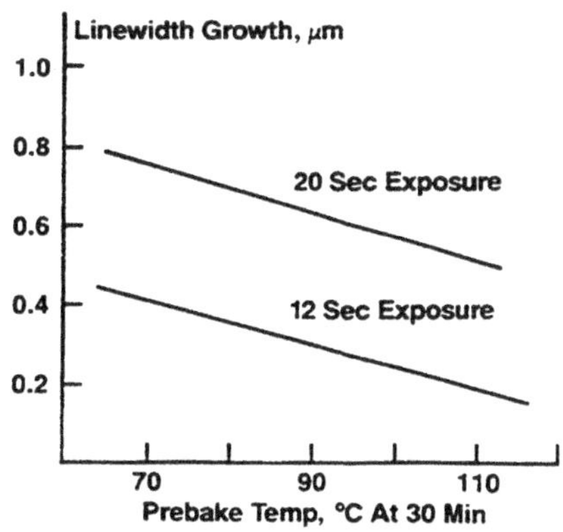

FIGURE 7-5-6. Linewidth changes for DQN resist prebaked and processed at two levels of exposure. From Ref. 50, copyright 1979 and courtesy of Eastman Kodak.

7-6. SUMMARY

Prebake is a prime lithographic example of the interaction of a process, a resist material, and process equipment. For example, the DQN resists require careful control of isothermal heating to avoid excess heating and loss of sensitivity. All of the heating systems (IR, convection, conduction, and microwave) rely mainly on the absorption of energy by the silicon substrate and the outward transport of the heat into the resist film. In the resist, the removal of the casting solvent is diffusion controlled in an exponential (Fickian) power function. The size of the solvent, the interaction of solvent and sensitizer with polymer, and the T_g of the resist, dominate the diffusion of solvent out of the film. Since most of the DQN films cannot be baked above their T_g, residual stress remains in the DQN resist which can cause adhesion failure during development. Suitable primers are required to seal the interface and relieve the stresses concentrated at the interface of the cast film. Accurate chemical and functional tests are needed to ascertain the differences, if any, between absorption, conduction, and convection heating.

For processing resists, the prebake temperature is more important than time, as long as steady-state temperature is achieved. The time should be chosen to remove most of the solvent to a constant level, or in the case of PMMA to remove all of the solvent in heat-insensitive resists. Prebake conditions should seek a maximum R/R_0 and good linewidth process latitude.

In the future, we can expect solventless vacuum- or plasma-deposited resists to gradually replace liquid-cast films. In this case, prebaking would only be used on a limited scale to relieve internal film stresses.

REFERENCES

1. W. Moreau, *Opt. Eng.* **22**, 878 (1983).
2. T. Yamazaki, Y. Suzuki, and H. Nakata, *J. Vac. Sci. Technol.* **17**, 1348 (1980).
3. P. Huggett, K. Frick, and H. Lehmann, *Appl. Phys. Lett.* **42**, 592 (1983).
4. E. Walker, *IEEE Trans. Electron Devices* **ED-22**, 464 (1975).
5. A. Ouano, in *Macromolecular Solutions*, edited by R. Seymour and G. Stahl, Pergamon Press, Elmsford, N.Y., 1982, p. 208.
6. T. Batchelder and J. Piatt, *Solid State Technol.* Aug. 1983, p. 211.
7. D. Meyerhofer, *IEEE Trans. Electron Devices* **ED-27**, 921 (1980).
8. M. Bowden, L. Thompson, S. Farenholtz, and E. Doerries, *J. Electrochem. Soc.* **128**, 1304 (1981).
9. S. Croll, *J. Appl. Polym. Sci.* **23**, 847 (1979).
10. R. Bond, S. Dzioba, and H. Naguib, *J. Vac. Sci. Technol.* **18**, 335 (1981).
11. S. Yamamoto, K. Kobayashi, and Y. Toyama, *Fujitsu Sci. Tech. J.* June 1978, p. 143.
12. L. White, *J. Electrochem. Soc.* **130**, 1543 (1983).
13. H. Fujita, in *Diffusion in Polymers*, edited by J. Crank and G. S. Parks, Academic Press, New York, 1968, p. 75.
14. N. Atoda, M. Komuro, and H. Kawakatsu, *J. Appl. Phys.* **50**, 3707 (1979).
15. J. Vossen and E. Davidson, *J. Electrochem. Soc.* **119**, 1708 (1972).
16. R. Eaton and F. Willeboordse, *J. Coat. Technol.* **52**, 63 (1980).
17. D. Robinson, A. Higginbotham, and P. Wankat, *I & EC Process Des. Dev.* **8**, 502 (1969).
18. D. Newman, C. Nienn, and J. Oliver, *J. Paint Technol.* **47**, 70 (1975).
19. R. Barrer, in *Diffusion in Polymers*, edited by J. Crank and G. S. Park, Academic Press, New York, 1968, p. 166.
20. H. Frisch, *Polym. Eng. Sci.* **20**, 2 (1980).
21. T. Gilbert, *J. Paint Technol.* **43**, 93 (1971).
22. F. Dill and J. Shaw, *IBM J. Res. Dev.* **21**, 210 (1977).

23. J. Bieron and R. Conley, *J. Appl. Polym. Sci.* **1**, 171 (1963).
24. U. S. Patent 3,898,350 (1975), IBM.
25. D. Meyerhofer, *IEEE Trans. Electron Devices* **ED-27**, 921 (1980).
26. S. Croll, *J. Appl. Polym. Sci.* **23**, 847 (1979).
27. M. Long and C. Walker, Kodak Microelectronics Seminar Proceedings, **G-102**, 56 (1979).
28. C. Deckert, Kodak Microelectronics Seminar Proceedings, 1977, p. 44.
29. L. Gavens, B. Wu, D. Hess, A. Bell, and D. Soony, *J. Vac. Sci. Technol.* **B1**, 481 (1983).
30. N. Atoda, M. Komuro, and H. Kawakatsu, *J. Appl. Phys.* **50**, 3707 (1979).
31. W. Prest and D. Luca, *J. Appl. Phys.* **50**, 6067 (1979).
32. W. Prest and D. Luca, *J. Appl. Phys.* **51**, 5170 (1980).
33. J. Greeneich, *J. Electrochem. Soc.* **122**, 970 (1975).
34. A. Morinaka and Y. Asano, *J. Appl. Polym. Sci.* **27**, 2139 (1982).
35. J. Anderson, R. Baker, and J. Forbes, *J. Colloid Interface Sci.* **31**, 372 (1964).
36. A. Rossi and G. Charland, *J. Paint Technol.* **44**, 65 (1972).
37. J. Shaw, M. Frisch, and F. Dill, *IBM J. Res. Dev.* **21**, 219 (1977).
38. J. Macbeth, Kodak Microelectronics Seminar, 1982, G-136, p. 87.
39. M. Hockey, *SPIE Proc.* **275**, 192 (1981).
40. J. Kolyer, F. Custode, and R. Ruddell, Kodak Microelectronics Proceedings, 1979, G-102, p. 150.
41. K. Kadota, Y. Taki, and S. Shimizu, *SPIE Proc.* **275**, 173 (1981).
42. H. Keller, *Solid State Technol.* June 1978, p. 45.
43. R. Newman and J. Prousnitz, *J. Paint Technol.* **45**, 33 (1973).
44. J. Walsham and G. Edwards, *J. Paint Technol.* **43**, 64 (1971).
45. J. Leers, *Solid State Technol.* March 1981, p. 90.
46. K. Wickersheum, *Electron. Packag. Prod.* Sept. 1981.
47. E. Egerton, A. Nef, W. Milkinsun, and D. Barr, *Solid State Technol.* Aug. 1982, p. 84.
48. A. Martynenko, B. Strizhov, and V. Nikolosku, *Russ. J. Phys. Chem.* **49**, 1310 (1975).
49. J. Pacansky and J. Lyerla, *IBM J. Res. Dev.* **23**, 42 (1979), *J. Electrochem. Soc.* **124**, 862 (1977).
50. E. Hryhorenko, Kodak Microelectronics Proceedings, G-102, p. 18.
51. M. Cagan, *Electrochem. Soc. Ext. Abstr.* **82-1**, 323 (1982).
52. T. Gupta, *Eur. Polym. J.* **17**, 1127 (1981).
53. M. Goldrick and L. Rankel Plauger, *Photogr. Sci. Eng.* **17**, 386 (1973).
54. I. Calder, R. Sue, and H. Naguib, *J. Electrochem. Soc.* **130**, 1390 (1983).
55. U. S. Patent 3,771,948 (1973), Nissho.
56. U. S. Patent 4,243,744 (1981), Exxon.
57. Anon, *Circuits Manuf.* Nov. 1974, p. 62.
58. A. Endo, *Jpn. J. Appl. Phys.* **23**, 795 (1984).
59. U. S. Patent 4,508,812 (1985), Hughes.
60. V. Marriot, Y. Lin, and G. Fuller, *SPIE Proc.* **469**, 65 (1984).
61. European Patent Application 114,126, *Chem. Abstr.* **102**, 123116 (1985).
62. M. Suzuki, Y. Ohnishi, and A. Furuta, *J. Electrochem. Soc.* **132**, 1390 (1985).
63. T. Pampalone and P. Zanzucchi, *J. Electrochem. Soc.* **133**, 1917 (1986).

8

Optical Exposure

8-1. INTRODUCTION

Optical "lightography" combines the physical disciplines of optics, mechanics, and photographic chemistry. Stevens, in his classic book[1] on microphotography, lists some of the interactive requirements (Table 8-1-1). Any form of printing results in some degradation of edge sharpness (Fig. 8-1-1). Projection of a two-dimensional circuit pattern results in the degradation of the edge slope of the pattern (Fig. 8-1-2). It is the resist which must restore the modulated sinusoidal intensity of the beam into an acceptable rectangular mask for subsequent image transfer by etching or lift-off.

If two slits (a) are placed a distance (b) apart, the unexposed space gradually receives exposure because of five possible limits (Fig. 8-1-3):

a. Diffraction
b. Focal limit of lens (out of focus)
c. Low-contrast resist
d. Standing waves (reflection from substrate)
e. Refraction in resist

The image of a point source formed in the focal plane of a perfect lens is never a true point but rather the light is spread out into a diffraction pattern known as an Airy disk (Fig. 8-1-4). Thus, in a positive photoresist, the unexposed space, for example, is partially exposed by the sinusoidal diffracted beam and by light reflected from the substrate (Fig. 8-1-5) resulting in a Gaussian-shaped profile and loss of film thickness. Relatively thick and steep wall profiles (Fig. 8-1-6) are required for the subsequent dry etching process because of the limited selectivity between the etch rates of the resist and the substrate. Due to the intraproximity effect (diffraction loss of forward intensity), isolated exposed lines do not receive sufficient radiation and must be exposed at a higher dose (distorting images larger than 3 μm or unexposed spaces smaller than 3 μm) or be developed with loss of unexposed (Fig. 8-1-7). A resist with a contrast $\tau > 3$ would not develop lines $< 2\ \mu$m.

Accordingly, it is the task of the optical exposure step to overlay and reproduce in the resist the two-dimensional features of photomasks to within $\pm 15\%$ of nominal feature size and to within $\pm 5\%$ of the required edge slope (Fig. 8-1-8). Each device layer must also be precisely aligned to within $\pm 25\%$ of the smallest printed feature. To achieve these objectives, device fabrication involves

TABLE 8-1-1. Interaction of Tool and Process in Emulsion Printing[a]

Requirements	Components	Limits
Size of original	Emulsion (λ, ASA, resolution)	Astigmatism
Size of final copy	Illumination system	Aberrations
Contrast of print	Depth of focus	Field curvature
Number of copies	Numerical aperture	Emulsion thickness
Overlay accuracy	f of lens	Gamma
Reproducibility		Flatness

[a] From Ref. 1.

multilevel registration and image transfer from the resist pattern into a film pattern. The root mean square of all the image and registration factors should be integrated to assess the contribution of the projection optics and alignment (Fig. 8-1-9).[3] The usable linewidth is about 4 × the registration accuracy.[4] Not only must the dimensional pattern fidelity and overlay be achieved, but a reasonable and reproducible process must be obtained to realize a good yield (Fig. 8-1-10) and an economical throughput from the million dollar optical exposure tool. Interacting in the optical arena are the physical limitations of light (diffraction, interference, refraction, and reflection), the mechanical limitations of the exposure tool, the thermally warped wafers and their undulating topography, and the resist response (including contrast and photospeed). Under such circumstances, a wide process window latitude (Fig. 8-1-11) is desired to reproduce a 1.0-μm pattern to within ± 0.1 μm.

Accordingly, a practical resist must be chosen in terms of its dose requirement, film retention, and response to a particular optical scan setting (Fig. 8-1-12). Exposure tools with high numerical aperture (NA > 0.4) are able to project sharper and smaller features to about 1.0 μm in size without the influence of the resist contrast response, but the field of exposure (Fig. 8-1-13) unfortunately shrinks and only one chip at a time can be printed. In addition, the depth of focus approaches the thickness of the resist (1.0 μm) and each field must be releveled and refocused before each exposure.

The exposure sources can be point sources such as lasers or extended Hg lamps (Fig. 8-1-14) which emit in three spectral regions,[2]

- Deep UV 100 to 200–300 nm
- Mid UV 300–360 nm
- Near UV 360–450 to 700 nm

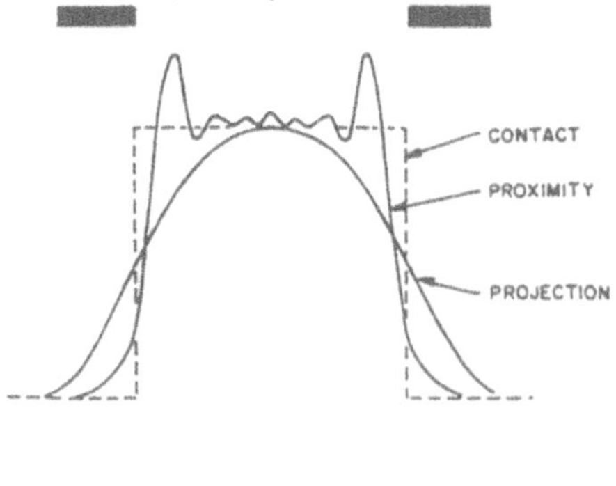

FIGURE 8-1-1. Incident profiles of masked pattern defined by contact printing, proximity printing, and projection. Adapted from J. Skinner, Kodak Microelectronics Proc. G-35, 54 (1974) by permission.

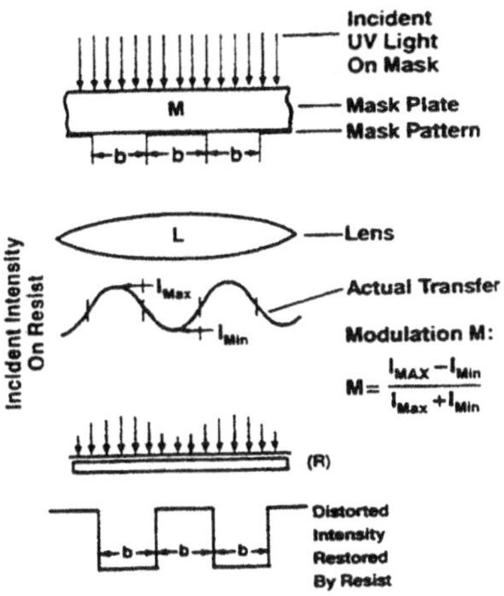

FIGURE 8-1-2. The distortion of the two-dimensional mask (M) by the MTF of the lens (L) must be restored by the resist (R). Adapted from D. Widmann and K. Stein, Solid State Circuits, Proc. 2nd Eur. Solid State Circuits Conf., 1976, pp. 29–47.

Because of the unavailability of transparent glasses of various refractive indexes, it is difficult to fabricate lenses for the deep- and even the mid-UV region. Ultimately, if the materials problems with deep-UV glasses are solved, it will be possible to fabricate lenses that have an NA of 0.5 and resolve 0.25-μm features. The mercury lamp (Fig. 8-1-15) emits strongly at 365 nm (I line), 405 nm (H line), and 435 nm (G line). The lamp output is weakest in the deep UV and new sources are required to maintain the same incident intensity as the 360- to 450-nm region.

Three basic types of optic tools are available:

- Shadow casting
- Refractive projection
- Reflective projection

Shadow casting involves the simple process of a 1 × mask in physical contact

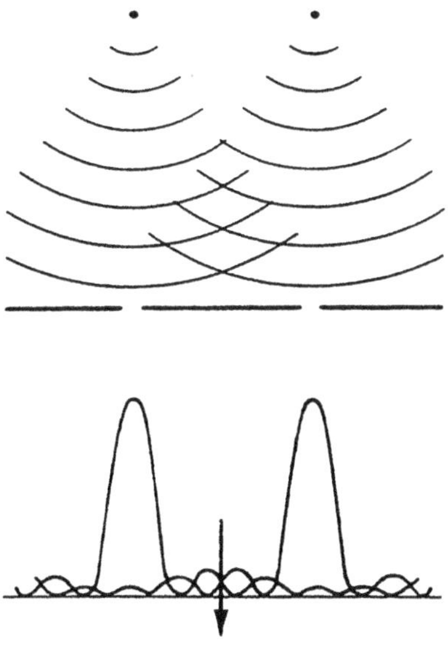

FIGURE 8-1-3. Two slits which are separated by a dark space are resolved. From Ref. 147, copyright 1980 and courtesy of Butterworths, Inc.

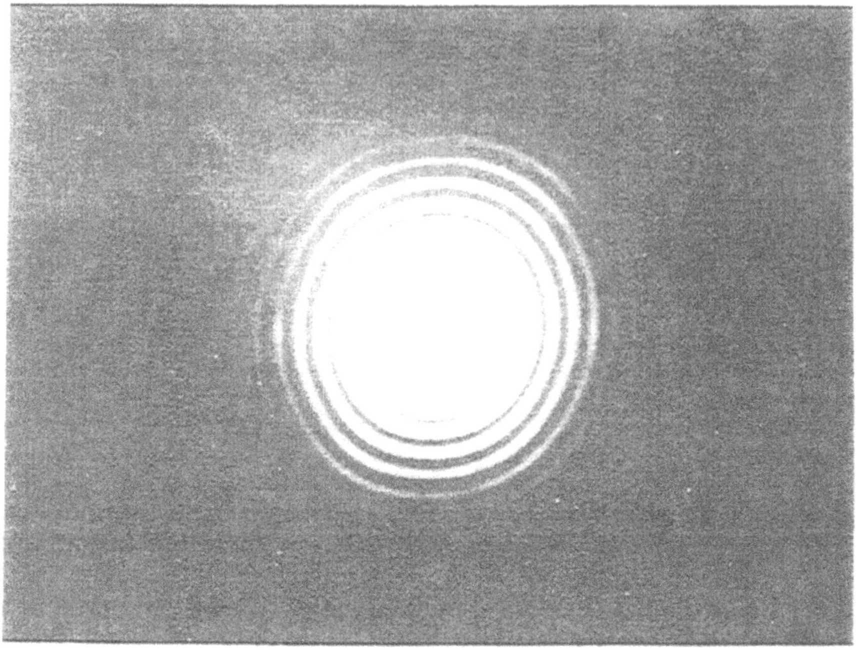

FIGURE 8-1-4. Airy disk at the focal point of a lens. Adapted from A. Van Heel and C. Velzen, *What is Light*, World University Press, London, 1968, p. 61.

or separated by a few micrometers as in proximity printing (Fig. 8-1-16). The constant mask deterioration and limited alignment are the major drawbacks of contact printing.

Projection optics uses lenses (Fig. 8-1-17) or mirrors (Fig. 8-1-18) to project a 10×, 5×, or 1× mask onto a 20-mm square field or a stripe (1.5 mm) which is scanned across the wafer. The five key attributes required of a step and repeat system are[230]: (1) good, every chip alignment to compensate for mask and wafer

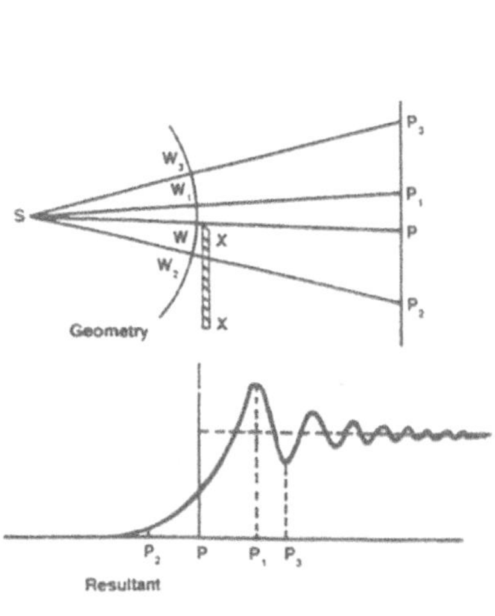

FIGURE 8-1-5. Diffraction of light by a straight edge and the resultant intensity distribution. From Ref. 147.

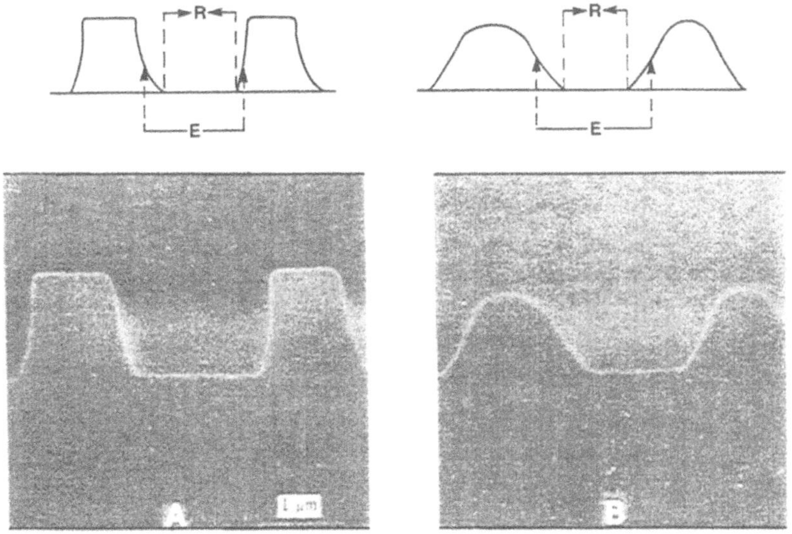

FIGURE 8-1-6. Examples of (A) acceptable steep profile in resist for image transfer by reactive ion etching (RIE); (B) unacceptable shallow profile. Adapted from V. Marriott, *SPIE Proc. Opt. Microlithogr. II* **394**, 146 (1983), copyright and courtesy of SPIE.

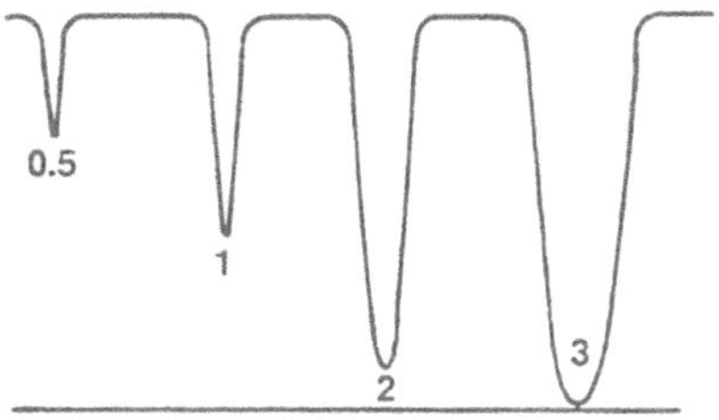

FIGURE 8-1-7. Trace profiles of energy deposited in positive resist of 3-, 2-, 1-, and 0.5-μm lines developed and measured from the top surface. The diffraction-limited lens does not expose the smallest features (intraproximty effect) with equidoses.

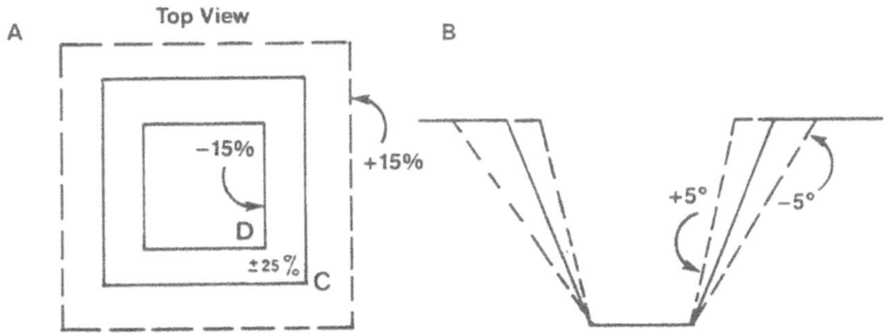

FIGURE 8-1-8. Goals of lithographic imaging. (A) Reproduction of mask dimensions within ±15%. (B) Reproduction of edge slope within ±5°. (C) Overlay of C to D with ±25%.

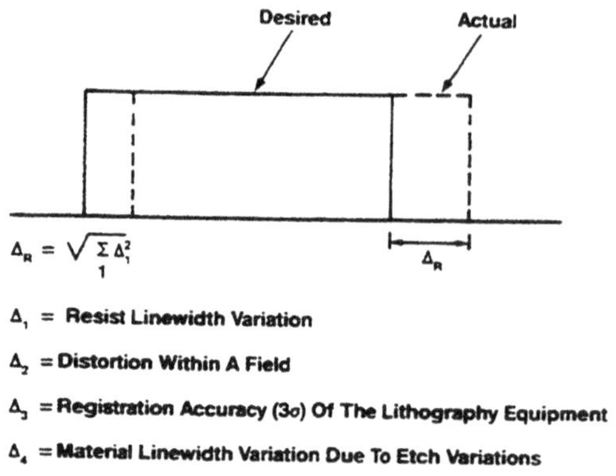

FIGURE 8-1-9. The "usable" linewidth composed of the resist linewidth, etched and registration accuracy. From Ref. 3, courtesy of SPIE.

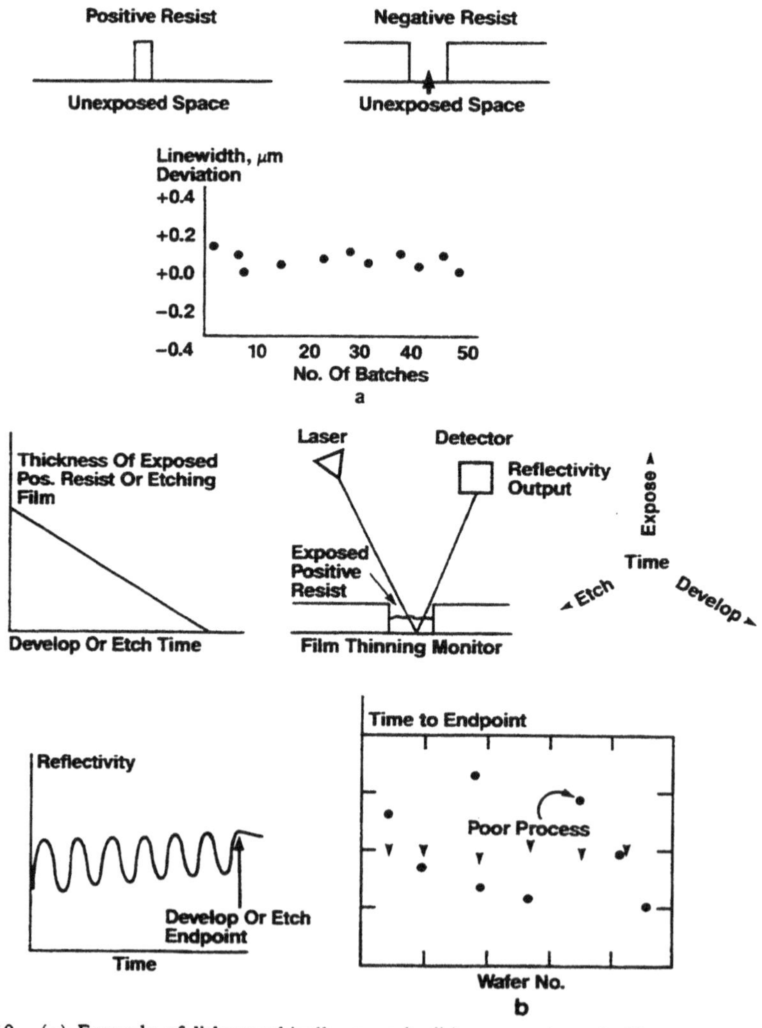

FIGURE 8-1-10. (a) Example of lithographically reproducible processing of silicon wafers to meet target linewidth objective. (b) Variation of develop endpoint time monitored by endpoint detect for a poor process (●) and a good process (▼). The variation in develop time is a result of exposure fluctuations.

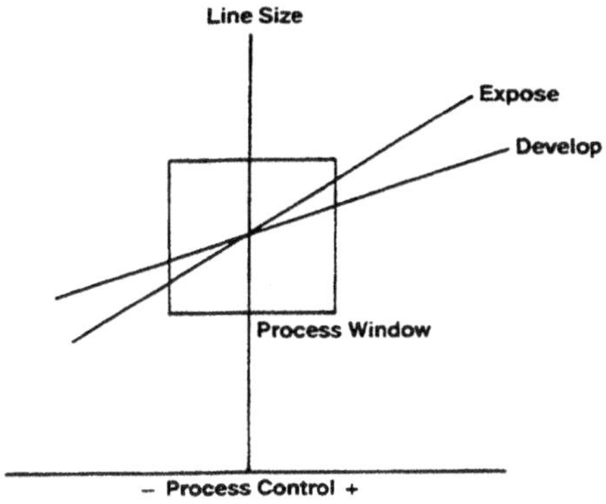

FIGURE 8-1-11. Process window in exposure and development steps of resist image. Adapted from Ref. 149.

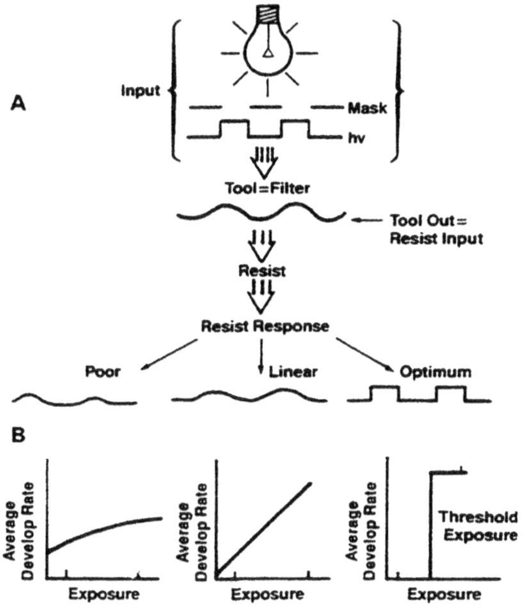

FIGURE 8-1-12. (A) Tool output and resist response. (B) Resist response by develop rate. After M. Hakey and J. Straub,[16] copyright 1982 and courtesy of *J. Electron. Mater.* **11**, No. 4, 1982, a publication of the Metallurgical Society of AIME, Warrendale, Pa.

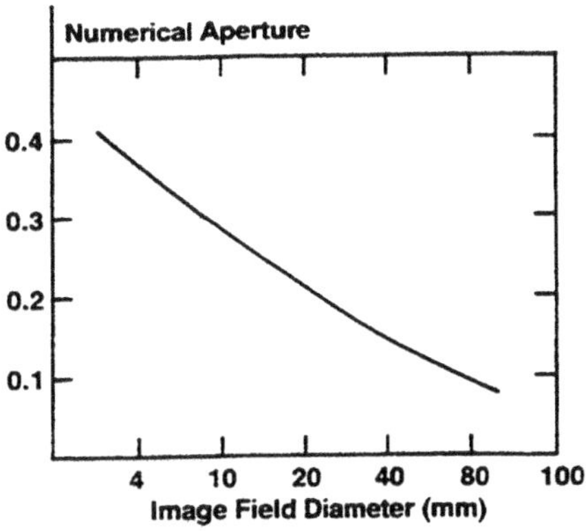

FIGURE 8-1-13. Projection lenses plotted with respect to numerical aperture and field size. From Ref. 5, copyright 1978 and reprinted with permission of *Solid State Technology*, published by Technical Publishing, a company of Dun and Bradstreet.

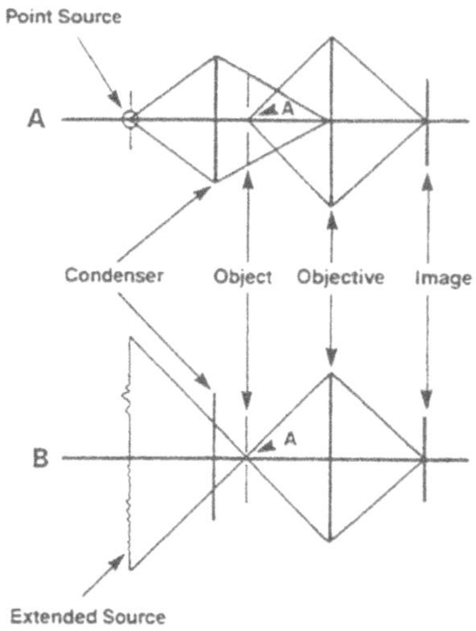

FIGURE 8-1-14. (A) Coherent projection point source and (B) incoherent extended source. From Ref. 2, reprinted with permission of *Solid State Technology*, published by Technical Publishing, a company of Dun and Bradstreet.

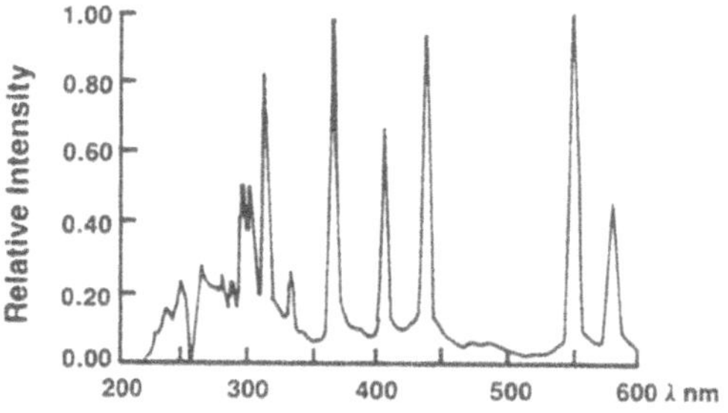

FIGURE 8-1-15. Output of high-pressure mercury lamp used in projecton printers. From Ref. 63, copyright 1983 and reproduced by permission of the American Chemical Society.

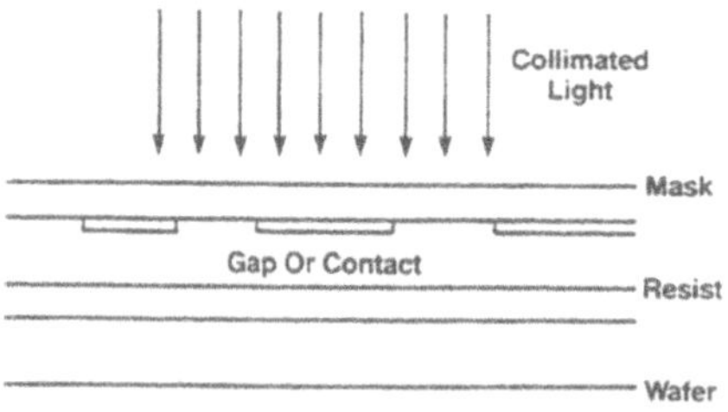

FIGURE 8-1-16. Schematic of shadow casting proximity printing.

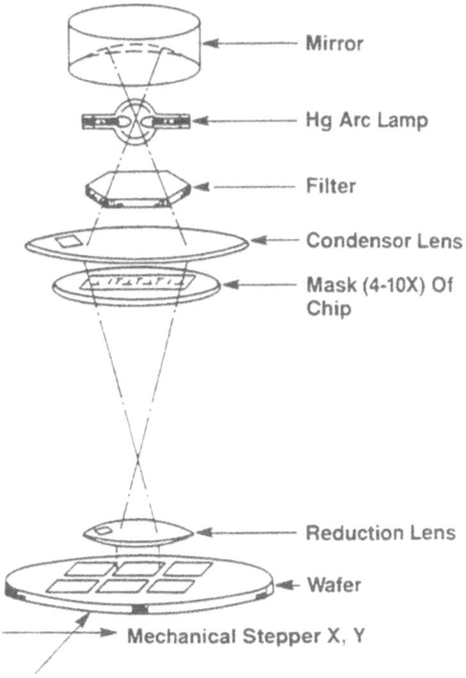

FIGURE 8-1-17. Schematic of refractive optics reduction projection printer. From Ref. 63, copyright and courtesy of the American Chemical Society.

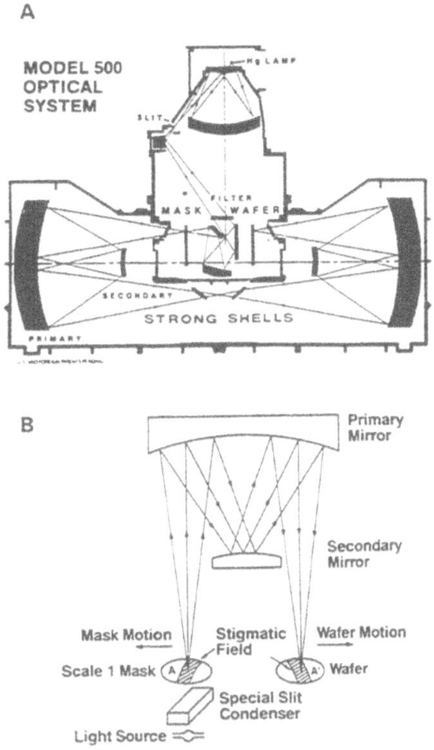

FIGURE 8-1-18. (A) Schematic of Perkin Elmer Model 500 reflective scanning optical exposure tool. From F. Zernike, H. Sewell, N. Barosi, and M. Gansfried, in *Microcircuit Engineering 83*, edited by H. Ahmed, J. Cleaver, and G. Jones, Academic Press, 1983, p. 211. (B) Optical arrangement of mirrors for simple scaning exposure tool. From Ref. 146, copyright 1980. Both figures reproduced by permission of Academic Press.

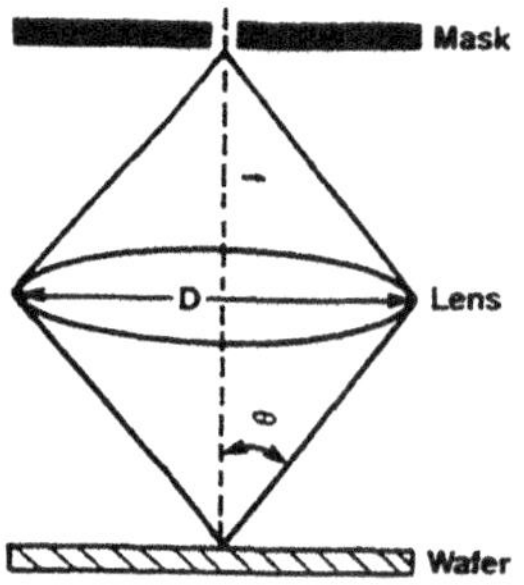

FIGURE 8-1-19. Relationship between the object, image, and focal length and the diameter of the lens to define the numerical aperture. From Ref. 150.

distortion; (2) an accurate laser table to take over from every chip alignment when the visibility of marks falls below an acceptable value; (3) the ability to change reticles without reducing throughput; (4) a small floor size to minimize the clean room area occupied by several exposure tools; (5) a throughput equivalent to about 40 (250-mm) wafers/hr. At present, no camera has all these attributes.

In simple imaging optics, the focal length (f) is a function of the lens diameter (D) (Fig. 8-1-19). The numerical aperture (NA) in a medium of refractive index n is given by

$$\mathrm{NA} = n \sin \theta = D/2f \tag{8-1-1}$$

The resolution (W) of a lens at the Rayleigh limit for two opaque objects that is barely detectable in the Airy disk (Fig. 8-1-4) is

$$W = K\lambda/\mathrm{NA} \tag{8-1-2}$$

K varies from 0.3 for top surface image resists, 0.5 for multilayer resists, 0.75 for single-layer resists, and 1.1 for resists over reflective surfaces like aluminum.[231] A practical resolution is three Rayleigh limits at an exposure wavelength, λ (Fig. 8-1-20):

$$W = 1.83\lambda/\mathrm{NA} \tag{8-1-3}$$

As a result, resolution is improved by using shorter wavelengths of exposure

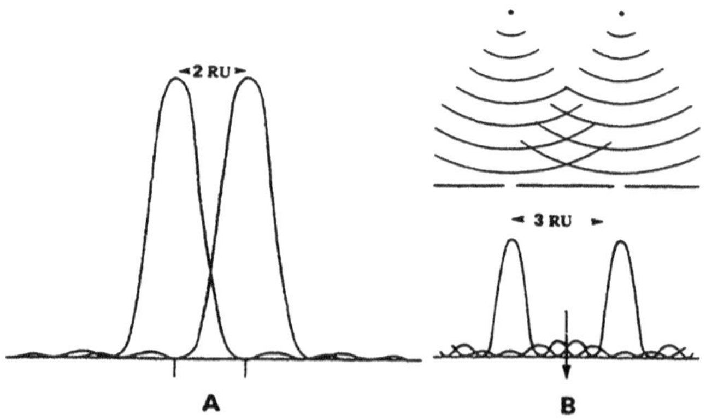

FIGURE 8-1-20. Images resolved (A) two Rayleigh units and (B) three Rayleigh units apart. From Ref. 24, copyright 1967 and reproduced by permission of Wiley & Sons. Inc.

TABLE 8-1-2. Numerical Aperture and Depth of Focus

NA	Depth of focus (μm) 400 nm	200 nm
0.2	5.0	2.5
0.3	2.0	1.0
0.4	1.5	0.8
0.5	0.8	0.4

(deep UV) and higher numerical aperture at the expense of smaller exposure fields.[157] Unfortunately, the depth of focus DF also decreases (Table 8-1-2) with higher NA and each chip site must be releveled and refocused,

$$DF = \lambda / \pm 2(NA)^2 \qquad (8\text{-}1\text{-}4)$$

For an NA of 0.35 and an exposure wavelength of 300 nm, the DF is less than 1.5 μm. The wafer flatness, uneven topographical features, and the resist thickness itself can exceed the in-focus condition for submicrometer imaging. Thus, processes such as top surface imaging are innovations necessary to keep the latent images in focus. Out-of-focus images rapidly lose image fidelity, especially in closely pitched features, and the intensity also decreases, requiring longer exposure times.

In the next section we will discuss some of the engineering optical principles which define the aerial (latent image factors) image. The resist contrast factor (τ) has already been discussed in Chapters 3 and 5.

8-2. OPTICAL PRINCIPLES

The primary physical factors of light which limit the latent image size and edge slope are:

- Coherency
- Diffraction
- Interference
- Astigmatism
- Chromatic aberrations

In coherent radiation (Fig. 8-2-1), light waves travel together so that their crests

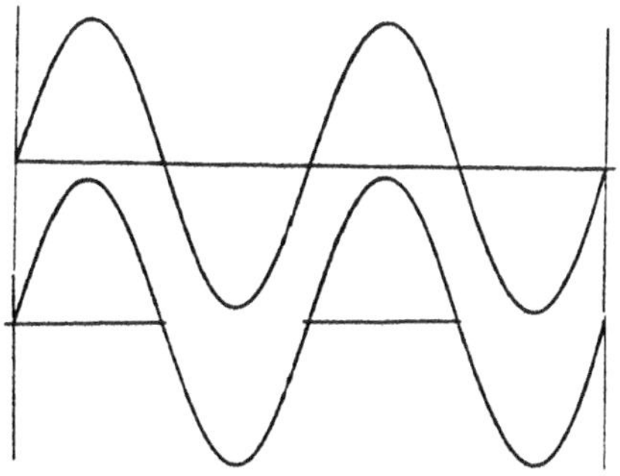

FIGURE 8-2-1. Coherent radiation: the monochromatic light waves travel in phase.

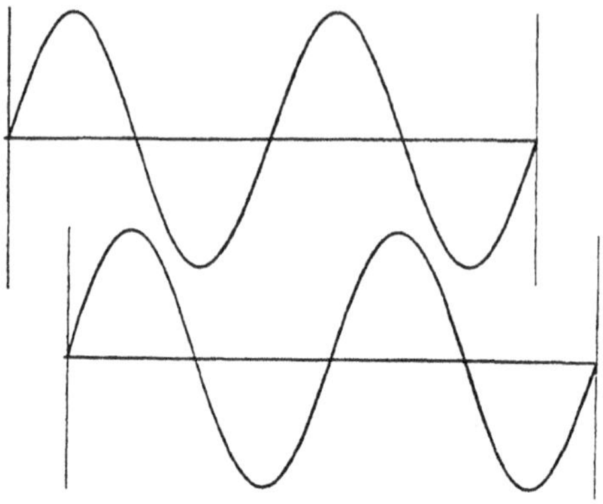

FIGURE 8-2-2. Incoherent or out-of-phase monochromatic light.

and troughs are matched in "phase" with each other. Monochromatic light waves that travel randomly so that their crests and troughs are not matched are out of phase (partial) and are said to be partially coherent or incoherent light (Fig. 8-2-2). The laser is a source of both spatially and temporally coherent radiation, which means that the emitted light is perfectly monochromatic. Conventional arc sources emit light in 4π steradians. Lasers are spatially coherent point sources of high brilliance which emit a narrow beam with an angular divergence θ limited by diffraction,

$$\theta = K\lambda/d \tag{8-2-1}$$

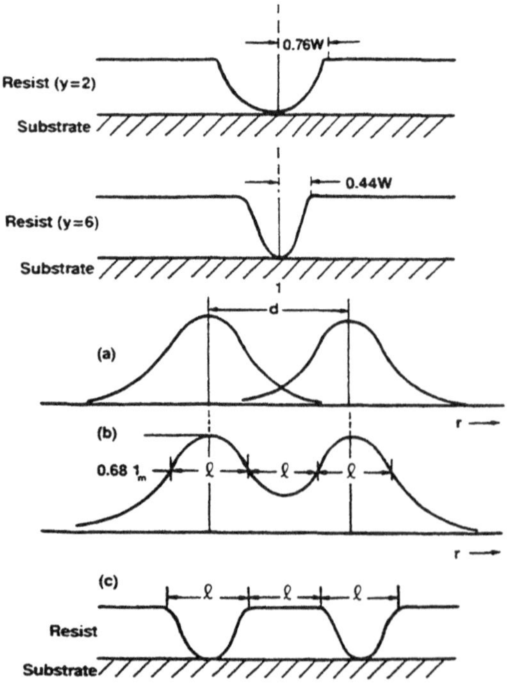

FIGURE 8-2-3. Resist profiles formed by a projected Gaussian laser beam. (a) Focused Gaussian beams with separation d. (b) Exposure profile in resist of contrast $\gamma = 6$ for $d = 1.836w$. (c) Developed profile in resist of contrast $\gamma = 6$ with lines and spaces of equal width at the resist surface. From Ref. 8.

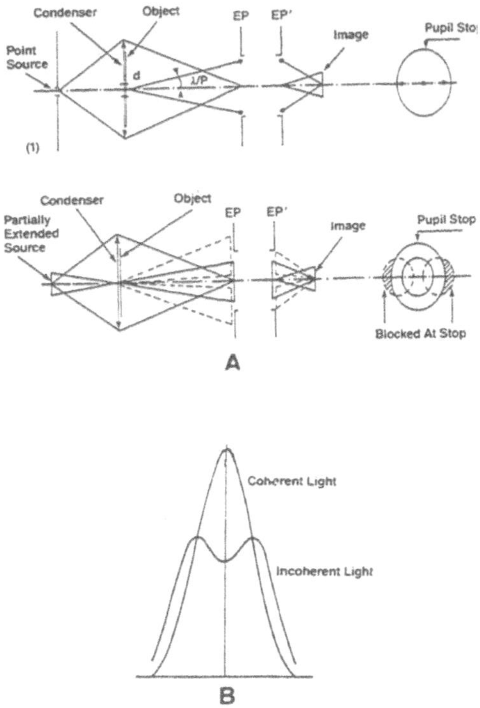

FIGURE 8-2-4. (A) Coherent point and extended source. From Ref. 2, copyright 1980 and reprinted with permission of *Solid State Technology*, published by Technical Publishing, a company of Dun and Bradstreet. (B) Resultant intensity for close objects. From Ref. 147, copyright 1980 and courtesy of Butterworths, Inc.

where λ is the wavelength, d is the output diameter through which the light radiates from the cavity; $K = 1.2$ for a uniform beam and $K = 2/\pi$ for a Gaussian beam.[(2)] In a laser beam, the collimation prevents focusing of all the emitted energy into a small spot the size of which is limited only by the diffraction of the focusing lens and the contrast (τ) of the resist. Johnson[(8)] (Fig. 8-2-3) printed 0.1-μm (100-nm) features in a resist using a Gaussian laser beam ($\lambda = 3250$ Å) with exposures of a line or grating just sufficient to reach the substrate.

The limiting diffraction is

$$\phi = Kf\lambda/D \tag{8-2-2}$$

where f is the focal length and D is the aperture. For an $F/1$ condenser, the spot size is q. With a laser, a perfect coherent point source can be achieved (Fig. 8-2-4A).[(12)] The main drawback to coherency is the severe image degradation due to diffraction[(11)] effects. To alleviate this problem, the illumination is typically decollimated by a few degrees[(12)] to reduce the lens diffraction. By partially filling

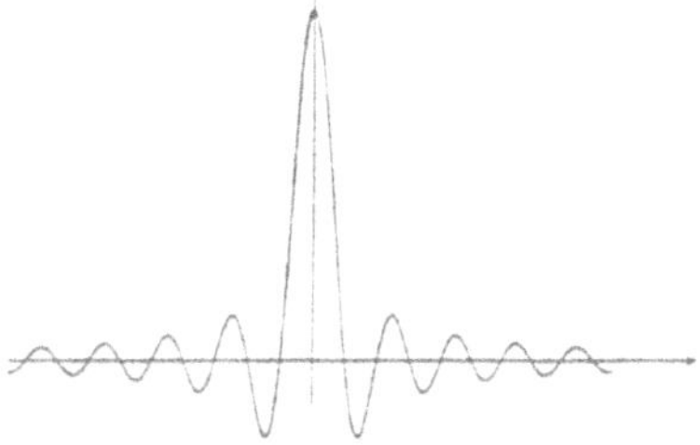

FIGURE 8-2-5. Partially coherent illumination and resultant intersity curve. Diffraction orders from objects increase in size. From Ref. 2.

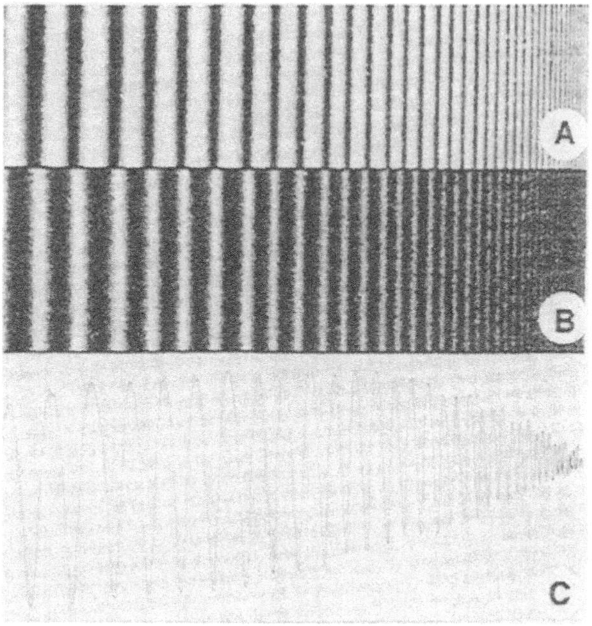

FIGURE 8-2-6. Influence of line frequency on image contrast. (A) Test bar target; (B) image reduction in emulsion; (C) decreasing depth profile in emulsion. From Ref. 1, copyright 1967 and reproduced by permission of John Wiley & Sons, Inc.

the entrance pupils of the lens, certain frequencies of the image are enhanced, sharpening the edge of the image. The ratio (σ) between the diameter of the effective source and that of the entrance pupil of the illuminator is often taken as a measure of the partial coherency.

When the illumination is changed from a coherent to a partially coherent source, the contrast of the lens (MTF), the effective resolution, and the depth of focus (Fig. 8-2-4B) of a projected image are lowered. In a Kohler source, the exposure source is imaged through the entrance pupil of the projection lens and the source image is the effective source size. If the source size increases, to an extended arc lamp source, the phase correlation decreases and diffracted orders from the

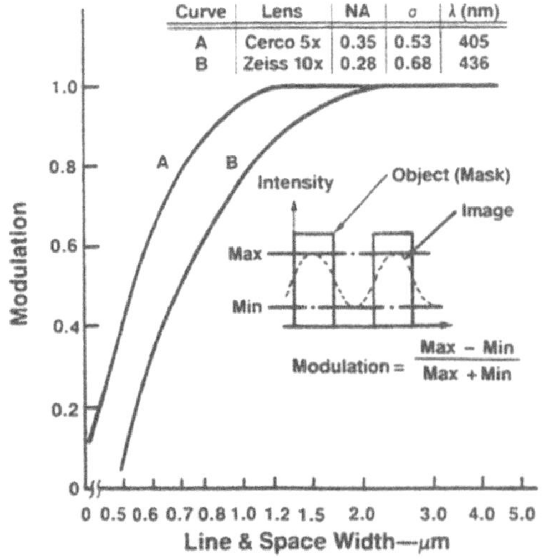

FIGURE 8-2-7. MTF of commercial lenses. From Ref. 13, courtesy of AIP.

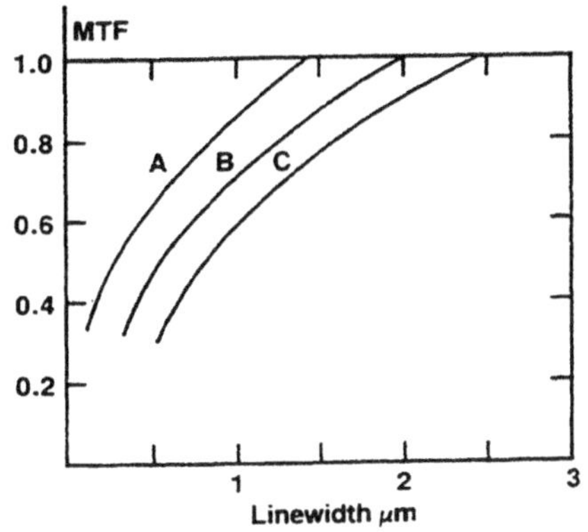

FIGURE 8-2-8. MTF as a function of resist thickness. (A) 0.4 μm; (B) 0.8 μm; (C) 1.2 μm. From Ref. 14, courtesy of AIP.

object increase in size (Fig. 8-2-5). Diffracted orders may be partially eliminated by the pupil stop.

The engineering definition of the resolution of a lens is the ability of the lens to resolve a (Fourier) series of equal lines and spaces (a diffraction grating) based on the linear recording in a silver emulsion (Fig. 8-2-6) of a two-dimensional pattern (transmitted and opaque). The modulation transfer function (MTF) is the expression of the object M_1 and image M_2 (Fig. 8-2-7) given by[1,9]

$$\text{MTF} = M_1/M_2 = (\text{Max} - \text{min})/(\text{Max} + \text{min}) \tag{8-2-3}$$

A high-quality lens will resolve lines and spaces equal to the thickness of the resist. The concept of resolving presumes no intensity loss of features but as can be seen in Fig. 8-2-6, the higher (of the closely spaced lines) the spatial frequency (lines/mm or cycles/mm), the lower is the image contrast. Once the MTF of the lens is determined, predictions[10] can be made regarding line spread functions, linewidth control, and sensitivity to exposure conditions. A modulation of 60% implies a difference of 80% and 20% in intensity of the maximum and minimum light transmitted by the diffraction elements of the lens. The minimum MTF of a lens has to be about 0.60 which allows for a 20% loss of unexposed resist (Fig. 8-2-7).[13] Mathematically, the MTF of a diffraction-limited incoherently illuminated system is identical to the Fourier transform of a circular aperture,[14]

$$\text{OTF} = 2/\pi(\cos^{-1})[f/2f_c - f/\pi f_c][1 - (f/2f_c)^2]^{1/2} \tag{8-2-4}$$

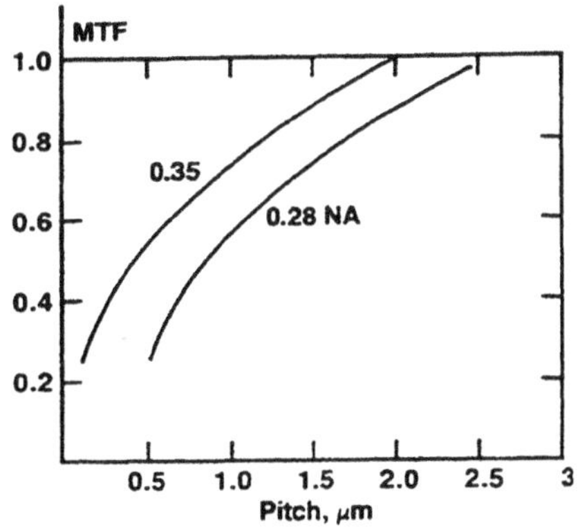

FIGURE 8-2-9. MTF as a function of numerical aperture. From Ref. 13, courtesy of AIP.

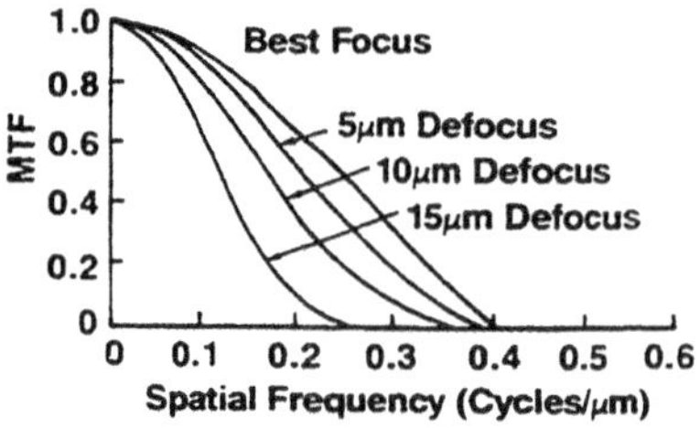

FIGURE 8-2-10. MTF curves as a function of defocus. 0.1 cycle/μm = 2.5 μm. From Ref. 15, courtesy of SPSE.

with the (coherent) spatial cutoff frequency given by

$$f_c = [M/(1+M)]\ 1/\lambda(\tan \sin^{-1} \mathrm{NA}) \tag{8-2-5}$$

where M is the system magnification and λ is the wavelength of light.

The finite resist thickness is taken into account by averaging the MTFs of the system in focus on top of the resist ($t=0$) and out of focus on the bottom of the resist (t). Defocusing is treated as an aberration. The defocused MTF is the product of the focused MTF and the Fourier transform of the Airy disk,

$$F(f) = (1/\pi Rf)^{J}(2\pi Rf) \tag{8-2-6}$$

where R is the disk radius and J is the Bessel function of first order. Therefore, for finite resist thickness (Fig. 8-2-8),

$$\mathrm{MTF}_t = (1 + F/2)\ \mathrm{MTF}_0 \tag{8-2-7}$$

The MTF of optical tools[14] (Fig. 8-2-8) falls off sharply in frequency space favoring the transmittal of pattern periodicity. A larger NA and shorter λ, increase the fidelity of pattern transfer (Fig. 8-2-9). Defocusing[15] can be treated (Fig. 8-2-10) as an aberration which reduces the higher spatial frequency. As a result, thinner films in multilayer or top surface imaging resists favor higher resolution, especially (due to diffraction) for closely spaced lines and features (Fig. 8-2-11) which experience a rapid falloff in MTF.

The MTF of a lens is the first estimate of the evaluation of one projector resolution versus another. The imagery of a pattern of arbitrary shape cannot be constructed on the basis of MTF alone.[17, 217] A true evaluation has to include the

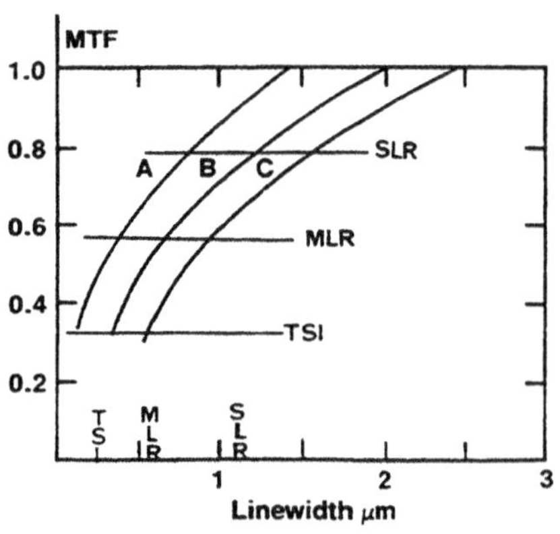

FIGURE 8-2-11. MTF as a function of wavelength for a numerical aperture of 0.41[14] and for a single film resist (SLR), a multilayer resist (MLR), and a top surface imaging (TSI). The top surface imaging resist offers the potential for submicrometer resolution.

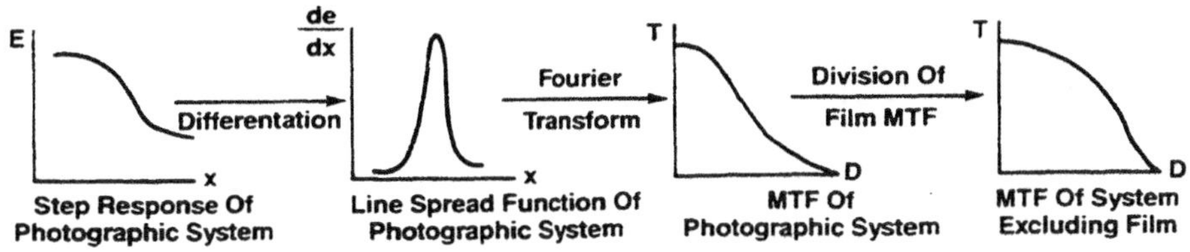

FIGURE 8-2-12. Schematic of reconstruction of MTF from the system step response of photographic film. From Ref. 16, courtesy of SPSE.

MTF_{lens}, the edge slope of the beam, and the MTF_{resist} of the resist (photographic system) (Fig. 8-2-12).[152]

In submicrometer imagery, the MTF_{lens} does not provide enough process latitude margin over the minimum MTF_{resist} required for image formation in polymer resists. Tai *et al.*[13] introduced the concept of the critical MTF_{resist} of a resist which is related to the gamma slope of a resist (Fig. 8-2-13). The minimum size feature using any combination of resist and lens is then determined by

$$MTF_{lens(optical)} > CMTF_{resist} \tag{8-2-8}$$

Experimentally, the resist contrast may be determined by using a thin layer of resist as a threshold detector.[229] Most resist have contrasts (τ) of 1–2 or a $CMTF_{resist}$ of 0.6. Therefore, the optical MTF must be about 0.8, which is about 1.0 μm in resolution (Fig. 8-2-14).[58] In order to achieve a higher optical MTF, deep-UV exposure tools using reflective optics have been engineered (Fig. 8-2-14) to provide an MTF of 0.6 at 1.0-μm dimensions.[6, 18] If a 0.6 optical MTF is just met for 1.0-μm images, then deep-UV resists with higher contrast ($\tau > 2$) or thin resists (< 0.4 μm) such as in multilayer resists are required for submicrometer resolution.

The translation of the MTF into actual resist[166] profiles integrates the edge slope of the beam (Fig. 8-2-15), the developer[98] interaction with the resist (contrast), and the MTF of the optical system into the differential expression for linewidth variation,[7, 19, 60, 210, 219]

$$dy/dx = (\delta y/\delta E)(\delta E/\delta x) \tag{8-2-9}$$

where E is the energy absorbed within the resist. Assuming a low-absorption resist and a low-reflectivity substrate, the first term $\delta y/\delta E$ is resist and resist processing

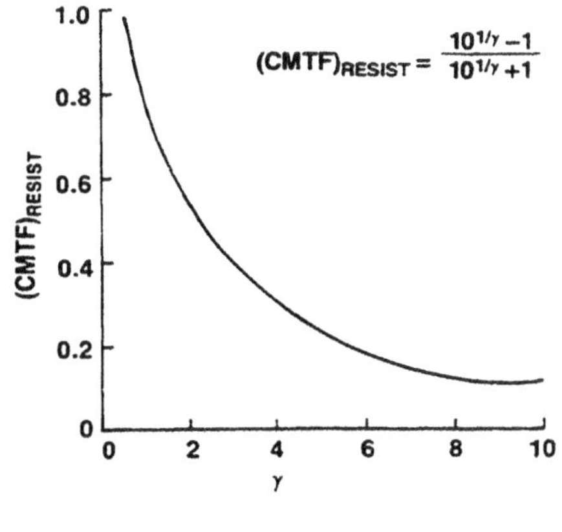

FIGURE 8-2-13. Critical MTF for resists. From Ref. 13, courtesy of AIP.

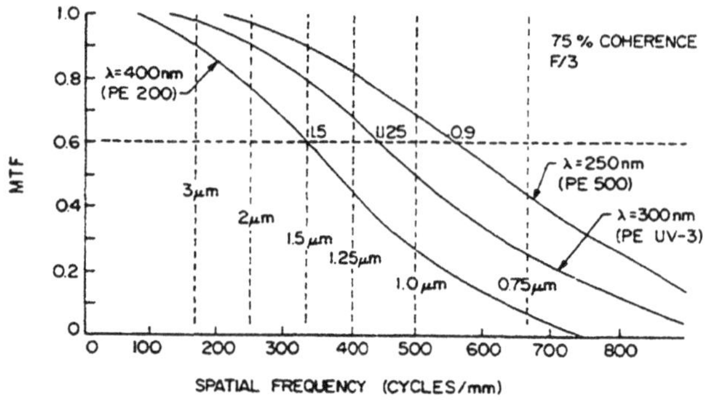

FIGURE 8-2-14. Effect of exposure wavelength on resolution for a reflective scanner. From Ref. 6, copyright and courtesy of *J. Electron. Mater.*, a publication of the Metallurgical Society of AIME, Warrendale, Pa.

dependent whereas the second term is dependent only on the optical system. The term $\delta E/\delta x$ is the intensity profile of the image and is affected by the wavelength of exposure, λ, the NA, the deviation (Δz) of the resist plane from the focal plane, and the uniformity of illumination,

$$\delta E/\delta x \approx 2\mathrm{NA}/\Gamma[1-k(\Delta z(\mathrm{NA})^2/\Gamma)]^2 \tag{8-2-10}$$

The parameter $k=1$ and varies slightly for different degrees of source partial coherence. Next, a positive resist is described by its contrast ratio,

$$\tau=[\log_{10}(E_0/E_1)]^{-1} \tag{8-2-11}$$

where E_1 is the exposure energy below which there is no resist removal in the developer and E_0 is the exposure energy at which no resist remains after development. In general, E_1 is independent of the thickness, t, whereas E_0 will vary approximately with the reciprocal absorbance $(-E=10^{\alpha t})$. Under these assumptions,

$$\tau=(b+\alpha t)^{-1}=2.303/\alpha \tag{8-2-12}$$

where β is a constant and α is the resist absorption. For an absorbance of 0.4, the light absorption is uniform throughout the resist film and an approximate resist contrast of 2.5 can be estimated.[213, 214] The process-dependent term then becomes

$$\delta y/\delta E=\tau/E_0 \tag{8-2-13}$$

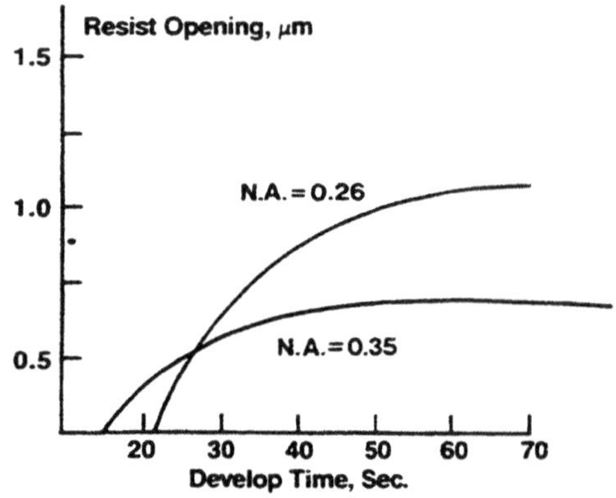

FIGURE 8-2-15. The change in linewidth of a DQN optical photoresist as a function of develop time. Exposure through a larger numerical aperture lens provides a larger process develop window. From Ref. 20, by permission of *Solid State Technology*, published by Technical Publishing, a company of Dun and Bradstreet.

TABLE 8-2-1. Exposure Changes and Process Latitude for Scanning Exposure at 3100 Å of DQN Resists[a]

Resist type	Linewidth thickness (μm)	Exposure change, %	
		ΔL for 0.1 μm	τ
AZ-2400	1	15	2.6
2500 Å	2	15	2.6
AZ-2400	1	11	2.4
5000 Å	2	11	
AZ-2400	1	7	2.1
10,000 Å	2	7	
HPR-204	1	4	1.7
10,000 Å	2	4	

[a] From Ref. 19.

which predicts the variation of the resist profile in identifiable parameters,

$$\frac{dy}{dx} \cong \frac{\text{NA}}{\lambda(\beta + \alpha t)\, E_0}\left[1 - k\left(\frac{\Delta z(\text{NA})^2}{\lambda}\right)\right]^2 \tag{8-2-14}$$

which states that a high-contrast resist (Table 8-2-1) with low absorbance will have a wider exposure latitude and tolerate a larger intensity variation in exposure lamp output. A 10% overexposure causes lines and spaces to change about 0.1 μm for a low-contrast resist. Secondly, the modeling of two lenses of different NA shows a sharper edge slope (Fig. 8-2-16) and wider development latitude for the higher NA system.[20, 211]

In the linewidth modeling and MTF derivations, we assume no contribution of the substrate to the image. Negative resists, for example, do not exhibit a gray scale and may actually wash off the resist in the underexposed condition. By the nature of the depthwise insolubilization, negative resists are always overexposed in order to obtain sufficient adhesion to the substrate. In general, the resist–substrate interaction can contribute significantly to image exposure and linewidth variations by reflection of destructive and constructive (interference) waves (Fig. 8-2-17). If a node occurs at a substrate surface, the resist may not be exposed leading to residues in positive photoresist images and delamination of negative resists. The most severe

FIGURE 8-2-16. Edge slope by cross sections of 1-μm optical beams and electron beam. A steeper edge slope in the case of optical beams is obtained with a larger numerical aperture lens. From Ref. 20, by permission of *Solid State Technology*, published by Technical Publishing, a company of Dun and Bradstreet.

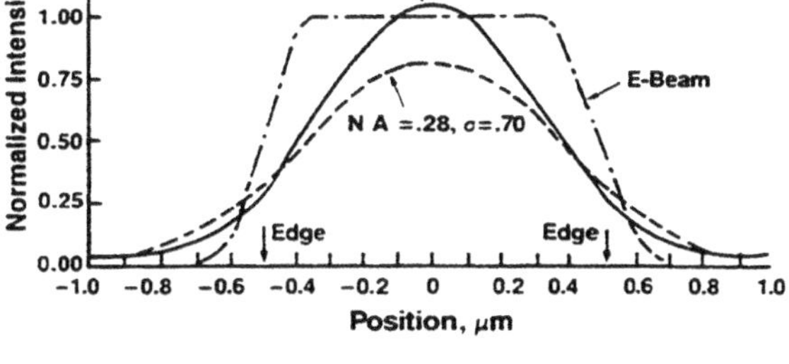

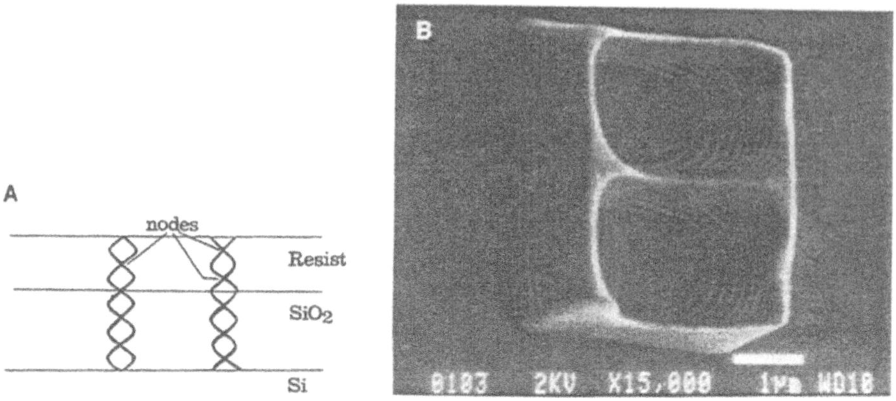

FIGURE 8-2-17. (A) Schematic of origin of standing waves in resist. From Ref. 87, copyright 1974 and courtesy of Benn Electronics Publ., Ltd. (B) SEM of standing waves recorded by resist. Courtesy of Jeol Inc., Medford, Mass.

problem occurs in resist images: overstepped features (Fig. 8-2-18). Due to the different refractive layers, interference occurs between the layers of resist, oxide, and other underlayers. Standing waves (Fig. 8-2-19) are present at $2N\lambda/2n$ and are replicated into single-resist films, producing edge distortion (Fig. 8-2-20) and halos around the periphery of patterns.

The difficulties associated with standing waves are as follows:

1. Small resist thickness variations (± 200 Å) can alter the intensity (I_{max}/I_{min}) ratio of the standing wave to a value of 2.5 for perfect reflection. The resist can be overexposed by about half of this factor depending on its absorptivity and bleaching rate.[13, 23]
2. Variations in oxide underlayer thickness of ± 300 Å can cause the energy deposited in a resist to vary by 50% and the linewidth to change by 0.5 μm for a resist with a τ of 1.2.[21]
3. Standing wave effects coupled with a step limit the resolution more severely. The linewidth variation ΔW,

$$\Delta W = 2h\left(\frac{1}{\tau t} + \alpha\right)\frac{I\lambda}{\mathrm{NA}f} \tag{8-2-15}$$

($f = 1.8$ for a degree of coherence of σ of 0.7[22]) over a step (h) increases with thin films (t), low-contrast (τ) resist, and high absorption, α. The right-hand factor reflects the influence of the optical characteristics of the projection system on linewidth variation; they increase with higher λ and lower NA.

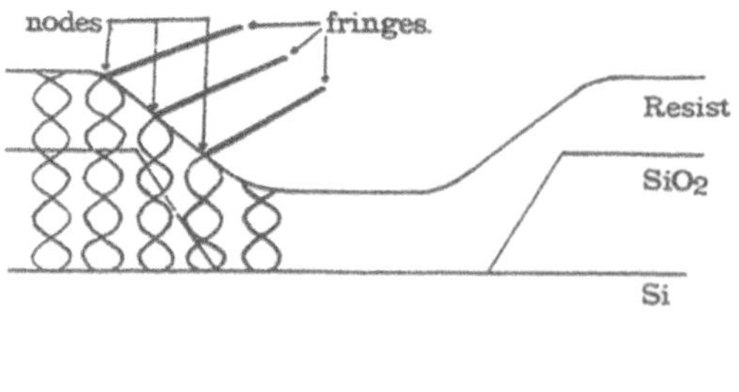

FIGURE 8-2-18. Standing wave patterns in resist coated over oxide steps causing edge fringes in resist image. From Ref. 87, copyright and courtesy of Benn Electronics Publ., Ltd.

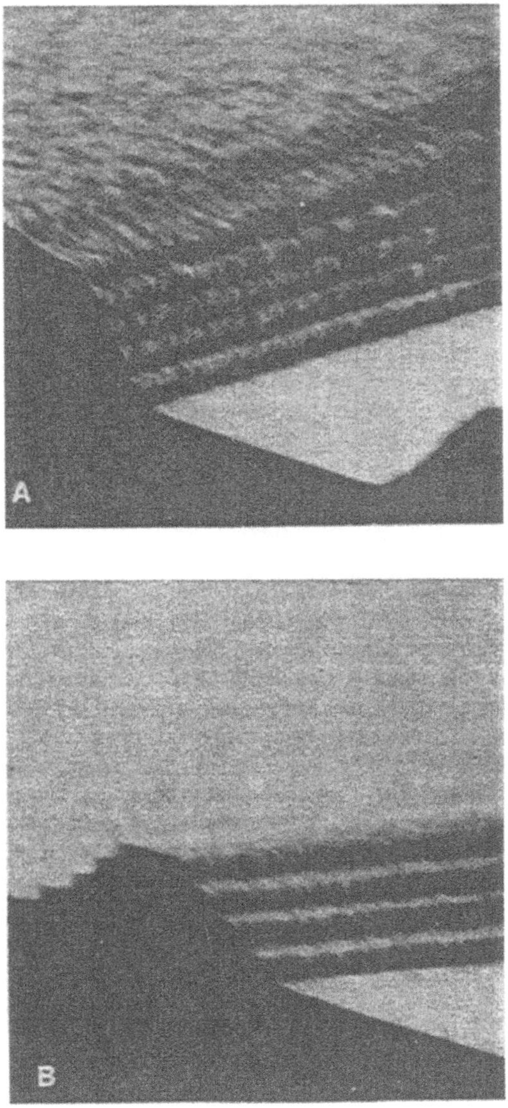

FIGURE 8-2-19. Standing wave patterns for a line and land in DQN resist. (A) Edge of positive DQN resist; (B) edges of unexposed land in positive DQN. Both ×20,000. Courtesy of M. Narisinham, IBM.

The last major (but not least) optical problems are the diffraction at the limit of a lens—and diffraction from the edge of a mask, in all forms of printing.

In Fresnel and Fraunhofer[24] type diffraction (Fig. 8-2-21), the projected forward intensity loss of a single aperture produces the interproximity effect at 2λ and with overexposure (Fig. 8-2-22) of unexposed spaces for closely pitched lines. If the mask is separated from the wafer, the diffraction effect acts to defocus the image from an anisotropic beam into an isotropic beam of twice the aperture size (Fig. 8-2-23). The edge slope of the projected pattern falls off severely at gaps $> 3\lambda$ of exposure. When a mask containing closely spaced slits is separated from a resist by even 40 nm, nodes occur in the forward intensity (Fig. 8-2-24) and the beams widen and show second-order interference pattern as the gap approaches 2 μm in

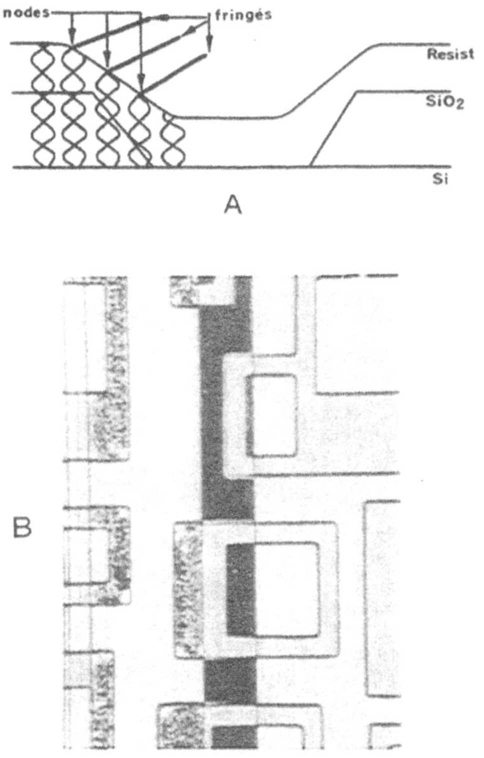

FIGURE 8-2-20. (A) Standing waves in DQN resist over oxide giving rise to unexposed resist. (B) Halos at edges of exposed resist. From Ref. 100, copyright 1976 and reprinted with permission of *Solid State Technology*, published by Technical Publishing, a company of Dun and Bradstreet.

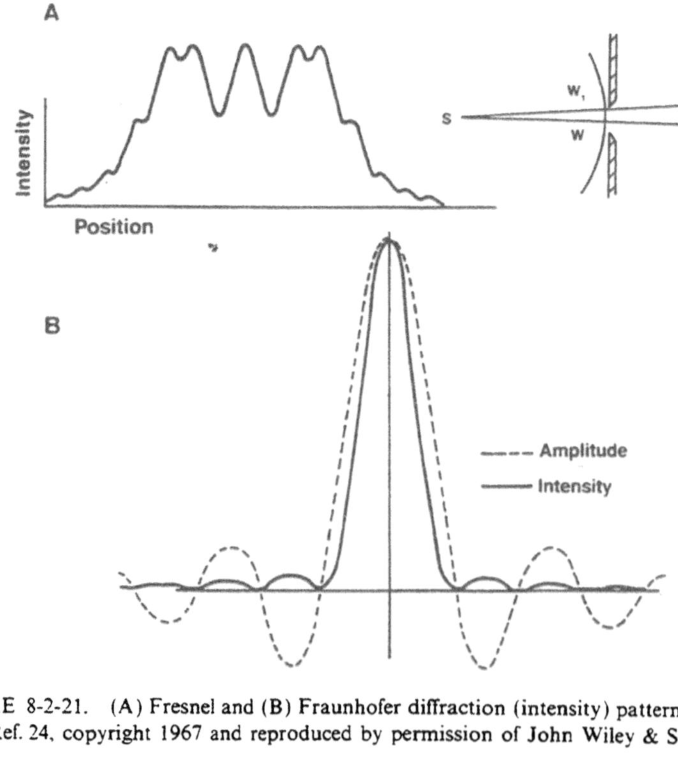

FIGURE 8-2-21. (A) Fresnel and (B) Fraunhofer diffraction (intensity) pattern from a single aperture. From Ref. 24, copyright 1967 and reproduced by permission of John Wiley & Sons, Inc.

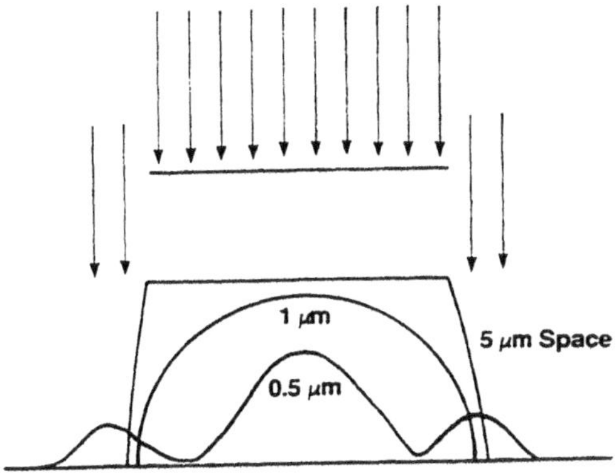

FIGURE 8-2-22. Interproximity effect in UV exposure. Underexposed space is exposed by diffracted light as space width approaches λ. Reproduced by permission of *Semiconductor International*, Cahners Publishing Co.

width.[25] As the image size approaches within 2λ, even on a nonreflecting substrate, for an NA in the 0.2–0.4 range, linewidth variations due solely to the intensity profile (Fig. 8-2-25) at each feature edge can contribute a ± 0.2-μm image bias for $\pm 10\%$ variation of exposure in a 1-μm-thick resist.[26]

In addition, due to diffraction effects, lines smaller than 2 μm do not develop out in positive photoresists at the same time as larger features. The smallest lines must be overexposed or overdeveloped since the mask cannot compensate for diffraction losses unless contrast-enhanced layers are present on the mask[27] or on top of the resist.[28] Overdevelopment or overexposure invariably causes image widening, poor edge slope, and loss of the unexposed space in a positive resist (Fig. 8-2-26) or a closed gap in a negative resist. These types of images would be unacceptable for subsequent anisotropic dry etching.

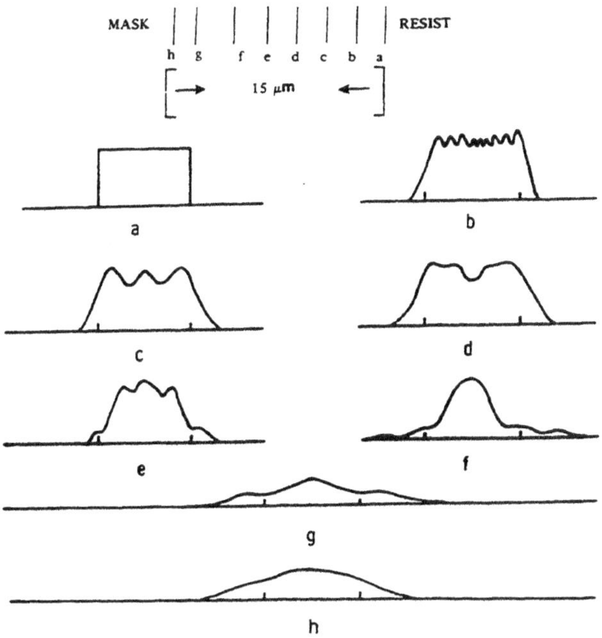

FIGURE 8-2-23. Intensity distribution of light falling on a resist after passing through a mask in perfect contact with a wafer and separated by various gaps. From Ref. 151.

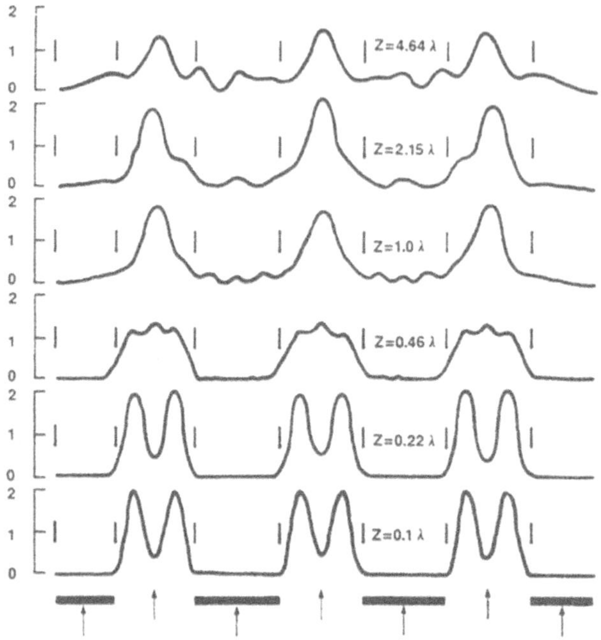

FIGURE 8-2-24. Plots of light intensity after passing through three slits of width 2λ. The resist is Z units away from the slit. From Ref. 25, copyright and courtesy of IEEE.

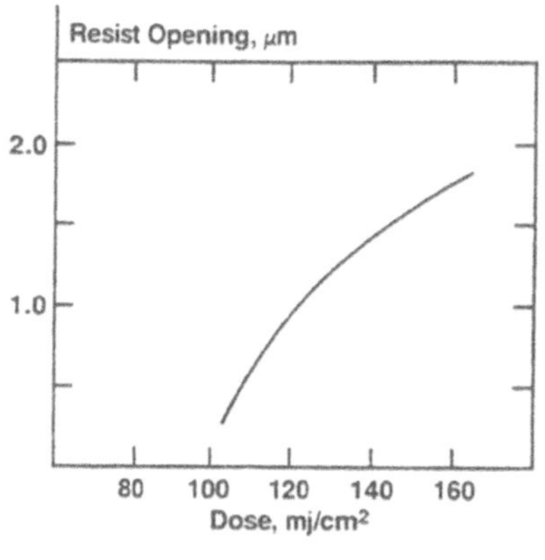

FIGURE 8-2-25. Resist opening versus exposure dose for 1-μm exposed line at numerical aperture of 0.28 and exposure wavelength of 436 nm for a DQN resist. From Ref. 26.

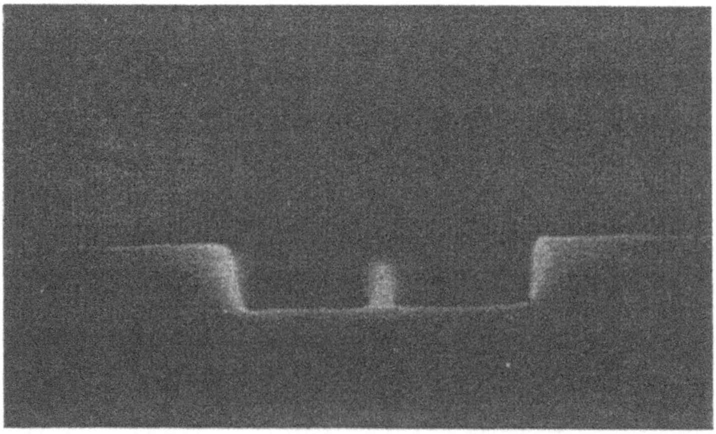

FIGURE 8-2-26. Loss of unexposed space in a positive resist printed by optical projection. Courtesy of G. Hefferon, IBM.

8-3. CONTACT AND PROXIMITY PRINT

By physically contacting the mask and wafer or by providing *in situ* molecular deposited masks on top of a resist, infinite resolution is possible. Dimensions as small as 10 Å have been contact printed in a photoresist.[41, 192] In practice, however, molecular contact is rarely achieved and damage to the mask during alignment and printing occurs after ten or more passes. To extend mask life, top-coatings and lubricants placed between the resist and mask have lengthened the useful life to about 500 passes. The movement of both the mask and the wafer during alignment causes operator errors and restricts alignment to about ± 1 μm. In the early decades of lithography, contact printing was the workhorse method of producing 3- to 10-μm images. Since for wet etching purposes the width of the resist image and not the profile was important, a print bias of ± 1 μm could be tolerated by the $\pm$ 1-μm wet etch bias.

For contact or near-contact printing, a collimated coherent (offset a few degrees to reduce diffraction) source is used. The exposure tool (Fig. 8-3-1) consists of a high-pressure mercury lamp, a high-efficiency ellipsoidal reflector, an infrared filter, a solenoid-operated shutter, an optical integrator (fly's eye), a collimation aperture plate, a turning mirror, and a collimating lens mounted in an air-cooled housing.[29]

The MTF of contact printing is very high (> 0.8) and images down to 0.1 μm in size can be printed with a perfect contact conformable mask[25] or bilayer resists.[30] If short deep-UV radiation is used, 1.0-μm images can be proximity

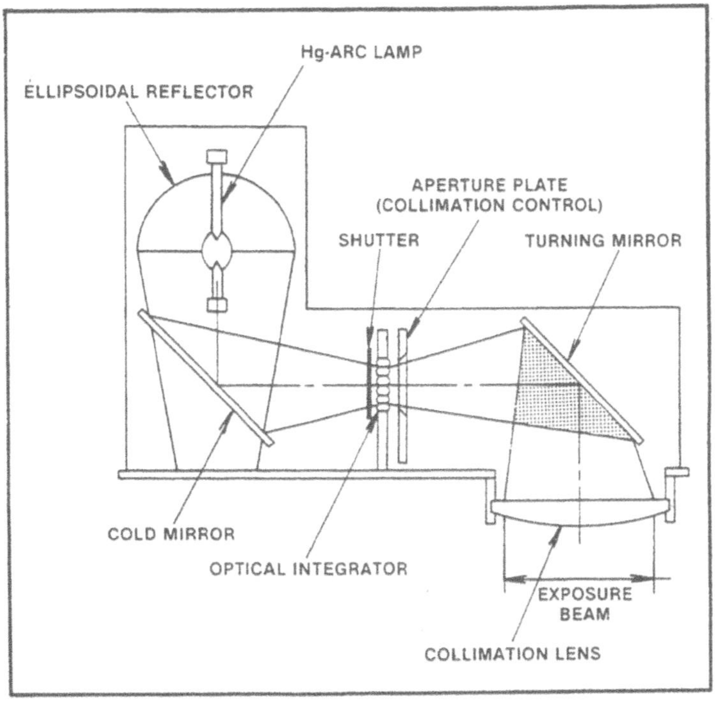

FIGURE 8-3-1. Refraction-collimated exposure source for contract printer. From R. Sheets, *Microelectron. Manuf.*, Oct. 1982, p. 19, reprinted by permission of Lake Publishing Co., Libertyville, Ill. 60048.

printed in PMMA as shown by Moreau and Schmidt in 1970.[31] As the gap (Z) between the mask and the wafer exceeds the Fresnel limit ($\pm 5\%$ intensity tolerance and a $\pm 20\%$ linewidth tolerance), the limit resolution W is about 1–2 μm for a gap of 5–10 μm,

$$W \sim \sqrt{0.7\lambda Z} \tag{8-3-1}$$

If the gap exceeds a first-order limit, the second- and third-order diffraction (Figs. 8-2-23 and 8-2-24) projected and a resist image is printed with a tapered profile (Fig. 8-3-2). Certain resist thicknesses accentuate the interference diffraction effects and edge slope distortion. Figure 8-3-3 illustrates the diffraction along the normal incidence axis for several resist thicknesses.[42] Filling the air gap with an index matching fluid improves the contact printing resolution.[225] For superior submicrometer contact printing, thin glass plates are used as conformable masks.[40] Very little pattern[41] distortion occurs (100 nm over a 2-cm^2 area). In addition, the high index of refraction of a DQN positive photoresist (1.6) and the nonlinear dissolution behavior of the resist suppress the diffraction broadening of the top surface of the images.

For deep-UV contact printing, highly collimated beams from excimer lasers operating at 157 nm[33] and at 193–250 nm[34] have been used. No reciprocity intensity effects were noted for nanosecond pulsed exposure of DQ resists.[34] Because of the low power output (microwatts/cm^2), the previous use of continuous-wave lasers as scanning spots or for exposure by nonlinear wavefronts[35] required long exposure times. Excimer lasers, in contrast to most other lasers, have the important property that their beams lack good spatial coherence due to the highly multimode nature of the laser emission. The poor spatial coherence eliminates speckle[11] spots, a problem that has hindered the use of lasers in resist holography.[34] Speckling is not a property of the laser beam itself but is present when phase variations are

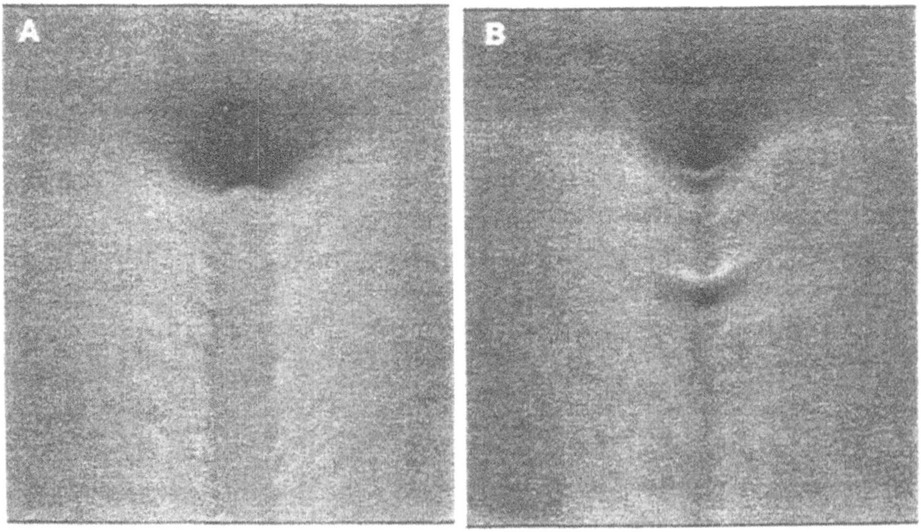

FIGURE 8-3-2. Proximity printing with (A) 1.5-μm line and (B) 1.0-μm line at a gap exceeding the third-order diffraction limit at 436 nm.

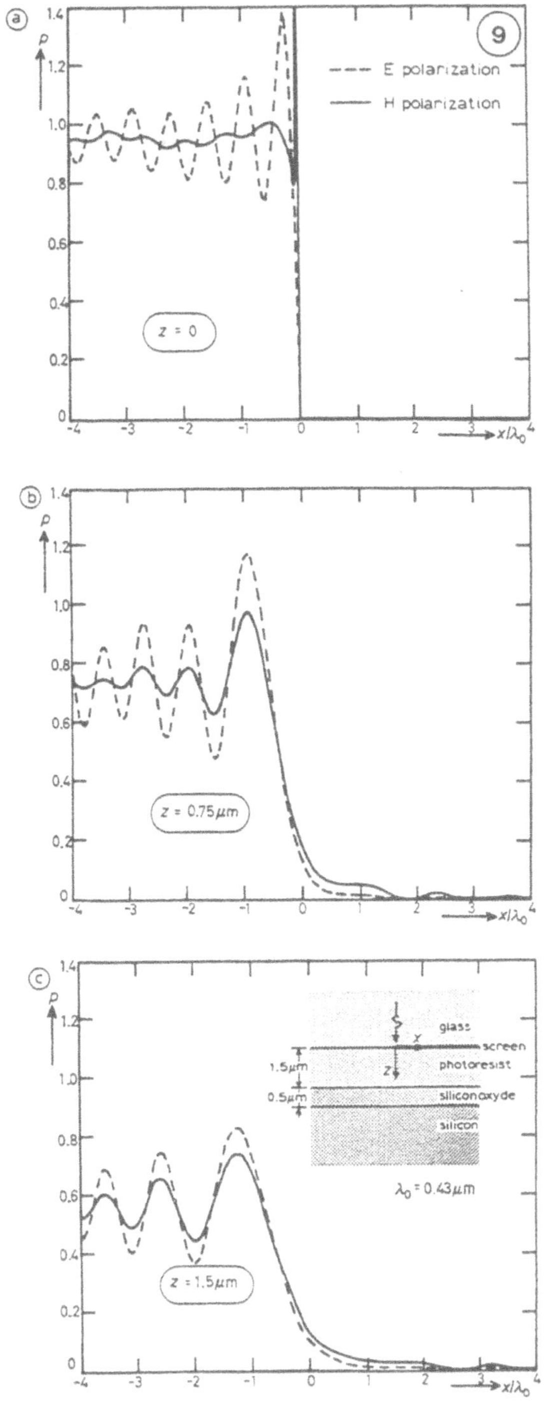

FIGURE 8-3-3. Diffraction by thin half plane placed between the glass of a mask and a photoresist on top of a SiO_2/Si substrate. Simulated results for proximity printing are depicted as a mark that is not in contact with the wafer. The solid line represents the normalized field intensity and the dotted line represents the normalized power flow density as the gap is varied from 0.05 μm to 1.55 μm in cases a–c. From W. Hestman and P. Van Den Berg, *Can. J. Phys.* **53**, 1310 (1975), courtesy of *Can. J. Phys.*

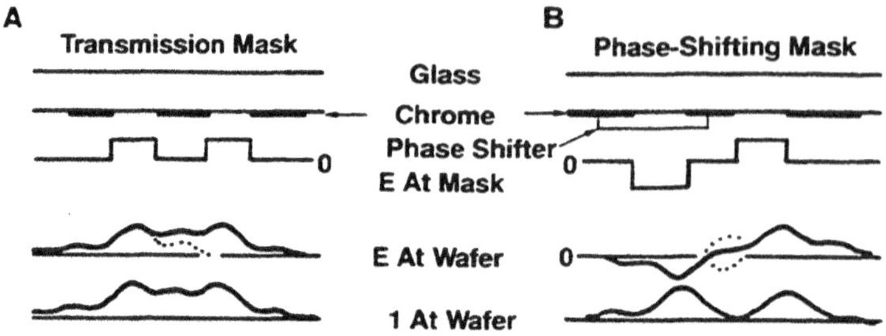

FIGURE 8-3-4. Comparison of diffraction of (A) an ordinary transmission mask with (B) a phase-shifting mask. From M. Levenson *et al.*, *IEEE Trans. Electron. Devices* **ED-29**, 1828 (1982), copyright and courtesy of IEEE.

introduced into wavefronts by reflection from a surface.(11) With excimer lasers, contact printing at 200–300 nm produces excellent-quality speckle-free images with 0.5-μm resolution.(34, 196) With a monochromatic excimer source, one also expects standing waves. Only a trace was found possibly due to interference node cancellation at the oxide interface or due to resist heating during exposure which contributes to a reduction in standing wave effects.(36) Ehrlich and Tsao(38) also noted with a pulsed laser that heating effects lowered the required dose from 1000 mJ/cm^2 to 8 mJ/cm^2 for Ge-Se inorganic resist. The temperature rose to 150°C. Lasers can also be used to dry develop ablate resist by heating a thin layer at a time.(38) The high extinction (10^5) coefficient of the polymeric films confines the radiation to a top surface 100-nm layer for outstanding contrast and submicrometer resolution.(39, 153–155)

Closely spaced features in contact or proximity printing are spread by constructive interference between waves diffracted by adjacent apertures (Fig. 8-3-4A).(43) If adjacent apertures are 180° out of phase with each other by imposition of a resist layer (Fig. 8-3-4B), destructive interference occurs at a thickness (t),

$$t = \lambda(2n - 1) \tag{8-3-2}$$

which minimizes diffraction and allows for a twofold larger gap.

The use of shorter wavelengths in contact or proximity printing can tolerate a larger gap. The maximum separation (or dirt particle between mask and resist)

TABLE 8-3-1. Maximum Gap in Proximity Printing at Various Wavelengths

W (μm)	Spectral	Maximum gap (μm)
2.5	Deep UV	3.8
	Near UV	2.4
2.0	Deep UV	2.3
	Near UV	1.4
1.0	Deep UV	0.9
	Near UV	0.3
0.5	Deep UV	0.26
	Near UV	0.20

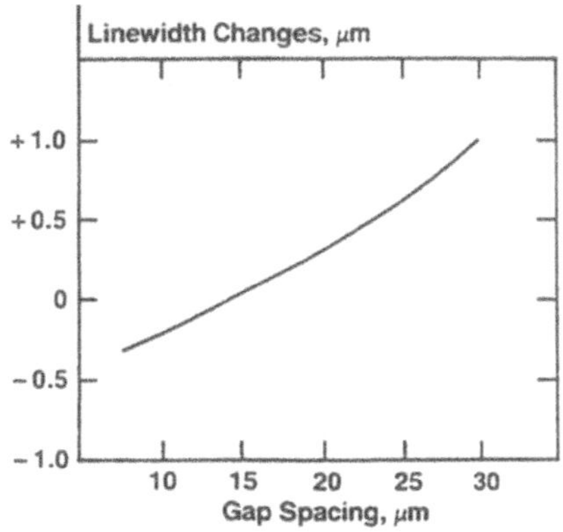

FIGURE 8-3-5. Proximity printing an image in PMMA as a function of gap width. From Ref. 44, copyright and courtesy of *RCA Review*.

which can be tolerated for less than ±10% image (W) distortion has been calculated by Lin[32] (Table 8-3-1).

PMMA has a very low extinction coefficient and absorbs only 10% of the light at 254 nm. The higher diffraction orders are not readily reproduced and larger gaps of 10–20 μm can be used with a ±0.2-μm deviation in a 1.5-μm line (Fig. 8-3-5).[44] Other resists such as AZ-2400, a DQ resist, required tighter (depth of focus) gap control and a smaller spacing.[44] Because of multiple diffractions, the gap setting is

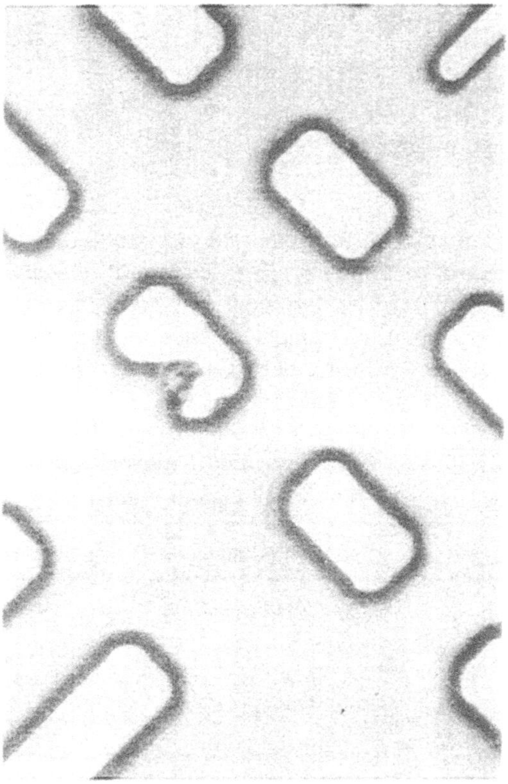

FIGURE 8-3-6. Rounding of rectangular images in proximity printing at a gap of 20 μm.

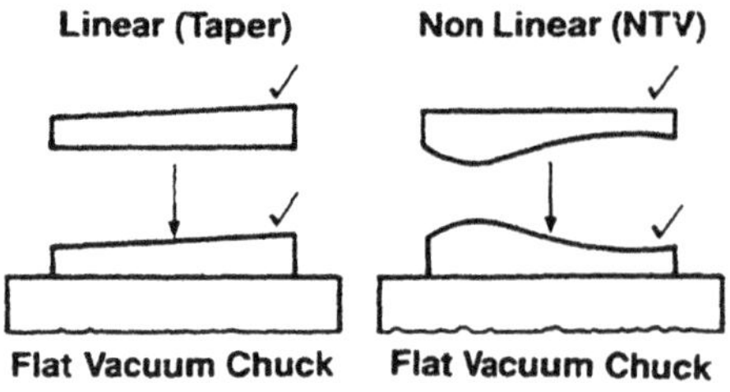

FIGURE 8-3-7. Different types of wafer flatness variation. From Ref. 55, reprinted with permission of *Solid State Technology*, published by Technical Publishing, a company of Dun and Bradstreet.

not the same for all shapes. Circular holes[45] reproduce easier than square features which tend to round corners due to interproximity diffraction (Fig. 8-3-6). Pinholes in a photomask do not print since they are sufficiently diffracted. In proximity printing, negative resists suffer since the forward intensity is lowered and the higher orders are printed. Positive resists with high τ and a long induction effect are required to reduce the printing of the diffracted orders at the top of the resist image.

Practical proximity printing is a compromise of resolution versus the cost of replacing masks (as in contact printing). Precise gap setting, alignment, and good collimation are required. A viewing microscope (for mask-to-wafer alignment) of at least 10-mm depth of focus is necessary for alignment to ± 1 μm. Precision gap tools are able to set a space within ± 0.1 μm.[46] Because of the significant amount of research in X-ray printing, alignment schemes have been improved and these new methods can be reapplied to deep-UV proximity printing. These techniques use the interference properties of light (λ/n) to align to ± 0.1 μm in proximity printing:

- Moire fringes[47, 175]
- Interference[48]
- Diffraction from gratings[49, 50]
- Zone plates[51]

For practical contact printing, the major problems are concerned with the poor mask-to-wafer contact and the accumulation of defects. Poor contact between the mask and wafer can be caused by linear[55] and nonlinear wafer distortions from processing (Fig. 8-3-7) epitaxial spikes, dirt particles, and edge lips from spin coating.[56] During contact printing with DQ resists, the inadvertent separation of

TABLE 8-3-2. Advantages and Disadventages of Contact and Proximity Printing

	Contact printing	Proximity printing
Advantages	High resolution, >0.1 μm	>1.0 μm Large pattern area High throughput Alignment to <0.5 μm
Disadvantages	Imperfect contacts Mask damage Printable defects Difficult alignment	Small depth of focus Wafer distortion factor Precise collimation

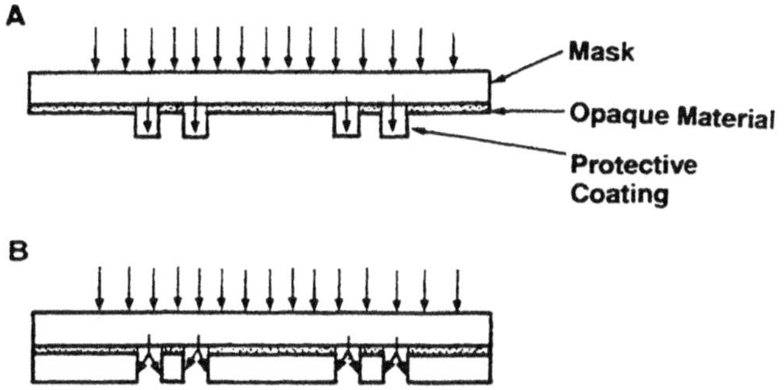

FIGURE 8-3-8. (A) Photoresist applied to photomask to serve the function of a protective layer and as an optical waveguide during proximity–contact printing. (B) The gap is maintained by a photoresist applied and exposed through the topside of the photomask. From Ref. 52.

mask from wafer can occur due to the pressure "popping" from the N_2 released by the DQ irradiation,[156]

As mask with channels to release the gas and to guide the light beam and reduce diffraction (high refractive index) can be fabricated by coating the chrome face with a resist followed by reverse exposure (Fig. 8-3-8).[52] The resist can also be used as a "pseudopellicle" to protect the mask during contact printing.

Protective strippable polymer coatings on chrome masks have found applications as a preservative and as a renewable surface to reduce[145] the number of opaque particle defects and to lower the amount of photoresist pickup.[53] Bleachable or photochromic dyes which block out large features in order for diffracted lines to be printed have been applied (Fig. 8-3-9) to masks[177] as contrast enhancement layers.[54, 177]

To sum up, the pros and cons of contact and proximity printing are reviewed in Table 8-3-2. The most useful forms of contact printing are the portable conformable mask made of imaged AZ-1350 on top of PMMA (the PMMA sublayer is exposed with deep UV; Chapter 12)[40] and the bilayer mask[30, 57] (also see Chapter 12). A suitable dry etch-resistant deep-UV resist of sensitivity $< 50\ \mathrm{mJ/cm^2}$ and high contrast ($\tau > 3$) is required for the survival of deep-UV proximity printing of 1- to 2-μm geometries.

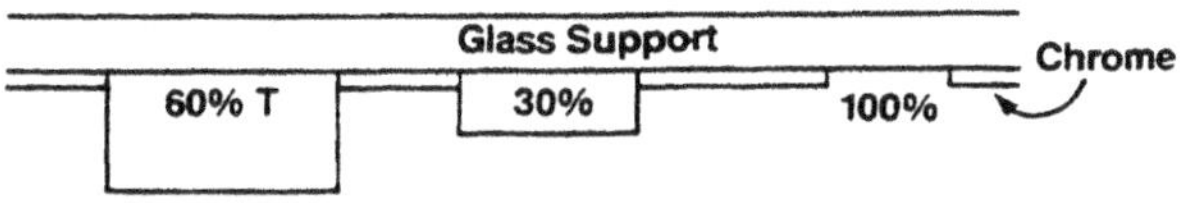

FIGURE 8-3-9. Variable transmission photomask with lower transmission in the larger features to overcome the intraproximity effect. The smaller features can be overexposed to size while the feature sizes of the larger widths are maintained. U.S. Patent 4,374,991 (1981).

8-4. PROJECTION PRINT

Scanning systems and stepping aligners use reflective and refractive optics (Figs. 8-1-17 and 8-1-18). With the scanning technique, the wafer is exposed in a succession of scans, and there is no opportunity to realign to local targets in midwater.[17, 19, 63] While the first-generation steppers used fixed steps with no reference to local targets in the middle of the wafer, modern steppers have the ability to align to each field and make adjustments to X, Y, rotation, focus, and tilt.[61, 62] In addition, scanners can readjust the focus on the fly. Both steppers[65–68] and scanners are driven by alignment accuracy, throughput, and resolution. A figure of merit has been suggested by Allsop[4] to compare the two types of exposure (Table 8-4-1). The figure of merit, M, is,

$$M = TY \times 10^6 / CL^2 \qquad (8\text{-}4\text{-}1)$$

where T is throughput in wafers per hour, Y is yield (see below), C is cost in thousands of dollars, and L, the achievable linewidth in μm, is given by $4\times$ the 3σ alignment. The yield is given by

$$Y = \exp(-ADN) \qquad (8\text{-}4\text{-}2)$$

where A is mask area (in.^{2}), D is defect density (defects/in.^{2}), and N is number of masking levels. On this basis, the following systems were rated for exposing a 125-mm wafer[4]:

Printing	Merit
Contact	0
Proximity	1
1:1 scanner	1
1:1 stepper	17
10:1 stepper	25

TABLE 8-4-1. Comparison of Reflective Scanner and Refractive Stepper[a]

Performance	Stepper	Scanner
Reduction	10:1, 5:1, 1:1	1:1
Resolution (μm)	>0.7	>1.0
NA	0.2–0.4	0.2–0.3
λ (nm)	365, 405, 435	Polychromatic
Aberration	Yes	No
Field size	5–20 mm square	100–150 mm wafer
Depth of focus (μm)	3–10	5–10
Align, λ (nm)	436, 546, 633	436, 546
Align, ±μm (3σ)	0.3–1.0	0.3–1.0
Resist sensitivity factor	Low	75% throughput
Mask defects[b]	Not too sensitive	Sensitive
Throughput, 100 mm/hr	15–40	30–60

[a] From Ref. 4.
[b] A 5:1 or 1:1 stepper will print 6–10× (more steps) more defects because of smaller field size than a 10:1 stepper.

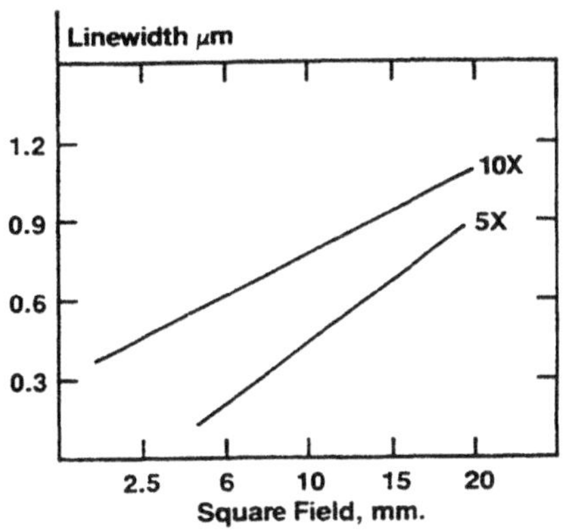

FIGURE 8-4-1. Limiting resolution versus field size for 10× and 5× lenses. From Ref. 59, courtesy of AIP.

Steppers offer higher alignment accuracy, use cheaper masks and the present positive photoresist exposing at 365–435 nm. However, they are lower in throughput than optical scanners. The major advantages of a 10:1 stepper are the reduction of reticle inaccuracies to unimportance and higher resolution than a 5:1 reduction lens.[172] On the other hand, a 5:1 line sacrifices resolution (Fig. 8-4-1) to obtain a substantially larger image field.[62, 62]

If a small chip can be tolerated, it is possible to projection print images as small as 0.1 μm in resist. Features that small were projection printed over a 7-mm field by Hennings and Schuetze[70] as early as 1966. For achieving high resolution[70] and throughput, an exposure tool can be expected to combine the best features of a scanner and a stepper.[230] A combination of a scanner with deep-UV optics and a 4× stepper with higher field overlay accuracy is contemplated for submicrometer lithography.[59, 171, 230] Assuming equal performance of a stepper and scanner at 1.5-μm resolution, the throughput (T) is dependent on[81]

$$T = [1/(1 + t_{\text{setup}})]\{3600/[t_{\text{OH}} + N(t_{\text{align}} + t_{\text{step}} + t_{\text{expose}} + t_{\text{prealign}})]\} \quad (8\text{-}4\text{-}3)$$

The overhead (OH) is composed of

- Expose
- Align
- Step
- Setup
- Prebleach at 435 nm

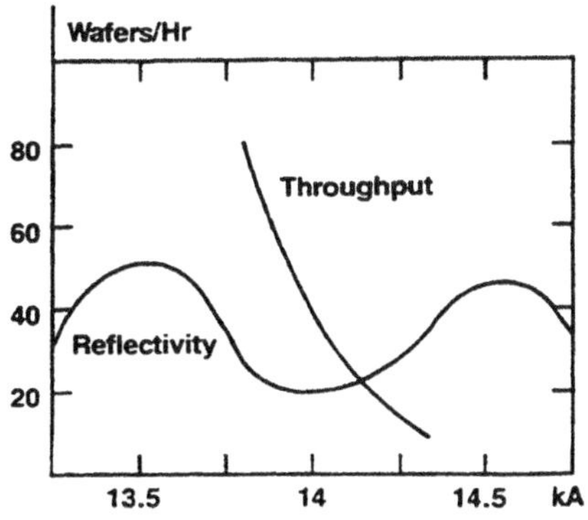

FIGURE 8-4-2. Effect of thickness of resist on oxide on reflectivity and throughput of a stepper. From Ref. 73, copyright 1983 and courtesy of Cahners Publishing Co.

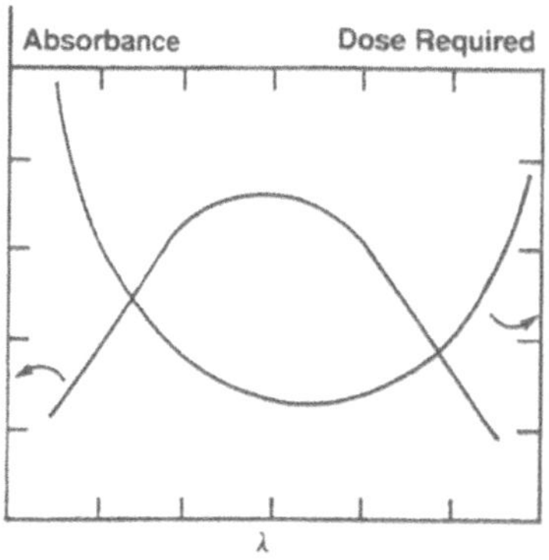

FIGURE 8-4-3. Actinic response of a resist for incident dose with a constant-intensity λ source.

The field of exposure (NA) determines the number of steps per size of wafer. The expose time is dependent on

- Thickness of resist
- Output λ of lamp
- Absorptivity of resist
- Step heights
- Reflectivity of substrate
- Presence of contrast enhance layer
- Intensity of source

In steppers, the move and align time dominates the throughput. If no reciprocity is present, higher-intensity mercury lamps or lasers for flash on the fly will reduce the exposure dependency to less than the align and move time (< 20 sec per 125-mm wafer).[171] The intensity of most mercury lamps is on the order of 300–500 mW/cm^2 when focused to a 10-mm square. For present resists, which require about 25–100 mJ/cm^2, the exposure time will be about 300 msec per 10- to 20-mm square field of steppers. Nonuniformities of resist thickness of $\pm$ 500 Å can

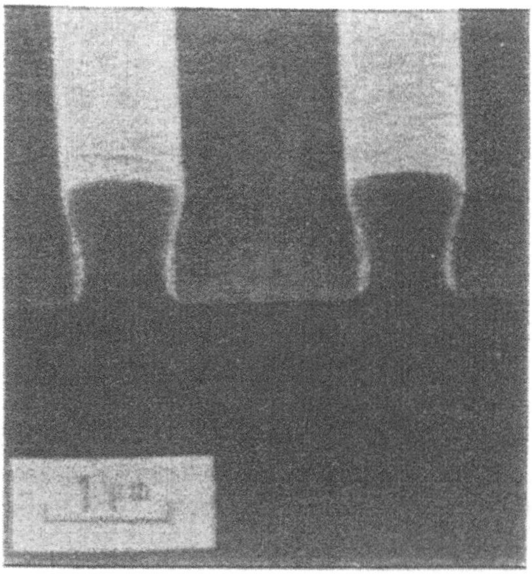

FIGURE 8-4-4. Image produced in azide phenolic resist with high absorbance in top of film. From T. Hashimoto, S. Nonogaki, S. Koibuchi, and D. Makino, *Polym. Eng. Sci.* **23**, 935 (1983), courtesy of SPE.

TABLE 8-4-2. Photoresist Thickness Factor on Throughput of Step and Repeat Aligner[a]

Film thickness (Å)	Wafers/hr
15,350	57
15,050	64
14,850	72
14,800	68
14,750	10
14,650	10

[a] From Ref. 73.

induce sufficient reflectivity (Fig. 8-4-2) so as to cause a stepper aligner which senses reflectivity to vary in throughput by 6× (Table 8-4-2).(73)

Other throughput factors are the thickness and absorbance of the resist. Since the absorbed dose is not measured, the exposure time increases with thicker films(226) or low-reflectivity substrates (Table 8-4-3). The absorptivity of the resist (Section 8-3) plays the important role of determining the resolution and speed of a resist. The response of a resist is called the actinic response and the exposure time is a minimum at the spectral absorption maximum for a polychromatic isointensity source (Fig. 8-4-3). The resist can absorb too strongly or too weakly. In a bleaching resist, high absorbance leads to overcut profiles (Fig. 8-4-4) which are useful for lift-off or top surface imaging. Too low an absorbance (< 0.4)(213, 214) leads to top surface thinning and interproximity effects(22, 74) (Fig. 8-4-5). An absorbance of about 0.4 is ideal(75) for image control in positive or negative resists. For azide or DQN resists(76, 80) exposed in the deep UV, the high absorbance of the novolak resin leads to overcut profiles (inner filter effect). Studies of addition of dyes to DQ resists to prevent the light scattering from grainy aluminum and reduce the reflective notching of resist over steps have been moderately successful.(92, 223, 224, 231) In the deep UV, the lack of a mercury lamp intensity at 254 nm equivalent to the 365-nm line output, requires the resist to be as sensitive as 10 mJ/cm^2 in order to maintain a similar throughput(76) as the near-UV exposure tools. By using the unfiltered exposure of the full mercury lamp, more light is absorbed by the resist and the exposure time of DQN is reduced by an order of magnitude in a scanner system.(77) On a scanner tool, careful spectral tuning to prevent unwanted novolak resin

TABLE 8-4-3. Effect of Thickness of Resist and Reflectivity of Substrate (1-μm-Thick Resist) on Exposure Time

		Exposure time (sec)
Thickness (μm)	0.5	6
	1.0	10
	1.5	14
	2.0	18
Reflectivity, 435 nm (%)	10	7
	30	5
	80	3

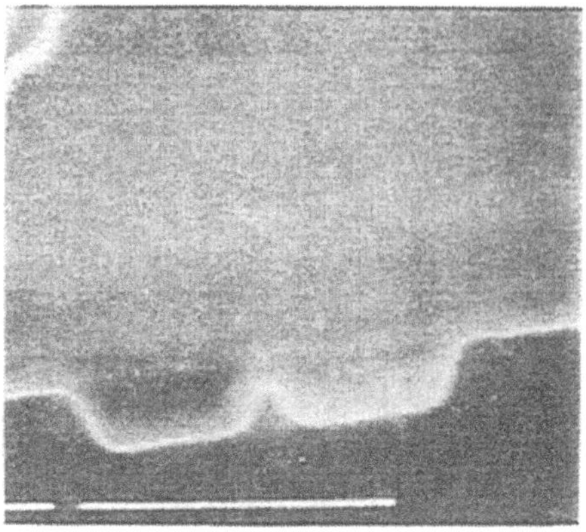

FIGURE 8-4-5. A positive resist with low absorbance (<0.25) leads to exposure and thinning of unexposed space.

absorbance can lead to better resist profiles for 300- to 330-nm mid-UV exposure.[79, 126]

Anti-halation layers on top of the resist, additives to the film, or underneath the resist are used to reduce interference waves. Any of these additives or coatings will absorb radiation and the resist will require up to twofold more exposure time (or dose) to compensate for the inner filter effect.

Diffraction and interference also modulate the intensity, exposure time, and, hence linewidth.[99] Monochromatic light reduces the Fresnel diffraction[83] and chromatic aberrations[82] but enhances standing waves (Fig. 8-4-6) Standing waves occur at quantized refractive index–wavelength nodes (Table 8-4-4). On flat surfaces, by making adjustments in the resist thickness and oxide layers,[97, 107] the interference can be reduced but not for the case of coatings over steps where the resist thickness is gradually conforming[85, 212, 231] to the substructure. Polychromatic light[83, 87] reduces the effect significantly due to destructive inter-

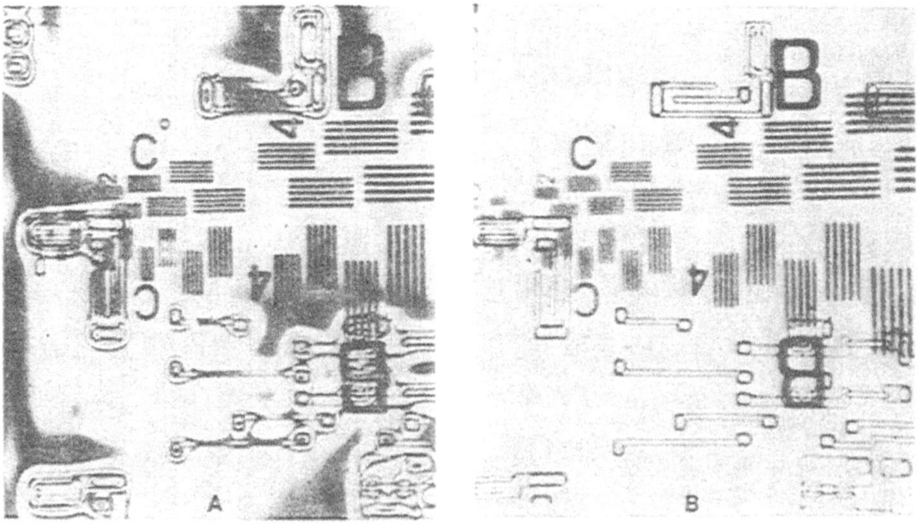

FIGURE 8-4-6. Interference halos and standing waves reproduced in resist with (A) monochromatic light and (B) polychromatic light. From Ref. 87, copyright 1974 and courtesy of Benn Electronics Publ. Ltd.

ference (Fig. 8-4-6 and 8-4-7) but leads to chromatic aberrations.[82] Interference is a result of refractive index differences between SiO_2 (1.4) and the novolak resist (1.6). Photoresists with the refractive index the same as the oxide have not been formulated but siloxane polymers with low refractive index have been used as topcoatings on resists to reduce the surface reflectivity.[220]

For proximity printing, the adjustment of the mask spacing will minimize standing waves at[194]

$$n_2 = (n_1 n_3)^{1/2} \tag{8-4-4}$$

$$h_2 = \lambda/4n_2 \tag{8-4-5}$$

where n_1, n_2, and n_3 are the refractive index of the resist, spacer, and substrate, and h_2 is the gap or thickness of the spacer. However, the distortion is even more severe with coherent optics.

Antireflection coatings on the chrome[89, 90] mask or substrate are important in reducing frustrated reflection.[88] Imperfect contact in contact printing also results in lateral image fringes caused by multiple reflections between the mask and substrate.

Three major effects influence linewidth variations over steps[110, 231]: (1) the bulk effect from the resist thickness variation in the vicinity of the steps; (2) the

TABLE 8-4-4. Variation of Haloing with Oxide Photoresist Thickness[a]

Oxide thickness d_2 (Å)	$\frac{n_2 d_2}{\lambda}$	AZ-1350 thickness d_1 (Å)	$\frac{n_1 d_1}{\lambda}$	$\frac{1}{\lambda}\sum n_1 d_1$ in 1/4	Halo	Reflectance (%)	Exposure time (sec)
1210	0.441 ~ 2/4	3200	1.19 ~ 5/4	7	No	65	7
		4000	1.49 ~ 6/4	8	Yes	44	5
		4800	1.79 ~ 7/4	9	No	49	9
		5800	2.16 ~ 9/4	10	Yes	32	9
2160	0.735 ~ 3/4	3200	1.19 ~ 5/4	8	Yes	48	5
		4000	1.49 ~ 6/4	9	No	38	8
		4800	1.79 ~ 7/4	10	Yes	36	7
		5800	2.16 ~ 9/4	12	Yes	48	6
2500	0.912 ~ 4/4	3200	1.19 ~ 5/4	9	No	53	8
		4000	1.49 ~ 6/4	10	Yes	63	6
		4800	1.79 ~ 7/4	11	No	28	9
		5800	2.16 ~ 9/4	13	—		—
3240	1.185 ~ 5/4	3200	1.19 ~ 5/4	10	Yes	40	5
		4000	1.49 ~ 6/4	11	No	33	9
		4800	1.79 ~ 7/4	12	Yes	45	6
		5800	2.16 ~ 9/4	14	Yes	54	6
4500	1.64 ~ 7/4	3200	1.19 ~ 5/4	12	Yes	51	5
		4000	1.49 ~ 6/4	13	No	16	9
		4800	1.79 ~ 7/4	14	Yes	59	6
		5800	2.16 ~ 9/4	16	Yes	52	6
5400	1.94 ~ 8/4	3200	1.19 ~ 5/4	13	No	42	6
		4000	1.49 ~ 6/4	14	Yes	67	5
		4800	1.79 ~ 7/4	15	No	32	8
		5800	2.16 ~ 9/4	17	No	17	8

[a] From Ref. 86.

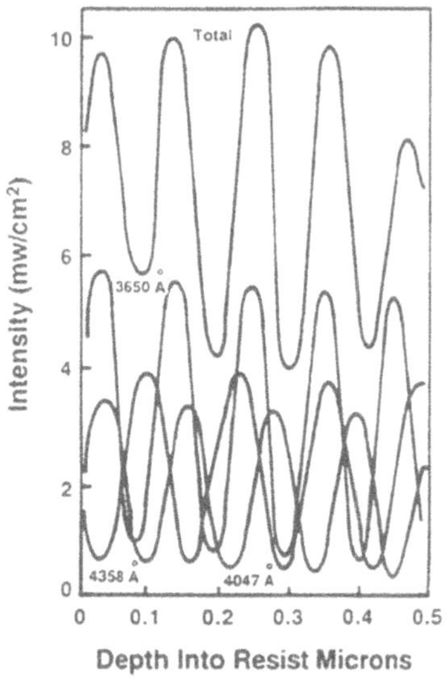

FIGURE 8-4-7. Intensities of exposing light with monochromatic light and total (365, 405, +435 nm) for 0.5 μm AZ-13500J on 0.5 μm oxide of silicon. From Ref. 83, courtesy of AIP.

interference effect from the reflective step; and (3) light scattering effects within the resist itself. These combined effects cause linewidths to vary in width[64, 91] as the line crosses over steps. Many material solutions to the interference problems have been disclosed:

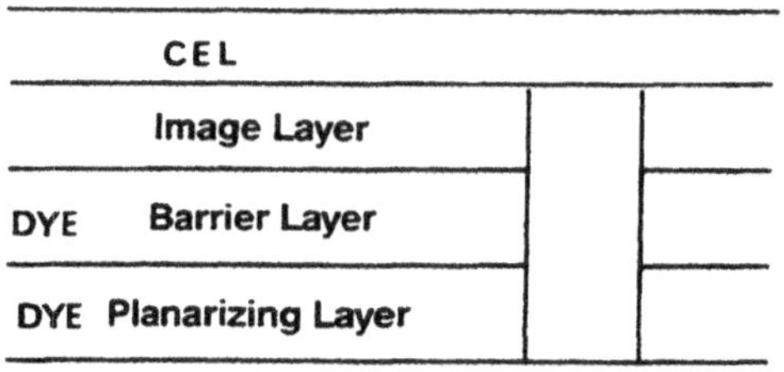

(8-4-6)

1. Coatings on top of resist (contrast-enhanced layer of azidonitrone)[28, 163, 164, 180, 185, 189, 191, 207–210, 221]
2. Dyes added to resist[92, 195, 218, 231–233]
3. Dyes incorporated in barrier layer[91, 181] (Fig. 8-4-8)
4. Dyes added to planarizing layer in multilayer resists[93, 94, 161, 162]
5. Dye layer underneath resist[95]
6. Baked DQN underlayers[96]
7. Antireflective oxide underlayers[176, 184]

With the dyed top layers, longer exposure times are required and the alignment at 436 nm can be severely distorted unless darkfield techniques are used.[61, 62, 72] Light scattering from interference effects over steps becomes the most severe at the resolution limit (MTF < 0.6) of a lens.[174, 178] Under such circumstances, the energy distribution of an optical image projected on a photoresist film is nearly sinusoidal (Fig. 8-1-2). The addition of dyes as absorbing layers reduces the light scattering within the resist or from the substrate and they also broaden the exposure process latitude (tolerate greater overexposure). Multilayer resists and optical proximity corrections[28] should also be considered. A thick resist of

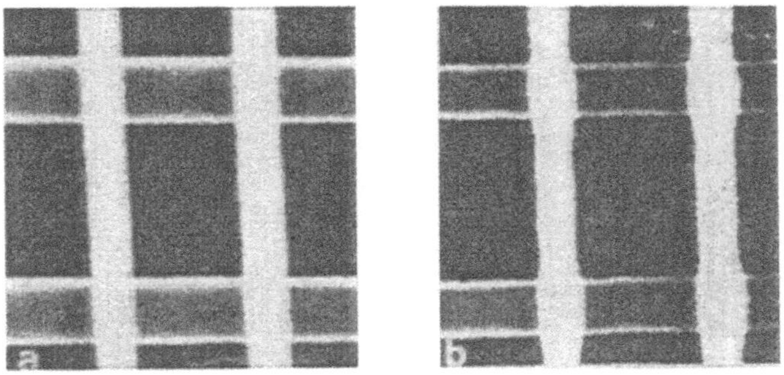

FIGURE 8-4-8. Linewidth variations of AZ-1450J over oxide step (a) with dyed underlayer in multilayer resists and (b) single film of resist. From S. Matsui and N. Endo, *Microelectron. Eng.* 1, 51 (1983), copyright and courtesy of Elsevier Science Press.

DQN[74] or a dyed DQN[195] can act as its own contrast enhancement layer. A substitute is to use a layer of a water-soluble diazonium or diazoquinone sulfonic acid in polyvinylpyrollidone (or water-soluble polymer),

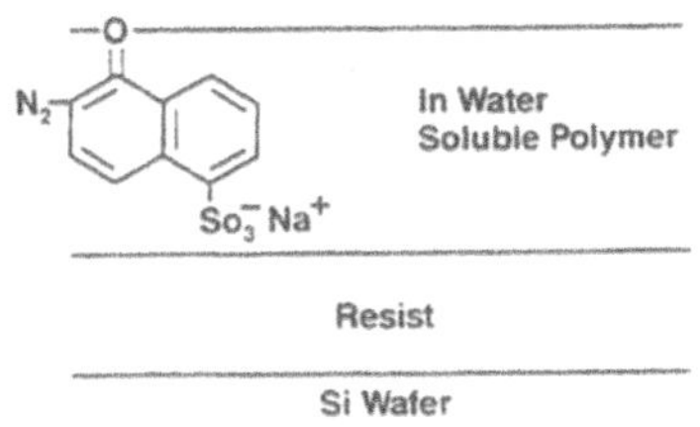

The top layer absorbs the second-order diffraction[207–210] and also refracts the incoming beam (defocuses higher diffraction orders). The top layer easily dissolves in the aqueous alkaline developer for the DQN underlayer. In the contrast layers that bleach upon exposure, the dyes absorb the second-order diffracted light emanating from the projected pattern. Finally, image reversal of a positive DQ

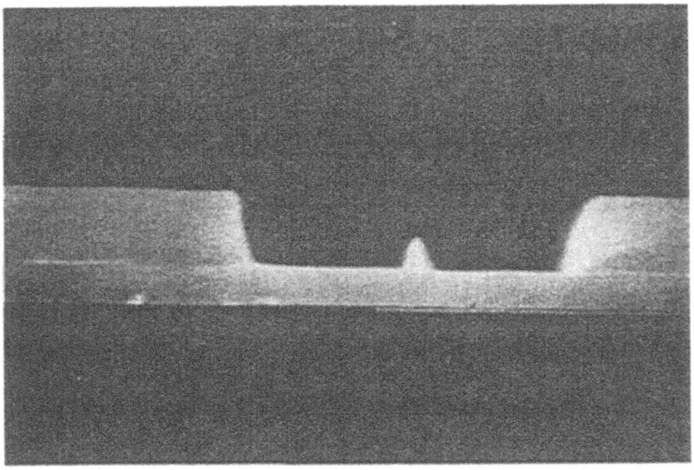

FIGURE 8-4-9. Interproximity efects of the loss of image profile in the unexposed space of a positive resist land. Courtesy of G. Hefferon, IBM.

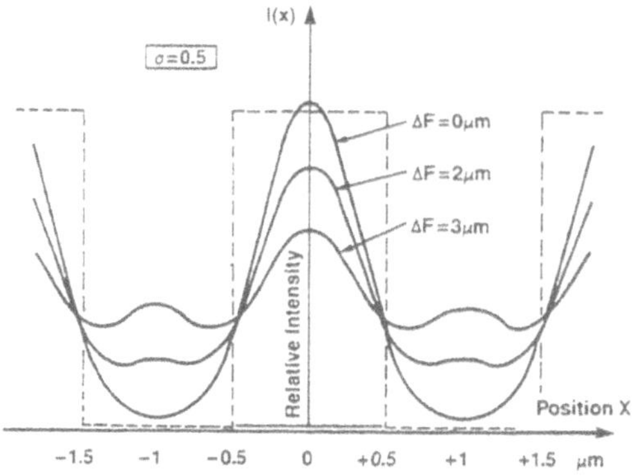

FIGURE 8-4-10. Intensity distribution changes as a function of out-of-focus distance F for an NA 0.28 lens of partial coherence. From Ref. 64, copyright 1979 and courtesy of IEEE.

resist itself is an excellent means of not reproducing the scattered and standing waves in the latent image.(215)

For a resist coated on top of metal,(167) a severe reflection problem can arise at the metal–photoresist interface where the tangential electric field is shortened by the conductive film and there is practically no exposure at all.(106, 176, 211) A technique(102) for reduction of the standing waves takes advantage of the change in polarization of reflected light. Exposure of a mask–resist sandwich at 45° incident beam angle produces a plane of incidence which is perpendicular to the plane of incidence and the standing waves are eliminated since the electric field vector of the incident and reflected waves are then orthogonal. A practical solution is to expose the resist-coated wafer at Brewster's angle which reduces the standing waves from 38 to 2. However, some features which are closely spaced are difficult to reproduce due to the interproximity(103, 105) effects (Fig. 8-4-9) and the loss of unexposed spaces. A mechanical solution to the interference problem is said to involve vibrating the wafer at the exposure frequency.(106)

For projection printing, the profile and linewidth control are the product of the exposure optics, the resist contrast, the substrate reflectivity, and the focal length of the exposure tool. Diffraction causes(103) 1.5-μm spaces to suffer the interproximity

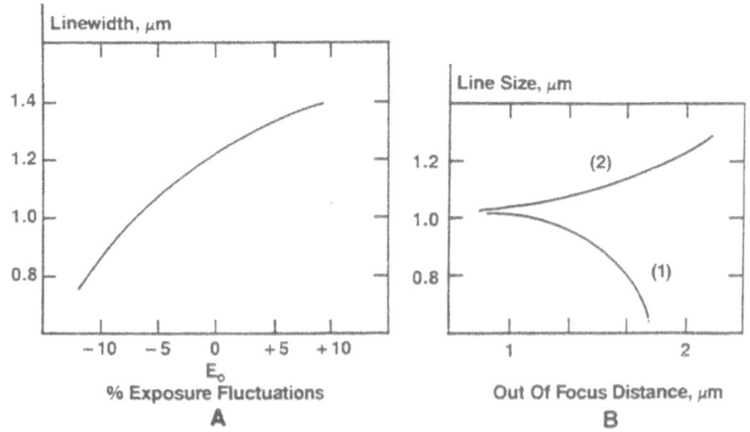

FIGURE 8-4-11. (A) Linewidth changes in DQN for under- and overexposure. (B) Linewidth changes for (1) a 1-μm line and space and (2) a 1-μm isolated line. From Ref. 64, copyright 1979 and courtesy of IEEE.

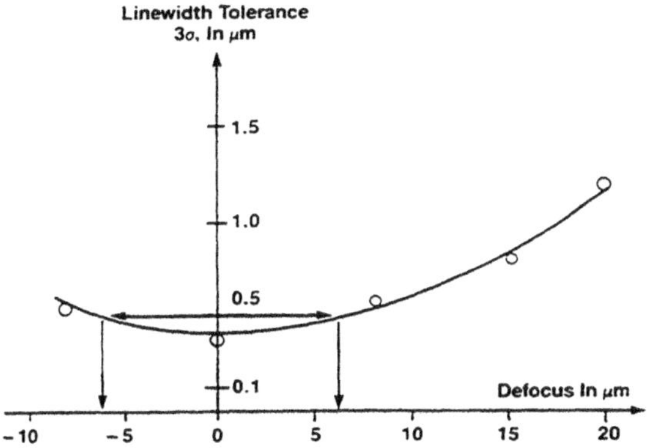

FIGURE 8-4-12. Linewidth tolerance versus defocus. From Ref. 111, copyright 1979 and courtesy of IEEE.

effect (Fig. 8-4-9).(109) In scanners, the astigmatism(108) can distort the images printed in the X, Y directions. Aberrations change the focal position of the lens and hence the shape of the incident beam.(82)

As the depth of focus of projection systems approaches ± 1–$2\ \mu m$, every field, even in a scanner, will require autofocusing.(119, 120) The change of the beam incident profile (Fig. 8-4-10) distorts the unexposed spaces of a positive photoresist more severely from adjacency effects than in isolated lines (Fig. 8-4-11). On processed wafers, reflection differences can occur to distort the autofocus signal and lead to a symmetrical spot. Double reflection spots over different areas are used to keep the lenses focused to $\pm 0.5\ \mu m$.(64, 68) Linewidth control of focus is equivalent

TABLE 8-4-5. Linewidth Tolerance in Optical Projection Lithography[a]

Exposure tool	NA = 0.28 $\sigma = 0.70$ $\lambda = 436$ nm	NA = 0.35 $\sigma = 0.65$ $\lambda = 436$ nm
Minimum feature	1-μm line/1-μm space	0.75-μm line/0.75-μm space
Process	1 μm AZ-1350J resist over Si 150 mJ/cm² 1:1 AZ developer 60 sec 300 Å plasma ashing	0.75 μm AZ-1350J resist over Si 130 mJ/cm² 1:1 AZ developer 60 sec 300 Å plasma ashing
Nominal process bias	+0.30 μm	+0.20 μm
Linewidth tolerance components		
Dose (± 5 mJ/cm²)	±0.03 μm	±0.03 μm
Development process (equivalent to ±5 sec develop time	±0.04 μm	±0.02 μm
Pattern line sensitivity (isolated exposed lines, equal line/space array, two large exposed lines separated by a small resist space)	±0.10 μm	±0.07 μm
Defocusing	±0.05 μm (±2.5 μm defocusing on top of resist)	±0.05 μm (±1.5 μm defocusing on top of resist)
Resist/substrate light coupling variance	±0.18 μm	±0.18 μm
Total linewidth tolerance	±0.21 μm	±0.21 μm

[a] From Ref. 78.

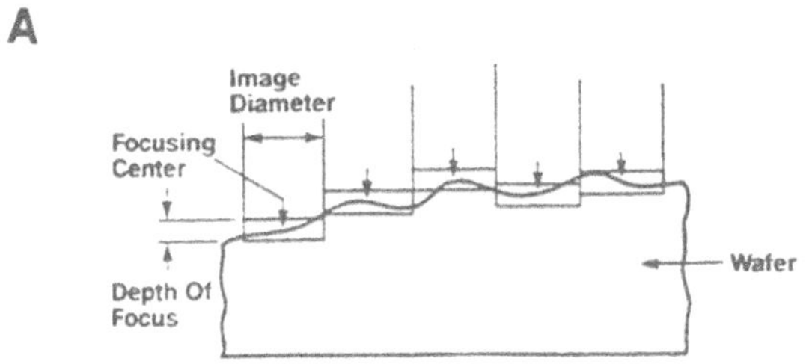

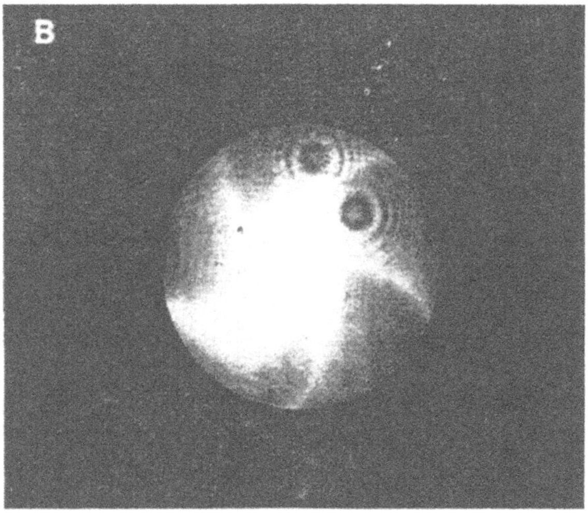

FIGURE 8-4-13. (A) Relationship between exposure depth of focus and "flatness" of wafer. Regions of imaged field not within focus will suffer poor resolution and poor linewidth control. (B) Interference photograph of wafer distortion. (A) From Ref. 113, copyright 1981 and courtesy of Cahners Publishing Co.

to printing in the proximity mode. A typical linewidth tolerance is $\pm 1\ \mu m$ for a ± 5-μm focus condition as shown in Fig. 8-4-12.[109] As the exposure tool drifts out of focus due to changes in substrate topography, astigmatism, and aberrations, the linewidth responds in a second-order compensation. The combination of focus errors, wafer distortion, intensity and development fluctuations, and interference effects have been modeled for micrometer and submicrometer images.[78] The total linewidth tolerance (1σ) is $\pm 0.21\ \mu m$ exclusive of the alignment errors (Table 8-4-5).

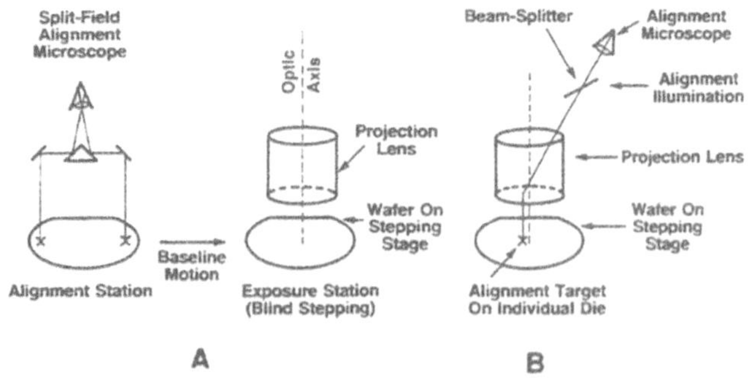

FIGURE 8-4-14. Schematic of (A) global alignment and (B) local alignment. From B. Heflinger, *SPIE Proc. Opt. Microlithogr.* **334**, 70 (1982), courtesy of SPIE.

Since optical scanners use global alignment (Fig. 8-4-14), the control of wafer flatness (Fig. 8-4-13) during hot processing[112] is paramount as the projected dimensions approach 0.8 μm in size.[122] Regions of the imaged field which are not within the depth of focus will suffer poor resolution and poor linewidth control.[113] Decreasing the coherence from 0.9 to 0.7 lowers the intensity by 50%. In the same manner, images out of focus due to steps or warped wafers suffer from a similar intensity loss. An out-of-focus condition of ± 1 μm corresponds to a 20% intensity loss or a ± 0.2-μm variation in linewidth,[78] while an intensity control of $\pm 5\%$ is required to keep linewidth variations below ± 0.1 μm.[19, 59, 78]

In order to take advantage of the small depth of focus of the projection system, dust particles are taken out of focus by applying a thick layer of Canadian balsam resin (10 μm) on top of the DQN resist (same principle to defocus particles on pellicles). The refractive index of the "defocus" layer matches that of the novolak resist.[114] In projection printing, dirt accumulates on the photomask. Printable defects have been reduced by pellicles which use a thin layer of plastic film to defocus the image of the dirt particle on the mask surface.[4, 69, 199]

8-5. OVERLAY

A circuit is built from several patterns which should ideally be printed sequentially in the exact relative positions specified by the circuit designer. Most IC design rules call for an overlay tolerance no greater than a small fraction ($\frac{1}{4}$) of the minimum feature size. From the standpoint of exposure tool design, overlay, not linewidth control, becomes the limiting characteristic of the exposure technology.[117]

Wafer alignment is done by two basic means, either remote from the projection lens (global) or through the projection lens (local alignment) (Fig. 8-4-14).[115] Global alignment signifies rotational and translational alignment of the whole wafer to the mask.[71] For full-field scanners, the whole wafer is exposed after a single leveling and global focus alignment. Local alignment is variously referred to as through the lens, site by site, die by die, or field by field alignment.[115] For step and repeat tools, the overlay depends on the ability to reposition the sample for each exposure field, and on the distortions of the image field by the wafer. Repositioning is accomplished by means of laser interferometry or by sensing the position of the wafer through the imaging lens, before the exposure is made.[230]

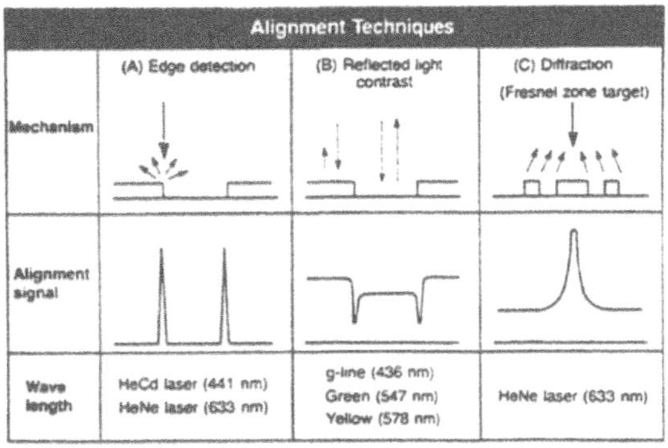

FIGURE 8-5-1. Basic die-by-die alignment techniques. From Ref. 193, courtesy of Cahners Publishing Co.

Alignment is based on the optics of the system, the flatness of mask[186] and substrate, the alignment mark, and the signal processing[160] (Fig. 8-5-1). Every field alignment has been plagued by difficulties with the visibility of alignment marks. Cases arise where the resist (and its process thickness variations, Fig. 8-4-2) and the wafer processing reduces the contrast of the alignment mark below an acceptable value. For example, the film of photoresist over the mark causes interference changes[158] as a result of different optical path lengths. For a signal requirement, a phase difference of $\lambda/2$ is necessary for a step height (h) with a resist of refractive index n,

$$h = (\lambda/2)[1/(1+n)] \tag{8-5-1}$$

Distortions in the alignment signal due to resist thickness (± 100 Å) fluctuations or vibrations[222] can have a severe impact on throughput (see Table 8-4-2).

The major factors that cause misalignment (Fig. 8-5-2) include lens-to-lens magnification-distortion, thermal or process runout[118] of mask–wafer, and alignment errors (Table 8-5-1).[227, 228] In order to adjust the positional accuracy of the exposure tool, mechanical vibration > 10 Hz (20 μm/sec) should be damped.

Mix-and-match tool strategy involves the patterning and alignment with stepper and scanning tools (Table 8-5-2).[116] Because the distortion in refractive lenses is on the same order as that of the full-field scanners, i.e., about 0.1 μm, there is no particular advantage to the smaller field step and repeat lenses. As a result, the overlay is only better for step and repeat systems when the wafer distortion, either local or due to poor temperature control, is the dominant factor.[230] For example, "hot" processes include the thermal distortion expansion of the mask and wafer (Table 8-5-3) and the wafer mounting errors which give an average of 10 ppm in the relative position of chips.[55] Thermal distortions of 1 °C will cause misalignment

FIGURE 8-5-2. Example of misalignment—"light rectangle" not centered in middle of darker rectangle.

TABLE 8-5-1. Factors of Registration Error[a]

Error	Stepper	Scanner
Align	Wafer align Reticle align Relative position[b]	Wafer to mask
Machine stability and compatibility	Stepping accuracy X/Y total pitch Stage orthogonality Line reduction/distortion field rotation	X/Y magnification Cross scan distortion Nonlinear distortion
Mask	X/Y total pitch Orthogonality Nonlinear pattern location error	
Thermal Expansion	Temperature of mask and wafer variation of mask temperature	
Residual	Flatness and bending error of wafer and mask Nonlinear wafer distortion	

[a] From Ref. 118.
[b] Relative position stability between align and projection optics.

due to thermal expansion (Table 8-5-3) and will cause the distance between two points 1 cm apart on a silicon wafer to change 0.023 μm. For two points, 10 cm apart corresponding to a full wafer printing, the change is 0.23 μm or well over the 0.1-μm limit for submicrometer (0.5-μm) technology. In addition, the silicon wafer is warped from the hot process deposition of thin films. Distortions as much as 0.15 μm in 1 cm have been reported.[4, 113, 118]

After consideration of all the alignment factors and the operational exposure tools, Allsop[4] (Table 8-5-4) concluded that the step and repeat cameras are able to print a usable linewidth (4 × the root mean square of all align errors) of 1.5 μm in single-resist films while full-field scanners are able to meet 2- to 3-μm device ground rules.

TABLE 8-5-2. Registration Accuracy and Factors for a GCA 4800 DSW Refractive Stepper and Perkin Elmer 140 Scanner[a]

	Registration error (μm)	Align error (μm)	Machine error (μm)	Mask error (μm)	Thermal expansion (μm)	Residual error (μm)
Stepper, same tool	0.28	0.28	0.07	0.02	0.02	0.02
Stepper, tool to tool	0.30	0.28	0.12	0.04	0.04	0.04
Scanner, same tool	0.43	0.38	0.10	0.12	0.14	0.04
Scanner, tool to tool	0.50	0.38	0.23	0.12	0.14	0.04
Stepper to scanner	0.35	0.28	0.17	0.12	0.06	0.06

[a] From Ref. 116.

TABLE 8-5-3. Contributing Factors to Alignment Overlay Errors[a]

Factor	Alignment error (μm)					
	10:1 stepper	1:1 stepper	Mirror projection (near UV)	Mirror projection (deep UV)	Contact proximity (near UV)	Contact proximity (deep UV)
1. Mask overlay	0.1	0.1	0.5	0.25	0.5	0.5
2. Alignment tolerance	0.3	0.3	0.3	0.3	0.5	0.5
3. Thermal expansion (1 °C)	0.1	0.1	0.1	0	0.5	0
4. Distortion	0.25	0.25	0.35	0.35	0	0
Total overlay errors	0.40	0.40	0.84	0.52	0.87	0.71
Usable linewidth (4 × alignment tolerance)	1.6	1.6	3.36	2.08	3.48	2.84
Resolution	1.0	1.6	2	0.8	2	0.8

[a] From Ref. 4.

Alignment errors (A) are one part of the total device ground-rule tolerances. Tolerance (T) or the ability to overlay spatially one masking layer to the previous etched or deposited pattern (Fig. 8-5-2) is given by the standard deviations of the independent variables of the mask (M), etch (E), and A errors,

$$T = (A^2 + M^2 + E^2)^{1/2} \tag{8-5-2}$$

Typical mask errors[187] range from 0.1 to 0.3 μm, etch errors from 0.2 to 0.4 μm, and alignment errors from 0.2 to 0.5 μm. Thus, the tolerances of current lithographic processing lie in the range of 0.5–0.8 μm. For the next generation of submicrometer ground rules, a fourfold reduction in alignment and other errors will be necessary.

To assess and monitor alignment errors, the metrology of alignment involves

TABLE 8-5-4. Misalignment Due to Thermal Expansion[a]

Material	Coefficient of thermal expansion (in./in.-°C)	Misalignment over 4 inches for 1 °C temperature rise	
		μin.	μm
Soda-lime glass	8.7×10^{-6}	34.8	0.87
Hoya low-expansion	3.7×10^{-6}	14.8	0.37
Corning 7059	4.7×10^{-6}	18.8	0.47
Fused silica (quartz)	5.5×10^{-7}	2.20	0.05
Silicon wafer	2.33×10^{-6}	9.38	0.23

[a] From Ref. 4.

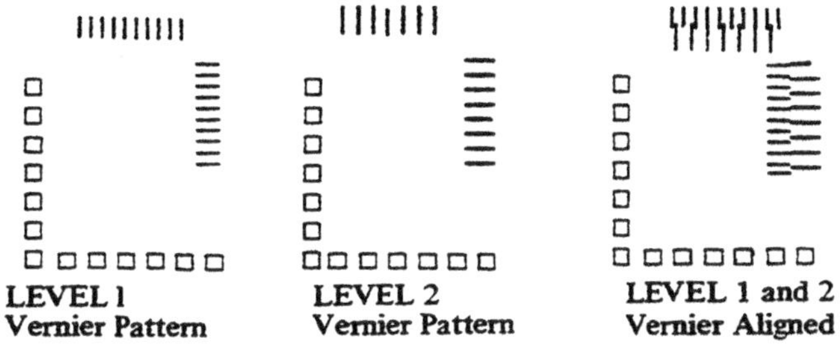

FIGURE 8-5-3. Standard optical vernier test pattern for overlay evaluation. From Ref. 123, reprinted with permission of *Solid State Technology*, published by Technical Publishing, a company of Dun and Bradstreet.

measurement of wafer flatness,[113] linewidths, and overlay using verniers (Fig. 8-5-3). Common verniers are as follows[121]:

1. Electrical verniers—conducting layers are patterned to form voltage dividers.[118, 165, 202]
2. Optical—interference displacement monitors; coherent laser light is diffracted and the amplitude measured.[193, 200]
3. Edge detection—by reflected laser beam.
4. Video—computer pattern recognition.
5. Microscopy—SEM.

Autoalign marks can serve as process control monitors.[123] The positive resist itself can be used as a process vernier since it can be developed, reexposed, and repatterned to test tool or mixed exposure tool performance. The metrology of alignment in photomask production is just as important as device alignment. For example, a major source of registration failure was found to be due to variations in optical field sizes defined by the dies caused by unsatisfactory process focus control.[124, 125]

8-6. MISCELLANEOUS EXPOSURE TECHNIQUES

With the discovery of holography, several attempts[126–129] have been made to project holographic images into resists. Some success has been achieved in printing diffraction gratings,[130] but laser speckling and the high degree of coherence have limited the practical resolution. For old but still active means of projection printing, these include fly's eye lenses and the pinhole camera.[131, 132] With a fly's eye lens, one $10\times$ or $1\times$ chip could be projected[133] over a whole wafer in one exposure.[203]

8-7. PHOTOMASKS

Progress in the fabrication of photomasks is essential for future optical lithography. In $1\times$ masks, windage corrections[56] to $-0.5\ \mu m$ and chrome edge control to $\pm 0.05\ \mu m$ are necessary for submicrometer optical lithography. As wafers increase in size, masks become more difficult to coat, image, and process.[134]

Today, mask fabrication is primarily done by electron-beam exposure (Chapter 9). The opaque layer in photomasks can be made of various films:

- Ag emulsion[136]
- Ion-bombarded resist[137]
- Diazonium polymers[138]
- Iron oxide[139]
- Germanium on glass[140]
- Chrome on glass[141]
- Heated polyacrylonitrile[142]
- Europium oxide[143]

A practical review of mask making has been published in SPIE and Kodak proceedings from 1970 to 1980.[144] In mask substrates, the specifications for ultraflat quartz masks have been tightened on flatness and antireflection coatings are now applied to minimize light scattering. Quartz substrates are necessary to keep thermal induced substrate expansion to a minimum.[135] In order to produce and maintain high-quality plates,[198] automated linewidth and defect analysis[204] have been complemented with ion-beam[197] and laser[206, 216] repair. Printable pinholes or pindots depend on the power transmitted or diffracted which falls off as the size (diameter) approaches $0.5\lambda/\mathrm{NA}$.[188] For the repair of missing chrome, opaque dabs,[201, 205] lift-off metal deposition, gallium ion implant and chemically vapor deposited (laser assisted) chrome and aluminum have been used. Negative opaque defects are removed inside a scanning ion microscope by the sputtering action of the ion beam or in a laser microscope by thermal vaporization.

After the production of the $10\times$ master reticle, optical photorepeaters take over on a precision stage for photoreduction to a $1\times$ mask. The accuracy of the stage feed pitch is highly important in obtaining accurate registrations between masks.[125] It is also important to align the reticle accurately in order to prevent chip rotation of misalignment of the target pattern.[134]

8-8. SENSITOMETRY

Photosensitometry is involved with the measurement of the incident dose (intensity–time product) and the response of the resist in terms of chemical or physical (dissolution) changes. In advanced measurements not involving the simplified monochromatic sources, the various output of the lamps, filters, resist absorption, and substrate reflectivity, are integrated into the actinic bandwidth response function.[182, 183] In all sensitometric treatments, it is assumed that the photoresists obey the Bunsen–Roscoe reciprocity relation,

$$E = It \tag{8-8-1}$$

The resist responds equally to a high-intensity (I) excimer laser[34] or with a long exposure time (t) with a low-intensity tungsten lamp. Those resists (e.g., DQN) based on unimolecular reactions are insensitive to the light intensity, while resists (e.g., azides) sensitive to oxygen (bimolecular reactions) exhibit reciprocity failure at low intensity due to an excess of oxygen quenching.

After measurement of the incident intensity with a thermopile or a calibrated silicon photodiode, the resist response can be intensity modulated with a step wedge (see Chapter 15) and its dissolution rate determined with film endpoint systems. Chemically, the bleaching of DQs can by monitored by IR or in the UV to even back-calibrate the output of full-field scanning exposure tools.[179] In the same manner, the quality control of incoming lots of resists can be analyzed with the calibrated exposure source.

8-9. OUTLOOK

The basic prognosis is that optical systems will last for the next two decades down to 0.25 μm dimensions. No other exposure systems can compete in throughput and large mask stability. It has been said that a resist is only as good as the camera or vice versa. The evolution of optical exposure from contact to proximity to projection was driven by mask damage and overlay requirements. At present, the economical production of submicrometer VLSI devices (over the 1-μm barrier) rests on the field size of steppers, the global alignment accuracy of scanners, and the flatness of wafers after hot processes. As the number of resolution elements within the chip field ($> 10^8$) exceeds focus limits, the chip field could be partitioned into subfields (to offset the higher NA of the lens) just as the wafer field has been partitioned for step and repeat printing.[59] Registration, alignment, and linewidth control must also continue to improve to meet the shrinking tolerances of new devices.

The intra- and interproximity effects are major problems for optical exposure. Diffraction and interference distort features over raised topographies. Will there be an exposure tool, process, or material solution to the problem? A process and/or material solution appears to be the plausible answer. The benefits of bilayers, trilayers, antihalation and contrast enhancement layers will outweigh the added costs. High-contrast single deep-UV resists will improve the linewidth control and process latitude.

Top surface imaging of resists or high-absorptivity materials is in need of further development for submicrometer imaging. In top surface imaging, the key will be to make the resist layer appear only a few nanometers thick and allow a low-optical-MTF [$K < 0.3$, Eq. (8-1-2)] exposure system to be used. Resists sensitive to 1 mJ/cm^2 would allow flash on the fly exposure of smaller fields and make throughput only dependent on the global align and step. Statistically, this creates a situation where there are less photons than resist molecules. Quantum gains of 10^2 will be required for very thin layers of resist to reduce the number of molecules. The three-prong approach of better lenses and alignment techniques, flatter wafers, planar films, and improved resists should keep optical exposure in the limelight of the next decade of device fabrication.

Devices sensitive to high-energy (*e*-beam or X-ray) damage may only be practical to fabricate using UV exposure. "Lightography" will remain as the workhorse of the high-volume–low-part-number lithographic technology.

REFERENCES

1. G. Stevens, *Microphotography*, Wiley, New York, 1967.
2. M. Lacombat, *Solid State Technol.* Aug. 1980, p. 115.

3. A. Tasch, *SPIE Proc.* **333**, 68 (1982).
4. B. Allsop, *Microelectron. Manuf. Test.* July 1981, p. 28.
5. J. Roussel, *Solid State Technol.* May 1978, p. 64.
6. M. Hakey and J. Straub, *J. Electron. Mater.* **11**, 813 (1982).
7. P. Blais, *Solid State Technol.* Aug. 1977, p. 76.
8. L. Johnson, *Appl. Opt.* **21**, 1892 (1982).
9. M. King and M. Goldrick, *Solid State Technol.* Feb. 1977, p. 37.
10. M. King, in *VLSI Electronics*, Vol. 2, edited by N. Einspruch, Academic Press, New York, 1980, p. 57.
11. P. Concidine, *J. Opt. Soc. Am.* **56**, 1001 (1966).
12. D. Doane, *Solid State Technol.* Aug. 1980, p. 101.
13. K. Tai, R. Vadimsky, C. Kemmer, J. Wagner, V. Lambertini, and A. Timko, *J. Vac. Sci. Technol.* **17**, 1169 (1980).
14. W. Stickel and G. Langner, *J. Vac. Sci. Technol.* **B1**, 1007 (1983).
15. M. Nakase and Y. Matsumoto, *Photogr. Sci. Eng.* **23**, 215 (1979).
16. R. Jones, *Photogr. Sci. Eng.* **11**, 102 (1967).
17. M. King, *IEEE Trans. Electron Devices* **ED-26**, 711 (1979).
18. A. Offner, *Photogr. Sci. Eng.* **23**, 374 (1979).
19. T. Shankoff, I. Bruning, and R. Johnston, *Polym. Eng. Sci.* **20**, 1105 (1980).
20. T. Chang, D. Kyser, and C. Ting, *Solid State Technol.* May 1982, p. 60.
21. M. O'Toole, R. Liu, and M. Chang, *IEEE Trans. Electron Devices* **ED-28**, 1405 (1981).
22. W. Arden, H. Keller, and L. Mader, *Solid State Technol.* July 1983, p. 143.
23. F. Dill, A. Neureuther, J. Tuttle, and E. Walker, *IEEE Trans. Electron Devices* **ED-22**, 445 (1975).
24. R. Longhurst, *Geometrical and Physical Optics*, Wiley, New York, 1967, p. 208.
25. H. Smith, *Proc. IEEE* **62**, 1361 (1974).
26. T. Chang, D. Kyser, and C. Ting, *Solid State Technol.* May 1982, p. 60.
27. U. S. Patent 3,511,653 (1970), American Cyanamid.
28. B. Griffing and P. West, *Polym. Eng. Sci.* **23**, 947 (1983).
29. R. Sheets, *Microelectron. Manuf. Test.* Oct. 1982, p. 19.
30. B. Lin, *J. Vac. Sci. Technol.* **12**, 1317 (1975).
31. W. Moreau and P. Schmidt, *138th Electrochem. Soc. Ext. Abstr.* 1970, p. 459.
32. B. Lin, in *Fine Line Lithography*, edited by R. Newman, North-Holland, Amsterdam, 1980, p. 141.
33. H. Craighead, J. White, R. Howard, L. Jackel, R. Behringer, J. Sweeney, and R. Epworth, *J. Vac. Sci. Technol.* **B1**, 1186 (1983); T. McGrath, *Solid State Technol.* **26**, 165 (1983).
34. K. Jain, C. Willson, and B. Lin, *IBM J. Res. Dev.* **26**, 151 (1982); *IEEE Electron Device Lett.* **EDL-2**, 52 (1982).
35. M. Levenson, K. Johnson, V. Hanchett, and K. Chiang, *J. Opt. Soc. Am.* **71**, 737 (1981).
36. E. Walker, *IEEE Trans. Electron Devices* **ED-22**, 464 (1975).
37. K. Polasko, *IEEE Electron Device Lett.* **EDL-5**, 24 (1984).
38. D. Ehrlich and J. Tsao, *J. Vac. Sci. Technol.* **B1**, 969 (1983).
39. R. Srinivasan and V. Mayne-Banton, *Appl. Phys. Lett.* **41**, 576 (1982).
40. H. Smith, F. Bachner, and N. Efremow, *J. Electrochem. Soc.* **118**, 821 (1971); *IEEE Trans. Electron Devices* **ED-22**, 496 (1975).
41. A. Voschenkov and H. Herrman, *Electron. Lett.* **17**, 61 (1980).
42. W. Hestman and P. Van Den Berg, *Can. J. Phys.* **53**, 1310 (1975).
43. M. Levenson, N. Viswanathan, and R. Simpson, *IEEE Trans. Electron Devices* **ED-29**, 1828 (1982).
44. D. Meyerhofer and J. Mitchell, *RCA Rev.* **43**, 608 (1982).
45. L. White, *RCA Rev.* **43**, 391 (1982).
46. D. Flanders and T. Lysczarz, *J. Vac. Sci. Technol.* **B1**, 1196 (1983).
47. M. King and D. Berry, *Appl. Opt.* **11**, 2455 (1972); U. S. Patent 3,690,881 (1972).
48. D. Flanders and H. Smith, *Appl. Phys. Lett.* **31**, 426 (1971).
49. H. Kinoshito, A. Une, and M. Iki, *J. Vac. Sci. Technol.* **B1**, 1276 (1983).
50. H. Kleinknecht, *SPIE Proc.* **174**, 63 (1979).
51. M. Feldman, A. White, and D. White, *SPIE Proc.* **333**, 125 (1982).
52. U.S. Patent 3,676,002 (1972), IBM.
53. D. Flowers and H. Hughes, *J. Electrochem. Soc.* **124**, 1599 (1977).
54. U.S. Patent 3,511,653 (1970), American Cyanamid.
55. W. Baylies, *Solid State Technol.* Feb. 1981, p. 132.

56. R. Bracken and S. Rizi, in *VLSI Electronics*, Vol. 6, edited by N. Einspruch, Academic Press, New York, 1983, p. 295.
57. B. Lin, *SPIE Proc.* **174**, 121 (1979).
58. L. Contente, *Appl. Opt.* **8**, 75 (1969).
59. J. Bruning, *J. Vac. Sci. Technol.* **17**, 1146 (1980).
60. J. Cuthlert, *Solid State Technol.* Aug. 1977, p. 59.
61. P. Burggraaf, *Semicond. Int.* April 1982, p. 57.
62. P. Burggraaf, *Semicond. Int.* Feb. 1984, p. 88.
63. M. Bowden and L. Thompson, in *Introduction to Microlithography*, edited by M. Bowden, L. Thompson, and C. Willson, American Chemical Society, 1983, p. 24.
64. H. Binder and M. Lacombat, *IEEE Trans. Electron Devices* **ED-26**, 698 (1979).
65. D. Doane, *Solid State Technol.* Aug. 1980, p. 101.
66. M. Nakase and T. Shinozaki, *IEEE Trans. Electron Devices* **ED-28**, 1416 (1981).
67. J. Bruning, *J. Vac. Sci. Technol.* **16**, 1925 (1979).
68. S. Wittekoek, *Solid State Technol.* June 1980, p. 82.
69. U.S. Patent 4,131,363 (1978), IBM.
70. K. Hennings and H. Schuetze, *SCP Solid State Technol.* July 1966, p. 31.
71. C. Ausschnitt, *SPIE Proc.* **394**, 64 (1983).
72. J. Wilcyznski, *J. Vac. Sci. Technol.* **16**, 1929 (1979).
73. D. Angel, *Semicond. Int.* April 1983, p. 72.
74. P. Tigreat, *J. Vac. Sci. Technol.* **16**, 1938 (1979).
75. A. Reiser and E. Pitts, *J. Photogr. Sci.* **29**, 187 (1981).
76. T. Iwayangi, *IEEE Trans. Electron Devices* **ED-28**, 1306 (1981).
77. R. Leonard and W. Cordes, *SPIE Proc.* **394**, 125 (1983).
78. T. Chang, C. Ting, and D. Kyser, *IEEE Trans. Electron Devices* **28**, 1299 (1981).
79. A. McCullough, *J. Electrochem. Soc.* **128**, 225 (1981).
80. A. McCullough and H. Sewell, *SPIE Proc.* **394**, 107 (1983).
81. H. Stover, *SPIE Proc.* **334**, 60 (1982).
82. P. Jain, A. Neureuther, and W. Oldham, *IEEE Trans. Electron Devices* **ED-28**, 1410 (1981).
83. S. Fujimori, *J. Appl. Phys.* **50**, 615 (1979).
84. S. Middelhoek, *IBM J. Res. Dev.* **14**, 117 (1970).
85. D. Widmann and H. Binder, *IEEE Trans. Electron Devices* **ED-22**, 467 (1975).
86. D. Ilten and K. Patel, Proceedings of SPSE Seminar on Applications of Photopolymers, p. 79; *Image Technol.* **2**, 9 (1971).
87. K. Clark and E. Juleff, *Microelectronics* **6**, 37 (1974).
88. H. Van Den Berg and J. Van Staden, *J. Appl. Phys.* **50**, 1212 (1979).
89. U.S. Patent 3,622,319 (1971), Western Electric.
90. U.S. Patent 3,647,566 (1972), Corning.
91. S. Matsui and N. Endo, *Microelectron. Eng.* **1**, 51 (1983).
92. A. Neureuther, *IBM Tech. Discl. Bull.* **16**, 334 (1973).
93. T. Brewer, R. Carlson, and J. Arnold, *J. Appl. Photogr. Eng.* **7**, 184 (1981).
94. M. O'Toole, E. Liu, and M. Chang, *IEEE Trans. Electron Devices* **ED-28**, 1405 (1981).
95. U.S. Patent 4,414,314 (1983), IBM.
96. C. Ting and K. Liauw, *J. Vac. Sci. Technol.* **B1**, 1225 (1983).
97. G. Thomas, *SPIE Proc.* **174**, 15 (1979).
98. P. Blais, *Solid State Technol.* Aug. 1977, p. 70.
99. D. Widdmann, *Appl. Opt.* **14**, 931 (1975).
100. K. Clark and K. Okutsu, *Solid State Technol.* April 1976, p. 82.
101. I. Korba, *Appl. Opt.* **9**, 969 (1970).
102. R. Lang and B. Smilowitz, *IEEE Trans. Sonics Ultrason.* **SU-27**, 134 (1980).
103. T. Liu and B. Lin, *IEEE Trans. Electron Devices* **ED-30**, 1259 (1983).
104. P. Robertson, A. Neureuther, C. Ting, and A. Nasi, *SPIE Proc.* **334**, 37 (1982).
105. H. Van Den Berg and J. Ruigrok, *Appl. Phys.* **16**, 279 (1978).
106. H. Van Den Berg and V. Zieren, *J. Opt. Soc. Am.* **70**, 110 (1980).
107. U.S. Patent 4,239,970 (1981), RCA.
108. S. Iwamatsu and K. Asanami, *Solid State Technol.* May 1980, p. 83.
109. D. Elliott and M. Hockey, *Solid State Technol.* June 1979, p. 53.
110. D. Fehrs, *Proc. Kodak Microelectron. Semin.* 1979, p. 135.

111. H. Moritz, *IEEE Trans. Electron Devices* **ED-26**, 705 (1979).
112. L. Yau, *IEEE Trans. Electron Devices* **ED-26**, 1299 (1979).
113. S. Morgan, Z. Sobezak, G. Lynch, and L. Reed, *Semicond. Int.* Oct. 1981, p. 109.
114. U.S. Patent 4,379,831 (1979), Censor.
115. B. Heflinger, *SPIE Proc.* **334**, 70 (1982).
116. S. Uolya, W. Wakiyama, A. Ale, and N. Nakato, *SPIE Proc.* **394**, 80)1983).
117. H. Stover, N. David, and T. Lewis, *Solid State Technol.* Oct. 1982, p. 126.
118. K. Nichols, I. Stemp, and N. Brockman, *J. Electrochem. Soc.* **128**, 609 (1981).
119. A. Wat and K. Chem, *SPIE Proc.* **394**, 58 (1983).
120. C. Ausschnitt, T. Bruner, and R. Cronin, *SPIE Proc.* **394**, 64 (1983).
121. C. Van Peski, *Solid State Technol.* May 1982, p. 25.
122. T. Bettes, *Semicond. Int.* Feb. 1981, p. 77.
123. H. David and H. Stover, *Solid State Technol.* June 1982, p. 131.
124. H. Rottmann, *IBM J. Res. Dev.* **26**, 553 (1982).
125. H. Rottmann, *IBM J. Res. Dev.* **24**, 461 (1980).
126. A. Voschenkov and P. Hanson, *IEEE Electron Device Lett.* **EDL-3**, 208 (1982).
127. H. Kiemle, *Int. Elektron. Rundsch.* **7**, 176 (1970).
128. M. Beesley, *Electron. Lett.* **4**, 49 (1968).
129. M. Levenson, K. Johnson, and V. Hanchett, *J. Opt. Soc. Am.* **71**, 737 (1981).
130. N. Efremow, *J. Vac. Sci. Technol.* **19**, 1234 (1981).
131. W. Rudge, W. Harding, and W. Mutter, *IBM J. Res. Dev.* April 1963, p. 146.
132. P. Newman and V. Rible, *Appl. Opt.* **5**, 1225 (1966).
133. J. Moran, *Appl. Opt.* **10**, 412 (1971).
134. Y. Hashimoto, *Jpn. Semicond. Technol. News* **2**(5), 18 (1983).
135. T. O'Neill, *Semicond. Int.* Aug. 1981, p. 81.
136. U.S. Patents 4,004,925, 4,225,659, 3,567,447, and 3,639,125.
137. B. MacIver, *J. Electrochem. Soc.* **129**, 827 (1982).
138. U.S. Patent 4,149,888, GAF.
139. U.S. Patent 4,027,052 (1977), Bell.
140. U.S. Patent 3,830,686.
141. U.S. Patent 3,647,566 (1972), Corning.
142. U.S. Patent 3,650,796.
143. U.S. Patent 3,895,147.
144. D. Elliott, *Integrated Circuit Fabrication Technology*, McGraw–Hill, New York, 1982, pp. 339–362.
145. T. Matsuzawa, H. Yanazawa, N. Hashimoto, and H. Mishimagi, *J. Electrochem. Soc.* **128**, 184 (1981).
146. J. Trotel and B. Fay, in *Electron Beam Technology in Microelectronic Fabrication*, edited by G. Brewer, Academic Press, New York, 1980, p. 309.
147. W. Fincham and M. Freeman, *Optics*, Butterworths, London, 1980, p. 306.
148. A. Van Heel and C. Velzel, *What is Light*, World University Library, London, 1968, p. 35.
149. K. Harris, P. Sandlund, and R. Singleton, *SPIE Proc.* **394**, 239 (1983).
150. I. Brodie and J. Muray, *Physics of Microfabrication*, Plenum Press, New York, 1982, p. 273.
151. G. Geikas and B. Ables, *Kodak Photoresist Semin.* 1968, p. 22.
152. M. Nakase, *Photogr. Sci. Eng.* **27**, 254 (1983).
153. M. Latta, R. Moore, S. Rice, and K. Jain, *J. Appl. Phys.* **56**, 585 (1984).
154. H. Jelinek and R. Srinivasan, *J. Phys. Chem.* **88**, 304B (1984).
155. U.S. Patents 4,414,059 (1984) and 4,417,948 (1984), IBM.
156. Japanese Patent 58,28,739, *Chem. Abstr.* **100**, 183203 (1983).
157. H. Stover, *SPIE Proc.* **470**, 22 (1984).
158. H. Ohtsuka, H. Funatsu, G. Kushibiki, and T. Kokideka, *SPIE Proc.* **470**, 62 (1984).
159. W. Trutna and M. Chen, *SPIE Proc.* **470**, 62 (1984).
160. H. Mayer and E. Loebach, *SPIE Proc.* **470**, 147 (1984).
161. F. Buiguez, *SPIE Proc.* **470**, 34 (1984).
162. D. Hofer, R. Miller, C. Willson, and A. Neureuther, *SPIE Proc.* **469**, 108 (1984).
163. B. Griffing and W. Lorenson, *SPIE Proc.* **469**, 102 (1984).
164. B. Griffing and R. West, *SPIE Proc.* **469**, 94 (1984).
165. R. Allen, M. Cagan, and M. Foster, *SPIE Proc.* **470**, 111 (1984).
166. J. Peterson and R. Kowalski, *SPIE Proc.* **469**, 46 (1984).

167. L. Hsia, *J. Electrochem. Soc.* **131**, 2133 (1984).
168. D. MacDonald, M. Nager, C. VanPeki, and T. Whitney, *SPIE Proc.* **470**, 212 (1984).
169. M. Prouty and A. Neureuther, *SPIE Proc.* **470**, 228 (1984).
170. A. Lewis, *Ultramicroscopy* **13**, 227 (1984).
171. D. Markle, *Solid State Technol.* Sept. 1984, p. 159.
172. T. Omata, *Solid State Technol.* Sept. 1984, p. 173.
173. U.S. Patent 4,474,864 (1984), IBM.
174. T. Matsuzawa, T. Iwayanagi, H. Obayashi, and H. Tomioka, *Microelectron. Eng.* **1**, 1185 (1983).
175. U.S. Patent 4,386,849 (1983), IBM.
176. M. Kim and P. Piacente, *Thin Solid Films* **119**, 75 (1984).
177. U.S. Patent 3,571,248 (1973), Bell.
178. V. Miller and H. Stover, *Solid State Technol.* Jan. 1985, p. 127.
179. U.S. Patent 4,474,864 (1984).
180. European Patent Application EP 110,165 (1984), *Chem. Abstr.* **102**, 15162 (1985), GE.
181. Japanese Patent 58,198,040, *Chem. Abstr.* **101**, 181319 (1984).
182. N. Gold, *SPIE Proc.* **334**, 34 (1982).
183. D. Novotony, *Photogr. Sci. Eng.* **21**, 351 (1977).
184. Japanese Patent 59,103,343, *Chem. Abstr.* **101**, 201560 (1984).
185. L. Halle, *J. Vac. Sci. Technol.* **B3**, 323 (1985).
186. T. Matsuzawa, H. Sunami, and N. Hashimoto, *Microelectron. Reliab.* **16**, 173 (1977).
187. C. Kim and W. Ham, *RCA Rev.* **39**, 565 (1978).
188. M. Rosenfield, D. Goodman, A. Neureuther, and C. Willson, *J. Vac. Sci. Technol.* **B3**, 377 (1985).
189. D. Hofer, R. Miller, A. Neureuther, and C. Willson, *SPIE Proc.* **469**, 108 (1984).
190. B. Griffing and R. Lorenson, *SPIE Proc.* **469**, 102 (1984).
191. B. Griffing, P. West, and E. Balch, *SPIE Proc.* **469**, 94 (1984).
192. A. Yoshikawa, S. Hirota, O. Ochi, A. Takeda, and Y. Mizushima, *Jpn. J. Appl. Phys.* **20**, 181 (1981).
193. W. Wakima and M. Nakajima, *Semicond. Int.* May 1985, p. 261.
194. U.S. Patent 4,498,775 (1985), RCA.
195. U.S. Patent 4,444,869 (1984), Fujitsu.
196. U.S. Patent 4,458,994 (1984), IBM.
197. U.S. Patent 4,503,329 (1985), Hitachi.
198. Japanese Patent 59,119,353, *Chem. Abstr.* **102**, 36789 (1985).
199. U.S. Patent 4,482,591 (1985), J. T. Baker.
200. U.S. Patent 4,443,911 (1984), Tokyo Shibaura.
201. U.S. Patent 4,444,801 (1984), Hitachi.
202. U.S. Patent 4,399,205 (1983), IBM.
203. G. Maclay, *J. Vac. Sci. Technol.* **B3**, 926 (1985).
204. S. Gupta, A. Bagchi, and N. Kundu, *Microelectron. J.* **16**, 22 (1985).
205. U.S. Patent 4,523,974 (1985), Perkin Elmer.
206. M. Opysko, M. Bernek, and P. Young, *IEEE Electron Device Lett.* **EDL-6**, 344 (1985).
207. M. O'Toole, *IEEE Electron Device Lett.* **EDL-6**, 282 (1985).
208. B. Griffing and P. West, *Solid State Technol.* May 1985, p. 152.
209. A. Neureuther, D. Hofer, and C. Willson, in *Microcircuit Engineering*, edited by H. Beneking and H. Beureuther, Academic Press, New York, 1985, p. 53.
210. A. Neureuther and W. Oldham, *Solid State Technol.* May 1985, p. 139.
211. K. Polaski and B. Griffing, *SPIE Proc.* **631**, 180 (1986).
212. K. Ismail, *Microelectron. Eng.* **1**, 295 (1983).
213. G. Thommes and V. Weber, *J. Imag. Technol.* **29**, 113 (1985).
214. S. Babu and V. Srinivasan, *IEEE Trans. Electron Devices* **ED-32**, 1896 (1985).
215. F. Vollenbroek, H. Kroon, J. Bartsen, and J. Del, in *Microcircuit Engineering*, edited by H. Beneking and H. Beureuther, Academic Press, New York, 1985, p. 556.
216. U.S. Patent 4,510,222 (1985), Hitachi.
217. L. Cescato, G. Mendes, and J. Freijlich, *Appl. Opt.* **24**, 1142 (1985).
218. M. Watts, D. D. Debruin, and W. Arnold, *Polym. Eng. Sci.* **26**, 1165 (1986).
219. C. Mack, *SPIE Proc.* **538**, 207 (1985).
220. Japanese Patent 60,38,821, *Chem. Abstr.* **103**, 113352 (1985).
221. German Patent 3,428,565 (1985), GE.

222. T. Brunner and R. Allen, *IEEE Electron Device Lett.* **EDL-6**, 329 (1985).
223. J. Bohland, H. Sanford, and S. Fine, *SPIE Proc.* **539**, 267 (1985).
224. A. Brown and W. Arnold, *SPIE Proc.* **539**, 259 (1985).
225. U.S. Patent 4,499,162 (1985), AT&T.
226. U.S. Patent 4,500,602 (1985).
227. K. Edmark and C. Auschnett, *SPIE Proc.* **538**, 91 (1985).
228. S. Murakani, T. Matsura, M. Ogawa, and M. Uehara, *SPIE Proc.* **538**, 9 (1985).
229. D. Chien, L. Chien, and M. Chen, *SPIE Proc.* **538**, 197 (1985).
230. F. Vollenbroek, W. Nyssen, H. Kroon, and B. Yelmaz, SPE RETEC Photopolymers, Ellenville, N.Y., 1985, p. 309.
231. M. Bolsen, G. Buhr, H. Merrem, and K. Van Werden, *Solid State Technol.* Feb. 1986, p. 83.
232. L. White, *RCA Rev.* **47**, 345 (1986).
233. T. Pampalone and F. Kunyan, *J. Electrochem. Soc.* **133**, 192 (1986).

9

Radiation Exposure

9-1. INTRODUCTION

The traditional form of exposure of resists involves pure uncharged ultraviolet photons. From actual microscopy and diffraction theory, the resolution is limited by the practical wavelength of observation. Thus, shorter-wavelength, higher-energy photoelectrons from electron beams, ion beams, and X rays have been used in microscopes to observe and analyze objects the size of atoms. These analytical tools formed the basis of high-energy radiation exposure systems to expose resists and build devices to a nanometer scale. In some cases, masks disappear and resolution with some forms of high energy becomes infinite, if one is willing to pay for the high capital and maintenance costs at low wafer exposure throughput. The primary application of these forms of exposure tools has been to fabricate, measure, and repair photomasks but gradually scanning exposure tools have been applied to direct silicon write and in the next decade, flood exposure will also enter into the production of memory circuits.

For high-energy exposure, the beam wavelength is now measured in angstroms and not nanometers. Every type of polymer, inorganic, or film-forming material becomes a "resist." Even "resistless" lithography is possible as inorganic films can be induced to etch, vaporize, or ion implanted to become semiconductors. Enough energy is available to excite atoms as well as rearrange any chemical bond. The new area of exposure is called radiolithography which encompasses all wavelengths below 100 nm (Table 9-1-1).

The price that is paid for all the extra energy is throughput (wafers per hour) or cost per exposure. High-energy exposure takes over when optical lithography cannot compete in:

- Alignment
- Low volume/high part number
- Depth of focus ($< 2\ \mu m$)
- Fabricating optical $10\times$ or $1\times$ masks
- Submicrometer linewidths
- Nanometer devices

Historically, the basic elements of scanning exposure, especially electron beams, were in the early 1960s already in commercial pieces of equipment for electron microscopy and cathode-ray tubes. Scanning *e*-beam exposure of photographic

TABLE 9-1-1. High-Energy Lithography

Radiation	keV	λ (Å)	Commercial	Area exposed
Electron beam	2–100	1	Yes	5 × 5 mm
Ion beam	10–300	1–5	No	1 × 1 mm
X ray	0.3–3	1–100	Almost	100 cm^2
Gamma ray	1000	0.1	Yes	1000 cm^2

film[1] was already a contemplated product for microfilm and television recording. The cathode-ray tube and electron gun (Figs. 9-1-1 and 9-1-2) were well established. In 1967, the cathode-ray tube was even contemplated as a rapid and economical photomask generator for the personalization of integrated circuits[2] (Fig. 9-1-3). The UV exposure mask was generated by exposure of a photographic (20) film attached to the faceplate of a cathode-ray tube (12). Discretionary wiring was accomplished by computer-aided design (22).

Instead of generating a hard-copy photomask, the electron beam was used in 1968 to directly expose a resist on a circuit board (Fig. 9-1-4). Pattern tracing was achieved by controlling the deflection of the beam in conjunction with the oscillatory motion of a substrate mounted on a stage with a grating (20) to feed back its position.[3] Later, direct silicon-write electron beams became available after the development of mask generators for writing on photographic plates[4] and later on resist-coated chrome plates.[5,6] Even though mask generators require up to 8 hr/plate using electron beams, they are considerably faster than optical reticle (flash on the fly) generators requiring many man-hours for exposure. The first manufacturing application for *e*-beam exposure directly on a silicon wafer was reported by IBM in 1976 for the personalization of the metal interconnections of bipolar chips.[8,9]

Ion-beam exposure systems for flood exposure of resists are based on the components of commercial ion implanters. X-ray exposure of photographic film was common in medicine and structural analysis when it was applied to expose resists to overcome deficiencies of UV proximity printing. Ion-beam and X-ray exposure are good contenders for step-and-repeat printing if durable masks and high-intensity sources can be developed. Electron beams generate masks for all other forms of lithography (X-ray, UV, and ions). Direct wafer write with electron beams will

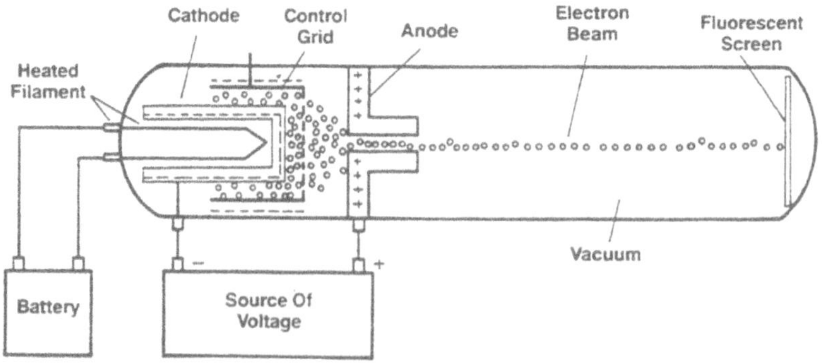

FIGURE 9-1-1. A simple electron-beam exposure unit, a cathode-ray tube. From Ref. 1, copyright and courtesy of Allyn & Bacon Publishing Inc.

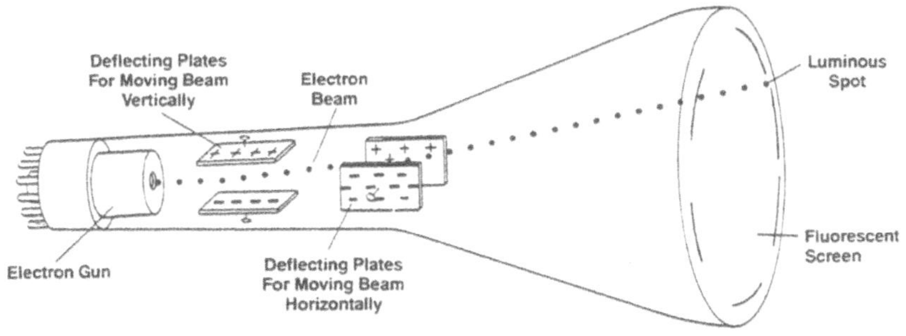

FIGURE 9-1-2. In a cathode-ray tube setup, the electrostatic deflection plates control the position of the beam to expose photographic film or resist deposited on the faceplate. From Ref. 1.

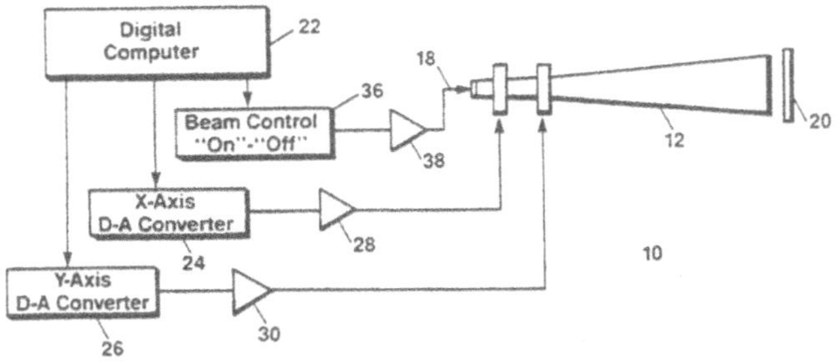

FIGURE 9-1-3. Schematic of an early *e*-beam mask maker for custom personalized circuits of low volume/high part number. A photographic plate (20) is placed against the faceplate of a cathode-ray tube (12). The circuit pattern drawn on the faceplate is concerted into an exposure photomask. U.S. Patent 3,581,385 (1971), Texas Instrument.

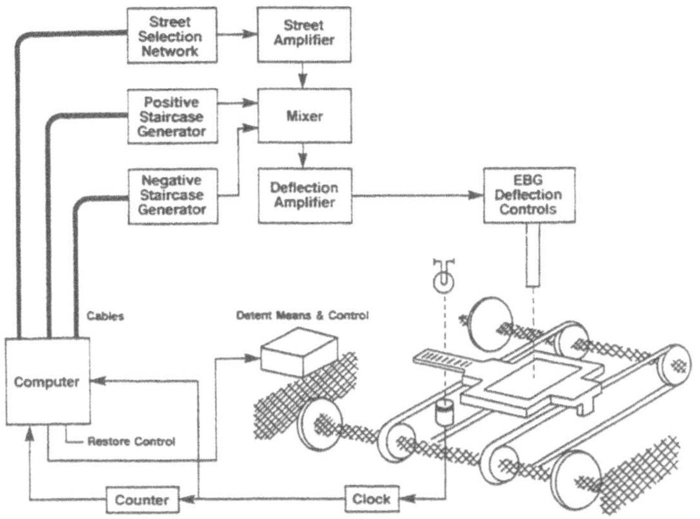

FIGURE 9-1-4. Schematic of an early *e*-beam exposure tool for directly exposing a photoresist-coated workpiece. U.S. Patent 3,575,588 (1971), IBM.

TABLE 9-1-2. Comparison of Mask and Maskless Exposure Techniques

	Maskless		Mask	
Exposure	Chip[a]	Wafers/hr[b]	Full wafer[c]	Wafers/hr[d]
UV	0.5–2.0	1	1.0–5.0	20–60
e-beam	0.1–2.5	1–5	3.5	*g*
X ray[e]	0.5–2.5	0.1	1.5–3.5	5
Ion beam[f]	0.5	0.1	5	10 (est.)

[a] Experimental production features (μm).
[b] Wafers per hour of 7.5-mm chips in 125-mm wafer.
[c] 125-mm wafer, exposure time in minutes
[d] 125-mm wafers/hr.
[e] In experimental stages.
[f] Highly experimental.
[g] For 125-mm wafer estimated at 36.(196)

slowly complement optical mix-and-match lithography and eventually dominate submicrometer lithography.

At present, a healthy competition exists between step-and-repeat (chip-by-chip) UV and UV scanning full-field exposure (Table 9-1-2) but perhaps in the near future flood X rays and *e*-beams will enter the marketplace.

The finest features which can be resolved are dependent on the shape of the incoming beam, its energy (keV), and the range of backscattered radiation. With all forms of radiation (Fig. 9-1-5), nanolithography is possible even with UV exposure if diffracted orders are used.(9,208,210,211,220) The smallest reported fabricated devices have been in the 1- to 10-nm range for electron beams and in one or two instances for X-ray or ion-beam exposure. Scanning exposures replicate features at about 5× the minimum(10) (Fig. 9-1-5) spot size. Since *e*-beam exposure is the dominant form of radiography, we will concentrate on this technology from a resist and

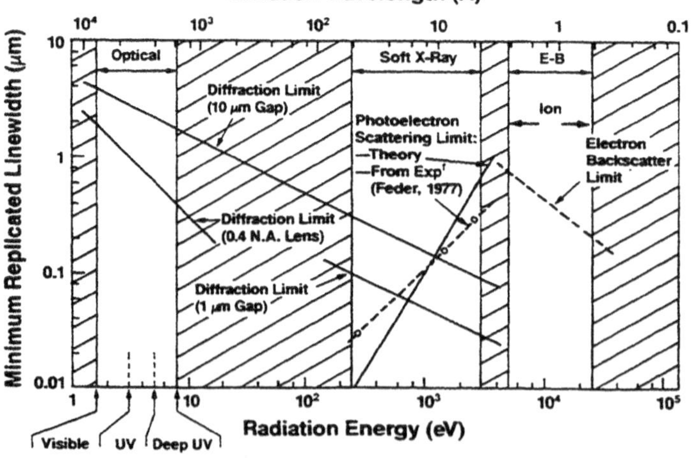

FIGURE 9-1-5. Minimum replicated linewidth = 5× minimum pixel versus radiation energy or wavelength. From Ref. 10, copyright 1980 and courtesy of Academic Press.

process viewpoint, examining the tool itself and the interaction with resists in the areas of contamination, registration, proximity effects, radiation damage, and nanolithography.

9-2. ELECTRON-BEAM PERFORMANCE

The next generation of submicrometer bipolar VLSI will likely utilize high-resolution *e*-beam lithography. Technically, the several key process and exposure tool issues are the machine performance, its resolution–alignment capability, and linewidth control (Table 9-2-1).[12]

The finest features which can be resolved are dependent on the edge slope of the beam just before it is absorbed by the resist. Two types of beam shapes are used: a round Gaussian beam and a square or rectangular-shaped beam (Fig. 9-3-3).[13] For a round Gaussian beam, the intensity distribution at the half-width of 50% equals the spatial resolution (d). To provide a true copy of the pattern, the half-width of the spot has to be at least five times smaller than the smallest element of the pattern. Hence, many more spots have to be written than a square-shaped spot. In a square beam, the size of the square spot is generally equal to the minimum pattern linewidth. To achieve equivalent resolution, the edge slope of the intensity distribution (defined as slope from 10 to 90% intensity) is made equivalent to the half-width of a Gaussian beam. The edge slope is on the order of 0.2 μm for a 1.5-μm square beam.[13]

In order to compare the performance of optical and *e*-beam resolution, Stickel and Langner[14] propose the edge slope contrast (C_{EC}) rather than the intensity modulation transfer function, C_{MTF}. The C_{MTF} is generally defined as the difference ratio of intensity maximum (Fig. 9-2-1). A value of 60% C_{MTF} has been declared as acceptable for processing based on an exposure process which allows for 25% exposure and loss of unexposed spaces. While resist thinning can be tolerated for wet etching, thinned resists do not provide sufficient protection in reactive ion etching wherein up to 50% of the resist is eroded. A 60% MTF also implies that the resist has a nonlinear response suppressing the unwanted exposure,

$$C_{EC} = (2/\pi) \tan^{-1}(W/E)\, C_{MTF} \tag{9-2-1}$$

where W/E is slope width in Fig. 9-2-2 at the 90 and 10% intensity points.

For comparison, the edge contrast of (coherent and incoherent, 254 nm) UV and *e*-beam exposure as a function of linewidth are shown in Fig. 9-2-3. In order to

TABLE 9-2-1. Electron-Beam Lithography—Key Issues[a]

Resist related	Machine dependent	Linewidth control
Resolution	keV	Bias
Sensitivity	Dose control	Proximity effects
Process	Alignment	Edge slope
Compatibility	Throughput	Address increment
Thickness	Radiation damage	

[a] From Ref. 12.

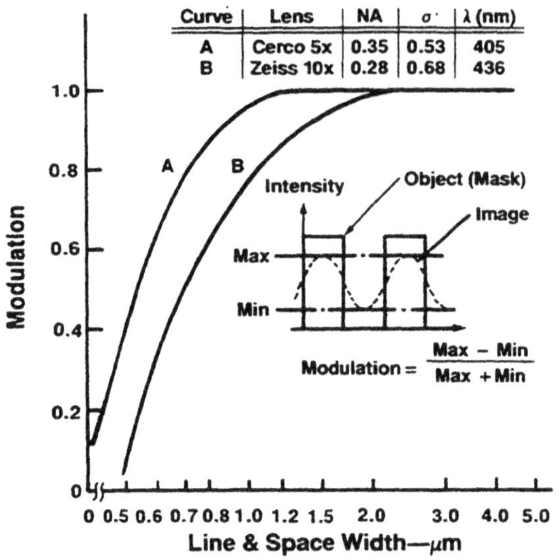

FIGURE 9-2-1. Modulation transfer function of the transfer efficiency of optical lens to convert the image of a mask and project the (intensity) pattern onto a resist without distortion. From Ref. 7, copyright and courtesy of AIP.

maintain an edge slope of one-fourth the maximum linewidth as well as maintain the alignment to the same accuracy, the edge contrast must stay above 70% contrast in the case of *e*-beams and 83% for optical exposure. Only an optical tool with coherent radiation (standing waves included) in the deep UV can match the image quality achievable by a 1-μm *e*-beam using a fixed or variable shaped beam. However, the edge contrast or energy deposited is only valid for negative[15] resists which are independent of the rate of development (development time). For a positive resist, the rate of development (R) or profile as a function of depth (Z) is

$$dR/dZ = (dR/dE)(dE/dZ) \qquad (9\text{-}2\text{-}2)$$

The response of dissolution to the absorbed energy (E) is the first term on the right side while the second term is the edge slope influence. Only at high dose (100% MTF) does the developer have a small influence depending on the resist contrast and the induction (nonlinear response of a resist).[16] At moderate to low doses, the developer significantly contributes to the shape of the image.[221] When a resist is sufficiently thin, the absorbed energy distribution can be considered to be uniform in a direction normal to the surface.[15,19,179] When the resist is thicker

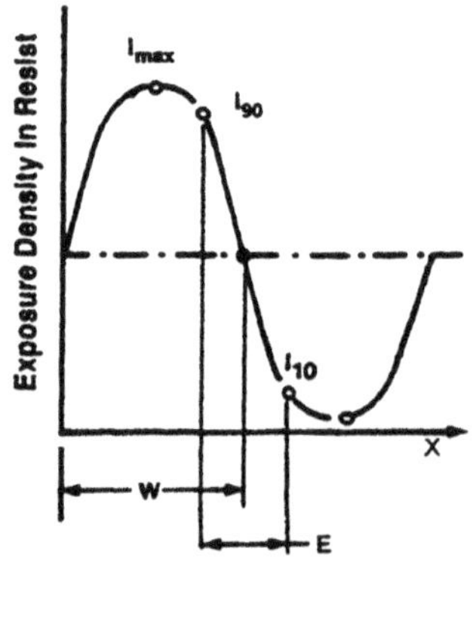

FIGURE 9-2-2. Exposure density profiles and quantities used in contrast definitions for an *e*-beam exposure. From Ref. 14, copyright and courtesy of AIP.

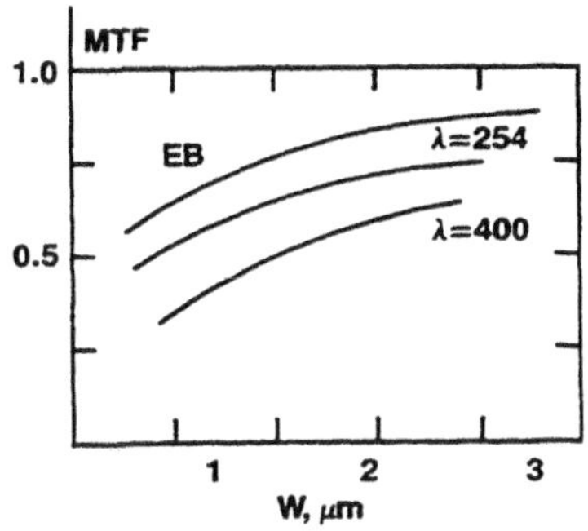

FIGURE 9-2-3. Edge contrast for 254-nm UV exposure compared to *e*-beam exposure of a 0.5-μm-thick film of resist on silicon. From Ref. 14, copyright and courtesy of AIP.

(> 1.5 μm), the absorbed energy distribution is not uniform and the developed process must be considered, wherein the absorbed energy distribution in the nearest layer to the substrate will determine the developed pattern dimension.[33,113] The combination of a low-contrast resist ($\gamma = 1$–3) and a beam with a steep edge slope can give a similar resist profile as a high-contrast resist at moderate doses, provided very tight process controls are exercised.[113,221]

Because of distortion of the edge contrast (slope) by backscattered electrons, the removal[17] of the substrate (Fig. 9-2-4A) will serve to eliminate backscatter from the silicon wafer.[23] Secondly, the use of thin[18] resist layers such as in multilayer resists (Fig. 9-2-4B) lowers the MTF and enhances the resolution of *e*-beam tools. Multilayer resists are the most practical way to reduce proximity effects and improve resolution.[15,247] Figure 9-2-5 is an example of a pattern made in a single 2.0-μm-thick *e*-beam resist and the same pattern made in a polysiloxane bilayer

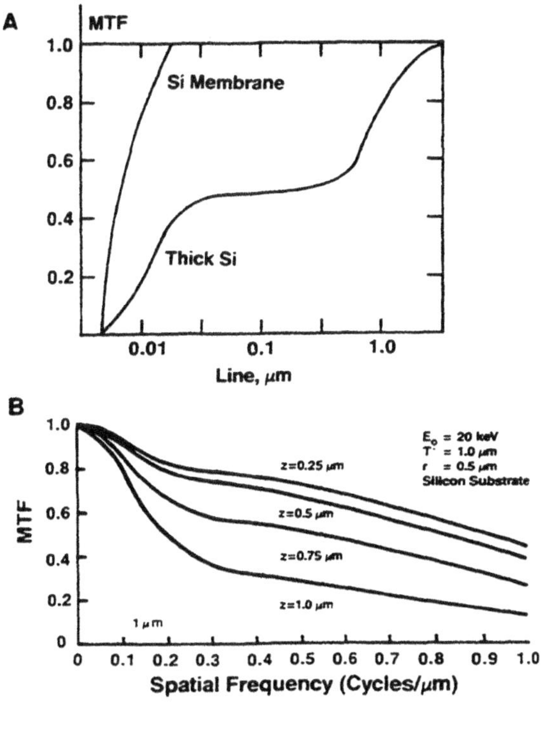

FIGURE 9-2-4. (A) MTF for *e*-beam on thick and thin substrates. From Ref. 17, copyright 1981 and courtesy of IEEE. (B) MTF for *e*-beam as a function of film thickness. Thinner films reduce proximity and improve resolution. From M. Nakase and Y. Matsumoto, *Photogr. Sci. Eng.* **23**, 215 (1979), copyright and courtesy of SPSE.

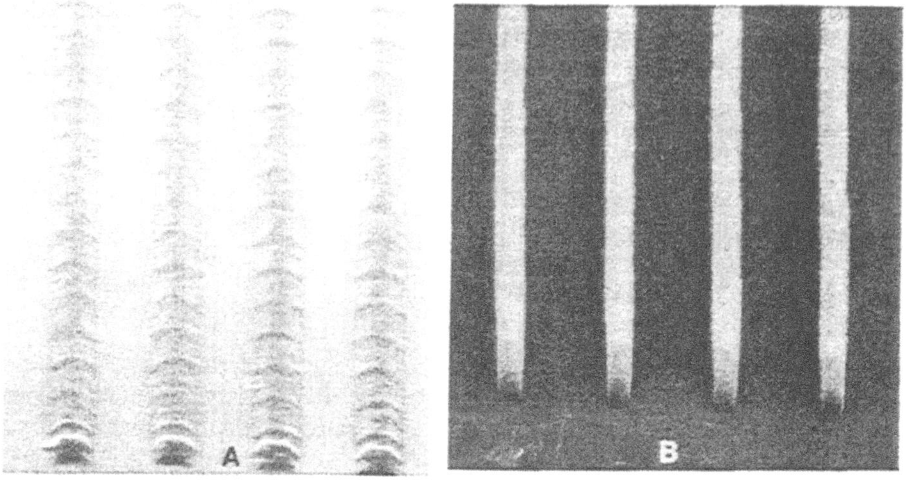

FIGURE 9-2-5. (A) Image in 2.0-μm-thick single film of resist distorted by backscattered radiation from silicon substrate and (B) the same image made by using a thin (0.5 μm) image layer in a trilayer resist (see Chapter 12) stack.

resist. The single film resist suffers from interproximity backscattered electrons. In a single resist film, backscatter effects can be put to work by top surface imaging (Fig. 9-2-6).[20] The backscatter covers the bottom of the positive resist image while the opening at the top of the film serves as a perfect lift-off mask.[21]

Along with high resolution, a reasonable silicon wafer throughput through the *e*-beam exposure tool has to be achieved. Some of the important throughput factors are listed in Table 9-2-2. When compared to optical exposure systems, *e*-beam exposure cost per wafer is an order of magnitude higher but the systems are justified on the basis of quick turnaround (no photomasks required) and where a multiple of chip exposures is required. Because of electron–electron interactions (aberrations)[22,63] in an *e*-beam of decreasing size, the beam current is lowered for

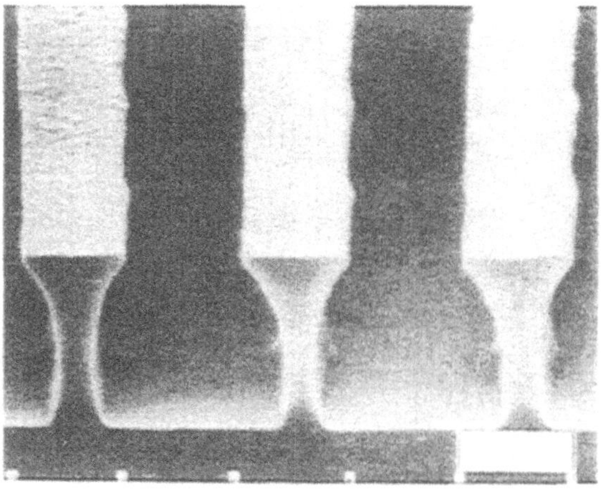

FIGURE 9-2-6. Example of top surface imaging of a single film of resist by *e*-beam exposure. The top surface image is defined by the forward scatter in the top 100-nm thickness of the resist. The profile is suitable for lift-off metal definition which is defined by the top width opening of the resist (see Chapter 12). From S. Gillespie, *IBM J. Res. Dev.* **28**, 457 (1984), copyright and courtesy of IBM.

TABLE 9-2-2. Electron-Beam Throughput Factors

Round spot versus square spot versus variable shaped spot
Beam intensity versus spot size (Fig. 9-2-7)
Resist dose
Proximity corrections (dose, spot size)
Overheat (align and step)
Size of chip and wafer
Pattern density
Vector or raster
Statistical dose limit

smaller exposure spots (Fig. 9-2-7) resulting in longer exposure times for submicrometer features. With the exposure time (t) given by

$$t = \text{Resist dose}/I(\text{A/cm}^2) \qquad (9\text{-}2\text{-}3)$$

the exposure dose required for resist ranges from 1 mC/cm^2 to 1 μC/cm^2. Exposure systems such as the variable shaped spot tool which use the largest spot possible can expose 1.0-μm features at the rate of 5–10 (125-mm) wafers/hr, while a Gaussian raster tool such as EBES has to use 0.025-μm-diameter beams at lower intensity and thus requires up to several hours to expose one wafer. Lowering the resist sensitivity by an order of magnitude to 1 μC/cm^2 increases the throughput by about a factor of 3. Exposure times per spot of 0.1 to several microseconds are common with W or LaB_6 guns emitting 0.1–50 A/cm^2 and field emission guns can reach[(11)] 10^6 A/cm^2 but are somewhat unstable. Because of lens aberrations and space charge effects, all the (10^6 A/cm^2) available source intensity cannot[(22)] be focused[(63)] into nanometer-sized spots, to make up for the larger (10^2–10^4) area to be exposed. About the same 1–100 A/cm^2 intensity is maintained for a 1-μm spot as a 1-nm spot with the same exposure time per spot.

Nanometer lithography uses beams of 1 nm at 50 keV and 10^{-11} A to expose monomolecular films[(26)] or 10^{-12} A at 100 keV to expose 60-nm-thick films of PMMA.[(11)] For nanolithography, we also encounter statistical problems of insufficient electron dose in a nanometer spot. For example, with a 100-nm spot we would like to use a more sensitive resist to enhance the throughput and as thin a resist as possible (100 nm) to increase the resolution[(242,248)] (Table 9-2-3).

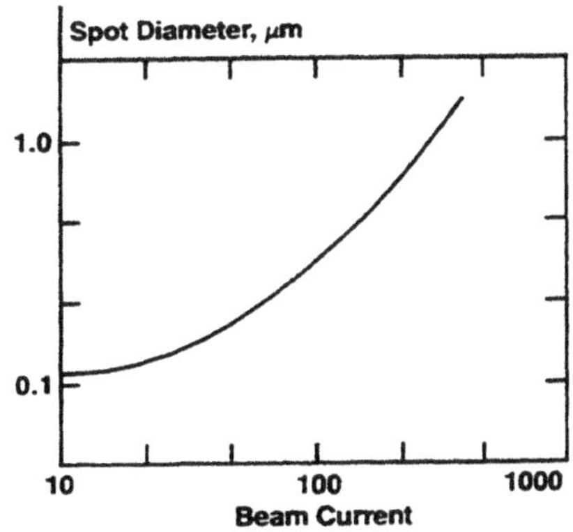

FIGURE 9-2-7. The current available in a focused e-beam system as a function of beam spot diameter. From Ref. 22, copyright and courtesy of IEEE.

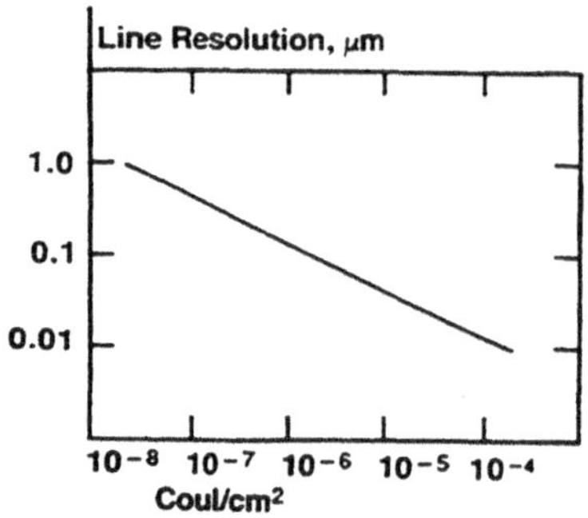

FIGURE 9-2-8. Statistical shot noise limit of *e*-beam exposure of higher dose needed for nanometer dimensions. From Ref. 25, copyright 1980 and courtesy of Academic Press.

FEATURE SIZE, μm			*ALIGN LEVEL, μm*
2.0 1.0 0.5	FULL FIELD EXPOSURE 1990	SUB FIELD EXPOSURE 2000	0.5 0.25 0.12
0.25	LIMITED BY DEVICE SIZE		0.06
0.125 0.062	NANOLITHOGRAPHY		0.03 0.01
0.03 0.01	MOLECULAR DEVICES		0.006 0.002

FIGURE 9-2-9. Linewidth and alignment requirements for future devices. Adapted from Ref. 172.

ELECTRON BEAM TOOL

Spot size Edge slope keV Dose Pattern factor Proximity

↓

SUBSTRATE

Quartz photomask Silicon wafer X-Ray mask Ion-beam mask

↓

RESIST

Positive Negative Thickness Multi- or Single-layer
Contrast Heating effects Postexposure reactions

↓

PHYSICS

Resist outgassing Heating effects Charging Registration

FIGURE 9-2-10. Considerations for selection of an *e*-beam resist.

TABLE 9-2-3. Dose and Electron Flux for a 320-nm Spot

Dose (μC/cm^2)	Number of electrons required in a 0.1-μm^2 area
50	30,000
5	3,000
0.5	300
0.05	30

The number of resist molecules present in a corresponding film (spot diameter = resist thickness) is calculated in Table 9-2-4. In a 100-nm spot there are about 10^2 chains to be broken or 10^3 molecules to be rearranged. At least 10^3 electrons or a dose $> 2\ \mu\text{C/cm}^2$ is required (Fig. 9-2-8). Because of the statistical nature of electron emission, the shot noise limit is around 200 electrons[24,25] and the limiting dose is given by

$$D = 1/(\text{line size})^2 \tag{9-2-4}$$

Because of the higher dose, nanometer lithography will require an inordinate writing time (Table 9-2-5) for a single 125-nm wafer, unless field emission[11] sources[26] can be invoked, or a resist which can be used that is image amplified at 0.01 μC/cm^2 doses.[27,28] In addition, there are fundamental physical factors in electron optics, analog circuitry, beam settling, circuit delays, and step time which limit the spot size, dose, exposure time, and deflection field. These factors, along with the above beam intensity, and edge slope, in turn limit the throughput to 1–10 wafers/hr for 100- to 150-mm-diameter wafers on the fastest *e*-beam exposure tools based on the variable shaped spot.[30,31]

Although high resolution and throughput are opposing or neutralizing factors in *e*-beam lithography, a stronger technical lithography driver for present and future device fabrication is the outstanding precision and accuracy of level-to-level overlay (Fig. 9-2-9). Electron-beam exposure systems have achieved automated alignment to $\pm 0.1\ \mu$m (3σ) (see also Table 9-5-2) and *e*-beam alignment schemes have even been proposed to improve the accuracy of X-ray proximity[246] and ion-beam proximity printers.[251]

To complement the *e*-beam tool performance, the choice of resist depends on many process and tool considerations (Fig. 9-2-10). Factors such as resist outgassing, charging, and susceptibility to the proximity effect are as important to the

TABLE 9-2-4. Molecules per Image Size Features

	Image size (Å) and thickness of resist (Å)			
	10,000	1000	100	10
Molecules/cm^2	10^8	10^6	10^4	10^2
Molecular chains/cm^2	10^6	10^4	10^2	10

TABLE 9-2-5. Estimated Exposure Time for a 125-mm Wafer at Various Beam Sizes

Beam size	Expose time	Dose
1 μm	10 min[20,29]	1 $\mu C/cm^2$
100 nm	20 hr	10 $\mu C/cm^2$
10 nm	4000 hr	100 $\mu C/cm^2$
1 nm	?	>100 $\mu C/cm^2$

tool performance as the process consideration of the durability of the resist to reactive ion etching. We will discuss some of the resist–tool interactions in the next three sections.

9-3. ELECTRON-BEAM EXPOSURE EQUIPMENT

The principal requirements of any lithographic system are (1) critical dimensional control, (2) alignment accuracy, (3) cost effectiveness, (4) flexibility, and (5) compatibility with other exposure tools. For a tool configuration, there are several options available for *e*-beam[32,197] scanners:

- Round Gaussian probe versus a shaped beam.
- Step-and-repeat table versus a continuously moving one.
- Electron source: W filament, LaB_6 emitter, field emission tip, a single source versus multiple sources or images.
- Proximity corrections (neighboring exposure influence) by dose, windage (beam bias), or a combination of linewidth and dose.
- Variable beam voltage (5–100 keV).

Besides scanning beams modeled after SEM, full wafer or full chip exposure have been explored. From work done at Westinghouse in the late 1960s, a full wafer photocathode exposure system has been developed into a potential 10-keV exposure tool (Fig. 9-3-1).[231] A CsI cathode pattern is irradiated by deep-UV light to emanate a stream of electrons which are collimated by an external magnetic field to expose a wafer. Poor cathode life has hampered commercialization of the tool. In close analogy to optical steppers, *e*-beam projectors of 10×, 4×, or 1× chips have also been examined as step-and-repeat exposure systems (Fig. 9-3-2).[33] Electron-beam proximity printers show (Table 9-3-1) the best promise as a tool capable of competing with X-ray proximity printing with the distinct advantage of the fact that the beam can be accurately aligned by electronically shifting or tilting and that it is easy to obtain millisecond per chip exposure times. The major disadvantages are the complementary two-level stencil masks needed for isolated patterns and the difficulty of correcting for proximity effects by adjusting the windage of the mask.

Several different scanning tools have been configured for mask making and for direct silicon write. For mask making, no registration is needed and throughput is not a major issue. Conventional optical reticle generators require 40 hr for 10× single segments (10^6 flashes) while an *e*-beam can produce the reticle in less than an hour.[34]

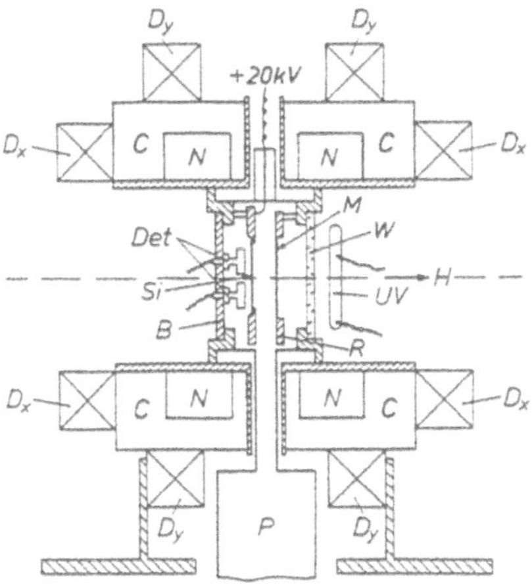

FIGURE 9-3-1. (A) Photocathode image projector. Schematic cross section of the electron-image projector. M is the mask, which is coated with a layer of cesium-iodide photoemitter and Si the silicon slice to be exposed, covered in an electron-sensitive resist. M and Si are situated in a vacuum chamber closed by the window W and the disk B. The chamber is pumped via a port P. Outside the vacuum, to the right of the window, is the ultraviolet lamp UV. The radiation from this lamp (184.9 nm) causes electrons to be emitted by the photoemitter behind the clear areas of the mask. The electrons are accelerated to S by an electric field (15 kV/cm) and focused by the highly uniform magnetic field H (about 1 kOe) provided by the split solenoid C, with correction notches N. The projected image is aligned in the x and y directions by means of currents through the deflection coils D_x, D_y; rotational alignment is provided by mechanical rotation of the mask holder R. Alignment signals are obtained by markers on the slice which emits X-ray signals which are detected by solid-state detectors (Det). From J. Scott, courtesy of *Philips Tech. Rev.* **37**, 347 (1977).

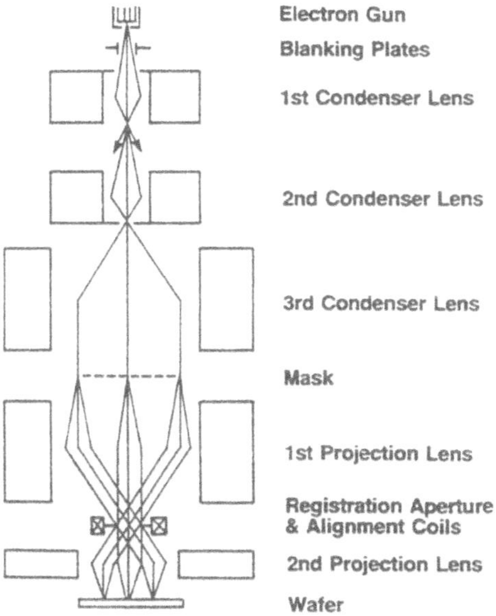

FIGURE 9-3-2. Electron foil (10×) projection system. From Ref. 54, courtesy of AIP.

TABLE 9-3-1. Type of Electron-Beam Exposure Systems

System	Wafers/hr, 82 mm	Reference
Round (Gaussian) raster	1–2	5, 35, 36
Round vector scan	2–3	37, 38
Square (rectangular) raster	10–20	13, 39
Variable shape, raster/vector	20–40	40–47
Multiple round, fly's eye	*a*	48–51
Proximity print	*a*	52
Projection print	*a*	53–55
Photocathode proximity	20–25	56–60

[a] Under development—wafers/hr of 10–40 estimated.

The basic elements of an *e*-beam exposure system are a source, a spot projection lens, a beam blanking system, a registration and deflection diode, a motorized stage, and a computerized interface. Apart from an electron gun and lenses, the important components are[32]:

- A beam blanking unit of either electrostatic or electromagnetic type to switch off the electron beam by deflecting it off the optical axis.
- A deflection unit of either the double deflection type housed inside the bore of the final lens or a single deflection type located after the lens. Although both electrostatic and electromagnetic deflection units can be used, the latter is generally preferred for reasons of lower aberrations and better immunity to surface charging effect.
- A dynamic focus correction unit to overcome deflection aberration difficulties.
- An electron detection system to collect electron signals for viewing of registration marks and other features on the target surface.
- A high-precision workstage with suitable motorized traverse to cover the entire wafer.
- A vacuum system.

To produce LSI and VLSI circuit patterns on silicon chips, an electron beam (A) (Fig. 9-3-3) is magnetically focused on a resist-coated circular wafer situated on an X–Y worktable (B), forming a specified image. This tiny image is then sequentially positioned and moved by a combination of magnetic and electrostatic controls (C, D) along areas on the wafer to be exposed (E). During this process, *e*-beam on–off control (F) is used to selectively expose specific patterns in areas on the silicon wafer, as determined by digital design data stored in a computer (G). An electron detector (H) monitors and coordinates the exposure patterns on each chip. A calibration (I) is used to establish and maintain the positioning and movement accuracy of the beam during exposure operations to less than 0.15 μm.[61]

In the early days of *e*-beam lithography (1960–1970), many commercial SEMs were modified with a beam blanking unit, a motorized stage, and computer interface.[35,36] The same setup is now available commercially for linewidth measurements of resist images below 2 μm. Instead of a digital or analog drive, early workers used flying spot scanners to trace a normal chrome or emulsion photomask onto a wafer.[30]

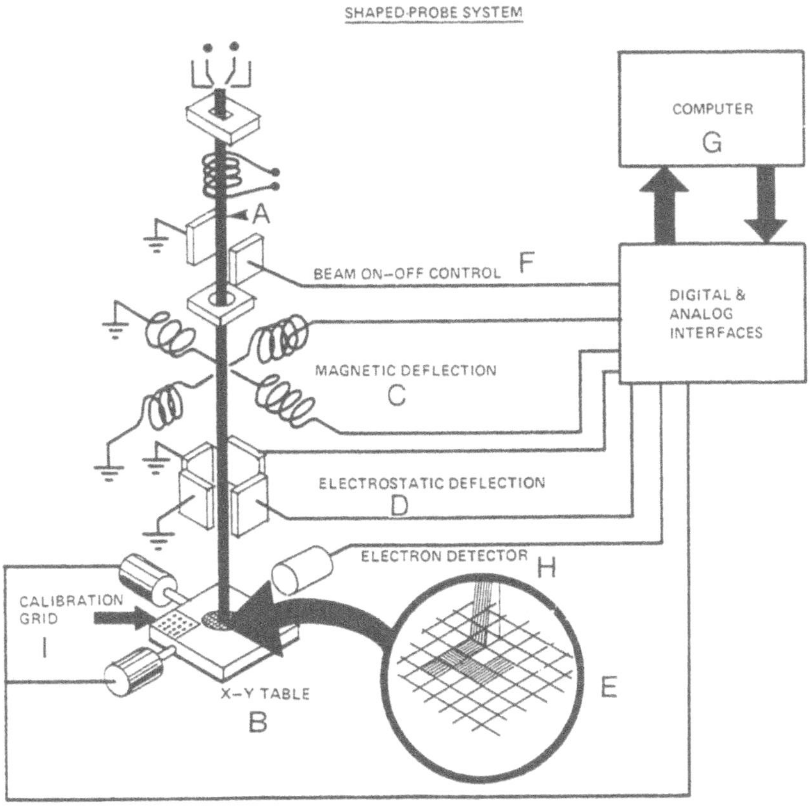

FIGURE 9-3-3. Schematic of square *e*-beam system. From Ref. 8, reprinted with permission of *Solid State Technology*, published by Technical Publishing, a company of Dun and Bradstreet.

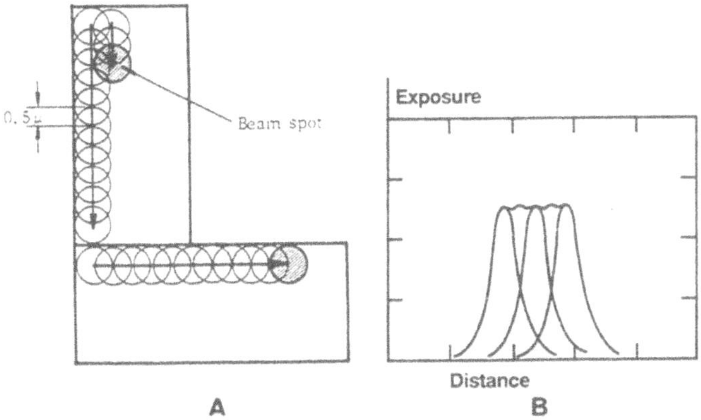

FIGURE 9-3-4. (A) Raster writing with round Gaussian beam. From Ref. 200. (B) Formation of a single circuit line by overlap of Gaussian beams. From Ref. 201, reprinted with permission of *Solid State Technology*, published by Technical Publishing, a company of Dun and Bradstreet.

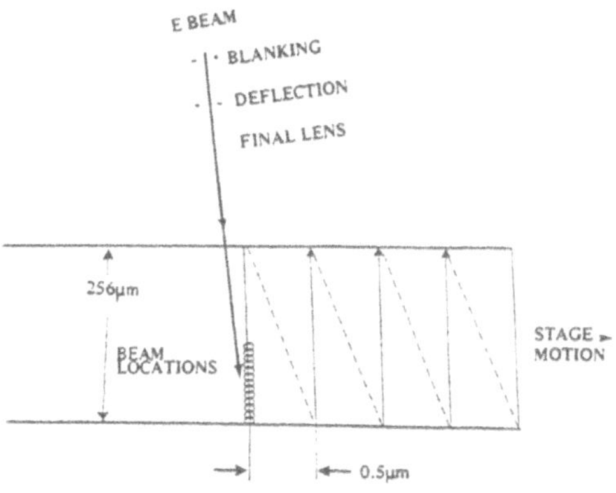

FIGURE 9-3-5. Principle of MEBES (moving electron-beam exposure system). From Ref. 34, reprinted with permission of *Solid State Technology*, published by Technical Publishing, a company of Dun and Bradstreet.

The raster scheme typically divides the chip pattern into subfields where each subfield is scanned in a TV raster.[81] The *e*-beam spot is blanked (off and on) along the serpentine path (Fig. 9-3-4a). A round Gaussian spot about one-quarter the finest dimension is overlapped to form a pattern (Fig. 9-3-4b). Some edge scallop from intraproximity effects is noted.

To speed up the raster scan for mask making, workers at Bell Labs used a moving stage to overcome the small deflection (0.3 μm) of the round *e*-beam. The *e*-beam is deflected along the *y* axis while the stage advances in the *x* direction. The beam voltage is 10 kV for reduced proximity effects (Fig. 9-3-5).[34] The curing of negative resists such as epoxy COP is performed in a cassette holder inside the vacuum system.

Another round beam[38] version uses a vector scan mode. The *e*-beam is deflected to only the areas requiring exposure (Fig. 9-3-6). For low-density patterns such as via holes, the throughput can be significantly enhanced. Comparison of raster and vector scanning is summarized in Table 9-3-2.

In order to speed up the (3×) throughput of raster round beams, a new field emission gun operating at 300 A/cm^2 and a fast D-A converter have been developed achieving 10–15 (100-nm) wafers/hr.[64] In further developments (Fig. 9-3-7), a high-current (300 A/cm^2) field emission round beam raster system has been reported[74] to achieve 10 wafers/hr at 2.0-μm features and 5 wafers/hr at 1.0-μm features.

Gaussian beams require overlap of four to five smaller spots to form a 1-μm square and suffer from loss of beam intensity (Fig. 9-3-7). Projection of a square[39] or variable rectangular shapes enhances the throughput by providing a square 1-μm spot with a flat cross-sectional current profile (Fig. 9-3-8).[13] Even higher

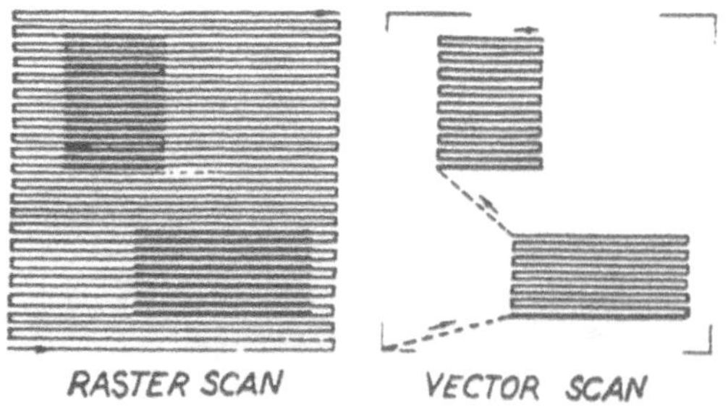

FIGURE 9-3-6. Writing path of raster and vector *e*-beam exposure. From Ref. 38, courtesy of AIP.

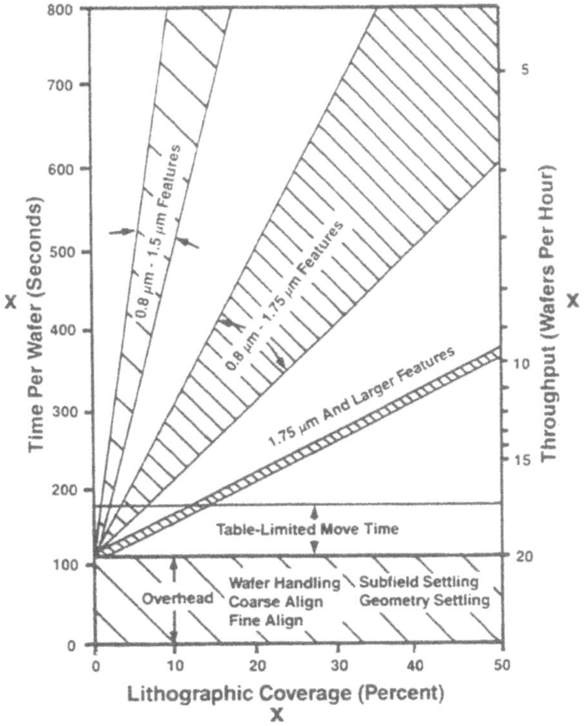

FIGURE 9-3-7. Wafer throughout achieved by variable spot size, high current density, Gaussian beam. From W. Livesay, *Microelectron. Manuf. Test.*, March 1984, p. 58, copyright and courtesy of Lake Publishing Co., Libertyville, Ill. 60048.

throughput can be obtained by projection of full characters(64) using a combined raster/vector mode (Fig. 9-3-9). The reduced writing time for a variable character projection is evident in Fig. 9-3-10 which compares a round, a fixed shape, a variable shape, and character projection. The larger the area projected by one exposure, the higher is the throughput. Enhancement of 16 to 100 in write time reduction over Gaussian beams has been achieved (Table 9-3-3) with variable shaped spot systems. Another significant advantage of the shaped beam approach is that throughput is independent of the address increment (lsb). The address increment is the precision with which the edge of a shape can be placed within the exposed area.(188) With a round beam raster scan system, the address and edge definition are set by the beam diameter. If the address has to be smaller than the edge definition, then a smaller round beam must be used at lower intensity (longer exposure time per spot and hence a significant throughput hit).

TABLE 9-3-2. Comparison of Raster and Vector Scanning Round Beams

Raster	Vector
Both positive and negative resist	Preferably positive
Lower cost	Need good deflection coils
Can be used at 1-nm spot	Spot limited
Slow spot overlap	Faster, direct write
Not readily proximity corrected	Ease of proximity correction
Need fast D/A	Ease of data compaction

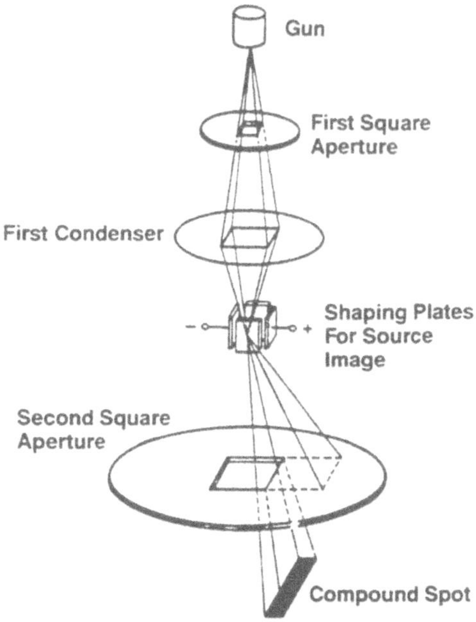

FIGURE 9-3-8. Compound rectangular-shaped *e*-beam exposure system. U.S. Patent 3,894,271 (see also Ref. 39).

The pattern density (factor) is very important in variable shaped beams (Fig. 9-3-11). The throughput for sparse patterns is independent of resist dose (sensitivity) or exposure time but for dense patterns the throughput is proportional to the writing time and hence faster resists (1 $\mu C/cm^2$) are required.[20]

The key internal element to the success of variable shaped beams is the final variable axis lens[65,66] or a moving[67–73] objective lens whose focus can be changed to correct for electron–electron aberrations. Some of the sacrifices for square and

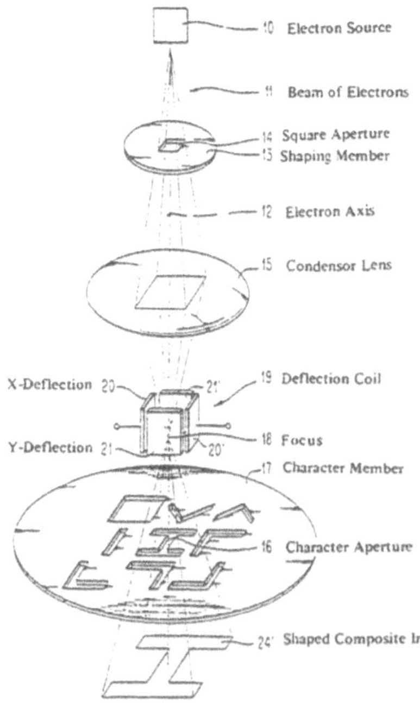

FIGURE 9-3-9. Full-character *e*-beam projection. U.S. Patent 4,213,053.

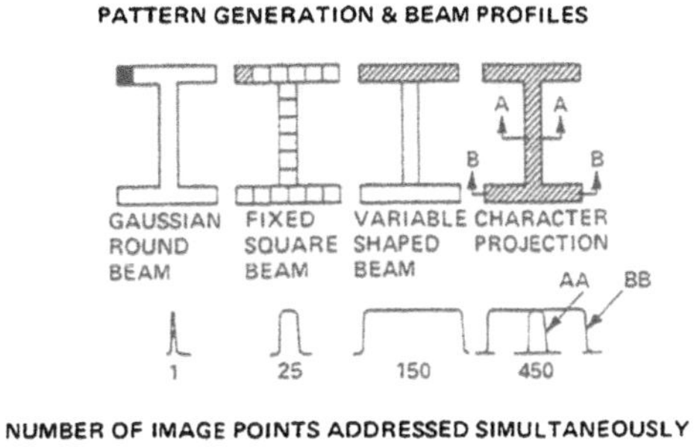

FIGURE 9-3-10. Number of image points written by a round, square, variable shaped spot, and character projection. From Ref. 8, reprinted with permission of *Solid State Technology*, published by Technical Publishing, a company of Dun and Bradstreet.

variable shaped beams lie in the electron optic complexity and a lower beam intensity as compared to Gaussian beams for spots > 1.0 μm since only a portion of the source output is used (Fig. 9-3-8). Since the beam current also decreases with the reduction[45] of field (chip) size (Fig. 9-3-12), faster resists (< 2 μC/cm^2, Fig. 9-3-13) are needed to compensate the loss of beam intensity.[46]

In a variable shaped spot system, the pattern density and minimum dimension can have a dramatic effect on exposure time per chip and throughput (Fig. 9-3-14).[20] For a low pattern density, a positive *e*-beam resist would obviously be used for a low-density via and a negative resist for high etching areas.

Writing schemes to overcome the pattern factor include the outline exposure of a pattern followed by metal deposition and electrolysis to remove excess metal (Fig. 9-3-15). This scheme has been termed "canyon" or "brushfire" lithography[75,204] with reference to outlining of a region and consumption by contraction. A combination of *e*-beam and optical exposure has also been used to speed up the writing of slow DQN resists.[76] The *e*-beam writes the smallest features and a complementary photomask is aligned and photoexposure completes the resist exposure of the larger features.

The throughput of *e*-beam tools is composed of writing, align, move, and

TABLE 9-3-3. Pattern Writing Time/Chip[a,b]

System	Type scan	Beam stepping rate (MHz)	Resist sensitivity (C/cm^2)	Beam current density (A/cm^2)	Beam addresses/ chip	Minimum geometry (μm)	Pattern writing time (sec)/chip
EBES	Raster	20	10^{-6}	20	64×10^6	2.5 (5 passes)	3.2
EBES	Raster	20	10^{-6}	20	256×10^6	1.25 (5 passes)	12.8
VSI	Vector	10	2.5×10^{-5}	300	16×10^6	2.5 (5 passes)	1.6
VSI	Vector	10	2.5×10^{-5}	300	64×10^6	1.25 (5 passes)	6.4
EBM III	Vector	20	5×10^{-5}	120	64×10^6	1.25 (5 passes)	3.2
EBM III	Vector	10	5×10^{-5}	60	16×10^6	2.5 (5 passes)	1.6
EBM VB	Vector	100	10^{-6}	120	64×10^6	1.25 (5 passes)	0.6
EBM VB	Vector	50	10^{-6}	60	16×10^6	2.5 (5 passes)	0.3
EL1	Raster	5	10^{-5}	50	4×10^6	2.5 (1 pass)	0.8
EL1	Raster	5	10^{-5}	50	16×10^6	1.25 (1 pass)	3.2
EL2	Vector	5	10^{-5}	50	4×10^6	1.25 (1 pass)	0.8
EL3	Vector	5	10^{-5}	50	1×10^6	1.25 (1 pass)	0.2

[a] From Ref. 82.
[b] Chip size, 200 × 200 mil^2; pattern area, 25%; variable shaped beam.

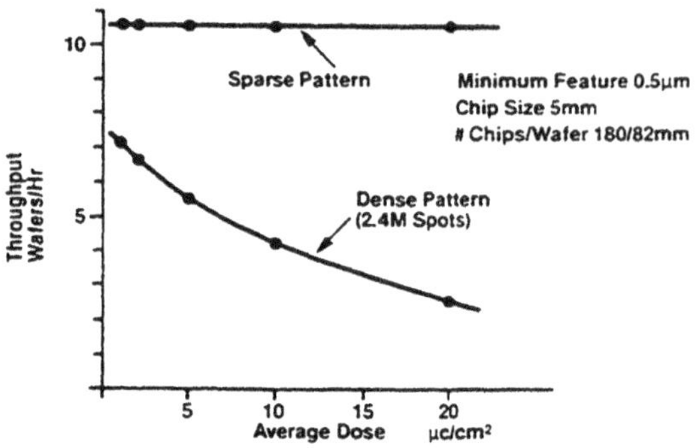

FIGURE 9-3-11. Throughput of a variable shaped beam as a function of pattern factor. From Ref. 20, courtesy of AIP.

overhead. A breakdown of the overall throughput rate of an *e*-beam system is shown in Fig. 9-3-16.[74]

The time (T) for drawing a pattern by *e*-beams is the sum of the exposure time, t_e, and the waiting time, t_w,

$$T = t_e + t_w \tag{9-3-1}$$

The t_w is the sum of times for data transfer, stage movement, position adjustment, and so on. Although t_w is never negligible, our present concern is mainly on t_e. If we use *e*-beams of current density j which expose an area a at the same time t, then the time t_e for exposing a field of area A is given by

$$t_e = k(S/j)(A/a) \tag{9-3-2}$$

where k is the ratio of actual scanning area to A and S is the resist sensitivity.

The value of k is determined by the pattern and the scanning scheme ($k = 1$ in the raster scanning scheme, and $k = 0.2$–0.4 in the vector scanning scheme). Therefore, in order to reduce the exposure time t_e, one must increase the beam intensity or ja, or the total beam current.[51] The overhead time consists of data processing,[80] and subfield and table settling. Data processing requires harmony between the shape of the circuit pattern design and the basic digital system formats used by the beam to pattern a configuration.[29]

Although many classes of pattern designs contain regular repeats of pattern

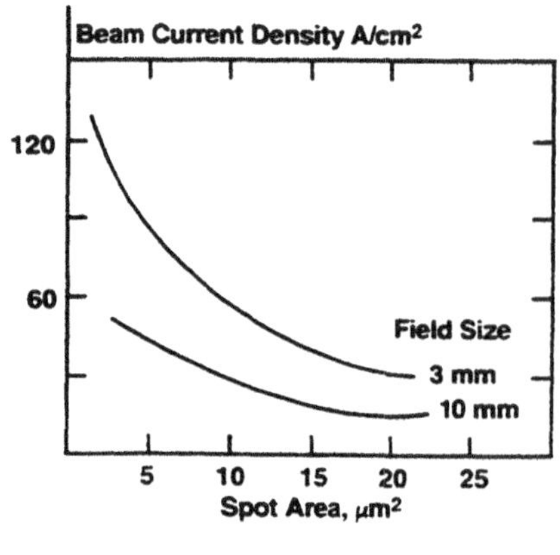

FIGURE 9-3-12. Beam current in variable shaped spot versus area and field size. From Ref. 45, courtesy of AIP.

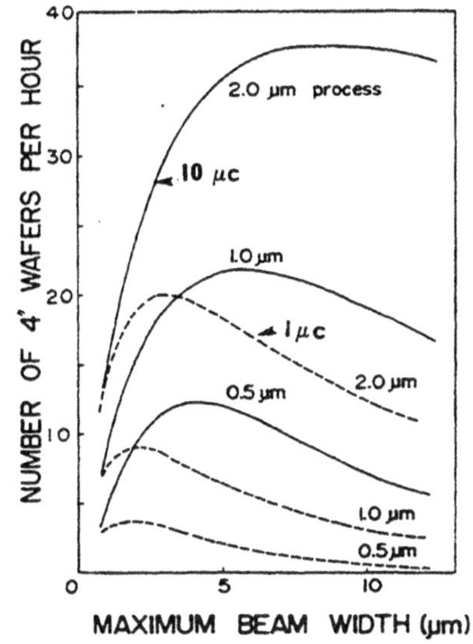

FIGURE 9-3-13. Throughput of a variable shaped spot system. Dashed lines, 10 μC; solid lines, 1 μC. From Ref. 46, courtesy of AIP.

shapes, design manipulations such as personalization,[77] image reversal for proximity correction,[78] or *e*-beam postprocessor functions (partitioning and dose correction for proximity) must be efficiently written to reduce writing times. Digital control of pattern writing can be formed in five basic formats[79,94]:

1. Dot matrix format—the mask pattern is divided into a dot matrix and stored in rectilinear coordinates.
2. Definition by lines—pattern is expressed as a series of vectors.
3. Raster scan—entire pattern is covered by a television-type raster.
4. Rectangular division—mask is divided into a series of constituent rectangles.
5. Pattern synthesis—mask is synthesized from basic patterns held in computer.

Each format must be able to accommodate negative or positive resists. For large dimensions ($> 2.0\ \mu m$), raster scan systems have the advantages of high beam

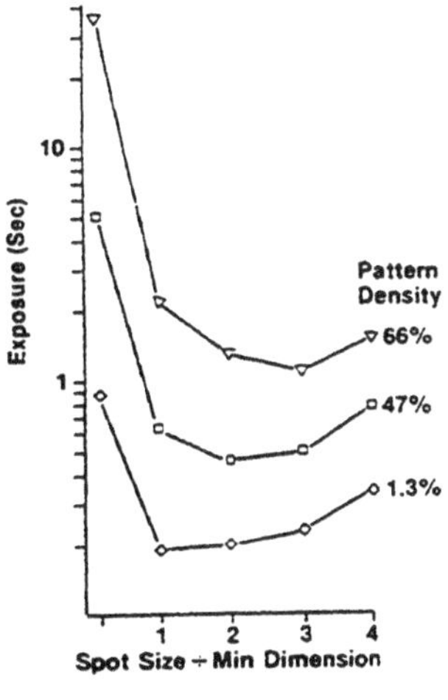

FIGURE 9-3-14. Maximum spot size effect on exposure time for various pattern densities. From Ref. 45, courtesy of AIP.

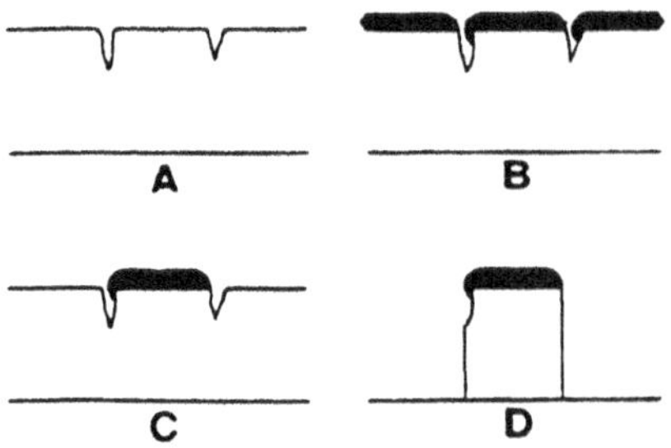

FIGURE 9-3-15. Canyon lithography; a cross section of the resist taken through a rectangular feature at various stages of forming the structure. (A) After canyon formation. (B) After oblique metal evaporation. (C) After toning by removing the metal of the feature. (D) After completion by, say, ion milling with oxygen. From Ref. 75.

current, accurate pattern definition, and short overheat time. However, the systems require a large amount of bit data to be transmitted to the writing interface with appropriate speed without ceasing while a work stage is continuously moving.[30] A vector scan system is limited in throughput by the overhead rate and exposure rate. The overhead rate is a function of the pattern generator, deflection amplifier, and electron optics.[31]

In variable shaped spots, the major problem lies in the writing[32] of nonrectangular shapes and in proximity corrections by dose partitioning. The partitioning of a large shape into smaller shapes for proximity correction leads to high data[32] handling problems and loss of throughput. Shaped beams will likely require multilayer resist structures to handle the proximity effects.[74] Varnell *et al.*[82] have assessed the writing time of commercial and reported round and variable shaped raster and vector systems (Table 9-3-3). The fastest system is the vector EL3 reported by workers at IBM.[20,40,82,83] For the EL3, a faster resist (1 $\mu C/cm^2$) would reduce the writing time per chip from 0.2 sec to 0.07 sec but even faster resists would not be useful since the settling time of the magnetic deflection and other electronic circuitry limits the beam spot time. Pfeiffer[192] has compared the exposure concepts of scanning systems including multibeam[229,230] (Table 9-3-4) and assessed the throughput of each tool. Variable shaped beams are the primary choice for direct silicon write. Other systems (raster) under development are based on field emission guns[74] emitting up to 10^2–10^6 A/cm^2. If long-term reliability can be achieved in gun stability and spot control, less sensitive but more stable inorganic resists (100 $\mu C/cm^2$) may be economical to process. Projection *e*-beam systems are

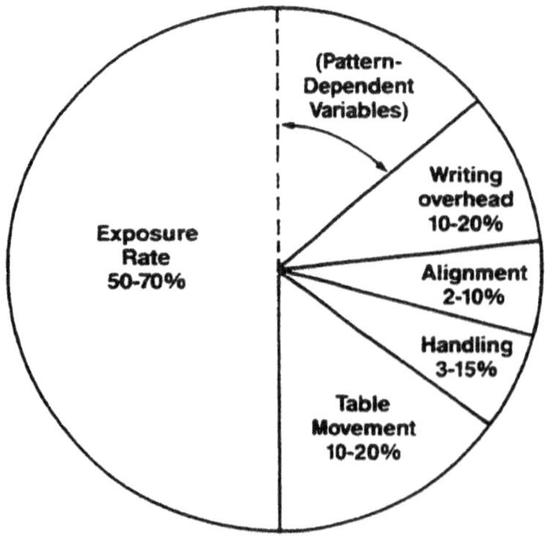

FIGURE 9-3-16. Breakdown of throughput factors on round beam *e*-beam tool. From Ref. 74, copyright 1984 and courtesy of Lake Publishing Co.

TABLE 9-3-4. Assessment of Electron-Beam Systems[a]

	Gaussian round beam	Multibeam	Shaped beam
Pro	Simplicity Flexibility Proven mask-making capability (MEBES)	Parallel processing (high exposure efficiency) Low brightness Slow electronics	Parallel processing (high exposure efficiency) Flexibility, variable shape Proven direct write and mask-making capability (EL3)
Con	Strict serial (1 pixel at a time) High-brightness source High-speed analog system	Limited flexibility Alignment restrictions Low current utilization	Complexity (high cost) Resolution varies with shape size (current dependency correctable)

[a] From Ref. 84.

constantly being improved to solve problems in mask fabrication and full-field overlay and these systems will be used for the high-volume production of low-part-number, submicrometer-size devices (memory).

9-4. ELECTRON-BEAM–RESIST INTERACTION

The resist material can interact with the components of the electron-beam system and cause problems such as contamination, charging, poor registration, and poor long-term reliability including periodic or more frequent maintenance. Problems in this area are sometimes difficult to diagnose. They may be the result of either poor machine performance or cumulative resist exposure.

9-4-1. Charging of Resist

Charge buildup during exposure or registration can occur in two forms: (1) due to thick dielectric underlayers and resist layers up to 3 μm thick for lift-off, and (2) contamination buildup from outgassing resists or other sources on the electrostatic lenses, deflection plates, or proximity printing masks. Full photocathode or multibeam exposure of resists may generate a large volume or product not readily handled by a vacuum system designed for serial exposure of resists. In fact, the lifetime of the photocathodes is proportional to the volume of outgassing product.[260]

To prevent charging such as in SEM work, gold is usually deposited over organic and low-dielectric-constant specimens.[199] For exposure of Ag emulsion on glass plates, a thin layer of tin oxide is used as a ground plane.

Because of multilayer resist structures or thick underlayers, up to 5 μm of dielectric may be present. At low energy (< 10 keV), the penetration of the beam is on the same order as the thickness of the dielectric layers. Since the backscattered electrons are lower in energy and do not escape the film, charge buildup can occur. Deflection errors are found to be proportional to the sample voltage.[83,84,223] When a dielectric coated metallic or conductive sample is charged to a potential V, an electric field is created between the sample and the nearest machine parts such as the holder for backscatter electron detectors. The sample voltage should be smaller than 1.6 V at 20 keV to keep the registration error below 0.1 μm.[85]

There are several ways to reduce charging and conduct the charge away:

1. Thin metallic,[207] Si_3N_4,[111,236] or carbon[84,186] coating on top of, in the middle of multilayer resists, or underneath the resist layer.
2. Sufficient beam voltage (> 10 keV), and good grounding of the silicon wafer by removal of native oxide.
3. Conductive resist by loading with nanometer-size carbon, metal, or electronic conductors. Ionic resists, solution conjugated,[86] or photoconductive resists[205] are still in the experimental stages.
4. Thin resist layers (< 1 μm). A 5-keV beam penetrates 0.7 μm and the 15-keV penetration range in organic resists is 5 μm.

During registration, bubbling of the resist (see Fig. 9-5-3) can foam and create molecular voids which may not dissipate the electrical charge along the path of dielectric breakdown.

9-4-2. Resist Outgassing

Positive resists such as PMMA and polyolefin sulfones can outgas hydrocarbons, CO_2, SO_2, and solvents during exposure. Some novel (Chapter 3) polyaldehydes develop *in situ* upon exposure. Hopefully, the mean free path is sufficient (several meters at 10^{-6} torr) to remove products at high pump speeds and prevent deposition or reaction with internal surfaces. In some cases, the lens components are continuously heated to prevent organic buildup. To prevent gun contamination, a separate gun pumping pathway and interlocks are usually built into the system. A recent innovation is a microseal *e*-beam orifice which allows the wafer to be exposed external to the system and avoid the vacuum system.[254,258]

LaB_6 guns are more susceptible to poisoning than tungsten filaments.[82] Avidenko and Malev[87] studied the poisoning of LaB_6 filaments and found that hydrocarbons poison the gun at the lowest vacuum pressure (Table 9-4-1). A synergistic lowering of the pressure limit was noted for a mixture of CO_2 and hydrocarbons. CO_2 gas can rejuvenate hydrocarbon-poisoned guns by the following reactions:

$$LaB_6 + C_nH_{2n} \rightarrow LaC_2 + H_2 + B_4C$$

$$B_4C + LaC_2 + XCO_2 \rightarrow LaB_6 + CO$$

TABLE 9-4-1. Pressure Limits of LaB_6 Poisoning[a]

Gas	Poison pressure (Torr at 1500 °C)
O_2	10^{-3}
CO_2	10^{-2}
H_2O	10^{-4}
Acetone	10^{-5}
Hydrocarbons	10^{-7}
CO_2 + hydrocarbons	10^{-5}

[a] From Ref. 87.

Outgassing from the resist may also be induced by beam heating of different substrates on a silicon wafer: a resist coating experiences only a 13°C rise while the same resist on quartz plates may be heated[88] to 160°C for a 1-μm spot (20 keV) and 50 A/cm^2 beam intensity (10^{-4} C/cm^2). If the resist experiences a temperature rise above its glass transition or ceiling temperature, the linewidths may be altered by the accelerated decomposition or diffusion reactions. For example, polyolefin sulfones undergo depropagation with activation energies of about 20 kcal/mole and a 10°C rise during exposure may cause a twofold enlargement of linewidth. However, high-sensitivity resists require low doses ($< 2\ \mu$C/cm^2) and the temperature rise on a silicon wafer (but not on quartz) may be negligible. Negative resists such as epoxy (COP) which exhibit postpolymerization linewidth growth (Chapter 5) are heat sensitive while polysiloxanes and polychlorostyrenes do not exhibit an active thermal susceptibility. Subtle reciprocity and proximity effects especially on quartz plates will also be susceptible to the beam writing pattern for localized heating effects.[89] Someday, preheated resists inside an *e*-beam tool may be used for resist sensitivity enhancement.

9-5. REGISTRATION

Registration, overlay, field-to-field alignment, is one component of the total design registration accuracy.[82] The design registration tolerance of a device is defined to be the registration tolerance of one etched device pattern level to another so that the device meets all performance objectives.[82] Common to all exposure systems are the level-to-level registration and the linewidth control in the imaged resist. Several other factors such as the mask set overlay accuracy,[217] slice flatness (chucking or electrostatic clamping during exposure),[82] temperature rises, and in-plane distortion affect the registration accuracy (Table 9-5-1) in some exposure tools more than others. Electron-beam direct write (maskless) is not concerned with mask placement accuracy.

TABLE 9-5-1. Design Registration Accuracy for Projection UV and Electron-Beam Writing on Silicon[a]

	Mask-based optical, UV, and *e*-beam projection print		Maskless *e*-beam writing	
	Status (μm)	Future (μm)	Status (μm)	Future (μm)
1. Registration accuracy (3σ)	± 1	± 0.3	± 0.25	± 0.1
2. Mask set overlay accuracy	± 0.5	± 0.3		
3. *e*-beam accuracy			± 0.2	± 0.1
4. Slice nonflatness, mask instability, temperature effects, in-plane distortion	± 0.1–0.2	± 0.1–0.2		
5. Mask set critical dimension	± 0.2	± 0.2		
6. Linewidth control on slice (resist and etching	± 0.5	± 0.3	± 0.3	± 0.1
$\text{rms} = \sqrt{1^2 + 2^2 + \cdots + 6^2}$	± 1.2	± 0.6	± 0.4	± 0.17

[a] From Ref. 82.

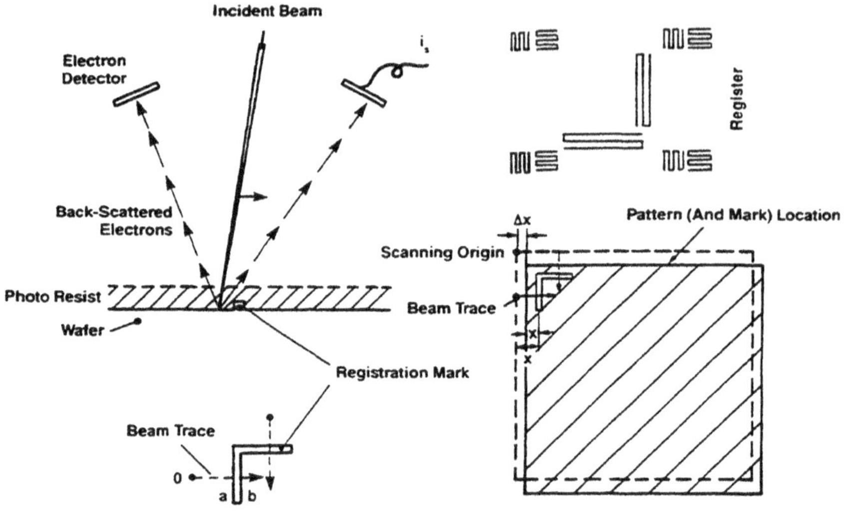

FIGURE 9-5-1. Schematic of a registration scheme for an *e*-beam exposure system. From Ref. 200.

In *e*-beam registration, alignment marks are typically fabricated as trenches,[98,250] pedestals of silicon,[240] or for more effective backscatter signals, from high-Z (atomic number) metals. While the *e*-beam scans over the mark[62] (Fig. 9-5-1), the backscattered electrons are detected and the scan field of the beam (Fig. 9-5-2) is corrected to overlay the field to the exact dimensions of the chip.[90] The registration signal is strongly influenced by the work function of the substrate, the incident beam voltage, the composition of the resist, and its film coverage. While detectors can be channel multipliers, scintillators, and diffuse diodes,[91] four components are essential:

1. An efficient and accurately placed detector[92]
2. A scattering and dithering beam[91] smaller than the mark[225]
3. A configured mark with the correct size, shape, and resist thickness
4. A high signal-to-noise data retrieval and a feedback loop to distort the scan field to overlay the chip[93]

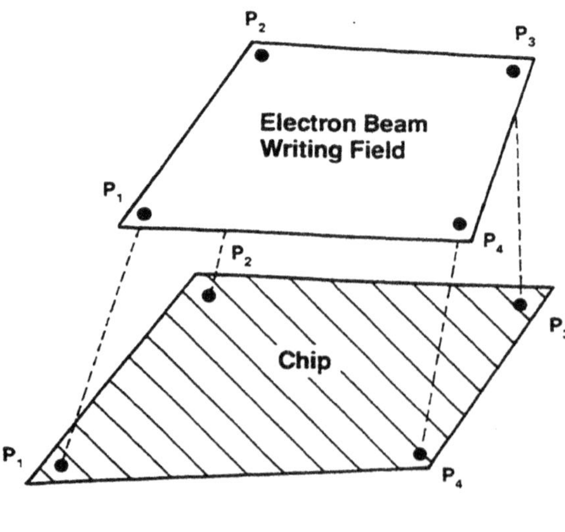

FIGURE 9-5-2. Electronic adjustment of *e*-beam exposure field to match chip overlay. Electron-beam field stretched to match chip. From Ref. 90, courtesy of IBM.

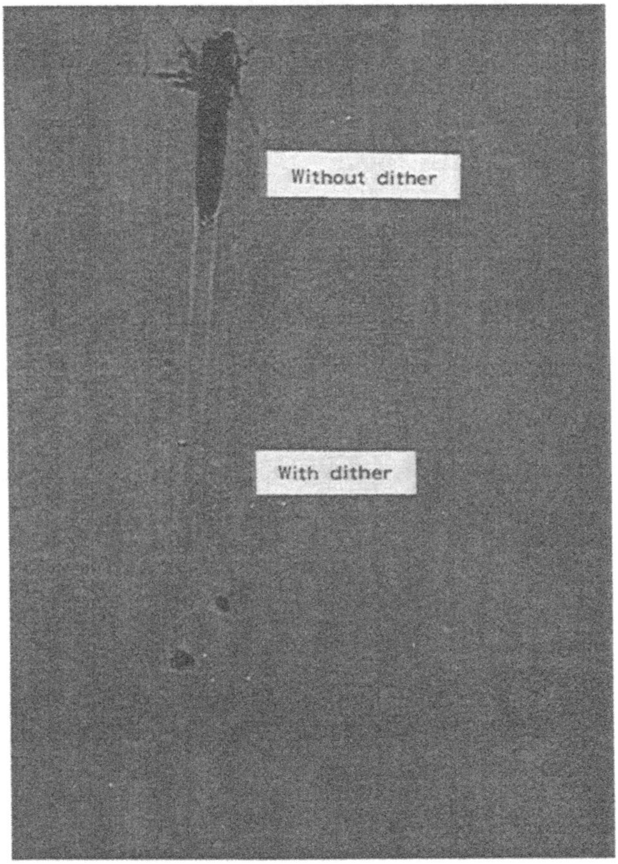

FIGURE 9-5-3. Bubbling of PMMA resist during *e*-beam registration without beam dither and with dither (spreading beam out and writing slower to allow gas by-products to diffuse out of the resist).

If the alignment mark is raised and composed of a high-Z (Au) metal, a 10× registration signal can be obtained versus a mark made in silicon. In fact, precise calibration grids of gold on silicon are used to map the deflection field and correct for beam position, rotation, and orthogonality.[90,102] The registration procedures are a sequence of events[90] involving

1. Mechanical align to 75 μm
2. Whole wafer align to adjust magnification and rotation to 2 μm
3. Precise align using four corner chip registration marks to ± 0.1 μm
4. Independent motion with laser interferometer X–Y table

Closely spaced registration marks increase the signal[99] and V-shaped marks with a detector set at 54° give maximum signal to noise.[95–97] Since the resist

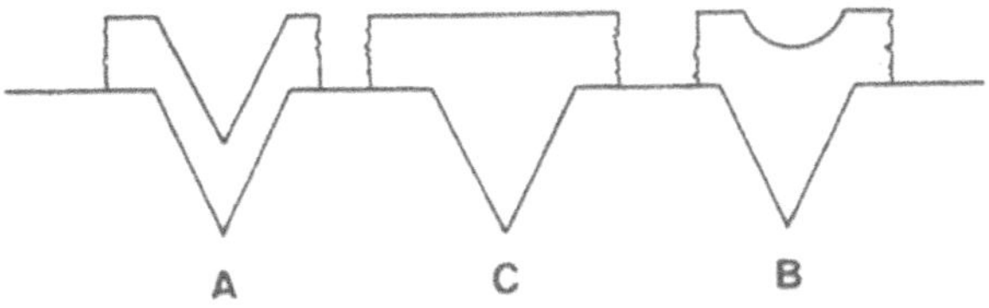

FIGURE 9-5-4. Typical cross sections of planarizing resist covering V-shaped groove registration mark of (A) conformal coating, (B) partial conformal coating, and (C) planarizing coating. Adapted from Ref. 105.

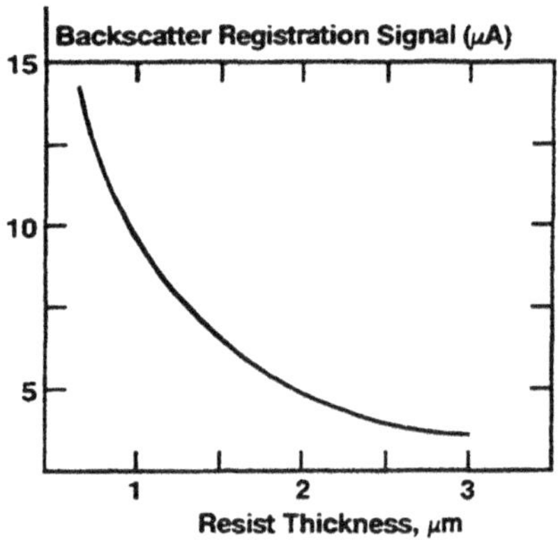

FIGURE 9-5-5. Registration signal at 25 kV in backscattered *e*-beam hitting diode after passing through various thicknesses of PMMA. From Ref. 90, courtesy of IBM.

absorbs part of the backscattered signal, the thicker the resist coating over the mark, the poorer is the signal.[224] During registration the bubbling and outgassing[245,259] of resists (Fig. 9-5-3) such as PMMA can be caused by heating and higher doses (10^{-3} C/cm^2). By spreading the registration beam in a "dithering" motion of elongation of the beam in a direction orthogonal to the scanning direction,[91,222] the signal distortion in PMMA resists is eliminated (Fig. 9-5-3). In some instances, resists have to be removed from marks by deep UV or other forms of exposure[103] to enhance the signal. Because of the limited range of 10-keV electrons (3 μm), higher-energy electrons (50 keV) will greatly improve the registration signal but the loss of resist sensitivity and radiation damage may have to be compromised back to 25 keV at least for 1-μm geometries. Over the alignment mark, the nature of the resist coating the mark (Fig. 9-5-4) will depend on

1. Thickness of step, hole, or pedestal[105]
2. Thickness[106,107] and viscosity of resist (Fig. 9-5-5)
3. Radial orientation of mark during spin coating[105]
4. Spacing between marks[105–107]
5. Planarization by resist[104]

TABLE 9-5-2. Overlay Accuracy in an Electron-Beam System[a]

Item	Parameter	Variation (μm)
Beam	Position drift	0.05 ~ 0.1
Alignment	Position ripple	<0.02
	Mark detection	~0.05
Stage	Alignment procedure	~0.05
	Stage position	~0.05
Deflection	Static	<0.02
	Dynamic	<0.02
Distortion	Drift	~0.05
	Distortion	<0.02
Wafer	Height variation	<0.02
	Temperature	<0.02
Environment	Vibration	<0.02
Total		0.113 ~ 0.142

[a] From Ref. 100.

V-shaped grooves can be obscured by a thick resist. The thickness uniformity also depends on the mark location on a wafer.[101,105] If the resist flows along the V-shaped grove, a symmetrical resist profile can be obtained; otherwise, an asymmetrical resist coating will result if the resist flows perpendicularly to the V-shaped groove.[104] The asymmetry in resist profile may lead to a detectable difference in alignment marks at different wafer positions.[250]

Since positive *e*-beam resists such as PMMA tend to thin volatize with repeated registration scanning, the signal and accuracy improve. If the marks, however, are left unprotected, they may be eroded in subsequent etching operations. Due to the thickness of the coatings, multilayer resists may present registration problems, especially over V-shaped grooves (up to 7 μm of resist including 3- to 4-μm-deep groove). In this case, a raised mark would have a thinner resist coating due to the planarizing nature of spin coating. Consequently, the planar the device, the better is the uniformity of coatings and the subsequent reduction of noise. Table 9-5-2 summarizes the detailed analysis of the *e*-beam registration for fabricating MOS devices.[100]

In the final analysis, the accuracy of *e*-beam registration exceeds device ground rules and provides the necessary precision for the present and future submicrometer lithography.

9-6. PROXIMITY EFFECTS

Proximity effects have been the bane of *e*-beam lithography. At 25 keV, the backscatter range of electrons for a 1-μm spot is 5 μm wide and becomes 12 μm wide at 50 keV (Fig. 9-6-1). The range of backscatter follows an $E^{1.7}$ dependency.[215] The proximity effect leads to undesired exposure in regions not addressed by the *e*-beam while lithographic exposed patterns in the absence or presence of near neighbors, respectively, give rise to the "intra-" (isolated exposure) or interproximity effect.[143] In the intraproximity effect, the falloff of the range of backscattered electrons (Fig. 9-6-2) only allows isolated linewidths below 2 μm to be exposed by one or two adjacent spots.

If a 0.5-μm line and a 2.0-μm line are exposed and developed for the same time, the smallest geometry only partially develops. Even overdevelopment of the 2.0-μm line will not force develop the small features (Fig. 9-6-3). The 0.5-μm wide line

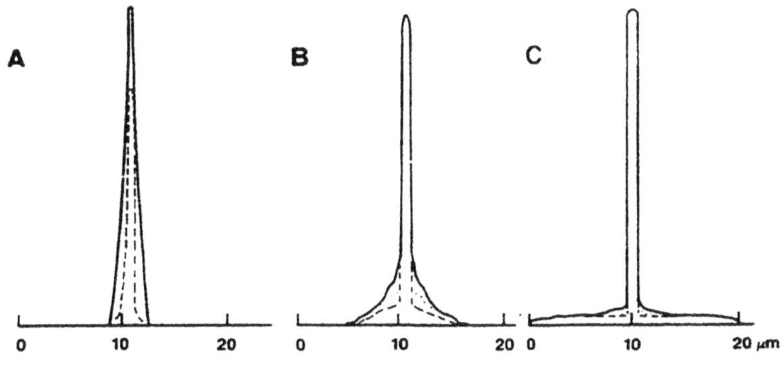

FIGURE 9-6-1. Energy deposited by a 1.0-μm square *e*-beam into a 1.0-μm-thick resist on a silicon substrate at (A) 10 kV, (B) 25 kV, and (C) 50 kV. Courtesy of N. Eib, IBM.

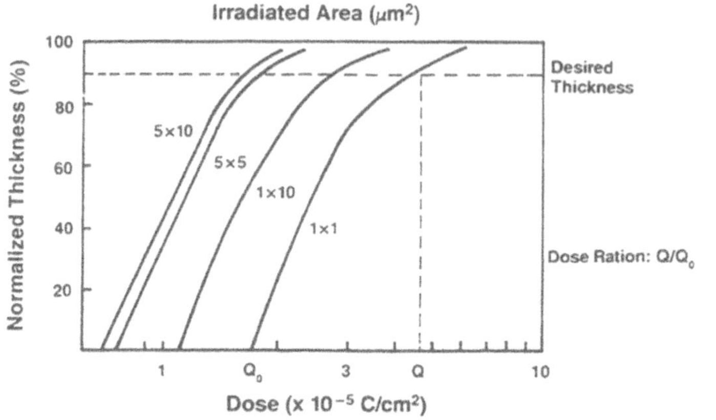

FIGURE 9-6-2. Falloff of backscattered electrons in the intraproximity effect for negative polydiallylisophthalate resist resulting in more dose needed for smaller line. From Ref. 73, of *Fujitsu Scientific and Technical Journal*, Fujitsu Ltd.

would require twice the dose of the 2.0-μm line to develop to the same bias and film retained. The intraproximity effect is due to the lack of backscatter (falloff of range) owing to the deeper penetration of the incident beam into the substrate and the lower resist absorption at the higher energy (Fig. 9-6-4).[108,109] When the voltage is low (1–10 keV), most of the forward energy is deposited in the resist (Fig. 9-6-5) and the image spread is minimal, but at higher energies, the exposure is dominated by backscatter.

When resists are exposed on other substrates or heavy metals (Fig. 9-6-6) such as gold[110] or tungsten[198] on Si in Josephson devices, the exposure intensity on

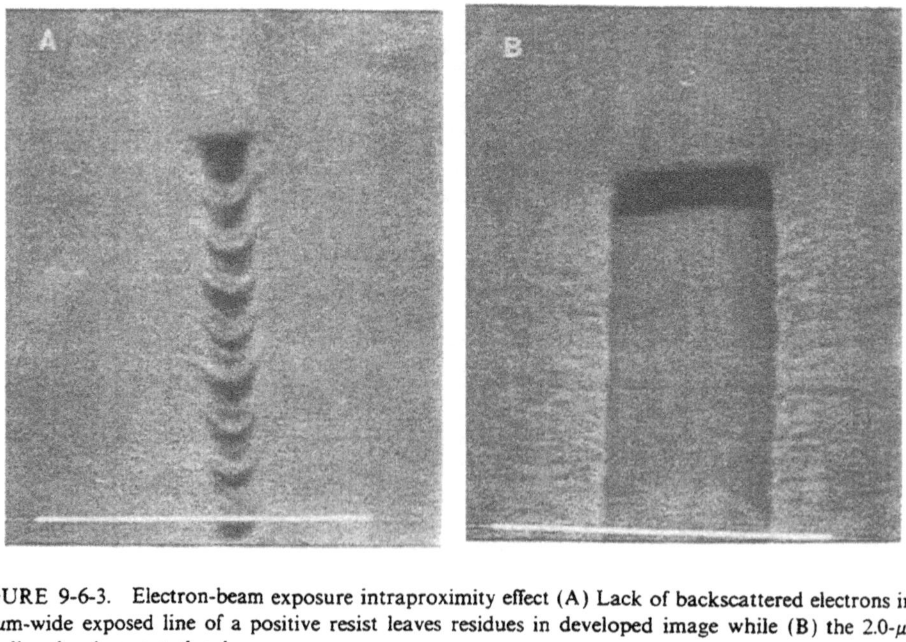

FIGURE 9-6-3. Electron-beam exposure intraproximity effect (A) Lack of backscattered electrons in a 0.5-μm-wide exposed line of a positive resist leaves residues in developed image while (B) the 2.0-μm-wide line develops out cleanly.

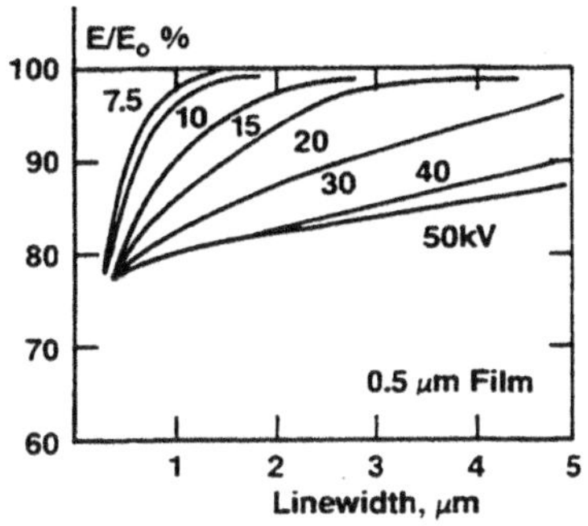

FIGURE 9-6-4. Energy absorbed at 0.5-μm depth for a 0.5-μm-thick PMMA film on silicon versus linewidth and beam voltage. From Ref. 108.

the metal film becomes greater in the vicinity of the *e*-beam incident point due to its large backscatter coefficient (Fig. 9-6-7). Interactively, this results in a twofold increase in resist dose (apparent sensitivity) and some intraproximity compensation. To take advantage of the backscatter effect, overcut lift-off profiles can be created in PMMA (Fig. 9-6-8) by a dose twice that of normal.

In order to compensate for the intraproximity effect, an extra dose must be programmed into the beam (longer spot time).(212) The thicker the resist, the greater is the dose required to reduce the bias. The thinner the resist (Figs. 9-6-17 and 9-6-18) such as in a multilayer structure, the less is the intraproximity effect.(147) Also, the lack of backscatter is revealed in spot placement error or spot butting due to the poor spot overlap in subfields (Fig. 9-6-9).(114)

In the scan field, boundaries result at the end of stitching.(29) Because of *e*-beam spot size limitations, very long lines must be broken into segments that are written sequentially. In order to prevent underlap or overlap (Fig. 9-6-10), it is necessary to keep the positioning accuracy several times higher than the edge resolution (± 0.2 μm). Gaps of 0.2× the spot size leave bridges and the beam must be enlarged at subfield boundaries to overcome the lack of extra backscattered electrons.(115) Other beam irregularities besides spot placement error such as beam defocus and beam tilt have been modeled for linear PMMA and nonlinear dissolving positive resists.(116)

The other main type of backscatter (interproximity effect) involves the exposure of near neighbors. Unexposed spaces between lines are exposed by backscattered electrons. In a photocathode(117) mask exposure (Fig. 9-6-11), mutual rays generate an electric field which reflects backscattered electrons beyond the limit of the range of first pass (1- to 10-μm reflection), up to several millimeters away.(114,118) The interproximity effect causes the unexposed space in positive resists

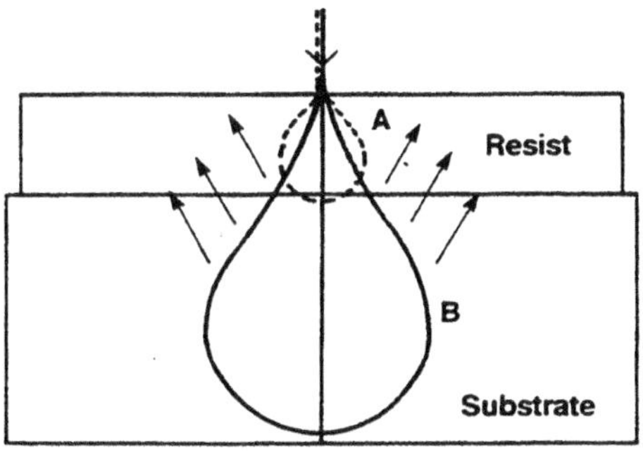

FIGURE 9-6-5. (A) Backscatter of electrons when the penetration range equals the resist thickness and (B) when the range is deep into the silicon wafer. From Ref. 179, courtesy of AIP.

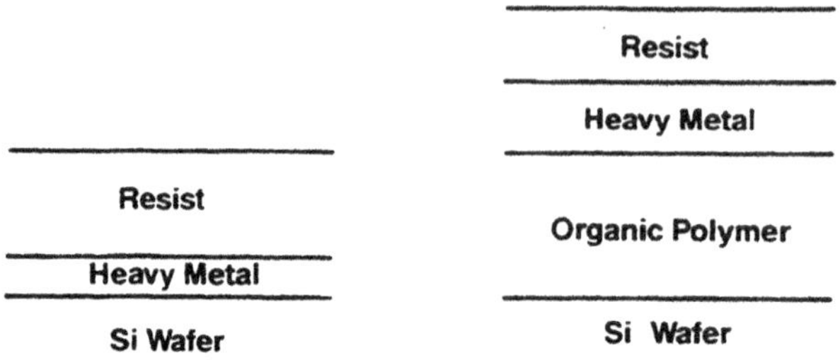

FIGURE 9-6-6. Resist on top of high-atomic-number (Z) substrate which enhances the resist sensitivity (less primary electrons are needed since there is more efficient backscatter on Au than Si). The metal interlayer in a trilayer (Chapter 12) stack also reduces the interproximity effect.

(Fig. 9-6-12) to thin and become narrow (unsuitable for reactive ion etch). In negative resists (Fig. 9-6-13), the unexposed space is gradually filled in with scumming resist. Typical troublesome areas in the interproximity effect are shown in Fig. 9-6-14.[119] Raster systems are more difficult[120] to correct than vector systems which can partition[119,121] the energy into dose or size corrections.[121–124,256,257] In controlling the extent of the proximity effect, the thickness of the resist[215,216] and the energy of the beam have the most impact. Low energy (10 keV) and thin-resist films (0.2 μm thick) exhibit minimum proximity effects.[125] For certain repetitive features, a 50-keV[203] beam shows a minimum proximity effect.[126–130,183,234,252] Yoshimi and co-workers[126] were able to image a 0.25-μm line and space combination in 1-μm-thick PMMA with practically zero bias (same as beam diameter) (Fig. 9-6-15).

The advantages[132] of high-voltage (50–100 keV) e-beam writing are largely due to the diffuse lateral distribution of backscattered electrons at the interface of a resist film and an Si substrate. On the other hand, the diffuse distribution may

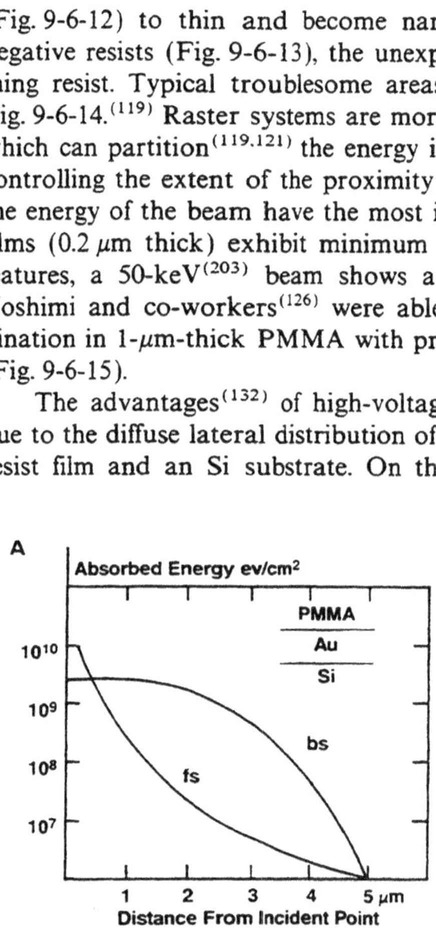

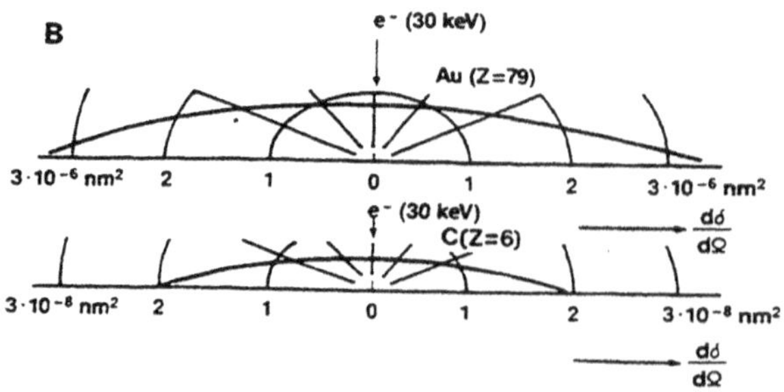

FIGURE 9-6-7. (A) Foward (fs) and backscatter (bs) contributions from Au/Si deposited under PMMA. From Ref. 110. (B) Polar diagram of Rutherford backscattering from gold and carbon film at 30 keV. From H. Niedrig, *J. Appl. Phys.* **53**, R23 (1982), courtesy of AIP.

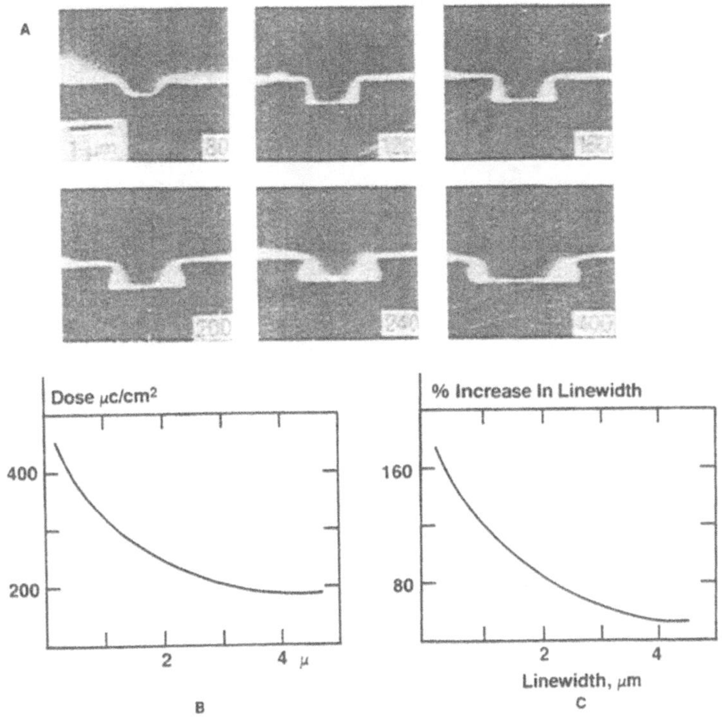

FIGURE 9-6-8. (A) Cross section of a 1.0-μm line in PMMA ranging in dose from 80 to 400 $\mu C/cm^2$. (B) Exposure dose used to generate undercut profiles in PMMA as a function of linewidth. (C) Resulting percent increase in feature size. From Ref. 112, courtesy of AIP.

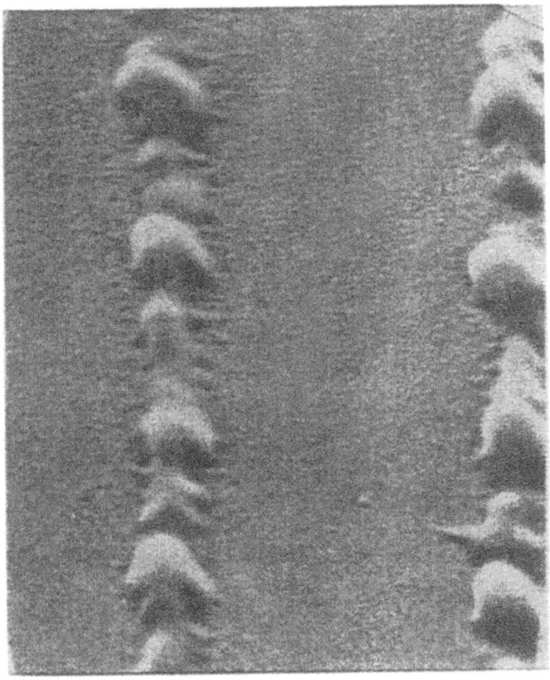

FIGURE 9-6-9. Poor spot overlap in raster *e*-beam as revealed in a negative *e*-beam resist.

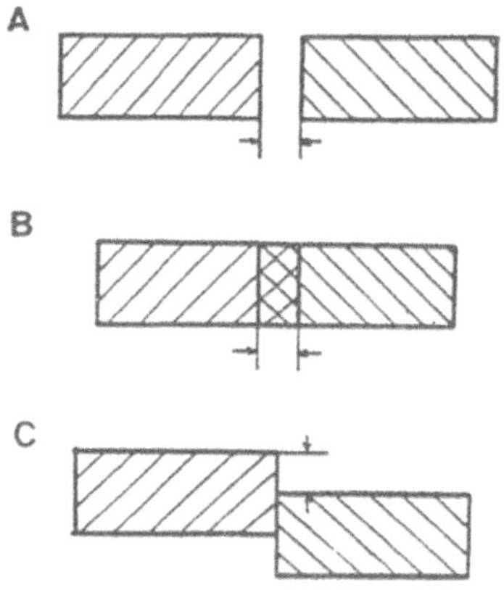

FIGURE 9-6-10. Schematic of potential spot overlap problems in *e*-beam exposure. (A) Spot underlap; (B) spot overlap; (C) misaligned square spots. Adapted from Ref. 115, courtesy of AIP.

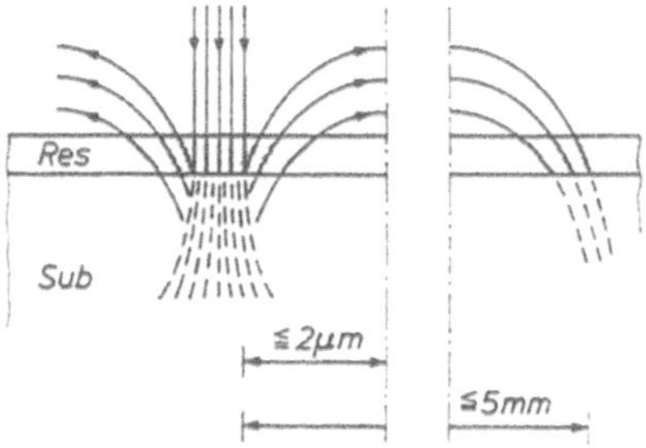

FIGURE 9-6-11. Proximity effects in full-field photocathode exposure. Backscattering of electrons in the resist (Res) and substrate (Sub) occurs in all electron-lithographic processes and limits the resolution obtainable. In the image projector the strong electric field also deflects the scattered electrons back to the substrate, up to 5 mm away from the scattering point. This affects the contrast and means that different exposures must be used for densely covered patterns, for which the backscattered signal is highest. From J. Scott, copyright and courtesy of *Philips Tech. Rev.* **37**, 352 (1977).

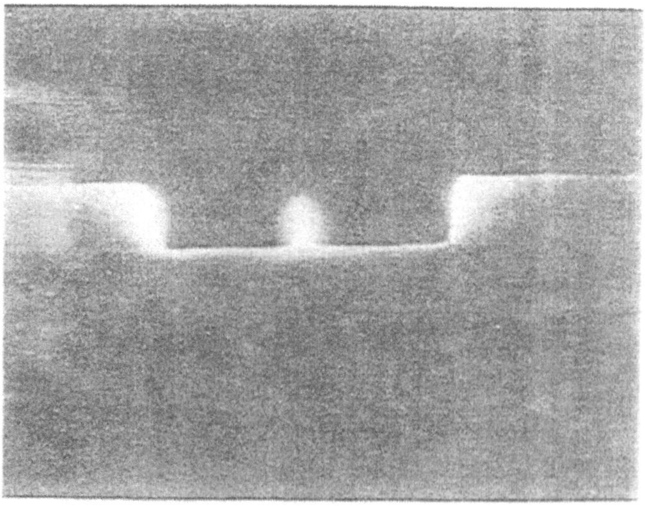

FIGURE 9-6-12. Interproximity effect in positive *e*-beam resist exposure. The unexposed space of 1.0 μm is thinned and profile is shallow due to exposure from adjacent 4-μm lines. Courtesy of G. Hefferon, IBM.

FIGURE 9-6-13. Interproximity effect in negative *e*-beam resist exposure. The unexposed space is scummed with residual partially exposed resist which must be ashed away.

cause a nonaddressed small island in an infinitely large exposed area to suffer from a large amount of background exposure by electrons backscattered from the surrounding exposed area. The background exposure results in reduction of the resist film thickness of the nonaddressed island and finally in its washing-out. The ratio of the amount of background exposure in the nonaddressed island to the amount of exposure at the point of incidence in the infinitely large exposed area is expressed as $\eta_E/(\eta_E+1)$, by using the reciprocity principle proposed by Chang,[131] where η_E is the ratio of total energy deposited by the backscattered electrons to that deposited by forward scattered electrons (Fig. 9-6-7). If η_E is unity, $\eta_E/(\eta_E+1)$ is equal to 0.5 and the nonaddressed island is given half the energy deposition of the point of incidence in the large exposed area. $\eta_E/(\eta_E+1)$ depends on acceleration voltage, when η_E depends on the acceleration voltage. The dependence of η_E on the acceleration voltage, however, has not been well established in the literature.[132]

The decrease of η_E with the increasing acceleration voltage is desirable for high-voltage *e*-beam lithography, because $\eta_E/(\eta_E+1)$ decreases with the reduction of η_E. The proximity function describes the forward (B_f) and backscatter (B_b) contributions as the sum of two Gaussians[131] for the energy absorbed in the resist (E) per depth (Z) as a function of radial distance (r) away from the beam,

$$dE/dZ = E(r) = k \exp(-r^2/B_f)^2 + \eta_E B_{f^2}/B_{b^2} \exp(-r^2/B_b)^2 \qquad (9\text{-}6\text{-}1)$$

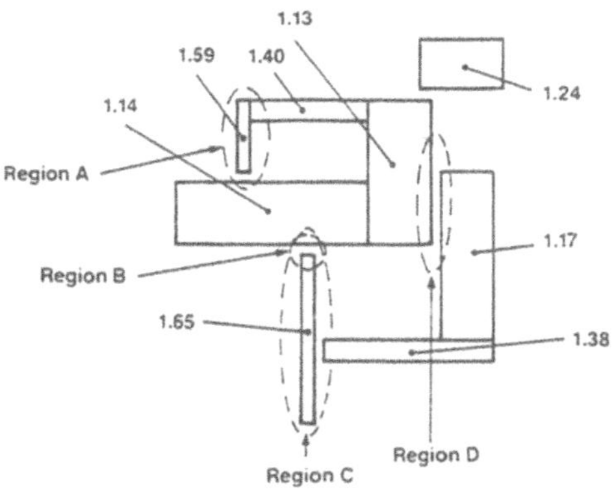

FIGURE 9-6-14. Troublesome (proximity effect) patterns. A pattern consisting of eight rectangles. Note regions where proximity corrections are necessary for complete dissolution of the resist as well as pattern fidelity. If this pattern is not proximity-corrected, a relative incident electron exposure value of unity is given to each rectangle. If this pattern is corrected via the self-consistent algorithm, incident exposure values are given to each rectangle as noted. From M. Parikh and D. Schreiber, copyright and courtesy of *IBM J. Res. Dev.* **24**, 531 (1980).

FIGURE 9-6-15. Reduction of proximity effect by using 50-kV exposure of PMMA. From Ref. 183, courtesy of Elsevier Science Press.

The coefficients are determined empirically[122,132–134] or modeled by Monte Carlo calculations.[133] Included in the energy deposited per depth term (dE/dZ) is the rate of resist dissolution (dR/dE) per energy absorbed for linear or nonlinear dissolving resist.[116] The rate (R) of film dissolution per depth is given by

$$dR/dZ = (dE/dZ)(dR/dE) \tag{9-6-2}$$

The rate of dissolution (dR/dE) is modeled by film dissolution studies as a function of a flood dose[135] (Chapter 10). The proximity correction parameters (B) depend on the energy (keV), the atomic number (Z) of the substrate, the resist contrast, and the *e*-beam current. Since negative resists do not involve the developer term (dR/dE), a quick and definitive test of proximity correction involves the use[177] of negative resists as a single film or on multilayer resist structures since all backscattered electrons[136,137] are recorded[15] in the resist (without developer erosion as in a positive resist). Etched chrome lines and electrical resistance measurements have also been used for an automated proximity test.[239]

In order to improve pattern accuracy, Machida and co-workers[227] investigated the limitations of the exposure tool and the correction method for defining 0.5-μm patterns in a 1.0-μm-thick polydiallylphthalate negative resist. They developed a pattern accuracy method based on the dose difference between the resist development energy and the exposure energy points at the edge of each shape.

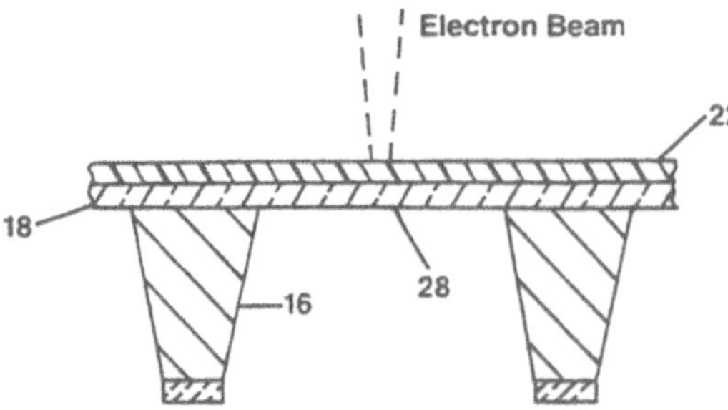

FIGURE 9-6-16. Elimination of back-scatter interproximity effects by reducing the thickness of the silicon (18) substrate to a thickness (etch back) of a membrane less than the pitch of the pattern. U.S. Patent 4,316,093 (IBM).

TABLE 9-6-1. Proximity Effect Factors and Engineering Controls

Dominant factors	Corrections
Thick resist (>0.5 μm)	Multilayer resist[145]
Energy (keV)	<12 or >40[184,203]
η, B_f, B_b, backscatter	Empirical, thin membrane
Pattern pitch (line > space)	Bias, dose, dose partitioning and exposure position, flood reverse tone[123,213,214,253]
Edge slope	Smaller spot, square[182] beam
Resist thinning	High-contrast[144] resist, $\gamma > 4$, non-cross-linking negative (no developer influence)
Backscatter	Top surface resist image[20]
	Au or W multilayer[185]
	Siloxane bilayer[228] (10 keV)

Also, instead of using the conventional resist contrast, γ, they defined a resist evaluation parameter, $K = Q/Q_0$, where Q is the exposure dose for the final desired resist thickness and Q_0 is the interface gel dose. To achieve 0.5-μm patterns with ±0.1-μm accuracy in a 1.0-μm-thick resist, the resist evaluation parameter K must be less than 2, the rounding error of the exposure dose must be less than 2.5%, and the beam addressing unit (lsb) must be less than 0.025 μm.

Substantial reduction, especially for nanolithography (Fig. 9-6-16), of the backscatter for positive[138,139] or negative resists[40] can be achieved by having a membrane 5–10× thinner than the dimension for the pattern to be formed.[141,232] The electronic control of the interproximity effect involves dose or beam shape partitioning, or windage[142] (bias) corrections which can be mathematically intractable for certain patterns. A summary of the programmable and process controls of proximity effect is given in Table 9-6-1.

Since the proximity effect is almost directly proportional to the resist thickness,[226,227,238] the thinner the resist (Fig. 9-6-17) whether in a single film or in a multilayer stack (Fig. 9-6-18), the smaller is the proximity effect (Fig. 9-6-3). Multilayer resists are the practical solution to the proximity effect.[145] By incorporating a heavy metal of tungsten as a plasma etchable barrier layer in a multilayer stack[198] (Fig. 9-6-6), the proximity effect and radiation damage to the device can be reduced. When using a thick enough (> 0.3 μm) tungsten layer, the ratio of the

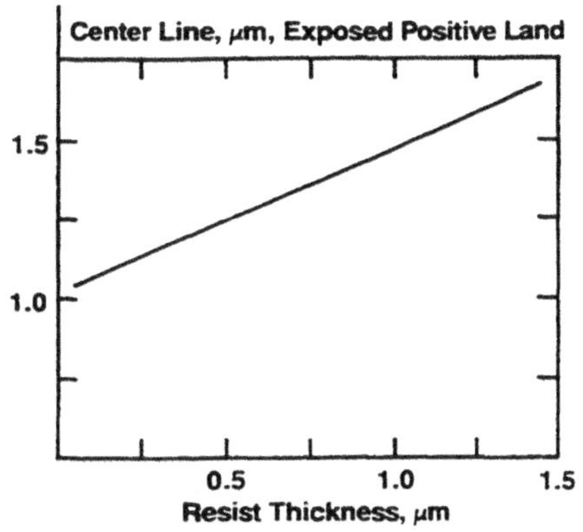

FIGURE 9-6-17. Interproximity effect on a pattern of 1.0-μm line and space versus thickness of resist on silicon wafer exposed at 25 kV. Courtesy of N. Eib, IBM.

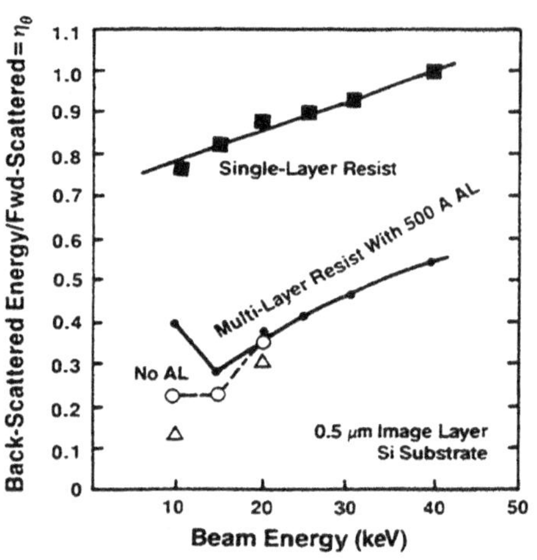

FIGURE 9-6-18. Ratio of backscattered to forward scattered electrons deposited in PMMA resist in a trilayer structure with an interlayer of Al film versus a single layer of PMMA on silicon wafer. From Ref. 145, reproduced by permission of the Electrochemical Society Inc.

electron dose (intraproximity effect) of an isolated line pattern, to that of the small space pattern between larger patterns is smaller than that of the silicon substrate. Therefore, the proximity effect correction decreases and it becomes easier to form the resist pattern accurately.[198] The proximity effect sets the lower pitch limit to *e*-beam lithography[146] to about 1.5 μm for a single-film resist and 0.5 μm for multilayer resists.

9-7. RADIATION DAMAGE

Electron beams, X rays, and high-energy ions penetrate to depths of 5–10 μm into the silicon device areas.[180] Plasma etching (Chapter 13) and plasma stripping (Chapter 14) also bombard devices with radiation and induce damage. For military applications, semiconductor devices have to be able to withstand nuclear bombardment and many yearly symposiums are held on radiation damage in devices.[148] The damage in MOS devices results from creation of electron–hole pairs, interface states at the SiO_2–Si interface, and additional neutral oxide traps.[149–153]

While positive oxide charges and interface states can be annealed at temperatures between 300 and 450°C, radiation-induced neutral traps cannot.[149,150,154] Since the damage is proportional to the absorbed dose,[151] low-energy (< 10 keV) beams and higher-sensitivity resists would minimize damage. At the device metal personalization level, a positive *e*-beam resist coupled with subtractive metal removal would eliminate direct irradiation of the critical device area.

Ion-beam exposure with protons does not penetrate deeply into silicon and little damage is apparent.[237] High-atomic-number metals in trilayer resists (as barrier layer) have been used to prevent radiation damage.[180,198] Critical metal contact layers may have to be done by image reversal (or negative resists) to prevent direct *e*-beam or X-ray irradiation of the active devices. Projection UV may also have to be used to overcome permanent irradiation damage.[155]

9-8. X-RAY AND ION-BEAM EXPOSURE

9-8-1. X Rays

In Chapter 3 we discussed the factors of resist sensitivity in X-ray and ion-beam exposure. In this section we will examine some of the physical-equipment factors.

In the simplest form, X-ray (Fig. 9-8-1) and ion-beam printing use a shadow mask, limited by the source and mask-to-wafer penumbra, similar to the inverse square law for a diffuse optical source.[208,220] Proximity printing requires four components for printing[1]:

1. A high-intensity collimated source
2. A mechanical, electrical, optical, backscatter, or transmission alignment of mask accurate to $\pm\frac{1}{4}\,\mu m$[162]
3. A precise gap control accurate to less than 0.2 μm (Fig. 9-8-2)
4. An inexpensive foil or membrane mask. Boron nitride is preferred as a membrane[156] mask

The distortion introduced by irradiating a mask with a divergent beam (Fig. 9-8-1) is $\Delta r/r = \Delta d/l$ with l the source-to-mask distance and Δd the mask-to-wafer gap. If a distortion of less than 0.1 μm is required, the source-to-mask distance must exceed 10 μm (Fig. 9-8-2). For submicrometer printing, the gap[17] must be on the order of 1 μm. Wafer buckling that arises during many standard integrated circuit processes makes this difficult to achieve.[233] The run-out error could be avoided by going to a step-and-repeat mode or if resist sensitivity was improved to 1 mJ/cm^2 so that exposure could be made at a point farther removed from the source.[233] Major difficulties with mask stability and heating by the exposure beam cause changes in lateral dimensions that produce overlay errors.

The continued problems with shadow printing are the mask cost, complexity, stability, and uniformity. Resist exposure times have been disappointing for X rays, since the absorption is low and times on the order of minutes even with step-and-repeat printing of one chip.[161] The first steps to improve X-ray proximity print are:

1. High-speed resists (Chapters 3 and 5)
2. High-brightness source

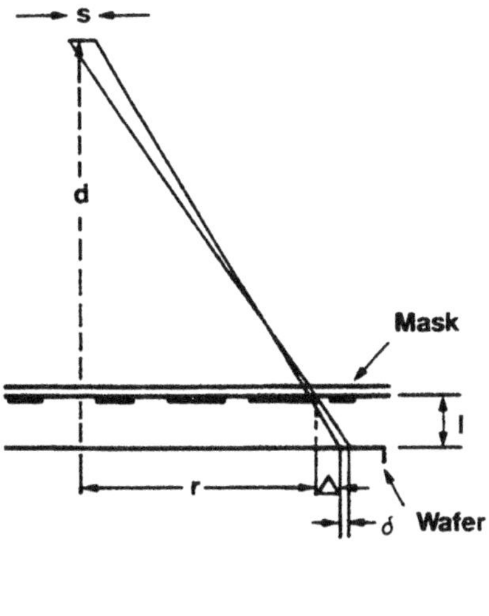

FIGURE 9-8-1. Penumbra runout effect in X-ray shadow printing due to falloff of collimation. From Ref. 157, courtesy of Morgan, Grampian Publishing Co. Inc.

3. Reliable and accurate alignment
4. A transparent and stable mask

The output of the X-ray source and the absorption bands of the resist must be actinically matched for minimum exposure time (Fig. 9-8-3). Precise alignment[159] and precise gap settings[160] are necessary because of Fraunhofer diffraction in conventional X-ray or Fresnel diffraction in synchrotron exposure.[158,195] Due to penumbral blurring, an intraproximity effect occurs in conventional X-ray printing.[188] In synchrotron exposure, penumbral blurring can be neglected but Fresnel diffraction is serious at larger than 1-μm gaps (Fig. 9-8-4). X-ray-induced secondary electron emission limits the ultimate resolution of X-ray lithography to 0.1 μm.[202,259]

Obtaining rapid (< 1 sec/chip) exposure times for a resist has been a technical impetus to develop high-intensity sources for X-ray technology. While resist sensitivity ranges from 10 to 1000 mJ/cm^2, the source intensities are

Electron-beam source	0.01–0.1 mJ/xm^2 sec
Plasma source	0.01–1 mJ/cm^2 sec
Synchrotron source	10–100 mJ/cm^2 sec

Exposure[164] for the best single-film resist such as a DQN requires about 0.5–2 J/cm^2. Faster X-ray resists are available at 0.05 J/cm^2 but they involve costlier trilayer processing. For a conventional novolak-based resist, the throughputs based on the above fluxes are for a 125-mm wafer (see also Fig. 3-1-8):

Conventional source	1 wafer/hr
Plasma source	3 wafers/hr
Storage ring	30 wafers/hr

A synchrotron source of a conventional storage[165] ring, or a compact smaller source,[166] requires about 2–10 sec/chip. The synchrotron source[249] is so intense that PMMA or other resists are dry developed.[176,178,255] Since some of the overhead is due to stepping, plane X-ray scanning mirrors have been investigated to reduce the step time.[167] With present costs, the ideal source is a small *e*-beam excited anode. In order for X-ray printing to be more cost effective than optical systems, exposure times (comparable to optical steppers) of milliseconds per chip

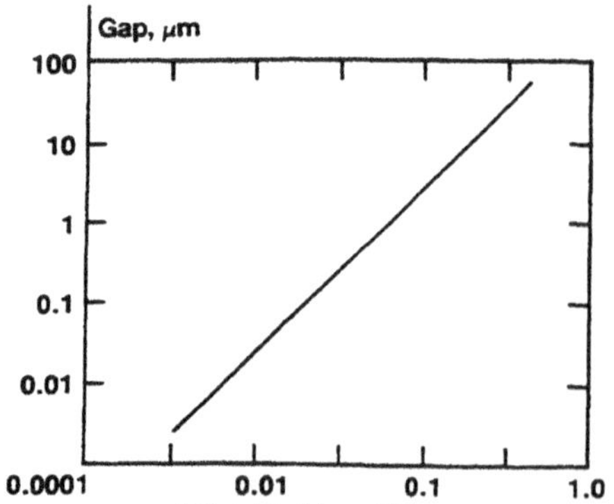

FIGURE 9-8-2. Printed linewidth versus spacing of X-ray mask and resist surface. From Ref. 172, courtesy of SPIE.

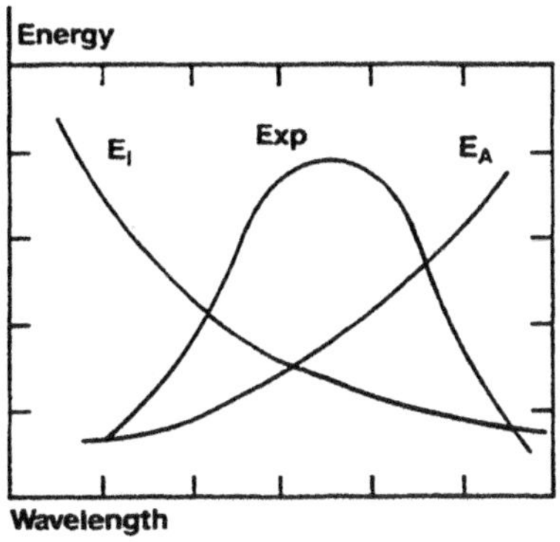

FIGURE 9-8-3. Actinic matching of X-ray absorption of resist and X-ray output of source. E_I—X-ray source output λ. E_A—absorption of resist. Exp—maximum—optimum for minimum exposure time. From Ref. 206, reproduced by permission of the Electrochemical Society Inc.

should be achieved. From a process viewpoint, it would be desirable to overcome the problems of optical printing of single films over steps. Because of the large depth of focus, a high aspect ratio of 10 to 1 of linewidth to resist thickness has been achieved in X-ray printing. Unfortunately, no good single-film X-ray resists of less than 5 mJ/cm^2 exist and at present we have to rely on more costlier multilayer resist[168] or swelling chlorostyrene negative resists. A good single-film RIE-resistant X-ray resist is needed.

Improved masks with smooth edges are also required for submicrometer lithography.[194,208,243] Field-to-field overlay for large wafers is a significant problem (Fig. 9-8-5).[172] Wafer flatness variations of ± 2 μm can create large overlay errors and it is unlikely that full wafer exposure can be accomplished for wafers larger than 7.5 cm.[233] Thus, X-ray steppers are likely to be the primary exposure tool. A new reflective optics X-ray exposure system is under investigation.[193] X-ray system throughput has not been able to compete with optical lithography. The major hope for X rays lies in submicrometer applications that do not involve precise alignment.[219]

9-8-2. *Ion Beams*

Ion-beam lithography shows promise in two forms: (1) as a step-and-repeat printer and (2) as a focused ion beam of H^+, Si^+, or P, B ions.[187,209–211,218] Focused ion beams are based on the many years of development of ion

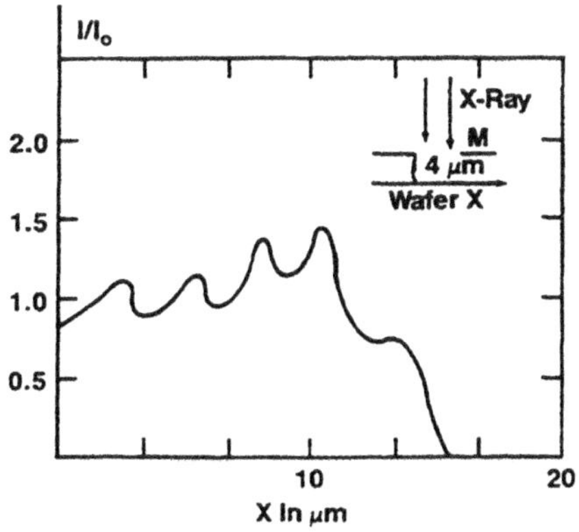

FIGURE 9-8-4. Limiting Fresnel diffraction by edge of X-ray mask as a function of gap between the X-ray mask and the wafer for synchrotron exposure. From Ref. 163, copyright and courtesy of AIP.

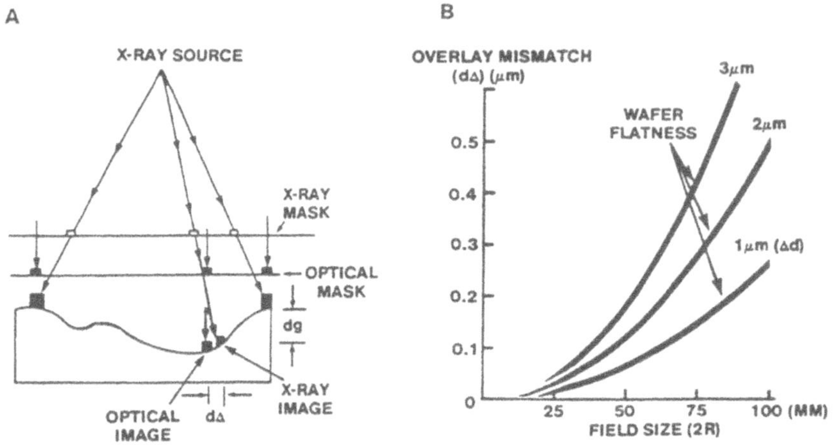

FIGURE 9-8-5. (A) The overlay error (Δd) induced by wafer nonflatness in mixed lithography applications. (B) The magnitude of the overlay error (Δd) for various wafer flatnesses versus the field size of exposure in an X-ray system. From Ref. 172, copyright and courtesy of SPSE.

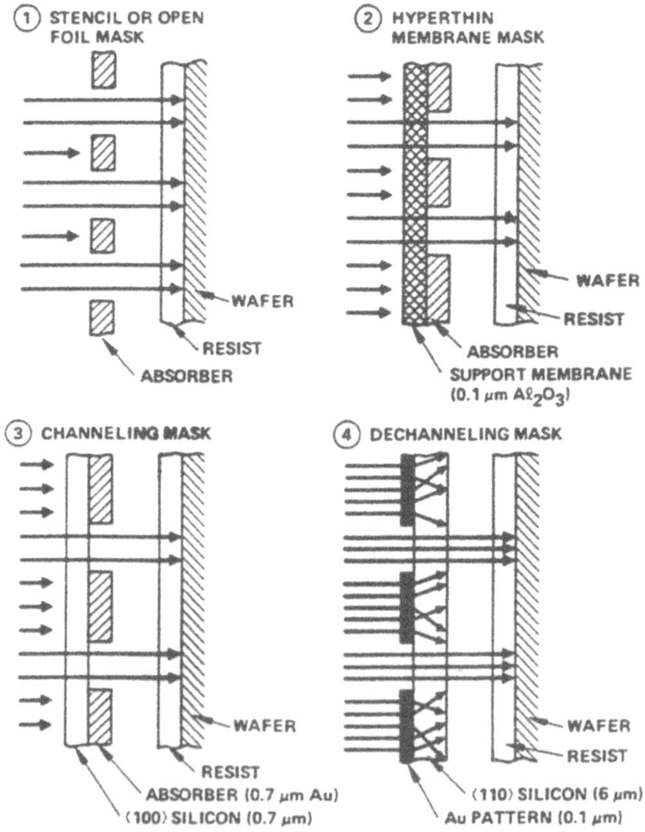

FIGURE 9-8-6. Mask types for flood ion-beam exposure. From J. Jensen and C. Slayman, *SPIE Electron, X-ray and Ion Beam Techniques for Submicron Lithography III* **471**, 53 (1984), copyright and courtesy of SPSE.

implanters[244] and some scanning ion microscopes.[169] For resist exposures, high-brightness metal ion sources of Ga and In have been used to expose top layers of the resist followed by O_2 plasma development.[170]

The main advantage of using ion beams over *e*-beams is the lack of backscattered and hence minimum proximity effect. Early experiments demonstrated that 40-nm lines could be formed in 400-nm-thick PMMA with a 60-keV H^+ beam and with only a 10-nm broadening at a dose of 1 $\mu C/cm^2$.[171] Heavier ions such as Ga^+ scatter and show proximity effects due to nuclear-ion scattering.[173,211] Since PMMA has about the same dose requirements as the DQNs, a single-film resist could be used if a suitable mask (Fig. 9-8-6) and whole wafer flood source were available.[181] Masked ion-beam lithography uses metal on silicon or channeling masks (Fig. 9-8-7).[175,176,191,241] In stencil masks, arbitrary geometries can be reproduced by partitioning the pattern into complementary masks.[52] A compact source is needed to be competitive with X-ray printing. Even poor-resolution swelling negative resists (Chapter 5) can be exposed and developed to form straight-wall profiles (Fig. 9-8-7).[175] The same alignment schemes found in an X-ray exposure tool can be used, but a large field exposure is not likely due to penumbra effects and wafer warpage.[233]

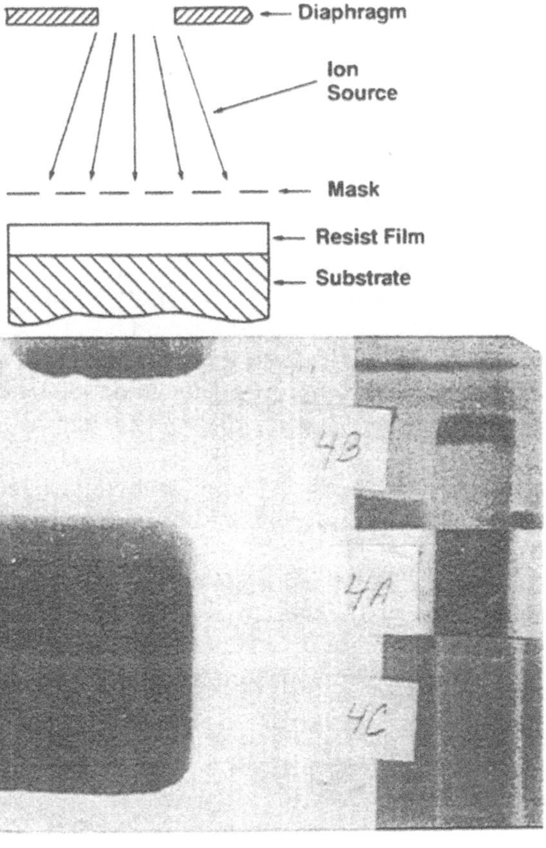

FIGURE 9-8-7. Proton ion-beam exposure of resist. (4A) Ion-beam stencil mask opening, (4B) image produced in positive PMMA resist, and (4C) image produced in negative KTFR resist. Both images were exactly reproduced in the two different resist types.

Focused ion beams can be used to expose resists, repair chrome mask defects,[235] or in resistless lithography (Chapter 16), expose silicon oxide and etch out the region directly. Ion sources have been improved but still do not produce high enough beam currents for adequate throughput[233] even with 0.1 $\mu C/cm^2$ resists. The difficulty is not the lack of source brightness, but the lack of adequate total flux from the source[233] due to the classical space charge spreading[174] (beam broadening) which is more serious in beams of particles of higher mass/charge (m/q) than in *e*-beams.[175] Compared to an estimate of 10 (200-mm) wafers/hr for *e*-beam exposure of 0.5-μm geometries, a scanning ion beam would require 50 hr to expose one wafer.[233] Therefore, scanning ion beams are restricted at present to photomask repair.[261]

9-9. OUTLOOK

From a resist and process viewpoint, we have to take a serious look at the pros and cons of radiation exposure:

1. No single-film thick (2–3 μm) resistant resist is able to be exposed rapidly enough to be competitive with optical exposure. Electron-beam exposure and X-ray exposure require multilayer resists to recover from the proximity effects in *e*-beam and the lack of speed in X rays.
2. Conventional X-ray sources are too weak for present single-film resists. Faster resists are required. At present, synchrotron sources are too expensive and better masks are needed.
3. Field emission *e*-beam sources have achieved sufficient brightness for resist exposure.
4. The proximity effect is a serious limitation for submicrometer *e*-beam lithography. The solution to the proximity effect will be a combination of tradeoff of throughput, process steps, and extra resist-underlayer materials.
5. Significant breakthroughs in lower cost, reliable, and stable masks and high-intensity sources are required for flood *e*-beam, X-ray, and ion-beam proximity printing to be competitive with optical scanners or projection systems.
6. In the year 2000, we can expect individual molecule (angstrom or molecular lithography) to be addressed by high-energy exposure and readout.[86]
7. Electron beams will be the premier mask-making systems for all other lithographies and will serve to handle custom circuit fabrication. Full device fabrication by *e*-beam will not likely occur until the year 2000 unless the overlay requirements of submicrometer devices cannot be met in optical systems.

REFERENCES

1. A. Tafel, *Physics*, Allyn & Bacon, 1981, p. 434.
2. U. S. Patent 3,581,385 (1971), TI.
3. U. S. Patent 3,575,588 (1971), IBM.

4. E. Samaroo, J. Raamot, P. Parry, and G. Robertson, *Bell Syst. Tech. J.* **49**, 2077 (1970).
5. U. S. Patent, 3,900,737 (1975), Bell.
6. J. Ballantyne, *J. Vac. Sci. Technol.* **12**, 1257 (1975).
7. K. Tai, R. Vadimsky, C. Kemmer, J. Wagner, V. Lambertini, and A. Timko, *J. Vac. Sci. Technol.* **17**, 1169 (1980).
8. E. Weber and R. Moore, *Solid State Technol.* May 1979, p. 61; E. Weber, *SPIE Proc.* **333**, 95 (1982).
9. L. Fried, J. Havas, G. Paal, J. Logan, P. Totta, and J. Lechaton, *IBM J. Res. Dev.* **26**, 364 (1982).
10. J. Trotel and B. Fay, in *Electron Beam Technology in Microelectronic Fabrication*, edited by G. Brewer, Academic Press, New York, 1980, p. 309.
11. A. Muray, M. Issacson, I. Aesida, and B. Whitehead, *J. Vac. Sci. Technol.* **B1**, 1019 (1983).
12. R. Havemann, R. Smith, S. Evans, L. Arledge, R. Love, and G. Varnell, *J. Vac. Sci. Technol.* **19**, 903 (1981).
13. H. Pfeiffer, *J. Vac. Sci. Technol.* **12**, 1170 (1975).
14. W. Stickel and G. Langner, *J. Vac. Sci. Technol.* **B1**, 1007 (1983).
15. J. Paraszczak, D. Kern, M. Hatzakis, J. Bucchignanao, E. Arthur, and M. Rosenfield, *J. Vac. Sci. Technol.* **B1**, 1372 (1983).
16. N. Eib and F. Jones, *J. Vac. Sci. Technol.* **B1**, 1327 (1983).
17. A. Broers, *IEEE Trans. Electron Devices* **ED-28**, 1268 (1981).
18. M. Nakase and Y. Matsumoto, *Photogr. Sci. Eng.* **23**, 215 (1979).
19. J. Shaw, M. Hatzakis, J. Paraszczak, J. Liutkus, and E. Babich, *Polym. Eng. Sci.* **23**, 1054 (1983).
20. D. Davis, S. Gillespie, S. Silverman, and W. Stickel, *J. Vac. Sci. Technol.* **B1**, 1003 (1983).
21. J. Phang and H. Ahmed, *J. Vac. Sci. Technol.* **16**, 1754 (1979).
22. J. Wolfe, *IEEE Trans. Electron Devices* **ED-27**, 1475 (1980).
23. A. Broers and M. Pomerantz, *Thin Solid Films* **99**, 323 (1983).
24. H. Ryssel, G. Prinke, H. Bernt, K. Harberger, and K. Hoffmann, *Appl. Phys. A* **27**, 239 (1982).
25. J. Greeneich, in Ref. 10, p. 245.
26. D. Stephani, E. Kratschner, and H. Beneking, *J. Vac. Sci. Technol.* **B1**, 1011 (1983).
27. H. Ito and C. Willson, *Polym. Eng. Sci.* **23**, 1012 (1983).
28. M. Gazard, C. Duchesne, J. Dubois, and A. Chapiro, *Polym. Eng. Sci.* **20**, 1069 (1980).
29. S. Moriya, K. Komatsu, K. Harada, and T. Kitayama, **B1**, 990 (1983).
30. D. Herriott and G. Brewer, in Ref. 10, p. 1.
31. B. Piurczk and A. Williams, *Solid State Technol.* June 1982, p. 74.
32. A. Wilson, *Proc. IEEE* **71**, 575 (1983).
33. N. Aizaki, *J. Vac. Sci. Technol.* **16**, 1726 (1979).
34. J. Doherty, *Solid State Technol.* May 1979, p. 83.
35. L. Lin and H. Beauchamp, *J. Vac. Sci. Technol.* **10**, 987 (1973).
36. B. Piwczk and K. McQuhal, *J. Vac. Sci. Technol.* **10**, 1016 (1973).
37. U. S. Patent 3,876,883 (1975), IBM.
38. A. Wilson, A. Speth, T. Chang, and A. Kern, *J. Vac. Sci. Technol.* **12**, 1235 (1975).
39. U. S. Patents 3,894,271 (1975) and 3,930,181 (1975), IBM.
40. R. Moore, G. Caccomo, H. Pfeiffer, E. Weber, and O. Woodard, *J. Vac. Sci. Technol.* **19**, 950 (1981).
41. H. Pfeiffer, *J. Vac. Sci. Technol.* **15**, 887 (1978).
42. Y. Sakokibara, T. Ogawa, K. Komatsu, S. Moriya, M. Kobayashi, and T. Kobayashi, *IEEE Trans. Electron Devices* **ED-28**, 1279 (1981).
43. G. Jones, V. Rao, H. Sun, and H. Ahmed, *J. Vac. Sci. Technol.* **B1**, 1298 (1983).
44. E. Goto, T. Soma, and M. Idesawa, *J. Vac. Sci. Technol.* **15**, 883 (1978).
45. E. Weber and R. Moore, *J. Vac. Sci. Technol.* **16**, 1780 (1978).
46. S. Moriya, *J. Vac. Sci. Technol.* **B1**, 990 (1983).
47. M. Fujinami, T. Matsuda, K. Takamoto, H. Yoda, T. Ihiga, N. Saitou, and T. Komoda, *J. Vac. Sci. Technol.* **19**, 941 (1981).
48. C. Lemmond, E. Buschmann, T. Klotz, and G. White, *IEEE Trans. Electron Devices* **ED-21**, 598 (1974).
49. D. Smith and K. Harte, *J. Vac. Sci. Technol.* **19**, 953 (1981).
50. I. Brodie, E. Westerberg, D. Cone, J. Murray, M. Williams, and L. Gasiorek, *IEEE Trans. Elect. Dev.* **ED-28**, 1422 (1981).
51. T. Sasaki, *J. Vac. Sci. Technol.* **19**, 963 (1981).

52. H. Bohlen, J. Greshner, and P. Nehmiz, *J. Vac. Sci. Technol.* **16**, 1834 (1979), **19**, 972 (1981), *IBM J. Res. Dev.* **26**, 568 (1982).
53. H. Koops, *J. Vac. Sci. Technol.* **12**, 1141 (1975).
54. M. Heritage, *J. Vac. Sci. Technol.* **12**, 1135 (1975).
55. M. Frosien, B. Liechke, and K. Anger, *J. Vac. Sci. Technol.* **16**, 1827 (1979).
56. U. S. Patent 3,519,873 (1970), Westinghouse.
57. T. O'Keefe, *IEEE Trans. Electron Devices* **ED-17**, 465 (1970).
58. T. Bril and J. Snyders, *J. Vac. Sci. Technol.* **16**, 1665 (1979).
59. R. Ward, *J. Vac. Sci. Technol.* **16**, 1830 (1979), **15**, 1917 (1978).
60. K. Nicholas, R. Ford, H. Brockman, and I. Stemp, *J. Vac. Sci. Technol.* **B1**, 1020 (1983).
61. U. S. Patents 3,644,700 (1972) and 3,894,271 (1975), IBM.
62. M. Larken and R. Matta, *Solid State Electron.* **10**, 491 (1967).
63. J. Eidson and R. Scudder, *J. Vac. Sci. Technol.* **19**, 932, 936 (1981).
64. U. S. Patent 4,213,053 (1980), IBM.
65. U. S. Patent 4,376,249 (1983), IBM.
66. H. Pfeiffer and G. Langner, *J. Vac. Sci. Technol.* **19**, 1058 (1981).
67. U. S. Patent 4,112,305 (1978), Rikagu.
68. U. S. Patent 4,117,349 (1978), Rikagu.
69. U. S. Patent 4,122,369 (1978), General Electric.
70. U. S. Patent 4,199,688 (1980), Hitachi.
71. U. S. Patent 4,199,689 (1980), Tokyo Shibaru.
72. H. Ohiwa, *J. Vac. Sci. Technol.* **15**, 849 (1978).
73. Y. Machida, S. Yamamoto, H. Hakayama, S. Furuya, and T. Hisatsuga, *Fujitsu Sci. Tech. J.* **19**, 21 (1983).
74. W. Livesay, *Microelectron. Manuf. Test.* March 1984, p. 58.
75. G. Dolan and T. Fulton, *IEEE Electron Device Lett.* **EDL-4**, 192 (1983).
76. T. Berker and D. Casey, *IEEE Trans. Electron Devices* **ED-29**, 524 (1982).
77. W. Grobman and T. Studwell, *J. Vac. Sci. Technol.* **16**, 1803 (1979).
78. G. Owen and P. Rissman, *J. Appl. Phys.* **54**, 3573 (1983).
79. J. Pasiecznek and J. Frey, *J. Vac. Sci. Technol.* **10**, 1012 (1973).
80. M. Sumi, F. Chiba, and M. Niomiya, *J. Vac. Sci. Technol.* **16**, 1809 (1979).
81. A. Dexiang, W. Jian-kun, and Q. Pei-yong, *J. Vac. Sci. Technol.* **19**, 975 (1981).
82. G. Varnell, P. Shah, and R. Havemann, *Proc. IEEE* **71**, 612 (1983).
83. M. Miyazaki, N. Saitou, and C. Munakata, *J. Phys. E* **14**, 194 (1981).
84. G. Langner, *Proc. Microcircuit Eng.* 1979, p. 261.
85. N. Saitou, C. Munakata, and A. Maekawa, *Jpn. J. Appl. Phys.* **10**, 351 (1971).
86. F. Carter, *Molecular Electronic Devices*, Dekker, New York, 1982, p. 1; *J. Vac. Sci. Technol.* **B1**, 959 (1983).
87. A. Avidenko and M. Malev, *Vacuum* **27**, 583 (1978).
88. H. Ralph, G. Duggan, and R. Elliott, *Electrochem. Soc. Ext. Abstr.* **82-1**, 514 (1982).
89. T. Lin, *IBM J. Res. Dev.* Sept. 1967, p. 527.
90. D. Davis, R. Moore, M. Williams, and O. Woodard, *IBM J. Res. Dev.* **21**, 498 (1977).
91. D. Davis, *IBM J. Res. Dev.* **24**, 545 (1980).
92. U. S. Patent 4,056,730 (1977), IBM.
93. U. S. Patent 3,644,700 (1972), IBM.
94. E. Wolf, P. Coane, and F. Ozedimir, *J. Vac. Sci. Technol.* **12**, 1266 (1975).
95. H. Friedrich, H. Zeitler, and H. Bierhenke, *J. Electrochem. Soc.* **124**, 627 (1977).
96. D. Stephani, *J. Vac. Sci. Technol.* **16**, 1739 (1979).
97. J. Lida and T. Everhart, *J. Vac. Sci. Technol.* **15**, 917 (1978).
98. R. Henderson, D. Mayer, and J. Nash, *J. Vac. Sci. Technol.* **16**, 260 (1979).
99. W. Stickel, *J. Vac. Sci. Technol.* **15**, 901 (1978).
100. S. Okazaki, F. Mular, J. Takeda, K. Mochyu, H. Kume, and S. Asar, *J. Vac. Sci. Technol.* **19**, 927 (1981).
101. L. White, *J. Vac. Sci. Technol.* **B1**, 1235 (1983).
102. C. Ting, R. Anderson, D. Saiki, and A. Kraft, *J. Vac. Sci. Technol.* **15**, 948 (1978).
103. Y. Lin, A. Neureuther, and I. Aesida, *J. Appl. Phys.* **53**, 899 (1982).
104. Y. Lin, A. Neureuther, and I. Aesida, *J. Electrochem. Soc.* **130**, 939 (1983).
105. Y. Lin and A. Neureuther, *Solid State Technol.* Feb. 1984, p. 117.

106. S. Yamamoto, H. Nakayama, S. Furuya, T. Hisatsugu, and Y. Machida, *Fujitsu Sci. Tech. J.* **14**, 143 (1978).
107. L. Rothman, *J. Electrochem. Soc.* **127**, 2216 (1980).
108. D. Kyser and C. Ting, *J. Vac. Sci. Technol.* **16**, 1759 (1979).
109. D. Kyser and R. Pyle, *IBM J. Res. Dev.* **24**, 426 (1980).
110. N. Aizaki, *J. Vac. Sci. Technol.* **16**, 1726 (1979).
111. B. Sohler, C. Snider, and R. Shuman, *J. Electrochem. Soc.* **131**, 868 (1984).
112. C. Dix, *J. Vac. Sci. Technol.* **19**, 911 (1981).
113. T. Chang, C. Codella, and R. Lange, *IEEE Trans. Electron Devices* **ED-28**, 1428 (1981).
114. R. Moore, G. Caccomo, H. Pfeiffer, E. Weber, and O. Woodard, *Electronics* Nov. 3, 1981, p. 142.
115. M. Idesawa, Y. Somo, and E. Goto, *J. Vac. Sci. Technol.* **19**, 983 (1981).
116. N. Eib and F. Jones, *J. Vac. Sci. Technol.* **B1**, 1327 (1983).
117. K. Nicholas, R. Ford, H. Brockman, and I. Stemp, *J. Vac. Sci. Technol.* **B1**, 1021 (1983).
118. H. Sewell, *J. Vac. Sci. Technol.* **15**, 927 (1978).
119. M. Parikh and D. Schreiber, *IBM J. Res. Dev.* **24**, 531 (1980).
120. M. Nakase and M. Yoshimi, *IEEE J. Solid State Circuits* **SC-15**, 525 (1980).
121. M. Parikh, *J. Vac. Sci. Technol.* **15**, 931 (1978).
122. J. Greeneich, *J. Vac. Sci. Technol.* **19**, 1269 (1981).
123. G. Owen and P. Rissman, *J. Appl. Phys.* **54**, 3573 (1983).
124. W. Grobman, A. Speth, and T. Chang, *IBM J. Res. Dev.* **24**, 537 (1980).
125. D. Kyser and C. Ting, *J. Vac. Sci. Technol.* **16**, 1759 (1979).
126. Y. Yoshimi, M. Yoshimi, K. Kawabuchi, T. Takigauna, M. Takahashi, and Y. Kato, *Electron. Lett.* **18**, 880 (1982).
127. T. Neil and C. Bull, *Electron. Lett.* **16**, 621 (1980).
128. A. Broers, *J. Electrochem. Soc.* **128**, 166 (1981).
129. R. Howard, H. Craighead, L. Jaeckel, and P. Mankiewich, *J. Vac. Sci. Technol.* **B1**, 1101 (1983).
130. D. Stephani, E. Kratschner, and H. Beneking, *J. Vac. Sci. Technol.* **B1**, 1011 (1983).
131. T. Chang, *J. Vac. Sci. Technol.* **12**, 1271 (1975).
132. P. Vettiger, T. Forster, and D. Kern, *J. Vac. Sci. Technol.* **B1**, 1383 (1983).
133. D. Kyser, *J. Vac. Sci. Technol.* **B1**, 1391, (1983).
134. C. Shaw, *J. Vac. Sci. Technol.* **19**, 1286 (1981).
135. J. Greeneich, *J. Electrochem. Soc.* **122**, 970 (1975).
136. H. Nokata, T. Kato, K. Murata, Y. Hirai, and K. Nagami, *J. Vac. Sci. Technol.* **19**, 1248 (1981).
137. N. Nakayama, Y. Machida, S. Furuya, S. Yamamoto, and T. Hisatsugu, *Electrochem. Soc. Ext. Abstr.* **82-1**, 509 (1982).
138. U. S. Patent 3,971,860 (1976), IBM.
139. I. Aesida and T. Everhart, *J. Appl. Phys.* **51**, 5994 (1980).
140. U. S. Patent 4,316,093 (1982), IBM.
141. R. Howard and D. Prober, in *VLSI Electronics*, edited by N. Einspruch, Academic Press, New York, 1982, p. 146.
142. M. Rosenfield, A. Neureuther, and C. Ting, *J. Vac. Sci. Technol.* **19**, 1242 (1981).
143. D. Chow, J. McDonald, D. King, W. Smith, K. Molnar, and A. Steckl, *J. Vac. Sci. Technol.* **B1**, 1383 (1983).
144. K. Harada, T. Tamamura, and O. Kogure, *J. Electrochem. Soc.* **129**, 2576 (1982).
145. J. Greeneich, *Semicond. Int.* April 1981, p. 159.
146. D. Kern, *Proc. 8th Electron Ion Beam Conf.* **80-6**, 327 (1980).
147. J. Greeneich, *Proc. 8th Electron Ion Beam Conf.* **80-6**, 293 (1980).
148. *IEEE Trans. Nucl. Sci.* **NS-30**, No. 6 (1983).
149. J. Tzou, *J. Appl. Phys.* **55**, 846 (1984).
150. J. Aitken, D. Young, and K. Pan, *J. Appl. Phys.* **49**, 3386 (1978).
151. M. Peckerar, *J. Vac. Sci. Technol.* **16**, 1658 (1979).
152. K. Galloway, *J. Electrochem. Soc.* **127**, 1862 (1980).
153. K. Galloway, S. Majo, and P. Roctman, *J. Electrochem. Soc.* **126**, 2245 (1979).
154. D. Brown, D. Ma, C. Dozier, and M. Peekerar, *IEEE Trans. Nucl. Sci.* **NS-30**, 4025 (1983).
155. K. Galloway, *Semicond. Int.* May 1979, p. 65.
156. D. Maydan, G. Coquin, H. Levenstein, A. Sinha, and D. Wang, *J. Vac. Sci. Technol.* **16**, 1959 (1979).
157. J. McCoy, *Circuits Manufacturing*, Nov. 1977, p. 57.

158. H. Aritome, *J. Vac. Sci. Technol.* **15**, 922 (1978).
159. S. Austin, H. Smith, and D. Flanders, *J. Vac. Sci. Technol.* **15**, 984 (1978), **16**, 1949 (1979).
160. D. Flanders and T. Lyszczarz, *J. Vac. Sci. Technol.* **B1**, 1196 (1983).
161. E. Hundt and P. Tescher, *J. Vac. Sci. Technol.* **15**, 1004 (1978).
162. K. Heinrich, H. Betz, A. Heuberger, and S. Pongrantz, *J. Vac. Sci. Technol.* **19**, 1254 (1981).
163. N. Atoda, H. Kawakatsu, H. Tanino, S. Ichimura, and H. Koh, *J. Vac. Sci. Technol.* **B1**, 1267 (1983).
164. J. Lane, J. Maldanado, A. Cleland, R. Haelbich, J. Silverman, and J. Warlamount, *J. Vac. Sci. Technol.* **B1**, 1072 (1983).
165. R. Haelbich, J. Silverman, W. Brobman, J. Maldanado, and J. Warlamount, *J. Vac. Sci. Technol.* **B1**, 1262 (1983).
166. W. Grobman, *J. Vac. Sci. Technol.* **B1**, 1257 (1983).
167. M. Beiber, H. Scheunemann, H. Betz, and A. Heuberger, *J. Vac. Sci. Technol.* **B1**, 1271 (1983).
168. D. Johnson, *Semicond. Int.* March 1984, p. 83.
169. T. Shiokawa, *J. Vac. Sci. Technol.* **B1**, 1117 (1983).
170. W. Brown, T. Venkatensan, and A. Wagner, *Solid State Technol.* Aug. 1981, p. 60.
171. L. Karapiperis and C. Lee, *Appl. Phys. Lett.* **35**, 395 (1979).
172. C. Fencil and G. Hughes, *SPIE Proc.* **333**, 100 (1982).
173. B. Siegel, G. Hanson, M. Szilazi, D. Thomas, R. Blackwell, and H. Park, *SPIE Proc.* **333**, 152 (1982).
174. Y. Yau, T. Groves, and R. Pease, *J. Vac. Sci. Technol.* **B1**, 1141 (1983).
175. C. Slayman, J. Bartelt, and C. McKenna, *SPIE Proc.* **333**, 168 (1982), *J. Vac. Sci. Technol.* **19**, 1166 (1981).
176. J. Randall, D. Flanders, N. Economu, J. Donnely, and E. Bromley, *J. Vac. Sci. Technol.* **B1**, 1152 (1983).
177. Y. Machida, H. Nakayama, S. Furuya, T. Hitsatsugo, and S. Yamamoto, *Fujitsu Sci. Tech. J.* **19**, 21 (1983).
178. S. Ichimura, M. Hirata, H. Tanino, N. Atoda, M. Ono, and K. Oh, *J. Vac. Sci. Technol.* **B1**, 1076 (1983).
179. A. Barraud, C. Rosillo, and A. Ruadel-Texrier, *J. Vac. Sci. Technol.* **16**, 2003 (1979).
180. M. Shiyama, O. Nakajima, C. Hashimoto, and Y. Sakabara, *J. Electrochem. Soc.* **131**, 1391 (1984).
181. M. Geis, J. Randall, T. Deutsch, N. Efremow, J. Donnely, and J. Woodhouse, *J. Vac. Sci. Technol.* **B1**, 1178 (1983).
182. R. Howard, H. Craighead, L. Jackel, and P. Mankewich, *J. Vac. Sci. Technol.* **B1**, 1101 (1983).
183. T. Takigawa, K. Kawabuchi, M. Yoshimi, and Y. Kato, *Microelectron. Eng.* **1**, 121 (1983).
184. D. Joy, *Microelectron. Eng.* **1**, 103 (1983).
185. U. S. Patent 4,456,677 (1984), U.S. government.
186. M. Kakuchi, M. Hikita, A. Sugita, K. Onose, and T. Tamamura, *J. Electrochem. Soc.* **133**, 1755 (1986).
187. I. Aesida, M. Zhang, and E. Wolf, *J. Electron. Mater.* **13**, 689 (1984).
188. C. Chen, *SPIE Proc.* **471**, 1 (1984).
189. Y. Suzuki, H. Itoh, and T. Shigura, *J. Vac. Sci. Technol.* **B3**, 1009 (1985).
190. R. Jaeger and B. Heflinger, *SPIE Proc.* **471**, 110 (1984).
191. J. Randall, D. Flanders, and J. Economu, *SPIE Proc.* **471**, 110 (1984).
192. H. Pfeiffer, in *Microcircuit Engineering*, edited by H. Ahmed, J. Cleaver, and G. Jones, Academic Press, New York, 1984, p. 3.
193. H. Matsumura and T. Takeshi, *Appl. Phys. Lett.* **45**, 3 (1984).
194. A. Neukermans, *Solid State Technol.* Sept. 1984, p. 185.
195. A. Shikiunas, *Solid State Technol.* Sept. 1984, p. 192.
196. H. Bohlen, P. Nehmiz, and J. Greshner, *Solid State Technol.* Sept. 1984, p. 210.
197. H. Pfeiffer, *Solid State Technol.* Sept. 1984, p. 223.
198. M. Suzuki, H. Namatsu, and A. Yoshikawa, *J. Vac. Sci. Technol.* **B2**, 665 (1984).
199. O. Wells, *Scanning Electron Microscopy*, McGraw–Hill, New York, 1974.
200. S. Miyauchi, K. Tanaka, and J. Russ, *IEEE Trans. Electron Devices* **ED-17**, 450 (1970).
201. B. Piwczyk and A. Williams, *Solid State Technol.* June 1982, p. 74.
202. K. Murata, *J. Appl. Phys.* **57**, 575 (1985).
203. L. Jackel, R. Howard, and P. Mankiewich, *Appl. Phys. Lett.* **45**, 698 (1984).
204. German Patent 3,342,319 (1984), *Chem. Abstr.* **101**, 201548 (1985).

205. Japanese Patent 59,116,745 (1984), *Chem. Abstr.* **101**, 15133 (1985).
206. W. Buckley, Proc. 7th Int. Symp. Electron Ion Beam Sci. Technol., Electrochemical Society, 1978, p. 462.
207. R. Ono, J. Sauvageau, A. Jain, D. Schwartz, K. Springer, and J. Lukens, *J. Vac. Sci. Technol.* **B3**, 282 (1985).
208. G. Atkinson and A. Neureuther, *J. Vac. Sci. Technol.* **B3**, 421 (1985).
209. D. Rensch, J. Chen, W. Clark, and M. Courtney, *J. Vac. Sci. Technol.* **B3**, 286 (1985).
210. T. Kato, H. Morimoto, K. Saitoh, and H. Nakata, *J. Vac. Sci. Technol.* **B3**, 50 (1985).
211. I. Aesida, E. Kratschner, E. Wolf, A. Muray, and M. Isaacson, *J. Vac. Sci. Technol.* **B3**, 45 (1985).
212. A. Chen, A. Neureuther, and J. Pavovich, *J. Vac. Sci. Technol.* **B3**, 148 (1985).
213. P. Rissman and G. Owen, *J. Vac. Sci. Technol.* **B3**, 159 (1985).
214. G. Owen, P. Rissman, and M. Long, *J. Vac. Sci. Technol.* **B3**, 153 (1985).
215. P. Mankewich, L. Jackel, and R. Howard, *J. Vac. Sci. Technol.* **B3**, 174 (1985).
216. K. Gamo, Y. Yamashita, F. Emoto, S. Namba, N. Samoto, and R. Shizumi, *J. Vac. Sci. Technol.* **B3**, 117 (1985).
217. L. Hsia and E. Weber, *J. Vac. Sci. Technol.* **B3**, 128 (1985).
218. H. Yamaguchi, A. Shimase, S. Haraichi, and T. Miyauchi, *J. Vac. Sci. Technol.* **B3**, 71 (1985).
219. K. Muller and H. Burghause, *J. Vac. Sci. Technol.* **B3**, 241 (1985).
220. L. Semenzato, E. Eaton, A. Neukermans, and R. Jaeger, *J. Vac. Sci. Technol.* **B3**, 245 (1985).
221. C. Dix, P. Flavin, P. Hendy, and M. Jones, *J. Vac. Sci. Technol.* **B3**, 131 (1985).
222. L. Hsia and E. Weber, *J. Vac. Sci. Technol.* **B3**, 128 (1985).
223. N. Saitou, H. Morishito, S. Nonogaki, H. Itoh, and A. Maikawa, *Jpn. J. Appl. Phys.* **10**, 1486 (1971).
224. T. Matsuda, K. Miyoshi, R. Yamaguchi, S. Moriya, T. Hosoya, and K. Harado, *IEEE Trans. Electron Devices* **ED-32**, 168 (1985).
225. M. Rosenfield, D. Goodman, A. Neureuther, and M. Prouty, *J. Vac. Sci. Technol.* **B3**, 337 (1985).
226. A. Chen, A. Neureuther, and J. Pavovich, *J. Vac. Sci. Technol.* **B3**, 148 (1985).
227. Y. Machida, N. Nakayama, S. Furyu, and S. Yamamoto, *IEEE Trans. Electron Devices* **ED-32**, 831 (1985).
228. J. Paraszczak, D. Kern, M. Hatzakis, J. Bucchignano, E. Arthur, and M. Rosenfield, *J. Vac. Sci. Technol.* **B1**, 1372 (1983).
229. U. S. Patent 4,467,211 (1984), Control Data.
230. U. S. Patent 4,465,934 (1984), VEECO.
231. U. S. Patent 4,460,831 (1984), Thermoelectric.
232. A. Broers, *Solid State Technol.* June 1985, p. 119.
233. A. Broers, in *Microcircuit Engineering*, edited by H. Beneking and H. Beureuther, Academic Press, New York, 1985, p. 1.
234. T. Kokigawa, K. Kawabuchi, M. Yoshimi, and Y. Kato, *Microelectron Eng.* **1**, 121 (1983).
235. P. Heard, J. Cleaver, and H. Ahmed, *J. Vac. Sci. Technol.* **A3**, 87 (1985).
236. B. Soller, C. Snider, and R. Shuman, *J. Electrochem. Soc.* **131**, 868 (1984).
237. R. Devine and C. Froe, *J. Appl. Phys.* **57**, 5162 (1985).
238. T. Kitayama, T. Tamamura, and K. Harada, *Microelectron. Eng.* **2**, 97 (1985).
239. D. Yen, L. Linholm, and W. Glendenning, *J. Electrochem. Soc.* **132**, 1726 (1985).
240. H. Beneking, *Microelectron. Eng.* **2**, 65 (1985).
241. K. Gamo and S. Namba, *Microelectron. Eng.* **2**, 74 (1985).
242. S. Mackie and S. Beaumont, *Solid State Technol.* Aug. 1985, p. 117.
243. K. Suzuki, K. Okada, J. Matsui, N. Endo, Y. Ieda, and M. Kukuina, *IEEE Electron Devce Lett.* **EDL-6**, 353 (1985).
244. H. Golen, N. Schell, and J. George, *Solid State Technol.* June 1985, p. 155.
245. N. Samato, R. Shimizu, and H. Hashimoto, *Jpn. J. Appl. Phys.* **24**, 482 (1985).
246. U. S. Patent 4,513,203 (1985), IBM.
247. S. Hasegawa and Y. Ieda, *IEEE Trans. Electron Devices* **ED-32**, 95 (1985).
248. H. Craighead, *J. Electron Microsc. Tech.* **2**, 147 (1985).
249. G. Atkinson and N. Cheung, *Nucl. Instrum. Phys. Res.* **B7-8**, 872 (1985).
250. U. S. Patent 4,503,334 (1985), U.S. Phillips.
251. H. Morimoto, Y. Sasaki, H. Onoda, and T. Kato, *Appl. Phys. Lett.* **46**, 898 (1985).
252. G. Jones, P. Sargent, T. Norris, and H. Ahmed, *J. Vac. Sci. Technol.* **B3**, 124 (1985).
253. G. Owen, P. Rissman, and M. Long, *J. Vac. Sci. Technol.* **B3**, 153 (1985).

254. P. Petrie and J. Schoeffel, *Proc. SPIE* **537**, 17 (1985).
255. K. Mochizu, T. Kimura, H. Obayashi, and H. Maezawa, *Proc. SPIE* **539**, 56 (1985).
256. U. S. Patent 4,520,269 (1985), IBM.
257. U. S. Patent 4,504,558 (1985), IBM.
258. U. S. Patent 4,425,508 (1985), GCA.
259. K. Murata, K. Nakami, and S. Namba, *IEEE Trans. Electron Devices* **ED-32**, 1694 (1985).
260. P. Whipps, SPE RETEC Photopolymers Proceedings, Ellenville, N.Y., 1985, p. 147.
261. M. Kakuchi, M. Hikita, and T. Tamamura, *Appl. Phys. Lett.* **48**, 835 (1986).

10

Developing Resist Images

10-1. INTRODUCTION

The exposure process induces chemical changes in polymers such as random scission or cross-linking and molecular rearrangements in small molecules added to polymers. The chemical products are distinguished from the unexposed reactants by a developer or process of development (such as heating) which attempts to maximize the readout:

$$\text{Developing readout} = \frac{\text{Exposed resist properties}}{\text{Unexposed resist properties}}$$

Various differential properties (Table 10-1-1) can be used to produce one-, two-, and three-dimensional images in polymers. For resists, the differential rate of solubility in positive resists or the insolubility in negative resists (liquid developer) have been the basic mechanisms for the formation of three-dimensional images. In positive resists, the kinetics of polymer dissolution determine the speed and contrast of a resist, while traditional negative resists rely upon the mechanical integrity of a cross-linked or insoluble (or nonvolatile) region that was radiation transformed.

The object of the development step is to remove the exposed resist (positive) region or unexposed (negative) region at a faster rate (R) than the background (R_0) and in a reproducible controlled manner. In the resist image, the suitable profiles for the various image transfer steps shown in Table 10-1-2 are generated by the appropriate resist dose and developer combination.

In the overcut profile, the shape is dominated by the dose and reflected exposure rays, while in the undercut profile, the developer dominates the resist profile and removes part of the unexposed film. Isolated images and profiles are easier to reproduce than closely spaced features (Fig. 10-1-1). For example, the unexposed land in a positive resist receives exposure from adjacent patterns (inter-proximity effect) and the image is considerably thinner than the background thickness. For positive resists, the concept of the kinetic ratio of exposed dissolution rate (R) to the unexposed rate (R_0) is an important one-dimensional parameter to assess the speed (dose requirement) of a resist (Table 10-1-2). A high ratio implies vertical profiles (unless swelling of the resist image occurs) and no loss of the unexposed (no pinholes or excess thinning over steps, higher yields, less defects) while a low ratio (< 6) reveals the isotropic image widening caused by a poor edge slope of the exposing beam (pattern) and the radiation scattering in the resist

TABLE 10-1-1. Image Differentiation in Polymers

	Type of image	
Imaging property	Positive	Negative
Solubility rate	inc.[a]	dec.
Volatility rate	inc.	dec.
Tackiness	inc.	dec.
Adhesion[b]	dec.	inc.
Adsorption	inc.	dec.
Viscosity	inc.	dec.
Solidification	dec.	inc.
Permeability	inc.	dec.
Transmission (opt.)	inc.	dec.
Refraction	dec.	inc.
Plasma erodability	inc.	dec.

[a] inc., increase in exposed region; dec., decrease in exposed region.
[b] W. Woodruff, W. Jeffers, and R. Snedeker, Image formation by photo-adhesion, *Photogr. Sci. Eng.* **11**, 93 (1967).

(forward and reflected). Because of this distortion, Gaussian-shaped images are formed and the spaces (unexposed region in a positive resist) are subject to inter-proximity effects, i.e., the lines continue to enlarge with overdevelopment and the spaces shrink (Fig. 10-1-2). Isolated images are also subject to enlargement (Fig. 10-1-3) if not properly developed.

The process control of isolated line dimensions is easier than the control of lines and spaces since the mask can be compensated by negative windage (smaller openings). If an isolated space is deliberately enlarged in the photomask (positive windage), the final printed and developed feature can be made less sensitive to variations in development.[1] While positive resists have tight process latitudes and one must be concerned about the unexposed thinning and image enlargement

TABLE 10-1-2. Profiles in Resist for Semiconductor Fabrication

Profile	Formation	Application	R/R_0[a]
1. Overcut 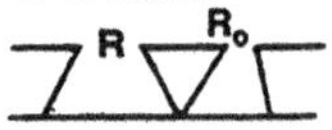	Overexposure High dose Multiple layers Positive resist Image reversal	Ion implant Lift-off	>10
2. Vertical 	High contrast ($\gamma > 3$) Positive resist Multilayer resist	Ion implant Plasma etch Wet etch	10
3. Undercut 	Low contrast ($\gamma < 2$) Positive or negative resist Low dose Resist thinning	Wet etch	<6

[a] Positive resist example.

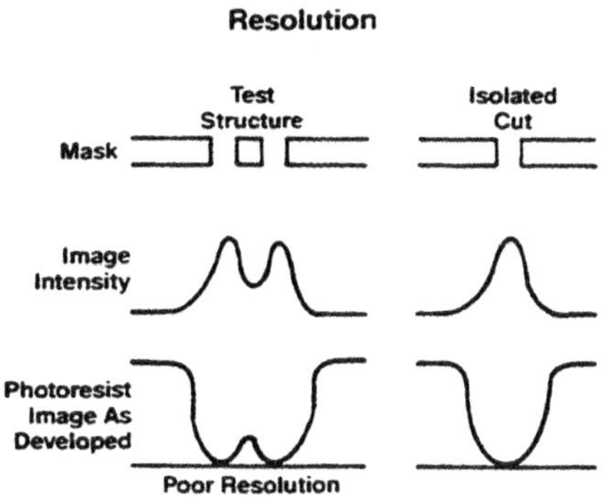

FIGURE 10-1-1. Resolution loss of 1-μm space versus good resolution in isolated 1-μm line. From P. Gwodz, *SPIE Semicond. Lithogr.* **275**, 156 (1981), courtesy of SPIE.

(Fig. 10-1-4), negative cross-linking resists are invariant to the development time and exhibit a wide process latitude.[2]

It is apparent from Figs. 10-1-3 and 10-1-4 that the control of image sizes in positive resists is very dependent on the combination of the exposure and develop time. In the various stages (Fig. 10-1-5) of developing an image, the positive resist reveals the underexposure in a depthwise (stage a) manner. In order to prevent gross adhesion failure (Fig. 10-1-6) proper adhesion by priming (Chapter 6) must be obtained prior to applying the resist film to prevent gross image failure or edge lifting by the lateral attack of the developer along the resist–substrate interface (stage d). By contrast, optically exposed negative photoresists do not reveal underexposure since the radiation (dose) has not penetrated the resist sufficiently to insolubilize the image (Fig. 10-1-7). However, in order to promote adhesion at the interface, negative photoresists are overexposed. In *e*-beam exposure, the negative resists are initially exposed at the interface by the backscattered radiation and for thinner images, they remain (underexposed) on the wafer (Fig. 10-1-8). Since backscatter is the predominant exposure mechanisms in negative electron resists (Fig. 10-1-9), the interproximity effect is a serious problem (Fig. 10-1-10) and the effect is also responsible for the "scumming" (extraneous resist remaining between spaces).

Although the development process for a negative cross-linked resist is time independent, all of the scattered radiation (Fig. 10-1-11) is revealed since the developer encounters a lightly exposed but swollen outer shell which impedes the removal of the sol phase. Only with non-cross-linking types of negative resists or with nonscattering (X-ray or ion-beam) exposure can one expect vertical profiles in negative resists.

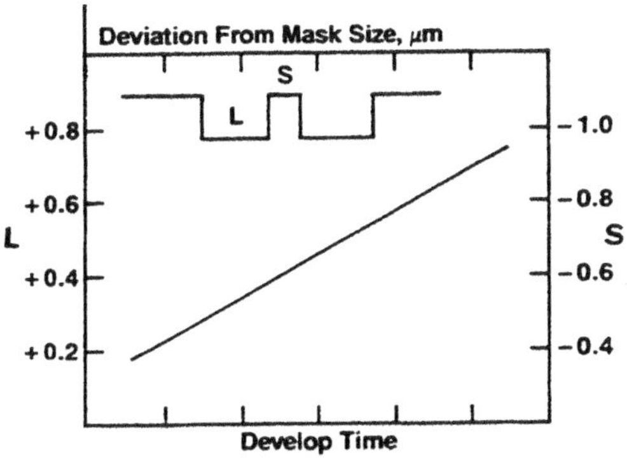

FIGURE 10-1-2. Bias of developed image deviation from mask size as a function of developed time. L = trench in positive resist; S = remaining resist stud.

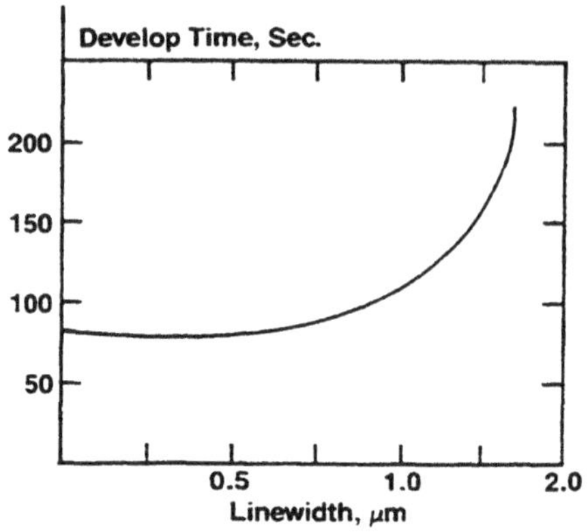

FIGURE 10-1-3. Linewidth versus development time for a nominal 1-μm line in AZ-1350 positive resist. From Ref. 1, copyright 1975 and courtesy of IEEE.

Since the develop time is a function of the exposure time or dose, the rate (R) is given by

$$R = R_0 e^{-\alpha k I t} \tag{10-1-1}$$

where I is the intensity of exposure, α is the energy absorption coefficient, and t is the exposure time. The dose is given by

$$E = It \tag{10-1-2}$$

The resist development rates are also subject to the prebake conditions, the concentration or type of developer, and the developer temperature.[2, 285, 356, 357] The rate, R, for a linear dissolving resist (Fig. 10-1-12) can be expressed by a power expression,

$$R = aM^k \tag{10-1-3}$$

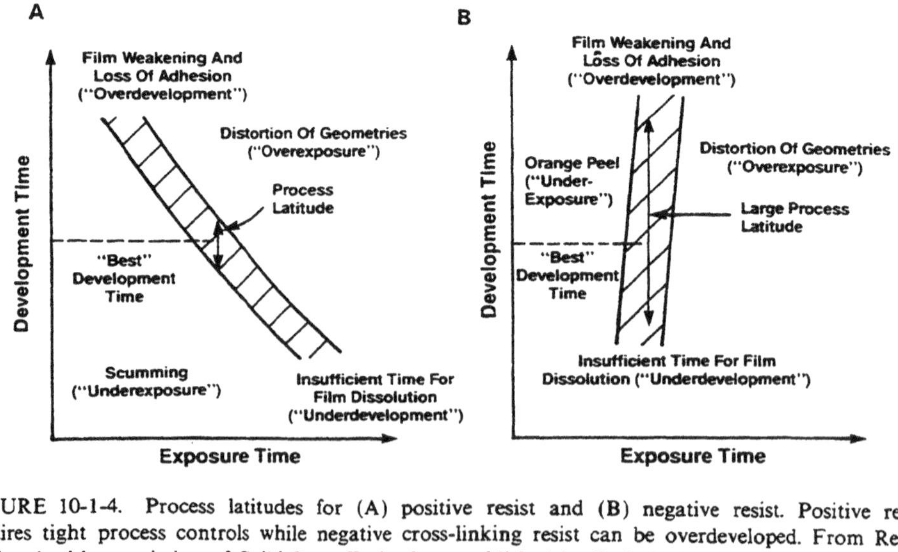

FIGURE 10-1-4. Process latitudes for (A) positive resist and (B) negative resist. Positive resist requires tight process controls while negative cross-linking resist can be overdeveloped. From Ref. 2, reprinted with permission of *Solid State Technology*, published by Technical Publishing, a company of Dun and Bradstreet.

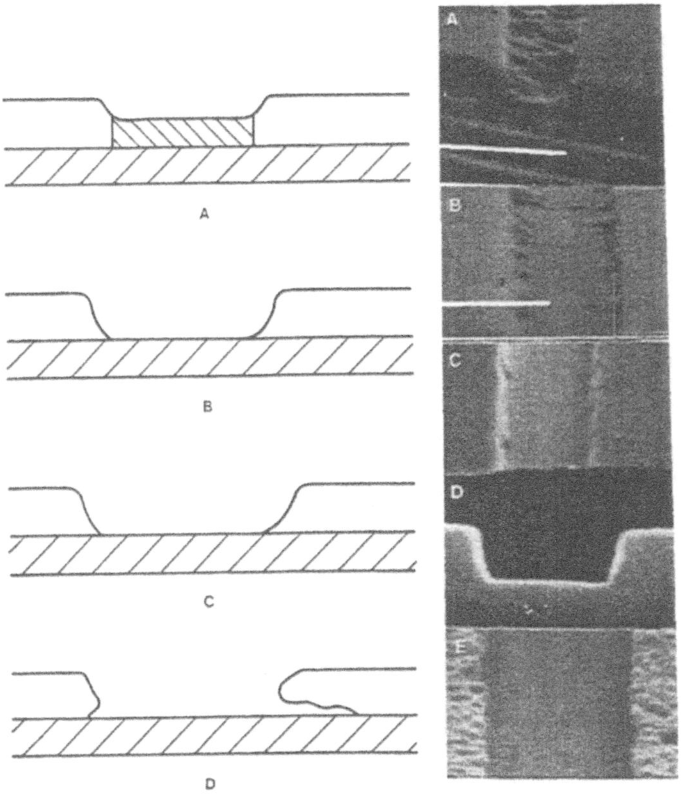

FIGURE 10-1-5. Stages of development of positive resist. (A) Underdeveloped or underexposed image; (B) developed image; (C) slightly overdeveloped; (D) overdeveloped with possible edge lifting and shallow profiles.

FIGURE 10-1-6. Adhesion failure on PMMA developed images.

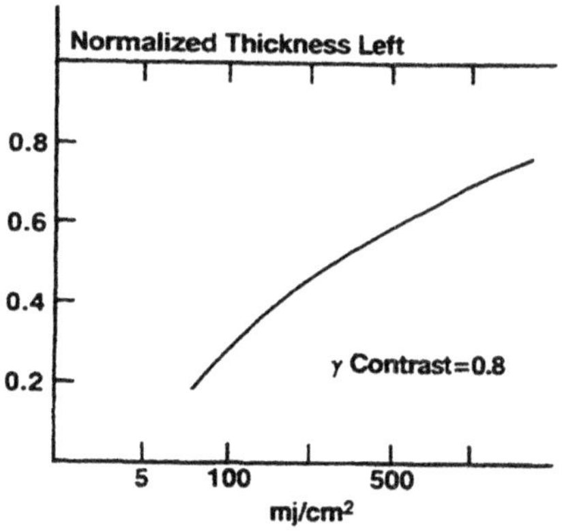

FIGURE 10-1-7. Film contrast for KTFR. From J. Frejilich, *J. Opt. Soc. Am.* **67**, 92 (1977).

where a is the moles of dissolving species.[7, 270] A plot of R versus dose on a log scale (Fig. 10-1-13) reveals that a 90°C prebake will be the least sensitive[285] (smaller k) to overexposure and in the cases of a larger k, tighter process controls are necessary to prevent linewidth growth per thickness (Z) of resist, which is given by

$$dR/dZ = (dR/dE)(dE/dZ) \tag{10-1-4}$$

where the first part (dR/dE) is the developer term dependent on the developer and inhibitor concentration and the second term is the energy absorbed dependent on the quantum yield (ϕ) or G value, the absorption (extinction coefficient), and the wavelength of exposure. At high doses, the developer has little influence and in some cases such as DQN resists, the inhibitor (Chapter 2) is completely destroyed

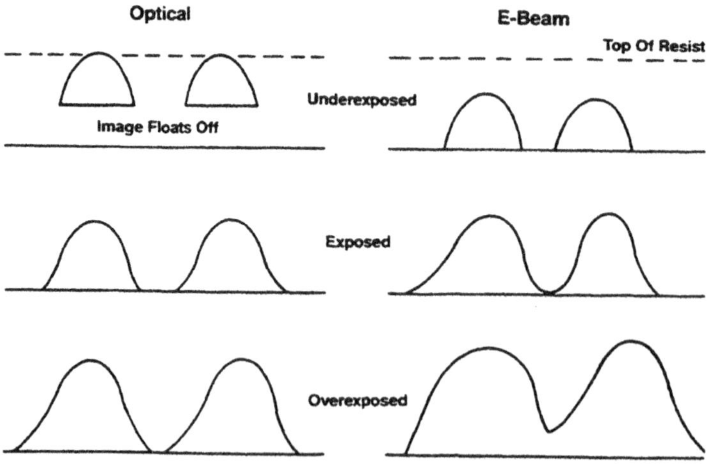

FIGURE 10-1-8. Characteristic response of cross-linking negative resist to optical and *e*-beam exposure. In optical exposure, radiation is absorbed in the forward mode and may be insufficient in the underexposed case to insolubilize the resist at the resist–substrate interface (image floats off in developer). For *e*-beam exposure, the backscattered radiation predominates and the resist records the absorbed dose outwardly from the substrate–resist interface.

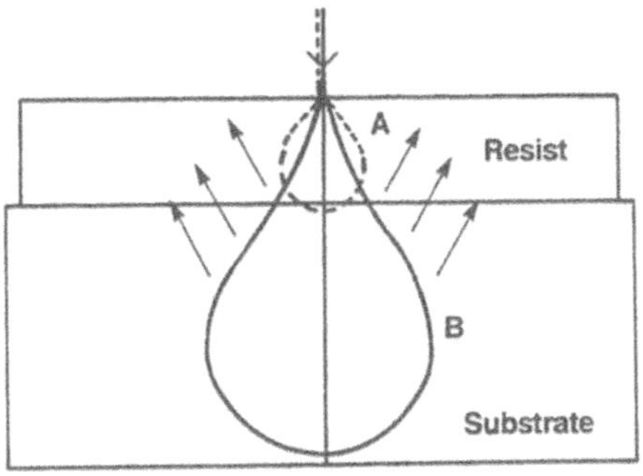

FIGURE 10-1-9. Backscattered electrons from deep in the silicon causing interproximity exposure beyond beam width. From Ref. 385.

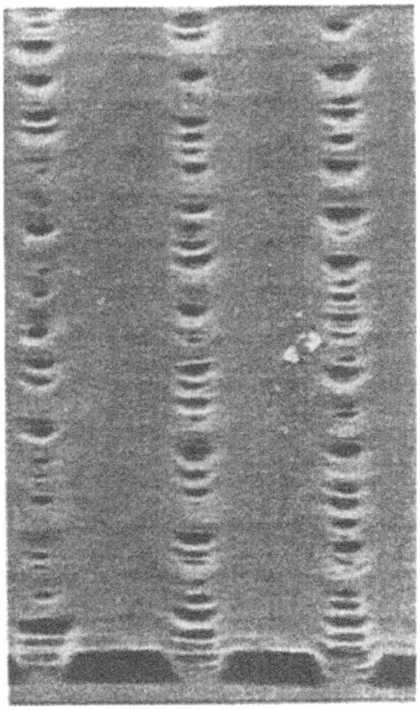

FIGURE 10-1-10. Scumming between unexposed negative resist due to proximity effect from backscattered electrons.

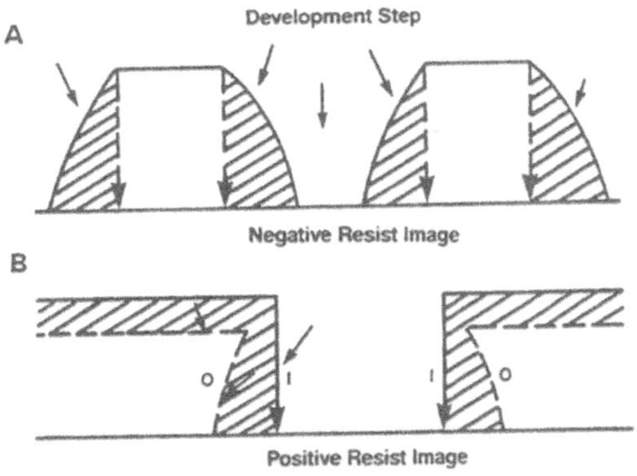

FIGURE 10-1-11. (A) Negative image developing inwardly from the outer light scatter zone to the heavily exposed region; (B) positive resist image developing outwardly from the heavily exposed zone into the light scatter zone. On a kinetic basis, the positive resist development can be stopped to subtend the development of the scatter zone, while a negative cross-linked image reveals (time independent) all of the light scatter.

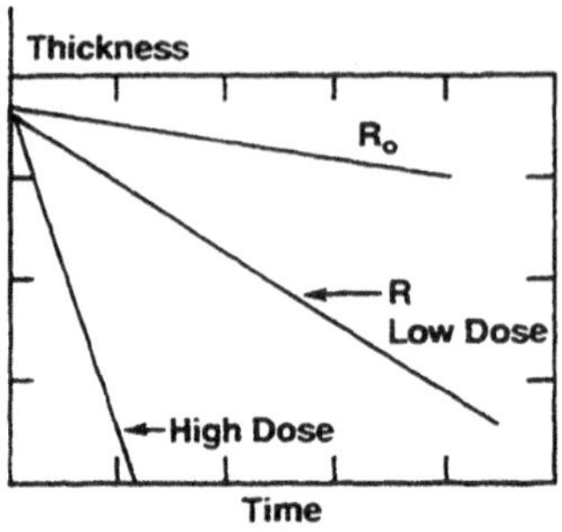

FIGURE 10-1-12. Linear dissolving positive resist.

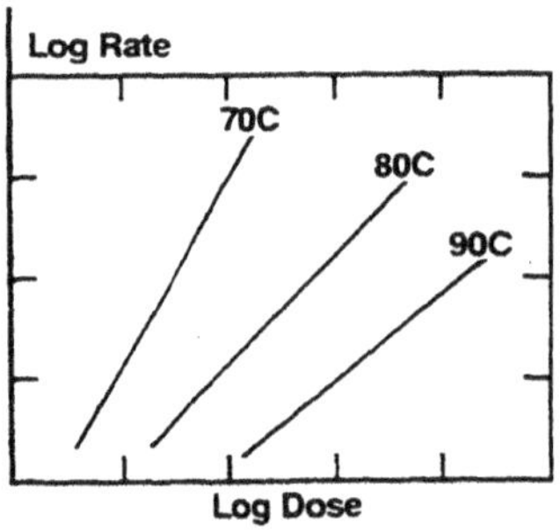

FIGURE 10-1-13. Determination of rate of dissolution power constant k as influenced by three prebake temperatures for DQN resist. The 90°C will be the least sensitive to overexposure and bottom of linewidth variations.

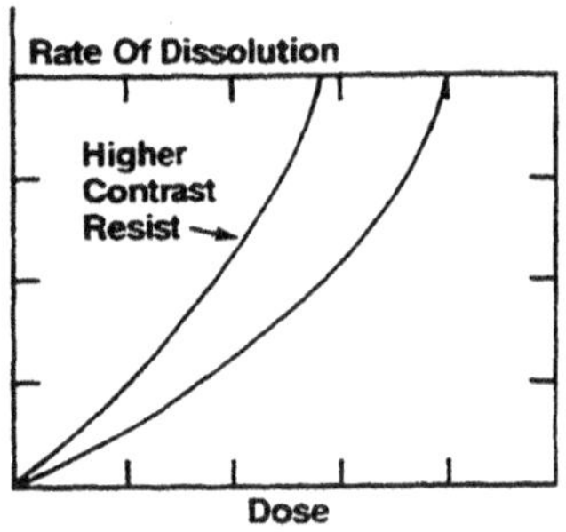

FIGURE 10-1-14. Plot of rate of dissolution of positive resist versus dose.

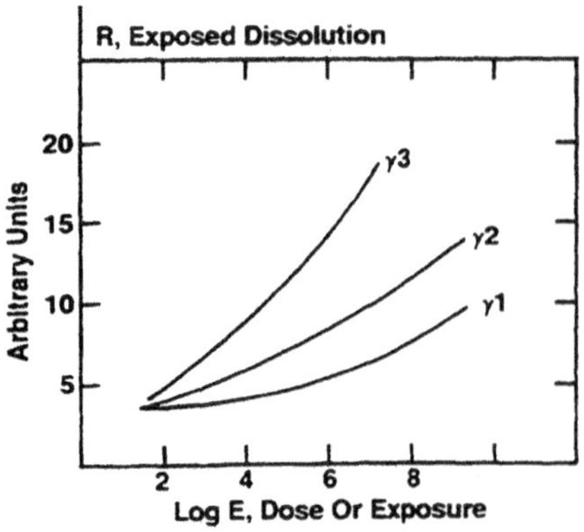

FIGURE 10-1-15. Effect of dose on rate of developing of positive resist and the effect of resist contrast γ: a higher-contrast resist dissolves more rapidly over a smaller dose range.

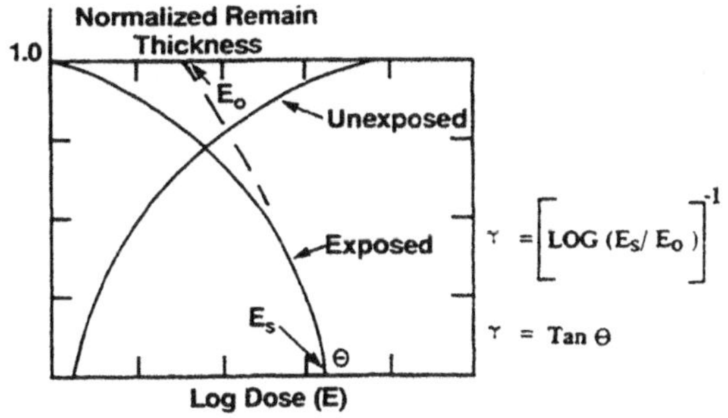

FIGURE 10-1-16. Contrast (γ) curve for positive resist. γ, slope of $E_0 - E_s$.

and the development rate (Fig. 10-1-14) is constant independent of dose. A high contrast ($\gamma > 3$) is achieved with high doses (Fig. 10-1-15).

The normal contrast (γ) is defined as the slope or tangent (θ) of the abscissa of a plot of normalized exposed thickness remaining (N) (Fig. 10-1-16) versus log dose (E), where N is

$$N = (T_i - T_f)/T_f \tag{10-1-5}$$

In the unexposed region of positive resists, a similar plot is used to define E_s (Fig. 10-1-17), the minimum dose at which no thinning occurs. The thinning of the resist from underexposure and overdevelopment ($R/R_0 < 6$) is a major problem on silicon wafers. Since spun resists planarize a step (Fig. 10-1-18), excess thinning does not provide sufficient resist (thickness) protection to compensate for plasma erosion. In addition, many pinholes are revealed ($> 10/\text{cm}^2$) for resist films that are deliberately thinned more than 50% of their original thickness by a combination of developer and reactive ion erosion.

Resists with low contrast tend to reveal top surface image distortion especially if used at low R/R_0. Fortunately, some diazonovolak formulations exhibit an altered outer surface layer which inhibits dissolution and causes an induction effect (Fig. 10-1-19) which retards the dissolution of the unexposed region. In comparison (Fig. 10-1-20A), a linear dissolving diazonovolak resist exposed at moderate to low doses will show shallow profiles, especially if exposed near the MTF falloff of the lens. Diazonovolak resists with nonlinear dissolution (Fig. 10-1-20B) will show steeper profiles, which are advantageous for reactive ion etching. In some cases, to form shallow profiles for via hole plasma etching, a developer and dose combination which provides low contrast and thinning (Fig. 10-1-21) is chosen for

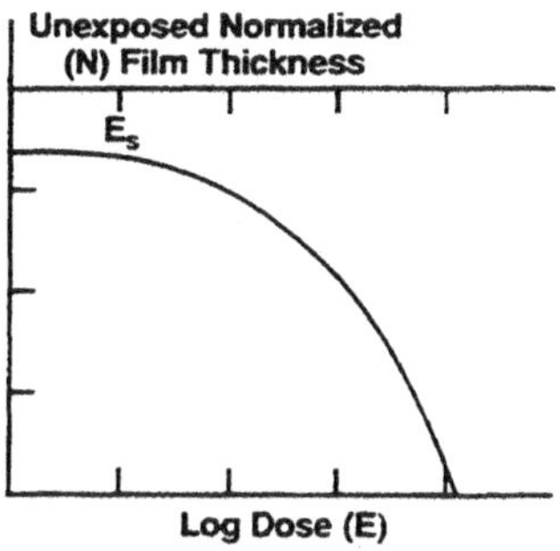

FIGURE 10-1-17. Finding minimum (E_s) dose in a positive resist at which no thinning occurs.

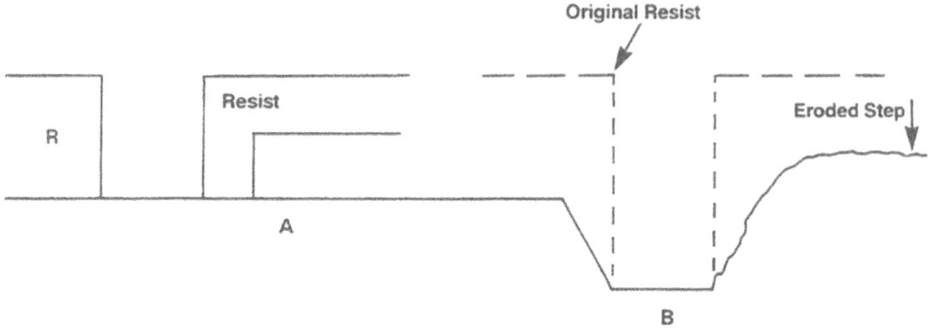

FIGURE 10-1-18. (A) Planarizing resist cast over a previous step developed image. (B) After reactive ion etching involving resist erosion. The step is eroded because of insufficient resist thickness over the top of the step.

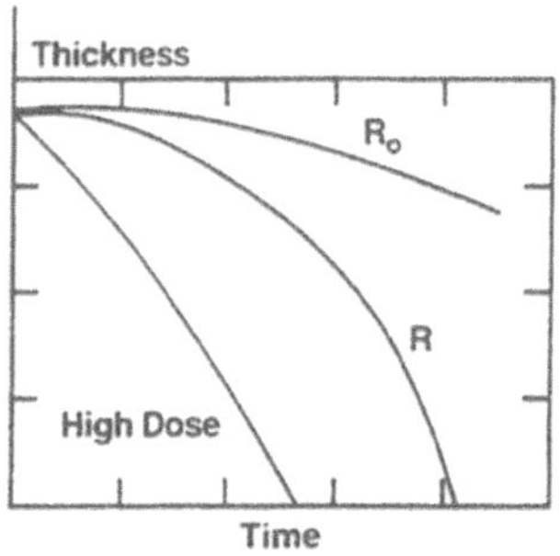

FIGURE 10-1-19. Nonlinear dissolving (induction effect) positive resist.

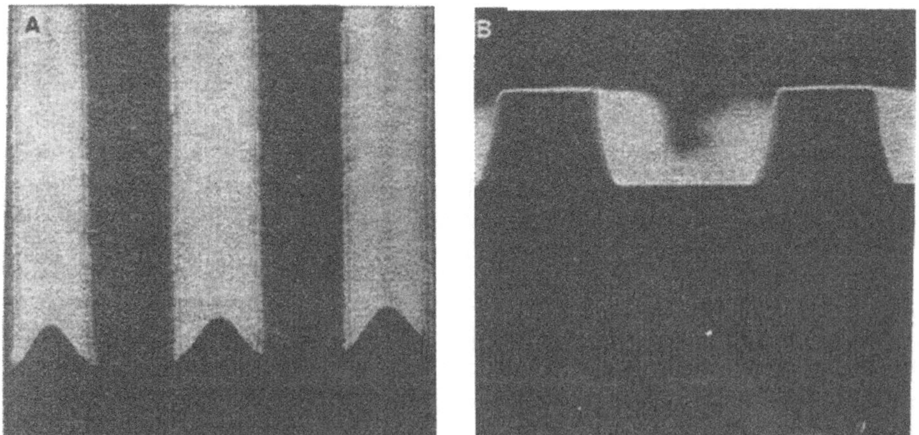

FIGURE 10-1-20. (A) Low-contrast DQN resist with linear dissolution. The top of the unexposed region is dissolved. (B) DQN resist with induction (little top surface thinning) exposed under same scan setting. From G. Willson, R. Miller, D. McKean, T. Tompkins, N. Clecak, and D. Hofer, *Polym. Eng. Sci.* **23**, 1004 (1983), courtesy of SPE.

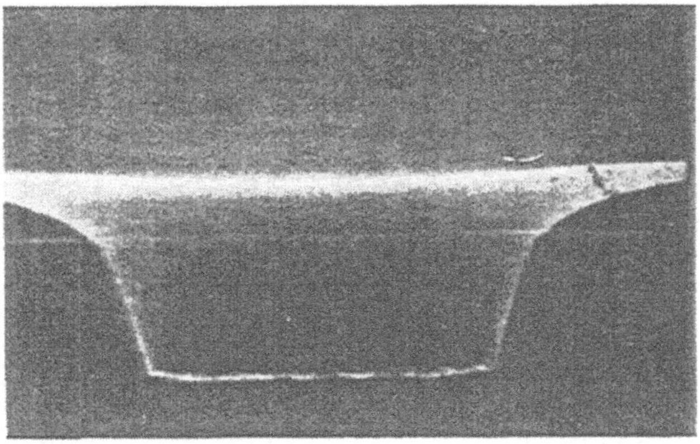

FIGURE 10-1-21. Profile in PMMA exposed at low dose and thinned 30% to develop image.

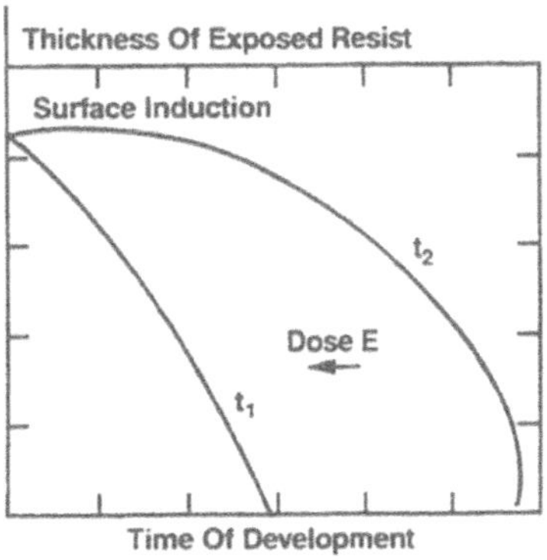

FIGURE 10-1-22. Determination of the development characteristics of a positive resist. First the dissolution profile of the unexposed region (t_2) is determined. Then various doses are used to determine correct dose and develop time (t_1) which avoids resist thinning (unless required for shallow profiles) and provides best image size and widest process latitude.

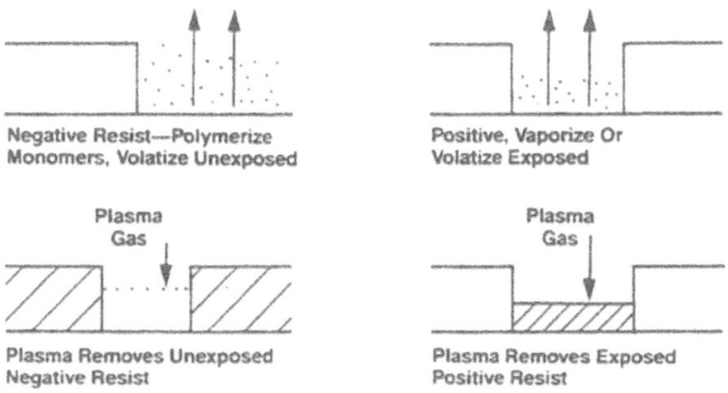

FIGURE 10-1-23. Dry processing resist.

TABLE 10-1-3. Liquid Developers for Resists

Resist	Developer
DQN	KOH + surfactant (unbuffered)
	KOH + Na silicate and surfactant (buffered)
PMMA	Ketones or acetate (solvent)
	Ketones + nonsolvent
Negative resists	"Poor" solvents
Inorganic resists	Oxidizing or reducing agent

profiles $< 60°$. At low doses with surface induction, the exposed region may also exhibit nonlinear dissolution (Fig. 10-1-22) and the rate (R) [Eq. (10-1-1)] cannot be determined from the final film thickness remaining. At high doses (t_1, Fig. 10-1-22), most of the dissolution is linear and the rate is constant, while at low doses (t_2), the rate increases deeper into the film. Nonlinear dissolution can also be due to nonuniform absorption of the exposure radiation due to too high of an absorbance (> 0.4) or backscatter energy which concentrates the absorbance at the interface. Thus, development rate monitors are essential in establishing the kinetics of dissolution and the correct endpoint times for different resist thicknesses and process conditions.

Besides the empirical nature of resist development kinetics, the role of the composition of the developer should also be considered, not only from the contrast or resist sensitivity viewpoint but with the practical problems of toxicity, safety, handling, shelf life, cost, and disposal. With these in mind, aqueous based developers are the safest and least costly to handle. The response of a resist to different developers is dependent on the nature of the radiation products and the selectivity of the developer to the products, based mainly on the kinetic removal of the products (positive resist case) or the reactants (negative resist). Developers can be chosen on the basis of the molecular size, polarity, surface tension or surface reactivity, and even chemical reactivity. Some representative developers for the popular PMMA and DQN resists and other resists are given in Table 10-1-3.

Besides liquid development, dry develop is being pursued for a total concept of dry resist lithography. The dry process depends on differences in volatility or differential removal in a plasma (Fig. 10-1-23). The formation of an organosilicon product which does not etch in an oxygen plasma is commonly used for negative working resists. Some examples of self-developing or plasma-assisted developing are given in Table 10-1-4.

TABLE 10-1-4. Dry Developing Resists

Resist	Developer	Tone	Reference
PMMA	CF_4, H_2[288]	Pos.	3
PMMA, Si^+	O_2	Neg.	4
Arylazide	CF_4	Neg.	5
Silicone graft	O_2	Neg.	6
Aryl monomer graft	CF_4	Neg.	7–9
As_2S_3	CF_4	Neg.	10
Polybutene SO_2	Volatile	Pos.	11
Polymerized monomers	Volatile	Neg.	12

In the next sections, the development of the three major types of resist—(1) PMMA–organic solvent developed, (2) DQN–aqueous alkaline develop, and (3) negative resists—will be modeled as well as dry development.

10-2. GENERAL MECHANISMS

The most general case of development is the Fickian diffusion or Bruner's Law of Solid Dissolution. In Fig. 10-2-1, a microscopic view of the surface of polymer dissolving without swelling is shown. The polymer continues to dissolve until a layer of liquid immediately adjacent to the polymer has become saturated. The dissolved polymer molecules will diffuse from the saturated layer to the less concentrated neighboring layers. Thus, the rate of dissolution reaction is essentially the rate of diffusion itself.

The diffusion coefficient of the developer D in units of cm^2/sec $(10^{-6}$–$10^{-8})$ is related to the viscosity (η) of the developer by the Stokes–Einstein relation,

$$D = RT/\eta = RT/KM^{-a} \qquad (10\text{-}2\text{-}1)$$

where KM^{-a} is the Mark–Houwink (M) molecular weight dependency on the solubility parameter, a. For undirectional dissolution in 1 sec,

$$DT = 10^{-3}\ \text{cm} \qquad (10\text{-}2\text{-}2)$$

Activation energies for diffusion-controlled development or etching (see Chapter 13) are on the order of 2–8 kcal/mole.

If the liquid developer is vigorously stirred, the diffusion of molecules away from the surface will be speeded, but a thin layer of concentration gradient 10^{-2}–10^{-3} cm thick (100–10 μm) adheres strongly to the surface. If Σ is the thickness of the static layer (Fig. 10–2–1). A_s represents the saturation concentration and C the concentration of the polymer dissolved made uniform by stirring,

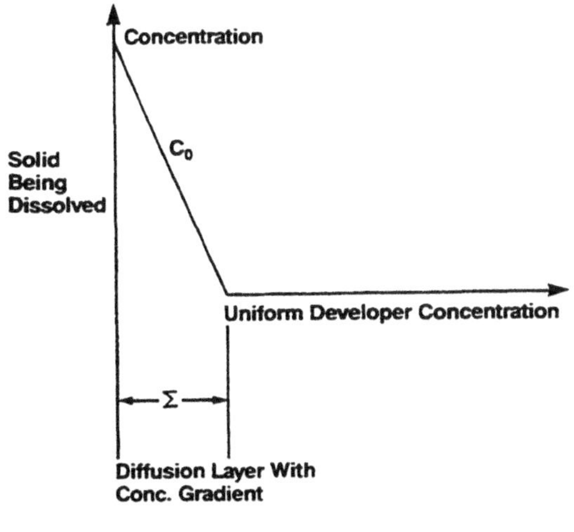

FIGURE 10-2-1. Concentration gradient at the surface of a developing resist.

Fick's law for differential mass m of polymer dissolving with time (dt) of surface area S is

$$dm = DS\,A_s - C/\Sigma\,dt \quad (10\text{-}2\text{-}3)$$

However, the thickness of the layer Σ will vary with stirring and the overall rate of dissolution will increase with stirring (Fig. 10-2-2). The concentration (C) is related to the volume (V) of the liquid and the molecular weight (M) of the polymer by

$$dm/MV = dC \quad (10\text{-}2\text{-}4)$$

Integration of Eq. (10-2-3) yields

$$\ln A_s - C/A_s = -DSt/MV\Sigma \quad (10\text{-}2\text{-}5)$$

which is equivalent to a first-order reaction. Substitution of the density p and thickness t for the mass or concentration C of dissolving polymer reveals that the rate of dissolution is an exponential function,

$$dt/dx = e^{-m} \quad (10\text{-}2\text{-}6)$$

While Fickian diffusion assumes that D of the developer is independent of the molecular weight of the polymers, this is not the case for many glassy polymers in which swelling occurs as a resistance to dissolution.[13–15, 362, 371, 378, 379]

The dissolution of glassy polymers occurs in principle like a solid composed of small molecules. A small molecule is dissolved from the solid when the intermolecular attractions are replaced by molecule–solvent interactions.[33] However, in the case of a monomeric solid, each solvated molecule is immediately free to move away from the solid and into the solvent. By contrast, a solvated group on a polymer has limited mobility because of its covalent bonding and after solvation, it can only move away to the extent of the length of the solvated portion of the chain. Only when all the monomer units of the same chain are completely solvated is the entire chain free to move away from the bulk polymer. Thus, the longer chains, as found in the highest-molecular-weight polymers, take a longer time to dissolve than lower-molecular-weight chains. In the initial stages of polymer dissolution, the

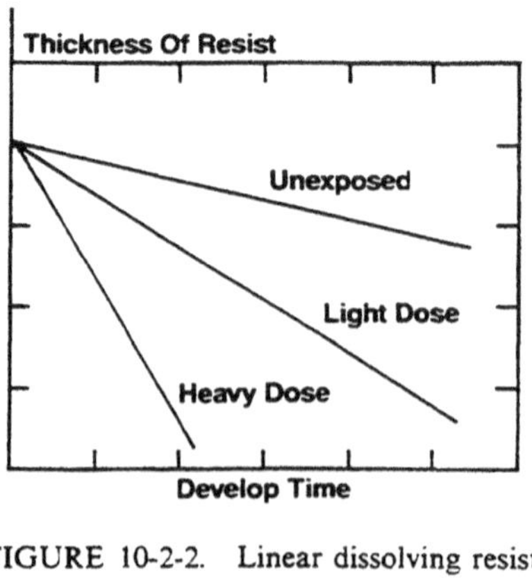

FIGURE 10-2-2. Linear dissolving resist.

solvent readily penetrates the holes (free volume), but deep solvent penetration is slowed down by the increase of the viscosity of the dissolving polymer at the surface of the polymer. At this point the dissolution represents a Fickian diffusion process which is concentration gradient controlled.

For the Fickian case, it can be seen in general that the rate of dissolution of a polymer is inversely proportional to its molecular weight,[16,17]

$$R = kM^{-a} \tag{10-2-7}$$

For resist dissolution, two general rates are observed:

- Case 1. The dissolution of exposed and unexposed are fairly linear assuming uniform absorption of light or energy throughout the film (Fig. 10-2-2). This is observed in DQN and PMMA of lower molecular weight ($< 10^4$). Because of higher absorption at the surface of DQN films ($> 1\ \mu$m thick, absorbance > 1.0), the rate of dissolution (R) is higher at the surface of the film and decreases exponentially with depth. For *e*-beam exposure, the energy deposited is uniform in thin films (< 3000 Å), but increases with depth in thicker films due to a stronger backscatter contribution. Case 1 is a simple etching case of one molecular film layer by layer.
- Case 2. Before the film dissolves, the dissolution of the unexposed is delayed by an induction period (Fig. 10-2-3).[14, 304] Case 2 involves the penetration of developer to fill the voids, but no immediate dissolution occurs because of the formation of a swelling gel layer or the adsorption of products of dissolution on the surface. The formation of several different physical layers on a dissolving polymer surface was modeled by Uebereiter and co-workers (Fig. 10-2-4).[16, 17] Dissolution can also occur by cracking and crazing to relieve the internal swelling stresses.[17]

With induction, the formation of a gel layer can be very beneficial since the dissolution of the unexposed resist (R_0) is inhibited for several minutes after the image has partially developed.

Several polymeric factors and the choice of developer influence the rate and mechanism of development. From a polymer materials viewpoint these include:

1. Molecular weight—M_w and M_n, weight- and number-average molecular weight
2. Molecular weight distribution, M_w/M_n

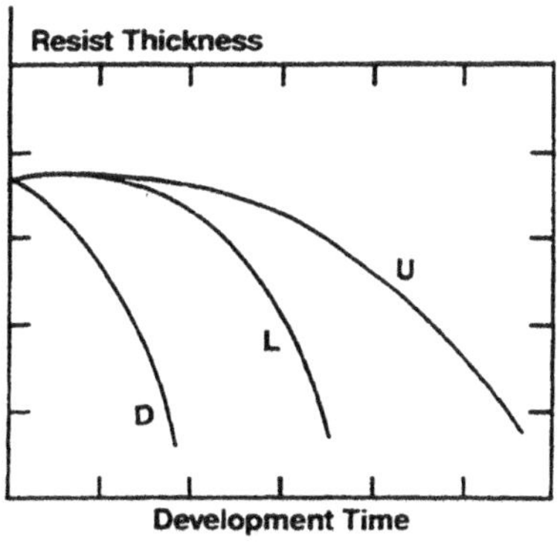

FIGURE 10-2-3. Nonlinear dissolving resist with induction effect. U, unexposed; L, light dose; D, heavy dose.

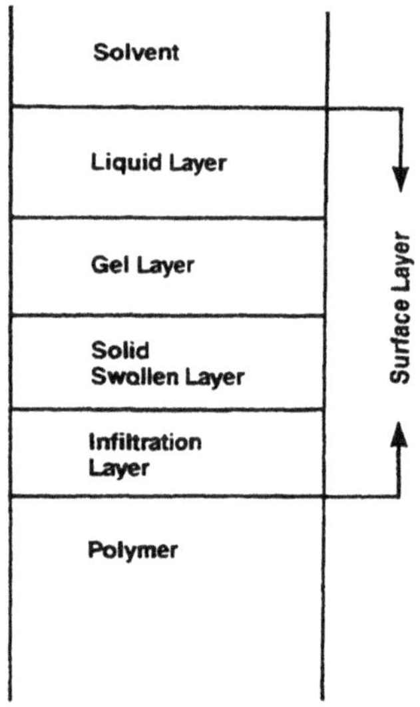

FIGURE 10-2-4. Schematic of various layers of dissolving polymer. From Ref. 16, reproduced by permission of John Wiley & Sons, Inc.

3. Bake temperature and residual solvents
4. Tacticity of polymer
5. T_g of polymer and solvent combination in gel layer, glass transition and developer temperature
6. Formation of gaseous products to microfoam exposed region and accelerated infusion of developer[317]
7. Plasticization by nonvolatile or trapped products

Studies of the activation energy of dissolution (E_D) give some clues to the structure of the polymer and the penetration mechanism. Meares[13] has introduced the concept that the diffusion of a solvent takes place along a cylindrical volume,

$$\pi d^2 g/4$$

where g is the length of a diffusion step and d is the diameter of the penetrant molecule. According to the theory, the liquid diffusivity should be inversely proportional to the cube root of the solvent's molar volume.[13, 378] The diffusion coefficient from Eq. (10-2-3) follows the Arrhenius equation,

$$D = D_0 \exp(-E_D/RT) \tag{10-2-8}$$

The E_D is related to the volume of the average unit diffusion step and the polymer cohesive energy density (CED) is given by

$$E_D = \pi/4 d^2 g N(\text{CED}) \tag{10-2-9}$$

Thus, in a homologous series of developers such as methyl-heptyl acetate for PMMA, a plot of activation energy of dissolution versus d^2 is fairly linear

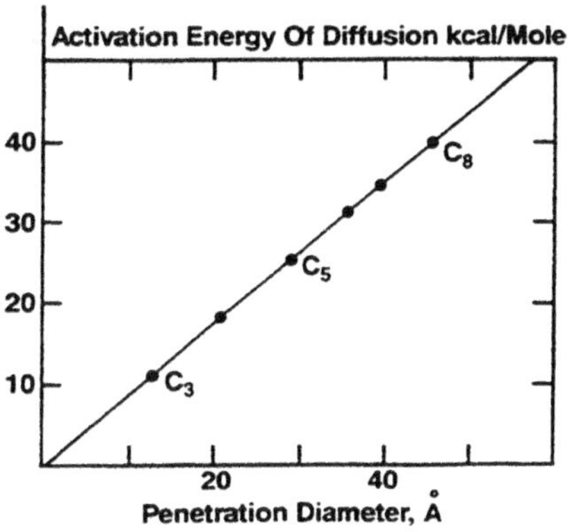

FIGURE 10-2-5. Activation energy of diffusion of alkanes into polycarbonate film. From Ref. 14, reproduced by permission of John Wiley & Sons, Inc.

(Fig. 10-2-5).[14] The higher alkyl solvent derivative (C_5) will provide a greater selectivity (a higher R/R_0). Choosing developers on the basis of molecular size and activation energy can greatly influence their ability to penetrate, distinguish, and dissolve the higher- and low-molecular-weight fragments. In methyl isobutyl ketone (MIBK) developer, the activation energy for developing PMMA is 24 kcal/mole, and 48 kcal/mole in a mixture of MIBK to 3 parts of isopropanol.[18, 378] Lapick[19] and Valko studied the dissolution of polyvinylchloride in cyclohexanone which proceeds with an induction period. They found that the activation energy of dissolution increased from 6 to 12 kcal/mole with a tenfold increase in M_n (Fig. 10-2-6). The dependence on molecular weight indicates that the chain ends make a greater contribution to swelling and dissolution than the interior of the polymer chain. Since the dissolution is sensitive to the number of chain ends, low-molecular-weight polymers dissolve faster and with lower activation energies (Fig. 10-2-6). In the cases of dissolution without swelling,[16, 378, 379] the rate-determining step is the control by a relaxation mechanism in which a sharp boundary exists between the glassy core (Fig. 10-2-4) and the solvent front.

Activation energy is necessary to uncoil the polymer chains and overcome the intramolecular and intermolecular forces. According to the free volume theory,[23] the movement of solvent molecules takes place through local density fluctuations or holes which are produced as a result of polymer segment mobility. The polymers are envisioned to uncoil as they dissolve in a "good" solvent, but not in a ther-

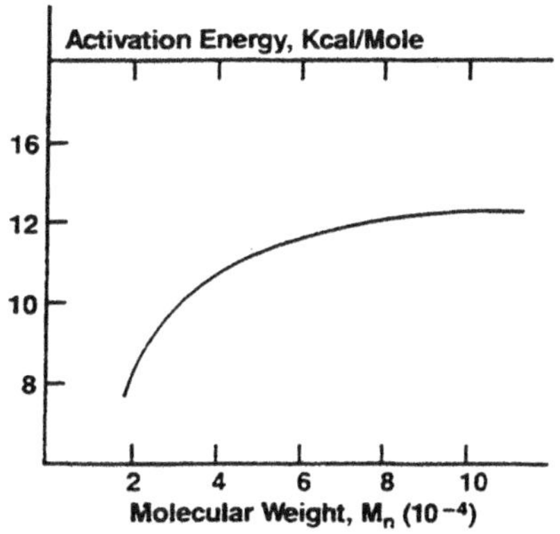

FIGURE 10-2-6. Activation energy for dissolution of polyvinylchloride in cyclohexanone as a function of molecular weight. From Ref. 19, reproduced by permission of John Wiley & Sons, Inc.

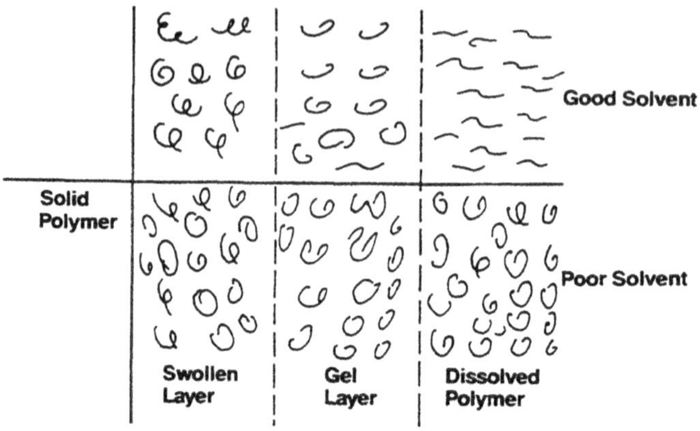

FIGURE 10-2-7. Dissolution of a polymer in both good and poor thermodynamic solvents. Uncoiling is evident in a good solvent.

modynamically poor solvent (Fig. 10-2-7). In a poor solvent, such as 3-heptanone for PMMA, the solvent breaks the intermolecular forces but not the intramolecular forces of the excluded volume.[19–22] A steady-state competition exists between the diffusion of the solvent (D) into the film and the diffusion of the dissolved polymer into the liquid developer.[24] In the case of glassy polymers with $M_n > 10^5$, they exhibit similar activation energies since the number of end groups remains essentially constant.

The rates of diffusion can be estimated for various solvents. Unfortunately, little data exist in the literature for resist polymers since they have no other applications in coating technology or the science of separation by osmosis technologies. For a homologous series, the choice of the effectiveness of a solvent increases to the swelling limit. This limit is associated with self-diffusion. For PMMA, the dissolution ceases after C_7 acetates.[25] The movement of large penetrants is independent of solvent size since the movement is not controlled by free volumes[371] or "holes" penetrant smaller than holes), but by the motions of polymer[22] molecules (penetrant larger than holes). Thus, C_4–C_5 solvents are the

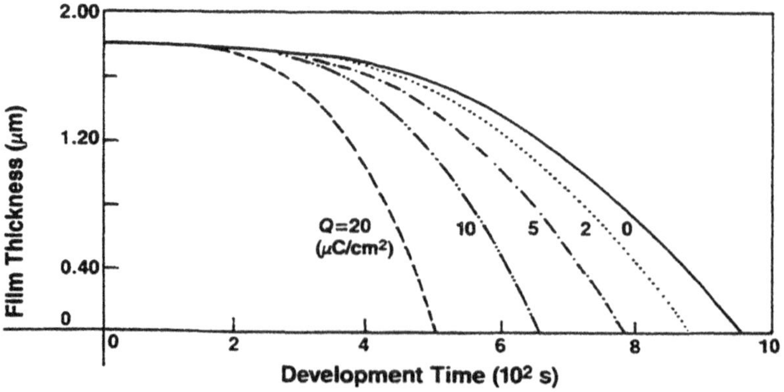

FIGURE 10-2-8. Dissolution curve of PMMA as a function of dose. From Ref. 32, courtesy of IBM.

practical upper limit ($R \geqslant 1000$ Å/min) for R of developers for most amorphous glassy polymers (positive type).

While swelling (Case 2) of the polymer complicates dissolution and limits the resolution of the image, it is useful in inhibiting the dissolution of the unexposed resist. Very-high-molecular-weight polymers show swelling on the order of Dt, where D is in the range of 10^{-8}–10^{-10} cm^2/sec. Figure 10-2-8 shows the dependence of swelling time (t) or induction period encountered in development of PMMA of M_n of 10^5 in heptanone developer.[32] The thickness of the swelling layer can be derived on the thermodynamic assumption that swelling can be represented by a coil of effective mean square radius of gyration.[19] The thickness of the layer (T) will be proportional to the molecular weight M_n and solubility parameter a,

$$T = kM^a/D \tag{10-2-10}$$

The velocity of dissolution, V, is inversely proportional to the thickness,

$$V = kD/T \tag{10-2-11}$$

of the swollen layer. Since the surface layer contains coils filled with solvent, Ueberreiter[16] related the thickness T to the coil diameter by

$$T \propto d \tag{10-2-12}$$

and d can be estimated from the Flory relation,

$$d = kM^a \tag{10-2-13}$$

A plot of log of dissolution (V) versus log M (molecular weight) gives a slope a. Solvents with high a can be gleaned from activation energy studies of the rate of dissolution. The developer with the highest activation (Fig. 10-1-17) and suitable induction will maximize the contrast. Thus, developers with high a values will

TABLE 10-2-1. Developer Solubility Parameter and γ

γ	a	Developer	Reference
1.7	1.4	MIBK	26
1.7	1.5	MIBK	18
1.7	1.5	MIBK	27
3.0	3.9	MIBK/IPA (1:1)	27
3.5	2.0	MIBK/IPA (1:3)	28
4.1	4.3	Amyl acetate	27
4.2	4.3	Isoamyl acetate	29
4.4	4.2	3-Heptanone	30
6–11	4.1	MIBK/isopropanol	31

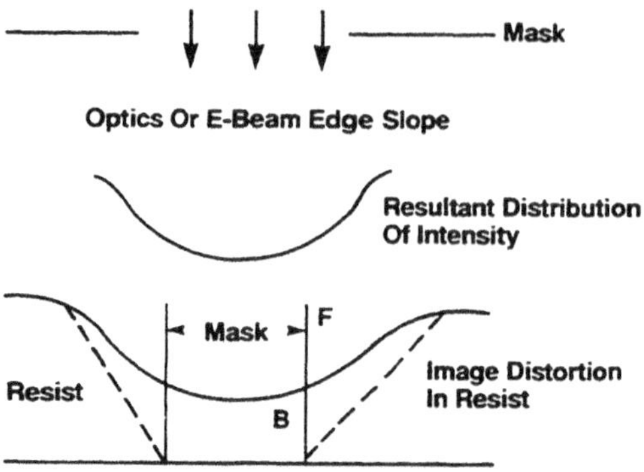

FIGURE 10-2-9. Image distortion in a resist due to forward (F) and backscattered radiation (B).

exhibit a large distinction between the unexposed (high molecular weight) and exposed (lower molecular weight) in a positive resist such as PMMA. Poor solvents tend to swell the polymer since the uncoiling is very slow for the high M_n and the excluded volume is filled with excess solvent. In actual practice, mixtures of solvents and nonsolvents provide the highest a and in turn the highest contrast (τ) and resolution for PMMA (Table 10-2-1). It is to be noted that many workers have reported a sensitivity of PMMA at 100 $\mu C/cm^2$ for MIBK/isopropanol mixtures. At this dose, the develop time is several minutes. Doses of 10 $\mu C/cm^2$ or less can be used if one can tolerate a longer (60 min) develop time. Other mixtures can be used if one varies the ratio and waits for a sufficient period of time to develop instead of using a fixed develop time (1–2 min) and increasing the dose to achieve a high R. Straight isopropanol can also be used as a high-contrast developer for PMMA but high doses (100 $\mu C/cm^2$) are required.[368]

The action of the developer actually traces out the edge slope of the beam and the image distortion by the resist. These major distortion factors are

1. The edge slope of the incoming e-beam or the diffracted optical beam (Fig. 10-2-9)
2. The forward and backscattered radiation in the resist medium (Fig. 10-2-9)

Once the modulated beam scatters in the resist medium such as PMMA (Fig. 10-2-10), various increasing molecular weights are created from the center of the beam outwardly to the fringes of scattering (left side of Fig. 10-2-10) to create the corresponding contrast, τ, and the receding profiles (right side of Fig. 10-2-10). Resists with a low contrast (1–2) reproduce the scattered radiation when the resist

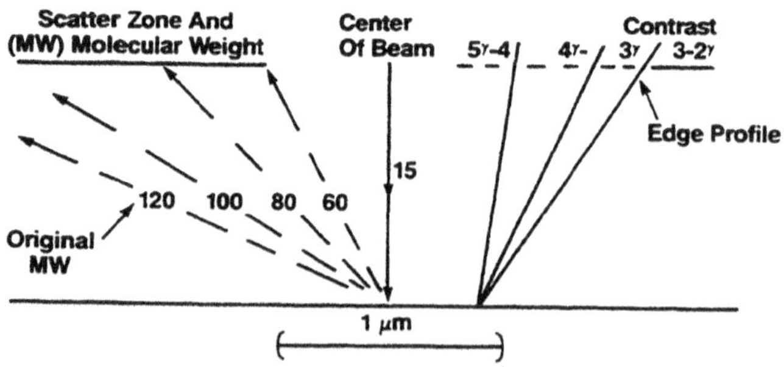

FIGURE 10-2-10. Simulated profile of an image in PMMA as a result of e-beam exposure at 25 keV. The contrast and resultant molecular weight decrease after exposure are shown for a 1-μm-wide beam incident on a 1-μm-thick film.

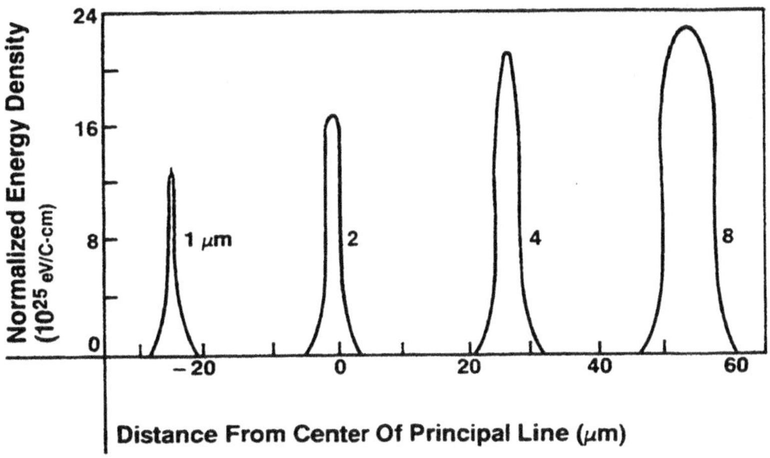

FIGURE 10-2-11. Lateral energy deposited in PMMA for isolated lines received same dose. From Ref. 32, courtesy of IBM.

is given a moderate dose and slightly overdeveloped. The contrast concept (τ) is necessary to visualize the response of a resist to the incoming beam and the distortion by the resist and substrate. An exposed, isolated line in a positive resist such as PMMA, is a family of isocurves of various molecular weights created by the scattered radiation, when the resist is given a moderate dose and slightly overdeveloped. Overdeveloping is necessary to compensate for the intraproximity effect (Fig. 10-2-11) due to the intensity loss by optical diffraction in deep-UV exposure or due to the loss of backscatter electrons in *e*-beam exposure of 1-μm lines.

In summary, we have discussed good and bad solvents and their interaction with glassy polymers. A good starting point for choosing solvents is the θ temperature or theta solvent[33] at which the polymer chain attains its unperturbed dimensions. In the excluded volume, the attractive and repulsive forces are equal. A good thermodynamic solvent depends on its chemical constitution and the interaction of atomic groups. The kinetic suitability is chiefly connected with molecular size and steric hindrance. In general, the positive resist polymer should be of low dispersity, $M_w/M_n < 2$, in order to avoid thinning of the unexposed positive area.

Swelling of the resist in the developer can be both a friend and a foe. While swelling can prevent background loss, in low-T_g resists such as polybutene sulfone[34] or fluorobutyl methacrylates[35] of films (> 0.5 μm thick), appreciable swelling can constrict submicrometer images and cause loss of adhesion from the stress force. In some polymers such as in polyolefin sulfones,[36] polycarbonates,[37] and poly-α-methyl styrene, swelling can cause surface crazing and deep cracking.

The choice of a good developer can be gleaned from activation energy and viscosity–molecular weight plots and θ conditions. Kinetic measurements on molecular weight fractions can also serve to establish the R/R_0 ratio. Solvent–developer combinations which maximize the a parameter as a function of molecular weight [Eq. (10-2-7)] assure that high contrast will be achieved.

In the next section, we will examine PMMA in terms of the primary factors (G_s, molecular weight) and secondary factors (M_w/M_n, tacticity, T_g, and prebake) which can be manipulated in combination with suitable developers to enhance the sensitivity and contrast.

10-3. PMMA DEVELOPER SENSITIVITY

Although PMMA was the first positive electron[38]-beam resist, it has also been used for deep-UV, X-ray, and ion-beam exposure. The primary radiation event is the scission of the main chain and ester group,

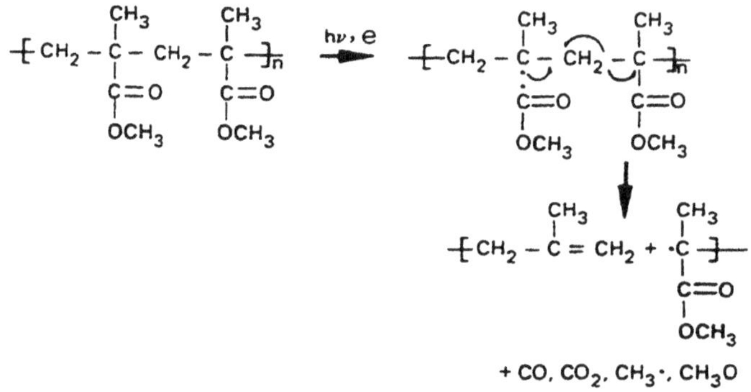

A measure of the chain scission efficiency is the G value,

$$G_{\text{scission}} = \text{Number of permanent scissions/100 eV absorbed} \qquad (10\text{-}3\text{-}1)$$

G values for γ radiation are 1–1.5 but for e-beam radiation the G_s is lower (0.75[28]) since the forward Grun range[18] and the backscatter contribution are properly included in the energy absorbed term [Eq. (10-3-1)]. Otherwise, the e-beam exposure values are close to the γ-radiation G values if only the forward absorption is used.

While permanent scissions imply no recombination,[62] more radicals than can be accounted for with a G_s of 1 are formed and recombination occurs once the exposed PMMA is in contact with the developer. The trapped radicals in PMMA are very stable and can be grafted[39] with monomers and organometalics[305] to amplify the image, even at the low dose of 10^{-7} C/cm^2. In the work of Geuskens[40] and Thompson,[41] the racemization of isotactic PMMA into atactic PMMA was used to demonstrate a G_{rad} value of 6 for radical formation,

(a) PMMA ⟶ PMMA· $\xrightarrow[\text{grafting}]{n(\text{monomer})}$ PMMA-graft $\xrightarrow{\text{develop}}$ IMAGE

(b) PMMA $\xrightarrow{e}$ PMMA fragments $\xrightarrow{\text{develop}}$ IMAGE

Isotactic ⟶ Atactic $\quad G = 6$

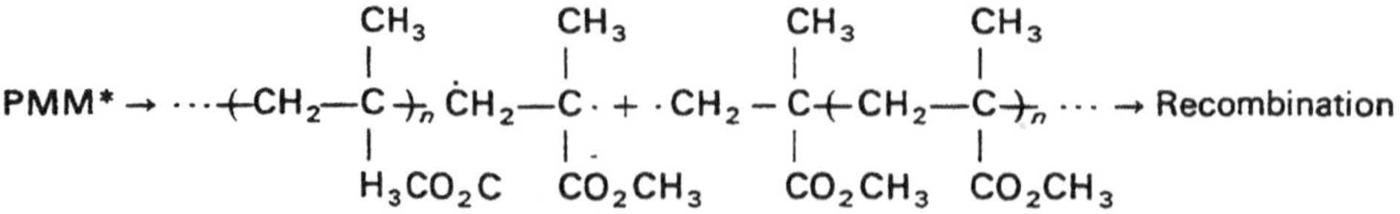

As a result of the isomerization, a negative image can be formed since the higher-T_g atactic form (104°C) dissolves slower than the isotactic form (T_g of

TABLE 10-3-1. Dose and M_0/M_e for PMMA

	M_0[b]			
Q[a]	10K	40K	400K	800K
100	8[c]	16	130	280
40	4	8	70	140
20	2	4	35	71
10	1	3	18	35
4	—	2	7	18
2	—	1	4	9

[a] Dose (μC/cm^2) at 20 kV.
[b] Number average molecular weight.
[c] Ratio of M_0/M_e.

42°C[25]). Thus, more scissions occur than anticipated, but recombination consumes most of the radicals.

In the permanent scission model of Charlesby, Ku and Scala[42] assumed that the ratio of the number of chain scissions M_0 (original molecular weight) to M_e (exposed molecular weight) is approximated at high doses by

$$M_0/M_e - 1 = M_0/M_e \tag{10-3-2}$$

and the dose Q is given by

$$Q = P_s(100q)\,pzN/EG_sM_0 \tag{10-3-3}$$

where P_s is the probability of permanent scission, q is the electron charge, p is the film density, z is the film thickness, N is Avogadro's number, and E is the energy absorbed in the film per incident electron. Since the solubility rate R [Eq. (10-2-7)] proportional to the molecular weight, at low doses,[43] substitution in Eq. (10-3-2) yields

$$R/R_0 = k(M_0/M_e)^{-a} = (1 + kQGM_0)^a \tag{10-3-4}$$

The effect of the initial molecular weight on the ratio of M_0/M_e can be seen for PMMA in Table 10-3-1. Starting with a higher initial molecular weight (M_0) achieves a greater ratio (R/R_0) at the same dose. Therefore, the maximum sensitivity is achieved using the highest initial molecular weight.[44–48] Polybutene[34, 44, 49] sulfone and other methacrylates such as tertiary butyl[51] and fluoromethacrylates[31, 50, 83, 334, 337] also exhibit higher sensitivity (lower dose required) at higher initial molecular weights. By fractionation[45] of the polymer prior to casting or by leaching out the lower molecular weights directly on a prebaked film (solvent dip), an improved sensitivity and lower pinholes can be obtained in PMMA.[52] To maximize the resist sensitivity, ultrahigh-molecular-weight PMMAs[46](16×10^6 and 5×10^6, M_n) were studied. The 16M PMMA was found to have an induction dissolution period in MIBK if baked at 200°C for 60 min (Fig. 10-3-1). Most other workers have used a prebake of 160°C or less and found no advantageous induction. However, Greeneich[18] found that dissolution

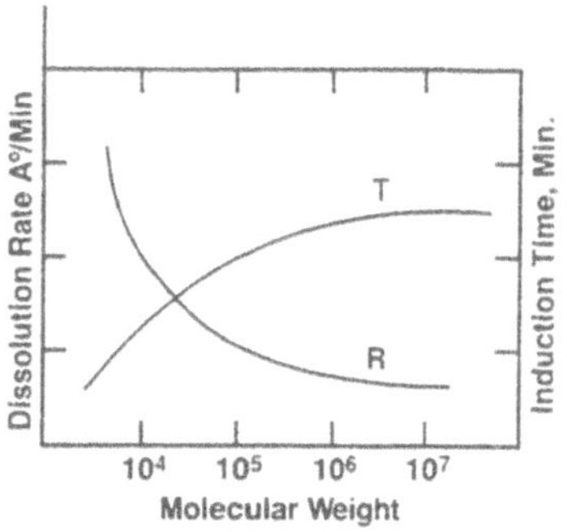

FIGURE 10-3-1. Dissolution rate (R) and induction time (T) as a function of PMMA molecular weight.

kinetics depended strongly on the casting solvent and prebake condition (160°C for 30 min; see Fig. 7-1-1). If PMMA is baked at 200°C, an induction effect on the unexposed dissolution is noted.(46) If the ultrahigh-molecular-weight PMMAs are cast as good films, doses as low as 2×10^{-6} C/cm^2 could be used but the images had some undercutting.(46) In addition, Hess(54) and co-workers experienced many problems with 5.4M PMMA. The swelling limited resolution to 10 μm (the prebake, however, was not at 200°C) and the contrast was extremely low ($\tau = 0.7$ at 10^{-5} C/cm^2 in methyl ethyl ketone/IPA developer). The major resolution problem was the swelling(286) of even the exposed image during development. The best way to take advantage of the induction effect and not experience undercutting or swelling is to use a dual layer resist of high-molecular-weight(53) PMMA on top of (faster dissolving) lower-molecular-weight PMMA (Chapter 12).

The modeling of the induction(32, 46) effect has been done to include inhibition in the exposed region(32) with the rate of change with thickness (z) given by

$$R = f(1 - e^{-z\alpha}) + R_s \qquad (10\text{-}3\text{-}5)$$

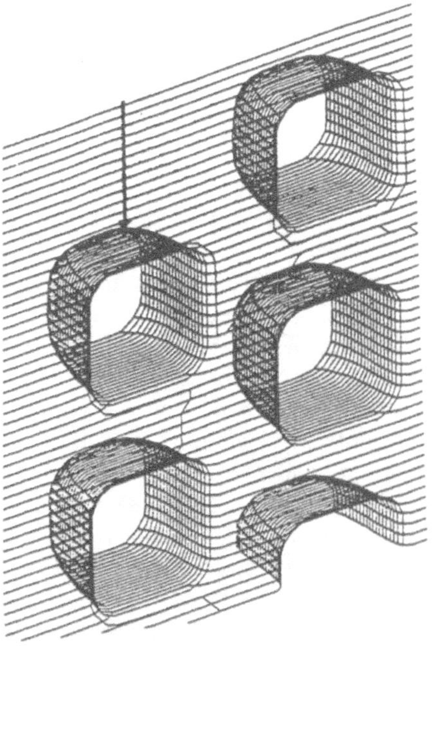

FIGURE 10-3-2. Three-dimensional profiles of image in PMMA. From Ref. 56, copyright and courtesy of IEEE.

where R_s is the rate at the surface and the first term is the rate after the induction where α is the depth,

$$z \gg 1/\alpha \qquad (10\text{-}3\text{-}6)$$

Using Eqs. (10-3-5) and (10-3-6), the modeling of profiles based on the *e*-beam exposure has been applied to PMMA resists for proximity corrections,[55] writing bias,[61] X-ray,[57] ion-beam,[58] deep-UV,[59] and 3-D profiles[56, 86] (Figs. 10-3-2 and 10-3-3) (2-D[87] and 3-D[56] profiles). For modeling, a convenient program called SAMPLE[60] was developed at the University of California by A. Neureuther and co-workers. The deposited energy profiles and developed profiles have proved accurate to ± 0.1 μm in 1.0-μm images.[123] When combined with kinetic endpoint measurements, modeling is a powerful lithographic tool to simulate the lithography needed for manufacturing future submicrometer devices.[118]

The last area of molecular weight effects is polydispersity (M_w/M_n). Upon exposure, polydisperse materials converge upon several scissions to a $M_w/M_n = 2$ (Fig. 10-3-4).[34] This fact is based on the probability that all chains rupture with the same efficiency. While the convergence has been demonstrated for polybutene

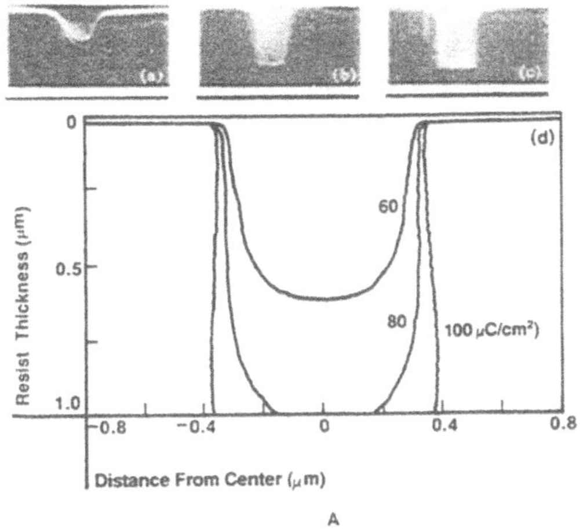

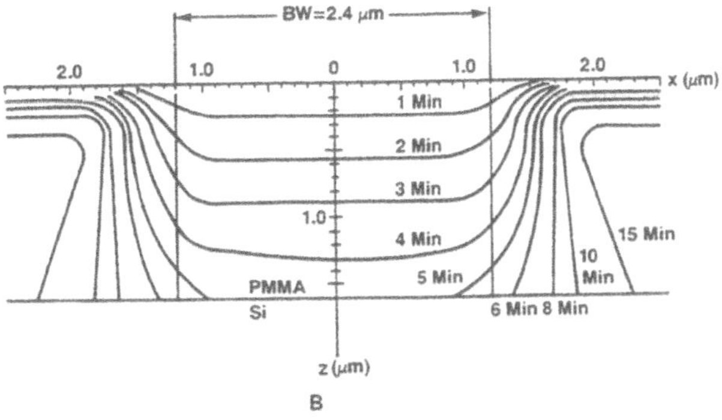

FIGURE 10-3-3. (A) Time evolution of contours in PMMA for doses of (a) 60, (b) 80, (c) 100 μC/cm², and (d) simulated profile. (B) Time evolution of profiles in PMMA. B from Ref. 32, courtesy of IBM.

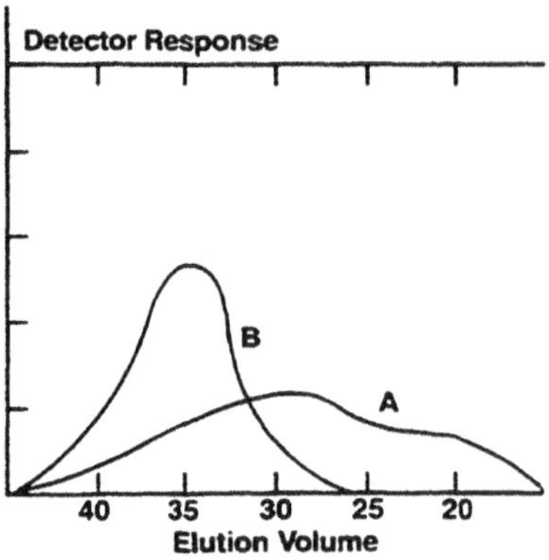

FIGURE 10-3-4. GPC traces of degrading polymer. (A) Dispersity of 24; (B) after exposure dispersity of 2.5. See Ref. 34.

sulfone,[34] PMMA,[63, 64] and polyisobutylene,[63] the rate of dissolution for polydisperse $M_w/M_n > 2$ has been found to be more dependent on M_w than M_n. In this case, intentional mixtures of polydisperse PMMA can be used to take advantage of the faster dissolution of the exposed resist with a converging M_w/M_n.[65] Also, mixtures of monodisperse syndiotactic and isotactic which form polymeric complexes, can be formed to manipulate the polydispersity.[66]

10-4. ENHANCEMENT OF R AND R_0 OF PMMA

The enhancement of the sensitivity of PMMA in Eq. (10-3-4) can be achieved by manipulation of the primary parameters of molecular weight and the developer solubility parameter. Other secondary factors to manipulate R or R_0 are given in Table 10-4-1 along with the primary factors. In the following section, we will briefly review the primary factors and examine the secondary contributions to high sensitivity in PMMA.

TABLE 10-4-1. Speed Factors for PMMA

	Major effect	
Enhancement	R	R_0
Primary		
Molecular weight		+
Substituents and copolymers	+	
Developer		+
Secondary		
Gaseous products	+	
Pre-cross-linked		+
M.W.D.[a]	+	
T_c	+	
Multilayers		+
T_g of scissioned	+	
Additives	+	
Stereo structures	+	
Prebake		+

[a] Molecular weight distribution (polydispersity).

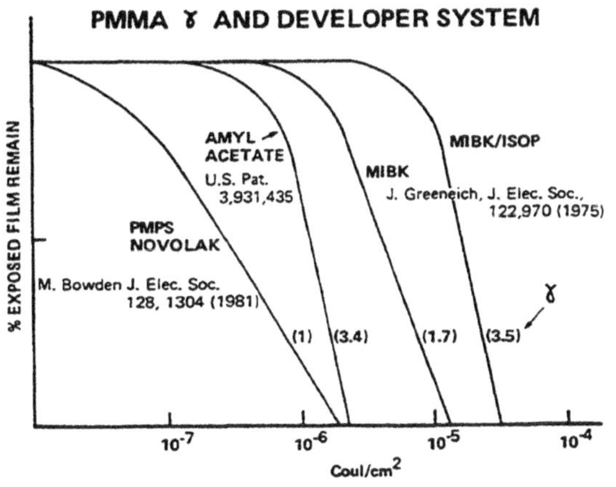

FIGURE 10-4-1. PMMA contrast and developer systems.

Primary Factors

1. Choose developers with high a (Table 10-2-1). In principle, one can study various molecular weight fractions of γ-exposed polymer coatings to simulate e-beam fractions. A developer with a higher a (Fig. 10-4-1) leads to a higher contrast, τ (Table 10-2-1). Table 10-8-4 at the end of the chapter lists common solvents broken down in polar and nonpolar components. θ solvents, θ conditions, fractionating solvents, and homologous series can be gleaned from the *Polymer Handbook*.[68]

2. Increase the initial molecular weight of the resist and maintain $M_w/M_n < 2$.[44, 183, 266]

3. Increase the G_{scission} with substituents in the α position or with copolymers of acid, halogens, and cyano groups[38, 69] (G increases to 7).

Secondary Factors

1. Release copious quantities of gaseous products which form microvoids to increase R. Ouano[70] using Eq. (10-3-4) calculated the final (M_e) molecular weights of PMMA and found experimentally higher rates for R in acetate developers. He concluded that micropores on the order of 10 Å in diameter may be formed (from permeability measurements) to facilitate developer penetration through the pores of similar size. Later work by Rodriguez and co-workers[328] found that irradiated resists dissolved faster than films of lower-molecular-weight synthetic polymers and concluded that micropores or trapped product may be responsible for the increased dissolution of the irradiated resist. Later, work on PMMA-methacrylic anhydride copolymers demonstrated a 4× enhancement of R. When compared to PMMA, even more gaseous products are formed[71] (5×) from the anhydride rupture (Fig. 10-4-2). Polybutene sulfone and similar sulfones have a G_{gas} of 20 which is higher than the G of scission.[11] For the formulation of two-component resists, a sensitizer or dissolution inhibitor which forms copious quantities of gas (polyolefin sulfones, diazo and azide compounds) induces sufficient disruption of the exposed region to allow the developer to penetrate into the exposed region.[317, 318] However, if large bubbles are formed, the edges of the image may become ragged.

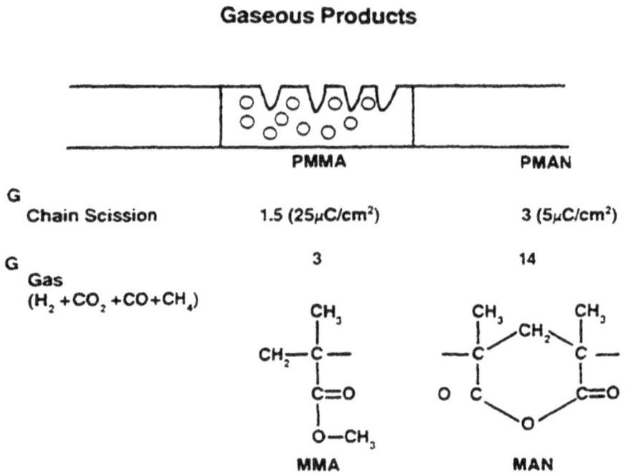

FIGURE 10-4-2. Gaseous products from exposure of PMMA and polymethacrylic anhydride on a per mole basis for the same incident dose of electrons.[38,71]

2. Use pre-cross-linked resists. Many[38] pre-cross-linked resists based on succinic anhydride[84] have been formulated (Fig. 10-4-3). Although the unexposed resist is insoluble in the developer, fairly high doses ($> 15\ \mu C/cm^2$) are required to scission the multiple cross-links. If only one or two cross-links could be synthesized and selectively broken by radiation, much lower doses could be used.

3. Molecular weight distribution. Intentional mixtures to manipulate the convergence of M_w (see above).

4. Ceiling temperature (T_c). Exposure of PMMA above[72] its T_c of 150°C results in a high G_s (10) due to the thermal depolymerization and release of fragment radicals out of the glassy cage effect. Both UV and *e*-beam exposure of resists at elevated temperatures have demonstrated enhanced sensitivity.[73] Dry development (see Section 10–8) or postexpose bake[82] (predevelop bake) may also amplify degradation and enhance *R*.

5. Multilayer resist (see Chapter 12[79–81]).

6. T_g of scissioned polymer. Since the excluded volume is larger in low-T_g polymers, the resists dissolve more rapidly.[37] If copolymers fragment to release low-T_g soluble components in the exposed region, higher sensitivity will be obtained. For example, PMMA copolymers of butyl-fluoro methacrylates and isobutylene[74] are examples of exposure fragments which enhance *R*.

7. Polar scission products. Copolymers of PMMA such as PMMA-acid[71] or anhydride[75] with higher polar (solvent) solubility than PMMA release polar fragments which dissolve in polar developers while the unexposed hydrophobic region is unaffected. An outstanding example of exposure-induced polar fragments is derived from a PMMA with diazoquinone side chains,

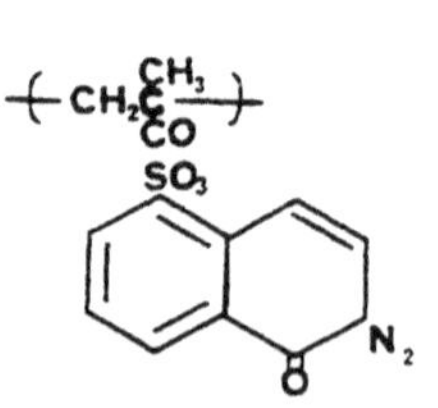

Photoexposure releases water-soluble alkaline (nonpolluting) indene acid products.[76]

8. Additives. Additives to enhance *R* or suppress R_0 can be small molecules or

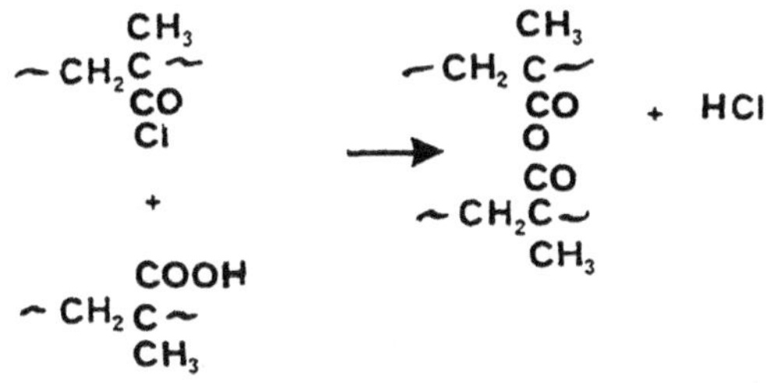

FIGURE 10-4-3. Formation of pre-cross-linked anhydride resist from PMMA acid and acid chlorides.

compatible polymer additives. Polyvinyl acetate has been used to inhibit R_0 and achieve higher sensitivity.(77) Free radical sensitizers to enhance chain scission may be useful. The G_{scission} of PMMA is high in a chain transfer solvent(78) (Table 10-4-2) since the solvent radicals attack the scissioned chain. Unfortunately in the solid state, the radicals are immobilized. In addition, the electronic energy transfer by the Forster mechanism(230) has not been successful in the PMMA system. Other additives, such as surfactant added to the developer, have been claimed to preferentially absorb and enhance PMMA sensitivity.(23, 83, 85) Treatment of PMMA with swelling agents followed by a deswelling developer has been claimed to suppress R_0 and enhance R/R_0.

9. Tacticity effects. Isotactic PMMA is threefold more sensitive than atactic PMMA and tenfold more sensitive than the syndiotactic form. Since isotactic PMMA has a lower T_g (60°C) than the other forms (atactic 104°C, syndiotactic 127°C), the radical chain scission intermediates can diffuse away from recombination effects.(67)

10. Prebake effects. Since the dissolution properties of PMMA (R_0) are sensitive to trapped solvent (Fig. 10-4-4), the resist must be baked at sufficiently high temperatures above the T_g to enhance R and suppress the unexposed dissolution with an induction effect.

In summary, the manipulation of R and R_0 requires a keen understanding of the radiation products, the kinetics of polymer dissolution, and the physical or chemical contribution of the products to alter the interaction with a liquid developer. Not only do efficient radiation transformations have to occur, but the liquid medium must fractionate between a chemical reaction that does not go to

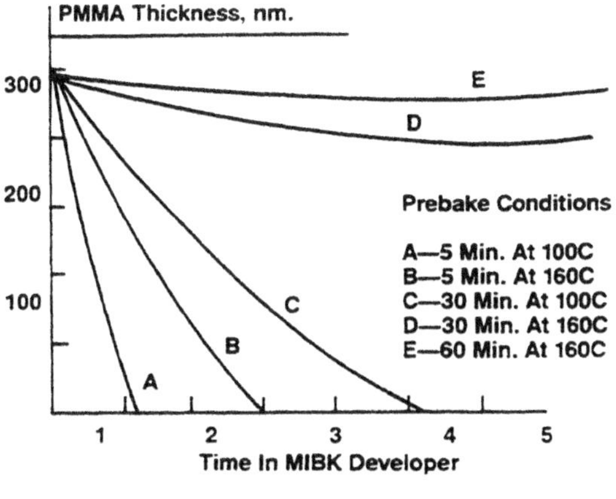

FIGURE 10-4-4. Effect of prebake on the dissolution of PMMA in MIBK developer for the unexposed region. From Ref. 18, reproduced by permission of the Electrochemical Society Inc.

TABLE 10-4-2. G_s of PMMA Irradiated in Liquids

G_s	Irradiation liquid
1.8	Benzene
3.2	Acetone
4.5	Ethylene dichloride
6.1	Chloroform

100% completion. Continued studies in other polymer disciplines of polymer permeability, diffusion, and morphology, along with phenomenological modeling, will assist in the selection of a developer to maximize a resist's sensitivity or resolution.

10-5. DIAZOQUINONE–NOVOLAK RESISTS

10-5-1. *Introduction*

The novolak-based resists contain a dissolution inhibitor such as diazoquinone[88](**A**) or a polyolefin sulfone[89] (**B**) dispersed in a phenolic-formaldehyde resin (**D**),

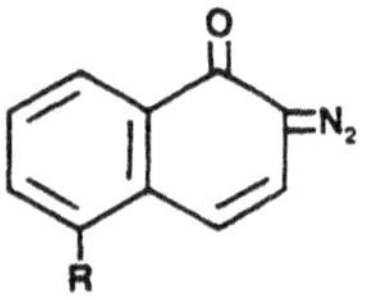

A

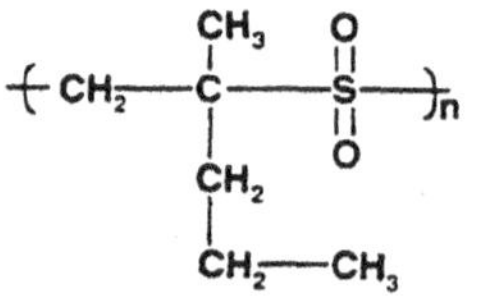

B

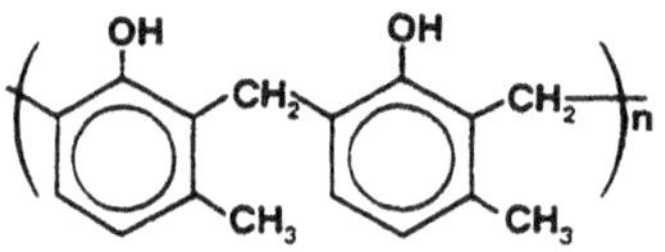

D

By adding a hydrophobic sensitizer at 5–25% by weight of resin, the dissolution of the unexposed resin in 0.1–0.3 N alkaline developer is temporarily inhibited. Upon exposure of **A** or **B** the dissolution inhibitor is removed. The exposure of **A** to radiation forms a soluble indene acid (**E**),

The moisture[90] is supplied by the water trapped in the film ($<$ 1.5% by weight) or by moisture present in the air. If the exposure is done in the absence of moisture (at

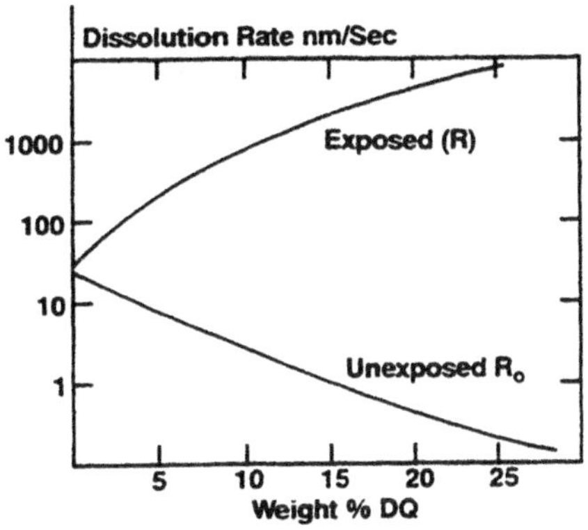

FIGURE 10-5-1. Effect of sensitizer DQ concentration on R and R_0 for DQN resist. From Ref. 93, copyright 1980 and courtesy of IEEE.

low humidity[144] or in a vacuum), the ketene reacts with the OH group of the resin to form ester cross-links,[91]

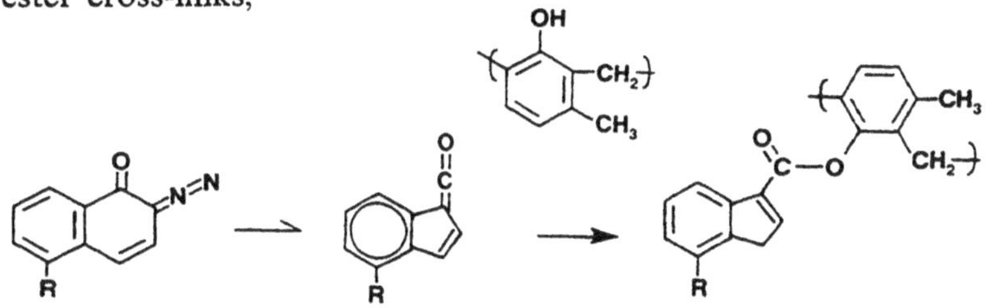

Thus, a negative image (reversal) can be formed by exposure with any radiation in the absence of moisture, followed by predevelop baking and/or flood exposure (300–450 nm) and mild alkaline development.[92, 252] In the exposed resist, the presence of the acid product (**E**) slightly enhances the rate R above that of the pure resin.[359] Figure 10-5-1 illustrates the effect of increasing the sensitizer concentration of diazoquinone (**A**) on the rate of dissolution of the exposed resist (R) and the inhibited region (R_0). With the addition of more sensitizer (inhibitor), the exposed region becomes more hydrophilic and the unexposed region is able to tolerate a stronger developer concentration. Increasing the concentration of developer enhances both R and R_0 (Fig. 10-5-2) but the relative rate ratio (R/R_0) remains constant.[93]

Depending on the substituents on the sulfonic acid ester of diazoquinone, up to 50% DQ by weight is miscible with resin. With the higher sensitizer loading, a higher contrast (Fig. 10-5-3) is obtained, since the unexposed dissolution is inhibited.[93]

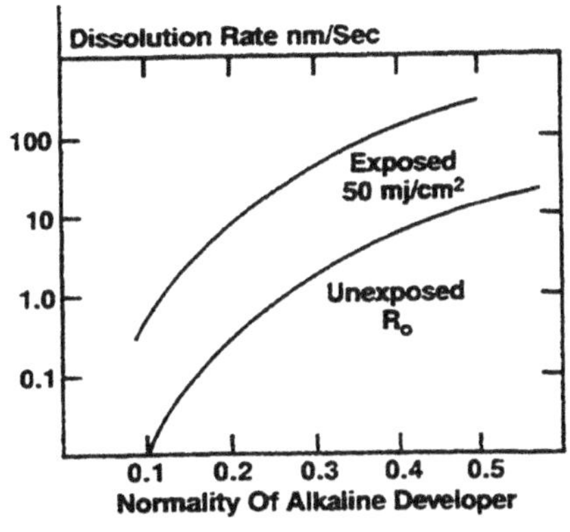

FIGURE 10-5-2. Dissolution rate of DQN as a function of normality of developer. From Ref. 93, copyright 1980 and courtesy of IEEE.

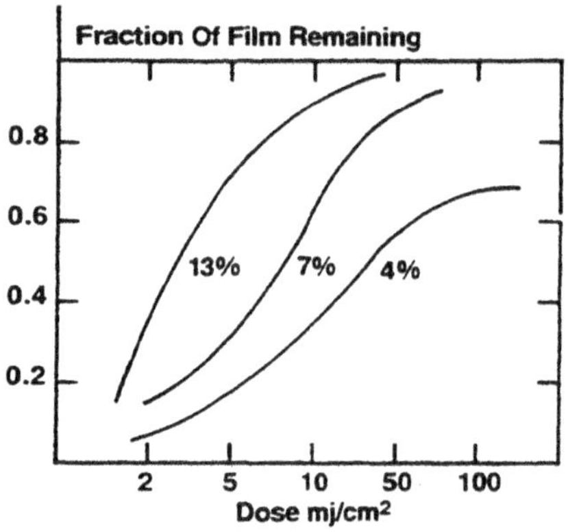

FIGURE 10-5-3. Contrast curves of DQN at different DQ loading. From Ref. 93, copyright 1980 and courtesy of IEEE.

The amount of indene acid (**E**) formed is governed by the quantum yield (ϕ) of the primary reaction (disappearance of DQ) and the uniformity or depth-dependence of absorbed light. Because of standing waves reflected from the underlayers of oxide (Fig. 10-5-4), the absorption of light is nonuniform.[94, 95] Some of the standing waves can be suppressed by the addition of inert absorbing dyes. With standing waves, the edges of a developed image reveal the faithful recording of the interference patterns (Fig. 10-5-5).[95] Postexpose (predevelop) bake (Fig. 10-5-6) mixes the bleached DQ with the partially bleached DQ to smooth out the edges of the profile.[96]

In the exposed resist, the rate ratio of development (R) and R_0 (unexposed) is proportional to the dose, Q, quantum yield, ϕ, and the developer concentration, z, by a similar expression derived for PMMA sensitivity,

$$R/R_0 = (1 + kQ\phi)^z \tag{10-5-1}$$

A higher quantum yield (> 0.20) provides a larger rate ratio at a lower dose. For optical exposure in the region of 300–400 nm, the actinic response is determined by the spectral actinic absorption[97] (μm^{-1}) or the extinction coefficient (ε, 7600 liters/mole-cm) of the DQ sensitizer (Fig. 10-5-7). Since the quantum yield (ϕ, 0.25) is essentially independent of the wavelength at 365-, 405-, and 435-nm

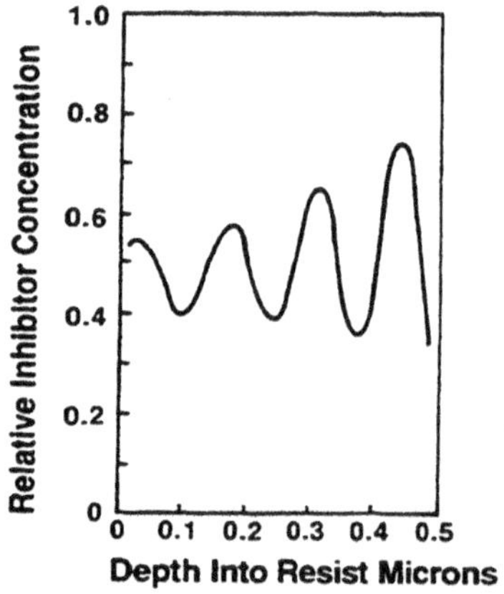

FIGURE 10-5-4. Absorption of light in DQN resist on oxide of silicon. From Ref. 94, courtesy of AIP.

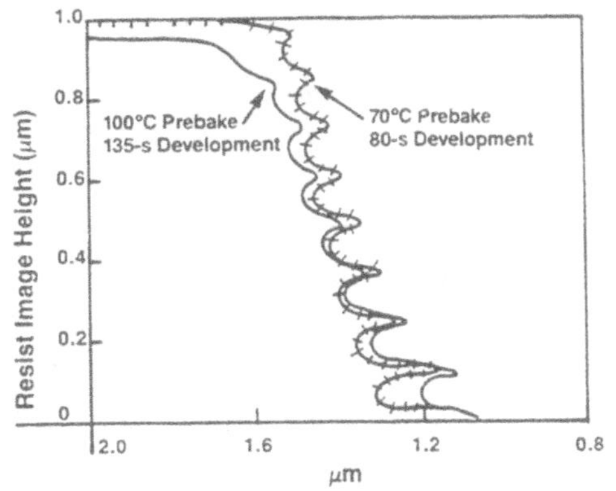

FIGURE 10-5-5. Standing waves reproduced in edges of DQN resist. From Ref. 90, courtesy of IBM.

exposure,[88, 98] the solubility rate, R, is a linear function of the fraction of inhibitor remaining (sensitizer destroyed) (Fig. 10-5-8).[88] In e-beam,[88, 102, 103] ion-beam,[100] X-ray,[99] and deep-UV[101] exposure, both the sensitizer and resin absorb radiation and bimolecular or independent reactions may occur to form inhibiting or insoluble products. Indeed, at high e-beam or high-energy doses, the resin or novolak resist insolubilizes into image-reversed negative images.[92, 102, 104, 252] On the other hand, chlorinated novolaks undergo decomposition in DQN formulations to accelerate the dissolution (R) of the exposed resist.[105]

10-5-2. *Modeling DQN Development*

The modeling of DQN involves the dose–developer interaction in three major parts:

1. The photokinetics of exposure
2. The kinetics of dissolution
3. The influence of composition and process parameters

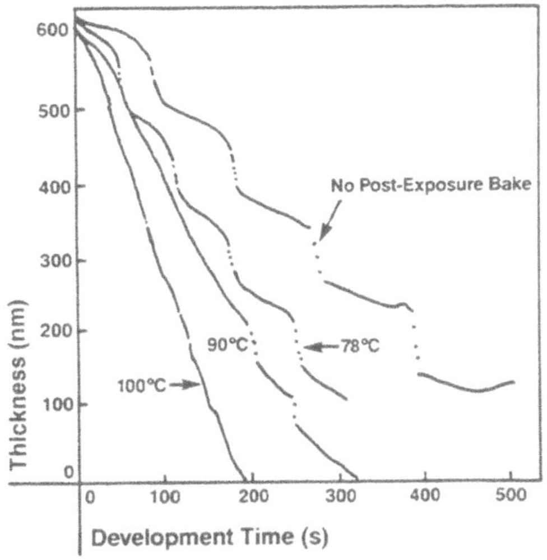

FIGURE 10-5-6. Reduction of the standing waves in AZ-1350J with predevelop (post-exposure bake). From Ref. 96, courtesy of IBM.

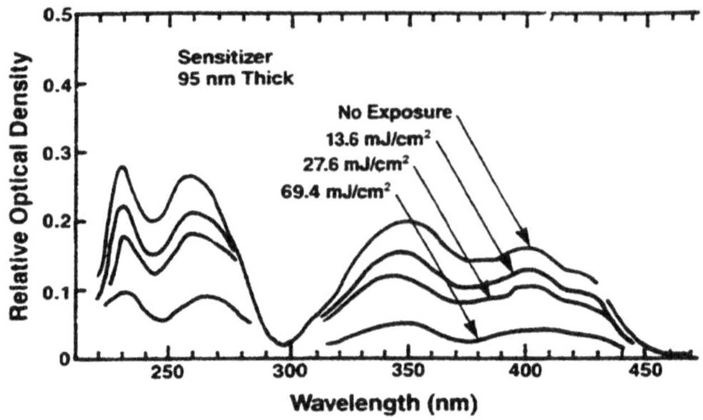

FIGURE 10-5-7. Bleaching in DQN resist as a function of dose. From Ref. 88a, copyright 1977 and courtesy of RCA Co.

10-5-2-1. *Photokinetics of Exposure.* In the photobleaching of the DQ sensitizer of AZ-1350 shown in Fig. 10-5-7, the residual absorbance is due to a soluble azo side product from coupling of the indene acid with DQ,

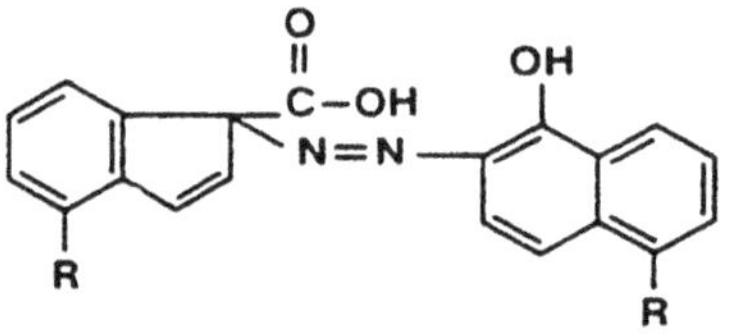

In the modeling of the bleaching, the kinetics is corrected for the absorbance of the azo photoproduct (acts as inner filter).[95, 106] In the simplest case, light is uniformly absorbed throughout the film (no edge slope or reflected wave distortion as in Figs. 10-5-4 and 10-5-5). Upon exposure, the concentration of diazoquinone (D_0) sensitizer decreases in the following manner,

$$dD(t)/DT = D_0 e^{-\phi It} \tag{10-5-2}$$

where $D(t)$ is the resist thickness, ϕ is the quantum yield or number of DQ bleached/absorbed photons, and I is the quantity of absorbed light in mole quanta/cm^2 sec (1 J/cm$^2 = 10^{18}$ photons or a mole per einstein) $I = 10^{-2}$ J/cm^2.

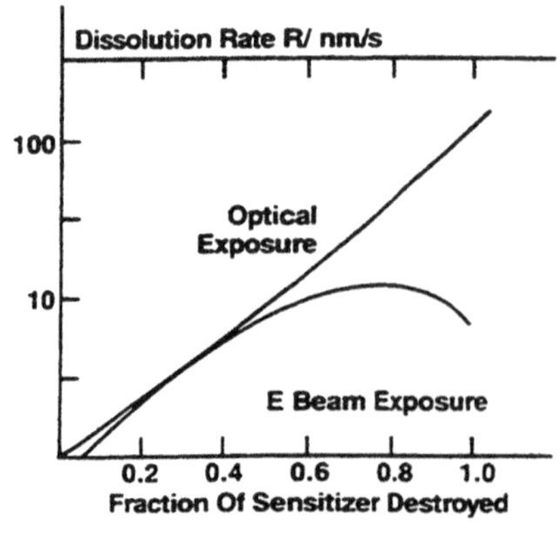

FIGURE 10-5-8. Dissolution rate of DQN resist as a function of inhibitor content. From Ref. 88, courtesy of SPE.

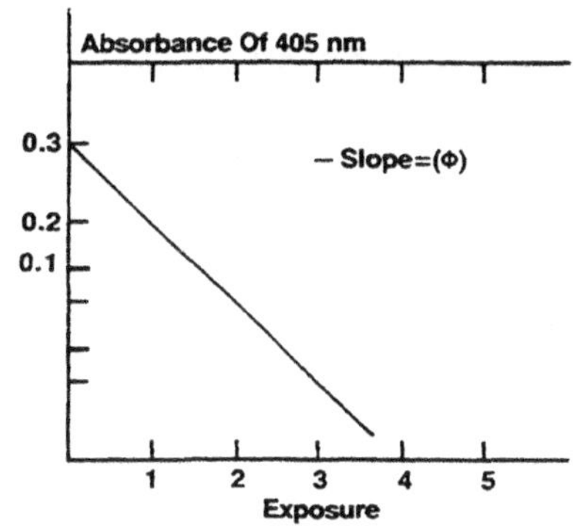

FIGURE 10-5-9. Quantum yield (ϕ) of bleaching of DQN resist. From Ref. 98, reproduced by permission of the Electrochemical Society Inc.

If the amount of incident light is I_0 and ε is the absorption coefficient, I is also related to the absorbance (A) of the film,[370]

$$I = I_0(1 - A) \qquad (10\text{-}5\text{-}3)$$

or

$$A = A_0 \exp[-\varepsilon D(t)]$$

Substitution of Eq. (10-5-3) into (10-5-2) leads to[108]

$$\ln A = \ln A_0 - k\phi t \qquad (10\text{-}5\text{-}4)$$

If the plot of log absorbance versus exposure time is linear, the slope is the quantum yield, ϕ (Fig. 10-5-9).[98] The expression (10-5-3) can also be integrated for polychromatic exposure and corrected for the back Fresnel reflection of light from silicon substrates.[107] The conversion of D_0 to products proceeds exponentially into the depth of the film.[98]

If we assume that the rate of development is proportional to the dose E,

$$E = I_{\text{abs}}\, t \qquad (10\text{-}5\text{-}5)$$

the depth of development, Δd, normal to the surface for develop time, Δt,

$$\Delta d = [(\Delta r)\, \Delta T]\, kE \qquad (10\text{-}5\text{-}6)$$

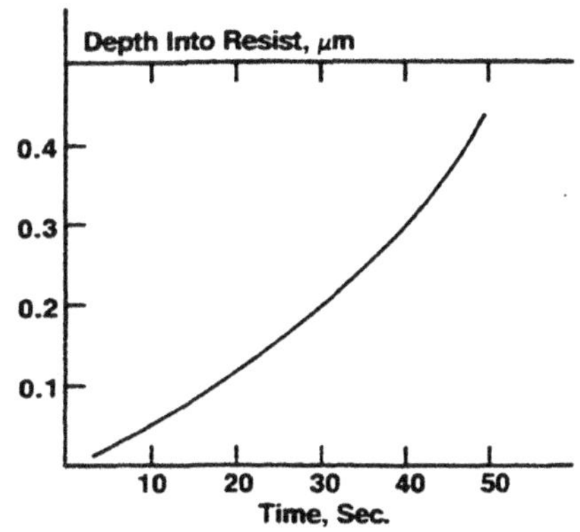

FIGURE 10-5-10. Development depth versus exposure at constant develop time for AZ-1350 resist. From Ref. 113.

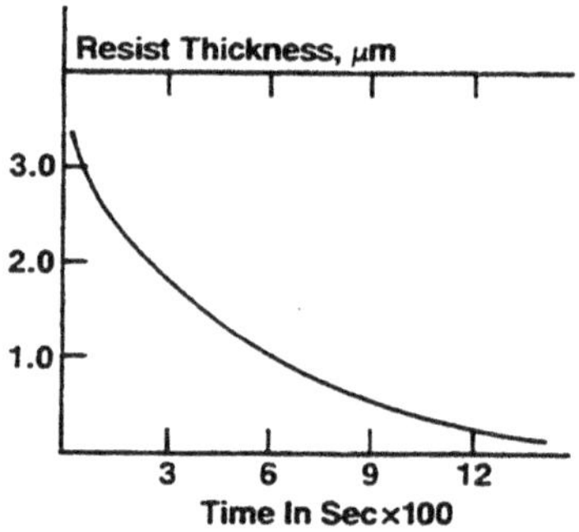

FIGURE 10-5-11. Thickness versus time of development of AZ-1350. From Ref. 111, copyright 1975 and courtesy of IEEE.

where Δr is the difference in resist etch (dissolution) rates of exposed and unexposed regions

$$\Delta r = R - R_0 \tag{10-5-7}$$

assuming no induction effects.[108]

Figure 10-5-10 illustrates the linearity of development per depth for AZ-1350J versus dose at constant develop time. At the surface, the absorption is linear since[112] $1 - e^{-x} = 1 - x$ for small x. Above an absorbance of 0.2 (with film thickness 0.5 μm), the absorbance of light is nonlinear and the exponential depth distribution of product follows a first-order kinetic dissolution[111] as the film dissolves (Fig. 10-5-11). In the absence of standing waves, the resist acts as an internal[113] step wedge (at constant develop time and varying dose) to monitor the dose (Fig. 10-5-12). Since the exposure of a DQ resist does not require 100% bleaching of D to form an image, about 40% bleaching is used with a developer to provide a differential image. The fraction of molecules bleached (D_b) is inhibited by the remaining fraction (D_r),

$$D_b + D_r = 1 \tag{10-5-8}$$

The bleached fraction is proportional,[109, 110]

$$D_b = \exp(-kE) \tag{10-5-9}$$

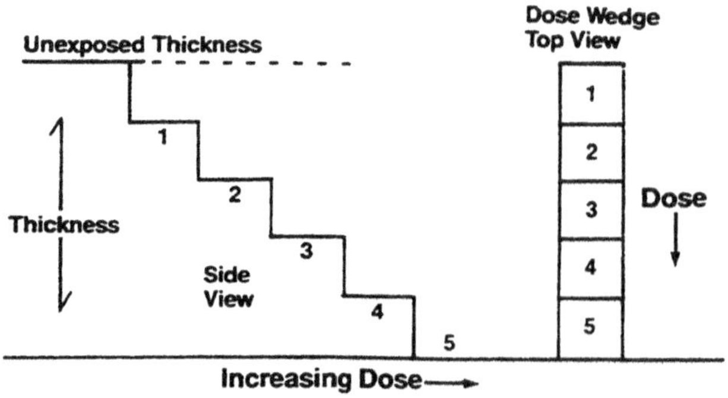

FIGURE 10-5-12. Cross section of response of positive resists to a step wedge exposure. The highest dose is at wedge 5 and the lowest dose (lowest % transmission of wedge) is at position 1. If the resist dissolves linearly (first order), then any depth position (number of fringes in laser endpoint apparatus) can be used to determine image cleanout.

therefore,

$$D_r = 1 - \exp(-kE) \tag{10-5-10}$$

The amount of photoresist developed at any depth Δd in the exponential bleaching case is given by

$$\Delta d = (D_b R + D_r R_0)\, T \tag{10-5-11}$$

or

$$\Delta d = \{D_b R + R_0[1 - \exp(-kE)]\}\, T \tag{10-5-12}$$

Since Δd is an exponential function of exposure (Fig. 10-5-10), at no exposure, $E = 0$, and

$$\Delta d = R_0 T \tag{10-5-13}$$

at a higher dose where

$$R/R_0 > 10 \quad \text{and} \quad kE \ll 1 \tag{10-5-14}$$

$$\Delta d = RT \tag{10-5-15}$$

or

$$\Delta d = (R - R_0)\, TkE + R_0 T \tag{10-5-16}$$

where k is about 0.05 cm^2/mJ for AZ-1350J.[114] These elementary bleaching kinetics point out the benefits of modeling the profiles in three dimensions[350] on the basis of the fraction of inhibitor remaining (D_r). If the inhibitor concentration

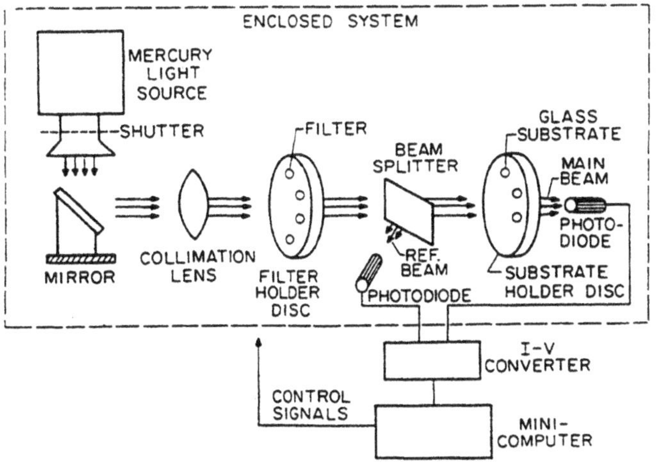

FIGURE 10-5-13. Schematic of automated exposure-bleaching system for DQN resist. From Ref. 259, courtesy of Dr. A. Neureuther, UCLA, Berkeley.

can be calculated from optical scattering and absorption coefficients, then k can be measured and the R calculated.[370] Dill[95, 106, 111] and later Neureuther[259] (Fig. 10-5-13) determined k by bleaching transmission measurements and correlated the rate of bleaching and the rate of film thinning. Although no absolute values of ϕ or ε could be measured since no density of photoactive sites was assumed,[88] the ASA speeds or k in Eq. (10-5-12) (C in Dill's work[95]) is proportional to the quantum yield in Eq. (10-5-4). Table 10-5-1 lists several DQN resists (examined by Shaw and Hatzakis[115]), which peak at the highest absorption band of the D sensitizer (405 nm).

All of the photokinetic models have assumed an exponential uniform absorption of light in three dimensions. Several factors, however, prevent this from occurring:

1. Strong optical absorbance in films of DQN greater than 1.0 μm thick, absorbance > 1.0.
2. The back reflection of light especially over steps which causes coherent interference effects (standing waves).[116, 117]
3. The edge slope of patterning beam.[119]
4. Striations in coated resist which cause inhomogeneous concentration of D sensitizer to be dispersed in the resist film[119] (Fig. 10-5-14). Striations are readily observed by thinning the unexposed resist in the developer.

Because of striations, overexposure is required to destroy the uneven sensitizer distribution in the cast film. To reduce striations, new casting solvents of diglyme[120] or silicone oil additives to AZ-1350 have been used.[327]

Due to standing waves, the local intensity variations can cause the local absorbed energy to vary by a factor of two (Fig. 10-5-4). Because of the interference effects, a variation in the rate of development of one order of magnitude can be detected which can result in a ragged edge profile [especially if the resist has a low[122] contrast ($\gamma < 2$)] (Fig. 10-5-14). The variation can be reduced significantly by overdevelopment[121] (Fig. 10-1-2) or by plasma ash descumming of the undeveloped resist. Besides standing waves, the incoming beam is also distorted by the lens. The inclusion of the intensity distortions due to light is treated by

TABLE 10-5-1. Absorption Rate Decay (k) of Inhibitor for Various λ

Resist	λ (nm)	k (cm²/mJ)
Kodak 809	365	0.011
	405	0.017
	435	0.010
Polychrome 129	365	0.010
	405	0.019
	435	0.009
AZ-2400	365	0.012
	405	0.011
	435	0.001
AZ-1350J	365	0.010
	405	0.020
	435	0.010

FIGURE 10-5-14. Striations in AZ-1350J resist film. From J. Margelin and D. Webb, *IBM J. Res. Dev.* **24**, 561 (1980), courtesy of IBM.

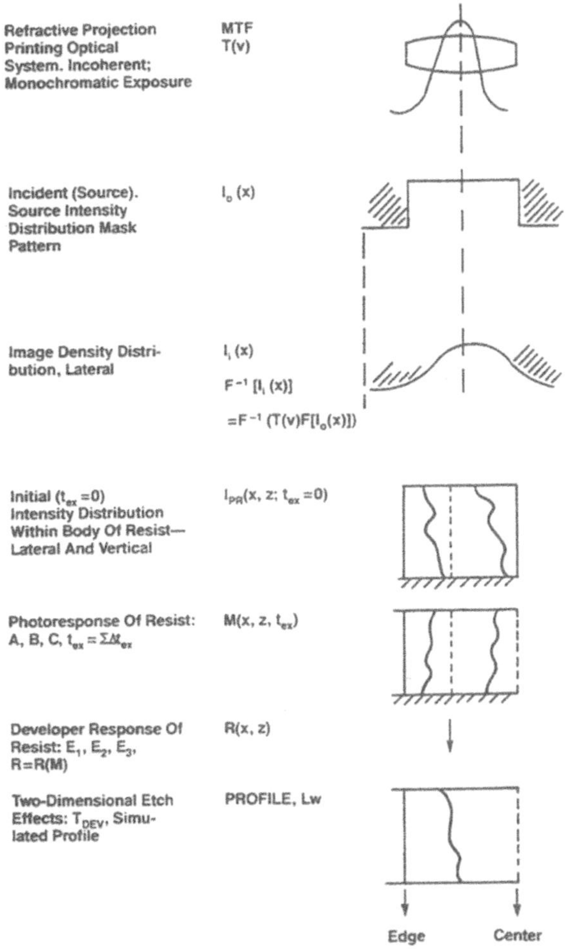

FIGURE 10-5-15. Schematic of the steps in algorithm calculations to determine the profile in photoresist as a function of incident beam shape, intensity, and developer response. From Ref. 124, copyright 1979 and courtesy of IEEE.

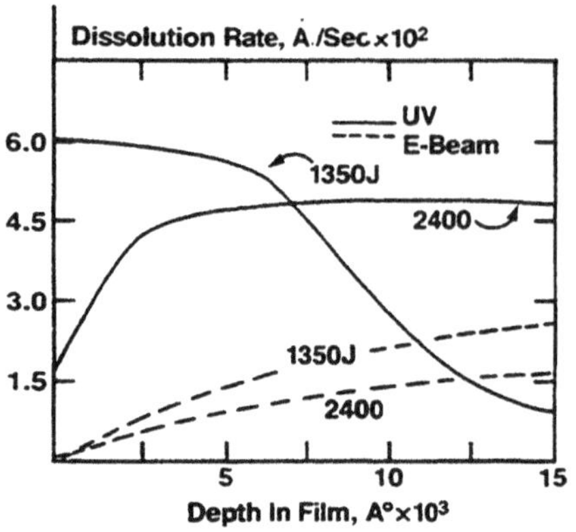

FIGURE 10-5-16. Development rates versus depth into film for UV- and *e*-beam-exposed AZ-1350J and AZ-2400. From Ref. 115, copyright and courtesy of IEEE.

Fraunhofer diffractions(116, 117, 123, 124, 346, 350) to calculate the lateral and vertical cross sections of an exposed and developed image in DQN resist. An outline of SAMPLE, the computation algorithm, is shown in Fig. 10-5-15.(124) The end result of the photoresponse treatment is the inhibitor content at every point in the resist image.(56) Once the inhibitor content is known, we can proceed to simulate and model developed profiles.

10-5-2-2. *Kinetics of Dissolution.* In the simplest version, the exposure to radiation removes or alters the inhibitor to free the phenolic resin which rapidly dissolves in the developer. The indene acid product also may increase the hydrophilicity of the exposed region (wetting by the aqueous polar developer) and the nitrogen gas released by the DQ may also microfoam the resist to create cavities to assist the developer infusion.(170, 359) In the positive photoresist of DQN, less light is absorbed deeper in the film, hence R decreases with depth (Fig. 10-5-16).(115) Such is the case for AZ-1350J but not for AZ-2400, which may have more radial(119) spun-cast striations to offset the exponential absorption. Due to the multiple backscattering (Fig. 10-5-17), the rate (R) of development increases with depth in *e*-beam exposure.(32) For UV-exposed AZ-2400 and Kodak 809, the top 2500 Å is slower dissolving than the remainder of the film (Fig. 10-5-18) because an induction effect retards the dissolution of the top of the film.(125) The inhibition is due to the dominance of R_0 in the following equation:

$$R = R_0 + R(\text{exposed}) \qquad (10\text{-}5\text{-}17)$$

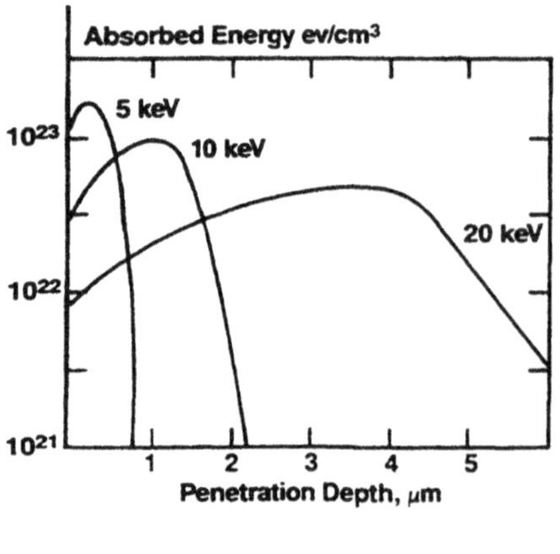

FIGURE 10-5-17. Energy deposited in PMMA for *e*-beam exposure at three energy levels. From Ref. 32, copyright and courtesy of IBM.

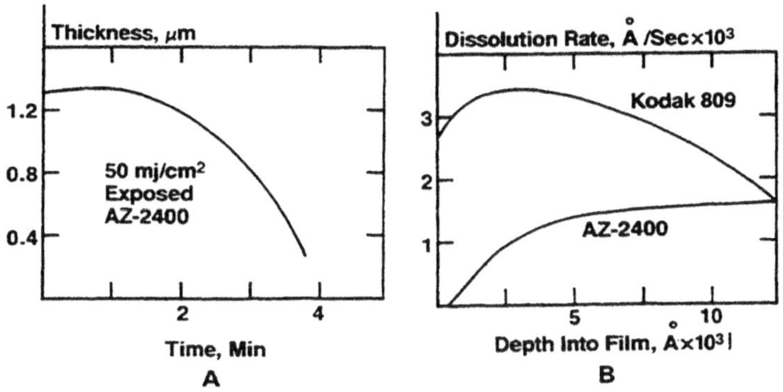

FIGURE 10-5-18. (A) Film thinning for AZ-2400 and (B) rate of development versus depth for UV-exposed ($50\ mJ/cm^2$) AZ-2400 and Kodak 809 positive DQN resists. From Ref. 115, copyright and courtesy of IEEE.

In the AZ-2400 resist, either the resin component has characteristics of dissolution different from other novolaks or a higher sensitizer concentration is used. During prebake, the oxidation of the resin may also cause an induction effect.[89, 93] Thus, the overall rate of dissolution R may be comprised of the contributions of many rates (k),[351]

$$R = k_{H_2O} + k_{\text{indene acid}} + k_{\text{gas}} + k_{\text{solvent}} + k_{\text{resin}} + k_{\text{surface}} \tag{10-5-18}$$

The presence of an induction effect has been previously noted for PMMA exposure[32] [Eq. (10-3-5)]. For AZ-2400 exposed by *e*-beam,[118, 352] the rate of development follows a Bose–Einstein type of curvature,[304, 320]

$$R(z, E) = (2.5 + 2.07 \times 10^{-19} E^{1.065})(1 - e^{-1.7z}) + (0.11 + 5 \times 10^{-32} E^{1.7}) \tag{10-5-19}$$

where z is the depth into the resist film (μm) and E is the absorbed energy density (keV/cm^3).

For optical exposure, a similar expression can be used.[125] The induction effect can enhance the profile and the contrast of a resist by limiting the lateral isotropic erosion of the scattered image at the top of the resist film (Fig. 10-5-19).[125, 304] In the scattered zone, the dose of $20\ mJ/cm^2$ is the average value of the scattered film

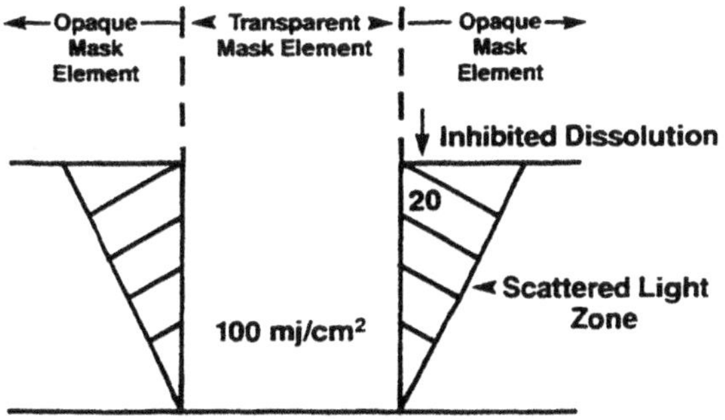

FIGURE 10-5-19. The induction effect at top of exposed positive resist inhibits profile rounding. Adapted from B. Grant *et al.*, *IEEE Trans . Electron. Devices* **ED-28**, 1300 (1981), copyright 1981 and courtesy of IEEE.

TABLE 10-5-2. Profiles and Contrast in DQN

Resist	Profile[a]	Contrast[b]
KTI Pos II	80	3.3
Kodak 809	77	
Polychrome 129	63	
AZ-1350J	75	3.2
AZ-2400	90	4.5

[a] Sidewall profile in degrees.
[b] Ref. 122.

near the top of the image. The induction effect retards lateral dissolution while the heavily exposed (100 mJ/cm^2) region develops in an anisotropic manner. In Table 10-5-2, the profiles of the resist shown in Fig. 10-5-18, along with the reported contrasts, are given.[115] AZ-2400 exhibits a strong induction effect while the other diazonovolaks without an induction effect,[285] dissolve isotropically especially near the substrate interface (last 20% of film remaining) due to the falloff of the MTF of the lens.

Several material and process factors influence the induction effect:

Composition Effects

1. Use high-molecular-weight narrow-dispersity novolak fractions[126]; *p*-vinylphenol[268] or other synthetic polymers may be easier to synthesize as narrow molecular weights. Presoak of DQN resist in chlorobenzene achieves fractionation[143] by removing low molecular weights. *Ortho*- and *para*-novolaks retard dissolution.
2. Substitute alkyl groups on phenols in the *para* position as *t*-butyl retards dissolution.[127]
3. Use higher DQ sensitizer loading > 20%.[88, 128, 129]
4. Esterification of phenol groups follow by silylation to decrease dissolution.[165]

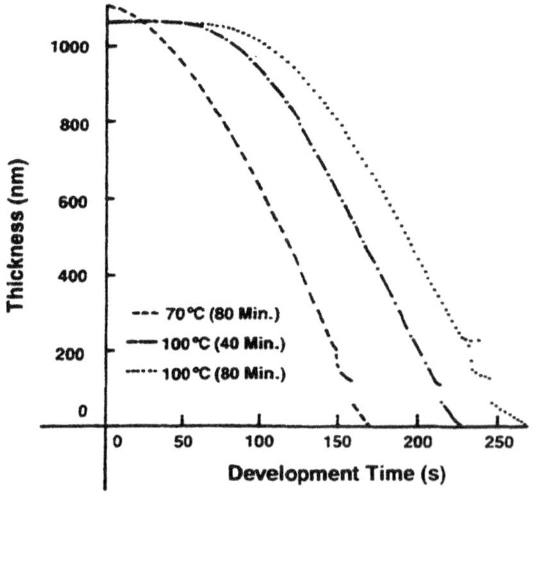

FIGURE 10-5-20. Dissolution rate of novolak resin after various prebake conditions. From Ref. 96, copyright and courtesy of IBM.

5. DQN cast from good solvents (diglyme) exhibits induction while films cast from poor solvents do not.[27, 131] Propylene glycol acetate gives a higher-speed resist than Cellosolve acetate (Table 2-2-1).

Processing Effects

1. Prebake resist > 90°C (preferably on a hot plate to reach steady-state conditions).[90, 287] Prebake of the top surface while keeping the silicon surface cold has been claimed to create a large induction effect sufficient for a lift-off profile.[130] At the high prebake, several physical and chemical reactions may occur:

   a. Removal of solvent by prebaking 10°C below DQ sensitizer decomposition.[90, 133, 135]
   b. Cross-link resist by ketene (E) phenol ester formation (see Section 10-6).[91, 132, 133, 252] The prebake may be irreproducible due to various sensitizer loadings in commercial formulations as well as different phenol (titer) groups in resin (D).
   c. Oxidize novolak resin by heat (Fig. 10-5-20) or deep UV expose in the presence of O_2,[89, 93, 96, 134]

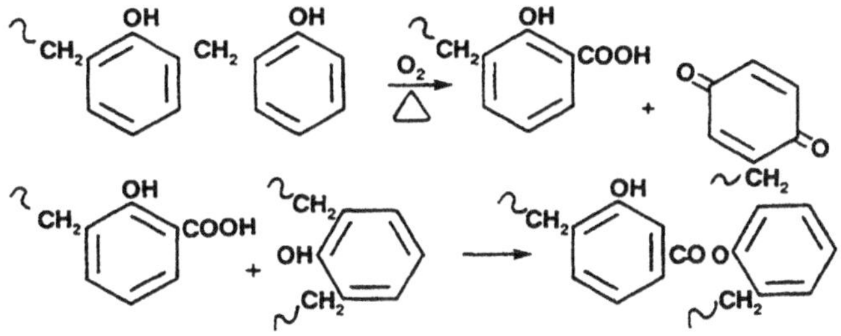

   Postexpose (predevelop bake) also oxidizes the resist on the surface in addition to mixing the exposed products that are isolated by standing waves.[133, 136, 294]

   d. The developer is also claimed to activate an azo dye formation between the resin in DQ and the original sensitizer which retards dissolution,[139, 140]

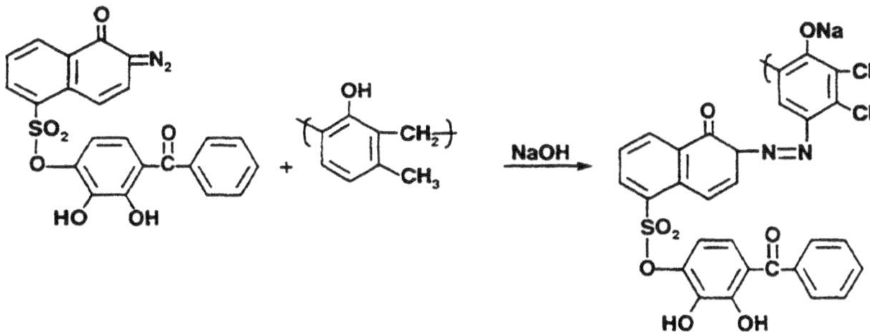

e. Di- and triesters of diazonaphthoquinones are more effective inhibitors than the monoesters, because two or more DQ groups can couple intermolecularly with phenolic chains and top surface involubilize the resist,

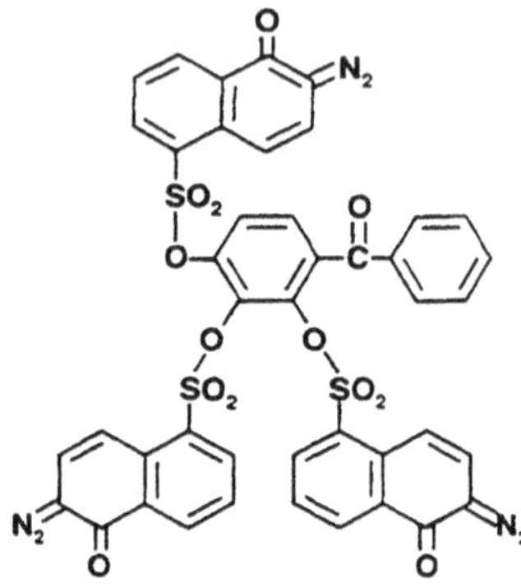

2. Concentration of developer. Concentrated developer (0.25 N) generally increases sensitivity but lowers contrast[353, 356, 357] and causes pronounced thinning ($R/R_0 < 8$).[138] The silicate content in buffered K_2O developers controls the contrast.[226] Although a weaker developer requires higher doses or a longer develop time, a significant induction effect is noticeable (Fig. 10-5-21).[137] In the literature there is some contradiction on the DQN contrast and sensitivity, since the normality and composition of commercial developers are not specified.

Surface-active agents such as polyethylene glycol[162, 358] added to the developer, may preferentially adhere to the unexposed novolak resin and retard the unexposed dissolution resulting in a high-contrast resist. Since the surface-active agents are used at trace concentrations (1 ppm), they are consumed in batch development processes. A dramatic demonstration of the effect of different modes of development in surface-active developers was given by Marriott and co-workers (Fig. 10-5-22). While puddle and dip development rapidly lose their potency due to irreversible surfactant

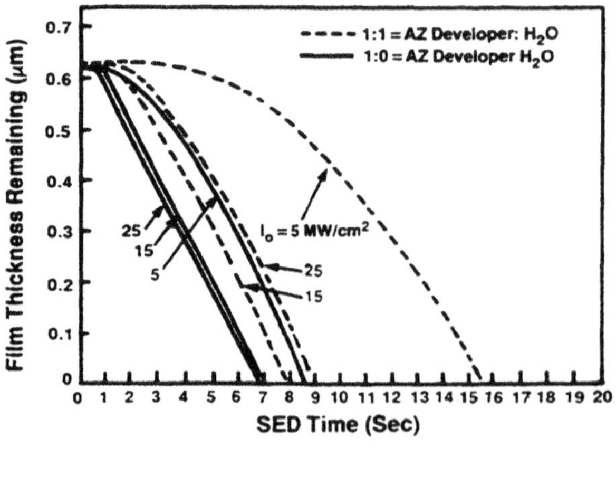

FIGURE 10-5-21. Dissolution rate of AZ-1350 at two developer concentrations, undiluted and diluted. From W. Tsang, *Appl. Opt.* **16**, 1918 (1977), courtesy of OSA.

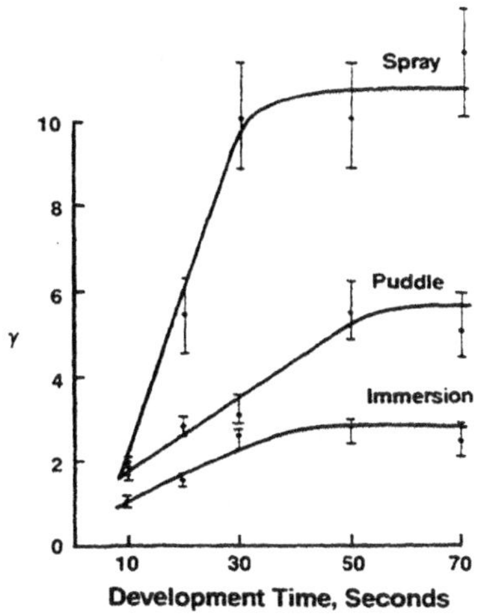

FIGURE 10-5-22. Effect of technique of developing on the contrast of DQN photoresist developed in alkaline solution. From Ref. 340, courtesy of SPIE.

adsorption, spray development provides a fresh supply of surfactant and allows the DQN resist to achieve the highest contrast (τ of 10) ever reported for an optical resist.

3. Presoak and interrupted developer. Presoak of the resist briefly in water or in the developer is also claimed to retard the unexposed dissolution.(141, 225, 297) The other presoaks or pretreatments are with NH_3 or amines to catalyze azo dye reaction.(142) A brief dip in chlorobenzene removes low-molecular-weight components and forms profiles suitable for lift-off.(143) Presoak in water(354) or storage in high humidity (> 35%) can also cause the induction effect (Fig. 10-5-23). If the humidity is not specified or controlled in processing DQN resists, various contrasts may be obtained. Control of the relative humidity at $35 \pm 5\%$ is recommended.(144) In the presoak in water, the absorption of moisture may cause strong H bonds with the resin. The effective strength of the alkaline developer is diluted by the 1–2 M concentration of water (1–5% by weight of resin) as it penetrates by a solubility mechanism.(145) The casting solvent is also removed by a water soak, increasing the induction effect.

Another variation of the presoak process is interrupted development. The resist is partially developed, quenched in water, and redeveloped. The induction effect prevents lateral erosion at the edge of the image at the top

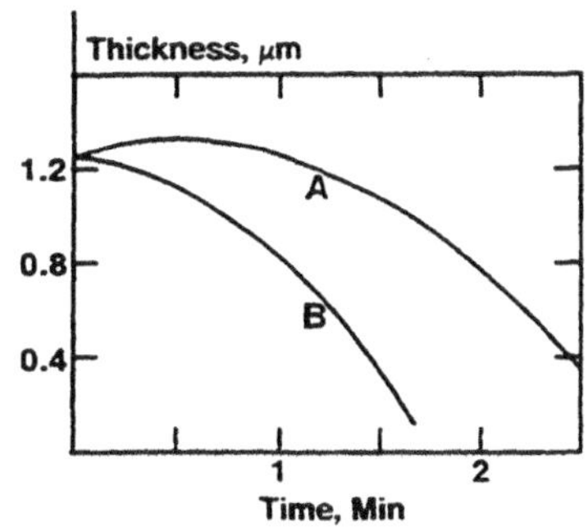

FIGURE 10-5-23. Dissolution rate of exposed DQN (A) stored after exposure for 1 hr at 50% relative humidity and (B) stored and developed at 5% relative humidity.

surface.[146] A similar step is to develop the resist for 90% cleanout in a weak developer and clean up the last 10% with a stronger developer especially in cases of underexposure or with nonuniform intensity across the wafer.[147] The two-step or higher multistep process is especially useful when combined with low-temperature (14°C) develop to extend the sensitivity of *e*-beam-exposed DQ resists.[148]

4. Chemical nature of the developer. The common developers are salts which can produce a pH of 12–13, and are derived from KOH, NaOH, Na phosphates, Na silicates, and tetraalkylammonium hydroxydes, (metal free) bicarbonates plus a surfactant to aid wetting.[149, 150] Alcohols and polyethylene glycols are also added to assist the removal of the exposed.[151] For high doses, no differences between buffered and unbuffered developers have been seen.[283] For lower doses, such as in pushing *e*-beam or optical exposure, the buffered developed resist exhibit no induction effect or loss of contrast. The common ion effect of Na^+ or K^+ prevails in buffered developers to drive the reaction to the right in favor of dissolution of the unexposed,

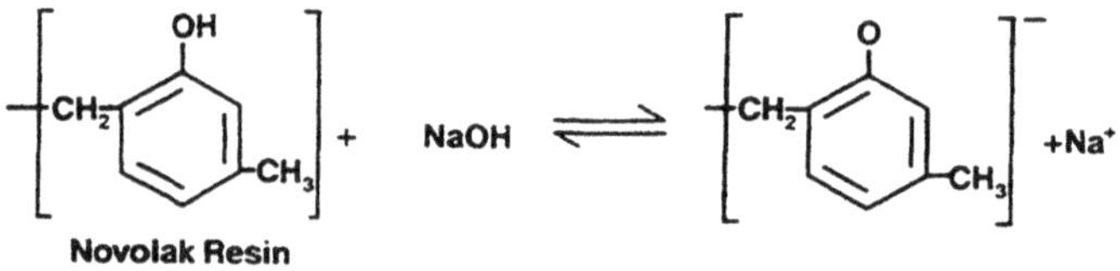

Thus, the addition of 0.2 g of novolak resin to a liter of developer is claimed to reduce thinning due to an induction effect caused by a reverse common ion effect (inhibits novolak dissolution).[158] For high doses of R/R_0 of 50 or more, NaCl is added to allow lower doses to be used up to R/R_0 of 10.[152] A summary of the various additives to the developer is given in Table 10-5-3. Traces or small amounts of surfactants in developer (ppm) can have a profound effect (2–4 ×) on R and R_0. The foaming of

TABLE 10-5-3. DQN Developers

Developer	Use	Reference
$HCONH_2$, EtOH	Metal free	153
$(NH_4)_2S_2O_3$	Longer shelf life of KOH	154
Triethanolamine	Forms holes in exposed, high R_0	155
$KMnO_4$, H_3PO_4	Increases resolution	156, 295
R_4NOH	Metal free	157, 315, 345
Add resin to developer	Inhibits thinning	158
Methyl triethanol NOH	Increases develop latitude	159
Triethanolamine	Improves contrast	160, 313
Surfactants	Improves cleanout	161, 162, 312, 314
NaCl, Na_2SiO_3	Improves sensitivity	152
Add surfactants or polyglycols	Improves cleanout and τ	163, 355, 358
R_4N silicate	Improves contrast	326
50 ppm acetone	Reduces adhesion loss	336

the developer bath with successive use may be due to buildup of a soaplike metal indene acid photoproduct which is suppressed by the addition of antifoaming agents.[338] Finally, developing of the DQN resist in an electric field is claimed to inhibit the unexposed dissolution.[228]

5. Temperature of the developer. The diffusion of the developer into the resist film involves three stages: (1) the diffusion of the developer from the bulk to the resist surface, (2) the diffusion of the developer into the film, (3) the diffusion of products out of the film. Because the diffusion coefficient of water intermixing within a polymer increases,

$$D = D_0 \exp(-\Delta E/kT) \tag{10-5-20}$$

exponentially with temperature (5–15 kcal/mole), the rate of development also increases. The activation energies for various resists are given in Table 10-5-4.[163] For a 10 kcal/mole activation, a 1°C change in temperature will change the rates R and R_0 by 20%. A 2°C fluctuation varies the linewidth in AZ-1370 by 0.2 μm.[285, 350] The Arrhenius plot (Fig. 10-5-24) exhibits a break at 15°C indicating a different dissolution mechanism below room temperature. For low-temperature (15°C) development, a major benefit in the resist of AZ-1350J is the inhibition of the unexposed dissolution. Suitable profiles can be produced for lift-off at *e*-beam doses of 10 μC/cm^2 and development at 14°C (Fig. 10-5-24).[158]

To summarize induction effects, the combination of composition, processing, and improved developers can produce conditions for high-resolution profiles (75–90°C) in novolak-based resists. For maximizing the induction effects, high-molecular-weight and narrow-dispersity resins along with more stable sensitizers are desirable synthetic factors.

For physical effects, the outer surface layer of a novolak resist may differ from the lower depths, due to combined effects of a lower solvent content and novolak oxidation at the surface of the prebaked film. In addition, the molecular weight distribution, the solvent used, method of coating, and drying can influence the film dissolution properties. Reproducing the skin on the surface, where the boundary layers offer greater resistance to permeation,[263] is the challenge of resist formulation and resist process engineering.

TABLE 10-5-4. Activation Energies for DQN

Resist	Activation energy (kcal/mole)
AZ-2400	
Unexposed	6
e-beam exposed	9
UV exposed	8
AZ-1350J	
Unexposed	13
e-beam exposed	15
UV exposed	9

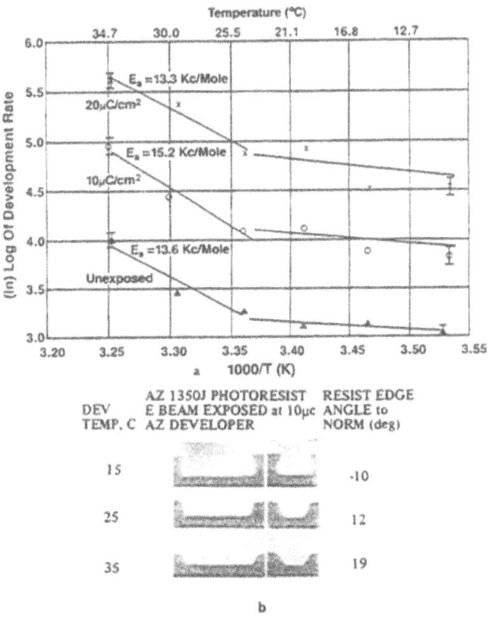

FIGURE 10-5-24. (a) Arrhenius plot of AZ-1350 development and (b) profiles in developed resist. From J. Shaw and M. Hatzakis, *J. Electrochem. Soc.* **126**, 2026 (1979), copyright 1979 and reprinted by permission of the Electrochemical Society, Inc.

10-6. DEVELOPMENT OF NEGATIVE RESISTS

Negative photoresists fall into three major categories:

	T_g(°C)
Polyvinylcinnamate	88
Cyclized rubber-azide	52
Novolak-azide	85

Negative *e*-beam, X-ray, or ion-beam resists are based on:

	T_g(°C)
Epoxy—glycidylmethacrylate epoxidized butadiene	<50
Allyl—allylphthalate or ether	<25
Vinyl—chlorostyrene	100
Siloxane–dimethylsiloxane	<0

Some of the negative resist polymers are thermoplastics, but most are gum rubbers of very low T_g that attract particulates (act as flypaper) and are easily swollen. The basic mechanism for exposure in a (nonchain polymerization) intermolecular cross-linking reaction is that gel formation occurs and the gel fraction is equivalent to the normalized film (Fig. 10-1-16) thickness remaining (> 50%) after development.

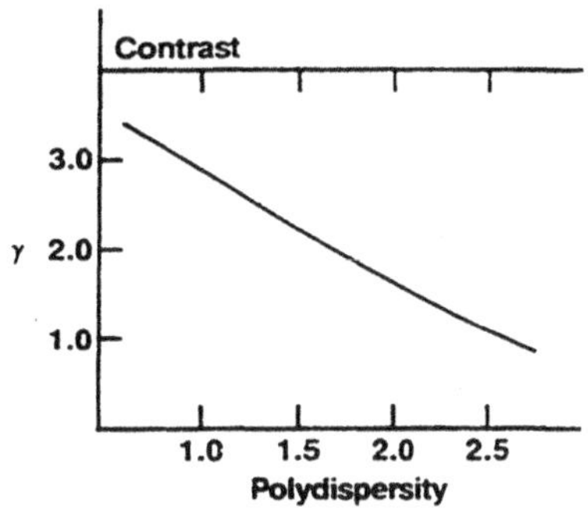

FIGURE 10-6-1. Contrast of PGMA negative resist versus polydispersity. From Ref. 168, courtesy of *Japanese Journal of Applied Physics.*

The cross-linking efficiency is given by a coefficient,

$$\delta = \delta_0[1 - \exp(-Kq/\delta_0)] \tag{10-6-1}$$

where δ_0 is the total number of cross-linkable sites per molecule and q is the dose. K is given by

$$K = G_x M_w E/50epN_a \tag{10-6-2}$$

where G_x is the number of cross-links per 100 eV of the absorbed energy, E is the energy absorbed per unit chain length, e is electronic charge, p is density, and N_a is Avogadro's number. In resists, δ_0 is on the order of 100 whereas the gel fraction saturates when δ is larger than 10.[164] The contrast of resists that insolubilize by gel formation is an exponential function of the polydispersity,[166]

$$\gamma = 2/E\exp(-M_w/M_n)^2 \tag{10-6-3}$$

Since the polydispersity of most commercial resists lies in the 2–4 range, the contrast ranges from 0.5 to 2.0. With a narrower dispersity (1–2), the contrast increases to 3–4 (Fig. 10-6-1).[168]

If resolution is to be improved, narrow-dispersity (1.1–1.3) polymers must be synthesized. Thus, the selection of the correct molecular polydispersity is the first criterion for developing reasonable images in negative cross-linking resists.

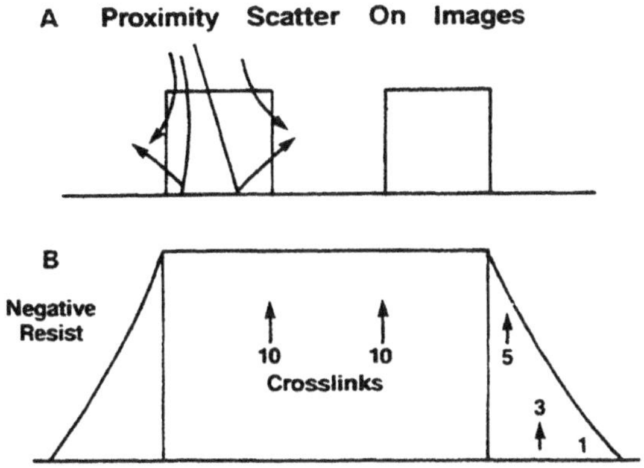

FIGURE 10-6-2. (A) Scatter of radiation in negative resist images and (B) the failure of the developer to remove the ligh scatter zone. The removal of the scatter zone is impeded by the swelling of the lightly cross-linked region.

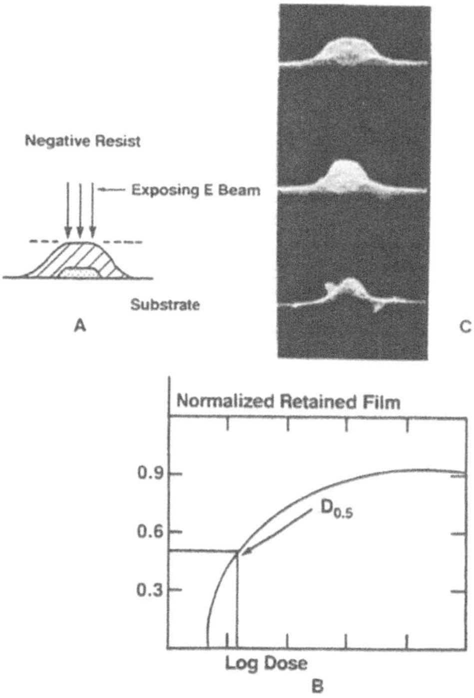

FIGURE 10-6-3. Electron-beam formed images in negative crosslinking KTFR resist. (A) Backscattered radiation (interproximity effects) subtended by using thinner resist.[26] (B) Contrast curve for negative resist with dose of $D_{0.5}$ (50% film loss.)[26] (C) Profiles in KTFR at various electron beam doses.[384]

In the latent image of cross-linked negative resists, the forward and backscatter radiation tend to form Gaussian-shaped images which overlap in closely spaced lines (pitch equal to 2 μm or less). In the developing process, the incoming developer encounters a lightly exposed but slightly swollen zone (Fig. 10-6-2) of a polydisperse polymer and fails to remove the gel fraction. Thus, thick resists (0.5 μm) are distorted and in order to subtend backscatter photons or electrons, thin films (Fig. 10-6-3) are used at a severe risk of pinholes and poor step coverage. Optical resists[169] are generally exposed until > 90% of the original film thickness remains, while negative *e*-beam resists are thinned to 50% film loss.[167] In a negative azide rubber resist which uses only a 2% sensitizer loading, film loss can also occur due to the oxygen effect (Fig. 10-6-4).[170] Into the top of the film, oxygen diffuses generally anisotropically, to produce a constant film loss, regardless of the initial film thickness. The oxygen effect in bisazide–rubber photoresists is due to the interception of the nitrene intermediate with oxygen to form nitroso compounds,[170]

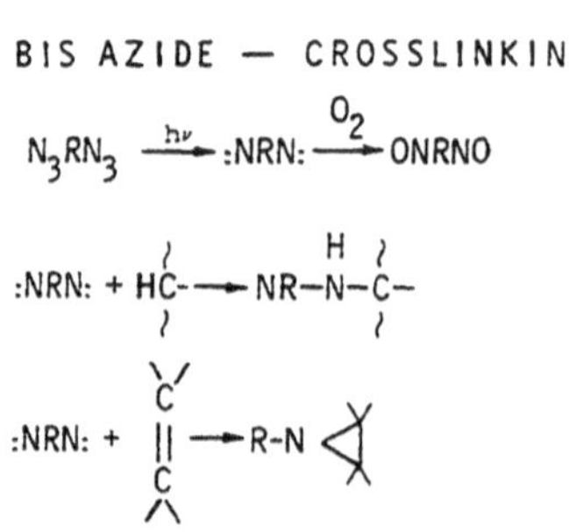

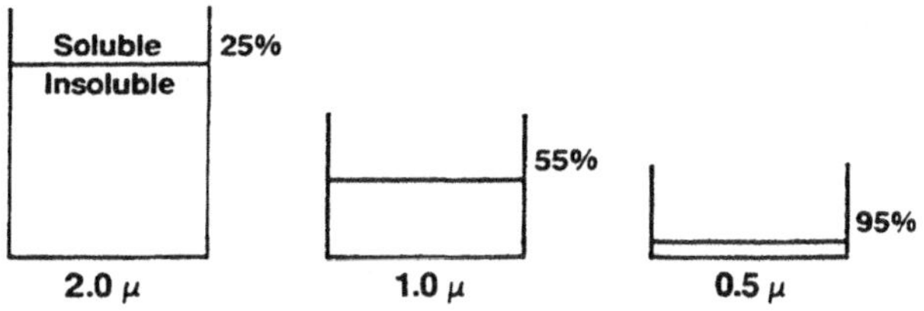

FIGURE 10-6-4. Oxygen effect in KTFR azide resist. The oxygen diffuses into the film during exposure in air, and quenches cross-linking. A constant thickness of film is lost independent of initial thickness.

Vertical film loss can be retarded by (1) exposure in the absence of O_2, (2) by topcoating with polymers such as polyvinylpyrollidone cast from water, or (3) by flash exposure in the millisecond range (faster than the rate of oxygen diffusion).[170] For a photopolymer system such as RISTON based on acrylate monomers, the oxygen trapped in the film actually improves the resolution by diffusing from the unexposed regions (Fig. 10-6-5).[172]

In *e*-beam exposure, a similar oxygen effect has been noted in *e*-beam negative resists of the epoxy or allyl type with low T_g and long side chains. Because of the low T_g, the cross-linking of the side chains continues after polymerization[173] and if the wafer is left in vacuum, the linewidth grows with time (Fig. 10-6-6).[171] The dark reaction increases the sensitivity at the expense of a time-dependent process curing. If the system is opened to air immediately after exposure, as shown by ESR studies the oxygen diffuses in the film and attacks the long-lived epoxy radical.[172] Polystyrene which does not exhibit the oxygen curing effect forms radicals which are short-lived at room temperature.[173, 227] By adding[171] a radical trap of DPPH (Fig. 10-6-7), the polymerization growth is reduced but with loss of sensitivity in the COP resist.[174, 235] Thus, the interproximity effect of scumming of the resist between exposed lines (Fig. 10-1-10) can be controlled to some extent by oxygen or radical inhibition. The addition of DPPH to limit the interproximity effect has been claimed for polydiallylphthalate *e*-beam resist[175] and KTFR bisazide photoresist.[176]

Another film loss of the negative resist image occurs, due to the intraproximity effect in *e*-beam or with optical exposure at the diffraction limit of a lens. The intraproximity effect self-limits resolution, by providing insufficient dose in micrometer lines (< 2 μm) (Fig. 10-6-8) while larger images in optical or uncorrected *e*-beam exposure retain more thickness, but at the expense of a larger base

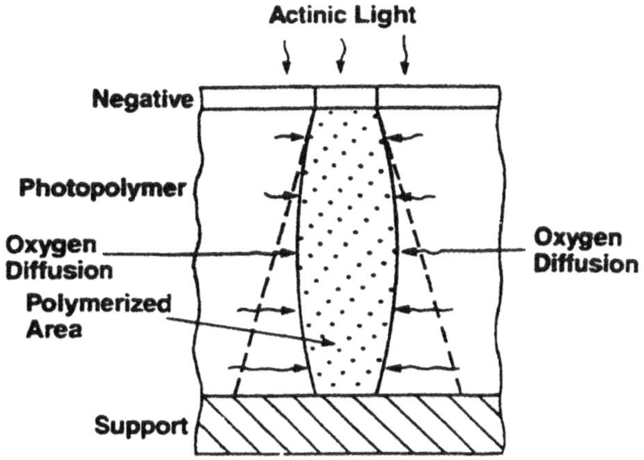

FIGURE 10-6-5. Enhanced resolution in photopolymer negative resist. The latent oxygen trapped in the film diffuses into the irradiated scattered zone and quenches free radicals to subtend the polymerization. From Ref. 172, courtesy of SPSE.

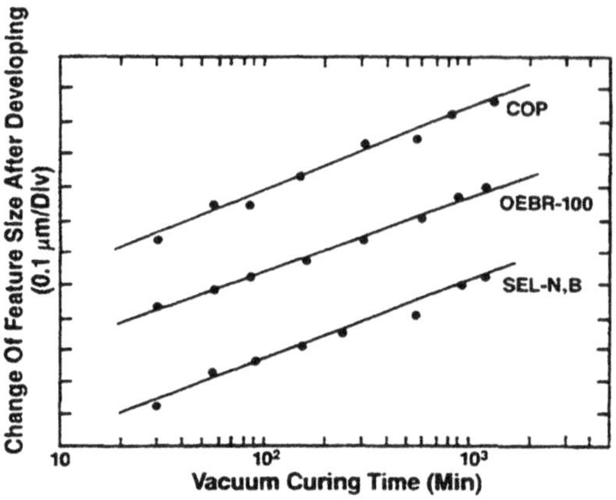

FIGURE 10-6-6. Change in linewidth for commercial negative *e*-beam resists left in *e*-beam vacuum for various times. From Ref. 341, copyright and courtesy of AIP.

width. Standing waves can also create layers of unpolymerized resist, which may cause orange peel reticulation undercut, and even loss of adhesion (Fig. 10-6-9).

Thus, in summary, the several major exposure factors of the oxygen effect, intraproximity effect, backscattered radiation and standing waves which alter the vertical or lateral negative resist image response, must be comprehended before the negative resist is placed in a developer.

Liquid Development of Negative Resists

Since cross-linking is a drastic change in solubility, practically any solvent could be used for development. Swelling, cracking, loss of adhesion, or excess thinning of the resist should be avoided. Redeposition or failure to completely remove the unexposed resist may be due to a high prebake or heavy overexposure. With cross-linked polymers, the swelling of the resist is a common problem. Swelling is a strong isotropic mechanical force which can overcome adhesion of the resist and produce wriggling lines or complete delamination (Fig. 10-6-10). The swelling is caused by osmotic pressure on the order of 50 kg/cm^2.[177] Once a solvent is absorbed in the cross-linked polymer, the bonds holding the chain are fully extended and swelling equilibrium is obtained. The highest degree of swelling occurs with solvents of the smallest molar volume.[371] For closely spaced lines in films thicker than about 1.5 μm, swelling may constrict narrow openings between large exposed pads

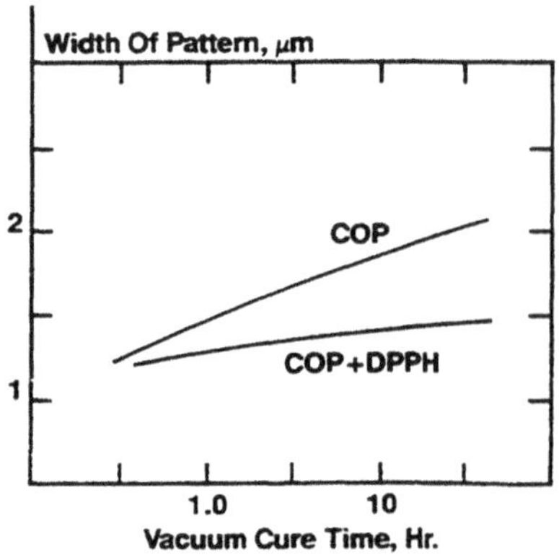

FIGURE 10-6-7. Postirradiation changes in *e*-beam-exposed COP with and without DPPH. From Ref. 171, copyright 1981 and courtesy of AIP.

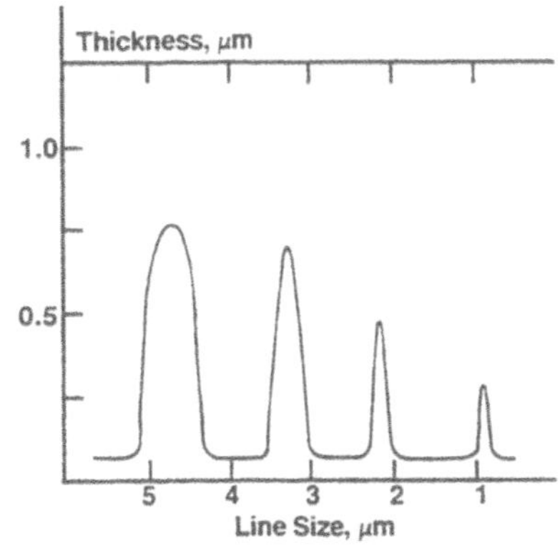

FIGURE 10-6-8. Film thickness remaining versus linewidth for an *e*-beam- or photon-exposed negative resist.

(Fig. 10-6-11). Spray development may assist in the more rapid removal of the unexposed line, before considerable swelling constricts the opening.

When spun cast, the film is also under stress on the order of 50 to 200 kg/cm^2.[177] Thus, swelling is a liquid phase extension of the residual stress. In Fig. 10-6-12, the profile of an image is shown for the cases of a dry, unconstrained image swollen with solvent and an image swollen but constrained to the substrate.[178] The degree of swelling is related to the physical properties of the polymer and solvent by the Flory equation,[179]

$$q^{5/3} = (nM_c/V_1)[(1.5\ \mathrm{M}_c/M)]^{-1}(1/2 - x) \qquad (10\text{-}6\text{-}4)$$

where q is the ratio of the volumes in the swollen and dry states, n is the specific volume of the polymer, C_1 is the molar volume of the solvent, M_c is the molecular weight between cross-links, M is the molecular weight of the polymer, and x is the polymer–solvent interaction parameter.

The influence of solvent on the extent of swelling can be estimated from x. A low x minimizes q and poor θ-solvents do not swell as much as good solvents, as, for example, in polystyrene,[347]

good solvent: toluene; poor solvent: decalin

due to less chain uncoiling.[182] For each polymer there are unique sets of solvent and θ temperature combinations,[33] which compress the mean dimensions to that predicted by models which neglect the excluded volume.[324]

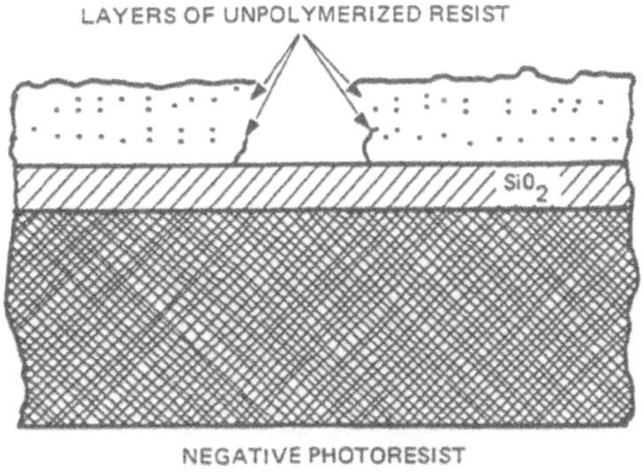

FIGURE 10-6-9. Standing wave effect in negative photoresist. From Ref. 298, courtesy of Morgan–Grampian Publ. Co.

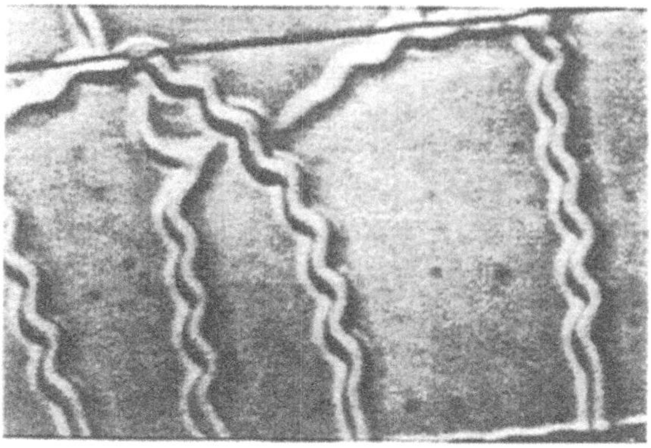

FIGURE 10-6-10. Swelling distortion of underexposed negative photoresist. From M. Long and C. Walker, *Kodak Microelectronics* **G-102**, 125 (1979), courtesy of M. Long.

The molecular weight between cross-links[180] is inversely proportional to the T_g of the polymer. Tightly cross-linked polymers such as styrenes cross-linked by vinyl backbones have a lower tendency to swell than glycidyl methacrylate with eight or more alkyl groups between the main chains. Thus, low-T_g elastomers tend to swell and elongate up to 300% of their volume (V) and length (l) (Fig. 10-6-13),[182]

$$3l/l_0 = V/V_0 \tag{10-6-5}$$

The shear modulus (G) is directly proportional to the number of cross-links, n,[198]

$$G = nRT \tag{10-6-6}$$

and the swelling is also a function of G, the density, ρ, and the specific volume, v,[181]

$$G/v = \rho RT/M_c \tag{10-6-7}$$

Polymers with a high shear modulus have lower-molecular-weight cross-links[198] and a lower degree of swelling.

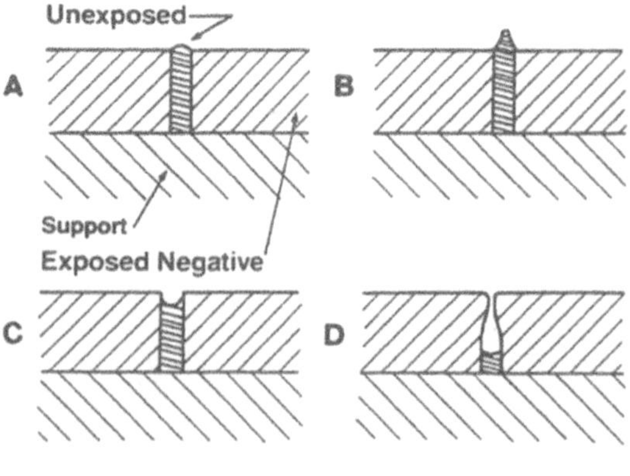

FIGURE 10-6-11. Swelling stages restricting development of micrometer line in negative photoresist. From Ref. 269, copyright 1961 and reproduced by permission of John Wiley & Sons, Inc.

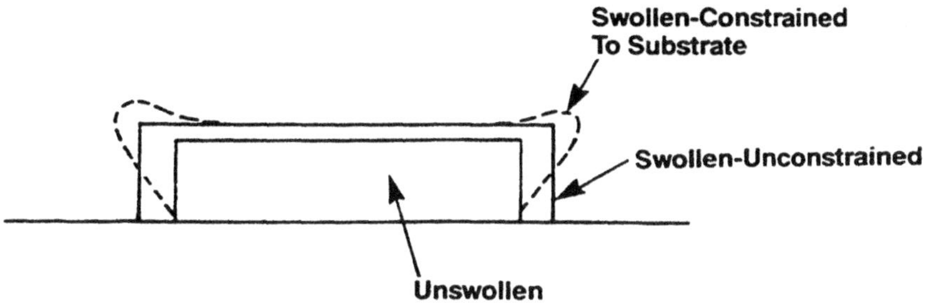

FIGURE 10-6-12. Swelling of negative resist constrained to a substrate upon immersion in a swelling developer. From Ref. 178.

Factors which reduce swelling and improve resolution of negatives are based on the following composition or process changes.

Composition

1. High-T_g polymers have a lower tendency to swell.[187, 377]
2. Narrow-dispersity polymers such as polystyrene,[184] chloromethylstyrene,[183] glycidyl methacrylate,[168] chlorostyrene,[185] and polyvinylcinnamate.[186]
3. Use θ or poor solvents such as chlorotoluene,[188] xylene, and butyl acetate[189, 192] for KTFR or MIBK + MEK for COP.[193]
4. Rinse with butyl acetate to reduce the swelling for KTFR.
5. Use water-based alkaline developers such as bisazide in phenol-formaldehyde (novolak) resins.[190]
6. Use non-cross-linking mechanism such as inorganic resists or donor–acceptor complexes of polythiafulvalene bromide.[191]
7. Image reverse diazonovolaks.[92]
8. Use dual tone poly-*t*-carbanatostyrene non-cross-linking resist.[332]

Negative resists are limited in resolution and contrast if elastomeric swelling polymers are used. Higher-T_g polymers based on chlorostyrenes have improved resolution due to a narrow dispersity and reduced swelling.[309]

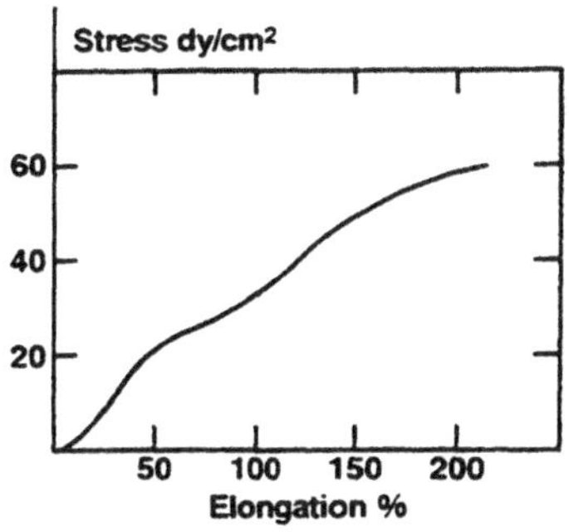

FIGURE 10-6-13. Stress elongation for cross-linked rubber in chloroform. From Ref. 182, reproduced by permission of John Wiley & Sons, Inc.

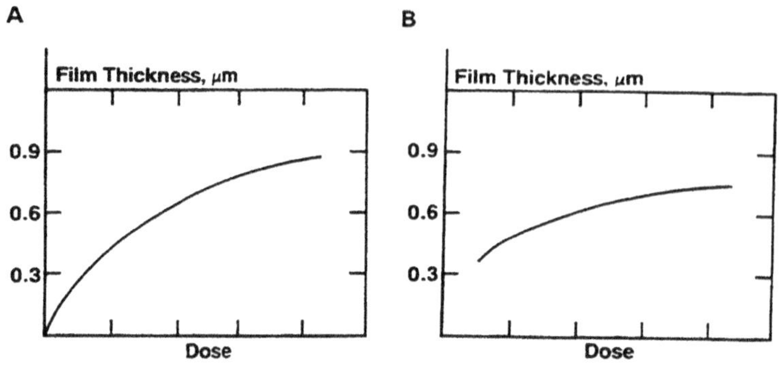

FIGURE 10-6-14. Film thickness of a negative resist after (A) *e*-beam and (B) optical topside exposure.

Process

1. Try thinnest film as possible.
2. Reduce standing waves in multilayer resist structure.
3. Use high-energy exposure and proximity correction.
4. Dry develop in O_2 (see Section 10-7).
5. Spray develop with surfactant in developer.
6. Graft monomers to top surface images.[39, 139, 196]

Although there are similarities in the chemical cross-linking of negative resists induced by either photons or electrons, because of the efficient electron backscatter, an underexposed electron negative resist (Fig. 10-6-14) retains the pattern in a stepwise response but an optical resist may wash off in the developer. Optical exposure (underexposure) is not as revealing and below 40% film loss, small features delaminate since the interfacial layers have not received a sufficient gelation dose. This is especially true for films thicker than 2 μm which contain optical sensitizers that do not bleach. Thus, underexposure of a negative photoresist may be more difficult to detect.

The development latitude for a negative resist is large (Fig. 10-1-4B) and the times on the order of 1–2 min. For most negative resists, residues from the interproximity effect (Fig. 10-1-10) remain between the lines. The residues can also be due to high prebake or ozone-induced cross-linking of coated resists such as in polyvinylcinnamate plates. If the residue between the lines is less than 1000 Å thick, plasma descum (ashing or etching) in an O_2 plasma[194, 195] has been routinely used (Chapter 14) to strip the residue with a slight increase in linewidth (0.1–0.2 μm). Since barrel ashers are not uniform for descum operations, an RF diode plate reactor is a more reproducible tool.

For improved resists, commercial bisazide–rubber resists require a new polymer base and/or developer combination to reduce swelling and enhance their submicrometer capability. To eliminate the swelling problems of conventional negative resists, other insolubilization or differential development mechanisms besides cross-linking such as ionic salt formation or molecular rearrangement, need to be further developed. For example, monomolecular layers which polymerize to a linear polymer may be useful if sufficient rigidity or water-based developers can be used.[197]

10-7. DRY DEVELOPMENT

Dry development is the process of solventless development by sublimation of the exposed resist brought about by direct irradiation or by a normal develop step using a reactive plasma gas to chemically differentiate an image. For example, the intense heat from a pulse of an excimer laser vaporizes polymer films.[330, 366, 367] An additive case of dry development (Chapter 16) is the direct deposition of resists from a gas-phase monomer.

The advent of dry development was fostered by the need to improve the performance of negative resists and resist processing in several ways:

1. As a nonpolluting process to replace the toxic organic solvents and the cost of disposal of developers for negative photoresists.
2. For improving the resolution of negative resists by overcoming the swelling.
3. An anisotropic development process using directed reactive plasma gas.
4. A promise of reduced pinholes and little loss due to adhesion failure.
5. A process for an all-dry vacuum technology in which the wafer never exits the vacuum.

Since plasma stripping and etching organic films were already in place in the 1970s, the tools were available for the dry development of resists. Secondly, several degradation and polymerization mechanisms, well established for liquid-based resists, could be applied to dry development:

1. Depolymerization of vinyl polymer to a monomer, especially if the exposure is sufficient and the temperature of the resist is above its ceiling temperature, T_c.
2. The differential plasma removal of exposure products such as lower-molecular-weight polymers.
3. The inhibited retention of exposed product and removal of the unexposed region.
4. The failure of oxygen plasma to strip silicones.

In the first case, direct image formation has not been too successful due to the high doses and restricted mobility of polymer fragments in a solid film. The second and third cases have been more successful, since monomer fragments are not required and lower doses can be used.

10-7-1. *Direct Image Formation*

In situ or direct volatilization of the image was first noted by Moreau and Schmidt during deep-UV (200 nm) proximity printing of PMMA.[199] In the reverse of polymerization,

$$\text{Polymer} \rightarrow \text{Monomer} \qquad \Delta F = + \qquad (10\text{-}7\text{-}1)$$

$$\text{Monomer} \rightarrow \text{Polymer} \qquad \Delta F = - \qquad (10\text{-}7\text{-}2)$$

$$\text{Monomer} \rightarrow \text{Polymer} \qquad \Delta F = 0 \qquad (10\text{-}7\text{-}3)$$

$$\Delta F = \Delta H - T_c \Delta S \qquad (10\text{-}7\text{-}4)$$

where ΔH is the heat of polymerization and ΔS is the entropy of polymerization. At the ceiling temperature, the free energy (ΔF) is zero, so

$$T_c = \Delta H/\Delta S \tag{10-7-5}$$

Above the ceiling temperature, the reversion to monomer is favored as long as the monomer product is physically removed from the latent image. Many polymers are stable, even above their ceiling temperature, because of the difficulty of initiating degrading centers on the polymer chain. Chain unzipping usually occurs through the rupture of foreign weak links at the polymer endgroups or occasionally in the main chain [more favorably in ultrahigh-molecular-weight ($M_w > 10^6$) polymers]. Since the activation energy for depolymerization, H, is quite high (10–26 kcal/mole), supplemental, preheated wafers speed up the direct volatilization processes. Some of the ceiling temperatures for resist polymers are listed in Table 10-7-1.[200] Thus, polyaldehydes and polysulfones are likely to depolymerize, if chain scission is initiated above the ceiling temperature. Unzipping from the random fragments or scissioned ends,

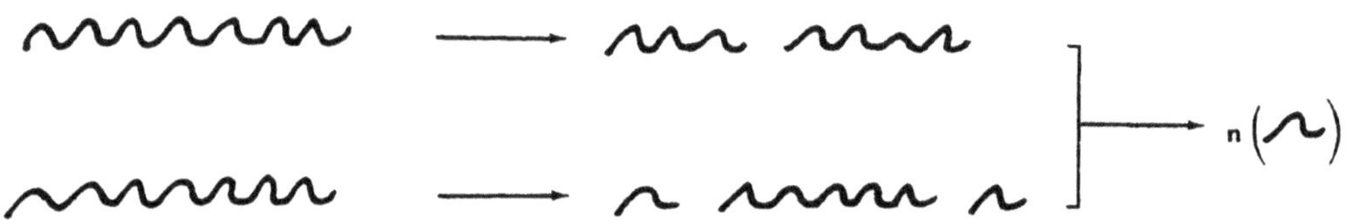

is accelerated by heat. A polymer of 100,000 molecular weight (M_n) contains 1000 monomer units which require 10^3 scissions for complete degradation to a volatile molecular weight of 100. Since high-energy radiation creates random scission, the longer the chain length, the greater is the number of potential fragments which can unzip into a monomer. The rate of film thinning of thickness z,

$$-dz/dt = 2kD_P z \tag{10-7-6}$$

is proportional to the degree of polymerization D_P (molecular weight) of the polymer.[201] Figure 10-7-1 shows the increased effectiveness of higher-molecular-weight polymethylpentene sulfone exposed at 100°C. In the polyolefin sulfone family of resists, the lowest boiling tertiary alkanes are the most sensitive.[260] At room temperature, doses of 10^{-4} C/cm^2 are required, but at elevated temperature, the G_s reaches 100 or more units and the unzipping reaction is an inherent chemical

TABLE 10-7-1. Ceiling Temperatures of Polymers: Resist Monomer

	T_c (°C)
2-Methyl-1-pentene-SO_2	−33
Cis-2-butene-SO_2	33
Acetaldehyde	−31
Methylmethacrylate	173
Isobutylene	50

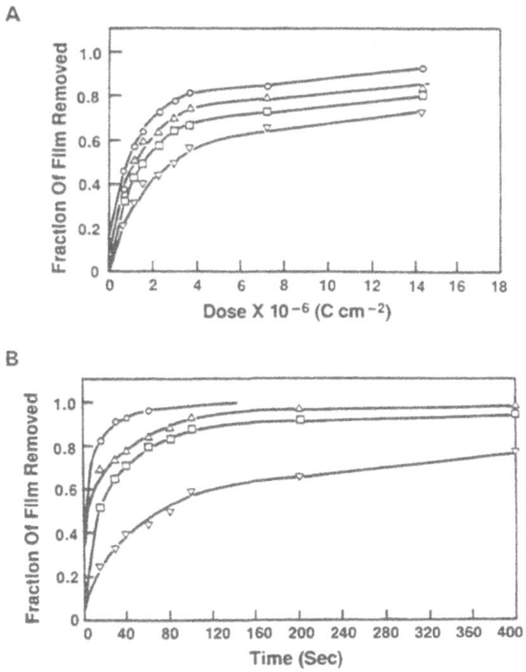

FIGURE 10-7-1. Fraction of film removed in polymethylpentene sulfone resist for direct *e*-beam irradiation (dry develop) of (A) differential initial molecular weights and (B) at different beam intensities for 500-nm-thick films. From Ref. 342, courtesy of SPE.

gain up to 10^4 monomer units. Unfortunately, the vapor development of olefin sulfones is not complete and residues remain in the image. The residues are thought to be due to the recombination of olefin fragments and cross-linking of the olefin units left after SO_2 expulsion.

In other depolymerizing resists such as polycarbonates (sensitized by onium salts),[332] polyaldehydes of alkyl[202, 316] or aryl aldehydes,[203] doses of only 1 $\mu C/cm^2$ are required to lose up to 60% of thickness. However, the contrast of the alkyl aldehydes is low (< 1) and plasma ashing is required to clean up the residues.[202] Since the polysulfones and polyaldehydes have low ceiling temperatures, they are also not very plasma resistant. They may be used for wet etching of chrome masks[260] or for etching chrome masks by reversal plasma etching.[208] The polysilanes with (–Si–Si–Si–) backbones are resistant to oxygen plasma and serve as useful bilayer masks (Chapter 12). Electron-beam exposure of heat-sensitive resists on quartz substrates for mask making, can also self-develop heating from heat buildup wherein temperatures in the images of over 100°C have been reached.[204] Even the postexpose bake of the latent image in polyolefin sulfones[89] or PMMA[307] can generate slight decomposition and partial depressions.

For deep-UV exposure, the complete depolymerization of polymers can be initiated at 200 nm or below using excimer lasers. PMMA,[205, 206, 273, 293] polysilanes,[306, 362] Mylar (polyethyleneterephthalate),[207] and nitrocellulose[265, 329] have been ablated with excimer exposure. Ablative recording may find useful applications for storing information in optical memories.[207]

In self-developing exposure, a major contrast or resolution benefit is the limited range of the isotropic scattering of the beam. In *e*-beam exposure with self-developing, the backscatter radiation is high and the contrast is low. However, for optical exposure at 193 nm, the extinction coefficient (ε) of polymers is $> 10^4$ and

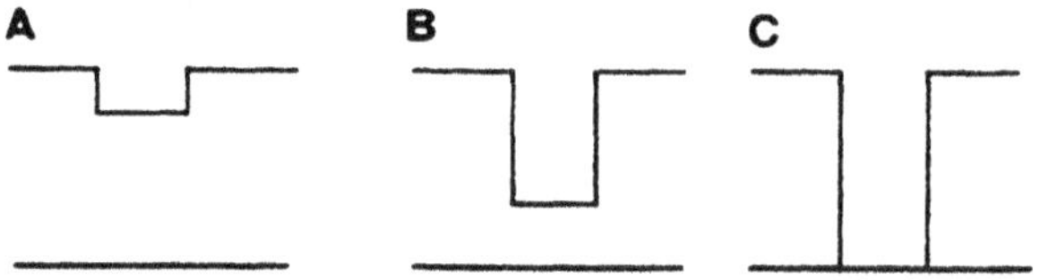

FIGURE 10-7-2. Limited scatter in high-absorbing volatizing (self-developing) positive resist. (A) Low dose; (B) partial dose; (C) fully exposed.

photon penetration is only 0.1–0.2 μm. As the resist anisotropically volatizes (Fig. 10-7-2), the forward path length decreases and the isotropic scatter is confined to a short lateral distance (< 500 Å).

Besides vapor development, several attempts have been made for a total dry process derived from PMMA or PMMA containing tin. The process has been coined vacuum lithography.(210) The PMMA is plasma deposited, exposed, and self-plasma developed. However, the plasma-deposited PMMA contains cross-links and very high doses are required for self-develop.(209) In the Sn-PMMA resist system, doses of about 100 μC/cm^2 are required to develop an image with an oxygen plasma.(240)

Self-developing negative resists are based on the polymerization of *in situ* vinyl or silicone monomers using *e*-beam or X-ray exposure. The surface-activated polymerization requires high doses to achieve solid films but no development is required, since the monomer condenses and is pumped away from the unexposed region.(211) For a negative plasma-developable resist, Lewis acids formed by exposure of onium salts can cationically induce top surface image grafting of silicone monomers (see Fig. 4-6-4).(381) Likewise, the same Lewis acids can be used in a self-developing positive resist by inducing acid-catalyzed main chain scission of polycarbonates into monomer fragments (see Section 2-8).(387)

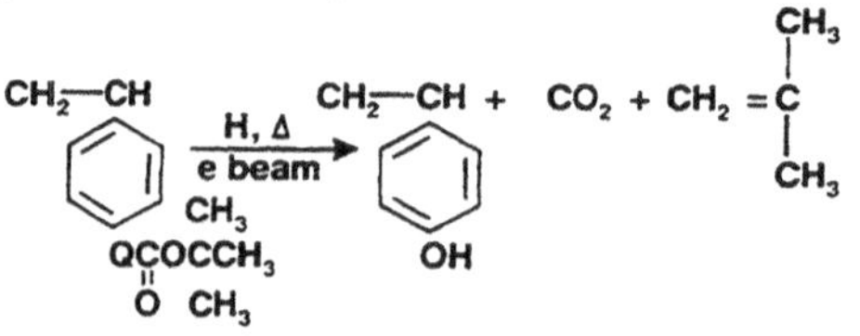

These new findings support the principle that reactions which form polymers:

1. Vinyl polymerization
2. Condensation reactions
3. Cationic polymerization
4. Anionic polymerization
5. Charge transfer

can be reversed to degrade or rearrange polymers into positive working and dry developable resists.

10-7-2. *Plasma-Developed Resists*

10-7-2-1. *Monomer Matrices for Plasma-Developable Resists.* In oxygen plasma, aromatic resists strip more slowly than aliphatic resists. Since the empirical formula of aromatic resins is on the order of C_8–C_{15}, more moles of oxygen are required than for C_3–C_8 aliphatic polymers. Secondly, organometallic polymers do not form volatile oxides and therefore do not etch in oxygen plasma. These two fundamental facts form the basic chemical factors in the formulation of plasma-developable resists by incorporating volatile silicone(299) or aromatic monomers

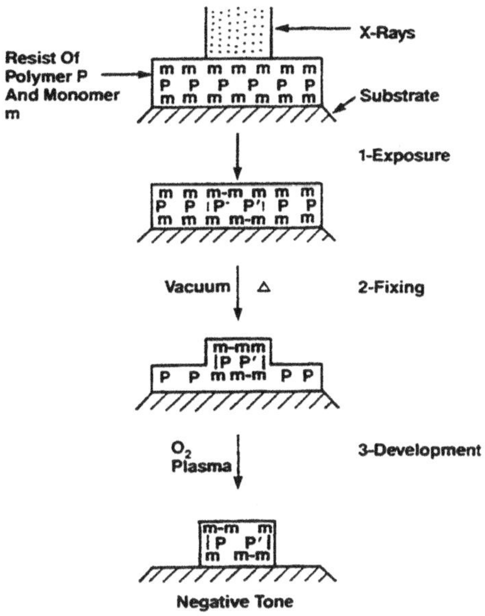

FIGURE 10-7-3. Schematic of negative dry develop resist based on a silicone or aryl monomer polymerizing in a resin matrix followed by heat fixing and plasma developing. From Ref. 215, reprinted with permission of *Solid State Technology*, published by Technical Publishing, a company of Dun and Bradstreet.

into a resist formulation (Fig. 10-7-3). The monomers are grafted or copolymerized upon exposure and the monomers are removed from the unexposed region by heating. A positive working system is based on the incorporation of metal into the unexposed resist such as the treatment of the KTFR azide with $SnCl_4$ followed by O_2 plasma developing.[271]

The incorporation of monomers into a polymer matrix for the purpose of polymerization has been extensively utilized in RISTON-type photopolymers for dry film resist and packaging applications. For microelectronic resists of the negative type, the plasma inhibitor is usually an aromatic or silane vinyl or allyl monomer (Table 10-7-2). Some of the initial formulations were based on early work by Horizons on iodoform sensitizers in dry-developable (heat) non-silver halide recording film.[213, 221] In principle, any resist can be converted into a dry plasma-developable system by the appropriate addition of a reactive monomer. Both classic

TABLE 10-7-2. Dry Development Compositions[a]

Exposure	Plasma inhibitor	Host polymer	Reference
UV	Vinyl carbazole	Vinylbutyral	212
UV	Vinyl carbazole	Vinylbutyral	213
X-ray	Vinyl carbazole	Chloroacrylate	214, 215, 217
X-ray	Vinyl silane	Chloroacrylate	216
e-beam	Diphenylacetylene	Trichloroacrylate	218
UV	Aryl azide	PMIPK[b]	219, 284
UV	Aryl azide	PMIPK	220
UV	Phenyl silane	PMMA, Cl-styrene	272, 275–278, 325
UV	Iodoazide	Polyvinylphenol	285
UV	Silazane	DQN	333

[a] O_2 plasma developed.
[b] PMIPK, polymethylisopropenylketone.

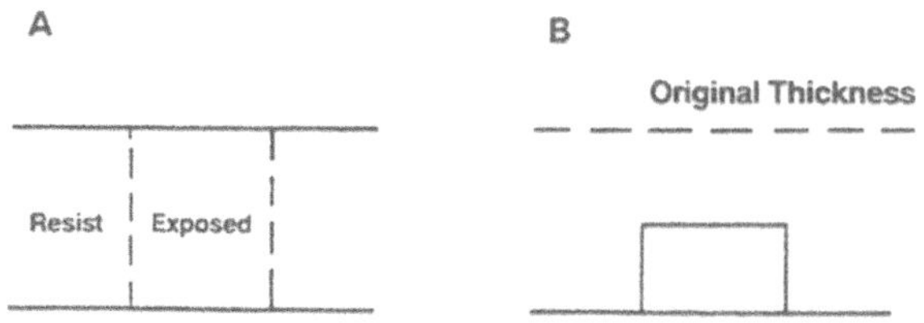

FIGURE 10-7-4. Plasma-developed resist (A) before developing and (B) after developing of negative system with loss of exposed original thickness.

diazoquinone (DQN)[331] and KTFR azide negative photoresists form such systems by the addition of volatile silicone compounds.[279, 280] In the actual negative monomer formulations, the inhibiting monomer[339] is removed by a predevelop bake,[365, 375] which results in background film thinning (Fig. 10-7-4).[310] A second film thinning, up to 59% of the exposed image, occurs, during the oxygen plasma development step. Precooling the O_2 plasma to $-20°C$ is said to prevent resist film loss.[335] Because of monomer outgassing inside the vacuum of an *e*-beam exposure unit, monomers such as vinylcarbazole cannot be used unless a polymer topcoat (polyvinylpyrollidone) is applied on top of the resist and removed before the predevelop baking and ashing.[218]

Because of excessive thinning and image spread, the vinyl-based resist is not suitable as a single-film resist and is confined to a multilayer resist as image and O_2 transfer layers.

Aromatic-based dry-developable negative resists are a concentration modification of KTFR azide photoresists, in which (10–30 ×) more azide sensitizers are loaded into PMMA or PMIPK resists.[214, 220, 281, 291, 302, 319, 363] Tsuda and coworkers[291, 302] propose that the azides form aromatic amines which quench the oxygen plasma singlet state. Although amines quench the excited state, azides have also been known to quench singlet oxygen.[210] Another factor to consider is the formation of an azo polymer from the self-polymerization of the bisnitrene intermediate,[231]

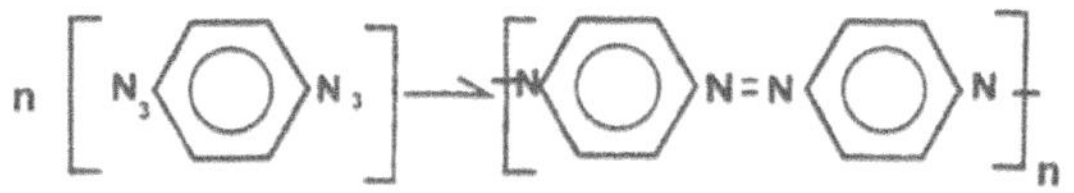

which would be very stable to oxygen plasma. Since oxygen etching is sensitive to cross-links[232, 233] and aromatic polymers, the background is removed at about a threefold faster etch rate. The addition of azides to polybutene sulfone positive resists is said to form an oxygen plasma negative resist.[234, 311] If polymethylpentene is loaded in a novolak resist, a dry-developable system is formed.[369] For a positive working system, a polyvinyl phenol was silylated and exposed in the presence of onium salts to remove the silyl ester, followed by dry O_2 etching of the exposed region,[380]

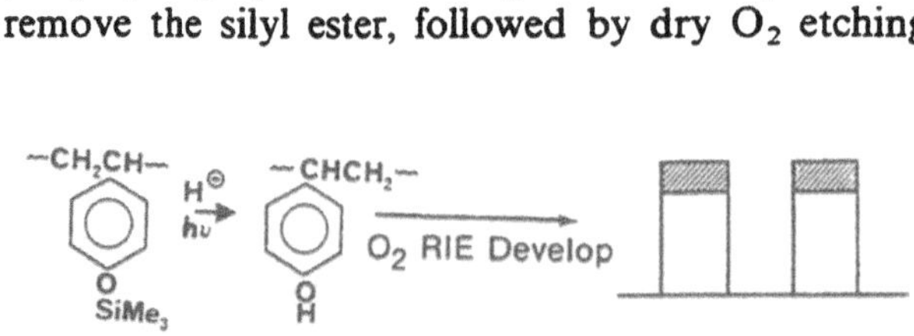

For the highest resolution resist capable of submicrometer resolution, top surface imaging has been used (Fig. 10-7-5).[224–226] The metal inhibitor is incorporated by

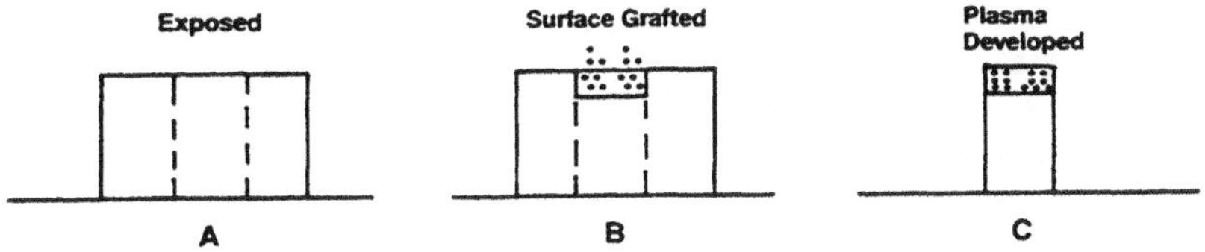

FIGURE 10-7-5. Surface grafting steps. (A) Resist exposed to generate graft sites at surface; (B) surface grafting with postreaction; (C) plasma developed in oxygen or other gas to erode ungrafted region.

direct implantation of a polymer with Si^+ or In^+[236, 237] or the exposure forms a product such as polyvinyl phenol,[372, 373]

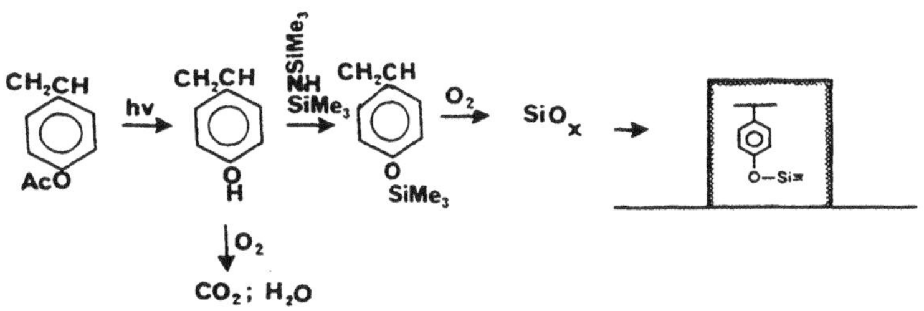

which can be silylated with hexamethyldisilazane and subsequently oxygen plasma developed. In other systems, KTFR is grafted with B_2H_6[271] or $TiCl_4$,[374] PMMA with vinyl silane,[242] and polyvinylacetate with vinylferrocene.[292] Although a top surface process confines the image to a thin layer, some isotropic image spread has been noted with vinyl postgrafting (runaway polymerization).[360] In this case, simple esterification reactions, as the above polyvinylphenol example, would confine the image differential to only the exposure confines.

10-7-2-2. *Polymer Matrices for Plasma Development.* The direct volatilization of PMMA requires an enormous dose of 10 J/cm^2 or about 10^3 or about 10^4 more dose than the liquid-developable system.[300, 301] A plasma-developable PMMA also relies upon a faster erosion of the exposed (lower molecular weight) region.[3, 238]

$$R/R_0 = (1 + kQM_n)^b \qquad (10\text{-}7\text{-}7)$$

Even the chain length and end groups have an effect on the rate of etching.[239] For plasma developing of an *e*-beam-exposed resist, a dose of 50 μC/cm^2 is required, and for self-developing in an *e*-beam, a dose of 400 μC/cm^2 is needed.

If 4% Sn is incorporated by copolymerization, a lower (25 μC/cm^2) dose can be used.[240] Polymethacrylonitrile plasma develops in O_2 at a 3 to 1 erosion rate after a dose of only 10 μC/cm^2.[241] The addition of diazoquinones to polymethacrylonitrile permits UV images to be dry developed after doses of 250 mJ/cm^2,[241] and conventional AZ-1350 can be oxygen dry developed after a dose of 500 mJ/cm^2.[243] In AZ-1350, image reversal can be accomplished by soaking the exposed resist in Mg acetate. The Mg ion complexes with the indene acid photoproduct and inhibits the exposed region from a dry-developing[244] resist. Soaking of polymethylacrylic acid or PMMA in metal salts also preferentially incorporates metals.[305] Some inorganic resists also form differential plasma-developable images.[348, 349, 364] Se-Ge resists doped with Ag can be developed in

CF_4 due to the formation of volatile GeF_4 and SeF_4, while the AgF_4 formed in the background inhibits plasma removal.[251] Finally, some silicone resists have been claimed to be dry developable in CF_4[245] and if imaged on top of a planarizing layer, a bilayer mask can be formed by switching to oxygen plasma. A variation of this process is to partially develop a resist image, overcoat with a silicone polymer, and reverse the image with consecutive CF_4 and O_2 plasma etching,[308, 382]

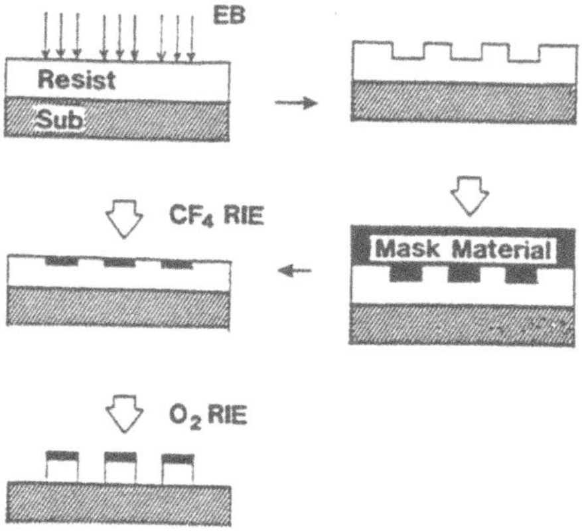

10-7-3. *Plasma Etching without Development*

The direct etching of an area beneath a latent resist image is an interesting application of using permeability to develop an image. Negative resists such as KTFR have been shown to be more permeable to HF vapors in the exposed region (image reversal).[246, 247] Polyvinylcinnamate photoresist was also shown to exhibit the same phenomenon.[248] For plasma etching with an exposed resist mask in place, an extension of this concept was applied to mask Cr etching with CCl_4[249] or in SiO_2 etching by CF_4.[250]

10-7-4. *Outlook for Dry Development*

The dry development of resists is an intriguing process which should make a significant cost impact in processing printed circuits, to replace the huge quantities of organic developers. For microelectronics, improved resolution of negative resists can be obtained by overcoming the traditional swelling problems of cross-linked negative resists, but at present the compositions suffer from excessive unexposed resist loss upon development. For superior resolution, more top surface metal-impregnated resists should provide higher differential plasma erosion rates.

The direct exposure development of a resist is a process of high resolution if the penetration of the beam is confined to top surface layers. Dry development of all forms has a good potential for low defect processing, since all of the steps including the resist deposition can be performed in a vacuum or in a plasma reactor. Because of the increasing investment of plasma tools, research in dry developing should be vigorously pursued, toward the goals of a total lithography (gas) processing inside class 0 clean conditions.

10-8. PRACTICAL DEVELOPMENT OF RESISTS

The major objective of process control in the development step is to minimize the exposure bias. Because of variations in exposure and resist thickness over steps, overdevelopment predominates with positive resists. A "process window" incorporates the mask width distortion, the exposure fluctuations, the resist thickness, and the development variations in time, temperature, and concentration. The process windows are narrow for positive resists, while cross-linking negative resists offer wide latitude but poor resolution. While a bias of 0.5 μm can be tolerated in a 3.0-μm resist image, it is not acceptable for a 1.0-μm image.

Depending on the resist contrast (Fig. 10-8-1), the edge slope of the beam, and the resist thickness, various profiles can be obtained for respective processing (Table 10-1-1). Although straight walls can be obtained with a low- ($\tau < 2$) and a high-contrast resist, if, in both cases, the dose is precisely controlled ($\pm 5\%$). The low-contrast resist also requires precise development monitors and controls ($\pm 3\%$).[352, 376] If not controlled in a low-contrast resist, the top of the image can exhibit lateral isotropic image distortion (Fig. 10-8-2). Overdevelopment, in addition to increasing the bias of a line or space, can lead to other random defects such as pinholes or unacceptable plasma-etched profiles. Because of the intense backscattering of electrons, linewidths (Fig. 10-8-3) can change more rapidly than in optically exposed and developed positive resists.

The ability of a lens to transfer an aerial image of a mask is controlled by the MTF of the lens (Fig. 10-8-4). The MTF of the lens optics should be at least 0.6 and exceed the CMTF of the resist,

$$\mathrm{MTF}_{\mathrm{lens}} > \mathrm{CMTF}_{\mathrm{resist}}$$

High-contrast resists are required for lowering the CMTF of the resist (Fig. 10-8-5) and for reducing the interproximity effect. Due to backscattered electrons or the falloff of the MTF of an optical projector, the spaces of a positive resist, narrower

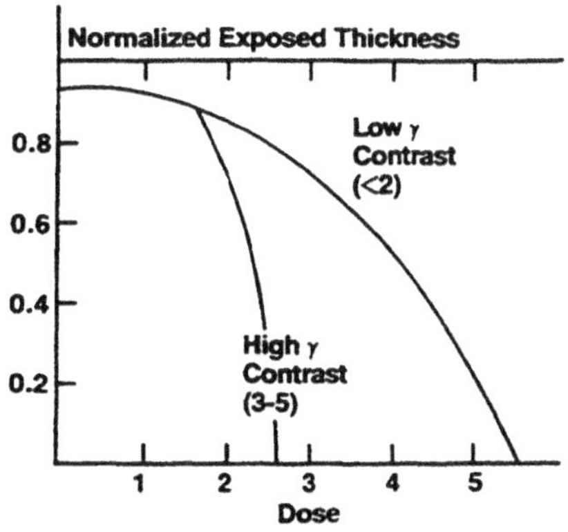

FIGURE 10-8-1. Contrast (γ) curves for positive resist determined by the tangent of the thickness loss with dose intersection.

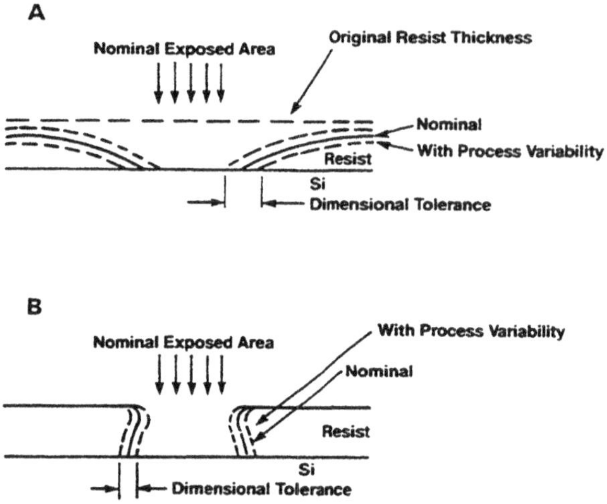

FIGURE 10-8-2. (A) Low-contrast resist process window and large image bias. (B) High-contrast resist. From Ref. 383.

than 2 μm in width, are exposed and thinned (Fig. 10-8-6). Measurement of R and R_0 using flood exposure overlooks the interproximity effect. The true R_0^* of the resist is longer (50%) than measured by the flood endpoint. An SEM examination of a 1 μm land exposed and developed in a DQN resist reveals loss (Fig. 10-8-7) of extra resist. Insufficient resist remains to protect the features in a subsequent reactive ion etch process.

Since the profiles and linewidths are a strong function of the exposure and dose conditions, there are several means of achieving proper geometries:

1. Choose a resist and developer of highest contrast.
2. Use minimum resist thickness (multilayer if necessary).
3. Add dye to resist or change conditions to minimize reflection.
4. Use highest NA lens or steep edge profile beams.
5. Apply proximity corrections and mask windages.[322, 323]
6. Choose correct exposure wavelength or energy (keV).
7. Vary dose at constant develop time.
8. Vary develop time at fixed dose (Fig. 10-8-8).
9. Find and determine maximum R/R_0 conditions.

At high doses (Fig. 10-8-8), the line size is essentially dominated by the absorbed energy and independent of the short develop time, but at low doses

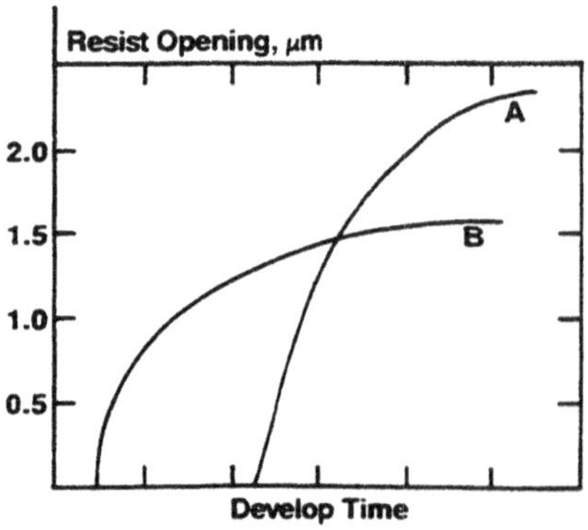

FIGURE 10-8-3. Positive resist opening as a function of develop time for a 1-μm line formed by (A) optical projection and (B) 20 keV electron beam on 1.0-μm-thick positive DQN resist. From Ref. 123.

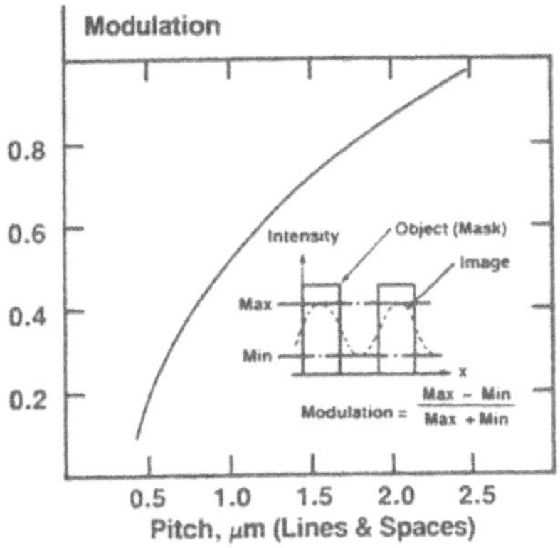

FIGURE 10-8-4. MTF for a 436-nm 10× stepper lens of 0.28 NA versus pitch resolution. From Ref. 343, courtesy of AIP.

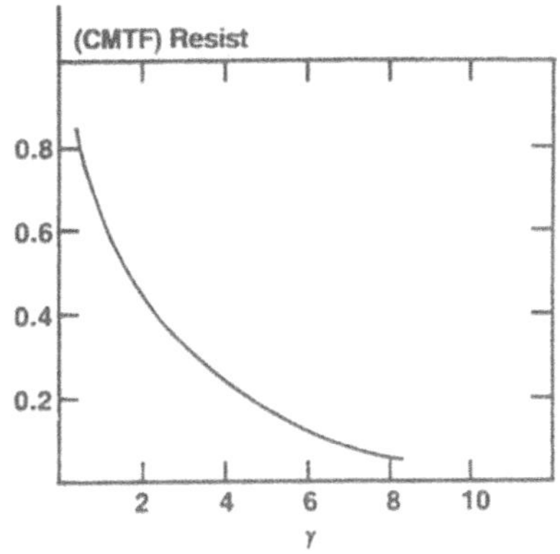

FIGURE 10-8-5. MTF of resist as a function of resist contrast. From Ref. 343, courtesy of AIP.

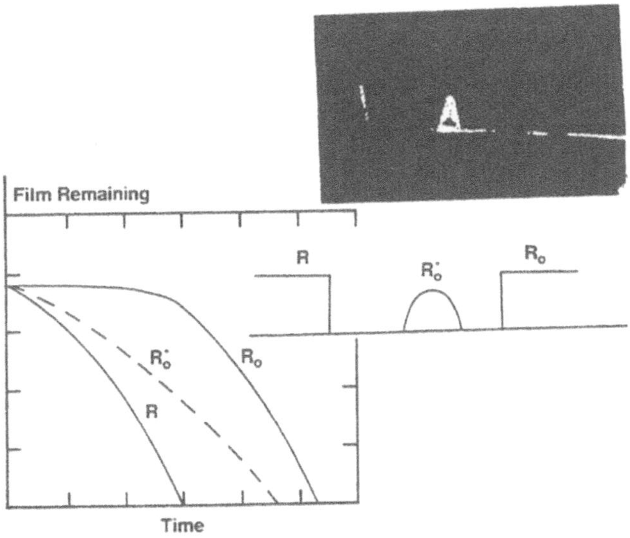

FIGURE 10-8-6. Typical film thinning curves for a positive resist showing R (exposed dissolution rate), R_0 (unexposed dissolution of large land), and R_0^* (dissolution rate of small 1-μm line in positive resist).

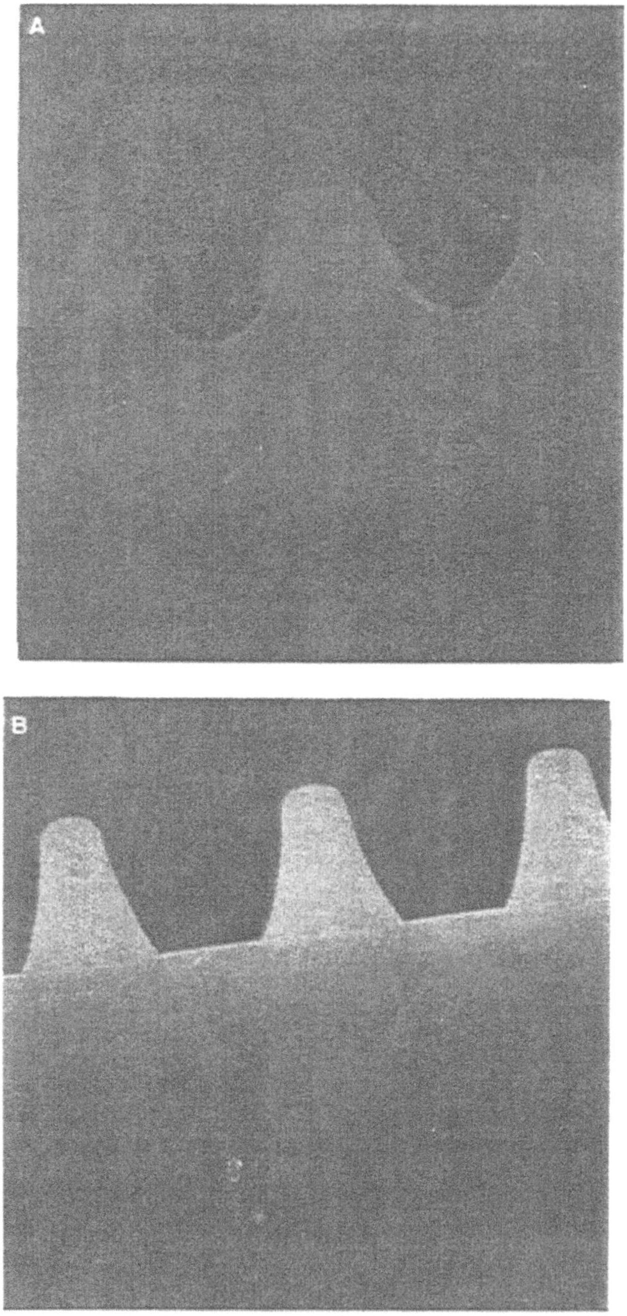

FIGURE 10-8-7. SEM of (A) partially developed 1-μm land in scanner optical exposure, (B) developed feature showing thinning of land. Courtesy of D. Merritt, IBM.

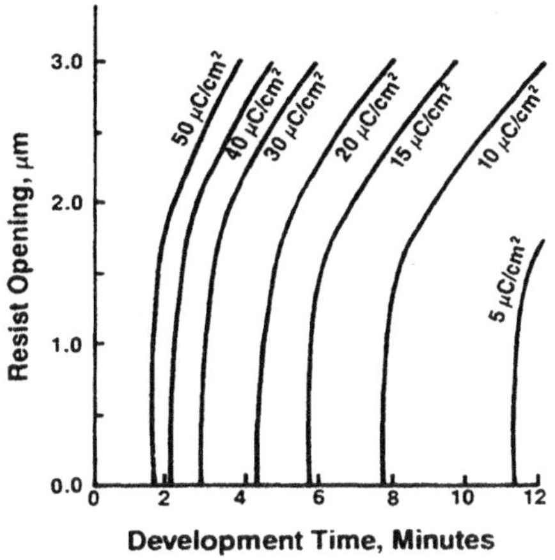

FIGURE 10-8-8. Line width versus develop time for a DQN resist exposed with different electron beam doses at 25 keV. From Ref. 118.

($< 20\ \mu C/cm^2$), overdevelopment leads to large biases as the developer dominates the absorbed energy profile.

For most exposures, we choose a minimum dose (E) that provides the minimum linewidth bias but maintains sufficient throughput. Experimentally, the exposure dose can be determined by a resist that obeys the reciprocity relation,

$$E(\text{dose}) = I(\text{intensity})t(\text{time}) \tag{10-8-1}$$

in which a resist responds equivalently in linewidth to a high-intensity (short exposure) or a low-intensity beam with a long exposure time. All of the resists with unimolecular decomposition or rearrangement (PMMA, DQN) exhibit no reciprocity failure. Reciprocity failure has only been noted in bimolecular reaction systems at low intensity such as KTFR (azide rubber) subject to the oxygen effect[170] and polymethylpentene sulfone–novolak *e*-beam resist.

For optical sensitometry, it is convenient to vary the dose with an intensity step wedge (Fig. 10-8-9). A step wedge is a series of variable transmission stripes of thin metal or an opaque resist. The wedge is placed over the photomask and with a dose of 20 sec at 40% transmission, the actual exposure time is 8 sec. In addition,

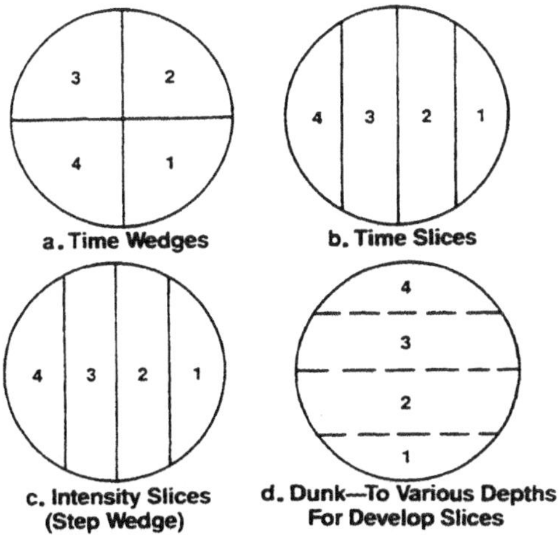

FIGURE 10-8-9. Various ways of using a single wafer to maintain prebake and expose constant to determine exposure (reciprocity) and develop time for images. (a) Vary exposure time in wedges and develop to same time. (b) Expose in slices to test exposure across wafer and develop at same time. (c) Vary intensity using step wedge instead of exposure and develop for constant time. (d) Expose at one dose and develop for various times by immersion to different depth.

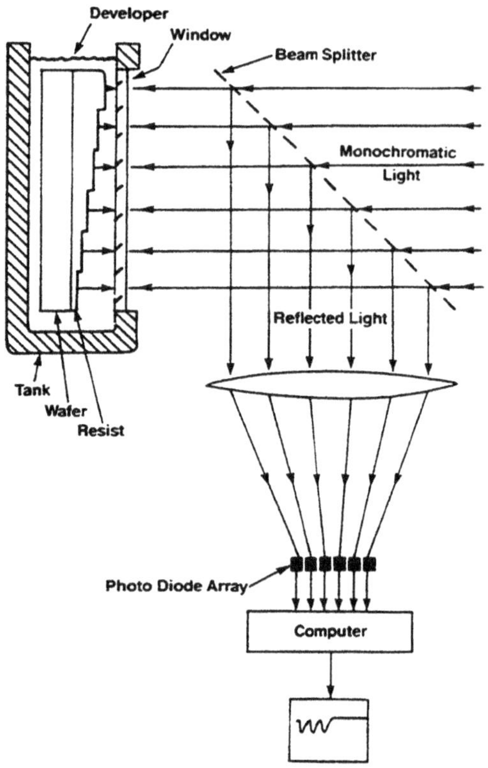

FIGURE 10-8-10. Multiple beam detector which simultaneously monitors the time to develop multiple levels of exposure in one area. From Ref. 266, copyright Perkin–Elmer Co. and courtesy of Dr. A. McCullough.

one can vary the dose by time slice exposure at fixed intensity. By immersing the wafer to different depths or times, the development time can also be determined on a single wafer which eliminates the prebake and exposure variables. A more sophisticated commercial system for scanning exposure tools, exposes a wafer with a step wedge, and simultaneously monitors the development time for each step (Fig. 10-8-10).[267, 386]

For a laboratory of process line monitor,[295, 303, 361] a setup of an inexpensive endpoint apparatus (Fig. 10-8-11) consists of a laser beam interferometer, which if beam split (Fig. 10-8-11C) can simultaneously monitor a large pad for R_0, a large space for R, and a diffraction grating[274] for the cleanout (proximity effect) of submicrometer features (R^*).[253–255] Because of the inter- and intraproximity effects, different line sizes clean out at different times beyond endpoint (EP); for example,

1-μm line	EP + 50%
2-μm line	EP + 25%
3-μm line	EP + 15%
1-μm line and space	EP + 15%

The endpoint trace (Fig. 10-8-13) can also be used to compute the initial film thickness from the refractive index (Table 10-8-1) of the resist (preferably on a bare silicon surface) and the instantaneous rate of development to test for any induction effects. If swelling of the resist occurs or if the surface roughens during development,

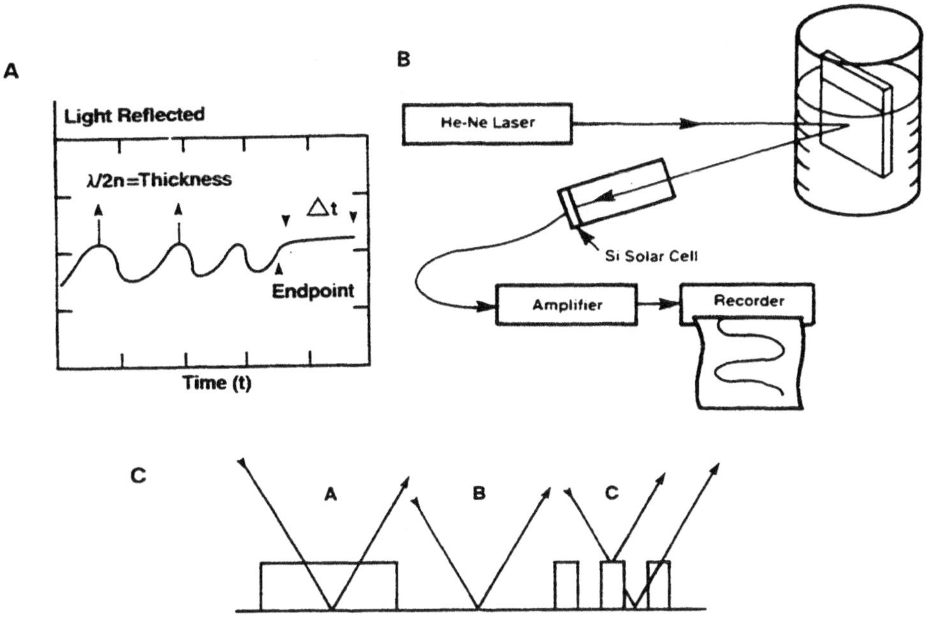

FIGURE 10-8-11. (A) Resist development endpoint film thinning signal. From Ref. 93, copyright 1980 and courtesy of IEEE. (B) Typical apparatus. From Ref. 229, copyright and courtesy of *RCA Review*. (C) Three-beam endpoint detect to determine endpoint of develop. Beam A—monitors large unexposed pad. Beam B—monitors large exposed pad. Beam C—1-μm diffraction grating. From Ref. 253, copyright 1978 and courtesy of IEEE.

the amplitude of the signal (Fig. 10-8-13) will collapse and the endpoint will be obscure. Poor signals and obscure endpoints can be improved by electronic digital filters.[321] If the signal is still distorted, an alternative is to use a microbalance based on an oscillating piezoelectric crystal.[359] Spray development can also be performed if the signal is electrically modulated to overcome the noise from the turbulent developer.[321]

In the future, the concept of single wafer monitored development will be important for submicrometer linewidth control in large (250-mm-diameter) wafers. In Fig. 10-8-14, an example of batch processing development versus single wafer developing is shown. The automatic process-controlled wafers exposed at high or lower doses, consistently developed to the desired pitch. Any changes in the develop time for a resist can be detected by the process monitors. The fluctuations in R could be due to such factors as:

- Dose (nonuniform intensity, time fluctuations, focus errors, or reflectivity from substrate)
- Concentration of developer
- Temperature of developer

Likewise, changes in the unexposed dissolution (R_0) could be due to:

- Prebake variations
- Nonuniform thickness
- Humidity of environment or during wafer storage
- Concentration of developer

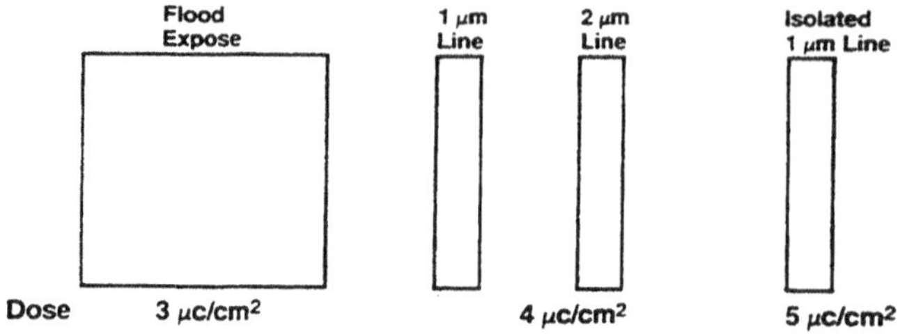

FIGURE 10-8-12. Electron-beam test pattern for determining correct dose and development time combination by taking advantage of intraproximity effect. An isolated line is exposed at higher dose and monitored at a lower flood exposure dose. The time to endpoint of a 1-cm^2 area exposed at 3 $\mu C/cm^2$ is the same as the time to clean out a 1-μm line exposed at 5 $\mu C/cm^2$.

Positive resists require tight process controls. Although cross-linked negative resists are tolerant of overdevelopment, because of image swelling and the overexposure required to achieve interfacial insolubilization, their resolution (Fig. 10-8-15) is limited to about twice their thickness, while positive resists can resolve their thickness.

On-line process controls (Table 10-8-2) to achieve a $\pm 0.2\ \mu m$ image bias in a positive resist, rely upon the prebake and exposure steps to achieve tight process specifications, reinforced by endpoint monitors on a send-ahead or production wafer. For a typical diazoquinone (DQN) positive resist with a contrast of 2.5, Table 10-8-3 lists the contribution of the exposure and development steps to the width of a trough, with a $\pm 10\%$ process variation. In DQN resists, the concentration of the alkaline developer is a dominant and sensitive variable. Figure 10-8-16 shows the variation in printed line size with exposure time for a DQN resist exposed at 405 nm on a 10× stepper. Because of the falloff of the lens MTF, the smallest features require a longer develop time. Consequently, the larger features are overdeveloped (Fig. 10-8-17). If a weaker developer (Fig. 10-8-17C) of 0.5 N is used, the large linewidths are less susceptible to enlargement. Unfortunately, this also requires a lower dose (longer exposure time). A compromise is the 0.20 N developer. However, at any concentration (Fig. 10-8-18), small changes in

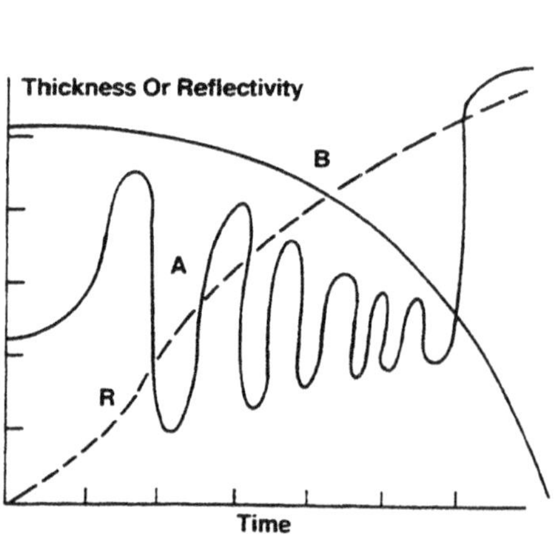

FIGURE 10-8-13. Endpoint development data for a positive resist. Raw data of (A) reflectivity versus time, (B) transformed into thickness versus time, and (R) instantaneous rate of development for an *e*-beam-exposed DQN resist. Courtesy of N. Eib, IBM.

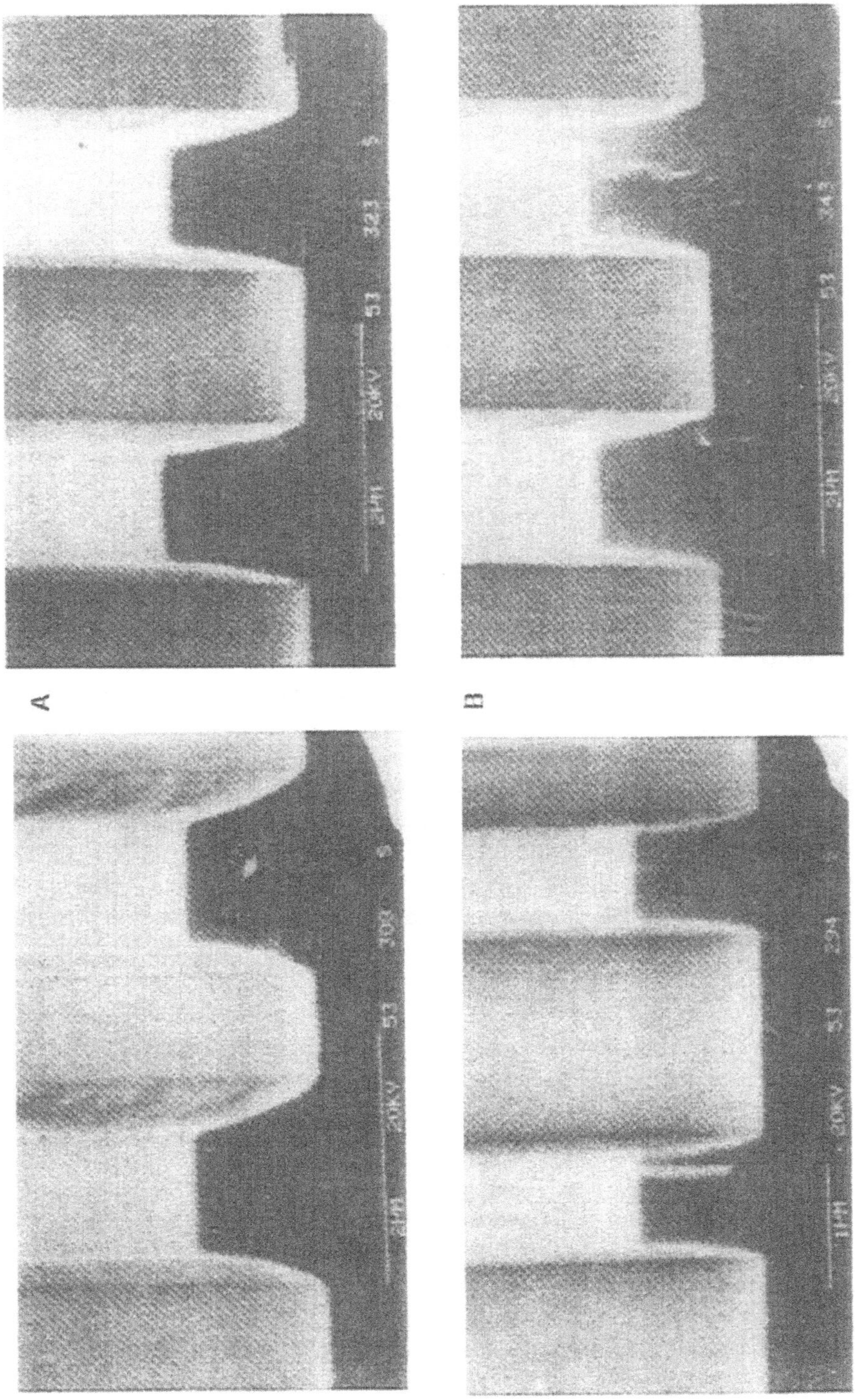

FIGURE 10-8-14. SEM showing the unchanged wall angle of resist of constant linewidth for 1.25-μm features under APC (automatic process control) contrasted with equivalent resist profiles obtained without APC control. (A) Low exposure dose; (B) high exposure dose. From Ref. 321, reprinted with permission of *Solid State Technology*, published by Technical Publishing, a company of Dun and Bradstreet.

TABLE 10-8-1. Refractive Index of Resists at 6238 Å

Resist	Refractive index
Kodak 820 (unexp.)	1.620
Kodak 820 (exp.)	1.608
AZ-1350 (unexp.)	1.621
PMMA (unexp.)	1.582

developer normality (± 0.001) can alter R by 10% and thus, endpoint monitors are required to guard against abrupt changes, or the aging (Fig. 10-8-19), of the developer. Upon continued exposure to air, the alkaline developer strength is degraded by CO_2 pickup and OH^- depletion,[258]

$$CO_2 + OH^- \rightarrow CO_3^{2-} + H_2O$$

A KOH solution left in air shows a depletion of 0.03N in 4 hr and the development rate decreases from 0.4 μm/min to 0.3 μm/min. Over the same period, tetramethyl ammonium hydroxide developer only loses 0.004 N.[258] The addition of sodium sulfite is claimed to stabilize an alkaline developer for up to 1 day in air.[154, 296] For safe processing, developer tanks should be purged with N_2, covered, and otherwise kept free of CO_2.

In a production environment, troubleshooting the development step is primarily attended to situations in which the resist does not develop out cleanly, or nonuniformly (Table 10-8-2). Residues can usually be traced to exposure conditions if the developer has been consistent. Redeposited resist (Figure 10-8-20) can be a problem in static development. Spray development can help prevent the redeposition of alkaline insoluble products, especially if dyes or other materials are added to the DQN resist. However, spray development can introduce problems of

TABLE 10-8-2. Troubleshooting Residues in Developed Images

Defect	Cause	Remedy
Does not develop at all	Dose wrong	Monitor intensity
	Prebake high	Monitor wafer
	Wrong developer	pH monitor
	Developer weak	Titrate, analyze
Does not develop cleanly	Underdevelop	Monitor rate
	Nonuniform dose	Profile intensity
	Variable agitation	Redesign tank
	Loading effects	Titrate
	Thicker film	Monitor
	Humidity	$35 \pm 5\%$
	Reflective substrate	Multilayer resist
	Prebake loading	Hot plate base

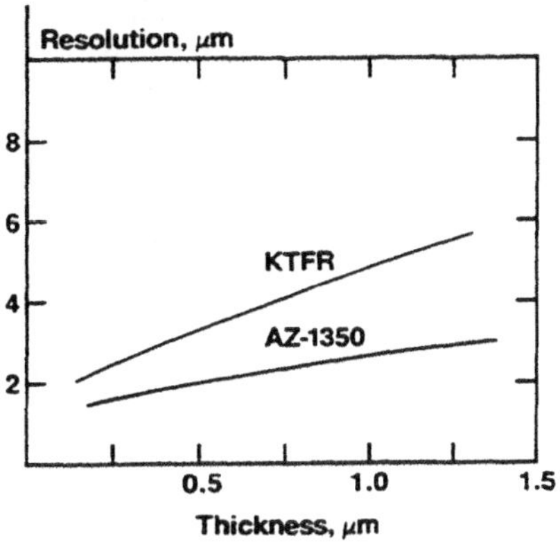

FIGURE 10-8-15. Resolution as a function of film thickness for negative KTFR and positive AZ-1350 photoresists. From Ref. 344, courtesy of SPSE.

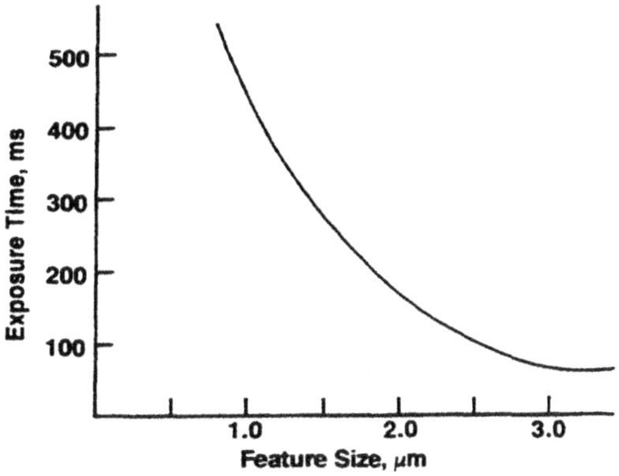

FIGURE 10-8-16. Exposure time required on a 405-nm optical stepper to print AZ-1350 features. From Ref. 257, courtesy of SPE.

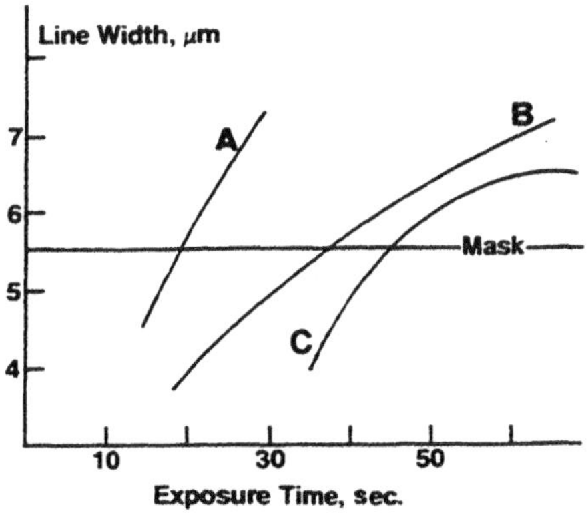

FIGURE 10-8-17. Exposure versus linewidth for (A) 0.25 N, (B) 0.20 N, and (C) 0.15 N AZ-1350 resist. From Ref. 256, reprinted with permission of *Solid State Technology*, published by Technical Publishing, a company of Dun and Bradstreet.

TABLE 10-8-3. Process Variable Contributions to a Positive Resist (1.0 μm Thick)

Variable	Image bias (μm)
Exposure time or intensity	0.2
Develop time	0.1
Developer normality	0.3
Resist thickness	0.05
Developer temperature	0.2
Humidity	0.1

poor temperature control and salt deposition from metallic alkaline developers. A variety of development methods exist:

- Spin
- Spin and spray
- Puddle and spin
- N_2 bubble
- Cascade
- Ultrasonic[279, 290, 298]
- Vapor jet spray[282, 289]

Agitation can double the rate of development (Fig. 10-8-21). The kinetics of development are governed by the surface diffusion reactions and thus are sensitive

TABLE 10-8-4. Solubility Parameter for Solvents[a]

	Components		
Solvent	Nonpolar	Polar	H-bond
Methanol	7.4	6.0	10.9
Isoropanol	7.7	3.0	8.0
Glycerol	8.5	5.9	14.3
Propylene glycol	8.2	4.6	11.4
Diacetone alcohol	7.7	4.0	5.3
Acetic acid	7.1	3.9	4.6
Dimethyl formamide	8.5	6.7	5.5
N-Methyl pyrollidone	8.8	6.0	3.5
2-Ethoxyethanol	7.9	4.5	8.0
Butyl acetate	7.7	1.8	3.1
Isoamyl acetate	7.5	1.5	3.4
MIBK	7.5	3.0	2.0
Chlorobenzene	9.3	2.1	1.0
MEK	7.8	4.4	2.5
Benzene	9.0	0.5	1.0
Toluene	8.8	0.7	1.0
Cyclohexane	8.2	0.0	0.0
Hexane	7.2	0.0	0.0

[a] From Ref. 262.

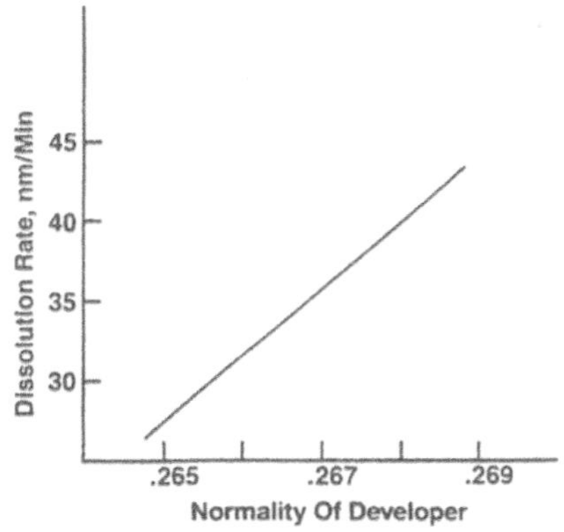

FIGURE 10-8-18. Effect of normality (alkalinity) of developer on DQN dissolution rate of exposed resist.

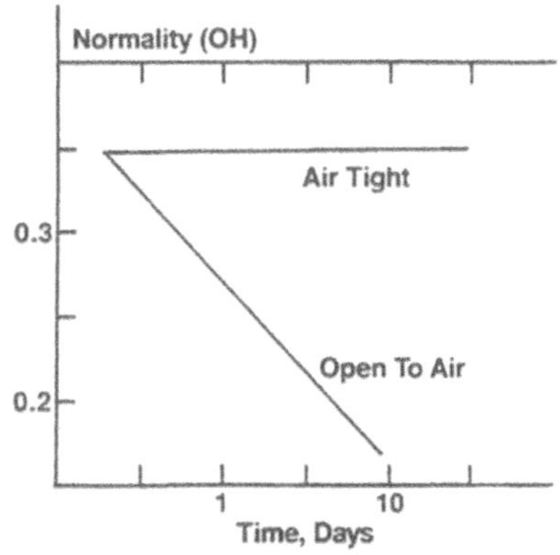

FIGURE 10-8-19. DQN alkaline developer shelf life. From Ref. 261, reprinted with permission of *Solid State Technology*, published by Technical Publishing, a company of Dun and Bradstreet.

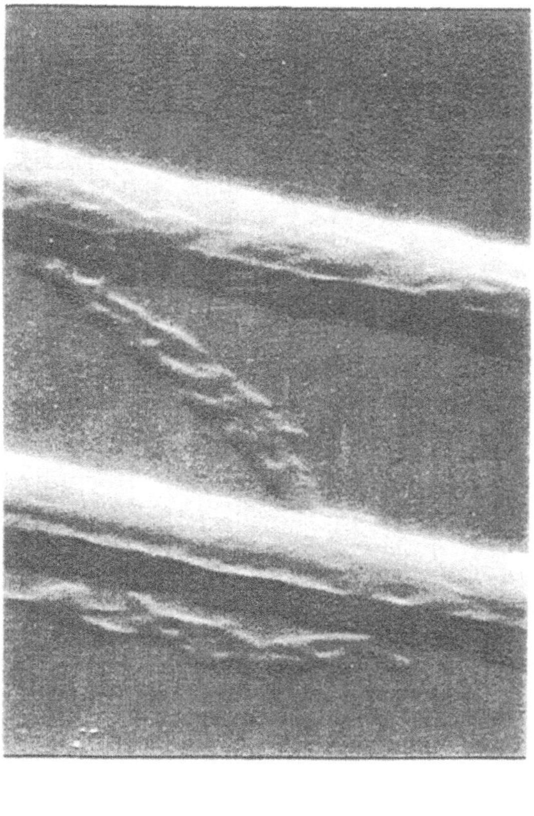

FIGURE 10-8-20. Redeposited resist on image after development in a static development tank. If adhesion failure of image occurs on a wafer, redeposits can be found on subsequent wafer.

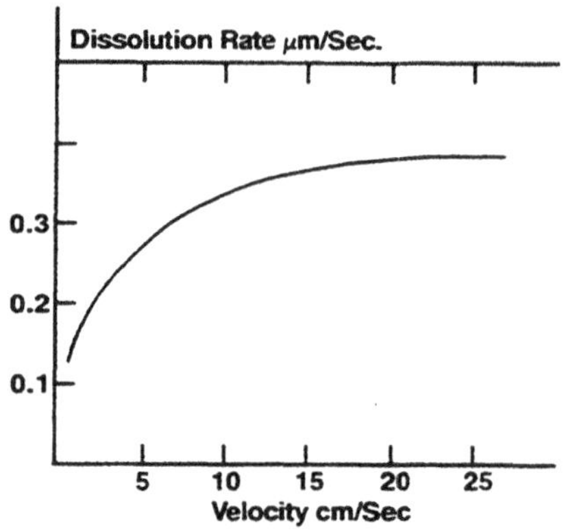

FIGURE 10-8-21. Dissolution rate of AZ-1470 as a function of developer rate flow. From Ref. 259, courtesy of Dr. A. Neureuther, UCLA, Berkeley.

to mass transport of the removal of products from the surface. Choosing an appropriate type of developer and the concentration of developer can be determined by measurement of R and R_0. If possible to arrive at the appropriate contrast (Fig. 10-8-22), trade-offs between throughput (dose) and contrast may be required to achieve acceptable linewidths at moderate costs per exposure. Statistical methods, such as response surface analysis routinely applied to process control plasma etching,(264) are appropriate for defining the process limits of prebake and exposure, congruent with developer conditions.

In summary, the practical development of a resist involves the choice of developer (concentration), an endpoint method or apparatus, a set of exposure conditions, and a selection of develop process conditions, all of which work together for maximum image profiles, minimum linewidth biases, and a reasonable developer process window. Endpoint development monitors will be necessary with positive resists and new nonswelling negative resists to override daily prior or *in situ* processing fluctuations, as well as aid in the establishment of proper developer or exposure conditions to achieve high throughput and/or high resolution.

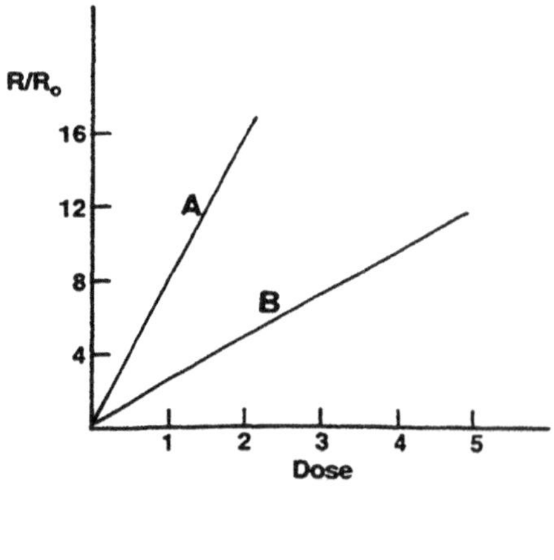

FIGURE 10-8-22. Film thinning ratio and dose. (A) High contrast; (B) low contrast (<2).

REFERENCES

1. F. Dill, A. Neureuther, J. Tuttle, and E. Walker, *IEEE Trans. Electron Devices* **ED-22**, 456 (1975).
2. C. Deckert and D. Peters, *Solid State Technol.* Jan. 1980, p. 76.
3. M. Yamada, S. Mattori, and S. Morita, *J. Electrochem. Soc.* **129**, 2598 (1982).
4. I. Adesida, J. Chinn, L. Rathbun, and E. Wolf, *J. Vac. Sci. Technol.* **21**, 666 (1982).
5. M. Tsuda, S. Orkawa, W. Kanai, A. Yokota, I. Hijikata, A. Uehara, and H. Nakane, *J. Cac. Sci. Technol.* **19**, 259 (1981).
6. G. Taylor, T. Wolf, and J. Moran, *J. Vac. Sci. Technol.* **19**, 872 (1981).
7. P. Blais, *Solid State Technol.* August, 1977, p. 78.
8. U.S. Patent 4,241,165 (1980), Motorola; U.S. Patent 4,278,753 (1981), Horizons.
9. G. Taylor and T. Wolf, *J. Electrochem. Soc.* **127**, 2665 (1980); U.S. Patent 4,232,110 (1980) Bell Labs.
10. M. Chang and J. Chen, *Appl. Phys. Lett.* **33**, 892 (1978).
11. M. Bowden and L. Thompson, *Polym. Eng. Sci.* **14**, 525 (1974).
12. H. Hiroaka, *Appl. Phys. Lett.* **31**, 503 (1977).
13. H. Frisch, *Polym. Eng. Sci.* **20**, 2 (1980).
14. S. Chen and J. Edin, *Polym. Eng. Sci.* **20**, 40 (1980).
15. G. Park, in *Diffusion in Polymers*, edited by J. Crank and G. Park, Academic Press, New York, 1968, Chapter 5, pp. 140–162.
16. K. Ueberreiter, in Ref. 15, pp. 219–257.
17. K. Ueberreiter and F. Asmussen, *J. Polym. Sci.* **52**, 75 (1957); **57**, 187 (1962); **57**, 199 (1962).
18. J. Greeneich, *J. Electrochem. Soc.* **122**, 970 (1975).
19. L. Lapcik and L. Valko, *J. Polym. Sci.* Part A-2, **9**, 633 (1971).
20. S. Ju, H. Lu, J. Duda, and J. Vrentas, *J. Appl. Polym. Sci.* **26**, 3735 (1981).
21. O. Aboul-Nasr and R. Huang, *J. Appl. Polym. Sci.* **23**, 1819 (1979).
22. J. Vrentas, H. Lu, and J. Duda, *J. Appl. Polym. Sci.* **25**, 1793 (1980).
23. M. Kumbar, *J. Macromol. Sci.* **A5**, 1301 (1971); L. Rebenfeld, *J. macromol. sci., Rev. macromol. chem.* **C15**, 279–393 (1976).
24. Y. Tu and A. Ouano, *IBM J. Res. Dev.* **23**, 131 (1977).
25. E. Gipstein, A. Ouano, D. Johnson, and O. Need, *IBM J. Res. Dev.* **21**, 143 (1977); *Polym. Eng. Sci.* **17**, 396 (1977).
26. M. Hatzakis, C. Ting, and N. Viswanathan, Proc. Electron Ion Beam Sixth Conf., San Francisco, Electrochemical Society, May 1974, p. 542.
27. A. Ouano, *Polym. Eng. Sci.* **18**, 306 (1978).
28. J. Greeneich, *J. Electrochem. Soc.* **121**, 1669 (1974).
29. U.S. Patent 3,987,215 (1977), IBM.
30. U.S. Patent 4,078,098 (1978), IBM.
31. K. Harada, T. Tamamura, and O. Kogure, *J. Electrochem. Soc.* **129**, 2576 (1982).
32. D. Kyser and R. Pyle, *IBM J. Res. Dev.* **24**, 426 (1980).
33. F. Billmeyer, *Textbook of Polymer Science*, Wiley, New York, 1971, pp. 33, 74, 84.
34. M. Bowden, L. Thompson, and J. Ballantyne, *J. Vac. Sci. Technol.* **12**, 1294 (1975).
35. K. Murase, M. Kakuchi, and S. Sugarwara, Proc. Intl. Conf. on Microlithography, Paris, 1977, p. 265; *J. Electrochem. Soc.* **126**, 1831 (1979); **127**, 491 (1980).
36. G. Chui, E. Gipstein, W. Moreau, and O. Need, *J. Appl. Polym. Sci.* **21**, 3477 (1977); U.S. Patent 3,916,036 (1976), IBM.
37. R. Kambour, E. Romagosa, and C. Gruner, *Macromolecules*, **5**, 335 (1972).
38. W. Moreau, *Opt. Eng.* **22**, 181 (1983).
39. M. Gazard, C. Duchnese, J. Dubois, and A. Chapiro, *Polym. Eng. Sci.* **20**, 1069 (1980).
40. G. Geuskens, E. Hellinck, and C. David, *Eur. Polym. J.* **7**, 87 (1971), *Makromol. Chem.* **160**, 135 (1972), **160**, 347 (1971).
41. E. Thompson, *Polym. Lett.* **3**, 675 (1965).
42. H. Ku and L. Scala, *J. Electrochem. Soc.* **116**, 980 (1969).
43. C. Ting, in *Record of the 11th Symposium on Electron, Ion and Laser Beam Technology*, edited by R. Thornley, San Francisco Press, 1971, p. 337.
44. U.S. Patent 3,996,393 (1976), IBM.
45. R. Harris, *J. Electrochem. Soc.* **120**, 272 (1973).
46. M. Atoda, M. Komuro, and H. Kawakatsu, *J. Appl. Phys.* **50**, 3707 (1979).

47. French Patent 2,304,933 (1979), Thompson CSF; *Chem. Abstr.* **91**, 220371 (1979).
48. Japanese Patent 79,41719 (1979), Cho LSI; *Chem. Abstr.* **91**, 999536 (1979).
49. M. Bowden, *J. Polym. Sci.* **12**, 499 (1975).
50. M. Kakuchi, S. Sugawara, K. Murase, and K. Matsuyama, *J. Electrochem. Soc.* **124**, 1648 (1977).
51. U.S. Patent 3,898,350 (1977), IBM.
52. Japanese Patent 77,93493 (1977), AGIST, *Chem. Abstr.* **86**, 198006 (1978); **88**, 56981 (1979).
53. U.S. Patent 3,934,057 (1976), IBM.
54. L. Gavens, D. Hess, B. Wu, A. Bell, and D. Soong, *J. Vac. Sci. Technol.* **B1**, 481 (1983).
55. R. Hawryluk, *J. Vac. Sci. Technol.* **19**, 1 (1981), and references therein.
56. F. Jones and J. Paraszczak, *IEEE Trans. Electron Devices* **ED-28**, 1544 (1981).
57. K. Heinrich, H. Betz, A. Heuberger, and S. Pongratz, *J. Vac. Sci. Technol.* **19**, 1254 (1981).
58. L. Karapiperes, I. Adesida, C. Lee, and E. Wolf, *J. Vac. Sci. Technol.* **19**, 1259 (1981).
59. H. Santini and N. Viswanathan, *Proceedings of Microelectronics Seminar*, Interface 82, Kodak Publication **G-136**, Eastman Kodak Co., Rochester, New York, 1982, p. 47.
60. SAMPLE (Simulation and Modeling Profiles in Lithography and Etching, Department of Electrical Engineering, University of California, Berkeley, California; P. Jain, A. Neureuther, and W. Oldham, *IEEE Trans. Electron Devices* **ED-28**, 1410 (1981).
61. M. Rosenfield, A. Neureuther, and C. Ting, *J. Vac. Sci. Technol.* **19**, 1242 (1981).
62. M. Ballauf and B. Wolf, *Macromolecules* **14**, 654 (1981).
63. N. Viswanathan, *J. Polym. Sci., Polym. Chem. Ed.* **14**, 1553 (1976).
64. A. Ouano, D. Johnson, B. Dawson, and L. Pederson, *J. of Polym. Sci.* **14**, 701 (1976).
65. Japanese Patent 80,105,244, Matsushita.
66. U.S. Patent 4,051,271 (1981), AGIST.
67. V. Sharma, R. Pethrick, and S. Affrossman, *Polymer*, **23**, 1732 (1982).
68. *Polymer Handbook*, edited by J. Brandup and E. Immergut, Wiley, New York, 1975, Chapter 4.
69. J. Helbert, B. Wagner, P. Caplan and E. Poindexter, *J. Appl. Polym. Sci.* **19**, 1201 (1975).
70. A. Ouano, *Polym. Eng. Sci.* **18**, 306 (1978).
71. W. Moreau, W. Moyer, I. Merritt, M. Hatzakis, L. Pederson, and D. Johnson, *J. Vac. Sci. Technol.* **16**, 1989 (1979); U.S. Patent 4,121,935 (1981), IBM.
72. A. Charlesby and R. Blackburn, *Nature*, **210**, 1036 (1966).
73. W. Moreau and P. Schmidt, *Electrochem. Soc. Ext. Abstr.* 1970, p. 459.
74. E. Gipstein, W. Moreau, and O. Need, *J. Electrochem. Soc.* **123**, 1105 (1976).
75. I. Haller, R. Feder, M. Hatzakis, and E. Spiller, *J. Electrochem. Soc.* **126**, 154 (1979); U.S. Patent 3,984,582 (1979).
76. U.S. Patent 4,193,797 (1981), DuPont.
77. U.S. Patent 3,931,435 (1977), IBM.
78. A. Chapiro, *Radiation Chemistry of Polymeric Systems*, Wiley, New York, 1962, p. 546.
79. U.S. Patent 3,934,057 (1976), IBM.
80. U.S. Patent 4,024,293 (1980), IBM.
81. M. Hatzakis, *Solid State Technol.* August 1981, p. 74.
82. Japanese Patent 80,144247; *Chem. Abstr.* **94**, 163479, (1980).
83. Japanese Patent 79,143232; *Chem. Abstr.* **92**, 155909 (1979).
84. U.S. Patents 3,981,985 (1977); 4,061,382 (1980), Phillips.
85. Japanese Patent 79,116227; *Chem. Abstr.* **92**, 320281 (1979).
86. J. Greeneich, *J. Appl. Phys.* **45**, 5264 (1974).
87. D. Kyser and N. Viswanathan, *J. Vac. Sci. Technol.* **12**, 1305 (1975).
88. M. Kaplan and D. Meyerhofer, *RCA Rev.* **40**, 166 (1979); *Polym. Eng. Sci.* **20**, 1973 (1980).
89. M. Bowden, L. Thompson, S. Farenholtz, and E. Doerries, *J. Electrochem. Soc.* **128**, 1304 (1981); U.S. Patent 4,289,845 (1981), Bell.
90. J. Shaw, M. Frisch, and F. Dill, *IBM J. Res. Dev.* **21**, 219 (1977).
91. J. Lyerla and J. Pacansky, *IBM J. Res. Dev.* **23**, 42 (1979).
92. W. Oldham, *IEEE Electron Device Lett.* **EDL-1**, 217 (1980).
93. D. Meyerhofer, *IEEE Trans. Electron Devices* **ED-27**, 921 (1980).
94. S. Fujimori, *J. Appl. Phys.* **50**, 621 (1979).
95. F. Dill, A. Neureuther, J. Tuttle, and E. Walker, *IEEE Trans. Electron Devices* **ED-22**, 456 (1975).
96. F. Dill and J. Shaw, *IBM J. Res. Dev.* **21**, 210 (1977).
97. D. Ilten and R. Sutton, *J. Electrochem. Soc.* **119**, 539 (1972).
98. D. Ilten, *J. Electrochem. Soc.* **119**, 537 (1972).

99. J. Lane, Abstracts of 1983 Conference on Ion, Electron and Photon Beams, Los Angeles, California, F-3.
100. Y. Wada, K. Mochiji, and N. Obayashi, *J. Electrochem. Soc.* **130**, 187 (1983).
101. Japanese Patent 81,92536, Fijitsu; *Chem. Abstr.*, **96**, 60866d (1981).
102. H. Hiroaka and A. Gutierrez, *J. Electrochem. Soc.* **126**, 860 (1979).
103. R. Morgan and C. Pollard, *Electron. Lett.* **1B**, 1038 (1982).
104. S. Farenholtz, *J. Vac. Sci. Technol.* **19**, 1111 (1981), U.S. Patent 4,173,470 (1979), Bell.
105. U.S. Patent 4,379,826 (1983), IBM.
106. F. Dill, W. Hornberger, P. Hauge, and J. Shaw, *IEEE Trans. Electron Devices* **ED-22**, 445 (1975).
107. J. Perez and A. Tobar, *J. Photochem.* **19**, 133 (1982).
108. R. Bartolini, *Appl. Opt.* **11**, 1275 (1972).
109. S. Austin and F. Stone, *Appl. Opt.* **15**, 1071 (1976).
110. S. Norman and M. Singh, *Appl. Opt.* **14**, 818 (1975).
111. K. Konnerth and F. Dill, *IEEE Trans. Electron Devices* **ED-22**, 452 (1975).
112. D. Novotony, *Solid State Technol.* March 1981, p. 83.
113. C. Livanos, A. Katzir, J. Shellan, and A. Yarw, *Appl. Opt.* **16**, 1633 (1977).
114. R. Bartolini, *Appl. Opt.* **13**, 129 (1974).
115. J. Shaw and M. Hatzakis, *IEEE Trans. Electron Devices* **ED-25**, 425 (1978).
116. S. Middelhoek, *IBM J. Res. Dev.* **14**' 117 (1970).
117. D. Windmann and H. Bender, *IEEE Trans. Electron Devices* **ED-22**, 467 (1975).
118. T. Chang, C. Codella, and R. Lange, *IEEE Trans. Electron Devices* **ED-28**, 1428 (1981).
119. P. Frasch and K. Saremski, *IBM J. Res. Dev.* **26**, 561 (1982).
120. W. Moreau and W. Moyer, *IBM Tech. Discl. Bull.* **23**, 2304 (1980).
121. F. Dill, *IEEE Trans. Electron Devices* **ED-22**, 440 (1975).
122. D. Leers, *Solid State Technol.* March 1981, p. 90.
123. T. Chang, C. Ting, and D. Kyser, *Solid State Technol.* May, 1982, p. 60.
124. M. Narasimham, *IEEE Trans. Electron Devices* **ED-22**, 478 (1975).
125. D. Hofer, C. Willson, A. Neureuther, and M. Halsey, *SPIE Proc.* **334**, 196 (1982).
126. U.S. Patent 3,639,184 (1972), IBM.
127. U.S. Patent 4,173,479 (1978), Bell.
128. U.S. Patent 4,266,001 (1982), American Hoechst.
129. U.S. Patent 4,174,222 (1982), Tokyo Ohko.
130. U.S. Patent 4,115,129 (1978), IBM.
131. A. Ouano, *Am. Chem. Soc. Org. Coat. Prepr.* **45**, 431 (1981).
132. Japanese Patent 80,88048; *Chem. Abstr.* 94, 22984 (1980).
133. T. Batchhelder and I. Pratt, *Solid State Technol.* August 983, p. 211.
134. R. Conley and J. Bieron, *J. Appl. Polym. Sci.* **1**, 171 (1963).
135. T. Gupta, *Eur. Polym. J.* **17**, 1127 (1981).
136. E. Walker, *IEEE Trans. Electron Devices* **ED-22**, 464 (1975).
137. W. Tsang, *Appl. Opt.* **16**, 1918 (1977).
138. T. Shaughnessy and R. Ruddell, *Semicond. Int.* May 1980, p. 79.
139. W. Deforest, *Photoresist Materials and Processes*, McGraw Hill, New York 1975, p. 54.
140. M. Dinaburg, *Photosensitive Diazo Compounds*, Focal Press, New York, 1964, p. 77.
141. U.S. Patent 3,961,000 (1976), RCA.
142. Brazilian Patent 78,07666, IBM; *Chem. Abstr.* **93**, 85225 (1978); U.S. Patent 4,359,520 (1982), IBM.
143. R. Halverson, W. MacIntyre, and W. Motsiff, *IBM J. Res. Dev.* **26**, 590 (1981), IBM.
144. L. Harriman, *Solid State Technol.* June 1983, p. 155.
145. J. Barrie, in *Diffusion in Polymers*, edited by J. Crank and G. Park, Academic Press, New York, 1968, Chapter 8.
146. U.S. Patent 3,961,101 (1976), RCA.
147. P. Hershey, *Kodak Microelectronics Seminar* **G-136**, 33 (1982).
148. E. Fredericks, *IBM Tech. Discl. Bull.* **20**, 2317 (1977).
149. U.S. Patent 3,402,044 (1968), Shipley.
150. U.S. Patent 3,649,283 (1972), Shipley.
151. U.S. Patent 3,859,099 (1975), Kodak.
152. U.S. Patent 4,379,830 (1983), Polychrome.
153. Japanese Patent 80,43537; *Chem. Abstr.* **93**, 141023 (1980).
154. U.S. Patent 4,253,888 (1982), Kodak.

155. S. Grigorovich, *Zh. Prikl. Khim.* (Leningrad) **48**, 1307 (1975); *Chem. Abstr.* **84**, 128719t (1975).
156. German Patent 2,447,225 (1976), IBM; *Chem. Abstr.* **85**, 102423m (1976).
157. Japanese Patent 75,158,280 Tokyo Shibaura; *Chem. Abstr.* **85**, 12374t (1975).
158. Japanese Patent 81,162746, Fujitsu; *Chem. Abstr.* **97**, 31276 (1981).
159. U.S. Patent 4,141,733 (1979), Kodak.
160. Russian Patent 608,850; *Chem. Abstr.* **89**, 120922 (1978).
161. Russian patent 691,794; *Chem. Abstr.* **92**, 102324a (1980).
162. U.S. Patent 3,868,254 (1975), GAF.
163. J. Shaw and M. Hatzakis, *J. Electrochem. Soc.* **126**, 2026 (1979).
164. N. Atoda and H. Hawakatsu, *J. Electrochem. Soc.* **123**, 1519 (1976).
165. Japanese Patent 78,135621, Hitachi; *Chem. Abstr.* **90**, 130686 (1978).
166. A. Reiser and E. Pitts, *J. Photogr. Sci.* **29**, 187 (1981).
167. T. Kobayashi and E. Arai, *J. Appl. Phys.* **52**, 4785 (1981).
168. Y. Taniguchi, Y. Hatano, H. Shiraishi, S. Horigome, S. Nonogaki, and K. Naroaka, *Jpn. J. Appl. Phys.* **18**, 1143 (1979).
169. A. Reiser and E. Pitts, *Photogr. Sci. Eng.* **20**, 225 (1976).
170. W. Moreau, SPE RETEC, Ellenville, New York, Photopolymer Conference Proceedings, October 1970, p. 143.
171. Y. Ohnishi, M. Itoh, K. Mizuno, H. Gokan, and S. Fujewara, *J. Vac. Sci. Technol.* **19**, 1141 (1981).
172. D. Smith, *Photogr. Sci. Eng.* **12**, 264 (1968).
173. T. Tada, *J. Electrochem. Soc.* **129**, 1070 (1982).
174. U.S. Patent 4,279,986 (1982), Nippon.
175. Japanese Patent 81,162746; *Chem. Abstr.* **97**, 31276 (1981).
176. Japanese Patent 79, 17104.
177. T. Bullett, *J. Adhes.* **4**, 73 (1972).
178. E. Davidson, SPE RETEC, Ellenville, New York Photopolymer Conference Proceedings, October, 1970, p. 141.
179. P. J. Flory, *Principles of Polymer Chemistry*, Cornell University Press, Ithaca, New York, 1953, p. 578.
180. Z. Rigbi, *J. Appl. Polym. Sci.* **12**, 242 (1968).
181. N. Taylor and E. Bagley, *J. Appl. Polym. Sci.* **21**, 113 (1977).
182. K. Gandhi and M. Williams, *J. Appl. Polym. Sci.* **16**, 2721 (1972).
183. S. Imamura, T. Tamamura, K. Harada, and S. Sugawara, *J. Appl. Polym. Sci.* **27**, 937 (1982); U.S. Patent 4,286,049 (1982), Nippon.
184. J. Lai, *J. Electrochem. Soc.* **126**, 697 (1979).
185. E. Feit, *Polym. Eng. Sci.* **20**, 1058 (1980); U.S. Patent 4,201,580 (1980), Bell.
186. B. Gong, Y. Ye, H. Gu, and Q. Chang, *J. Vac. Sci. Technol.* **16**, 1980 (1979).
187. German Patent 3,041,261, JSR; *Chem. Abstr.* **94**, 22989 (1980).
188. German Patent 2,817,256, Allied; *Chem. Abstr.* **90**, 195637 (1978).
189. Japanese Patent 80,155353, Tokyo Ohko; *Chem. Abstr.* **94**, 200872.
190. U.S. Patent 4,148,655 (1979), Oji Paper.
191. F. Kaufmann, A. Schroeder, E. Engler, and V. Patel, *Appl. Phys. Lett.* **36**, 423 (1980); **37**, 314 (1980), U.S. Patent 4,338,392 (1981), IBM.
192. Japanese Patent 79,140,535; *Chem. Abstr.* **92**, 119732 (1979).
193. Japanese Patent 82,178239, Hitachi; *Chem. Abstr.* **98**, 117127 (1982).
194. C. Tan and J. Rauner, *J. Vac. Sci. Technol.* **19**, 1348 (1981); U.S. Patent 4,289,842 (1982), Kodak.
195. O. Kogure, *Jpn. J. Appl. Phys.* **21**, 206 (1982).
196. East German Patent 179164; *Chem. Abstr.* **86**, 81717 (1978).
197. A. Barraud, *Thin Solid Films* **85**, 77 (1981).
198. R. Kaiser, G. Miller, D. Thomas, and L. Sperling, *J. Appl. Polym. Sci.* **27**, 957 (1982).
199. W. Moreau and P. Schmidt, *Electrochem. Soc. Ext. Abstr.* **138**, 459 (1970).
200. H. Sawarada, *J. Macromol. Sci., Rev. Macromol. Chem.* **C3**, 313 (1969).
201. M. Bowden and L. Thompson, *Polym. Eng. Sci.* **14**, 525 (1974).
202. K. Hatada, *Polym. Bull.* **8**, 469 (1982).
203. G. Willson, H. Ito, J. Frecht, and F. Houlihan, Proceedings of 28th IUPAC Macromolecular Symposium, July 1982, International Union of Pure and Applied Chemists, Amherst, Massachusetts, p. 448.
204. H. Raply, G. Duggan, and R. Elliott, *Electrochem. Soc. Ext. Abstr.* **82**-1, 514 (1982),

205. N. Ueno, S. Koniski, K. Tanimoto, and K. Sugita, *Jpn. J. Appl. Phys.* **20**, L709 (1981).
206. R. Srinivasan and V. Mayne-Banton, *Appl. Phys. Lett.* **41**, 576 (1982); B. Garrison and R. Srinivasan, *Appl. Phys. Lett.* **44**, 849 (1984).
207. M. Mitsuya, M. Terao, Y. Taniguchi, T. Kabu, K. Shigematsu, and M. Akayai, *J. Appl. Phys.* **54**, 3710 (1983).
208. T. Yamazaki, K. Tanaka, and H. Nakata, *J. Electrochem. Soc.* **127**, 1860 (1980); *Jpn. J. Appl. Phys.* **21**, 1518 (1982).
209. S. Morita, J. Tamano, S. Hattori, and M. Ieda, *J. Appl. Phys.* **51**, 3938 (1980).
210. J. Tamano, S. Hattori, M. Ieda, and S. Morita, *Plasma Chem. Plasma Process.* **1**, 261 (1981).
211. H. Hiroaka, *Appl. Phys. Lett.* **31**, 503 (1977).
212. U.S. Patent 4,241,165 (1980), Motorola.
213. U.S. Patent 4,278,753 (1981); 4,292,384 (1981), Horizons.
214. U.S. Patent 4,232,110 (1980), Bell.
215. G. Taylor and T. Wolf, *J. Electrochem. Soc.* **127**, 2668 (1980): G. Taylor, and T. Wolf, *Solid State Technol.* Feb. 1984, p. 145.
216. G. Taylor, T. Wolf, and J. Moran, *J. Vac. Sci. Technol.* **19**, 82 (1981).
217. G. Taylor, T. Wolf, and M. Goldrick, *J. Electrochem. Soc.* **128**, 361 (1981).
218. U.S. Patent 4,386,152 (1983), Honeywell.
219. M. Tsuda, S. Oekawa, W. Kanai, A. Yokoto, I. Hiijita, J. Uekara, and H. Nakane, *J. Vac. Sci. Technol.* **19**, 259 (1981).
220. M. Tsuda, *J. Vac. Sci. Technol.* **19**, 1351 (1981).
221. M. Issacson and A. Muray, *J. Vac. Sci. Technol.* **19**, 1117 (1981).
222. J. Lane, J. Maldonado, A. Cleland, R. Haelbich, J. Silverman, and J. Warlaumont, *J. Vac Sci. Technol.* **B1**, 1072 (1983).
223. N. Samoto, R. Shemizu, H. Hasimoto, I. Adesida, E. Wolf, and S. Namba, *J. Vac. Sci. Technol.* **B1**, 1367 (1983).
224. S. Gillespie, *IBM J. Res. Dev.* **28**, 454 (1984).
225. V. Nagawasan and R. Carlson, *J. Electrochem. Soc.* **131**, 1369 (1984).
226. U.S. Patent 4,452,880 (1984), Konishiroku Photo.
227. S. Imamura, T. Tamamura, and O. Kogure, *Polym. J.* **16**, 391 (1984).
228. Japanese Patent 58,04143, Fujitsu; *Chem. Abstr.* **100**, 200947 (1983).
229. PCT Int. Patent Appl. WO. 83,04,320 (1983), *RCA*; *Chem. Abstr.* **100**, 604 (1983).
230. N. Turro, *Modern Molecular Photochemistry*, Benjamin, Reading, Massachusetts, 1978, p. 590.
231. R. Cox, N. Clecak, and W. Moreau, *Polym. Eng. Sci.* **14**, 491 (1974).
232. H. Gokan, S. Scho, and Y. Ohnishi, *J. Electrochem. Soc.* **130**, 143 (1983).
233. G. Taylor and T. Wolf, *Polym. Eng. Sci.* **20**, 1087 (1980).
234. Japanese Patent 81,77844; *Chem. Abstr.* **96**, 26852 (1981).
235. J. Moran and G. Taylor, *J. Vac. Sci. Technol.* **16**, 2020 (1979).
236. I. Adesida, J. Chen, L. Rathburn, and E. Wolf, *J. Vac. Sci. Technol.* **21**, 666 (1982).
237. T. Venkatesan, G. Taylor, A. Wagner, B. Wilkens, and D. Barr, *J. Vac. Sci. Technol.* **19**, 1379 (1981).
238. B. Wu, D. Hess, D. Soong, and A. Bell, *J. Appl. Phys.* **54**, 1725 (1983).
239. M. Yamada, S. Hattori, and S. Morita, *J. Electrochem. Soc* **129**, 2598 (1982).
240. M. Yamada, J. Tamano, K. Yoneda, S. Morita, and S. Hattori, *Jpn. J. Appl. Phys.* **21**, 768 (1982); *J. Electrochem. Soc.* **130**, 1962 (1983).
241. H. Hiroaka, *J. Electrochem. Soc.* **128**, 1065 (1981).
242. D. Follett, K. Weiss, J. Moore, A. Steckel, and W. Liu, *Electrochem. Soc. Ext. Abstr.* **82**-2, 321 (1982).
243. T. Yamazaki, K. Tanaka, and H. Nakata, *Jpn. J. Appl. Phys.* **20**, 2191 (1981).
244. U.S. Patent 4,307,178 (1981), IBM.
245. Japanese Patent 82,50430; *Chem. Abstr.* **97**, 48231 (1982).
246. P. Holmes and J. Snell, *Microelectron. Reliab.* **5**, 337 (1966).
247. D. Weston and R. Mattox, *J. Vac. Sci. Technol.* **17**, 466 (1980).
248. M. Jun-Ru, S. Kuo-Hsung, E. Wolf, and T. Everhart, *J. Vac. Sci. Technol.* **19**, 1385 (1981).
249. Japanese Patent 77,2259; *Chem. Abstr.* **87**, 193575 (1977).
250. Japanese Patent 77,12235; *Chem. Abstr.* **88**, 82744 (1977).
251. S. Yoshikawa, O. Ochi, and Y. Mizushi, *Appl. Phys. Lett.* **36**, 197 (1980); P. Huggett, *Appl. Phys. Lett.* **42**, 592 (1983).

252. U.S. Patent 4,307,176 (1981), Hitachi.
253. B. Lin, *IEEE Trans. Electron Devices* **25**, 419 (1978); U.S. Patent 4,142,107 (1979), IBM.
254. U.S. Patent 4,039,379 (1977), IBM.
255. H. Kleinknecht and H. Meier, *J. Electrochem. Soc.* **125**, 798 (1976); *J. Electrochem. Soc.* **130**, 655 (1983).
256. D. Elliott, *Solid State Technol.* Sept. 1977, p. 66.
257. M. Borovicka, SPE RETEC Photopolymers, Ellenville, New York, 1979, p. 20.
258. A. Johnson, *Kodak Microelectronics Seminar*, **G-136**, 60 (1982).
259. D. Kim, W. Oldham, and A. Neureuther, *Kodak Microelectronics Seminar*, **G-136**, 100 (1982).
260. U.S. Patent 3,935,331; 3,935,332 (1977), RCA.
261. H. Keller, *Solid State Technol.* June 1978, p. 45.
262. L. Rebenfeld, P. Makarewicz, H. Weigmann, and G. Wilkes, *J. Macromol. Sci., Rev. Macromol. Chem.* **C15**, 279 (1976).
263. F. Sacher, *J. Polym. Sci., Polym. Lett.* **21**, 111 (1983).
264. S. Bergeron and B. Duncan, *Solid State Technol.* August 1982, p. 98, and references therein; G. Box and J. Hunter, *Ann. Math. Stat.* **23**, 195 (1957).
265. M. Geis, J. Randall, T. Deutsch, P. DeGraff, K. Krahn, and L. Stern, *Appl. Phys. Lett.* **43**, 74 (1983).
266. H. Deckman and J. Dunsmuir, *J. Vac. Sci. Technol.* **B1**, 1166 (1983).
267. A. McCullough, SPE RETEC Photopolymers, Ellenville, New York, 1982.
268. U.S. Patent 4,439,516 (1984), Shipley.
269. G. Stevens, *Microphotography*, Wiley, New York 1966, p. 189.
270. G. Pannetier and P. Souchay, *Chemical Kinetics*, Elsevier, New York, 1967, p. 150.
271. G. Taylor, L. Stillwagon, and T. Venkatesan, *J. Electrochem. Soc.* **131**, 1658 (1984).
272. U.S. Patent 4,464,455 (1984), Fujitsu Ltd.
273. U.S. Patents 4,414,059 and 4,417,948 (1984), IBM.
274. Z. Pelzbauer and R. Wagner, *J. Appl. Polym. Sci.* **29**, 1427 (1984).
275. Japanese Patent 59,12,433; *Chem. Abstr.* **102**, 165450 (1984).
276. Japanese Patent 57,202,536; *Chem. Abstr.* **100**, 112257 (1982).
277. Japanese Patent 57,202,532; *Chem. Abstr.* **100**, 148509 (1982).
278. Japanese Patent 57,211,143; *Chem. Abstr.* **100**, 12988 (1982).
279. Japanese Patent 57,168,246; *Chem. Abstr.* **100**, 430474 (1982).
280. Japanese Patent 59,02,042; *Chem. Abstr.* **102**, 31147 (1984).
281. Japanese Patent 59,02,037; *Chem. Abstr.* **102**, 46313 (1984).
282. German Patent DE 3,315,395; *Chem. Abstr.* **101**, 129934 (1983).
283. W. Hinsberg and M. Guiterrez, *SPIE Advances in Resist Technology*, **469**, 57 (1984).
284. U.S. Patent 4,468,447 (1984), Tokyo Ohko.
285. M. Watts, *J. Vac. Sci. Technol.* **B3**, 434 (1985); U.S. Patent 4,465,768 (1984).
286. W. Flack, *J. Electrochem. Soc.* **131**, 2200 (1984).
287. V. Marriott, *SPIE Advances in Resist Processing* **469**, 65 (1984).
288. M. Hori, *Plasma Chem. Plasma Process.* **4**, 119 (1984).
289. D. Dimter and M. Hanson, *SPIE Proc. on Optical Lithography* **470**, 203 (1984).
290. W. Bonivert, *Plat. Surf. Finish.* **71**, 58 (1984).
291. M. Tsuda, S. Oekawa, M. Yabuta, A. Yakota, H. Nakane, in *Microcircuit Engineering*, edited by J. Cleaver, H. Ahmed, and G. Jones, Academic Press, New York, 1984, p. 371.
292. D. Fauval and S. Gourrier, in *Microcircuit Engineering*, edited by J. Cleaver, H. Ahmed, and G. Jones, Academic Press, New York, 1984, p. 371.
293. G. Koren and J. Yeh, *J. Appl. Phys.* **56**, 2120 (1984).
294. U.S. Patent 4,015,986 (1977), IBM.
295. U.S. Patent 4,462,860 (1984), Bell.
296. U.S. Patent 4,464,461 (1984), Kodak.
297. Japanese Patent 59,84,426; *Chem. Abstr.* **101**, 161274 (1984).
298. Anonymous, *Circuits Manufacturing*, November 1974, p. 20.
299. Japanese Patent 58,108,529; *Chem. Abstr.* **101**, 141083 (1983).
300. R. Srinivasan and B. Braren, *J. Polym. Sci.* **22**, 2601 (1984).
301. J. Andrew, *J. Appl. Phys.* **43**, 717 (1983).
302. M. Tsuda, S. Oekawa, M. Yabuta, A. Yakota, and H. Nakane, *Jpn. J. Appl. Phys.* **23**, 259 (1984).
303. T. Pampalone, M. Hannifan, S. Jain, and C. Krieger, *J. Electrochem. Soc.* **131**, 2670 (1984).

304. D. Kim, W. Oldham, and A. Neureuther, *IEEE Trans. Electron Devices* **ED-31**, 1730 (1984).
305. U.S. Patent 4,103,073 (1978), DIOS.
306. Anonymous, *Indust. Chem. News.* November 1985, p. 5.
307. V. Sharma, S. Affrossman, and R. Pethrick, *Br. Polym. J.* **16**, 73 (1984).
308. Japanese Patent 59,119,276, *Chem. Abstr.* **102**, 15135 (1984).
309. Japanese Patent 59,05,245: *Chem. Abstr.* **102**, 15119 (1984).
310. Japanese Patent 59,05,245; *Chem. Abstr.* **102**, 15120 (1984).
311. Japanese Patent 59,44,046, *Chem. Abstr.* **102**, 201541 (1984).
312. Japanese Patent 58,09,143; *Chem. Abstr.* **101**, 201527 (1983).
313. East German Patent 211,192 (1984), *Chem. Abstr.* **102**, 219864 (1984).
314. Japanese Patent 59,24,847; *Chem. Abstr.* **102**, 219845 (1984).
315. Japanese Patent 58,15,0949; *Chem. Abstr.* **102**, 181997 (1984).
316. Japanese Patent 59,925,32; *Chem. Abstr.* **102**, 238180 (1984).
317. B. Grant, N. Clecak, R. Twieg, and G. Wilson, *IEEE Trans. Electron Devices* **ED-28**, 1300 (1981).
318. U.S. Patent 4,359,820 (1982), IBM.
319. M. Tsuda, S. Oekawa, M. Yabuta, A. Yakota, H. Nakane, K. Yamashito, K. Gano, and S. Namba, *J. Vac. Sci. Technol.* **B3**, 481 (1985).
320. N. Eib, *J. Vac. Sci. Technol.* **B3**, 425 (1985).
321. L. Hauchlan, K. Sautter, and T. Batchelder, *Solid State Technol.*, April 1985, p. 333.
322. P. Rissmann and G. Owen, *J. Vac. Sci. Technol.* **B3**, 159 (1985).
323. C. Dix, P. Flavin, P. Hendy, and H. Jones, *J. Vac. Sci. Technol.* **B3**, 131 (1985).
324. Japanese Patent 59,125,729; *Chem. Abstr.* **102**, 15153 (1985).
325. Japanese Patent 59,125,729; *Chem. Abstr.* **102**, 15153 (1985).
326. Japanese Patent 59,62,850; *Chem. Abstr.* **102**, 15127 (1985).
327. Japanese Patent 59,155,836; *Chem. Abstr.* **102**, 36776 (1985).
328. F. Rodriguez, P. Krasciky, and R. Groele, *Solid State Technol.* May 1985, p. 125.
329. D. Henderson, J. White, H. Craighead, and I. Adesida, *Appl. Phys. Lett.* **46**, 900 (1985).
330. G. Gorodetsky, T. Kazyaka, R. Melcher, and R. Srinivasan, *Appl. Phys. Lett.* **46**, 828 (1985).
331. Japanese Patent 59,180,545; *Chem. Abstr.* **102**, 53980 (1985).
332. Japanese Patent 59,165,952; *Chem. Abstr.* **102**, 87659 (1985).
333. British Patent 2,139,777; *Chem. Abstr.* **102**, 876668 (1985).
334. Japanese Patent 50,176,743; *Chem. Abstr.* **102**, 87661 (1985).
335. Japanese Patent 59,72,441; *Chem. Abstr.* **102**, 123107 (1985).
336. Japanese Patent 59,182,444; *Chem. Abstr.* **102**, 123112 (1985).
337. Japanese Patent 59,29,853; *Chem. Abstr.* **102**, 36767 (1985).
338. Anonymous, *Semicond. Int.* May 1985, p. 30.
339. U.S. Patent 4,464,455 (1985), Fujitsu.
340. V. Mariott, *SPIE Proc. on Optical Microlithography* **394**, 194 (1983).
341. K. Sogo, Y. Tanaka, and K. Uchiho, *Kodak Microelectronics Seminar* **G130**, 44, (1980).
342. M. Bowden and L. Thompson, *Polym. Eng. Sci.* **14**, 525 (1974).
343. M. Kato and H. Nakanse, *Photogr. Sci. Eng.* **23**, 209 (1984).
344. U.S. Patent 4,103,073 (1985), Dion.
345. U.S. Patent 4,530,895 (1985), Hoechst.
346. C. Mack, *SPIE Proc. on Optical Microlithography* **538**, 207 (1985).
347. U.S. Patent 4,535,054 (1985), Hughes.
348. H. Fuchs and J. Petermann, *J. Appl. Phys.* **58**, 1056 (1985).
349. U.S. Patent 4,506,005 (1985), GCA.
350. T. Matsuzawa, T. Ito, and M. Tanuma, *IEEE Trans. Electron Devices* **ED-32**, 1781 (1985).
351. W. Waldo, *Semicond. Int.* October 1985, p. 116.
352. H. Yamashita and Y. Todokoro, *J. Vac. Sci. Technol.* **B3**, 1004 (1985).
353. J. Peterson and M. Stan, *Microelectron. Manuf. Test.* January 1986, p. 23.
354. F. Sebillote, A. Weill, and P. Paniez, *Macromol. Chem.* **186**, 1695 (1985).
355. German Patent 3,346,979 (1985), Merck; *Chem. Abstr.* **103**, 113376 (1985).
356. M. Watts and R. Hannifan, *SPIE Opt. Eng.* **539**, 21 (1985).
357. M. Watts, *J. Vac. Sci. Technol.* **B3**, 434 (1985).
358. European Patent Appl. 129,106 (1985), Allied; *Chem. Abstr.* **102**, 22949 (1985).
359. W. Hinsberg, C. Willson, and K. Kanazawa, *SPIE Opt. Eng.* **539**, 6 (1985).
360. Japanese Patent 6028881, Sumimoto; *Chem. Abstr.* **103**, 75809 (1985).

361. F. Rodriguez, R. Groele, and P. Krasicky, *SPIE Opt. Eng.* 539, 14 (1985).
362. D. Soong, *SPIE Opt. Eng.* **539**, 2 (1985).
363. M. Tsuda, S. Oikawa, M. Yabuta, A. Yakota, H. Nakane, K. Yamashita, K. Gamo, and S. Namba, *J. Vac. Sci. Technol.* **B3**, 481 (1985).
364. A. Muray, *J. Vac. Sci. Technol.* **B3**, 1773 (1985).
365. U.S. Patent 4,481,279 (1984), Fujitsu.
366. J. Branon, J. Lankard, A. Baise, F. Burns, and J. Kaufman, *J. Appl. Phys.* **58**, 2036 (1985).
367. R. Srinivasan and S. Lazare, *Polymer* **26**, 1247 (1985).
368. H. Yamashita, and Y. Todokoro, *Electron. Lett.* **21**, 646 (1985).
369. U.S. Patent 4,497,891 (1985), IBM.
370. S. Babu and V. Srinivasan, *SPIE Advances in Resist Technology* **539**, 36 (1985).
371. L. Thomas and J. Windle, *Polymer* **23**, 529 (1982).
372. U.S. Patent 4,552,833 (1985), IBM.
373. S. MacDonald, H. Ito, H. Hiroaka, and C. Willson, SPE RETEC Photopolymers, Ellenville, New York, 1985, p. 177.
374. L. Stillwagon, P. Silverman, and G. Taylor, SPE RETEC Photopolymers, Ellenville, New York, 1985, p. 87.
375. U.S. Patent 4,500,628 (1985), AT&T.
376. M. Flannagan and R. Wake, *SPIE Resist Technol.* **539**, 45 (1985).
377. G. Bendikt, *SPIE Resist Technol.* **539**, 242 (1985).
378. U.S. Patent 4,587,205 (1986), USA.
379. V. Starov, *J. Imag. Technol.* **30**, 74 (1986).
380. U. S. Patent 4,587,203 (1986), Hughes.
381. U.S. Patent 4,551,418 (1985), IBM.
382. T. Ishu, *Appl. Phys. Lett.* **47**, 123 (1985).
383. E. Weber and R. Moore, *Solid State Technol.* May 1979, p. 61.
384. M. Hatzakis, *J. Vac. Sci. Technol.* **12**, 1276 (1975).
385. A. Barraud, C. Rosilio, and A. Ruandel-Texier, *J. Vac. Sci. Technol.* **16**, 2003 (1979).
386. S. Grindle and E. Pavelchek, *Test Meas. World* May 1986, p. 102.
387. J. Frechet, F. Bouchard, F. Houlihan, B. Kryczka, E. Eichler, N. Clecak, and C. Willson, *J. Imag. Sci.*, **30**, 59 (1986).

11

Postbake

11-1. INTRODUCTION

Postbake or hardbake, initially used to harden printing plate resists, is the process of applying heat to enhance the performance of a developed resist image. Later the procedure was applied to deswell developed images in negative rubber azide resists. More recently, the process has been applied to improve the performance of DQN resists.

Involving both physical and chemical reactions in resists, postbake is to be distinguished from postexpose (predevelop) bake (Chapter 10) which is used to improve the resolution of mainly positive novolak-based resists. Predevelop baking is also used to reverse the image of DQN resists. For some moderate-temperature (< 100°C) processes such as in mild dry etching, postbaking is not necessary, because no liquid is present to attack and weaken the resist–substrate interface.

For wet etching, postbaking may restore the interfacial adhesion weakened by developer penetration along the resist–substrate interface or the adhesion weakened by the swelling of the resist in negative images. The practical limit of postbake conditions is the excessive distortion of the image caused by physical melting or by chemical degradation. In the extreme case, some positive resist images are intentionally flowed to produce tapered images for the dry etching of via holes[1] or for better planarization of the steps coated with resist (Chapter 12).[2]

Postbake process conditions are dependent upon many factors:

- Thickness of resist
- Type of resist and developer
- Exposure dose
- Etchant and conditions
- Image profile and size of land
- Subsequent process requirements (temperature, volatility, ease of stripping)
- Presence or absence of air
- Thermochemical reactivity of the resist

Postbaking involves the thermochemical (thermolysis) reactions of the resin, sensitizer (if present), and residual solvents with heat and air. Hardening can also be achieved by radiation and chemical agents such as melamines, epoxies, strong acids, azides, and formaldehyde.

Postbaking has been a common process with both negative azide and DQ

TABLE 11-1-1. Postbake Reactions in Resists

Reaction	Chemical	Physical	Example
Cross-linking	x		DQN
			Azide rubber
			Negative vinyl
Oxidation	x		Novolak resists
Volatilization		x	Casting solvents
			Developer removal
			Sensitizer decomposition
Wetting–adhesion		x	Plastic flow
Pinholes		x	Shrinkage due to plastic due
Image reversal[57]	x		DQN

positive resists. In the ideal case, the resist thermosets prior to plastic flow. Plastic flow is most evident in novolak-based resists and flow occurs at temperatures in excess of 150°C. Base-soluble resins of maleic imide copolymers tolerate 200°C[3] and have been introduced as replacements for novolak resins in DQN.

Some of the physical and chemical reactions which can be induced by postbaking are given in Table 11-1-1.

The latent flow of the resist for positive materials is dependent on the glass transition temperature (T_g) of the composition. Since amorphous polymers are mixtures, the melting points are not sharp transitions: in passing from a solid to a liquid.

The surface adhesive forces, surface tension, and the internal modulus of the resist counteract the plastic flow. The T_g of polymers is determined by the intermolecular bonds between polymer chains. Polymers with polar groups have strong intermolecular H-bonds and have a high T_g. Weakly interacting hydrocarbon polymers are elastic and have a low T_g (< 50°C). Cross-linking can raise the T_g by 100°C or more. Tightly cross-linked polymers exhibit no flow and only ablate at elevated temperatures. The physical properties of polymers exhibit a major change in passing through the T_g. Table 11-1-2 gives the T_g of common resist polymers.

The thermal transitions as measured by differential thermal analysis are shown in Fig. 11-1-1 for PMMA, AZ-1350, and KTFR. For AZ-1350, a melt endotherm is followed by an exotherm for N_2 evolution from DQ sensitizer decomposition. Polymers with wide polydispersity such as novolak resins exhibit a wide glass transition phase change (Fig. 11-1-2). On a microscopic scale of polymer motion, the T_g is about 30°C lower than the macroscopic plastic flow (melting) region of a resist polymer. Figure 11-1-3 shows the flow and rounding of a profile edge for DQN resist after postbaking to 100–160°C.[4] Rounding may be tolerated in wet etching

TABLE 11-1-2. Glass Transition of Resists

Resist	T_g (°C)
PMMA	105
DQN	70–110
Polybutene sulfone	62
Cyclized rubber	52

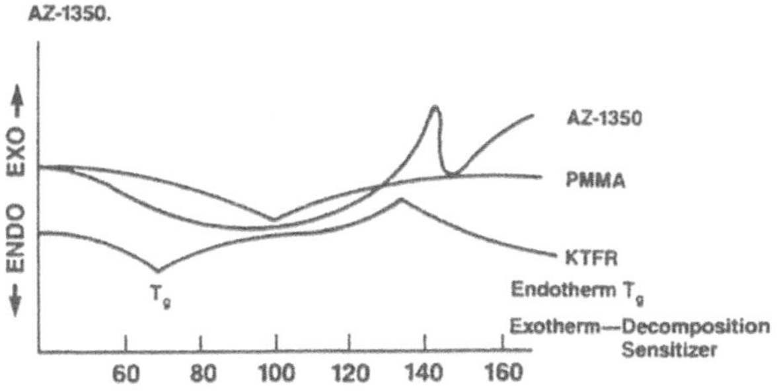

FIGURE 11-1-1. Differential thermal analysis of prebaked resist films.

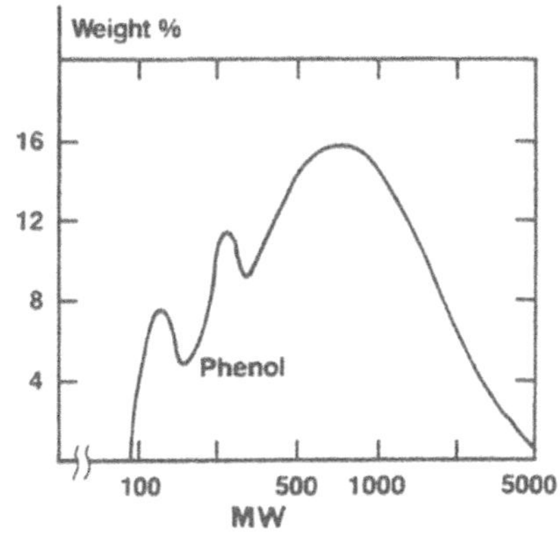

FIGURE 11-1-2. Molecular weight distribution in a typical phenol–formaldehyde resin made by condensation polymerization. From Ref. 2.

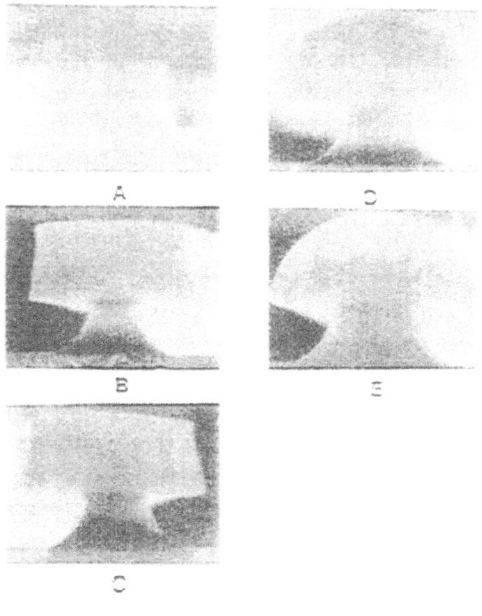

FIGURE 11-1-3. Profiles in DQN. SEM photography of the cross section of a 4-μm-wide line after a period of 20 min after baking. (A) No baking—the magnification is different from others and no etching was done; (B) 100°C; (C) 120°C; (D) 140°C; (E) 160°C. From K. Jinno *et al.*, *Photogr. Sci . Eng* . **21**, 290 (1977), copyright Soc. Photogr. Sci. Eng., reproduced by permission.

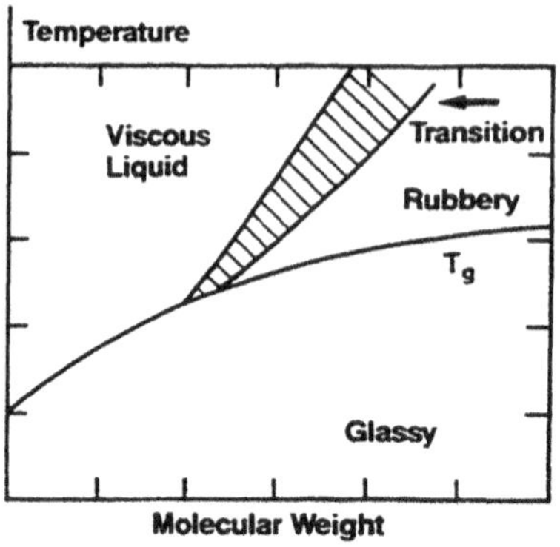

FIGURE 11-1-4. Phase transitions of typical polymer materials.

but not in dry etching. Dry etching depends critically on obtaining a vertical resist profile whereas wet etching is a bottom surface process. The width of the image at the resist–substrate interface determines the size of the final etched image. Thus, slight top surface rounding may be acceptable for aqueous etching.

The molecular weight (M_w) of the polymer can exhibit some influence on the melting point. Low-molecular-weight novolaks melt at about 60°C for a M_w of 1000 and at 120°C for a M_w of 10^4. The typical dependence of the melting point on the molecular weight of amorphous polymers is shown in Fig. 11-1-4.

Above a certain limit, the melting point is independent of the molecular weight. Furthermore, branching can also raise the melting point of novolak resins.[5] For example, an image in PMMA of M_w of 60k will soften at about 100°C and become quite rubbery. On further heating, a shear stress will cause viscous flow to predominate. If a M_w of 10^6 is used, flow would not be evident until 140°C. In addition, cross-linking the PMMA with anhydride groups will raise the flow point to 200°C or more.[6,62]

Postbaking is a trade-off in many cases between resist stabilization and melting. Treatments of resist with radiation or chemical agents are alternative ways of hardening (postbaking) at room temperature. The following sections will discuss the chemical reactions which can occur during postbake for various resists as well as other hardening techniques and the practical aspects of image flow, monitoring, and postbaking equipment.

11-2. PHYSICAL CHEMISTRY OF POSTBAKE

After developing, rinsing, and drying, the resist is ready for postbake. An exposed negative resist of the bisazide rubber type consists of cross-linked rubber. In the exposed region, most of the unreacted bisazide has been leached out by the developer. The swollen resist then contains the developer and the rinse solvent.

In a positive DQN resist, the edges of the image may contain exposure products (Fig. 11-2-1) due to the fact that the resist develops anisotropically from the inside-out. The highly exposed zone initially dissolves until the walls of the pattern isotropically begin to dissolve. Some indene acid product may also be present in the walls of the DQN image. In PMMA, the lower-molecular-weight (lower

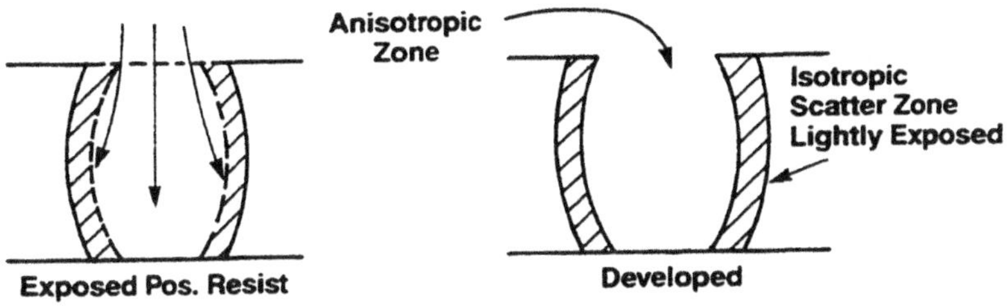

FIGURE 11-2-1. In a positive resist, a gradual zone of reaction products can exist at edges of image due to isotropic exposure. The layer of products may have a lower melting point than the melting point of the bulk polymer.

melting) products are present in the walls to a depth of 0.1–0.5 μm depending on the dose, the development time, and the angle of forward and backscattered radiation. The lower-molecular-weight products depress the melting point of PMMA and edge distortion may occur at 10–15°C lower than the normal 130°C flow point. Continued heating of PMMA will eventually cause thermal depolymerization reversion.

The unexposed region may also contain water absorbed by the resist during development and rinsing as well as a few percent of casting solvent left after prebake. For a prebake of 80°C for 30 min, up to 10% of the casting solvent is retained[7,60] in the DQN resist (Fig. 11-2-2). The solvent remaining, as shown by dissolution studies,[8] is distributed with a maximum concentration at the middepth of the film of DQN. Baking a resist film below its T_g does not remove all of the casting solvent. Therefore, PMMA is baked above 130°C to remove the casting solvent[7] (Fig. 11-2-3).

Since water is absorbed by the prebaked DQN resist films after developing and rinsing with water, postbaking removes most of the water. The subsequent immersion of a postbaked film in water causes very little absorption of moisture (Fig. 11-2-4). If the absorbed water outgasses, however, during lift-off processing (Chapter 12), the postbaked films should be stored in a dry box and vacuum baked inside the deposition chamber prior to metal deposition.

Baking a resist film at 100–150°C for DQN, at 125–150°C for KTFR, and at 100–140°C for PMMA can cause physical and chemical changes such as:

1. Volatilization of the casting solvent, the absorbed water, developer, and rinse

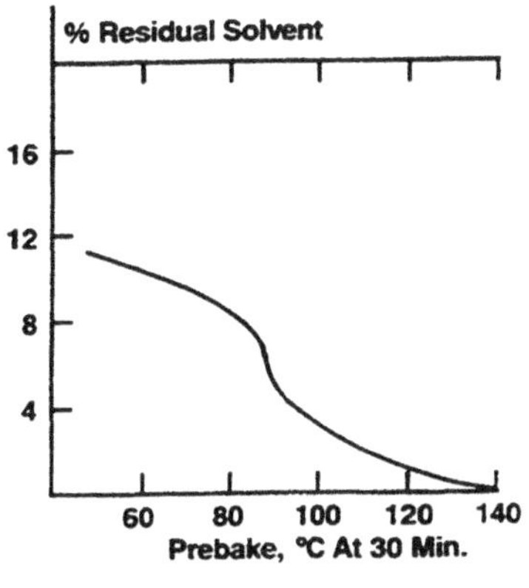

FIGURE 11-2-2. Solvent remaining in a 1-μm-thick film of DQN after prebaking at various temperatures. From Ref. 7.

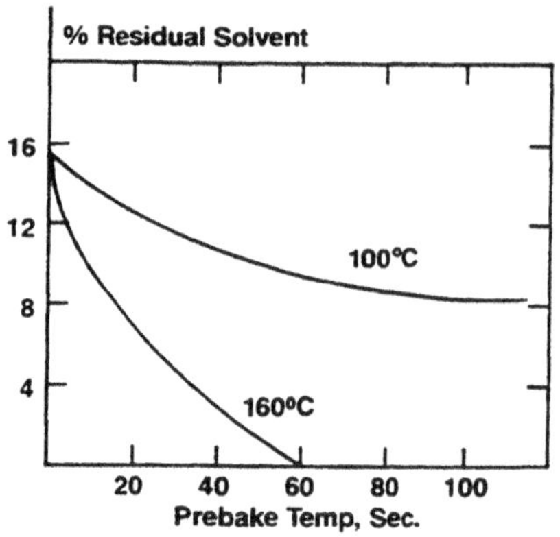

FIGURE 11-2-3. Solvent remaining in a 1.5-μm-thick film of PMMA after prebaking for various times and temperatures. From Ref. 7.

2. Volatilization and decomposition of sensitizer
3. Shrinkage and flow of images
4. Inter- and intramolecular resin and sensitizer reactions
5. Oxidation of resin

These effects occur together, not independently, and they can be significant factors in the chemical and physical resistance of the resist. Image distortion, especially of the profile, is undesirable for dry etching. Moreover, the removal of solvents, moisture, and low-molecular-weight species which compete for adhesion sites, improves the bonding and reduces side effects in dry etching and in ion implantation. Finally, thermal stabilization can be induced by intermolecular reactions between sensitizer and resin.

The popular DQN resists may contain mono-, di-, or trisubstituted DQ sulfonates, as well as polymeric DQs which can cross-link at different rates. A basic test of the thermal reactions of the DQN sensitizer can be performed by baking and monitoring the loss of the diazo band in the infrared or the DQ bleaching at 435 nm until baking has destroyed all of the sensitizer. In the resin component, different alkyl substituents can raise or lower the melting range by approximately 25°C.

11-2-1. Adhesion

The improved adhesion resulting from postbake has been noted for DQNs[9–12] at bake temperatures of 120–150°C (Fig. 11-1-3). Under these conditions, solvents

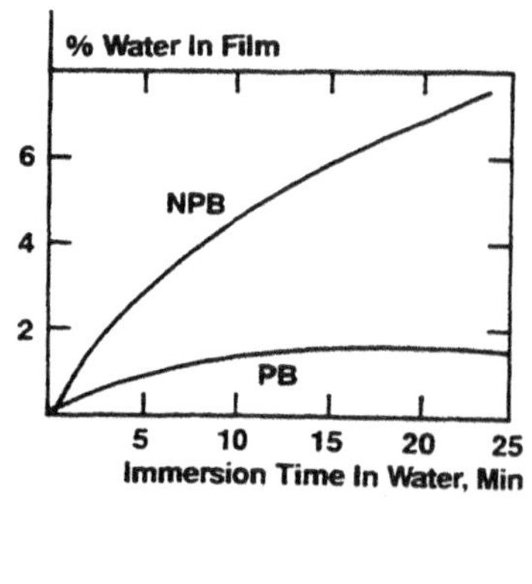

FIGURE 11-2-4. Absorption of water by film of DQN (1.5 μm thick) postbaked (PB; 1 hr at 140°C) or unbaked (NPB).

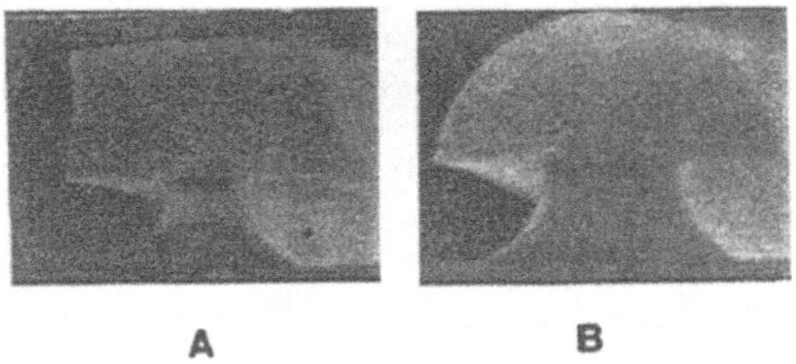

FIGURE 11-2-5. (A) DQN resist with no postbake. The underlying oxide layer is etched in buffered HF. Note large undercut of land. (B) DQN postbaked at 160°C and etched in HF. Note less undercut of land of oxide. From Ref. 4, courtesy of SPSE.

and adsorbed water are removed and the bonding of the resin to the substrate is improved. The slight melting of the resin acts as a hot melt adhesive to seal the interface between the resist and substrate and reduce edge lifting(9,10) during wet etching (Fig. 11-2-5). The flow point for thicker resists will occur at higher temperatures. The plastic flow conditions may induce interdiffusion between a silylated surface primer and the overlayer of resist which improves adhesion. Maximum adhesion is not obtained in a thermoplastic until all of the casting solvent has evaporated. This does not occur until the boiling point of the solvent is reached.(13) In DQN resists, the 2-ethoxyethyl acetate casting solvent would be released at temperatures above 160°C. At this temperature, however, considerable image rounding of the commercial DQN may occur. Another consequence of plastic flow is the decrease of the surface tension of the polymeric resist. For PMMA baking above the T_g, improved wetting of the substrate is obtained (Fig. 11-2-6).

Baking also induces thermomechanical stresses in films due to the mismatch of thermal expansion between the resist and substrate. Thermal stress δ_t, caused by the difference between the thermal expansion coefficients for the film and the substrate, is expressed by(15)

$$\delta_t = (\delta_f - \delta_s)[E_f/(1 - v_f)](T - T_d) \tag{11-2-1}$$

where δ_f and δ_s are the thermal expansion coefficients for film and substrate and E_f and v_f are Young's modulus and Poisson's ratio for the film, respectively. T is the temperature at which the stress is measured, and T_d is the film deposition temperature.

The stress(14) for AZ-1350 is on the order of 10^8 dynes/cm^2 while for the silicon(15) wafer the stress is 10^9 dynes/cm^2. The stress is independent of the coating thickness and resist solution concentration.(16) Narrow-molecular-weight polymers without self-plasticization are likely to crack. Novolaks become brittle upon heating in excess of 125°C due to cross-linking and oxidation. Thus, high postbake may induce enough internal stresses in novolaks to cause edge lifting in aqueous

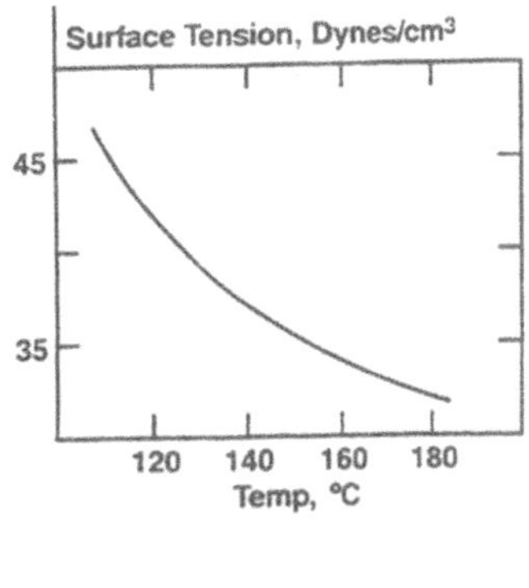

FIGURE 11-2-6. Surface tension of PMMA at elevated temperatures. From Ref. 61.

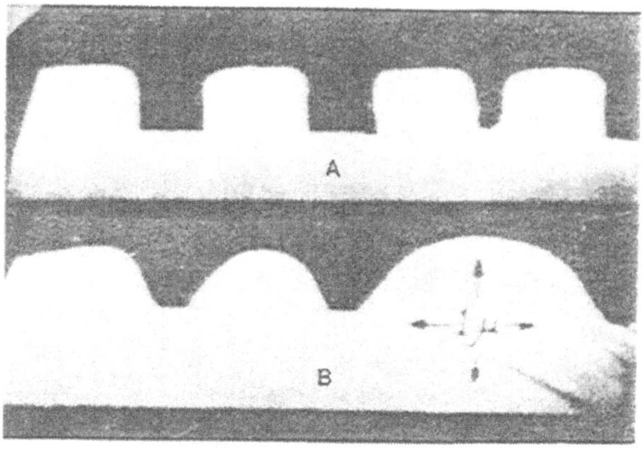

FIGURE 11-2-7. PMMA images (A) before postbake and (B) after baking at 135°C. From M. Hatzakis, *J. Electrochem. Soc.* **116**, 1033 (1969), copyright and reproduced by permission of the Electrochemical Society, Inc.

etchants. The mismatch of the thermal coefficients of expansion and the wide difference in surface energies between the substrate surface tension and the resist can give rise to internal stress. Good adhesive materials are elastic and have a low modulus to counteract thermal induced stresses. Thus, resists designed for high-temperature dry etching or lift-off have a high T_g and are not as suitable for wet etching as elastomeric rubber-based negative resists. High-T_g polymers suffer from edge lifting during wet etching. The edge lifting due to internal stresses, causes severe undercutting or even delamination of small resist islands.

11-2-2. Melting

In diazonovolaks (DQN) and PMMA, melting occurs above 135°C. In KTFR, melting and film loss by oxidation occur above 180°C. Some DQN formulations which use high-molecular-weight cresol novolaks, *p*-vinyl phenol,[17,54] maleimide,[3,75] polyimide,[18] tris siloxane additives,[19] or polymeric diazoquinones[20] are claimed to be stable to at least 200°C. These new formulations are most suitable for dry etching or ion implantation. No postbake or image hardening is necessary.

In melting or softening, surface tension minimizes the surface area, limited by such factors as the difference between the resist and substrate surface tension, gravity, and the rheological melt–flow properties of the resin (unless chemical reactions change the structure of the polymer resin). Large lands of resist are more resistant to flow than micrometer-size islands. In the fabrication of self-aligned vias

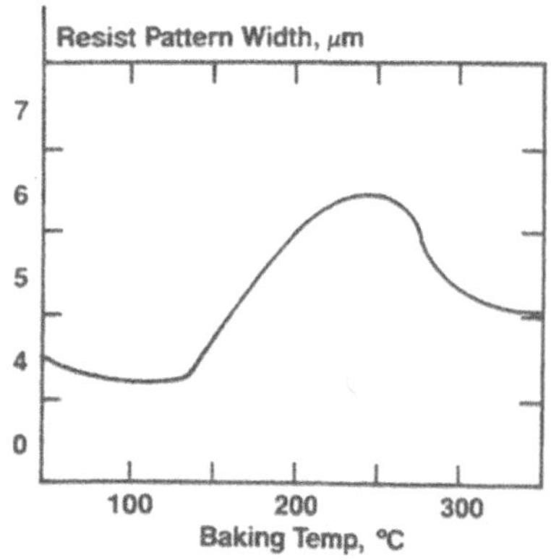

FIGURE 11-2-8. Postbake temperature versus linewidth for a DQN resist. From Ref. 4, courtesy of SPSE.

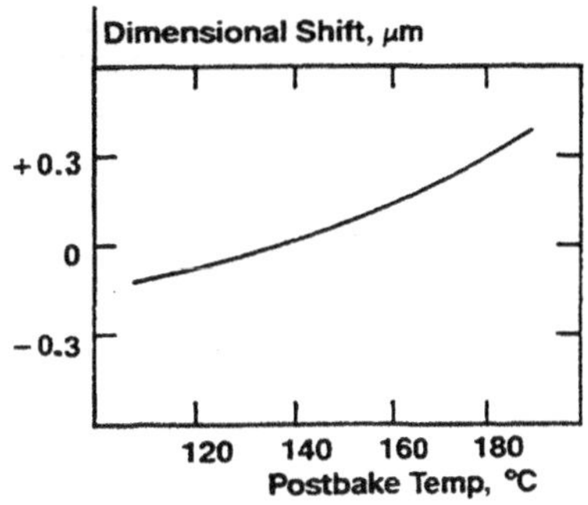

FIGURE 11-2-9. Dimensional changes of a (2.5 μm) line in DQN resist after postbaking at elevated temperatures.

by dry etching, the resist is deliberately melted to form 45° profiles prior to isotropic etching.[21,67,74] For PMMA, the image flow occurs at temperatures above 135°C (Fig. 11-2-7). In melting, there is a tendency to round sharp contours and gradually smooth edges.[12] The resist softens sufficiently so that surface and interfacial tension forces modify the image shape (Fig. 11-2-7) to minimize the energy associated with them. Figure 11-2-8 shows the lateral widening (vertical shrinkage) of a resist land in OFPR-2 DQN resist[4] as a result of the phase change at 150°C.

For Kodak 820 resist, a dimensional widening in a line of 3.75 μm of 0.2 μm is noted for baking above 150°C. The 0.2-μm negative bias is not independent of line size and must be determined for each size and shape (Fig. 11-2-9). Because of structural differences in the resin and the sensitizers, various commercial DQN resists show different image flow points and linewidth distortions at high postbake conditions (Fig. 11-2-10).[22] Upon high postbake of 170°C, a thickness loss of 8–20% is noted for Kodak 820 resist (Fig. 11-2-11). This could be a problem in subsequent dry etching due to the insufficient resist remaining to protect the top of planarized steps.

The weight loss as noted by mass spectrometry of AZ-1350 corresponds to the following volatiles[13]:

Component	Condition	Percent Weight loss
Cellosolve acetate	Prebake	11
Water	Postbake	2
N_2	Postbake	1
Phenols	Postbake	3
Cellosolve acetate	Postbake	3

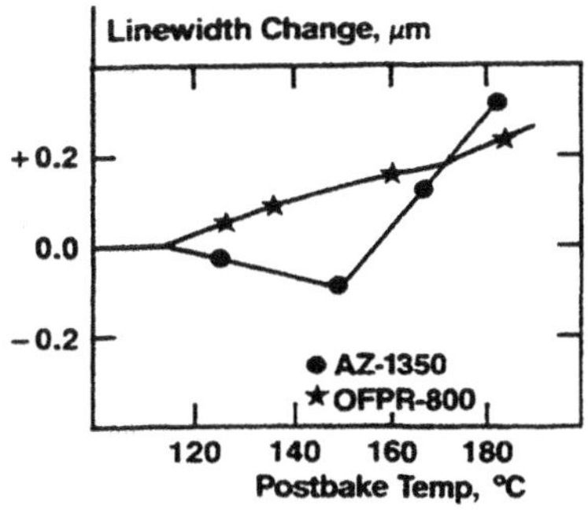

FIGURE 11-2-10. Linewidth changes after postbaking a DQN resist. From Ref. 4, courtesy of SPSE.

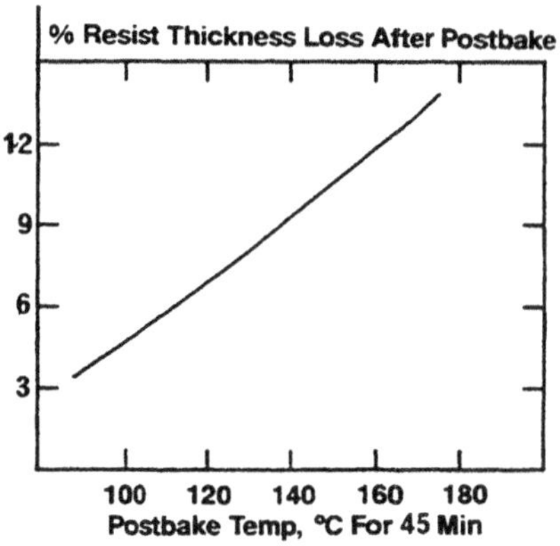

FIGURE 11-2-11. Thickness loss of a DQN resist after postbaking at various temperatures.

In DQN postbake, heating causes the exterior of the image to form an impervious oxidized crust. The N_2 gas evolved from the DQ decomposition cannot readily diffuse through the crust. Consequently, the stress is later relieved by edge lifting in wet etching.[64] A flood UV exposure prior to postbake allows the gas to slowly escape.

One of the side benefits of the physical melting of the resist is the closure of the smallest pinholes. A trade-off between image flow and defect closure has to be made.

11-3. CHEMICAL REACTIONS IN POSTBAKE

Novolak resists baked in air generally become insoluble due to oxidation and cross-linking. If not baked above 170°C, they remain strippable in acetone (organic solvents). Thus, postbaking can lead to complications in solvent stripping (Chapter 14).

For resole-based DQN resists, the following reactions are believed to occur[4]:

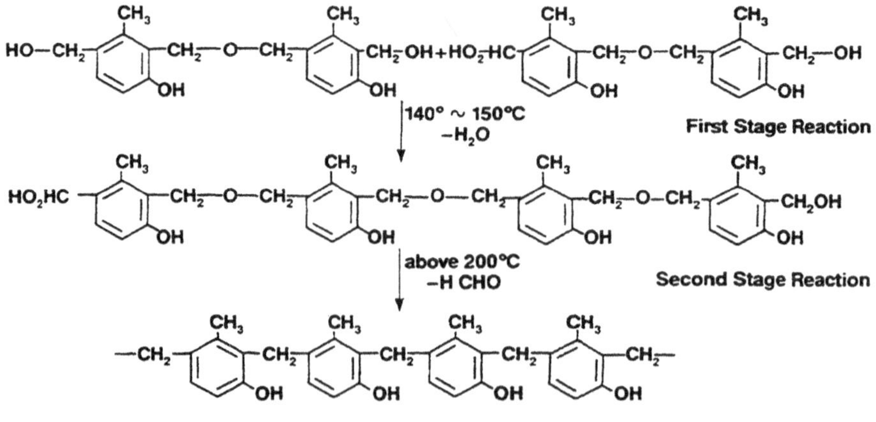

The methyl end groups react to form branches and cross-links. The chemical reactions, however, compete with physical melting of the image. For novolak-based resists, the thermal oxidation of the methylene bridge occurs as evidenced by infrared studies (Fig. 11-3-1).[23] The oxidation also occurs with deep-UV hardening in air[24,63] (O_2) and even in the absence of air since oxygen is supplied by the OH groups,

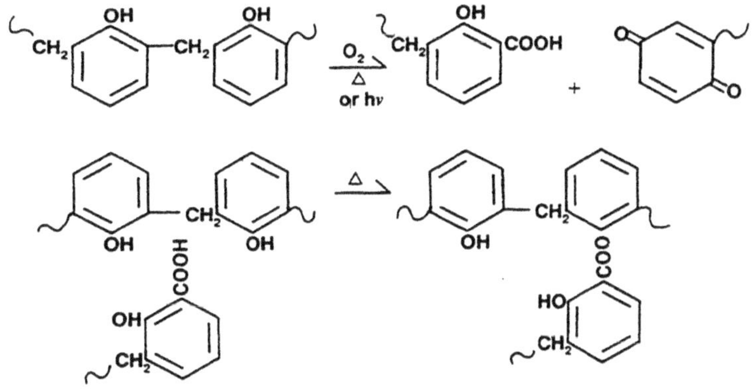

(11-3-1)

An activation energy of 40 kcal/mole has been estimated.[23] Even at the 70–100°C prebake, a slight surface oxidation of novolaks[25,63] in DQN or polyolefin sulfone–novolak resists occurs. The surface oxidation is an important mechanism in establishing an induction effect of the inhibited dissolution of the unexposed resist (Chapter 10) in alkaline developers.

To eliminate standing waves at 436 nm during exposure,[26] the hard postbake of AZ-1350 has also been used to create a dark absorbing underlayer in a multilayer resist (Fig. 11-3-2). The dark color is likely due to quinode formed in reaction (11-3-1). Pure novolaks are water white but commercial resins are orange-red. In the large-scale synthesis of novolak resin, the excess formaldehyde is removed by baking thick resin melts. During the baking, considerable oxidation of the novolak resin occurs.

The presence of DQ in the novolak resists leads to some self-hardening. Pacansky and Hiroaka[27] achieved 150°C stability in AZ-1350 by generating the

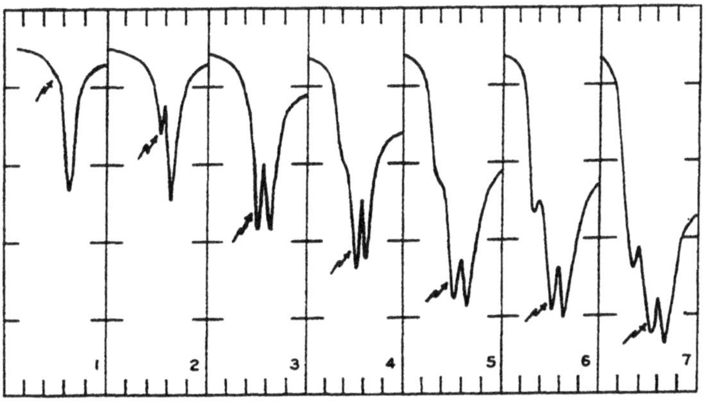

FIGURE 11-3-1. Growth of carbonyl band at 1650 cm^{-1} monitored in thermal oxidation of novolak resin. (1) No oxidation; (2) after 2 min at 170°C; (3) after 12 min; (4) after 25 min; (5) after 100 min; (6) after 175 min; (7) after 300 min. From Ref. 23, copyright 1963 and courtesy of John Wiley & Sons, Inc.

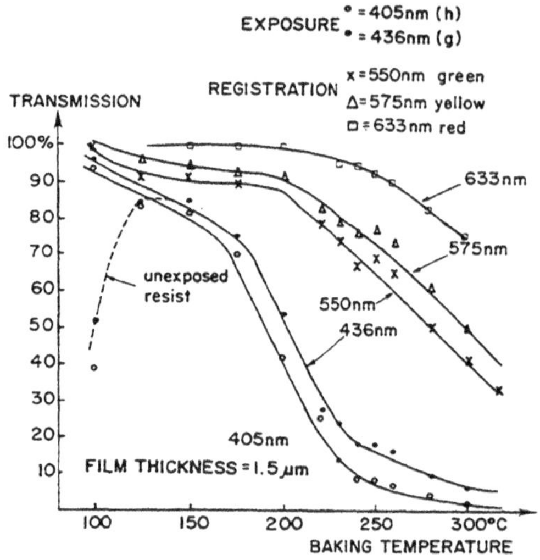

FIGURE 11-3-2. UV transmission properties of baked AZ-1350 photoresist. From Ref. 26, reproduced by permission.

ketene at 435 nm by UV flood hardening of the developed resist. The bisketene adds to the phenol and cross-links the resin (in the absence of water):

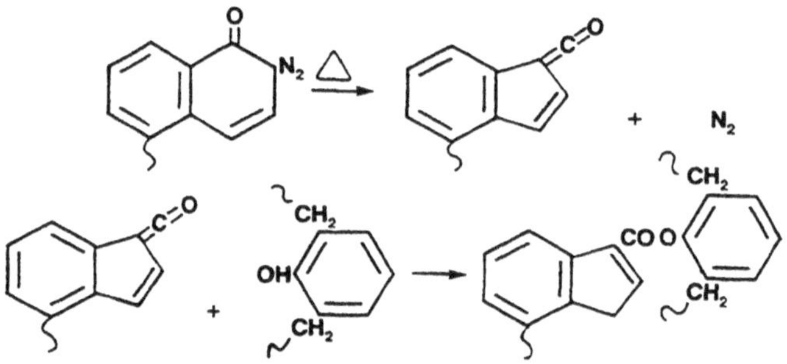

Baking the DQ sensitizer in the presence of water in air forms the indene acid.[55,56] The bis acid product couples and cross-links the resin. Nitrogen, lactones, and azo dimers[28] are the other decomposition products of the DQ sensitizer.

The DQ in AZ-1350 decomposes at 130°C with an activation energy of 40 kcal/mole. At 120°C, 90% remains undecomposed for a 30-min bake. The DQ in novolak is less thermally stable (33 kcal/mole) than in the isolated pure DQ[13] the addition of the ketene intermediate, dependent on the mobility of the ketenes

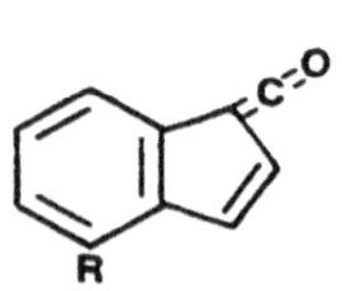

and novolaks. The mobility of the novolak increases with resin melting. The cross-linking reaction is also responsible for the image reversal of AZ-1350 exposed to electrons or deep-UV photons.[29,30] Polymeric DQs also act as efficient cross-linkers since two side chain reactions result in multichain cross-linking:

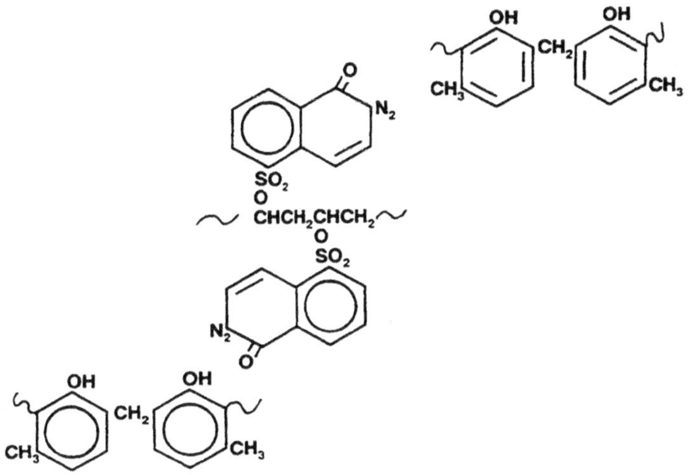

Although the melting[73] of DQN competes with cross-linking, the latter can be effected at 120°C if postbake is performed for 24 hr in air. Without excessive melting, the slow oxidation and ketene reactions harden the resist. The addition of melamine resins to DQN formulations is said to induce cross-linking before the melting of the resin occurs during postbake.[69]

11-4. OTHER METHODS OF HARDENING DQN

Radiation or chemical treatments prior to postbake can stabilize the novolak-based resists. The developed image is subject to a treatment which hardens the outer shell (Fig. 11-4-1A) and prevents lateral flow (Fig. 11-4-1B), or hardens the whole matrix (Fig. 11-4-1C). A physical shell can be formed by overcoating with a thin metal such as Al or by spin coating a layer of PMMA over the developed DQN image.[70] Inside the shell, the normal cross-linking reactions of the resin and sensitizer occur. The flow is prevented by the rigid outer shell. The normal resin–resin oxidation or resin–DQ reaction will harden the interior resist. The outer shell prevents outward flow. Deep hardening throughout the matrix of resist can be achieved by exposure with penetrating ion, electron, X-ray beams or by chemical treatment with formaldehyde[31] or silanes.[71] A summary of the hardening techniques is given in Table 11-4-1.

Deep-UV hardening was first noted by Moreau and Schmidt who reported negative images in AZ-1350 upon 2537-Å exposure.[45] Deep-UV hardening was applied as early as 1966 to stabilize polyvinylcinnamate to 200°C[46] and to extend

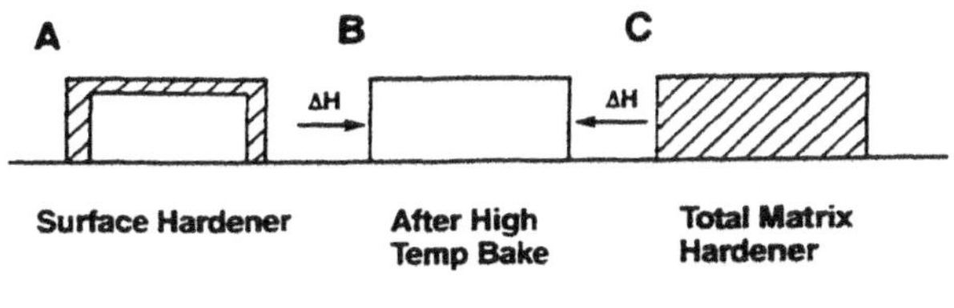

FIGURE 11-4-1. Hardening of DQN resist prior to reactive ion etching. (A) Surface hardened such as with deep-UV irradiation or plasma overcoat. (B) Stabilized DQN image after curing in subsequent postbake. (C) Chemical hardening of DQN such as with formaldehyde and curing (B) in postbake.

the printing plate durability. Since novolak resins absorb strongly in the deep UV, films thicker than 0.3 μm are not thoroughly hardened. Exposure in oxygen requires 1–5 J/cm^2 while exposure in vacuum requires 7–12 J/cm^2. Flash exposure is limited by the rates of diffusion of excited oxygen (or ozone) into the film. Plasma treatment with singlet oxygen also induces cross-linking. Further hardening of the interior occurs during high postbake. Highly intense microwave-powered lamps which emit strongly in the deep UV are used to reduce the flood exposure time to under 10 min.(58) Exposure of a heated resist also reduces the required deep-UV dose.

The highest stability achieved with surface treatments is about 200°C. However, matrix hardening can achieve 300°C stability. The hardening process reduces plasma erosion rates by 2–5×.(31,32) For stabilization of DQN resists to chlorine reactive ion etchants, the resist is treated with hexamethyldisilazane.(71) In deep-UV hardening, ester formation from two reactions is possible:

1. The formation of an ester from DQ and phenol via ketene reaction,(47)

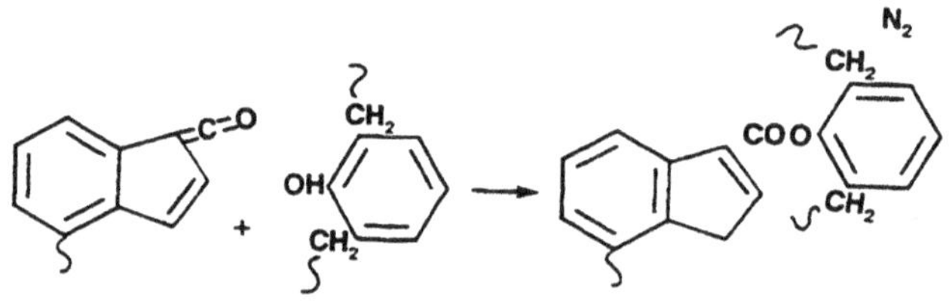

2. The formation of an ester by reaction of photooxidized products,

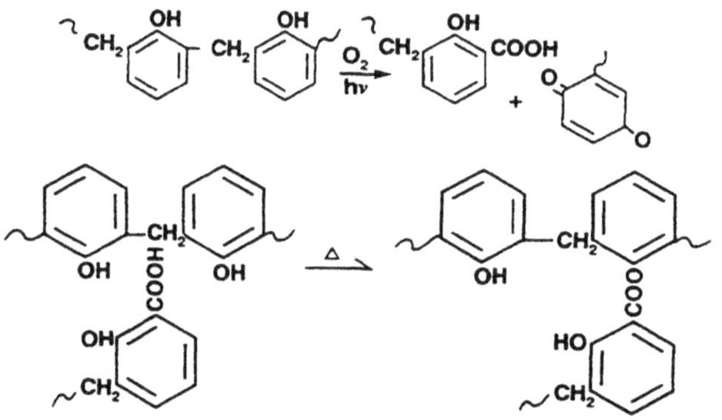

The addition of bisazides to the DQN formulation lowers the dose required for deep-UV hardening.(65) The azide absorbs in the 200- to 365-nm region. A dual-tone negative or positive resist can also be formed whereby exposure at 435 nm generates a normal positive resist image in 0.25 N alkaline developer. Exposure at lower wavelengths and development in stronger alkaline developer forms a negative resist image.

The thermal stability achieved by the various treatments is shown in Table 11-4-2. The highest stability is achieved by soaking of the resist in a formaldehyde bath. A thermal gravimetric comparison of deep-UV and HCHO hardening of DQN is shown in Fig. 11-4-2. The treatment with HCHO involves an intermolecular condensation reaction of HOCH$_2$OH and novolak resin whereby

TABLE 11-4-1. Thermal Stabilization of DQN Resists

Treatment	Dose	Application	Stable temperature (°C)	Reference
1. Deep UV	$>2\ J/cm^2$	RIE Al	200	32, 58, 66
		Dry etch	100	24, 27
		Lift-off	150	33
2. Electron beam	$>10^{-3}\ C/cm^2$	Dry etch	200	34, 59
3. Ion beam	$>10^{15}$ ions/cm	Ion implant	200	35
4. Plasma (CF_4, N_2)		Dry etch	180	36, 37, 39
		Ion implant	150	38
5. Al overcoat		Dry etch	180	40
6. Soak in DQS[a]		Dry etch	180	41
7. Soak in KMnO		Dry etch	180	42
8. Soak in HCHO		Dry etch	300	43
9. Add peroxides		Dry etch	180	44
10. Add tetraamines				68

[a] DQS, 1-oxo-2-diazo-5-naphthalene sulfonic acid.

the $HOCH_2OH$ is derived from a reaction of 37% formaldehyde in water (formalin) in an acid medium (1 M H^+). The wafer coated with DQN is immersed for 5 min at 85°C (37 kcal/mole activation energy). The resin is highly cross-linked in the *para* position with methylene bridges,[(48)]

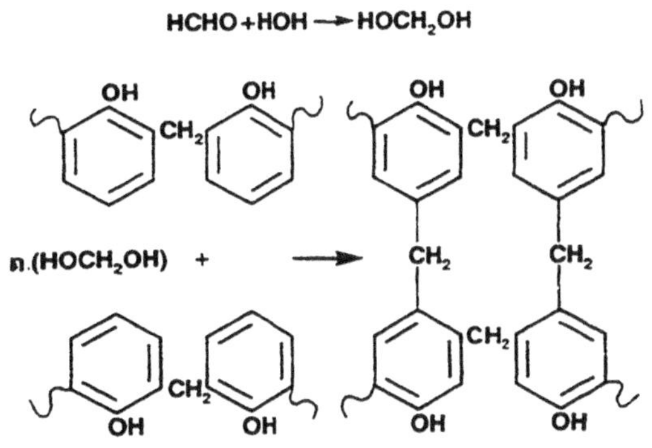

Monosubstituted *ortho*-phenol and *para*-phenol react with formaldehyde to produce novolaks (ON, *ortho* novolak; PN, *para* novolak). The *ortho*- and *para*-

TABLE 11-4-2. Thermal Stability of DQN Resists

Method of hardening	Onset of weight loss (°C)
Peroxide additive	180
Plasma	190
Oxidation	210
Deep UV	220
Electron beam	225
HCHO	300

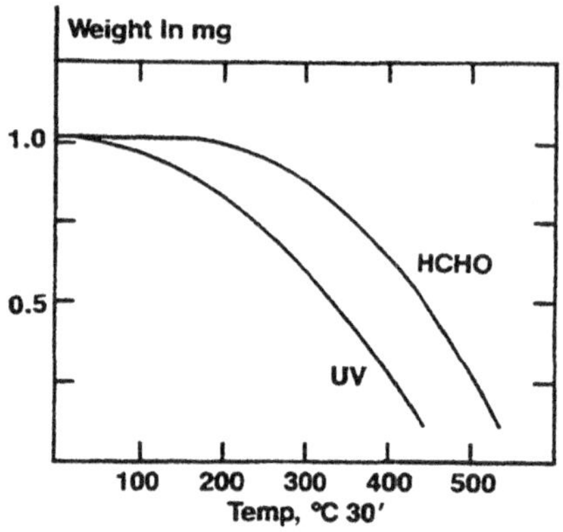

FIGURE 11-4-2. Thermal gravimetric curve of deep-UV-cured and formaldehyde-cured films of DQN.

substituted novolaks cannot further react with formaldehyde to form cross-linked resins,

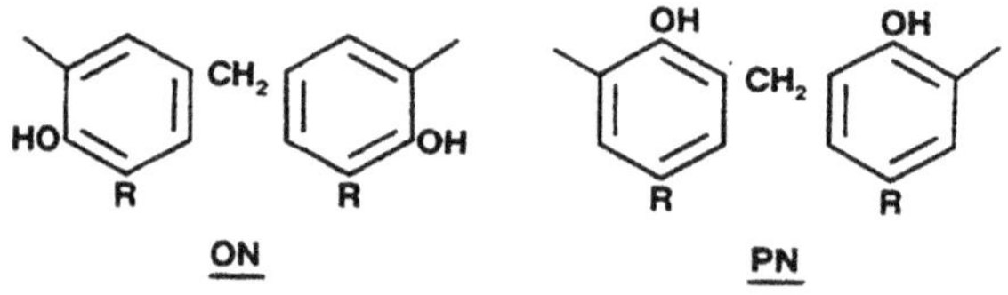

Meta-substituted phenols, however, have available *ortho* and *para* positions to further react with formaldehyde to generate cross-linked resins,[48]

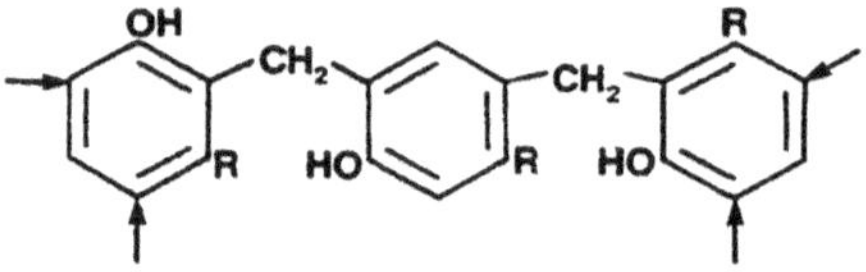

The HCHO-stabilized DQNs are melt resistant and thermally stable to 300°C. In addition, the hardened resists are also more stable to plasma etchants (Table 11-4-3).

The high-temperature (HCHO stabilized) stability of the DQN resist before and after heating to 300°C is shown in Fig. 11-4-3. An additional advantage of

TABLE 11-4-3. Plasma Etch Rates of Treated DQN Resist

Resist[a]	Erosion rate (Å/min)	Etch diode gas
T	300	CCl_4
U	1250	CCl_4
T	70	CF_4
U	700	CF_4
T	500	O_2
U	3000	O_2

[a]T, HCHO treated; U, untreated.

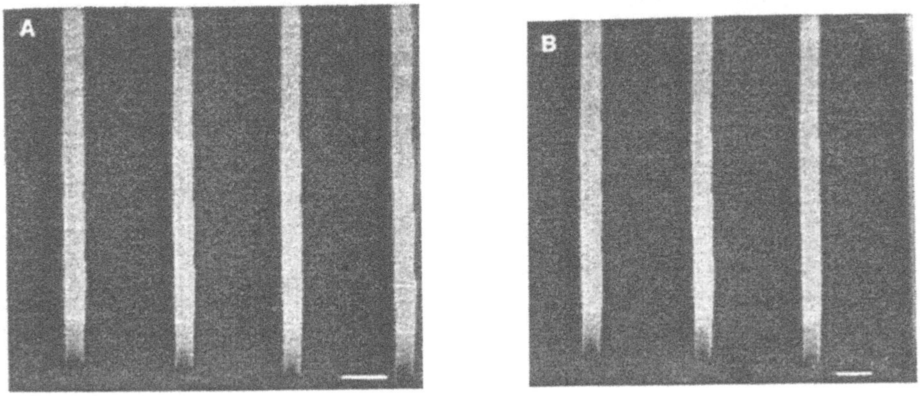

FIGURE 11-4-3. DQN resist stabilized by formaldehyde treatment (A) before heating and (B) after heating for 2 hr at 300°C. Scale bar: 1 μm.

HCHO treatment is the insolubility of the DQN resist in any solvent. Treatment of the OH groups of the treated image with salts of Mg^{2+} or Zn^{2+} incorporates metals which form nonvolatile halides.[(49)] The impregnated DQN resist exhibits minimal thinning in a plasma reactor. Cross-linking reactions similar to HCHO can be used to introduce metallic reagents or other cross-linkers. Bissilylamines, amines, and epoxy reagents react with the OH group of the novolak resin. In addition, vapor-phase reagents avoid solvent swelling. Incorporation of metals such as Si, Al, Ti, Sn, B, and Li, which form nonvolatile metal oxides in an oxygen plasma, allows the resist to be used in bilayer processing for subsequent reactive ion etching or lift-off applications.

11-5. PRACTICAL POSTBAKING

The practice of postbaking involves:

1. A choice of conditions determined by the T_g of the particular resist polymer
2. Conditional limits to avoid image flow or distortion
3. A thermal energy source and detection[(50)] system to monitor the heat absorbed and the temperature achieved
4. A monitor of image profile and a functional test to ascertain the effectiveness of the postbake treatment

The postbake equipment is the same as that used in prebake and is divided into the three heating categories of conduction, convection, and absorption of electromagnetic radiation (infrared or microwave). Hot plate heating is slow but achieves isothermal conditions while the other techniques generally do not achieve steady-state conditions. Baking is usually performed in a batch reactor. Compensations in batch mode have to be made for loading effects.

For wet etching with DQN, a mild postbake of 100–140°C is used to remove interfacial solvents and improve adhesion. A higher bake coupled with a hardening process provides high-temperature stability for subsequent dry processing. Postbake typically causes changes of 0.1–0.3 μm in linewidths.[53]

The process definition of postbaking involves:

1. Functional analysis of etch undercut (Chapter 13)
2. SEM of cross sections
3. Analytical polymer characterization such as differential thermal analysis (T_g, T_m) and thermal gravimetric analysis (weight loss)
4. Linewidth and film thinning measurements
5. Measurement of the IR absorption constants of the substrate for IR bake or thermal constants of the wafer and the boat for conduction and convection baking

On-line process monitors are:

1. A thermocouple bonded to a wafer to ascertain the time–temperature conditions. Noncontact fluorooptic measurements are also very accurate techniques.[50]
2. Flow or image distortion by monitoring the superposition of a grating pattern.[51]
3. Gas chromatography of residual solvent or reaction products.

From an analytical and kinetic viewpoint, the decomposition of DQN can be monitored by changes in the ketone band, and by the diazo loss in IR or UV (365–435 nm). Similar procedures are used in prebake to avoid thermal decom-

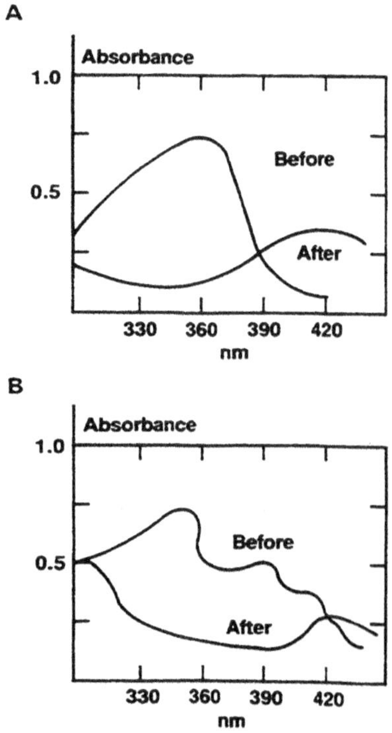

FIGURE 11-5-1. UV spectrum of (A) 150°C postbaked KTFR and (B) 150°C postbaked AZ-1350.

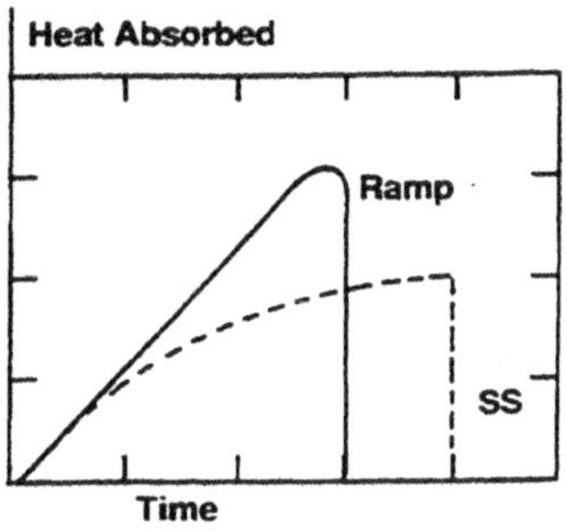

FIGURE 11-5-2. Integrated heat absorbed in ramp heating versus steady-state heat input into a resist during postbake.

position. Figure 11-5-1 shows the disappearance of the azide (negative) or DQ (positive) group after baking above the decomposition temperature of the sensitizer.

Since an activation energy of 40 kcal/mole is operative for DQN postbake, the temperature can have a profound effect on the rate of chemical conversion in postbake. A change of $\pm 10°C$ corresponds to a $\pm 10\times$ change in the extent of conversion. A temperature control of $\pm 1°C$ is recommended. Volatilization during prebake or melting absorbs about 20 kcal/mole. Thus, a control of $\pm 2°C$ for "physical" postbake reactions is sufficient. The time of postbake should commence at the point an isothermal condition has been achieved. Therefore, on-line temperature monitors are usually required. Table 11-5-1 gives some of the typical thermal characteristics of DQN positive and negative azide rubber resists. Most organic-based resists will have similar thermal constants.

In IR bake, several parameters such as substrate absorption, belt speed, and lamp emitter temperature, influence[52] the shape of the time–temperature ramp heating curves (Fig. 11-5-2). Since heat can be rapidly absorbed, the IR bake time and/or belt speed is very critical. Steady-state (hot plate) heating requires a longer ramp time of bake (Fig. 11-5-2) but easily achieves isothermal conditions.

The heat balance equation[52] for heat gain and loss during prebake/postbake has been modeled for an IR cure. Modeling requires accurate optical constants some of which are not available and have to be measured.[72] The thickness of the resist has a slight effect on the rate since most of the heat is absorbed in the substrate (Fig. 11-5-3). The silicon wafer absorbs heat more effectively than a glass or metal substrate. After the absorption of IR heat, the wafers are converted into miniature hot plates.

High postbake conditions, such as baking DQN in air above 150°C, induce insolubilization and difficulty of removal in liquid strippers. Dry plasma stripping is

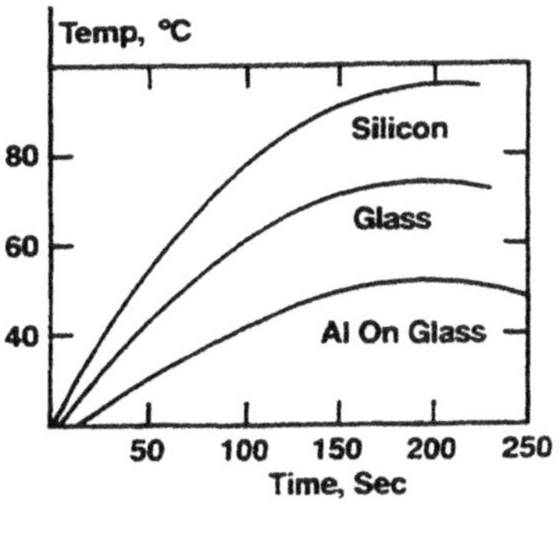

FIGURE 11-5-3. Time–temperature profiles for IR baking of various films of silicon wafer and of glass substrate with and without Al-deposited film. From Ref. 52, copyright 1983 and reproduced by permission of the Electrochemical Society, Inc.

TABLE 11-5-1. Thermal Properties of Resist (CGS Units)

	DQN	Azide rubber
Specific heat	0.36	0.37
Thermal conductivity	1.04	1.02
Density	1.21	1.19
Coefficient of heat expansion	1.0×10^{-4}	3×10^{-4}

insensitive to postbake conditions unless chemical hardening has incorporated metallic elements that do not erode in oxygen plasma. Other stripping gases or wet methods may have to be used to strip the treated resist.

11-6. SUMMARY AND OUTLOOK

Postbake especially for wet etching or aqueous processing is necessary in most resists to reduce or overcome the loss of interfacial adhesion due to swelling, film stresses, or imbided solvent and developer penetration along the resist–substrate interface. The process is also used to reduce the defects of pinholes by melting (slightly) and to chemically convert the resist to a material which is able to withstand dry processing at elevated temperatures.

The process time and temperature are chosen to be adequate and not cause excessive phase melting and linewidth distortion. Alternate chemical and physical hardening techniques are available for stabilizing DQN resists for subsequent dry etching, lift-off, or ion implantation. The deep-UV hardening of DQN is now a routine commercial process.

In the future, resists with higher thermal stability are likely to be commercialized.[75] These resists based on thermoset or higher-T_g polymers will be designed to eliminate the traditional flow problems of novolak-based resists.

With the trend to multilayer resist structures, the underlayer is chosen from melt-resistant polymers. The image layer resist is therefore not required to be postbaked or withstand temperatures above 100°C. Cooled reactive ion etch reactors also reduce the thermal stability requirements of resists. Dry-deposited resists also eliminate prebake and hence postbake if dry development is used. Thus, thermal postbake is likely to be eliminated from future resist processes. Radiation or chemical posttreatments of the developed image will continue to harden resists for harsh process environments.

REFERENCES

1. L. Fried, J. Havas, G. Paal, J. Lechaton, J. Logan, and P. Totta, *IBM J. Res. Dev.* **26**, 362 (1982), U. S. Patent 4,201,800 (1981), IBM.
2. U. S. Patent 3,976,524 (1975), IBM.
3. L. White, *J. Electrochem. Soc.* **130**, 1543 (1983).
4. K. Jinno, Y. Matsumoto, and T. Shinozaki, *Photogr. Sci. Eng.* **21**, 290 (1977).
5. A. Knob and W. Scheib, *Chemistry and Applications of Phenolic Resins*, Springer-Verlag, Berlin, 1979, p. 28.
6. W. Moreau, *Opt. Eng.* **22**, 181 (1983).

7. A. Ouano, in *Macromolecular Solutions*, edited by R. Seymour and G. Stahl, Pergamon Press, Elmsford, N.Y., 1982, p. 208.
8. T. Batchelder and J. Piatt, *Solid State Technol.* Aug. 1983, p. 211.
9. G. Parisi, S. Hazko, and G. Rozgoni, *J. Electrochem. Soc.* **124**, 917 (1977).
10. M. Andrasi, *Thin Solid Films* **67**, 229 (1980).
11. R. Brandes and R. Dudley, *J. Electrochem. Soc.* **120**, 140 (1973).
12. F. Dill and J. Shaw, *IBM J. Res. Dev.* **21**, 210 (1977).
13. J. Shaw, M. Frisch, and F. Dill, *IBM J. Res. Dev.* **21**, 219 (1977).
14. C. Deckert, Kodak Microelectronics Seminar Proceedings, 1977.
15. A. Morinaka and Y. Asano, *J. Appl. Polym. Sci.* **27**, 2139 (1982).
16. S. Croll, *J. Appl. Polym. Sci.* **23**, 847 (1979).
17. Japanese Patent 76,120,712, *Chem. Abstr.* **86**, 13177 (1976).
18. German Patent 2,631,535; U. S. Patent 3,868,254 (1975), GAF; U. S. Patent 4,339,521 (1981), Siemens.
19. Japanese Patent 78,135,621, *Chem. Abstr.* **90**, 130686 (1978).
20. U. S. Patent 4,139,384 (1981), Fuji.
21. U. S. Patent 4,311,533 (1981), Thompson CSF.
22. K. Kodota, Y. Taki, and O. S. Shimzu, *SPIE Proc.* **275**, 1373 (1981).
23. J. Bieron and R. Conley, *J. Appl. Polym. Sci.* **1**, 171 (1963).
24. R. Allen, M. Foster, and Y. Yen, *J. Electrochem. Soc.* **129**, 1379 (1982).
25. M. Bowden, L. Thompson, S. Farenholtz, and E. Doerries, *J. Electrochem. Soc.* **128**, 1304 (1981).
26. C. Ting, I. Avigal, and B. Lu, Kodak Microelectronics Seminar, 1982, G-136, p. 139.
27. J. Pacansky and H. Hiroaka, *J. Vac. Sci. Technol.* **19**, 1132 (1981).
28. A. Martynenko, B. Strizhov, and V. Nikolsku, *Russ. J. Phys. Chem.* **49**, 1310 (1975).
29. W. Oldham, *IEEE Trans. Electron Devices* **ED-29**, 554 (1982).
30. T. Berker and D. Casey, *IEEE Trans. Electron Devices* **ED-29**, 524 (1982).
31. W. Moreau, in *Microcircuit Engineering*, edited by J. Cleaver, H. Ahmed, and G. Jones, Academic Press, New York, 1984, p. 327.
32. R. Mattox and C. Tracy, *Solid State Technol.* June 1982, p. 83.
33. B. Cantos and D. Dobkin, *IEEE Electron Device Lett.* **EDL-2**, 222 (1981).
34. R. Morgan and C. Pollard, *Electron Lett.* **18**, 1038 (1982).
35. K. Mochiyi, H. Obayashki, and K. Wasa, *J. Electrochem. Soc.* **130**, 187 (1983).
36. U. S. Patent 4,187,331 (1980), IBM.
37. J. Moran and G. Taylor, *J. Vac. Sci. Technol.* **19**, 1127 (1981).
38. U. S. Patent 3,920,483 (1975), IBM.
39. U. S. Patent 4,253,888 (1981), Hitachi.
40. U. S. Patent 4,259,369 (1981), IBM.
41. U. S. Patent 4,125,650 (1978), IBM.
42. German Patent 2,447,225 (1978), IBM.
43. U. S. Patent 4,289,573 (1981), IBM.
44. U. S. Patent 4,007,047 (1979), IBM.
45. W. Moreau and P. Schmidt, *Electrochem. Soc. Ext. Abstr.* 1970, p. 459.
46. U. S. Patent 3,506,440 (1970), Mitsui.
47. J. Lyerla and J. Pacansky, *IBM J. Res. Dev.* **23**, 42 (1979).
48. H. Adabbo and R. Williams, *J. Appl. Polym. Sci.* **27**, 893 (1982).
49. U. S. Patent 4,289,573 (1981), IBM.
50. E. Egerton, A. Nef, W. Milkinsun, W. Cook, and D. Barr, *Solid State Technol.* Aug. 1982, p. 84.
51. H. Kleinecht and H. Meir, *J. Electrochem. Soc.* **125**, 799 (1978).
52. I. Calder, R. Sue, and H. Naguib, *J. Electrochem. Soc.* **130**, 1390 (1983).
53. E. Hryhorenko, Kodak Microelectronics Seminar, 1979, p. 18.
54. U. S. Patent 4,439,518 (1984), Shipley.
55. D. Johnson, *SPIE Proc.* **469**, 72 (1984).
56. V. Marriott, Y. Lin, and G. Fuller, *SPIE Proc.* **469**, 65 (1984).
57. M. Long, *SPIE Proc.* **469**, 189 (1984).
58. J. Matthews and J. Wilmott, *SPIE Proc.* **470**, 194 (1984).
59. U. S. Patent 4,476,216 (1984), Amdahl.
60. J. Koyler, F. Custode, and R. Rudell, *Kodak Microelectron. Semin. Proc.* **140**, 150 (1979).
61. S. Wu, *Polymer Interfaces and Adhesion*, Dekker, New York, 1982, p. 69.

62. W. Moreau, *Opt. Eng.* **22**, 181 (1983).
63. H. Hiroaka, *Am. Chem. Soc. Symp. Ser.* **266**, 358 (1984).
64. H. Yanazawa, N. Hasegawa, and K. Douta, *J. Appl. Polym. Sci.* **30**, 547 (1985).
65. German Patent 3,337,315, *Chem. Abstr.* **101**, 181207 (1984).
66. U. S. Patent 4,435,533 (1984), Xenon.
67. U. S. Patent 4,487,652 (1984), Motorola.
68. *Res. Discl.* 242 (1984), *Chem. Abstr.* **101**, 60461 (1984).
69. W. Feely, J. Imhof, and C. Stein, SPE RETEC Photopolymers, 1985, p. 49.
70. B. Lin, V. Chao, K. Petrillo, and B. Yang, SPE RETEC Photopolymers, 1985, p. 75.
71. Japanese Patent 85,4365, *Chem. Abstr.* **103**, 1485 (1985).
72. R. Bond, S. Szioba, and H. Naguib, *Electrochem. Soc. Ext. Abstr.* **81-1**, 729 (1981).
73. U. S. Patent 3,976,524 (1975), IBM.
74. L. White, *J. Electrochem. Soc.* **130**, 1543 (1983).
75. S. Turner, R. Arcus, C. Houle, and W. Schleigh, SPE RETEC Photopolymers, 1985, p. 35.

12

Additive Processes

12-1. INTRODUCTION

In additive processes, a solid or precursor is deposited through a resist mask onto a substrate. The solid is deposited from a liquid, gas, or solid state. In some cases, the solid is added to the resist and the resist is removed to deposit the solid. Metal films are most frequently deposited by lift-off or electroplating techniques. In Fig. 12-1-1, a panorama of additive processes using resists is shown. Some of the processes such as electroplating, fill and fire resists, and lift-off originated from printed circuit (mil-size) processes. By themselves, the resists find applications as light waveguides in optical information transmission and storage. For semiconductor fabrication, ion-beam implantation and the lift-off of metal interconnections have been the major uses of additive processing.

In the lift-off process (Fig. 12-1-2), the resist acts as a vertical standoff evaporation mask to separate the desired metal from the extraneous material.[1] Generally, an aspect ratio of twice the resist thickness to metal thickness is required. The undesired metal is lifted off by dissolution or by swelling (adhesion failure) of the resist underlayer. The main objective of additive processes is to minimize the "etch bias" by confining the deposited dimensions to the printed bias and, thus, reduce the bias usually lost in subtractive isotropic etching. Since Al alloys and multiple layers of metals can be deposited without having to worry about selective liquid or plasma etchants and resist failure, the lift-off process is a key metal patterning process. In some cases, materials are not etchable by reactive ion etching, since they do not form volatile products. Since the early 1960s, the lift-off process has been used for image reversal and mask repair of chrome photomasks. IBM has reported[2] its use since 1980, for the personalization of the metal interconnections of bipolar logic chips.[3]

In the lithography of lift-off, positive resists such as DQN and PMMA are frequently used since they are soluble in organic lifting solvents, while negative resists are rarely used since they remain insoluble and are difficult to lift even in organic phenol or sulfonic acid strippers. However, because most positive resists are thermoplastic polymers of moderate T_g (50–125°C), the metal deposition temperatures for single-film resists are restricted to lower than 150°C. Deposits of insulators and other higher temperature films can be produced, if inorganic or selectively etchable lift-off layers are used, instead of polymeric organic resists.

The additive processes are critically dependent (Table 12-1-1) on the three-

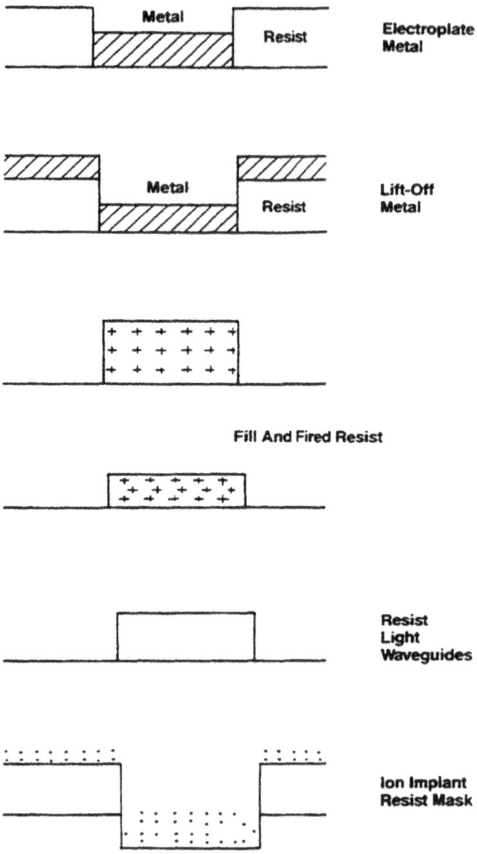

FIGURE 12-1-1. Additive techniques using the resist as a deposition mask or as a depositing material.

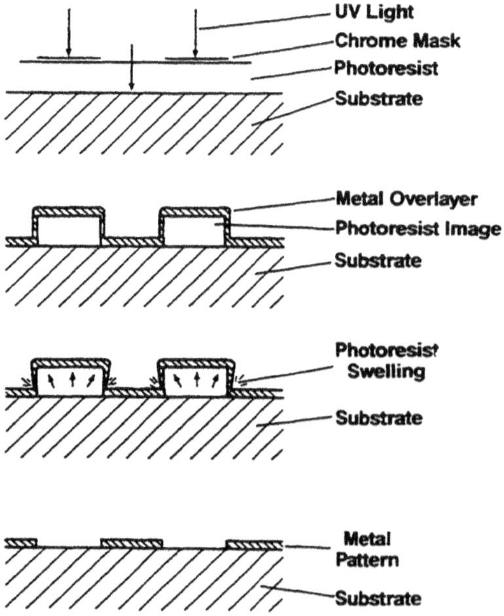

FIGURE 12-1-2. Schematic of metal deposition and lift-off technique. From Ref. 1, reproduced by permission of the Electrochemical Society Inc.

TABLE 12-1-1. Profiles for Lift-off

Profile	Application	Exposure lithography
 a. 45–75° Normal	Fair for lift-off and for thin metal ($\frac{1}{4}$ resist thickness), could be used as image layer in MLR.	Resist with contrast <3, at moderate dose with unexposed ≤20% positive resist loss
 b. 75–95° Vertical	Suitable for lift-off and plasma etching up to 80% resist thickness	Resist contrast <4, at high dose no resist thinning
 c. 95–110° Overcut	Excellent for lift-off but not plasma etching and also for ion implantation	High resist contrast (>6) at very high dose and backscatter radiation

dimensional nature of the resist profile, since the walls define (as in electroplating) or separate, the deposited metal (as in lift-off) from the undesirable metal. Very unusual profiles are generated in resists for separating the deposited metal in the vertical direction from the extraneous metal.

For the specific profiles in shown in Table 12-1-1, shallow profiles (case a) are normally achieved with photon or electron exposure. Vertical profiles (case b) are normally achieved with ion-beam and X-ray (nonscattering) exposure. At high electron doses, the backscattered electrons (case c) dominate the exposure process, to produce overcut profiles. Thus, profiles in single layers of resist suitable for lift-off are done at the expense of higher dose and lower wafer throughput (higher cost per exposure). In photon exposure, many physical-tool and optical factors limit the resist profiles to <75°. These include[7,34,362,369,414]:

- Diffraction
- Multiple reflection of light over steps
- Variable resist thickness from nonplanarization
- The image distortion by the MTF falloff
- The small depth of focus of optical projection systems
- The contrast of the resist
- Standing waves

The print bias is usually on the order of 25% of the resist thickness (3σ of $\pm$0.2–0.4 μm) measured at the bottom of the image and 0.4–1.0 μm measured at the middle and 0.4–1.0 μm at the top of a slope profile.[4] Thus, unless process tricks or changes can be produced to form top surface images with small biases, single-film lift-off resists are limited to about 2- to 3-μm-wide by 0.5-μm-high metal geometries.

For *e*-beam, X-ray, or ion-beam exposure, the resist sensitivity and thermal stability have been the limiting manufacturing economic factor. Resists with *e*-beam sensitivities of 1 $\mu C/cm^2$ and X-ray of 1 mJ/cm^2 are required for vertical profiles. Only polyphthaldehydes and polyolefin sulfones have approached these sen-

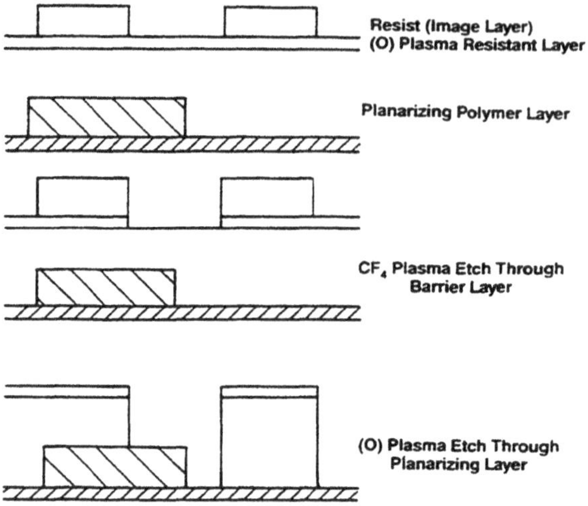

FIGURE 12-1-3. Steps in fabrication of a trilayer resist structure suitable for subsequent RIE or lift-off.

sitivities, but they are low-melting ($< 100°C$) polymers and not practical lift-off layers. In addition, polyolefin sulfones thicker than 0.7 μm crack upon development. Because of the thermal limitations, Al metal could not be deposited above 150°C with suitable grain size for bipolar devices.[2,6] Secondly, since the upper resolution of single films is limited by the film thickness, multilayer resists evolved in the early 1970s to overcome the thermal, dose, and profile limits of single-film resists. Although Al plasma etching was known, many postcorrosion problems limited its implementation. Fortunately, O_2 plasma (stripping) etching of polymers was widely accepted for stripping resists. Thus, a suitable lift-off mask could be formed if an O_2-resistant mask could be (sandwiched) created *in situ* on a polymer relief.[2,159,430] The direct application of dry reactive ion O^+ etching to transfer the image from a thin planar layer (Fig. 12-1-3) to a thicker planarizing layer derived from a soluble high-temperature polymer, such as an imidized polyimide, was a significant lithography breakthrough which started a whole series of two-layer, three-layer, and even four-layer composite resist structures of metallic, inorganic, and organic layers. Besides plasma transfer, other means of forming additive resist profiles (Fig. 12-1-4) include liquid developers, deep-UV exposure, or the high-energy exposure with electrons, X rays, and ion beams.

Indirectly, plasma etching has benefited from the generation of suitable lift-off profiles because of the high thermal and halogen plasma resistance of the planarizing layers. Today, Al lines can be fabricated by lift-off or direct plasma etching

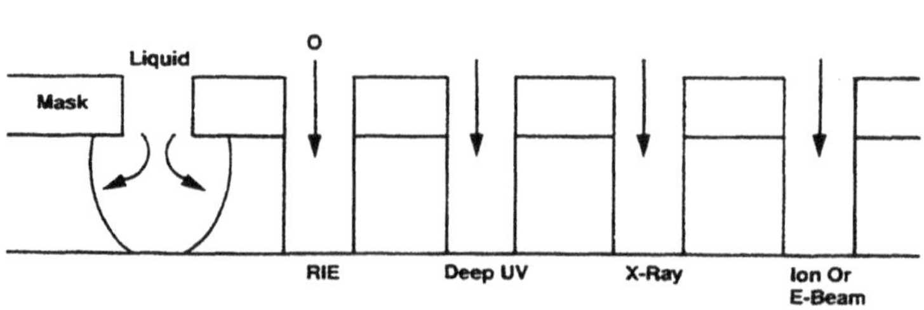

FIGURE 12-1-4. Image transfer methods of forming a final image in a multilayer resist. Both wet and radiation processing including gas or liquid development can be used.

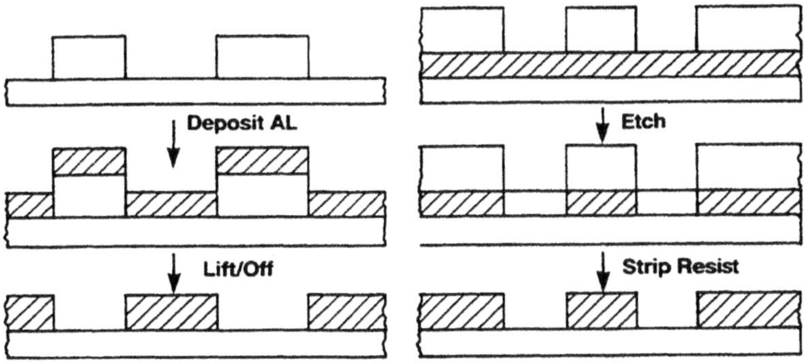

FIGURE 12-1-5. Lithographic fabrication of an Al metal interconnection by additive lift-off or subtractive etching.

(Fig. 12-1-5) using the same multilayer resist mask. The multilayer resists offer high thermal stability, excellent resolution (because of submicrometer planar image layers), and low plasma erosion. Negative resists could also be used in this scheme as image layer and plasma etch underlayer, but not as a lift-off underlayer (unless an inert stripper can be used). The multilayer resists have evolved to overcome the resolution limitations of single-film resists and the inherent physical limitations of organic resists. We will now examine the processes and materials that comprise single- and multiple-layer resists in additive processes.

12-2. SINGLE-LAYER RESIST FOR RESIST PROCESSES

Of all the resist structures for additive processes, the single-layer resist represents the lowest cost application in lithography. The first production use of single-layer metal lift-off involved the fabrication of chrome photomasks[3] using DQN resists. Later in 1969, Al interconnections were formed with vertical profiles by the lift-off of patterns generated in PMMA at an aspect ratio of 2 to 1.[5]

Of the profiles shown in Table 12-1-1, the vertical and overcut profiles are suitable for lift-off. The normal positive resist profile ($<75°$) requires selective back-etching or strict deposition conditions to be operational with deposited films thicker than 0.4 μm. In the case of vertical and overcut profiles, the metal should be deposited close to normal incidence or shadowing will inhibit lifting (Fig. 12-2-1).

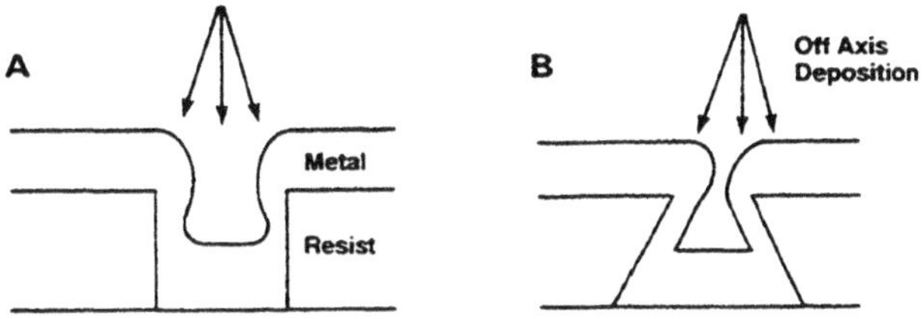

FIGURE 12-2-1. Metal deposition for lift-off at off-axis angle for (A) vertical and (B) overcut resist profiles.

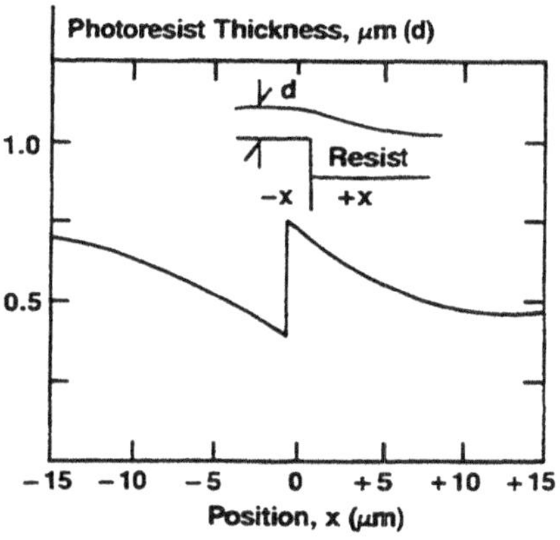

FIGURE 12-2-2. Profile of resist traversing a step of 0.5-μm height. From Ref. 7, copyright and courtesy of IEEE.

The creation of vertical profiles in thick resists is easily achieved with contact printing. However, projected images suffer from inter- and intraproximity effects due to diffraction, reflectivity, attenuation by variable resist thicknesses, image distortion by the lens, and standing waves. These limitations constantly challenge the implementation of single-layer lithography. The desirable characteristics of single layers for lift-off are as follows:

1. A soluble 200°C melt-resistant and stable resist (no volatilization or melting) to allow Al–Cu deposition at elevated temperatures to minimize electromigration[6] and promote good adhesion at the silicon or silicide contacts.
2. The steps over which the conductors are to be deposited must not be too steep or the deposited metal may short due to necking in the images of the planar resist[7] (Figs. 12-2-2 and 12-2-3). The resist should also be free of striations.[4]
3. The top surface opening of a single film of resist defines the evaporated metal profile on the substrate. Top surface imaging can be achieved by using a fast resist at higher *e*-beam doses.[8]
4. The aspect ratio of the resist thickness to the deposited metal thickness should be at least 1.5.[9]

We will now examine the two popular positive resists of DQN and PMMA as single-layer lift-off additive resists.

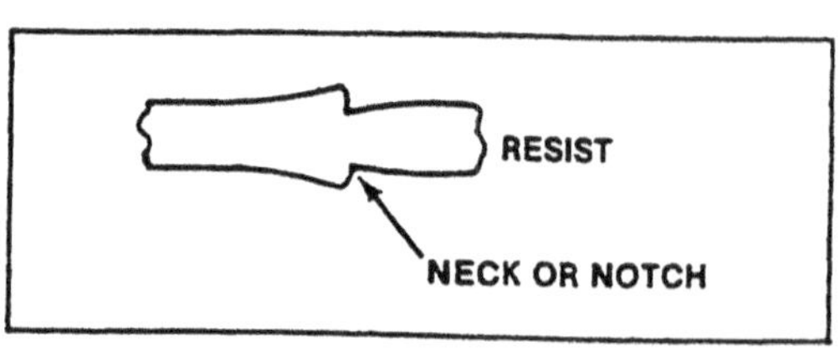

FIGURE 12-2-3. Top view of developed resist image traversing a step. From Ref. 311.

12-3. DQN FOR LIFT-OFF

The DQN resists are the most versatile of all resists and they have been applied for every type of additive processing (Table 12-3-1) with the primary interest directed to metal patterning. At present, metal interconnections are about 1 μm thick with 2- to 3-μm linewidths on 7-μm centers. In the next decade, 1-μm × 1-μm × 3-μm lines and spaces will probably be required. The ability of single-film resists to form a high aspect ratio of greater than 1.5 to 1 will be dependent on the MTF and NA of the stepper-scanner projection system (probably operational at about 250 nm with flash laser exposure). In the meantime, it is difficult to produce vertical or overcut profiles in positive resists on present topographies. Resist images traversing stepped topography suffer from nonuniform reflectivity (Fig. 12-2-2) and the images are constricted at the step junction (Fig. 12-2-3). The addition of dyes to the resist has reduced, in some cases, the internal reflections and has improved the linewidth control.[36]

The response of the resist to the absorbed forward and reflected radiation depends on the incident dose, the quantum yield or G value, and the dissolution rates of the resist for a particular developer. The resist contrast is a function of the dose. At a contrast of > 2.5, one can expect a vertical profile if the images in small features do not suffer from the intraproximity effect. A basic attempt to ensure that all lines receive equivalent dose, has been the inclusion of an antihalation layer. In photographic emulsion processing, an antihalation layer can be present above, below, or inside a silver emulsion to cutoff the nonlinear radiation. Figure 12-3-1 illustrates the loss of the forward intensity of a 1-μm projected image due to lens diffraction.

Griffing and West[35,314,408] applied the concept of a contrast-enhancing layer on top of AZ-1470J (Fig. 12-3-2) which is a variable transmission filter layer to absorb the second-order lens diffraction of the projected image. The pattern is

TABLE 12-3-1. DQN Resists in Additive Processes

Resist	Process	Reference
AZ-1350B	Lift-off overcut profile created by exposure of backside of glass	10, 11
DQN	Electroplate Cr, deposit Permalloy, etch Cu, and lift-off of Permalloy	12
DQN	Overcut profile, expose, treat image H^+, flood expose, bake, develop, image reverse	13
DQN	Expose, chlorobenzene soak, develop, overcut profile	14–19
DQN	Electron-beam, or deep-UV vacuum expose, bake, flood expose develop, overcut image reverse	20
AZ-1350	Add amines, bake, flood expose, overcut, image reverse	21, 289
AZ-1350	Xylene soak prior to expose develop and O_2 descum	22
AZ-1350	Prebake top of film with hot air while wafer cooled, overcut	23
DQN	Arsenic sulfide etched with DQN masks, use as 400 °C lift-off	24
AZ-1350B	Deposit Ti at 30° angle for 0.15-μm line	25
AZ-1370	Pattern, sputter deposit SiO_2, HF etch off resist walls, lift-off	26
AZ-1350	Contact print 1-μm lines × 2 μm thick, lift-off	1
AZ-1350J	Electron-beam, 10 μC/cm^2, low-temperature develop, overcut profile	27
AZ-1350	Postexpose, predevelop bake to remove standing wave edge raggedness	67
DQN	Develop in isoamyl acetate	394

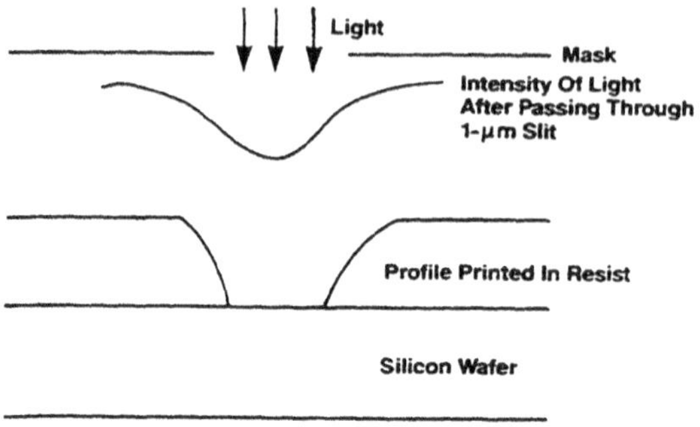

FIGURE 12-3-1. Diffraction-limited printing in resist.

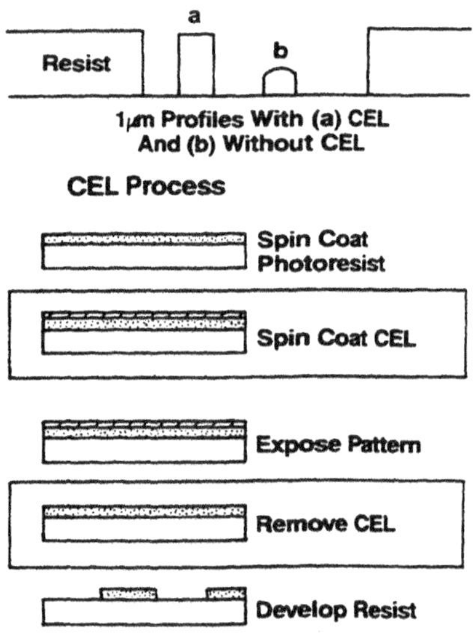

FIGURE 12-3-2. Contrast-enhanced lithography (CEL) using a bleachable topcoat to modulate diffracted lines. From Ref. 35, courtesy of SPE.

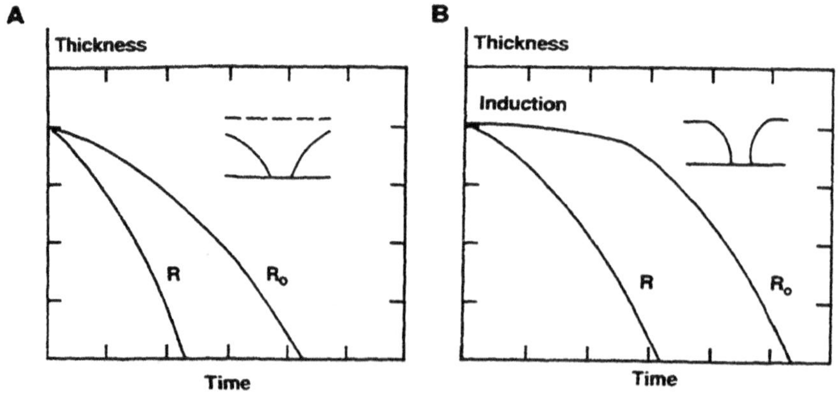

FIGURE 12-3-3. Dissolution curves of R and R_0 (A) without and (B) with induction.

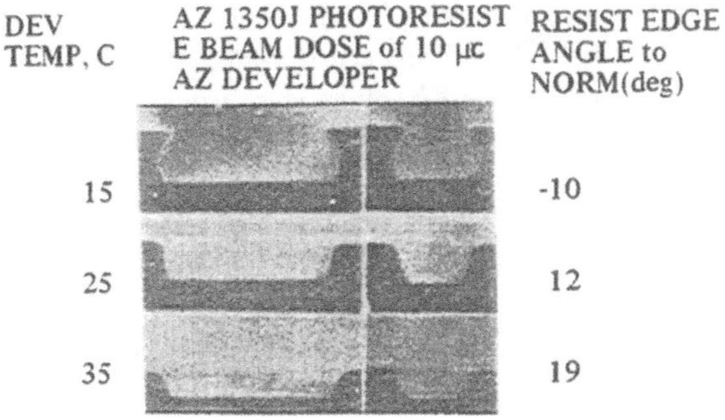

FIGURE 12-3-4. Effect of developer temperature on the profiles of *e*-beam-exposed AZ-1350J. From Ref. 27, reproduced by permission of the Electrochemical Society Inc.

exposed at 3–4× higher dose to offset the inner filter absorption and the bleached dyed layer is removed prior to the AZ-1470J development. Lines and spaces as small as 0.5 μm were resolved.

The dissolution characteristics of the unexposed DQN resist greatly affect the vertical profile in single-film resists. Some DQN resists, because of differences in novolak structure, molecular weight, and sensitizer loading, exhibit no induction effect while others have large induction effects (Fig. 12-3-3). In addition, DQN developers, comprised of buffers or other common ions, tend to eliminate the induction period. If induction is present, overcut features suitable for lift-off can be produced. Shaw and Hatzakis found that low developer temperatures (15°C) (Fig. 12-3-4) increased the induction effect and even created overhang (overcut profiles) in AZ-1350J, *e*-beam exposed at 10 $\mu C/cm^2$.[(27)]

Presoak Single-Film Lift-off Process

Since the novolak resins are formed by a condensation reaction, the molecular weight dispersivity is very broad, ranging from 3 to 20. If the exposed film is soaked in a fractionating solvent, low molecular weights and photoactive components are preferentially removed.[(15–19,416)] The absorption of solvent at the top surface also retards dissolution and contributes to the induction and overhang. Workers at IBM

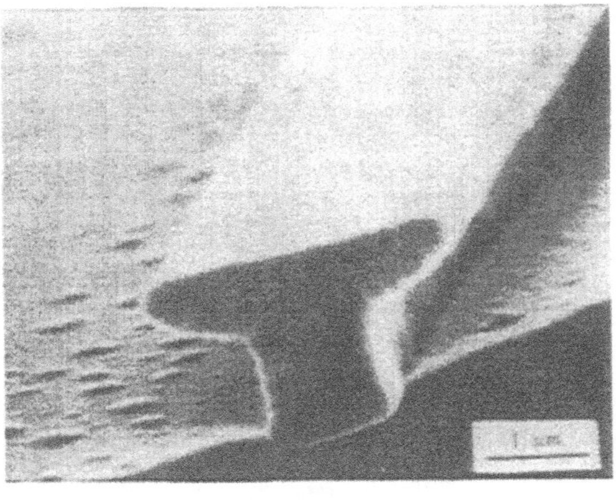

FIGURE 12-3-5. Profile in AZ-1350J after predevelop soak in chlorobenzene and processing. From Ref. 19, copyright and courtesy of AIP.

used chlorobenzene soak after exposure to create the profiles shown in Fig. 12-3-5.[15–19] The chlorobenzene diffuses into the top layer forming a swollen gel layer. The depth of penetration is controlled by the soak time, solvent content after prebake, and impurities in the chlorobenzene.[16,392,416] In production, tight process controls include the prebake, the soak time, the develop, and the expose time. In Fig. 12-3-6 are shown a series of overhangs as a function of soak time.[17] The soak process is limited to pitches of 4–5 μm and will vary from one commercial DQN to another. The top crust after soak is grainy. If the crust is not sufficiently thick, it curls up and can be removed in small geometries[22] by the action of the developer. If the gel layer is made thicker by a longer soak time, 1-μm geometries exhibit ragged edges or they fail to clear altogether. Batchhelder[22] recommends only a brief soak in xylene to create 75–90° profiles from a stepper exposure. To form 0.06-μm patterns, Speidel[19] evaporated aluminum at a 30° incidence angle (Fig. 12-3-7) to shadow the deposition (Fig. 12-3-8). Another example of the pileup technique is the low-angle (15°) (Fig. 12-3-9) evaporation of Cr/Au metals to form a submicrometer Schottky gate.[18,347–351]

Presoaking of the diazonovolaks in KOH,[37,282,397] alcohols,[38] NH_3, water,[39] and amines[41] can also cause an induction effect on R_0 and an overcut or vertical profile. Developing DQNs with a water rinse–interrupt cycle up to four times, can also produce vertical profiles for top surface imaging.[8,40] Lastly, the reversal of the images in DQN with overcut profiles (Table 12-3-1) can be formed by e-beam[424] or deep-UV exposure followed by flood normal UV expose and develop.[14,400] In the absence of moisture, the heavy dose of the e-beam or deep UV causes the bisketene generated from the DQ sensitizer to react with the OH groups

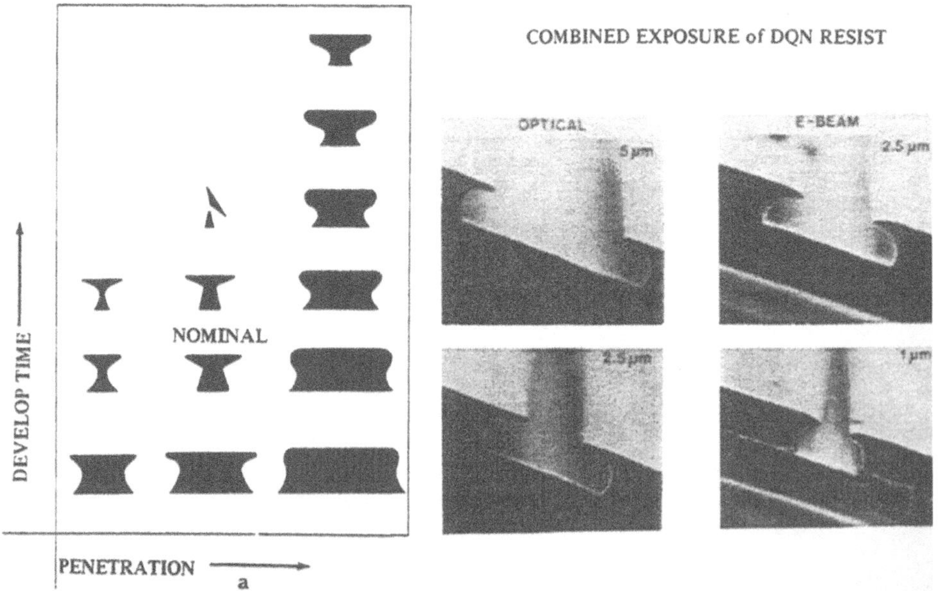

FIGURE 12-3-6. (a) Changes in profiles in AZ-1350J after soaking for various times in chlorobenzene. From Ref. 17, copyright 1982 and courtesy of IBM. (b) Examples of lift-off profiles created by chlorobenzene soak of diazo resist. From Ref. 294, copyright 1983 and courtesy of Academic Press.

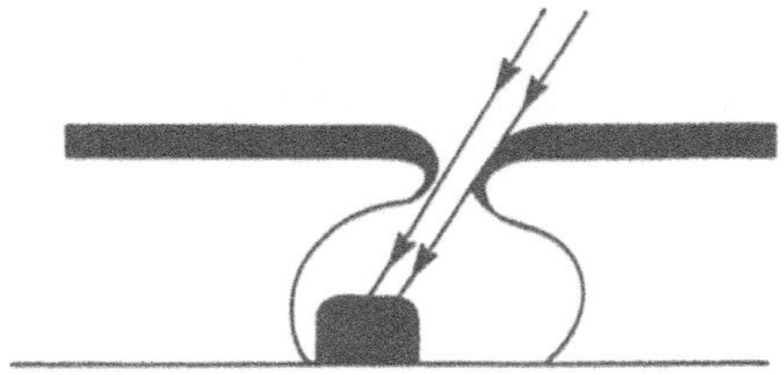

FIGURE 12-3-7. Shadow deposition of submicrometer metal using lift-off structure created by chlorobenzene soak prior to development of AZ-1350J. Metal is deposited off-axis to create 100- to 200-nm lines. From Ref. 19, courtesy of AIP.

of the phenolic resin to form intermolecular cross-links and negative insoluble images,[20,72,363,391]

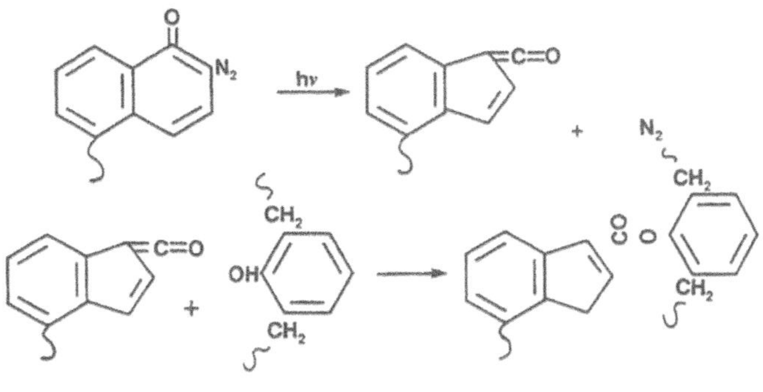

If water reacts with the ketene to form the acid, a predevelop bake at 100°C will induce a similar esterification with the phenolic resin.[73] Since the phenolic resin component of DQN absorbs strongly in the region of 200–300 nm,[240] deep-UV exposure of DQN produces an overcut profile due to insufficient light penetration to the silicon surface[120] (Fig. 12-3-10). Image reversal can also occur by the addition of organic bases such as imidazole to the DQN resist, followed by a

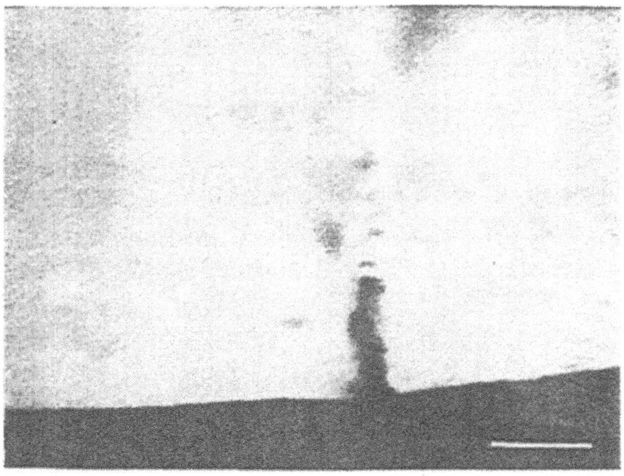

FIGURE 12-3-8. Metal lines of 0.1 μm deposited by shadow deposition (Fig. 12-3-7). From Ref. 19, copyright 1981 and courtesy of AIP. Scale bar: 0.5 μm.

postexpose (predevelop) bake at 100°C and a flood UV[21,46,393,397,398] exposure (Fig. 12-3-11). The intermediate salt thermally decomposes,[45,47]

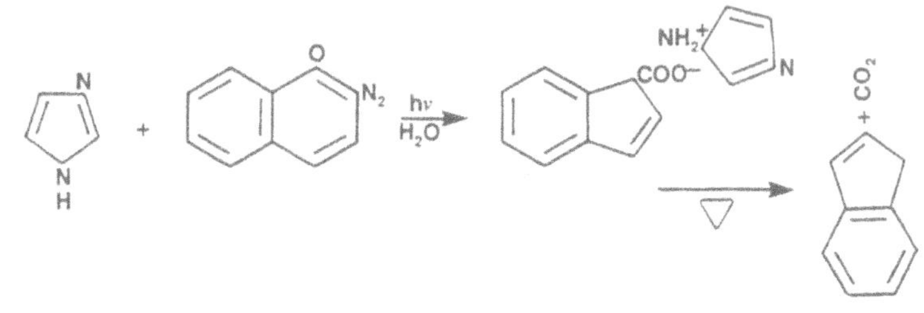

to form indene which is insoluble in NaOH. The images formed by high-dose exposure or by chemical reversal do not flow below 140°C, but require strong strippers for lift-off. However, DQN formulations with added imidazole, exhibit poor shelf life due to the base-catalyzed azo-coupling reaction of the DQ sensitizer. Substitution of a nonaromatic DQ of Meldrum's acid instead of the conventional DQN[51,52]

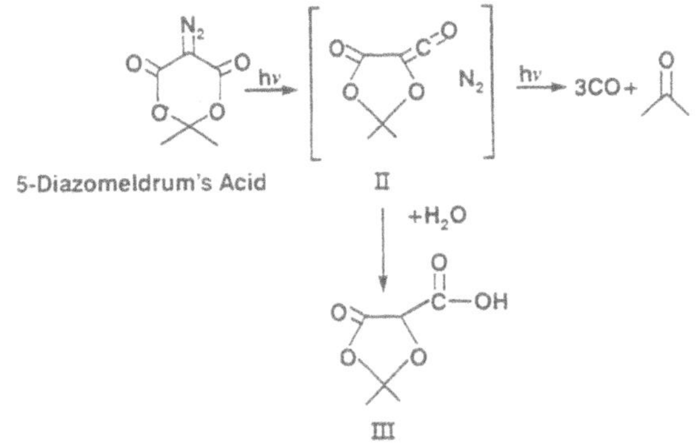

produces a coatable sensitizer which can be used as a self-developable film or as a dispersion in a novolak resin for deep-UV lithography and lift-off.[241]

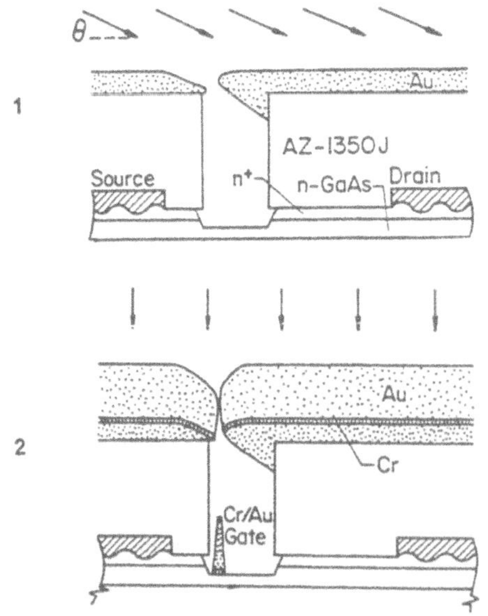

FIGURE 12-3-9. (1) Low-angle deposition to build up narrow constriction. (2) Normal-incidence evaporation to deposit submicrometer gate. From P. Chao, W. Ku, and C. Lowe, *IEEE Electron. Device Lett.* **EDL-3**, 286 (1982), copyright and courtesy of IEEE.

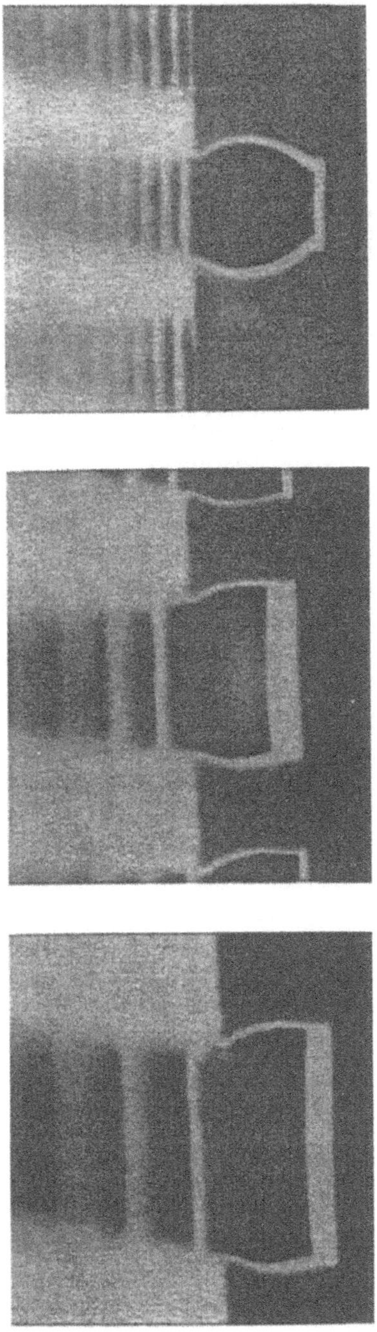

FIGURE 12-3-10. Profiles developed in deep-UV-exposed DQN resist. From Ref. 120, courtesy of IBM.

FIGURE 12-3-11. Inverted profiles formed in DQN by imidazole image reversal. Courtesy of M. Skinner, IBM.

Inverted profiles in negative resists such as Hitachi RD-200ON can be formed by the exposure of a bisazide in polyvinylphenol[48,233] (Fig. 12-3-12). Although the DQ bleaches at 313 nm, the resin absorbs strongly and light barely reaches the substrate. The scattered zone is etched out by the alkaline developer to produce (Fig. 12-3-12B) an inverted profile. The images are stable to about 100°C and can be lifted in organic solvents.[48] Instead of an azide–novolak system, a single-component DQ sulfonic acid ester of a novolak resin has been used with deep-UV inner filter effects to create overhang profiles for lift-off.[425]

Since images in PMMA and DQN resists are limited to about 120°C deposition conditions, polyimides with DQ sensitizers have been used as high-temperature (200°C) resists.[53,54] As more deep-UV exposure tools are developed, more high-temperature resists are possible, since photon energies (>5 eV) are sufficient to decompose polymer chains. Exposure with ultradeep UV (180 nm) self-develops PMMA, PMIPK (polymethylisopropenyl ketone), AZ-1350, polyether sulfones, and polyimides to produce vertical profiles in more thermally stable resists.[49,50] In addition, X rays, and ion beams can create any aspect (vertical)

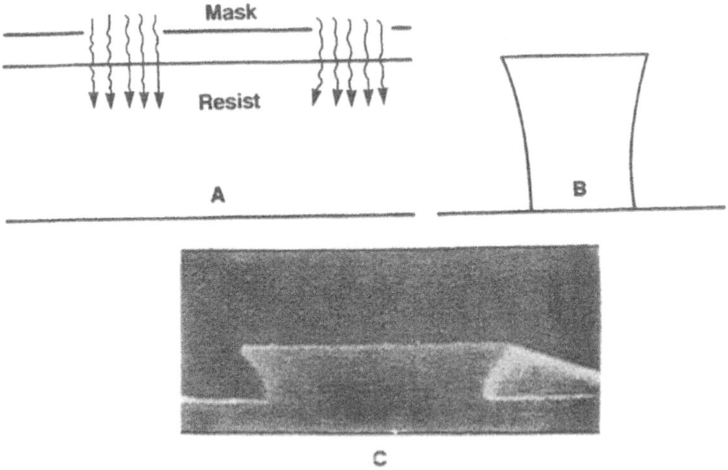

FIGURE 12-3-12. Imaging in highly absorbing azide–novolak resist. (A) Light absorption; (B) developer undercut; (C) actual profile. From Ref. 233, copyright and courtesy of IEEE.

ratio in DQN since the radiation is nonscattering. Basic improvements in overlay (alignment < $\pm 0.3\ \mu m$) and throughput [> 10 (125-mm diameter) wafers/hr] are required before these exposure tools become economical and applicable.

12-4. LIFT-OFF WITH PMMA FILMS

PMMA was the first lift-off *e*-beam resist and continues to be used for nanometer metal lines (Table 12-4-1). Unfortunately, the doses required for inverted profiles are considerable (> 100 $\mu C/cm^2$) and the images are distorted by the interproximity effect, unless 10-keV exposure is used.[55] To reduce the proximity effect, PMMA is best used in a multilayer resist structure at reduced film thickness (0.5 μm).[56] Figure 12-4-1 illustrates the interproximity effect in PMMA at a 25-keV dose in a 2.0-μm pitch of a 1-μm array, as a function of film thickness. While a 1.5-μm-thick film has an image of 1.0 μm in width enlarged with a bias of $\pm 0.7\ \mu m$, a 0.5-μm-thick film in a multilayer lift-off sandwich would have a bias of only $\pm 0.25\ \mu m$.

Since PMMA images flow above 125°C, several attempts to increase the thermal stability of PMMA have used polymethylacrylyl chloride (MAC) and polymethacrylic acid to form anhydride cross-links,[32,57]

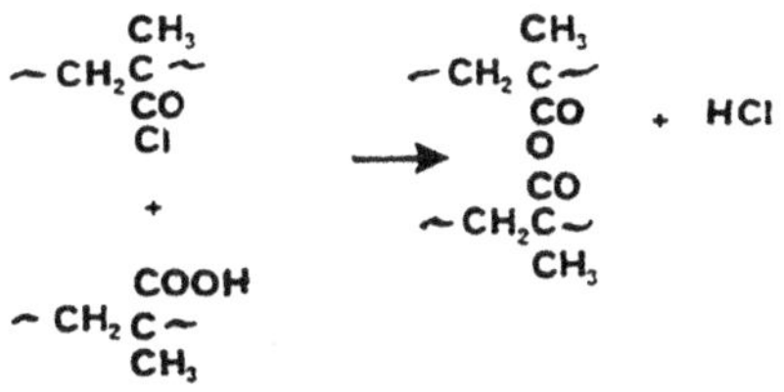

The cross-linked polymer is stable to 200°C but is not soluble in organic solvents and the lift-off requires pressure cooker treatment with ammonia.[32,60] Other deep-UV-exposable and cyclized polymers include soluble polyglutaramides,[433,434] and polymethacrylic anhydrides which can be derived from poly-t-butyl methacrylate,[58] polymethacrylic acid,[88] or polyitaconic acid.[59]

TABLE 12-4-1. PMMA Resist in Single-Film Additive Processes

Process	Reference (year)
1. First lift-off, 2-μm vertical profiles	5 (1969)
2. Cross-linked, lift in NH_4OH	32 (1975)
3. Lift-off on Si_3N_4 membrane, 250-Å metal	31 (1978)
4. X ray, 10 to 1 aspect ratio, 0.1 μm	28 (1976)
5. X ray, lift 175 Å	29 (1980)
6. Lift-off on thin C foils, 17 nm	30 (1981)
7. Expose, develop step, deposit, electrochemically remove metal underlayer (sacrificial layer), RIE, O_2, Al remain acts as mask	33 (1983)

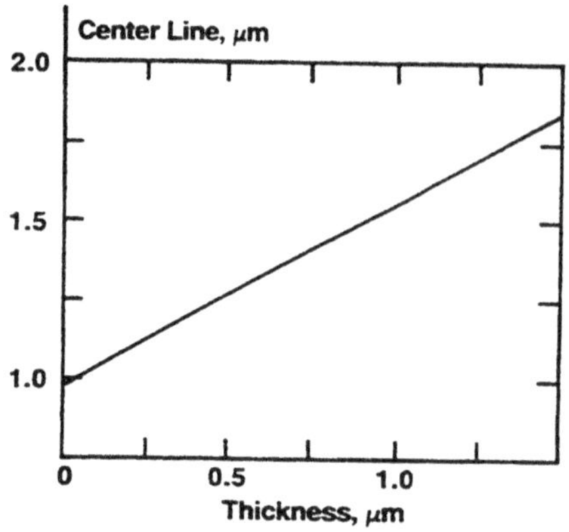

FIGURE 12-4-1. 1.0 μm by 1.0 μm in width line and space array, center resist line (trench) opening as a function of resist thickness for PMMA exposed by 25-keV electrons. Courtesy of N. Eib, IBM.

12-5. SINGLE-FILM LIFT-OFF WITH SHALLOW-PROFILE RESISTS

Single layers of resists with profiles less than 75° at an aspect ratio of 1 to 1 can be used for lift-off, if the sidewall deposits of Al[61] and SiO_2[62,116,411] are etched back with liquid etchants (Fig. 12-5-1) or plasma etchants. The process involves the HF removal of the thinner sidewall deposits with an accompanying sacrificial removal of the deposited (vertical) thickness. With an aluminum deposit, the resist melting is retarded by the Al overcoat, but deposition temperatures over 150°C cannot be tolerated, since the DQN resist will outgas, cross-link, and be difficult to lift. If a higher-temperature phenolic or maleimide resin is substituted for the novolak resin in the DQN formulation, temperatures as high as 200°C can be tolerated. In the etch back of Al or SiO_2, careful etching controls are needed and even ion-beam etching at acute angles can also be used to selectively remove the sidewalls.[63] To reduce sidewall deposition, the sputter deposition of metals or insulators, as opposed to the traditional line-of-sight boat evaporation, is claimed to reduce the sidewall deposition. This allows shallow single-resist profiles to be used and metals of Al alloys,[64,66] Al_2O_3,[65] and SiO_2 can be sputtered deposited at thicknesses greater than the resist thickness! Al_2O_3 up to 10 μm thick was deposited and lifted off with a 2-μm-thick resist. The secret of this process is the

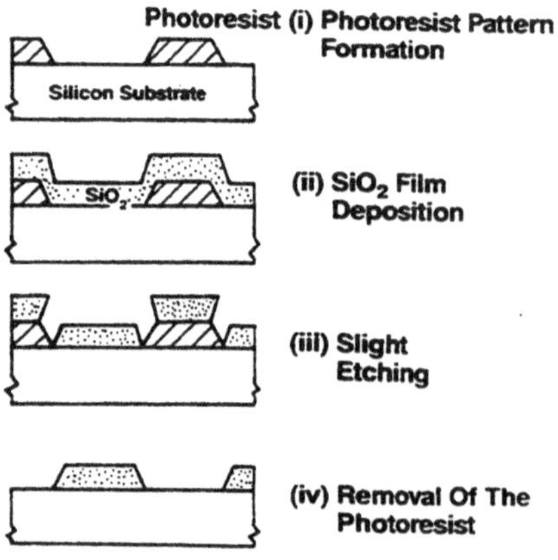

FIGURE 12-5-1. Lift-off profiles and process for depositing and lifting sputtered quartz. An etch back with HF is used to create gap (iii). From Ref. 62, reproduced by permission of the Electrochemical Society Inc.

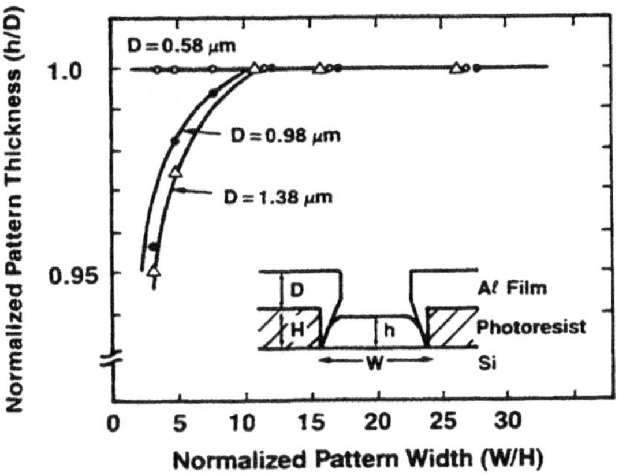

FIGURE 12-5-2. Lift-off of Al using metal deposition thicker than the resist thickness. The height-to-width ratio and normalized deposited metal thickness (D) are shown. From Ref. 66, reproduced by permission of the Electrochemical Society Inc.

development of a stress-induced microcrack at the conformal coating cusp (Fig. 12-5-2).

The distribution of Al atoms ejected by an Ar sputtering system obeys the cosine law.[65] Because the mean free path is 50 mm at the sputter pressures (1 Torr), the sputtered atoms experienced many collisions before reaching the wafer surface. The films grow on top of the resist and substrate, but do not fuse together because of the difference of thermal expansion of the resist and the substrate. Because of the low-deposition conditions, microcracks are created. Al films of 2 μm wide by 4 μm thick were lifted successfully and the slope of the metal images are only about 45–60° (isotropic wet etch profiles) but with smaller biases (approx. 0.5 μm) than wet etch (1- to 2-μm bias).

12-6. SUMMARY AND OUTLOOK OF SINGLE FILMS

Single films of resist reveal all the defects of physical scattering of projected lithographic radiation, especially in resist films overlying uneven topographies. The interproximity effects in thick films hamper the placement of closely spaced features for successful metal lift-off. In order to produce vertical profiles, high-speed resists with contrasts of 3–5, are required to offset the modulation loss in 1- to 2-μm-thick resist films. High-speed positive (1 $\mu C/cm^2$) *e*-beam or (1 mJ/cm^2) optical and X-ray resists are needed to fulfill the throughput requirement of the more sophisticated exposure tools. Nevertheless, the lower cost and simplicity of single-film lift-off remain a technical advantage over multilayer resist processing, provided the linewidths are greater than 1 μm (and film thicknesses < 0.5 μm) and that high-temperature (> 150°C) deposition conditions are not needed.

The limited thermal melt stability (100°C) of DQN and PMMA resists remains as a drawback to the minimum deposition temperature requirements of 150–200°C for Al–Cu–Si metallurgy, but replacement resists stable to 200°C are imminent with maleimide or glutaramide-based polymers. Confining the image to the top surface of overcut profiles requires many process tricks which trade off cost, tight process controls, and the ever-present process irreproducibility.

12-7. DOUBLE-LAYER RESIST

The objective of a double-layer resist process is to separate the image layer (I) from the planarizing lift-off layer (L),

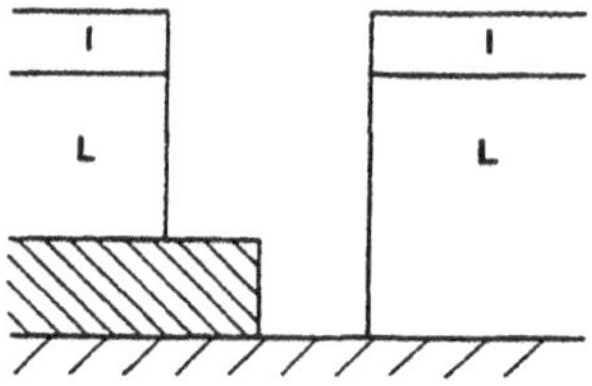

and thus overcome the resolution and thermal deficiencies of single-layer resists. Since the image layer can be used at about one-half the thickness of a normal resist layer (< 0.5 μm thick), class 10 or better clean room process conditions are required to reduce the pinholes (Fig. 12-7-1). Patterning of a 2-μm-thick film over 1-μm steps by optical exposure is subject to multiple reflections (Fig. 12-7-2)[74] and resist necking in the images (Fig. 12-2-3). The variable resist thicknesses require different exposures for optimum linewidth control (vertical intraproximity effect). If one assumes the acceptable linewidth variation to be ±10%, then the achievable resolution will be lower than the capability of the lens. It has been reported that if a lens of NA of 0.3 capable of resolving 1-μm features is used to print in a 1-μm-thick resist over a 0.5-μm oxide step, the minimum feature size is 1.75 μm.[75] For a 1-μm step in Al, the minimum feature size is 3.5 μm.[75] By using a higher NA lens [0.42, but with a smaller field (7 mm), and 1.2-μm depth of focus], the distortion over a step can be reduced to about 0.5 μm.[177] As an approximation, the effective optical thickness of a 2-μm-thick resist is the thickness divided by the refractive index of a resist (1.6) or about 1.25 μm.[74] In a single-layer resist cast over topography, a variation of 25 to 50% in thickness can occur, which gives rise to considerable standing waves (Fig. 12-7-3) and image distortion. To reduce the standing waves, several techniques have been used which later were applied to bilayer resists: (1) polychromatic exposure light[76]; (2) an antireflective underlayer (a form of a bilayer resist) comprised of a base etchable dyed polyimide[77]; (3) the control of oxide thickness.[76] Secondly, the depth of focus of some step and repeat systems is close to 2 μm which sets a practical upper limit of a single-film resist thickness to about 1.0 μm and a resolution of about 1.5 μm over normal device topographies.

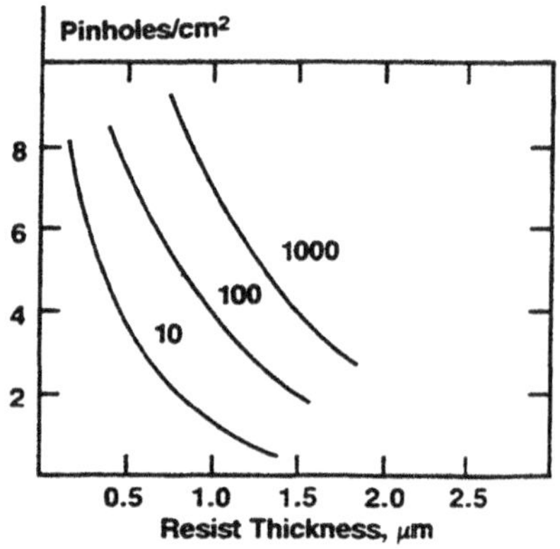

FIGURE 12-7-1. Pinholes (>0.3 μm in diameter) in resist films cast in clean rooms of different classes at various resist film thicknesses.

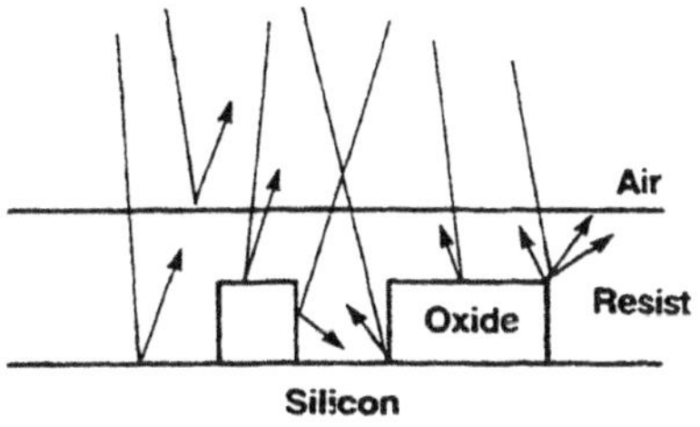

FIGURE 12-7-2. Multiple reflections of light from the substrate and resist interfaces.

To overcome the basic resolution limits of single-film resists, various structures can be created in a double-layer resist. The image layer serves both as an image layer and as an anisotropic image transfer (barrier) layer inert to O etch. The underlayer can also selectively absorb standing waves, which exposes the lift-off layer. For example, the bilayer resist can even be composed of two familiar resists, DQN and PMMA. Since DQN (Fig. 12-7-4) absorbs strongly in the deep UV (200–300 nm) and PMMA only moderately absorbs (<254 nm), an image layer of DQN masks a PMMA sublayer for image transfer by flood UV exposure. Even a 0.2-μm-thick film (Fig. 12-7-5) will absorb enough radiation to mask PMMA for deep-UV (254 nm) exposure. Figure 12-7-5 shows the optical dual-layer resist (DQN/PMMA) of a 0.75-μm line and space transferred from a thin edge of standing waves created in the top DQN image layer.

Various structures can be created in double-layer resists by a combination of exposure, plasma etching, or develop–etch transfer which mimic single-layer profiles, but at lower doses. Table 12-7-1 illustrates the many profiles and conditions for producing the single- and double (or triple)-layer profiles suitable for additive processes.

Overhang profile 9 or extended profile 6 are difficult to obtain for closely pitched features because backscattered electrons obscure the spacing. Vertical profiles 2, 5, and 8 are desirable for dry etching and electroplating applications. Profiles 3, 6, and 9 are liable to collapse or curl with small pitches (2–3× film thickness) while profiles 2, 3, 5, 6, and 9 are suitable for lift-off since the metal deposit can be defined by the opening in the image layer and sufficient vertical cavities (gaps) are present to offset the contact metal from the excess layer.

Three basic configurations have evolved for dual-layer resists. The initial

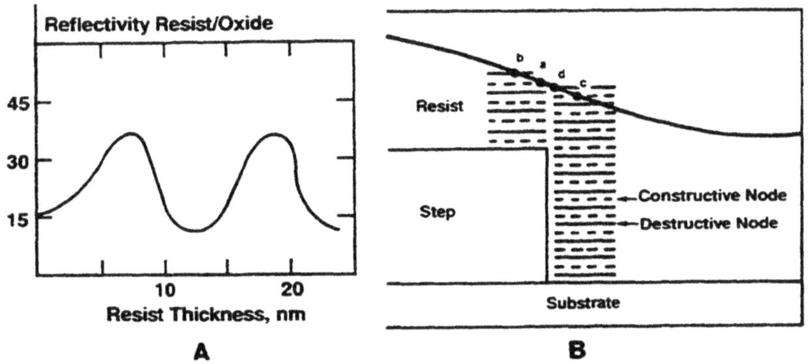

FIGURE 12-7-3. (A) Reflectivity from resist oxide interference. (B) Schematic of interference of light in resist traversing a step. (B) from Ref. 78, courtesy of Dr. C. Ting, Intel.

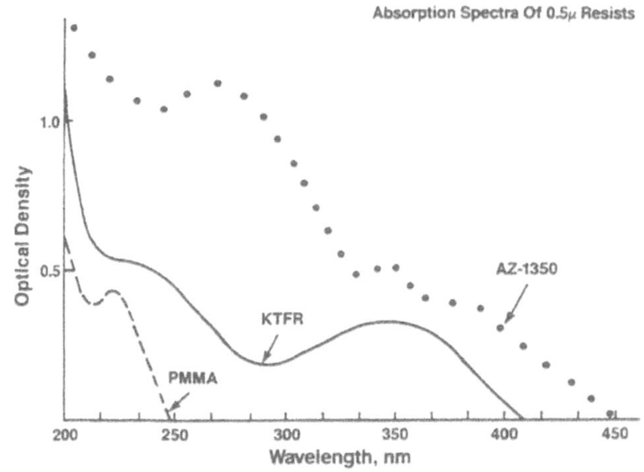

FIGURE 12-7-4. UV absorption spectra of 0.5-μm-thick films of AZ-1350, KTFR, and PMMA.

systems (1976) were layers of similar resists with single or mutually exclusive developers. The next generation (1979) was the conformable photomask—DQN on PMMA. At present, plasma etching is used with dual-purpose layers of inorganic, polysiloxane, silicone-containing resist, and postsilylated resist serving as both the barrier and image layer on top of high-temperature but soluble polyimides or other stable polymers. Not only are these structures popular for lift-off, they have become the dominating structures for plasma etching and high-resolution lithography with all forms of exposure (deep UV, *e*-beam, ion beam, and X rays).

12-7-1. Developer Transfer Bilayer Resist

The use of single-layer resists is limited by the inherent sensitivity of the film and the tolerable or suitable profile for the subsequent process. As a single-film

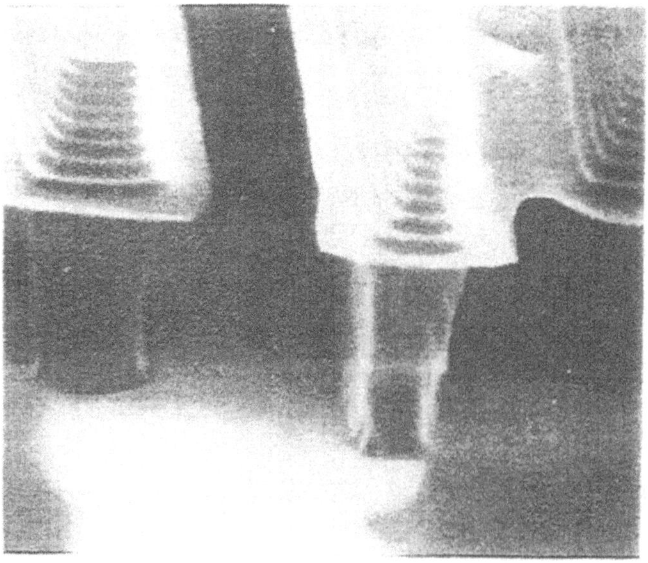

FIGURE 12-7-5. Standing waves printed in AZ-1350 portable conformable mask on top of film of PMMA. The waves can be reduced by appropriate absorbers in the PMMA underlayer. From Ref. 78, courtesy of Dr. C. Ting, Intel.

TABLE 12-7-1. Creation of Single- and Double-Layer Resist Profiles

Number	Profile	Exposure and transfer conditions	Reference
1		Single film, optical or *e*-beam moderate dose	5
2		Contact print ion or X rays, and high-dose *e*-beam	1
3		High-dose *e*-beam	5
4	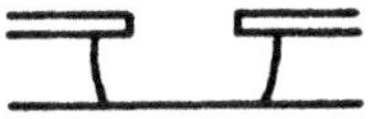	Dual-layer resist, *e*-beam developed	79
5		Plasma or deep-UV transfer	75, 80
6	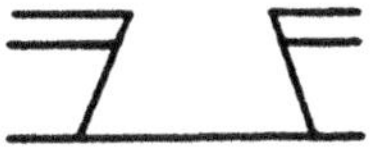	High-dose *e*-beam, two developers	82, 83
7	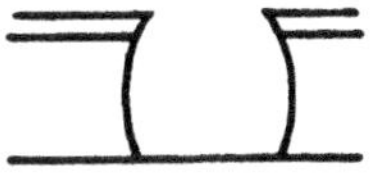	Plasma *e*-beam, single developer	81, 105
8		RIE, O_2 + overetch	81
9	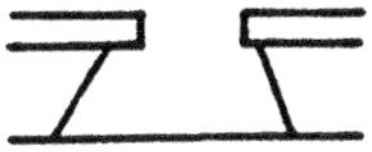	High-dose *e*-beam	86

e-beam resist, the sensitivity of PMMA is about 20–100 $\mu C/cm^2$. If a higher-molecular-weight PMMA is overcoated on top of a lower-molecular-weight PMMA (Fig. 12-7-6), a 2–4 × increase in sensitivity can be obtained. For a PMMA of 500k M_n, a dose of 10 $\mu C/cm^2$ produces an R/R_0 of 6 while a lower layer of 100k M_n has an R/R_0 of 2,

$$R/R_{0(\text{layer 1})} = 600/100 = 6$$

$$R/R_{0(\text{layer 2})} = 2000/1000 = 2$$

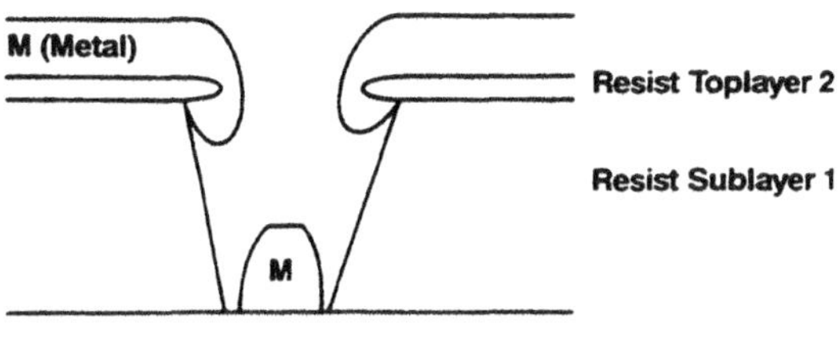

FIGURE 12-7-6. Bilayer resist image formed by high-molecular-weight PMMA deposited on top of lower-molecular-weight PMMA. The exposed underlayer dissolves faster than the top layer creating an appropriate undercut for lift-off.

If we form a bilayer,

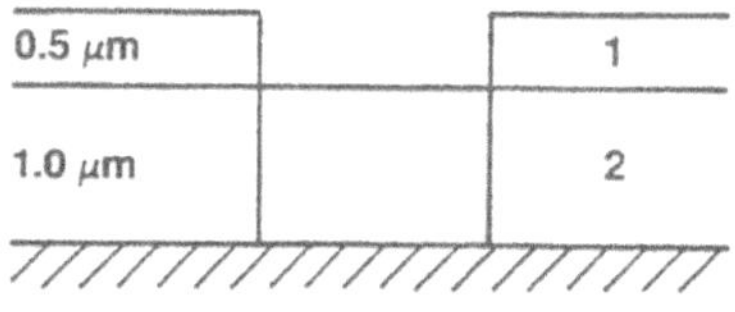

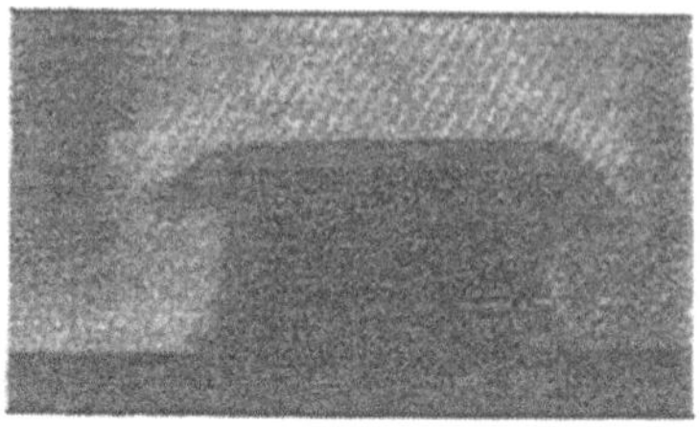

the top layer (1) will undergo a film loss of 1300 Å,

$$R_0 = 100(8' + 5') = 1300\ \text{Å/min}$$

to develop the bottom of layer 2,

$$R = 600(8') + 2000(5') = 15{,}000/13 = 1200\ \text{Å/min}$$

therefore,

$$R/R_0 = 1200/100 = 12$$

and a gain in sensitivity and a better lift-off profile are obtained.[84] If two mutually exclusive developers are used, a further increase in sensitivity can be obtained. In structures in which the top layer of PMMA is replaced by a copolymer of PMMA-methacrylic acid (MA)[85,86] or MAN, PMMA-MA-methacrylic anhydride,[87,88] the top layer is developed in alcohols, while the bottom PMMA layer is developed in aromatic solvents which do not erode the image layer. The acrylate double layer has been used to make submicrometer devices at moderate *e*-beam doses (Table 12-7-2) with thin ($< 0.5\ \mu$m thick) metal deposits. In forming dual layers, acrylate polymers which are not glassy ($T_g < 75°$C) are difficult to coat without interfacial mixing.

TABLE 12-7-2. Acrylate Dual-Layer Electron-Beam Resists

Image layer	Planarizing layer	Reference
PMMA (high M_w)	PMMA (low M_w)	84
PMMA-MA	PMMA	89
PMMA-MA (high M_w)	PMMA-MA (low M_w)	90, 121
PMMA	PMMA	91
PMMA-MA-MAN	PMMA	87
XLPMMA	PMMA	89
PMMA	PMMA-MA	92
PMMA	PMMA-MACl	89
PMMA	PMMA-MA	93–96
PTBMA	PMMA	97

12-7-2. Optical Dual-Layer Resist

The popular bilayer structure of DQN on PMMA originated from the work of Lin[34,236] (1976), who combined the deep-UV sensitivity of PMMA reported in 1970[98] with the high absorbance of the phenolic resin (Figs. 12-7-4 and 12-7-5) in the deep mid-UV (200–320 nm).[99] This basic concept evolved into many PCMs (portable conformable masks) (Table 12-7-3) consisting of a highly absorbing, usually aromatic-based resist on top of a deep-UV acrylate-based resist.

In forming the DQN/PMMA sandwich, the major problem[78] associated with DQN over PMMA is the interfacial polymer miscibility, which leaves a PMMA developer-resistant scum layer that prevents the PMMA underlayer development. Since the prebaked DQN contains trapped solvent and is a lower-T_g polymer than PMMA, the novolak resin diffuses into the PMMA layer. The immiscible layer has been studied by Lin as a function of empirical parameters, such as spin coat time and prebake, but no prevention of intermixing was found.[34,74]

From a materials viewpoint, the intermixing can be addressed by choosing an immiscible top or underlayer. The top layer could be a variety of resists such as:

1. A polymeric diazoquinone one-component system[115,228]
2. A polyvinyl phenol-azide (Hitachi RD-2000N)[48]
3. Low-molecular-weight diazoquinone ester novolak[328,364]
4. KTFR azide-rubber
5. Negative Cl-styrene resists (CMS)[234]
6. Polytetrathiafulvalene[111,343]
7. Use of a trilevel resist to prevent layer mix[100]
8. A dyed polymethylpentene-novolak resist[373]
9. A dyed photoresist[257,361,413,421]
10. A polyvinylalcohol barrier layer[345]

Since the negative resists are solvent resistant, they are more useful. The DQN top layer is capped (insolubilized) by 1 J/cm^2 deep-UV exposure due to resin–sensitizer cross-links[72] or by overcoating DQN with a conformable mold of PMMA, followed by baking and stripping.[356,415] Another process trick is to treat the PMMA[235] layer with a CF_4 plasma prior to coating the DQN.[118,278,320] Finally,

TABLE 12-7-3. DQN and Aromatic Dual-Layer Resists

Imaging layer	Planarizing layer	Reference
AZ-1350J	PMMA	74, 75, 80, 99–102, 232
DQN	Exposed and baked DQN	103, 251, 297, 404
AZ-1350J	AZ-1350J saturated sensitizer	104
DQN	Polybutene sulfone	105
AZ-1350J	PMMA + β-carotene	106
KTFR	DQN	107
KTFR	Polyimide	108
RD2000N	PMMA	109, 110, 317
PTSF	DQN	111
DQN	PMMA	112–114
RE500P	PMMA	339
CMS	PMMA	339
AZ-1350	Polyglutaramide	433, 434

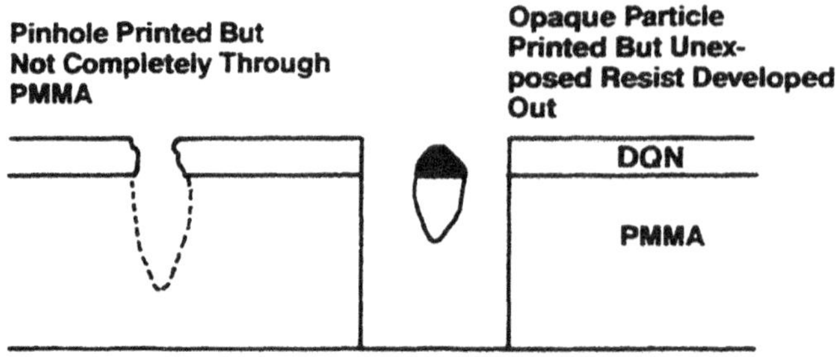

FIGURE 12-7-7. Nonharmful defects in bilayer portable conformable mask.

an aqueous developed resist such as PMMA-glutaramide or methacrylic acid anhydride deep-UV resist eliminates the interfacial layer by dissolution in the subsequent alkaline developer. If an interfacial layer persists, it is removed by an O_2 plasma prior to blanket deep-UV exposure and develop. Initially, the blanket deep-UV exposure of PMMA underlayer required a dose of several hours from H_2 or Hg lamps, but recently high-intensity Cd lamps[113] or microwave-powered Hg lamps[114,119] emitting in the deep UV have reduced the flood exposure time to minutes.[113,242,243]

The resolution of the DQN–PMMA is better than the total film thickness and resolution to 0.5 μm has been achieved in a 1.0-μm sandwich.[123] For device processing, the other major performance aspect of the DQN–PMMA system is the pinhole level in the 0.3-μm AZ top layer. As expected (Fig. 12-3-11), the pinhole level is about 6 to 8 defects/cm^2.[34] The defects are usually a reflection of the cleanliness of the coating environment and not the resist itself. In the developed DQN image (Fig. 12-7-7), some of the pinholes or opaque particles, due to diffraction effects, may not be transferred completely into the PMMA.[34]

The usefulness of the DQN–PMMA is limited by the reactive plasma or thermal stability of the PMMA underlayer. The deep-UV exposure serves to transfer the image in an efficient form of contact printing.[266] Higher-temperature underlayers of acrylate or other deep-UV resists are needed for more manufacturing applications. In this regard, recent work on cyclized polyglutaramides,[301,366,370,409]

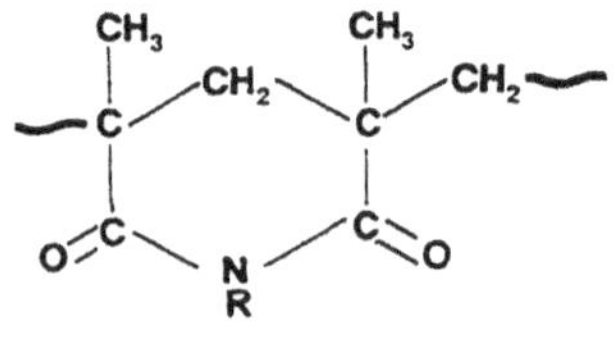

has shown deep-UV sensitivity, melt stability to 175°C, and alkaline developability. Another limitation of a DQN resist as the top layer is the gray zone of absorbance at edges of the images of the resist. Other polyaromatic resins, of higher extinction coefficient in the deep UV, and anthracene-type dyes added to DQN, provide higher absorbance than the 0.4-μm-thick novolak resists.[373]

12-7-3. Dual-Layer Plasma Etch Transfer

The transfer of an image by reactive oxygen plasma etching increases the possibility of using high-melting thermoplastic soluble[229] polyarylethersulfones and polyimides as planarizing layers and subsequently for plasma etching or lift-off. Molding of the dual properties of an image resist and O_2 barrier resistance into one resist is achievable by selected inorganic and silicone polymers. Even the DQN resist can be postsilylated with silanes to incorporate at least 7% Si[419] into the resist,[420] to serve as an oxygen plasma transfer barrier,

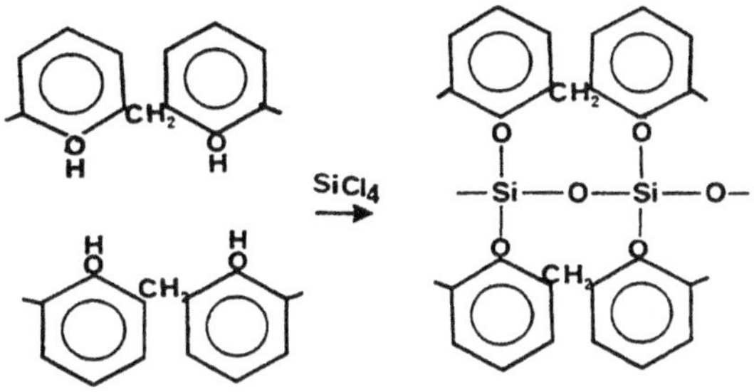

A plasma-resistant top image layer can be readily derived from inorganic[290] resists, polysiloxanes,[179,223] or silicon-doped PMMA resists (Table 12-7-4).

The bottom layer can be any carbonaceous polymer, since O^+ plasma etching is used to transfer the pattern.[336] Among the top layers which are resistant to CF_4 or O_2 are As_2S_3, polyiodostyrene,[341] AgBr,[387] dry-developed Ge-Se,[274] silicone resin,[132] PMMA-Sn MMA,[132,133,375] and any silicone monomer–polymer combination.[261,376] For an oxygen plasma barrier, a minimum amount of Sn or Si of 7% by weight is sufficient to act as an oxygen etch barrier.[149,150,256,335,373,386,418] Images can be confined to 1000-Å layers using ion-beam exposure (Si^+, In^+), followed by plasma O_2 etching of the *in situ* mask.[69,379–381] Stripping of the

TABLE 12-7-4. Dual-Layer Plasma Transfer Resists

Imaging/barrier layer	Planarizing layer	Reference
Metal films, ion exposed	PMMA	91
AgBr	PMMA	124, 387, 395
Ag_2Si/SeGe	DQN	75, 290
Ag/SeGe	Polyimide	125, 406
SeGe	Baked DQN	126, 127
Silicone monomers	DQN	128
Si^+ implanted	PMMA	129, 142
In^+ implanted	PMMA	68, 334
Silane-treated PMMA	PMMA	70, 264, 300
Polysiloxane	DQN	130, 131, 269–270
Polysiloxane	Polyethersulfone	135
PMMA silylyloxime	Polyimide	245
Polysilsesquioxanes	PMMA	246, 316
Silicone grafted	Polystyrene	255
Polysilane	PMMA	265, 302, 313, 315, 371, 407
DQSi-novolak	DQN	331, 426
Polysilyl-MMA	DQN	342, 385
Ga implant SiO_2	DQN	352
Iodostyrene	PMMA	341
Siloxane-MMA	DQN	353
Azide-PMIPK	PMMA	322

Sn-based resists will require a hydrogen plasma to form volatile tin hydrides, rather than the traditional oxygen plasma.[200,201]

A plethora of silicon-containing resists have originated ever since polydimethylsiloxane was shown to be able to create submicrometer images in bilayer stacks as thick as 4 μm.[130,131] Almost every resist including DQN has undergone "silicosis" of the sensitizer or polymeric component. Some of the examples of silicon incorporation are as follows:

Positive resists:	Polysilanes, PMMA-silicone[377,382,389]
	Polysilylsulfones[281]
Negative resists:	Polychloromethylsiloxanes[328,368,378,390,396]
	Polytriallylphenylsilanes-azide[322]
	Polydimethylsiloxane-copolymers[360]
Polymerize silicones:	Vinylsilanes[128]
Graft silicone:	PMMA-graft vinylsilanes[71,330]
Silicon photosensitizer:	Diazoquinone Si-ester[426]
Silicone DQN:	DQN silylated novolak[325,331,374,388,417]
Postsilylate:	Silylate OH, COOH groups[300,384,387,421,427]
Silicone additives:	PMMA and triphenylsilanes[128]
Ion implant:	Si^+ irradiate resists[379–381]
Posttreat with B_2H_6[357,385]:	Boron reacts with OH, COOH

Positive working silicone resists have been derived from the onium salt decomposition of silylesters of polyvinylphenol,[412]

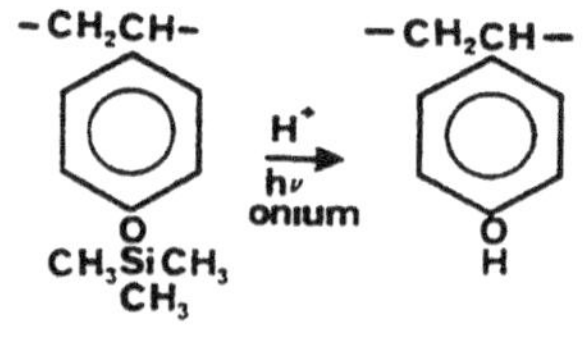

or from the posttreatment of polyvinylphenol formed from the onium salt decarbonylation,[421]

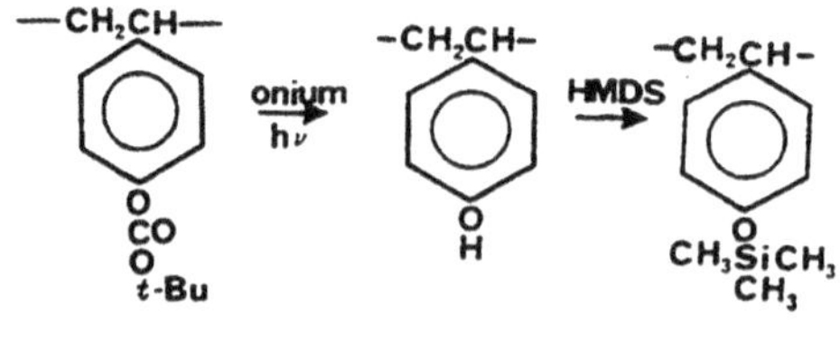

A dual-purpose deep-UV resist-dry-developable and oxygen barrier layer was synthesized from a silyl-oxime methacrylate,[245]

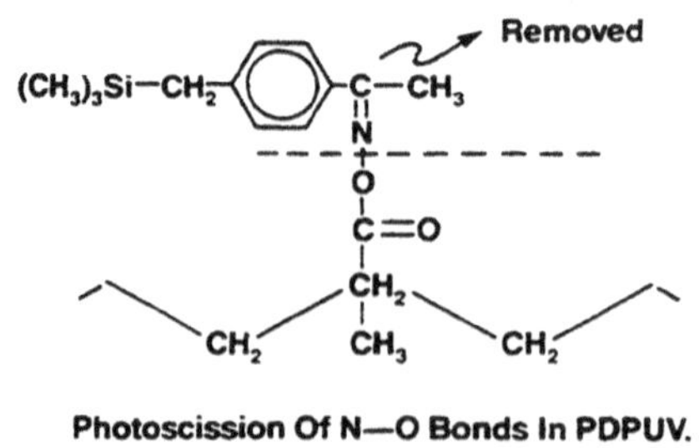

Photoscission Of N—O Bonds In PDPUV.

Deep-UV exposure releases silicone fragments and the exposed image is dry developed in oxygen. Similarly, polysilylsulfones are exposed by *e*-beam and developed to serve as oxygen barrier resists,[358,359]

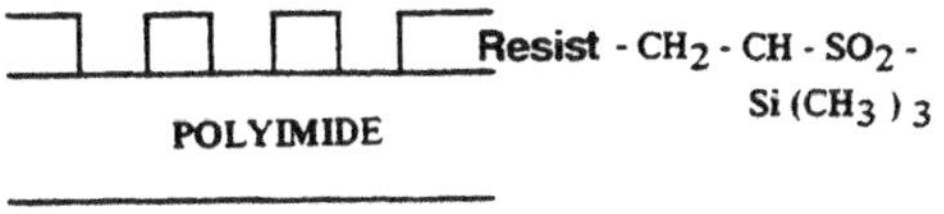

U.S. Patent 4,357,369 (1982), RCA

For device metal definition, the dual-layer O_2 image transfer process has an excellent chance of achieving minimum print and/or etch lift-off bias on the order of 0.1 μm, if thin pinhole-free image layers (< 0.3 μm) can be deposited. Fortunately, by using thin polysiloxanes of low T_g, the pinholes are closed by prebaking.[249] Thin image layers (< 0.3 μm thick) overcome the interference and proximity effects, and even top surface imaging is possible.[421] Photosensitive silicone[134]-based resists such as the deep-UV chloromethylsiloxanes, and methacrylated siloxanes[418] are a good investment for deep-UV and mid-UV scanning systems. Unfortunately, most of the siloxanes are rubber elastomers and may curl if used for a lift-off resist. Recently, chloromethylphenylsiloxanes,[269,329]

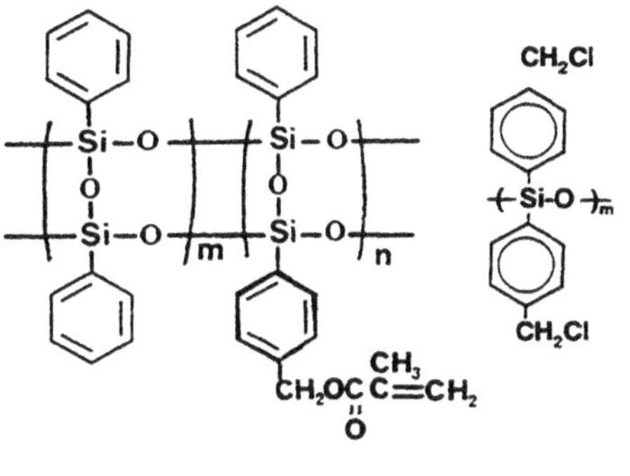

have been synthesized with a T_g of 150°C.

If acetylene is added to the oxygen etchant, the etching of the underlayer using a polysiloxane mask produces anisotropic or undercut patterns (Fig. 12-7-8).[122] In the O^+ transfer, the ratio of vertical anisotropic etching transfer to the horizontal undercut is strongly dependent on the pressure in the diode reactor.[123,250,268] Many profiles can be created by adjustment of the reactor conditions[142,143] but batch processing uniformity can vary as much as ±25–50% and must be improved for profile control. Contamination from sputtering of siloxane or other barriers must be monitored and corrected by postetch cleanup in liquid HF dip. In addition, the failure of the bottom underlayer to planarize, leads to image widening and too much undercutting. The slightly undercut pattern can be used for high-temperature (200°C) lift-off, if a high-temperature polyarylether sulfone (PES),

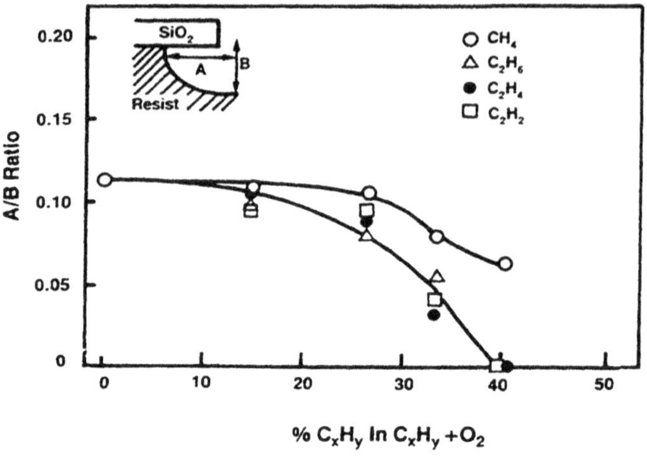

FIGURE 12-7-8. Undercutting of resist under SiO_2 nonerodable mask by addition of alkanes to O_2 plasma etchant. From Ref. 122, copyright 1983 and reproduced by permission of the Electrochemical Society Inc.

is used as the planarizing layer[135,144–147] since the PES have a high T_g (190–230°C). The PES or polyimide[174] are also resistant to common halogen plasma etchants.

12-8. HIGH-TEMPERATURE LIFT-OFF

Instead of using an organic polymer to serve as the planarizing and lift-off layer, another auxiliary layer such as an etchable Si_3N_4 can be used as a release layer (Fig. 12-8-1) or as a planarizing underlayer.[403] The high-temperature lift-off layer is patterned by dry etching and lifted by wet etching.[344] The lift-off layer can also be deposited by lift-off processes (MgO or As_2S_3) or by electroplating (Table 12-8-1).

The sacrificial layers are probably useful for exotic metallurgy, refractory layers,[238] and special devices.[312] The profiles in the layers should be close to vertical and the aspect ratio should be two or greater.

TABLE 12-8-1. Lift-off Underlayers for High Temperatures

Layer	Reference
Cu	238
As_2S_3	136, 248
Si_3N_4	137, 297
Mo	138, 258, 259
Cu electroplate	139
SiO_2	140, 254, 295, 296, 304, 410
Poly Si	140
MgO	141
AlN	231
Water soluble	237
Carbon	425

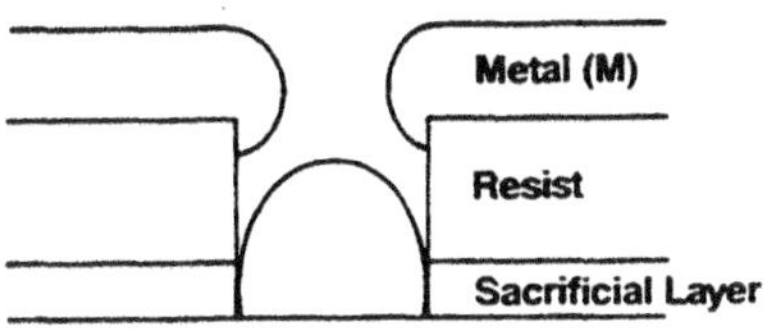

FIGURE 12-8-1. Resist lift-off layers with sacrificial bottom layer etchable in volvent or acid to facilitate lift-off.

12-9. SUMMARY AND OUTLOOK OF DOUBLE-LAYER RESISTS

Double-layer resists are capable of high aspect ratio (3- to 4-μm-thick films) and submicrometer features, if the oxygen etching is carefully controlled. The double-layer systems separate the image layer from the planarizing layer.

Since the image layer can use conventional DQN resists, or silicone-containing (postsilylation or silylated novolak) resists,[419,427] submicrometer images can be extended by projection printing in the region of 300–400 nm. The two-layer structure can be used with a PMMA underlayer for moderate lift-off conditions or with improved glutaramide derivatives. New top layer or more intractable planarizing layers are required to prevent interfacial mixing. For higher deposition temperatures ($>100°C$) or deep trench plasma etching conditions, oxygen transfer plasma etching is required to pattern the stable lift-off layer.

The recent application of inorganic resists on Sn- and Si-containing resists as the image layer, is a major advance for a reasonable cost with few additional steps.

Major improvements in inorganic Ag-based resists are required for consideration as deep-UV or optical resists. For X-ray, ion-beam, and *e*-beam exposure, bilayer resists reduce the proximity effect and allow resists to be used at lower dose, $R/R_0 < 6$. New high-temperature bilayer materials are required to expand the bifunctional lift-off or plasma etch capabilities of dual-layer resists.

12-10. TRILAYER RESISTS

Trilayer resists were created to overcome the thermal and plasma limitations of single-layer positive resists of PMMA and DQN. PMMA and DQN were originally used in the late 1960s for wet etching of oxides and metal films. As circuit pitches decreased to 5 μm and linewidths shrunk to 2–3 μm, the need for smaller print biases (< 0.5 μm) and vertical profiles induced workers to turn to lift-off and reactive plasma etching. The inherent deficiencies in the physics of printing in films

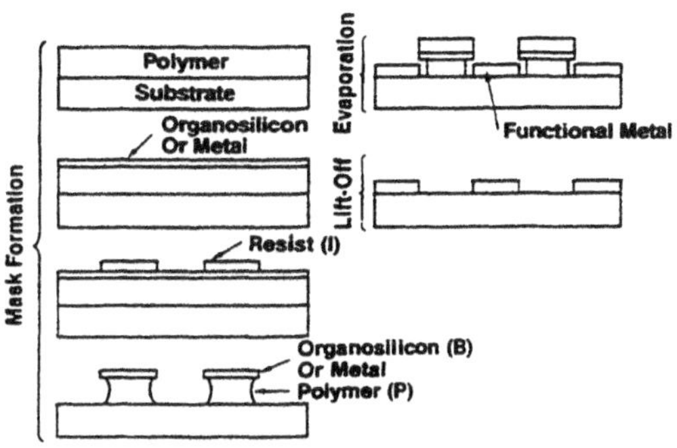

FIGURE 12-10-1. Trilayer resist production steps. The image layer is on the order of 0.2–1.5 μm in thickness, the oxygen plasma barrier layer is 100–200 nm, and the planarizing bottom layer is about 1–2 μm. From Ref. 2, courtesy of IBM.

TABLE 12-10-1. Planarizing Layers for Multilevel Resists

Planarizer	Reference
Polymethylisopropenyl ketone	151
PMMA	123, 126, 129, 152–157
Baked 200 °C DQN	130, 131, 146, 158–167, 232
Dyed DQN	161
DQN	168, 169
Exposed DQN	117, 149
Polyimide	142, 170, 171, 174
Polyarylethersulfone	135, 144–147, 229, 403
Polyvinylformal	99, 172

thicker than 1 μm (over steps) prompted the development of planar image layers.[266] Since oxygen plasma-resistant image layers were not known in the early 1970s (although polysiloxanes were known as *e*-beam resists in 1968),[32] inorganic or metal films were patented as O^+ barrier layers. The basic trilayer resist (Fig. 12-10-1) consists of an image (I) layer (0.2–1.0 μm), a barrier (B) layer (0.1–0.3 μm), and a planarizing (P) layer (0.5–3.0 μm thick), which can serve as a lift-off or plasma etch mask. Some representative planarizing and barrier layers are listed in Tables 12-10-1 and 12-10-2. The definition of a resist image in a thin planar layer separated from the planarizing sublayer produces superior resolution independent of the substrate topography (Fig. 12-10-2).

12-10-1. Planarizing Layers

The original underlayer was baked AZ-1350 (200°C), which is difficult to solubilize and lift-off in organic solvents. To improve the lifting, a skin release layer

TABLE 12-10-2. O_2-Resistant Barriers for Multilayer Resist

Barrier layer	Reference
Plasma deposited SiO_x	148, 154, 164, 170, 171, 174
Si_3N_4	161, 162, 263, 265, 402
Polysiloxanes	130, 135, 142, 143, 159, 163, 247, 270, 321
Polysilanes	253
Silylated novolak	417
Si	100, 160, 405
Ge, Ge-Se	99, 126, 175
Al	124, 156, 165–167, 259
As_2S_3	259, 305, 327
Mo, MoO_3	138, 432
W	300
Polystannylacrylates or styrene	149, 150, 236
Si^+-doped PMMA	99, 125, 129, 153
Silicone-polyimide	291
Silica gel + dye	244, 318
Polyvinylsilylsulfone	292
Polyhexafluorbenzene	293
Spun-on TiO_2	401
Spun-on Al_2O_3	401
Spun-on SiO_2	262, 326, 372

FIGURE 12-10-2. Trilayer resist of optical resist/SiO_2 barrier layer/resist over 0.4-μm steps. The 1.6-μm spatial period grating has lines 0.8 μm in width. Patterns have been overexposed, so that lines are narrower than spaces. However, there is no linewidth variation across steps. From F. Buiguez, P. Parrens, and B. Picard, "Submicron Electron Beam and Optical Lithography using a Trilevel Resist Scheme," Proceedings of SPIE Symposium on Electron-Beam, X-ray and Ion Beam Techniques for Submicron Lithographies II **393**, 192 (1983), copyright and courtesy of SPIE and Dr. F. Buiguez, LETI-CEA.

of a polyarylether sulfone is precoated before the application of the baked AZ underlayer.[300] Recently, a soluble high-T_g (325°C) polyimide (Ciba–Geigy XU-218) has been successfully used to deposit metal at 300°C, followed by lift-off in chlorinated solvents.[173,174] For elimination of standing waves, optically absorbing dyes are added to the underlayer. These findings increase the versatility of multilayer resists.

12-10-2. Barrier Layers

In the transfer of the image from layer I to layer P (Fig. 12-10-1), the image layer is removed by O etching to reveal the barrier layer and the lift-off metal is deposited on B. If B is polymeric in nature and not melt stable ($>$ 150°C), the layer may curl or flop down,

due to the difference in thermal expansion between the metal and the barrier layer. The curling and cracking may distort the profile of the metal deposited on the wafer contact hole (similar to edge lifting of resist during wet etching; Chapter 13). Inorganic barrier layers have lower coefficients of expansion (10^2–10^4 lower) to offset the expansion of the planarizing layer. However, they may be too opaque to the alignment wavelengths. Furthermore, the deposition of inorganic or metallic barrier layers is not without its share of problems. If the DQN (Table 12-10-1) layer is not baked above 150°C to oxidize–cross-link the film and to remove solvents, outgassing may form occlusions in deposited film or cause stress cracking. Sometimes the film can be perfect before deposition but will crack after deposition and prebake of the image layer.[176] Low baked DQN will also outgas N_2 and cause pinholes to form in the depositing barrier layer.[81] Although the metal barrier layers also reduce *e*-beam charging,[303] they may hamper optical alignment due to high absorbance in the visible region. A separate process may be required such as metal lift-off[174,299] to deposit and reveal the alignment marks.

Another function of the barrier layer is to prevent intermixing of the image layer and the planarizing layer.[277] The planarizing layer should be cured and not contain extractables which mix with the spin-coated polysiloxane barrier layer. The polysiloxane barrier layers may be derived from spin-on dopants[163,244,264,275,276] which generally exist as siloxane cyclics with pendant OH groups,[160,270]

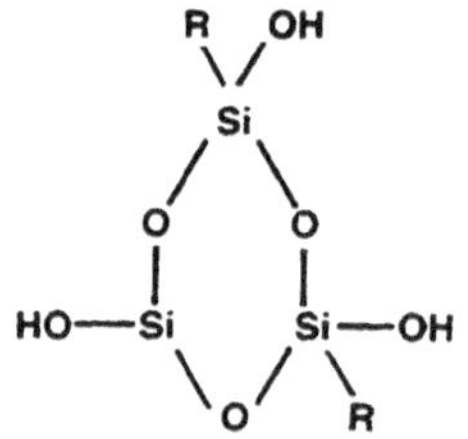

The terminal OH groups cure by condensation reactions by heating in the presence of trace catalyst. If the siloxane layer is not properly cured, the silicone polymer may diffuse into the image layer or planarizing layer and not be removed during O etch transfer. Thus, the barrier polymer layer must generally be highly cross-linked, free of mobile metal ion, and alkaline resistant if used under diazonovolak resists. Furthermore, the proper choice of barrier layers and the reliable processing of incompatible organic layers in a trilayer structure are necessary to eliminate defects and rework.

Finally, barrier layers can be fabricated from films opaque to various resist exposure radiation such as UV, X rays, and ion beams (Table 12-10-3).

TABLE 12-10-3. Radiation Barrier Layers for Multilayer Resists

Masking layer	Radiation masked	Reference
Au	H^+	152
Au	X ray	154, 281, 324
Al	$h\nu$	151, 168, 169, 280, 301
DQN	$h\nu$, 254 nm	Table 12-7-3
RD-2000N Hitachi	$h\nu$, 254 nm	178

TABLE 12-10-4. Image Layers for Multilayer Resists

Image layer	Reference
DQN	113, 153–162, 169, 171, 174
PMMA	89, 153, 160, 162, 175
Polybutene SO_2	181
Polychlorostyrene	81
Polychloroacrylate and phenyl monomers	99
Ag-Se	124, 126

12-10-3. Image Layer

The final deposition step involves the image layer, which functions as an image layer and a reactive etch CF_4 mask to transfer the image through the barrier layer. The most popular resists are the DQN resists such as American Hoechst, Hunt, and Kodak DQN which have been used in all optical exposures, while PMMA has been the most popular resist for *e*-beam MLR exposures (Table 12-10-4). The aromatic and inorganic resists have sufficient plasma resistance to survive the plasma transfer, while PMMA requires a CF_4–H_2 polymerizing etchant to survive the transfer step.[333] A spin-on glass barrier etches 20× faster than DQN in CHF_3/CO_2 plasma.[163]

The transfer of the image into the barrier layer can be accomplished by wet etching, dry etching, or lift-off itself, to create an image-reversed barrier layer (Table 12-10-5).

Since the image layers are less than 1.0 μm thick, to reduce *e*-beam inter-proximity effects and enhance the optical depth of focus, a selectivity (resist plasma erosion to barrier erosion) of at least 0.5 is required to prevent excess thinning and pinhole formation (Fig. 12-3-11). The selectivity factors which reduce resist erosion such as the etchant and tool parameters are discussed in Chapter 13. If the image layer is as thin as 0.4 μm, proximity *e*-beam effects are reduced to a ± 0.2-μm print bias.[56,172] Although the use of a multilayer resist (MLR) reduces the proximity

TABLE 12-10-5. Methods of Transfer from Image Layer to Barrier Layer

Barrier layer	Plasma transfer	Reference
Si	CHF_3	158
SiO_x	CF_4/O_2	159, 181
Poly SiO_x	CF_4, H_2	100
Si_3N_4	CF_4	100, 161, 162, 164, 170
Spin-on SiO_2	CHF_3/CO_2	163
	Other methods	
Al	Wet etch	99, 152, 153, 165–168, 182
Au, Mo	Sputter etch	138, 152, 238
MgO	Wet etch	183
Al	Lift-off	154
Ti	Lift-off	184
PMMA	Electron-beam, develop	155

effect by the use of thin image layers, the carbonaceous planarizing layers have to be very thick (5–10 μm for 10–25 keV) to completely absorb the backscattered electrons. At these thicknesses, charging, poor registration, and possible resist heating may offset any gain in the reduction of proximity effects.

The outcome of all the materials and process considerations for trilayer resists is the superb high-aspect-ratio images (Fig. 12-11-2) even over steep topography.

12-11. PRACTICAL MLR PROCESSING

Trilayer resists were fabricated to promote high-temperature (150–200°C) lift-off structures,[159] but later adopted[150] as excellent resists for plasma etching. For lift-off, the aspect ratio can be as high as 3–10, since anisotropic vertical images can be formed in the planarizing layer (Table 12-11-1).

The basic applications of trilayer resist structures are many and range from additive to subtractive processes (Table 12-11-2).

Three major problems associated with trilayer lift-off and other uses are: (1) contamination from O_2 plasma etching, (2) defects in barrier layer or stresses in composite layers, and (3) incomplete planarization by the bottom layer.[369] Reactive ion etching with O^{2+} can lead to high contact resistance prior to lift-off, due to the sputter redeposition of Al_2O_3[160] from the Al barrier layers and the Al electrodes[81] of the plasma tool or SiO_2[100,283] from the Si barrier layers. A brief dip in HF or phosphoric acid is used to clean the Si contacts prior to lift-off.[2,100] Although argon sputter clean can also be used to clean the contacts, the PMMA underlayer may cross-link and become insoluble in the lift-off solvent. In this case, a thin Al barrier layer was used to stop the ion-induced cross-linking.[154] Secondly, a milder H_2 plasma etchant has also been used to transfer the image from the barrier layer into the planarizing layer.[251] Deposition of the Si barrier on top of the planarizing polymer layer can lead to considerable blistering. The stresses[189] from linear expansion and the defects in one layer can propagate throughout the sandwich layers. While silicon deposited at low temperature loses adhesion, the deposition of silicon followed by annealing, prior to topcoat deposition, alleviated silicon blisters.[100] Spun-on glass barrier layers may be susceptible to attack by the alkaline DQN developers. Cracking of the barrier layers has been noted to occur due to developer infusion and from the thermal mismatch of thermal expansion of all trilayer films.[367] Spun-on TiO_2 has been claimed to be less susceptible to film defects than SiO_2 deposits.[401] Finally, a brief dip of the spun-on SiO_2 barrier in HF is claimed to reduce cracking.[412]

In the process of oxygen RIE transfer, the undercut profile is highly dependent on the pressure of oxygen.[283–286] Irregular vertical profiles and undercutting result at low pressure, due to the wandering radicals created in the plasma. Although high-pressure etching in magnetron fields achieves fast rates of 1–2 μm/min for single-wafer processing, the nonuniformity of etching increases rapidly with slight pressure fluctuations.[287] The uniformity of etching is improved by oscillating the wafer during the plasma transfer process.[287] Another curious process problem has been noted during the oxygen etching of isolated images. Large openings of > 3 μm in diameter, etch 1.2–1.5× faster than submicrometer holes.[284,288] A plausible explanation for this intraproximity effect is the restriction of the diffusion of oxygen into the smallest orifice.[279]

TABLE 12-11-1. Trilayer Lift-off Composite Films and Method of Transfer

Image layer	Barrier layer	Method of transfer	Planarizing layer	Method of transfer	Lift-off solvent	Lift-off material	Reference
DQN	SiO	CF_4 gas etch	Polyimide	O_2 gas etch	$CHCl_3$	NbSi 200 °C	175
DQN	SiO_2	CF_4 gas	Polyimide	O_2 etch	CH_2Cl_2	Al-Cu 300 °C	174
DQN	Polysiloxane	CF_4 gas	Polyimide	O_2 etch	NMP	Al-Cu	159
DQN	Polysiloxane	CF_4 gas	DQN	Expose and develop	Acetone	Al	185
DQN	Si_3N_4	DQN + dye	O_2 etch	Used for CF_4 etch		Al	161
DQN	Al	Wet etch	DQN	$h\nu$ develop	Acetone	Al	168, 182
DQN	Si	CF_4 gas	PMMA	$h\nu$ develop	Acetone	Al	100
PMMA high M_w	PMMA medium M_w	*e*-beam develop	PMMA low M_w	*e*-beam develop	Acetone	Al-Cu	155
PPMA	Ge	CF_4	PPMA-MA	O_2 etch	Acetone	Al	175

TABLE 12-11-2. Applications of Trilayer Resists

Application	Reference
Deposition of metals Al-Cu and NbSi	173
Ion implantation mask	153, 186, 190
Reduce *e*-beam proximity effect	160
Reduce *e*-beam charging	160, 161, 425
Reduce *e*-beam radiation damage	252
Reactive ion etch mask	158
17-nm lines by lift-off	30
3-mm-thick metal by lift-off	164
Lift-off by electrolysis	138
Zero-overlap vias by etch lift-off	164
Increase resist sensitivity	76, 86, 88, 158
Two-level interconnect metal	138
Coplanar metal and insulators	167
Image reversal	187
Reduce optical standing waves	161
Electroplate mask	154
X-ray *in situ* mask	154, 158
Deep-UV *in situ* mask	157, 100
Ion-beam *in situ* mask	152
Electron-beam *in situ* mask	154

Planarization

The ability of the bottom layer to planarize over steps or cover trenches is dependent on the rheological flow properties of the spin-cast film. If the spinning film dries too quickly and becomes rigid, little plastic flow occurs to level the disparities. Since the surface tension of the liquid is the driving force that causes a liquid coating to level, the priming treatment of the surface to lower the surface tension, will enhance leveling. For liquid resists, their surface tension is about 35 dynes/cm (Chapter 13) for nonpolar solvents and 40–50 dynes/cm for polar solvents (i.e., polyimide in casting solvents). The leveling action of the casting solvent is inversely proportional to the liquid coating surface tension.[188]

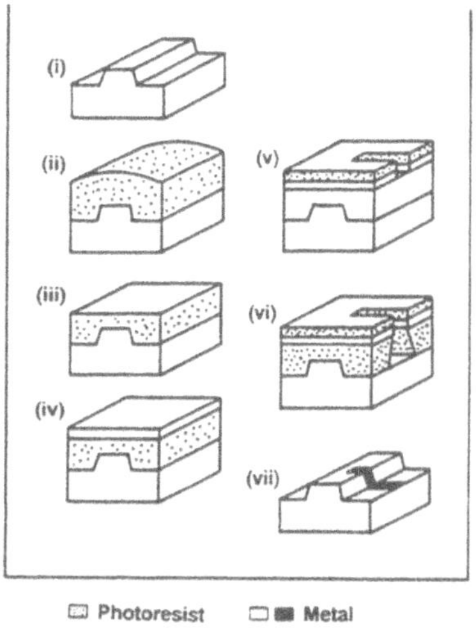

FIGURE 12-11-1. Schematic representation of the method used to delineate a fine pattern of the metal electrode (Schottky gate) across a bevel on the substrate. (i) Etching of the substrate to form a bevel; (ii) coating of the substrate with a positive resist layer followed by prebaking; (iii) thinning and smoothing of the resist layer; (iv) evaporation of a metal layer on the whole surface of the resist layer; (v) pattern formation on the metal layer by chemical etching with a mask produced by a negative photoresist layer using usual lithographic techniques; (vi) overexposure of the positive resist layer followed by development; (vii) evaporation of a metal film through the window in composite layers and subsequent removal of the layers by immersion in acetone, leaving the desired pattern of the metal electrode. From O. Wada, S. Yanagisawa, and H. Takanashi, *J. Electrochem. Soc.* **123**, 420 (1976), reproduced by permission of the Electrochemical Society Inc.

FIGURE 12-11-2. Planarizing features of trilayer resists over steps. From J. Moran and D. Maydan, "Device Patterning Using the Trilevel Technique," SPE RETEC on Photopolymers, Ellenville, N.Y., 1979, p. 248, courtesy of SPE.

Since there are many topography differences in logic and memory surfaces produced by the various oxidations and depositions, it is necessary to form metal patterns over these beveled steps (Fig. 12-11-1).[202] Multilayer resists planarize the step and permit metal interconnections to traverse the step with a conformal coating. Figure 12-11-2 illustrates the image features of a trilayer resist showing the planarized resist crossing the oxide steps on the wafer. Due to the fact that spin coating proceeds from a leveling liquid state to a centrifuged dried film, the coatings, if properly applied, tend to planarize, except in the vicinity of the step (Fig. 12-11-3). When a patterned surface is spin coated with a polymer solution and dried, the resulting surface is not completely flat due to the nonuniform film shrinkage (differential volume changes in the vicinity of the step). If the planarizing layer is wavy, the image layer may also fail to planarize. Consequently, this causes incomplete exposure on optical steppers due to the interference effects from the ±100-Å variation in film thickness. It has been determined that the degree of planarization depends on many of the following factors:

- Thickness of the polymer coating
- The shrinkage of the polymer solution[355]
- The flow properties of the polymer during the bake cycle
- The spacing, step height, and radial orientation of the step on the wafer[334]

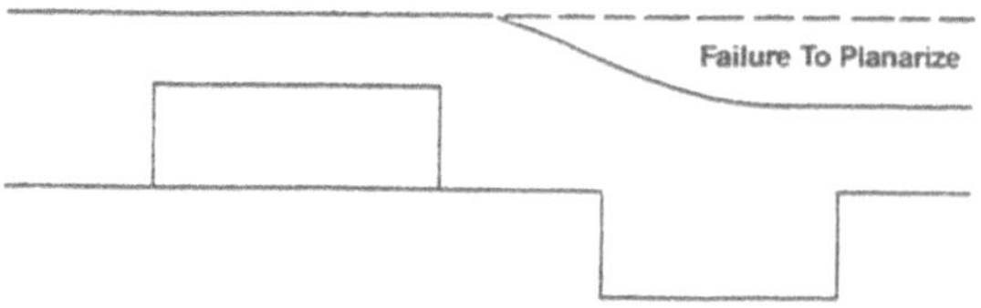

FIGURE 12-11-3. Failure of resist to planarize steps.

At the rising edge of the step, the resist film tends to conform under the influence of the coating thickness. Rothman[192,193] studied the planarization of polyimide films (2 μm thick) for second-level metallization over an Al step. Figure 12-11-4 illustrates the degree of planarization and planarization as a function of linewidth and space (pitches). Failure to planarize a 1.0-μm step can range from 10 to 50% of step height.[192]

In the spacing between features, a proximity effect is noted. Isolated lines are more difficult to planarize than closely spaced lines or isolated spaces. As the pitch increased from 2 to 25 μm for 0.8-μm steps, PMMA and polyimide films suffer wave effects and exhibit conformal, rather than planarizing film properties.[100] A maximum planarization of 80% was achieved for steps 14 μm apart in linewidth[100] and complete planarization was noted for thick resists (2–3 μm) and for pitches less than 20 μm.

The degree of planarization is affected by many physical variables:

1. Method of spin coating including puddle, solvent atmosphere, ramp acceleration.[100,194]
2. The thicker the resist coating, the smoother is the film.[191,195,196]
3. The lower the T_g, the smoother is the coating. Novolaks (DQN) < PMMA < polyimide are lower in T_g and hence DQN provides more planar coatings.[191]
4. Thin lines and narrow spaces (pitch < 20 μm) planarize smoothly while larger pads or spaces do not.[100,191,193,196]
5. Lower-viscosity lower-molecular-weight PMMA planarizes smoother films than higher-molecular-weight PMMA.[100,196]

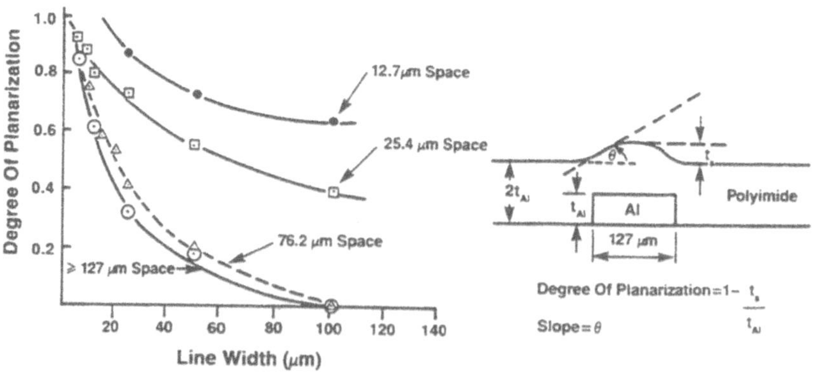

FIGURE 12-11-4. Planarization as a function of linewidth and spacewidth of patterns in the substrate spin coated with polyimide films. From Ref. 192, reproduced by permission of the Electrochemical Society Inc.

6. High-solid-content solutions exhibit less shrinkage upon drying and planarize better than low-solid-content polymer solutions.[335]
7. Height of step and orientation of step with respect to center of wafer.[340,362,369]

For all geometric topographies, complete planarization is not achievable with current resist or polymer multilayer underlayers.[431] In many cases, acceptable planarization can be achieved by several materials and process changes:

1. Apply/bake/reapply multiple thin coats (PMMA, polyimide).[100,193]
2. Use lowest-molecular-weight, broadest-polydispersity PMMA[100,196] or polyimide.[192]
3. Choose striation-free commercial resists.[4]
4. Short spin times and low rpm tend to smooth film before drying at higher rpm.[197]
5. Exposure of prebaked film to solvent atmosphere during or after coating has been claimed to smooth (flow) DQN.[198]
6. Melting of cast film with a high bake of the planarizing layer above T_g tends to planarize film.[191,195]
7. Dip[199] or other coating technique may produce more planar films.

If the bottom layer fails to completely planarize, stratified distortions may be transmitted to the uppermost image layer. Variations of ±500 Å in the step planarization can cause a 50% energy difference in absorbed energy, due to optical interference of the monochromatic exposure light used in optical steppers.[161] The variations can be overcome by the addition of a dye to the planarizing layer,[429,435] or by a resist topcoating of a bleaching–masking or contrast enhancement layer,

CEL layer (ABCD)

resist/barrier layer

dyed underlayer

Si wafer

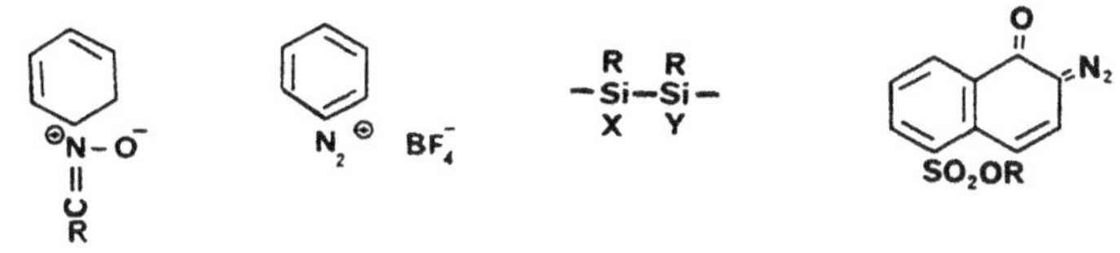

A B C D

Figure 12-11-5 illustrates the reflectivity (interference loss) as a function of film thickness with and without an absorbing dye incorporated into the underlayer of

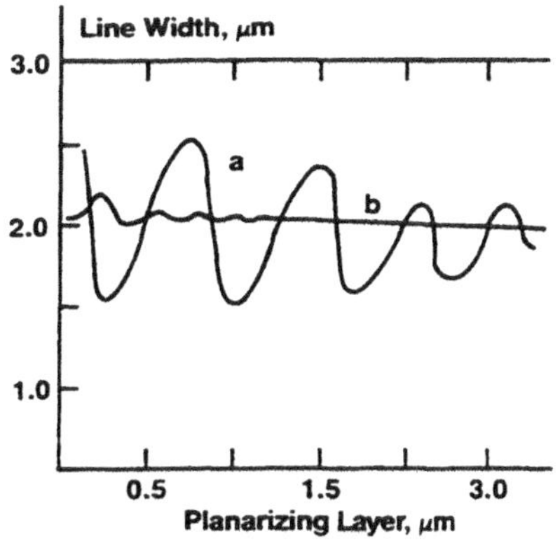

FIGURE 12-11-5. Simulated linewidth in trilayer resist (A) with bottom layer transparent to 436 nm and (B) with dyed bottom layer. From Ref. 161, copyright 1981 and courtesy of IEEE.

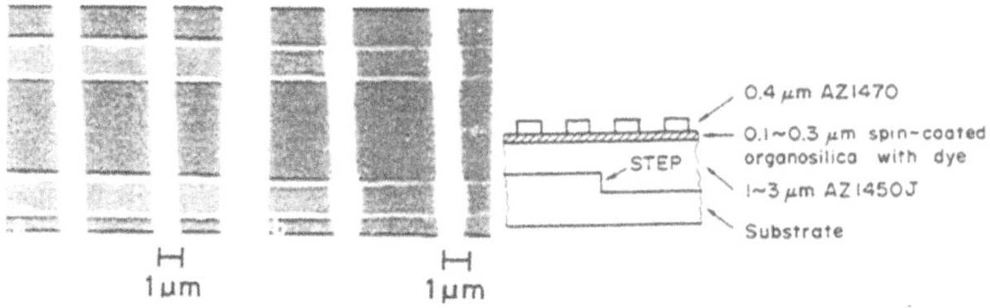

FIGURE 12-11-6. (a) Lines of 1 μm in width printed across steps using a trilayer resist structure with antireflective absorbing dye incorporated in the barrier layer. (b) Same lines printed using a single DQN resist showing distortion of image traversing over steps. From Ref. 244, copyright 1983 and courtesy of Elsevier Science Press. Scale bar: 1 μm.

DQN resist. Dyed layers reduce the line biases to ±0.4 μm[306–308] providing the underlayer planarizes to sufficient thicknesses over steep topographies. In multilayer resists, the actinic absorbing dye can also be incorporated into the organosilicone barrier layer, to ensure a constant uniform absorptivity (Fig. 12-11-6).

Electron-beam exposure tools have much larger (10–20 μm) depth of focus than optical exposure tools (1–5 μm) and are not sensitive to nonplanar surfaces. However, the proximity effect is still present and thin image layers must be used or lower-energy (10 keV) electrons (with registration schemes) to minimize thickness-induced variations in linewidths.

To avoid the step coverage problems, planar devices can be built by backfilling a preetched via hole directly with Al metal and avoid the mask alignment step (Fig. 12-11-7). In addition, the SiO_2 layer (Fig. 12-11-8) or Si_3N_4 insulating layer can also be deposited over an aluminum land by lift-off.[298]

For future devices, there are basic needs to fabricate more planar devices to ease the current exposure problems of the decreasing depth of focus in optical

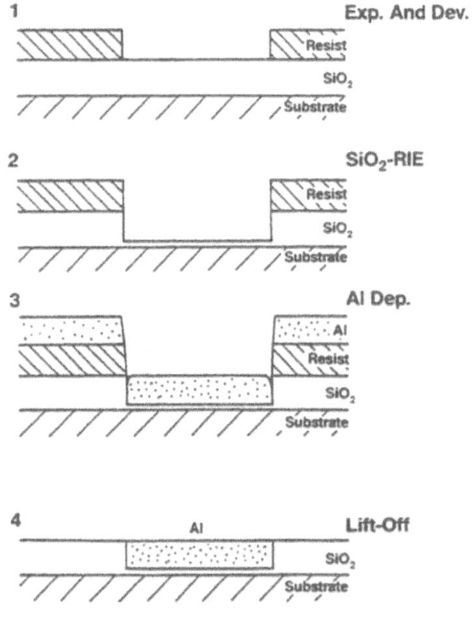

FIGURE 12-11-7. Process sequence of planar lift-off. From Ref. 271, copyright 1980 and reproduced by permission of the Electrochemical Society Inc. (Figure originally presented at the spring ECS meeting held in St. Louis, Mo.)

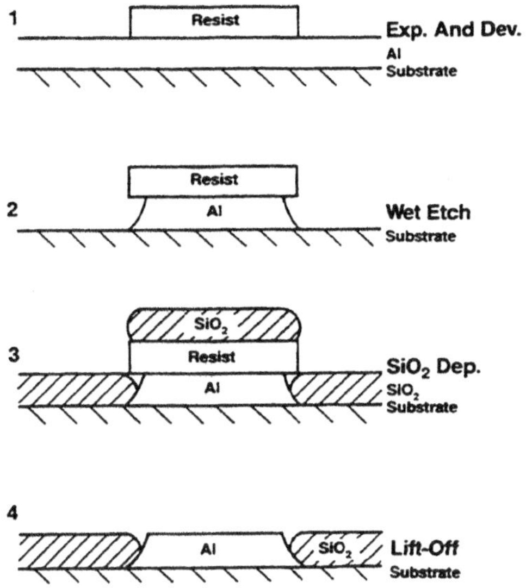

FIGURE 12-11-8. Process sequence for planar SiO_2 lift-off. From Ref. 272, copyright 1983 and courtesy of Academic Press.

exposure tools and the inherent difficulties in obtaining perfectly planarized coatings over all topographies. Unfortunately, high-temperature ($T_g > 150°C$) thermoplastic and thermoset polymers used for planarization, become progressively worse in planarization as their melt stability increases. New coating precursors and polymeric materials are essential for improving the step coverage. Accordingly, in underlayer materials, a trade-off of thermal melt resistance or plasma resistance with planarizing properties, may be required in balancing lithographic performance and polymeric rheological properties.

Once the planar coatings have been achieved and the images totally transferred to a clean contact, the MLR structure is ready for metallization and lift-off.

12-12. METAL DEPOSITION AND LIFT-OFF

The top opening of the MLR serves as the metal deposition mask (Fig. 12-12-1). A high aspect ratio of resist to metal thickness ensures that the metal deposit will not become a continuous film. If the resist profile is tapered (Fig. 12-12-2), protruding feet may form at the top of the metal land to cause subsequent problems with the sputtered quartz topcoat.[155] For deposition of the metal over a sharp step (Fig. 12-12-3), consideration has to be given to the step height and profile. Sharp steps will require some beveling of the step to minimize the microcrack at the top of the substrate step (Fig. 12-12-4). Predeposition of a metal blanket increases the metal step coverage,[260] and the worst possible step coverage occurs when the metal is deposited at an angle nearly normal to the wafer surface and at low pressure.[22,399] In general, problems occur in metal lift-off under the following conditions:

1. Metal deposit as thick as lift-off thickness and the metal deposits walls of resist causing "burrs."
2. Large pitches between isolated lines. Large real estate of resist to be swollen and long times for solvent permeation (Fig. 12-12-5).

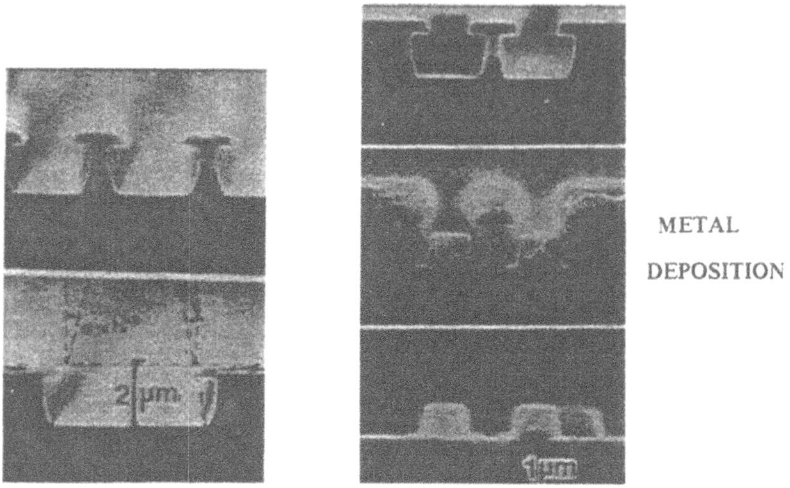

FIGURE 12-12-1. Metal deposition angle α in bilayer resist followed by metal deposition and lift-off. From Ref. 86, courtesy of AIP.

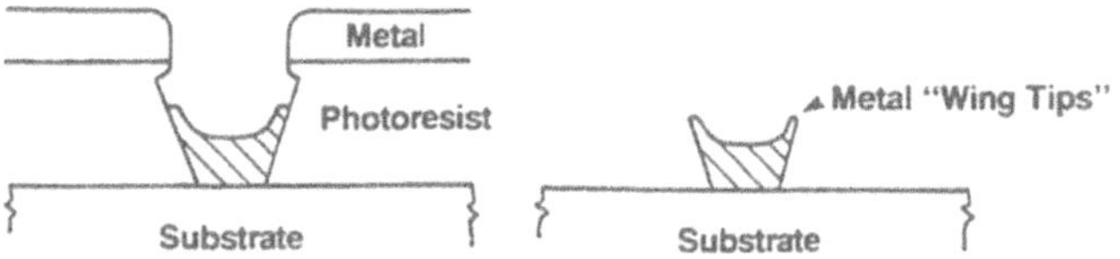

FIGURE 12-12-2. Formation of metal "wing tips" in lift-off due to resist foot in planarizing layer. From J. Frary and P. Seese, "Lift-off Techniques for Fine Line Metal Patterning," *Semiconductor International*, Dec. 1981, p. 72, copyright and courtesy of Cahners Publishing Co.

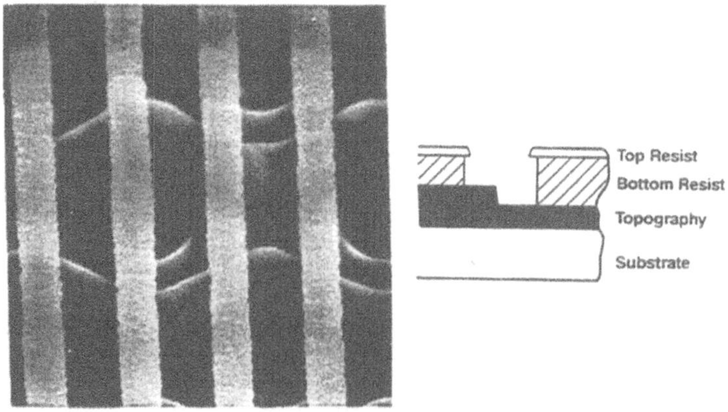

FIGURE 12-12-3. Lines of 1 μm made by bilayer resist over 1-μm oxide step prior to metal deposition. Courtesy of C. Lyons, IBM.

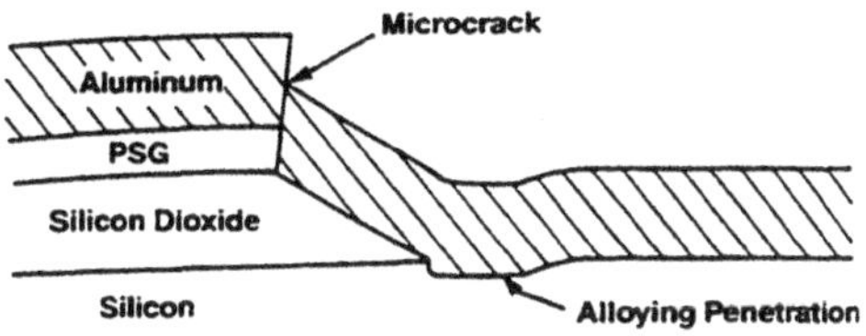

FIGURE 12-12-4. Schematic of microcracks in Al deposited over steps. From Ref. 211, copyright 1974 and courtesy of *Journal of Electronic Materials*, a publication of the Metallurgical Society of AIME, Warrendale, Pa.

3. Poor metal adhesion due to low temperature, sharp substrate steps, or contaminants on contact hole surface.
4. Redeposition of metal in lift-off solvent tank or wafer surface. Glycerine added to acetone is claimed to eliminate this problem.[203]
5. Curling of the metal and barrier layer in the MLR (Fig. 12-12-6).
6. Outgassing from resist causing pyramidal profiles (see Fig. 12-13-1).

The mechanical strength of the deposited metal and the expansion from the heat of metal deposition and elasticity can also limit the rigidity of the metal/resist composite. For example, the curling of the metal deposited on top of PMMA has been noted.[155] More rigid and resistant lift-off layers such as polyimide[174] or inorganic[26,273] SiO_2 may be required for depositing various new metals. The curling (Fig. 12-12-6) can be modeled by a thin plate buckling under differential expansion. Chromium, for example, has a Young's modulus of 5×10^{10} dynes/cm while PMMA's is about 10^9 dynes/cm. The PMMA layer or the barrier layer in a trilayer resist must be about 0.7–1.0 μm thick to offset the intrinsic Cr stress. Thin (0.2 μm) polysiloxane[159] barrier layers may also be too thin to prevent metal stress, and more rigid barrier layers as Si_3N_4[161,162] may be required to offset the stressing metals.

The metal deposition process is similar to the ray tracing a beam of energy deposited into a resist. In modeling, studies of metal deposition have focused on the evaporation angles of deposition to produce good step coverage (no metal breaks, Fig. 12-12-4), but with rectangular cross sections, if possible, for maximum current capacity. Blech[204,205] and co-workers and Wasik and Gneiwik[206] have modeled the general problem of depositing aluminum on any surface. Their program was refined by Widmann[209] and later by Neureuther and Oldham[208] (SAMPLE) to account for the surface migration effect from a heated boat source and for a sputter deposition source.[9] For maximum step coverage (Fig. 12-12-4), the deposition angle is 40°. The studies and models of metal deposition angles and conditions for maximum cross section and hence resolution, have attempted to estimate:

1. Maximum pitch of lift-off as a function of resist vertical thickness[9]
2. Maximum deposited metal thickness[209]
3. Deposited metal profiles as a function of deposition angle[9] (Fig. 12-12-7)
4. Source evaporation opening angle and throughput[399]

In recent studies by Lehman and co-workers, it was concluded that requirements of good step coverage and linewidth control are mutually exclusive:

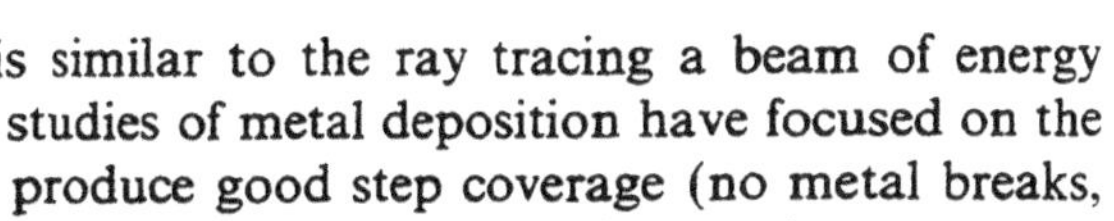

FIGURE 12-12-5. Large real estate of metal (M) over resist land. Long lift-off times or powerful strippers are required to lift metal.

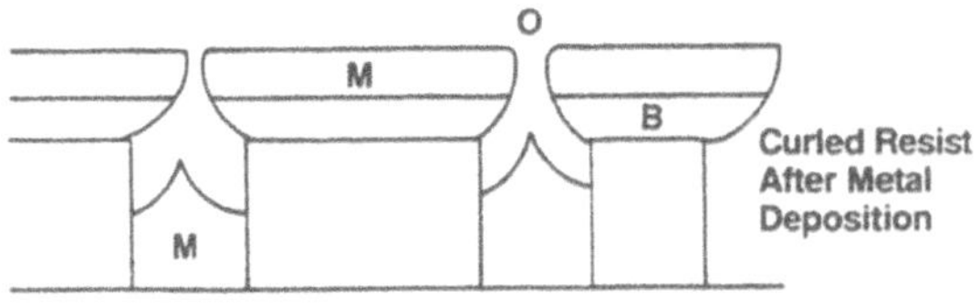

FIGURE 12-12-6. Curling of barrier (B) layer caused by melting of resist or mismatch of metal–polymer linear expansion. The deposition of metal (M) constricts the lift-off opening (0).

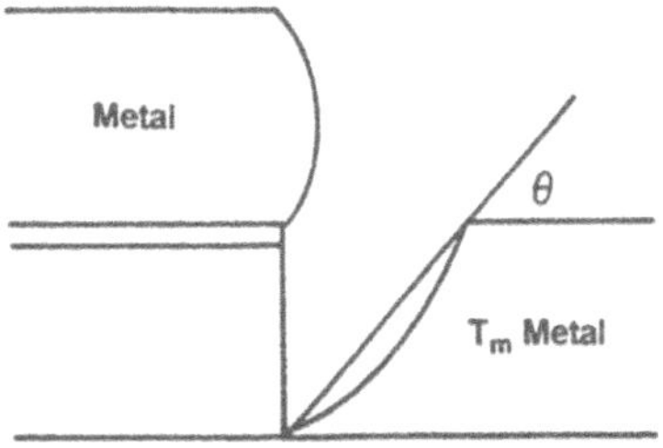

FIGURE 12-12-7. Change in slope of deposited metal as thickness (T_m) of metal builds up on lift-off profile. From Ref. 210, copyright 1982 and courtesy of IEEE.

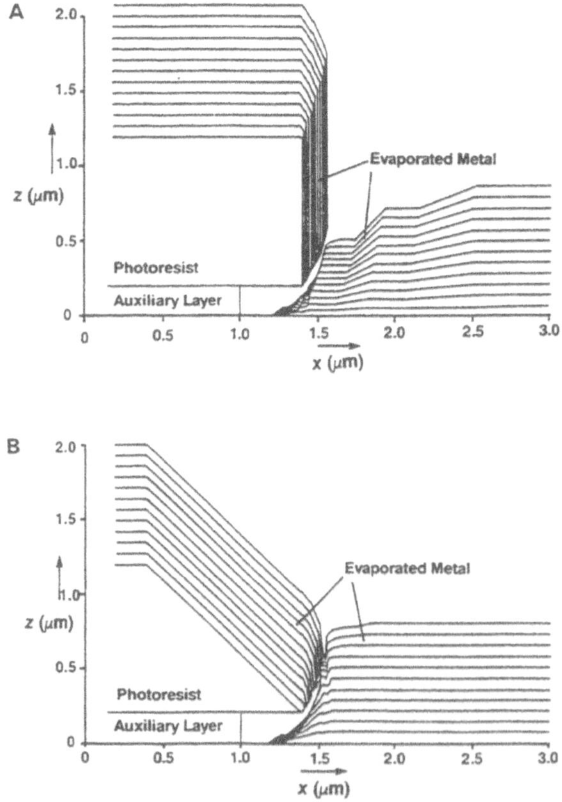

FIGURE 12-12-8. Computer simulation of metal deposition (A) over vertical profile and (B) over 45° profile to form boundary crack at step junction. From Ref. 209, copyright 1976 and courtesy of IEEE.

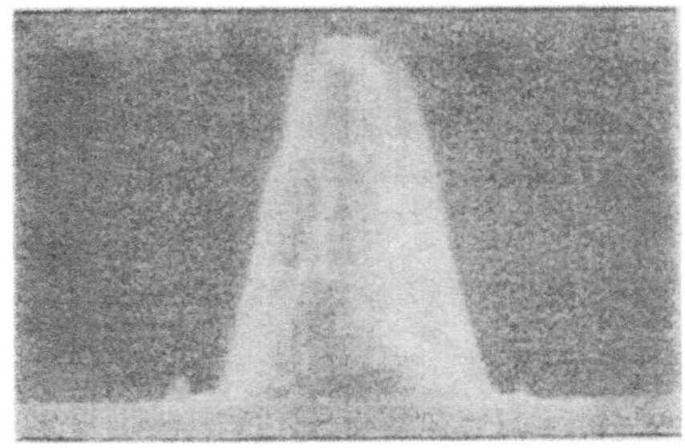

FIGURE 12-12-9. Lift-off profile of 2.5-μm line by 1 μm high. From Ref. 19, courtesy of AIP.

while an acceptable step coverage can be achieved by tapering the step and using a vapor source with a large source angle, good linewidth control can only be obtained by using a vapor source with a narrow source angle.[399]

12-12-1. Resist Profile and Lift-off

Assuming that the deposited metals undergo no diffusion after deposition, a planetary source was modeled.[207] The angular distribution of incident atoms in the center of the substrate is symmetrical, with respect to the substrate axis. The evaporated film profiles are the same for any orientation of photoresist geometries, and the mean free path of evaporated atoms is assumed to be larger than the source-to-substrate distance, or the height of steps in the substrate surface. Widmann's results for the two cases of a vertical resist profile and a 45° resist profile are shown in Fig. 12-12-8. Accordingly, the main result of the calculation is that a microcrack remains between the metal deposited on the resist and over the contact hole, and the maximum slope of the resist profile after lift-off is about 70°. Furthermore, the actual profile shown in Fig. 12-12-9 with a cross section of lift-off metal of 1 μm × 2 μm high is close to Widmann's prediction. In the case of a 45° resist profile, the metal film is separated from the deposited film, only if the metal layer is twice the auxiliary layer thickness. However, if the metal is sputter deposited, Sakurai and Serikawa[66] have shown that a resist with 45° profiles (Fig. 12-12-8B) can still be used, even with metal deposits thicker than the resist. A microcrack occurs along the sidewall of the resist pattern and that of the metal on the Si surface (Fig. 12-12-4). For the vertical case (Fig. 12-12-8A), the etch-back technique of removing the sidewall metal with H_3PO_4 can be used successfully[22,209] (Fig. 12-12-10) provided that the sidewall thickness is 10% of the vertical metal thickness.

In the usual production mode, some variation from the normal incidence of deposition is bound to occur. The lateral growth on the resist surface pinches off

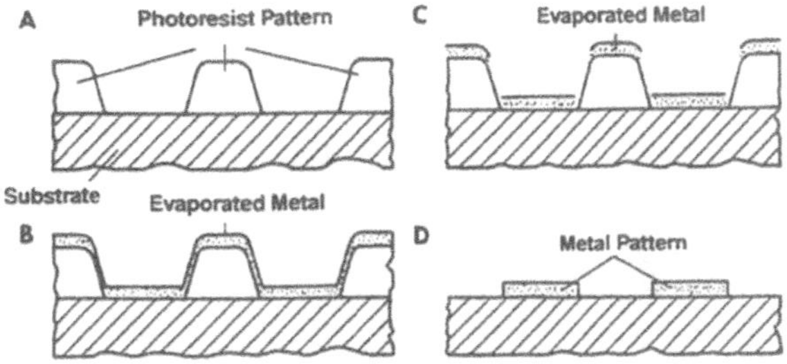

FIGURE 12-12-10. Schematic of lift-off using sloped (A) walled resist (60°). The evaporated metal in (B) deposited on the sidewall is etched back with liquid acid (C) and lifted (D). From Ref. 209, copyright 1976 and courtesy of IEEE.

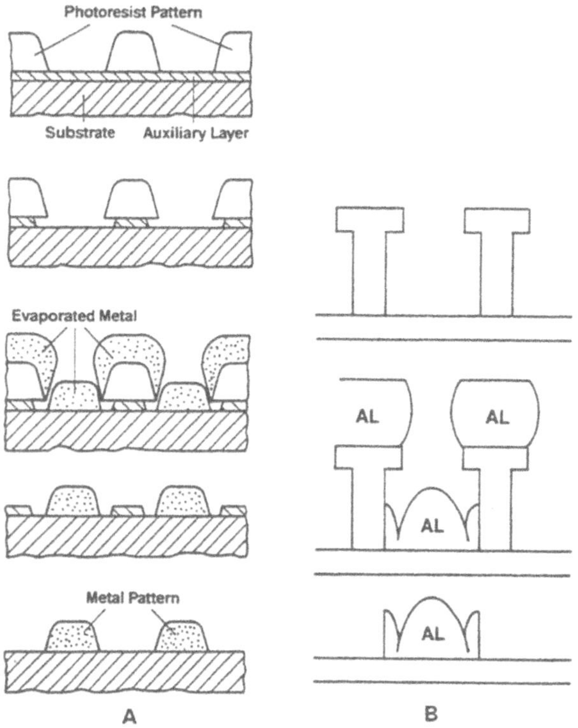

FIGURE 12-12-11. (A) Lift-off process using sacrificial layer. From D. Widmann, *IEEE J. Solid State Circuits* **SC-11**, 466 (1976), copyright 1976 and courtesy of IEEE. (**B**) Schematic showing off-axis metal deposition on lift-off profiled resist. From Ref. 22, reprinted with permission of *Solid State Technology*, published by Technical Publishing, a company of Dun and Bradstreet.

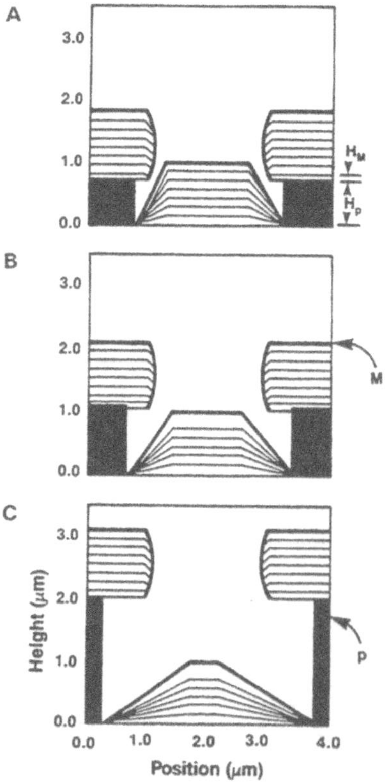

FIGURE 12-12-12. Effect of polyimide (p) thickness on deposited metal (M) profiles for polyimide thickness of (A) 0.7 μm, (B) 1.0 μm, and (C) 2.0 μm. From Ref. 210, copyright 1982 and courtesy of IEEE.

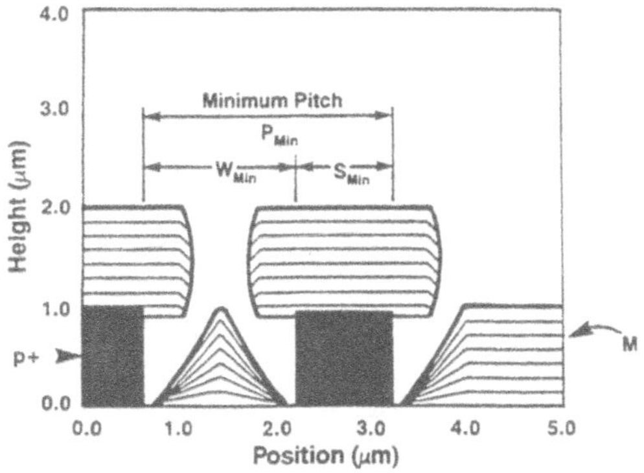

FIGURE 12-12-13. Calculated minimum pitch of deposited metal thickness. From Ref. 210, copyright 1982 and courtesy of IEEE.

the deposition. This results in tapered profiles or edge skirts[22] (Fig. 12-12-11). In some cases, a predeposit of a thin blanket of metal to planarize the substrate improves step coverage. Overall, a bias of about one-fourth the vertical thickness of the metal can be added to the base of the metal land. Tapered deposits for Josephson devices can be achieved by oblique deposition with a rotating substrate.[168]

12-12-2. Resist Thickness and Lift-off

The effects of resist thickness in a trilayer (resist/MO/polyimide) resist were studied by Homma and co-workers.[210] Increasing the polyimide thickness from 0.7 to 2.0 μm increases the base width of the profile (Fig. 12-12-12). Thus, an MLR over steps requires a planarizing layer thicker than the height (>1 μm) but not too thick to cause isotropic deposition (2.0-μm case in Fig. 12-12-13). In one sense, the limit is the height of the underlying topology. The minimum pitch for a 1.0-μm-thick deposited metal was also measured for varying polyimide thicknesses. For a 1.0-μm-thick metal film, the minimum pitch will be 2.6 μm, with a metal land width of 1.6 μm and a space (horizontal) of 1.0 μm (Fig. 12-12-13).

12-13. LIFT-OFF PROCESS CONTROL

Since the uniformity of metal deposition can vary as much as $\pm 25\%$, single-wafer deposition with *in situ* monitoring may be required. Secondly, variations in metal thicknesses across the wafer or from batch to batch depositions can result in poor lifting and added biases of 0.2–0.4 μm During the metal deposition, the fluctuations of the vacuum pressure are evidence for resist outgassing from such components as the decomposing N_2 in the DQN sensitizer, the phenol in the novolak resin of DQN, trapped water, and casting solvents. Outgassing from walls of the resist (Fig. 12-13-1) act as a streaming source to isotropically divert the depositing metal into forming pyramidal profiles. To minimize outgassing, careful bake conditions, low metal deposition temperatures as well as stable resist materials are required. Monitoring of the metal lift-off by optical reflectance or other means has not been reported, but endpoint systems could be easily constructed based on reflectivity differences.

In processing the lift-off step, common lift-off solvents include acetone for

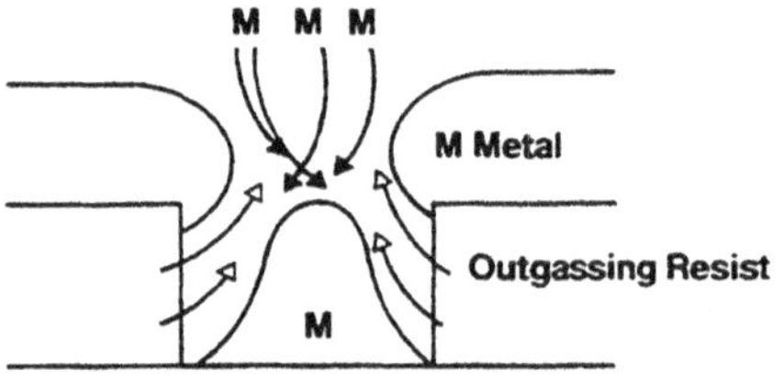

FIGURE 12-13-1. Outgassing resist (at elevated temperatures and stray electron bombardment during *e*-beam metal deposition) distorting angle of metal deposition.

PMMA and DQN, $CHCl_3$ and NMP for some polyimides, hydrazine for cross-linked polyimides, and aqueous etchants for inorganic layers. Organic-based strippers which do not attack Al (Chapter 14) can also be used. Underneath the resist, polyethersulfone polymers are also used as thin release layers[2,310] (under baked DQN) to assist in lift-off. In addition, a high-temperature aqueous stripping layer would be useful in reducing the problems of the safe handling and the high cost of disposal of organic solvents. In this regard, a "dry" lift-off process has been developed using the depolymerization of PMMA by heat, to rupture the metal film prior to lift-off.[240]

After metal lift-off, the frequent filtering and changes of the lift-off tank are necessary to prevent polymer and metal redeposition. Glycerin[203] is added to acetone lift-off solvent to prevent metal redeposition. Heating of the lift-off solvent facilitates permeation, and swelling and ultrasonic agitation are also used to assist in stressing the undesired metal.[346] If the solvent does not dissolve the lift-off layer, adhesion failure must occur at the polymer–substrate interface. To encourage abhesion of the lift-off underlayers, strong surfactants and penetrants (Chapter 14) are added to the resist solvents. The permeation of sulfonic acids[319] or polar solvents with higher surface tension than the lift-off polymer, should induce adhesion failure.

If the MLR layer is used for lift-off, the metallurgy lines will be subject to developer plasma contaminants in the lift-off line, prior to metal deposition. These contaminants will prevent metal deposition or opens in the circuit. For example, the bridging across metal lines can occur through cracks in the lift-off layers. Frequently, a sputter clean or preetching of the contacts is used to remove contaminants prior to metal deposition, but surface analysis should be made, to ensure that considerable resist distortion does not occur. Pinholes in the trilayer stack are not a major problem in the lift-off process, since the deposited stalactites are not connected to active Si. Nicks in resist images are more of a problem as they lead to metal shorts.

Heating of the metal/resist prior to lift-off may expand the resist and assist in solvent penetration. Recent soluble high-temperature polyimides[174] or polyether aryl sulfones[147] are more suitable soluble lift-off underlayers than the insoluble baked DQN underlayer.[2,159] More suitable (soluble) lift-off layers are required for practical and reproducible lift-off times.

Summary of Lift-off of MLR

The MLRs have evolved to overcome the planarizing deficiencies of spun-cast films, the image distortion over steps, and the poor reactive ion or thermal melting of DQN and PMMA resists. In Table 12-13-1, the comparative analysis of the advantages and disadvantages of MLRs is given. From the needs of the resist stabilization, the separation of the image layer from the masking layer enabled the

TABLE 12-13-1. Comparative Anatomy of Multilayer Resist Masking

Advantages	Disadvantages
Reduce proximity effects	More process steps
Reduce standing waves	Higher tooling costs
Improve resolution	More difficult registration
RIE/thermal stable masks	Mixing of interlayers
Multiuse RIE or lift-off	Stresses in layers
Can use lower dose	Higher pinholes (thinner films)
Reduce *e*-beam charge	Precision metal evaporation
Image reversal	Precise O_2 + etching (RIE step)
Zero-overlap via contacts	Limited high-temperature polymers
Multiple metal deposit, Al-Cu	
Planarization	
Wide spectrum of materials	

production of more ideal resist structures. A composite of materials ranging from metal and silicone barriers, silicone and metal–organic resists, inorganic resists, and inorganic lift-off layers have been engineered into bi- and trilayer resists. The most useful system is the posttreatment of DQN resists with silanes to impregnate the resist and convert the resin into an oxygen plasma mask.(419) With this advantage, conventional steppers can be extended to submicrometer geometries.

For MLRs, the trade-off lies in the extra process costs versus higher yields and higher circuit density. For the definition of metal patterns,(222) the competitive alternative is reactive ion etching (Chapter 13), ironically, using an MLR structure made by plasma processing at double the costs. Personalized metal conductors using *e*-beam exposure and lift-off are cost-effective for low-volume–high-part-number logic circuitry (Fig. 12-13-2)(2) in which the break-even point is about 100 wafers per part number. Otherwise, optical exposure will suffice. However, because of the increasing cost of *e*-beams (> \$4000 K), high-speed resists with 1 $\mu C/cm^2$ (*e*-beam) or 1 mJ/cm^2 (X-ray) are needed to maintain reasonable beam writing schemes in MLRs [> 10 (125 mm) wafers/hr]. The lift-off process can be performed with various configuration considerations:

	One-layer	Two-layer	Three-layer
Best candidates	DQN PMMA	Si-DQN/UL Siloxane-UL	Many with appropriate barrier
Steps	$\leqslant 6$	$\leqslant 11$	$\leqslant 15$
Limits	Profile	Underlayer performance	Barrier cost
	Steps	Top layer curling	RIE cost
	Dose	Pinholes image layer	
Improved resists	Thermal	Faster inorganic	Soluble lift-off layer
	Conformable	Positive O^+ resist	Stable lift-off layer
	Higher speed		Cheap barrier

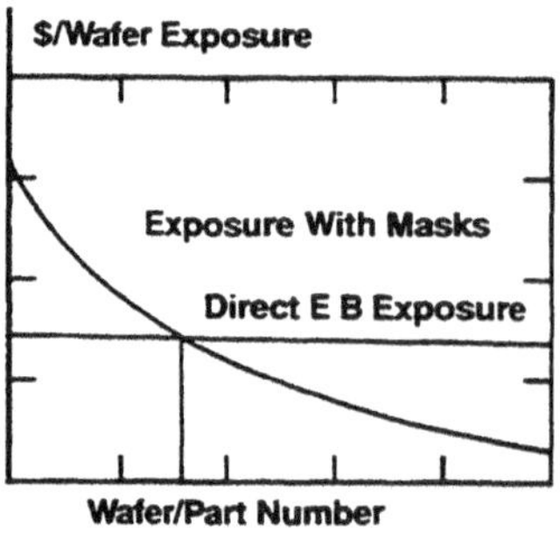

FIGURE 12-13-2. Personalization cost factor of *e*-beam exposure to do metallurgy lithography in custom circuits. For low volume, high part numbers, *e*-beam exposure is less expensive than making a photomask for a limited printing.

The additive processes are competitive with metal etching.[222] Etching of Al–Cu plasma has been a major problem. Likewise, the deposition of Al–Cu requires elevated ($>150°C$) temperatures[2,211] for suitable electrical performance. A comparison of subtractive and additive metal definition for 1-μm-thick metal indicates that reactive ion assisted etching of metals can provide a smaller bias than lift-off, if product redeposition and postcorrosion can be eliminated (Chapter 13). In Table 12-13-2, a perspective comparison of the three metal patterning processes is given.

Single- or double-layer resists should be initially tried to offset costs. However, if dry etching is required to form the MLR structure, then the extra cost must be offset by the smaller biases and closer pitches of 1-μm metal lines.

MLRs appear to offer the dual purpose of applicability to either metal lift-off or metal etching. Although both processes are maturing into the 1980s, both lift-off and RIE appear to be required for submicrometer circuitry. Acquiring basic experience in lift-off processing is necessary, before selection of suitable lift-off or RIE mask layers and before selection of the primary metal process. A soluble high-temperature reactive ion layer is required to complement the option of lift-off or reactive subtractive etching with the MLR structures. More suitable materials are required before the tooling costs and the number of steps can be reduced.

TABLE 12-13-2. Comparison of Metal Lithography

	Wet etch	Dry etch	Lift-off
Advantages	Simple Clean contact High-temperature deposit Profile insensitive	High resolve Clean contact Low bias	Step insensitive Tight tolerance
Disadvantages	Large bias Pitch limit Harsh chemicals Liquid waste	Resist erosion Profile control Redeposition	Deposit temperature Residue sensitive Profile sensitive Multiple cost
Bias and tolerance in 1-μm-thick film	1 ± 1 μm	0.2 ± 0.5 μm	0.5 ± 0.3 μm
Cost	Lowest (excluding waste cost)	Intermediate	Highest

12-14. MISCELLANEOUS ADDITIVE PROCESSES

Besides metal lift-off, the deposition of metals by electroplating or the deposition of dopants (ion implant) using a resist mask, followed by strip and oxidation, have been selected uses of resists. Light optical waveguides and resists filled with additives (Chapter 16) are specialized patterns formed by resists.(227) We will discuss electroplating and then ion implantation as pertinent uses of resists in additive processing.

12-15. ELECTROPLATING WITH RESISTS

Electroplating with resists has a strong foundation in printed circuit technology. Resists have evolved from silk-screened "resists" to dry film resists (positive and negative) for electroplating,

Silk screen → Negative cinnamate → Positive diazoquinone → Dry film acrylate

Additive electroless plating of Cu has been used for the plating of conductors and holes.(211) The inability of liquid resists to cover holes has been solved by dry film resists (Fig. 12-15-1)(213) in which the dry film resist bridges and protects a plate trough hole.

In other applications, the resist can act as a nonconducting mask to allow electroplating of exposed conductive underlayers.(214) The resist can also contain a catalyst to create a resistless process (Chapter 16). These commercial processes are called Physical Development and Reduction (Phillips Co.) and Photoformation (Photocircuits).(215) The substrate is coated with Cu acetate in glycerin, UV exposed to create Cu seeds, and Cu is electrolessly plated.

Another variation of electroless plating involves sensitizing the AZ-1350 resist pattern with Sn^{2+} (Fig. 12-15-2). The Sn^{2+} adheres selectively to the substrate and not to the resist surfaces and the adsorbed Sn^{2+} acts as a site for electroless plating.(216)

The effectiveness of plating metal lines as in lift-off is dependent on the resist profiles and resist thickness. Since electroplating is the inverse of lift-off, the bottom profile at the resist–substrate interface defines the cross section (Fig. 12-15-3) instead of the top of the resist as in lift-off.(212)

Thick photoresists of AZ-1350 (range 2–16 μm in thickness) have been studied as an electroplating mask.(217) Difficulties in producing uniform films by roller coating, spray or dip without rim buildup, were eliminated by spin coating. The steepness of AZ-1350H, AZ-340, AZ-119, and AZ-2400 resists was investigated at

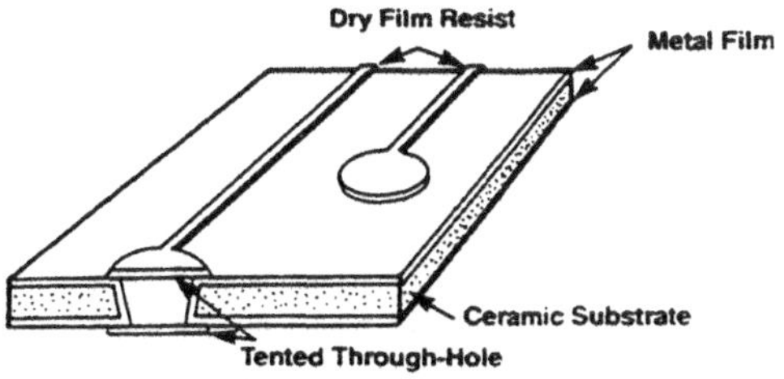

FIGURE 12-15-1. Dry film resist covering (tenting over hole) for printed circuit plating. From Ref. 213, reprinted with permission of *Solid State Technology*, published by Technical Publishing, a company of Dun and Bradstreet.

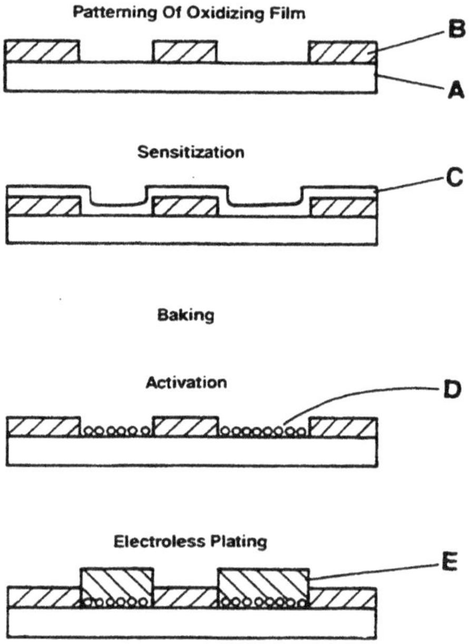

FIGURE 12-15-2. Steps of the stannous ion sensitized electroless plating of metals. The resist selectively masks the Sn catalyst to activate desired regions. From Ref. 216, copyright 1971 and courtesy of RCA.

different postbake temperatures of 90–110°C. To withstand the plating bath chemicals, postbaking is necessary to obtain sufficient adhesion and plating resistance to prevent undercut plating at the resist–substrate interface (Fig. 12-15-4). However, after postbaking, the wall profile drops to 40° for AZ-1350H (4 μm thick) baked at 110°C. The flow is probably due to the considerable casting solvent trapped in the film at the 70° prebake. AZ-340 was more elastic and could be used at 7-μm thickness without cracking during plating. Furthermore, overplating produces mushroom profiles (Fig. 12-15-5) which leads to major shorting.

In microelectronics, electroplating additive processes have not been used on silicon due to the fact that Al cannot be readily deposited. Electroplating has been used for fabrication of magnetic bubble memories[218] and X-ray exposure masks.[28] Permalloy and gold are electroplated through masks of AZ-1350J and lines of 1 μm in width have been plated. In addition to the lateral dimensional control, good thickness and composition uniformity are required in the plated films. These parameters are directly related to the local current density which in turn is influenced by the mask design, cathode holder construction, plating bath formulation, and the thickness of evaporated plating base. Alkali in the plating bath attacks PMMA

FIGURE 12-15-3. Comparison of additive lift-off and additive electroplating. (A) The resist mask in lift-off defines the metal width by the opening at the top of the resist. (B) In electroplating the resist width at the surface of the metal (bottom of resist image) defines the width of the deposited metal.

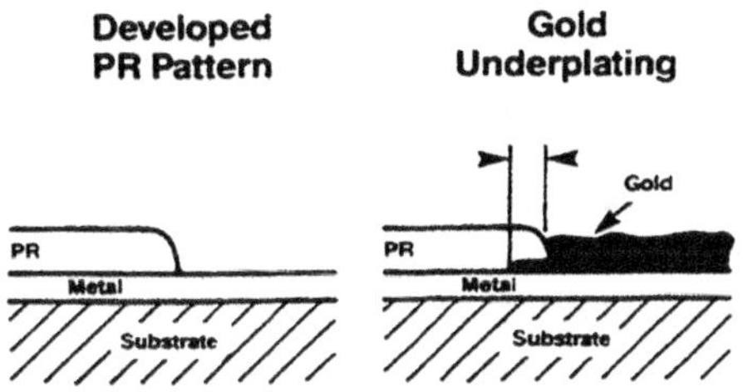

FIGURE 12-15-4. Plating of gold with a resist mask. Overplating of opening under resist due to poor resist adhesion or resistance to chemicals in electroplating bath. From Ref. 309, reprinted with permission of *Solid State Technology*, published by Technical Publishing, a company of Dun and Bradstreet.

and DQN resists. Since the electroplated metal reproduces the walls (even standing waves) of the resist, vertical profiles produced by the MLR are required to produce vertical plated features. The production of X-ray[(219)] masks relies on lift-off of a base metal and subsequent Au electroplating (Fig. 12-15-6). A high-temperature lift-off profile can be produced by electroplating Cu mushrooms, depositing the desired metal, and etching the Cu lift-off layer.[(220)]

The ultimate resolution of electroplated lines depends primarily on the resolution of the resist and to a much lower extent on the electroplating. Aspect ratios of 2 to 1 of height to width are possible[(221)] (usually 25% thicker metal than resist thickness) as compared to the 70% of resist thickness limit for the lift-off technique. With X-ray exposure, resist profiles with aspect ratios of 5 to 1 were produced as small as 1 μm wide by 5 μm thick separated by 0.1-μm space.[(28)]

After the electroplating has been completed, the resist and plating base, if necessary, are stripped in organic solvents and etchants which do not corrode the plated metal. Sometimes the base has to be removed by physical milling since etchants also attack the plated metal.[(218)] Overall, the additive electroplating process is limited to the more noble metals and will find much utility in microelectronic devices, but will continue to find more applications in chip packaging.

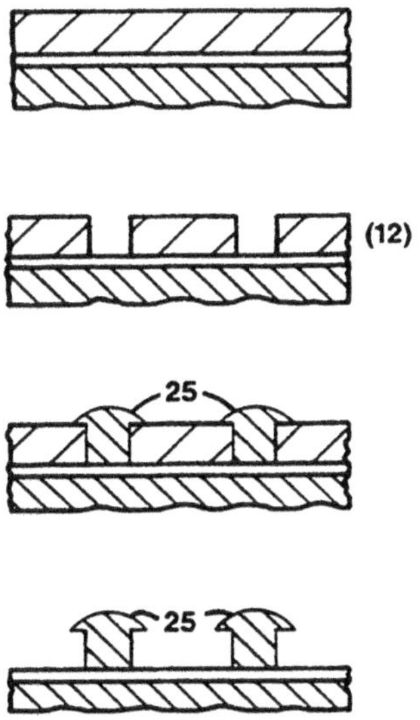

FIGURE 12-15-5. Overplating of metal (25) on top of resist produces characteristic "mushroom cap." U.S. Patent 4,224,361 (1980).

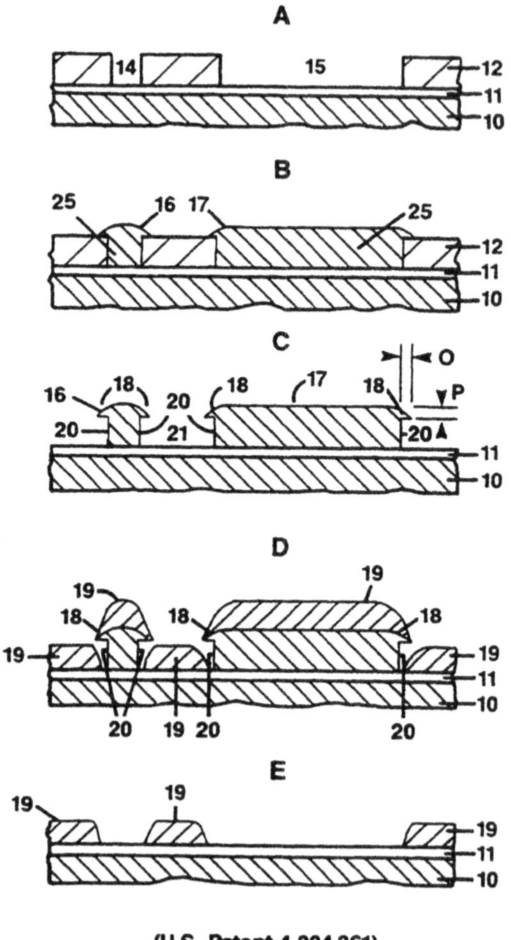

FIGURE 12-15-6. Steps in high temperature lift-off. (A) Pattern (15) formed in resist (12); (B) overplate metal (25) on top of resist; (C) strip resist (12); (D) deposit Al (19) on top of Cu (18); (E) etch of Cu (18) to leave Al (19).

12-16. ION IMPLANTATION RESIST MASKING

Instead of (traditionally) thermally doping transistors with P, B, and As through resist-formed holes in SiO_2, ions have been implanted directly through resist masks.[224,225] The stopping power of resists for B and P ions from an ion

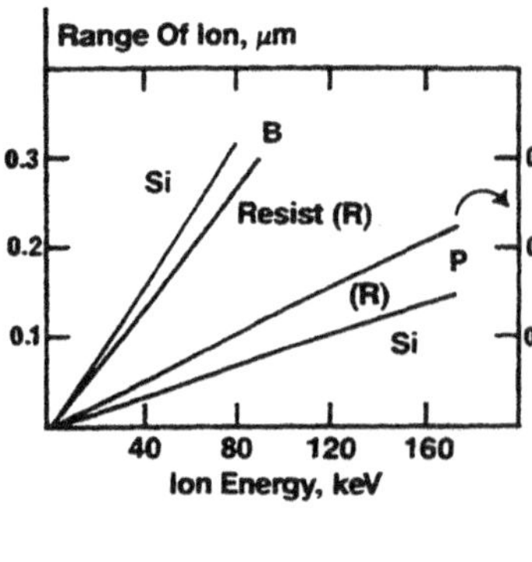

FIGURE 12-16-1. Range of P or B ions in Si and in organic C-based resist (R).

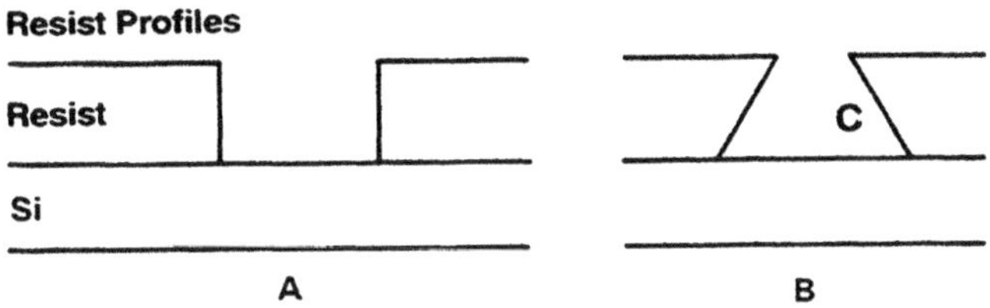

FIGURE 12-16-2. (A) Vertical and (B) overcut profile in resist for masking ions. Overcut profile prevents DQN resist from hardening. Only the top of the overcut profile hardens, leaving the bottom half for liquid stripping.

implanter is shown in Fig. 12-16-1.[337] An organic resist thickness of about 1.0 μm will mask all dopant ions.[239] During ion implantation, the resist of DQN self-hardens due to ion bombardment and cross-linking.[226] To permit the removal of DQN in organic strippers, an inverted profile (Fig. 12-16-2) similar to an overcut (high dose) profile for single-film lift-off, has been proposed to prevent insolubilization of DQN along the sidewalls.[186] The overcut profile acts as a top surface mask to absorb the incident ions and prevent sidewall hardening. Subsequently, the resist can be more readily stripped in liquid strippers. Perhaps in the future resistless lithography (Chapter 16) will implant patterns into Si directly with ion beams.

Outgassing from a DQN resist during implantation has been solved by various resist hardening techniques (see Table 13-12-3). Metal-on-polymer masks, such as trilayer resists, have also been used, due to the high stopping power of metals for B and P ions. If the resist that is used to etch to the silicon is left in place as an implant mask, self-aligned holes can be generated.[105,186]

Dopants can also be introduced from B-, As-, or P-containing polysiloxane ladder polymers. The tris-diorganosilylene diphosphate is coated, exposed to an *e*-beam, developed, and fired in O_2 to selectively drive in P, B, or As (Fig. 12-16-3).[32,180] Thus, P-doped glass or patterns of insulating SiO_2[230] can be formed in this fashion. Perhaps in the future submicrometer optical circuits will be fabricated directly on silicon or other substrates by patterning and firing.[227]

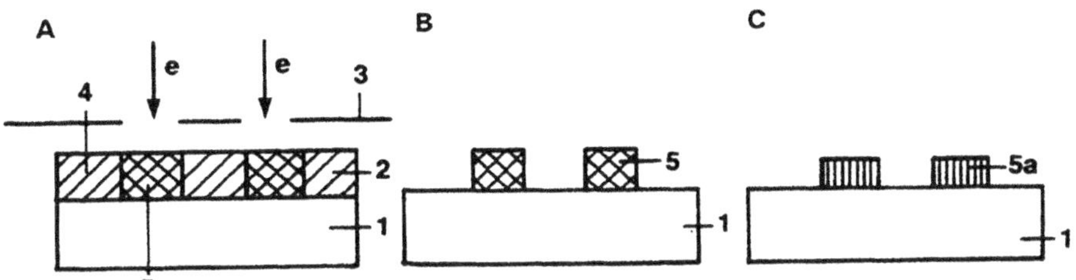

FIGURE 12-16-3. Doping resist process. Dopant of P introduction from phosphorus-containing siloxane resist. The resist (2) is patterned by *e*-beams to create a doping pattern (5) on top of the silicon. In step (C), the dopant is thermally driven into silicon and phosphosilicate glass is formed. From Ref. 180.

REFERENCES

1. H. Smith, F. Bachner, and N. Efrenov, *J. Electrochem. Soc.* **118**, 821 (1971).
2. L. Fried, J. Lechaton, P. Totta, J. Logan, J. Havas, and G. Paal, *IBM J. Res. Dev.* **26**, 362 (1982).
3. T. Maple, *Solid State Technol.* Sept. 1966, p. 60.
4. P. Frasch and K. Saremski, *IBM J. Res. Dev.* **26**, 561 (1981).
5. M. Hatzakis, *J. Electrochem. Soc.* **116**, 1033 (1969).
6. A. Learn, *J. Electron. Mater.* **3**, 531 (1974).
7. D. Widman and H. Binder, *IEEE Trans. Electron. Devices* **ED-22**, 467 (1975).
8. D. Davis, S. Gillespie, S. Silverman, and W. Stickel, Abstract D-2, Int. Symp. on Electron, Ion, and Photon Beams, Los Angeles, May 1983; *Solid State Technol.* Sept. 1983, p. 174; *J. Vac. Sci. Technol.* **B1**, 1003 (1983).
9. Y. Homa, A. Yajima, and S. Harada, *IEEE J. Solid State Circuits* **SC-17**, 142 (1982).
10. M. Feuer and D. Prober, *IEEE Trans. Electron Devices* **ED-28**, 1375 (1981).
11. U. S. Patent 4,154,613 (1980), Am. Hoechst.
12. U. S. Patent 4,224,361 (1980), IBM.
13. U. S. Patent 4,007,047 (1977), IBM.
14. U. S. Patent 4,212,935 (1980), IBM.
15. M. Hatzakis, B. Canavello, and J. Shaw, *IBM J. Res. Dev.* **24**, 452 (1980).
16. R. Halverson, M. MacIntyre, and W. Motsiff, *IBM J. Res. Dev.* **26**, 58 (1982).
17. G. Collins and C. Halstead, *IBM J. Res. Dev.* **26**, 595 (1982).
18. Japanese Patent 80,008,014, Fujitsu; *Chem. Abstr.* **92**, 224306 (1980).
19. J. Speidel, *J. Vac. Sci. Technol.* **19**, 693 (1981).
20. W. Oldham and E. Hieke, *IEEE Electron Device Lett.* **EDL-1**, 217 (1980).
21. U. S. Patent 4,104,070 (1979), IBM.
22. T. Batchhelder, *Solid State Technol.* Feb. 1982, p. 111.
23. U. S. Patent 4,115,120 (1978), IBM.
24. U. S. Patent 3,622,365 (1971), Fairchild.
25. E. Jelks, G. Kerber, and H. Wilcox, *Appl. Phys. Lett.* **36**, 93 (1980).
26. T. Serikawa and T. Yachi, *J. Electrochem. Soc.* **128**, 918 (1981).
27. J. Shaw and M. Hatzakis, *J. Electrochem. Soc.* **126**, 2029 (1979).
28. E. Spiller, M. Heritage, R. Feder, and T. Topalian, *Solid State Technol.* April 1976, p. 62.
29. D. Flanders, *Appl. Phys. Let.* **36**, 93 (1980).
30. S. Beaumont, P. Bower, T. Tamamura, and C. Wilkinson, *Appl. Phys. Lett.* **38**, 436 (1981).
31. A. Broers, J. Harper, and W. Molzen, *Appl. Phys. Lett.* **33**, 392 (1978).
32. E. Roberts, *Philips Tech. Rev.* **35**, 41 (1975).
33. G. Dolan and T. Fulton, *IEEE Electron Device Lett.* **EDL-4**, 178 (1983).
34. B. Lin, in *Introduction to Microlithography*, edited by L. Thompson, M. Bowden, and G. Willson, American Chemical Society, Washington, D.C., 1983, pp. 287–349.
35. B. Griffing and P. West, *IEEE Electron Device Lett.* **EDL-4**, 14 (1983).
36. U. S. Patent 4,287,289 (1981), Tokyo Ohko.
37. C. Hamel and E. Symula, *IBM Tech. Discl. Bull.* **22**, 1399 (1979).
38. C. Hamel and J. Kristoff, *IBM Tech. Discl. Bull.* **24**, 215 (1981).
39. U. S. Patent 3,961,000 (1977), RCA.
40. U. S. Patent 3,961,101 (1977), RCA.
41. Brazilian Patent 78,077666, IBM; *Chem. Abstr.* **93**, 85225 (1981).
42. W. Oldham and A. Neureuther, *IEEE Trans. Electron Devices* **ED-27**, 1455 (1980).
43. T. Berker and D. Casey, *IEEE Trans. Electron Devices* **ED-29**, 524 (1982).
44. Japanese Patent 81,92,536, Fujitsu; *Chem. Abstr.* **96**, 60866 (1982).
45. Japanese Patent 80,32,088, Fuji; *Chem. Abstr.* **93**, 195530 (1980).
46. U. S. Patent 4,104,070 (1979), IBM.
47. S. MacDonald, R. Miller, C. Willson, G. Feinberg, R. Gleason, R. Halverson, M. Macintyre, and W. Motsiff, Interface 82, Kodak Proceedings of Microelectronics Seminar, 1982, p. 114.
48. T. Matsuzawa and H. Tomika, *IEEE Trans. Electron Devices* **ED-28**, 1284 (1981).
49. N. Veno, S. Konishi, K. Tanimoto, and K. Sugita, *Jpn. J. Appl. Phys.* **20**, L709 (1981).
50. Y. Kawamura, K. Toyoda, and S. Namba, *J. Appl. Phys.* **53**, 6489 (1982).
51. U. S. Patent 4,339,522 (1980), IBM.
52. G. Willson, N. Clecak, B. Grant, and R. Twieg, *Electrochem. Soc. Ext. Abstr.* **80-1**, 696 (1980).

53. U. S. Patent 4,339,521 (1981), Siemens.
54. German Patent 2,631,535, GAF; *Chem. Abstr.* **86**, 198007 (1978).
55. D. Kyser and C. Ting, *J. Vac. Sci. Technol.* **16**, 1759 (1979); M. Yoshimi, *Electron. Lett.* **78**, 880 (1982).
56. J. Greeneich, *Electrochem. Soc. Ext. Abstr.* **80**, 261 (1980).
57. U. S. Patent 4,061,832 (1980), Mullard; Y. Mimura, *J. Electrochem. Soc.* **126**, 1881 (1979).
58. A. Miura and S. Hideyama, *J. Vac. Sci. Technol.* **16**, 1996 (1979).
59. U. S. Patent 4,276,365 (1981), Fujitsu.
60. W. Moreau and R. Lange, *IBM Tech. Discl. Bull.* **25**, 2725 (1982).
61. D. Widmann, *IEEE J. Solid State Circuits* **SC-11**, 466 (1976).
62. T. Serikawa and T. Yachi, *J. Electrochem. Soc.* **128**, 918 (1981); N. Yamauchi, *Jpn. J. Appl. Phys.* **22**, L595 (1983).
63. U. S. Patent 4,214,966 (1980), Bell.
64. W. Curry, G. Galyon, J. Giddings, and A. Lee, *IBM Tech. Discl. Bull.* **22**, 4511 (1980).
65. R. Burkhart, *IBM Tech. Discl. Bull.* **24**, 2081 (1981).
66. T. Sakurai and T. Serikawa, *J. Electrochem. Soc.* **126**, 1257 (1979).
67. E. Walker, *IEEE Trans. Electron Devices* **ED-22**, 64 (1975).
68. T. Venkatesan, G. Taylor, A. Wagner, C. Wilkens, and D. Barr, *J. Vac. Sci. Technol.* **19**, 1379 (1981).
69. I. Aesida, J. Chinn, L. Rathbun, and E. Wolf, *J. Vac. Sci. Technol.* **21**, 666 (1982).
70. D. Follett, K. Weiss, J. Moore, A. Steckl, and W. Lu, *Electrochem. Soc. Ext. Abstr.* **82-2**, 321 (1982).
71. M. Gazard, C. Duchesne, J. Dubas, and A. Chapiro, *Polym. Eng. Sci.* **20**, 1069 (1980).
72. J. Pacansky and J. Lyerla, *IBM J. Res. Dev.* **23**, 42 (1979).
73. A. Martyneko, B. Stizhov, and V. Nikolskii, *Russ. J. Phys. Chem.* **49**, 1310 (1975).
74. B. Lin, *Solid State Technol.* May 1983, p. 105.
75. K. Tai, R. Vadimsky, C. Kemmerer, J. Wagner, V. Lamberti, and A. Timko, *J. Vac. Sci. Technol.* **17**, 1169 (1980).
76. S. Middelhoek, *IBM J. Res. Dev.* March 1970, p. 117.
77. T. Brewer, *J. Appl. Photogr. Eng.* **1**, 184 (1981).
78. C. Ting, I. Avigal, and B. Lu, Proceedings of Kodak Microelectronics Seminar, Oct. 1982, G-136, p. 139; *J. Vac. Sci. Technol.* **B1**, 1225 (1983).
79. U. S. Patent 3,934,057 (1976), IBM.
80. B. Lin, E. Bassous, V. Chao, and K. Petrillo, *J. Vac. Sci. Technol.* **19**, 1313 (1981).
81. H. Namatsu, Y. Ozaki, and K. Hirata, *J. Vac. Sci. Technol.* **21**, 672 (1982).
82. U. S. Patent 3,996,393 (1978), IBM.
83. U. S. Patent 3,987,215 (1977), IBM.
84. U. S. Patent 3,934,057 (1976), IBM.
85. U. S. Patent 4,024,293 (1978), IBM.
86. M. Hatzakis, *J. Vac. Sci. Technol.* **16**, 1984 (1979).
87. D. Webb, *IBM J. Res. Dev.* **24**, 554 (1980).
88. W. Moreau, W. Moyer, D. Merritt, M. Hatzakis, L. Pederson, and D. Johnson, *J. Vac. Sci. Technol.* **16**, 1990 (1979).
89. Y. Todokora, *IEEE Trans. Electron Devices* **ED-27**, 1443 (1980).
90. Japanese Patent 80,140,836, Fujitsu; *Chem. Abstr.* **94**, 183462 (1980).
91. H. Hiroaka, *SPIE Proc.* **469**, 127 (1984).
92. R. Howard, E. Hu, and L. Jackel, *IEEE Trans. Electron Devices* **ED-28**, 1378 (1981).
93. R. Howard, E. Hu, and L. Jackel, *Appl. Phys. Lett.* **36**, 141 (1980).
94. P. Hammel and R. Richardson, *Physics* **107B**, 611 (1981).
95. M. Yamada, J. Tamano, K. Yoneda, S. Morita, and S. Hattori, *Jpn. J. Appl. Phys.* **21**, 768 (1982).
96. British Patent 2,064,152, Bell; *Chem. Abstr.* **96**, 13668f (1982).
97. U. S. Patent 3,898,350 (1975), IBM.
98. W. Moreau and P. Schmidt, *Electrochem. Soc. Ext. Abstr.* No. 187 (1970).
99. T. Venkatesan, *J. Vac. Sci. Technol.* **19**, 1368 (1981).
100. E. Bassous, L. Ephrath, G. Pepper, and D. Mikalsen, *J. Electrochem. Soc.* **130**, 478 (1983).
101. U. S. Patent 4,214,966 (1980), Bell.
102. B. Lin, *J. Electrochem. Soc.* **127**, 202 (1980).
103. U. S. Patent 3,873,313 (1974), IBM.

104. U. S. Patent 4,238,559 (1981), IBM.
105. Y. Todokoro, *Trans. IEEE Jpn.* **E-65**, 23 (1982).
106. German Patent 3,036,710, Siemens; *Chem. Abstr.* **97**, 48247 (1982).
107. U. S. Patent 3,982,943 (1976), IBM.
108. U. S. Patent 4,092,442 (1980), IBM.
109. A. McCullough and E. Pavelche, Symp. Electron, Ion, Laser Beam Technol., San Francisco, 1983, Abstr. K-5; *J. Vac. Sci. Technol.* **B1**, 1241 (1983).
110. E. Ong, Ref. 109, Abstr. K-6; *J. Vac. Sci. Technol.* **B1**, 1247 (1983).
111. B. Lin, V. Chao, F. Kaufman, and S. Kramer, *Electrochem. Ext. Abstr.* **82-1**, 450 (1982).
112. H. Santini and N. Viswanathan, Ref. 78, p. 47.
113. B. Griffing, *J. Vac. Sci. Technol.* **19**, 1423 (1981).
114. T. Batchelder and C. Takemoto, *Semicond. Int.* July 1981, p. 7.
115. W. Moreau, *Opt. Eng.* **22**, 181 (1983).
116. U. S. Patent 4,139,384 (1981), Fuji.
117. D. Bobkin and B. Cantos, *IEEE Electron Device Lett.* **EDL-2**, 222 (1981).
118. G. Chui and W. Ma, *IBM Tech. Discl. Bull.* **21**, 3623 (1979).
119. M. Vry, J. Matthews, and C. Wood, *SPIE Proc.* **334**, 241 (1982).
120. K. Jain, C. Willson, and B. Lin, *IBM J. Res. Dev.* **26**, 151 (1982).
121. C. Horwitz, *Appl. Phys. Lett.* **32**, 803 (1978).
122. H. Namatsu, Y. Ozaki, and K. Hirata, *J. Electrochem. Soc.* **130**, 523 (1983); H. Gokan, M. Itoh, and S. Esho, *J. Vac. Sci. Technol.* **B2**, 34 (1984); I. Watanabe, *Jpn. J. Appl. Phys.* **23**, 487 (1984).
123. B. Lin, E. Bassous, V. Chao, and K. Petrillo, *J. Vac. Sci. Technol.* **19**, 1313 (1981).
124. J. Lavine, J. Masters, G. Goldberg, and A. Das, *IEEE Trans. Electron Devices* **ED-28**, 1311 (1981).
125. T. Venkatesan, G. Taylor, A. Wagner, B. Wilkens, and D. Barr, *J. Vac. Sci. Technol.* **19**, 1379 (1981).
126. K. Tai, W. Sinclair, R. Vadimsky, J. Moran, and M. Rand, *J. Vac. Sci. Technol.* **16**, 1977 (1979).
127. A. Wagner, D. Barr, T. Venkatesan, W. Crane, V. Lamberti, K. Tai, and R. Vadimsky, *J. Vac. Sci. Technol.* **19**, 1363 (1981).
128. G. Taylor, T. Wolf, and J. Moran, *J. Vac. Sci. Technol.* **19**, 872 (1981).
129. I. Aesida, J. Chinn, L. Rathburn, and E. Wolf, *J. Vac. Sci. Technol.* **19**, 875 (1981).
130. M. Hatzakis, J. Paraszczak, and J. Shaw, *Microcircuit Eng.*, Grenoble, France, Sept. 1981, p. 396.
131. M. Hatzakis, J. Shaw, J. Paraszczak, J. Liutkus, and E. Babich, SPE RETEC Photopolymers, 1982, Ellenville, N.Y., p. 285.
132. Japanese Patent 82,50,430, *Chem. Abstr.* **97**, 48231 (1982).
133. M. Yamada, J. Tamano, K. Yoneda, S. Morita, and S. Hattori, *Jpn. J. Appl. Phys.* **21**, 768 (1982).
134. U. S. Patent 3,782,940 (1974), Dai Nippon Print.
135. K. Beyer and J. Logan, *IBM Tech. Discl. Bull.* 17, 1600 (1974).
136. U. S. Patent 3,622,365 (1977), Fairchild.
137. U. S. Patent 3,519,584 (1976), IBM.
138. Y. Homma, H. Nozaiva, and S. Harada, *IEEE Trans. Electron Devices* **ED-28**, 552 (1981).
139. U. S. Patent 4,224,361 (1980), IBM.
140. U. S. Patent 4,315,984 (1982), Hitachi.
141. L. Rothman, *Electrochem. Soc. Plasma Processes* **81-1**, 192 (1981).
142. Y. Todokoro, *Electron. Lett.* **18**, 543 (1982).
143. H. Namasu, Y. Ozaki, and K. Hirata, *J. Vac. Sci. Technol.* **21**, 672 (1982).
144. L. Rothman, *Electrochem. Soc. Ext. Abstr.* **82-1**, 582 (1982).
145. U. S. Patent 4,132,568 (1981), IBM.
146. U. S. Patent 3,985,597 (1976), IBM.
147. L. Gregor, W. Moreau, J. Zingerman, and L. Kaplan, *IBM Tech. Discl. Bull.* **24**, 3837, 5538 (1982).
148. H. Namatsu, Y. Ozaki, and K. Hirata, *J. Vac. Sci. Technol.* **21**, 672 (1982).
149. S. Macdonald, G. Willson, H. Ito, and R. Miller, Int. Symp. Electron, Ion, Photon Beams, Los Angeles, May 1983, Abstr. I-4.
150. K. Tanigaki, *Am. Chem. Soc. Org. Coat. Prepr.* **48**, 179 (1983).
151. P. Van Pelt, *Soc. Photogr. Inst. Eng. Proc.* **275**, 150 (1981).
152. L. Karapiperis and C. Lee, *Appl. Phys. Lett.* **35**, 395 (1979).
153. B. Hunt and R. Burhman, *J. Vac. Sci. Technol.* **19**, 1308 (1981).
154. M. Hatzakis, D. Hofer, and T. Chang, *J. Vac. Sci. Technol.* **16**, 1631 (1979); U. S. Patent 4,267,259 (1980), IBM.

155. J. Kitcher, *J. Vac. Sci. Technol.* **16**, 2030 (1979).
156. U. S. Patent 4,283,483 (1981), Hughes.
157. B. Lin and T. Chang, *J. Vac. Sci. Technol.* **16**, 1669 (1979).
158. J. Moran and D. Maydan, *J. Vac. Sci. Technol.* **16**, 1620 (1979); U. S. Patent 4,244,799 (1981), Bell.
159. J. Havas, *Electrochem. Soc. Ext. Abstr.* **76-2**, 743 (1976); U. S. Patents 3,873,361 (1973) and 4,004,044 (1977), IBM.
160. J. Kruger, P. Rissman, and M. Chang, *J. Vac. Sci. Technol.* **19**, 1320 (1981).
161. M. O'Toole, E. Liu, and M. Chang, *IEEE Trans. Electron Devices* **ED-28**, 1405 (1981); U. S. Patent 4,370,405 (1983); U. S. Patent 4,102,683a (1978), RCA.
162. U. S. Patent 4,202,914 (1980), IBM.
163. G. Ray, S. Peng, D. Burriesi, M. O'Toole, and E. Liu, *J. Electrochem. Soc.* **129**, 2153 (1982).
164. S. Lyman, J. Jackel, and P. Liu, *J. Vac. Sci. Technol.* **19**, 1325 (1981).
165. U. S. Patent 3,873,361 (1976), IBM.
166. U. S. Patent 3,982,943 (1976), IBM.
167. U. S. Patent 4,035,276 (1977), IBM.
168. U. S. Patent 4,256,816 (1981), Bell.
169. K. Grebe, J. Ames, and A. Ginzberg, *J. Vac. Sci. Technol.* **11**, 485 (1974); U. S. Patent 3,849,136 (1974), IBM.
170. P. Grabbe, E. Hu, and R. Howard, *J. Vac. Sci. Technol.* **21**, 33 (1982).
171. U. S. Patent 4,367,119 (1983), IBM.
172. T. Venkatesan, *J. Vac. Sci. Technol.* **19**, 1368 (1981).
173. D. Tenant, *J. Vac. Sci. Technol.* **B1**, 494 (1983).
174. A. Milgram, *J. Vac. Sci. Technol.* **B1**, 490 (1983).
175. L. Jackel, R. Howard, E. Hu, D. Tenant, and P. Grabbe, *Appl. Phys. Lett.* **39**, 268 (1981).
176. Ref. 34, p. 335.
177. W. Arden, H. Keller, and L. Mader, *Solid State Technol.* July 1983, p. 143.
178. P. Burggraaf, *Semicond. Int.* June 1983, p. 53.
179. J. Dubois and M. Gazard, *Electrochem. Soc. Ext. Abstr.* **72**, 332 (1972).
180. U. S. Patent 3,877,980 (1975), U. S. Phillips.
181. R. Watts, W. Fitchner, E. Fuls, L. Thilbant, and R. Johnston, *IEEE Trans. Electron Devices* **ED-28**, 1338 (1981).
182. L. Dunkelberger, *J. Vac. Sci. Technol.* **15**, 88 (1978).
183. U. S. Patent 4,132,586 (1977).
184. M. Itoh, H. Gokan, S. Esho, and K. Asakawa, *J. Vac. Sci. Technol.* **20**, 21 (1982).
185. British Patent 1,450,509 (1976), IBM.
186. U. S. Patent 4,144,101 (1979), IBM.
187. W. Grobman, H. Luhn, T. Donohue, A. Speth, A. Wilson, M. Hatzakis, and T. Chang, *IEEE Trans. Electron Devices* **ED-26**, 360 (1979).
188. T. C. Patton, *Paint Flow and Pigment Dispersion*, Wiley, New York, 1979, p. 356.
189. S. Croll, *J. Appl. Polym. Sci.* **23**, 847 (1979).
190. D. Tenant, *J. Vac. Sci. Technol.* **B1**, 494 (1983).
191. L. White, *J. Electrochem. Soc.* **130**, 1543 (1983).
192. L. Rothman, *J. Electrochem. Soc.* **127**, 2216 (1980).
193. L. Rothman, *J. Electrochem. Soc.* **130**, 1131 (1983); U. S. Patent 3,985,597 (1975), IBM.
194. W. Daughton and F. Givens, *J. Electrochem. Soc.* **129**, 173 (1982).
195. A. Adams and C. Caparo, *J. Electrochem. Soc.* **128**, 423 (1981).
196. S. Yamamoto, K. Kobayashi, and Y. Toyama, *Fujitsu Sci. Tech. J.* **14**, 143 (1978).
197. E. Mondou and P. Schmidt, *IBM Tech. Discl. Bull.* **18**, 391 (1975).
198. U. S. Patent 3,982,943 (1976), IBM.
199. B. Kuang and C. Chang, *J. Vac. Sci. Technol.* **16**, 2025 (1979).
200. U. S. Patent 4,210,579 (1980), Motorola.
201. S. Irving, *Solid State Technol.* June 1971, p. 47.
202. O. Wada, S. Yamamoto, K. Kobayashi, A. Taguchi, and Y. Toyama, *J. Electrochem. Soc.* **124**, 959 (1977).
203. U. S. Patent 4,042,387 (1979), Rockwell.
204. I. Blech, D. Fraser, and S. Maszko, *J. Vac. Sci. Technol.* **15**, 13 (1978).
205. I. Blech, *Thin Solid Films* **6**, 113 (1970).
206. C. Wasik and J. Gneiwik, *J. Vac. Sci. Technol.* **8**, 441 (1971).

207. A. Neureuther, C. Ting, and C. Luv, *IEEE Trans. Electron Devices* **ED-27**, 1449 (1980).
208. W. Oldham, A. Neureuther, C. Sung, J. Reynolds, and S. Nandgoankar, *IEEE Trans. Electron Devices* **ED-27**, 1455 (1980).
209. D. Widmann, *IEEE J. Solid State Circuits* **SC-11**, 446 (1976).
210. Y. Homma, A. Yajima, and S. Harada, *IEEE Trans. Electron Devices* **ED-29**, 512 (1982).
211. A. Learn, *J. Electron. Mater.* **3**, 531 (1974).
212. R. Ryan, T. McCurdy, and L. Wolff, *RCA Rev.* Dec. 1968, p. 582.
213. D. Keleman, *Solid State Technol.* Aug. 1976, p. 37.
214. S. Spitz, *Circuit Manuf.* Jan 1981, p. 69.
215. U. S. Patent 3,993,802 (1976), Photocircuits.
216. N. Feldstein and T. Lancsek, *RCA Rev.* **32**, 306 (1971).
217. J. Maes, A. Van Nie, and G. Hut, *Microelectron. Reliab.* **17**, 325 (1978).
218. L. Romankiw, *IEEE Trans. on Magn.* **MAG-10**, 828 (1974).
219. R. Feder, M. Heritage, E. Spiller, and J. Topolian, *Solid State Technol.* April 1976, p. 62.
220. U. S. Patent 4,224,361 (1980), IBM.
221. L. Romankiw, M. Hatzakis, and E. Castellani, *Electrochem. Soc. Ext. Abstr.* **74-1**, 489 (1974).
222. L. Romankiw, in *Proc. Symp. Etching Pattern Definition*, edited by H. Hughes and M. Rand, Electrochemical Society, 1976, p. 161.
223. A. Sugita, M. Morita, and A. Tamamura, Extended Abstracts, 16th International Conference on Solid State Devices, pp. 19–22 (1984).
224. Y. Irda, *Jpn. J. Appl. Phys.* **16**, 1313 (1977).
225. D. Lee, *Proc. IEEE* **62**, 1241 (1974).
226. Y. Okuyama, T. Hashimoto, and T. Koguich, *J. Electrochem. Soc.* **125**, 1293 (1978).
227. M. Gazard, J. Dubois, and C. Dubois, *Am. Chem. Soc. Org. Coat. Abstr.* **33**, 372 (1973).
228. U. S. Patent 4,315,984 (1982), Hitachi.
229. U. S. Patent 4,367,119 (1983), IBM.
230. U. S. Patent 3,962,004 (1976), RCA.
231. U. S. Patent 4,030,942 (1977), IBM.
232. U. S. Patent 4,093,942 (1977), IBM.
233. T. Matsuzawa and H. Tomioka, *IEEE Electron Device Lett.* **EDL-2**, 90 (1981).
234. H. Shirashi, Y. Tanaiguchi, S. Horigime, and S. Nonogaki, *Polym. Eng. Sci.* **20**, 1954 (1980).
235. D. Dobkin and B. Cantos, *IEEE Electron Device Lett.* **EDL-2**, 222 (1981).
236. B. Lin, *IBM J. Res. Dev.* May 1976, p. 213; U. S. Patent 4,211,834 (1980).
237. U. S. Patent 4,387,145 (1983), Fairchild.
238. R. Howard, *Appl. Phys. Lett.* **33**, 1934 (1978).
239. K. Pickar, *Solid State Electron.* **15**, 239 (1977).
240. U. S. Patent 4,456,675 (1984), IBM; K. Jain, C. Wilson, and B. Lin, *IBM J. Res. Dev.* **26**, 155 (1982).
241. N. Clecak, B. Grant, C. Wilson, and R. Twieg, *IEEE Trans. Electron Devices* **ED-28**, 1300 (1981).
242. J. Andrew, P. Dryer, D. Foster, and P. Key, *Appl. Phys. Lett.* **43**, 717 (1983).
243. K. Sugita, N. Ueno, S. Konishi, and Y. Suzuki, *Photogr. Sci. Eng.* **27**, 146 (1983).
244. S. Matsui and N. Endo, *Microelectron. Eng.* **1**, 51 (1983).
245. W. Meyer, B. Curtis, and H. Brunner, *Microelectron. Eng.* **1**, 29 (1983).
246. U. S. Patent 4,427,713 (1984), RCA.
247. Japanese Patent 57,141,642, *Chem. Abstr.* **99**, 222415 (1983).
248. B. Singh, *Appl. Phys. Lett.* **45**, 74 (1984).
249. F. Jones, J. Paraszczak, and A. Speth, *J. Appl. Phys.* **55**, 3092 (1984).
250. B. Soller, R. Shuman, and R. Ross, *J. Electrochem. Soc.* **131**, 1353 (1984).
251. F. Robb, *J. Electrochem. Soc.* **131**, 1670 (1984).
252. M. Shimaya, O. Nakajima, C. Hashimoto, and Y. Sokakibara, *J. Electrochem. Soc.* **131**, 1391 (1984).
253. U. S. Patent 4,464,460 (1984), IBM.
254. K. Ehara, T. Morimoto, S. Muramoto, and S. Matsuo, *J. Electrochem. Soc.* **131**, 419 (1984).
255. S. Gillespie, *IBM J. Res. Dev.* **28**, 454 (1984).
256. S. Macdonald, C. Wilson, R. Miller, and H. Ito, *Polym. Mater. Sci. Eng.* **49**, 104 (1983).
257. European Patent Application 98,992 (1984), IBM (dual layer positive or negative resist based on dyeing the developed top layer and exposing bottom layer).
258. U. S. Patent 4,448,800 (1984), Nippon Tel. (refractory metal in lift-off).

259. U. S. Patent Application 524,828 (1984), U. S. Navy (trilevel of resist/metal/polyimide for lithography on ceramic substrates).
260. Anon, *Res. Discl.* **239**, 106 (1984); *Chem. Abstr.* **100**, 148411 (1984) (deposit blanket metal film before lift-off).
261. M. Morita, S. Imamura, T. Tamamura, O. Kogure, and K. Murase, *J. Electrochem. Soc.* **131**, 653 (1984).
262. S. Gupta and C. Audain, *SPIE Proc.* **469**, 179 (1984).
263. European Patent Application 98,318, *Chem. Abstr.* **100**, 18347 (1984).
264. D. Hofer, R. Miller, and C. Willson, *SPIE Proc.* **469**, 16 (1984).
265. M. Watts, *SPIE Proc.* **469**, 1 (1984).
266. D. Meyerhofer and L. White, *SPIE Proc.* **469**, 11 (1984).
267. C. Ting and K. Liauw, *SPIE Proc.* **469**, 24 (1984).
268. R. Castellano, *J. Electrochem. Soc.* **131**, 2340 (1984).
269. M. Morita, S. Imamura, A. Tanaka, and T. Tamamura, *J. Electrochem. Soc.* **131**, 2402 (1984).
270. C. Rosilio, A. Rosilio, and F. Buiguez, *Microelectron. Eng.* **1**, 197 (1983).
271. T. Bril, R. deWert, and P. Willemse, *Electrochem. Soc. Ext. Abstr.* **80-1**, 203 (1980).
272. R. Verhaar and W. Hoek, in *Microcircuit Engineering*, edited by J. Cleaver, H. Ahmed, and G. Jones, Academic Press, New York, 1983, p. 405.
273. J. Reekstin and R. Kowalchuck, *IEEE Trans. Magn.* **MAG-9**, 485 (1973).
274. P. Hugget and H. Lehman, in *Microcircuit Engineering*, edited by J. Cleaver, H. Ahmed, and G. Jones, Academic Press, New York, 1983, p. 363.
275. K. Harada, K. Myoshi, H. Namatsu, and S. Moriya, in *Microcircuit Engineering*, edited by J. Cleaver, H. Ahmed, and G. Jones, Academic Press, New York, 1983, p. 313.
276. S. Uoya, D. Stephani, and M. Bolsen, in *Microcircuit Engineering*, edited by J. Cleaver, H. Ahmed, and G. Jones, Academic Press, New York, 1983, p. 199.
277. East German Patent 205,287, *Chem. Abstr.* **101**, 161264 (1983).
278. Japanese Patent 59,84,428, *Chem. Abstr.* **101**, 161267 (1983).
279. J. Kruger, M. O'Toole, and P. Rissman, in *VLSI Electronic Microstructure Science*, Vol. 8, edited by N. Einspruch, Academic Press, New York, 1984, p. 91.
280. Japanese Patent 58,169,910, *Chem. Abstr.* **101**, 120498 (1983).
281. H. Gozdz, H. Craighead, and M. Bowden, *J. Electrochem. Soc.* **132**, 2809 (1985).
282. Japanese Patent 58,105,140, *Chem. Abstr.* **101**, 120489 (1983).
283. Japanese Patent 59,84,426, *Chem. Abstr.* **101**, 161274 (1983).
284. Y. Todokoro, *Electron. Lett.* **15**, 543 (1982).
285. H. Nanamasu, Y. Ozaki, and K. Hirata, *J. Vac. Sci. Technol.* **21**, 672 (1982).
286. J. Paraszczak, E. Babich, J. Shaw, M. Hatzakis, and J. Liutkis, *SPIE Proc.* **393**, 8 (1983).
287. J. Yeh, K. Grebe, and M. Palmer, *J. Vac. Sci. Technol.* **A2**, 1292 (1984).
288. P. Degraff and D. Flanders, *J. Vac. Sci. Technol.* **16**, 1906 (1979).
289. U. S. Patents 4,378,383 and 4,377,633 (1983), IBM.
290. U. S. Patent 4,373,018 (1983), Bell.
291. U. S. Patent 4,430,153 (1984), IBM.
292. U. S. Patent 4,396,702 (1983), RCA.
293. U. S. Patent 4,452,65 (1984), IBM; H. Hiroaka, *J. Electrochem. Soc.* **131**, 2938 (1984).
294. P. Vettiger, K. Daetwyler, and D. Webb, in *Microcircuit Engineering*, edited by J. Cleaver, H. Ahmed, and G. Jones, Academic Press, New York, 1983, p. 295.
295. U. S. Patent 4,489,146 (1984), IBM.
296. U. S. Patent 4,489,101 (1984), Hitachi.
297. Y. Takasu and Y. Todokoro, *Electron. Lett.* **20**, 1013 (1984).
298. K. Tsuji, M. Sasgo, and K. Kugimiya, *IEEE Trans. Electron Devices* **ED-31**, 1861 (1984).
299. I. Aesida, M. Zhang, and E. Wolfe, *J. Electron. Mater.* **13**, 689 (1984).
300. U. S. Patent 4,289,573 (1981), IBM.
301. U. S. Patent 3,964,908 (1976), IBM.
302. U. S. Patent 4,357,369 (1982), RCA.
303. M. Suzuki, H. Namatsu, and A. Yoshikawa, *J. Vac. Sci. Technol.* **B2**, 665 (1984).
304. K. Arai, F. Yanagawa, and S. Kurosawa, *J. Vac. Sci. Technol.* **B2**, 669 (1984).
305. I. Lauks, *Appl. Phys. Lett.* **45**, 74 (1984).
306. Y. Lin, V. Marriott, K. Orvek, and G. Fuller, *SPIE Proc.* **469**, 30 (1984).
307. M. Livistan, *SPIE Proc.* **470**, 85 (1984).

308. Y. Lin, *J. Appl. Phys.* **55**, 1110 (1984).
309. H. Keller, *Solid State Technol.* June 1978, p. 45.
310. E. Fredericks, *IBM Tech. Discl. Bull.* **20**, 989 (1977); P. Carr, *IBM Tech. Discl. Bull.* **18**, 1396 (1975).
311. D. Buckley, *Microelectron. Manuf. Test.* Feb. 1982, p. 19.
312. U. S. Patent 4,496,648 (1985), Sperry Rand.
313. M. Ishikawa, *J. Polym. Sci.* **22**, 669 (1984).
314. European Patent Application 110,165, *Chem. Abstr.* **102**, 15162 (1984).
315. U. S. Patent 4,464,460 (1984), IBM.
316. Japanese Patent 59,105,637, *Chem. Abstr.* **101**, 238186 (1984).
317. Japanese Patent 59,135,793, *Chem. Abstr.* **101**, 238190 (1984).
318. Japanese Patent 58,190,040, *Chem. Abstr.* **101**, 181319 (1984).
319. *Res. Discl.* **246**, 498, *Chem. Abstr.* **101**, 219627 (1984).
320. U. S. Patent 4,470,871 (1984), RCA.
321. N. Endo and S. Matsui, *Jpn. J. Appl. Phys.* **22**, L109 (1983).
322. K. Saigo, Y. Ohnishi, M. Suzuki, and H. Gokan, *J. Vac. Sci. Technol.* **B3**, 331 (1985).
323. A. Faithmulla, *J. Vac. Sci. Technol.* **B3**, 25 (1985).
324. R. Ono, J. Sauvageau, A. Jain, D. Schwartz, K. Springer, and J. Lukens, *J. Vac. Sci. Technol.* **B3**, 282 (1985).
325. C. Wilkins, E. Reichmanis, T. Wolf, and B. Smith, *J. Vac. Sci. Technol.* **B3**, 306 (1985).
326. N. Gellrich, H. Beneking, and W. Arden, *J. Vac. Sci. Technol.* **B3**, 335 (1985).
327. B. Singh, G. Chern, and I. Lauks, *J. Vac. Sci. Technol.* **B3**, 327 (1985).
328. Y. Yamashita, R. Kawazawa, K. Kamura, S. Ohno, T. Asano, K. Kobayashi, and G. Mamagatsu, *J. Vac. Sci. Technol.* **B3**, 314 (1985).
329. U. S. Patent 4,507,384 (1985), Nippon Tel.
330. Y. Kamakami, T. Aoki, and Y. Yamashita, *Macromolecules* **18**, 580 (1985).
331. Y. Saotome, H. Gokan, K. Sargo, M. Suzuki, and J. Ohnishi, *J. Electrochem. Soc.* **132**, 909 (1985).
332. M. Tsuda, S. Oekawa, M. Yabuta, A. Yakota, H. Nakane, K. Yamashita, K. Ganco, and S. Namba, *J. Vac. Sci. Technol.* **B3**, 481 (1985).
333. Japanese Patent 59,119,276, *Chem. Abstr.* **102**, 15135 (1985).
334. B. Stangl, J. Mitteraurer, F. Reudenaurer, and G. Marawsky, *J. Vac. Sci. Technol.* **B3**, 477 (1985).
335. N. Chou, C. Tang, J. Paraszczak, and E. Babich, *Appl. Phys. Lett.* **46**, 31 (1985).
336. Japanese Patent 59,119,276, *Chem. Abstr.* **102**, 15135 (1985).
337. D. Tenant, H. Dayem, R. Howard, and E. Westerwick, *J. Vac. Sci. Technol.* **B3**, 458 (1985).
338. Japanese Patent 59,31,976, *Chem. Abstr.* **102**, 36770 (1985).
339. Y. Takasu and Y. Todokoro, *J. Vac. Sci. Technol.* **B3**, 869 (1985).
340. L. White and M. Miskowski, *J. Vac. Sci. Technol.* **B3**, 862 (1985).
341. T. Ueno, H. Shiraishi, T. Iwazanagi, and S. Nonogaki, *J. Electrochem. Soc.* **132**, 1168 (1985).
342. E. Reichmanis and G. Smolinsky, *J. Electrochem. Soc.* **132**, 1178 (1985); U. S. Patent 4,481,049 (1984).
343. European Patent Application 113,034, *Chem. Abstr.* **102**, 140887 (1985).
344. U. S. Patent 4,497,684 (1985), Amdahl.
345. Japanese Patent 59,175,725, *Chem. Abstr.* **102**, 140887 (1985).
346. Japanese Patent 59,126,686, *Chem. Abstr.* **102**, 70242 (1985).
347. R. Dean and R. Matarese, *IEEE Trans. Electron Devices* **ED-22**, 358 (1975).
348. D. Prober, *IEEE Trans. Electron Devices* **ED-28**, 1368 (1981).
349. M. Feuer and D. Prober, *IEEE Trans. Electron Devices* **ED-28**, 1375 (1981).
350. G. Dolan, *Appl. Phys. Lett.* **31**, 337 (1977).
351. R. Howard and D. Prober, in *VLSI Electronics*, Vol. 5, edited by N. Einspruch, Academic Press, New York, 1982, pp. 146–185.
352. A. Milligram and J. Puretz, *J. Vac. Sci. Technol.* **B3**, 879 (1985).
353. A. Tanaka, M. Morita, and K. Onose, *Jpn. J. Appl. Phys.* **24**, 1112 (1985).
354. M. Suzuki, Y. Ohnishi, and A. Furata, *J. Electrochem. Soc.* **132**, 139 (1985).
355. D. Lavergne and D. Hoger, *SPIE Proc.* **539**, 115 (1985).
356. B. Lin, V. Chao, K. Petrillo, and B. Yang, SPE RETEC Photopolymers, Ellenville, N.Y., 1985, p. 75.
357. L. Stillwagon, P. Silverman, and G. Taylor, SPE RETEC Photopolymers, Ellenville, N.Y., 1985, p. 87.

358. A. Gozdz, H. Craighead, and M. Bowden, SPE RETEC Photopolymers, Ellenville, N.Y., 1985, p. 157.
359. U. S. Patent 4,357,369 (1982), RCA.
360. M. Hartney and A. Novembre, SPE RETEC Photopolymers, Ellenville, N.Y., 1985, p. 211.
361. F. Vollenbroek, W. Nyssen, H. Kroon, and B. Yilmaz, SPE RETEC Photopolymers, Ellenville, N.Y., 1985, p. 309.
362. L. White, SPE RETEC Photopolymers, Ellenville, N.Y., 1985, p. 271.
363. E. Alling and C. Stauffer, *SPIE Proc.* **539**, 194 (1985).
364. U. S. Patent 4,306,005 (1985), GCA.
365. J. Underhill, V. Nguyen, M. Kerbaugh, and D. Sundlez, *SPIE Proc.* **539**, 83 (1985).
366. M. de Grandpre, D. Vidusek, and M. Legenza, *SPIE Proc.* **539**, 103 (1985).
367. J. Wijdenes and M. Geomini, *SPIE Proc.* **539**, 97 (1985).
368. Y. Ohnishi, M. Suzuki, K. Saigo, Y. Saotome, and H. Gokan, *SPIE Proc.* **539**, 621 (1985).
369. L. White, *SPIE Proc.* **539**, 29 (1985).
370. M. Legenza, D. Vidusek, and M. de Grandpre, *SPIE Proc.* **539**, 250 (1985).
371. J. Zeigler, L. Harrah, and A. Johnson, *SPIE Proc.* **539**, 166 (1985).
372. J. Tomeoka, *SPIE Proc.* **539**, 158 (1985).
373. Japanese Patent 60,84,541 (1985); *Chem. Abstr.* **103**, 132419 (1985).
374. U. S. Patent 4,521,274 (1985).
375. S. MacDonald, H. Ito, and C. Willson, *Microelectron. Eng.* **1**, 269 (1983).
376. Japanese Patent 60,47,419 (1985); *Chem. Abstr.* **103**, 113354 (1985).
377. E. Reichmanis, G. Smolinsky, and C. Wilkins, *Solid State Technol.* Aug. 1985, p. 130.
378. Japanese Patent 60,08,839 (1985); *Chem. Abstr.* **102**, 19521 (1985).
379. U. S. Patent 4,507,331 (1985), IBM.
380. H. Miligram and J. Puretz, *J. Vac. Sci. Technol.* **B3**, 879 (1985).
381. R. Brault, R. Kubena, and R. Metzger, *SPIE Proc.* **539**, 70 (1985).
382. Japanese Patent 60,35,727 (1985); *Chem. Abstr.* **102**, 229481 (1985).
383. T. Matsuda, T. Ishii, and K. Harada, *Appl. Phys. Lett.* **47**, 123 (1985).
384. U. S. Patent 4,552,833 (1985), IBM.
385. European Patent 136,130 (1985), AT & T; *Chem. Abstr.* **103**, 79509 (1985).
386. F. Buiguez, J. Gilbert, C. Rosilio, A. Rosilio, F. Schue, R. Gagnes, B. Sevres, L. Geral, W. Abou-Madi, and C. Montgenoul, in *Microcircuit Engineering*, edited by H. Beneking and H. Heuberger, Academic Press, New York, 1985, p. 471.
387. J. Enerva, K. Skarbova, and S. Kitora, *J. Imag. Technol.* **11**, 174 (1985).
388. U. S. Patent 4,521,274 (1985), AT & T.
389. U. S. Patent 4,481,049 (1984), AT & T.
390. A. Tanaka, M. Morita, and K. Onose, *Jpn. J. Appl. Phys.* **24**, L112 (1985).
391. Japanese Patent 60,45,246 (1985); *Chem. Abstr.* **103**, 62591 (1985).
392. Japanese Patent 60,32,047 (1985); *Chem. Abstr.* **103**, 212712 (1985).
393. H. Moritz, *IEEE Trans. Electron Devices* **ED-32**, 672 (1985).
394. European Patent 124,265 (Oki); *Chem. Abstr.* **102**, 229486 (1985).
395. Japanese Patent 60,59,734 (1985); *Chem. Abstr.* **103**, 113366 (1985).
396. U. S. Patent 4,507,384 (1985), Nippon Tel.
397. B. Lin, K. Petrillo, and V. Chao, in *Microcircuit Engineering*, edited by H. Beneking and H. Heuberger, Academic Press, New York, 1985, p. 32.
398. H. Moritz, in *Microcircuit Engineering*, edited by H. Beneking and H. Heuberger, Academic Press, New York, 1985, p. 45.
399. H. Lehman and R. Widmer, in *Microcircuit Engineering*, edited by H. Beneking and H. Heuberger, Academic Press, New York, 1985, p. 493.
400. F. Vollenbroek, H. Kroon, J. Bartsen, and J. Dil, in *Microcircuit Engineering*, edited by H. Beneking and H. Heuberger, Academic Press, New York, 1985, p. 555.
401. H. Umezaki, N. Koyama, Y. Maruyama, Y. Sugita, and R. Suzuki, *J. Electrochem. Soc.* **132**, 2440 (1985).
402. V. Nguyen, J. Underhill, S. Fredmann, and P. Pan, *J. Electrochem. Soc.* **132**, 1925 (1985).
403. European Patent 132,585 (1985), IBM; *Chem. Abstr.* **103**, 4584 (1985).
404. J. Gobrecht and M. Rossinelli, *Proc. Electrochem. Soc.* **85-1**, 235 (1985).
405. P. Poppert, S. Novak, and P. Wright, *SPIE Proc.* **538**, 461 (1985).
406. Y. Usuzi, A. Yoshikawa, and T. Kitayama, *Microelectron. Eng.* **2**, 281 (1984).

407. European Patent 98,992 (1984), IBM; *Chem. Abstr.* **100**, 219054 (1984).
408. M. O'Toole, *IEEE Electron Device Lett.* **EDL-6**, 282 (1985).
409. U. S. Patent 4,524,121 (1985), Rohm & Haas.
410. U. S. Patent 4,535,053 (1985).
411. P. Sheldon, J. Dick, and R. Hayes, *J. Vac. Sci. Technol.* **A3**, 883 (1985).
412. F. Buiguez, J. Giubert, M. Tacussel, C. Rosilio, and A. Rosilio, in *Microcircuit Engineering*, edited by H. Beneking and H. Heuberger, Academic Press, New York, 1985, p. 485.
413. K. Li and M. Oprysko, *Appl. Phys. Lett.* **46**, 997 (1985).
414. M. Kaplan, D. Meyerhofer, and L. White, *RCA Rev.* **44**, 135 (1983).
415. B. Lin, F. Lai, and Y. Vladimirsky, *J. Vac. Sci. Technol.* **B4**, 426 (1980).
416. Y. Mimura, *J. Vac. Sci. Technol.***B4**, 15 (1986).
417. R. Kawazu, Y. Yamashita, T. Ito, K. Kawamura, S. Ohno, T. Asano, K. Kobayashi, and G. Nagamatsu, *J. Vac. Sci. Technol.* **B4**, 409 (1986).
418. M. Morita, A. Tanaka, and K. Onose, *Jpn. J. Appl. Phys.* **24**, L112 (1985).
419. R. Morgan, Plasma Etching, in *Semiconductor Fabrication*, Elsevier, 1985, p. 302; British Patent 2,154,330 (1985); *Chem. Abstr.* **104**, 43197 (1986).
420. M. Hayashi, T. Ueno, H. Shiraishi, T. Nishida, M. Trumi, and S. Nonogaki, *Polym. Mater. Sci. Eng.* **55**, 611 (1986).
421. S. MacDonald, H. Ito, H. Hiroaka, and C. Willson, *SPE RETEC Photopolymers*, Ellenville, N.Y., 1985, p. 177.
422. S. MacDonald, L. Pederson, A. Patlach, and C. Willson, *Polym. Mater. Sci. Eng.* **55**, 611 (1986).
423. R. Allen, S. MacDonald, and C. Willson, *Polym. Mater. Sci. Eng.* **55**, 290 (1986).
424. F. Watanabe and Y. Ohnishi, *J. Vac. Sci. Technol.* **B4**, 422 (1986).
425. M. Kakuchi, M. Hikita, A. Sugita, K. Onose, and T. Tamamura, *J. Electrochem. Soc.* **133**, 1755 (1986).
426. E. Babich, J. Shaw, M. Hatzakis, J. Paraszczak, D. Witman, and B. Grenon, *Microelectron. Eng.* **5**, 299 (1986).
427. F. Koopmans and R. Bruno, *SPIE Proc.* **633**, 262 (1986).
428. L. Myers and M. Spencer, *J. Vac. Sci. Technol.* **B4**, 1259 (1986).
429. U. S. Patent 4,609,614 (1986), RCA.
430. U. S. Patent 4,599,137 (1986), Nippon Telephone.
431. L. Bushnell, L. Gregor, and C. Lyons, *Solid State Technol.* June 1986, p. 133.
432. H. Namatsu and T. Shibata, *Jpn. J. Appl. Phys.* **24**, L790 (1985).
433. U. S. Patent 4,524,121 (1985), Rohm & Haas.
434. A. McCullough, D. Vidusek, M. Legenza, M. de Grandpre, and J. Imhof, *SPIE Proc.* **631**, 316 (1986).
435. K. Kilichowski and L. White, *RCA Tech. Notes*, No. 136, January 23, 1985, p. 1.

13

Subtractive Etching

13-1. INTRODUCTION

Dragon's blood is a sidewall banking agent (Fig. 13-1-1) that is dusted on lithographic plates to reduce lateral undercutting. The banking agent is an early attempt at anisotropic etching of amorphous materials. Since isotropic etching involves neutral species and generally proceeds to etch laterally and vertically in regions defined by a resist mask, the width of the desired feature can be widened by isotropic subtractive etching by as much as twice the thickness of the etched substrate film. On the other hand, anisotropic directional etching involves charged species and can achieve almost perfect reproduction of the opening in the resist mask. Etching is the most critical step in the lithography process since images are irreversibly transferred into a permanent film.

Subtractive etching involves the selective (chemical) patterning of diffusion masks, insulators, or conductors whereby the etched region is chemically transformed into a soluble or volatile species. In lithography, the etching processes are mainly used to define diffusion masks in thermal oxide of silicon and via holes in insulators for metal contacts. Metallic conductors are patterned by selective removal (image reversal) of gaps, and photomasks are also made of etched metal films. The task of the process engineer is to transfer the resist image into a substrate film with a minimum etch bias (E) and process tolerance ($\pm$T) (Fig. 13-1-2). In Fig. 13-1-2, we see that the resulting etched product (E) contains the image distortion of the mask (± 0.1 μm), the resist bias (± 0.5 μm), and the final etched bias of $\pm$ 1.0 μm with a tolerance of $\pm$ 1.0 μm.

Depending on the crystallinity of the film, and the resist integrity (no lifting during wet etching or erosion during plasma etching), the etch bias can be as large as or larger than D, the thickness of the film. Although the thicknesses of semiconductor films have remained the same for the last 10 years, the horizontal images have shrunk from 4–5 μm to 1–2 μm. Isotropic liquid etching with a large lateral etch component (L) had to be replaced by gas-phase anisotropic etching ($D/L \gg 1$) (Fig. 13-1-3).

Isotropic etching refers to an etching process which proceeds randomly at a uniform rate in all spatial directions, L and D. Anisotropic etching results from some deviation from isotropic etching. Preferably, the depth of etching (D) is much greater than the lateral etch (L). Overetching refers only to a lateral etching rate since the vertical etching has ceased. The degree of isotropy can be classified by the

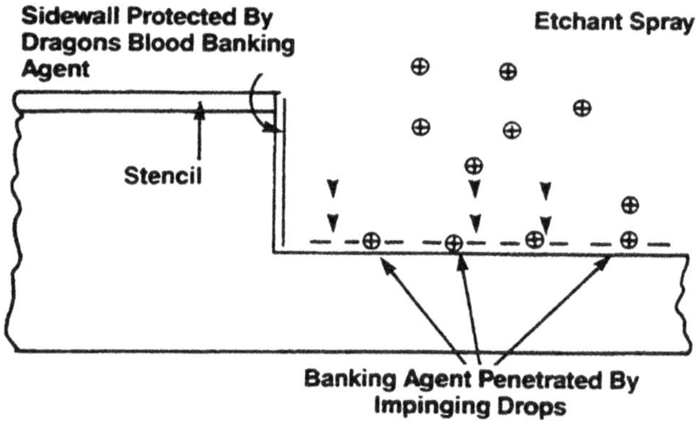

FIGURE 13-1-1. Sidewall banking agent of dragon's blood used in lithographic plate manufacturing.

ratio of L to D (Fig. 13-1-4). Many physical parameters influence the degree of isotropy. In wet etching, the static parameters consist of the resist adhesion, postbake conditions, and the composition of etchant. In dry etching, the dynamic parameters such as power, pressure, and resist erosion strongly influence the degree of isotropy. In wet etching, the degree of undercutting is dependent on previous steps in the process such as surface preparation and postbake, while dry etching has more *in situ* parameters such as power, pressure, and flow rate which control the directionality of the etchant and its concentration. If we plot D versus L as in Fig. 13-1-5, the angle at the origin approximates:

- The beveled shape (α) of the negative[1] undercut
- The centroid cusp (β) of the isotropic[2] etch
- The steep profile (γ) of the positive[1] anisotropic

The etched image is said to be naturally undercut in the normal isotropic etch ($D/L \leqslant 1$) and heavily undercut in the negative anisotropic etch. The primary objective of etching is to reduce L and to limit the total etch bias (E) (Fig. 13-1-2).

Various shapes[3] can be configured in films using wet (primarily electrochemical reactions) liquid etching or the recent and prevalent plasma-assisted dry etching (Fig. 13-1-6). The highly anisotropic shapes are etched with sputtering ions of noble gases in conjunction with reactive halogens (C, D, and F in Fig. 13-1-6)

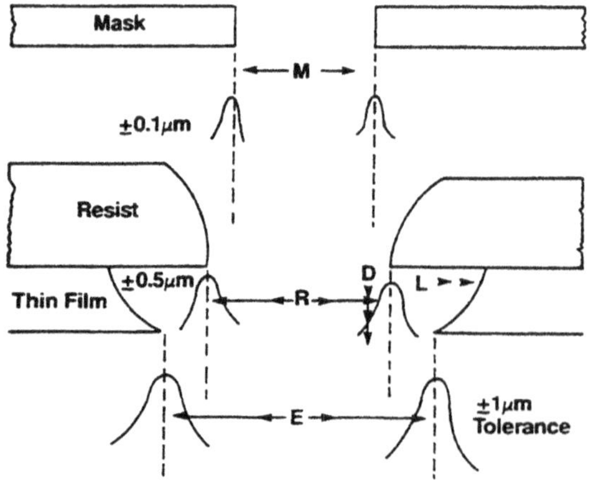

FIGURE 13-1-2. Etch bias in transfer of the image from the resist into the substrate by isotropic etching.

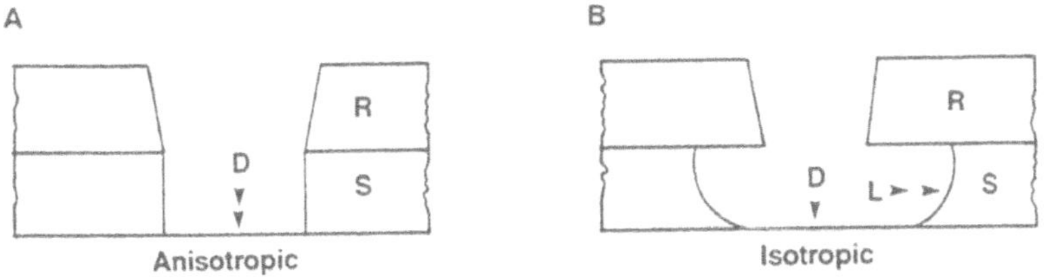

FIGURE 13-1-3. (A) Anisotropic and (B) isotropic etching. R, resist; S, substrate.

while neutral plasmas or liquid etchants produce isotropic or beveled profiles (A, B, and E in Fig. 13-1-6). A beveled edge is preferred for better (subsequent) metal step coverage.[4, 5]

The metal conductor patterns are the most important lithography step, especially in the personalization of "masterslice" logic circuits. The metal line and its nearest neighbor (space) determine the pitch. In Table 13–1–1, the trend in the pitch of metal interconnections is shown. As circuits decrease in size, more closely spaced metal lines are required. The pitch is more important than the absolute line size in the densification of circuitry. The conductivity of the metal line is sensitive to the profile (cross-sectional area). Metal lines can be produced by both additive

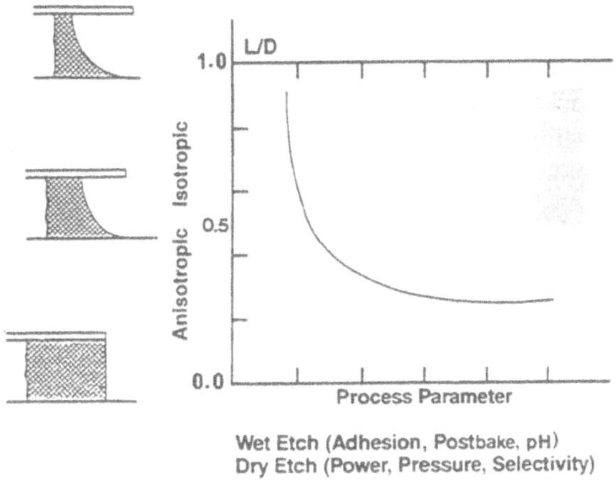

FIGURE 13-1-4. Degree of isotropy (L/D) of lateral etch rate (L) and vertical etch rate (D). From Ref. 1, reproduced by permission of the Electrochemical Society Inc.

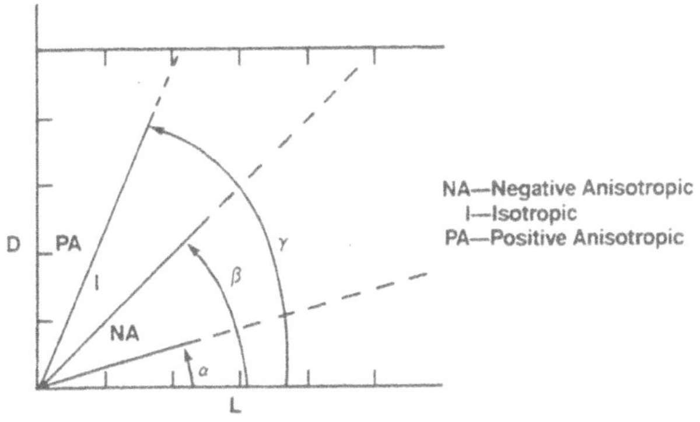

FIGURE 13-1-5. Plot of depth of etching (D) versus lateral etch rate (L) for various isotropy of etching.

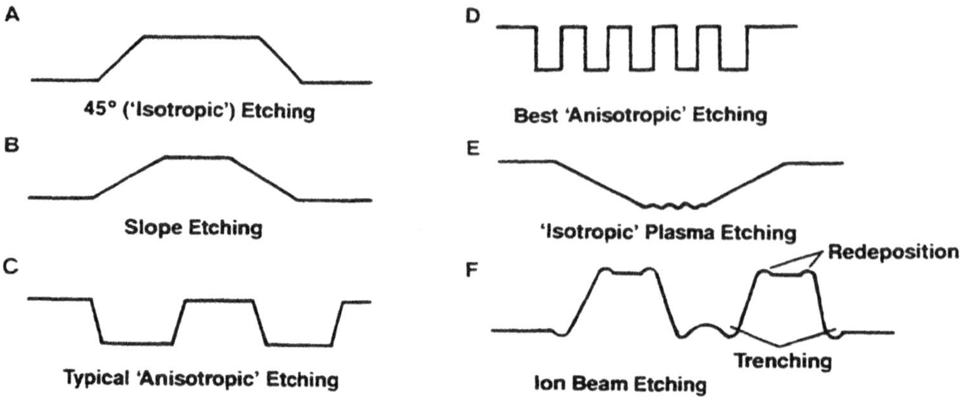

FIGURE 13-1-6. Examples of etched profiles obtained by wet and dry gas etching. From Ref. 3, courtesy of Institute of Physics.

(electroplate and lift-off, Chapter 12), and subtractive (etch) processing. A comparison of profiles is shown in Fig. 13-1-7. The linewidth is 2.5 μm and the thickness of the metal is varied. In the comparison it is assumed that the metal can be deposited at less than 100°C (lift-off and electroplate). Only the thinnest metal film (0.5 μm) can be reproduced closer to desired dimensions.

Metals can be deposited at 200°C, if special multilayer resists are used.[7] Wet etching produces the worst undercut at the thickest films but dry sputter and ion etching require an inorganic or intermediate metallic mask since organic resists erode at too rapid a rate. To minimize undercutting, wet etching is limited to thin films ($\leqslant$ 0.5 μm). Electroplating is only practical with the transition elements (Ag, Cu, Ni, and Au).

In 1976, the transition from wet to dry etching occurred after the discovery that silicon could be etched in a dry asher by introducing fluorocarbon gases. In the period of 1970–1985, a combined literature search of the Abstracts of Chemistry, Science and Engineering revealed a total of 4700 references to all types of electronic device etching.[8] Histograms of references to microelectronic etching for the periods of 1970–1976 and 1976–1984 are shown in Fig. 13-1-8.

In the period of 1970–1984, the minimum linewidths have diminished from 6 to 2 μm. A wet etch bias[9] of 1–2 μm could not be tolerated with the dimensional shrinkage of linewidths. The acceptance of plasma etching and the enforcement of laws pertaining to the disposal of aqueous toxic inorganic etchants prompted the switch to dry etching. Following barrel isotropic etching was the introduction of CF_4 gas to an ion sputter system which paved the way for anisotropic reactive ion etching.[10–13]

TABLE 13-1-1. Metallurgy Pitch for 1-μm Thickness

	Year			
	1976	1978	1984	1986
Pitch	11 μm	8 μm	5 μm	4 μm
Al line	4.5 μm	3.5 μm	2.5 μm	2 μm

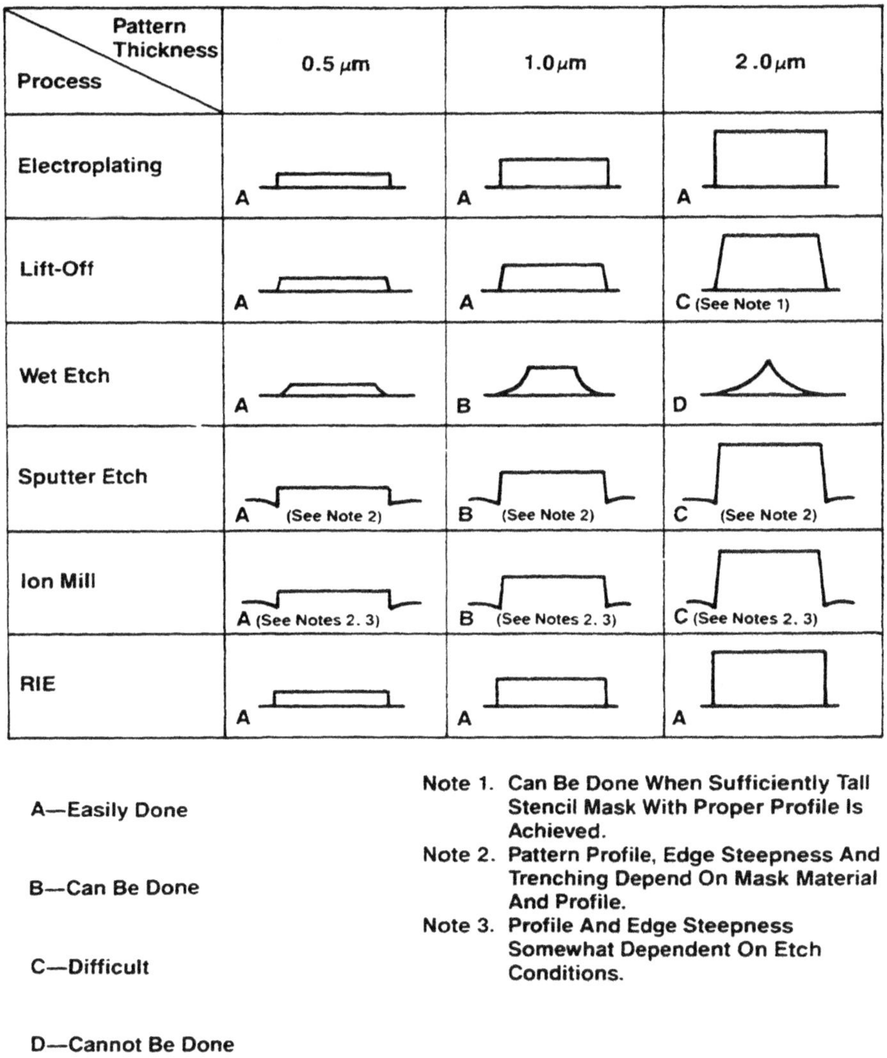

FIGURE 13-1-7. Additive and subtractive techniques for defining patterns in metal films. Modified from Ref. 6; figure originally presented at the 1979 ECS Los Angeles meeting, reproduced by permission of the Electrochemical Society Inc.

The introduction of ion implantation overcame the isotropy of thermal diffusion and allowed the dimensional shrinkage of the size of transistors.

Wet etching consumes silicon real estate by the lateral nature of the etch undercut and the stress failure of the resist during etching. If we assume an etch bias of 1.0, 0.5, and 0.25 μm (which approximate the wet and several cases of dry etching) and a practical overlay and dimensional tolerances of the same magnitude, we can make an estimate of the nonusable real estate of a device consumed by the process bias and tolerance. The results of the analyses are shown in Fig. 13-1-9. The limiting linewidth for wet etching is 2.0 μm for films on the order of 0.5–1.0 μm thick.

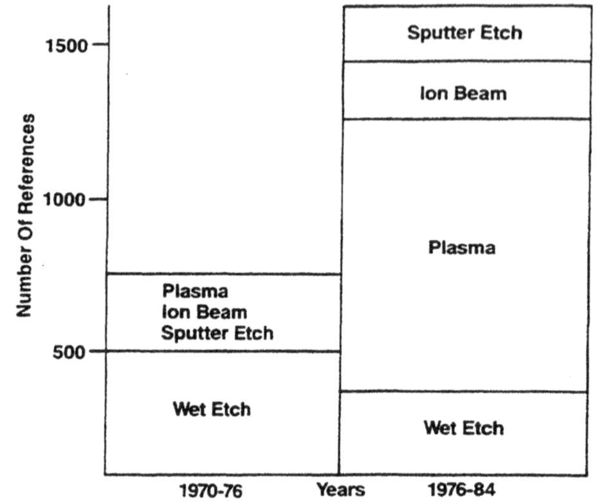

FIGURE 13-1-8. Cumulative references to semiconductor etching.

In wet etching, the larger the etch bias, the lower is the yield, and higher is the cost of large-scale integration. Some of the lost real estate can be regained by choosing an etching process with a high degree of anisotropy, but in some cases with a more costly multilayer resist process. In order to compensate for the undercutting of isotropic wet etching, the photomask would have to be shrunk. In Fig. 13-1-10, the windage compensation of the mask is shown for various degrees of

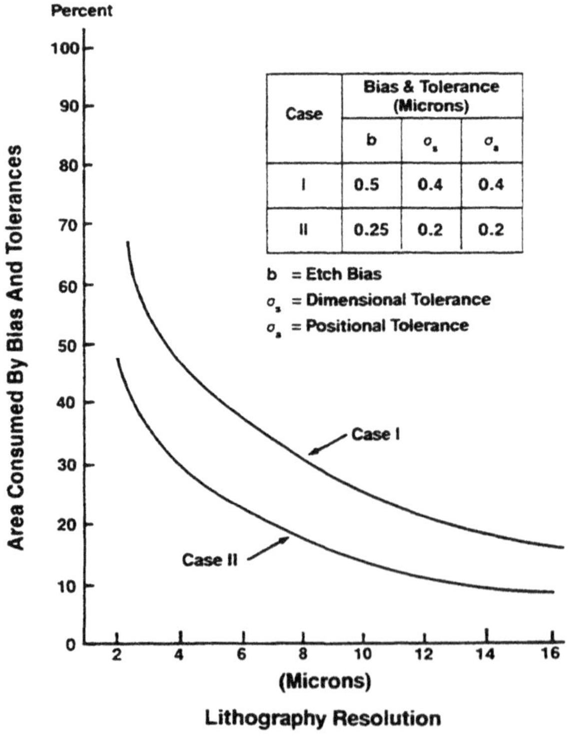

FIGURE 13-1-9. Area consumed by isotropic wet etch bias (see Fig. 13-1-2) as a function of lithography resolution. Below 1 μm, the undercutting consumes all of the device real estate. Courtesy of J. Lounsberry, IBM.

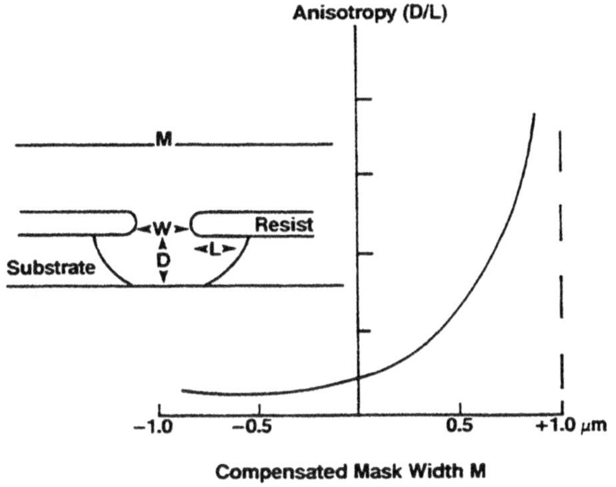

FIGURE 13-1-10. Compensation (shrinkage of opening in mask) required for isotropic ($D/L > 2$) versus anisotropic etching. From Ref. 14, reprinted with permission of *Solid State Technology*, published by Technical Publishing, a company of Dun and Bradstreet.

anisotropic etching. For normal isotropic etch, the D/L is 1 with no resist attack and good adhesion. To produce a line of width w_e, the resist image, w_w, would have to be shrunk by an amount equal to twice the lateral undercut (L),

$$w_r = w_e - 2L \tag{13-1-1}$$

For a moderately anisotropic etch, $D/L = 3$, the resist image would have to be shrunk to 0.2 μm in order to produce a 1.0-μm line and the mask image (M) would have to be about 0.05–0.1 μm in width to compensate for the resist print bias. If $D/L = 10$, a 1-μm line could be etched from a compensatory 0.7-μm-wide resist pattern. The difference in the compensation of the resist image size can be clearly seen in Fig. 13-1-11 for dry etching and wet etching of Si_3N_4. Accordingly, the limits of lithography lie in the bias of the etching process. The etch biases of today are the images of tomorrow. In the etch transfer step, the challenges of the etching process, materials, and equipment are threefold:

1. To consistently etch an image uniformly within a wafer and in a batch of wafers

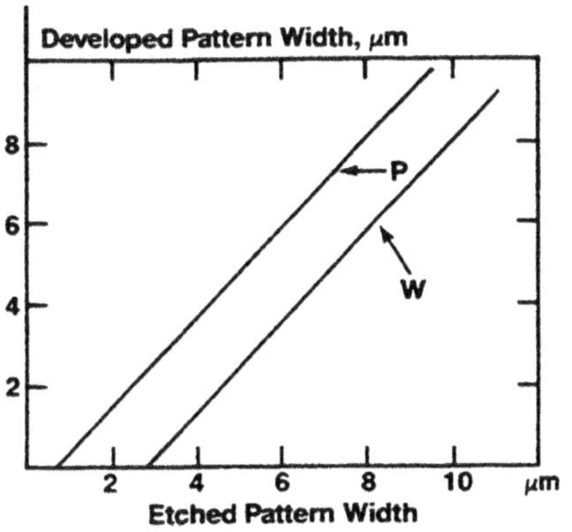

FIGURE 13-1-11. Wet etch (W) versus plasma etch (P) of 0.25-μm-thick Si_3N_4 using a patterned mask in both cases. From Ref. 392, courtesy of the *Japanese Journal of Applied Physics*.

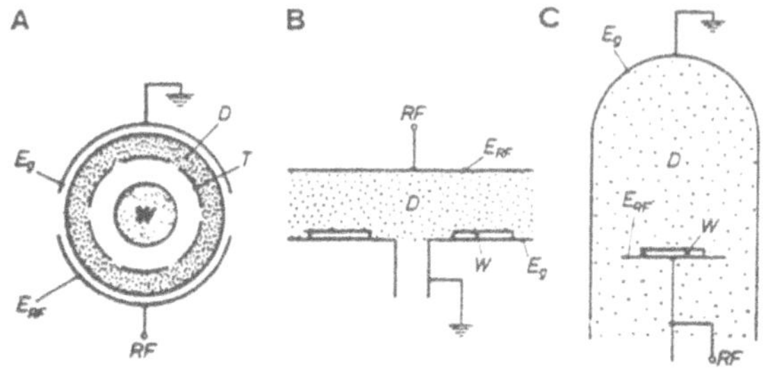

FIGURE 13-1-12. Three main types of dry etchers. (A) Barrel reactor which contains a perforated aluminum tunnel (T) surrounding the wafers to be etched (W). E_{RF} electrode is connected to the radio frequency generator (RF). E_g, grounded electrode. Since the wafers are not in direct contact with the discharge (D), the main contribution to the etching process comes from the long-lived radicals. (B) Planar reactor with the wafers on the grounded electrode. Since the wafers are now in direct contact with the discharge, short-lived radicals now contribute to the etching process. Ions also make a limited contribution. (C) Reactor for reactive ion etching. The wafers are located on the electrode connected to the RF generator. This electrode is much smaller than the grounded electrode. The ion contribution to the etching is considerable. From H. Kalter and E. Van de Ven, *Philips Tech. Rev.* **38**, 202 (1978).

2. To limit the loss of excessive material inside the walls of an etched image
3. To economically etch large wafers (200 mm) with reasonable throughput

Whether the anisotropic etching process occurs in the gaseous or liquid state, depends on ionic species and the direction of the electric field surrounding the etchant. Both electrolytic liquid (electroetching) and reactive ion etching (and some high-power plasma etching) are diffusion-controlled reactions (<6 kcal/mole activation energy). However, the electric field in liquid etching is weak and the ions lose direction and strike the surface in an oblique angle distorting the anisotropy. In the gaseous state, the high electric field between the capacitors accelerates the dissociated gases to strike the silicon wafer at normal incidence. Both neutral radicals and ions are created in the activated gas plasma etching. The schematics of three popular reactor configurations—a barrel radical reactor, a planar radical reactor, and an ion parallel plate dry etch reactor—are shown in Fig. 13-1-12.

In the etching chemistry, the solid reactants,

$$\text{Solid} + \text{etchant} \rightarrow \text{Products}$$

consist of silicon, its oxides and nitrides, and many metals. Aluminum and alloys with Si and Cu are the major metals used for chip-level interconnections, with Al the major first-level material (Table 3-1-2). The oxides of silicon can be thermally grown, chemically or sputter deposited, or doped with phosphorus or boron. The metals can exist as pure films, passivated films, alloys, sandwiches, and intermetallics. While silicon can exist as crystalline or polysilicon films, the crystalline materials exhibit both short- and long-range order. Since etching rearranges ordered structures into a random disorder, entropic ($+\Delta S$) changes as well as

TABLE 13-1-2. The Materials of Semiconductors

Conductors	Ag, Al, Au, Cr, Cu, Mo, Ni, Pb, Pt, Ta, Ti, W
Semiconductors	Si, Ge, GaAs
Insulators	SiO_2, Si_3N_4, resist, polyimide

enthalpic (ΔH) (heats of solution or vaporization) are the basic thermodynamic considerations for the free energy (ΔF) driving force,

$$\Delta F = \Delta H - T\Delta S \qquad (13\text{-}1\text{-}2)$$

In the reaction of glass etching,

$$SiO_2(s) + 6HF(1) \rightarrow H_2SiF_6 + 2H_2O \qquad (13\text{-}1\text{-}2a)$$

the reaction is endothermic, $\Delta H = +11$ kcal/mole. The entropy assists in rearranging the short-order forces in the amorphous solid. Slight defects such as stress, strain, doping levels can alter the etch rate. In crystalline silicon, the low index planes exhibit[13, 723] different etch rates proportional to the number of free bonds (Table 13-1-3). The availability of high-quality silicon crystals and precise lithography has enabled many three-dimensional structures such as ink jet nozzles, diffraction gratings, and X-ray masks to be fabricated.[13, 723] Anisotropic etching occurs preferentially in liquid- and gas-phase etching[685] along one plane of the silicon crystal.

The conversion of a metal or silicon to an aqueous species involves the ionization of the metal (ionization potential) and the transfer of an electron to a suitable reducing agent of high electron affinity:

$$M(s) \rightarrow M^{n+}(aq.) + ne \qquad (13\text{-}1\text{-}3)$$

The reaction consists of three steps:

$$M(s) \rightarrow M(g) \quad \text{sublimation} \qquad (13\text{-}1\text{-}4)$$

$$M(g) \rightarrow M^{n+}(g) + ne \quad \text{ionization} \qquad (13\text{-}1\text{-}5)$$

$$M^{n+}(g) + H_2O \rightarrow M^{n+}(aq.) \quad \text{hydration} \qquad (13\text{-}1\text{-}6)$$

The enthalpy changes for sublimation and ionization are positive (endothermic) while hydration is exothermic (negative). In the gas phase, etching energy (momentum) has to be absorbed by metals in order to sputter them by sublimation. When a metal is dipped into a solution of its own ions [Eq. (13-1-3)], the metal

TABLE 13-1-3. Orientation Effects in Silicon Etchants

Orientation	Relative number of free bonds	Relative etch rate
111 plane	0.58	0.62
110 plane	0.71	0.89
100 plane	1.00	1.00

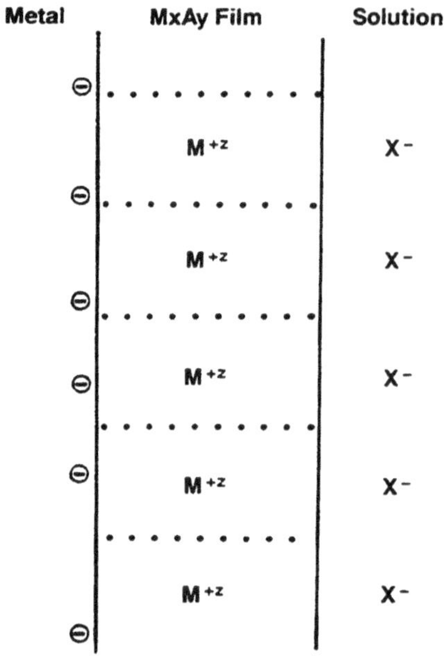

FIGURE 13-1-13. Helmholtz double layer at the interface of a metal in equilibrium with its ionic liquid state (M^+) and its anion (A^-). From Ref. 152, copyright 1981 and courtesy of Butterworths, Inc.

becomes negatively charged as the ions pass into solution (Fig. 13-1-13). The metal forms its own anode and the metal M^{n+} ions are attracted to the metal forming an electric double layer (Helmholtz layer). The potential difference is called the absolute electrode potential. The standard oxidation or reduction potential is tabulated in texts on electrochemistry. At the cathode, a counteroxidation occurs as a cathodic reaction in solution:

$$ne + X^{n-} \rightarrow X_n \tag{13-1-7}$$

The magnitude of the electrode potential and hence the net free energy, ΔF, is (Fig. 13-1-14)

$$\Delta F = -n \mathscr{F} \Delta E \tag{13-1-8}$$

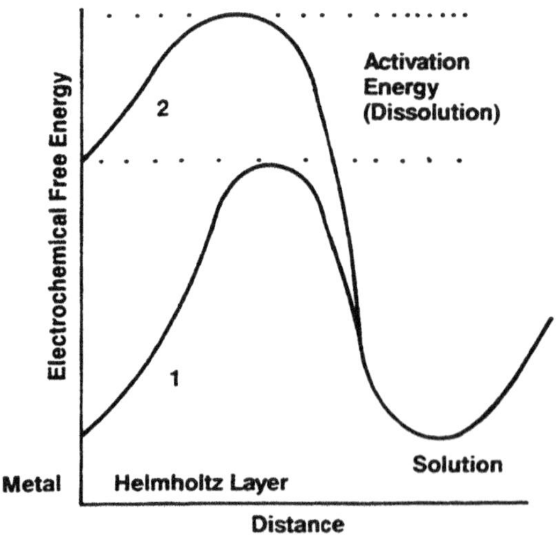

FIGURE 13-1-14. Schematic of free energy relationships at metal–solution interface (1) in equilibrium and (2) anodically polarized. From Ref. 9.

TABLE 13-1-4. Chemical Transformations in Etching

	Rates of etching (μm/min)		Mechanism[a]		Activation energy	
	Liquid	Plasma	Liquid	Plasma	Liquid	Plasma
Metal	0.01–0.2	0.1–1.0	ET	AET	6–20	1–6
Semiconductor	0.01–0.2	0.1–1.0	ET	AET	6–20	1–6
Insulator	0.01–0.2	0.1–1.0	AS	AS	6–20	1–6

[a] ET, electron transfer and ion formation; AS, atomic substitution; AET, atom electron transfer, e.g., $Al + 3Cl \rightarrow AlCl_3$.

where ΔE is the difference between the anode and cathode potential and $\mathscr{F}$ is Faraday's constant. F is affected by such factors as:

1. Metal—purity, crystallinity, stress, method of deposition, and alloy content
2. Activity of M^{n+} ions in solution
3. Ionic strength of electrolyte
4. Temperature
5. Solvent media

The etching of insulators involves no electron transfer but an acid–base reaction:

$$SiO_2 + 6HF \rightarrow H_2SiF_6 + 2H_2O \qquad (13\text{-}1\text{-}9)$$

$$SiO_2 + CF_{4(gas)} \rightarrow SiF_4 + CO_2 \qquad (13\text{-}1\text{-}10)$$

The Si–O bond is replaced by a Si–F bond. Since the Si–O and Si–F bond energies are comparable, entropy is the driving force.

A thermodynamic comparison of the liquid and plasma gas etching of a metal, a semiconductor, and an insulator is shown in Table 13-1-4. Liquid etching proceeds with higher activation energy than plasma etching and is controlled by the rate of reaction. Plasma reactions occur at low concentrations (nonsteady state) and are controlled by the diffusion of the excited gas to the solid surface (4 kcal/mole).

13-1-1. *General Principles of the Rate of Etching*

Heterogeneous solid-phase reactions involve many disciplines of chemistry, mechanics, and physics. A typical rate process involves a sequence of reactions[14 722]:

1. Transport of the reactant
2. Adsorption of the reactant ΔH_{ads}
3. Reaction on the surface ΔF
4. Desorption of the products ΔH_{vap}
5. Transport of the products

The slowest step controls the rate of reaction. Reactions can be of simple order:

$$\text{Rate} = k \qquad \text{zeroth order} \qquad (13\text{-}1\text{-}11)$$

$$\text{Rate} = kE \qquad \text{first order} \qquad (13\text{-}1\text{-}12)$$

dependent on the concentration of etchant (E) for the first-order case. In etching, we seek a reaction of the least variable, preferably linear rates of removal with variable anisotropy (controlled by physical variables) and a high degree of selectivity (i.e., no attack of the resist or the film underneath the etching film). For example, in a zeroth-order reaction we do not have to worry about a small depletion of the etchant. However, in plasma reactions of the first order, the etchant is not in sufficient excess and can be depleted by the load of ten or more wafers. In a simple order, a plot of etched thickness or log thickness versus time is linear. The reaction endpoint can be monitored and correctly determined by extrapolation.

We will now examine the transport mechanisms for the two basic types of reactions: diffusion controlled and reaction rate controlled. All three phases of matter can be involved in etching:

1. Solid—the latent chemical energy and physical structure of the film
2. Liquid—the transport of ions in a liquid dielectric of high viscosity
3. Gas—the chemisorption, recombination, ionization, and mean free path of gas at reduced pressure

A careful kinetic study of the rate of production of reactants and products coupled with the dynamics[14, 722] of mass transport will ultimately lead to a uniform and reproducible manufacturing process. The prerequisite is a uniform and consistent substrate film.

13-1-2. *Phenomenological Mechanism of Etching*

The conversion of a solid to a liquid or gas,

$$\text{Solid film} + \text{etchant} \xrightarrow{k} \text{Products} \tag{13-1-13}$$

depends on the diffusion of the participants:

$$SiO_2(s) + 6HF(1) \rightarrow H_2SiF_6 + 2H_2O \tag{13-1-14}$$

$$SiO_2 + CF_4 \rightarrow SiF_4 + CO_2 \tag{13-1-15}$$

If we let r be the ratio of molar volumes,

$$r = (m/d)/(M/D) \tag{13-1-16}$$

where m and M are the molecular weight of the product and etchant and d and D are the respective densities, then when $r > 1$ (as in the etching of glass) the product does not cover the solid surface completely (Fig. 13-1-15).

Since the product does not hamper the etchant permeation, the rate of etching is controlled by the rate of reaction [k in Eq. (13-1-13)] of the etchant with the solid surface. Activation energies are on the order of 7–20 kcal/mole (Table 13-1-4).

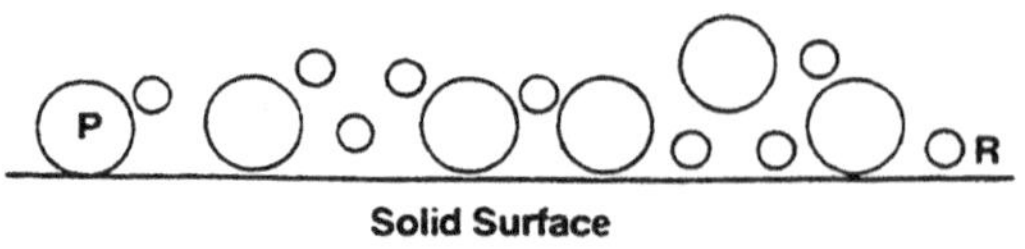

FIGURE 13-1-15. The product (P) covers the solid (reacting) surface poorly and the surface is accessible to the reactant (R).

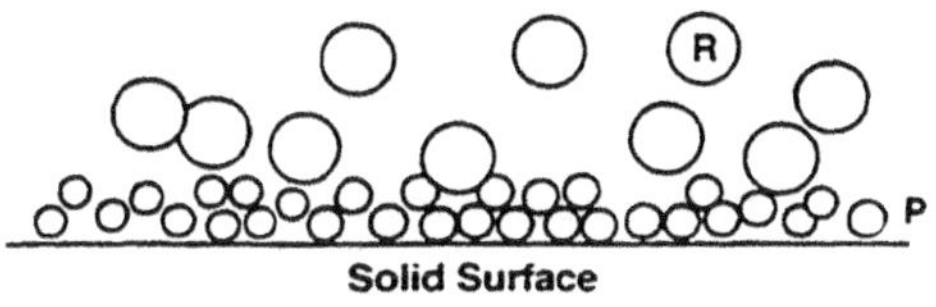

FIGURE 13-1-16. The product (P) covers the etching surface completely and blocks the access of the etchant (reactant, R) to the etching surface.

When $r < 1$, the etchant no longer has free access to the surface and must diffuse through the barrier layer (Fig. 13-1-16) and the Helmholtz layer (Fig. 13-1-13). In electrochemical etching, an electric field is also present (Fig. 13-1-13).

The basic models of diffusion were elucidated by Fick. The fundamental assumption is that diffusion is subject to the same laws as the conduction of heat. At the surface of the solid a concentration gradient exists (Fig. 13-1-17).

The amount dM of a substance which diffuses through a cross-sectional area S during time dt is proportional to S, to the time dt, and in terms of dM to the concentration gradient, dc/dx, at a point on the surface of the solid of surface area S,

$$dM = -DS\, dc/dx - dt \tag{13-1-17}$$

where D is the diffusion coefficient in units of cm^2/sec (same as thermal diffusivity). The cross-sectional opening (S) is assumed to remain constant during etching. However, in wet etching, lifting of the resist usually occurs, widening the opening. In sputter etching or reactive ion etching, erosion of the resist may occur. The coefficient, D, has been shown by the Einstein–Stokes relation to be dependent on the viscosity (η) (same as thermal resistance to heat flow; see Section 16-5),

$$D = RT/\eta \tag{13-1-18}$$

$$dM/dt = \text{Rate of etching} \approx D \approx 1/\eta \tag{13-1-19}$$

$$\eta D = \text{const } T \tag{13-1-20}$$

$$\eta = e^{E_{vis}/RT} \tag{13-1-21}$$

$$E_{vis} = \mathrm{E}_{etch} \tag{13-1-22}$$

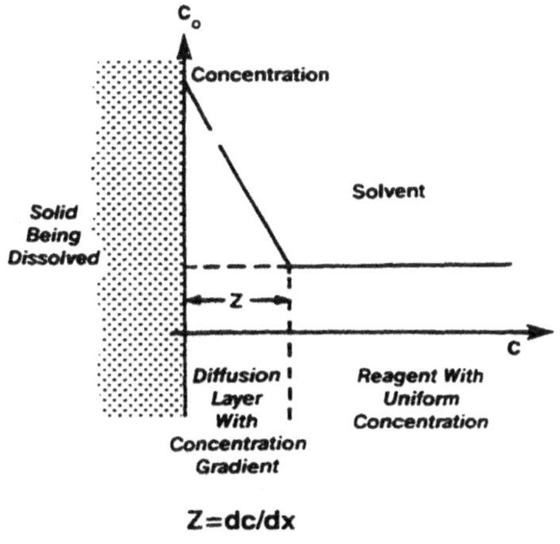

FIGURE 13-1-17. Dissolution of a solid layer in a liquid. The dissolved solid molecules diffuse through the saturated layer to the less concentrated zone. From Ref. 537.

The viscosity of the medium can lower the rate to that of a diffusion-controlled (1–6 kcal/mole) reaction. Frequently, glycols are added to increase the viscosity of the aqueous etchants.

Three general types[16] of solid etching can occur:

- Case 1—The chemical process at the surface is slow and the observed rate is that of the surface process [$r > 1$, Eq. (13-1-16)].
- Case 2—The chemical process at the surface is so fast that convection and diffusion cannot maintain an appreciable concentration of reactant, $r > 1$. The observed rate is that of transport (diffusion) to the surface.
- Case 3—The diffusion and reaction rates are of the same magnitude of diffusion and reaction, but the concentration does not fall to a negligible value at the surface.

The simplest rate equation for Case 1 is

$$dM/dt = k_1\, SC \tag{13-1-23}$$

where S is the surface area and C is the etchant concentration. This equation assumes the rate to be first order with respect to the concentration of etchant. It ignores intermediate adsorption and surface roughness.

For Case 2 it is necessary to assume an effective thickness (σ) layer of concentration gradient (Fig. 13-1-17) and apply Fick's law [Eqs. (13-1-17) and (13-1-18)],

$$dM/dt = DSC/\sigma = k_2SC \tag{13-1-24}$$

For Case 3, it is considered that the concentration falls to C_s (s for surface) at the interface,

$$dM/dt = k_1SC_s = k_2S(C - C_s) \tag{13-1-25}$$

If the difference in effective areas is incorporated into k_1,

$$dM/dt = k_1k_2SC/k_1 + k_2 = k_3SC \tag{13-1-26}$$

Since Eqs. (13-1-23), (13-1-24), and (13-1-26) are formally the same, it is necessary to find experimental criteria to distinguish the three cases. Some distinctions are as follows.

The attributes of diffusion-controlled reactions [Eq. (13-1-24)] are:

1. The activation energy is viscosity controlled, 1–6 kcal/mole [Eq. (13-1-22)].
2. The reaction rate increases with agitation. An exception[17] is the autocatalysis effect of NO on etching of Si in HNO_3. The NO product of etching promotes the reaction. Enhanced removal by stirring results in a decreased rate of reaction.
3. All substances and crystal orientations etch at the same rate.
4. The activation energy increases with stirring. An exception[18] to this is found in etching silicon in HNO_3($\Delta H = 100$ kcal/mole) in which an exothermic reaction liberates a considerable amount of heat and causes an increase in the rate of diffusion and the rate of etching. Stirring in this case would decrease the rate of etching by dissipating heat.

TABLE 13-1-5. Dissolution Rate of Metals in HCl[a]

Metal	Rate (cm/min)
Cd	0.88
Pb	0.88
Sn	0.88
	1.71 (concd HCl)
Cu	0.24

[a] From Ref. 16.

The attributes of rate-controlled reactions [Eq. (13-1-23)] are as follows:

1. The rate changes with etchant concentration.
2. The rate is not sensitive to agitation.
3. The activation energy is 8–20 kcal/mole.

Case 3 [Eq. (13-1-26) requires extensive experimentation to vary k_1 and k_2 and to define k_3. For example, King[(16)] found the etch rates for metal dissolution in HCl shown in Table 13-1-5. The rate at 0.88 represents a diffusion-controlled [k_2, Eq. (13-1-24)] reaction while the others were rate controlled or complicated by product blocking the surface.

Thus, it is important in establishing the basic kinetics and the subsequent process controls in etching or resist development (dissolution of phenolic resins in base, Chapter 10) to study many kinetic parameters (Table 13-1-6). The kinetic study should try to establish the order of the reaction and whether the reaction is rate or diffusion controlled. Secondly, the role of the stability of the resist also has to be integrated into the surface area (S) dependency of the reaction.

For gas-phase etching, the importance of understanding fluid and gas dynamics (and electrical fields in plasma) is evident in the profiles generated in plasma reactions at low and high pressures. At low pressure, the etchant can diffuse randomly over a long mean free path (cm) before deactivation. At higher pressures, considerable deactivation occurs and the processes become diffusion controlled. If the etch products are adsorbed on a vertical surface, they may induce anisotropic etching since the etchant has to diffuse through a thick boundary layer. On the horizontal surfaces, the bombarding ions clear away the reaction products.

TABLE 13-1-6. Kinetic Variables in Etching

Activation energy
Agitation
Viscosity
Concentration (power in plasma)
Thickness
Crystallinity of film
Ratio of etchant/product (3rd body effect in gas)
Rate of product (not easy in liquids)
Surface effects (oxides, wetting, surface energy)

The influence of the mass transport of etching is evident in the distinction by metallurgists[19] for different etchants:

1. *Preferential*—etchant produces pit whose facets have a definite crystal orientation.
2. *Polish*—etchant produces a smooth surface with everything obliterated except grain boundaries.
3. *Nonpreferential*—etchant not only produces a polished surface but also reveals dislocations.

Turbulence and agitation are very important in controlling the undercutting of metal foils, circuit boards, and in deep metal etching.[20]

We have elucidated the general principles of etching and the key kinetic factors. We will now proceed to examine liquid etching and gas phase dry etching in considerable detail with major emphasis on the role and the interaction of the resist with the etching chemicals, process, and equipment.

13-2. LIQUID ETCHING

Liquid etching involves oxidation–reduction reactions for metals and displacement (acid–base) reactions for inorganic oxides. Strong oxidants are used in acidic media. In strongly ($pH > 12$) alkaline etchants, the medium attacks positive diazonovolak resists. In the next sections, we will discuss the etching of SiO_2, Si_3N_4, Si, and Al since they are the primary semiconductor materials.

13-2-1. SiO_2 *Etching*

The etching of glass is an ancient art form.[4, 599] Basically, glass exists as a crystalline material of a silicon atom tetrahedrally surrounded by oxygen atoms (Fig. 13-2-1). Both crystalline and amorphous phases can exist in glass materials. Sputtered quartz exists as the amorphous material of SiO_2 in the SiO_4 tetrahedron. For etching reactions, elemental fluorine can easily replace the O atom in SiO_2 since fluorine has a smaller ionic radius (1.40 Å) than Si–O (1.60 Å) The bond energy of Si–F is one-half that of Si–O.[20] The important uses of glasses as semiconductor films are:

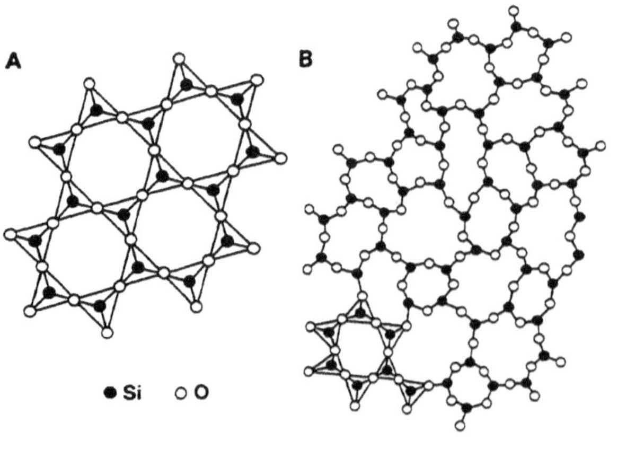

FIGURE 13-2-1. (A) Structure of crystalline tetrahedron SiO_2; (B) structure of amorphous SiO_2. From W. Vogel, *Structure and Crystallization of Glasses*, 1971, p. 14, copyright and courtesy of Pergamon Press.

- Good dielectric insulation
- Barrier for ion diffusion and implantation
- Low intrinsic stress
- High degree of structural perfection and film uniformity
- Step coverage and conformality
- High purity, uniform density, and pinhole free

Glasses of different forms are deposited by chemical vapor diffusion, sputtering, or grown by steam oxidation of silicon. Both thermal and anodic oxidation of silicon result in amorphous films that have little Si–Si bonds.[21, 22] While their properties are similar to those of vitreous fused silica, their surfaces, however, may contain silanol (–SiOH) groups due to the reaction of broken siloxane (O–Si–O) groups with water. Silanol groups are favorable sites for forming primary silylesters with silane treatments (primers) to promote resist adhesion.

Because of the internal stress, the structures of oxides deposited by various methods possess short-order bonds which influence the etch[22] rates (Table 13-2-1). Thus, the microheterogeneity of glass influences the etch rate in liquid etching (and later in plasma-assisted etch). Lopez[23] demonstrated that thermally oxidized silicon of 3000-Å thickness is not the same as the oxide that was formed from etching 6000 Å and more pinholes are present in the thinned oxide. In spite of its long use, the microstructure of thermally oxidized silicon is still questionable.[24] Therefore, empirical etch conditions must be established for each type of glass film. For consistent etching, process monitors to ensure batch-to-batch uniformity and endpoint detect[25] are necessary.

The aqueous HF etching of SiO_2 with a photoresist mask proceeds in an isotropic manner due to undercutting accented by partial resist delamination midway through the depth of etching. Almost anisotropic vertical profiles can be produced with a stable and stress-free Si_3N_4 mask (Fig. 13-2-2). Beveled edges are also produced with a 30 to 1 solution by weight of NH_4F to HF.[26] The three different profiles are due to the resist adhesive failure or retention (Si_3N_4) of the nitride resist–SiO_2 interface (the mechanism will be discussed later).

The chemistry of etching SiO_2 involves a nucleophilic attack by a fluorinated species on the Si–O network. In buffered HF (7 parts of 40% NH_4F to 1 part of concentrated HF), two species are prevalent[27, 28]:

$$\mathrm{HF} \xrightarrow{k_1} \mathrm{H^+} + \mathrm{F^-} \qquad k_1 = 10^{-3} \tag{13-2-1}$$

$$\mathrm{HF} + \mathrm{F^-} \xrightarrow{k_2} \mathrm{HF_2^-} \qquad k_2 = 10^{-1} \tag{13-2-2}$$

TABLE 13-2-1. Etch Rate in Buffered 7:1 HF of SiO_2

Method of formation	Relative etch rate (μm/min)
Steam grown[a]	1.0
Anodic grown	8.5
Pyrolytic	3–10
Sputtered	0.5
Doped oxide	3–5

[a] Approximately 0.1 μm/min (20°C).

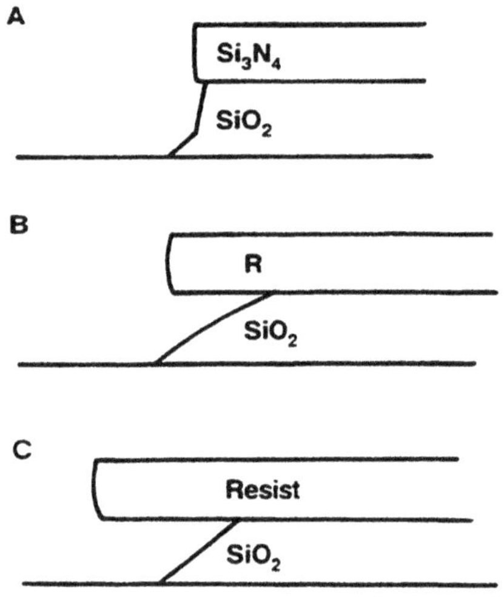

FIGURE 13-2-2. Etched profiles obtained with liquid etchant of 6:1 NH_4/HF with (A) mask of Si_3N_4 and (B) photoresist mask. In (C) an etchant of 30:1 NH_4F/HF was used with a photoresist mask. From Ref. 26, reproduced by permission of the Electrochemical Society Inc.

HF_2^- is the major species in buffered HF (Fig. 13-2-3). The linearity[29] of dissolution in static and agitated buffered HF are shown in Fig. 13-2-4. The system is sensitive to agitation and is likely to be diffusion controlled. Figure 13-2-5 shows the linear dependence[28] of dissolution as a function of HF_2^- and HF. Thus, the rate of thinning of SiO_2 is

$$d(SiO_2)/dt = A(HF) + B(HF_2^-) + C \qquad (13\text{-}2\text{-}3)$$

where A, B, and C are the constants of 2, 5, and 9.7, respectively, at 25°C. Unbuffered HF is dissociated to only 10^{-3} and is about 4× slower in the rate of etching (0.925 μm/min). Since unbuffered HF is also a powerful penetrant, it diffuses readily through the resist, causing pinholes and eventual delamination.

The permeability-sensitivty of exposed KTFR and KPR to vapors of HF has been used as a form of "dry develop–etch." Images[30–34] are etched into oxide without a development step by immersion in vapors of HF at 100°C. The difference in permeability of the cross-linked region versus the linear unexposed polymer may be responsible for image differentiation. High *e*-beam or optical doses (10–100×) are required to augment this distinction.

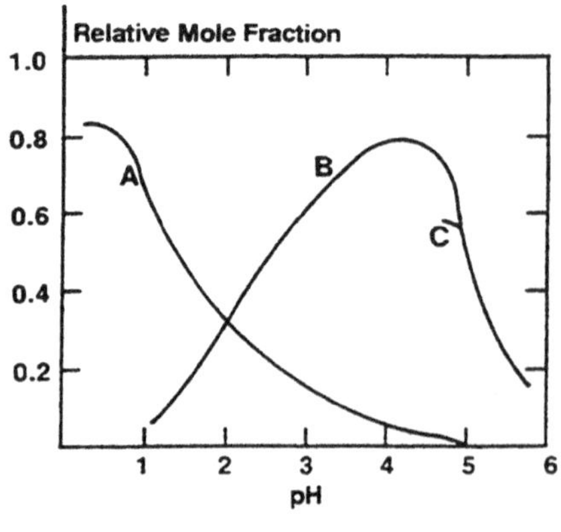

FIGURE 13-2-3. Mole fraction of fluoride as HF and HF_2^- calculated for 15 M total [F]. (A) Mole fraction HF; (B) mole fraction HF_2^-; (C) 7:1 buffered etch of $NH_4F = 10.50F$ and $HF = 3.53F$. Data from Refs. 27 and 28.

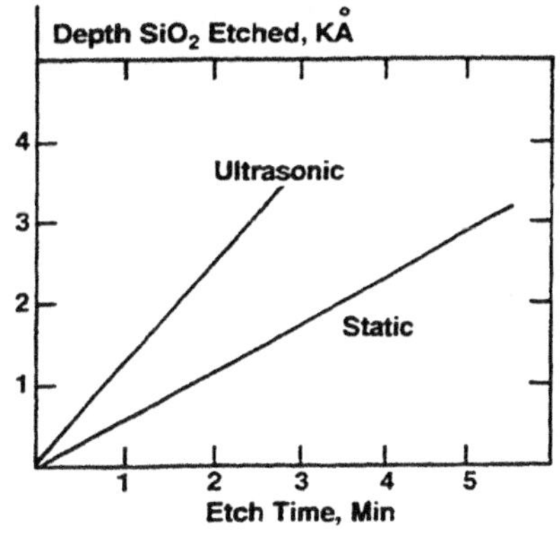

FIGURE 13-2-4. Effect of agitation on etching SiO_2 in buffered 7:1 HF. From Ref. 29.

The attack of the bifluoride ion on a silicon dioxide surface may be envisioned as involving the transition state

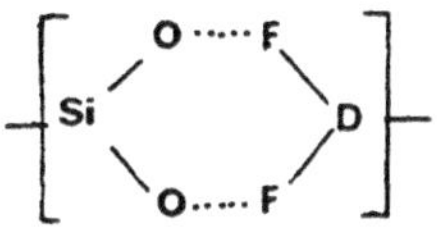

The attack of HF involves the possible surface states[34]

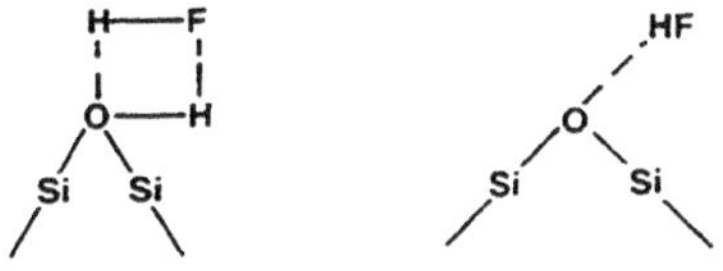

Eventually, the fluorine replaces oxygen,[34, 601]

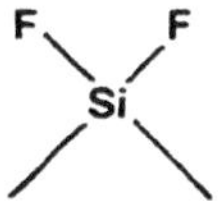

The hydrogen ions attach themselves to the oxygen atom on the SiO_2 and the

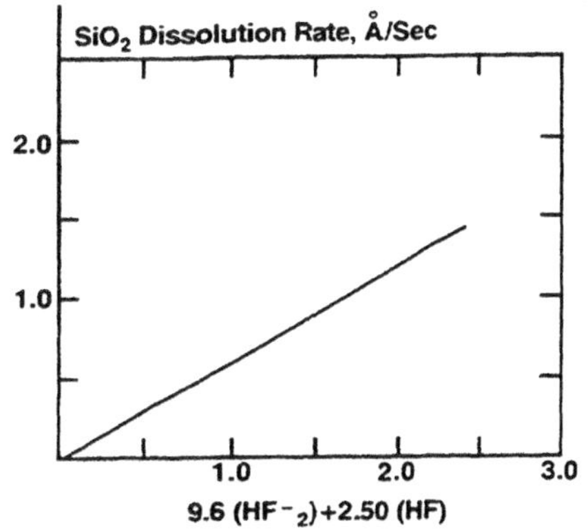

FIGURE 13-2-5. Linear dissolution rate of SiO_2 at 23°C. From Ref. 28, reproduced by permission of the Electrochemical Society Inc.

SiF_4 coordinates two or more fluoride ions to form $S_iF_6^-$ in solution. The overall etching may be represented by

$$6HF + SiO_2 \rightarrow H_2SiF_6 + 2H_2O \qquad (13\text{-}2\text{-}4)$$

Harrell[35] added NH_4F and H_2F_6 to buffered HF and found that the etch rate increased due to HF_2^- formation. The addition of H_2SiF_6 is countered by the precipitation of $(NH_4)_2SiF_6$,

$$H_2SiF_6 + NH_4F \rightarrow (NH_4)_2SiF_6 + HF \qquad (13\text{-}2\text{-}5)$$

White[36] studied the mechanism of etching SiO_2 using DF (deuterated) to examine the nature of the transition state by the kinetic isotope effect. By comparing the rates k_{HF} and k_{DF}, one can examine the response of the transition state. Substitution of DF for HF would affect the DF vibrational mode in the transition state, and change the rate,

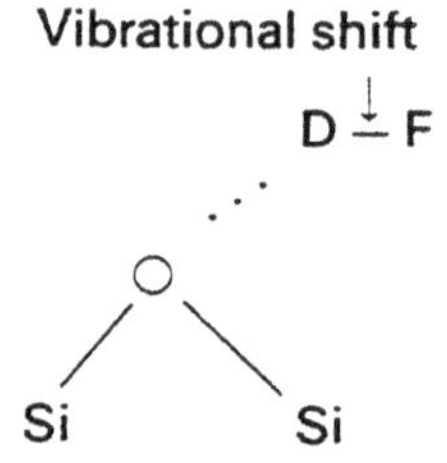

Since no change in rate was observed, the predominant etching species is DF_2^- (HF_2^-) and involves the transition state,

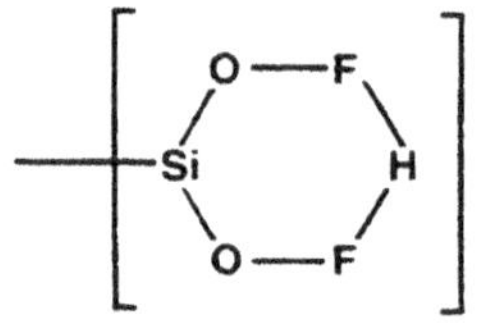

The addition[34] of stronger nucleophiles as NH_4Cl, –Br, and –I increased the rate (Table 13-2-2), indicating the etching proceeds via nucleophilic displacement. Insight into the mechanism can also be gleaned from activation energy studies of

TABLE 13-2-2. Effect of Halogen on SiO_2 Etch Rate[a]

Buffer ion	Etch rate (Å/sec)
F^-	10
Cl^-	20
Br^-	23
I^-	33

[a] From Ref. 34.

TABLE 13-2-3. Activation Energies of Etching SiO_2 and Si

Etch composition	ΔE (kcal/mole)	Reference
NH_4F/HF		
7:1	10	29
6:1–30:1	9–11	37
6:1–30:1	10	26
7:1	10	36
7:1	10	27
$NH_4F/(NH_4)_2HPO_4$, PSG	11	38
HF at 130 °C	5	31
HF/NHO_3, Si	5	39
Alkali, Si	5	40

etching of SiO_2 and Si (Table 13-2-3). In spite of the high activation energies, Lawrence[37] presented an argument for a diffusion-controlled rate [also supported by Harrell's results (Fig. 13-2-4) of the increased rate due to ultrasonic agitation]. Figure 13-2-6 shows the profiles depicted by Lawrence for 6:1 and 30:1 NH_4F/HF ratios. His results for the 30:1 composition indicate a beveled edge profile with a 30–40° wall and he ascribed the severe undercut to an accelerated lateral diffusion (L) rate. Later, Parisi[26] and co-workers repeated this work and ascribed the undercut to resist failure (lifting during etching) (Fig. 13-2-7). No beveled profiles were obtained with a Si_3N_4 mask.

Since agitation effects were not evident in the vertical component (D) of etching, it is unlikely (Fig. 13-2-6) that a lateral component (L) would have faster rates, unless directional forces (spraying) were impinging on the walls of the oxide surface. The beveling effect was later noted by Andrasi,[41] who found that the thinnest film of AZ-1350 of 0.5-μm thickness produced severe undercutting while a thicker film of resist of 1.5 μm produced less undercutting. In addition, the residual stress is less with the thicker film and a smaller degree of delamination during etching is likely to occur.

13-2-2. *Wetting, Adhesion, and Undercutting*

The considerable range of undercutting that is observed in etching SiO_2 (Fig. 13-2-2) has been the result of the failure of the resist–SiO_2 interface. The

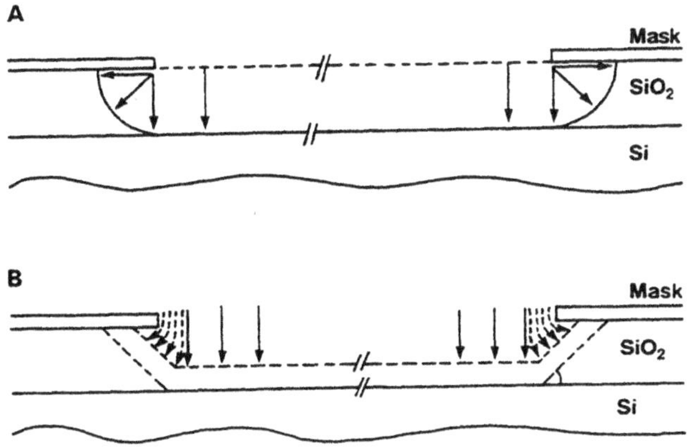

FIGURE 13-2-6. Etching models for HF liquid etching of SiO_2 films: (A) activation energy-controlled reaction and (B) diffusion-controlled reaction. From Ref. 37; figure originally presented at the 1972 Fall Electrochemical Society meeting, reproduced by permission of the Electrochemical Society Inc.

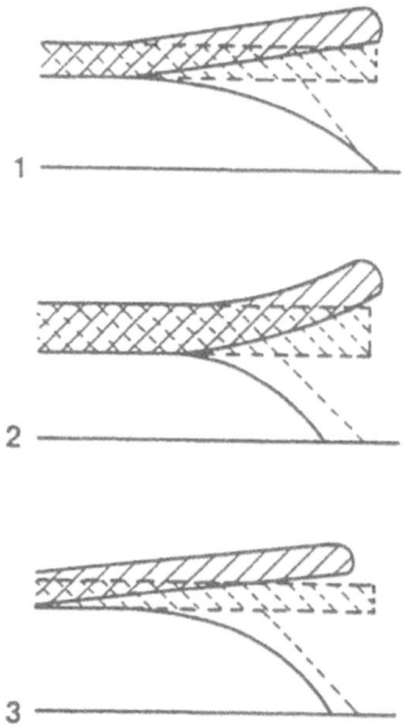

FIGURE 13-2-7. Edge lifting during etching (1) of 0.2-μm-thick resist, (2) at 50% overetch of 0.5-μm-thick resist, and (3) of 0.2-μm-thick resist and 50% overetch of SiO_2. From Ref. 41, copyright 1980 and courtesy of Elsevier Sequoia S.A.

attainment of good adhesion of a resist film has long been a desired objective but a goal which has been measured only by functional performance. In the study of thin film adhesion, the major problem is the lack of susceptibility to measurement.[42] I will now present some of the thermodynamic arguments for wetting and resist adhesion and the practical results achieved by changing (priming) the SiO_2 surface, the resist, or the process conditions.

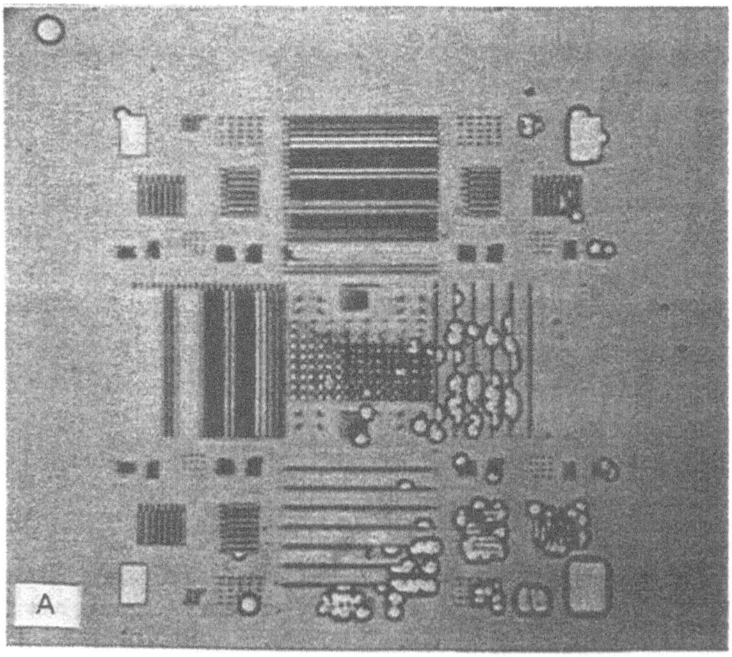

FIGURE 13-2-8. (A) Center dewets of resist on Si wafer.

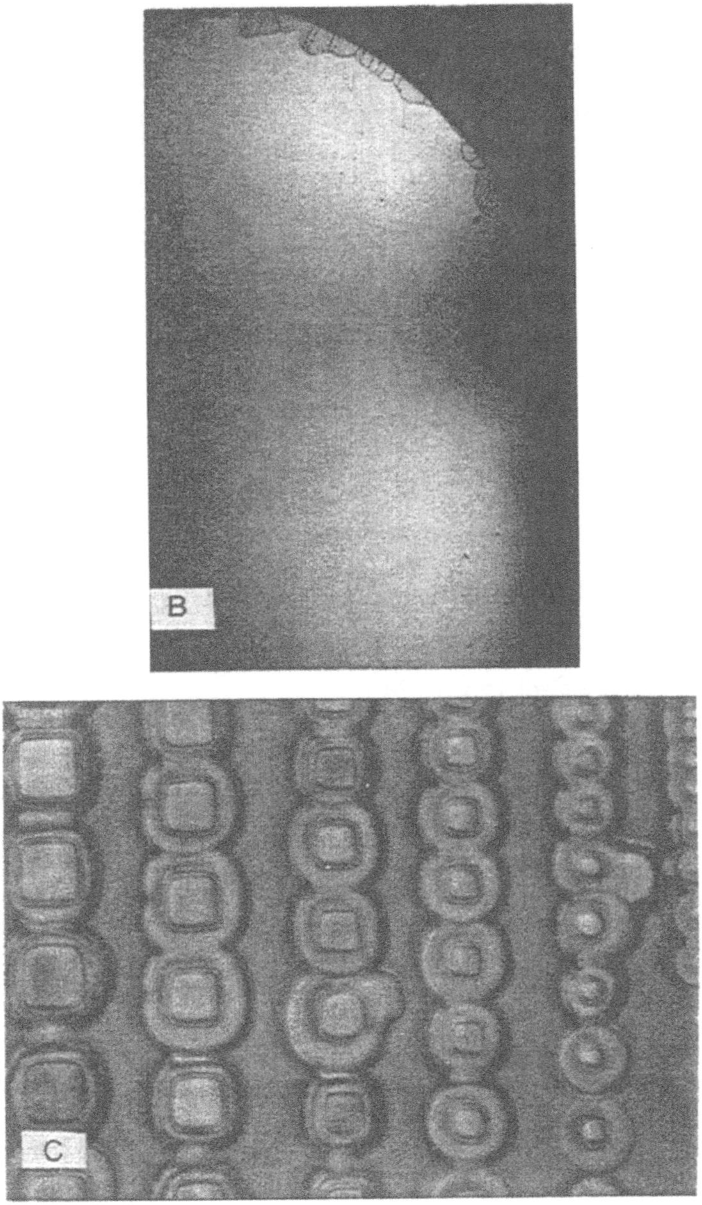

FIGURE 13-2-8. (*Continued*) (**B**) edge dewets of resist on Si wafer; (C) severe undercutting of sputtered quartz in HF etching.

When a resist liquid is spun on a wafer and spun dry, a continuous film should be formed. In some instances, the film is discontinuous and exhibits large holes or contractions from the edge. The resist liquid or film is said to be dewet. Figure 13-2-8 shows the center and the edge dewet of a resist cast on a silicon wafer. Since the surface tension of the casting resist is stronger than the adhesive bond, dewetting occurs. In some cases, spots of foreign contaminants are present on the wafer surface and the defect sites have a repellent surface free energy. No wetting occurs with a hydrophobic resist on a hydrophilic surface and the resist liquid "beads up" due to the fact that its surface tension is larger than that of the solid. When surface tension acts on the circumference of the drop of liquid, a minimum volume is sought. During spin coating, the contraction[734] of the planarized resist film (over topographical features) as it dries, is also responsible for the failure of spin-coated films to completely planarize steps. Contraction of the film by 25% is not uncommon.

The basic criterion for wetting is a low contact angle (θ) at the liquid–solid interface,

$$W_a = \tau_{LV}(1 + \cos\theta) \qquad (13\text{-}2\text{-}6)$$

where W_a is the work of adhesion and τ_{LV} is the liquid–vapor interfacial energy. The value[44] of τ_{LV} for AZ-1350J is 26 dynes/cm. The critical surface tension (τ_c) is associated with the wettability or the repellent qualities of a surface. Liquids with a surface tension below the critical surface tension will wet the surface, i.e., show a contact angle of θ, $\cos\theta = 1$ (Fig. 13-2-9),

$$\tau_{LV} - \tau_{SV} = \cos\theta \qquad (13\text{-}2\text{-}7)$$

Some examples of the critical surface tensions of solids including those treated with silanes are given in Table 13-2-4. It is generally conceded[44] that if the surface tension of the liquid is 20 dynes/cm larger than the surface tension of the substrate, wetting will not occur. The silane treatment of high-energy surfaces lowers the surface tension of the solid to the 30–40 dynes/cm range to match the surface tension of the organic resist and promote bonding. Precoating the wafer by just spinning on the resist casting solvent prior to the application of the liquid resist is said[533] to improve the wetting of the resist and the adhesion. Rubber-based resists applied in 100-nm thicknesses are also used to promote the adhesion of DQN positive resists.

The concept that good wetting and good adhesion go hand in hand is not

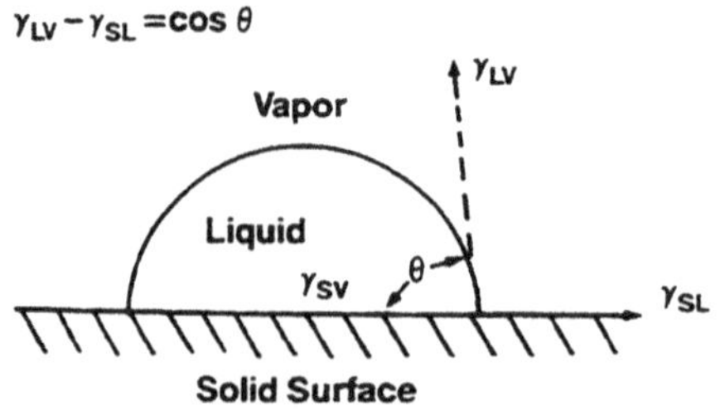

FIGURE 13-2-9. Contact angle of a wetting liquid on a solid surface.

TABLE 13-2-4. Critical Surface Tensions[a,b]

	γ_c (dynes/cm)
Polytetrafluoroethylene	18.5
Methyltrimethoxysilane	22.5
Vintriethoxysilane	25
Paraffin wax	25.5
Ethyltrimethoxysilane	27.0
Propyltrimethoxysilane	28.5
Glass, soda lime (wet)	30.0
Polychlorotrifluoroethylene	31.0
Polypropylene	31.0
Polyethylene	33.0
Trifluoropropyltrimethoxysilane	33.5
3-(2-aminoethyl)-aminopropyltrimethoxysilane	33.5
Polystyrene	34
Cyanoethyltrimethoxysilane	34
Aminopropyltriethoxysilane	35
Polyvinylchloride	39
Phenyltrimethoxysilane	40.0
Chloropropyltrimethoxysilane	40.5
Mercaptopropyltrimethoxysilane	41
Glycidoxypropyltrimethoxysilane	42.5
Polythyleneterphthalate	43
Copper (dry)	44
Aluminum (dry)	45
Iron (dry)	46
Nylon 6/6	46
Glass, soda lime (dry)	47
Silica, fused	78
Titanium dioxide (anatase)	91
Ferric oxide	107
Tin oxide	111

[a] From Ref. 54.
[b] Critical surface tensions for silanes refer to treated surfaces.

necessarily true. The spreading coefficient (τ_{rs}) and work of adhesion (W_a) between the two surfaces of resist (r) and substrate (s) are given[(44)] by

$$\tau_{rs} = 2\phi(\tau_r \tau_s)^{1/2} - 2\tau_r \tag{13-2-8}$$

$$W_a = 2\phi(\tau_r \tau_s)^{1/2} \tag{13-2-9}$$

Both are maximized when the interaction parameter ϕ has its maximum value. If we have a substrate with a given surface tension, τ_s, and polarity component, X_s^P, what should the value of τ_r and X_r^P be? If we let

$$\delta\phi/\delta X_r^P = 0 \tag{13-2-10}$$

to find the maximum, the maximum spreading occurs when the X_P of the two surfaces are equal. The total τ_r or τ_r is composed of the polar X^P and dispersion, X^d, van der Waals (nonpolar) forces. Some typical values of the components of the surface tension of solids encountered in semiconductor lithography are given in

TABLE 13-2-5. Surface Tension (dynes/cm) of Lithographic Solids

Surface	X^p	X^d	Reference
SiO_2	35	15	45
SiO_2 treat[a]	8	25	48
SiO_2 treat[b]	25	10	48
Al	38	25	45
KTFR	2	28	47
AZ-1350	5	23	43
PMMA[d]	12	20	46
Polystyrene	6	23	46
H_2O	60	10	49
Cellosolve acetate[c]	4	33	49

[a] Contact of 6 minutes with HMDS liquid.
[b] Contact of 1 minute with HMDS liquid.
[c] Trademark of Union Carbide.
[d] Polymethylmethacrylate.

Table 13-2-5. Since resists do not have a large X^P, they do not wet more polar surfaces. Treatment with silanes (Tables 13-2-4 and 13-2-5) lowers the surface tension to allow the X^d forces to maximize the adhesion.

The other basic question for adhesion is the optimum surface tension of the adhesive (resist) having a given τ_s and X_r^P component. The resist structure is usually dictated by radiation (exposure) chemistry requirements although copolymers may be built into the structure for adhesive purposes such as PMMA–polyisobutylene[(50)] without reducing radiation sensitivity.

Increasing the adhesive surface tension τ_r will decrease the spreading coefficient [Eq. (13-2-8)] but increase the work of adhesion [Eq. (13-2-9)]. Since we have antagonistic predictions, the optimum condition must then be a compromise between the two mechanisms. The optimum condition is when spontaneous spreading occurs,[(44)]

$$X_r^P = X_s^P \tag{13-2-11}$$

when

$$\tau_r = \phi^2 \tau_s \tag{13-2-12}$$

and the work of adhesion is

$$W_a = 2\phi^2 \tau_s \tag{13-2-13}$$

Thus, it is very important to match the polar forces in the adhesive bond. While a high τ_s implies a high surface energy[(45)] with a substrate such as SiO_2 ($\tau_s = 50$ dynes/cm), a high polar X_s^P substrate in practice is not desirable because of its high affinity for water, $X_s^P = 60$ dynes/cm (Table 13-2-5). Water will preferentially be adsorbed and easily rupture a resist bond by displacement. Secondly, the water may not be desorbed during prebake leading to edge attack and lifting during etching in aqueous baths. In addition, the aqueous alkali developer used in

TABLE 13-2-6. Work of Adhesion on SiO_2[a]

Solvent	W_a (ergs/cm^2)
H_2O	388
Isopropanol	158
Acetone	133
Benzene	11
Heptane	79

[a] From Ref. 51.

phenolic-based resists may also penetrate along the adsorbed water interface and cause lifting of the resist during the development step. The high affinity of water for a SiO_2 surface can be seen in the work of adhesion measured for various liquids deposited on SiO_2 at 25°C (Table 13-2-6). Strikingly, water has such a high affinity for polar surfaces that even as little as 0.02% by weight (200 ppm) of water in benzene will give almost the same heat of wetting [(51)] as pure water. Thus, 10 ml of photoresist which contains usually[(52)] up to 0.5% water by weight (5000 ppm) will wet polar surfaces but fail to adhere in the development step [small (1–2 μm^2) islands will fail first]. The nonpolar solvents will fail to displace the adsorbed water. If the surface contains adsorbed water, additional adhesive work has to be expended,

$$W_{\text{observed}} = W_a - W_{\text{water desorption}} \tag{13-2-14}$$

Laird[(51)] showed that the rate of diffusion of water along a glass resin interface of a bonded epoxy was 450 times faster than permeation through the resin. Accelerated aging of resist images at high relative humidity and heat would provide insight into the propensity to fail of a particular processing condition (e.g., preclean, primer, prebake, and postbake conditions). In production, surfaces such as fresh thermal oxide should be immediately primed or coated in dry conditions to avoid water adsorption during storage. Cleaning and rinsing in water only adds surface water. Primers which react with surface water are ideal since they form a hydrophobic barrier to further water adsorption during storage. Any adsorbed water can be easily removed by vacuum treatment.

Because of the competition of high-energy surfaces for water, it is difficult to raise the surface energy or use a surface with a surface tension > 30 dynes/cm. Some surfaces such as Si_3N_4 and polysilicon are low-energy ($\tau_s < 40$ dynes/cm) surfaces to begin with. Oxidation[(53)] in a plasma may produce a SiO_2 surface which absorbs enough water for silane treatment with a cationic polymerizable primer such as bistrifluoroacetamide (BTSA). If necessary, there are some silanes with pendant amino or polar groups which can raise the surface tension (see Table 13-2-4).

13-2-3. *Surface Primers and Treatments*

Resists are generally not formulated on the basis of being an adhesive material, although rubber-based materials are fortuitously good adhesive materials and sufficiently elastomeric to counteract the latent stresses induced by spin coating and rapid drying-form contraction. Silylation is the replacement of an active hydrogen

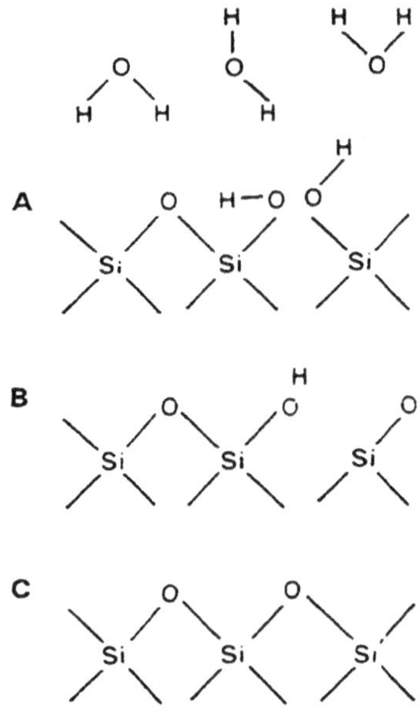

FIGURE 13-2-10. Schematic representation of thermal SiO_2 surface structures. (A) Type 1, with adsorbed molecular water referred to as a hydrated surface; (B) type 2 containing silanol groups with no adsorbed molecular water; (C) type 3, predominately a siloxane structure with no adsorbed molecular water. From Ref. 84, reproduced by permission of the Electrochemical Society Inc.

of a protic material with a substituted silicon atom. Thermal oxide exists in three forms (Fig. 13-2-10).

The silanes belong to the general formula

$$R_nSiX_{(4-n)}$$

where X is a hydrolyzable group of halogen, alkoxy, acyloxy, or amine. While the

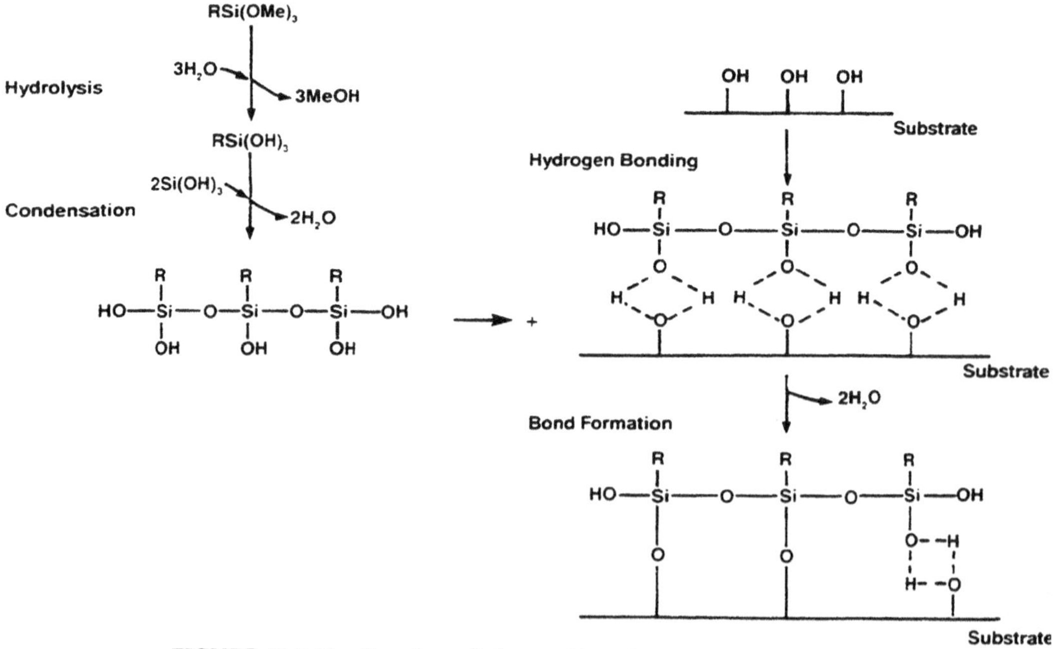

FIGURE 13-2-11. Reactions of silanes with surface water. From Ref. 54.

TABLE 13-2-7. Adhesion Promoters for Resists

Primer[a]	Resist	Reference
TMSC	KPR	55
HMDS	KTFR	56
MAPTMS	KPR	57
BTSA	AZ-1350	58
CrAc	AZ-1350	59
HMCTS	AZ-1350	60
DMClS	KTFR	61

[a] TMSC, trimethylsilylchloride; HMDS, hexamethyldisilazane; MAPTMS, γ-methacryloxypropyltrimethoxysilane; BTSA, bistrimethylsilylacetamide; CrAc, chromium acetylacetonate; HMCTS, hexamethylcyclotrisilazane; DMClS, dimethylchlorosilane.

halo and amino silanes can react directly with the silanol substrates, the silane is subject to hydrolysis with the adsorbed surface water (Fig. 13-2-11). Some of the most popular silanes for priming applications on SiO_2 are shown in Fig. 13-2-12 and Table 13-2-7. While TMCS reacts slowly with water or silanol groups, HMDS reacts with silanol and water,

$$2SiOH + [(CH_3)_3Si]_2NH \rightarrow 2SiOSi(CH_3)_3 + NH_3 \quad (13\text{-}2\text{-}15)$$

$$H_2O + HMDS \rightarrow [(CH_3)_3Si]_2OH + NH_3 \quad (13\text{-}2\text{-}16)$$

Although HMDS is a primer that does not polymerize on the surface, TMCS or BTSA (Fig. 13-2-11) reacts with surface water and polymerizes on the surface. HMDS does react with surface water to form ammonia and hexamethyldisiloxane which should be removed or the surface tension will fall below 30 dynes/cm (Fig. 13-2-14). If the same spin bowl is used to spin on HMDS and then apply a resist as AZ-1350, the HMDS or ammonia vapors may react with the spun resist film and cause insolubilization of the resist due to diazocoupling and silylation of the phenolic ester. In fact, this same process is used to silylate DQN into a bilayer resist mask. Vapor-phase application of HMDS in a reduced vacuum is the most

Me_3SiCl
TMCS
(Trimethylsilyl chloride)

$Me_3SiNHSiMe_3$
HMDS
(Hexamethyldisilazane)

$CF_3C[=NSi(CH_3)_3]\,OSi(CH_3)_3$
BSTFA
(Bistrifluorosilylacetamide)

$NH_2N(CH_2)_3Si(OCH_2CH_3)_3$
A-1100
(Aminopropyltriethoxysilane)
Union Carbide product

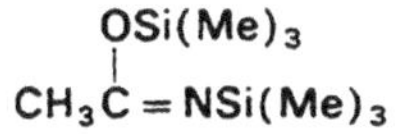

BTSA
(Bistrimethylsilylacetamide)

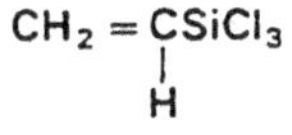

Vinyltrichlorosilane
(Si_3N_4 primer)

FIGURE 13-2-12. Popular silane derivatives for priming semiconductor surfaces for resist coatings.

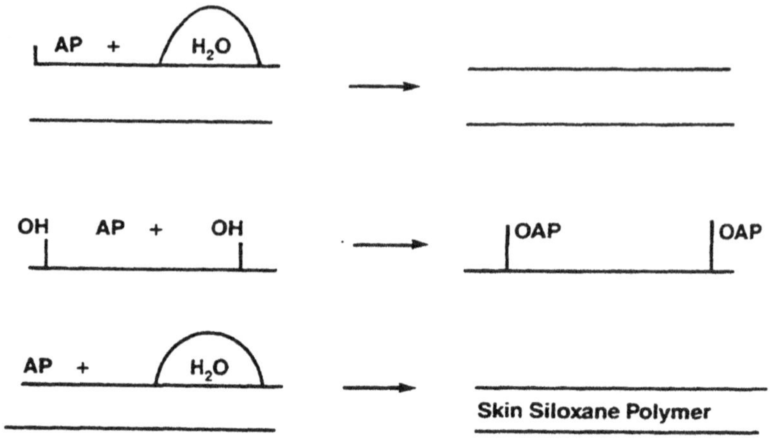

FIGURE 13-2-13. Schematic of three major silylation reactions of priming a surface of SiO_2 with adhesion promoter (AP).

effective and contamination–free method of priming wafers. The reaction time and conditions should be determined to achieve about 40 dynes-cm (Fig. 13-2-14).

The silanol groups of the substrate are esterified to lower the τ_s^P (surface tension of the substrate). Thus, the two general types of surface treatments are:

1. *Primary*—HMDS which reacts to silylate silanol groups.
2. *Polysiloxane formation*—BTSA, A1100, HMCTS, and DMClS which rely on surface water to effect a monomolecular polymerization.[63, 64] A continuous adhesive layer is formed.

Priming a surface will depend on the number of reactive sites, the reaction time, and the temperature. Various degrees of surface tension can be achieved[62] (Fig. 13-2-15). For subsequent process steps, the primed surface should resist the action of organic solvents, (negative and some positive resist developers) alkaline developers, and in some cases, possess reactive groups (vinyl) for enhanced bonding of *e*-beam negative resists. While vinyl groups are thought to polymerize and react

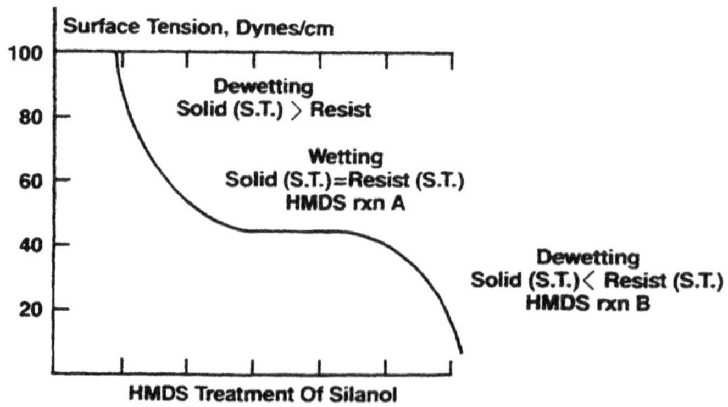

FIGURE 13-2-14. Wetting of silanol surfaces by sufficient priming reaction (A). Reaction B leads to dewet by excessive layer buildup of siloxane to lower surface tension. Dewetting occurs when surface tension (S.T.) of substrate and resist are mismatched.

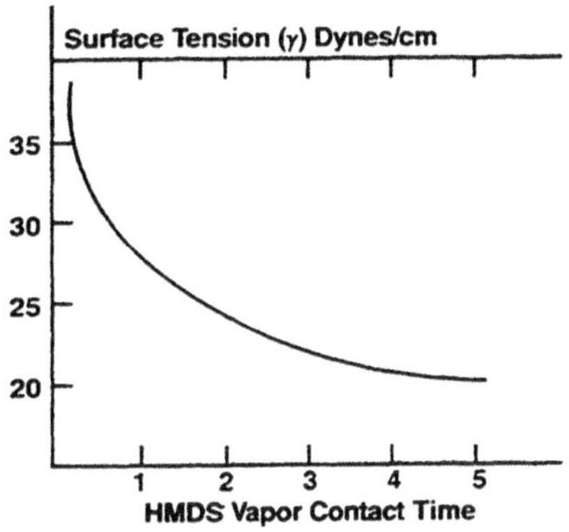

FIGURE 13-2-15. Prime time of HMDS with thermal SiO_2.

with surface primer groups, the primed surface should, however, be removed by the etchant (gas or liquid) and not inhibit etching.

The success of the reactions with silanes depends on the available reaction sites. Quantitative surface analysis is necessary to establish the reproducibility of the process and, hence, the lithographic process. The low surface area of silicon wafers is too small to permit direct analysis by conventional IR although Fourier transform IR(63) has been used to show by difference spectrum that treatment of silica with vinyltriethoxysilane is equivalent to a coating obtained by directly depositing a film of polyvinylsiloxane.(64) Polyvinylsiloxane further cross-links upon heating or electron irradiation.

Wagner(87) and co-workers have used electron spin resonance (spin labeling silanol groups) to measure the silanol content of the surface and they showed that heating to 200°C removes only surface water and not silanol groups. Silica surfaces cleaned and baked will rapidly absorb contaminants. The spin labeling technique allows determination of the cleaning efficiency of precleaning steps (Chapter 6).

Since the primers are in contact with the semiconductor surface, they should be of highest purity (< 1 ppm of any residues or metals). On the surface, vapor-phase treatment minimizes inorganic and polymeric residues. In the polymer formed by grafting onto the silanol groups, the stability to etchants and alkaline developers is in reverse order of their chemical reactivity(54):

$$\text{BTSA} > \text{HMDS} > \text{TMSC}$$

The silicone layer should be slower etching than the oxide layer or beveling (lateral undercut beyond $D/L = 1$) will occur. Beveling agents or films are based on

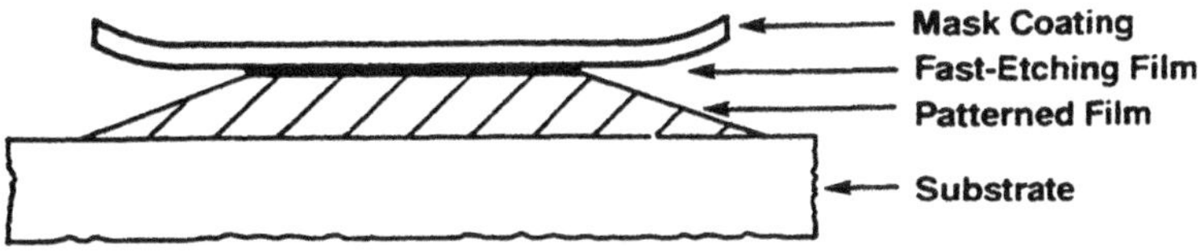

FIGURE 13-2-16. Use of a fast-etching underlayer to bevel SiO_2 in wet isotropic etching. From Ref. 65, copyright 1972 and courtesy of Elsevier Sequoia S.A.

TABLE 13-2-8. Undercutting–Priming of SiO_2 in Buffered 7:1 HF

Resist	Primer	Undercut (μm)	Reference
AZ-1350	None	2.1	66
AZ-1350	HMDS	1.2	66
KTFR	None	1.9	66
KTFR	HMDS	0.9	66
KTFR	None	2.5	58
KTFR	BTSA	1.0	58
KTFR	DMCIS	1.5	58
AZ-1350	None	2.8	67
AZ-1350	HMDS	1.6	67
AZ-1350	BTSA	1.0	67
AZ-1350	MONA[a]	1.0	67
OMNR	Alcohols[b]	1.5	182

[a] 1-hydroxymethylimidazoline (Monazoline, Mona Industries) added to AZ-1350.
[b] Autoclave treatment of OMNR Tokyo Ohko Resist.

a faster etching layer[65] between the resist and oxide substrate (Fig. 13-2-16). They are used for producing 30° profiles in oxide for subsequent stress-free metal deposition over the steps. In Table 13-2-8, a comparison of the effectiveness of adhesion promoters for semiconductor surfaces of SiO_2 is given. The polymerizing primer of BTFA is more effective than the primer of HMDS. Changing the surface properties (τ_r of the resist) by additives is also an effective treatment. Four basic approaches have been used to improve adhesion:

1. Use a resist subbing layer of rubber or baked AZ-1350 as a double layer. Immiscible polymer layers applied as skin layer (< 1000 Å) serve to reduce pinholes and contamination. Oxygen ash may be necessary to develop images to the substrate.
2. Use a polymerizable monomer sublayer such as cyanoacrylate which reacts with surface water in an anionic polymerization. Silanes as we have discussed serve in a similar function.
3. Add elastomeric adhesives or (Monazoline imidazoline soaps) surface-active agents to the resist.
4. Overcoat the resist image with a cross-linking resist and ash off the excess. The liquid resist creeps into the interface between the resist and substrate but is not ashed away.

All of the treatments or extra steps have to be compromised in some cases due to losses of sensitivity or extra process costs. A comprehensive summary of the above adhesive improvements of DQN resists of the AZ-1350 type is given in Table 13-2-9.

In the wet etching of SiO_2 with HF, there are several major factors which can influence the amount of undercut. The major factors are summarized in Table 13-2-10. Unprimed surfaces (Fig. 13-2-17) and low postbake of DQN positive resists (Fig. 13-2-18) are the major physical contributors to severe undercut. However, postbaking above 120°C can lead to 50% more undercut than at lower postbake.[627] At the higher postbake, oxidation of the surface of the novolak resin

TABLE 13-2-9. Adhesion Treatments or Additives to Promote Adhesion of DQN Resists

		Reference
Treatment	a. Treat by oxidizing in $KMNO_4$ (add to developer)	68
	b. Use rubber resists OMNR (Tokyo Ohko) under AZ-1350	69
	c. Baked and exposed AZ-1350 sublayer under AZ-1350 (plasma ash after)	70, 71
	d. Cyanoacrylate treat surface	72
	e. Electron-beam polymerize silanes	183
Additive to AZ-1350	a. Alkylimidazolines	67
	b. Butyric acid	73
	c. Alkyl acids	74
	d. Polyhydroxystyrene	75
	e. Polyvinylmethylether	76
	f. t-Butyl novolaks	77
	g. Epoxy silanes	78
	h. Epoxy polymers	79
	i. Silane	80

forms a husk and the N_2 evolving from the decomposition of the DQ sensitizer cannot espace from the resist. This buildup of pressure is released by adhesive failure during etching. An undercut of 50% more than without high postbake was noted. To slowly release the N_2 without stressing the film, a flood UV exposure is used prior to postbaking.

For the chemical composition factors in wet etching, the addition of nitric acid to buffered HF or the use of high NH_4F to HF mole ratios (> 10) in buffered HF can lead to severe undercutting. Combinations of factors may interact to increase the undercut.

Surfaces devoid of silanol groups such as Si_3N_4 and PSG are severely prone to adhesive failure. The dot test of measuring the adhesive failure of micrometer-size dots of resist[62] was used to assess various HMDS treatments on various surfaces

TABLE 13-2-10. Undercut Factors for Etching SiO_2 with Buffered HF

Factor	Reference
1. Low prebake or high postbake[627]	81
2. No or low postbake	82 (Figs. 13-2-17 and 13-2-18)
3. Phenolic resist versus rubber resist	26, 41, 66, 83
4. No surface promotion	26, 47, 62, 66, 81, 82
5. 1 μm^2 versus 10 μm^2 pattern	62, 66
6. Spin versus dip coat	83
7. Siloxane > silanol > hydrated silanol	84–86
8. NH_4F/HF ratio >10	26
9. PSG (phosphosilicate glass)	81
10. $\gamma_s^p > 3$	48
11. No mask	65
12. Add NHO_3 to etchant (lift agent)	87

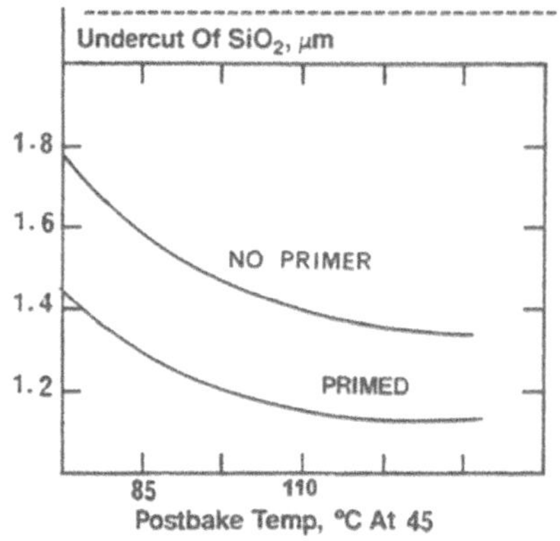

FIGURE 13-2-17. Effect of HMDS primer applied to a SiO_2 film on the undercut of an etched (HF) image masked by a Kodak 830 positive photoresist. From Kodak technical literature on 820 photoresist.

in which the processed wafers were agitated in water (Table 13-2-11) and the time for the island of resist to delaminate was noted. The HMDS treatment preserved the integrity of the dot for a longer period of time than untreated surfaces. For good adhesion to Si_3N_4, an oxygen-rich surface of silicon nitride is required for reaction with silanes. On Si_3N_4, silanol groups are formed by plasma oxidation and reaction with moisture.[678] In practice, a more durable or effective "subbing" layer is needed for the (PSG, Si_3N_4) nonsilylating surfaces to later withstand aqueous etching or development. Effective treatments for the more difficult layers are given in Table 13-2-12. The most effective treatments are subbing layers which have to be processed as multilayer resists (Chapter 12).

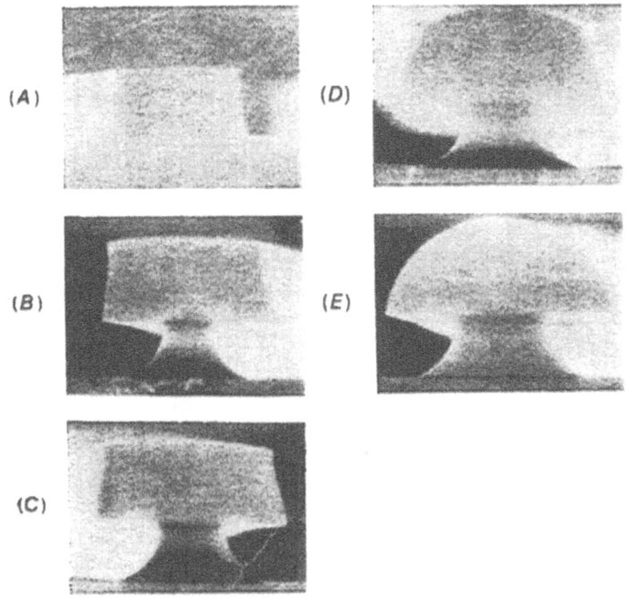

FIGURE 13-2-18. Effect of postbake of Tokyo Ohko OFPR-2 positive photoresist on HF wet etching of SiO_2 layer. (A) No bake, no etch; (B) 100°C bake and etch; (C) 120°C bake and etch; (D) 140°C bake and etch; (E) 160°C bake and etch. All postbakes were 20 min. From K. Jinno, Y. Matsumoto, and T. Shinozaki, Photog. Sci. Eng. 21 290 (1977), courtesy of SPSE.

TABLE 13-2-11. Adhesion Time to Fail Test[a]

Surface	HMDS treated	No treatment
SiO_2	9 min	3 min
Si_3N_4	4 min	1 min
PSG	3 min	<10 sec
HF-etched SiO_2	1 min	<10 sec

[a] From Ref. 62.

13-2-4. *Stress in Resist Films*

One of the most important findings in all the studies of undercut was reported by Andrasi[(41)] who showed that lateral undercut of AZ-1350 on oxidized silicon or aluminum was in the order of the resist film thickness:

$$0.5\ \mu m > 1.0\ \mu m > 1.5\ \mu m$$

Previously, Deckert[(66)] had pointed out the importance of stresses in the adhesion of resist film. A stress of 10^8 dynes/cm^2 was measured for KTFR and AZ-1350 using a basic stress (σ) formula,[(89, 90)]

$$\sigma = E_s T_s^2 / 6(1 - V) R T_r \tag{12-2-17}$$

where E_s is the modulus of elasticity (Young's) of the substrate, V is Poisson's ratio, R is the net radius of curvature of the substrate disk [R is negative for convex curvature (compressive stress) and positive for concave curvature (tensile stress)], T_r is the resist thickness, and T_s is the substrate thickness. Thinner resist films induce more stress in the coating.

Since all of the work of adhesion [W_a, Eq. (13-2-9)] is accomplished in the initial wetting, spreading of the resist, displacement of water (if any), and adsorption of the fluid on the solid surface, the subsequent solidification of the resist cannot increase the attractive forces. In the transition from a spun film to a prebaked solid, the resist films shrink in volume. Over 40% of the solvent is lost from the final film. In AZ-1350, the solvent retention[(91)] is about 5–10%. Residual solvent also competes with the resist polymer for bonding sites to the substrate. The compaction of the film results in stress development.

TABLE 13-2-12. Adhesive Treatments for Other Dielectric (Non-SiO_2) Surfaces

Layer	Treatment	Reference
Si_3N_4	Plasma oxidize	53
Si_3N_4	AZ-1350 sublayer	70, 71
Polysilicon	Plasma oxidize	53
Al_2O_3	Cr complexes	88
PSG	AZ-1350 sublayer	70, 71
PSG	Polyvinylmethylether	76
Gold	S, P, CO-silanes	581

Since the coefficient of thermal expansion of the resist film ($2–5 \times 10^{-5}$ cm/deg) and the substrate (2×10^{-6} cm/deg for Si) are not equivalent, heating or cooling will produce additional stresses. If we use a value of E of 10^6 dynes/cm^2(modulus of elasticity) of glassy phenolic resists,[92] the stress can be estimated to be 7.5×10^{-3} for a 125°C excursion in postbake. A stress of 50 to 200 kg/cm^2 can be anticipated for solvent-cast films.[92]

The W_a of polymers[92] is in the range of 100–300 kg/cm^2. The true or observable adhesion [Eq. (13-2-14)] should be modified to include the stress:

$$W_{\text{obs.}} = W_a - W_{\text{water des.}} - W_{\text{stress}} \tag{13-2-18}$$

Although stress is a force[93] and W_a is energy and not of the same units, it is instructive to consider the forces working against adhesion. Edge lifting of the thinnest resist was observed by Andrasi who studied the undercut of AZ-1350 on SiO_2.[41] Since the stress (σ) is proportional to $1/T_r$ [Eq. (13-2-17)], edge lifting will occur in the thinnest film when the adhesive bond is broken between the resist and etched oxide (Fig. 13-2-19). The lifting will result in beveled profiles. Strong HF_2^- concentrations[26] in 30 to 1 NH_4F/HF produce 30° profiles due to rapid interface penetration and edge lifting.

In a negative cross-linked resist, a force due to swelling may exist. Narrow 1-μm lines form serpentine patterns due to delamination caused by swelling upon development (see Fig. 10-6-10). The swelling is caused by osmotic pressure[92] on the order of 50 kg/cm^2. In stripper formulations, small molecules (CH_2Cl_2 and CH_3OH; Chapter 14) are powerful swelling agents which can induce forces of 200 kg/cm^2. Finally, even the casting solvent can influence the residual stress in a film. Films of nitrocellulose cast from acetone had a stress of 135 kg/cm^2 while butyl acetate-cast films had a stress of 50–70 kg/cm^2. In the film cast from butyl acetate, the lower stress may be due to the retention of the slower evaporating butyl acetate.[94]

Some reduction in the intrinsic stress in spin casting can be achieved by using

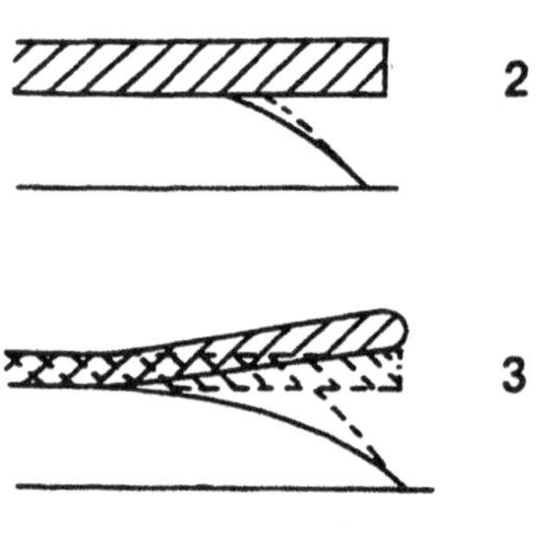

FIGURE 13-2-19. Undercut profiles of SiO_2 etched in HF liquid at resist thickness of (1) 1.5 μm; (2) 0.5 μm; (3) 0.23 μm. All etched at 50% overetch for a 0.7-μm-thick SiO_2 layer. The severest undercutting occurs with the thinnest resist. From Ref. 41, courtesy of Elsevier Sequoia, S.A.

TABLE 13-3-1. Passivating SiO_2 Etches

Sandwich	Etchant	Reference
SiO_2/Al	BHF + glycol + NaCl	97
SiO_2/Al	BHF + glycol + FC134 (surfactant, 3M Co.)	98
SiO_2/Al	BHF + isopropanol	99
$SiO_2/Si/Al$	Diamine + isopropanol	100
SiO_2/Al	BHF + glycerine (color change)	186

elastomeric resists of (KTFR) rubber, low-E, or elastomeric additives and plasticizers (Table 13-2-9) such as butyl phthalate. Heating a resist polymer above its glass transition temperature (T_g) will relieve some casting stresses.[95] In novolak-based resists, it is difficult to heat them above the T_g, because oxidation and DQ desensitization occur in the prebake. Annealing of the film in a solvent atmosphere has been shown to reduce radial stresses from spin coating.[96]

13-3. PASSIVITY IN ETCHING SiO_2

Hopefully, the etching of thermal oxide with buffered HF shows selectivity and ceases at the Si or Al substrate. Al is attacked by HF. Some additives (Table 13-3-1) have been used to passivate the Al substrate to further attack. Alcohols bond strongly and form Al alkoxides which are strong bases not displaced by HF (Table 13-3-1).

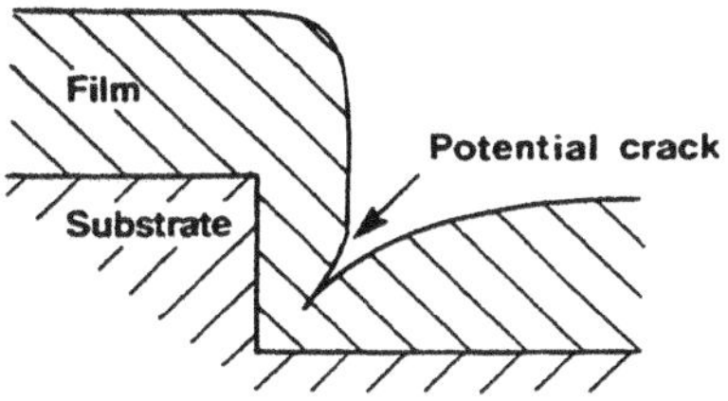

FIGURE 13-3-1. Profile of metal deposited over step subject to stress cracking. From Ref. 102, courtesy of Elsevier Sequoia, S.A.

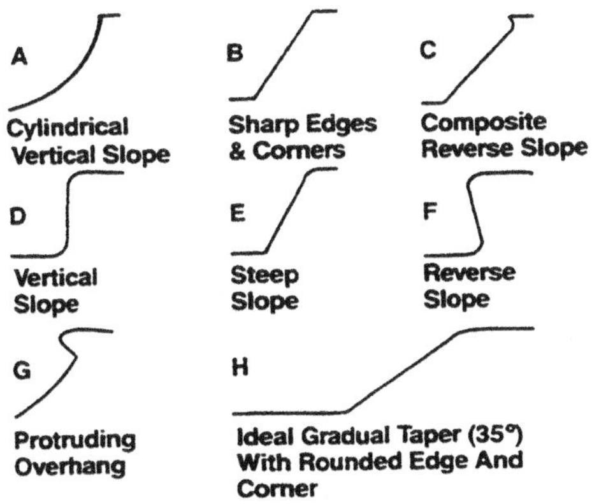

FIGURE 13-3-2. Profiles of metal edges subject to reliability problems compared to ideal case (H). From Ref. 4.

TABLE 13-3-2. Beveling SiO_2

Process	Reference
1. Resist lifting with 20:1 BHF	26
2. Lifting agents, NHO_3 in BHF	87
3. Thin resist (0.5 μm)	41
4. PSG toplayer	65
5. Sputtered quartz toplayer	103
6. Ion-implanted SiO_2	104

Beveled Etching

The gradual tapered profiles of etched SiO_2 or polysilicon are important in reducing the stress in metals deposited in contact holes.(101) Metal deposited over steps is subject to stress cracking(4) (Fig. 13-3-1) and long term reliability problems. Beveled profiles (Fig. 13-3-2) have been obtained by resist lifting or by the inclusion of a rapid dissolving (Fig. 13-2-16) layer between the resist and etching substrate. Several of the techniques are listed in Table 13-3-2.

13-4. SILICON ETCHING

The etching of silicon involves the oxidation step:

$$Si + [O] \rightarrow SiO_2 + 14 \text{ kcal/mole} \tag{13-4-1}$$

followed by the etching of SiO_2,

$$SiO_2 + 6HF \rightarrow H_2SiF_6 + 2H_2O - 11 \text{ kcal/mole} \tag{13-4-2}$$

In the HF/HNO_3 system,(39)

$$Si + 2HNO_3 + 6HF \rightarrow H_2SiF_6 + 2HNO_3 + 2H_2O + 125 \text{ kcal/mole} \tag{13-4-3}$$

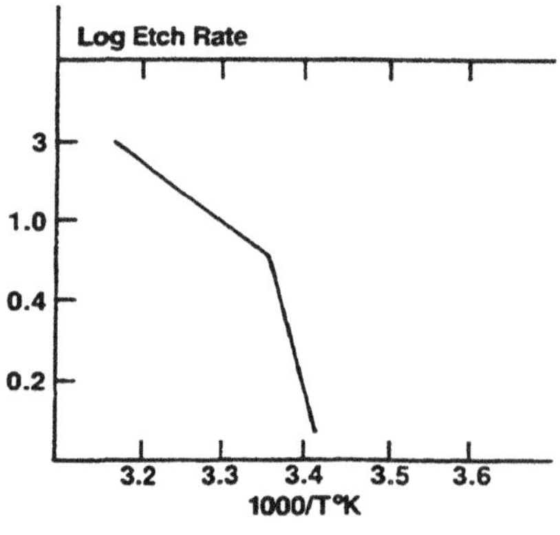

FIGURE 13-4-1. Plot of etch rate versus 1000/T for etching Si in HNO_3/HF. From Ref. 105, reproduced by permission of the Electrochemical Society Inc.

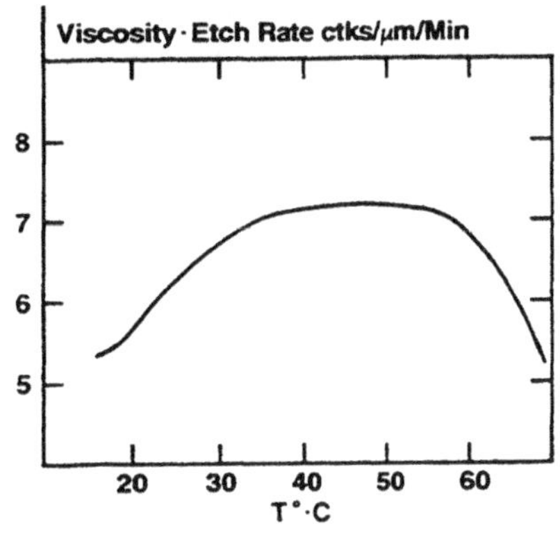

FIGURE 13-4-2. Product of viscosity × etch rate versus temperature for etching silicon using glacial acetic acid as viscuant. From Ref. 106, reproduced by permission of the Electrochemical Society Inc.

Each atom of Si dissolving uses up to 2 moles of HNO_3 and 6 moles of HF. If the reaction is diffusion controlled, the maximum etch rate should occur when the mole ratio of HNO_3 to HF is 1 to 3. Examination of the Arrhenius plot of etching Si in HNO_3/HF reveals a break (Fig. 13-4-1) corresponding to a change from a diffusion-controlled to a reaction rate-controlled process. The diffusion-controlling step (6 kcal/mole) is the HF diffusion through the product layer and the rate-controlling step (4 kcal/mole) is the oxidation of the silicon. In a diffusion-controlled reaction, the product of viscosity × rate is a constant [Eq. (13-1-20)]. Glacial acetic acid is added as a viscosity moderator (Fig. 13-4-2).[106]

The isotropic etchants for silicon use a mask of nonetching metal,[146] Si_3N_4 or SiO_2 (sometimes for shallow etch). Rarely is a resist used since HF–HNO_3 permeates rapidly through film. Alkaline etchants have also been used to etch silicon,

$$Si + 2OH^- + H_2O \rightarrow SiO_3 + 2H_2 \qquad (13\text{-}4\text{-}4)$$

Ethylenediamine, hydrazine, and OH^- act as oxidants, and pyrocatechol and alcohols as complexing agents for SiO_3^+. In addition, the hydrogen gas can inhibit etching on the surface of polysilicon. Surfactants[122] are added to displace the H_2.

The alkaline etchants are preferential[623, 723] anisotropic etchants for the low index planes. The free bonds[3] (dangling crystal defects) are available in a relative proportion of 1.00 to 0.71 to 0.58. The reason for choosing the 100-oriented silicon

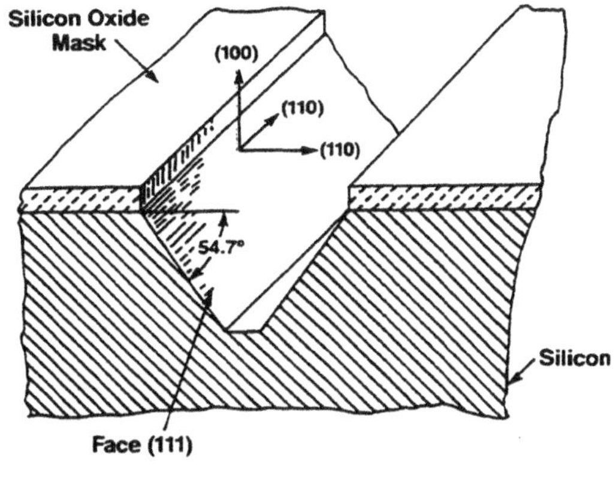

FIGURE 13-4-3. Preferential etching of the 100/110 crystal axis of silicon. U.S. Patent 3,765,969.

TABLE 13-4-1. Isotropic and Anisotropic Silicon Etching

	Etch rate (μm/min)			Undercut (μm/edge)[a]		
Etching solution	PS	ES	BS	PS	ES	BS
Isotropic[b]	3	4	4	1.5d	1.5d	1.5d
Isotropic[c]	0.8	0.6	0.5	1.0d	1.0d	1.0d
Anisotropic[d]	0.7	0.9	1.1	(0.1–1.0)d	<0.1d	<0.1d

[a] d, depth of etching.
[b] HNO_3(65%)/HF (40%)/$NaNO_2$ = 95 ml/5 ml/1 g.
[c] HNO_3(65%)/H_2O/HF (40%) = 100 ml/40 ml/6 ml.
[d] KOH/H_2O/*n*-propanol = 15 g/50 ml/15 ml.

is that this is the only one of the major planes(723) in which the 110, 111, 100, and 211 planes intercept with regular symmetry and hence the orientation most likely to be used in the etching of deep trenches in silicon. It should be noted that the geometry of a surface produced by anisotropic etching will depend on the geometry of the original surface, since the convex surfaces will limit on fast-etching planes while slow-etching planes will be bound on the concave surface.(40) In the 100 direction, the etch rate is up to 100 times faster than in the 111 direction. Figure 13-4-3 is an example of the preferential etching of a 54° groove in a 110/100/111 intersection using KOH and isopropanol at 85°C. KOH and isopropanol are popular etchants with a 55 to 1, 100/111 etch rate ratio.

Kaminsky(723) has made an extensive investigation of the methods and liquid etchants for crystalline silicon. The alkaline etchants with inorganic masks are the most effective. During etching, the silicon wafer is placed in a holder and vigorously agitated.

The slope of the sidewall of the 111 orientation forms an angle with the slice surface determined by the crystallographic surface. The value of the angle is expressed from a geometric analysis to be the arc tan of $\sqrt{2}$ which equals 54.7.(123) Alcohols are added to the etchant(40) to be preferentially(124) adsorbed on the 111 plane to allow anisotropic etching to occur in the other directions. The rate of etching is diffusion controlled with an activation energy of 4 kcal/mole(40) because the alkali must diffuse through a barrier of complexed silicon.

TABLE 13-4-2. Anisotropic Silicon Etchants

Etchant	Application	Reference
EDA, PYROC, H_2O	100	128
EDA, PYROC	SiO_2, Si_3N_4, pinhole detect	129
Hydrazine, IPA, H_2O	100, Al mask	40
KOH, *sec*-alcohols	100	130–132
KOH, ethyleneglycol	Texture solar cells	133
Diamines, KOH, IPA	No Al attack	134
KOH, IPA, H_2O	100	135
R_3N^+OH, IPA, H_2O	100	136
R_3N^+OH, surfactant	Surfactant stops H_2	122
R_3N^+OH	Avoid Na^+ in etch	137
H_3PO_4 + trace As_2O_3	*n*-type	138
CuF_2, AZ-1350 mask	Electroetch	139

TABLE 13-4-3. Silicon Isotropic Etchants

Etchant	Application	Reference
HF, HNO_3, CH_3COOH	All Si	107–110
HF, HNO_3, CH_3COOH	Low-resistivity Si	111
HF, $KMnO_4$, CH_3COOH	Epitaxial Si	112
HF, HNO_3; $H_2O_2 + NH_4OH$	Remove Cu impurities	113
HF, HNO_3, CH_3COOH	PNP multilayers	114
HF, HNO_3	PNP multilayers	115
NH_4F, H_2O_2	Minimum undercut	116
HF, HNO_3, I_2	General etch	117
HF, HNO_3, CH_3COOH	100 plane undercut	118
HNO_3, HBF_4, NH_4BF_4	AZ-1350 mask	119
NH_4F, H_2O_2, NH_4HPO_4	Si/PSG 2 to 1 (etch rate)	120
KOH + alcohols	Polycrystalline	121

The anisotropic or isotropic etching polysilicon for gate electrodes was studied by Raetzel and co-workers.[125] A summary of their results of etching poly-, epi-, and bulk silicon is shown in Table 13-4-1. Etchants which have slower rates have less undercutting since the endpoint can be more easily sensed. A summary of all the anisotropic etchants for silicon is given in Table 13-4-2.

The most popular etchant for etching single and polycrystalline Si is ethylene diamine-pyrocatechol (EDA) with an activation energy of 8 kcal/mole,

$$2NH_2(CH_2)_2\,NH_2 + Si + 3\phi(OH)_2 \rightarrow 2H_2 + Si(\phi O_2)_3 + 2NH_2(CH_2)_3NH_3 \quad (13\text{-}4\text{-}5)$$

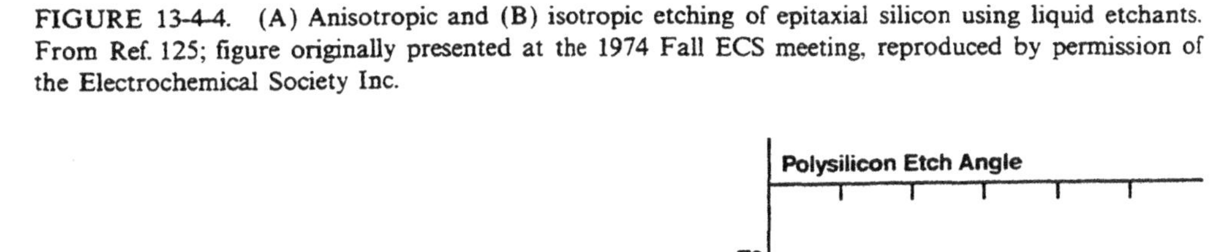

FIGURE 13-4-4. (A) Anisotropic and (B) isotropic etching of epitaxial silicon using liquid etchants. From Ref. 125; figure originally presented at the 1974 Fall ECS meeting, reproduced by permission of the Electrochemical Society Inc.

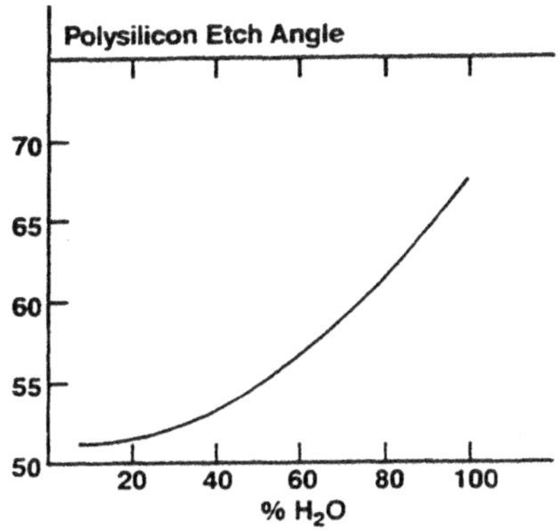

FIGURE 13-4-5. Polysilicon etch angle versus water content of KOH/alcohol/H_2O etchant. U.S. Patent 3,909,325.

An increase in activation energy to 11 kcal/mole and an increase in selectivity from 10 to 20 over 100 of the 111 plane have been achieved by the addition of 1000 ppm of aromatic pyrazine(126) to the reaction. For diagnostic investigations, silicon etching is used to highlight the SiO_2 pinholes. Boron-doped silicon etches at a slower rate than undoped Si.(543)

The beveling of polysilicon in KOH and alcohol is a function of the water content of the etchant. Isotropic images are formed in anhydrous alcohol. The degree of anisotropy is a function(121) of the water content of the etchant (Fig. 13-4-5). In Table 13-4-3, a summary is given of the isotropic etchants for silicon.

13-5. SANDWICH ETCHING

Sandwich etching consists of etching through various layers with a single or multiple etchants. Single etchants are preferred. The major task is to formulate an etchant which etches all layers at equal rates in order to prevent the "Christmas tree" profile. Overall, the most intensive studies have focused on etching Si_3N_4/SiO_2 sandwich in which equal etch rates are required for smooth tapered holes. Films of Si_3N_4 are only etchable in HF or boiling H_3PO_4 at 180°C. At 180°C, no organic resist can withstand the severe conditions. Etching(140) with HF follows the same rate law found by Judge(27) for SiO_2 etching,

$$\text{Rate} = A(\text{HF}) + B(\text{HF}_2^-) + C \qquad (13\text{-}5\text{-}1)$$

Kern and Deckert have extensively reviewed the etching of Si_3N_4.(110) A comparison of the Si_3N_4 and SiO_2 etch rates in HF indicates that equal but slow rates of 100 Å/min can be achieved by:

1. Varying temperature
2. Varying HF/HF_2^- ratio

The etch rate of the oxide is reduced to 100 Å/min by diluting a tenfold dilution of HF. At the low HF concentration, the dissolution of SiO_2 changes from a rate-controlled to a diffusion-controlled (4 kcal/mole) reaction.(141, 142) Another composition used phosphoric acid–fluoroboric acid and adjusted the temperature to attain

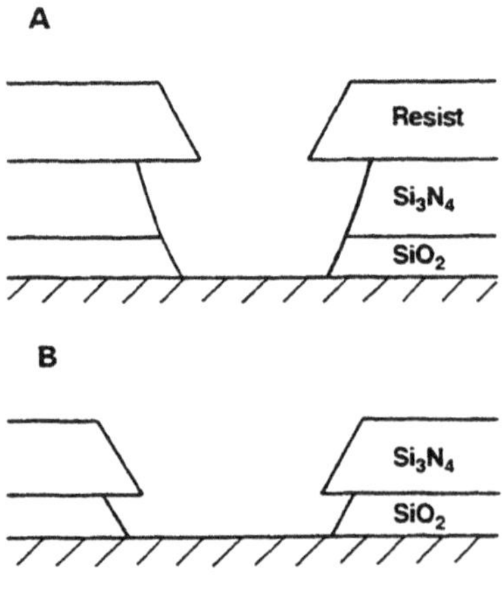

FIGURE 13-5-1. (A) Sandwich etching of Si_3N_4/SiO_2 with faster SiO_2 etch rate; (B) isotropic etching of sandwich with equal etch rates.

100 Å/min for SiO_2 and Si_3N_4.[143] However, the phosphoric acid attacked the Al and Si underlayer. Also, sulfuric acid has been added to slow the attack.[147] The addition of diols also prevents the attack of the underlayer of Al.[97, 98, 186]

Faster but equal etch rates have been achieved by manipulating the viscosity of the etchant. Glycerol or other high-viscosity alcohols are added up to 50% by weight to replace the water.[140, 144] NH_4F is also added to moderate the HF.[145] Typical tapers etched in Si_3N_4/SiO_2 are shown in Fig. 13-5-1.

Another approach which uses reverse etching is to use tungsten patterned by resist as a mask on top of SiO_2/Si_3N_4. The oxide layer is undercut, the W stripped, and the Si_3N_4 is profiled since the oxide layer has attained the desired configuration.[149] A compromise etchant has been devised for etching a resist/Si_3N_4/borosilicate glass sandwich in which a mixture of 70% H_3PO_4, 29% glycerin, and 1% HBF_4 at 103°C has been used. Higher concentrations of HBF_4 attack KTFR or AZ-1350 resists.[148]

13-6. ALUMINUM ETCHING

In the liquid state, metal etching involves many electrochemical principles. The metal acts as its own anode, a chemical effect which induces passage of current:

$$M \rightarrow M^{n+} + ne \qquad (13\text{-}6\text{-}1)$$

Since most metals are covered by a native oxide, a secondary etchant is added to etch the passivating layer. In solution, a cathode reduction of the oxidizing agent occurs. Practically all first-level metallurgy uses Al, Al-Cu, Al-Si, and Al-Cu-Si. Al-Cu can form precipitates and grain boundaries (Fig. 13-3-1). At the boundaries, precipitates of $CuAl_2$ occlude and etch very slowly to form residues or edge hillocks. Generally, a single level of metal is used but as logic circuitry density increases, some times up to four levels of metal interconnections are used involving multiple etchants. For example, a multiple layer of Al/W/Ni/Cu has been etched with HF/HNO_3, H_2O_2, and H_2SO_4 using a resist as mask.[150]

Many oxidizing agents of the inorganic type are known. The most common agent is H^+. The main task of the process chemist is to choose a suitable oxidizing etchant to satisfy the thermodynamic requirement,

$$\Delta F = -n\mathcal{F}\Delta E \qquad (13\text{-}6\text{-}2)$$

TABLE 13-6-1. Metals and Oxidizing Agents

Metal	Oxidizer
Al	F_2
Zn	H_2O_2
Cr	MnO_4^-
Ni	$Cr_2O_7^{2-}$
Sn	Br_2
Cu	HNO_3
Ag	Fe^{2+}
Au	I_2

TABLE 13-6-2. Metal Line Definition

	Year			
	1976	1978	1984	1986
Pitch	11 μm	8 μm	5 μm	4 μm
Al line	4.5 μm	3.5 μm	2.5 μm	2 μm

where ΔF is the free energy and ΔE is the difference in the oxidation and reduction potentials (Table 13-6-1).

Although the metal etching of large mil-sized features has been practiced for years in chemical milling, electropolishing, and corrosion, no studies have been published of the Al wet etching of metal interconnections on silicon devices. Romankiw[(151)] has published the only critical review of Al etching and cited only one brief paper[(156)] on the mechanism.

The patterning of metal is one of the major limiting steps in lithography. Metal conductor sizes can be projected to decrease (Table 13-6-2) but pitches smaller than 2 μm are severely limited by capacitance.[(559)]

Al wet etching is isotropic and subject to an etch bias (Fig. 13-1-2) of 1–2 μm and a batch etch process reproducibility of ± 1 μm. Patterns are generally overetched since the metal films are subject to defects such as grain size, precipitates, and stress. Excellent photographs and examples of defects in Al etching can be found in the review by Romankiw.[(151)] The electrochemistry of etching can be found in treatises on corrosion.[(152)] As such, MacArthur[(153)] has shown how to use polarographic current voltage data in selecting an oxidizing agent. Metals in contact with other metals (galvanic effect) are altered in their electrochemical potential [Eq. (13-6-1)] and accelerate their own etch rate. Consequently, severe undercutting generally occurs of the submetal layer. For example, Al alone will etch in 7 min in NaOH. In contact with Pt/Au, etching is complete in 1 min.[(154)]

The amount of current donated by the metal [Eq. (13-6-1)] can widely vary by complexation, electrolytic polarization, crystallinity, and product adsorption.[(152)] Thus, each metal will require an empirical determination of its electrode potential.

13-6-1. *Etchants for Aluminum*

The etching of aluminum can be performed in alkaline or acidic medium. A popular formulation consists of concentrated H_3PO_4 (76%), glacial acetic acid (15%), concentrated nitric acid (3%), and water (5% by volume). According to polarographic studies by Harvilchuck and Judge,[(156)] the reaction consists of a rate (Al^{3+} formation)-controlled and an $AlPO_4$ product-controlled step,

$$Al_2O_3 \xrightarrow{\text{slow}} Al \xrightarrow[HNO_3]{-3e} Al^{+3} \xrightarrow{\text{fast}} \text{Film} \xrightarrow{\text{slow}} \text{Soluble } AlPO_4$$

In phosphoric acid, water impedes the dissolution of Al_2O_3 but is essential for solubilizing the secondary product of $AlPO_4$. The current flow is proportional to the etch rate. If a current is applied to the aluminum, anisotropic etching has been

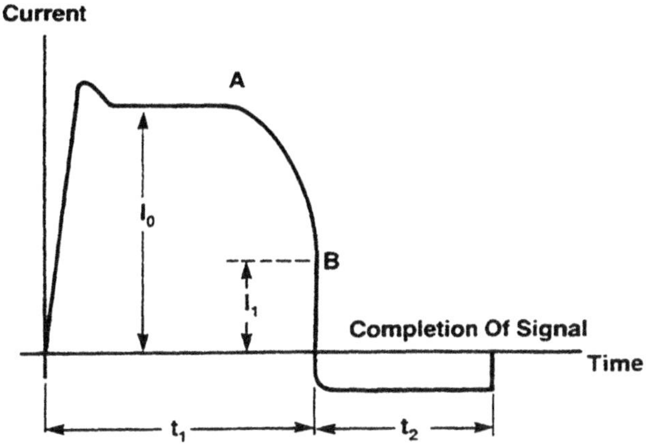

FIGURE 13-6-1. Operation of an automatic electrochemical Al etcher monitoring current I to endpoint t_2. From Ref. 179.

noted. As early as 1971, a commercial electrochemical etcher was reported.[179] The wafers are the anode of a cell and a bias voltage is applied (Fig. 13-6-1). The cell current rises rapidly to a peak as the electrolyte is polarized and then settles to a steady-state value of I_0. The endpoint of etching is a fixed percentage drop of the current.

The activation energy[156] for Al etching in H_3PO_4/HNO_3 is 13.2 kcal/mole suggesting that the rate-controlled process is the H_3PO_4 dissolution of the Al_2O_3. The gas evolved is a mixture of H_2, NO, and NO_2. On the Al surface, the adsorption of the gases is a persistent problem with high-viscosity etchants. Bubbles can inhibit the etching by bridging small-pitch gaps, resulting in shorts. Various techniques have been tried to reduce or displace the bubbles produced in Al etching:

1. Interrupt etching
2. Ultrasonic agitation[157]
3. Addition of surfactant[158, 159]
4. Oxidation of AZ-1350 by $KMnO_4$[160] to change contact angle of resist from 80° to 30°
5. Apply vacuum over etch tank

Adventitiously, the adsorption of bubbles has been used to bevel images of a nickel–iron film[161] etched in HNO_3 (Fig. 13-6-2). As the etching process commences, the nitrogen oxide bubbles collect along the liateral edge. The etching accelerates in the lateral direction since the adsorbed intermediate, NO_2, acts as a rapid oxidant for metal etching. The adsorption of gases on the etching sidewall (Fig. 13-6-2) has also been used to reduce the lateral undercut in Al etched in H_3PO_4.[740] An evacuated etch bath at a pressure of 9 Torr was used to reduce the

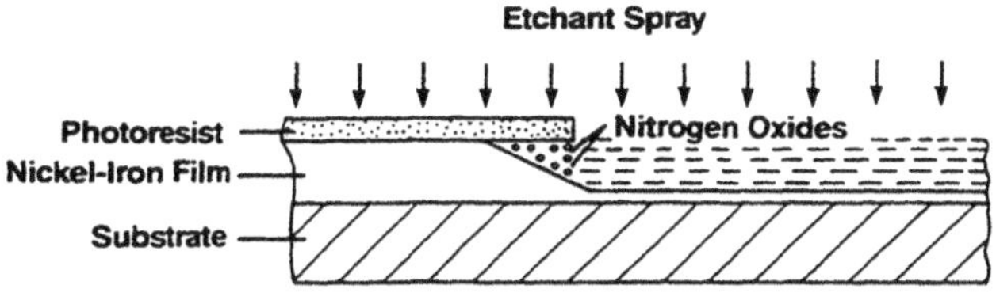

FIGURE 13-6-2. Formation of bubbles during the course of wet etching of a film of nickel–iron alloy. The preferential adsorption of the gaseous product on the sidewall limits lateral undercutting. U.S. Patent 3,808,070.

TABLE 13-6-3. Aluminum Etchants

Etchant	Resist[a]	Application	Reference
1. Na_3PO_4, Na_2CO_3, K_3FeCN_6	DQN, AN	Reduce undercut to 0.5 μm	162
2. K_3FeCN_6	AN	Minimum bubbles	163
3. H_3PO_4, $HClO_4$, H_2O, surfactant	DQN, AN	Less undercut	164
4. H_3PO_4, HNO_3, surfactant, sucrose	AN	0.25 μm undercut	165
5. HCl	AN	All galvanic metals, charge Al, spray, eliminate undercut	166
6. HCl, HNO_3, $Cu(NO_3)_2$		Al/Cu eliminate nonuniform etch	167
7. Alkaline, isopropanol		General etch	168
8. DQN developer		Al attack at some concentration	

[a] AN, azide cyclized rubber, KTFR type; DQN, diazoquinone–novolak, AZ-1350 type.

undercut from 0.8 μm at 1 atm to 0.4 μm. An effective diffusion barrier was formed by the adsorption of small hydrogen bubbles on the sidewall.

Several etchants (Table 13-6-3) consisting of additives of sucrose (polyalcohol) and surfactant have been claimed to reduce undercut from 1.0 μm to 0.25 μm in Al etch! Potassium ferricyanide etchant does not produce significant bubbles. If these systems are combined with spray or monitored and controlled dip etching, the undercut factor can be as good as that achieved by reactive ion etching or lift-off.

The uniformity of Al-Cu etching depends on controlling the size of grain boundaries and $CuAl_2$ precipitates (Fig. 13-6-3).[187] The metal deposition conditions determine the size of the grains.[731] Excessive formation of $CuAl_2$ creates regions or occlusions which etch slowly and leave residues. By overetching, the residues are cleaned up at the expense of further undercut etch biases.

The adhesion of resists to Al has not been critically studied. Aluminum surface tension is reported[45] to be in the range of 45–60 dynes/cm with an X^P of

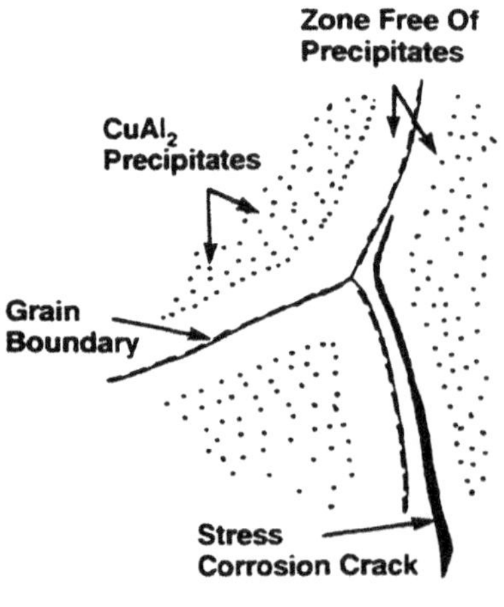

FIGURE 13-6-3. Stress etch cracks in Al–4% Cu alloy. From Ref. 187.

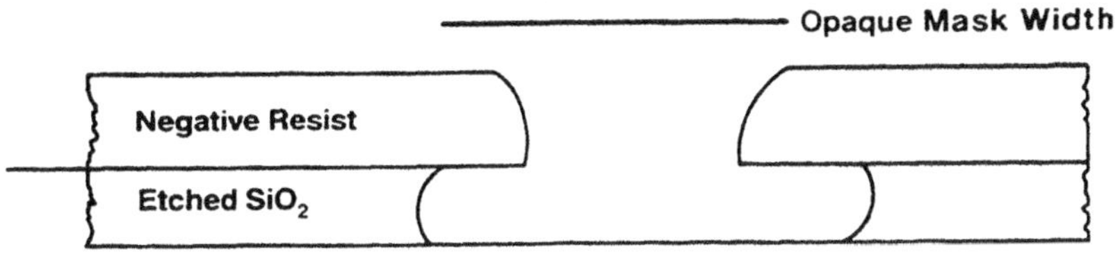

FIGURE 13-6-4. Compensation of the lateral undercutting with a mask of a negative resist. The unexposed opening in the resist is undersized from the exposure step.

38 dynes/cm and X^d of 10–20 dynes/cm. Even though the surface is polar, both positive and negative resists (AZ-1350 and KTFR) of 30–35 dynes/cm appear to adhere adequately. Although negative azide photoresists were popular Al etch masks, they were difficult to strip without the use of hot acids. The acids also attacked the aluminum. Positive DQ resists made their first impact as replacements for negative resists since they were strippable in mild reagents which did not harm the Al. With the advent of hydrocarbon sulfonic acid strippers (Chapter 14), negative resists could be reinstated. The pickup of water in the stripper bath should be avoided or the sulfonic acid strippers will ionize and corrode Al. In wet etching (Fig. 13-6-4), negative resists self-compensate the etch bias since the image is the reversal (smaller than the opening) of the mask and the wet etching enlarges the geometry to the desired size.

Pretreatment of Al with Cr_2O_3 etchant(169) is claimed to improve the adhesion of the resist to the substrate. Aluminum stearate was also added to KTFR to improve its adhesion to Al.(170) The contact angle of the etched sample decreases with the extent of etching.(171)

Since the undercutting of Al is on the order of 1–2 μm, several attempts have been made to invert an isotropic etching into a positive anisotropic etch (Fig. 13-1-5). The interrupt–resist collapse method(172) has been tried to displace the isotropic undercut (Fig. 13-6-5). In this technique, the etching is stopped about one-quarter of the depth of the thickness of the metal. Next, the resist is heated to flow the cantilever onto the sidewalls. The etching proceeds with a new locus for undercut. The process is similar to an earlier version(173) of a process for deep etching germanium. A gold layer is undercut during etching (Fig. 13-6-6) and the gold foil collapses and banks the sidewall to limit lateral isotropy.

The beveling of aluminum is important for second-level and subsequent metal connections. The quality of the deposited quartz over the metal pattern is determined by the quartz conformality(4) to the metal slope. Two approaches have been studied. The first is the controlled photoresist(4, 26, 41) undercut which was found to be an irreproducible process. The second is the fast (top layer) etching layer. For example, anodic oxide (Al_2O_3) is formed in a thin layer of 700–2500 Å by electrochemical oxidation in oxalic acid. The anodic aluminum dissolves laterally under the photoresist.(174) The formation of metal oxides can be accelerated by exposure to an oxygen plasma. Metals such as Si, Al, Cr, and Pt can be oxidized to a state which inhibits etching. A photoresist mask is used in conjunction with a plasma to selectively oxidize the metal. Next, the resist is stripped and the metal is etched. Platinum was etched in this manner with aqua regia.(556) Anodic aluminum oxide is also used to planarize first level metallurgy for further metal contacts.(175) Metal cross-overs of intersecting aluminum lands are converted to oxides and then selectively etched to form passivating layers.(176–178)

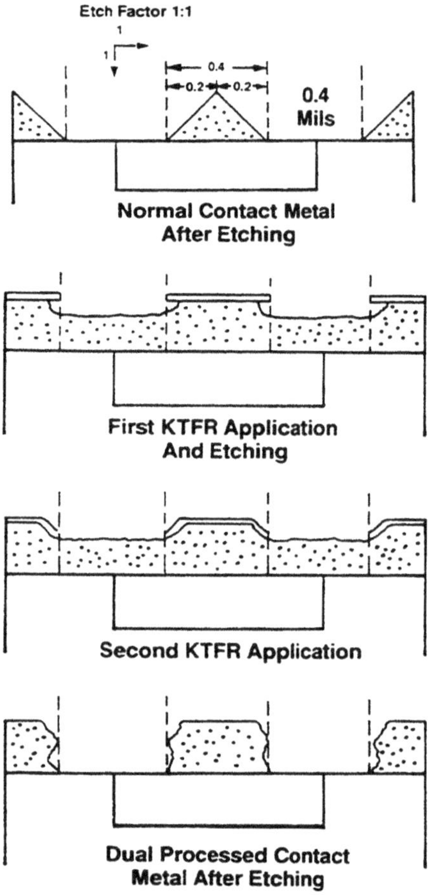

FIGURE 13-6-5. Sidewall banking agent created by interrupting wet etching halfway down, baking resist to collapse onto sidewall, and continuing etching to obtain pseudoanisotropic features. From Ref. 172.

The monitoring of Al etching has been attempted in order to reduce the large tolerances ($\pm 2\ \mu m$) on batch etching. The ideal monitor is the electrochemical current (electrolytic etcher) which was an early commercial venture (Fig. 13-6-1).[179]

Inconsistent Al etching can be attributed to several factors:

1. Underdeveloped resist
2. Thickness variations
3. Stress in films over steps[4]
4. Galvanic acceleration due to Al-Cu precipitates[151]
5. Variable oxide thickness
6. Poor temperature control ($> \pm 1°C$)

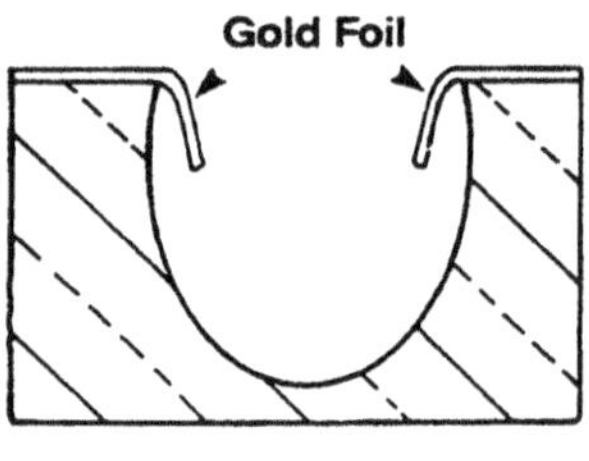

FIGURE 13-6-6. Collapse of gold foil mask during the course of wet etching to act as a sidewall protective agent in metal foil etching. U.S. Patent 3,544,401.

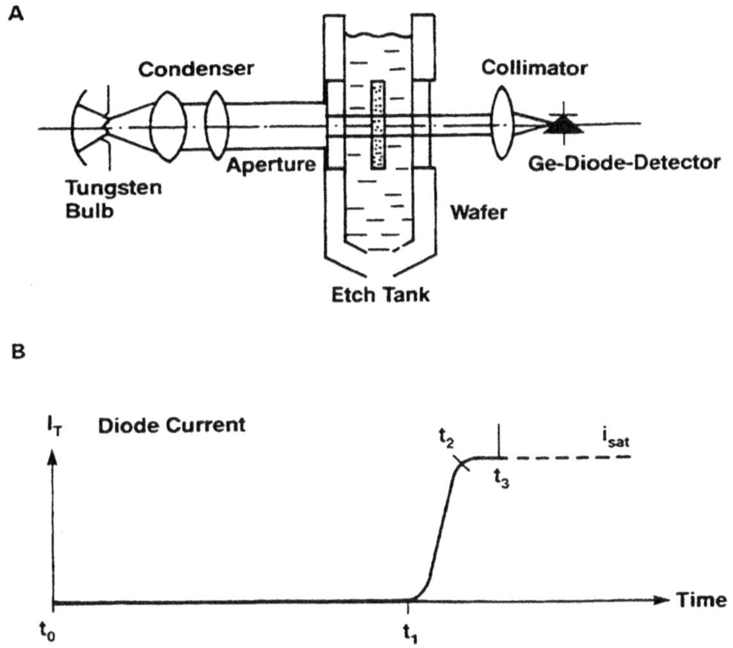

FIGURE 13-6-7. (A) Etch endpoint detect setup for aluminum wet etching monitored by IR ray passing through Al film. (B) Endpoint detect trace; t_0 is start of etching, t_1 is where the first areas of Al break through, t_2 is at the last area of Al removal, and t_3 is when the acid is drained. From Ref. 180, reprinted with permission of *Solid State Technology*, published by Technical Publishing, a company of Dun and Bradstreet.

These factors result in overetching and short circuits. Moritz[180] described an aluminum etch monitor based on the infrared ray system of transmission monitoring (Fig. 13-6-7). Although tedious, monitoring each wafer reduced the etch bias and tolerance from $2 \pm 1\ \mu m$ to $1 \pm 0.3\ \mu m$ (3σ) at an excellent throughput of 25 wafers/hr for 1-μm-thick films. Excursions due to temperature and compositional changes are automatically compensated by the process monitor.

Other metals such as Cr, W, Mo, Ti, and Au are important to mask (UV and X-ray) making, metallurgy, and as evaporation or ion-beam masks. The etchants for these systems have been tabulated by Kern and Deckert.[110] The etchants are usually composed of an acidic solution of a strong oxidant such as the cerate ion. Au requires complexing agents of cyanide ion.

Chromium is the second most etched metal. Cr is used extensively in the fabrication of optical exposure masks. Frieser[193] has studied the morphology of thin chromium films used on glass for mask making. Ceric sulfate/HNO_3 was used as the etchant and the etching rate was observed by ultraviolet spectroscopy of the thinning film. In Fig. 13-6-8, a generalized etch curve is shown depicting the influence of the inhibiting (induction effect) surface oxide on the total rate of etching. The oxide layer varies from sample to sample. Because of the induction effect, the film etches in a nonlinear fashion and the initial film thicknesses cannot be used to estimate the endpoint of etching. An *in situ* etch monitor corrects for the variable induction due to the topside Cr_2O_3 layer.

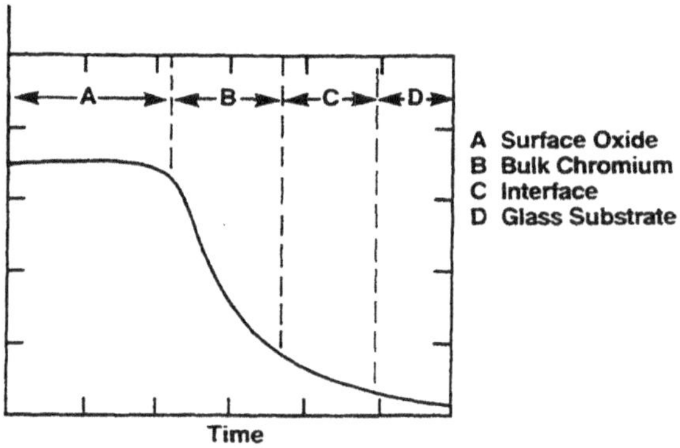

FIGURE 13-6-8. Generalized etch curve showing inhibition of chrome etching by layer of Cr_2O_3. From Ref. 192, reproduced by permission of the Electrochemical Society Inc.

13-6-2 *Electroetching*

Instead of using equilibrium-driven etching, applying a potential to a resist-patterned metal can delineate patterns in milder etchants. Platinum,[181] for example, is usually etched in hot aqua regia (HCl/HNO_3), which strips most resists. Applying a potential of 1.0 V allows platinum to be etched with dilute HCl. The electroetching process is desirable because it offers distinct advantages in sensing completion (Fig. 13-6-1) and can be easily automated with electroplating equipment. Contact resistance has to be low (< 1 ohm-cm) to assure proper current flow to the wafer. Anisotropic images in tungsten[195] were etched in 2.5-μm-thick films by electroetching. Submicrometer images have not been made due to the high resistance or due to isolated images ("donut-in-hole problem"), i.e., loss of electrical contact. In this case, a solution to this problem is to use an etchant capable of acting as electroetchant and a chemical etchant (for the last 1000 Å of film). Electroetching is analogous to reactive (plasma) ion-beam etching since the metal ions are directionally removed. More work is required for this promising process.

13-6-3. *Summary*

The wet etching of aluminum remains stagnated at 3-μm linewidth and 8-μm pitch. Process monitoring is definitely needed to keep the batch tolerance under control. Some etchant formulations claim to hold the etch bias to 0.25 μm. A consistent and uniform film of Al is required for reproducible lithographic etching. Alone or in combination with a chemical etchant, electroetching can be readily monitored and has good potential for anisotropic etching. Dry processes (lift-off and reactive ion etch) loom as strong replacements for the wet etching process. More critical studies of the kinetics and processes such as spray etching are needed to extend wet etching into the 2-μm range for the remainder of the 1980s and reliable etch monitors are needed to reduce the etch biases.

13-7. PRACTICAL ASPECTS OF WET ETCHING

The practical aspects of wet etching are concerned with the static and dynamic factors and the end results:

Static	Dynamic	Results
Film uniformity	Spray	Undercut
Film composition	Agitation	Pinholes
Etchant molarity	Rate	Shorts, opens
Etchant composition	Resist erosion	Bias
Resist image size	Adhesion failure	Tolerance
Temperature	Endpoint	Selectivity
Static bath	Exhaust etchant	Beveling

For practical processing, a kinetic monitor of thickness versus time (Figs. 13-6-1, 13-6-4, and 13-6-8) is a basic requirement for reproducible etching and developing (positive novolak resists). In wet etching, the three major process variations are the thickness of the film to be etched, the control of the temperature, and the time to endpoint. Agitation of the liquid etch bath is not important in rate-controlled processes [except in dry etching, where the flow rate is important since the processes are diffusion controlled (±4 kcal/mole)]. Most of liquid etching (HF) is rate controlled. Typical variations in these variables can result in overetching. Table 13-7-1 is a computed estimate of the overetching for etching an oxide film of 5400-Å thickness. If a thickness varies ±10%, for example, within a batch, the endpoint is set to handle the thickest film (+10% thicker). This results in overetching of the thinnest film (−10% thinner). Underetching or poor cleanout is fatal not from a dimensional override but from the inhibition of a subsequent step such as diffusion or high contact (metal) resistance. The thicker the film to be etched, the worse is the undercut bias and process tolerance. In the results of overetching in Table 13-7-1, the poor temperature control is the largest contributor

TABLE 13-7-1. Estimate of Etch Bias in Isotropic Etch

		Thickness of film (μm)			
Variable (V)	ΔV	Overetch[a] of: 0.1	0.5	1.0	2.0
		Etch bias (μm)			
Time					
60 sec	±20	0.25	1.0	2.0	4.0
180 sec	±15	0.16	0.8	1.7	3.2
300 sec	±10	0.15	0.7	1.5	3.0
600 sec	±10	0.15	0.7	1.5	3.0
Thickness					
0.1 μm	±100 Å	0.15			
0.5 μm	±400 Å		0.8		
1.0 μm	±0.1 μm			1.7	
2.0 μm	±0.1 μm				3.2
Temperature					
Diffusion control[b]	±1 °C	0.18	0.8	1.8	3.6
Rate	±1 °C	0.21	1.0	2.1	4.1

[a] In overetch, we assume that the time is derived for the thickest film, e.g., 5400 Å with an undercut of 1.5 times the thickness of the film (see Table 13-1-4).
[b] +5 kcal/mole, 1 °C change is 8% in R rate; +10 kcal/mole, 1 °C change is 20% in R rate.

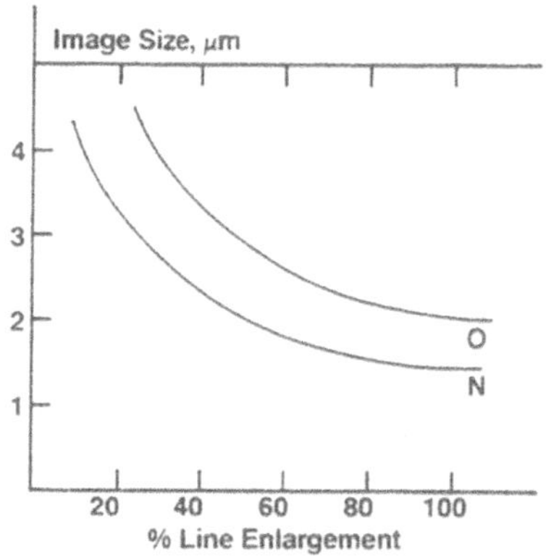

FIGURE 13-7-1. Isotropic wet etching of a 1.0-μm-thick film. N, normal etched profile; O, typical etched profile in batch etching (overetched).

to etch image bias. The bias is independent of nominal image size except for square lands etched on all four sides. Temperature control should be better than ±0.5°C or less, if necessary, for images of 2.0-μm feature size.

The effect of the natural undercut and "overetching" can be seen in Figs. 13-1-10 and 13-7-1, on the line size-enlargement in a wet etching process. Most semiconductor processes attempt to keep the etch distortion under 30%.[14] As the nominal image size shrinks, tighter process controls (Table 13-7-1), single wafer-endpoint processing, and anisotropic etching will be essential tool and process *modus operandi.*

An experimental indicator of undercut is the etching of various size dots. Both adhesion and undercutting are tested for the smallest features that will stand the "dot test." The undercut has to be corrected for the undercut of opposite sides. If the 1-μm dots disappear, the undercut is at least 0.5 μm (Fig. 13-7-2).

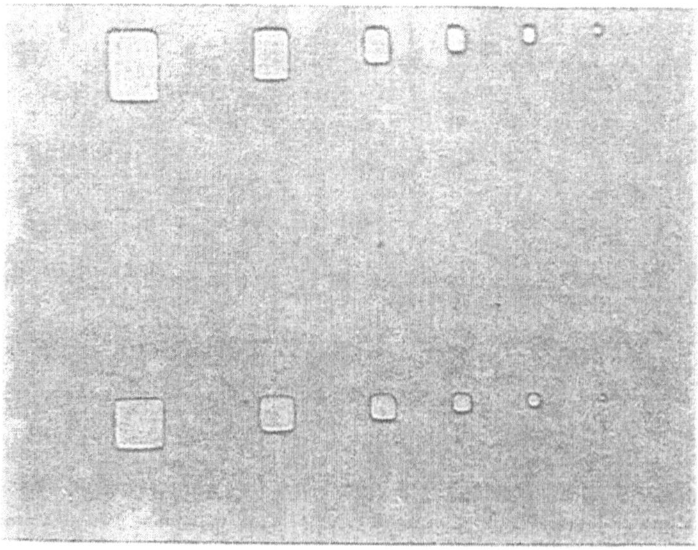

FIGURE 13-7-2. Etching squares (0.5–500 μm on a side) as a dot test of etch resistance due to adhesion failure and undercutting of the resist image.

Pinholes (Chapters 6 and 15) are yield losses, dependent on the circuit susceptibility to the size, location, and population of defects. The etch process transfers the resist pinhole into the device pattern. If the peak size of pinholes is on the order of the etch bias (0.4–1.0 μm), then a high potential exists for opens in closely pitched arrays. The etching process itself should be tested to reveal the limit, if any, to the size of pinholes which will be transferred. In liquid etching, surface tension and capillary action may obstruct the etching of submicrometer defects. Dry etching may also be inhibited, due to electrical charge buildup on the resist surface and deflection of the positive etchant ion.

13-8. TROUBLESHOOTING ETCHING

Troubleshooting is a daily test of one's experiences and intuition in lithographic processing. The cause and effect may not be a single predictable factor related to a tool, process, or materials defect, but a series of random failures of several steps. Hopefully, the problem is not confined to a small area of chip ($\leqslant 10\ \mu m^2$) but is distributed throughout the wafer batch. Figure 13-8-1 illustrates a localized defect of a pattern, in which three out of four diffusion holes are cleanly etched. One of the holes is only partially etched, due to blocking (scum in the developed resist image), or some artifact during etching. Several other etch defects are shown in Fig. 13-8-2, which shows opens, ragged etching, and severe undercutting. In diagnosing process defects, the task is to define the problem, in terms of the location and frequency of the defect, and hopefully correct the tool, process, or materials problem. Is the defect due to etching or some previous step? A troubleshooting list of symptoms is given in Table 13-8-1. Some of these will also apply to dry etch.

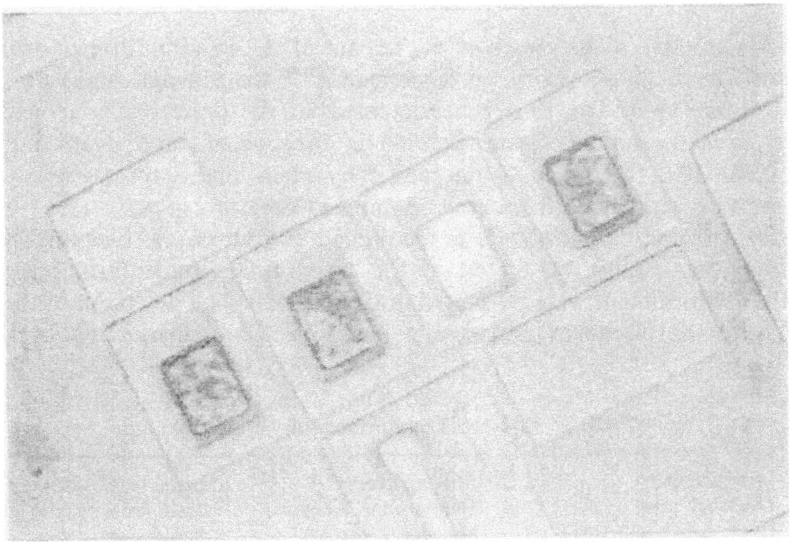

FIGURE 13-8-1. Localized defects in wet etching of oxide of Si. Three out of four holes did not each properly due to random residues.

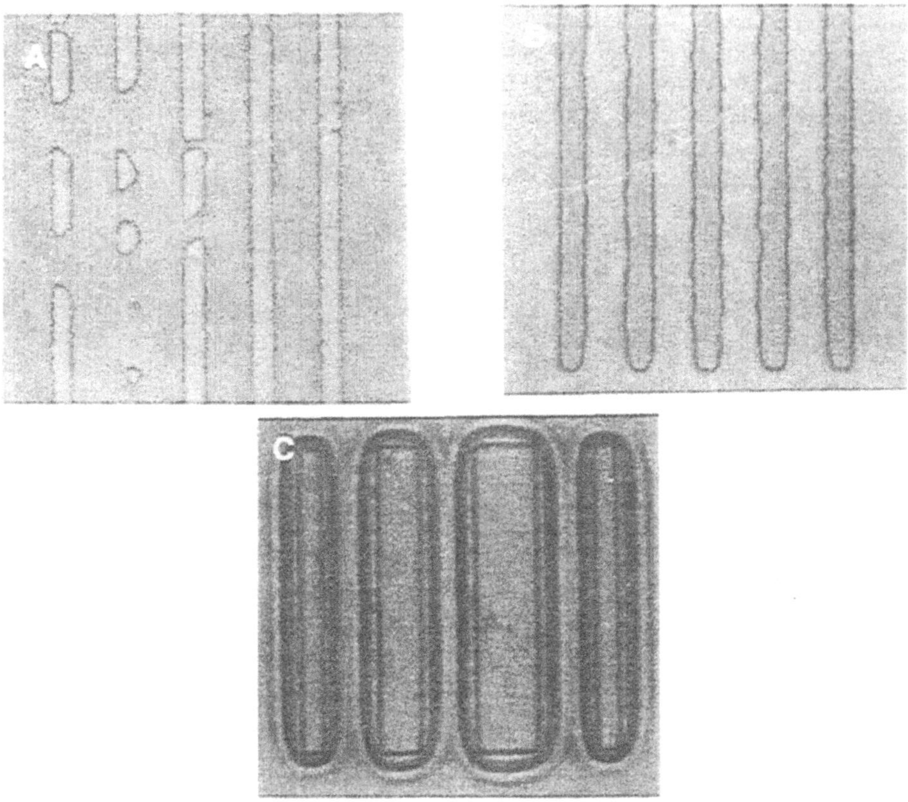

FIGURE 13-8-2. Defects in wet etching of SiO_2. (A) Opens in lines; (B) ragged edges due to localized undercutting; (C) severe undercutting.

It is instructive to be aware of the impact of the random defects on reliability and yield. Figure 13-8-3 shows an assessment,[184] from a mask-making viewpoint (but applicable to devices in a general sense), of the defects of pinholes, pindots, bridging, breaks, and nicks, encountered in the contact and projection printing modes. Generally, these defects are caused by mask defects or are resist induced, and have to be segregated from etching-induced defects.

Systematic etch failures, such as incomplete inhibition or "blocking" (complete inhibition), can usually be traced to the resist image-underdevelopment. Resist image and development were discussed in Chapter 10. The detection of the invisible film left after development underlies a basic need for physical tools capable of

TABLE 13-8-1. Troubleshooting Wet Etching

Does not etch	Missing images	Longer etch time
Unetched areas	Halo around images	Bath discoloration
Large patterns etch only	Notches in pattern	Bubbles on surface
Deposits on surface	Tapered images	Particles in bath
Resist lifting	Skewed images	Pinholes and pindots

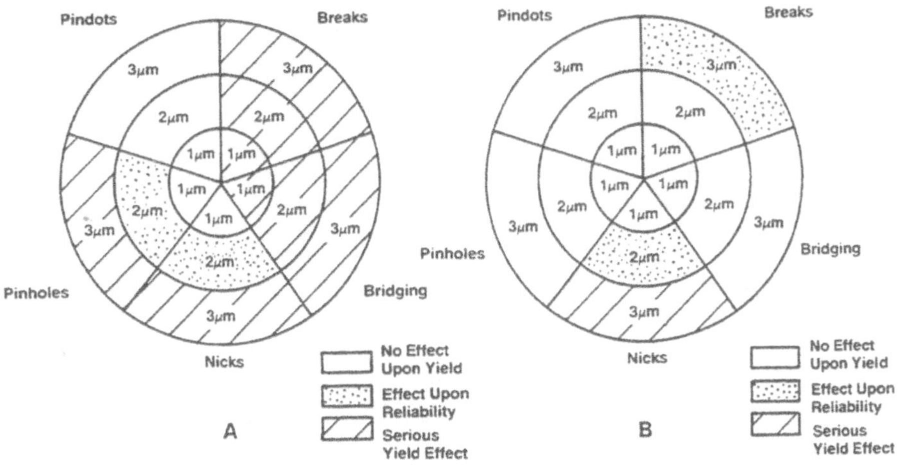

FIGURE 13-8-3. The effect of printed mask defects on devices of (A) 3-μm in size and (B) 1.5 μm in size. From Ref. 184, copyright 1980 and courtesy of Cahners Publishing Co.

detecting resist or etch residues in areas of 25 μm^2. Some of the potential causes and remedies for poor or inhibited liquid etching are listed in Table 13-8-2.

Many types of chemical, physical, and engineering analysis must be brought to bear upon liquid etching. Since it is difficult to monitor the products, the reactants are monitored for the correct composition and etch rate. Hopefully, no extraneous impurities are introduced from the wafer, boat, or materials of the etch tank. For example, nitric acid vapors can contaminate the HF etch bath.[194]

13-8-1. *Spray versus Static Etching*

If a process is rate controlled, spray or agitation should have no effect on the process. Many artisans maintain that spray etching or developing ensures that the reaction products are transported into the bulk of the liquid. If necessary, spraying

TABLE 13-8-2. Defective Wet Etching

Problem	Remedy
Thicker film	Measure, specify, monitor
Impure film	Compositional assay and monitor
Wrong etch	Compositional assay and monitor
Exhausted etch	Assay, batch life, renew
Moisture absorption	Closed system, assay
Evaporative loss	Closed system, assay
Nonuniform spray	Clogged nozzle, smaller orifices
Etchant dewets	Add surfactant
Temperature drop	Bath thermostat
Invisible film	Surface primer etch off
Invisible film	Correct, bake, expose, develop
Wafers touching	Boat, orientation
Deposits residues	Filter etchant

TABLE 13-8-3. Advantages and Disadvantages of Spray Etching

Advantages	Disadvantages
Cleanout of submicrometer geometries	Nozzle clogging
Displace adsorbed bubbles	Separation of etch comp.
Self-contained	Nonuniform wetting
Potential control of anisotropy	Rapid etch, overetch
Applicable to rinsing, developing	Poor endpoint control
Temperature controlled	Low throughput

or uniform agitation may be useful for diffusion-controlled reactions. Although no decisive studies of the advantages, principles, and disadvantages of spray etching have appeared in the literature of semiconductor processing, some advantages have been noted by equipment technologists.[185] Spray etching has been used successfully for printed circuits and printing plate. Some of the advantages and disadvantages are cited in Table 13-8-3. For wafer processing, most wet etch tools use a spray, with a wafer spinning during etching. The endpoint of spray etching of chrome photomasks is detected by transmitted light.[668]

If the etchants can maintain a charge, such as in electrostatic spray painting, then spray etching of metal films (with grounded wafer) may offer some anisotropic features. Some potential control of anisotropy has been claimed by adjustment of the angle at which the spray impinges on the wafer.[185] More basic chemical engineering is needed to foster spray etching, as a bridge between isotropic static (liquid) etching and anisotropic dry etching. Spray development will also benefit from work in spray etching.

13-8-2. *Other Etching Considerations*

Agitation is important in maintaining uniform etch bath temperatures. During etching, constant ultrafiltration removes resist particles and other solids, which may block etching. The etch tank should be closed and separated from all other stations to prevent contamination. To avoid leaks or contamination, etch compatible tank and equipment components have to be carefully chosen. Most semiconductor etchants are sensitive to trace metals and anions, such as Cl^-, NO_3^-, and PO_4^{3-}, which act as powerful complexing agents. Regardless of the physical state of matter, contamination can be transported as a gas (HCl, HNO_3), liquid, or solid. Extraneous heat or traces of surfactants (mold release agents from plastic parts) can also alter the etch rates. An *in situ*, etch rate monitor[25] (Fig. 13-8-4) is the best indicator of process changes. For image endpoint, the simultaneous etching of diffraction gratings, or features smaller than the desired pattern, is a good indicator of overetching. The batch life of an etchant can be estimated from the stoichiometry of the chemical reaction. In SiO_2 etching, 1 mole of SiO_2 requires 6 moles of HF. If we assume a 25% etch area of a 100-mm wafer in 7:1 buffered 2 M HF (see Fig. 13-2-5) and an 8-liter tank or 16 moles of HF,

$$SiO_2 + 6HF \rightarrow H_2SiF_6 + 2H_2O$$

Assuming a drop in etch rate of 20% is the limit, this drop corresponds to a

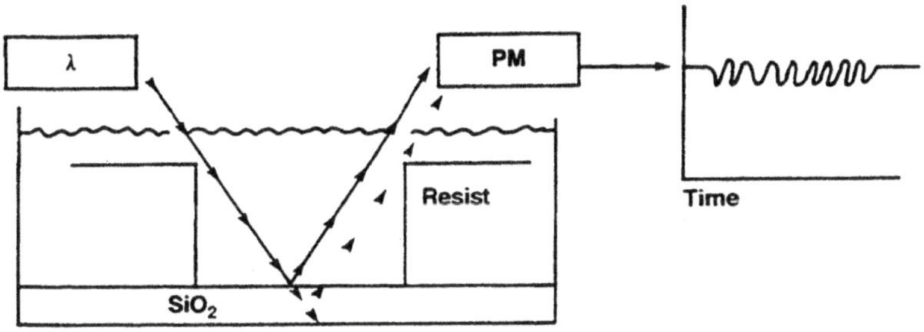

FIGURE 13-8-4. Setup of an etch endpoint detect. A laser beam is reflected off the etching area into a photomultiplier (PM). Etching ceases when the interference pattern disappears.

20% decrease (Fig. 13-2-5) in "HF" concentration. For a 0.5-μm-thick SiO_2 (t) of density D, the moles of SiO_2 consumed,

$$\text{Moles SiO}_2 = \pi r^2 tD/\text{SiO}_2 = 3.14(25\ \text{cm}^2)5 \times 10^{-5}\ \text{cm}(2.3\ \text{g/cm}^3)/60$$

Since 4.8×10^{-4} moles of HF is consumed per wafer, a drop of 20% or 3.2 moles of HF, will sustain up to 25,000 wafers. If 500 wafers/day are processed, an etch bath life of 50 days is possible (if there are no material losses or contamination).

13-9. MODELING OF WET ETCHING

Two-dimensional modeling of wet chemical and dry etching has been pioneered by Neureuther[188] and Viswanathan,[189] who adopted the work of Dill on resist developing.[190] The etching is a function of $R(x, y, C)$, where x and y are directions and C are the usual process conditions, such as temperature and concentration. The etching proceeds with a cross section, as shown in Fig. 13-9-1.

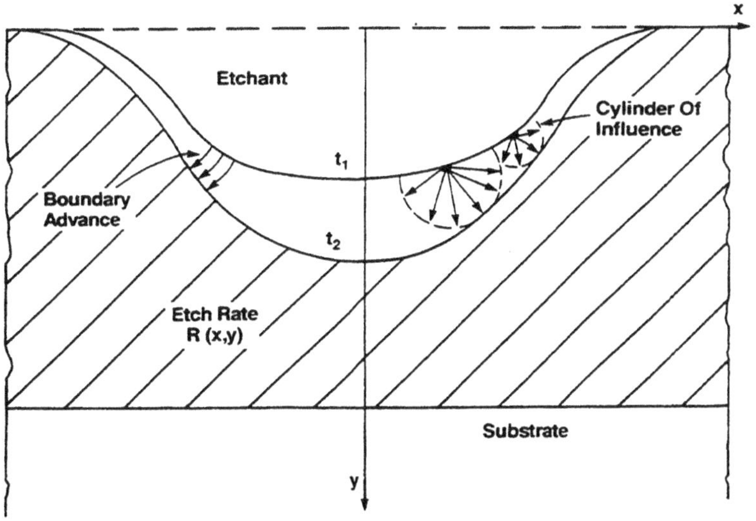

FIGURE 13-9-1. Computer modeling of isotropic wet etching. From Ref. 191, courtesy of Dr. A. Neureuther, University of California, Berkeley.

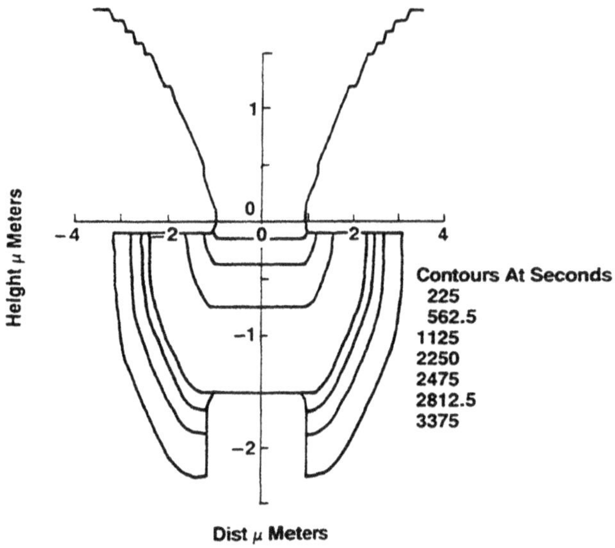

FIGURE 13-9-2. Modeling of the depth of isotropic etching with a non-erodable mask at an etch rate of 40 nm/min. From Ref. 189, courtesy of AIP.

Exact mathematical models[(636)] for diffusion-controlled etching involve equations from heat transfer (Stefan) of continual displacement of a boundary in time. Since the diffusion field is faster than the etch rate, closed-form equations predict a bowl-shaped image for isotropic etching. For submicrometer images, special consideration may be necessary for concentration gradients building up in small orifices.

Since chemical kinetics cannot be readily measured in some etching processes, the phenomenological simulations are invaluable aids in predicting the limits of microminiaturization, due to etching losses (Fig. 13-1-9). Compensatory changes in resist image size, due to mask windage changes, can be predicted. The severest tests and greatest needs for etch modeling will involve the corrections for the proximity effects in *e*-beam exposure.

The linear etch rate is determined by a measurement of thickness versus time. Various computer methods[(603)] of cell removal, string and ray tooling, are used as

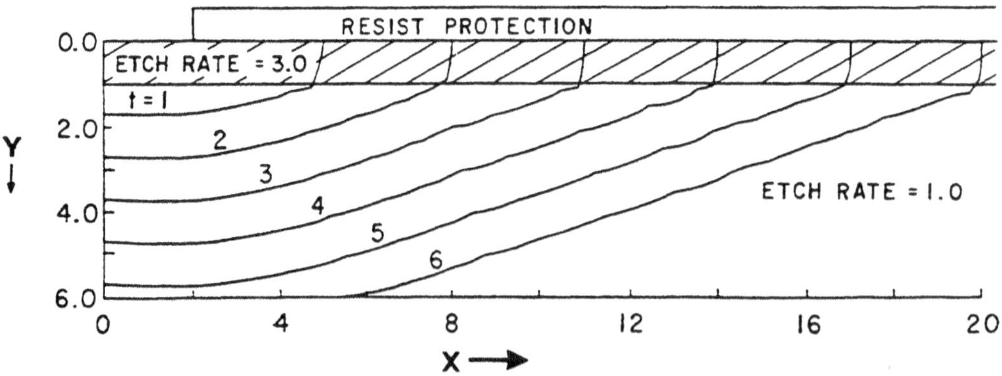

FIGURE 13-9-3. Modeling of the etching of a fast-etching intermediate layer underneath the resist (phosphosilicate glass) to produce a beveled SiO_2. From Ref. 191, courtesy of Dr. A. Neureuther, University of California, Berkeley.

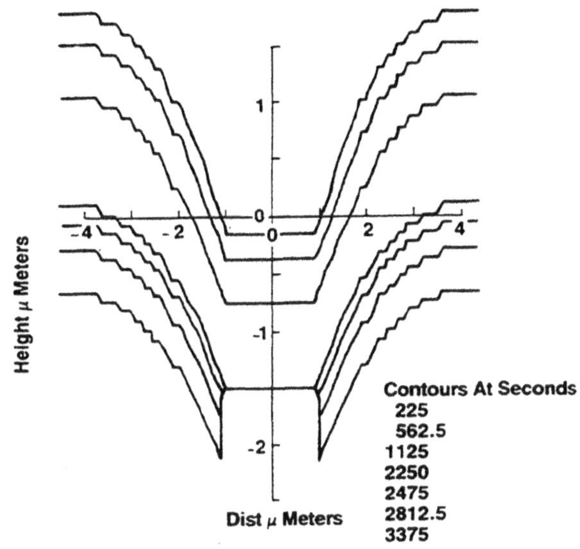

FIGURE 13-9-4. Computer modeling of directional etching of SiO_2 in a plasma of CF_4. The resist erosion rate is 50 nm/min and the oxide etch rate is 40 nm/min. From Ref. 189, courtesy of AIP.

miniature etching cells to track the progress of undercut. A simulation of modeling of profiles of lithography (SAMPLE) has been used to predict profiles, after wet and dry etching, in developed images exposed by *e*-beam and optics, proximity effects, and after metal evaporation.[189, 190] Examples of modeling[604] are the via-hole profiles for various depths of etching[190] (Fig. 13-9-2) and the beveling of CVD oxide, using PSG glass as a fast-etching sublayer[191] (Fig. 13-9-3).

In Fig. 13-9-4, the biases of an etched image can be seen. The image at the bottom surface does not widen as much as does the top of the image. Comparison of the profile of the actual etched image (2 μm deep) with the simulated image, can indicate such factors as overetching, changes in etch rate, or interfacial resist (lifting during etching) adhesion failure. The simulated profiles in sandwich layers of metal can indicate if accelerated etching occurred due to galvanic action. The modeling of plasma etching, to simulate wet etching of via holes, has proven invaluable to process development wherein the model predicted the correct erosion rate needed for the resist[190] (Fig. 13-9-4). Thus, the plasma barrel etching could obtain the same sloping tapers as formed in wet etching.

13-10. OUTLOOK OF WET ETCHING

Etching is a critical step in the lithographic process. As far as the resist performance is concerned, the etching step severely tests the functional adhesion, the impermeability, the defect level, and the chemical inertness of a resist. In defining a resist process, it may be more important to rank the etch resistance and adhesion of a resist, above all other performance criteria. The selection of a resist based on excellent etch performance will ensure minimum risks in the image transfer from the resist pattern. The practical limit of wet etching is 1.5- to 2.0-μm width and etch biases of ± 0.2–0.5 μm.

Although the etching of thermal oxides and other insulators has received detailed attention, the etching of metals has been inadequately studied. Another area of continued interest and investigation is the adhesion of the resist during etching. For example, the determination of the kinetic factors of adhesion failure during etching is more important than the thermodynamic assumptions for equilibrium conditions. In addition, primers and adhesive treatments are vital aids to resist technology in the wet etching step.

The process control of etching requires a knowledge of the static and dynamic variables. In wet etching, the control of the thickness, time, and temperature are the most important process variables. For process control, an etch endpoint detect system is a vital step toward reproducible etching, as well as a signal for variations in the chemical etch composition. If single wafers could be controlled by endpoint monitors, the etch biases could be held within $\pm 0.25\ \mu m$ for a 1.0-μm-thick film.

The simplicity of wet etching has enabled the process to survive the test of time. However, the trends are to use more dry etching of silicon nitride and polysilicon in devices, and wet etching for thin thermal oxides. The densification of circuits results mainly from a decrease in line pitch. For the wet etching of metals, the interproximity effects of undercutting and overetching will limit the pitch to about 6 μm. Except for beveled profile etching, wet etching with its high disposal costs of materials, will be of lesser importance in future processing. Electrochemical etching shows good potential for anisotropic imaged profiles. Finally, better endpoint detectors, operational at the micrometer linewidth, are vitally needed to engineer wet processing to compete for the 1- to 2-μm linewidth challenges of VLSI circuitry.

13-11. INTRODUCTION TO GAS ETCHING

In gas-phase etching, activated fluorine- or chlorine-based etchants are created by an electrical discharge. It has long been recognized that reactions which proceed at high temperatures ($>300°C$), such as the etching of silicon with chlorine gas, will proceed at room temperature if the etchant is an activated atom, radical, or ion. In the early 1960s, the ion-beam etching or sputter etching of semiconductor films was attempted with argon ions excited by an RF field. If ions of greater than 50 eV strike a surface, momentum transfer (sputtering), bond breakage, and lattice damage can occur. The damaged atoms or molecules are very susceptible to attack by free radicals or ions. Unfortunately, organic resists eroded at more rapid rates and no mask of resist survived the sputtering action of the beam. Secondly, the etch rates of silicon, for example, were only 200 Å/min, and considerable heating of the substrate occurred. What was needed was an etchant with a faster etch rate of silicon and without significant erosion of the organic resist. Traditionally, the photoexcitation of gaseous molecules is used to create "hot reactants" but similar species can be created by microwave (RF) plasma excitation. In this manner, the stripping of resist using excited O_2 was accomplished.

Plasma etching is said to be the chemistry of the light bulb. At low pressures, the electrical discharge in a gas creates an ionized gas of positive cations, negative anions, radicals, and vibrationally excited polyatomic molecules. The plasma is formed by internal or external RF excitation of fluorine or chlorine etchant precursors (Fig. 13-11-1) and the plasma is sustained at low pressure by infrequent recombination of wall collisions.

In excited gas etching, both physical and chemical components are necessary

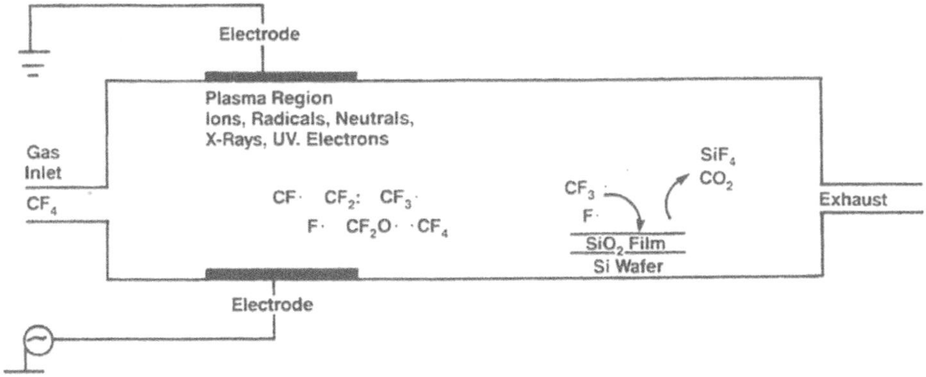

FIGURE 13-11-1. Schematic of simplified reactions occurring in plasma reactor. From *Electronics*, July 21, 1977, p. 89, copyright by McGraw–Hill, reproduced by permission.

to impart anisotropic features.[539] Since the resist is a form of anisotropic masking, uncharged radicals or nondirectional ions in a "barrel" reactor strike the wafer surface and proceed to etch in an isotropic manner similar to liquid-phase etching (Fig. 13-11-2). Inside the barrel etcher (Fig. 13-11-1), the wafer surfaces are at the floating potential of the glow discharge and would be subject to only low-energy ($\leqslant 30$ eV) ion bombardment. In some instances, a metal mesh (etch tunnel) is placed around the wafers to exclude energetic charge particles and allow only radicals to etch the surfaces.[195]

The most popular etchant for films of Si, SiO_2, and Si_3N_4 has been CF_4. For aluminum etching, CCl_4 has been widely used. At least 200 etchants, mixtures, and multistep processes have been investigated.[195–197]

The plasma excitation of CF_4 primarily involves dissociative electron capture,

$$CF_4 + e \rightarrow CF_3^+ + F^* + 2e$$

Neutral radicals such as F^* cannot be directed toward a target except in a magnetic field. Ions (CF_3^+) are easily directed if a capacitor plate configuration is used

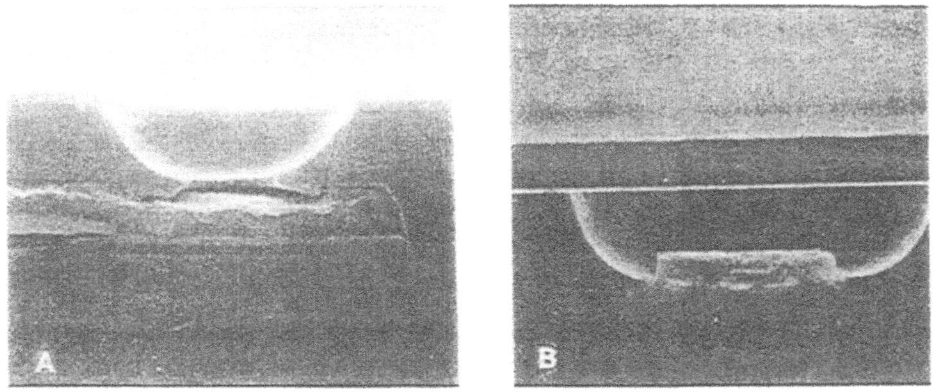

FIGURE 13-11-2. Comparison of etched images in SiO_2 generated (A) by isotropic (barrel plasma etching) and (B) by HF wet etching. From Ref. 189, courtesy of AIP.

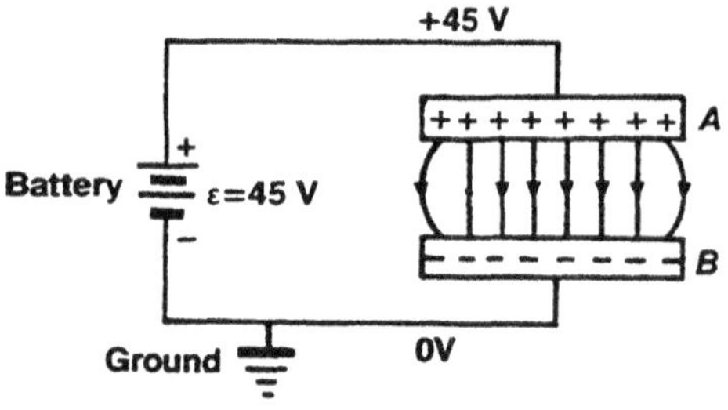

FIGURE 13-11-3. A parallel plate capacitor showing uniform field between plates. From F. Miller, *College Physics*, Harcourt–Brace–Jovanovich, 1982, p. 421, reproduced by permission of Harcourt–Brace–Jovanovich.

(Fig. 13-11-3). In a capacitor plate arrangement, the electrical field lines of force are parallel to the grounded electrode surface. The field serves to collimate the beam such as in an ion microscope or ion projection exposure tool.[198] Thus, a broad beam of ions is created over a several square meter area to bombard the surface with ions of energy > 100 eV and etch a masked surface in a process sometimes known as reactive ion etch, reactive sputter etch, or reactive ion beam etch. Collectively, all three are known as "plasma-assisted etching." If the ions are not chemically reactive, such as the rare gases of argon, only sputter or ion beam damage occurs. In RIE, the ions of fluorocarbons are chemically reactive as well as capable of momentum sputtering. Thus, etch rates in RIE are orders of magnitude greater than sputter etch rates. RIE is a flood exposure process as opposed to ion projection lithography which uses a proximity[198] mask and inert ions to damage a silicon surface (the patterned surface is subsequently taken out of the reactor and wet etched). Ultimately (Chapter 16) a "resistless" process will evolve, combining a reactive beam projected over a whole wafer with a circuit pattern. The pattern is formed directly on the surface by reactive beam etching.

While plasma etching primarily involves free radicals generated at higher pressures (> 10 mm) than ion etching, ion etching can take many different forms depending on the electrode configuration, the power input, and the ion selector:

Sputter etching Ion milling	Physical ions
Reactive sputter etching Reactive ion etching Reactive ion beam etching	Chemically reactive ions, radicals, and physical ions

Physical sputtering refers to the formation of a volatile or sublimed product without change in the chemical structure of the film, while chemical sputtering refers to the formation of a new chemical compound under the influence of gaseous ion bombardment. In Fig. 13-11-4, a schematic illustrating the operating range of ion and neutral gas processes is shown.

A plasma may be described as a conductive "gas" with equal densities of positively and negatively charged particles. However, the plasma shares precious few properties with the gaseous state. A plasma may be created by generating a strong electric field between oppositely charged plates by coupling, through an impedance matching network, the RF power to a gas at very low pressure. Inside, the evacuated chamber exhibits an extremely small, though measurable, conduc-

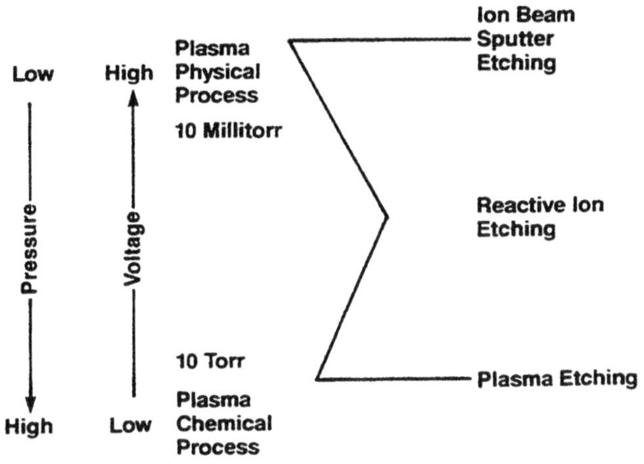

FIGURE 13-11-4. Transition in plasma etching from chemically assisted reactions to physical sputtering by changing voltage applied and pressure of reactor. From C. Mogab and W. Harshberger, *Electronics* August 31, 1978, p. 117, copyright 1978 and courtesy of McGraw–Hill.

tivity due to residual ionization caused by cosmic rays from outer space, γ rays and the like. The electrons arising from this ionization can be accelerated to high enough energies within an electric field so as to strip the electrons from neutral atoms and molecules by collision. At high frequencies, the electrons oscillate rapidly causing randomness through collision and converting most of the energy received into kinetic energy.[12]

Plasma radical etching generally produces isotropic features, unless sidewall[200] recombination or sidewall banking agents (polymers)[199] are present to limit the lateral undercut. The major contribution of plasma etching has been the elimination of the cost of disposal of large quantities of toxic liquid etchants. However, since plasma etching is a nonequilibrium condition with a limited quantity of etchant, more reactor controls are required than in a liquid etch tank.

Various barrel reactors are used for isotropic etching with external cylindrical electrodes which are coupled inductively (A), capacitively (B), or coaxially (C) to a 13.56-MHz (FCC allowed) RF field (Figs. 13-11-5 and 13-11-6). The major advantages of the barrel reactor are simplicity, low cost, good selectivity, and radiation damage-free etching. The disadvantages are the isotropic etching "bull's-eye" clearing of wafers from the outer edge, poor batch reproducibility and uniformity, poor temperature control, residues, and loading susceptibility.

Placing the electrodes inside the discharge in a parallel diode configuration provides ion directionality. The wafers can sit on a cathode-powered or anode-

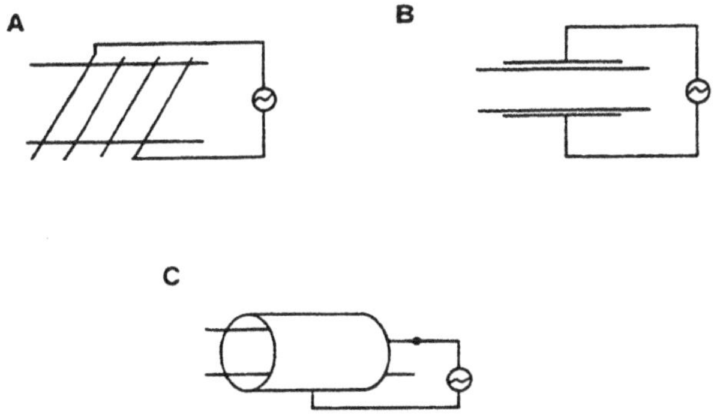

FIGURE 13-11-5. Schematic of a barrel etcher (A) inductively, (B) capacitively, and (C) coaxially coupled to the RF field. U.S. Patent 3,705,091.

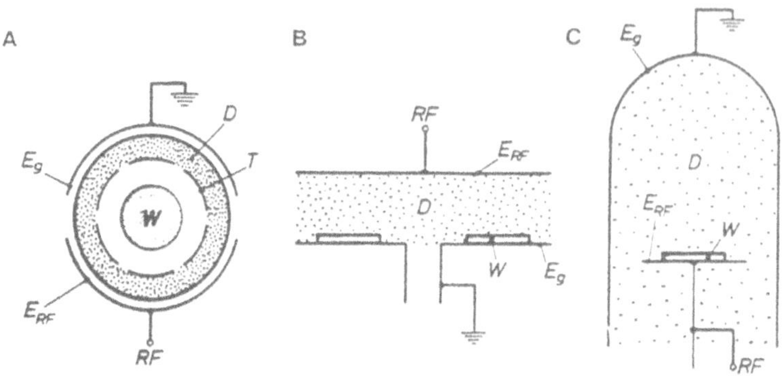

FIGURE 13-11-6. Schematic of the three main types of reactors used for gas-phase plasma etching. (A) Barrel reactor containing a perforated Al tunnel T, outer electrode E_{RF}, and grounded electrode E_g. Since the wafers are not in direct contract with the discharge, the radicals produced by the plasma discharge are primarily responsible for the isotropic etching. (B) Planar reactor with the wafers in the grounded electrode. Since the wafers are in contact with the discharge, both ions and radicals contribute to the semi-isotropic etching. (C) Reactor for reactive ion etching. The wafers are located on the electrode connected to the RF generator. Since ions are mainly produced, the etching is primarily anisotropic. From H. Kalter and E. Van de Ven, *Philips Tech. Rev.* **38**, 200 (1978).

powered electrode (Fig. 13-11-7). Generally, the wafers sit on a grounded plate. In this configuration, the resist and masked wafer are exposed to a broad-beam low-energy (100–400 eV) ion and free radical bombardment. In addition, photons, electrons, and X rays also strike the wafer. The advantages of planar plasma etching are: (1) good selectivity, (2) a degree of anisotropy, (3) better uniformity, and (4) some wafer temperature control since the electrode is cooled. The disadvantages are a smaller wafer capacity, more radiation damage than a barrel reactor, and etch residues.

If pressures lower than 1 Torr are used and the extracted ion energy is increased up to 1500 eV by RF biasing of the substrate electrode, the chemically reactive ions are attracted toward the cathode. The increased voltage drop and reduced pressures with a longer mean free path enable reactive ions such as CF_3^+, CCl_3^+, Cl^+, and CF_2^{+2} to strike the wafer as a broad collimated beam. The advantages of RIE are good anisotropic etching and reduced residues, while the disadvantages are lower selectivity, increased resist erosion, radiation damage, and poor control over process parameters such as the ion flux which affect the etch process reproducibility.

Ion-beam milling is a purely physical process[(201)] in which inert (Ar^+) ions are extracted from a broad-beam ion source. Etching is accomplished by momentum transfer (sputtering). Since etching is a sputtering process, the etch products are nonvolatile and redeposition and trenching are common problems[(201)] (Fig. 13-11-8). Although the beam milling process is anisotropic, the poor selectivity and high rate of resist erosion have limited the ion milling to thin ($<0.2\ \mu m$) films or exotic material etching not accessible by chemically active etchants.

Substitution of a chemically[(200)] reactive ion such as CF_3^+ or Cl^+ for argon in an ion miller (Fig. 13-11-9), has enabled greater process control of ion flux than in RIE and hence has obtained good uniformity with anisotropic features.[(500)] Moreover, low resist erosion and low radiation damage are other advantages. The

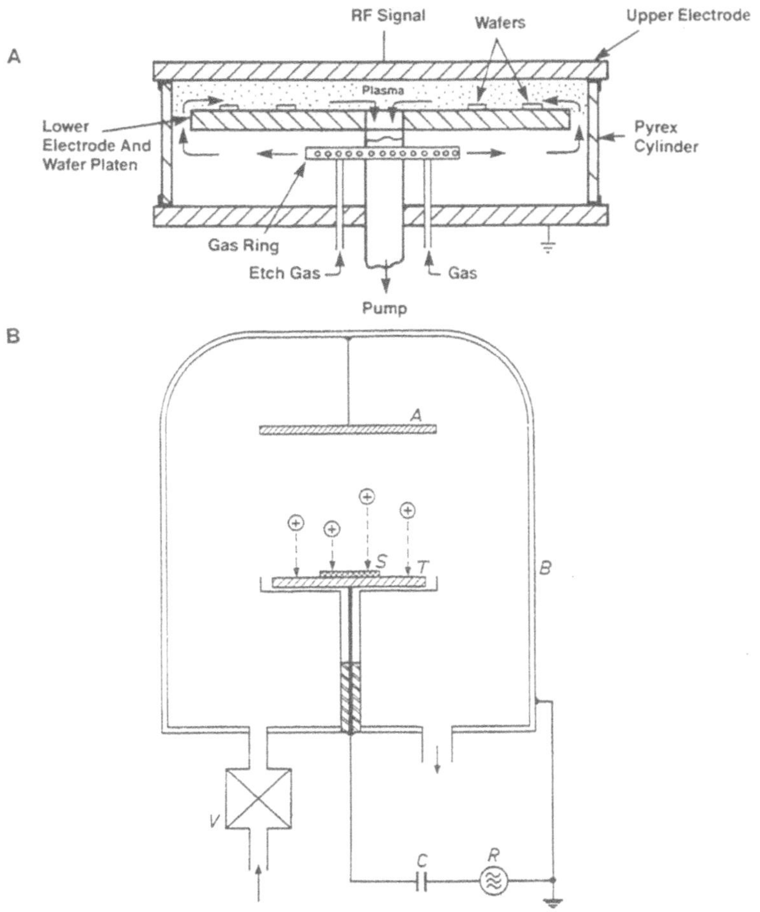

FIGURE 13-11-7. (A) Schematic of parallel plate reactor. After Reinberg, U.S. Patent 3,757,733. (B) Schematic of low-pressure cathode sputterer which can be used for etching. B, bell jar of conducting material with vacuum pump attached. Argon is admitted via valve V until a pressure of 1 Pa is reached. R, RF source; C, capacitor; A, anode in electrical contact with bell jar; T, target cathode with silicon wafer S attached. From H. Dimigen and H. Luthje, *Philips Tech. Rev.* 35, 199 (1975).

main disadvantages are lower throughput and lower selectivity in etching materials which require a physical sputter or ion damage component.

The rapid implementation of anisotropic plasma etching has enabled dry etching to significantly reduce the etch bias. Low etch biases are needed when the ratio of feature width to layer thickness drops below 3 to 1. Table 13-11-1 a

TABLE 13-11-1. Etch Bias in Pattern Transfer[a]

Thickness of material	Wet bias	Dry bias
0.5 μm	0.4 μm	0.1 μm
1.0 μm	1.0 μm	0.3 μm
1.7 μm	5.0 μm	0.4 μm

[a] From Ref. 202.

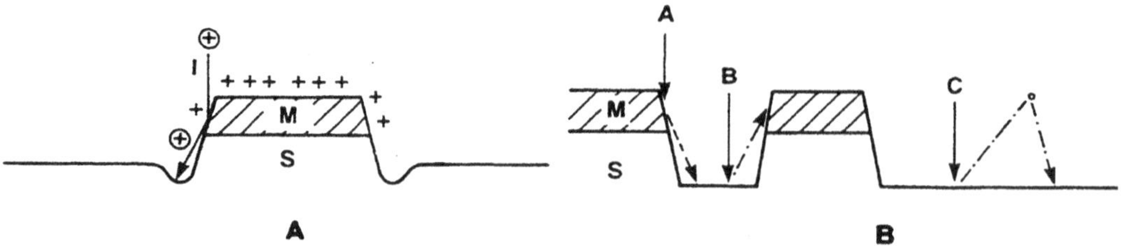

FIGURE 13-11-8. (A) Formation of ditches in "physical ion etching." M, photoresist masking; S, specimen to be etched. In this form of etching there is some etching of the mask as well. The mask is etched most rapidly in the neighborhood of the transitions between the upper surface of the mask and the side. The slope of the sides of the mask with respect to the surface to be etched then becomes smaller, and ions are no longer incident parallel to the sides. Some of them collide with the sides of the mask before they arrive at the etch surface. There is consequently an increase in the number of incident ions close to the edges. This gives a local increase in the etch rate, so ditches are formed. (B) Diagrammatic representation of some undesirable redeposition processes that can arise in physical ion etching. In process A, particles from the photoresist masking M are scattered onto the surface of the specimen S; this delays the etching process. In process B, particles from the specimen are sputtered onto the sides of the mask which can cause the formation of sharp edges after the resist is removed. In process C, particles that were originally sputtered from the specimen arrive back at the etch surface after colliding with gas atoms. This also delays the etching process. The effect increases in magnitude with increasing gas pressure. From H. Kalter and E. Van de Ven, *Philips Tech. Rev.* **38**, 200 (1978).

presents comparison of wet and dry etching. Anisotropic etching can easily achieve submicrometer etch biases. Another benefit is the selectivity of etching silicon nitride which required hot liquid phosphoric acid as a liquid etchant. In this regard, the versatility and selectivity of dry etching over the wet etching of a Si_3N_4/SiO_2 sandwich are shown in Fig. 13-11-10. Liquid-phase etching required a multiple step lithography to define a mask of pyrolytic oxide followed by an H_3PO_4 etch for the Si_3N_4. Using HF etching for the next layer, the SiO_2 layer under the nitride was severely undercut by liquid isotropic etchants. On the other hand, with a proper gas

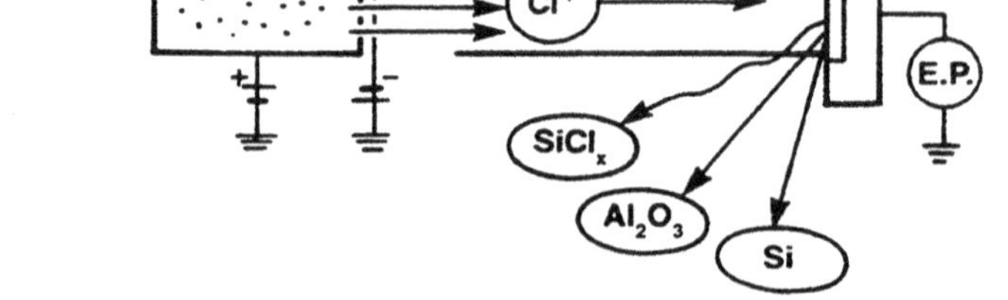

FIGURE 13-11-9. Chlorine reactive ion etching of Al–Cu films. From Ref. 201, reprinted with permission of *Solid State Technology*, published by Technical Publishing, a company of Dun and Bradstreet.

A

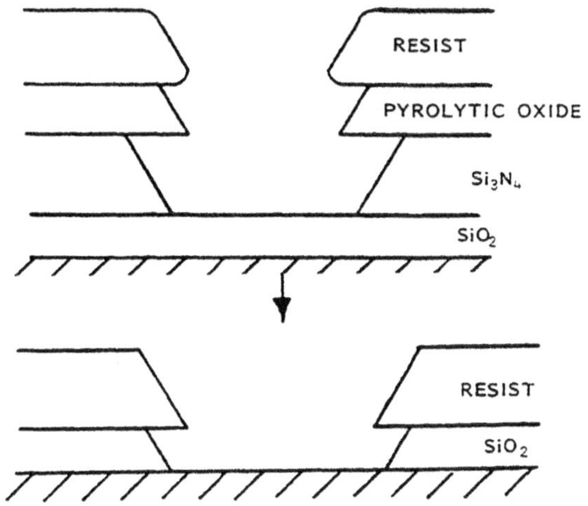

B

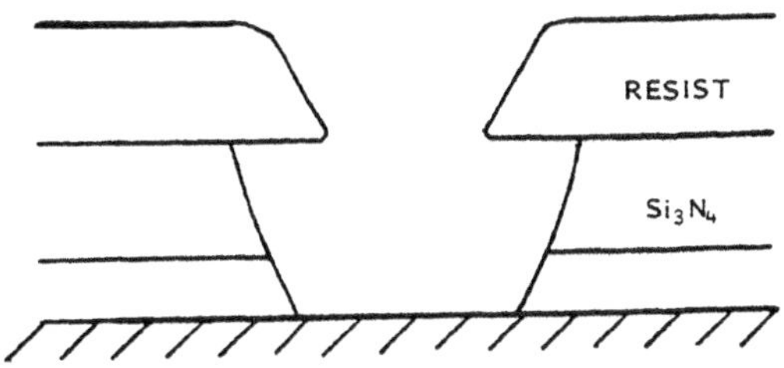

FIGURE 13-11-10. Comparison of (A) the wet etching of a Si_3N_4/SiO_2 sandwhich which produces a Christmas tree undercut and (B) under plasma etch conditions of equal etch rate for Si_3N_4 and SiO_2. U.S. Patent 3,880,684.

etchant, both the Si_3N_4 and SiO_2 layers were etched at equal rates to provide a smooth tapered hole.

In wet etching, the passivity of the substrate is very high. The etching ceases at the surface of the sublayer while in plasma etching, erosion of the substrate can occur such as in an Al/Si sublayer (Table 13-11-2). The selectivity is defined as the ratio of the etch rate of the etching film to the etch rate of the substrate. With high

TABLE 13-11-2. Selectivity of Gas Plasma Etch[a]

Film	Substrate	Selectivity[b]	Resolution
Si_3N_4	SiO_2	4	Isotropic
Si_3N_4	Si	7	Anisotropic
SiO_2	Si	6	Anisotropic
Al	Si	2	Anisotropic
Poly Si	SiO_2	8	Anisotropic
Poly Si	SiO_2	30	Isotropic

[a] From Ref. 208.
[b] Selectivity = etch rate ratio of film/substrate.

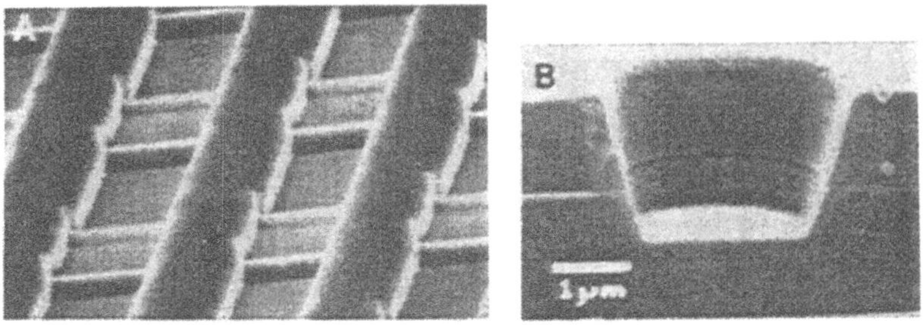

FIGURE 13-11-11. Representative profiles in reactive ion etching of (A) Al lines 1μm in width using CCl_4 and Ar ions; (B) sputtered quartz via hole etched by CF_4 and Ar. Copyright and courtesy of Veeco, Inc.

selectivity (> 10), excellent etched profiles with little underlayer attack can be obtained. Representative RIE profiles in films of Al and sputtered quartz are shown in Fig. 13-11-11.

The plasma[(204)] is a dynamic gaseous state and does not supply an excess of etchant as in the liquid etch processes. The plasma loses energy to its environment mainly by radiation and conduction to the walls, by etching (reactive mechanisms), and by changes of state (vaporization at the walls of the reactor). Therefore, to maintain the plasma state, energy and mass must be supplied as fast as they are lost. In dry etching, the dynamic variables such as power input, pressure, flow rate, and surface area of wafer to be etched constantly must be controlled.[(594)]

In a plasma etcher, low pressures (5–200 mm) are necessary to prevent ion and radical collisional deactivation. At the low pressure, the lifetime of the F* radical is as long as 1 sec and the mean free path is a few centimeters (sufficient to diffuse to the wafers in a barrel reactor). Other ions such as CF_3^+ last only 10 μsec, but in a parallel plate reactor the ions are accelerated by the electric field to strike the wafer surface.

In the following sections, I will discuss metal etching, silicon etching, and silicon insulator etching, followed by an examination of the chemical etch mechanisms, the role of the resist, and the practical aspects of dry etching.

13-12. ALUMINUM ETCHING

13-12-1. *Introduction*

The dry etching of aluminum and its alloys (2% Si and 5% Cu) are critical processes for high-density circuits. Aluminum dry etching was not seriously addressed until the late 1970s since the need (Table 13-6-2) for metal interconnections smaller than 3 μm or pitches smaller than 4 μm was not required until mid-1980. Meanwhile, the etching of Al and especially the Cu alloy, has proven to be a more formidable task than selectively etching Si and SiO_2. The reaction products of $AlCl_3$ are absorbed by the resist and upon contact with moisture form acids to corrode the metal patterns. Due to its low volatility, the $CuCl_2$ product requires sputter etch removal that in turn distorts the resist and the etched patterns.

The dry etching of Al represents the gas-phase answer to the neglected liquid electroetching of Al (Section 13-6-2). The bias associated with isotropic liquid etching is on the order of 0.5–2.0 μm for a 1.0-μm-thick Al film. Historically, an unsuccessful attempt at anisotropic liquid etching has been the spray HCl etching of a charged Al film.(166) In addition, electrolytic etching has achieved some anisotropy when sidewall banking agents (dragon's blood) of sucrose or polyalcohols(166) are added to the H_3PO_4–HNO_3 etch. A reduction of etch undercut bias to the 0.25- to 0.50-μm range was claimed.

In dry etching, a glow discharge form of "dragon's blood" is formed by the *in situ* polymerization of chlorocarbon etchant fragments on the walls of the etching Al pattern. In electroetching or liquid etching, it is necessary to have the H_3PO_4 component to remove the Al_2O_3 passivation layer and an oxidant (HNO_3) to etch the Al film. Similarly, in dry etching, the physical components damage or sputter Al_2O_3 to allow Cl radicals and ions to oxidize Al to a volatile product of $AlCl_3$. For Al plasma etching, the choice of the etchant and the reactor should meet the following objectives:

1. Etch bias—2.0 μm
2. Etch tolerance—$\leqslant 0.3$ μm
3. Anisotropic features—$>80°$ profile
4. Selectivity—Al/resist >3
   Al/Si >4
   $Al/SiO_2 > 10$
5. Operating temperature—$<130°C$
6. Postcorrosion—None
7. Loading effects—None
8. Etching uniformity—$>90\%$ wafer-to-wafer

As a prerequisite, the integrity, uniformity, and reproducibility of the deposited Al film will influence the final process bias, tolerance, yield, and uniformity. If oxygen inadvertently is leaked into the evaporator or sputter deposition system for Al, Al_2O_3 will become incorporated into the bulk of the deposited aluminum. Al_2O_3 does not readily etch in chlorine plasmas. Secondly, if the Al-Cu alloy is not sputtered under slow deposition conditions, precipitates of $CuAl_2$ form which also do not readily etch in the chlorine plasma.(731)

13-12-2. *Mechanism of Al Etching*

The gas-phase etching of Al involves six basic reactions:

$$Al_2O_3 + 12\ \text{eV} \rightarrow \text{Lattice damage}$$

$$2Al_2O_3 + 6''Cl_2'' \rightarrow 4AlCl_3 + 3O_2$$

$$\text{Al–Al} > 6\ \text{eV} \rightarrow \text{Lattice damage}$$

$$Al(s) + Cl(g) \rightarrow \text{Chemisorption(ad)}$$

$$Al(s) + 3Cl(ad) \rightarrow AlCl_3(s)$$

$$2AlCl_3(s) \rightarrow Al_2Cl_6(g)\text{—Sublimation}$$

In order for Al etching to commence, the passivating Al_2O_3 layer must be removed by chemical reduction or vaporized by sputtering before the fresh Al etches. The passivating Al_2O_3[205] layer etches at ≈100 Å/min in CCl_4 plasma and induces an induction effect before the 1000 Å/min Cl "radical attack" on the clean Al film. The activation energy for the removal of Al_2O_3 or Al in CCl_4 is 4 kcal/mole[206] for a typical diffusion-controlled reaction (Section 13-1-2). An unusually low activation energy of 0.5 kcal/mole is noted for BCl_3 etching of Al_2O_3 or Al. Since a polymer also forms in CCl_4 and not in BCl_3 etching, a different mechanism or activated (rate limiting) step is evident. For example, the addition of BCl_3 to chlorine-based etchants facilitates the removal of Al_2O_3, while BCl_3 alone etches Al very slowly.[737] Furthermore, pure Cl radicals will not attack Al_2O_3[206] and hence barrel etching is not successful. The Cl radical in the low pressure of a barrel etcher (1 mm) has only a millisecond lifetime and will not diffuse to the wafer surface.

On the fresh Al surface, the chemisorption of chlorine[207] occurs rapidly at the low pressure of 10^{-5} mm. Chlorine can interact with Al in two states, as a physisorbed molecule $(Cl_2)_{ads}$ and after dissociation, as a Cl–Al complex:

$$Cl_1 \xrightarrow{k_1} (Cl_2)_{ads} \qquad (13\text{-}12\text{-}1)$$

$$(Cl_2)_{ads} \xrightarrow{k_{-1}} (Cl_2)_{desorp} \qquad (13\text{-}12\text{-}2)$$

$$(Cl_2)_{ads} \xrightarrow{k_2} (Cl\text{–}Al)_s \qquad (13\text{-}12\text{-}3)$$

$$(Cl\text{–}Al)_s \xrightarrow{k_{-2}} (Cl_2)_{ads} \qquad (13\text{-}12\text{-}4)$$

$$(Cl\text{–}Al)_s \xrightarrow{k_3} (AlCl_3)_{gas\ desorp} \qquad (13\text{-}12\text{-}5)$$

The activation energy for chlorine adsorption [reaction (13-12-1)] is 4.2 kcal/mole. This is a reasonable value for physisorption. For the desorption step [k_3, Eq. (13-12-5)], a value of 22 kcal/mole has been reported.[207] For the adsorption of Cl radicals, a sticking coefficient is defined as the ratio of the rate of adsorption to the rate of impingement, V, as related by

$$V = P_{Cl}/(2\pi mkT)^{1/2} \qquad (13\text{-}12\text{-}6)$$

where m is the mass of Cl_2 and k is the Boltzmann constant. The sticking coefficient is 10^{-2} which can be compared to 0.03 for O_2 on Al.[207] If we define the adsorption coefficient b as the ratio of k_1/k_{-1}, the fraction of surface covered by the adsorbed etchant is

$$\sigma = bP/1 + bP \quad \text{(Langmuir adsorption)} \qquad (13\text{-}12\text{-}7)$$

Using this equation, the Langmuir–Hinshelwood equation can be derived (as shown by Battey[208] for O_2 stripping of resists, Chapter 14) for the rate of Al etching (Fig. 13-12-1) as a function of pressure,

$$\text{Rate of etching} \propto K_R P_R/1 + K_P P_P \qquad (13\text{-}12\text{-}8)$$

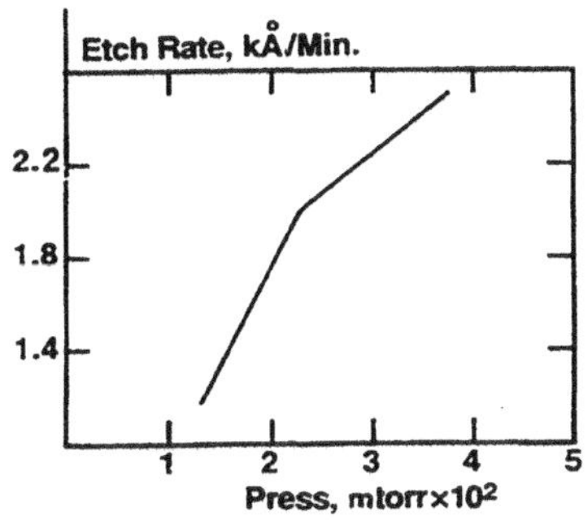

FIGURE 13-12-1. Etch rate of Al as a function of pressure. From Ref. 209, reprinted with permission of *Solid State Technology*, published by Technical Publishing, a company of Dun and Bradstreet.

where K_R and K_P are first-order rate constants and P_R is the pressure of the reactant etchant (R) and P_P is the product pressure (concentration). At lower pressure (Fig. 13-12-1), the rate will be proportional to P_R, but at a higher pressure where the product covers the reactant sites and makes them unavailable (slows down the K_R reaction), the rate will be proportional to P_R/P_P.[209]

If the Al is simultaneously damaged by ion bombardment in the reactor, then k_2 [Eq. (13-12-3)] will be enhanced at a lower activation energy. The overall activation energy [Eq. (13-12-8)] for CCl_4 etching is 4 kcal/mole and 0.5 for BCl_3.[206] While BCl_3 etching only involves Eqs. (13-12-1)–(13-12-8), it can be assumed to be controlled by the ion bombardment and desorption of product (Fig. 13-12-3a). If the sidewalls are not protected with an inert layer, Cl radicals or activated Cl* (ion or neutral) will readily undercut the pattern.

In Al etching, anisotropy is obtained by the formation of polymeric deposits[639] on the sidewall. Since vertical etching does not occur due to the low ion bombardment (etch) dependency for normal incidence, lateral undercutting is diminished by factors of ten. Furthermore, no mechanism pertaining to ion-damaged Al has been observed in the Al chlorine etched system.[671, 741]

Since a protective polymer layer forms on all surfaces during CCl_4 etching[210, 211] (Fig. 13-12-3B),

$$CCl_4 \rightarrow CCl_3^* \rightarrow CCl_2^* \rightarrow \text{Polymer} \qquad (13\text{-}12\text{-}9)$$

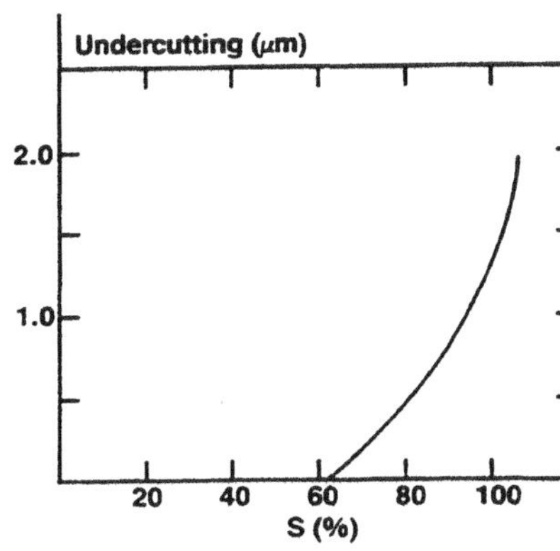

FIGURE 13-12-2. Undercutting of an Al film on top of a phosphosilicate glass during the course of plasma etching. The oxygen radicals derived from the phosphosilicate glass attack the plasma polymer deposit which forms on the Al sidewalls. The degree of undercutting is a function of *S*, the etchable Al area/whole wafer area. Fom M. Oda and K. Hirata, *Jpn. J. Appl. Phys.* July 1980, L405, courtesy of *Japanese Journal of Applied Physics*.

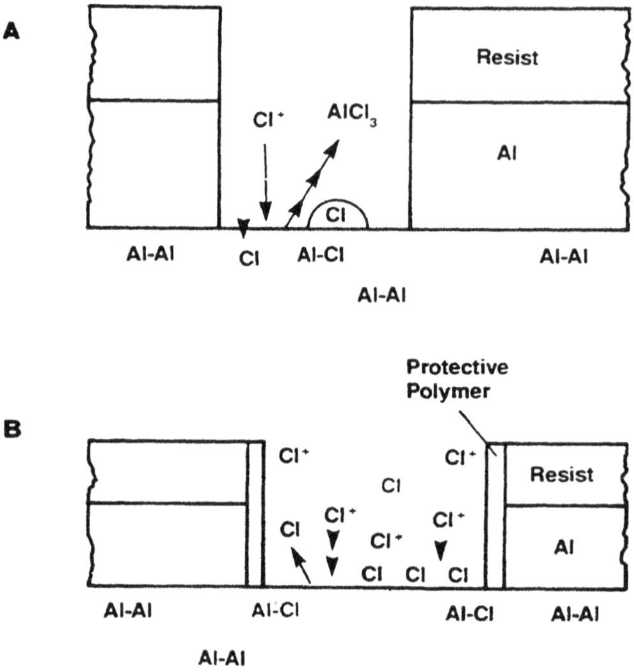

FIGURE 13-12-3. (A) Al plasma etching and chlorine ion desorption of Al–Cl etch product and (B) sidewall protection of lateral undercutting by the deposition of a plasma polymer during the course of etching.

an additional step is required to remove the polymer deposited on the etching surface,

$$CCl_3^+ + \text{polymer} \rightarrow \text{Sputter removal} \tag{13-12-10}$$

During Al etching, oxygen radicals readily attack the sidewall coating. This leads to edge pitting and undercutting (Fig. 13-12-2), The sources of oxygen are the electrode material, the underlying SiO_2 layer, and the resist. The undercutting is a function of the etched area. If more SiO_2 underlayer is available for attack by the etchant, larger etch biases are noted (Fig. 13-12-2). BCl_3 present in the etchant rapidly getters oxygen and at endpoint reduces the undercutting.[737]

If the Al is deposited over steps, anisotropic etching does not etch the vertical sidewall of Al metal. This is because the effective metal thickness to be removed over a vertical step, as seen by the anisotropic etchant, is the step height plus the metal thickness. With long overetches and the risk of undercut, residual metal can be left between conductors on steep sides and cause electrical shorting. Planarized insulators are necessary to reduce the step height and provide planar metals for future 1-μm interconnections.[747]

The etch temperature and activation energies play a critical role in Al etching and polymer formation. During etching, the counterelectrode can become covered with $AlCl_3$ and the rate of etching diminishes. Additional heat is required to sublime the reaction products.

Zarowin[212] has examined the temperature effects for etching gold with chlorine plasma. The Gibbs–Helmholtz equation for activation energy was modified to include ΔG, the free energy contribution,

$$\ln K = -(\Delta G - U/RT) \tag{13-12-11}$$

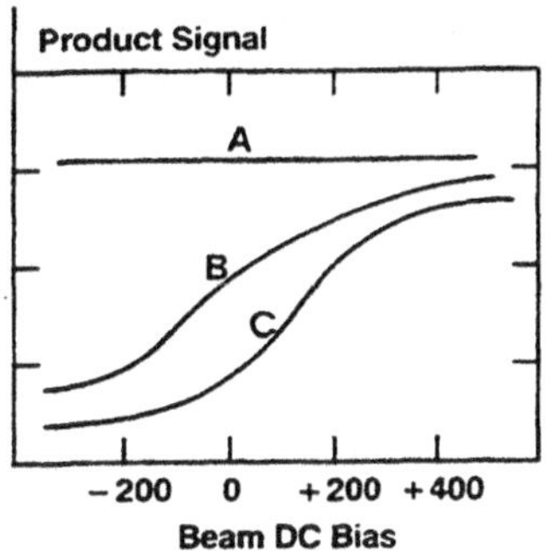

FIGURE 13-12-4. Dependence of desorbing etch product signal. (A) Etching of Si in CF_4/25% O_2 mixture; (B) etching of Si in Cl_2; (C) etching of Al in Cl_2. From Ref. 232, reproduced by permission of the Electrochemical Society Inc.

for the radical attack of Cl and U for the ionic contribution of etching. This equation is appropriate for Al etching if we lump the radical attack of Cl[207] and CCl_3^+ attack into G and include the ionic contribution of CCl_3^+. The temperature effects of etching Al have been studied empirically by Schwartz and Schaible.[216] During RIE, the wafers placed on an uncooled cathode are heated by ion bombardment. The temperature rise will be dependent on the emissivity of Si, Al, and resist, the power density, the cathode temperature, and the thermal heat-sink. They observed a maximum in etch of Al/4% Cu in CCl_4/Ar at 180°C (Fig. 13-12-4). As the reactor temperature rises to 180°C, the increase in etch rate is attributed to the increased volatility of the $AlCl_3$. $AlCl_3$ has a low vapor pressure of 10^{-4} mm at 25°C and 85 mm at 150°C.[217] At 180°C, $AlCl_3$ sublimes to form a low-volatile dimer,

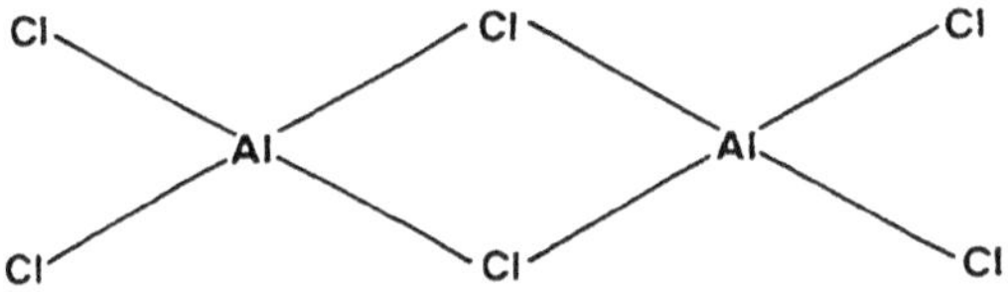

and the reaction slows down due to the adsorption of the dimer and desorption of the Cl radical [Eq. (13-12-2)]. If copper is present as a 5% alloy constituent, Cu_2Cl_2 and Cu_3Cl_3 form the copper chlorides which have a low vapor pressure of 10^{-4} mm at 200°C.[218] Argon ion-bombarded copper exposed to Cl plasma even at 200°C will not etch due to the formation of Cu_3Cl_3. In this case, sputtering is necessary to remove the copper chlorides formed in etching Al–Cu alloys. In

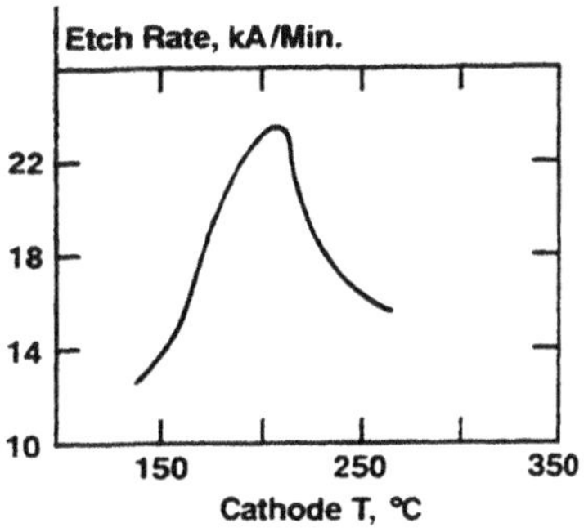

FIGURE 13-12-5. Etch rate of Al in CCl_4 plasma versus cathode temperature. From Ref. 216, reprinted with permission of *Solid State Technology*, published by Technical Publishing, a company of Dun and Bradstreet.

addition, reactor temperatures as high as 250°C are necessary to prevent redeposition of the sputtered copper chlorides.[220] Thus, high-temperature resists are required for some Al etchants. Al can be etched below 100°C (40°C in $BCl_3/CCl_4/O_2$).[219] Higher temperatures are required to remove the copper constituents.[220] However, severe photoresist degradation is noted in slow etching[221](< 300 Å/min) CCl_4 due to the resist heating. Above 130°C, on unhardened diazonovolak resist is susceptible to flow and degradation.[222] Thus, etchants and thermal conditions must be compromised if high-stability resists are not used.

In noncarbonaceous etchants such as BCl_3 and $SiCl_4$, some degradation (10%) of the resist is necessary to provide a polymer sidewall agent for anisotropic etching. Undissociated excited molecular CCl_4 will also attack pure Al[212, 213] and the SiO_2 underlayer[536] at 100°C:

$$\begin{aligned} &2Al + 6CCl_4 \rightarrow 2AlCl_3 + 3C_2Cl_6 \\ &2AlCl_3(g) + \tfrac{3}{2}SiO_2(s) \rightarrow Al_2O_3(s) + \tfrac{3}{2}SiCl_4(g) \end{aligned} \qquad (13\text{-}12\text{-}12)$$

In Al etching, the inclusion of both radical or molecular attack and the low ion contribution is based on the following observations of the uniqueness of Al etching in chlorine-type etchants:

1. Sato and Nakamura[214] studied Al and Si etching in $SiCl_4$ and CCl_4 and found by the results of self bias voltage measurements that the ion energy toward the surface decreases monotonically as the pressure increases. Films other than aluminum (silicon, Si_3N_4) depend primarily on the ion energy of the etchant. In Al, the etch rate shows little dependency on ion energy. A threshold of 400 eV is necessary to sustain the removal of Al_2O_3. It was concluded that Al can be etched by chlorine radicals. The effect of ion bombardment is to clear the Al_2O_3 and polymeric deposits, whereas silicon is etched primarily by ion bombardment (Cl radical does not etch Si at 25°C.[290]
2. Smith and Saviano[215] found, in contrast to Si etching in chlorine, that the Cl ion energy has a constant effect on the etching rate of Cl_2 in clean Al and that the etching continues, after the ion beam is extinguished, by radical or molecular etching. Figure 13-12-4 shows the major dependence of Si and independence of Al on the ion bombardment energy.
3. The anisotropic features of etching in Al result primarily from the ion milling of the polymer layer which builds up in chlorocarbon plasmas. Argon ion bombardment of chemisorbed CCl_4 had no effect on the etch rate.[689]
4. In a reactive ion-beam etcher, the etch rate of Al–Cu–Si in CCl_4 is linear[201] over the range of 600 to 1000 eV. Below 600 V, the ion components cannot sputter Al_2O_3 and leave Cu and Si residues. O_2, if present, retards the etching of Al by forming Al_2O_3. CCl_4 itself is a poor scavenger of O_2,[206]
5. In addition, Schaible and Schwartz[223] studied Al etching in CCl_4 with an inert (nonresist) MgO mask and found an increased lateral undercut at 225°C but not on an unheated wafer. The inhibiting polymer layer is not present at the higher temperature. These facts support the uniqueness of the polymeric protective agent as the primary anisotropic mechanism in Al, Cl, and CCl_4 plasma etching.

TABLE 13-12-1. Al Plasma Etchants

Etchant	Reference
CCl_4, $HCl+Cl_2$, CBr_4	U.S. Patent 3,994,793 (1976), IBM
BCl_3	U.S. Patent 4,030,967 (1977), Bell
$CCl_4+15\%\ NH_3$	U.S. Patent 4,073,669 (1978), ITT
BCl_3 80% + Cl_2	U.S. Patent 4,256,534 (1981), Bell
BCl_3	U.S. Patent 4,267,013 (1981)
CCl_4/H_2	223 (remove $AlCl_3$)
CCl_4/Cl_2	224 (no residues)
CCl_4/N_2	225 (low residues
CCl_4/Ar	217 (Al_2O_3 etching)
CCl_4/Xe	226
CCl_4	227 (selectivity–frequency)
Cl_2/H_2	228 (magnetron etcher)
Cl_2/Ar	217
Cl_2	229 (variable frequency)
$BCl_3/CCl_4/O_2$	219 75/20/5 (anisotropic, no residue)
BCl_3/O_2	219 (lateral overetch)
BCl_3/Cl_2	224, 545, 637
BCl_3/BBr_3	230
BCl_3/PCl_3	230
BCl_3/CCl_2F_2	231 (reduce postcorrosion)
$BCl_3/CHCl_3$	232 (lower resist attack)
$BCl_3/Cl_2/CHCl_3/He$	224, 233 (anisotropic)
BCl_3/CF_4	230 (low corrosion)
PCl_3	234
$SiCl_4$	214, 235 (low resist attack)
CCl_2F_2	236 (low corrosion)
$HCl/CHCl_3$	232

Although CCl_4 has been the initial choice of Al etchants, the B and Si chlorides and mixtures recently have been more successful at avoiding postcorrosion and residues. In Table 13-12-1, an extensive tabulation of Al etchants is given. BCl_3[590–592] and $SiCl_4$[585] offer the unique benefits of lower residues, less resist erosion, and better selectivity than Cl_2 or CCl_4. In BCl_3, polymers do not form.[206] Only B_2O_3 was found as an electrode residue.[205] Although BCl_3 removes the inhibiting Al_2O_3, it etches Al very slowly (3 nm/min). The rate-limiting step in etching with BCl_3 (0.5 kcal/mole activation energy) is the inefficient generation of Cl radical,

$$BCl_3 + e \rightarrow BCl_2 + Cl^* \tag{13-12-13}$$

Because of a low supply, the chlorine radical proceeds to slowly etch the Al film. Because of the low supply of Cl radicals, mixture of CCl_4 or Cl_2 with BCl_3 has been used. The Cl_2 and CCl_4 supply the Cl radical etchant species. Since BCl_3 is a more efficient scavenger of O_2 and moisture than CCl_4, a small amount of inhibiting Al_2O_3 is re-formed.[206, 737] The etch rate in BCl_3 is electrode dependent. With an electrode of anodized Al, the etch rate of Al is a factor of 2 higher due to the Cl radical recombination on steel surface.[206] At 20°C, a mixture of $BCl_3/CCl_4/O_2$ etches Al anisotropically.[219]

If there is a requirement to etch Al over steep SiO_2 sidewalls, or if there are problems in removing large bumps or hillocks under long overetch conditions (with excessive linewidth loss), Saia and Gorowitz[219] used a second overetch of BCl_3/O_2 to rapidly (isotropically) remove the bumps, and the sputtered chlorocarbon residues.

As a substitute Al etchant, $SiCl_4$ has been shown[214] to provide excellent resist protection (10% loss), high selectivity, little residues, and low postcorrosion effects at high anisotropy. The Al etching was examined with a resist and nonresist masks (Si_3N_4). With a resist mask, the organic species generated from the degraded resist deposit on sidewalls during vertical etching (Fig. 13-12-3b). As films on the sidewall cannot be easily removed by chlorine radicals, lateral etching under the mask is suppressed. With higher pressures, the concentration of radicals is increased. Frequent radical attacks on the sidewalls inhibit growth of organic films which leads to lateral etching of aluminum (a little resist goes a long way!). With an $Si_3N_4/Al/SiO_2$ sandwich, no undercut was observed at 3 Pa (3×10^{-2} mm) pressure. In the emission spectra, silicon atoms were detected which were thought to deposit on the sidewalls to inhibit lateral etching. Above 3 Pa, lateral etching was observed. In both masked cases, the vertical etch rate is not affected by the presence of resist. It was found that the ion etch rate of the plasma-deposited film in the vertical direction is larger than the deposition rate. Therefore, the growth of organic films is suppressed. At higher (> 3 Pa) pressures, the mean free path of ions is on the same order as the width of the ion sheath.[237] Ions collide with neutral etchant and impinge on the surface isotropically so that patterns are laterally undercut.

13-12-3. *Selectivity of Al Etching*

The erosion of the resist can be detrimental to the image size control and the electrical device yield. For example, pinholes can form on top of steps and the erosion of silicon contacts can lead to poor electrical contacts. Selectivity, which is defined as the ratio of the etch rate of the Al to the etch rate of the underlayer film or resist, lies in the range of 2–25 (Table 13-12-2). The erosion of the resist or SiO_2

TABLE 13-12-2. Selectivity of Al Etchants

Film	CCl_4	$SiCl_4$	BCl_3	Cl_2/H_2
Resist	2[209]	10[214]	5[224]	7[201]
	4[224]	15[235]		
	6[220]			
	8[238]			
SiO_2	2[201]	50[235]	30[224]	14[201]
	4[224]			
	7[209]			
	12[240]			
	25[227]			
	30[224]			
Poly Si	2[209]	2[214]		
	4[224]			
	10[238]			
100 Si	2[216,229,238]			
Si_3N_4	7[220]	10[214]		
PSG	10[235]	20[235]		

is a continual problems. In RIE, some erosion always occurs due to ionic bombardment and sputtering. The least resist attack and the lowest SiO_2 erosion occur in $SiCl_4$. Preferably, a resist selectivity ratio of Al/resist of at least 5 is necessary to control the profiles and prevent tapered profiles. The positive diazonovolak photoresists can achieve a good etch resistance ratio if the resist is properly hardened (see Table 13-12-3, Section 13-12-5). Treatment of the diazonovolak resist with $SiCl_4$, for example, silylates the hydroxyl (phenolic) group and incorporates silicon atoms into the polymer backbone,[757]

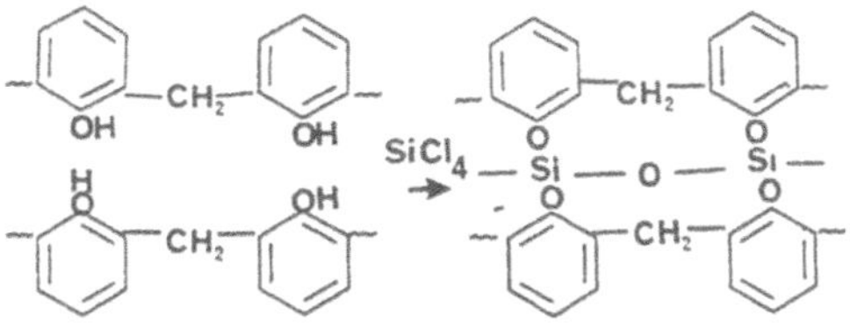

The resist erosion in chlorine plasma is substantially reduced.[704] Hardening is also necessary to thermally stabilize the resist image against flowing under the hot ($> 150°C$) reactor conditions.

Since Si is readily attacked by chlorine etchants, a selectivity of only 2 has been routinely obtained. A higher selectivity of 8 can be obtained if the power is dropped from 80 watts to 30 watts once the Al_2O_3 barrier has been etched.[238] In Al etching, overetching of up to 40% can occur due to the slow vertical etching of Al over steps,

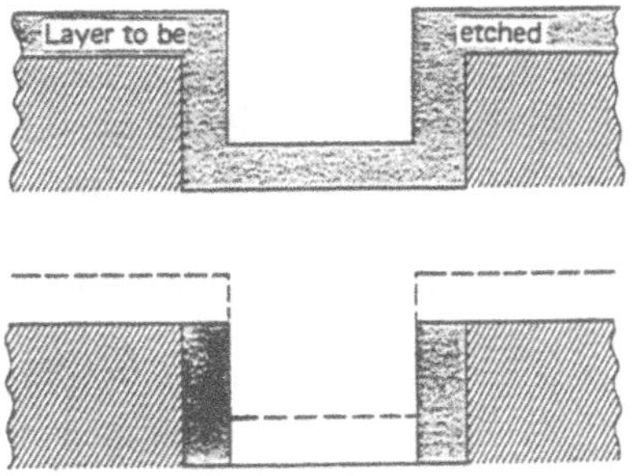

Vertical etch process
(no overetching)

the sublayer of silicon is invariably etched. In addition, because of poor etch uniformity ("bull's-eye clearing") or Al hillock residues,[219] the sublayer of silicon is invariably etched by additional overetching. The frequency[227, 229] of excitation of the RF field and the surface area[209] of the etchable wafers can also influence the Al/Si selectivity ratio. Thus, the prevention of the erosion of silicon contacts requires uniform etching and a precise endpoint determination.

13-12-4. *Residue and Postetch Corrosion*

In Al and alloy etching, pitting, corrosion, polymeric residues, deposits on the reactor walls, sidewall etching, and hillocks[731] are common problems. The $AlCl_3$ is not a gaseous product and can be absorbed by the resist and counterelectrodes. $AlCl_3$ reacts with moisture to form

$$AlCl_3 \cdot (H_2O)$$

$$Al_2O_3 \cdot H_2O$$

$$Al(OH)_3$$

In addition, HCl, which is corrosive to Al, is liberated upon reaction with water,

$$AlCl_3 + 3H_2O \rightarrow Al(OH)_3 + 3HCl \qquad (13\text{-}12\text{-}14)$$

$AlCl_3$ also forms Lewis adducts with aromatic groups in diazonovolaks. These adducts are precursors in the Friedel–Crafts alkylation of benzene.(241) If the resist, impregnated with $AlCl_3$, is stripped in O_2, Al_2O_3 is formed.(222) In addition, HCl can also be formed by the reaction of moisture with the impregnated $AlCl_3$. Also, $AlCl_3$ reacts with SiO_2 to form Al_2O_3.(536)

In the termination reaction of the carbon radicals and the chlorine radicals, the deposited polymer appears to be a polyvinylchloride,(210)

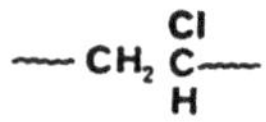

The polymer formed from CCl_4 etching is promoted by the use of active(26) electrodes of stainless steel, large wafer loads (more organic carbon from resist), and a rapidly consumed electrode. Furthermore, a polymer deposit is also formed with BCl_3 or $SiCl_4$ etchants. In these cases, the resist supplies the carbon-polymer backbone species. Polymer formation is necessary for anisotropic etching since the deposited polymer protects the sidewall against lateral etching (Fig. 13-12-3b). Higher temperatures ($< 180°C$) do not favor polymer formation.(216) Coating of the reactor walls with Al_2O_3 is claimed to reduce polymeric residues from forming on the reactor.(545)

Disadvantageously, oxygen plasma intermediates can be derived from the Al_2O_3 electrodes, the SiO_2 sublayer,(558) residual gases, or the resist. The oxygen radical will retard polymer formation by intercepting vinyl precursors(205, 206)

$$\text{"}Cl_n nC^{*}\text{"} + O^{*} \rightarrow COCl + CO \qquad (13\text{-}12\text{-}15)$$

Many physical or chemical steps have been tried to prevent, inhibit, getter, or remove the polymeric and inorganic residues. The treatments or procedures are as follows:

1. *Load locks.* Small-volume reactors (single-wafer processors) with graphite electrodes have been claimed(706) to reduce inorganic residues by effectively removing water and O_2.(242) Prior to etching, load locks reduce the gaseous O_2 and H_2O level. Moisture present during etching can inhibit etching by re-forming Al_2O_3.(250) Water reacts with the adsorbed $AlCl_3$ and forms corrosive HCl. Stripping of the resist directly in the reactor has been claimed to be helpful in reducing postcorrosion effects.(222)

2. *Choice of etchant.* $SiCl_4$ forms very little polymeric residues.[214] BCl_3[206] getters residual O_2 and H_2O more effectively than CCl_4 etchant.[589] The addition of H_2 to Cl_2[228] or to CCl_4[223] has been noted to reduce inorganic and organic residues. CF_4 has been used to harden novolak resist[249] prior to plasma etching. The fluoridated resist forms a passivation layer of AlF_3 on top of the resist (to prevent $AlCl_3$ pickup by the resist). Posttreatment of the etched Al with CF_4 followed by water,[246, 247] acid,[197, 248] or wet etchant rinse[195] has been used to remove residues. Removal of the resist in gases (H_2), other than O_2, can be used to prevent Al_2O_3 from being redeposited on the surface,

$$AlCl_3 + O_2 \rightarrow Al_2O_3 + Cl_2 \qquad (13\text{-}12\text{-}16)$$

By direct inclusion of fluorine in the etching gas such as in CCl_2F_2[236] (Table 13-12-1) or BCl_3/CCl_2F_2,[231] a dual-purpose etchant is obtained to etch and passivate (fluoridate the resist with AlF_3). It can be expected that various steps or etchant combinations will be used to contain and suppress the levels of $AlCl_3$ residue.

3. *High-temperature treatment.* Although high temperatures ($> 150°C$) during etching volatilize $AlCl_3$ and Cu_2Cl_2 (to some degree), isotropic etching may occur due to reduced polymer sidewall deposition.[216] A high-temperature bakeout[209] with stabilized resist[222] has been used to outgas the $AlCl_3$ trapped in the resist. Posttreatment of the etched image in O_2 at 200–450°C for $\frac{1}{2}$ hr has been claimed to getter corrosive products.[251, 252] Fok and co-workers[248] found that the O_2 treatment was irreproducible because the $AlCl_3$ on the sidewalls of the etched Al image was converted to Al_2O_3. They recommended a CF_4 postetch and an HF rinse.

4. *Stripping of resist.* Since the resist in CCl_4 etching becomes coated with a polymer that getters $AlCl_3$, stripping in a nonoxidizing medium would prevent Al_2O_3 deposits on the Al image. The use of water of water-based strippers presents problems of HCl formation. Organic-based strippers may be useful if the resist is not hard baked. If a multilayer resist composed of an image layer and an underlayer of organic soluble polyarylethersulfone is used,[253] the resist can be stripped in *n*-methyl pyrollidone. The polysulfones or soluble polyimides are stable to 220°C and are very plasma resistant.

13-12-5. *Resist Hardening*

Since Al–Cu etching may require $> 100°C$ temperatures to remove the chloride products and other process conditions such as ion implantation or deep Si trench etching may drive the resist temperature to 200°C, organic resists as stable as "rocks" (inorganic resists) have always been a major lithographic performance objective.

As 200°C materials, some negative polyimide, or acrylate pre-cross-linked resists[219] have been successful but they generally lack speed or high resolution. The workhorse resist has been the optically (Chapter 2) exposed positive diazoquinone phenolic-based resists. Recently, some DQ resists have been produced which use more stable phenolic or maleimide resins. Melt stability of about 150°C has been achieved.[292] However, the majority of commercial diazonovolaks melt

TABLE 13-12-3. Thermal Stabilization Treatments for Diazonovolaks

Treatment	Application	Reference
Deep (<320 nm) UV	RIE (Al)	222
	Dry etch	234, 255
	Lift-off	236
Add azides to DQN	RIE	548
Soak in HMDS	RIE (Al)	704
PMMA overcoat	Lift-off	742
Plasma	Dry etch	257, 258, 260, 261, 263
N_2	Ion implant	259, 263
O_2	HF vapor etch	259, 262
CF_4–O_2	RIE	261
CF_4	RIE	257, 261
CO	RIE	649, 664
Electron-beam	Dry etch	264, 265
Ion beam	Ion implant	266
	Dry etch	267
Soak in Na_2 DQ[a]	Dry etch	268
Soak in $KMnO_4$	Dry etch	269
Soak in HCHO–H^+	Ion implant	270
Soak in Mg acetate	Dry etch	271
Al overcoat, bake	Dry etch	272
Add peroxides	Ion implant	273

[a] DQ, 1-diazo-2-naphthoquinone-5-sulfonic acid, Na^+.

below 135°C. A host of physical and chemical posttreatments have evolved to stabilize the novolak resin to temperatures as high as 250°C (Table 13-12-3).

For mild hardening, the easiest batch technique[222, 257] is the low-pressure CF_4 plasma treatment. A stabilizing mold effect of the top surface is achieved by a plasma overcoat (Fig. 13-12-6B). The plasma treatment stabilizes the resist to about 160°C. However, the surfaces become reticulated because the N_2 gas escaping during the subsequent high postbake cannot readily permeate the topcoat. A PMMA mold overcoat is more permeable to the outgassing. The plasma, UV, and

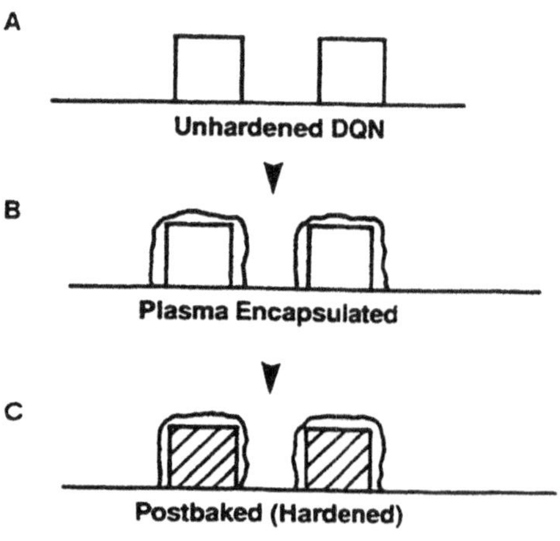

FIGURE 13-12-6. (A) Resist hardening by (B) plasma deposition of topcoat, followed by (C) slow baking of DQN at elevated temperatures.

metal overcoat cross-link or encapsulate the surface to allow the interior (Fig. 13-12-6C) of the novolak resin to undergo further cross-linking reactions,[274]

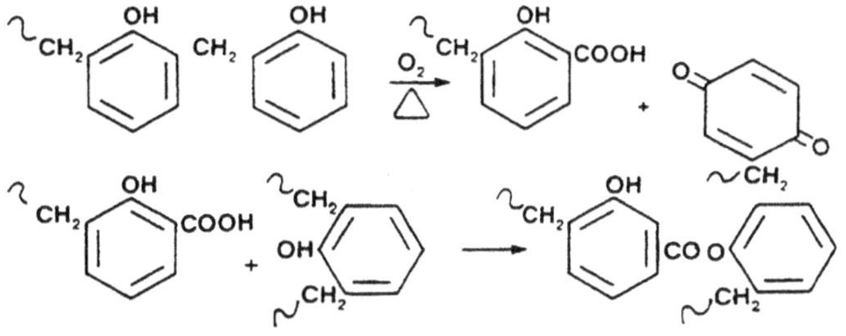

Other high-energy radiation such as *e*-beam, X-ray, and ion beam can achieve deeper penetration and stabilize the resist to 200°C (Chapter 11). Oxidation also contributes to hardening of the resin. In addition, treatment of the novolak image with formaldehyde[270] (HCHO) in an acidic medium continues the original novolak condensation reaction (between phenols and formaldehyde) to form a highly cross-linked (thermally stable to 250°C) insoluble polymer,

$$HCHO + HOH \longrightarrow HOCH_2OH$$

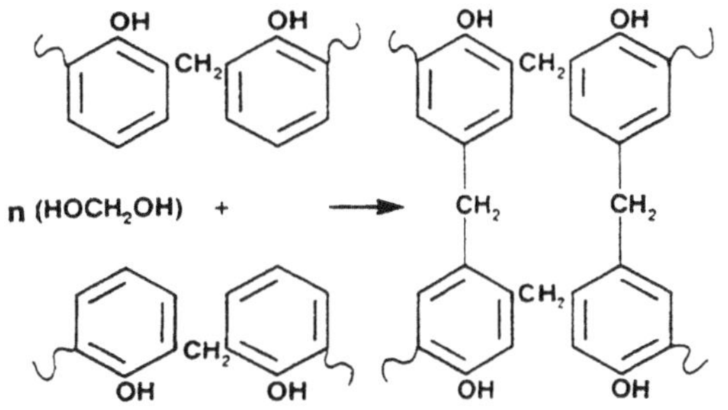

The HCHO-hardened resist can be soaked in NaF or other inorganic salts to further stabilize the resist against reactive ion erosion (formation of nonvolatile AlF_3). Treatment of the imaged novolak resist with HMDS incorporates (by silylation of the phenol) Si atoms into the resist. The silylated resist is resistant to CCl_4 etchant.[704] If the hardened resist does not survive the Al etching conditions, a metal resist TiW deposited by lift-off can be used as a stable etching mask.[621, 622]

A careful study of Al etching using masks of plasma and UV-hardened AZ-1350 was done by Tracy and Mattox.[222] They found that too much hardening provided no passivating sidewall polymer deposits for the eroding resist and that negative undercut images were formed (Fig. 13-12-7c). The undercut images were very similar to that images formed with an inert Si_3N_4 mask (Fig. 13-12-7d). The untreated resist provided too much erosion and loss of anisotropy. Thus, a controlled degree of resist loss is necessary to achieve anisotropic features. Furthermore, a trilayer resist mask with a baked diazonovolak underlayer was found to be superior than a single diazonovolak resist layer.[567]

Tracy and Mattox also found that stabilized photoresist masks may lead to negative anisotropic profiles (Fig. 13-12-7d) but that addition of H_2 or a H source to CCl_4 gas can be used to restore the anisotropy. In the H_2–CCl_4 etchant, the

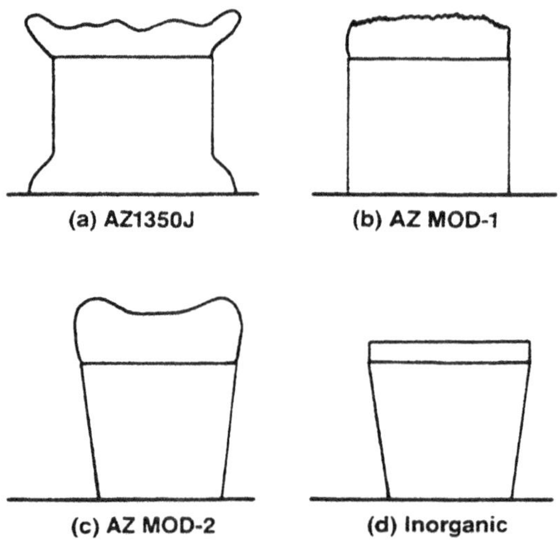

FIGURE 13-12-7. Schematic of profiles etched in Al using resist masks of DQN pretreated to harden resist. (a) Untreated resist leads to resist distortion during the middle of plasma etching resulting in protrusions. (b) Plasma-treated resist survives etch condition and anisotropically masks Al. (c) Deep-UV hardening of DQN leads to little resist distortion and undercutting. (d) Inorganic mask provides no carbon for sidewall banking agent and leads to isotropic undercutting at feet. From Ref. 222, reprinted with permission of *Solid State Technology*, published by Technical Publishing, a company of Dun and Bradstreet.

plasma deposits a passivating sidewall layer instead of obtaining H from the eroding resist. Meanwhile, the highly stabilized resist (Fig. 13-12-7c) was difficult to strip without Al_2O_3 residues since the resist absorbed more $AlCl_3$ than the partially stabilized resist used in the shown in Fig. 13-12-7a, b.

The work of Tracy and Mattox highlights the importance of studying the mechanism of etching with two types of masks, inert and erodable. The erodable mask may be critical to the formation of a sidewall protective coating. With an inert mask, the etching gases may or may not form a protective layer. In the case of an inert mask, ion bombardment mechanisms may be more critical. Many of the interpretations of the results of studies of plasma etching reported in the literature are based on a sole resist. Thorough interpretations of anisotropy should consider both erodable and nonerodable masking.

13-12-6. *Practical Al Etching*

Practical Al etching is concerned with choosing a resist, a reactor, and etchant(s), performing the etching to an endpoint, and assessing the final etch bias. In addition, the prevalent postcorrosion effects should be evaluated. The choice of a reactor is dependent on throughput and the efficient control of postcorrosion. For single-wafer processing, the rapid etching of a single wafer at high pressure or with a magnetron[(609)] source can provide the convenience of small-volume vacuum pumpout (less corrosive $AlCl_3$ deposits). In addition, the endpoint control of individual wafers will produce better reproducibility. Larger batch wafer etchers have to rely on single-wafer monitoring.

Some of the practical aspects of choosing anisotropic etchers have been surveyed by Broydo[(275)] and Weiss.[(276)] Representative and advanced Al etchers are listed in Table 13-12-4. For a parallel plate reactor, high rates of 6000 Å/min (normal: 0.1 μm/min) were achieved, if $CHCl_3$ was added to the Cl_2/BCl_3 etchant.[(233)] Prior to etching, the removal of moisture and O_2 is critical in limiting the re-formation of an inhibiting layer of Al_2O_3 and, secondly, the formation of corrosive HCl from the reaction of water with resist-impregnated $AlCl_3$.

Single-wafer etching[(278)] provides better uniformity ($>90\%$) and a lower etch

TABLE 13-12-4. Al Plasma Etchers

Configuration	Type	Reactant	Reference
1. Parallel plate	Reactive diode	CCl_4/He	224
2. Parallel plate	Reactive diode	$Cl_2/CHCl_3/BCl_3$	233
3. Hexagonal anode, cylindrical cathode, anode to cathode area >2	Reactive ion	CCl_4	276, 277
4. Multiple chamber, individual electrodes	Reactive ion	CCl_4	278
5. Parallel plate, magnetron(546,628)	Reactive ion	CCl_4	219, 228
6. Ion source, 500–1000 eV	Reactive ion	CCl_4	201
7. C electrode, 400 eV	Reactive ion	CCl_4	279
8. Ion mill		Ar	
9. Projection ion beam	100 keV	Ar	198
10. Projection *e*-beam	Radiation damage		281

bias (± 0.1 μm) than batch etching (± 0.5-μm etch bias). Since Al can be etched by both radicals and ions, the correct power, frequency, and pressure have to be established to provide sufficient but not excessive resist erosion (Fig. 13-12-3B) for sidewall protection. The undercut(293) of Al etching, as a function of CCl_4 pressure, has been as large as 0.5 μm.

Etching Al over SiO_2 can lead to undercutting since the O_2 product,

$$SiO_2 + 4Cl \rightarrow SiCl_4 + O_2$$

attacks(282, 558) the protective sidewall (plasma deposited) film. No undercutting of Al was found for oxygen-deficient Si or Si_3N_4 sublayers. For the process etching of Al films, the monitoring of the etch rate of the film is very important to overcome to adjust for the variations in Al_2O_3 inhibiting layer, the nonuniform Al thickness, grainy compositions,(538, 731) and the heating effects of the ion bombardment. Endpoint measurements of metal etching are more difficult than monitoring transparent films since there are no reflective nodes of interference.

Physical monitoring such as pressure changes, etching film thickness,(283) conductivity, or impedance(284) is simpler than chemical measurement by optical emission,(219, 285, 286) mass spectrometry,(215) or gas chromatography.(287) Etch monitors are necessary to compensate for the loading effect(230, 288) on the etch rate E,

$$E \propto (F/A)^{1/2}$$

where F is the flow rate and A is the etchable area of Al. If not compensated, during the final stage of etching, accelerated lateral etching will occur.(290)

Endpoint monitoring of the reactant (thickness) is the most exact method of measurement. Transparent films (resist, SiO_2) provide a natural interference pattern where one period of oscillation corresponds to a thickness change,(718, 719)

$$n\lambda/2$$

where n is the refractive index at the laser 6238-Å wavelength, λ. An optical monitor of the etching of a diffraction grating in SiO_2, can indicate the initial film thickness and the cleanout time of micrometer-sized features.[750] For opaque films,[283] an interferometer arrangement is used to monitor the resist erosion and Al etching. The difference in path length of one period of oscillation corresponds to the removal of an opaque layer,

$$\lambda/2 = \Delta X(n-1)$$

where ΔX is the thickness change of the resist (Table 13-12-3) during the same period of time. The disappearance of a preetched grating can also be used to monitor Al etching.

If we consider that up to 50% of the resist is lost, simultaneous monitoring of the etching of Al and the resist can provide a safeguard from excessive pinhole formation. The allowable thickness of resist lost (ΔT_{PR}) can be expressed as

$$\Delta T_{PR} = \frac{T_P E_{PR}}{E_P} + \left[\sum (\sigma_{EP} + \sigma_{TP}) \frac{T_P}{E_P} + \Delta_{EP} + S \right] E_{PR}$$

where T_P = polycrystalline silicon* film thickness, E_{PR} = photoresist etch rate, E_P = polycrystalline silicon* etch rate, σ_{EP} = overall percent standard deviation in polycrystalline silicon* etch rate, σ_{TP} = overall percent standard deviation in polycrystalline silicon* film thickness, Δ_{EP} = endpoint detection uncertainty time, and S = safety factor time (asterisks also refer to aluminum etch rate).

For a resist line with a 50–60° initial profile, Booth and Heslop[239] have shown that an etch bias of 0.3-μm widening, was achieved for a 1-μm line, etched 1 μm deep, using AZ-135OH, with a 50% resist loss (CCl_4). Etchants such as $SiCl_4$ (Table 13-12-2) only erode 10% of the resist and require less stringent monitoring.

Another physical measurement, which does not involve expensive equipment, is discharge impedance monitoring.[284, 720, 721] In RIE, it is found that the CCl_4 discharge impedance is reduced during aluminum etching and changes markedly, at the final aluminum interface. The discharge impedance Z_d is expressed as:

$$Z_d = (E_d)^2/P_d$$

E_d is the voltage between the electrodes and P_d is the RF power supplied to the electrode. When P_d is kept constant, the impedance change varies as the square of E_d. The impedance change is caused by product radicals, such as AlCl*, which cause a different voltage (conductivity) amplitude across the capacitor electrodes. The endpoint of impedance compared favorably with the 261.6-nm emission spectrum of AlCl.[286]

13-12-7. *Summary*

Al and alloy etching, although operational, is still troubled by residues, nonvolatile Cu products, and postreactive corrosion. Although CCl_4 has been an anisotropic etchant which forms a passivating layer, the unfavorable resist erosion and sensitivity to residual O_2 and water, have forced investigators to new mixtures of etchants. Recent findings appear to be that the $SiCl_4$ etchants offer low residue, little resist attack, and good selectivity. More extensive residue studies are necessary, before the dry etching of metals will be ready for submicrometer circuitry.

The etched bias of dry etching has achieved an order-of-magnitude (0.1–0.3 μm) reduction in linewidth over classical wet etching, and, in addition, has achieved smaller pitches than the lift-off technique. The low volatility of Cu or other additives may eventually require an inorganic or multilayer resist mask, to survive the higher reactor outgassing temperatures. Future development of projected, broad reactive ion beams (without a resist mask) may eventually supersede the present methods of resist masking.

13-13. SILICON ETCHING

13-13-1. *Introduction*

The subtractive removal of silicon dioxide, polysilicon, and silicon nitride has gradually switched from isotropic, wet etch conditions producing sloped images with 1-μm bias, to anisotropic, dry etch conditions. The minimum linewidth (μm) requirements for devices projected to 1990 are as follows:

Substrate	1978	1984	1990
Silicon dioxide	3	2	1
Polysilicon	4	3	1.5
Silicon nitride	5	4	2
Aluminum[a]	5(10)	4(8)	2(4)

[a] Values in parentheses are for pitch (line and space).

Single-wafer etching of large (200 mm) wafers is the predominant[(725)] method of etching. In fact, plasma etchers are the second most expensive lithographic tools, approaching \$500,000 in cost. The etcher is evolving into an expensive exposure tool. By 1990, the throughput of commercial tools will be the dominant competing factor. The present biases ($\pm$0.25–0.5 μm) of isotropic etching become the active real estate of future images. Anisotropic images with $<$ 25% final etch bias are a minimum requirement for 1990.

Nondirectional isotropic images are formed by neutral charged atoms, free radicals, and molecular species. Species of halogens are formed by electron impact dissociation or two-body collisions,

$$CF_4 + e \rightarrow CF_3\cdot + F\cdot \quad (13\text{-}13\text{-}1)$$

$$C_2F_4 + e \rightarrow 2CF_2 \quad (13\text{-}13\text{-}2)$$

$$CF_3OO\cdot + O\cdot \rightarrow COF_2O_2 + F \quad (13\text{-}13\text{-}3)$$

The neutrals strike the surface at random angles (Fig. 13-13-1) at pressures of $> 10^{-3}$ Torr. In a plasma reactor, the etch rate is controlled by three factors: the neutral atom and free radical concentration, the ion concentration, and the energy of the ion. The ion and radical concentration (density) control the reaction rate, while the ion energy provides the necessary activation and controls the degree of

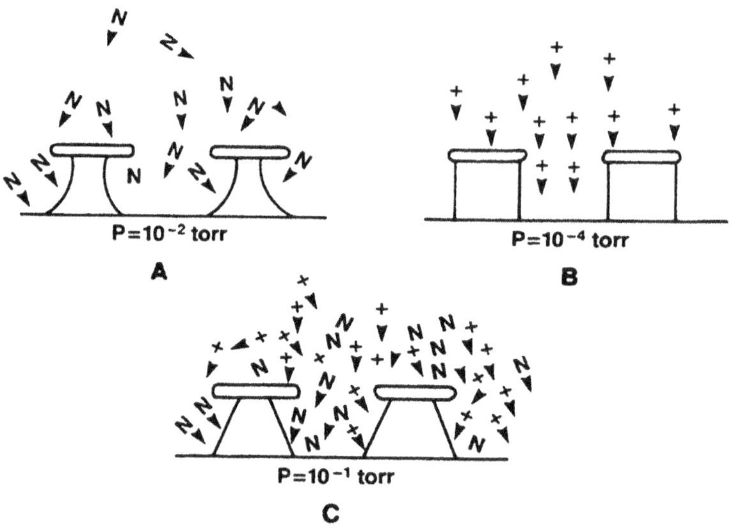

FIGURE 13-13-1. Plasma etching. (A) Neutral etchant molecules strike the surface at random angles to create isotropic images. (B) Positive charged ions strike the etching surface at normal incidence to vertical etch and produce anisotropic etched images. (C) Neutrals and ions collide at higher pressure to strike the surface obliquely and produce undercut images. From Ref. 327.

anisotropy. If anisotropic ($> 75^\circ$) image wall profiles are required, positive ions must be created at low pressure ($< 10^{-2}$ Torr), to strike the surface at normal incidence,

$$e + CF_4 \rightarrow CF_3^+ + F\cdot + e$$

$$CF_3 + Si \rightarrow SiF_4 + C^+$$

In RIE, ion bombardment (< 1000 eV) can induce many physical and chemical reactions[295]:

- Transfer energies to surface[625]
- Silicon bond fracture, Si · (to create reactive sites)
- Silicon ionization, Si^+
- Desorb products, SiF_x[531]
- Activate neutral etchants, F, XeF_2

By itself, sputter etching of silicon can yield about 0.4 atom/ion impact (< 100 Å/min of Si), but combined with any of the other chemical mechanisms, can achieve 30 atoms of Si removed per ion.[296]

Because of higher pressure ($> 10^{-2}$ Torr) in some plasma-induced reactions, the mean free path (λ) is shortened (at 10^{-4} Torr, $\lambda \approx 2$ cm; at 10^{-1} Torr, $\lambda \approx 10^{-2}$ cm), and the ions collide with neutrals, and impinge on the substrate, to form intermediate profiles.[297]

Selectivity, or the ability to stop the etching at the underlayer, is as important as the ability to transfer the image from the resist into the underlayer. Figure 13-13-2 gives examples of good and bad etching of polysilicon over SiO_2.[218] Conditions must be chosen to increase selectivity, and limit the isotropic undercut.

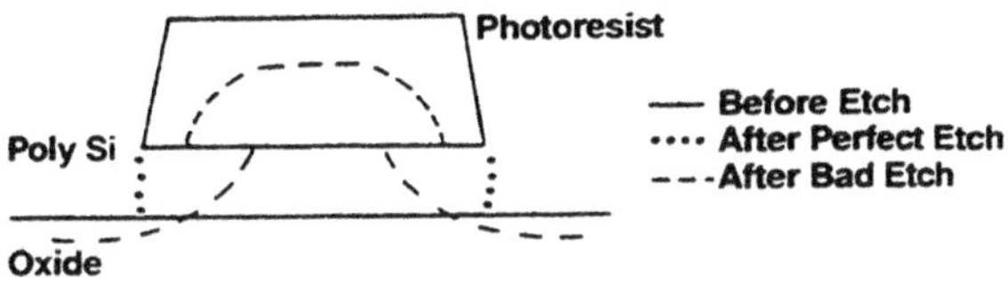

FIGURE 13-13-2. Examples of good anisotropic etching of polysilicon and poor etching with undercutting due to resist erosion. From Ref. 218, reprinted with permission of *Solid State Technology*, published by Technical Publishing, a company of Dun and Bradstreet.

The basic requirements for dry plasma etching, necessary for current and future devices, include[298]:

1. Profile control—selectable wall shapes with minimal etch bias
2. Process uniformity—less than 5% etch variation across the wafer and from wafer to wafer
3. Controlled selectivity—maximum etch rate ratio to silicon and variable ratio to resist
4. Minimum etch bias—less than 20% of the depth of etching
5. No contamination—less than 1–2 pinholes/cm^2 (one-third of minimum device linewidth) and no residues from metals, carbon products, including no permanent radiation damage
6. Production throughputs—high-speed single wafer (> 1 μm/min etch rate) and batch capability (> 25 wafers/hr)
7. Low toxicity—no corrosion of reactor, long pump life, and no harmful exposure to personnel and environment

A plethora of etchants have evolved, which attempt to satisfy some of the above requirements. The most popular etchants,[735] for silicon and for combinations, with silicon nitrides and oxides are:

CF_4	CF_4–H_2
CF_4–O_2	Cl_2
CCl_4	$SiCl_2$
C_2F_6	SF_6
CHF_3	

All of the above etchants must perform in suitable reactors, from a simple barrel surrounded by an RF coil, to complex ion-beam etchers. Some of the fundamental reactor and process characteristics are shown in Table 13-13-1. The choice of etchants, the reactor conditions, and the mechanisms of their interactions are subject to specific interpretations, because each reactor produces a ubiquitous distribution of excited states. In fact, plasma physicists need a periodic table of excited states to comprehend the fourth (plasma) state of matter.

On the silicon surface, chemical physicists have concentrated their efforts on understanding the consequences of ion bombardment. IBM workers such as Coburn,[531, 671] Kay, Chuang, and Winters[195, 300–302] have supported the chemisorption theory (ion activation of etchant or desorption of product) while Bell workers[691] such as Mogab, Donnelly, and Flamm have supported the ion lattice damage mechanism to enhance anisotropic product formation.[196, 197, 303] Both mechanisms are operable with silicon and its derivatives. Al anisotropic etching is an exception which does not conform to the above theories.

TABLE 13-13-1. Characteristics of Dry Etching Processes[a]

	Barrel reactor	Planar reactor	Reactive ion etcher	Reactive ion-beam etcher	Ion-beam miller
Location of substrate	Immersed in plasma	In plasma on grounded electrode	In plasma on powered electrode	Remote from plasma, in beam	Remote from plasma, in beam
Pressure (Torr)	$n \times 10^{-1}$	$n \times 10^{-1}$	$n \times 10^{-2}$	1×10^{-4}	1×10^{-4}
Flow rate (SCCM)	$n \times 10^{2}$	$n \times 10^{2}$	$n \times 10$	n	n
Ion energy (eV)	0	n–$n \times 10^{2}$	$n \times 10^{2}$	$n \times 10^{2}$	$n \times 10^{2}$
Active species	Atoms, radicals	Atoms, radicals, reactive ions	Reactive ions, radicals	Reactive ions	Ar ions
Products	Volatile	Volatile	Volatile	Volatile	Nonvolatile
Mechanism	Chemical	Chemical–chemical/physical	Chemical/physical	Chemical/physical	Physical
Selectivity	Usually very good; insignificant for SiO_2/Si	Usually good; fair for SiO_2/Si	Often excellent; good to very good for SiO_2/Si	Good for SiO_2/Si	Insignificant
Edge profile	Isotropic	Often isotropic; anisotropic for SiO_2	Usually anisotropic for SiO_2	Anisotropic	Anisotropic
Radiation damage	Slight	Moderate	High	Slight	Slight

[a] From Ref. 299.

Several centimeters above in the gas phase, workers such as Bruce, Reinberg, and Zarowin at Perkin–Elmer[304–307, 625] and Suzuki at Hitachi[297] have concentrated on the plasma sheath electric field[1] and the ion transport to the surface. Without proper ion impact to the surface, the chemistry of etching proceeds isotropically.

The interaction of a beam of energetic particles and the chemical reaction[605] on a surface is a multidisciplinary problem, which has critical technical implications for present and future lithography. In the next sections, we will review some of the present chemical–physical mechanisms of etching, and then discuss the practical aspects of profile control, etch bias, uniformity, process control, and contamination, as they pertain to etching silicon and its nitride and oxide layers.

13-13-2. *Gas-Phase Etching*

13-13-2-1. *Thermal Gas-Phase Etching.* Historically, gas-phase etching at high temperatures was used in the early 1960s to polish silicon wafers and for deep trench groove etching using oxide masks. An early reactor, including an RF coil for heating, is shown in Fig. 13-13-3. The gas-phase etching is not dependent on the silicon resistivity. Some etchants are preferential (pitting along crystal planes), while others produce smoother surfaces. A historical list of etchants is given in Table 13-13-2. Although gas-phase etching can occur at room temperature with XeF_2, it generally requires high temperature, to dissociate the etchant into radicals. Gas-phase etching has been proposed to follow Langmuir–Hinshelwood kinetics[310] in which three major steps are proposed for the reaction

$$Si + Cl_2 \rightarrow SiCl_2 \qquad (13\text{-}13\text{-}4)$$

1. Transport of the reactants to the reaction goal
2. Diffusion of the reactants and products to and from the surface

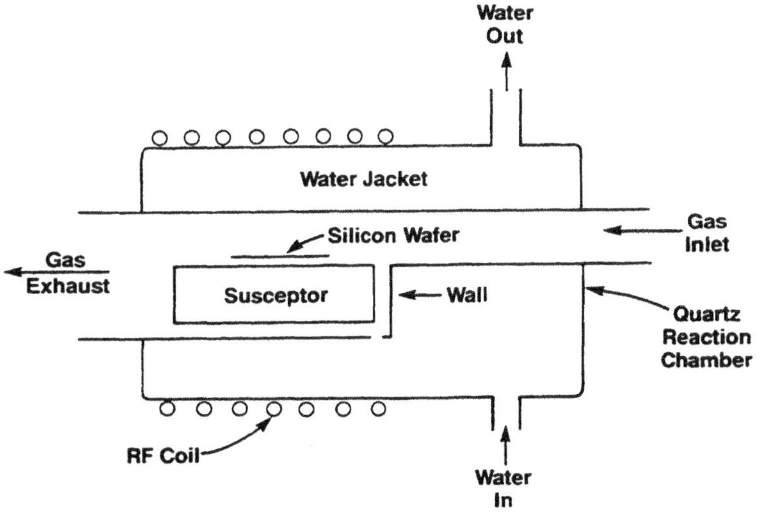

FIGURE 13-13-3. Schematic of gas-phase etcher for silicon. From Ref. 75b, reproduced by permission of the Electrochemical Society Inc.

3. Chemical reaction which includes the separate steps of adsorption,

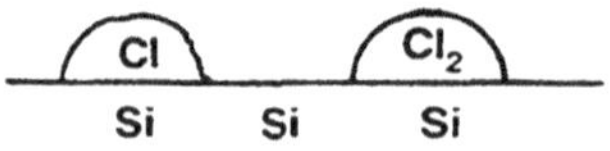

reaction on the surface,

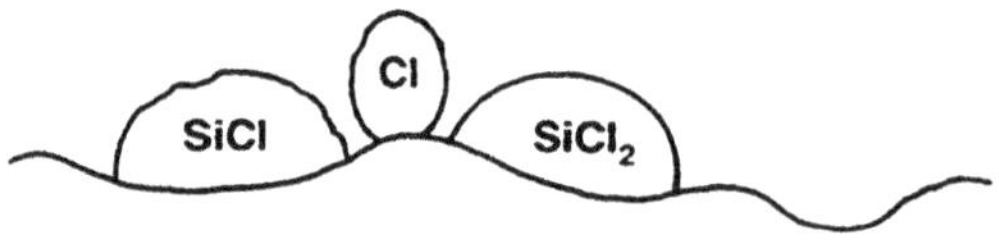

followed by desorption,

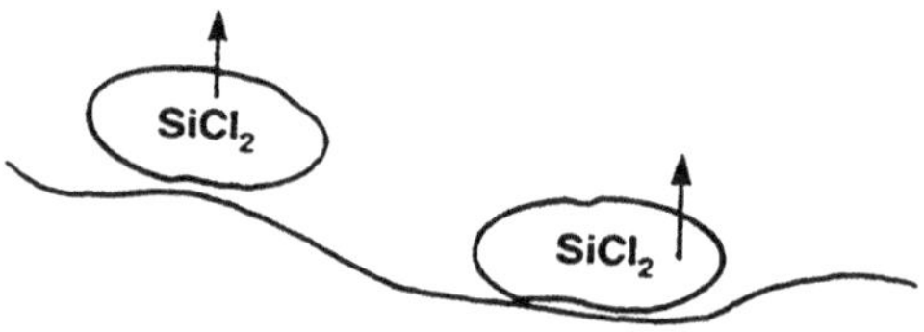

The three stages of the reaction can be examined for the reaction of Cl_2 with Si at 1100°C at an etch rate of 1 μm/min:

1. *Transport*—transport was not important since a flow system was used.
2. *Diffusion*—since the rate of gas diffusion increases with T^n and n is between 1.5 and 2.0,(319) the activation energy of 10 kcal/mole is for a diffusion-controlled reaction (see Section 13.1). The rate of Cl_2 etching varies as T^4 between 900 and 1100°C.
3. *Chemical reaction*—the rate of chemisorption appears to be the limiting step.

TABLE 13-13-2. Gas-Phase Silicon Etching by Thermal Activation

Etchant	Reference
H_2	U.S. Patent 3,243,323 (1964), TI
HI	U.S. Patent 3,366,520 (1968), IBM
$SiCl_4$	U.S. Patent 3,392,069 (1968), Siemens
ClF_3	U.S. Patent 3,511,727 (1970), Motorola
SF_6	U.S. Patent 3,546,036 (1971), Rockwell
HF	U.S. Patent 3,669,774 (1972), RCA
HCl	308, 309
Cl_2	310
HBr	311
HI	312
H_2S–H_2	313
H_2O	314
XeF_2	315, 316
F_2	317, 318
BrF_3	568

13-13-2-2. *Plasma Etching.* By contrast, in a plasma barrel etcher, the low activation energy[323] (1–6 kcal/mole) of etching of Si, such as with CF_4, indicates that the rate of etching is diffusion limited, and is controlled by the rate of diffusion of halogen neutrals to the wafer surface.

The diffusion coefficient (D) is inversely proportional to the gas pressure (10^{-4}–10 Torr). For atomic fluorine, D is about 500 cm^2/sec at 0.1 Torr and 150°C. The Γ (lifetime) is about 0.3 sec at 0.1 Torr and the distance the radical travels is $2\,Dt$, or about 25 cm.[323]

For Si etching with ion bombardment, the rate of diffusion[722] of a chemical etchant, such as HCl, is predominant. For example, HCl adds at room temperature to form Si–H by chemisorption,[309] but no Si–Cl is found. Heating to 500°C forms Si–Cl by displacing Si–H. If the silicon surface is previously bombarded by Ar^+ ions and placed in an HCl atmosphere, a rapid chemical reaction (with no energy barrier) occurs to form SiCl. Likewise, if the impinging[318] ion is Cl^+ and sputtering occurs,[687, 693]

$$\text{Si–Si} + 2\text{Cl}^+ \rightarrow \text{SiCl}_2 \qquad (13\text{-}13\text{-}5)$$

to remove Si and $SiCl_2$, then the diffusion of ions (limited step) predominates.

In the halogen family of etchants, the reaction of silicon with F_2 is exothermic at 86 kcal/mole, while $SiCl_2$ is formed with an exotherm of 49 kcal/mole.[310] Atomic fluorine etches Si with an activation energy of 2.8 kcal/mole[318] while molecular F_2 etching has been reported as requiring 8 or 9.6 kcal/mole.[317, 318] Typical activation energies are given in Table 13-13-3. Since the reported activation energies are subject to a loading effect, care must be exercised in interpretation. Since resists have higher activation energies, poor temperature control and high wafer surface temperatures will induce resists to erode faster than Si. Etching of Si and SiO_2 by CF_4/O_2, SiF_4, SF_6, and NF_3 proceeds in a plasma with low activation energies (1–4 kcal/mole). The atomic or neutral etchants produce isotropic images.[318]

13-13-2-3. *Plasma Etch Rates of Radical Systems.* Since plasma reactors operate at an upper pressure (P) of 10 Torr, the residence time (Γ) of radicals in a reactor of volume V is given by[324]

$$\Gamma = PV/Q \qquad (13\text{-}13\text{-}6)$$

TABLE 13-13-3. Activation Energies of "Si" Etching

Substrate	Etchant	Condition	kcal/mole	Reference
Si	F	Plasma	3	318
Si	F_2	Thermal	8	317
Si	F_2	Thermal	10	320
Si	Cl	Plasma	1	321
Si (polycrystalline)	CF_4/O_2	Plasma	2.5	322
SiO_2 (thermal)	CF_4/O_2	Plasma	2.4	322
SiO_2 (sputtered)	CF_4/O_2	Plasma	2.4	322
Si_3N_4	CF_4/O_2	Plasma	3.6	322
Resist	O	Plasma	9	323

where Q is the flow rate. Typical flow rates for reagent gases entering a plasma reactor range from 50 to 1000 ml/min. A flow rate of 100 ml/min corresponds to about 2×10^{21} molecules/min.[325] About 13% of the CF_4 molecules are converted into radicals,[326] so the wafers are subject to about 2×10^{20} radicals/min. Etching a silicon dioxide layer, on twenty-five 3-inch wafers to a depth of 5000 Å, involves the removal of 10^{21} silicon atoms from the batch in about 5 min.[325]

The number of radical etchants is comparable to the number of Si atoms. In wet etching, the number of etchant molecules is 10^5 higher (Section 13-8-2). Since the etching rate in a barrel reactor depends on the number of radicals and the number of Si atoms, the rate is markedly dependent on the wafer loading, RF power, flow rate, and pressure. These become important process variables in determining the rate of etching. Typical neutral–radical assisted etch rates are in the 100–3000 Å/min range, while conventional plasmas can only generate enough radicals to sustain 10^3 Å/min. However, if more effective power sources such as cyclotron or magnetron discharge[534, 726] are used, etch rates of 10^4 Å/min can be achieved.[327] The high-rate etchers are useful for uniform single-wafer etching at high throughputs.

Bombardment of Si with atomic radicals (1000 Å/min) is not as effective as chemisorption. While molecular or atomic fluorine has a low sticking coefficient of 10^{-5} of Si, XeF_2 absorbs strongly (sticking coefficient of 0.25) and rapidly etches Si at 25°C at 7000 Å/min. The generation of huge quantities of etchant radicals ($> 10^{23}$/min) by photons[739] (IR and UV laser) can produce rapid etching (see Chapter 16). Etch rates of 1200 μm/min were achieved for Si etching, using IR excitation of CF_3Br (Table 13-13-4). Since light sources cannot achieve a high power over $> 4\ cm^2$ area, their present usefulness is restricted to step-and-repeat operation. Laser-assisted plasma RIE has been used to augment Si etching. The laser creates electron–hole pairs on the semiconductor surface to accelerate etching by reactive ions.[738, 739]

13-13-2-4. *Plasma Ion Etching.* Ions such as CF_x^+ can transfer kinetic and ionization energy[683, 687, 693] to an Si surface, resulting in sputtering ionization, or sublimation of a new chemical species (SiF_x). Since Si is covered by native oxide, ions sputter away the oxide, to allow complementary neutral radicals and other ions to react with the fresh Si surface. Coburn and co-workers also investigated the interaction of CF_3^+ with SiO_2 and suggested that the ion bombardment raised the sticking coefficient of CF_x radicals.[300] Trifluoryl intermediates are more preferably chemisorbed on Si than SiF_2. In addition, kinetic plasma etching models overlook the structure[579]-dependent adsorption.

TABLE 13-13-4. Comparison of Radical Etching with IR, UV, and RF Sources for Etching Si

Excitation	Etchant	Rate (μm/min)	Reference
CO_2 (10.6 μm)	CF_3Br	1200	328
CO_2	SiF_6	12	315
Ar (546 nm)	Cl_2	5	314
RF (13.56 MHz)	Cl_2	0.1	321

TABLE 13-13-5. Ion-Assisted Si Etching[a]

Etch component	Etch rate (Å/min)
Ar^+	20
Cl^+	70
$Ar^+ + Cl^+$	730
$Ar^+ + Cl_2$	720

[a] From Ref. 333.

Ion bombardment can induce many physical and chemical actions[(671)]:

1. Desorb SiF_x products[(300–302, 330)]
2. Raise sticking probability of radicals[(302)]
3. Damage Si to increase radical attack[(303, 554)]
4. Sputter C and polymeric deposits[(290, 331, 332, 682)]
5. Sputter Si[(322, 333)]
6. Activate Cl_2 to form Cl on Si[(322)]
7. Transfer energy to activate Si[(625)]

In a mixture of etchants, the species with the lowest ionization potential will form the abundance of ions.[(334)] The etch rates of Si in Ar/Cl_2 can be separated into the contributions of each component (Table 13-13-5).

In order to provide high etch rates ($> 1\ \mu m/min$) such as in single-wafer etchers,[(631)] high plasma ion densities are created by the short mean free paths (planar diode reactor), auxiliary electron source[(728)] (triode reactor), or by increased electron path length in a magnetron source.[(631–634, 726, 727)]

Even if $Ar^+ + Cl^+$ bombards the surface or Cl_2 is added to Ar^+ (the Ar^+ dissociates the chemisorbed Cl_2), the ion-assisted etch rate is an order of magnitude higher than the Ar^+ sputter etch rate. For SiO_2 etching, the components of CF_4 ionization were analyzed (Table 13-13-6). Direct argon ion milling shows little selectivity and low-atomic-number species etch at the same rate (Table 13-13-7). Very large yields (25 Si atoms/1000-eV Ar^+ ions) have been measured.[(337)]

The larger the mass of the ion, the more effective the ion will perform SiO_2 removal.[(629)] In a CF_4 discharge, about 2% ionization occurs.[(297)] At low pressures ($< 10^{-2}$ mm), the etch rate is dependent on the ion flux, since the surface is

TABLE 13-13-6. Ion-Assisted SiO_2 Etching[a]

Etch component	Etch rate (Å/min)	S^b
CF_3^+	600	6
CF_2^+	360	3
CF^+	200	2
F^+	40	0.5
C^+	40	0.5
Ar^+	40	0.5[(336)]
Cl	30	0.7[(335)]

[a] From Ref. 322.
[b] Sputter atoms/ions.

TABLE 13-13-7. Etch Rates for Argon Ion Etching[a]

Film	Å/min
Si	360
SiO_2	400
KTFR	390
AZ-1350	600
PMMA	840
Al	444

[a] From Ref. 412.

saturated with materials.(322) Bombarding ions "clear" fluorinated products from the surface.(544) CF^+ is detected in the removal of Teflon polymers.(710)

In general, Si and SiO_2 etch rates of 500–1000 Å/min are achieved with Cl- and F-containing gases, with beams typically of 1000 eV and 0.5–1.0 mA/cm^2. These rates represent etching yields of about 3 atoms/ion, consistent with superposition of physical sputtering, and desorption from the ion flux itself.(295)

By itself, CF_4 does not absorb on SiO_2 but CF_2 does.(579) CF_2 is generated in an ion source discharge plasma, in addition to the CF_x (Table 13-13-6) fragments present in the beam as ions.(338) Additional increases in etch rates are achieved when a reactive molecular gas adsorbs and is activated, such as in etching Si/SiO_2 with Ar^+/CCl_4.(339)

13-13-2-5. *Role of Plasma-Deposited Polymers.* Besides the active radicals and ion-assisted etching, a side reaction occurs with some carbon-based etchants and organic resists. Plasma-deposited polymers have been suggested to polymerize by classical addition polymerization. Adsorbed species can polymerize by secondary electrons generated by ion bombardment.(340) Some catalytically active surface contaminants serve as chain initiators for polymer growth.(532) Most probably, chain initiators are nonvolatile metal fluorides, such as AlF_3 and CuF_2 (from mask and electrode) and NaF from the resist process.(532)

Table 13-13-8 compares the polymerization and etching tendency of fluorocarbon Si etchants. At 1-mm pressure, CF_4 does not readily polymerize from the carbene state, but instead an unsaturated fluorocarbon readily forms polymers. The plasma etching process involves removal of the fluorine by Si prompting a lower

TABLE 13-13-8. Etch-Polymer Tendency of Si Etchants as Function of F/C[a]

F/C	Etchant	Polymerization	Etch rate
4	CF_4	Low	High
2	C_2F_4	High	Low
3	C_3F_8	Moderate	Moderate
2	CH_2F_2	High	Low
3	CF_3Br	High	High

[a] From Ref. 341.

F/C ratio and an increase in the polymer deposition.[332] CF_4/H_2 will also form polymers via loss of F by HF formation. In a unique reaction, the CF_3Br etchant polymerizes since CF_3 removes Br[331] from the ends of monomers to promote additional polymerization. Once a skin layer of polymer forms, the growth of the new layers is suggested to occur by the activated growth model,[342]

$$CF_x\cdot + \text{adsorbed } P_n \rightarrow P_{n+1} \tag{13-13-7}$$

$$e, I^-, I^+ + P_{n+1} \rightarrow (P_{n+1})^* \tag{13-13-8}$$

$$CF_x\cdot + (P_{n+1})^* \rightarrow P_{n+2} \tag{13-13-9}$$

The adsorbed polymer can inhibit etching, unless preferentially removed by vertical ion bombardment[610–612, 666, 698, 707] (Fig. 13-13-4). From argon bombardment, the sputter yield is 10^2 per CF_2 group, for fluorocarbon polymers per ion. This is two orders of magnitude greater than the removal of Si ion.[354, 355] Thus, the inhibitor is easily removed while the unbombarded sidewalls act as a "dragon's blood" to inhibit lateral (sidewall) undercutting.

The F/C ratio can also be lowered by erosion of the organic resist, or by using polymeric electrodes. The passivation of Si occurs if CF_4/H_2 is used to protect SiO_2[343] and achieve a high resist or SiO_2 to Si etch rate ratio of > 20. Since the resist erosion consumes the F · to form CF_x, in a limited supply of reactants, the desired products exhibit a "loading of wafers effect." Mogab and Levenstein[290] and later Norstrom and co-workers[528] showed a resist pattern dependency, as well as a silicon batch dependency, on etch rates and selectivity. Thus, etch rates measured without resist masks are subject to reinterpretation in the presence of organic resist erosion. Highly eroding acrylate or polyolefin sulfones can contribute many polymerizable intermediates. In addition, the rate of polymer deposition must be greater than the etching rate or the total etching will be retarded. O_2 addition to CF_4 is envisioned to strip the deposited carbon or polymer layer.[300, 349] Thus, plasma etching in CF_4 or CF_4/O_2, even in a planar bed reactor, produces isotropic images at the usual 10^{-2} mm pressures, since the polymeric intermediates are intercepted by O radicals. Because F radicals do not adsorb strongly on Si, CF_4 undergoes little polymerization. A nonpolymeric deposit of $SiBr_4$, created by the addition of HBr to fluorine-based Si etchants,[705] has been used as a banking agent.

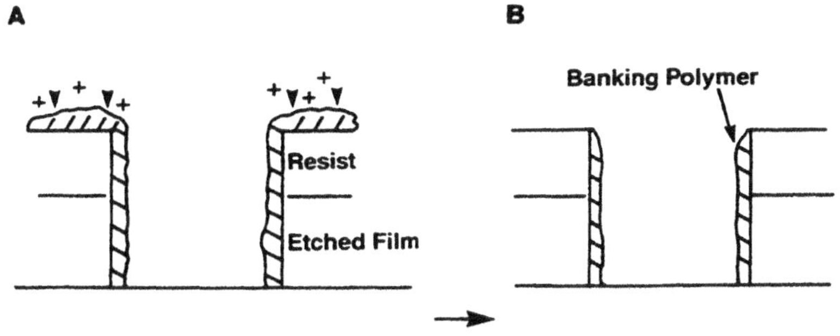

FIGURE 13-13-4. Plasma-produced banking agent deposits on (A) all etching surfaces or (B) is removed from horizontal surfaces by ions but not from vertical sidewalls.

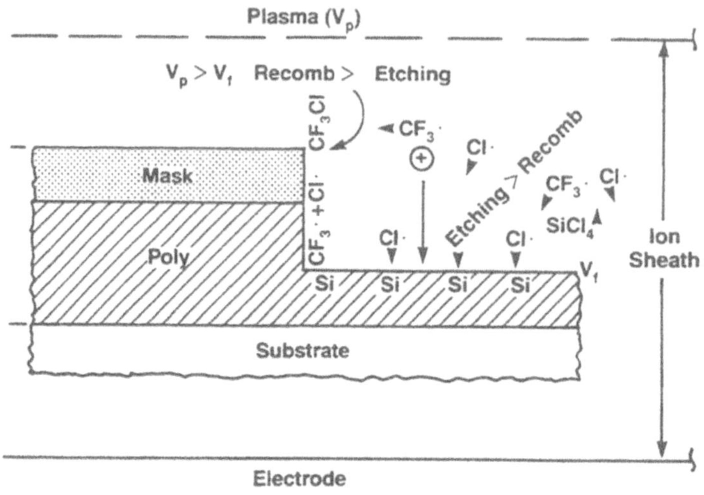

FIGURE 13-13-5. Recombination of radical etchants on sidewall to inhibit lateral etching. From Ref. 290, courtesy of AIP.

13-13-2-6. *Gas-Phase Recombination Model.* Chlorine-based etchants such as C_2F_6–Cl_2 produce anisotropic images, while CF_4 does not. Mogab and Levenstein(290) suggested that since chlorine-based etchants do not form polymers, they rely on a diffusion termination mechanism to inhibit lateral etching (Fig. 13-13-5). The surfaces of the sidewall are not directly bombarded by ions (except at high pressures due to nonuniform sheath potential, see Section 13-13-4-2. Recombination of the etchant can occur to terminate etching. In the vertical direction, the etchant can recombine, but is dissociated to components which etch silicon. Some gases (e.g., CF_3Cl, C_2Cl) produce both an etchant (Cl) and recombinants (CF_3, CF_2) in the plasma.(197) Adams and Capio(345) carried out a detailed study of the C_2F_6–Cl_2 ratio and concluded that a maximum (balance) concentration exists at a 4:1 mixture (see Section 13-13-4-1) for anisotropic features.

13-13-2-7. *Role of Electrode in Etching.* Any chemical rate will be dependent on the lifetime, concentration, and energetics of a particular atomic or molecular species. The breakdown of a gas in a plasma can generate a multitude of excited species. Flamm and Donnelly(197) have tried to categorize 12 etchants and 20 substrates into a table of etchants, each of which in a plasma can unscramble (hopefully) to produce hot species such as CF_x, F, and Cl atoms, neutrals, or ions. For example, the wide variation in Si etch rate, for a reactive sputter etcher, is shown in Table 13-13-9. Although there are many etchants, the choices considerably trimmed to one or two compositions after considering the selectivity, anisotropy, toxicity, and contamination issues (these will be discussed in the next section).(735)

In a chemical reaction, the reaction vessel is assumed to be inert. In ion-assisted etching, sputtering and recombination, or etching can occur on either electrode. The electrical extraction of ions, whether at a floating, grounded, or RF powered electrode, can strongly influence the ion concentration and, hence, the rate in a second-order plasma etching reaction. An example of the influence of electrode

TABLE 13-13-9. Etchants and Si Etch Rate[a]

Etchant	(Å/min)
Ar	124
C_2HCl_2	330
$CHCl_2F$	410
CF_4	900
$(CCl_2F)_2$	1280
$CHClF_2$	1430
CCl_3F	1670
$(CBr_2F)_2$	1850
CCl_2FCClF_2	2015
CCl_2F_2	2200

[a] From Ref. 346.

material is given in Table 13-13-10.The erosion of the electrode can lead to changes in F/C ratio and polymerization. Contamination of the wafer surface from sputter deposits, as well as gas phase–surface interactions, can influence the ultimate selectivity of an etching process.

In addition to the electrode material, the doping of silicon and silicon dioxide can also significantly influence the etch rate(551) (Table 13-13-11). *n*-type doping increases the etch rate while *p*-type doping inhibits it.(630, 711) No differences in etch rates for P or As dopant ($10^{18}/cm^3$) were noted. Matsumoto(347, 348) suggested that the more volatile P, As, or B halides are responsible for increased etching; Mogab and Levenstein(290) used a free electron model to explain the enhancement. The presence of extra electrons, rather than holes, lowers the Si band gap (bond energy). The concentration of free electrons, rather than the chemical nature of the impurity, determines the enhanced etch rate. In this case, the doping reduces the energy barrier to charge transfer from silicon to bound Cl atom. The halogen atoms become more anionic (Cl^-) and more mobile in the silicon lattice. Consequently, incident Cl atoms can more readily penetrate the surface layer, and increase the etch rate.(351)

TABLE 13-13-10. Electrode and Si Etch Rate (RF sputter electrode)

		Etchant (Å/min)	
Electrode	Ar	CCl_2F_2	CF_4/O_2
Silicon	124	2200	
Quartz	159	533	
Pyrex	103	347	
Al	166	1624	
Mo	185	836	
Steel	154	522	
Phenolic coated	185	410	
Al (grounded)			200(341)
Al (RF)			6000(341)

[a] From Ref. 346.

TABLE 13-13-11. Dopant Effects on Plasma Etching

Dopant	Å/min	Etchant	Reference
PSG	200	CF_4	347
AsSG	100	CF_4	347
BSG	110	CF_4	347
CVD SiO_2	75	CF_4	347
Thermal SiO_2	50	CF_4	347
P-poly Si ($10^{20}/cm^3$)	300	CF_4	348
P-poly Si ($10^{21}/cm^3$)	230	CF_4	348
Undoped poly Si	1300	C_2F_6–Cl_2	290
Undoped Si	600	C_2F_6–Cl_2	290
Thermal SiO_2	100	C_2F_6–Cl_2	290

13-13-2-8. *Summary of Plasma Etching Mechanisms.* A summary is given in Table 13-13-12 of the many influences on rate.[544] I have highlighted the significant physical and chemical role of the plasma neutrals, ions, and radicals in determining the isotropy of the direction of the etchant. Directed ions are required for anisotropic etching. The rate of etching is also influenced by the chemical structure of the etchant, the integrity of the electrode, and the reactor configuration. In most cases, the rate of etching is usually linear, except for an inhibition by native oxide or deviation caused by large thermal excursions (> 50°C).

The separation of chemical and physical factors that influence the isotropy or anisotropy of etching is given in Table 13-13-13. The chemical nature of the ion dominates the directionality and rate of etching. Both radical and ionic species are in some cases necessary for plasma etching. Reactive ion beam etching offers better control of the measurable concentrations of ionic etchant. A heavier mass of ion etchant results in greater enhanced etching.[629]

The passivation of the sidewalls with a polymer deposit (dragon's blood) is a critical inhibited mechanism for any plasma etcher, regardless of the pressure or ion flux. This is the *in situ*, resist chemical approach to anisotropy. The main physical approach for anisotropic etching is a collimated beam of ions, achievable by

TABLE 13-13-12. Major Chemical and Physical Influences on Rates

Chemical	Physical
Chemisorption	Sticking coefficient
Ion concentration	Adsorption isotherm
Radical concentration	Pressure
Atom concentration	Mass of ion
Resist products	Temperatures[640]
Wafer load	Area of electrodes
O concentration[a]	Power-frequency
C/halogen (etchant)	Dopant concentration in Si

[a] Can be formed from resist, electrode, or substrate (e.g., $C_6H_9O_2$, SiO_2).

TABLE 13-13-13. Isotropic versus Anisotropic Etching

Factor	Isotropic	Anisotropic	Reference
Chemisorption	Low	High	350
Radical concentration	High	Low	290
Ion concentration	Low	High	330
Ion mass	Low	High	352
Polymer formation	Low	High	303
Resist erosion	High	Low	345
Temperature	High	Low	321
Sputtering	Low	High	322
Loading	Low	High	290
Electrode spacing	Wide	Narrow	298
Volume of reactor	High	Low	290
Voltage	Low	High	330
Pressure	High	Low or high[a]	303, 353

[a] To form polymer deposits on sidewall.

parallel electric fields. Plasma gas-phase assisted etching can provide any degree of directional etching. For anisotropic etching, positive ions play three major roles:

1. Ions damage the silicon or other materials and break covalent bonds. The damaged material etches rapidly.[682–691]
2. The ion itself such as a fluorine or chlorine derivative is the principal etchant.[11, 685–691, 724]
3. The ion impact clears the horizontal surface of the protective polymer or nonetching deposits. The coated vertical sidewalls do not receive direct ion bombardment.[11, 682]

All three mechanisms can be found in Si, SiO_2, Si_3N_4 etching under a variety of conditions. Each etchant, apparatus, and set of conditions generates its own specific kinetics.[671] At present, the complete control of all plasma parameters, in order to achieve high selectivity and profile control, remains more of an empirical science, since most parameters produce a multitude of effects on the chemical and physical nature of the plasma environment.[735]

Current trends in research suggest that with more mechanistic studies, reactor conditions and etchants can be chosen for custom tailoring of any profile.

13-13-3. *Selectivity*

Selectivity is defined as the ratio of etch rates of the resist to the substrate film and to the sublayer. Preferably the etch rate of the film is high and the resist or sublayer does not etch at an appreciable rate. Wet etching offers infinite selectivity with respect to the resist or underlayer Si attack. However, the resist image may fail by adhesion-stress loss (Section 13-2-4). In plasma etching, the resist adhesion is unimportant during etching, but the resist erosion rate, because of the high yield of 10^2 CH_x units per ion impact,[354, 355, 673, 674] can be higher than Si or SiO_2 etch rates.

Selectivity is a main criterion in the performance of plasma etching. When needed, selectivity is available, such as in etching multiple layers of SiO_2/Si_3N_4

(Fig. 13-9-1) at equal rates. High selectivity ratios are required for $Si/SiO_2/Si_3N_4$ combinations, in order to etch active gate areas and contact windows. However, after etching upper layers, sometimes an etch stop layer is needed to protect the bottom layer from overetch.[613] The first requisite in high-selectivity anisotropic etching is a low-erosion-rate resist. In anisotropic etching, the resist loss of more than 20% leads to isotropic undercut.[345]

Organic resist erosion occurs readily in certain polymer structures. A very erodable resist is poly-1-butene sulfone (PBS) which strips at about $10\times$ the rate of DQN resists,[356] Nonaromatic structures, which are thermally degradable, erode at faster rates than resists composed of polyaromatic backbone polymers. The major instability of organic resists is the high yield of molecules per sputter ion impact. Over 10^2 C fragments are formed per ion bombardment.[295, 354, 355, 674] Hydrogen atoms are also abstracted by Cl or F radicals. Overall, the sputtered fragments are oxidized into small molecular species:

$$CH_x + O_2 \rightarrow CO_2 + H_2O \tag{13-13-10}$$

$$CH_x + F_2 \rightarrow CF_4 + HF \tag{13-13-11}$$

In a plasma system, the moles of etchant is limited from a mass viewpoint. Polymers with more C atoms require more moles of etchant[521, 674] and exhibit greater plasma stability (Table 13-13-14) since insufficient moles of etchant is present. Aromatic polymers, because of their higher empirical mass, require more moles of etchant than aliphatic polymers. Since short UV (200–300 nm) radiation exists in a plasma, the aromatic group also stabilizes the resist (by fluorescence) against chain scission. Once chain scission occurs, the PBS, polystyrene, and PMMA addition polymers can revert back to monomers by unzipping (see Chapter 14 on resist stripping) at thermal conditions above their ceiling temperatures, T_c,

$$\text{Polymer} \rightarrow \text{Monomer} \qquad > T_c \tag{13-13-12}$$

PBS reverts to monomer at $T_c > 60°C$.[522] Thus, a high-temperature excursion in the reactor (due to plasma bombardment or poor cooling) should be limited to less than the T_c and not above the glass transition temperature, T_g.[744] At low temperatures (< 45°C) on a cooled reactor plate, PBS has been used successfully to mask the etching of Si_3N_4 and poly Si, with a 10 to 1 resist/substrate selectivity.[356] Horike etched SiO_2 at 1 μm/min (25°C) with no AZ-1035J loss.[389] Etchants such

TABLE 13-13-14. Resist Mole Requirements

Resist	Moles (O_2, F_2)	Relative erosion, CF_4–O_2
$C_4H_8SO_2$–PBS	6	8.2
C_5H_8–P-isoprene	7	5.7
$C_5H_6O_2$–PMMA	6	4.2
C_8H_9–P-styrene	10	1.2
C_8H_8O–novolak	11	1

as C_3F_8 which self-polymerize tend to provide some deposits to show down the erosion of the photoresist. A DC bias applied to the wafer also reduces the resist erosion.(703)

The etch and thermal resistance of polymeric resists go hand in hand. Polymeric resists, stabilized to heat flow, are also more resistant to main chain fracture. Condensation polymers with aromatic groups(596) such as polyarylethers, sulfones, and carbonates offer heat and plasma resistance.(357) These resists are popular masks for use in multilayer resists. The model aromatic polymers for the best plasma resistance are the phenolic-based resists, or polystyrene-based negative resists. Only (2 ×) sputtered carbon(674) or inorganic resists show better plasma resistance. The addition of magnetic particles to AZ-1350 is claimed to slow down its erosion in a magnetron etcher.(709)

The posttreatments to thermally stabilize DQN (phenolic resist) were given in Table 13-12-3. Stabilized novolak resins are necessary, such as in Al etching, to prevent resist foaming and swelling.(401) Treatments to stabilize polyacrylate (PMMA) *e*-beam resists are given in Table 13-13-15. Incorporation of phenyl groups by direct substitution or postimage reaction(715, 745) can reduce PMMA erosion by 3 ×. Similarly, the incorporation of metals such as Na,(360)Pb,364) Sn,(362) or Ba(363) by copolymerization or postdiffusion into the developed image stabilizes the resist by the supposed formation of nonvolatile fluorides. In addition, silicon can be incorporated into novolak resists by posttreatments with HMDS or $SiCl_4$. The resist etches slowly in CCl_4 or O_2 plasma.(704) Recent studies have shown that PMMA can be used to etch Si and insulators if oxygen- and fluorine-deficient halocarbons are used as etchants.(365) The PMMA acts as a substrate for halocarbon polymer banking agent formation, which successfully competes with the resist erosion. Watanabe and co-workers(366) have added polymerizing CH_4 to an Ar sputter etcher and successfully etched, using a PMMA mask. A resist overcoat formed, to counter the reduction in resist pattern width (Fig. 13-13-6).

Negative deep-UV and *e*-beam chlorostyrene resists(521) offer 150°C thermal stability and plasma resistance equivalent to diazonovolaks. Practically all the deep-UV, X-ray, or *e*-beam negative resists, containing aromatic rings, will offer good plasma resistance. Iodopolystyrene etches 10 × slower in oxygen plasma due to the formation of I_2O_5.(660, 673–676) Since the erosion of organic resists is limited by the high sputter yield for polymeric carbon,(355, 674) inorganic resists or metallic (secondary) masks(295, 327) are the ultimate nonerodable mask. Basic enhancements in their exposure sensitivity are required before inorganic resists become popular.(373–376)

TABLE 13-13-15. Plasma Resistance Improvements of Acrylates

Resist	Remarks	Reference	Resistance factor
PMMA	Add aromatic antioxidants	358	2 ×
PMMA	Soak in epoxy monomer	359	2 ×
PMMA-MA	Cross-link with silanes, soak in NaOH	360	10 ×
PMMA-phenyl MA		361	3 ×
PMMA + polystyrene		*a*	2 ×

[a] U.S. Patent 4,405,708 (1983), Phillips.

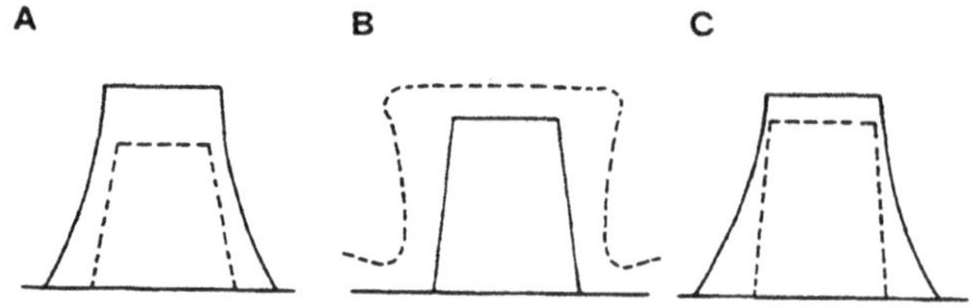

FIGURE 13-13-6. Changes of resist profiles with various plasma etchants (A) after sputter etch with Ar gas, (B) after hydrocarbon film deposit with CH_4 gas, and (C) with a mixture of Ar gas and CH_4 gas. From Ref. 366.

Metal masks have been used for all forms of ion-beam etching.[377] The metals are formed as the second layer in a trilayer (multilayer resist, Chapter 12) or are lifted off and patterned directly on Si,[661] silicon nitride, or silicon oxide. The metal masking requires additional process steps but offers the benefits of thin masks which overcome sputter redeposition[378] problems (Fig. 13-13-7) in ion-beam etching. By themselves, multilayer resists offer excellent thermal and plasma resistance, since the underlayer can be chosen from any halogen-resistant polymer. The bottom layer is patterned by reactive ion O_2 etch. Figure 13-13-8 illustrates common bilayer and trilayer resists which can be used for lift-off or etching with suitable polymeric underlayers, sputtered carbon, or inorganic films. For this application, the AZ-1350 underlayer can be thermally hardened at 150°C prior to topcoating with multilayers. Without proper baking, residual solvents left in the diazonovolak resist can give rise to erosion or popholes (pinholes), from solvent outgassing during plasma etching.[379, 380] In this case, ramped postbake is necessary to drive out the residual solvents.[746]

Some representative selectivities of the substrate (F)/resist (R) etching ratio are given in Table 13-13-16 for reactive ion-assisted etching. The single most important factor for selectivity is the choice of etching gas. Low resist erosion and low

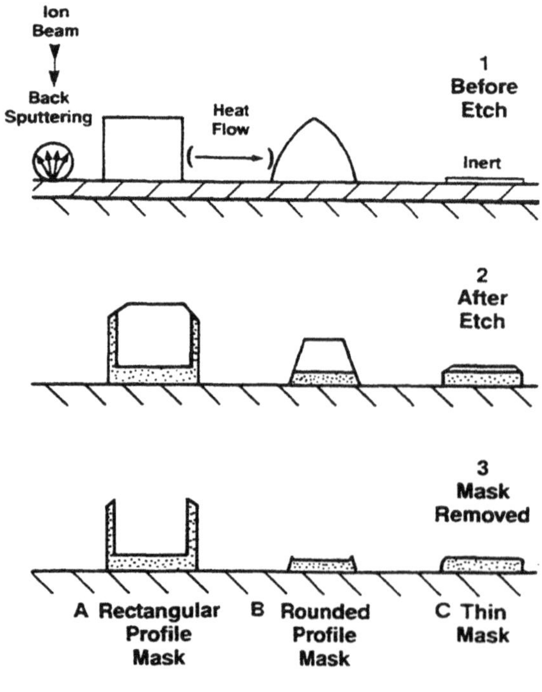

FIGURE 13-13-7. Masking during ion-beam etching. (A) Erodable rectangular mask; (B) rounded resist mask; (C) non-erodable thin metal mask. From Ref. 412, courtesy of AIP.

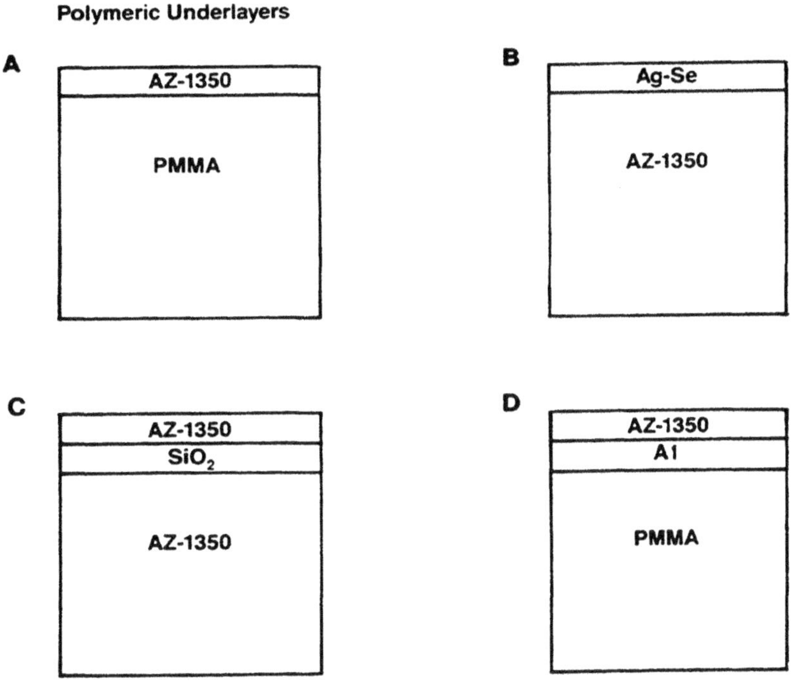

FIGURE 13-13-8. Some examples of bilayer and trilayer resist images that can be used for reactive ion etching. (A) From Ref. 758. (B) From Ref. 759. (C) From Ref. 760. (D) From Ref. 761.

underlayer erosion are desired. Furthermore, etch rate monitors are absolutely essential to terminate etching at each film sublayer.

By varying the power frequency and pressure, optimization of the throughput can be achieved.[385] Since sputter yields for Si, Si_3N_4, and SiO_2 are about equivalent for the same ion bombardment,[354] the chemical reactivity of Si, SiO_2, and Si_3N_4 to halogens, halocarbons, or halo etchants determines the selectivity ratio of film (F) to underlayer (U), (Table 13-13-17). For example, Kushner varied the Cl/F ratio in favor of a high Si/SiO_2 selectivity.[386] The high selectivity of SiO_2/Si is the major objective of the plasma etch process (to mimic HF wet etch). In order to accomplish high selectivity, the H/F ratio is manipulated in CF_4–H_2,[382] C_2F_6–C_2H_4,[394] and CHF_3.[395] There are three basic mechanisms of

TABLE 13-13-16. Resist-Plasma Selectivity

Film (F)	Resist (R)	F/R	Etchant	Reference
Si_3N_4	AZ-2400	10	CF_4	381
SiO_2	AZ-1350	5	C_2F_6	294
SiO_2	AZ-1350	10	CF_4/H_2	382
SiO_2	PMMA	4	CF_4/H_2	365
Poly Si	AZ-1470	5	NF_3	383
Poly Si	PBS	15	CF_4	356
PSG	AZ-1350	10	SF_6	384

TABLE 13-13-17. Selectivity (F/U) in Plasma Etching of Silicon

Film (F)	Underlayer (U)	F/U	Etchant	Reference
SiO_2	Si	20	H_2	387
SiO_2	Si	30	CF_4/H_2	388
SiO_2	Si	30	CF_3Cl	351
SiO_2	Si	8	HF	326
SiO_2	Si	5	C_2F_6	294
SiO_2	*p*-doped Si	20	CHF_3	389
Si	SiO_2	30	SF_6	390, 391
n-doped Si	Si	15	C_2F_5Cl	351
Poly Si	SiO_2	10	CF_4	392
Poly Si	SiO_2	14	NF_3	383
Poly Si	SiO_2	10	CCl_4	383
Si_3N_4	SiO_2	30	C_2F_6	393
Si_3N_4	SiO_2	12	CF_4/O_2	323
Si_3N_4	SiO_2	50	NF_3	398

SiO_2/Si selectivity: (1) a sidewall banking agent of polyvinyl fluoride film forms on the Si by growth from adsorbed C[698]:

$$CF_3(ads) + Si \rightarrow C(ads) + SiF \tag{13-13-13}$$

$$C(ads) + CF_x \rightarrow Poly(CF_x) \tag{13-13-14}$$

Excess H_2 generates carbon and a polymer deposits on all surfaces,[395] which requires a subsequent Cl_2/Ar cleanup.[396, 682] CHF_3/SiF_6 is recommended as a gas that forms little residue.[384, 707] In deep trench CF_4 etching of 0.2-μm-wide by 6-μm-deep holes in silicon, deposits of polymer build up on the orifice and eventually stop the etching.[751] (2) The second mode involves adjustment of the concentration of F* and CF_3^+. While the etch rate of Si will depend on the F* concentration, the etching of SiO_2 mainly involves the sputter-assisted reactions of CF_3^+. (3) Differences in the chemisorption of the etching gases have also been proposed as a basic mechanism for selectivity.[302] Br generated[523] from CF_3Br is suggested to adsorb preferentially on SiO_2, and inhibit F attack by inducing F atom dimerization.[397]

For active devices, the etchant attack of the Si layer underneath the SiO_2 remains a problem for shallow junction devices. A combination of mixed etching techniques serves to prevent the Si attack.[524] Initially, the plasma etching of the oxide to the last 500-Å thickness is performed. Wet etching with HF removes the remaining film of SiO_2, and any contamination, such as heavy metal left on the silicon surface.

Choosing selective etching gases and conditions remains a basic challenge to the plasma chemist, especially for future reactors using rapid single-wafer etching,[389] where overetching must be counteracted by a high selectivity ratio (>20).

13-13-4. *Etch Profiles and Image Bias*

The image etch bias is the loss of sidewall material beyond the edge of resist (Fig. 13-13-9). The biases of wet etching were as large (2–4 μm) as the dimensions

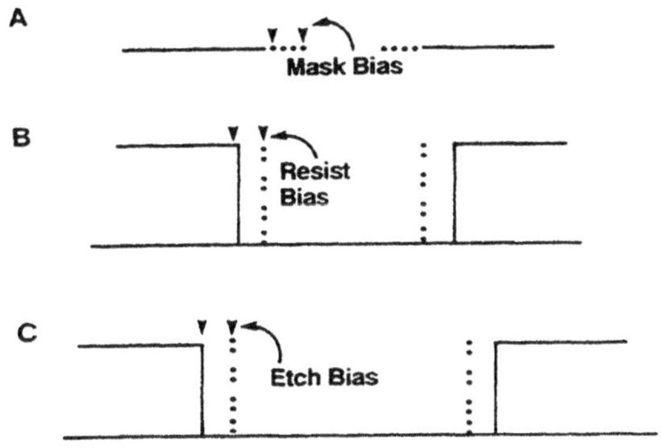

FIGURE 13-13-9. (A) Image distortion due to mask bias, (B) resist bias in printing image, and (C) etch bias in image transfer. The distortion $= (a^2 + b^2 + c^2)^{1/2}$.

in present circuits. Initially, plasma etching was performed with neutral reactants to mimic isotropic wet etching (Fig. 13-9-4). In this case, the etch biases were substantially reduced and the tolerances remained about equivalent to wet etching (Table 13-13-18). Since dry etching was 100 times cheaper than wet liquid baths, the economic cost factor was a major driving force for the rapid implementation of dry etching.[399]

As VLSI emerges toward higher density circuitry, the shrinkage of horizontal dimensions attempts to reclaim by using anisotropic reactive ions, the real estate lost in wet etching (Fig. 13-1-9). The biases of isotropic etching increase with the depth of film etched. Table 13-13-19 illustrates the typical bias associated with increasing etchable film thicknesses. The thicker the substrate(s) to be etched, the greater is the etch bias.

As the etching process encounters another sublayer, the controlling factor is the selectivity (protection) offered by the resist and the reactor conditions. A high selectivity (s/r) of substrate etch rate (s) to resist etch rate (r) is needed to cover up the poor wafer to wafer film uniformity, or print bias, by allowing a high percentage ($>$ 25%–50%) of overetching.[403] Since endpoint systems monitor large areas, overetching is a way of life and the process engineer estimates a fixed percentage beyond endpoint. Someday endpoint systems based on the e-beam scanning of a

TABLE 13-13-18. Isotropic Plasma[a] versus Wet Etch Biases and Tolerances[b]

Film	Method	Bias and 3σ tolerance (μm)
Si_3N_4	Wet	0.25 ± 0.75
	Plasma	0.0 ± 0.5
SiO_2	Wet	1.2 ± 0.75
	Plasma	0.35 ± 0.9
	Reactive sputter	0.0 ± 0.2
Si	Wet	0.9 ± 1
	Plasma	0.0 ± 0.8
Al	Wet	1.5 ± 3.0
	Plasma	0.5 ± 2.0

[a] 30% resist erosion.
[b] From Ref. 400.

TABLE 13-13-19. Plasma Etch Bias and Etch Depth

	Bias (μm)			
Selectivity (s/r[a]): Condition: s (μm thick)	(a) > 10/1 Anisotropic	(b) < 5/1 Anisotropic	(c) < 10/1 Isotropic	(d) < 5/1 Isotropic
0.1	0.0–0.1	0.1–0.2	0.1–0.2	0.1–0.2
0.5	0.0–0.1	0.1–0.3	0.3–0.4	0.5–0.7
1.0	0.1–0.2	0.3–0.4	0.4–0.5	0.7–0.9
2.0	0.2–0.4	0.4–0.5	0.5–0.6	0.9–1.0

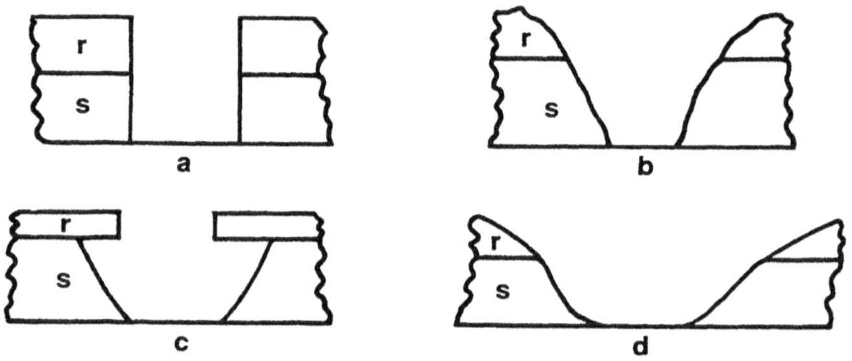

[a] Resist.

≤ 1-μm etching line will be available to precisely cease etching at the desired geometry.(719)

Selectivities of > 10 for *s/r* have been achieved with high-power microwave and magnetron sources.(404) The stabilization of the resist to heat distortion and erosion (Tables 13-12-3 and 13-13-15), which can be encountered in plasma etching, encourages higher selectivity. The resist erosion is further minimized by isothermal etching with rapid etching,(725) such as in high-pressure(353) or magnetron systems(389) (1 μm/min) using water-cooled electrodes.(434) For many of the commercial reactors, the improved batch-to-batch and wafer-to-wafer uniformities(406–9) of < ± 10% have reduced the process etch bias to 0.1–0.4 μm and the tolerances to ±0.2–0.5 μm (Tables 13-13-18 and 13-13-19).(406–409)

13-13-4-1. *Etch Profile Factors.* A conceptual understanding of the surface mechanisms of anisotropic reactions was presented earlier in this chapter. We will first examine some of the typical etch profiles in Si etching, and then examine the ion physics,(728) which controls the anisotropic features of the incoming beam of ionic and neutral species.

A comprehensive study of poly-Si etch profiles in a parallel plate reactor was performed by Adams and Capio(345) using anisotropic (A) and isotropic (I) conditions (Fig. 13-13-10). While the linewidth loss (bias) is independent of the feature size, it is primarily due to erosion of the resist, which proceeds faster in the lateral direction (isotropic) than in the vertical direction (Fig. 13-13-11). The profiles obtained by Adams and Capio are consistent with the simulations of Viswanathan(189) (Section 13-9, Fig. 13-9-4).

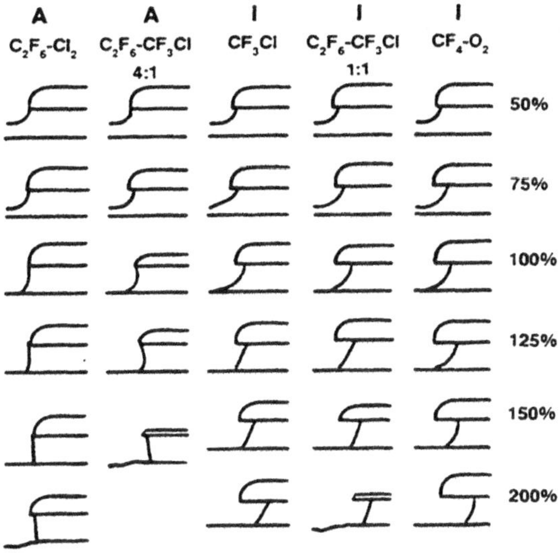

FIGURE 13-13-10. Etching profiles from dry etching of polysilicon produced with an AZ-1350 DQN photoresist and various etchants under anisotropic (A) and isotropic (I) conditions. From Ref. 345, reproduced by permission of the Electrochemical Society Inc.

The foot persistent at the 100% endpoint in Fig. 13-13-10 has been attributed to a charge buildup on the resist (insulator) during etching. Similar problems have been encountered in ion implantation wherein a secondary electron gun was used to neutralize the surface[408] charges (future need for a conductive resist). Figure 13-13-12 illustrates the charge deflection proposed by Bernacki,[409] wherein the incoming ions are deflected by the charge buildup on the resist. A charged mask (100 V) can be used to deflect the ions and form V-shaped grooves in Si.[410, 411] Positive charges on the sidewall can be built up on the insulating polymers, deposited during etching, and the charge layer can serve to protect the walls from direct ion bombardment and by combining with neutrals, diffusing to the walls (Fig. 13-13-13). This is similar to the Helmholtz potential buildup encountered in electrochemical liquid metal etching (Section 13-1-2 and Fig. 13-1-13).

The lateral erosion of the resist patterns (Fig. 13-13-10) depends on the initial profile angle of the resist with the substrate, and the isotropy of etching [vertical rate = horizontal rate (isotropic) or vertical = 10 × horizontal (anisotropic)]. For anisotropic etching, the rate of resist erosion is given by $R/\tan\theta$ (Fig. 13-13-14), and for isotropic etching by $R/\sin\theta$, where R is the linear etch rate of the resist. In Fig. 13-13-14, the rate of resist erosion divided by the resist etch rate (R) is shown as a function of the resist profile from the vertical. If the photoresist profile is ver-

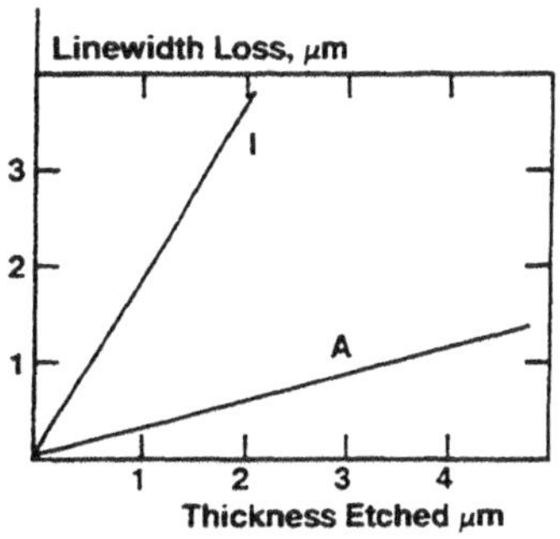

FIGURE 13-13-11. Linewidth loss for etching polysilicon with isotropic (I) and anisotropic (A) conditions. From Ref. 345, reproduced by permission of the Electrochemical Society Inc.

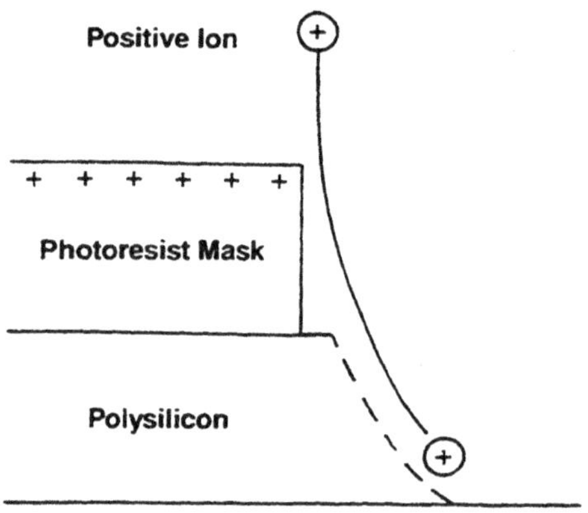

FIGURE 13-13-12. Charge buildup on resist mask and deflection of incoming reactive ion causing protruding feet in etched polysilicon. From Ref. 409; figure originally presented at Spring 1982 ECS meeting, reproduced by permission of the Electrochemical Society Inc.

tical, the erosion rate is zero for anisotropic etching. For precise RIE transfer, the control of the profiles in resist by the exposure and the development process is as important as obtaining the nominal etched linewidth. For example, to form 45° etched via holes in films of sputtered quartz, the 90° profiles in AZ-1350 are intentionally flowed in high postbake to 65° prior to RIE.(695, 701)

In the computer simulation of dry etching, the anisotropic etching component is proportional to the ratio of the local angle to flux incidence. A combination of anisotropic and isotropic conditions can be used in simulation, and be compared to published data.(544) Accordingly, experimental comparisons verify the accuracy of simulated results for both neutral and ion-assisted etching(526) in the original cast resist film. Upon development of a positive photoresist with an alkaline developer of NaOH or KOH, the silicon surface can adsorb metal ions which accelerate the silicon etching in a plasma.(468, 681)

Some unusual profiles can be created by incoming ions diverted from normal incidence, or by the resist swelling during plasma etching. Figure 13-13-15 is a

FIGURE 13-13-13. Positive charge buildup to deflect ions from eroding resist.

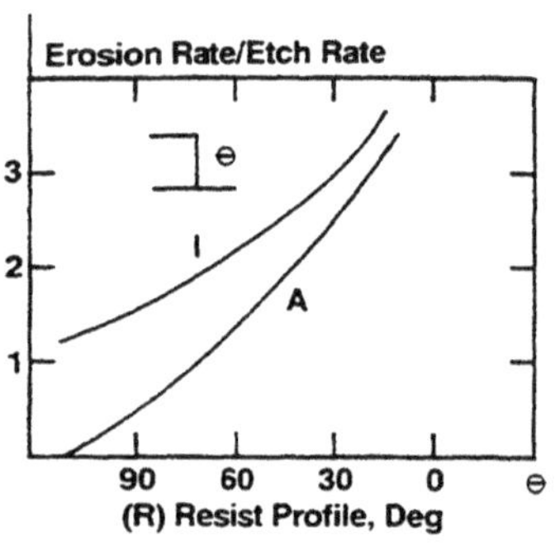

FIGURE 13-13-14. Ratio of resist erosion rate to substrate film etch rate for initial anisotropic (A) and isotropic (I) etch conditions. From Ref. 345, reproduced by permission of the Electrochemical Society Inc.

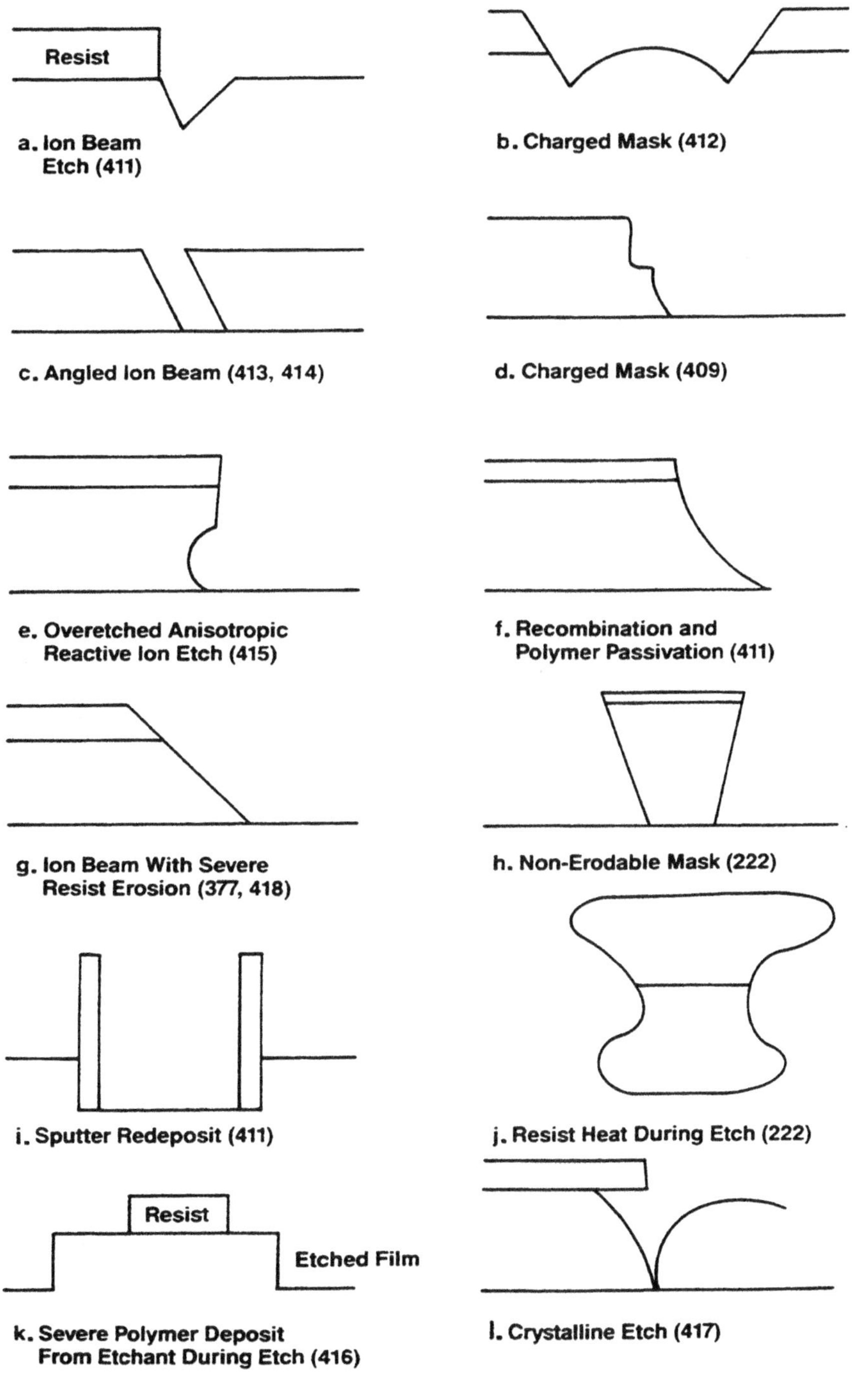

FIGURE 13-13-15. Examples of unusual etch profiles produced under dry etch conditions (references in parentheses).

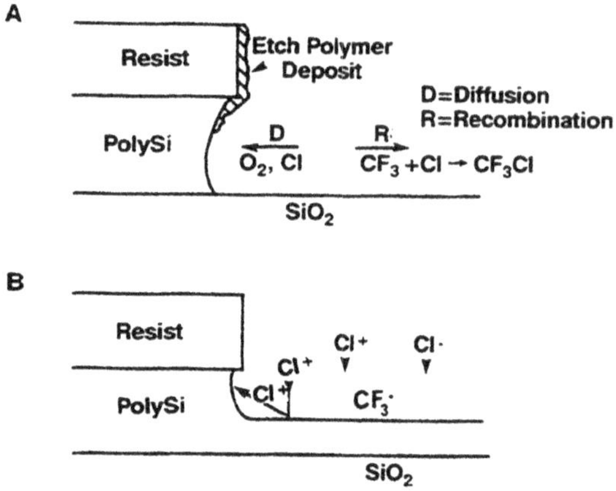

FIGURE 13-13-16. (A) Surface diffusion of oxygen or plasma radicals to attack polymer banking deposit built up during plasma etching causing isotropic undercut. (B) Backscatter of ions from substrate during plasma etching which also leads to sidewall attack and undercut. From Ref. 415, courtesy of SPIE.

portfolio of some (non-"cubical") unusual resist, or final resist-etched profiles. The reader is asked to consult the references for detailed explanations. Cases a, c, g, and i result from the extreme dependence of etching and topography on the pure ion-beam etching and sputter redeposition (i), while case k has been explained by a change in the sheath potential midway during anisotropic etching (see next section). With high-pressure (1–10 Torr) reactors becoming popular, it is expected that non-linear etching (variable ion velocity components) will produce unusual artifacts, due to ion–neutral collisions. Cases e and h bring out the importance of the overetch state, and the surface diffusion which can occur, due to the cessation of vertical etching.

In surface diffusion, neutral radicals or excited neutrals strike the lateral sidewalls. The model in Fig. 13-13-16 illustrates the role of surface diffusion in the etching of polysilicon over SiO_2. Selbrede[(415)] has proposed additional diffusion, up the sidewall, to produce a further cusp (Fig. 13-13-16). He noted that oxygen radicals, posibly from leaks (or from the SiO_2 underlayer), diffuse along the SiO_2 surface, and isotropically react with CF_3* to produce F*, which etches a notch. When isotropic ions remove the sidewall (protective polymer) deposits, this causes resist faceting in silicon etching.[(620)] Another explanation may be that of ion backscatter from SiO_2. The ions (Fig. 13-13-16) that are not involved in chemisorption or surface excitation on SiO_2, are available for sputtering. Finally, combinations of anisotropic and isotropic etching at endpoint, have been used to clean up[(409, 415)] the protruding feet (Fig. 13-13-15d, f) in the profiles and to remove the sputter or corrosion products.[(219)]

The anisotropic and isotropic profiles are the result of solid–gas interface reactions, and the "random" collision path of ions in the electric field above the wafer. We will now examine some of the parameters which control the anisotropy of the ions in the gas phase.

13-13-4-2. *Ion Transport Model.* The concentrations of positive n^+ and negative n^- ions in the plasma can be described by a Boltzmann distribution:

$$n^+ = n_p e^{-eV/kT} \qquad (13\text{-}13\text{-}15)$$

$$n^- = n_p e^{eV/kT} \qquad (13\text{-}13\text{-}16)$$

where n_P is the density of charged particles on the order of $10^{11}/cm^3$, and kT is 3–5 eV at 10^{-4} Torr and $10^9/cm^3$, 8–12 eV at 10^{-1} Torr.(327) The degree of ionization, α, is

$$\alpha = n_P/(n_0 + n_P) \tag{13-13-17}$$

where n_0 is the density of neutral particles, which is about 2% or less for plasma etching.(327)

In the plasma, the velocity direction of all the components is isotropic, as illustrated in Fig. 13-13-17.(199) A plasma ion sheath(659) forms a floating potential, V_P, to screen the plasma from the electrode potential, V_e. When the substrate is electrically isolated, the floating potential is given by(327)

$$V_P = kT_e/2e \ln 2\pi m/e'M = kT_e/2e \ln C/A \tag{13-13-18}$$

where e = absolute value of electric charge, m = electron mass, e' = natural log base, $kT_e = 4$ eV, M = ion mass, A = ion mass number, and $C = 1.26 \times 10^{-3}$. The value of V_P ranges from 11 to 25 V, while the sheath potential is approximately 5 V higher.(423) The T_e, electron temperature can be measured by the rotational Doppler effect of the plasma emission.(421) The width of d_{is} of the ion sheath is given by the Langmuir–Child equation,(327, 692)

$$d_{is} = 2^{5/4}/3\ (i_0/\varepsilon_0)^{-1/2}(M/e)^{-1/4}(-V_P)^{3/4} \tag{13-13-19}$$

where $i_0 = en_P(kT_e/e'M)^{1/2}$, i_0 = ion saturation current, and ε_0 = dielectric constant of vacuum. Since n_P is about $11^{11}/cm^3$, i_0 is 14 mA/cm^2 in a typical system and the

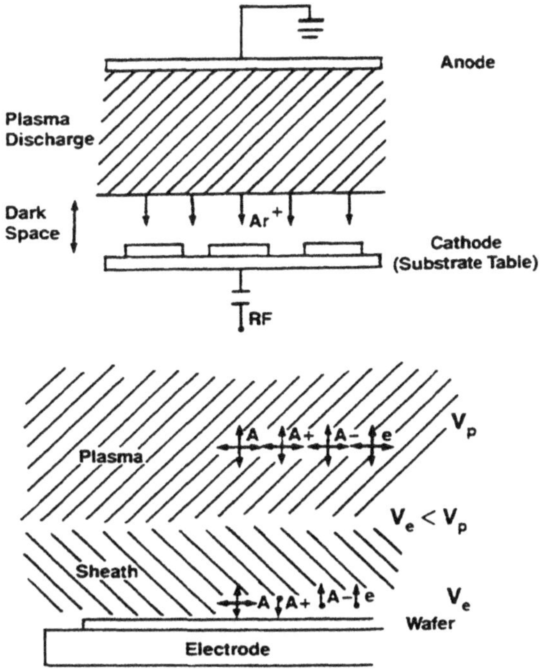

FIGURE 13-13-17. Schematic of plasma sheath and dark space near cathode electrode. V_p is plasma potential, V_e is electrode potential, A^+ are positive ions, A^- are negative ions, and e are free electrons. From Ref. 199, reprinted with permission of *Solid State Technology*, published by Technical Publishing, a company of Dun and Bradstreet.

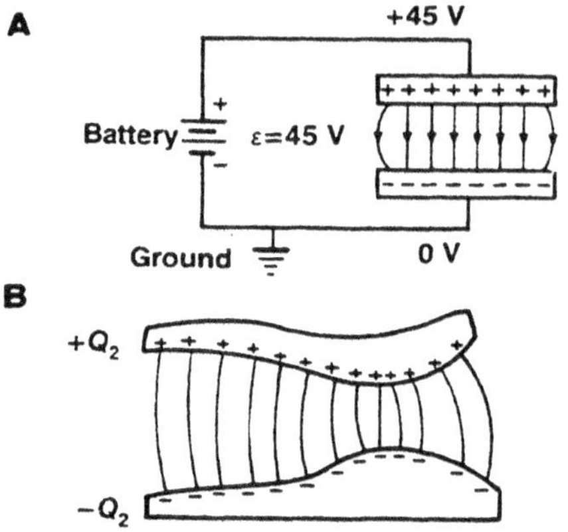

FIGURE 13-13-18. (A) Parallel plate capacitor showing uniform field between plates and (B) nonuniform field for uneven plates. From F. Miller, *College Physics*, Harcourt–Brace–Jovanovich, 1982, p. 421, reproduced by permission.

d_{is} is about 100 μm or 10^{-2} cm. The mean free path of ions between collisions, Γ_i, is a few centimeters at 10^{-4} Torr, and 10^{-1} cm at 1 Torr.

During the RF cycle, the V_e, electrode potential, is lower than the floating potential, and the ions are accelerated across the sheath without collisions since Γ is larger than d_{is} at 10^{-4}–10^{-1} Torr. The field lines are electrically equivalent to a capacitor with parallel plates (shunted by a resistor). Figure 13-13-18 illustrates a uniform field in a parallel and nonparallel plate arrangement (nonuniform field). At high pressures greater than 10 Torr, wafers or etched surfaces could disturb the field and cause isotropic ion collisions. The larger the ion sheath, the lower is the possibility of disturbance by the resist[531] or surface roughness ($\approx$ 1–5 μm). Since the sheath field is also inversely proportional to the electrode area, small reactors allow higher pressure (larger sheath field) to be used.[422] The ion response to the applied potential in the sheath governs the ion transport.[538]

When the mean free path of ions is the same order as the width of the ion sheath at higher pressures, the ions collide with neutral particles and impinge on the wafer isotropically (Fig. 13-13-19). In the plasma sheath, electric fields produce

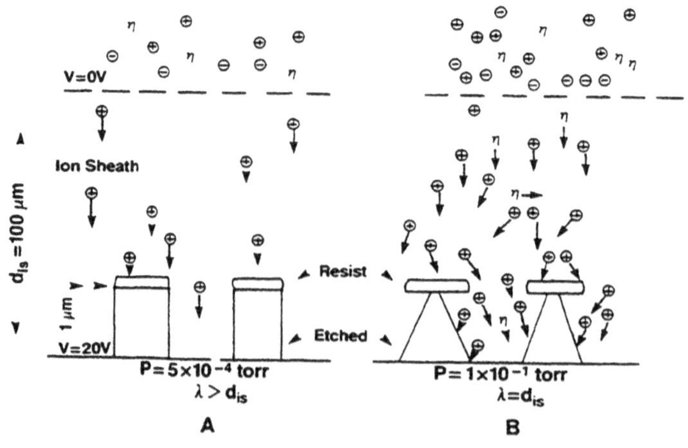

FIGURE 13-13-19. (A) Anisotropic ion pathway at pressure of 5×10^{-4} Torr. (B) Isotropic collisions at pressure of 1×10^{-1} Torr. λ = mean free path of ion; d_{is} = ion sheath width; η = neutral etchant; ⊕ = positive ion; ⊖ = negative ion. From Ref. 327, courtesy of *Japanese Journal of Applied Physics*.

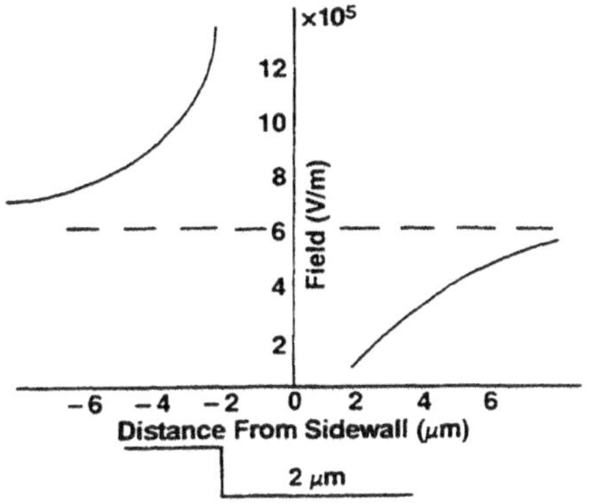

FIGURE 13-13-20. Field map potential calculated for a cathode length of 1 mm. The cathode fall potential is 300 V and the pattern height is 2 μm. From Ref. 424, courtesy of AIP.

anisotropic ion transport to the etch surface and induce chemical anisotropy.(577) Basically, the surface roughness is generally neglected. Ukai and Hanazawa(424) placed a wafer in the field, and the potential map on the sidewall was calculated (Fig. 13-13-20). The narrower the pattern, the stronger the field on the top is under the influence of both sidewalls. Figure 13-13-21 shows the pattern width dependence on etching poly-Si, 0.6 μm thick, were the smaller features disturb the field to a greater degree.

Another test of the vertical influence of ion etching was made by Matsuo,(426) who placed the wafers on a carbon disk 0.5 cm thick in a plasma with an ion sheath of 10^{-2} cm, and a dark space of 1–2 cm. As a result of the field distortion, the etching rate of SiO_2 is larger at the edge of a disk, where the ion flux concentrates (Fig. 13-13-22). By inclining the wafer at an angle θ, the flatness of the wafer and its influence can be tested. The etch rate will be proportional to $k \cos \theta$.(427) Recessing the wafers on the cathode(616) plate has been used to improve the poor (> 35%) uniformity of wafers in a batch. The etch rate is usually fastest at the outer edge of the wafer.(428)

Suzuki and co-workers(427, 692) also studied the role of ions and neutrals in microwave(615) plasma etching, by varying the angle of inclination, to gain insight into the z component. They found that at high pressures (> 10^{-2} Torr), Si is etched primarily by neutral species. Overetching can cause large undercutting and when the impinging ions hit the underlayer, they lose the opportunity to contribute

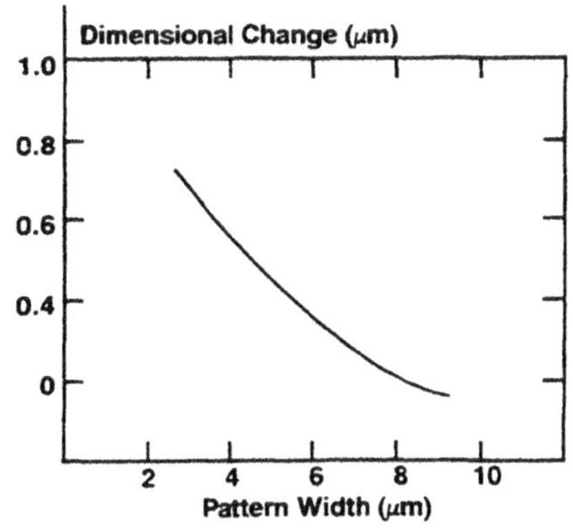

FIGURE 13-13-21. Pattern width dependence of a dimensional change for a 0.6-μm-thick crystalline silicon film placed in a parallel plate etcher. From Ref. 424, courtesy of AIP.

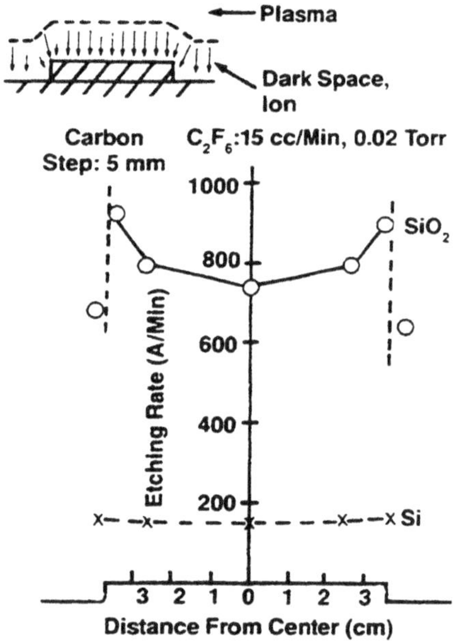

FIGURE 13-13-22. Etching rate disturbance when SiO_2 is etched on a 7-cm-diameter carbon disk, 0.5 cm thick. From Ref. 426, courtesy of AIP.

to the reaction. The ions diffuse along the surface (Fig. 13-13-16) and can accelerate sidewall etching (unless a polymer sidewall slows down the attack). At lower pressure, the ions are the primary etching species.

The ion sheath is also influenced by other factors, such as the dc bias, excitation frequency, and RF power density. Together, they act to control the ion current and directionality.[410] While the sheath field will tend to realign the ion velocities normal to the surface, the ion–neutral collisions produce ion velocity components parallel to the wafer surface. Excitation frequency besides 13.56 MHz can affect the [Eq. (13-13-17)] chemical (ion and neutral) composition and polymer formation. For example, CCl_4 readily forms polymers at high frequency.[232] The application of a dc bias will increase the sheath potential at the grounded electrode, where the higher sheath potential will realign the ions, which have suffered collisions at the higher pressure.

Operation of plasma reactors at high pressures (< 1 Torr) is desirable, since it can provide good etch results, at much higher etch rates (> 1 μm/min) to allow single-wafer processing.[410, 725] Zarowin[307] has extensively modeled the 1000-Pa (1–10 Torr) high-pressure region and its influence on plasma etch anisotropy. At high pressures, anisotropic images can be produced, by using an asymmetric electrode, and providing high sheath electric fields (100–300 V). Because of the trend to large single-wafer etching at rapid rates for sufficient throughput, the high-pressure reactor will be a critical area of practical and scientific investigation.[736]

13-13-5. *Uniformity of Plasma Etching*

A basic objective of any etching process is reproducible vertical and lateral etching (isotropic if required) over all surfaces within a batch, or from batch to batch. Table 13-13-20 shows the excellent results of Itakura and co-workers[429] of the 1σ process uniformity and reproducibility of a single-wafer (multichamber)

TABLE 13-13-20. Uniformity and Reproducibility of Etch Results

Uniformity of etched depth	Single wafer	Within ±2.7%	Within ±1.8%	Within ±4.2%
	Among wafers	Within ±2.3%	Within ±2.3%	Within ±8.0%
Reproducibility of etched depth	Single chamber	Within ±3.7%	Within ±1.7%	—
Uniformity of pattern width	Among wafers	Within ±0.11 μm	Within ±0.11 μm	Within ±0.1 μm

reactive diode etcher. As the wafer diameter continues to increase to 250 mm, batch processing will merge into single-wafer process.(725)

For etching in a batch parallel plate reactor, as shown in Fig. 13-13-23, variations of 25% in etch rate are not uncommon.(428) A realistic objective for etching anisotropic images may be a ±10% (3σ) uniformity. In a tubular reactor, even a ±25% uniformity is difficult to achieve.(403) As the linewidth shrinks into the submicrometer regime, uniformity across the wafer within a few percent may be required. Nonuniformity in etching can be broadly classified into deviations due to:

1. Film thickness variations and compositional differences
2. Isotropic ion diffusion(722) in the reactor and nonuniform etchant concentration(542)
3. Poor temperature uniformity (bonding wafers to cooled electrode assures low resist erosion)(434)
4. Excursions of power, frequency, or pressure during etching
5. General buildup of contamination on the electrodes and walls

Typically, nonuniformity of ±35% has been observed in batch processing a dozen or more wafers in a flat bed reactor.(430) One explanation for this nonuniformity is that although the etchant species is generated uniformly over the entire

FIGURE 13-13-23. Variations of etch rate of silicon wafer areas in a parallel plate reactor. Wafers are of 3-inch diameter. U.S. Patent 4,350,578.

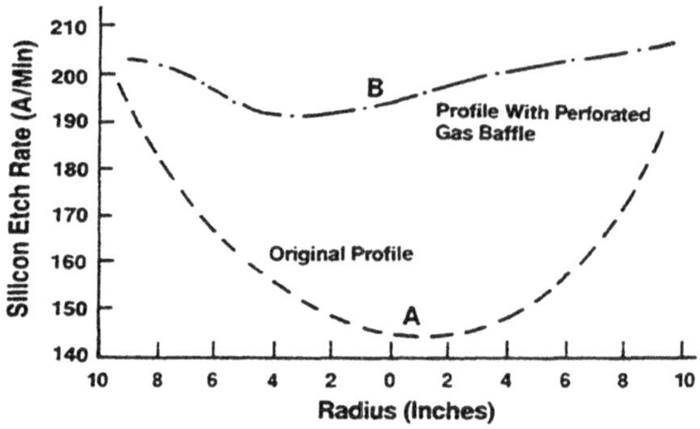

FIGURE 13-13-24. (A) Nonuniformity of etching silicon in a parallel plate reactor of 10-inch radius. (B) Improved uniformity of etching by insertion of a grid screen between anode and cathode. U.S. Patent 4,430,461.

chamber volume/surface, the dominant etchant loss occurs at the target. This creates a concentration gradient of etchant species which leads to an enhanced etch rate at the outer edge of the target. An example of the nonuniformity in a parallel plate reactor is shown in Fig. 13-13-24.[430] Vibrating the wafer during etching is claimed to improve the uniformity of etching.[430]

In the etching of individual wafers placed on the cathode, the etching has been noted to occur in a "bull's-eye pattern" clearing from the outer edge into the center of the wafer.[542] At the outer edge of the wafer, the stronger capacitance coupling of the RF field is said to induce a higher localized etchant concentration.[542] In order to improve the etch uniformity in a parallel plate reactor, a perforated gas baffle (Fig. 13-13-24) was conductively coupled to the anode to create an effective anode area larger than the cathode. The apertures have openings greater than the plasma Debye length. In another arrangement (Fig. 13-13-25), for reactive sputter etching, a hexagonally shaped cathode, a cylindrical anode, and a grid screen for uniformity are electrically floated between the anode and cathode.[431] The grid increases the selectivity of polysilicon to silicon from 6:1 to 20:1 and reduces resist erosion.

During the etching operation, temperature excursions in the power tuning of the RF field can cause localized hot spots in the plasma. Monitoring the gas temperature of the plasma with a pressure transducer has been used with variable RF power to limit fluctuations in etching.[432] The perforated aluminum shield originally used in tubular etchers to keep the nonuniform plasma away from the wafers also smoothes out thermal nonuniformities via its high thermal conductivity.[432] Another thermal gradient can arise at the wafer itself.[640] Nonuniform heating and cooling caused by poor wafer electrode contact is one cause of wafer-to-wafer variations in batch etching. Preheating of the chamber with hot N_2 has been used to reduce the heating effects.[433] Since the etch rate of the resist and substrate is proportional to E_a/kT, except for loading effects,[290, 352] the resist with a higher activation energy of 10 kcal/mole erodes (for a 10°C rise) at a rate double that of silicon which has a lower (3 kcal/mole) activation energy. In addition, the loading effect can lower the activation energy for a silicon area of 4 mm^2 to 4500 mm^2 from 4 kcal/mole to 0.4 kcal/mole[435] (as well as the selectivity). Thus, the resist erosion rate dominates as the temperature rises in a loaded batch reactor. In a barrel etcher, a temperature rise of the silicon wafer to 120°C at 200 W[436] and to 150°C at 600 W[437] were noted while in a parallel plate etcher, a rise of 40°C was measured.[436] Figure 13-13-26 illustrates the increased erosion of resists as the temperature rises during long etching times in a plasma reactor.[357] To prevent or

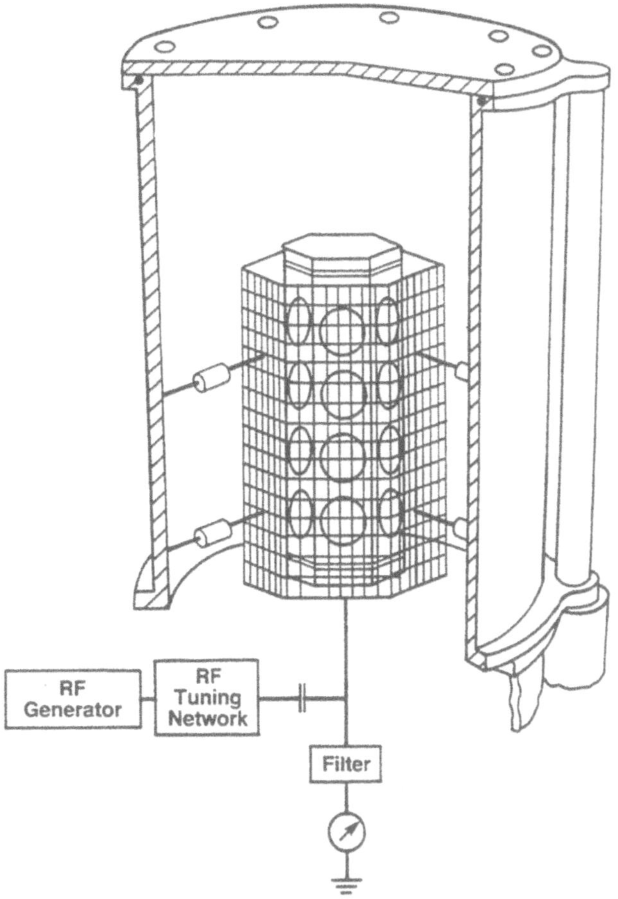

FIGURE 13-13-25. Insertion of wire grids in hexagonal reactor to improve batch etching uniformity. U.S. Patent 4,298,443.

reduce resist erosion and hence large etch biases, the importance of cooling or maintaining uniform temperatures cannot be overlooked.

Using a new and accurate thermometer based on rare earth emission (function of T), Egerton and co-workers studied the effect of good wafer contacts on a water-cooled electrode.(434) Wafers just resting on the electrode lost 5000 Å of resist while wafers thermally bonded lost only 500 Å. Earlier measurements of wafer temperatures such as by pyrometry(223) were subject to errors due to surface roughening.(434) Blackening the wafers on the back has been shown to reduce temperature rises.(302) Furthermore, He gas cooling has also been used to provide low and uniform temperatures.(434) At lower reactor temperatures, the loading effect is also suppressed for the film selected to have the small surface reaction rate constant.(496) The control of the temperature excursions will ensure that selectivity is maintained between the resist and underlayers. Greater overetching can then be tolerated to remove the underetched films in a batch reactor without the danger of resist degradation.

Besides temperature controls, other factors can dominate the concentration gradient of etchant. In commercial plasma etchers, the microprocessor control of

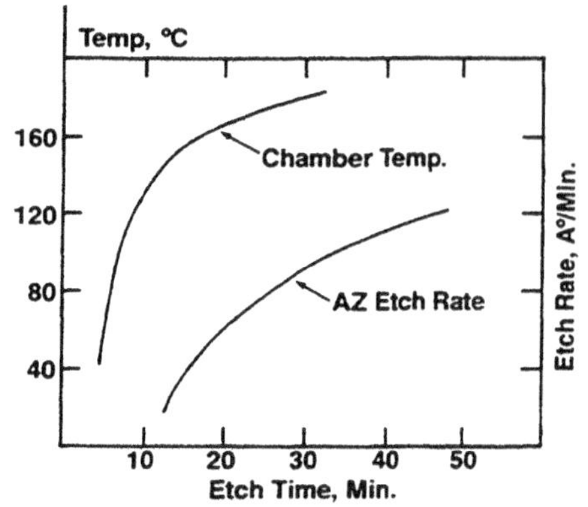

FIGURE 13-13-26. AZ-1350 etch rate versus chamber temperature profile which heats up during etching causing more resist loss for deeper etching (larger image loss). From Ref. 357, reproduced by permission of the Electrochemical Society Inc.

the flow, endpoint, power, and pressure are standard components in the growing number of plasma etchers. Flow rate effects and proper mixing(433) of etchants are also important in controlling the temperature and concentration of reactants. At low flow rates, etching can cease due to an inadequate supply of gas. In the active silicon area (A) being etched, there are 5×10^{14} Si atoms/cm²/A and 2.7×10^{14} SiO_2 atoms. A flow rate of 1 cm³/min is equivalent to 2.7×10^{14} molecules of silicon removed per min. The utilization factor U(438) is

$$U = \text{Etch rate of Si/input flux of } CF_4 \qquad (13\text{-}13\text{-}20)$$

$$= 1.9 \times 10^{-5}\ A_s E/Q \text{ for Si}$$

$$= 9.9 \times 10^{-6}\ A_s E/Q \text{ for } SiO_2$$

where A_s = substrate area (cm²), Q = flow rate (cm³/min), and E = etch rate (Å/min). A U factor of $< 70\%$ indicates a strong flow dependency (Fig. 13-13-27). In a radial flow reactor, the chemical composition of gases changes markedly.(440) The average residence time of the etchant is inversely proportional to the flow velocity obtained by integration over the volume of the reactor. The viscosity and laminar flow characteristics will change due to the changing gas composition across the reactor.(722) For a simple approximation, the average residence time is

$$t = \pi h r^2/Q \qquad (13\text{-}13\text{-}21)$$

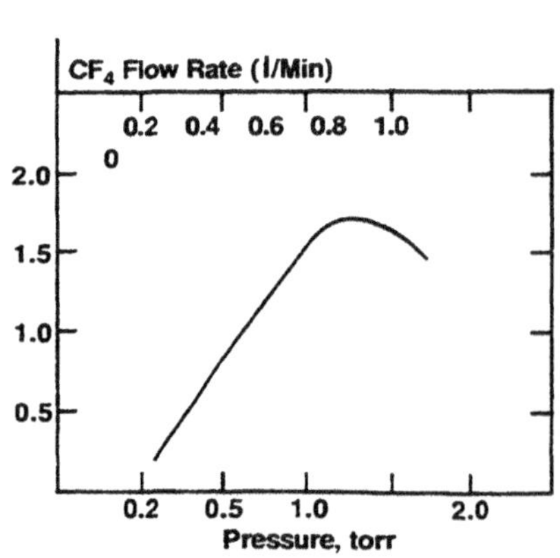

FIGURE 13-13-27. Variations in resist erosion in CF_4 gas with increase in flow rate and pressure in plasma reactor. From Ref. 438, courtesy of AIP.

where r is the radius of the discharge and h is the electrode separation. As long as a steady state is reached, the radicals created at the periphery have enough residence time to react. Mayer[440] found an etch rate uniformity of $\pm 10\%$ at a radial distance of 18 cm at C_2F_6 flow rates of 20–50 cm^3/min. At 82 cm^3/min, the etch rate fell by 40% toward the outer edge of the etching zone. At high flow regions, reactive species can be pumped away before they have an opportunity to react. For high flow rates,

$$t = V/S \tag{13-13-22}$$

where V is the volume and S is the pumping speed.[438] In a diode etcher at 10^{-3} Torr of 50-liter capacity, a flow of 20 cm^3/min means a residence time of 4 sec. Both the incoming flow and pumping speed have to be balanced to improve the uniformity, even at the loss of some throughput or selectivity.

Some of the above factors such as concentration, temperature, and flow uniformity have been examined. Once critical insight into the gas-phase kinetics continues to improve, the design of uniform reactors and built-in monitors (across the reactor) will be a natural extension of the physical–chemical mechanisms and modeling. Uniform etching is important not only for opening consistent holes or patterns to another layer, but also for planarization[441, 442] and in cases where the film is thinned rather than etched completely.[443] Consequently, uniform etching indicates that many factors interplay and there is no single reactor or set of processing parameters which can satisfy all etching needs.[729, 730] Rather, each reactor must be characterized and modeled for each new etching requirement.[444]

13-13-6. *Plasma Etching Contamination*

Contamination and damage in plasma etching[658] is more predominant than in liquid etching since ion and physical energies are sufficient to remove the metallic, inorganic, and organic films from one surface and redeposit them on a wafer surface. Plasma can implant foreign ions into semiconductor surfaces or oxidize, for example, P and As dopants.[445] Sometimes, it is required to dip plasma-etched wafers into HF or other liquid etches to remove residues and improve contact resistance.[445]

There are four major categories of contamination: (1) sputtered metals, (2) insulating deposits, (3) etch residues, and (4) radiation damage.[730] Residues and damage from resist stripping in a plasma are covered in Chapter 14. Physical sputtering, redeposition of the electrodes, and deposition of glow discharge polymers are the major mechanistic sources of contamination and radiation damage (Table 13-13-21). Except for downfield reactors (Chapter 14), all plasma reactors bombard the wafer with ions capable of inflicting ionization and temporary or permanent damage. Each type and sources of contamination and damage will be individually examined.

13-13-6-1. *Sputtered Metals.* Metals such as Fe, Al, Ni, Cu, and Zn are components of anodes, cathodes, plumbing, screens, and the wall fixtures of a reactor. Metals can also be present if the reactor was previously used for metal etching.[446, 657, 667] Metals not converted to volatile halides or oxides could deposit in a reactor or on etching surfaces (inducing pitting due to selective masking).[541] Under the influence of the electric field, metals such as Na^+, Zn, and Cu are dif-

TABLE 13-13-21. Factors Influencing Deposition of Etchant and Reactor Materials[a]

Reactants	Mechanism: Physical sputtering	Glow polymers	Redeposition
Electrode and reactor	Strong (increase with atomic weight)	No effect	No effect
Reacting etchant	Weak influence (due to decomposition of fragment ions)	Strong (due to vinyl polymerization)	Strong (formation of involatile products of substrate)
Substrate material	No effect	No effect	Strong (formation of involatile products of substrate)
Increase in current material	Strongly increasing	Reaches saturation level at gas pressure	Weakly increase
Increase in gas pressure	Decrease (due to redeposition of sputter particles on target)	Reaches saturation level at current density	Increase (due to reduced diffusion of etch particles)
Increase in gas flow	No effect	Reaches optimal decomposition and nonsaturation	No effect
Increase in substrate temperature	No effect	Weakly decrease	No effect

[a] From H. Poll, J. Meichsner, and A. Steinrucken, Film Deposition in Plasma Etching, *Thin Solid Films* **112**, 369 (1984).

fused into silicon and alter transistor action. During contact hole etching, heavy metal contaminants increase junction leakage.(447, 448) Heavy metals introduced by plasma etching are not removed by surface cleaning probably because the metals reside below the silicon surface. In this case, acid etching is required to remove the implanted metal.(677) Stripping of the resist can also introduce metals if present as Na^+ in the resist film. When alkaline developers (NaOH) are used in conjunction with diazonovolak resists, the adsorbed metal ion can accelerate the Si plasma etching.(468) A summary of the metal contamination studies in plasma etching is given in Table 13-13-22. Nonvolatile metal carbides are formed from sputtering resists.(379) Metal that is at a substantial negative voltage, relative to the plasma, is sputtered. In the reactive ion configuration, only the material on the cathode is sputtered. A suitable designed catcher, in combination with a cathode material with low sputtered yield, can be installed.(453) A substrate holder, which forms volatile Si species with the etchant, can reduce sputter redeposition.(344) Without catchers, metals and carbon are impregnated below the silicon surface. Backside ion implantation is not effective as a getter, unless the wafers are heated to high temperatures ($> 600°C$) after etching.(447) A Si coating on all parts of a reactor reduce metal contamination.(612)

Instead of alloys, pure Al substrate holders are recommended.(447) Silicon sub-

TABLE 13-13-22. Metal Contamination during Plasma Etching

Study	Metal	Result	Reference
1. Multiuse reactor	Au on Si	Higher resist erosion	446
2. RIE cathode sputter (C-V, MOS) 99.999% Al versus alloy	Al	Ni, Fe, Cu contaminants	447
3. Sputter etch, 10^{-5} Torr	Cu, Au, Ti	Ti getter O_2	449
4. Parallel plate Al versus SiO_2 electrode	Al	Al pitted	450
5. Parallel plate C_3F_8	Al	Metal incorporated into polymer deposits	451
6. RIE diode	Fe	Si stack faults	452
7. Sputter etch	Many	Higher resist erosion	379

strates exposed to various oxide plasma etching in stainless steel reactors, develop a large density of stacking faults on subsequent oxidation.(452) Al electrodes minimize stacking faults. However, Yabumoto and co-workers point out that regardless of electrode material [C, Fe, or Teflon (DuPont Co.)], in a reactive sputter etcher, C_2F_6 forms C precipitates 100 Å below the surface,

$$3Si + 2C_2F_6 \rightarrow 3SiF_4 + 4C \qquad (13\text{-}13\text{-}23)$$

to cause subsequent oxidation defects. The carbon defects can be removed by brief etching in Cl_2/Ar.(454) Besides junction device contamination,(455, 456) redeposited metals cause surface pitting. The deliberate etching of Si pits was applied to fabricate solar cells which absorb infrared (1 μm) radiation.(457) In some cases, the surface pitting can be eliminated, if a SiO_2 electrode is used in place of a metal electrode.

Heavy metals are incorporated into polymerizable etchants (CF_4/H_2, C_3F_8), which produce sidewall banking films.(451) These can be driven into the substrate during O_2 plasma strip. The resist erosion rate increases, due to the formation of metal carbides.(379) If a silicon cathode is used in place of a SiO_2 cathode, the etch rate of the resist is reduced by a factor of 2, which has been ascribed to a lower wafer temperature.(458) Similarly, metal electrodes offer good thermal conductivity to dissipate heat. Al appears to be an electrically acceptable and a plasma strippable material. Coating electrodes with a passivating SiO_2 film can prevent sputter contamination but O_2 formation may inhibit polymer sidewall formation. Electrodes coated with Si do not exhibit contamination, since the Si is converted to Si halides. Teflon has also been suggested, but not Al_2O_3, since the AlF is nonvolatile.(294)

13-13-6-2. *Insulation Deposits.* Insulating deposits of polymers,(556) derived from plasma etchants, are involved as resist sidewall banking agents in anisotropic etching. As long as the vertical erosion rate of the polymer deposit exceeds the deposition rate, silicon will exhibit high selectivity over SiO_2 (> 15/1). Fluorocarbons with a C/F ratio > 2 readily form polymers (Table 13-13-7). Electrodes which consume F radicals also alter the C/F ratio and induce plasma deposits. If uniform deposits are formed, no major problems are encountered. In a study of polymer-prone CF_4/H_2, Eprath and Petrillo found that a deposit was left in the etched contact hole.[448] A deposit of F and C about 20 Å deep was detected by Auger and ESCA on all surfaces, etched in CF_4.[439] An etch inhibitor of NH_4F deposits during CF_4 etching of Si_3N_4.[578] Sputter deposits of Si_3N_4 are cleaned in H_3PO_4.[700]

Reactive sputter etching is more susceptible to carbon contamination since the C atoms formed on the surface by C_2F_6 etch,

$$3Si + 2C_2F_6 \rightarrow 3SiF_4 + 4C \quad (13\text{-}13\text{-}23)$$

are driven into the Si subsurface. As noted by Yabumoto and co-workers, the surface[552, 635] damage can cause four categories of defects.[454] Figure 13-13-28 illustrates the four defects and the damage to the Si surface. In the CF_4/H_2 etching, H atoms implant into Si.[556, 598, 669] This can cause high resistance for Al/Si contacts and oxidative-induced stacking faults (OSF) in the Si-substrate. Oxygen sputter clean, prior to Al metallization, can lower the contact resistance.[396] Further, the removal of 500 Å of damaged silicon by a Cl_2/Ar plasma was found to minimize the OSF defects.[396]

Dry etching in Freon (DuPont trademark) leads to polymer formation by CF_x radical adsorption, initiated by secondary electrons from ion bombardment. The thickness of the polymer formed is a power-dependent function of pressure.[459, 460] By depositing during some phase of etching and producing nonlinear etch rates, polymer formation can affect etch reproducibility and uniformity. Polymer deposition is very sensitive to the reactive plasma beam ion energy. Indirectly, the polymer deposition on the interior of the reactor can affect the next run. If the reactor walls are not subsequently cleaned in O_2 plasma, Lehman and Widmer showed that a total encapsulation of the resist occurs.[462] After stripping of the resist, hollow outlines of $(C_xF_n)_n$ of the original AZ-1350 image were left. The Si etched

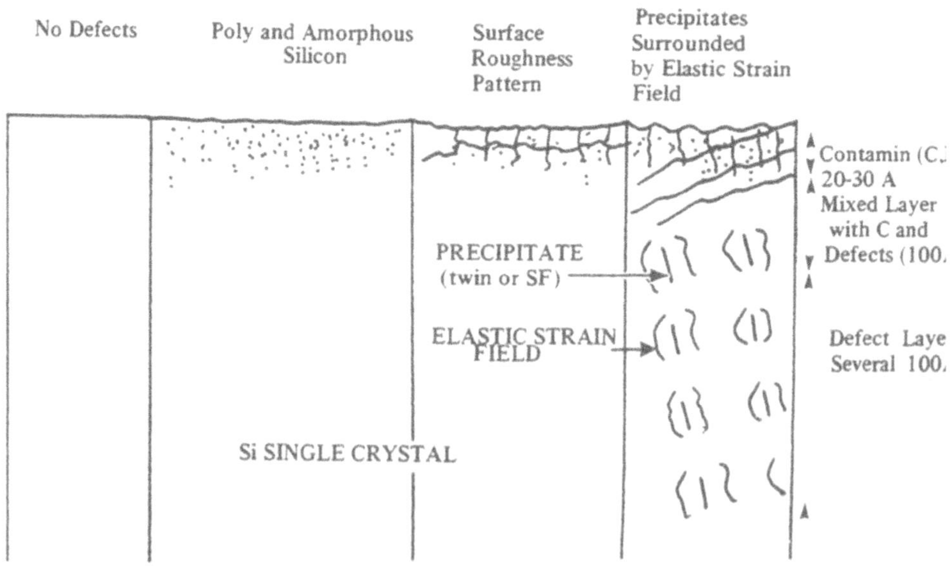

FIGURE 13-13-28. Radiation damage to silicon after RF sputter etching. From Ref. 454, courtesy of *Japanese Journal of Applied Physics.*

surfaces were pitted, due to a heavy rate of deposition, which was greater than the ion erosion rate in the vertical direction. Likewise in the CF_4 etching of phosphosilicate glass, columnar residues were formed.[463]

Sputtered metal can act as the site for growth on a wafer surface. Mullins observed that CF_x polymers grew as stalagmites and eventually toppled under their own weight.[298] In many cases, some of the polymer deposits are not effectively stripped by O_2 and may require liquid caustic strippers. Some electrodes, made of SiO_2, provide O radicals to inhibit or strip polymers, while Teflon electrodes will increase polymerization on the opposite electrodes. Cooled substrates will also tend to form polymer deposits.

Plasma conditions which lead to polymer formation include:

1. A high C/F ratio in the etchant
2. A degrading resist (acrylate)
3. Low resist prebake (trapped solvents)
4. Thick resist (> 2–$3\ \mu m$)
5. Low plasma power density
6. Low etch gas flow rate
7. A high etch gas pressure

Some of the above conditions are used for plasma hardening by resist encapsulation.[238] The formation of a polymer acts to protect the resist against thermal distortion. for example, heating a resist such as AZ-1350 above 120°C will produce 45° rounded profiles (suitable for taper via hole etching).[464] A polymer overcoat stabilizes the diazonovolak resists against flow.

The use of non-carbon-containing etchants, such as F_2, NF_3, and SF_6, has been pursued as a means of high selectivity and carbon-free contamination and hopefully no sulfur deposits. Negative halogenated ions have also been used to etch Si, but with low rates and no residues.[525]

In order to achieve the minimal deposition of polymers from CF_4/H_2 and to maintain high SiO_2/Si selectivity (> 20), requires tight process controls. For example, a 3% change in composition of CF_4/H_2 can cause a 44% change in selectivity and bring about heavy polymer deposits. Consequently, a microcomputer for adaptive control of flow, power, pressure, and etch rate was implemented.[467] To avert CF_4/H_2 polymer deposits, a mixture of CHF_3/SF_6 with a SiO_2/Si of selectivity of 6, has been suggested.[384] The contamination from polymer formation can be helpful or harmful to process anisotropy and uniformity. The simple avoidance of polymer formation in the first place, is probably a better choice, unless sidewall protection is necessary for anisotropic etching.

13-13-6-3. *Etch Residues.* Residues can be present from the prior resist process such as polysilicone adhesion promoter, photoresist, or resist developer ions. Oxide layers are present on silicon, Al, and most metals. In turn, these layers or residues may inhibit etching.

Positive DQN resists use alkaline NaOH and KOH developers. Improper rinsing can leave deposits of metal ions on polysilicon surfaces. In this case, the contaminated surface of polysilicon etches an order of magnitude faster in CF_4/O_2.[468, 681] As with many residues, a brief dip in HF or etchant will generally remove the residue or contaminant.[469]

13-13-6-4. *Radiation Damage.* Since ion bombardment is a component of plasma etching and X rays, deep UV, and electrons are formed, the rupture or ionization of silicon atoms or oxide bonds is an inevitable consequence of plasma etching. The damaging particles are (1) positive ions, (2) negative ions, (3) electrons, (4) energetic neutral particles (CF_3^*), (5) metastable atoms, (6) photons, and (7) X rays (Table 13-13-23).[471] There are several distinct types of degradation[596] which occurs in SiO_2 films exposed to plasma etching[730] or sputter deposition: (1) dielectric leakage, (2) SiO_2/Si interface fixed charge buildup, (3) SiO_2/Si system mobile charge buildup, and (4) SiO_2/Si interface state buildup.[471] Absorption of energy greater than the bandgap of silicon (>8 eV) can give rise to conductivity by generation of electron–hole pairs. Based on studies of MOS devices,[472,473] plasma-etched surfaces are generally free of damage due to electron irradiation. When the damage does occur, the process of thermal annealing must introduce electrons into the oxide, to neutralize the positive space charge. Furthermore, thermal annealing at 500°C can restore most of the interface states. Some of the neutral traps require baking above 500°C, which causes silicide formation.[474] Vossen has pointed out some alternative metal interconnections which allow lower-temperature annealing.[473] Unprotected gate oxides are very susceptible to permanent damage.[475]

The radiation damage is the highest in systems with high ion flux, such as reactive ion-beam etchers, ion-beam etchers, and reactive sputter etchers.[730] FET devices are more susceptible to damage than highly doped, bipolar logic devices[475] and the extent of damage in the depth range of 300 Å is proportional to the ion accelerating voltage.[476] Protection of *p–n* junctions and MOS devices by thin (200–350 Å thick) oxide layers will stop ion damage. Photon damage can only be stopped by opaque (metal) films. Decreasing the DC cathode bias reduces damage in magnetron RIE.[626] However, lower power or lower RF bias are trade-offs of the

TABLE 13-13-23. Energetic Particles in Diode Reactors[a]

Type of system	Type of particle	Typical energy current and intensity parameters
DC low-voltage triode (argon plasma)	Electrons	Energies up to 20 eV in column Energies up to 40 eV at substrate
	Positive ions (Ar^+, Ar^{2+}, CF_3^+)	Few eV maximum energy in column, 200 eV target, 40 eV substrate
	Photons	Energies up to 40 eV from electron striking metal etc. and from $Ar^{2+} \rightarrow Ar^0$ radiation
	Neutral atoms	Up to 100 eV (from target)
DC diode	Electrons	0–6 eV, up to V_A (maximum voltage in system)
	Positive ions (Ar^+, Ar^{2+})	Up to 1120 eV
	Photons	Up to maximum energy in system
	Neutral atoms	Up to V_A eV
RF diode	Electrons	Up to ±RF voltage
	Positive ions	Few eV–1200 eV
	Neutral atoms	Up to 200 eV
	Photons	Up to ±RF voltage; also at $Ar^{2+} \rightarrow Ar^0$ energy

[a] From Ref. 471.

throughput versus device reliability. In reactive ion etchers, materials used as cathodes can be sputtered, deposited, and ion implanted.[473, 677] Sodium ions or metals present in the resist can be sputtered and implanted into oxide or silicon surfaces. Imposition of a grid to lower electron energies has been used for RF powered diode etchers.[448] Indirectly, damage (Chapter 14) from plasma stripping of resists may occur, due to sodium ion implantation into silicon. As alternatives, lasers and *e*-beams[648, 651, 659, 665] have been used to form plasmas with lower radiation background. Downfield reactors in which the plasma-activated radicals are transported to the wafer surface are being pursued as radiation-free systems. However, the etch rates are slow in comparison to reactive ion bombardment.

The potential damage from plasma etching always exists. Constant testing, annealing methods, and new barriers to stop ions are always required. New circuit designs, barriers such as PtSi,[473] and protective layers are required, not only for plasma, but any high-energy exposure sources, such as X-ray, ion-beam, and *e*-beam technologies. Ironically, the RF plasma itself has been used to anneal *e*-beam damage.[477]

13-13-7. *Applications of Plasma Processing*

Subtractive etching or thinning of silicon, Si nitride, Si oxide, and Al are the primary functions of plasma etching. Anisotropic etching (ion assisted) is now as prevalent as barrel isotropic neutral etching. Figure 13-13-29 illustrates the application of plasmas to resist–organic polymer processing. Plasma can be used to deposit, treat, etch, and remove resists or other organic or inorganic films. As an outgrowth, anisotropic etching is responsible for the dry development of resists (Chapter 10) and for the formation of submicrometer straight-walled images in multilayer resists (Chapter 12).

The plasma processing of polymers involves polymerization, developing, hardening, and removal. A noteworthy achievement has been a complete lithography dry processing with plasma polymer deposited, developed, etched, and stripped PMMA resist. Table 13-13-24 gives examples of the ever-increasing use of plasma processing in semiconductor lithography. For multilayer lithography, a very important application for multilayer metal interconnections is the plasma planarization of conformal coatings. Sputtered quartz (as opposed to spun-on planarizing resist) conforms to the surface topography (Fig. 13-13-30). A primary cause of metal interconnection failures is the incomplete metal coverage over steep steps.[415] The abrupt steps in sputter-deposited P-glass are covered with a spun-on resist. However, the resist does not completely planarize the step (partial conformality)

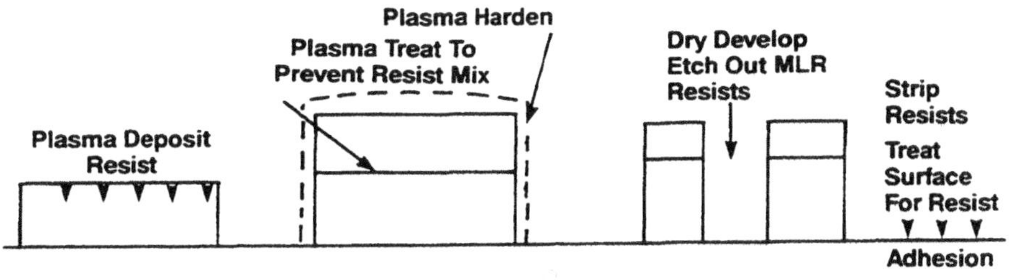

FIGURE 13-13-29. Applications of plasma to process or improve resist performance.

TABLE 13-13-24. Applications of Plasma to Semiconductor Lithography

Application	Reference
1. O^+etching of thick (10 μm) polymers	479
2. Planarization of SiO_2 layers	441, 463, 661
3. Planarization of sputtered quartz	480, 562, 662
4. Evaluation of pinholes in SiO_2	481
5. RF anneal of electron damage	477
6. Planarization of metal over quartz	482, 490
7. Multilayer resist etching for lift-off	483
8. Multilayer resist etching for reactive ion mask	484
9. Insoluble resist layers for fabricate lift-off	485
10. Improve resist adhesion	486
11. Dry develop resists	487
12. Plasma harden resists	488
13. Plasma deposit resists	478, 564
14. Plasma strip resists	Chapter 14
15. Dry etch Cr without developing resist	489
16. Plasma deposit adhesion promoter	563
17. Submicrometer gates by angular etch rate factor	666–667, 670
18. Slope via hole etching	671

(Fig. 13-13-30). Baking the resist at 150°C causes softening and viscous flow to smooth the surface. For planarization, etch conditions are chosen such that the resist and the P-layer etch at the same rate. The sacrificial resist layer erodes to expose the step and nearly planarize the P-glass layer.(441) Direct etchback of PSG without a sacrificial resist layer can be achieved if the underlying step is thick enough.(662)

In another application, the image reversal process using an etchback of polyimide is an interesting innovative combination of the planarization and dif-

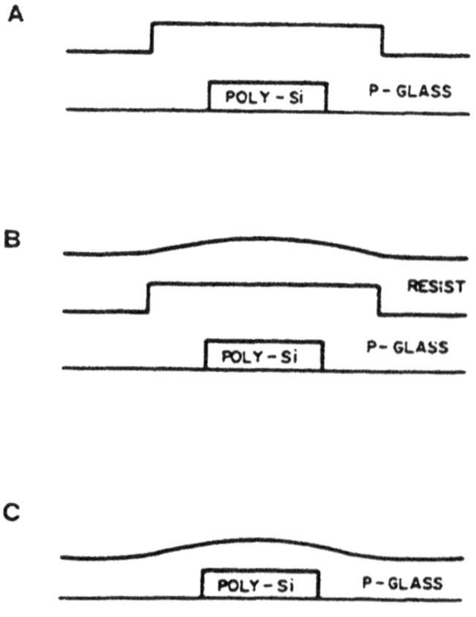

FIGURE 13-13-30. Schematic of steps of planarization of phosphosilicate glass deposited over polysilicon step. (A) Deposition of phosphosilicate (P) glass; (B) coating of partially conforming polymer; (C) after plasma etching to planarize P-glass using C_2F_6–Cl_2. From Ref. 441, reproduced by permission of the Electrochemical Society Inc.

ferential etch rates of two resists.[597] A positive image in a poor plasma-resistant material of PMMA is overcoated with polyimide and etched back with O_2. For top surface imaging, (SIEL),

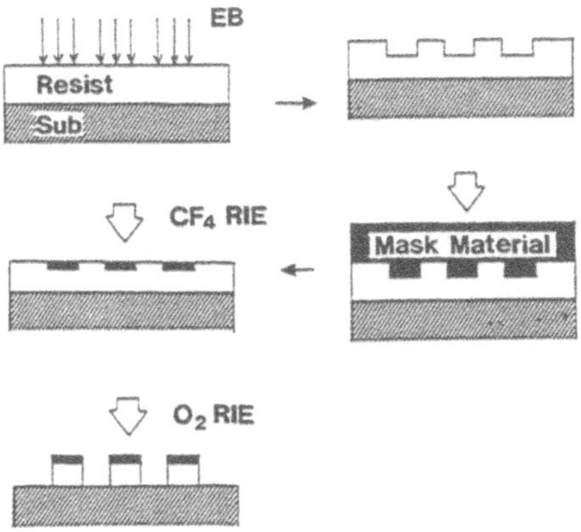

a partially developed image in PMMA is overcoated with polysiloxane, CF_4 etched back to the PMMA surface, and then RIE etched with O_2 to form 0.1-μm images.[748] Likewise, polyimide[656] and other polymers can be used to planarize the sputtered quartz by etching back prior to forming the second-level metallurgy (Fig. 13-13-31).[574] The polyimide is coated over the bumpy sputtered quartz. Next, the planarized polyimide is reactive ion etched at the same rate as the quartz layer. The process leaves no polyimide on the circuit and planarizes the sputtered quartz layer. Spun-on glass layer can also be used to replace the sputtered quartz and circumvent the etchback process.

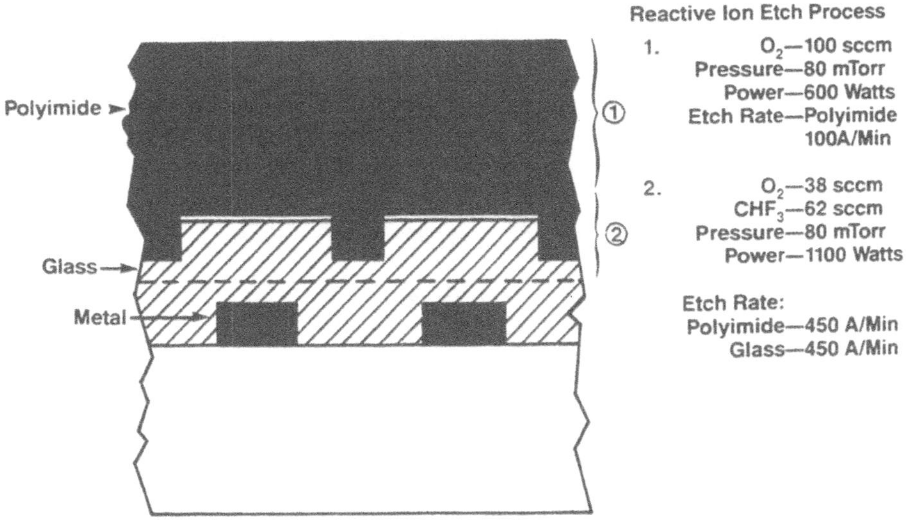

FIGURE 13-13-31. Planarizing and etchback process which leaves no polyimide on circuit prior to opening via holes for second metal contacts. From Ref. 574, reprinted with permission of *Solid State Technology*, published by Technical Publishing, a company of Dun and Bradstreet.

13-13-8. *Loading Effects*

Depletion of the active etchant can be expected, since the etchant is formed by a chemical reaction. In wet etching, the etchant is available in large excess. Dry etching is equivalent to a consecutive step reaction:

$$A \rightarrow B^* \qquad k_1 \tag{13-13-24}$$

$$B^* + C \rightarrow D \qquad k_2 \tag{13-13-25}$$

If k_1 exceeds k_2, a barely sufficient or minimum amount of B is available to react with substrate C. If C exceeds B, then mass-deficient (concentration) problems arise and the etch rate of C declines. The mean lifetime of B is dependent on recombination reactions with itself, or some other radical (Fig. 13-13-5). The loading effect was first noticed in plasma stripping of wafers[208] or in barrel etchers.[491] In flat bed reactors, etching with CF_3–O_2 decreases the rate from 4000 Å/min to 1000 Å/min as the number of 3-inch wafers increases from one to ten in the reactor.[496] In the practical sense, the overetching of silicon is needed to compensate the film or etch variations that consume the active etchant, before it reacts with the remaining SiO_2 layer.

Mogab[288] treated the second-order equation (13-13-25) and derived a mass balance to explain the etch rate (R) dependency,

$$R = \beta t G/(1 + \beta t d \text{Aw} N/V) \tag{13-13-26}$$

where $\beta = k_2$ rate constant of etching; t = mean lifetime of etchant [Eq. (13-13-24)]; G = rate of generation of active species, k_1; N = number of wafers of area A; V = volume of plasma; $d = bpN_0/M$ (where M = molecular weight of substrate of density p; b = preexponential frequency rate factor, k_2; N_0 = Avogadro's number). Loading effects occur when the denominator in Eq. (13-13-26) is $\ll 1$, such as in increasing the volume V or adjusting the flow rate [Eq. (13-13-22)]. The etchant lifetime, t, should be no longer than 0.1 msec.[493] Since the reciprocal of the etch rate is proportional to N in Eq. (13-13-26), a plot of reciprocal etch rate for silicon as shown in Fig. 13-13-32 yields a loading effect in CF_4–O_2 but not in CF_3Cl or CF_3Br. The etchant of CF_4–O_2 exhibits a loading effect due to the short CF_3^* lifetime in which the CF_3^* is consumed by oxygen radicals,

$$CF_3 + O \rightarrow COF_2 + F \tag{13-13-27}$$

A short inherent etchant lifetime can be realized by introducing fragments of CF_3 to recombine with fragments of Cl or Br derived from CF_3Cl or CF_3Br. No recombinant is present, since the CF_3^* fragment in CF_4–O_2 reacts with the O_2.[331] The loading effect is dependent on the pressure and reaction conditions, which involve chemical reactions or neutrals (radicals). At high power, for example, Mauer and co-workers have shown that etching in CF_4 is independent of the Si wafer load, since the etching mechanism is entirely due to sputtering.[495] In radical chemical etching, the loading effect can have a serious effect on the selectivity ratio of Si to nitride or oxide. Selectivity ratios in the literature (Tables 13-13-16 and 13-13-17) can vary by a factor of 4. These fluctuations may be due to the loading effect.[496] Since undercutting is a lateral etching attack of the free radicals on the walls of an

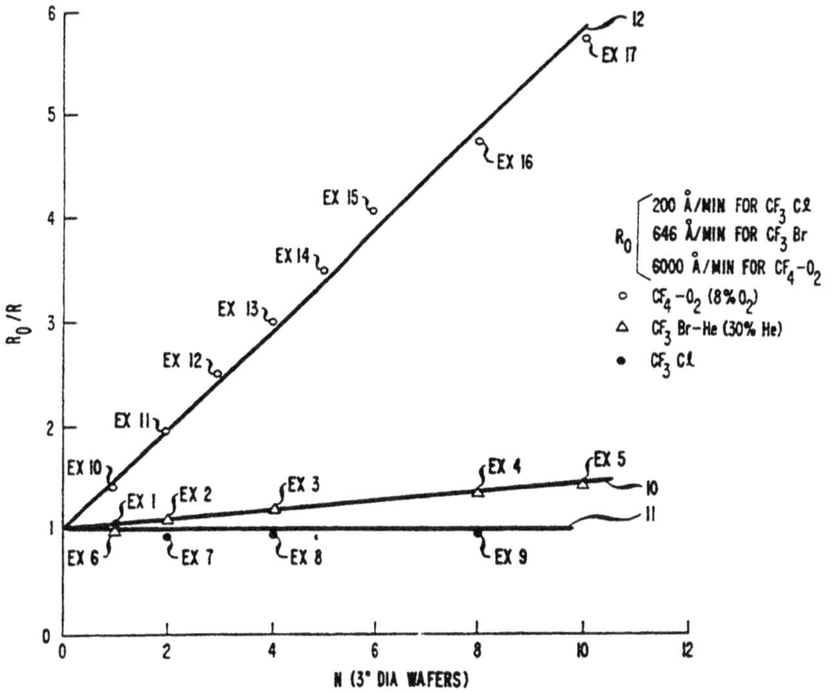

FIGURE 13-13-32. Etch rate ratio of single-wafer rate, R_0, to loading rate, R, versus the number of loadings for different etchants in plasma reactor. U.S. Patent 4,226,665.

etching pattern, a serious undercutting (3×) can occur if the wafers are etched under the loading effect conditions (Fig. 13-13-33).[496] Secondly, erosion of the resist, which may involve hydrogen abstraction by F or Cl radicals, will also be subject to the competitive consumption by silicon. Polymer deposition involving CF_x radicals will be altered under load conditions.

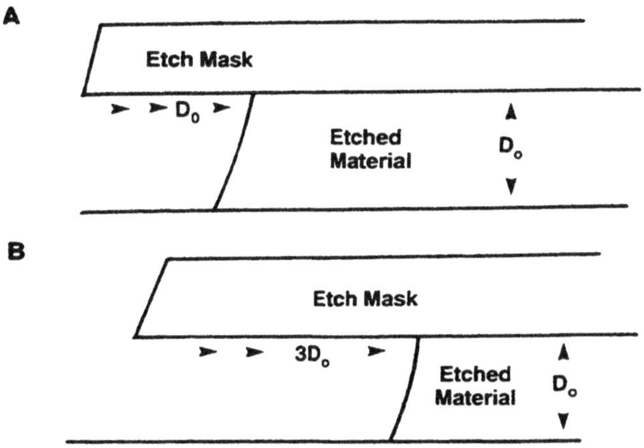

FIGURE 13-13-33. Model of undercut profiles in barrel plasma etching. (A) Lateral etch rate equals vertical etch rate; (B) lateral etch rate is 3× vertical etch rate. Both under identical loading conditions. From Ref. 496, courtesy of *Japanese Journal of Applied Physics*.

Under loading effect conditions, the activation energy for silicon etching can be inconsistent and the values vary from 4 kcal/mole at no load, to 0.4 kcal/mole in loaded region. The lowering of the activation energy is consistent with a transport-limited reaction (diffusion controlled). Zarowin[(422)] has separated the contribution of ions (I) and radicals (R) into the exponential sum $[\exp - (I+R)/kT]$ of the two components. Thus, lowering of the sample temperature, T, will also reduce the loading effect.[(496)] The detection of the loading effect can be performed by monitoring of the etch rate as a function of load (Fig. 13-13-32).

13-13-9. *Etch Rate and Endpoint Monitor*

Monitoring of the etch rate and the endpoint completion are an internal check on the reproducibility of many parameters that interact to produce a uniform etch process. In any chemical reaction, one can measure the decrease in reactants, or the increase in product. Endpoints and rate of etching monitors are necessary to prevent excessive overetching, as well as a means of sensing any changes in flow, pressure, power, and loading effects. Some of the methods include optical absorption and emission, mass spectrometry, and physical changes, such as pressure, thickness, RF peak voltage, and impedance. Table 13-13-25 summarizes the popular techniques and their limitations. Many workers have used the fluorine emission at 704 nm as an absolute indicator of etch rates of Si in CF_4. Coburn and Chen[(511)] have noted that the etch rate is not directly proportional to the fluorine intensity since at various power levels, the silicon etch rate decreases, while the intensity of F increases. They propose an internal standard of argon emission:

$$CF_4 + e \rightarrow CF_3 + F \qquad (13\text{-}13\text{-}28)$$

$$e + F \rightarrow F^* \qquad (13\text{-}13\text{-}29)$$

$$e + Ar \rightarrow Ar^* \qquad (13\text{-}13\text{-}30)$$

$$Ar^* + F \rightarrow F^* + Ar \qquad (13\text{-}13\text{-}31)$$

TABLE 13-13-25. Endpoint or Etch Rate Monitors

Technique	Limitation	Area sensed	Reference
Mass spectroscopy	Recombination of radicals	$>cm^2$	497
Optical emission	Detects atoms not molecules, second-order reactions	$>cm^2$	497, 614, 618, 650
IR absorption	Long-lived species only, molecules not atoms	$>cm^2$	498
Gas chromatography	Stable products only	$>cm^2$	499
Interference	Transparent films: texture problems	4 mm^2	500–505
DC voltage	Secondary method	Reactor	506
Impedance	Secondary method	Reactor	507
Pressure	Secondary method	Reactor	508
Quartz crystal	Secondary method	Reactor	509
Flow rate	Secondary method	Reactor	510
RF peak voltage	Secondary method	Reactor	517
Fluorescence (dyes in resist)	Signal/noise	cm^2	539

to counteract the variation of F*, as a function of power.(512) Therefore, the F* emission is a convenient, but a second-order monitor. Pulsed laser-induced emission can provide an order of magnitude more emission signal than the plasma-induced emission.(752)

The optical emission and mass spectra techniques require a minimum ($>1\ cm^2$) etched area.(514) More effectively, the mass spectra of ions, such as SiF_3^+, are stronger signals than the weak emission from Si in the SiO_2/Si etching requirement.(514) Because of weak signals near the endpoint for loss of Al or Si peak(513) (Fig. 13-13-34), light pipes have been patented to capture the diminished intensity.(515) In addition, improved emission techniques have been claimed by downstream excitation and monitoring of F* atoms, followed by chemiluminescent titration with Cl_2.(517)

For direct solid-phase etching,(565) monitors rely upon interference techniques.(508, 718, 719) For a transparent oxide on silicon, there are no major optical problems. The reflectance formula for Fresnel(653) reflection(517) has been applied for up to four layers such as resist/polysilicon/oxide/silicon.(518) For submicrometer

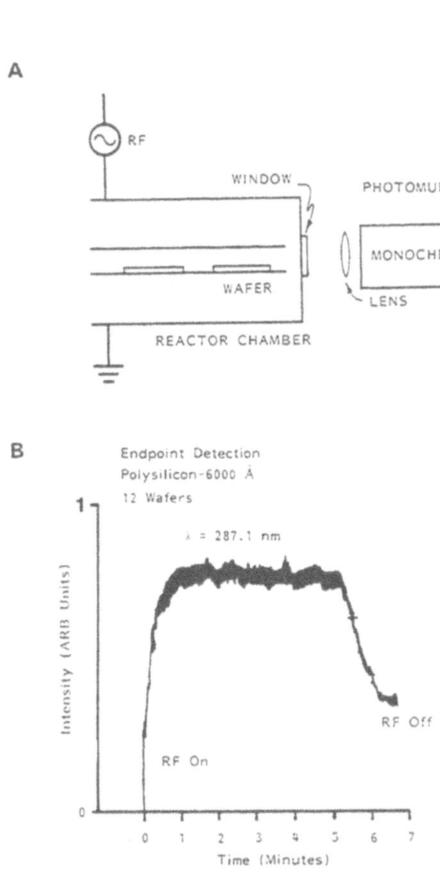

FIGURE 13-13-34. (A) Setup of optical emission monitor to determine endpoint of plasma etching. From Ref. 582, reproduced by permission of SPIE. (B) Output signal monitoring SiF_4 emission at 287 nm in polysilicon plasma etching. From Ref. 513, reprinted with permission of *Solid State Technology*, published by Technical Publishing, a company of Dun and Bradstreet.

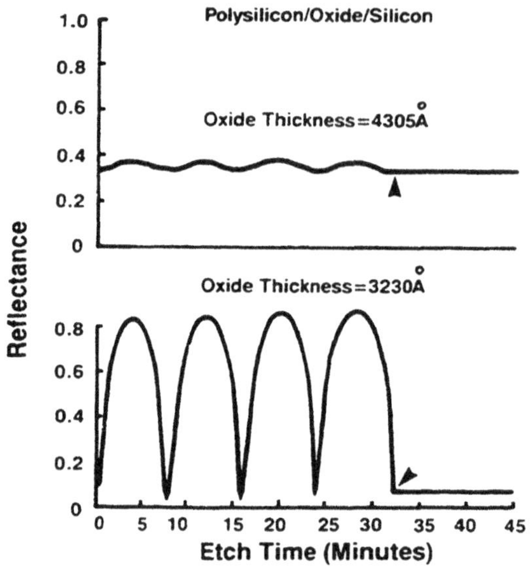

FIGURE 13-13-35. Endpoint detect of plasma etching of SiO_2 by laser reflectivity. From Ref. 518, reproduced by permission of the Electrochemical Society Inc.; figure originally presented at the Spring 1983 ECS meeting.

etching, diffraction gratings with 2-μm lines and space have been used to access the percent beyond endpoint that is needed to clean out micrometer geometries.[508, 750] A typical reflectivity curve is shown in Fig. 13-13-35. Each cycle corresponds to a change in film thickness, d, given by

$$d = \lambda/2n \tag{13-13-32}$$

where λ is the wavelength of light and n is the refractive index of the etching film.

For monitoring opaque substrates such as silicon, a two-beam interferometric method is used. A single beam is split to strike a reference nonetching surface and a surface being etched. A change in path length (thickness) can cause a phase shift and sinusoidal output.[519, 520]

Although present endpoint techniques are useful process monitors, more sophisticated optical-scanning electron beam techniques are needed in the future to monitor the actual etching of submicrometer dimensions.

TABLE 13-13-26. Resist Performance in Wet and Dry Etching

Requirement	Wet etch	Dry etch
Planar coating	No effect	Too thin over hills
Conformal coating	No effect	Desired
Step protection	No effect	Poor if >30% erosion
Thermal stable	<50 °C	100–250 °C
Adhesion	Basic requirement	Only to developer
Profile	No dependency	>75 °C or isotropic required
Pinholes, size	<1/5 linewidth	Same
Erosion	No attack	Require anisotropic
Erosion rate	No attack	Reproducible

13-14. SUMMARY AND OUTLOOK OF ETCHING

I have spent a long, but necessary amount of time describing wet and dry etching processes, mechanisms, and applications. In the last 10 years, the reactive ion anisotropic etching technique has made the most significant contribution to semiconductor lithography processes, in the ability to transfer resist images, with biases smaller than the print bias. As we progress to smaller submicrometer geometries, our understanding of both the theoretical and applied aspects of etching is still on a learning curve. There are three major components to etching—materials, process, and equipment. All three must be controlled, manipulated, and comprehended as to their pitfalls and limitations. For example, the critical role of the resist in wet and dry etching is compared in Table 13-13-26. The adhesion demands on the resist are higher for wet etching than for dry processing. For anisotropic dry etching, a high resist erosion resistance ($< 20\%$ loss) is required.

Exact profile control of the resist image is also required from the printing process since dry etching transfers without much distortion the shape and width of the resist image. Since diazonaphthoquinone–novolak resins were already in popular use in wet etching with exposure by scanning or projection tools when dry etching was introduced in 1975, they proved to be as applicable as dry etch resists and paved the way for the rapid introduction of dry etching. Improved resists have resulted from posttreatments of the basic diazonovolak resists.

Subtractive etching as opposed to additive techniques is the simplest image transfer process. Single-wafer processing, with self-monitoring of etching submicrometer geometries, is a realistic goal for the next generation of lithography. A more critical understanding of the chemical kinetics and mass transport of ions and radicals in the gas phase (or in solution) is required for integrating the choice of etchants, reactor configuration, and process conditions, into uniform and reproducible manufacturing process. Yes, there is still some magic in etching, and many more factors to unravel. Dry etching has to overcome the concentration-limited kinetics and have the luxury of wet etching, but derive the benefits of anisotropic control.

All submicrometer processing will rely upon the angular dependencies of charged particles striking a surface. Thus, in the future, physical scattering of ions will become a resolution limit. New resists for nanometer resolution of nanometer thickness will also be required to exhibit zero etch rates to limit etch biases to less than the thickness of the imaged resist layer. In this respect, vacuum-deposited inorganic resists and metal organic resists used in top surface imaging are the next generation of reactive ion resists.

REFERENCES

1. C. Zarowin and R. Horwath, *J. Electrochem. Soc.* **129**, 2541 (1982).
2. R. Brandes and R. Dudley, *J. Electrochem. Soc.* **120**, 140 (1973).
3. J. Curran, *J. Phys. E* **14**, 343 (1981).
4. W. Kern, J. Vossen, and G. Schnable, IEEE 11th Ann. Proc. Rel. Phys., 1973, pp. 214–223.
5. J. Vossen, G. Schnable, and W. Kern, *J. Vac. Sci. Technol.* **11**, 60 (1974).
6. L. Romankiw, S. Krongelb, M. Blakstey, A. Pfeiffer, and O. B. Stoeber, *Electrochem. Soc. Abstr.* **79-2**, 1190 (1979).

7. U. S. Patent 3,873,361 (1976), IBM.
8. A. Agajanian, *Semiconductor Devices: A Bibliography of Fabrication Technology, Properties, and Applications*, Plenum Press, New York, 1976. After 1976, internal search by author.
9. L. Romankiw, Proc. Symp. Etching Pattern Definition, 1976, pp. 161–193.
10. A. Reinberg, Proc. Symp. Etching Pattern Definition, 1976, pp. 91–110.
11. J. Coburn, *Plasma Chem. Plasma Process* **2**, 1 (1982).
12. B. Chapman, *Glow Discharge Processes, Sputtering and Plasma Etching*, Wiley, New York, 1980.
13. R. Frieser and C. Mogab, *Plasma Process, Electrochemical Society*, 1981.
14. P. Parry and A. Rodde, *Solid State Technol.* April 1979, p. 125.
15. P. Holmes, *Electrochemistry of Semiconductors*, Academic Press, New York, 1962.
16. H. King, in *Surface Chemistry of Metals and Semiconductors*, edited by H. Gatos, Wiley, New York, 1959, p. 358.
17. M. Cretella and H. Gatos, *J. Electrochem. Soc.* **105**, 487 (1958).
18. *Progress in Semiconductors*, edited by J. Gibson and H. Burgess, Vol. 9, Temple Press, London, 1965, p. 10.
19. W. Tegart, *The Electrolytic and Chemical Polishing of Metals in Research and Industry*, Pergamon Press, Elmsford, N.Y., 1959.
20. W. Harris, *Chemical Milling*, Oxford University Press (Clarendon), London, 1976.
21. W. Kern, *Semicond. Int.* March 1982, p. 89.
22. A. Revesz and K. Zaininger, *RCA Rev.* **29**, 22 (1968).
23. A. Lopez, *J. Electrochem. Soc.* **113**, 89 (1966).
24. W. Tiller, *J. Electrochem. Soc.* **128**, 689 (1981).
25. H. Kleinknecht and H. Meier, *J. Electrochem. Soc.* **125**, 799 (1978), U. S. Patent 4,039,370.
26. S. Haszko, G. Parisi, and G. Rozgonyi, *J. Electrochem. Soc.* **124**, 917 (1977).
27. J. Judge, *J. Electrochem. Soc.* **118**, 1772 (1971).
28. J. Judge, Proc. Symp. Etching Pattern Definition, 1976, pp. 19–36.
29. J. Dey, M. Lundgren, and S. Harrell, *Proc. Kodak Photoresist Semin.* **192B**, 4 (1968).
30. M. Jun-Ru, S. Kuo-Hsuing, E. Wolfe, and T. Everhart, *J. Vac. Sci. Technol.* **19**, 1385 (1981).
31. D. Wetson and R. Mattox, *J. Vac. Sci. Technol.* **17**, 466 (1980).
32. R. Bersin and R. Reichhelderfer, *Solid State Technol.* April 1977, p. 78.
33. P. Holmes and J. Snell, *Microelectron. Reliab.* **5**, 337 (1966).
34. N. Duffek and D. Pilling, *Electrochem. Soc. Ext. Absdtr.* Spring 1965, p. 244.
35. S. Harrell and J. Peoples, *Electrochem. Soc. Ext. Abstr.* Spring 1965, p. 247.
36. L. White, *Thin Solid Films* **79**, L73 (1981).
37. J. Lawrence, *Electrochem. Soc. Ext. Abstr.* Fall 1972, p. 466.
38. R. Esch and W. Pliskin, Proc. Symp. Etching Pattern Definition, 1976, p. 37.
39. B. Schwartz and H. Robbins, *J. Electrochem. Soc.* **123**, 1903 (1976).
40. D. Lee, *J. Appl. Phys.* **40**, 4569 (1969).
41. M. Andrasi, *Thin Solid Films* **67**, 229 (1980).
42. B. Chapman, *J. Vac. Sci. Technol.* **11**, 106 (1974).
43. T. Gupta, *Thin Solid Films* **75**, 319 (1981).
44. S. Wu, *Polymer Interfaces and Adhesion*, Dekker, New York, 1982, p. 371.
45. J. Klaubek, *J. Adhes.* **6**, 293 (1974).
46. C. Hammermesh and L. Crane, *J. Appl. Polym. Sci.* **22**, 239 (1978).
47. E. Davidson and G. Lee, *Electrochem. Soc. Ext. Abstr.* **71-1**, 79 (1971).
48. C. Deckert and D. Peters, *Thin Solid Films* **68**, 417 (1980).
49. J. Panzer, *J. Colloid Interface Sci.* **44**, 144 (1973).
50. E. Gipstein, O. Need, and W. Moreau, *J. Electrochem. Soc.* **123**, 1105 (1976).
51. A. Laird and F. Nelson, *SPE Trans.* **4**, 120 (1964).
52. D. Elliott, *Integrated Circuit Fabrication Technology*, McGraw–Hill, New York, 1982, p. 316.
53. U. S. Patent 4,176,003 (1979), NCR.
54. B. Arkles, "Technique for Silylation," Silicon Compounds, Register and Review, Petrach Systems, Levittown, Pa., 1979.
55. U. S. Patent 3,405,017 (1968).
56. U. S. Patent 3,549,368 (1970, IBM.
57. U. S. Patent 3,520,683 (1970), Bell.
58. U. S. Patent 3,586,554 (1971), IBM.
59. U. S. Patent 3,716,390 (1973), Bell.

60. U. S. Patent 4,103,045 (1978), Rhone–Poulene.
61. U. S. Patent 3,482,977 (1968), Sylvania. 62. H. Yanazawa, T. Matsuzawa, and N. Hashimoto, Proceedings of Kodak Microelectronic Seminar, 1977, p. 1.
63. H. Ishida and J. Koenig, *J. Colloid Interface Sci.* **64**, 555 (1978).
64. H. Calderon and L. St. Pierre, *J. Colloid Interface Sci.* **64**, 403 (1978).
65. L. Hall and D. Crosswaichte, *Thin Solid Films* **9**, 447 (1972).
66. C. Deckert, Proceedings of Kodak Microelectronic Seminar, 1977, p. 44.
67. U.S. Patent 3,827,908 (1974), IBM.
68. German Patent 2,447,225, IBM; *Chem. Abstr.* **85**, 102423 (1976).
69. U.S. Patent 4,142,636 (1981), Mitsubishi.
70. U. S. Patent 3,873,313 (1977), IBM.
71. U. S. Patent 4,332,881 (1982), Bell.
72. Eastman Kodak Technical Literature on Applications of Eastman 910 Adhesive.
73. *Kodak Res. Discl.* **149**, 27 (1976); *Chem. Abstr.* **86**, 362590 (1977).
74. U. S. Patent 4,036,644 (1979), IBM.
75. Japanese Patent 75,127,619; *Chem. Abstr.* **85**, 12367 (1976).
76. U. S. Patents 3,634,082; 4,148,654 (1982), Shipley Co.
77. S. Farenholtz, *Am. Chem. Soc. Org. Coat.* **35**, 306 (1975).
78. Japanese Patent 77,126,221; *Chem. Abstr.* **89**, 120920 (1978).
79. Japanese Patent 80,129,341, Hitachi; *Chem. Abstr.* **94**, 148396 (1981).
80. Japanese Patent 78,135,621, Hitachi; *Chem. Abstr.* **90**, 130686 (1979).
81. K. Kawase and T. Enomoto, Proceedings of Kodak Interface Resist Seminar **6-35**, 1973, p. 24.
82. Eastman Kodak Technical Literature on Kodak Micropositive 820 Resists.
83. M. Long and C. Walker, Kodak Microelectronics Seminar Proceedings **6-102P**, 1979, p. 125.
84. R. Lussow, *J. Electrochem. Soc.* **115**, 660 (1968).
85. A. Bergh, *J. Electrochem. Soc.* **112**, 457 (1965).
86. K. Mittal and R. Lussow, in *Adhesion and Adsorption of Polymers*, edited by L. Lee, Plenum Press, New York, 1980.
87. B. Wagner, J. Helbert, E. Poindexter, and D. Bates, *Surf. Sci.* **67**, 251 (1977).
88. U. S. Patent 3,716,390 (1973), Bell.
89. A. Sinha, T. Smith, and R. Severstein, *J. Appl. Phys.* **49**, 2423 (1978).
90. R. Scheuerman, *Electrochem. Soc. Ext. Abstr.* **74-1**, 341 (1974).
91. A. Ouano, *Am. Chem. Soc. Org. Coat.* **45**, 431 (1981).
92. T. Bullett, *J. Adhes.* **4**, 73 (1972).
93. K. Mittal, *Electrocomponent Sci. Technol.* **3**, 21 (1976).
94. S. Shreiner, P. Zubov, and A. Arwa, *Kolloidn. Zh.* **23**, 334 (1961).
95. W. Moreau, R. Cox, and N. Clecak, *Polym. Eng. Sci.* **14**, 491 (1974).
96. C. Cortellino, *IBM Tech. Discl. Bull.* **13**, 795 (1970).
97. D. Herman, *Electrochem. Soc. Ext. Abstr.* May 1971, p. 168.
98. U. S. Patent 4,040,867 (1977), Signetics.
99. U. S. Patent 3,518,135 (1967), Sylvania.
100. U. S. Patent 3,649,395 (1972), Phillips.
101. W. Kern, J. Vossen, and G. Schnoble, *IEEE 11th Annu. Rel. Phys.* **11**, 214 (1973).
102. I. Blech, *Thin Solid Films* **6**, 113 (1970).
103. G. Koel and T. Bertens, *Electrochem. Soc. Ext. Abstr.* **79-2**, 205 (1979).
104. J. Gotzlich and H. Ryssel, *J. Electrochem. Soc.* **128**, 617 (1981).
105. G. Schwartz and H. Robbins, *J. Electrochem. Soc.* **106**, 365 (1961).
106. A. Bogenschutz, W. Krusemark, K. Locherer, and W. Mussinger, *J. Electrochem. Soc.* **114**, 970 (1967).
107. U. S. Patent 3,749,619 (1973), Tokyo Shibaura.
108. U.S. Patent 3,756,877 (1974), Tokyo Shibaura.
109. W. Kern, *RCA Rev.* **39**, 278 (1978).
110. W. Kern and C. Deckert, *Chemical Etching in Thin Film Processes*, edited by W. Kern and J. Vossen, Academic Press, New York 1978.
111. U. S. Patent 3,767,494 (1974), Tokyo Shibaura.
112. M. Teunissen, J. Apples, and W. Verkuylen, *J. Electrochem. Soc.* **117**, 959 (1970).
113. U. S. Patent 3,891,983 (1975), GMBH.
114. U. S. Patent 4,050,979 (1977), TI.

115. U. S. Patent 4,029,542 (1977), RCA.
116. French Patent 1,266,612 (1961).
117. U. S. Patent 3,677,846 (1972), Phillips.
118. U.S. Patent 3,977,925 (1976), Siemens.
119. U. S. Patent 4,004,957 (1977), Rockwell.
120. U. S. Patent 4,171,242 (1979), IBM.
121. U. S. Patent 3,909,325 (1975), Motorola.
122. U. S. Patent 4,113,551 (1978), IBM.
123. U. S. Patent 3,765,969 (1973), Bell.
124. U. S. Patent 3,738,881 (1973), Bell.
125. C. Raetzel, S. Schild, and H. Schlotterer, *Electrochem. Soc. Ext. Abstr.* **74-2**, 336 (1974). Figure presented at Fall ECS Meeting (1974), St. Louis.
126. A. Reisman, M. Berkinbert, S. Chan, F. Kaufmann, and D. Green, *J. Electrochem. Soc.* **126**, 1406 (1979).
127. U. S. Patent 3,909,325 (1975), Motorola.
128. U. S. Patent 3,728,179 (1973), Rad. Inc.
129. R. Finne and D. Klein, *J. Electrochem. Soc.* **114**, 965 (1967).
130. U. S. Patent 3,506,509 (1970), Bell.
131. U. S. Patent 3,738,881 (1973), Bell.
132. U. S. Patent 3,765,969 (1973), Bell.
133. U. S. Patent 4,137,123, (1979), Motorola.
134. U. S. Patent 3,649,395 (1972), Phillips.
135. U. S. Patent 3,725,160 (1973), TI.
136. U. S. Patent 3,738,881 (1973).
137. U. S. Patent 4,172,005 (1979), Tokyo Shibaura.
138. U. S. Patent 3,759,968 (1973), Bell Northern.
139. W. Zurcker and S. Kurtz, in *Semiconductor Silicon*, edited by H. Huff and R. Burgess, Electrochemical Society, 1973, p. 315.
140. C. Deckert, *J. Electrochem. Soc.* **127**, 433 (1980).
141. V. Harrap, Ref. 139, p. 354.
142. U. S. Patent 3,607,480 (1971), TI.
143. U.S. Patent 3,811,974 (1974).
144. U. S. Patent 4,269,254 (1981), RCA.
145. U. S. Patent 3,979,241 (1976), Fujitsu.
146. U. S. Patent 3,519,504 (1970), IBM.
147. U. S. Patent 3,715,249 (1973), Bell.
148. U. S. Patent 3,979,238 (1976), RCA.
149. U. S. Patent 3,808,969 (1974).
150. U. S. Patent 4,220,706 (1980).
151. L. Romankiw, Proc. Symp. Etching Pattern Definition, 1976, pp. 161–193.
152. J. Moore, *Chemical Metallurgy*, Butterworths, Boston, 1981, pp. 172–184.
153. D. MacArthur, Proc. Symp. Etching Pattern Definition, 1976, p. 132.
154. J. Kelly and C. de Minjer, *J. Electrochem. Soc.* **122**, 931 (1975).
155. T. Agatsuma, A. Kikuchi, K. Nakada, and A. Tomozawa, *J. Electrochem. Soc.* **122**, 825 (1975).
156. J. Harvilchuck and J. Judge, Electrochem. Soc., Oct. 1969, Recent News Paper.
157. R. Hines, *Microelectron. Reliab.* **10**, 59 (1971).
158. U. S. Patent 4,080,246 (1978), GAF.
159. W. Motsiff, R. Malone, and C. Halsted, Int. Conf. Microlithogr. 1977, p. 333.
160. C. Halsted and M. Haller, *Electrochem. Soc. Ext. Abstr.* **76-2**, 748 (1976).
161. U. S. Patent 3,808,070 (1974), GE.
162. U. S. Patent 4,097,394 (1977), Phillips.
163. U. S. Patent 3,532,569 (1970), GM.
164. U. S. Patent 4,080,246 (1978), GAF.
165. U. S. Patent 3,715,250 (1973), GM.
166. U. S. Patent 3,738,879 (1973), Siemens.
167. U. S. Patent 3,859,149 (1975), Rolls Royce.
168. M. Ehman and J. Faust, *J. Appl. Phys.* **41**, 3169 (1970).
169. J. Campbell, E. Currier, and P. Hauth, *IBM Tech. Discl. Bull.* **13**, 948 (1970).

170. L. Martinson, Proceedings of Eastman Kodak Meetings, 1969, p. 34.
171. A. Smith, *J. Elec. Soc.*, **120**, 1551, (1977).
172. L. Blakely, Proceedings of Kodak Photoresist Seminar, 1969, p. 68.
173. U. S. Patent 3,544,401 (1970).
174. T. Agatsuma, *J. Electrochem. Soc.* **122**, 825 (1975).
175. G. Schwartz and V. Platter, *J. Electrochem. Soc.* **122**, 1516 (1975).
176. U. S. Patent 4,003,772 (1977), Hitachi.
177. U. S. Patent 3,825,454 (1974), Hitachi.
178. U. S. Patent 4,082,604 (1978), Motorola.
179. B. James, Proc. 1971 Semicond. Process Prod., p. 111.
180. H. Moritz, *Solid State Technol.* Aug. 1975, p. 54.
181. R. Frankenthal and D. Eaton, *J. Electrochem. Soc.* **123**, 703 (1976).
182. N. Hashimoto, H. Yanazawa, M. Ashikawa, and K. Douta, *Electrochem. Soc. Ext. Abstr.* **74-1**, 145 (1974).
183. J. Helbert, P. Seese, A. Gonjales, and C. Walker, *SPIE Proc.* **333**, 24 (1982).
184. D. Angel and P. Johnson, *Semicond. Int.* March 1980, p. 59.
185. P. Burggraaf, *Semicond. Int.* Feb. 1983, p. 49.
186. U. S. Patent 3,642,549 (1972), IBM.
187. J. Scully, *Fundamentals of Corrosion*, Oxford University Press, London, 1971, p. 165.
188. A. Neureuther, W. Oldham, S. Nandagonkar, and M. O'Toole, *IEEE Trans. Electron Devices* **ED-26**, 717 (1979), **ED-27**, 1455 (1980); *J. Vac. Sci. Technol.* **19**, 1242 (1981).
189. N. Viswanathan, *J. Vac. Sci. Technol.* **16**, 338 (1979).
190. K. Konnerth and F. Dill, *IEEE Trans. Electron Devices* **ED-22**, 452 (1975).
191. A. Neureuther, Kodak Microelectronics Seminar Proceedings, **6-47**, 1976, p. 81.
192. R. Frieser, *J. Electrochem. Soc.* **119**, 360 (1972).
193. Ref. 52, p. 285.
194. W. Kern and J. Shaw, *J. Electrochem. Soc.* **118**, 1699 (1971).
195. J. Coburn, *Plasma Chem. Plasma Process.* **2**, 1 (1982).
196. C. Melbiar-Smith and C. Mogab, in *Thin Film Processes*, edited by J. Vossen and W. Kern, Academic Press, New York, 1978, p. 497.
197. D. Flamm and V. Donnelly, *Plasma Chem. Plasma Process.* **1**, 317 (1981).
198. G. Stengl, R. Kartna, H. Loscher, R. Reider, P. Wolfe, and R. Sachen, *Solid State Technol.* Aug. 1982, p. 104.
199. P. Parry and A. Rodde, *Solid State Technol.* April 1979, p. 125; U. S. Patent 4,253,907 (1981), Western Electric.
200. C. Mogab and H. Levenstein, *J. Vac. Soc. Technol.* **17**, 721 (1980).
201. D. Downey, W. Bottoms, and P. Hanley, *Solid State Technol.* Feb. 1981, p. 121.
202. R. Porter, W. Harshberger, J. Clemens, and J. Cuthbert, *Electrochem. Soc. Ext. Abstr.* **78-2**, 515 (1978).
203. W. Harshberger and C. Mogab, *Electrochem. Soc. Ext. Abstr.* **78-2**, 511 (1978).
204. B. Chapman, *Glow Discharge Processes*, Wiley, New York, 1980.
205. K. Tokunaga and D. Hess, *J. Electrohem. Soc.* **127**, 928 (1980).
206. K. Tokunaga, F. Redeker, D. Danner, and D. Hess, *J. Electrochem. Soc.* **128**, 851 (1981).
207. T. Smith, *Surf. Sci.* **32**, 527 (1972).
208. J. Battey, *J. Electrochem. Soc.* **124**, 147 (1977).
209. L. Kammerdiner, *Solid State Technol.* Oct. 1981, p. 79.
210. A. Nagy and D. Hess, *J. Electrochem. Soc.* **129**, 2530 (1982).
211. F. Swift, R. Sung, J. Doyle, and J. Stille, *J. Org. Chem.* **30**, 3114 (1965).
212. C. Zarowin, *Thin Solid Films* **85**, 33 (1981).
213. W. Archer and K. Harter, *Corrosion* **34**, 159 (1978).
214. M. Sato and H. Nakamura, *J. Vac. Sci. Technol.* **20**, 186 (1982).
215. D. Smith and P. Saviano, *J. Vac. Sci. Technol.* **21**, 768 (1982); D. Smith and R. Bruce, *J. Electrochem. Soc.* **129**, 2045 (1982).
216. G. Schwartz and P. Schaible, *Solid State Technol.* Nov. 1980, p. 85.
217. N. Herman, V. Minkiewicz, and B. Chapman, *J. Vac. Sci. Technol.* **17**, 731 (1980).
218. S. Broydo, *Solid State Technol.* April 1983, p. 247.
219. R. Saia and B. Gorowitz, *Solid State Technol.* Aug. 1982, p. 95.
220. P. Schaible, W. Metzger, and J. Anderson, *J. Vac. Sci. Technol.* **15**, 334 (1978).

221. A. Chambers, *Solid State Technol.* Aug. 1982, p. 95.
222. C. Tracy and R. Mattox, *Solid State Technol.* June 1982, p. 83.
223. P. Schaible and G. Schwartz, *J. Vac. Sci. Technol.* **15**, 334 (1979).
224. R. Reichelderfer, *Solid State Technol.* April 1982, p. 160.
225. M. Szato and H. Nakamura, *J. Electrochem. Soc.* **129**, 2522 (1982).
226. K. Hirobe, Y. Kureishi, and T. Tsuchimoto, *J. Electrochem. Soc.* **128**, 2686 (1981).
227. L. Percoud, P. Laporte, M. Chevaliey, and K. Fried, *Electrochem. Soc. Ext. Abstr.* **81-2**, 712 (1981).
228. H. Okano, T. Yamazaki, and Y. Horike, *Solid State Technol.* April 1982, p. 170.
229. R. Bruce, D. Flamm, V. Donnelly, and B. Duncan, *Electrochem. Soc. Ext. Abstr.* **81-2**, 631 (1981).
230. M. Nakamura, M. Itoga, and Y. Ban, in *Plasma Processing*, edited by R. Frieser and C. Mogab, Electrochemical Society, 1981, p. 223.
231. Japanese Patent 78,140,240, *Chem. Abstr.* **90**, 208746 (1979).
232. D. Smith and R. Bruce, *J. Electrochem Soc.* **129**, 2045 (1982).
233. R. Bruce and G. Malsky, *Electrochem. Soc. Ext. Abstr.* **81-2**, 703 (1981).
234. Japanese Patent 80,85,670, *Chem. Abstr.* **94**, 23904 (1981).
235. G. Herb, R. Porter, P. Cruzan, J. Guerna, and B. Soller, *Electrochem. Soc. Ext. Abstr.* **81-2**, 710 (1981).
236. N. Hosokawa, R. Matsizaki, and T. Asamaki, *Jpn. J. Appl. Phys. Suppl.* **2**, Part 1, 435 (1974).
237. K. Suzuki, S. Okudaiara, and I. Kanamata, *J. Electrochem. Soc.* **126**, 1024 (1979).
238. R. Bersin, *Solid State Technol.* April 1978, p. 14.
239. R. Booth and C. Heslop, *Thin Solid Films* **65**, 111 (180).
240. Coberley Assoc., *Solid State Technol.* April 1978, p. 14.
241. L. Fieser and M. Frieser, *Advanced Organic Chemistry*, Reinhold, New York, 1961, p. 650.
242. J. Kiddon, T. Brewer, J. Sellerg, S. Bolgers, and T. Coutmas, *Electrochem. Soc. Ext. Abstr.* **81-1**, 746 (1981).
243. F. Custode, W. Fewer, M. Splinter, and D. Downey, *Electrochem. Soc. Ext. Abstr.* **81-1**, 748 (1981).
244. W. Winkler, *Electrochem. Soc. Ext. Abstr.* **81-1**, 253 (1982).
245. K. Donahue, *Electrochem. Soc. Ext. Abstr.* **81-2**, 706 (1982).
246. T. Herdon and R. Burke, Kodak Microelectronics Symposium, 1977.
247. R. Reichelderfer, D. Vogel, and R. Bersin, *Electrochem. Soc. Ext. Abstr.* **77-2**, 414 (1977).
248. T. Fok, *Electrochem. Soc. Ext. Abstr.* **80 – 2**, 310 (1980).
249. R. Ranadive, R. Khosta, and D. Losee, *Electrochem. Soc. Ext. Abstr.* **81-2**, 707 (1981).
250. T. Tsukada, *Electrochem. Soc. Ext. Abstr.* **81-2**, 715 (1981).
251. U. S. Patent 4,183,781 (1980), IBM.
252. W. Lee, J. Eldridge, and G. Schwartz, *J. Appl. Phys.* **52**, 2994 (1981).
253. L. Gregor, *IBM Tech. Discl. Bull.* **24**, 2994, 5538 (1982).
254. R. Allen, M. Foster, and Y. Yen, *J. Electrochem. Soc.* **129**, 1379 (1982), and references therein.
255. H. Hiroaka and J. Pacansky, *J. Vac. Sci. Technol.* **19**, 1132 (1981).
256. D. Dobkin and B. Cantos, *IEEE Electron Device Lett.* **EDL**-2, 222 (1981).
257. U. S. Patent 4,253,388 (1981), IBM.
258. W. Ma, *SPIE Proc.* **333**, 19 (1982).
259. U. S. Patent 3,920,483 (1975), IBM.
260. U. S. Patent 4,201,800 (1980), IBM.
261. U. S. Patent 4,187,331 (1980), IBM.
262. D. Weston and R. Mattox, *J. Vac. Sci. Rechnol.* **17**, 466 (1980).
263. J. Moran and G. Taylor, *J. Electrochem. Soc.* **19**, 1127 (1981).
264. R. Morgan and C. Pollard, *Electron. Lett.* **18**, 1038 (1982).
265. B. Canvello, V. DiMilla, and M. Hatzakis, *IBM Tech. Discl. Bull.* **22**, 2134 (1979).
266. Y. Wada, K. Mochyi, and H. Obayashi, *J. Electrochem. Soc.* **130**, 187 (1983).
267. Y. Wada, H. Okabashi, and H. Suzuki, *Jpn. J. Appl. Phys.* **16**, 1313 (1972).
268. U. S. Patent 4,125,650 (1978), IBM.
269. German Patent 2,447,225 (IBM), *Chem. Abstr.* **85**, 102423 (1978).
270. U. S. Patent 4,007,046 (1979), IBM.
271. European Patent Application 38,967, *Chem. Abstr.* **96**, 133203 (1982).
272. U. S. Patent 4,259,430 (1981), IBM.
273. U. S. Patent 4,007,047 (1977), IBM.
274. K. Jinno, Y. Matsumoto, and T. Shinozaki, *Photogr. Sci. Eng.* **21**, 290 (1977).
275. S. Broydo, *Solid State Technol.* April 1983, p. 159.

276. A. Weiss, *Semicond. Int.* Oct. 1982, p. 69.
277. U. S. Patent 4,298,443 (1981), Bell.
278. H. Okano, T. Yamazaki, and Y. Horike, *Solid State Technol.* April 1982, p. 166.
279. Commonwealt Sci., *Solid State Technol.* Oct. 1981, p. 100.
280. R. Bollinger and R. Fink, *Solid State Technol.* Oct. 1980, p. 79.
281. T. O'Keefe and R. Handy, *Solid State Electron.* **11**, 261 (1968); U. S. Patent 3,615,953 (1971), Westinghouse.
282. M. Oda and K. Hirato, *Jpn. J. Appl. Phys.* **7**, 1405 (1980).
283. U. S. Patent 4,260,259 (1981), IBM.
284. K. Uhai and K. Hanazawa, *J. Vac. Sci. Technol.* **16**, 385 (1979).
285. S. Dzioba, G. Este, R. Bond, and H. Naguib, *Electrochem. Soc. Ext. Abstr.* **82-1**, 259 (1982).
286. B. Curtis and H. Bruner, *J. Electrochem. Soc.* **125**, 829 (1978); C. Korman, *Solid State Technol.* April 1982, p. 115.
287. H. Tiller, D. Berg, and R. Mohr, *Plasma Chem. Plasma Process.* **1**, 247 (1981).
288. C. Mogab, *J. Electrochem. Soc.* **124**, 1262 (1977).
289. J. Hayes and T. Pandhumsoporon, *Solid State Technol.* Now. 1980, p. 71.
290. C. Mogab and H. Levinstein, *J. Vac. Sci. Technol.* **17**, 721 (1980).
291. W. Moreau, *Opt. Eng.* **22**, 181 (1983).
292. Eastman Kodak, Rochester, N.Y., Micropositive 820 resist; MacDermid Inc., Waterbury, Conn., Ultramar PR74 resist; Dynachem Co., Santa Fe Springs, Calif., OFPR-800 resist.
293. K. Ukai and K. Hanazawa, *J. Vac. Sci. Technol.* **15**, 330 (1978).
294. D. Tolliver, *Solid State Technol.* Nov. 1980, p. 99.
295. J. Harper and J. Cuomo, *J. Vac. Sci. Technol.* **21**, 740 (1982).
296. M. Geis, G. Lincoln, N. Efremow, and W. Pracentini, *J. Vac. Sci. Technol.* **19**, 1390 (1981).
297. K. Suzuki, S. Okudaira, N. Sakudo, and I. Kanomoto, *Jpn. J. Appl. Phys.* **16**, 1979 (1977).
298. C. Mullins, *Solid State Technol.* April 1982, p. 115.
299. B. Heath, *Solid State Technol.* Oct. 1981, p. 77.
300. J. Coburn and H. Winters, *J. Vac. Sci. Technol.* **16**, 391 (1979).
301. J. Coburn and E. Kay, *IBM J. Res. Dev.* **23**, 339 (1979).
302. Y. Tu, T. Chuang, and H. Winters, *Phys. Rev.* **B23**, 823 (1981).
303. D. Flamm, V. Donnelly, and D. Ibbotson, *J. Vac. Sci. Technol.* **B1**, 23 (1983).
304. R. Bruce and A. Reinberg, *J. Electrochem. Soc.* **129**, 393 (1982).
305. D. Smith and R. Bruce, *J. Electrochem. Soc.* **129**, 2045 (1982).
306. R. Bruce, *Solid State Technol.* Oct. 1981, p. 64.
307. C. Zarowin, *J. Electrochem. Soc.* **130**, 1144 (1983).
308. G. Lang and T. Stavish, *RCA Rev.* **24**, 488 (1963).
309. K. Miyamura, Y. Sakisaka, N. Nishijima, and M. Onchi, *Surf. Sci.* **72**, 243 (1978).
310. J. Dismukes and R. Ulmer, *J. Electrochem. Soc.* **118**, 634 (1971).
311. L. Gregor, P. Ballo, and F. Campagna, *IBM J. Res. Dev.* **9**, 327 (1965).
312. A. Reisman and M. Berkenblitt, *J. Electrochem. Soc.* **112**, 812 (1965).
313. P. Rai-Choudhury and A. Norieka, *J. Electrochem. Soc.* **116**, 539 (1969).
314. T. Chu, *J. Electrochem. Soc.* **113**, 156 (1966).
315. T. Chuang, *IBM J. Res. Dev.* **26**, 145 (1982).
316. H. Winters and J. Coburn, *Appl. Phys. Lett.* **34**, 70 (1979).
317. M. Chen, V. Minikiewicz, and K. Lee, *J. Electrochem. Soc.* **126**, 1946 (1979).
318. D. Flamm. V. Donnelly, and J. Micha, *J. Appl. Phys.* **52**, 3633 (1981).
319. M. Jost, *Diffusion in Solids, Liquids, and Gases*, Academic Press, New York, 1952, p. 425.
320. J. Micha, V. Donnelly, and D. Flamm, *J. Phys. Chem.* **85**, 3529 (1981).
321. G. Schwartz and P. Schaible, *Solid State Technol.* Nov. 1980, p.85.
322. T. Mayer and R. Bruce, *J. Electrochem. Soc.* **129**, 585 (1982).
323. H. Kalter and E. Van DeVen, *Philips Tech. Rev.* **38**, 204 (1978).
324. B. Chapman and V. Minikiewicz, *J. Vac. Sci. Technol.* **15**, 329 (1978).
325. A. Jacob, *Solid State Technol.* May 1976, p. 32.
326. G. Smolinsky, T. Mayer, and E. Truesdale, *J. Electrochem. Soc.* **129**, 1170 (1982).
327. K. Suzuki, S. Okudaira, N. Sakudo, and I. Kanomoto, *Jpn. J. Appl. Phys.* **16**, 1979 (1977).
328. J. Steinfeld, *J. Electrochem. Soc.* **127**, 514 (1980).
329. P. Ehrlich, R. Osgood, and T. Deutsch, *Appl. Phys. Lett.* **38**, 1018 (1981).
330. R. Barker, T. Mayer, and W. Pearson, *J. Vac. Sci. Technol.* **B1**, 37 (1983).

331. D. Flamm, P. Cowan, and J. Golovchenko, *J. Vac. Sci. Technol.* **17**, 1341 (1980).
332. M. Millard and E. Kay, *J. Electrochem. Soc.* **129**, 160 (1982).
333. H. Okano and Y. Horuke, *Jpn. J. Appl. Phys.* **20**, 2429 (1981).
334. W. Libby, *J. Vac. Sci. Technol.* **16**, 414 (1982).
335. D. Smith and R. Bruce, *J. Electrochem. Soc.* **129**, 2045 (1982).
336. J. Harper, J. Cuomo, P. Leary, G. Summa, H. Kaufmann, and F. Bresnock, *J. Electrochem. Soc.* **128**, 1077 (1981).
337. U. Gerlach-Meyer, J. Coburn, and E. Kay, *Surf. Sci.* **103**, 177 (1981).
338. T. Mayer, R. Barker, and L. Whitman, *J. Vac. Sci. Technol.* **18**, 349 (1981).
339. U. Gerlach-Meyer, J. Coburn, and E. Kay, *J. Appl. Phys.* **51**, 3362 (1980).
340. L. Holland, *Br. J. Appl. Phys.* **2**, 767 (1969).
341. J. Coburn and E. Kay, *Solid State Technol.* April 1979, p. 117.
342. A. Dagastino, F. Cramorasa, V. Colaprico, and R. Etole, *J. Appl. Phys.* **54**, 1284 (1983).
343. L. Eprath, *J. Electrochem. Soc.* **128**, 123 (1981).
344. H. Norstrom, R. Buchta, F. Runove, and P. Wiklund, *Vacuum* **32**, 737 (1982).
345. A. Adams and C. Capio, *J. Electrochem. Soc.* **128**, 366 (1981).
346. U. S. Patent 3,984,301 (1976), Varian.
347. Y. Matsumoto, H. Kinoshita, and Y. Matsumoto, *J. Electrochem. Soc.* **124**, 1258 (1977).
348. K. Jinno, H. Kinoshita, and Y. Matsumoto, *J. Electrochem. Soc.* **125**, 827 (1978).
349. D. Flamm, *Plasma Chem. Plasma Process.* **1**, 37 (1981).
350. H. Winters, *J. Appl. Phys.* **49**, 5165 (1978); *J. Vac. Sci. Technol.* **B1**, 469 (1983).
351. V. Donnelly and D. Flamm, *Solid State Technol.* April 1981, p. 161.
352. C. Zarowin, *J. Electrochem. Soc.* 130, 1144 (1983).
353. A. Reinberg, J. Dallave, G. Steinberg, and R. Bruce, *Electrochem. Soc. Ext. Abstr.* **81-2**, 669 (1981).
354. J. Harper, J. Cuomo, and H. Kaufman, *J. Vac. Sci. Technol.* **21**, 745 (1982).
355. C. Weissmantel, *Jpn. J. Appl. Phys. Suppl.* **2**, Part 1, 439 (1974).
356. K. Jinno, *Jpn. J. Appl. Phys.* **17**, 1283 (1978).
357. L. Pederson, *J. Electrochem. Soc.* **129**, 205 (1982).
358. K. Harada, *J. Electrochem. Soc.* **128**, 491 (1981).
359. Japanese Patent 80,142,337, *Chem. Abstr.* **94**, 112566 (1981).
360. U. S. Patent 4,289,573 (1981).
361. K. Harada, O. Kogure, and K. Murase, *IEEE Trans. Electron Devices* **ED-29**, 516 (1982).
362. M. Yamada, J. Tamano, K. Yoneda, S. Morita, and S. Hattori, *Jpn. J. Appl. Phys.* **21**, 768 (1982), **20**, 1969 (1981).
363. I. Haller, R. Feder, M. Hatzakis, and E. Spiller, *J. Electrochem. Soc.* **126**, 154 (1979).
364. M. Hatzakis and D. Webb, *J. Vac. Sci. Technol.* **16**, 2008 (1979).
365. J. Chinn, I. Adesida, E. Wolf, and R. Tiberio, *J. Vac. Sci. Technol.* **19**, 1418 (1981).
366. I. Watanabe, H. Yasihara, Y. Sartoh, and S. Matsuo, *Jpn. J. Appl. Phys.* **20**, L804 (1981).
367. S. Inamura, *J. Electrochem. Soc.* **126**, 1628 (1979), U. S. Patent 4,286,049 (1980), Nippon.
368. E. Feit and L. Thompson, *J. Electrochem. Soc.* **126**, 1699 (1979), U. S. Patent 4,262,081 (1980), Bell.
369. O. Kogure, *J. Appl. Phys.* **21**, 776 (1982).
370. H. Nakata, T. Kato, K. Murata, Y. Hira, and K. Nagami, *J. Vac. Sci. Technol.* **19**, 1248 (1981).
371. Y. Ohnishi, M. Itoh, K. Mizuno, H. Gokan, and S. Fujiwara, *J. Vac. Sci. Technol.* **19**, 1136 (1981).
372. S. Imamura and S. Sugawara, *Jpn. J. Appl. Phys.* **21**, 776 (1982).
373. J. Lavine, J. Masters, G. Golding, and A. Das, *IEEE Trans. Electron Devices* **ED-28**, 1311 (1981).
374. K. Tai, *Electrochem. Soc. Ext. Abstr.* Montreal, 1982; 16 papers on inorganic resists.
375. M. Chang, *J. Vac. Sci. Technol.* **16**, 1973 (1979).
376. A. Yoshikawa, O. Ochi, and Y. Mizushima, *Appl. Phys. Lett.* **36**, 107 (1980).
377. M. Cantagrel, *IEEE Trans. Electron Devices* **ED-22**, 483 (1975).
378. P. Gloerson, *Solid State Technol.* **19**, 68 (1976).
379. J. Vossen and E. Davidson, *J. Electrochem. Soc.* **119**, 1708 (1972).
380. R. Bond, S. Dzioba, and H. Naguib, *Electrochem. Soc. Ext. Abstr.* **81-1**, 729 (1981).
381. Y. Koguri, *Electrochem. Soc. Ext. Abstr.* **77-2**, 373 (1977).
382. L. Eprath and E. Petrillo, *J. Electrochem. Soc.* **129**, 2282 (1982).
383. D. Bower, *J. Electrochem. Soc.* **129**, 796 (1982).
384. E. Whitecomb, *Electrochem. Soc. Ext. Abstr.* **82-1**, 339 (1982).
385. Y. Yoyoda, M. Tobinaga, and H. Komiya, *Jpn. J. Appl. Phys.* **20**, 681 (1981).

386. M. Kushner, *J. Appl. Phys.* **53**, 2923 (1982).
387. R. Chang, C. Chang, and S. Darack, *J. Vac. Sci. Technol.* **20**, 45 (1982).
388. L. Eprath, *J. Electrochem. Soc.* **128**, 123 (1982).
389. H. Okano, T. Yamazaki, and H. Horike, *Solid State Technol.* April 1982, p. 166.
390. K. Eislele, *J. Electrochem. Soc.* **128**, 324 (1981).
391. R. Agostino and D. Flamm, *J. Appl. Phys.* **17**, 235 (1978).
392. H. Abe, Y. Sonabe, and T. Ecomato, *Jpn. J. Appl. Phys.* **12**, 154 (1973).
393. S. Matsuo, *Jpn. J. Appl. Phys.* **17**, 235 (1978).
394. S. Matsuo and Y. Tabehara, *Jpn. J. Appl. Phys.* **16**, 175 (1977).
395. H. Lehman and R. Widmer, *J. Vac. Sci. Technol.* **15**, 319 (1978).
396. Y. Horike, T. Sugawara, H. Okano, M. Shibagaki, and Y. Ueda, *Jpn. J. Appl. Phys.* **20**, 803 (1981).
397. F. Sanders, J. Dielman, H. Peters, and J. Saunders, *J. Electrochem. Soc.* **129**, 2559 (1982).
398. E. Surikin, *Electrochem. Soc. Ext. Abstr.* **82-1**, 341 (1982).
399. A. Bell and J. Hollahan, editors, *Techniques and Applications of Plasma Chemistry*, Wiley, New York, 1974, p. 356.
400. J. Bondur, *J. Vac. Sci. Technol.* **13**, 1023 (1976).
401. Y. Horike, T. Yamazaki, M. Shibagaki, and T. Kwisaki, *Jpn. J. Appl. Phys.* **21**, 1412 (1982).
402. Y. Kurogi, M. Tagima, K. More, and K. Sugawara, *Electrochem. Soc. Ext. Abstr.* **77-2**, 373 (1977).
403. M. Hutt and W. Class, *Solid State Technol.* March 1980, p. 92.
404. H. Itakura, H. Komiya, and K. Ukai, *Solid State Technol.* April 1982, p. 213.
405. F. Egitto, D. Wang, D. Maydan, and D. Benzing, *Solid State Technol.* April 1982, p. 213.
406. T. O'Neill, *Semicond. Intl*, April 1981, p. 67.
407. T. Mayer and J. McConville, Proc. IDEM Meet., 1979, p. 44.
408. U. S. Patent 3,507,709 (1970), Hughes.
409. S. Bernacki, *Electrochem. Soc. Ext. Abstr.* **82-2**, 345 (1982).
410. R. Bruce, *Solid State Technol.* April 1981, p. 64.
411. R. Bruce and A. Reinberg, *J. Electrochem. Soc.* **129**, 393 (1982).
412. R. Lee, *J. Vac. Sci. Technol.* **16**, 169 (1979).
413. G. Boyd, L. Coldren, and F. Storz, *Appl. Phys. Lett.* **36**, 503 (1980).
414. M. Geis, G. Lincoln, N. Efremow, and W. Pracentini, *J. Vac. Sci. Technol.* **19**, 1390 (1981).
415. S. Selbrede, *SPIE Proc.* **221**, 53 (1980).
416. H. Abe, *Jpn. J. Appl. Phys.* **14**, 1835 (1975).
417. H. Kinoshita and K. Jinno, *Jpn. J. Appl. Phys.* **16**, 381 (1977).
418. H. Itakura, H. Komiya, and H. Yoda, *Jpn. J. Appl. Phys.* **9**, 1429 (1980).
419. L. Mader and J. Hoepfner, *J. Electrochem. Soc.* **123**, 1893 (1976).
420. S. Bergeron and B. Duncan, *Solid State Technol.* Aug. 1982, p. 98; M. Jenkins, M. Mocella, K. Allen, and H. Sawin, *Solid State Technol.* April 1986, p. 175.
421. M. Oshima, *Jpn. J. Appl. Phys.* **17**, 1157 (1978).
422. C. Zarowin, *J. Electrochem. Soc.* **130**, 1144 (1983).
423. M. Vasile, *J. Appl. Phys.* **51**, 2503 (1980).
424. K. Ikai and K. Hanazawa, *J. Vac. Sci. Technol.* **15**, 338 (1978).
425. K. Suzuki, S. Okudura, N. Sakudo, and I. Kanomoto, *Jpn. J. Appl. Phys.* **16**, 1979 (1977).
426. S. Matsuo, *J. Vac. Sci. Technol.* **17**, 587 (1980).
427. K. Suzuki, S. Okudura, and N. Sakudo, *J. Electrochem. Soc.* **126**, 1027 (1979).
428. U. S. Patent 4,350,578 (1982), IBM.
429. H. Itakura, H. Komiya, and K. Ukai, *Solid State Technol.* April 1982, p. 209.
430. U. S. Patent 4,430,461 (1982), IBM.
431. U. S. Patent 4,298,443 (1981), Bell.
432. U. S. Patent 4,115,184 (1978), Northern Telecom.
433. S. Chung, *Solid State Technol.* April 1978, p. 114.
434. E. Egerton, A. Nef, W. Milikin, W. Cook, and D. Baril, *Solid State Technol.* Aug. 1982, p. 84.
435. T. Enomoto, M. Denda, A. Yasauka, and H. Nakata, *Jpn. J. Phys.* **18**, 155 (1981).
436. K. Ersell, *Rev. Phys. Appl.* **13**, 701 (1978).
437. R. Bond, S. Dzioba, and H. Naguib, *J. Vac. Sci. Technol.* **18**, 335 (1981).
438. B. Chapman and V. Minikiewicz, *J. Vac. Sci. Technol.* **15**, 329 (1978).
439. R. Heinecke, *Solid State Electron.* **19**, 1039 (1976).
440. T. Mayer, *J. Electron. Mater.* **9**, 513 (1980).
441. A. Adams and C. Capio, *J. Electrochem. Soc.* **128**, 423 (1981).

442. K. Hirata, Y. Ozaki, M. Oda, and M. Kimizuka, *IEEE Trans. Electron Devices* **ED-28**, 23 (1981).
443. M. Doken and I. Myata, *J. Electrochem. Soc.* **126**, 2235 (1979).
444. W. Ma, J. Bondur, and T. Krawiec, *Electrochem. Soc. Ext. Abstr.* **83-1**, 259 (1983).
445. J. Chang, *Electrochem. Soc. Ext. Abstr.* **83-1**, 280 (1983).
446. S. Muraka and C. Mogab, *J. Electron. Mater.* **8**, 763 (1979).
447. L. Eprath and R. Bennet, *J. Electrochem. Soc.* **129**, 1823 (1982).
448. L. Eprath and K. Petrillo, *J. Electrochem. Soc.* **129**, 2282 (1982).
449. H. Demminger and H. Luthge, *Thin Solid Films* **27**, 155 (1975).
450. J. Curran, *Thin Solid Films* **86**, 101 (1981).
451. E. Kay and A. Dilks, *J. Vac. Sci. Technol.* **18**, 1 (1981).
452. R. Booth and C. Heslop, *Thin Solid Films* **65**, 111 (1981).
453. L. Eprath, *J. Electron. Mater.* **7**, 415 (1981).
454. N. Yabumoto, *Jpn. J. Appl. Phys.* **20**, 893 (1981).
455. J. Bondur, *J. Electrochem Soc.* **126**, 227 (1979).
456. J. Vossen, *J. Electrochem. Soc.* **126**, 323 (1979).
457. J. Gittleman, E. Sichel, H. Lehmann, and R. Widmer, *Appl. Phys. Lett.* **35**, 742 (1979).
458. L. Rothmann and G. Schwartz, *Electrochem. Soc. Ext. Abstr.* **77-2**, 372 (1977).
459. M. Atik, *J. Appl. Phys.* **50**, 6567 (1979).
460. Y. Segui and B. Arg, *J. Appl. Polym. Sci.* **20**, 1611 (1971).
461. J. Harper and J. Cuomo, *J. Electrochem. Soc.* **128**, 1077 (1981).
462. H. Lehman and R. Widmer, *J. Vac. Sci. Technol.* **17**, 1177 (1980).
463. K. Hirata, Y. Ozaki, M. Oda, and M. Kimizuka, *IEEE Trans. Electron Devices* **ED-28**, 1323 (1981).
464. U. S. Patent 3,839,177 (1974).
465. Y. Kurogi, *Thin Solid Films* **92**, 33 (1982).
466. H. Hiroaka and L. Welsh, *Electrochem. Soc. Ext. Abstr.* **81–2**, 345 (1981).
467. F. Bresnock and T. Stumpf, *J. Vac. Sci. Technol.* **19**, 762 (1981).
468. T. Makino, H. Nakamura, and M. Asano, *J. Electrochem. Soc.* **128**, 103 (1981).
469. L. Eprath, *Solid State Technol.* July 1982, p. 87.
470. D. McCaughan and R. Kushner, *Proc. IEEE* **62**, 126 (1974).
471. E. Snow, A. Grove, and D. Fitzgerald, *Proc. IEEE* **55**, 1163 (1967).
472. K. Galloway, S. Mayo, and P. Roctman, *J. Electrochem. Soc.* **126**, 2245 (1979).
473. J. Vossen, *J. Vac. Sci. Technol.* **19**, 762 (1981).
474. L. Eprath, *IEEE Trans. Electron Devices* **ED-28**, 1315 (1981).
475. L. Eprath, *Solid State Technol.* July 1982, p. 65.
476. H. Matsumoto and T. Sugano, *J. Electrochem. Soc.* **129**, 2823 (1982).
477. T. Ma, G. Scoggan, and R. Leone, *Appl. Phys. Lett.* **27**, 61 (1975).
478. J. Tamano, K. Yoneda, S. Morita, and S. Hattori, *Plasma Chem. Plasma Process.* **1**, 261 (1981).
479. I. Goldstein and F. Kalb, *J. Vac. Sci. Technol.* **19**, 743 (1981).
480. U. S. Patent 4,025,411 (1977), Hitachi.
481. T. Maguro, H. Kunita, and T. Itoh, *J. Electrochem. Soc.* **128**, 1379 (1981).
482. U. S. Patent 4,029,766 (1978), IBM.
483. H. Namatsu, Y. Ozaki, and K. Hirata, *J. Electrochem. Soc.* **130**, 523 (1983).
484. J. Moran, *Polym. Eng. Sci.* **20**, 1097 (1980), *J. Vac. Sci. Technol.* **16**, 1620 (1978).
485. D. Bobkin and E. Cantos, *IEEE Electron Device Lett.* **EDL-2**, 222 (1981).
486. German Patent 2,703,738, *Chem. Abstr.* **89**, 13937 (1978).
487. P. Huggett, K. Frick, and H. Lehman, *Appl. Phys. Lett.* **42**, 592 (1983).
488. U. S. Patents 4,187,331 (1980), IBM; 4,253,888 (1981), Hitachi.
489. T. Yamezaki, K. Tanaka, and H. Nakata, *Jpn. J. Appl. Phys.* **20**, 2191 (1981).
490. U. S. Patent 4,307,179 (1981), IBM.
491. R. Poulsen, *J. Vac. Sci. Technol.* **14**, 266 (1977).
492. J. Hayes and T. Pandhumsoporon, *Solid State Technol.* Nov. 1980, p. 75.
493. F. Sanders, A. Kolfschoten, J. Dieleman, R. Haring, A. Haring, and A. de Vries, *J. Vac. Sci. Technol.* **A2**, 487 (1984).
494. H. Kawata, T. Shibano, K. Murata, and K. Nagami, *J. Appl. Phys.* **54**, 2720 (1983).
495. J. Mauer, I. Sogan, S. Zielinski, and G. Schwartz, *J. Vac. Sci. Technol.* **15**, 1737 (1978).
496. T. Enomoto, M. Denda, A. Yasuoka, and H. Nakata, *Jpn. J. Appl. Phys.* **18**, 155 (1979).
497. H. Kawata, T. Shibano, K. Murata, and K. Nagami, *J. Electrochem. Soc.* **129**, 1325 (1982).
498. M. Otsima. *Jpn. J. Appl. Phys.* **20**, 683 (1980).

499. H. Tiller, D. Berg, and R. Mohr, *Plasma Chem. Plasma Process.* **1**, 247 (1981).
500. H. Kleinknecht and H. Meir, *J. Electrochem. Soc.* **125**, 799 (1978).
501. H. Busta, R. Lajos, and D. Krewitt, *Solid State Technol.* Feb. 1979, p. 61.
502. U. S. Patent 4,367,044 (1983).
503. P. Herman and R. Schutz, *Electrochem. Soc. Ext. Abstr.* **83-1**, 257 (1983); *J. Electrochem. Soc.* **131**, 518 (1984).
504. U. S. Patent 4,208,240 (1980), Gould.
505. M. Sternheim and W. Geler, *Electrochem. Soc. Ext. Abstr.* **81-2**, 717 (1981).
506. F. Martinet, *Electrochem. Soc. Ext. Abstr.* **81-2**, 723 (1981).
507. H. Ukai and K. Hawazawa, *J. Vac. Sci. Technol.* **16**, 385 (1979).
508. M. Heichman and V. Eichenberger, *J. Vac. Sci. Technol.* **17**, 1378 (1980).
509. A. Mayeish and M. Baker, *Vacuum* **32**, 305 (1982).
510. U. S. Patent 4,362,068 (1982), IBM.
511. J. Coburn and M. Chen, *J. Appl. Phys.* **61**, 3134 (1980).
512. S. Griffin and J. Verdeyen, *IEEE Trans. Electron Devices* **ED-27**, 603 (1980).
513. C. Korman, *Solid State Technol.* April 1982, p. 115.
514. M. Oshima, *Jpn. J. Appl. Phys.* **20**, 1255 (1981).
515. U. S. Patent 4,328,068 (1982), RCA.
516. D. Darner, D. Flamm, and J. Mucha, *J. Electrochem. Soc.* **130**, 905 (1983).
517. P. Berning, *Phys. Thin Films* **1**, 69 (1963).
518. P. Herman and R. Schutz, *Electrochem. Soc. Ext. Abstr.* **83-1**, 257 (1983).
519. U. S. Patent 4,147,435 (1979), IBM.
520. U. S. Patent 4,367,044 (1983), IBM.
521. H. Gokan, S. Eshi, and Y. Ohnishi, *J. Electrochem. Soc.* **130**, 143 (1983), **131**, 1105 (1984).
522. T. Bowmer and J. O'Donnell, *J. Macromol. Sci.* **A17**, 243 (1982).
523. U. S. Patent 4,314,873 (1981), Bell.
524. Y. Kamoto and T. Kure, *Electrochem. Soc. Ext. Abstr.* **83-1**, 276 (1982).
525. H. Hiroaka and L. Welsh, *Electrochem. Soc. Ext. Abstr.* **81-2**, 345 (1981).
526. J. Reynolds, A. Neureuther, and W. Oldham, *J. Vac. Sci. Technol.* **16**, 1772 (1979).
527. H. Lehmann and R. Widmer, *Microcircuit Eng.* **1**, 34 (1983).
528. H. Norstrom, R. Buchta, F. Runovic, and P. Wiklund, *Vacuum* **32**, 737 (1982).
529. Y. Sakai, J. Reynolds, and A. Neureuther, *J. Electrochem. Soc.* **131**, 627 (1984).
530. J. Malehma, *Vacuum* **34**, 437 (1984).
531. J. Coburn, *Vacuum* **34**, 157 (1984).
532. A. Van Roosmalen, *Vacuum* **34**, 429 (1984).
533. C. Steinbruckel, *J. Vac. Sci. Technol.* **B1**, 38 (1983).
534. P. Revell and G. Goldspink, *Vacuum* **34**, 455 (1984).
535. U. S. Patent 4,426,274 (1984), IBM.
536. J. Spencer, *Solid State Technol.* April 1984, p. 203; A. Keaton and D. Hess, *J. Vac. Sci. Technol.* **A3**, 962 (1985).
537. A. Zettlemoyer and H. Hsing, *J. Colloid Interface Sci.* **55**, 637 (1976).
538. R. Gottscho, R. Burton, D. Flamm, V. Donnelly, and G. Davis, *J. Appl. Phys.* **55**, 2207 (1984).
539. P. Koldner, A. Katzir, and N. Hartsough, *J. Vac. Sci. Technol.* **B1**, 501 (1983).
540. H. Gokan and S. Esho, *J. Electrochem. Soc.* **131**, 1105 (1984).
541. M. Valente and G. Querirolo, *J. Electrochem. Soc.* **131**, 1132 (1984).
542. A. Nagy, *J. Electrochem. Soc.* **131**, 1871 (1984).
543. N. Raley, Y. Sugiyama, and T. VanDuzer, *J. Electrochem. Soc.* **131**, 161 (1984).
544. D. Eddelson and D. Flamm, *J. Appl. Phys.* **56**, 1522 (1984); N. Us, R. Sadowski, and J. Coburn, *Plasma Chem. Plasma Process.* **6**, 1 (1986); I. Plumb and K. Ryan, *Plasma Chem. Plasma Process.* **6**, 11 (1986).
545. U. S. Patent 4,419,120 (1984), Bell.
546. U. S. Patent 4,431,473 (1984), Toshiba.
547. U. S. Patent 4,431,477 (1984), Matheson.
548. *Chem. Abstr.* **100**, 129771 (1983).
549. Japanese Patent 59,04,025, *Chem. Abstr.* **100**, 129907 (1983).
550. Japanese Patent 59,33,830, *Chem. Abstr.* **100**, 201987 (1983).
551. Japanese Patent 59,33,830, *Chem. Abstr.* **101**, 46318 (1984).
552. S. Pan, *Solid State Technol.* April 1984, p. 249.

553. M. Toukhy and J. Jech, *SPIE Proc.* **469**, 159 (1984).
554. V. Donnelly, D. Flamm, W. Smith, and D. Werder, *J. Appl. Phys.* **55**, 242 (1984).
555. A. Kolfschoten, R. Haring, A. Haring, and A. de Vries, *J. Appl. Phys.* **55**, 3813 (1984).
556. M. Kim, Z. Gruenke, R. Saya, and S. Cohen, *Appl. Phys. Lett.* **44**, 462 (1984).
557. G. Koren and J. Yeh, *J. Appl. Phys.* **56**, 2120 (1984).
558. M. Oda and K. Hirata, *Jpn. J. Appl. Phys.* **19**, L405 (1980).
559. S. Mastrianni, *Solid State Technol.* May 1984, p. 155.
560. U. S. Patent 4,454,200 (1984), Varian.
561. T. Chow and A. Steckl, *J. Electrochem. Soc.* **131**, 2325 (1984).
562. U. S. Patents 4,389,281, 4,385,975, 3,976,524, 4,025,411, 4,073,054, 4,076,860, and 4,089,766.
563. U. S. Patent 4,371,565 (1983), IBM.
564. U. S. Patent 4,283,482 (1981), Nikon.
565. U. S. Patent 4,479,848 (1984), Hitachi.
566. G. Oehrlien, *Appl. Phys. Lett.* **45**, 420 (1984).
567. S. Chang, *J. Electron. Mater.* **13**, 955 (1984).
568. D. Ibbotson, J. Mucha, D. Flamm, and J. Cook, *J. Appl. Phys.* **56**, 2939 (1984).
569. V. Desmukh, T. Cox, and J. Benjamin, *Phys. Technol.* **15**, 301 (1984).
570. R. Walkup, *Appl. Phys. Lett.* **45**, 372 (1984).
571. S. Muraka, *J. Vac. Sci. Technol.* **B2**, 693 (1984).
572. G. Mendes, L. Cescato, J. Freijlich, E. Braga, and A. Mamanna, *Thin Solid Films* **117**, 107 (1984).
573. V. Kaliev and V. Danilov, *Sov. Microelectron.* **12**, 101 (1984).
574. S. Mastrionni, *Solid State Technol.* May 1984, p. 155.
575. T. Mele, J. Nulman, and J. Krusis, *J. Vac. Sci. Technol.* **B2**, 684 (1984).
576. N. Yamauchi, T. Yachi, and T. Wada, *J. Vac. Sci. Technol.* **A2**, 1552 (1984).
577. C. Zarowin, *J. Vac. Sci. Technol.* **A2**, 1537 (1984).
578. J. Brewer and G. Miller, *J. Vac. Sci. Technol.* **B1**, 932 (1983).
579. F. McFeely, I. Moran, N. Shinn, G. Landgren, and E. Himpkel, *Phys. Rev. B* **30**, 764 (1984).
580. W. Brandt and T. Honda, *J. Appl. Phys.* **57**, 119 (1985).
581. J. Helbert, *J. Electrochem. Soc.* **131**, 451 (1984).
582. P. Marcoux and P. Foo, *SPIE Proc.* **276**, 170 (1981).
583. Japanese Patent 59,11,895; *Chem. Abstr.* **101**, 201544 (1984).
584. B. Curtis, H. Bruner, and M. Ebenoether, *J. Electrochem. Soc.* **130**, 2242 (1983); U. S. Patent 4,421,593 (1983), RCA.
585. U. S. Patent 4,370,196 (1983), RCA.
586. U. S. Patent 4,373,990 (1983), Bell.
587. U. S. Patent 4,372,806 (1981), RCA.
588. U. S. Patent 4,442,388 (1984), RCA.
589. U. S. Patent 4,426,246 (1984), Bell.
590. U. S. Patent 4,412,885 (1983), Applied Mtls.
591. U. S. Patent 4,444,618 (1984), GE.
592. R. Light, *J. Electrochem. Soc.* **130**, 2225 (1983).
593. H. Kuiken, *Proc. R. Soc. London* **392**, 199 (1984).
594. H. Boehm, *Automatica* **18**, 665 (1982).
595. D. Allen and I. Routeledge, *IEEE Proc.* **130**, 49 (1983).
596. Y. Toduka, H. Yamane, and A. Usami, *J. Phys. D* **18**, 11 (1985).
597. Japanese Patent 59,103,337, *Chem. Abstr.* **101**, 201566 (1984).
598. R. Fieser, *J. Electrochem. Soc.* **130**, 2237 (1983).
599. J. Vondeling, *J. Mater. Sci.* **18**, 304 (1983).
600. N. Endo and S. Matsui, *Jpn. J. Appl. Phys.* **22**, L109 (1983).
601. H. Nielsen and D. Hackleman, *J. Electrochem. Soc.* **130**, 708 (1983).
602. C. Zarowin and R. Horwath, *J. Electrochem. Soc.* **129**, 3541 (1982).
603. W. Fitchner, in *VLSI Technology*, edited by S. Sze, McGraw–Hill, New York, 1983, p. 385.
604. S. Oh, Y. Choi, Y. Kwon, and C. Kim, *Electron. Lett.* **17**, 227 (1981).
605. D. Fricke, *Chem. Phys. Lett.* **94**, 421 (1983).
606. U. S. Patent 4,496,425 (1985), ATT & T.
607. U. S. Patent 4,430,547 (1984), Mitsubishi.
608. U. S. Patent 4,431,898 (1984), Perkin Elmer.
609. U. S. Patent 4,431,473 (1984), Tokyo Shibaura.

610. U. S. Patent 4,436,584 (1984), Sperry Rand.
611. U. S. Patent 4,432,132 (1984), Bell.
612. U. S. Patent 4,427,516 (1984), Bell.
613. U. S. Patent 4,447,824 (1984), IBM.
614. U. S. Patent 4,430,151 (1984), Anelva.
615. U. S. Patent 4,430,138 (1984), Hitachi.
616. U. S. Patent 4,400,235 (1983), Bell.
617. U. S. Patent 4,417,945 (1983) Shin-Estu.
618. U. S. Patent 4,415,402 (1983), Perkin Elmer.
619. U. S. Patent 4,448,637 (1984), Daicel.
620. A. Nagy, *J. Electrochem. Soc.* **132**, 689 (1985).
621. E. Sirikin and H. VanderPlass, *J. Vac. Sci. Technol.* **B2**, 152 (1984).
622. G. Schwartz and P. Schaible, *J. Electrochem. Soc.* **132**, 730 (1985).
623. R. Aikre and D. Economu, *J. Electrochem. Soc.* **132**, 684 (1985).
624. E. Palik, V. Bermudez, and O. Glembocki, *J. Electrochem. Soc.* **132**, 684 (1985).
625. S. McNevin and G. Becker, *J. Vac. Sci. Technol.* **B3**, 485 (1985).
626. K. Horobe and H. Azuma, *J. Electrochem. Soc.* **132**, 938 (1985).
627. H. Yanazawa, N. Hasegawa, T. Douta, and N. Hashimoto, *J. Appl. Polym. Sci.* **30**, 547 (1985).
628. U. S. Patent 4,417,947 (1983), Signetics.
629. J. Chinn and E. Wolff, *J. Vac. Sci. Technol.* **B3**, 410 (1985).
630. L. Baldi and D. Beardo, *J. Appl. Phys.* **57**, 2221 (1985).
631. S. Mathad, *Solid State Technol.* April 1985, p. 221.
632. M. Hill, *Solid State Technol.* April 1985, p. 243.
633. H. Sawin, *Solid State Technol.* April 1985, p. 211.
634. S. Schulttheis, *Solid State Technol.* April 1985, p. 233.
635. S. Fonash, *Solid State Technol.* April 1985, p. 201.
636. H. Kuiken, *Proc. R. Soc. London Ser. A* **392**, 199 (1984).
637. D. Choe, C. Knapp, and A. Jacob, *Solid State Technol.* March 1985, p. 165.
638. C. Tang and D. Hess, *J. Electrochem. Soc.* **131**, 115 (1984).
639. S. Bernacki and B. Kosicki, *J. Electrochem. Soc.* **131**, 1926 (1984).
640. M. Chen and Y. Lee, *J. Electrochem. Soc.* **131, 2118** (1984).
641. T. Hosoya, S. Ohfuji, and T. Shibata, *J. Electrochem. Soc.* **131**, 1135 (1984).
642. S. Uekkusa, K. Oigawa, and M. Tacano, *J. Electrochem. Soc.* **132**, 671 (1985).
643. E. Palik, H. Gray, and P. Klein, *J. Electrochem. Soc.* **130**, 149 (1983).
644. H. Kuiken and R. Tjiburg, *J. Electrochem. Soc.* **130**, 225 (1983).
645. Y. Harada, F. Matsumoto, and T. Nakakada, *J. Electrochem. Soc.* **130**, 149 (1983).
646. P. Smith, *J. Electrochem. Soc.* **130**, 225 (1983).
647. M. Kimiizuka and K. Hirata, *J. Vac. Sci. Technol.* **B3**, 16 (1985).
648. European Patent Application 127,188, *Chem. Abstr.* **102**, 54771 (1985).
649. French Patent 2,534,040, *Chem. Abstr.* **102**, 70237 (1985).
650. U. S. Patent 4,493,423 (1985), AT & T.
651. U. S. Patent 4,496,423 (1985), Colromm.
652. H. Lee, M. Chen, K. Ohn, and A. Bugert, *Thin Solid Films* **118**, 149 (1984).
653. U. S. Patent 4,496,425 (1985), AT&T.
654. Japanese Patent 59,132,626, *Chem. Abstr.* 102, 70244 (1985).
655. G. Corn and D. Baldwin, *J. Vac. Sci. Technol.* **B3**, 909 (1985).
656. F. Egitto, F. Emmi, R. Horwath, and V. Vukanovic, *J. Vac. Sci. Technol.* **B3**, 893 (1985).
657. J. Ma and B. Halon, *J. Vac. Sci. Technol.* **B4**, 822 (1986).
658. C. Mu, S. Fonash, and G. Oehrlin, *J. Appl. Phys.* **59**, 2958 (1986).
659. G. Fortuna, *J. Vac. Sci. Technol.* **A4**, 744 (1986).
660. T. Ueno, H. Shiraishi, T. Iwayangi, and S. Nonoyaki, *J. Electrochem. Soc.* **132**, 1168 (1985).
661. S. Suyama, T. Yachi, and T. Serikawa, *J. Vac. Sci. Technol.* **B3**, 905 (1985).
662. J. Mercier, H. Naguib, V. Ho, and H. Nentwich, *J. Electrochem. Soc.* **132**, 1219 (1985).
663. T. Mogami, M. Morimoto, H. Okabayashi, and E. Nagawa, *J. Vac. Sci. Technol.* **B3**, 857 (1985).
664. U. S. Patent 4,504,574 (1985), U.S. Phillips.
665. U. S. Patent 4,478,677 (1984), IBM.
666. U. S. Patent 4,358,340 (1982), TI.
667. R. Pinto, R. Babu, and P. Bhattachara, *Appl. Phys. Lett.* **48**, 1427 (1986).

668. F. Mizuno, M. Kata, T. Koizumi, Y. Takehana, and K. Sugawara, *J. Electrochem. Soc.* **132**, 1495 (1985).
669. G. Oehrlin, R. Tromp, J. Isang, Y. Lee, and E. Petrillo, *J. Electrochem. Soc.* **132**, 1441 (1985).
670. U. S. Patent 4,502,914 (1985), IBM.
671. J. Coburn and H. Winters, *Appl. Surf. Sci.* **22**, 63 (1985).
672. D. Suryanarayama and K. Mittal, *J. Appl. Polym. Sci.* **30**, 3107 (1985).
673. H. Gokan, K. Tanigaki, and Y. whnishi, *Solid State Technol.* **28**(5), 163 (1985).
674. H. Gokan, Y. Ohnishi, and K. Saido, *Microelectron. Eng.* **1**, 251 (1983).
675. T. Ueno, H. Shiraishi, T. Iwayanagi, and S. Nonogaki, *Proc. Electrochem. Soc.* **85-1**, 158 (1985).
676. M. Tsuda, *J. Imag. Technol.* **11**, 158 (1985).
677. T. Hosoya, Y. Ozak, and K. Hirata, *J. Electrochem. Soc.* **132**, 2436 (1985).
678. R. Narichania, J. Bruce, and S. Fridmann, *J. Electrochem. Soc.* **132**, 2700 (1985).
679. G. Oehrlein, R. Tromp, J. Tsang, Y. Lee, and K. Petrillo, *Proc. Electrochem. Soc.* **85-1**, 178 (1985).
680. K. Hirobe and H. Azuma, *J. Electrochem. Soc.* **132**, 938 (1985).
681. L. White and J. Maa, *Appl. Phys. Lett.* **46**, 1050 (1985).
682. D. Thompson and C. Helms, *Appl. Phys. Lett.* **46**, 651 (1985).
683. D. Oostra, A. Haring, A. deVries, F. Sanders, and K. Myaki, *Appl. Phys. Lett.* **46**, 1166 (1985).
684. B. Roop, S. Joyce, J. Shulta, N. Shinn, and J. Steinfeld, *Appl. Phys. Lett.* **46**, 1187 (1985).
685. R. Alexander, D. Heffelfinger, K. Padmanabhan, and J. Buclolz, *J. Appl. Phys.* **57**, 5268 (1985).
686. D. Chen, S. Chong, and G. Long, *J. Electrochem Soc.* **132**, 1705 (1985).
687. R. Nossen and H. Savin, *J. Vac. Sci. Technol.* **A3**, 881 (1985).
688. H. Winters, *J. Vac. Sci. Technol.* **A3**, 700 (1985).
689. S. Park, L. Rathbun, and T. Rhoden, *J. Vac. Sci. Technol.* **A3**, 791 (1985).
690. H. Winters, *J. Vac. Sci. Technol.* **A3**, 786 (1985).
691. F. McFeely, *J. Vac. Sci. Technol.* **A3**, 179 (1985).
692. K. Suzuki, K. Ninomiya, S. Nishimatsu, and S. Okudaira, *J. Vac. Sci. Technol.* **B3**, 1025 (1985).
693. S. McNevin and G. Becker, *J. Vac. Sci. Technol.* **B3**, 485 (1985).
694. J. Rembetski, S. Babu, N. Lu, and J. Hoffarth, *Proc. Electrochem. Soc.* **85-1**, 184 (1985).
695. G. Valco and V. Kapoor, *J. Vac. Sci. Technol.* **A3**, 1020 (1985).
696. I. Huang, T. Bril, D. Bernard, and B. Westlund, *Proc. Electrochem. Soc.* **85-1**, 482 (1985).
697. U. S. Patent 4,502,914 (1985), IBM.
698. U. S. Patent 4,528,066 (1985), IBM.
699. U. S. Patent 4,512,841 (1985), IBM.
700. E. Degenkolb, K. Park, J. Shorter, and M. Tabasky, *J. Electrochem. Soc.* **132**, 2027 (1985).
701. U. S. Patent 4,522,681 (1985), GE.
702. U. S. Patent 4,511,430 (1985), IBM.
703. G. Csanky and S. McCluskey, *Proc. Electrochem. Soc.* **85-1**, 464 (1985).
704. Japanese Patent 60,43,651 (1985), Matsushita; *Chem. Abstr.* **103**, 14585 (1985).
705. U. S. Patent 4,490,209 (1984), TI.
706. D. Choe, C. Knapp, and A. Jacob, *Solid State Technol.* **28**(3), 165 (1985).
707. H. Wang, *Proc. Electrochem. Soc.* **85-1**, 316 (1985).
708. P. Chang and H. Chao, *Proc. Electrochem. Soc.* **85-1**, 12 (1985).
709. Japanese Patent 60,06,942 (1985), Toshiba; *Chem. Abstr.* **1** 1995111 (1985).
710. T. Ogama, H. Kurokawa, and I. Karino, *Appl. Surf. Sci.* **22**, 82 (1985).
711. Y. Lee, M. Chen, and A. Bright, *Appl. Phys. Lett.* **46**, 260 (1985).
712. U. S. Patent 4,484,979 (1984), AT & T.
713. J. Vigoroux, D. Griscom, J. Durand, A. Le Moel, and C. Legressis, *J. Appl. Phys.* **57**, 5139 (1985).
714. E. Degenkolb, K. Park, J. Shorter, and M. Tabasky, *J. Electrochem. Soc.* **132**, 2027 (1985).
715. U. S. Patent 4,454,200 (1984), Varian.
716. T. Mizutani, C. Dale, W. Chu, and T. Mayer, *Nucl. Instrum. Methods Phys. Res.* **B7/8**, 825 (1985).
717. G. Chang, J. McVittie, J. Walker, and R. Dutton, *IEEE Electron Device Lett.* **EDL-5**, 514 (1984).
718. J. Canteloup and M. Lepinasse, *Vacuum* **35**, 83 (1985).
719. J. Roland, P. Marcoux, G. Ray, and G. Ranken, *J. Vac. Sci. Technol.* **A3**, 631 (1985).
720. W. Mlynko and D. Hess, *J. Vac. Sci. Technol.* **A3**, 499 (1985).
721. A. Van Roosmalen, W. Van den Hoek, and H. Kalter, *J. Appl. Phys.* **58**, 653 (1985).
722. R. Alkire and D. Economou, *J. Electrochem. Soc.* **132**, 648 (1985).
723. G. Kaminsky, *J. Vac. Sci. Technol.* **B3**, 1015 (1985).
724. U. S. Patent 4,501,061 (1985), AMD.

725. G. Mathad, *Solid State Technol.* **28**(4), 221 (1985).
726. S. Schultheis, *Solid State Technol.* **18**(4), 233 (1985).
727. R. Hill and D. Hinson, *Solid State Technol.* **28**(4), 243 (1985).
728. E. Bogle-Rohweir, D. Gates, L. Haylei, H. Kuraski, and B. Richardson, *Solid State Technol.* **18**(4), 251 (1985).
729. H. Sawin, *Solid State Technol.* **28**(4), 211 (1985).
730. S. Fonash, *Solid State Technol.* **28**(4), 201 (1985).
731. J. Maliham, *Vacuum* **34**, 437 (1984).
732. P. Revell and G. Goldspink, *Vacuum* **34**, 455 (1984).
733. D. Field, A. Hydes, and D. Klemperer, *Vacuum* **34**, 563 (1984).
734. U. S. Patent 4,476,217 (1985), Honeywell.
735. J. Mucha, *Solid State Technol.* March 1985, p. 123.
736. S. Fonash, *Solid State Technol.* Jan. 1985, p. 150.
737. D. Choe, C. Knapp, and A. Jacob, *Solid State Technol.* **28**(3), 165 (1985).
738. W. Holber, G. Reksten, and R. Osgood, *Appl. Phys. Lett.* **46**, 210 (1985).
739. P. Brewer, G. Reksten, and R. Osgood, *Solid State Technol.* **28**(4), 273 (1985).
740. T. Hara, T. Hirayama, H. Ando, and M. Furukawa, *J. Electrochem. Soc.* **132**, 2973 (1985).
741. H. Winters, *J. Vac. Sci. Technol.* **B3**, 9 (1985).
742. B. Lin, K. Petrillo, B. Yang, and V. Chao, SPE RETEC Photopolymers, Ellenville, N.Y., 1985, p. 75.
743. D. Lavergne and D. Hofer, *SPIE Proc.* **539**, 115 (1985); R. Wilson and P. Piacente, *J. Electrochem. Soc.* **133**, 981 (1986).
744. M. Tsuda, *J. Imag. Technol.* **11**, 148 (1985).
745. E. Roberts, *SPIE Proc.* **539**, 124 (1985).
746. U. S. Patent 4,532,001 (1985), AT & T.
747. T. Herndon, *Proc. Electrochem. Soc.* **85-6**, 87 (1985).
748. T. Matsuda, T. Ishii. and K. Harada, *Appl. Phys. Lett.* **47**, 123 (1985).
749. M. Hori, T. Muira, S. Hattori, and S. Morita, *Plasma Chem. Plasma Process.* **4**, 119 (1984).
750. G. Mendes, L. Ascato, J. Freijlich, and E. Braga, *Thin Solid Films* **117**, 107 (1984).
751. D. Chen, D. Long, and G. Long, *J. Electrochem. Soc.* **132**, 1705 (1985).
752. R. Walkup, P. Avoures, R. Dreyfus, J. Jasinki, and G. Seluryn, *Appl. Phys. Lett.* **45**, 372 (1984).
753. M. Park, M. Rhodin, and L. Rathbun, *J. Vac. Sci. Technol.* **A4**, 170 (1986).
754. U. S. Patent 4,582,581 (1986), Allied.
755. A. Weiss, *Semicond. Int.* June 1983, p. 144.
756. J. Dismukes and R. Ulmer, *J. Electrochem. Soc.* **118**, 634 (1971).
757. British Patent 2,154,330 (1985).
758. B. Ling, E. Bassous, V. Chao, and K. Petrillo, *J. Vac. Sci. Technol.* **19**, 1313 (1981).
759. K. Tai, W. Sinclair, R. Vadimsky, J. Moran, and M. Rand, *J. Vac. Sci. Technol.* **16**, 1977 (1979).
760. S. Lyman, J. Jackel, and P. Liu, *J. Vac. Sci. Technol.* **19**, 1325 (1981).
761. J. Kruger, P. Rissman, and M. Chang, *J. Vac. Sci. Technol.* **19**, 1320 (1981).

14

Stripping of Resists

14-1. INTRODUCTION

The stripping or removal of resists completes the total lithographic process. In a sense, resist removal is the reverse of the original coating process, since the resist and primer must be stripped to the original bare surface. Selective removal of the resist without substrate damage or residuals is necessary prior to ion implantation, metal or insulator deposition, or oxidation of silicon. In addition, the stripping of the resist is a vital step in the lift-off process (Chapter 12).

Stripping can be effected by physical or chemical means. Because the weak cohesive bonds of a polymer (0.1 eV) are easily broken by a solvent molecule, resist dissolution in a solvent is the simplest process.[1] However, a 1-nm interfacial film usually remains. The adhesive forces at the interface of the resist and substrate film are much stronger (1–10 eV) than the resist cohesive bonds (see Section 6-2-3-3). Therefore, the remaining resist requires physical volatilization or aggressive chemical reactions to remove all traces of material.

Volatilization involves energetic chemical reactions, but produces very clean surfaces. The choice of stripper and process conditions will depend mainly on the following practical considerations:

1. Susceptibility of the substrate to permanent electrical damage (oxidation, corrosion, ionic contaminants, complete dissolution, radiation damage)
2. Cost of stripping, including storage, handling, replenishing, and disposal
3. Type of resist—cross-linked, insoluble, inorganic, silicone, soluble positive, single layer, multilayer, or organometallic
4. Resist process history, including primer, postbake, hardening, etching, and ion implantation conditions

Strippers are available in a variety of liquid compositions or as a gas-phase process involving activated oxygen. Positive resists that belong to the thermoplastic family are easily removed by organic solvents, but positive resists of the thermoset class are usually post-cross-linked and are removed by stronger liquid compositions or by gaseous oxidation. Likewise, cross-linked negative organic resists require similar oxidative strippers but inorganic resists require specialized strippers. Moreover, multiple layers of resists may require consecutive stripping operations and consecutive rinsing steps may also be necessary to remove initial residuals.

Liquid strippers have been formulated from many types of acids (inorganic and organic), alcohols, surfactants, and solvents (mainly, chlorinated solvents). Due to the increasing cost of storage, handling, and disposal, gaseous plasma stripping has succeeded in replacing liquid strippers.

Before discussing the basic chemistry of liquid and gaseous strippers, it is essential that the process history of the resist be reviewed. The choice of stripper and its performance will depend heavily on the type of resist and its specific process history. In Fig. 14-1-1, a flow chart is presented as a reminder of the differential processing a resist may undergo. Several process factors influence the choice of stripper based on the solubility or volatility of the resist.

1. *Type of resist.* Positive photoresists (e.g., DQN) are thermoplastic materials, which remain soluble if not thermally oxidized or insolubilized during high ($>170°C$) process conditions. Most thermoplastic positive acrylic or olefin sulfones remain soluble, unless exposed to high *e*-beam doses ($>10^{-4}$ C/cm^2) (such as during registration in the *e*-beam). Negative resists of all types are insoluble, due to cross-link network formation. Some positive-acting *e*-beam (Chapter 3) or photoresists are pre-cross-linked (acrylic anhydride or epoxy based), and are insoluble in organic liquids. Most cross-links are of the imine type (nitrene cross-linked rubber) or involving carbon–carbon links. Yet, some of the pre-cross-linked acrylics involve anhydride or imide linkages, which can be stripped in alkaline

Resist type

Positive Negative

Inorganic Aromatic Aliphatic Organometallic Additives

Susceptibility of substrate

Corrosion Oxidation Alkali attack Adsorption of stripper Displacement Diffusion

Radiation damage Reduction

Resist thickness

Single layer Adhesion promoter Multilayer

Prebake

Exposure type and dose

Developer absorption (by resist)

Postbake

Time Temperature Thickness Atmosphere

Postharden

Chemical treatment UV Physical treatment

Ion-implanted resist

Etch or lift-off conditions

Choice of stripper

Plasma solvent Aqueous oxidizer Alkaline Swelling type

Strip conditions

Time Temperature Reagent Apparatus Batch size

Rinse and condition

Final test of effectiveness

Optical SEM ESCA SIMS Spectroscopic Wetting Electrical

FIGURE 14-1-1. Resist process factors that influence stripping.

(NH_3) medium. In addition, the polymer may possess functional groups such as ester, alcohol, or alkoxy, which can be saponified. As a last resort, the C–C or C–H linkages can be oxidized by liquid or plasma oxidizers.

Multilayer resists may contain a mixture of silicone polymers (resistant to oxidants) or other oxygen etch barriers such inorganic oxides or nitrides. Therefore, a fluorine-based stripper as well as consecutive operations may be required for each specific layer. The resist may contain trace metals of Si, S, I and elements which do not form volatile oxides. For example, residual S left after oxygen stripping is removed by a subsequent F plasma.[142]

Finally, in plasma stripping, the moles of oxygen is limited. Thus, the stoichiometric amount required will depend on the empirical formula of the resist composition, and polymers with larger numbers of carbon atoms (especially arranged in an aromatic ring) will be more resistant to removal.

2. *Thickness of resist coating.* Although the thickness of resists used in microelectronics ranges from 0.3 μm (masks of chrome) to 4.0-μm films for final metallurgy, resists on printed circuits or packaging layers are up to 250 μm thick. Resists cast over topography vary in cross-sectional thickness. In general, the thicker films require a longer strip time or a stronger strip formulation.

3. *Type of substrate.* Metals such as copper, chrome, and aluminum may be attacked by oxidizers, mineral acids, or alkaline strippers. Strong organic acids or bases may induce similar corrosion. Inorganic ions (Na^+, Cl^-) may be strongly adsorbed on SiO_2 surfaces. Oxygen plasma will oxidize silicon and some metals. In addition, the chemical products of stripping must also be inert or be diluted to be compatible with the substrate. Finally, the electrical damage to the substrate from the plasma radiation fields or ionic diffusion, as encountered in a plasma strip, should be investigated. Other radiationless stripping processes can be substituted if the damage cannot be annealed.

4. *Primer.* Surfaces primed with silanes may form layers resistant to oxygen plasma. Sometimes, adhesive polymers used as primers are inert to simple solvent attack.

5. *Curing of resist film.* Positive DQN resists are partially insolubilized by oxidation at 130–170°C. Further condensation reactions occur at higher temperatures, such as those encountered in plasma etching [thick (> 1 μm) films or in ion implantation]. The DQN resists have also been hardened by UV treatments, plasma, or formaldehyde, and are most difficult to strip.

Formulators of strippers should familiarize themselves with the full range of processes (Fig. 14-1-1). If the resist and process conditions are proprietary, different stripper formulations or conditions may have to be implemented. The final choice of stripper and its processing condition rests heavily with the user, who must determine the inertness of the device to the action of the stripping process.

Subsequent sections examine liquid strippers, gaseous plasma stripping, the mechanistic steps, and the process applications of stripping on semiconductor devices.

14-2. LIQUID STRIPPING FORMULATIONS

Liquid strippers have been widely used since the late 1960s. The formulations have been based upon strippers or cleaners used in the metal paint industries. Since the constituents of paints are similar to negative resist compositions, many paint strippers can be used as resist strippers. Some of the liquid strippers (cold solvent type) operate with physical agitation, requiring up to several hours' immersion. Hot solvent strippers are more effective solubilizers or emulsifiers of resists.

The compositions of resist strippers have been heavily influenced by the paint industry. Since resist manufacturers did not offer a complementary line of strippers, independent chemical firms evolved to market liquid strippers for negative polyvinylcinnamate (KPR, Eastman Kodak Co., Rochester, N.Y.) and cyclized rubber (KTFR, Eastman Kodak Co.). Since the negative resists are insoluble, the stripper should contain agents to break the adhesive bond and degrade the resist to an emulsifiable product.

14-2-1. Organic Solvent Strippers

Popular paint strippers contain a primary solvent, plus a cosolvent, an activator (penetrant), a thickener (polymeric coagulant to prevent solvent evaporation), an evaporation retarder, and an emulsifier (for aqueous rinse).

14-2-1-1. Simple Solvent Strippers. Simple solvents can be used on positive resists. DQN can be readily stripped in simple solvents, including the casting solvent. Negative resists, which require rework (prior to exposure), can also be stripped in the casting solvent or thinner. The most effective "thermodynamic" stripper matches the solubility parameter of the polymer with the solvent solubility parameter. The choice of a simple cold or hot solvent for dissolving resists is dependent on cost, purity (distilled—residue-free), toxicity, flammability, and disposal cost. Miscibility with water and/or alcohols (e.g., diglyme stripper) aids the subsequent rinse–dry step. The evaporation rate, flash point, and storage of organic solvents are hazardous constraints on their use in large-scale operations. Recycling of solvents by distillation can be done to save the costs of high-purity strippers.

The batch life of solvent strippers is limited. Since the resist dissolves in the stripper, eventually a resist buildup may occur, which will reprecipitate on the wafer during the rinsing operation. Spray stripping or vapor-phase degreasing are more effective techniques for residue-free processing. Azeotropes are effective stripping compositions, especially when combined with fluorocarbons to reduce flammability.

The effectiveness of solvent strippers has been examined. Bergin and Kaplan[2] found that acetone did not remove the last 100 Å of commercial positive diazoquinone resists. Similar results were noted by Deckert and Peters.[3] They found that the adhesive bonds at the substrate interface are stronger than the cohesive bonds of the uppermost layers. Consequently, the removal of the last microscopic layers probably requires oxidative techniques.

14-2-1-2. Multiple Component Strippers. These strippers have been formulated for negative resists or insoluble diazoquinone positive resists. The multicomponent strippers closely resemble paint formulations for removing oil-based paints and the multicomponent strippers are formulated from the major ingredients of solvent, activator, and wetting agent.

1. Primary solvent. The most popular ingredient for the primary solvent has been methylene chloride (dichloromethane, CH_2Cl_2). The effectiveness[4] of chlorinated solvents can be estimated by the "wrinkling rating" for fixed immersion

TABLE 14-2-1. Wrinkling[a] Effectiveness of Paint Films

Primary solvent	Percent effectiveness
Methylene chloride	100
Chloroform	69
Ethylene dichloride	45
Trichloroethylene	36
Monochlorobenzene	36
o-Dichlorobenzene	22
Trichlorobenzene	10

[a] From Ref. 4.

time in the respective solvent (Table 14-2-1). The swelling creates a strong force (200 kg/cm^2) to rupture the adhesive bonds (50 kg/cm^2) of the polymer film.

Methylene chloride has the smallest molar volume of the solution (cross section) and can readily permeate polymer films. In a homologous series of solvents such as alcohols, amines, and esters, the small methyl derivatives will permeate more rapidly into the polymer film.

Besides chlorinated solvents, other solvents which soften polymer films (in approximate decreasing order of effectiveness) are ketones, esters, aromatic hydrocarbons, alcohols, and aliphatic hydrocarbons.

The chlorinated hydrocarbons, as shown in Table 14-2-2, have been the major component of patented strippers. The primary solvent, in conjunction with sulfonic acid wetting agents (see below), also serves to reduce the viscosity of the formulation and render the sulfonic acid more emulsifiable in the aqueous rinse step. In addition, when phenol (a penetrant) is used in conjunction (50/50) with alkyl sulfonic acids, an insoluble reaction product may form. The primary solvent maintains both agents in solution.

If the stripper additives (penetrant and wetting agents) produce obnoxious fumes at elevated operating temperatures (> 70°C), a primary solvent (up to 75% by weight), such as carbon tetrachloride,[5] can serve as a protective mantle of vapors over the hot stripper. Since elevated temperatures are commonly used to

TABLE 14-2-2. Primary Solvents in Liquid Stripper

Primary solvent	U.S. patent
CH_2Cl_2	3,072,579 (1963)
CH_2Cl–CCl_3	3,335,088 (1967)
CCl_4	3,582,401 (1971)
Cl-toluene	3,650,969 (1972)
Cumol (hydrocarbon)	4,070,203 (1978)
α-Cl ethanol	4,087,370 (1978)
Tridecylbenzene	4,165,294 (1979)
Dimethyl sulfoxide	4,304,681 (1981)

increase the rate of removal, the decomposition of chlorinated primary solvent may occur to form phosgene,

$$C_xH_yCl_2 + O_2 \rightarrow COCl_2 \tag{14-2-1}$$

Hydrochloric acid can also form if water is present.

The major problems with the primary solvents have been their toxicity, flammability, and the cost of disposal of aqueous emulsified stripper-rinse. One proposal to lower the disposal cost is to use a solvent rinse instead of an aqueous rinse. A mixture[6] of methylene chloride (65%) and *t*-butanol (35%) was previously employed as an alternative to a pure alcohol rinse (flammable) or an aqueous rinse which tended to form two phases and left a surface film. The current trend in primary solvents (Table 14-2-2) is to avoid chlorinated hydrocarbons and replace them with less toxic and nonvolatile organic solvents.

TABLE 14-2-3. Activators in Stripper Formulation

Activator	Reference
Benzonitrile	U.S. Patent 4,221,674
Phenols	U.S. Patent 3,582,401
HF	U.S. Patent 3,335,088
	U.S. Patent 3,650,969
HCOOH	U.S. Patent 2,507,984
	U.S. Patent 2,507,985
	U.S. Patent 3,391,085
	U.S. Patent 4,087,370
	Res. Discl. **103**, 9 (1977)
FSO_3H	U.S. Patent 3,432,347
CH_3COOH	U.S. Patent 3,871,929
	U.S. Patent 4,087,370
Lactic acid	U.S. Patent 2,418,138
Chloroacetic acid	French Patent 2,072,172
Trichloroacetic acid	U.S. Patent 3,072,579
Glycerine	U.S. Patent 4,042,387
Benzaldehyde	U.S. Patent 4,078,102
$(Me)_4N^+OH^-$	U.S. Patent 4,202,703
NH_4OH	U.S. Patent 3,980,587
Dimethyl formamide	Japanese Patent 73,27,938
	Chem. Abstr. **79**, 6907 (1973)
Triethanolamine	Russian Patent 761,976
	Chem. Abstr. **93**, 248269 (1980)
Iodine	U.S. Patent 3,650,969
H_2O_2	Japanese Patent 75,55,405
Benzyl alcohol	Japanese Patent 74,14,681
	Chem. Abstr. **82**, 49904 (1975)
NaOH	U.S. Patent 3,980,587
KOH	British Patent 1,452,622
Tetrahydrothiophene dioxide	U.S. Patent 4,403,026
Dipropyl sulfone, NMP	U.S. Patent 4,401,747
Methyl formate	U.S. Patent 4,483,917
	U.S. Patent 4,438,192
MeEt ketone peroxide	Japanese Patent 60,39,647
	Chem. Abstr. **102**, 195220 (1985)

2. Cosolvents. When combined with methylene chloride, methanol forms a powerful low-cost stripper. The polar cosolvent reduces strip time, and is operable at lower temperatures. The cosolvents are effective at even 10% by weight.[6,7]

The cosolvents in conjunction with sulfonic acids[8] also prevent the resist polymer from precipitating or forming oily beads.[8] The formation of oily beads may lead to an undesirable oil layer remaining on the substrate.[9]

3. Activators. Activators enhance the penetration of the primary solvent into the resist coating. Phenols, amines, and organic acids have been the popular choices. A comprehensive list of activators is given in Table 14-2-3.

Since some of the activators are alkaline (OH^-), acidic (H^+), or ionic (Na^+) in stripper formulations, they should be carefully selected to avoid the potential of ion attack of the semiconductor surfaces. Because the strippers are in contact with all device surfaces for longer periods of time (and at elevated temperatures) than resist developers or liquid etchants, the strippers pose the greatest danger as device contamination or for permanent electrical damage. Not all strippers can be used on all surfaces. For example, organic sulfonic acids will attack aluminum. The addition of HF or dimethyl formamide was found[10] to inhibit the sulfonic acid corrosion.

14-2-1-3. Surface Wetting Agents. Wetting agents such as amines, alkyl or aryl sulfonic acids and esters can function in dual roles as activators and surface displacers. These additives promote the wetting of the hydrophobic resist, aid the solvent penetration, and rupture (displace) the adhesive bond of the resist to the substrate. A comprehensive list of surface agents is given in Table 14-2-4.

The hydrotropic aromatic sulfonic acids[9,133–135] themselves are effective stripping agents. Moreover, they prevent organic solvents from "oiling out" and staining upon rinsing[141] with water. However, organic surfactants present a major disposal problem, since they increase the biological oxygen demand of waste water. Some of the surfactants are also sodium salts and should be avoided. Water ($<1\%$ by weight) pickup by stripper containing sulfonic acids should also be avoided, since metal corrosion may be induced.

TABLE 14-2-4. Surface Wetting Agent Components of Strippers

Surface agent	Reference
Ethoxylated nonylphenol	Czechoslovakian Patent 173,324
	Chem. Abstr. **89**, 223,998 (1978)
Triton X-100 (Rohm and Haas Co., Philadelphia)	U.S. Patent 4,202,703
Dodecylbenzene sulfonic acid	U.S. Patent 4,070,203
	U.S. Patent 4,246,130
Ethyl morpholine surfactant	Czechoslovakian Patent 173,324
p-Toluene sulfonic acid	U.S. Patent 3,391,085
	U.S. Patent 3,582,401
	U.S. Patent 3,988,256
	U.S. Patent 4,070,203
	U.S. Patent 4,165,294
	U.S. Patent 4,221,674
	U.S. Patent 4,304,681
Benzene sulfonic acid	U.S. Patent 4,165,294

TABLE 14-2-5. Examples of Liquid Strippers[a]

	Percent by weight	Resist	Reference
a. *p*-Toluene sulfonic acid	25	KPR	U.S. Patent 3,582,401
Phenol	25	KTFR	To Mallinckrodt (1971)
o-Dichlorobenzene	45		
Tetrachloroethylene	5		
b. Formic acid	30	Riston	U.S. Patent 4,087,370
2-Chloroethanol	30		To ITT (1978)
Water	40		
c. Benzene sulfonic acid	30	AZ-1350	U.S. Patent 4,165,294
Xylene sulfonic acid	20	HR-100	To Allied (1979)
Tridecyl benzene sulfonic acid	50	KTFR, KPR	
d. Dodecyl benzene sulfonic acid	35	KPR	U.S. Patent 4,070,203
Cumol (aromatic solvent)	65	KTFR	To Merck (1976)
e. *p*-Toluene sulfonic acid	10	AZ-1350	U.S. Patent 4,304,681
Dimethyl sulfoxide	90		To Shipley (1981)

[a] The stripper formulations have evolved from toxic phenolchlorinated solvents to simpler less toxic systems. Negative resists are difficult to remove and required strong stripper additives. Positive phenolic-based resists in use today permit milder strip compositions.

14-2-2. Typical Liquid Strippers

Typical formulations which can be found in the patent literature are shown in Table 14-2-5.

The corrosion of metals such as copper, aluminum, and titanium has been noted with phenol–sulfonic acid strippers. On the other hand, the addition of HF at trace levels (ppm) inhibited metal corrosion.[10] In order to supply a constant amount of fluoride ion, a complex with organic nitriles, such as DMF (dimethyl formamide) or acetonitrile, has been used.[11-13] Thus, the addition of DMF reduces the loss of HF by evaporation, inhibits corrosion, and does not impede the stripping action of sulfonic acids.

In summary, the solvent strippers were developed as an answer to the poor batch life of oxidizing acids and their corrosive attacks on metallurgy. They have been an improvement over oxidizing acids, since they have a larger batch capacity, longer shelf life, are free of metal ions (in some cases), and are inert to metal surfaces (in some cases). The stripper formulation themselves have evolved into less toxic compositions. Since positive phenolic resists and plasma stripping have become popular, their continued improvement may be limited by the shrinking marketplace and high disposal cost.

14-3. WATER-BASED STRIPPERS

14-3-1. Acidic Strippers

Aqueous based strippers are composed of strongly acidic or alkaline agents which chemically convert resist polymers into emulsified or solubilized material. Main polymer chains, cross-links, and side chains are degraded, oxidized, or con-

TABLE 14-3-1. Oxidants in Acidic Strippers

Oxidant–acid	Reference
H_2O_2–H_2SO_4	U.S. Patent 3,900,337
H_2O_2–H_2SO_4	U.S. Patent 3,787,239
Cr_2O_3–H_2SO_4	U.S. Patent 3,962,108
$HClO_4$–H_2SO_4	U.S. Patent 3,962,108
$KMnO_4$–H_3PO_4	U.S. Patent 4,015,986
$K_2S_2O_8$–H_2SO_4	Def. Publ. 973,008 *Chem. Abstr.* **90**, 131801 (1979)
$(NH_4)_2S_2O_8$–H_2O_2–H_2SO_4	*J. Electrochem. Soc.* **126**, 1428 (1979)
HNO_3–H_2SO_4	*J. Electrochem. Soc.* **127**, 386 (1980)
$(NH_4)_2S_2O_8$–H_2SO_4	*J. Electrochem. Soc.* **127**, 386 (1980)

verted into polar components. Because strong acid–oxidant mixtures attack the adhesive bond at the interface, they are commonly used to clean glassware in chemical laboratories. They were also popular for the removal of insoluble negative resists or highly insolubilized (baked) positive phenolic resists. In the acid bath, the cyclized rubber backbone of the negative resist is oxidized into oxygen-bearing fragments which become emulsified or become soluble n the acidic medium. Table 14-3-1 indicates the variety of oxidants that have been used in conjunction with the strongly acidic medium of H_2SO_4.

The strongly acidic solutions (especially with water present at greater than 1% by weight) may corrode metallurgy. Therefore, test samples with different surfaces should be evaluated. The strongly oxidizing strippers have been used[2,132] with some success on highly baked (> 150°C) or ion-implanted (> 10^{15} ions/cm^2) positive DQN resists.

Because the oxidizing power of acidic solutions is limited, solutions such as Cr_2O_3 in concentrated sulfuric acid become discolored with use. In addition, the lifetime of a mixture of concentrated H_2SO_4 (70% by volume) and 30% H_2O_2 (30%) is 4–6 hr.[14] The peroxide is decomposed by the buildup of organic acids.[15] Because of their violent reactions with organic materials, the acidic oxidants require special handling, storage, and disposal techniques. Moreover, the Na^+ adsorption from dichromate–H_2SO_4 is very severe on glass surfaces[16] and the adsorption of Na^+ can lead to many electrical problems in *p–n* junctions.[16]

14-3-2. Alkaline Strippers

Strong bases are used to strip DQN resists. Even undiluted developers will strip DQN resists. If the resist is flood exposed, it will readily strip in normal developer.

For both DQN and polyvinylcinnamate (KPR) negative, a saturated solution of KOH and benzaldehyde at 70°C was very effective.[17] An alkaline solution of KOH, 2-ethoxyethanol, and $NaIO_4$ (1% by weight) was used for polyvinylcinnamate resist.[18] Anhydrous hydrazine was also used as a potent stripper.[138]

Polyvinylalcohol ($K_2Cr_2O_7$ sensitized) negative resists can be stripped in alkali. The alkaline materials are powerful strippers because (1) they readily penetrate the resist film, (2) they chemically saponify ester groups, phenolic groups, and ether linkages, and (3) they attack the interfacial resist–substrate adhesive bond. Because of their strong chemical potency, alkaline strippers should be avoided on silicon,[16] aluminum, copper, and glass.[16] In addition, cations such as Na^+ irreversibly adsorb on glass surfaces and may cause the surface state depletion of MOS devices.[16]

Oxidizing agents have also been added to alkaline strippers. Agents such as $KMnO_4$, $Fe_2(CN)_6$, and H_2O_2[3] in ammonia are compositions that have been used both as silicon precleaners and as photoresist strippers. An ammoniacal solution of H_2O_2 at 80°C was more effective for stripping baked AZ-1350, than an acidic oxidizing stripper.[3] The alkaline ion readily attacks the phenolic portion of the resin and assists in the solubilization. Again, as with acidic media, the safe handling and disposal of the spent liquids should be practiced.

14-4. PRACTICAL ASPECTS OF STRIPPING

With any liquid stripping process, there are various practical aspects which require process attention.

Safe Handling and Disposal of Strippers

Since the strippers may contain oxidizing agents, flammable solvents, corrosive materials, or nonbiodegradable additives, failsafe chemical practices should be mandatory. The transfer and disposal of reagents must be performed in vented areas. Because sludge buildup may occur in the strip tank and can block pumps, continuous filtration of the resist stripper is necessary.

Potency of the Stripper

The kinetics of liquid stripping depends on the history and type of resist, the time and temperature of stripping. Physical agitation may be necessary if emulsified products inhibit surface dislodgement. The effectiveness of stripping can be monitored by endpoint detect,[19] if the stripper liquid is transparent in some spectral region (6238 Å helium–neon laser).

An exemplary list of stripping problems and remedies is given in Table 14-4-1.

TABLE 14-4-1. Troubleshooting Liquid Stripping

Problem	Cause	Remedy
Poor stripping	Thicker film	Measure
	Low strip temperature	Adjust
	Low strip concentration	New batch
	Overbaked resist	Process control
	Air-baked resist	Adjust
	Low strip concentration	Cool vapor
Nonuniform stripping	Spray clogging	Clean
	Gels in tanks	Filter
	Corrosion of boat	Substrate
	Water absorption	Cover

14-5. PLASMA GAS STRIPPINGS OF RESISTS

Some basic problems persist with the liquid stripping of negative resists. Often, the insoluble film is not completely solubilized by the liquid stripper. Strip times are difficult to reproduce since the process relies on random adhesion failure. Some of the formulations will attack metal films. In addition, the storage and disposal costs of liquid strippers are very high. Although switching to positive resists made stripping easier (if the resist was not cross-linked), the disposal costs remained high. Lastly, the hard bake necessary for maintaining positive resist adhesion made some liquid strippers ineffective.

The introduction of oxygen plasma stripping in the early 1970s greatly reduced the cost of disposal since common combustion products (CO_2 and H_2O) were formed. The stripping equipment was already in use in analytical chemistry for the ashing of organic samples. Plasma cleaning and treatments of polymeric fibers were also in commercial stages when the plasma stripping of negative resists was recognized. The acceptance of plasma stripping paved the way for further use of plasmas in lithography such as etching, resist developing, and film deposition including resists.

In the next section, plasma stripping will be examined from the viewpoint of the basic mechanism, reactants, side effects, and applications of oxygen plasma.

14-5-1. Reactants of Plasma Stripping

The basic object of the gas–solid stripping reaction is to generate a volatile product. In general, the degradation of polymers into volatile products can be accomplished or accelerated by:

- Ultraviolet radiation (100–300 nm)
- Ozone
- Heat
- Free radicals (hydrogen abstraction)
- Ion bombardment (atomic displacement)
- Sputtering (atomic displacement)
- High-energy radiation (*e*-beam, X ray, γ ray, ion beam)

Plasma stripping involves one or a combination of radiation and chemical reactants, which attack polymeric chains. The plasma itself is a state within a gas where there are substantially an equal number of positively and negatively charged particles. The positive species are ions of atoms or molecules and the negative species are electron-rich atoms or molecules. Separation and identification of the most probable reagents is a formidable task for physical chemists and spectroscopists.

14-5-2. Polymeric Reactants

The overall phenomenological chemical reaction of oxygen plasma stripping is analogous to the combustion of polymers,

$$\text{Heat} + \text{radiation} + O_2^* + \text{resist} \rightarrow CO_2 + H_2O \qquad (14\text{-}5\text{-}1)$$

The polymers when burned release heat in an exothermic reaction. The total number of moles of O_2 required for combustion can be calculated from the stoichiometric reaction:

$$C_aH_bBr_cCl_dF_eI_fN_gO_hS_i + (a + b/4 + i/2 - h/2)\,O_2 \rightarrow CO_2 + H_2O \qquad (14\text{-}5\text{-}2)$$

Note that if a polymeric resist has oxygen in its repeat unit, we assume that the oxygen is available to aid in combustion. Consequently, the oxygen concentration demand is proportionately reduced. The stoichiometric amount of oxygen required to combust different types of resists based on their empirical formulas is calculated and given in Table 14-5-1.

From a stoichiometric viewpoint, aromatic polymers require more moles of oxygen than aliphatic resists. Phenolic resists strip about a factor of two slower than cyclized rubber. However, aromatic polymers such as polycarbonates, polystyrenes, and polyoxazoles require in excess of 10 moles of oxygen. Since a plasma stripper operates at low oxygen concentrations, sufficient flow of oxygen should be present for the various stoichiometric requirements of the resist. The heat evolved when polymers strip can be estimated from heat of combustion data[20] (Table 14-5-2).

Unless the exothermic heat is dissipated, self-heating will occur. The alteration of the rate of removal can occur in a barrel reactor because many wafers stacked in a boat will not reach as high an operating temperature as a single wafer.[29,30]

In the combustion of polymers, a sustained burning after ignition has been used to measure the limiting oxygen index (LOI), a ratio of oxygen required to the total amount of all gases present in the atmosphere. Representative LOI[21] numbers for various polymers are:

Polymer	LOI
Polyacetaldehyde	15
Polymethylmethacrylate	17
Polystyrene	18
Phenol–formaldehyde	36
Polytetrafluoroethylene	95

Polymers which require lower stoichiometric moles of oxygen (Table 14-5-1) for combustion will also sustain a combustion (a lower LOI). Aromatic polymers are more difficult to combust. The consensus of studies of combustion, flam-

TABLE 14-5-1. Oxygen Combustion Requirement of Resists

Resist	Empirical unit	Moles O_2
Phenol formaldehyde	$C_{15}H_{12}O_2$	17
Polyvinylcinnamate	$C_{10}H_{10}O_2$	12
Polychloromethylstyrene	C_9H_9Cl	11
Polymethylmethacrylate	$C_5H_6O_2$	6
Cyclized rubber	C_5H_8	7
Poly-1-butene sulfone	$C_4H_8SO_2$	6

TABLE 14-5-2. Heat of Combustion of Resists

Resist	$-\Delta H$ (kcal/g mole)
Phenol–formaldehyde	1496
Polystyrene	1033
Cyclized rubber	737
Polymethylacrylate	637
Poly-1-butene SO_2	619

mability, and photooxidation of polymers is that aromatic polymers are more resistant (2–10×) to degradation than aliphatic polymers.

The polymeric resists as reactants exist in the form of thin amorphous films. Since the films have a low surface-to-volume ratio, no apparent effects due to size, shape, or crystallinity of the polymers are present. Polymer plasma oxidation is dependent on the surface area (e.g., rods versus powders) but does not appear to be influenced by crystallinity.[(22)]

Polymeric variables such as molecular weight, additives, and type (addition versus condensation polymers) have been studied in the areas of photooxidation,[(23)] plasma treatment,[(24)] and in combustion.[(21)] In studies of plasma oxidative treatment, lower molecular weights are found in the treated polymer. Furthermore, in studies of the plasma oxidation of polymers, Hansen and co-workers[(25)] concluded that there is some correlation between thermal oxidation of polymers and plasma oxidation although the stepwise degradation may be different. As a result, antioxidants were found to have no effect on the plasma oxidized polymers but they inhibited the thermal degradation.

The random chain scission model of Charlesby[(26)] applied to radiation scission of vinyl polymers is applicable to the plasma degradation of polymers. The higher molecular weights of a polymer will suffer a more rapid decrease in molecular weight and the lower initial weights will be more resistant. In general, oxygen, attacks sites formed by the plasma radiative scission. More detailed kinetics are required to decipher the primary plasma processes. Variations in polymer structure of a resist due to high postbake or ion implantations may influence the initial surface kinetics of plasma strippings.

14-5-3. Oxygen Plasma

Oxygen has been the primary choice of plasma gas. Hydrogen, nitrogen, ammonia, water vapor, air, and halogens have been examined as secondary gases or in conjunction with devices sensitive to plasma oxidation. Oxygen gas has provided the fastest rates of stripping and oxygen with trace impurities provides a higher concentration of active species than pure oxygen.[(67)] The microwave excitation of oxygen provides a variety of neutral and ionic species,

- O_3
- O^+
- O^{2+}
- O^-
- O^{2-}
- $O(^3P)$—atomic oxygen
- $^1O_2(^1\Delta g)$—singlet oxygen

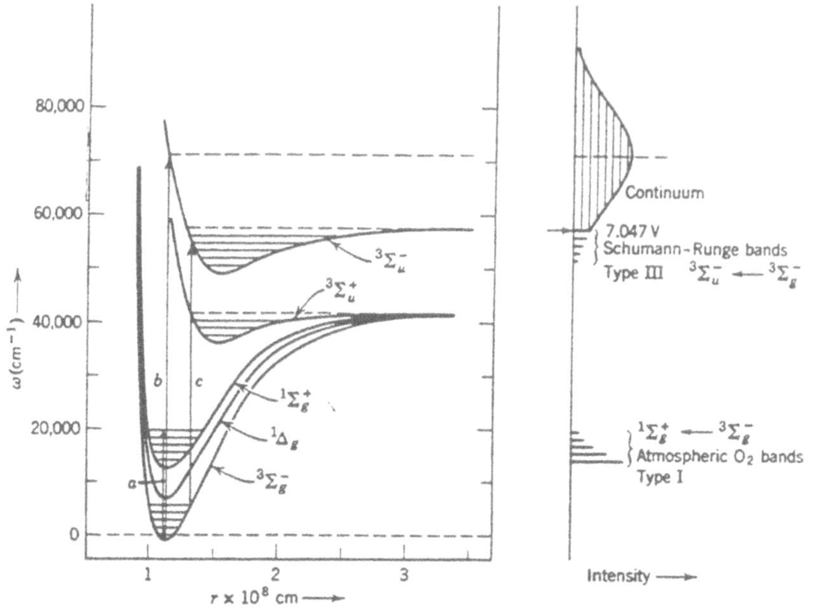

FIGURE 14-5-1. Potential energy diagram for the spectroscopic levels in oxygen. Adapted from Herzberg.[139]

Singlet oxygen is the most persistent and predominant neutral species. The ionic species have not been detected in a plasma stripper. McTaggart and Bell[27] have reviewed the plasma dissociation of oxygen. The dissociation of molecular oxygen can occur by one of two processes involving collisional excitation from the ground state to either $A^3\Sigma u^+$ or $B^3\Sigma u^-$ in the potential energy diagram shown in Fig. 14-5-1. Oxygen radicals in barrel reactor stripping are detected by ESR.[130] O_2^+ is also found[144] in a diode reactor,

$$e + O_2 O_2^*(A^3\Sigma u^+) \rightarrow O(^3P) + O(^3P) + e \quad (14\text{-}5\text{-}3)$$

$$e + O_2 O_2^*(B^3\Sigma u^-) \rightarrow O(^3P) + O(^1D) + e \quad (14\text{-}5\text{-}4)$$

Loss of atomic oxygen (O^*) can occur by heterogeneous recombination with the wall or other gases,

$$O^* + O^* \rightarrow M + O_2 + M^* \quad (14\text{-}5\text{-}5)$$

$$O^* + 2O_2 \rightarrow O_3 + O_2 \quad (14\text{-}5\text{-}6)$$

The low pressure of a plasma[28] in a stripper consists of 10–20% atomic oxygen and a similar quantity ($10^{16}/cm^3$) of singlet oxygen. The concentration of ions and electrons is in the range of 10^{10}–$10^{12}/cm^3$ and does not persist for very long outside the microwave excitation zone. Ozone[24] is also present in small amounts and can also act as a stripper. In a barrel reactor, the gases flows around the wafers Fig. 14-5-2. The reactive oxygen gases are transported to the wafers by diffusion. The diffusion coefficient[24] (D) and lifetime (T) of oxygen are of importance. For atomic oxygen, the value of D is about 500 cm^2/sec at 100 mTorr and

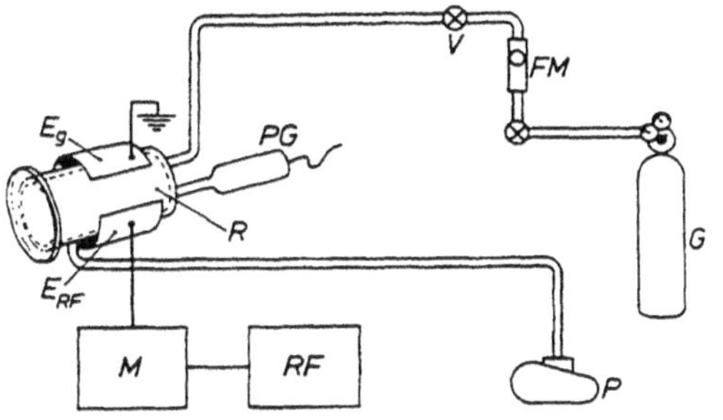

FIGURE 14-5-2. Barrel reactor for oxygen stripping of resists. R, reactor; E_g, grounded electrode; E_{RF}, microwave-powered electrode; M, matching network; RF, microwave generator; P, vacuum pump; V, valve; FM, flow meter; G, gas supply; PG, pressure gauge. Courtesy of H. Kalter and E. Van de Ven, *Philips Tech. Rev.* **38**, 200 (1978).

150°C. The lifetime is about 0.3 sec.[28] The distance the oxygen atoms can travel is $\sqrt{DT}$, which is about 25 cm. This is larger than the diameter of the barrel, which implies that the oxygen atoms strike the wafer surfaces. A partial vacuum is used in a plasma stripper to reduce the number of atoms in a chamber to avoid the self-quenching effect,[30]

$$O^* + O^* \rightarrow \text{Quench} \tag{14-5-7}$$

14-5-4. Plasma Radiation Field

In the physical layout of a barrel reactor (Fig. 14-5-2), the plasma and resist solid are physically in contact, electrically isolated, and have a steady-state interaction. Physical contact is produced by placing a resist in the volume where the plasma is generated or downstream from the glow. The resist film has an electrical potential which floats relative to the plasma. In a planar diode system,[31,50] the wafer can be grounded and electrical current is drawn from the plasma through the resist film. A current flow through the resist can alter the reactions at the resist surface, by producing an electric field which can change the plasma radiation field as well as the plasma gas composition.

The physical radiation field generated in a plasma can ablate a resist by bombardment with degrading UV (50–300 nm), neutral particles (O_2), and ionic fluxes. Resists containing aromatic polymers or sensitizers may shield the resist resin from direct UV absorption[32] (aromatic groups dissipate singlet energy by fluorescence).

The ions which impinge on ungrounded wafers in a barrel reactor possess excess kinetic, vibrational, and electronic energy. For a surface which floats relative to the plasma potential, the ions arrive at the surface with up to 5 V of energy.[31] Energy is dissipated by ablation of the resist and by electrical (ion) charging of the wafer surface. For a biased grounded surface (after most of the resist has been removed), the ions are accelerated by a cathode fall voltage, and hit the surface with several hundred volts of energy. The kinetic energy is dissipated by resist heating and sputtering. In some instances, small fragments ablate away forming large pits ($> 5\ \mu m$ in diameter).

The radiation field in a plasma can induce electrical damage in the SiO_2/Si interface (see Section 13-13-6-2). Radial damage on the wafers is evident from current–voltage studies of the wafer surface.

14-6. KINETICS OF STRIPPING

Before discussing the basic kinetics inside a barrel reactor, it is necessary to briefly list some of the important experimental observations reported since plasma stripping was first introduced in 1967[33]:

1. Lower removal rates at higher reactor pressure[34] (100 Torr) or increasing flow rates.[35] Faster rates at the lowest pressure.[31]
2. Strip rate is flow limited.[29,30,36]
3. Rates are generally nonlinear.[37] Thicker films require longer strip time.[37]
4. Wafer position affects rate of removal.[29,30] Wafers strip from the outer edge,[29,34] and exhibit greater inhomogeneity with larger-diameter wafers.[36] The rate of stripping decreases with increased wafer loading.[29]
5. Rates increase with self-heating[29,38,39] or with preheated wafers.[30,35]
6. Inorganic residues may remain behind.[40–42] Oxidation of silicon occurs.[41,43] Charges are induced in the oxide layers.[3,5,41,42,44]

14-6-1. Gas–Solid Interactions

Many reactions involving gases are known to occur more rapidly as a heterogeneous process on the surface of a solid (which acts as a catalyst) than as homogeneous reactions in the gas phase. The primary function of the solid is to adsorb by "chemisorption" the reacting molecules. The heat of adsorption is on the order of 0.5 to 4 eV/mole.[45] As the temperature of the reaction is raised, the rate of activated adsorption should increase and its extent should become more appreciable since adsorption is an exothermic process.

The stripping of resists involves a bimolecular reaction between a molecule or atom of the plasma gas and a molecule of the resist solid:

(G)	(S)	⇄	GS*	⇄	G–S	(14-6-1)
$\frac{1}{2}$ Gas	+ resist surface		Activated state		Adsorbed state	

The physical chemistry may be compared to the chemistry of flame combustion.[21] The resist surface reaction may be separated into five stages, the slowest of which will determine the rate of the overall process:

1. Transport of reactant (atomic oxygen) to the surface of the resist
2. Adsorption of the reactant
3. Reaction on the resist surface (oxidation)
4. Desorption of the products (CO_2, CO, H_2O, and radicals)
5. Transport of the products from the resist surface into the gas phase

Battey[30,36] has made an intensive study of steps 4 and 5. Water and carbon monoxide were deliberately added to a barrel plasma asher. He determined that when CO and H_2O products (P) are in excess, the law of mass action prevails,

$$\text{Resist} + \text{O} \rightarrow CO_2 + CO + H_2O \qquad (14\text{-}6\text{-}2)$$

Equation (14-6-2) predicts a shift in favor of a reduction in rate. The products cover the reaction sites (see Section 13-1).

The addition of CO or H_2O in amounts equal to atomic oxygen (O) (determined by NO titration) reduces the stripping rate by a factor of two. Addition of both CO and H_2O vapor reduces the rate by another factor of two without any reduction in atomic oxygen concentration. Presumably, the added gases are covering half of the reaction sites, making them unavailable for further reactions. The stripping rates are very slow at 100 Å/min at room temperature. This may be due to the low heat of adsorption (0.5 eV/mole) of atomic oxygen.

The results of the product inhibition study imply that:

$$\sim CH\cdot\ (\text{resist fragments}) + 2O \rightarrow CO + H_2O \tag{14-6-3}$$

$$\text{Rate} = k[O]^2 \tag{14-6-4}$$

Besides the consideration of chemical transport,[140] physical constraints on radial flow[147] in a plasma reactor can lead to localized inhomogeneity. If a boat load of 2-inch-diameter silicon wafers spaced at 3/16-inch intervals is stripped, the removal of the photoresist occurs inwardly from the outside of the wafer.[29,30,34] Since the diameter of the quartz barrel stripper is several times that of the wafers, the plasma gas flow is almost entirely external to the region between the wafers. Atomic oxygen must diffuse into the interwafer region and the by-products of the reaction must diffuse out. It is also observed that a boat load of 50 wafers with 3/16-inch spacing will take three to four times as long to strip clear as the interior wafer of a three-wafer array spaced 3/16 of an inch apart.

Possible explanations are the nonuniform heating of wafers by plasma microwave energy, the inhibition of the reaction of products, or the nonuniform diffusion of oxygen between the wafers. Battey[29,30] found that the etch rate at the edge is essentially independent of wafer diameter and spacing, whereas the etch rate at the center of the wafer increases linearly with wafer spacing and is inversely proportional to the square of the wafer diameter. Thus, increasing the wafer diameter by a factor of f leads to a greater edge to center inhomogeneity in strip rates which can be corrected by increasing the wafer spacing by a factor of f^2.

At the edge of the stacked wafers, the gaseous oxygen combines in a three-body reaction with the surface. The reaction products diffuse away and are not adsorbed. Moreover, the strip rate at the edge (10 kÅ/min) is temperature sensitive (100–200°C) with an activation energy of 6.5 kcal/mole (0.3 eV/mole). In the center of the wafer, however, the inwardly diffusing oxygen concentration is lower by consumption along the path. Furthermore, large wafers strip more inhomogeneously. The strip rate (Fig. 14-6-3) at the center (2500 Å/min) is independent of the temperature in the range of 100–200°C. At lower temperatures, the strip rate at the center is much lower. Consequently, wafers should be preheated or time alloted to attain a wafer temperature of 120°C.[35]

Bersin[34] observed that stripping times could be reduced by running the plasma at high pressure (10 Torr) and high powers rather than at lower powers and pressures of 1 Torr. Presumably, the higher pressure and powers serve to heat the wafers (desorb products). These pressures, however, had to be reduced to enable atomic oxygen to diffuse between the wafers. (Preheating the wafers would eliminate the two pressure–power cycling steps.)

Under conditions where there is no product inhibition, the chemical stripping rate is determined by the atomic oxygen concentration (function of flow rate,

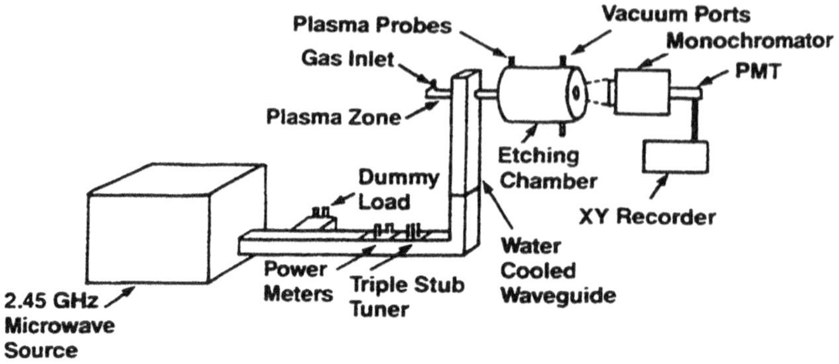

FIGURE 14-6-1. Schematic of downfield etching microwave-powered oxygen plasma stripper which avoids direct wafer bombardment. From Ref. 129, reproduced by permission of the Electrochemical Society Inc.

power, and pressure) and the temperature of the silicon substrate and reactor walls. The loading of more wafers reduces the thermal mass and decreases the rate of stripping.

Higher strip rates of 6 μm/min of resists in oxygen can be attained by creating a tenfold higher concentration of atomic oxygen. Microwave energy is used to create an abundance of long-lived atomic[131] oxygen in a downfield reactor (Fig. 14-6-1). The wafers are not in contact with the plasma (as encountered in barrel ashers) and radiation damage from stripping (Section 14-7-1) is averted.[129] Wang and Gelerent[47] found that in their downfield reactor, resist stripping did not take place in pure oxygen unless a small amount of CF_4 was added. The difference may be caused by differences in the composition of reactor wall materials.[129]

14-6-2. Activation Energy of Stripping

Temperature plays an important role in determining the rate and linearity or nonlinearity of the process. At constant wafer and reactor temperature and with a large excess of atomic oxygen ($> 10^{-3}$ moles/liter), the time of removal should be linearly dependent on the amount of resist to be stripped[29] (Fig. 14-6-2). Unfortunately, the temperature is not constant and the strip time is a function of position

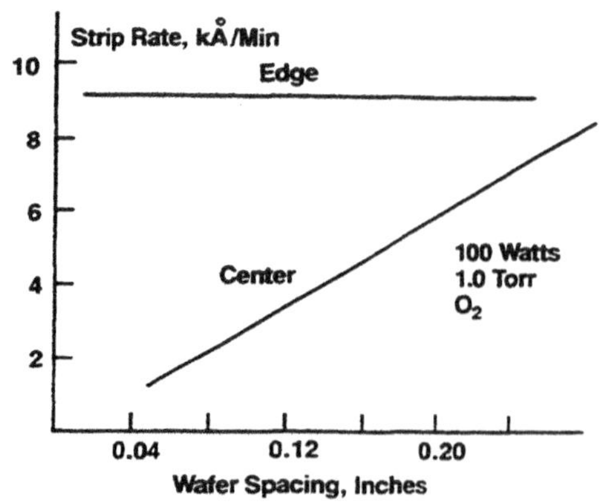

FIGURE 14-6-2. Strip rate at the wafer edge and center as a function of wafer separation under conditions of constant molecular oxygen flow rate, pressure, and power. The edge strip rate is linear in separation. From Ref. 29, copyright 1977 and reproduced by permission of the Electrochemical Society Inc.

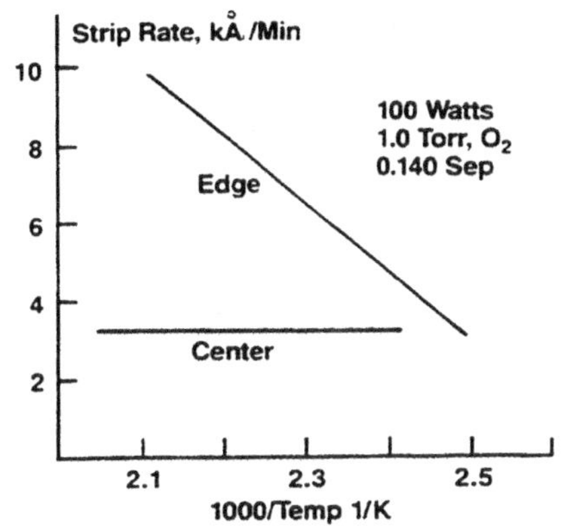

FIGURE 14-6-3. Strip rate at the wafer edge and center as a function of temperature under conditions of constant molecular oxygen flow rate, pressure, and power. The edge strip rate varies exponentially as $-E/RT$, where E is approximately 6.5 kcal/mole. The center strip rate is approximately independent of temperature. From Ref. 29, copyright 1977 and reproduced by permission of the Electrochemical Society Inc.

of the wafers in the chamber. In a barrel reactor with no tunnel, the wafer edges are quickly heated to 200°C or more because of the direct contact with the plasma and microwave input.

The activation energies for stripping reported in the literature are listed in Table 14-6-1. High activation energies of 12 kcal/mole are reported for wafers preheated to 200°C.(35) In a planar reactor at lower temperatures (<100°C), a lower activation energy of 6.5 kcal/mole was measured.(31) Shielding the wafers in a plasma (to reduce radiation damage) with a Faraday(35) cage, lengthens the heating, so the effect of temperature is much more noticeable. In this process the wafers can be preheated to restore the plasma self-heating effect.

More precise measurements of the activation energies by laser(39) monitoring of the wafer surface have revealed a nonlinear Arrhenius behavior(38) (Fig. 14-6-4). This behavior of the strip rate was attributed to a chemical gradient of trapped solvent in the various (oven baked) thicknesses of the films (Fig. 14-6-5). In a downfield reactor, the activation energies(131) of O_2-Freon increase with F content due to recombination of F and O.

If average stripping rates are used, there is general agreement in the literature of a value of 6.5 kcal/mole for the activation energy, even in a diode planar(31) stripper which can achieve rapid rates of micrometers per minute. Since the wafers are exposed to a laminar flow of atomic oxygen, the plasma strip rates are much higher

TABLE 14-6-1. Activation Energies of Stripping in Oxygen Plasma

Resist	Activation energy (kcal/mole)	Reference
AZ-1350J (2.0 μm)	20	46, 130
AZ-1350J (1.0 μm)	10	46
AZ-1470	1.2	47
Waycoat III	12	35
KTFR	6	29
Carbon film	6.5	48
Carbon film	10	49
KTFR	6.5	31

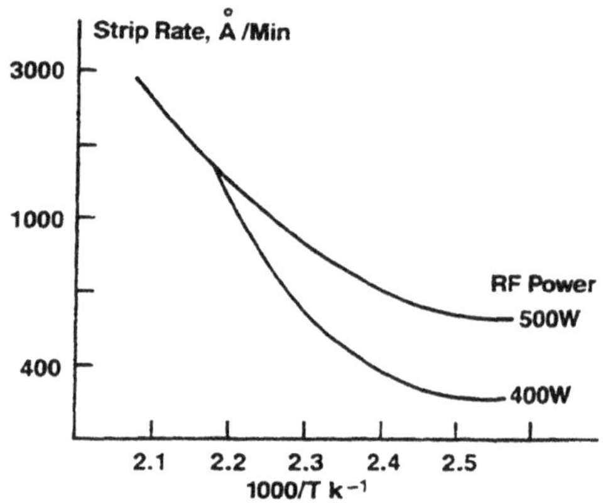

FIGURE 14-6-4. Arrhenius plots for two single stripping in O_2 runs at 400 and 500 W for 2.3-μm-thick AZ-1350. From Ref. 38; figure orginally presented at the 1981 Spring ECS meeting, reproduced by permission of the Electrochemical Society Inc.

(10×) in a planar diode system. For example, in order to prevent decomposition of the organic resist during plasma etching or resist plasma development, the resists are held at room or lower temperatures by cooling plates[51] (< 100°C). In addition, rapid stripping by magnetron[52] or high-pressure activated oxygen, minimizes the self-heating of the substrate. Even poly-1-butene sulfone (Table 14-6-4), which erodes 10× faster than most resists, can be used to dry etch chrome if the resist is kept below 50°C.[53] By contrast, in inorganic materials, an activation energy of only 2.5 kcal/mole is noted for etching Si, SiO_2, or polysilicon in CF_4 or $CF_4/8\,\%\ O_2$.[35] Heating effects are not noticeable in the plasma etching of inorganics wherein a ±10°C change is a ±1.2× change in rate. For an activation energy of 6.5–12 kcal/mole, a 10°C rise corresponds to a 2× change in removal rate. In order to prevent resist degradation during plasma etching of the substrate, conditions should be chosen in which the resist has the lowest activation energy (< 6 kcal/mole).

The effect of heat on the stripping of vinyl addition polymers,

$$\underset{\text{(Monomer)}}{CH_2{=}\overset{X}{\underset{Y}{C}}} \underset{T_c}{\rightleftarrows} \underset{\text{(Polymer)}}{-CH_2-\overset{X}{\underset{Y}{C}}-} \qquad (14\text{-}6\text{-}5)$$

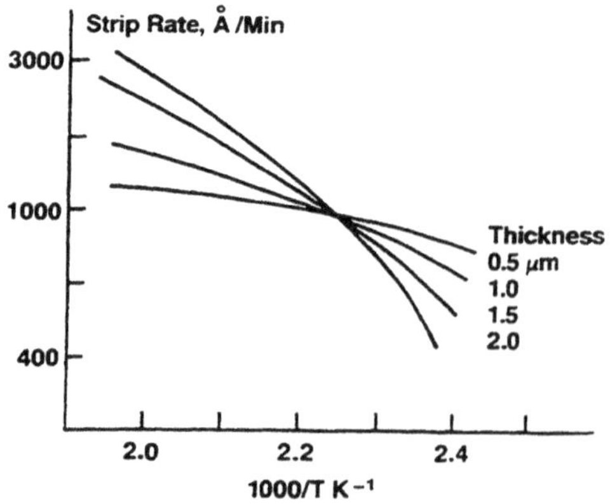

FIGURE 14-6-5. Arrhenius plots for various film thicknesses of AZ-1350 stripped in plasma. From Ref. 38; figure originally presented at the 1981 Spring ECS meeting, reproduced by permission of the Electrochemical Society Inc.

TABLE 14-6-2. Resist Stripping Monitors

Component	Monitor	Reference
Atomic oxygen	NO titrate	28, 36, 56
	Emission	37
Resist	Thickness	38
	Mass	30
	Contact angle	34, 40
CO_2	Mass spectroscopy	57
	Gravimetric	58
CO	Emission	56, 59
OH	Emission	56, 59
H_2O	Mass spectroscopy	57
H	Emission	60
CH	Chemiluminescence	47

can be appreciated in terms of T_c, the ceiling temperature.[54] Heating a polymeric resist such as PMMA above its glass transition[55] (melting) temperature, 105°C (T_g), and its T_c of 160°C, will amplify degradation of the scissioned radicals by end chain depolymerization. Thus, more rapid stripping can be accelerated by plasma stripping at elevated temperatures. The reaction [Eq. (14-6-5)] is driven in favor of the monomer above T_c.

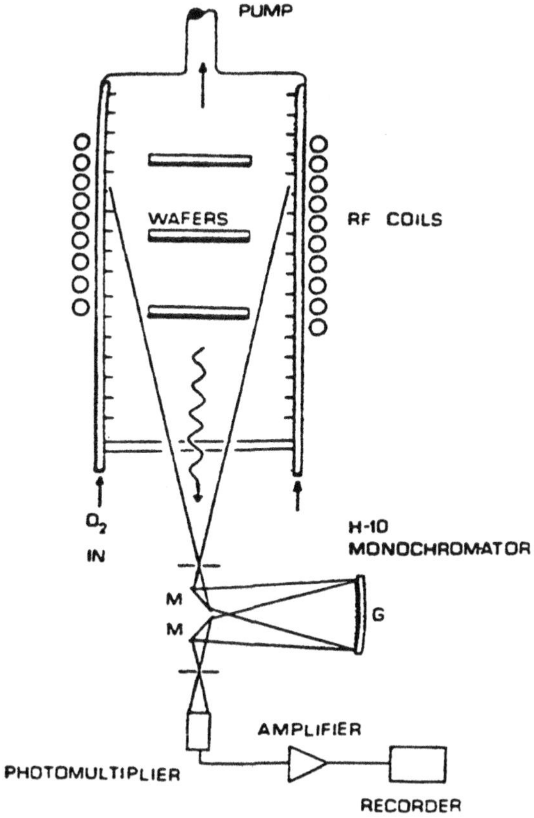

FIGURE 14-6-6. Experimental apparatus to monitor CO emission in plasma stripping. From Ref. 60, reproduced with permission.

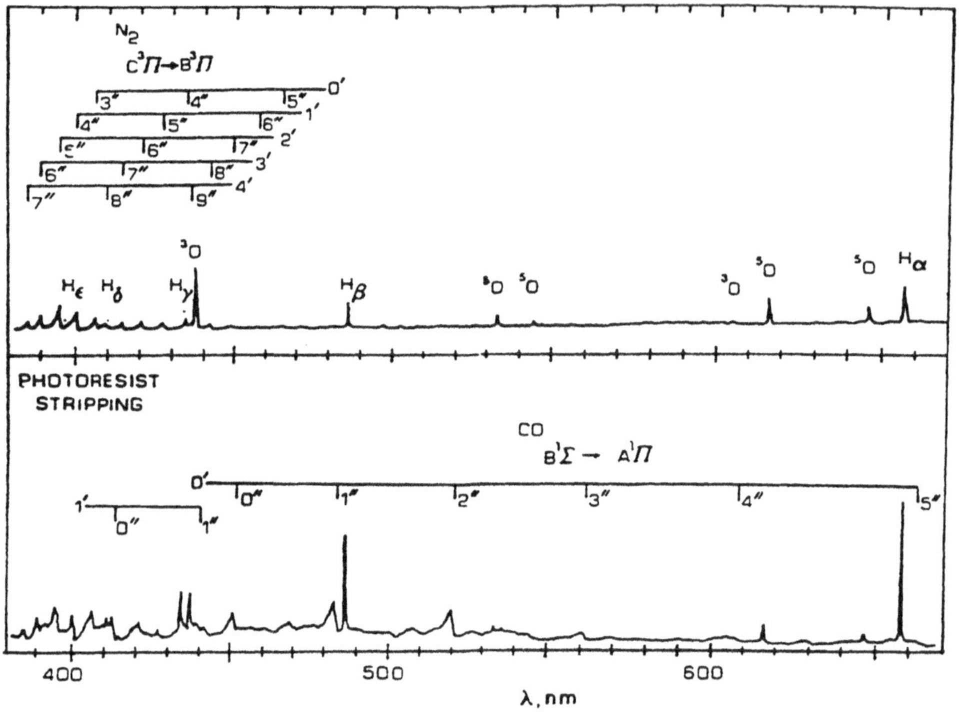

FIGURE 14-6-7. Emission spectrum from O_2 plasma stripping of resist. From Ref. 60, reproduced with permission.

14-6-3. Monitoring Plasma Stripping

The kinetics of resist stripping can be monitored by measuring the reactants or products. In Table 14-6-2, a summary of the spectroscopic and chemical techniques for examining reactants and products of the reaction is given.

$$O + \text{resist} \rightarrow CO_2 + CO + CH\cdot + OH\cdot + H_2O \qquad (14\text{-}6\text{-}6)$$

A typical spectroscopic setup(60) and emission spectrum for CO are shown in Figs. 14-6-6 and 14-6-7, respectively. The application(56) of a CO monitor in the effect of temperature of the wafer on strip rate is shown in Fig. 14-6-8.

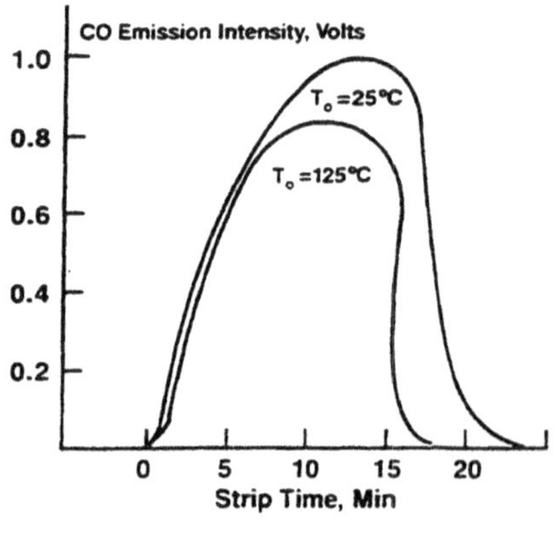

FIGURE 14-6-8. CO emission intensity (519 nm) versus time for stripping 1-μm-thick Waycoat 3 negative resist at two different start temperatures. From Ref. 57, reprinted with permission of *Solid State Technology*, published by Technical Publishing, a company of Dun and Bradstreet.

The spectroscopic monitoring of the reaction products is a convenient process control to decide the endpoint of stripping. For example, the loss[47] of emission of CH or neutral CO species is an indication of near completion. As plasma etching and stripping advance toward larger-diameter (125–150 mm) and single-wafer processing, endpoint monitors will be used to enhance throughput and prevent overstripping or etching.

14-6-4. Resist Stripping Rates and Polymer Structure

As previously discussed, the reactions of atomic oxygen, singlet oxygen, and ozone[137] have been studied for many years in the field of photochemistry and polymeric weathering.[23] Antioxidants and UV absorbers are added to stabilize the polymers against thermal and photon degradation.

Excited oxygen can attack polymers by three main reactions:

1. Hydrogen abstraction

$$\sim\mathrm{CH} + \mathrm{O} \rightarrow \sim\mathrm{C}\cdot + \mathrm{OH}\cdot \qquad (14\text{-}6\text{-}7)$$

2. Insertion

$$\sim\mathrm{CH} + \mathrm{O} \rightarrow \sim\mathrm{COH} \qquad (14\text{-}6\text{-}8)$$

3. Oxidation

$$-\mathrm{C{=}C} + \mathrm{O} \rightarrow -\underset{\diagdown\;\mathrm{O}\;\diagup}{\mathrm{C}\text{———}\mathrm{C}}- \rightarrow \mathrm{C{=}O} \qquad (14\text{-}6\text{-}9)$$

Oxygen also attacks radical intermediates, which can be formed by thermal or radiation-induced chain scission:

$$\sim\mathrm{C}\cdot + \mathrm{O_2} \rightarrow \mathrm{COO}\cdot \qquad (14\text{-}6\text{-}10)$$

The OH· radical has been detected by emission spectroscopy. In reactions involving atomic oxygen attack on polymers, the interception of chain fragments by oxygen prevents recombination to the original starting polymer. Moreover, the thermal reactions involving oxygen favor the abstraction and insertion steps; the attack by atomic oxygen[25] is favored by radical quenching [Eq. (14-6-10)]. Cventanovic studied the influence of polymeric (monomeric models) structure on the attack by atomic oxygen[63,64] (Table 14-6-3).

TABLE 14-6-3. Relative Rates of Reactions of Atomic Oxygen with Organic Compounds

Compound	Rate of oxidation
$H_2C{=}CH_2$	100
CH_4	0.5
CH_3CH_3	0.1
Benzene	0.05

TABLE 14-6-4. Plasma Oxygen Stripping of Polymers

Polymer	Relative rate of strip
Poly-1-butene SO_2	7.1
Polyisoprene	5.6
Polyisobutylene	3.6
Polymethylmethacrylate	2.4
Polyphenylmethacrylate	1.3
Phenolic resin	1.3
Polystyrene	1.0
Polydimethyl siloxane	0.0

Again, because of possible quenching of the excited oxygen by aromatic benzene, and because of the high heat of reaction, the aromatic molecules are very resistant to oxidation. Taylor and Wolf[65] found the relative rates of attack of plasma oxygen on polymers given in Table 14-6-4.

They proposed that aromatic groups (even as phenyl in methacrylate polymer) adsorb excited oxygen and quench the excited states by radiationless deactivation (ϕ is phenyl):

$$\sim\phi + O^* \rightarrow \sim(\phi\text{–}O) \rightarrow \phi^* + O \qquad (14\text{-}6\text{-}11)$$

Polymers which are resistant to attack or other plasma etchants such as CF_4 and CF_4–8% O_2 are also in most cases resistant to removal by oxygen. Pederson[66] reported similar results in CF_4 and CF_4–8% O_2 plasma erosion. Aromatic polymers were the most resistant. As previously noted (Tables 14-5-1 and 14-5-2), polymers containing aromatic groups such as polystyrene require 10–20 moles of O_2 per mole of compound, whereas polyolefins, polyolefin sulfones, and acrylates require 2–8 moles. Since the supply of atomic oxygen is limited by flow, power, and pressure, the rates of stripping are (mole dependent) concentration inhibited by the bimolecular adsorption reaction (14-6-1).

In general, silicone polymers are not attacked by O_2. Presumably, nonvolatile SiO_2 is formed. Likewise, resists containing elements such as Sn, I, Ti, B, Al, Cs, and Fe derivatives may form nonvolatile products which inhibit further degradation by oxygen. CH_4 is used to strip silicones and H_2 for Sn organics. Afterwards, a brief HF dip cleans up residues.

Variations in physical polymer structure which stabilize the polymer against thermal lability such as cross-links and branching will impede removal. The higher molecular weights of vinyl polymers will undergo more scissions per chain than lower molecular weights. Condensation polymers will tend to be more stable because of the low tendency to unzip.

14-7. SIDE EFFECTS OF PLASMA STRIPPING

The use of a plasma to strip resists is not without side effects on the electrical and surface states of the semiconductor device. When resists are removed in an oxygen plasma, there may be some radiation damage to the wafers by ions, elec-

TABLE 14-7-1. Concentrations (wt ppm) of a Number of Elements in Three Types of Photoresist[a]

Resist	Na	K	Cu	Zn	As	Br	Ag
Waycoat III	2.8	0.2	0.40	5.6	0.03	0.17	0.4
Shipley AZ-1350H	7.8	0.54	0.32		0.013	0.14	0.4
Shipley AZ-111	13	0.65	0.30	11	0.02	0.15	0.4

[a] The concentrations were determined by radiochemical analysis after excitation by a neutron flux of $2.2 \times 10^{11}/cm^2$-sec for 3 hr (from ref. 35).

trons, heating, and UV radiation. Not only physical instabilities such as charge states in MOS transistors, but also chemical reactions at the surface, such as oxidation, can occur.

However, in some cases, direct benefits of plasma cleaning can also be found. A higher breakdown voltage and much lower pinhole levels were noted in oxygen plasma-cleaned silicon (prior to gate oxidation of a MOS capacitor).[44] Moreover, the adhesion to silicon nitride is improved by plasma oxidation.

The evolution of plasma stripping was based on the plasma ashing of carbonaceous material to retain metal impurities. It is not surprising that in stripping, some resists synthesized by cyclization of rubber using Sn reagents were found to leave SnO residues. Phenolic positive resins can contain high Na^+ (>25 ppm) levels due to the gettering adsorption by the phenol group. Other metals such as K, Zn, Si, Fe, and P can be present at ppm levels. The results[35] of a radiochemical metal assay of a negative Hunt Waycoat and positive Shipley resists are shown in Table 14-7-1.

The presence of more than 2 ppm can lead to potential diffusion of mobile ions, such as Na^+ into the SiO_2 or Si substrate. Subsequent side effects of residues may cause problems[41] such as:

1. Crystallization of the oxide and breakdown of its passivating properties, leading to extraneous diffusions.
2. Precipitations in the semiconductor acting as getters for dopants, leading to uncontrollable sheet resistivity.
3. Isolated individual charges in the oxide giving rise to electrical instabilities, junction leakage, lower dielectric strength, uncontrollable drifts, and variable thresholds on MOS devices. In the case of MOS devices, the ionizing radiation[68] can induce a shift or degradation of device characteristics. More permanent damage can result from the presence of sodium during ion bombardment.[68,69]

Besides field-induced damage, oxidation of metals such as Si, Al, and Cu can occur[43] to a minor extent (50–100 Å). The oxide layer which is formed is deleterious to subsequent diffusion steps. Since it serves as a partial diffusion barrier for junction formation, an additional wet or dry etch cleaning step is required to remove the skin oxide prior to diffusion.

14-7-1. Detection of Damage

The capacitance voltage method of determining mobile ionic charges in SiO_2 has been used extensively to study the side effects of plasma stripping. The

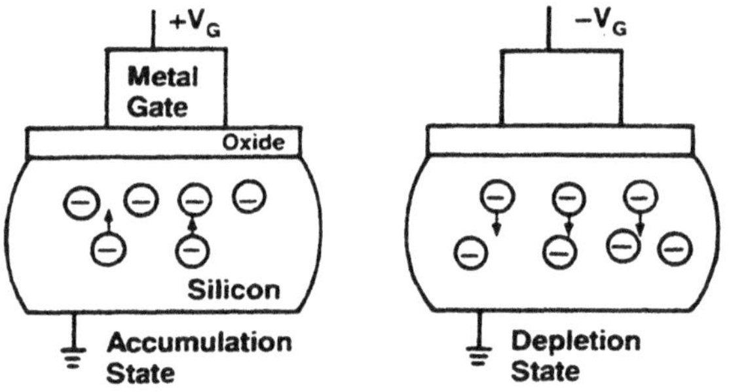

FIGURE 14-7-1. Schematic of capacitance voltage measurements for assessing plasma radiation. From Ref. 70, reprinted with permission of *Solid State Technology*, published by Technical Publishing, a company of Dun and Bradstreet.

schematic of a CV measurement is shown in Fig. 14-7-1. Zainiger and Heiman describe the CV technique in full detail.(71) Secondly, its application to MOS device studies of plasma stripping was shown by Irving,(41) Kalter,(35) Weston,(74) Maddox,(70) Bell,(73) and Szekeres.(31,72)

Ion bombardment of SiO_2 films on Si leads to interface charges in the oxide films located at the SiO_2 interface and on the outside surface.(58) The charge center is mobile under the bias–temperature stress conditions of 275°C, $\pm 10^6$ V/cm. Furthermore, sodium is the most mobile impurity which can be driven into the SiO_2 during and after plasma bombardment. The sodium ion transport can be traced by neutron activation analysis.(35,68,75)

Figure 14-7-2 shows the distribution of sodium ions after stripping. A pileup occurs at the SiO_2/Si interface. The driving force for this migration is the potential gradient in the oxide resulting from plasma interaction.(68) In a MOS transistor, such variations cause instabilities, because of the occurrence of charge variations in the oxide layer.

When a Hunt Waycoat(35) resist was removed in an oxygen plasma, more than half of the sodium was located in the SiO_2 layer and could no longer be removed by rinsing with HF (Table 14-7-2). A phosphosilicate barrier layer has been used as a barrier which getters sodium ions and prevents mobilization.(76)

The amount of sodium ions left on the surface is proportional to the thickness of the film. It is estimated that 40% of the Na^+ are volatilized (sputtered) during plasma stripping and of the remaining 60%, 10% are left on the surface and the rest diffuse into the bulk of the SiO_2 film.(75) Furthermore, rinsing with water removes only the surface ions.(75)

The transport of sodium into the SiO_2 is accelerated by the thermal gradient built up from wafer heating during plasma stripping (100–200°C). The ionic field bombarding the surface induces mobile ion diffusion under the thermal gradient.

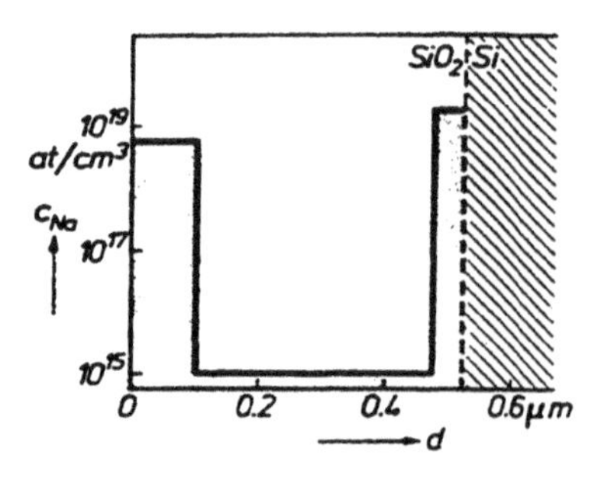

FIGURE 14-7-2. Schematic diagram of the variation in concentration of sodium (C_{Na}) of the radioisotope of ^{24}Na in a silicon dioxide layer as a function of the depth below the surface. The ^{24}Na originated in the photoresist (Waycoat III), which was first doped with the ^{24}Na, then removed from the oxide layer in a barrel reactor with no etch tunnel. A considerable number of sodium ions are concentrated close to the boundary (dashed line) with the silicon layer beneath the oxide. From Ref. 35.

TABLE 14-7-2. Residue of ^{24}Na on and in SiO_2 and SiO_2 Coated with 30 nm of Phosphosilicate Glass (PSG), after Removal of the Photoresist (Waycoat III) in an Oxygen Plasma, Both with and without Tunnel[a]

Material	Percentage of ^{24}Na	
	Without tunnel	With tunnel
SiO_2		
Without rinsing	100	100
After rinsing in H_2O or HNO_3	55	2
After etching for 30 sec in 1% HF	52	1
SiO_2 PSG		
Without rinsing	100	100
After rinsing in H_2O or HNO_3	45	20
After etching for 30 sec in 1% HF	2	1

[a] Values are percentages of the amount of ^{24}Na (about 10 ppm) initially used for doping the resist (from ref. 35).

McCaughan[68,69] proposes an ion-neutralization process to explain the means of low-energy ion bombardment of the SiO_2/Si interface. According to this model, when low-energy ions approach the SiO_2 surface, electron transfer from the surface sodium and/or from the SiO_2 valence band occurs to neutralize the incident ions. Then, ionized sodium is transported by a resultant drift field toward the SiO_2/Si interface where it piles up at the interface.

The results agree with the ion-neutralization model. The ionization[75] potentials of the incident ions, O_2^+ and O^+, are 12.2 and 13.6 eV, which are large enough to cause ion neutralization. Metal shielding of the wafers (Faraday screen) reduces direct ion bombardment. The addition of CF_4 gas in small amounts to the oxygen gas, induces Na^+ to form nonmobile NaF.[73]

Besides CV techniques, other physical techniques to study plasma stripping or oxidation are listed in Table 14-7-3.

In a separate study of Na^+ left after development with a positive novolak resist, X-ray photoelectron spectroscopy (XPS) was used to detect Na^+.[77] The

TABLE 14-7-3. Surface Characterization of Plasma Stripping

Technique	Application	Reference
Electron diffraction	Si oxidation	33, 43
Auger	Resist residues	41, 70
ESCA	Polymer treatment	78
X-ray photo electron	Na residues	77
Ellipsometry	Nb oxidation	78
Ellipsometry	Pb oxidation	79
Ellipsometry	Resist residues	2
Contact angle	Resist residues	3, 43
Radioactive I_2	Resist residues	81
Ion scan spectroscopy	Oxidation	70
Radioactive Na	Cleaning	16

residue of Na^+ left after development accelerated the etching of polysilicon in CF_4/8% O_2 by a factor of two. Rinsing the wafers in dilute acid eliminated the catalytic Na^+ effect.

Oxidation of the silicon surface occurs to the extent of 100 Å.[80] Electron diffraction[33] was used to monitor the disappearance of the silicon[111] surface. The low-temperature plasma oxidation of silicon can produce some desirable electrical features over thermally grown oxide. Higher dielectric strength and lower pinholes in SiO_2 are noted especially if *in situ* RF sputter cleaning is used prior to plasma oxidation.[82]

Resist residues left after wet and dry stripping were examined by methods of classical surface chemical wetting and ellipsometry. Plasma stripping left no residues while all wet methods left 5–50 Å of contaminants.[2,3]

14-7-2. Prevention and Control of Damage

Since metal ions are left and driven into the semiconductor surface during and after stripping, a limit on the metallic content of photoresists and/or developer should be enforced (< 2 ppm). Electrophoretic treatment of the liquid negative rubber resists has been suggested as a means of gettering ionic impurities.[83]

Tin has been a prevalent impurity in cyclized rubber resist.[40,41] Sodium has been noted in positive DQN resists.[31,35,70] The addition of organohalides (trichloroethylene) to the oxygen stripping gas induces formation of volatile tin chlorides[84] and the slow removal of some resists in an H_2 plasma is thought to form volatile tin hydrides.[41,85] The addition of trace amounts of CF_4 to O_2 stripping gas suppresses Na^+ diffusion into SiO_2 by forming immobile NaF.[73]

Both the direct shielding of the wafers from O^+ bombardment by a Faraday screen[74,76] and grounding the electrodes in a parallel plate[31] to reduce the ion energy (< 30 eV) have been noted to minimize MOS damage. Downfield plasma reactors avoid direct radiation damage to silicon.[129,130] The protection of the insulator surface by an Al gate electrode prevents MOS damage.[31] Likewise, phosphosilicate glass was used as a diffusion barrier to Na^+.[76] Preheating the wafer to 200°C prior to stripping enhances the strip rate and minimizes the plasma-induced diffusion of mobile ions.[35,76]

Postdip or cleaning to remove the trace oxide formed or metal residues has been practiced. In order to remove the oxide interlayer, a wet HF dip[74] or a brief CF_4 etch[31,84] has been used. Contaminants left on silicon have been cleansed with pyrocatechol,[86] KOH for Sn,[40] H_2O,[43] or HF.[31,41,43]

When the metallic ions are driven in by the plasma, postanneal in HCl[70] and H_2[72] have been used. Prior to lift-off or metal deposition over plasma-stripped surfaces, an additional RF sputter clean[74] may been used to reduce ohmic resistance.

Since metal ions such as Na^+ readily diffuse into the SiO_2/Si interface during plasma stripping, the resist formulation should be analyzed and specified to contain tolerable levels. Devices which are sensitive to the radiation field or mobile ions should be cautiously fabricated by plasma stripping. Treatments, techniques, and annealing to prevent or minimize damage have been discussed in Chapter 13. In the future, oxygen will be used to develop resist images as well as strip the resist. In the stripping case, a parallel plate reactor will replace the traditional barrel stripper. The parallel diode plate reactor can be a multipurpose tool since Al etching can also be performed. The parallel reactor (Fig. 14-5-2) exposes the wafer to direct damage by ionizing plasma radiation. In this case, low power, grounding, and soft conditions have to be used with proper controls to reduce radiation damage.[72] In addition, the resist itself will have to be pure and free of diffusing metal contaminants. Plasma-deposited resists, for example, should be free of metal impurities.[61]

The extent of damage depends on the time of exposure[72] to the plasma and the temperature rise[39] during stripping.[72] The temperature rise of wafers in an oxygen plasma can be monitored by refractive index changes.[39,46] At 300°C, the temperature is sufficient for some annealing to occur during a plasma exposure.[72] The radiation damage in MOS devices occurs during the first 10 min of stripping.[72,75] More studies of time–temperature are required to establish resist stripping with minimum device damage.

14-8. PLASMA STRIPPING APPARATUS

The major developments in the plasma apparatus can be traced in the patent literature. In Table 14-8-1, a compilation of patents is referenced. Originally, the plasma ashing equipment used in biological sciences was later applied to the processing of silicon wafers and chrome photomasks. The latest equipment in plasma stripping uses a downfield reactor to strip resists without direct radiation bombardment of the silicon wafer.[129,131]

14-9. MISCELLANEOUS STRIPPING TECHNIQUES

In addition to liquid and plasma stripping, other ancillary methods have been used. Volatilization by heat, light, or ozone have been physical means of stripping resists.

UV light (1849–3000 Å) has been used to strip cyclized rubber,[91] polyacetylene,[92] and polyacrylate resists.[94] If ozone and oxygen are present at

TABLE 14-8-1. Patents on Plasma Stripping

Application	Patent
Ashing organics	U.S. 3,410,776 (1968), LFE
Ashing organics	U.S. 3,428,548 (1969), LFE
Ashing polymers	U.S. 3,647,676 (1971), LFE
Ashing polymers	U.S. 3,671,195 (1972), Int. Plasma
Resist strip	U.S. 3,806,365 (1974), LFE
Resist strip	U.S. 3,837,856 (1974), Signetics
Resist strip residues	U.S. 3,867,216 (1975), LFE
Descum resist images	British 1,419,034 (1975), Std. Tel.
Postclean surface	U.S. 3,930,913 (1975), LFE
Chrome etch enhance	U.S. 3,951,709 (1976), LFE
Low-temperature ashing	U.S. 4,017,404 (1977), LFE

270°C, high strip rates of 1 μm/min could be obtained.[93] No residues were left as evidenced by Auger spectroscopy.[93]

A combination of UV/ozone has also been used[95,96,145,146] to clean wafers and surfaces. The advantage of the process is that wafers can be processed at room temperature and atmosphere. Radiation damage is not apparent as is the case for plasma stripping. The rates, however, are slow since each wafer must be totally exposed by UV. Ozone itself has also been used to strip cyclized rubber resists.[97,98]

Mechanical means of stripping thick resists or printing plates include high-pressure air incident at an acute angle,[99] mechanical scraping,[100] and a high-temperature stream of water.[101] Superheated steam[102] has been used to clean wafers and strip resists.[103] The removal of high postbaked resists such as a novolak can be facilitated by a combination of steam treatment followed by normal strippers.[103]

High-temperature flames[87,88] have been used in the direct stripping of Kodak KTFR rubber resists. The flame is a type of plasma which contains ionic particles and monoatomic oxygen. The KTFR stripped in 10 sec while Shipley AZ-1350 required 30–60 sec. No damage to a RAM device was noted since the surface dwell time–temperature is very short. The addition of ozone to the environment at 250°C speeded combustion.[88]

The direct pyrolysis of carbon films proceeds more rapidly in the presence of O_2 and ozone at elevated temperatures.[89] Phenolic resins themselves as powders or thick material do not pyrolize completely, but form graphite chars.[90] Micrometer films, however, will burn in an oxygen atmosphere.

The high-temperature pyrolysis of polymethylmethacrylate or polyolefin sulfone resist proceeds *in vacuo* if the polymers are heated above their ceiling temperature [Eq. (14-6-5)]. Flood exposure by *e*-beam, UV, or X rays of positive resists or pre-cross-linked acrylate resists will solubilize the resist for stripping in the developer.

14-10. APPLICATIONS OF OXYGEN PLASMA TO SEMICONDUCTOR LITHOGRAPHY

Oxygen plasmas have been used not only to strip resists but also in a variety of related processes. In Table 14-10-1, representative applications are given.

Descumming of negative *e*-beam[107,108] resists to overcome the intra- and interproximity effects in mask or device fabrication has been the most popular use. An early use was to descum the ozone-induced polymerization of Kodak KPR cinnamate resist.[104–106]

After plasma etching with CF_4, a thin veil of "Teflon"-type polymer may be present. Oxygen plasma was used to strip the inhibiting layer.[110] Some plasma gases add 8% O_2 to CF_4 to gradually erode the fluorocarbon[127] polymer on the etching[128] substrate. Unfortunately, the introduction of oxygen erodes the resist. Other nonfluorine, but chlorinated hydrocarbons have been tried to prevent polymer residues with minimum resist erosion.[110]

Residues from the RIE etching of Al by CCl_4 are cleaned in an oxygen plasma.[111] Cleaning of wafers by plasma oxidation is said to improve the adhesion of resists[117] to that surface.

TABLE 14-10-1. Applications of Oxygen Plasma

Application	Resist or system	Reference
1. Plasma descum of resist images	KPR	104–106
	Electron-beam negative	107, 108
	X-ray negative	109
2. Removal of plasma (RIE) etch residues	CF_4 polymer	110
	Al	111
3. Oxygen glow discharge	Clean metals	112–115
4. Improve adhesion of polymers	"Casing"	116
5. Oxidation of polysilicon	Resist adhesion	117
6. Fabrication of semitransparent masks	Polyvinyl-ferrocene	118
7. Dry develop resists	PMMA	119
	Silicone monomer	120, 143
	Aryl monomer	121
	Ion implant	122
8. Multilayer resist	Lift-off	123
	RIE	124
9. Stencil (mask) etch polymers	100 mm thick	125
10. Preclean wafers	Si	126

The perfection of the dry development of resists[61] involving oxygen, or the RIE etching of multilayer resists with oxygen[62] will require additional studies of kinematics of polymer removal, as well as process monitors.

14-11. OUTLOOK

The introduction of plasma ashing equipment paved the way for plasma etching. In turn, the study of the attack of oxygen on polymers led the way for dry development of resists. Soon, the dry development of single-layer or multilayer resists will be vital for future submicrometer lithography. The advances in this area will depend on critical studies of the nature of plasmas and the chemical and physical interactions with polymers and semiconductor surfaces.

REFERENCES

1. *Polymer Handbook*, edited by J. Brandup and E. Immergut, Wiley, New York, 1975, p. 337.
2. B. Bergin and L. Kaplan, *J. Electrochem. Soc.* **127**, 386 (1980).
3. C. Deckert and D. Peters, *J. Electrochem. Soc.* **126**, 883 (1979).
4. *Technology of Paint, Varnishes and Lacquers*, edited by C. Martens, Reinhold, New York, 1968, p. 613.
5. U. S. Patent 3,582,401 (1971), Mallinckrodt.
6. U. S. Patent 3,988,256 (1976), Allied.
7. Japanese Patent 80,64,233, *Chem. Abstr.* **93**, 177312 (1980).
8. U. S. Patent 4,070,203 (1978), Merck.
9. U. S. Patent 4,165,294 (1979), Allied Chem.; Japanese Patent 59,225,323, Allied; *Chem. Abstr.* **103**, 14572 (1985).
10. U. S. Patent 4,165,295 (1979), Allied Chem.
11. U. S. Patent 3,600,321 (1971), Air Prod.
12. U. S. Patent 4,215,005 (1980), Allied Chem.

13. U. S. Patent 4,221,674 (1980), Allied Chem.
14. F. Pintchoviski, J. Price, P. Tobin, J. Peavey, and K. Kobold, *J. Electrochem. Soc.* **126**, 1428 (1979).
15. D. Swern, *Chem. Rev.* **45**, 1 (1949).
16. W. Kern, *RCA Rev.* June 1970, p. 208.
17. U. S. Patent 4,078,102 (1978), IBM.
18. Japanese Patent 78,39,127, *Chem. Abstr.* **89**, 120921 (1978).
19. H. Kleinknecht and H. Meier, *J. Electrochem. Soc.* **125**, 798 (1978).
20. R. Joshi, in *Encyclopedia of Polymer Science and Technology* **13**, 788 (1968).
21. C. Cullis and M. Hivschler, *The Combustion of Organic Polymers*, Oxford University Press (Clarendon), London, 1981, p. 54.
22. E. Lawton, *J. Appl. Polym. Sci.* **16**, 1857 (1972).
23. B. Ranby and J. Rabek, *Photo-Degradation, Photo-Oxidation and Photo-Stabilization of Polymers*, Wiley, New York, 1975, p. 98.
24. J. Mollahan and A. Bell, Editors, *Techniques and Applications of Plasma Chemistry*, Wiley, New York, 1974.
25. R. Hansen, J. Pascale, T. DeBenedictis, and P. Rentzepis, *J. Polym. Sci. Part A1* **3**, 2205 (1965).
26. A. Charlesby, *Atomic Radiation and Polymers*, Pergamon Press, Elmsford, N.Y., 1960, p. 159.
27. B. McTaggart and R. Bell, *Plasma Chemistry and Electrical Discharges*, Elsevier, Amsterdam, 1967.
28. A. Bell and K. Kwong, *AICHE J.* **18**, 990 (1972).
29. J. Battey, *J. Electrochem. Soc.* **124**, 437 (1977).
30. J. Battey, *IEEE Trans. Electron Devices* **ED-24**, 140 (1977).
31. A. Szekeres, K. Kirov, and S. Alexandrova, *Phys. Status Solidi* **63**, 371 (1981).
32. Ref. 23, p. 362.
33. S. Irving, *Electrochem. Soc. Ext. Abstr.* J-6, No. 180, p. 55 (1967).
34. R. Bersin, *Solid State Technol.* June 1970, p. 39.
35. H. Kalter and E. Van de Ven, *Philips Tech. Rev.* **38**, 200 (1978).
36. J. Battey, *J. Electrochem. Soc.* **125**, 147 (1977).
37. E. Degenkolb and J. Griffiths, *Appl. Spectrosc.* **31**, 40 (1977).
38. R. Bond, S. Dzioba, and H. Naguib, *Electrochem. Soc. Ext. Abstr.* **81-1**, 729 (1981).
39. R. Bond, S. Dzioba, and H. Naguib, *J. Vac. Sci. Technol.* **18**, 335 (1981).
40. H. Hughes, W. Hunter, and K. Ritchie, *J. Electrochem. Soc.* **120**, 99 (1973).
41. S. Irving, *Solid State Technol.* June 1971, p. 47.
42. H. Akiya, K. Saito, and K. Kobayashi, *Jpn. J. Appl. Phys.* **20**, 647 (1981).
43. Int. Plasma Co., *Solid State Technol.* March 1970, p. 92.
44. S. Iwamatsu, *J. Electrochem. Soc.* **129**, 224 (1982).
45. A. Gaydon, *Spectroscopy of Flame*, Wiley, New York, 1957, p. 187.
46. R. Bond, S. Dizoba, and H. Naguib, *Electrochem. Soc. Ext. Abstr.* **81-1**, 729 (1981).
47. C. Wang and B. Gelerent, *Solid State Technol.* Nov. 1981, p. 121.
48. American Chemical Society, *Adv. Chem. Ser.* **80**, 232 (1969).
49. H. Marsh and T. O'Hare, *Trans. Faraday Soc.* **61**(2), 274 (1965).
50. U. S. Patent 3,757,733 (1973), TI.
51. E. Egerton, A. Nef, W. Millikim, W. Cook, and D. Baril, *Solid State Technol.* Aug. 82, p. 84.
52. Y. Horike, *Jpn. J. Appl. Phys.* **20**, L817 (1981).
53. K. Jinno, *Jpn. J. Appl. Phys.* **17**, 1283 (1978).
54. J. MacCallum and C. Schoff, *J. Chem. Soc. Faraday Trans.* **67**, 2372 (1971).
55. K. Harada, *J. Appl. Polym. Sci.* **26**, 1961 (1981).
56. B. Stafford and G. Gorin, *Solid State Technol.* Sept. 1977, p. 51.
57. G. Bunyard and B. Raby, *Solid State Technol.* Dec. 1977, p. 53.
58. R. Reichelderfer, J. Weitz, and J. Battey, *J. Electrochem. Soc.* **125**, 1926 (1978).
59. E. Degenkolb, C. Mogab, M. Goldrick, and J. Griffiths, *Appl. Spectrosc.* **30**, 520 (1976).
60. J. Griffiths and E. Degenkolb, *Appl. Spectrosc.* **31**, 134 (1977).
61. M. Yamada, J. Tamano, S. Morita, K. Yoneda, and S. Hattori, *Plasma Chem. Plasma Proc.* **1**, 261 (1981).
62. S. Hyman, *J. Vac. Sci. Technol.* **19**, 1325 (1981).
63. R. Cventanovic, *J. Chem. Phys.* **30**, 19 (1959).
64. R. Cventanovic, *Can. J. Chem.* **39**, 2444 (1961).
65. G. Taylor and T. Wolf, *Polym. Eng. Sci.* **20**, 1087 (1980).

66. L. Pederson, *J. Electrochem. Soc.* **129**, 205 (1982).
67. U. S. Patent 3,837,856 (1974), Signetics.
68. R. Krisher, D. McCaughan, V. Murphy, and J. Herlig, *Phys. Rev. B* **10**, 2632 (1974).
69. D. McCaughan, R. Kusher, and J. Murphy, *Phys. Rev. Lett.* **30**, 614 (1973).
70. R. Maddox and H. Parker, *Solid State Technol.* **21**, 107 (1978).
71. H. Zaininger and F. Heiman, *Solid State Technol.* June 1970, p. 46.
72. A. Szekeres, S. Alexandrova, and K. Kirov, *Phys. Status Solidi A* **62**, 727 (1980).
73. G. Bell and R. Stokan, *Electrochem. Soc. Ext. Abstr.* Fall 1977, p. 383.
74. D. Weston and J. Keller, *Electrochem. Soc. Ext. Abstr.* Fall 1977, p. 381.
75. H. Akuya, K. Saito, and K. Kobayashi, *Jpn. J. Appl. Phys.* **20**, 647 (1981).
76. H. Kalter and E. Van de Ven, *Electrochem. Soc. Ext. Abstr.* Spring 1976, p. 335.
77. T. Makino, H. Nakamura, and M. Asano, *J. Electrochem. Soc.* **128**, 103 (1981).
78. H. Yasuda, *Am. Chem. Soc. Polym. Prepr.* **16**, 57 (1975).
79. K. Korr and J. Leslie, *J. Electrochem. Soc.* **121**, 805 (1974).
80. J. Greiner, *J. Appl. Phys.* **45**, 32 (1974).
81. K. Heiner and G. Larrabee, *Solid State Technol.* April 1969, p. 44.
82. D. McCaughan and J. Heilig, *Int. J. Electron.* **34**, 737 (1973).
83. K. Hirobe and S. Iwanatsu, *J. Electrochem. Soc.* **126**, 1426 (1979).
84. U. S. Patents 3,806,365 (1974), 3,867,216 (1975), 3,951,843 (1976), LFE.
85. U. S. Patent 4,201,579 (1980), Motorola.
86. G. Das and F. Montillo, *IBM Tech. Discl. Bull.* **23**, 4490 (1981).
87. U. S. Patent 4,296,146 (1981), TI.
88. T. Penn, *IEEE Trans. Electron Devices* **ED-26**, 640 (1979).
89. U. S. Patent 2,443,373 (1948).
90. *High Temperature Polymers*, edited by C. Segal, Dekker, New York, 1967, p. 86.
91. U. S. Patent 3,664,899 (1972), GE.
92. U. S. Patent 3,890,176 (1975), GE.
93. D. Bolon and C. Kunz, *Polym. Eng. Sci.* **12**, 109 (1972).
94. Japanese Patent 8,059,458 (1980), *Chem. Abstr.* **93**, 177309d (1980).
95. R. Sowell, R. Cuthrell, D. Mattox, and R. Bland, *J. Vac. Sci. Technol.* **11**, 474 (1974).
96. J. Vig and J. LeBus, *IEEE Trans. Parts Hybrids Packag.* **12**, 365 (1976).
97. German Patent 2,635,066, *Chem. Abstr.* **86**, 149647 (1977).
98. P. Burrage, *IBM Tech. Discl. Bull.* **10**, 1260 (1968).
99. U. S. Patent 4,197,126 (1980), W. Grace.
100. U. S. Patent 4,116,715 (1978).
101. U. S. Patent 4,186,032 (1980), RCA.
102. U. S. Patent 3,515,607 (1970), West. Elec.
103. W. Moreau and R. Lang, *IBM Tech. Discl. Bull.* **24**, 1615 (1982).
104. British Patent 1,419,034 (1975), Std. Tel.
105. B. Leon, *Solid State Technol.* April 1977, p. 71.
106. U. S. Patent 3,705,055 (1972), West. Elec.
107. H. Choong and F. Kahn, *J. Vac. Sci. Technol.* **19**, 1348 (1981).
108. Z. Tan, C. Petropoulous, and F. Rauner, *J. Vac. Sci. Technol.* **16**, 2000 (1979).
109. G. Taylor, *J. Vac. Sci. Technol.* **16**, 2014 (1979).
110. J. Chinn, I. Aesida, E. Wolf, and R. Tiberio, *J. Vac. Sci. Technol.* **19**, 1418 (1981).
111. U. S. Patent 4,183,781 (1980), IBM.
112. A. Horgan and I. Dalins, *J. Vac. Sci. Technol.* **10**, 523 (1973).
113. A. Jones, E. Jones, and E. Williams, *Vacuum* **23**, 227 (1973).
114. D. O'Kane and K. Mittal, *J. Vac. Sci. Technol.* **11**, 530 (1974).
115. R. Lambert and C. Comrie, *J. Vac. Sci. Technol.* **11**, 530 (1974).
116. M. Hudis, in *Techniques and Applications of Plasma Chemistry*, edited by J. Hollahan and A. Bell, Wiley, New York, 1974.
117. U. S. Patent 4,176,003 (1979), NCR.
118. U. S. Patent 4,027,052 (1978), Bell.
119. M. Yamada, J. Tamano, K. Yoneda, S. Morita, and S. Hattori, *Jpn. J. Appl. Phys.* **21**, 768 (1982).
120. G. Taylor, T. Wolf, and M. Goldrick, *J. Electrochem. Soc.* **128**, 361 (1981), *J. Vac. Sci. Technol.* **19**, 872 (1981).
121. G. Taylor and T. Wolf, *J. Electrochem. Soc.* **127**, 2665 (1980).

122. T. Venhatesan, G. Taylor, A. Wagner, B. Wilkens, and D. Barr, *J. Vac. Sci. Technol.* **19**, 1379 (1981).
123. U. S. Patent 3,873,361 (1976), IBM.
124. J. Moran, *J. Vac. Sci. Technol.* **16**, 1620 (1979).
125. I. Goldstein and F. Kalk, *J. Vac. Sci. Technol.* **19**, 743 (1981).
126. D. Peters and C. Deckert, *J. Electrochem. Soc.* **126**, 883 (1979).
127. U. S. Patent 3,795,557 (1974), LFE.
128. U. S. Patent 4,314,873 (1982), Bell.
129. S. Dzioba, G. Este, and H. Naguib, *J. Electrochem. Soc.* **129**, 2537 (1982).
130. J. Cook and B. Benson, *J. Electrochem. Soc.* **130**, 2459 (1983).
131. J. Hannon and J. Cook, *J. Electrochem. Soc.* **131**, 1164 (1984).
132. P. Van Zant, *Semicond. Int.* April 1984, p. 109.
133. U. S. Patent 4,438,192 (1984), Dow.
134. U. S. Patent 4,403,029 (1983), J. T. Baker.
135. U. S. Patent 4,401,747 (1983), J. T. Baker.
136. U. S. Patent 4,395,479 (1983), J. T. Baker.
137. L. Zafonte and R. Chiu, *SPIE Proc. Opt. Microlithogr.* **470**, 164 (1984).
138. U. S. Patent 4,346,125 (1982), Bell (anhydrous hydrazine resist stripper).
139. G. Herzberg, *Spectra of Diatomic Molecules*, Van Nostrand, Princeton, N.J., 1950.
140. R. Alkire and D. Economu, *J. Electrochem. Soc.* **132**, 648 (1985).
141. U. S. Patent 4,491,530 (1985), Allied (sulfonic acid stripper).
142. U. S. Patent 4,501,061 (1985), Microdevice.
143. U. S. Patent 4,500,628 (1985), AT & T.
144. B. Stoller, R. Shuman, and R. Ross, *J. Electrochem. Soc.* **131**, 2285 (1985).
145. J. Vig, *J. Vac. Sci. Technol.* **A3**, 1027 (1985).
146. H. Norstrom, M. Ostling, R. Buchta, and C. Peterson, *J. Electrochem. Soc.* **132**, 2285 (1985).
147. A. Nagy, *J. Electrochem. Soc.* **131**, 1871 (1984).

15

Process Controls

15-1. GENERAL PROCESS CONSIDERATIONS

Lithographic process controls are vital to the cost-effective production of high-quality semiconductor circuits. In semiconductor lithography, good yields, reproducible processing,[1] and high productivity[2–4] are production objectives that offset the high cost of clean rooms and lithography process equipment. In semiconductor device fabrication, the final electrical yields of good chips versus the total number of fabricated chips can be as low as a few percent. For yield considerations, the factors (Table 15-1-1) include the density of circuits, the number of lithography operations, the circuit complexity, batch versus single wafer process controls, and the degree of contamination in processing.

The benefits of dedicated process controls in the lithography sector are higher yields, fast turnaround, and the early diagnosis of material or processing problems. Hopefully, the process defects can be traced to problems with the process tools, materials, or clean room environment and rectified before they become a persistent yield loss. Included in the controls are on-line monitors and the off-line quality control of the incoming resist materials and associated chemicals. The lithographic fabrication of semiconductor devices has three basic goals:

1. High productivity at the lowest costs
2. High yield production of critical circuit dimensions to within acceptable biases and tolerances
3. Overlay of circuit device layers to within one-quarter of the critical linewidth

The basic overlay (misalignment) factors are the translation errors, chip and wafer rotation, and the expansion of the mask or wafer processing errors (Fig. 15-1-1).[5,6] For example, in an optical stepper, distortion in the x axis can be $0.20 \pm 0.5\ \mu m$ (2σ) and $0.13 \pm 0.60\ \mu m$ (2σ) in y. Concerning the same positive resist, different chips were used to evaluate 25 different exposure tools by aligning in one wafer site and printing in the resist. The resist image was etched into chrome "resistors" and the linewidths were electrically measured.[7–9] The average distortion (Fig. 15-1-2) was ± 0.2–$0.3\ \mu m$ (2σ) with the outer edges having the largest distortion. The process control of overlay is concerned with the prevention of disfigured alignment marks and the reduction of in-line wafer warpage. Precision optics and a

TABLE 15-1-1. Yield Factors in Device Fabrication

Density of circuits	Active area Inactive area Linewidth Chip size Wafer size
Number of lithography steps and processes	Circuits Diffusion Implantation Isolation Metallurgy Passivation
Circuit complexity	Planarity Vertical integration Nonplanarity Thickness of films
Process bias and tolerances	Resist Image size Alignment (overlay) (Expose, develop, etch, ion implant, lift-off, image transfer)
Process	Handling Controls Contamination

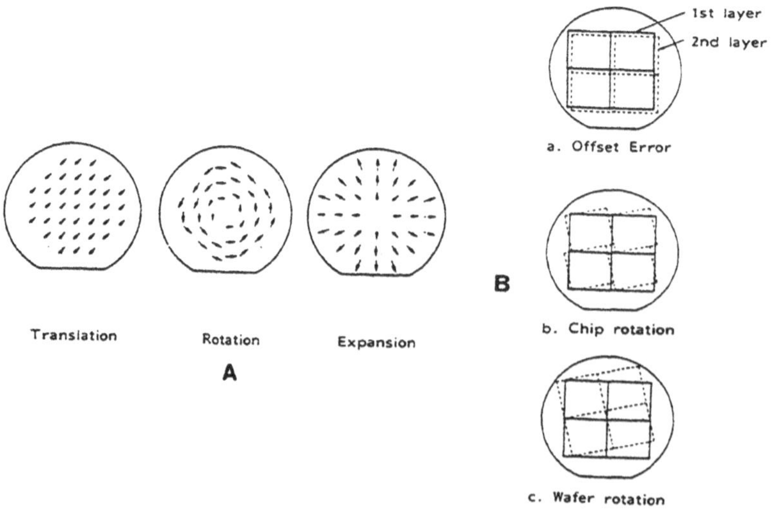

FIGURE 15-1-1. (A) Basic components of misalignment errors of translation–rotation–expansion. From Ref. 5, courtesy of SPIE. (B) Errors in off-axis alignment scheme. From Ref. 6, copyright 1981 and courtesy of IEEE.

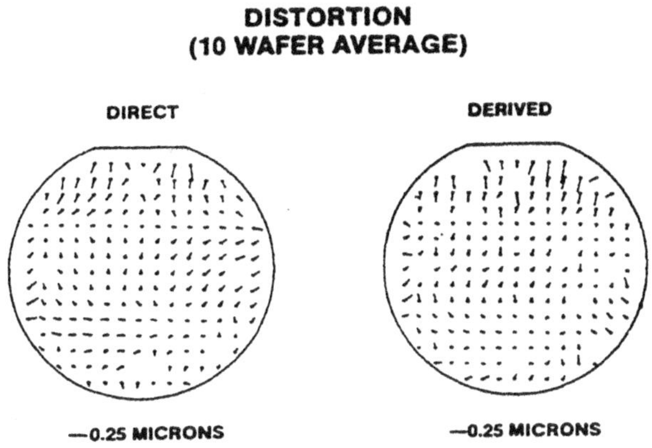

FIGURE 15-1-2. Distortion vectors of machine-to-machine overlay errors for optical projection printing averaged over ten 5-inch-diameter wafers. From Ref. 7, courtesy of SPIE.

refined signal analysis of the reflected light are key elements in the optical–electronic aspects of good alignment.

Lithography process controls are mainly concerned with keeping the final etched or deposited line dimension within acceptable limits (Fig. 15-1-3). At each appropriate step in the resist process, the size of the resist image and the final (etched or deposited) image should be measured to ascertain if the process controls are keeping the biases and tolerances within the device ground rules. The biases and tolerance of the process include the mask (M) distortions or beam spot placement, the resist image (R), the etched (E) or lift-off size, and the overlay errors (Fig. 15-1-4). For example, a series of process variables of prebake and exposure conditions (Fig. 15-1-5) can be varied, and statistically analyzed. Then the process controls can be defined in a process window for acceptable linewidth control.

The shape of a final etched line is a measure of the process integrity. An isotropic Gaussian-shaped profile implies a simplified process with natural scattering of radiation or the diffusion (translational)-controlled removal of material. Anisotropically formed images in amorphous materials imply precise processing with a higher order of a force (an external field) to overcome the random motion of molecules or energetic particles.

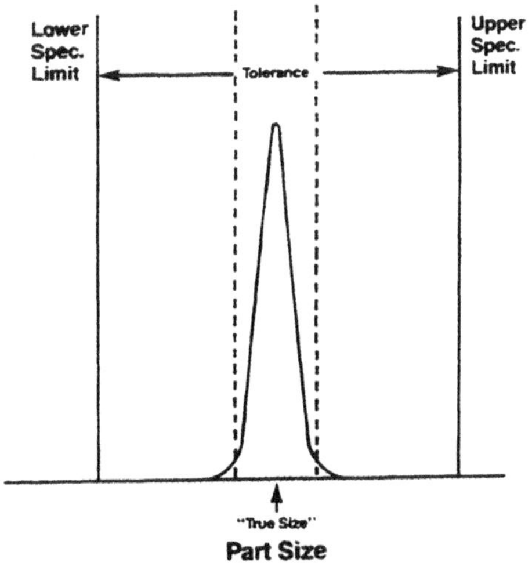

FIGURE 15-1-3. Accuracy (true size) and precision (tolerance of repeatability) desired in reproducing linewidths in a lithographic process. From Ref. 1, copyright 1980 and reproduced by permission of Eastman Kodak Co.

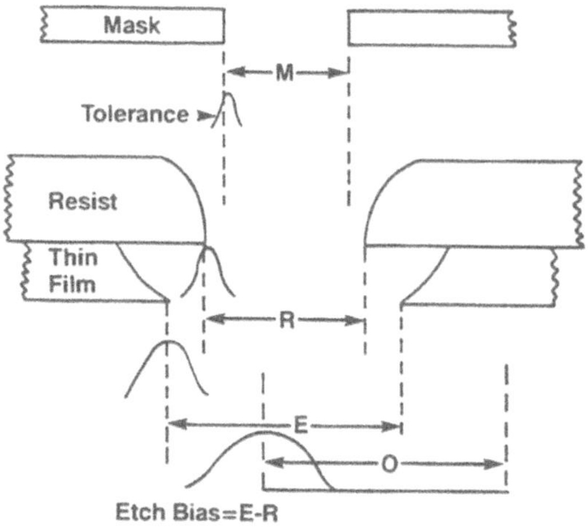

FIGURE 15-1-4. The lithographic biases and tolerance from mask (M) dimensions reproduced in final etched (E) pattern.

Resist lithography is faced with the production of a variety of profiles in a resist or in the final device film. Figure 15-1-6 shows examples of resist profiles acceptable for some types of images or image transfer, such as in wet etching or lift-off. For wet etching of lines > 3 μm, only the width at the base of the resist trough is important, not the upper features or profile. Because of plasma erosion, more three-dimensional demands on the resist profile for anisotropic reactive ion etching have emerged. Lift-off and electroplating almost duplicate the resist profile in additive processing and consequently are very important to control.

In the early days of lithography, it was sufficient to optically measure a resist linewidth. Today, profile measurements by SEM are important in qualifying a resist and process, in establishing operating limits, and in diagnosing process problems such as resist or plasma etch residues.

Process control involves the reproducible interaction of materials, process steps, and production equipment (Fig. 15-1-7). Before establishing process controls, process capability has to be established. Process capability is the measured inherent

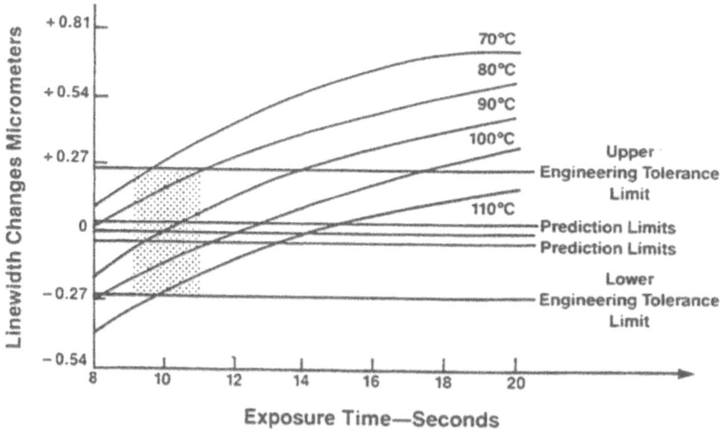

FIGURE 15-1-5. Linewidth changes for Kodak 820 photoresist as a function of exposure time and prebake time. From Ref. 1, copyright 1980 and reproduced by permission of Eastman Kodak Co.

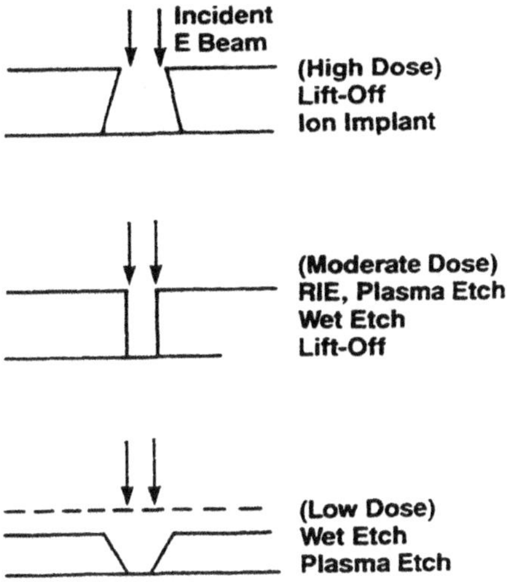

FIGURE 15-1-6. Resist profiles and their applications in subtractive and additive processes.

reproducibility of a product turned out by a process. A process is in a state of statistical control in the absence of assignable (specific and identifiable) causes of variation.[2]

Have all the major variables in resist processing been identified? Hopefully, I have pointed out the major variables in each of the prior chapters. Defects are either of the additive type (opaque or extra material) or of the subtractive type (missing material, i.e., pinholes). From a process viewpoint, we expect the resist material to be consistent in performance with a minimum of material concerns. The resist material should be reproducible, stable (prior to use), and of high purity. For all device levels, a sole resist (except for different thicknesses) would reduce errors in processing and prompt earlier diagnosis of problems (by shortening the learning curve of process definition). Secondly, the process equipment and process controls could be operated around a smaller, more consistent data base of material characteristics.

The most critical choice in resist processing is the selection of an exposure tool and the marriage of the resist with the particular exposure tool. Over 50 commercial resists and 100 resist formulations have been marketed or disclosed over the last 10 years. At least five different types of exposure tools have been marketed. In

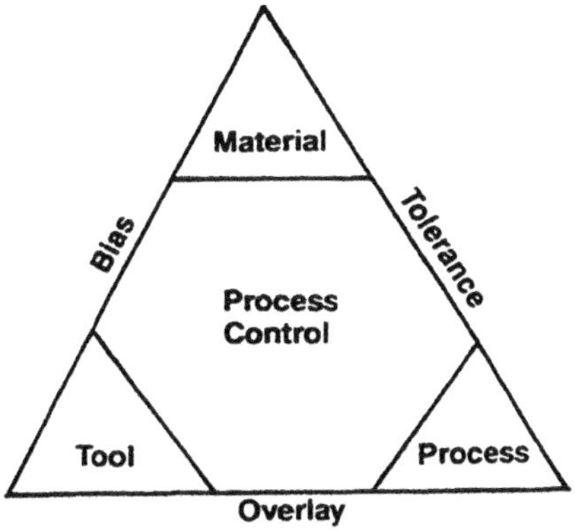

FIGURE 15-1-7. Interaction of resist materials, processes, and process equipment in lithography to determine the degree of process control.

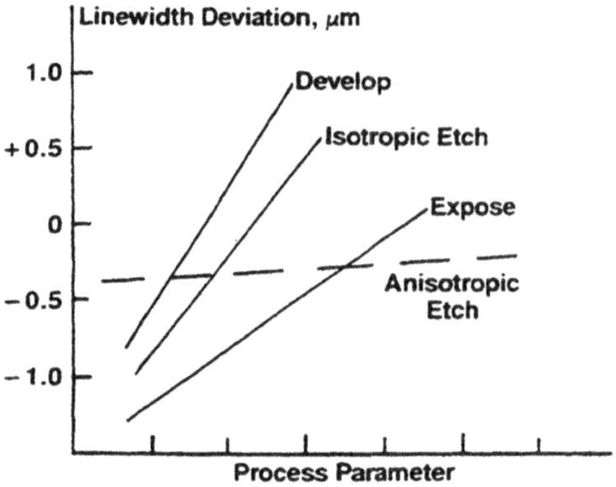

FIGURE 15-1-8. Critical process parameters for each lithographic step. The steeper the slope, the greater is the degree of process control.

the midst of the rapid technical improvisation, process definition and control are very difficult tasks, unless commitments are made for the long-term use of one resist with a specific exposure tool or dry etch tool. In most cases, the resist and associated processes have to be tuned to each specific exposure tool characteristic, such as MTF of the lens or edge slope of the beam. Optical exposure models such as SAMPLE and PROLITH assist in the design of optimum exposure controls.[12]

Three major steps or factors determine the width of a semiconductor pattern: the exposure tool (Chapters 2, 8, and 9), the developer and develop process (Chapter 10), etch tool (Chapter 13), and/or lift-off process (Chapter 12). A change in exposure and, moreover, development brings about a steep proportionate change in linewidth (Fig. 15-1-8).

The basic production objectives of resist processing and process controls are to produce and regulate:

- Linewidths: $\pm 1/4$ nominal size
- Overlay: $\pm 1/5$ linewidth
- Defect size: $< 1/4$ linewidth
- Defect density: $< 1/\text{cm}^2$
- Thickness uniformity: $\pm 5\%$

The process steps for lithography can be broadly classified into anisotropic and isotropic processes (Table 15-1-2). Accordingly, the extent of the control of the isotropic components of etching, developing, and exposure, determines the process

TABLE 15-1-2. Isotropic and Anisotropic Processes in Lithography

Isotropic	Anisotropic
Priming (adhesion promoter)	Spin coating
Exposure	Reactive ion etch
Wet develop	Lift-off
Plasma etch	Multilayer resists
Film defects	Chemisorption
Temperature	
Pressure	
Chemical concentrations	

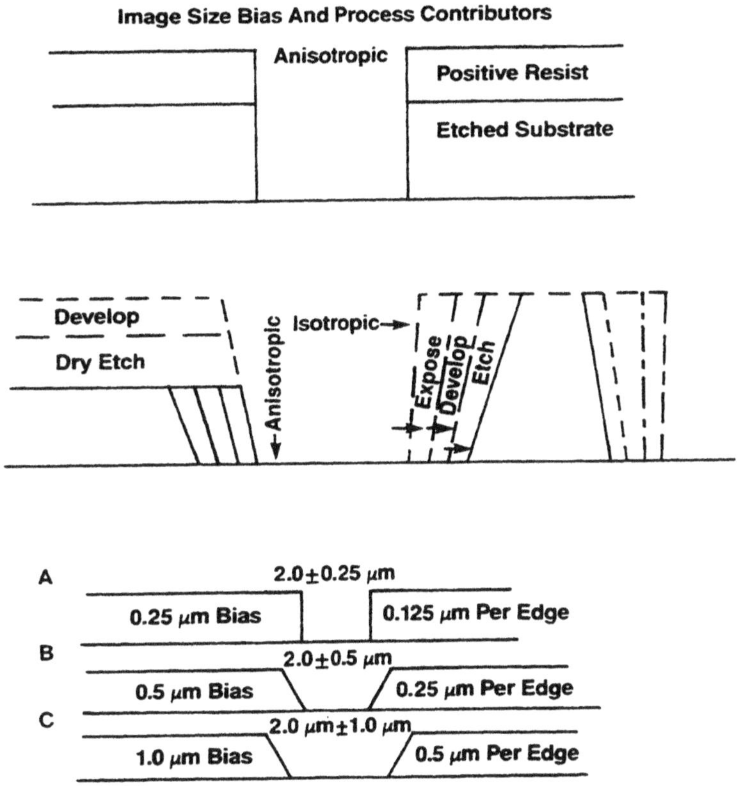

FIGURE 15-1-9. Image biases in lithography for process ground rules which require the final etched image bias to be within (A) 12% of nominal, (B) 25% of nominal, and (C) 50% of nominal. The isotropic contributions of exposure, development, and etching determine the final image biases.

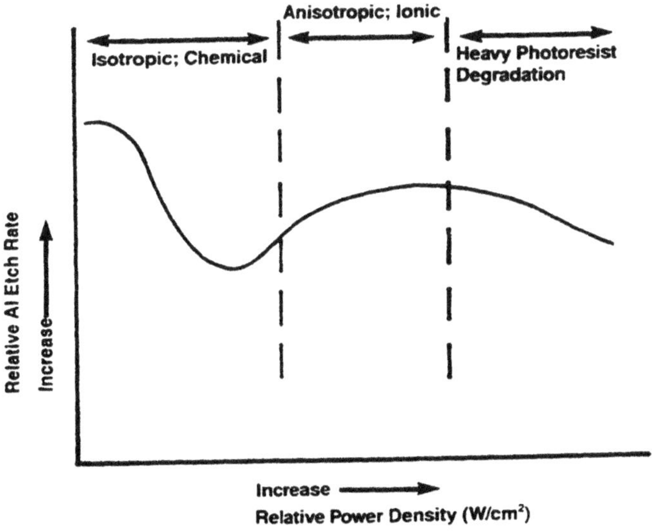

FIGURE 15-1-10. In gas plasma etching, the process changes from isotropic etching to anisotropic etching as the power increases. From Ref. 3, reproduced by permission of *Solid State Technology*, published by Technical Publishing, a company of Dun and Bradstreet.

bias of defining a specific linewidth and profile. Figure 15-1-9 illustrates the profiles of line etch biases of ±0.25, ±0.50, and ±1.0 μm, respectively. The *biases* of *today* are the *linewidths* of *tomorrow*.

Keeping a lithographic process under controls involves deciphering the interaction of several dependent variables. An etching process, for example, may suddenly interconvert from isotropic to anisotropic, such as in reactive ion etching (Fig. 15-1-10), by changing the power and ion concentration. Various degrees of process control are necessary to meet the design ground rules (Fig. 15-1-11) of

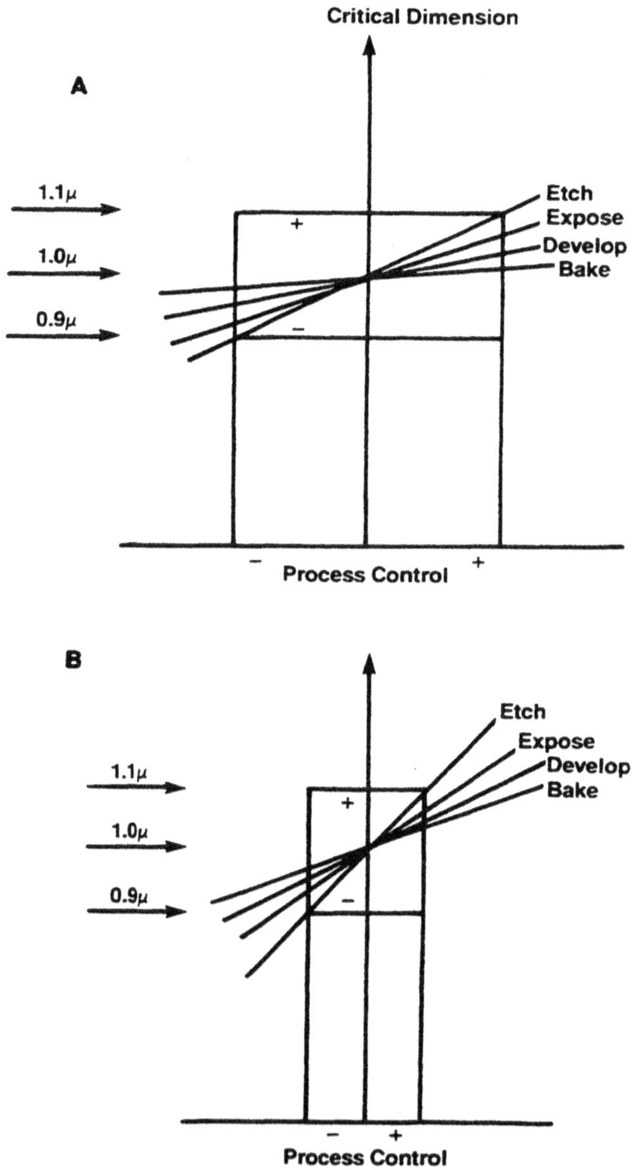

FIGURE 15-1-11. Examples of (A) wide process latitude to achieve 0.1-μm linewidth control; (B) tight process control necessary to achieve 0.1-μm control. Adapted from Ref. 63.

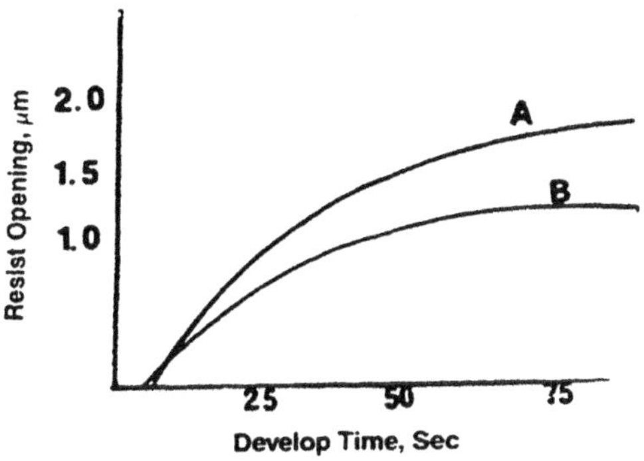

FIGURE 15-1-12. Simulations of openings in DQN resist in projection printing with (A) a numerical aperture of 0.28 and (B) an aperture of 0.35. From Ref. 4, reprinted with permission of *Solid State Technology*, published by Technical Publishing, a company of Dun and Bradstreet.

critical device dimensions. In case a, the process has a wider etch and exposure latitude for meeting 1.0-μm ground rules than in case b which requires tighter controls. Exposure latitude can be widened by a trade-off of throughput for the field size of exposure. In this case, the diffraction limit of a lens may be reduced by a higher numerical aperture and consequently, a smaller field size of exposure. At a larger line size (Fig. 15-1-12), the linewidth becomes invariant to development. As a precautionary point, specifying linewidth alone, does not account for the shape of the resist or for any loss of resist in the unexposed (positive resist) or exposed negative resist (Fig. 15-1-13). The loss of resist or incorrect profile may be fatal, such as in RIE processes over topography. Insufficient protection is afforded by the resist which loses more than 25% of its original thickness during RIE.

The thickness of the resist, the underlying topography, and the thickness and uniformity of the film to be etched, determine the degree of isotropy of a process or pattern. The thinner the etchable substrate, the smaller is the lateral etch deviation. Another concern with processing is the position of the chip or line on the wafer. Most lithographic processes show a radial dependency on yield (Fig. 15-1-14).

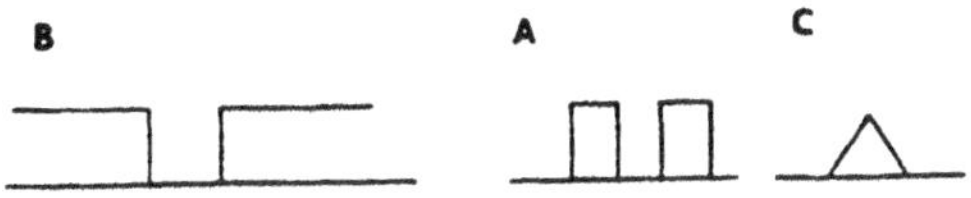

FIGURE 15-1-13. Three examples of resolution. (A) Equal lines and spaces in a positive or negative resist. (B) Unexposed space in a negative resists or exposed line in a positive resist. (C) Unexposed line in a positive resist or an exposed space of a negative resist but with film loss of original cast film thickness. (Case C would only be useful for wet etching.)

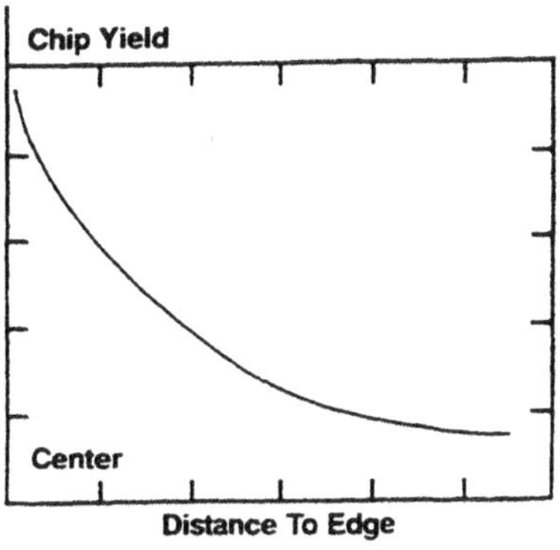

FIGURE 15-1-14. Radial dependency on position of chips in semiconductor wafer. Yields are lower at the outer edges.

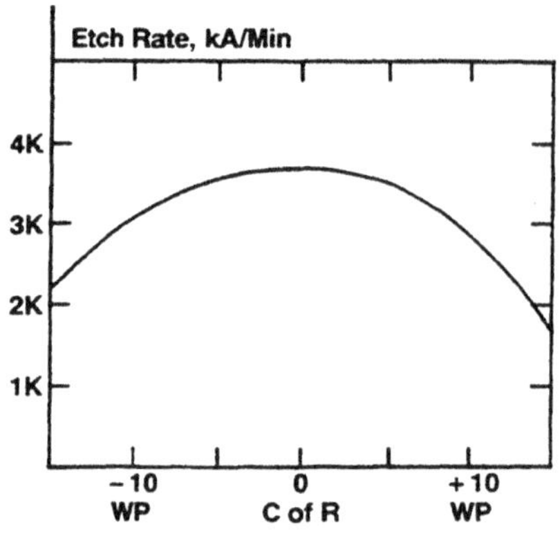

FIGURE 15-1-15. Etch uniformity of wafers positioned in a parallel plate reactor at center of reactor (C of R) and for wafer positioned radially toward outer edge of reactor (WP).

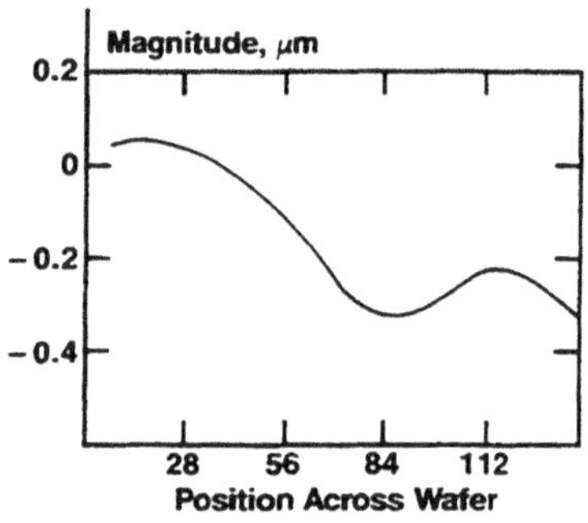

FIGURE 15-1-16. Linewidth deviation measured across processed wafer by electrical resistor method. From Ref. 8, copyright 1982 and reproduced by permission of the Electrochemical Society Inc.

Good chips can usually be found in the center of a wafer.[71,72] Defects predominate near the edges of masks. Batch processing in radial etchers or evaporators also tends to have a Gaussian distribution of nonuniformity (Fig. 15-1-15) which increases near the edges of a wafer or in a wafer positioned at the outer edge of the batch reactor plate. Overall, the linewidths across a wafer tend to vary in a nonlinear fashion (Fig. 15-1-16).

In addition to satisfying the basic linewidth requirements, process controls are necessary for maintaining sufficient throughput at lower processing costs. Resist thickness variations, for example, in excess of ± 100 Å can seriously slow down the throughput of a step-and-repeat alignment tool (Table 15-1-3).[10]

TABLE 15-1-3. Throughput of an Optical Stepper and Resist Thickness[a]

Thickness (Å)	Wafers/hr
13,000	25
13,100	35
13,200	40
13,300	29

[a] From Ref. 10.

15-2. LITHOGRAPHIC PROCESS AUTOMATION

Automation of fabrication processes is the dominant contributor to improved device yields (Fig. 15-2-1). Cleaner rooms and particulate-free operations are the next major yield enhancers.[78] Accompanying the successful automation of a lithographic process line are the necessary on- and in-line process monitors. Batch monitors tend to compensate for substrate thickness variations by greatly overetching or overdeveloping. A single-wafer process monitors the correct process conditions to come closer to the truer linewidth value (Fig. 15-2-2). Interactive process controls on the equipment, such as light intensity modulators on the exposure tool, and the dynamic power and pressure monitored by optical emission in a plasma etcher, are necessary to reduce the number of process variables and ensure process reproducibility. Since in many cases one is dealing with both dependent and independent variables, computer microprocessor controls and computer modeling are important in controlling and understanding process variables. Computer modeling, such as SAMPLE[12] for imaging and etching, grew out of studies of standing waves and *e*-beam proximity effects.[13] These models can simulate the resultant resist profiles and the dependent exposure variables of the coherence, the aberrations, and the modulation of optical tools. Etching,[75] *e*-beam exposure, and all process steps have been modeled.[14,15] The separation of the process variables, from resist and tool variations, is useful in diagnosing processes which exhibit large variations.

Once process controls are established, automation is possible on many levels (Fig. 15-2-3). Automation may be regarded as one means of increasing productivity and yield. Moreover, automation of the exposure tool, the alignment step and reticle changes are the most critical needs for productivity and overlay improvements. In-line and on-line computer control will improve production control, eliminate operator errors, and reduce operator contamination. Automation can be the simple interconnection of all process tools with automatic wafer handlers. Higher degrees of automation include microprocessor control[16] systems

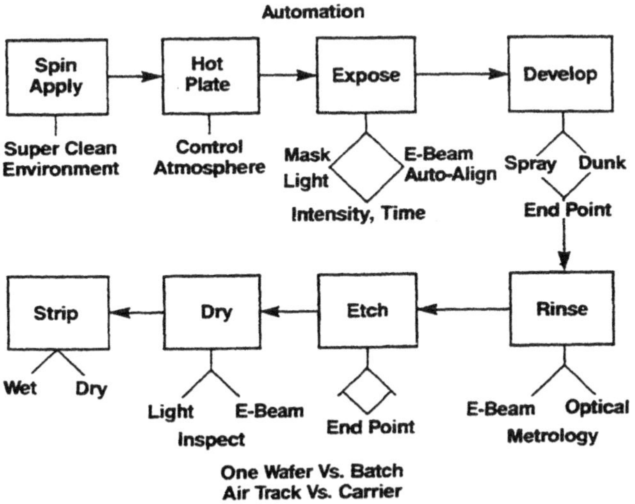

FIGURE 15-2-1. Schematic of an automated lithography process line.

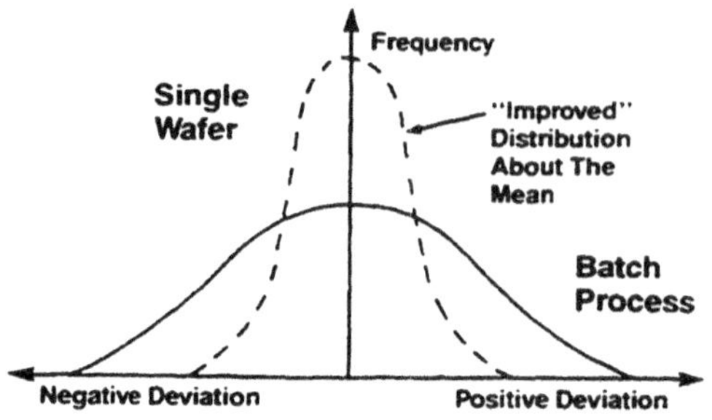

FIGURE 15-2-2. Single-wafer process control versus batch process deviation. Courtesy of H. Denk, IBM.

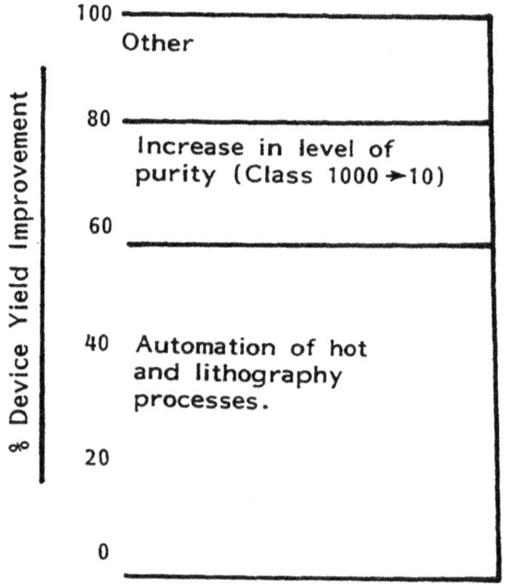

FIGURE 15-2-3. Factors contributing to improved yield. From Ref. 78.

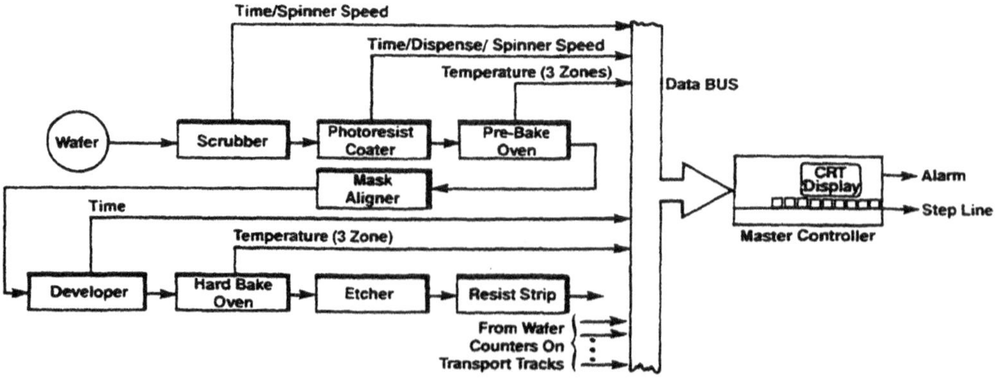

FIGURE 15-2-4. Microprocessor-controlled resist processing line. From Ref. 16, reprinted with permission of *Solid State Technology*, published by Technical Publishing, a company of Dun and Bradstreet.

TABLE 15-2-1. Status of Total Semiconductor Processing Including Cassette-to-Cassette Transfer (C-to-C) Assessment[a]

Process	Equipment	Batch (B) or single (S)	Status of automation[b]			Material handling	Technical restrictions	Cost performance
			C-to-C	Process	Monitoring			
Wet chemical treatment	Cleaning Wet etching	B B(S)	 △	○ △	 X	Transfer PFA-quartz	Endpoint detection	Good
Oxidation and annealing diffusion	Electric furnaces	B	△	○	X	Transfer PFA-quartz	Essentially no problem	Lower power consumption required
Ion implantation	Medium current High current	S B	○ ○	○ ○	○ ○	Loading/unloading mechanism	Uniformity	Not good—higher throughput required
Thin film deposition	LPCVD Atmospheric CVD Plasma CVD Epitaxy Sputtercoater	B B(S) B B B(S)	△ △ △ △ △	○ ○ ○ ○ ○	X X X X ○	Loading/unloading to quartz cassette Transfer mechanism Transfer mechanism Transfer mechanism	Figure conformability Uniformity	Not good—higher throughput required
Lithography	Coater/developer UV X ray Electron beam	S S S S	○ ○ ○ ○	○ ○ ○ ○	X ○ ○ ○		Mix-matching, resolution, and precise alignment	Not good—higher throughput required
Etching	Dry etching Ashing	B(S) B	○ ○	○ ○	○ ○	Loading/unloading to hexode Transfer PFA-quartz	Endpoint detection, selectivity, uniformity, precise gas flow rate control	Not good—higher throughput required

[a] From Ref. 19.
[b] ○, currently available; △, technically feasible; X, not feasible.

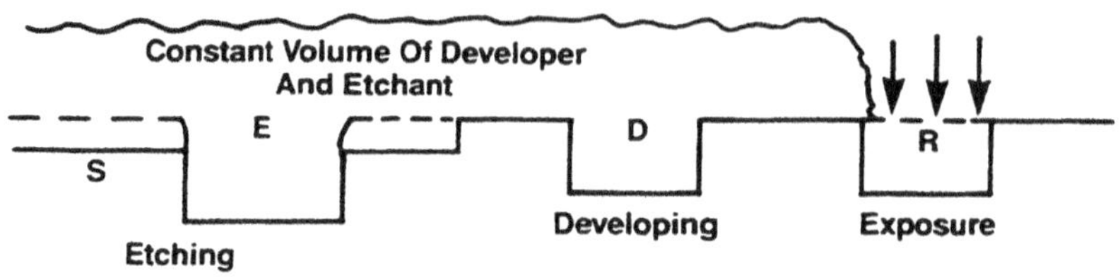

FIGURE 15-2-5. Schematic of a constant concentration (moles) of reactant at the surface of a resist or wafer during the chemical phases of resist processing (E, etchant; D, developer; S, substrate; R, resist).

(Fig. 15-2-4), air tracks,[79] and automatic linewidth[17] and defect measurement tools.[18] The status of automated lithographic tools is shown in Table 15-2-1.

The high capital cost of exposure tools must be offset by high throughput of > 50 wafers/hr.[19] In process control systems, the development endpoint monitor, or automatic thickness measurement after spin coating, have not been implemented on commercial coat/develop tools. For many years, batch processing has been the standard of the semiconductor industry. The batch processes were based on wet processing with wide process tolerances ($\pm 1\ \mu$m). The major process controls available to a process engineer are the three "T"s, and the concentration of reactants (M). The three "T"s are:

- Thickness (Z)
- Time (t)
- Temperature (T)

Chemical reactions like to be run with constant (M) moles of reactant at isothermal (T) conditions. The degree of thermal control depends on the activation energy for the process, ranging from 5 kcal/mole for diffusion-controlled reactions to as high as 40 kcal/mole for the rate-controlled decomposition of a sensitizer in the high postbake of diazoquinones (Chapter 11). The moles of chemical reactants should be as constant as possible. These include a constant volume of exposed resist (R), developer (D), etchant (E), and substrate to be etched (S) (Fig. 15-2-5).

Process control is concerned with the intrinsic (on-line) and external (off-line) parameters which require monitors, feedback, or stabilizing (isothermal) conditions. Since defects and yields are proportionate to the number of lithographic levels, devices such as bipolar may require more extensive process monitors and defect analyzers to control the level of yields, than in the production of FET memory circuits.

We will now examine the process controls for each related resist step with major emphasis on the material and "process" aspects. The tool–equipment aspects of process control are akin to many internal factors of commercial or in-house configurations and will be dependent on the manufacturer's particular recommendations.

15-3. RESIST MATERIAL CONTROLS

Resist material controls are concerned with the integrity of the as-received resist material and associated chemicals, and the maintenance of a steady-state concentration during processing. The latter factors are dependent upon the concen-

TABLE 15-3-1. Chemical Criteria and Controlling Factors in Lithography

Chemical criterion	Controlling factor
Developer concentration	Concentration of additives (surface active), flow rates, evaporation, CO_2 pickup, batch exhaustion, dilution (errors), normality
Etchant concentration	Plasma loading effects (flow rate, surface area, power), diffusion controlled, batch exhaustion
Reaction rates	Diffusion rates, temperature, chemisorption, ion flux, product removal, exposure dose, particle size

tration of reactants and their consumption in diffusion- or reaction rate-controlled (Table 15-3-1) reactions. Foreign contamination has to be eliminated by proper filtration, storage, and dispensing systems.

In material-oriented processing, batch-to-batch wafer tolerances have not been required to be reduced to $\pm 0.5\ \mu m$ or lower. Larger wafers have reduced cost, but batch processing can be a double-edged sword.[16] Small processing errors can ruin a whole batch of wafers valued at $1000 or more. If the temperature of a developer fluctuates beyond limits and remains unnoticed over an 8-hr shift, thousands of wafers may have to be reworked. In-line endpoint monitors safeguard against temperature or depletion effects. Concentration effects, such as the depletion of the developer or etchant (loading effect), should be avoided as they negate steady-state kinetics. The buildup of product deposits on the walls of a plasma etcher can lead to outgassing and alteration of the rates of etching of the next batch of wafers.[11] For a long shelf life, an excess of reactants guarantees reproducible rates, until replacement is necessary after the consumption by many batches of wafers.

Monitoring of the chemical reactions that occur in developing and etching is essential for reproducing, interactively, the rate and endpoint by:

- Correcting for over- or underexposure
- Resist thickness variations
- Prebake variations
- Etchant (plasma) depletion
- Substrate thickness nonuniformity
- Etch stop
- Temperature fluctuations

The classic example of concentration effect is the loading effect in plasma etching, where the rate is inversely proportional to the number of wafers,[22]

$$\text{Rate} = (\text{Area to be etched})^{-1} \qquad (15\text{-}3\text{-}1)$$

For developing with surface-active agents in an alkaline-based system of the DQN resist, a remarkable increase in contrast (γ) occurs with spray development (Fig. 15-3-1).[23] The contrast of the resist approaches 10 with spray developing. In

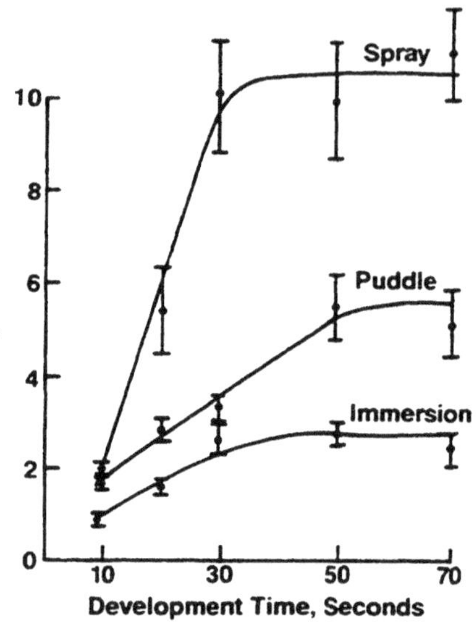

FIGURE 15-3-1. Contrast versus development time for three methods of development of a positive DQN resist. From Ref. 23, courtesy of SPIE.

batch immersion, the developer surface-active component is consumed after a few wafers are processed. In spray development, the surface-active component is constantly supplied (at ppm level) to freshly complex the unexposed resist region and inhibit its dissolution. A wide process latitude with low sensitivity to overexposure or overdevelopment are the process benefits of this unique developer (concentration-dependent system).

Most developer setups operate in the batch mode and should be insensitive to exhaustion or agitation effects. In general, the process concentrations of reactants are usually in great excess and only the rates of removal of the substrate have to be monitored and not the consumption of developer or liquid etchant. However, spray development reduces resist redeposition and should be considered for residue sensitive series circuits and metal interconnections sensitive to opens in lift-off or shorts in etched metals.

The integrity of the resist materials is based on the quality analytical and functional evaluation (Chapter 2) (Table 15-3-2) of incoming lots. Filtration of the dispensed resist (Chapter 6) should be as fine as one-fifth the resist film thickness ($< 0.2\ \mu m$) to safeguard against particulates. On- or off-line tests of resist assist in reducing or controlling material variations during processing (Table 15-3-3).

The testing of the developer is as important as the incoming evaluation of the resist. Confusion exists in the industry, since one alkaline-type developer may be used, even with different resist types. The developer labeling is not synonymous with the resist trade name. The correct dilution is sometimes a problem, since the final end-use composition, such as normality, is usually unspecified. Testing of diazoquinone developers would involve titration against a "standard" grade of (titered) acid and the measurement of the resist dissolution rates with a lot of previously qualified resist. In the case of a positive resist, such dose driven tests are:

- R—exposed dissolution rate
- R_0—unexposed dissolution rate
- Contrast (γ) and resist profile
- Resolution test on exposure tool

TABLE 15-3-2. Analytical and Functional Tests of Resists

Analytical	Functional
Molecular weight	R, R_0
Dispersity	Striations
Percent sensitizer	Pinholes (environmental)
Solid content	Melt point
Solvent content	Adhesion (surface dependence)
Viscosity	Plasma erosion (tool dependence)
Inorganic impurities	Resolution (tool dependence)
Percent water	Stripping (tool and process development)
Filterability	Contrast[80]
Gel or particle count	Development time
UV absorption	Reciprocity failure
Shelf life	Thickness uniformity
IR, NMR, Mass spectroscopy	
Gel permeation chromatography	
Thermal (T_g), DTA	

The laser reflective endpoint detect and film thinning monitor is an invaluable aid in quality control when coupled with a resist (SAMPLE) modeling. A monochromatic exposure source should be used and the dose accurately integrated. Simultaneously, the unexposed thinning rate or endpoint time for a 1.0-μm-thick film is a good secondary test of the resist consistency.

Supplementary polymers such as dyed antireflective layers, contrast enhancement layers, lift-off underlayers, multilayer resist barrier and planarizing layers, require the quality control of thickness, T_g, and viscosity. In solid cast resist films,

TABLE 15-3-3. Material Process Control

Process step	Off-line	On-line
Inventory	Handling Storage Disposal Toxicity	Dispensing Disposal Lighting Replenish
Priming	Assay Contact angle	Humidity Dehydrate bake
Apply	Viscosity	Thickness Humidity
Prebake/postbake	T_g of resist	Atmosphere
Expose	Quantum yield "G" value Extinction coefficient	Humidity Atmosphere
Develop/rinse	Normality Composition	R, R_0 endpoint pH, conductivity
Etchant	Concentration pH	Emission (gas etch) Mass spectroscopy
Stripper	Composition pH	Emission (gas)

the resist coatings on chrome plates or unprocessed wafers may deteriorate upon storage. The storage of DQNs in dry atmospheres usually requires conditioning at 35% humidity, to reabsorb the necessary moisture for the ketene conversion to the indene acid and for reproducible development rates. Films stored at low humidities dissolve faster than films stored at high humidities.

The dispensing of resists has been improved with pressurized containers with inner transfer tubes, instead of directly pouring from open glass bottles. Point-of-use filtration may also be useful to safeguard against particulate contaminants. In-line liquid particle counters watch out for potential residues depositing from unfiltered developer tanks.

15-4. RESIST PROCESSING CONTROLS

Resist process controls involve the four major variables of time (t), temperature (T), radiation flux (I), and concentration (previously discussed). Off-line and on-line controls are summarized in Table 15-4-1. For most liquid and gas processing, temperature (T) should be kept under control with isothermal conditions. The activation energy (1 eV = 24 kcal/mole or a 4× rate increase for a 10°C rise) ranges from 3 to 40 kcal/mole in resist processing (Table 15-4-2). The development and wet etch tanks should be controlled to within ±0.2°C.

15-4-1. Exposure Controls

Test chips, for monitoring the expose–develop endpoint monitor with step wedge or diffraction grating, are vital for exposure monitor and control. Ideally, one would like to monitor the latent resist image prior to development, such as on a scanning reflectometer. Some resists such as positive diazoquinone resists and contrast enhance layers bleach upon exposure and the increase in transmission can be used to detect a latent image. Precise optics are required to sense the exposed regions and the alignment marks.

TABLE 15-4-1. Resist Processing Controls

Step	On-line	Off-line	Critical
Preclean	Particle count	Surface analysis	Yes
Dehydrate bake	Time, temperature	Vacuum	No
Prime	t, T	Chemicals	Yes
Spin apply	t, acceleration	T, humidity	Yes
Prebake	t, T	Z, pinholes	No[a]
Expose	t	Intensity, T	Yes
Develop	t, T endpoint	Chemicals	Yes
	Agitation	SEM, profile	
Postbake	t, T	Atmosphere	No[b]
Etch	t, T, power	Chemicals	Yes
	Pressure, flow endpoint	Linewidth	
Strip	t, T, power, pressure, flow	Residues	Yes

[a] If executed properly, should not be critical.
[b] Unless melting occurs.

TABLE 15-4-2. Activation Energy of Resist Processing and Degree of Control Required

Process step	Activation energy (kcal/mole)	Rate change/1 °C (%)
1. Prime (silanes)	24	40
2. Apply	5 (viscosity)	1.0 of thickness
3. Prebake	6 (diffusion)	Negligible
4. Expose	15	Minor systems[a]
5. Develop	10–15	20[b]
6. Postbake	30–40 (decompose)	50
7. Etch		
Wet	15	30
Dry	6	10
8. Strip		
Wet	12	15
Dry	6	10

[a] Electron beam may heat resist on glass to > 100 °C.
[b] Develop control to ±0.1 °C is recommended.

The choice of expose and develop conditions is sometimes a compromise between the throughput on reflective scanners and the process latitude. Exposure with an MTF of 0.6 leaves little process latitude and even allows for a 10–20% thinning of unexposed positive resist, especially at spaces equivalent to thickness (Fig. 15-4-1).[23] Overexposure or overdevelopment will rapidly deteriorate the profile of the unexposed space.

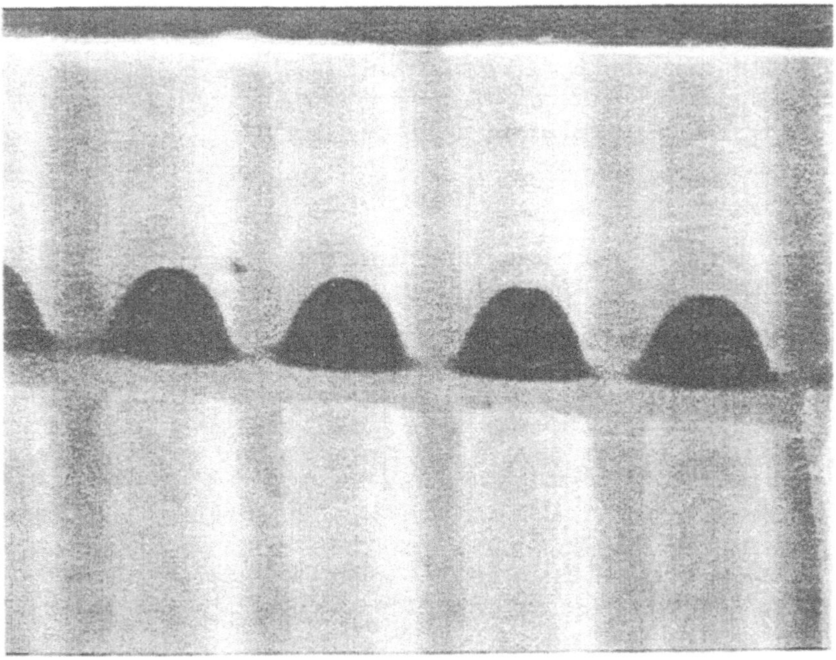

FIGURE 15-4-1. Line and space limit in photoresist (1.25 μm) produced in a reflective scanner. Courtesy of A. Scaduto, IBM.

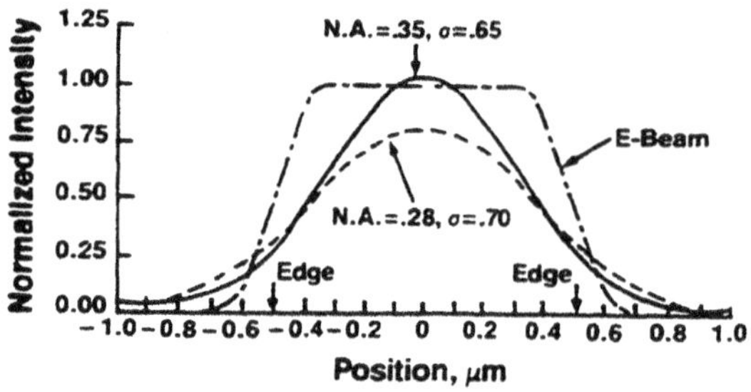

FIGURE 15-4-2. Incident beam shape of an *e*-beam and optical beam of two different numerical apertures and degrees of coherency. From Ref. 4, reproduced by permission of *Solid State Technology*, published by Technical Publishing, a company of Dun and Bradstreet.

Resist profiles are a strong function of the shape of the incident beam. The incident illumination profile for a 1-μm-wide *e*-beam and two optical beams of normal aperture are shown in Fig. 15-4-2.(4) Viewed upside, the energy deposited closely matches the profile in a positive resist. The edge slope of the resist image is a function of the dose E, and the rate of removal R in developer. From the contrast curve of γ (Fig. 15-4-3), the linewidth increases by W,

$$W = Z/\tan\theta = \gamma + Z_0/\text{constant} \tag{15-4-1}$$

where γ is the absorption coefficient of the resist, and θ is the incident angle of contrast at E_0 and initial thickness, Z_0. An optical absorption coefficient of 0.4 provides the most uniform deposition of light throughout a 1.0-μm-thick film of resist. The edge slope is given by

$$dR/dZ = (dR/dE)(dE/dZ) \tag{15-4-2}$$

As the nominal line (W) width decreases, $\Delta W/W$ should remain constant. The exposure term dE/dZ is the intensity profile of the image determined by the wavelength of exposure, the numerical aperture (d), the depth of focus, and uniformity of illumination across the wafer using partially coherent radiation.(25) The dR/dZ term is proportional to the energy deposited,(14)

$$R = (1 + kE_0)^a \tag{15-4-3}$$

where a is the solubility development parameter. The edge slope has been shown to be related to tool and $(\beta + \alpha t)$ contrast by(25)

$$\frac{dR}{dZ} = \frac{\text{NA}}{\lambda(\beta + \alpha t)\,E_0}\left[1 - k\left(\frac{\Delta z(\text{NA})^2}{\lambda}\right)\right]^2 \tag{15-4-4}$$

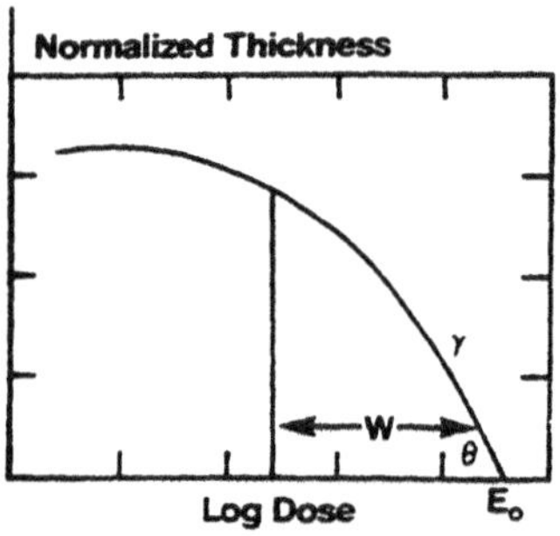

FIGURE 15-4-3. Contrast and edge slope effect on linewidth (process) profile and image widening (W) increasing with dose. From Ref. 25, copyright 1980 and courtesy of SPE.

TABLE 15-4-3. Exposure Control in Positive DQN Resists[a]

Resist	Thickness (μm)	Percent exposure change for 0.2 μm	γ
AZ-2400	0.25	—	2.6
AZ-2400	0.5	22	2.4
AZ-2400	1.0	14	2.1
HPR-204	0.50	13	2.1
On MLR resist[b]	0.15	—	3.6
HPR-204	1.0	8	1.7

[a] From Ref. 25.
[b] MLR—multilayer resist.

Again, a high NA, thin (z) films, and shorter wavelengths improve the edge contrast. The percent exposure change needed to increase ΔW by 0.2 μm is tabulated (Table 15-4-3) as a gauge of linewidth control latitude. Thin films of resist are relatively invariant to overexposure–develop conditions. Multiple exposures with a step wedge on the same test site, can be used as a simultaneous monitor to simulate the cleanout of fine lines. An endpoint monitor, plus a fixed point beyond, can handle both linear dissolving and nonlinear dissolving (Fig. 15-4-4) resists by determining the setting of the development time for a particular line, or line and space combination. Nonlinear dissolution requires dose–depth correction factors which can be calculated by a SAMPLE model.[14]

In a conventional stepper, exposure at 435 nm of a 1.0-μm line requires about 1.5× more dose due to the falloff of the MTF of the projection lens (Fig. 15-4-5).[31] Because of the overexposure or overdevelopment, a 3.0-μm line will have a bias of ±0.3 μm. If the line were exposed at a normal shorter time (100 msec), the bias would be reduced to ±0.15 μm. Thin imaging resists in multilayer structures have wide process latitudes and tolerate overexposure or overdevelopment conditions. Small features of 1.0 μm, however, require 1.5× more exposure than a 2-μm window.

As the device dimension decreases into the submicrometer region, a higher NA lens will be used. A lens of aperture of 0.4 at 436 nm, has a depth of focus (F),

$$F = \pm \text{wavelength of exposure}/2(\text{NA})^2 \tag{15-4-5}$$

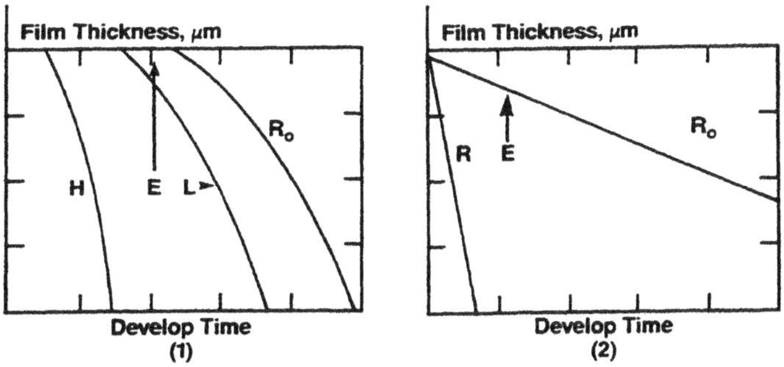

FIGURE 15-4-4. (1) Example of nonlinear dissolving resist at no dose (R_0), high dose (H), low dose (L), and endpoint (E). (2) Linear dissolving positive resist at high dose (R), no dose (R_0), and endpoint (E).

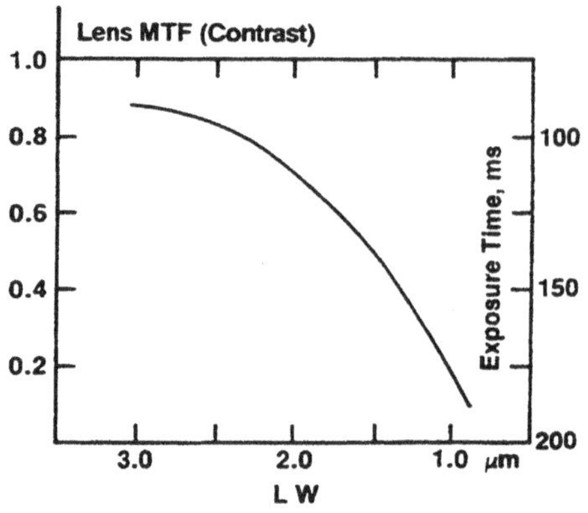

FIGURE 15-4-5. Exposure of DQN resist on a stepper (10 to 1). The effective MTF of the lens falls off as the linewidth decreases. From Ref. 31, copyright 1979 and courtesy of SPE.

of only 1.25 μm. The focus point may not be sufficient to cover the required resist thickness(21) and may well exceed the interplane wafer warpage.(20) Resists as thin as 0.5 μm in multilayer structures will be required and the wafer flatness to better than ±1 μm will be needed.(20,71,72) In Table 15-4-4, the modeling of optical projection lithography, by Chang and co-workers,(4) gives a good insight into the tool and process interaction. A process bias of ±0.2 μm can be expected.

Linewidths in a single resist of 1.0-μm width running over 0.5-μm-high steps, suffer from interference effects, which contribute a ±0.2-μm bias variations.(24) Moreover, striations in the resist are a persistent problem on silicon or chrome surfaces (Table 15-4-5) as they cause line notching and incomplete image cleanout. Striations in AZ-1350J on chrome mask caused image variations of ±0.2 μm.(26) Fortunately, striation-free resists based on added surfactants and improved casting solvents are now commercially available.

TABLE 15-4-4. Linewidth Tolerance in Optical Projection Lithography[a]

Exposure tool	NA = 0.28 σ = 0.70 λ = 436 nm	NA = 0.35 σ = 0.65 λ = 436 nm
Minimum feature	1 μm line/1 μm space 150 mJ/cm² 1:1 AZ developer 60 sec 300 Å plasma ashing	0.75 μm line/0.75 μm space 150 mJ/cm² 1:1 AZ developer 60 sec 300 Å plasma ashing
Nominal process bias	+0.30 μm	+0.20 μm
Linewidth tolerance component		
Dose (±5 mJ/cm²)	±0.03 μm	±0.03 μm
Development process (equivalent to ±5 sec dev. time)	±0.04 μm	±0.02 μm
Pattern sensitivity (isolated exposed lines, equal line/space array, two large exposed lines separated by a small resist space)	±0.10 μm	±0.07 μm
Defocusing	±0.05 μm (±2.5 μm defocusing on top of resist)	±0.05 μm (±1.5 μm defocusing on top of resist)
Resist/substrate light coupling variance	±0.21 μm	±0.21 μm

[a] From Ref. 4.

TABLE 15-4-5. Effect of Resist Striations on Resist Linewidths[a]

Resist	Striations	3σ, LW (μm)
AZ-1350J	Yes	±0.55
AZ-1470	Yes	±0.37
Hunt 204	No	±0.26

[a] From Ref. 26.

In the case of the exposure tool, several optical factors can affect linewidth control. The use of Mylar pellicles to defocus dirt particles on the photomask has been shown to cause loss of transmission and the generation of microscopic light scatter centers.(26) The distortion of 2-μm features on a Perkin Elmer III were:

Parameter	Variation (3σ, LW in μm)
Mask without pellicle	±0.31
Mask with pellicle	±0.42

In optical tools, incorrect changes of lamps and setup, in optical exposure have been noted to cause ±0.35-μm linewidth variations.(26)

The exposure and development process go hand in hand.(28) A low-contrast resist responds almost in direct proportion to the dose (scan setting), while a higher-contrast resist developer combination is tolerable of a ±10% dose (intensity × time) modulation (Fig. 15-4-6).

For *e*-beam exposure, the thickness of the resist has even a more dramatic effect, due to the rapid falloff of the beam MTF (Fig. 15-4-7). Backscattered electrons(29) scatter up to 4 μm from a 1-μm spot. The backscatter results in nonlinear resist development (Fig. 15-4-8). Likewise, thinner resist films reduce the proximity effect in *e*-beam exposure. Proximity effects dominate *e*-beam exposure and must be precisely corrected for accurate linewidth biases. A resist process window for *e*-beam exposure of DQN resists, using a dose-corrected algorithm (SPECTRE), is shown in Fig. 15-4-9.(14) In this case, a compromise in edge slope, throughput, and spot placement must be made to achieve the desired linewidth con-

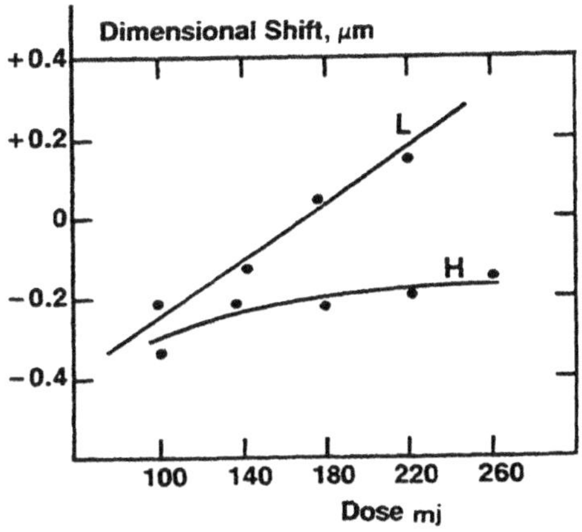

FIGURE 15-4-6. Linewidth changes in a DQN resist printed with a reflective scanner using a lower-contrast developer (L) and a higher-contrast developer (H).

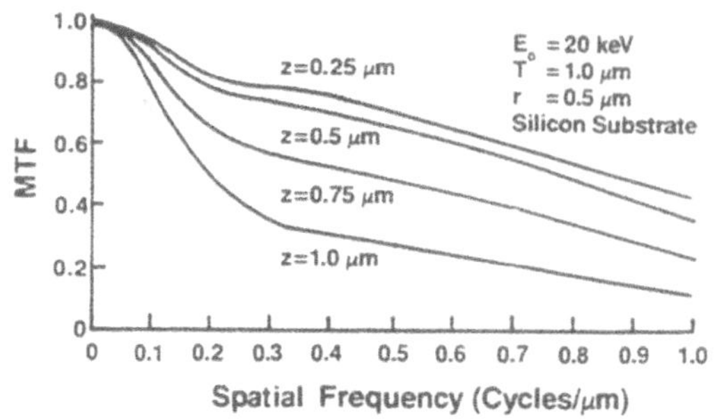

FIGURE 15-4-7. Enhanced MTF in *e*-beam resist with thinner films. (0.2 cycle/μm = 2.5 μm.) From Ref. 29, courtesy of SPSE.

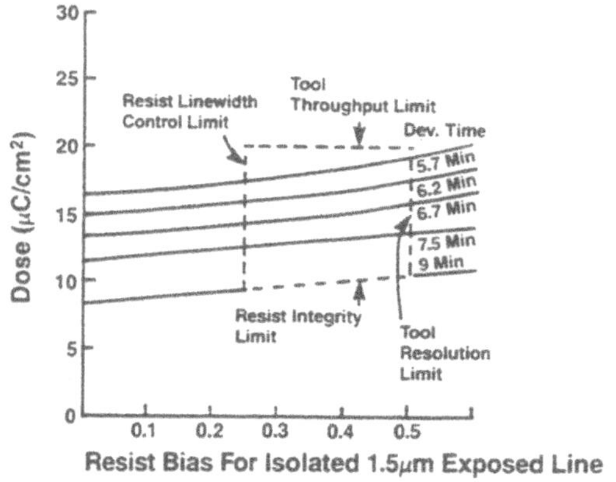

FIGURE 15-4-8. Resist process window for *e*-beam exposure and processing of a DQN resist. From Ref. 14, copyright 1981 and courtesy of IEEE.

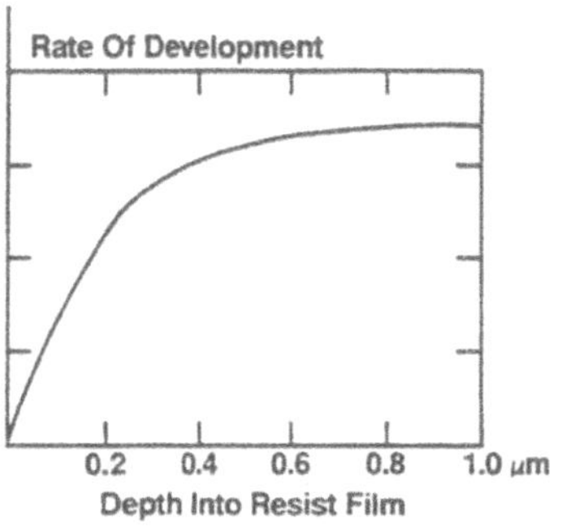

FIGURE 15-4-9. Rate of development into the depths of a film of positive resist exposed to electrons. The rate is slower at the top of the film due to the lower absorption of electrons and the lack of backscattered electrons. The latter are of lower energy and are more efficiently absorbed at the resist–substrate interface creating a faster rate of dissolution. Courtesy of N. Eib, IBM.

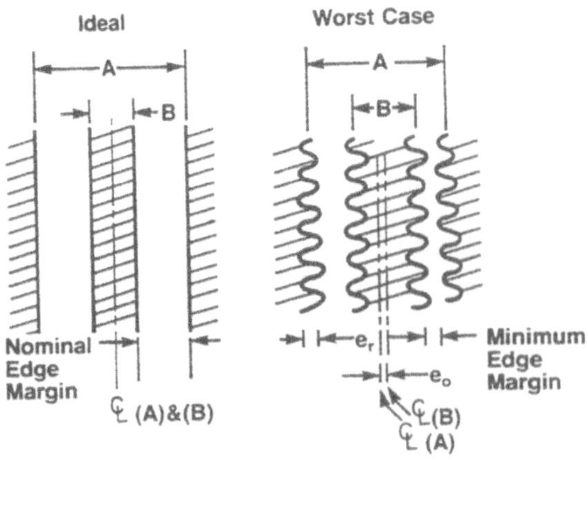

FIGURE 15-4-10. Edge margin reduction in *e*-beam 1× mask-making due to overlay errors and beam edge overlap (roughness). From Ref. 30, reprinted with permission of *Solid State Technology*, published by Technical Publishing, a company of Dun and Bradstreet.

TABLE 15-4-6. Linewidth Tolerance in a 1.5-μm Electron-Beam Direct Write Process[a]

Exposure tool	25 keV, 0.375 μm edge width/edge
Process	1 μm "nonlinear" resist over silicon 16.875 μC/cm² for 1.5 μm isolated exposed line
Develop process bias	+0.30 μm
Linewidth tolerance components	
Dose (±3%)	±0.10 μm
Beam edge width (±20%)	±0.05 μm
Development process (equivalent to ±0.1 min dev. time)	±0.5 μm
Resist thickness (±500 Å)	±0.15 μm
Residual proximity effect (SPECTRE, 1.5 μm edge framing), $\beta_f = 0.1$ μm, $\beta_b = 3.5$ μm, $\eta = 0.95$	±0.2 μm
Total linewidth tolerance	±0.30 μm

[a] From Ref. 4.

trol. A linewidth tolerance of ±0.3 μm was achieved for a 1.5-μm process (Table 15-4-6)[14] with a 0.375-μm beam edge slope. The development process contributes the largest bias (0.5 μm) and should be reduced by develop endpoint detect monitors. If a finer edge slope of 0.1–0.2 μm is used, a linewidth tolerance of the edge slope of ±0.1 to ±0.2 μm can be achieved (Table 15-4-7).[4]

The production of masks by *e*-beam is crucial to optical, X-ray, and ion-beam technology. Projection lithography requires an edge sharpness better than ±0.15 μm. Edge roughness and overlay in 1× masks made by *e*-beam have been

TABLE 15-4-7. Linewidth Tolerance in Electron-Beam Direct Write Lithography[a]

Exposure tool	25 keV 0.2 μm edge width	20 keV 0.1 μm edge width
Minimum feature	1 μm line/1 μm space	0.5 μm line/0.5 μm space
Process	1 μm nonlinear resist over Si 20 μC/cm² for 1 μm isolated exposed line 1:3.5 AZ-2401 developer; H_2O 5.3 min	0.50 μm nonlinear resist over Si 20 μC/cm² for 0.5 μm isolated exposed line 1.3:4 AZ-2401 developer; H_2O 3.3 min
Nominal process bias	+0.30 μm	+0.15 μm
Linewidth tolerance components		
Dose (±1 μC/cm²)	±0.10 μm	±0.05 μm
Development process (equivalent to ±0.1 min dev. time)	±0.05 μm	±0.03 μm
Residual proximity effect	±0.15 μm (SPECTRE, 1 μm edge sleeving, $\beta_f = 0.1$ μm, $\beta_b = 2.75$ μm, $\eta = 0.9$)	±0.07 μm (SPECTRE, 0.05 μm edge sleeving, $\beta_f = 0.1$ μm, $\beta_b = 2.50$ μm, $\eta = 0.7$)
Electron-beam edge width (EW)	±0.08 μm ΔEW = ±0.2 μm	±0.04 μm ΔEW = ±0.1 μm
Resist thickness	±0.04 μm	
Total linewidth tolerance	±0.21 μm	±0.11 μm

[a] From Ref. 4.

TABLE 15-4-8. Parameters of Electron-Beam Mask-Making[a]

Parameter	Sensitivity	± linewidth (μm) feature size contribution
Chrome thickness	1000 ± 100 Å	0.04
Resist type and thickness	3000 ± 100 Å (COP)	0.01
	3000 ± 200 Å (PBS)	
Beam current	±5%	0.03
Beam placement	Many parameters	0.01–1.0
Prebake	—	0.02
Cure time	COP 1-hr write, 0.5-hr cure	0.1
Development	PBS—humidity sensitive	0.03
Postbake	—	0.01
Descum	0.06 μm/min	0.02
Etch	Wet	0.03
		0.1 μm

[a] From Ref. 30.

reviewed by Cowan.(30) The edge margin reduction due to overlay error and edge roughness is shown in Fig. 15-4-10. The minimum edge margin (M) is seen to occur when A and B lines approach each other in size, where

$$M = A - B/2 - l_0 + l_e + l_r = <0.3\ \mu\text{m} \tag{15-4-6}$$

A summary of the parameters for *e*-beam mask-making can be found in Table 15-4-8.(30) Linewidth tolerances to ±0.1 μm should be possible in spite of poor contrast (< 1.0) and in resists which continue to polymerize after exposure (COP). Multilayer resists with a 0.1-μm-thick image layer will be needed to reduce the proximity corrections and *e*-beam problems, such as spot placement error, which is the largest contributor to noise in resist patterns.(15) Other contributions are:

- Beam defocus
- Beam tilt
- Dose spot variations
- Beam heating of the resist

Heating of the resist, by exposure of resist on chrome quartz photomask plates by *e*-beam dose > 2 μC/cm^2, can cause resist line scallops and linewidth differences across the plate.(35) Low-dose (high speed) resists are required to minimize extraneous line alterations.

Overall, the minimum usable linewidth has been suggested by Broers(32) to be 10× the standard edge deviation and that device linewidths must be able to tolerate the 3σ variation in linewidth. The production of chrome masks using optical steppers and plasma etching, achieves about ±0.2-μm linewidth edge control.(33) The measurement of resist thickness and the reflectivity of the chrome film on each plate are vital to production control.(33) Plasma etchers have been developed for etching whole plates with uniform etch rates of 220 ± 10 Å/min. For films of resist of 0.1- to 0.3-μm thickness on chrome, the best linewidth variance would be ±0.05 μm, while ±0.3 μm would be the worst case.

15-4-2. Development Process Controls

The rate of exposed resist[14] development is usually a power function of dose E,

$$R = R_0 + R_0(1 + kE)^a$$

where R is on the order of 5000–10,000 Å/min for DQN resists. Doubling of the dose increases R by 1.5–2.5×. Errors of 5–10% in exposure or develop time will produce a ±0.2-μm linewidth change. Resists with long develop time (> 3 min) will not be subject to time fluctuations. Resists with finite R_0 or $R/R_0 < 8$ will be subject to top surface image widening,

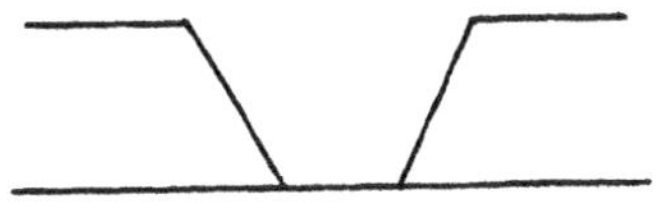

A typical DQN resist response upon development is shown in Fig. 15-4-11. Lands of the DQN resist disappear as the exposed lines widen with continued development.

The activation energy for development of DQN is on the order of 10–15 kcal/mole[73] and 24 kcal/mole for PMMA.[74] A 1°C change in temperature of the developer corresponds to a 40% change in rate or a ±0.2-μm linewidth variation. An unnoticed drop in temperature induces image residues, which require a longer development or removal by plasma ashing. Likewise, an unnoticed increase in temperature creates problems, since the thinnest resist with a ±5% thickness variation or overexposed (+5% of dose) will be overdeveloped by 40% or about ±0.2 μm for a 1°C rise. Thus, temperature control to ±0.1°C will maintain isothermal conditions and produce negligible linewidth fluctuations (±0.03 μm) for ⩾1.0-μm geometries.

Different DQN resists were reported by Frasch and Saremski[26] to produce linewidth fluctuations of ±0.7 μm on a Perkin Elmer III scanning printer (Table 15-4-9). Most of the variations were traced to striations in the resist film. Striations in the resist cause incomplete opening of a line and force one to overdevelop the film to clean out regions with localized deposits of sensitizer (Fig. 15-4-12).[27] Striation-free DQN resists are now commercially available.

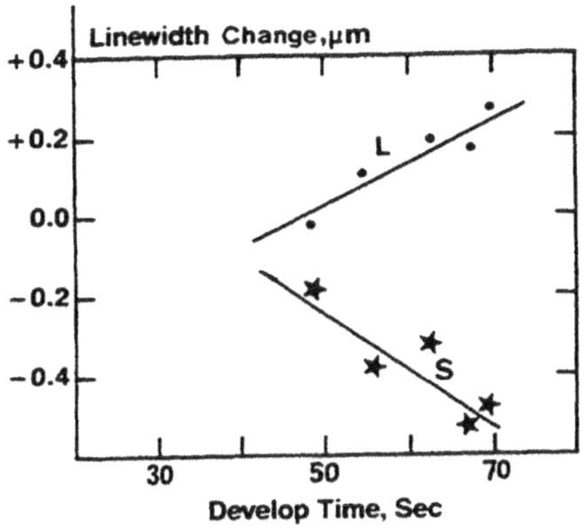

FIGURE 15-4-11. Typical response of a DQN resist to develop time, of an equal line and space configuration. The line L enlarges and the space S decreases from a nominal value.

TABLE 15-4-9. Chip-to-Chip Feature Size Variations with Different Resists[a]

Resist system	Developing technique	Chip-to-chip size variations (3σ) (μm)
AZ-1350J	AZ 1:1 developer, tank	±0.90
AZ-1470	AZ 1:1 developer, tank	±0.53
HPR-204	AZ 1:1 developer, tank	±0.60

[a] From Ref. 26.

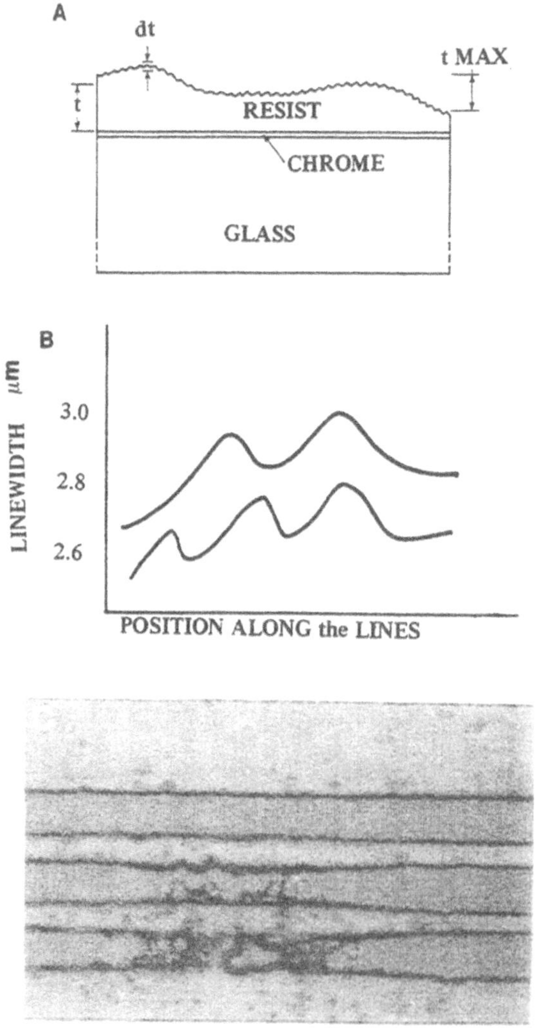

FIGURE 15-4-12. (A) Striations in AZ-1350 on chrome. (B) Impact of striations on lines printed perpendicular to striations. Both from Ref. 27, copyright 1982 and courtesy of SPSE. (C) Photo of lines printed in DQN with radial striations; lines are not fully open in striated areas due to higher sensitizer concentrations.

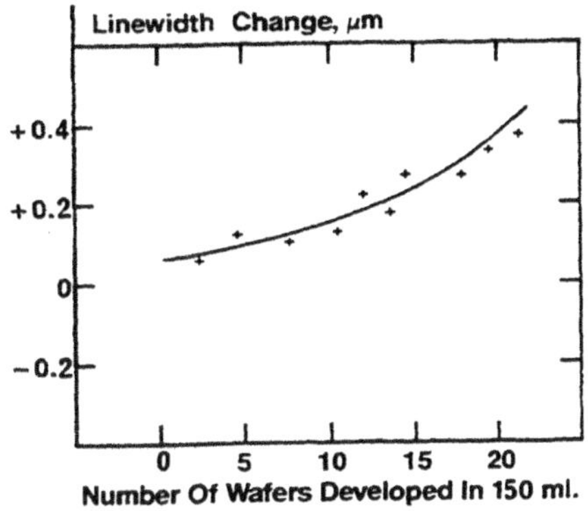

FIGURE 15-4-13. Linewidth changes in DQN resist as a function of the number of 100-mm wafers processed one at a time in 150 ml of developer.

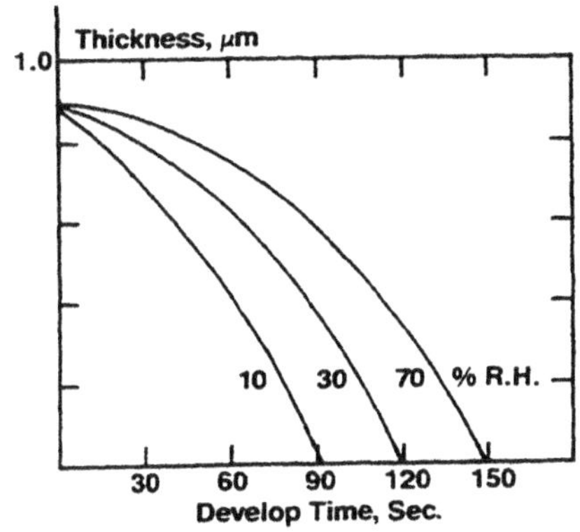

FIGURE 15-4-14. Develop time for AZ-1350J processed in lab with room relative humidities of 10, 30, and 70%.

After the exposure and resist considerations, the concentration of the developer should be monitored. The shelf life of alkaline developer deteriorates, if the liquid is exposed to air or in some cases (Fig. 15-4-13) to a loading effect (exhaustion of developer with increase in linewidth). A change of ±5% in developer normality changes the linewidth of a DQN resist by ±0.21 μm.[1] Rottmann[27] noted the variations given in Table 15-4-10 using AZ-1350J for chrome mask-making. Linewidth production of 1× photomasks requires better controls than in device fabrication since the required biases are a factor of 4× more stringent.

The processing of DQN in different humidities can alter the rates of development. Figure 15-4-14 shows a 1.0-μm film of AZ-1350J processed at 10, 30, and 70% relative humidity. The absorbed water dilutes the action of the developer and slows down the rate of removal. If the exposed wafer is acclimated at different

TABLE 15-4-10. Developer-Induced Linewidth Variations on Chrome with DQN[a]

Factor	Δ linewidth (μm)
Developer exhaustion per 8-hr use	±0.05
Delay of development after exposure	±0.03
Change of develop time of 1 sec/120 sec	±0.05

[a] From Ref. 27.

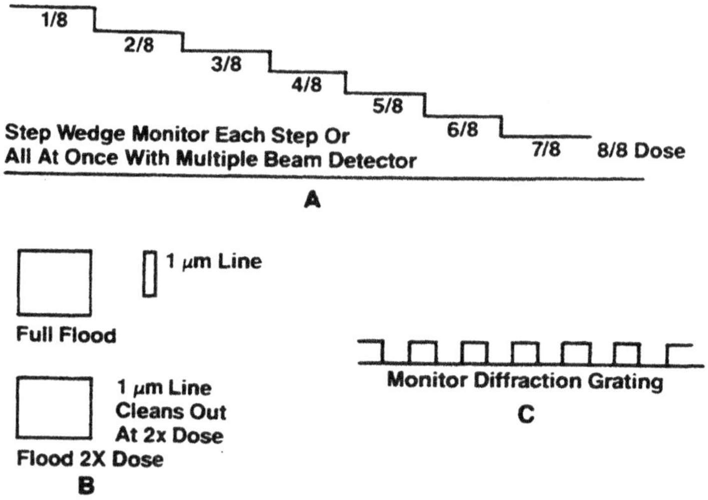

FIGURE 15-4-15. Various endpoint determination principles. (A) Step wedge monitor with multiple beam detector to measure finite depth into resist. (B) Flood expose monitor exposed at normal and higher dose to compensate for intraproximity effect in 1.0-μm line. (C) Monitor of diffraction grating to determine endpoint.

humidities and reprocessed, the rates of development are reversible. For batch processes and consistent development times, constant humidity should be maintained in the development sector.

If wafers are developed in batches, the fluctuations in rates of development and, hence, linewidth will remain undetectable. Development rate monitors are essential to process control (Chapter 10). The three basic patterns[36–38] consisting of a flood region, a corresponding dose wedge, and a diffraction grating, can be monitored (Fig. 15-4-15). Both the exposed and unexposed regions should be monitored. Because of the proximity effect in *e*-beam exposure or the diffraction distortion of images in optical exposure, a 1-cm^2 exposed area develops 50% faster than a 1-μm^2 pattern:

Area	Dose	Develop time
1 cm^2	75 mJ/cm^2	68 sec
1 μm^2	75 mJ/cm^2	139 sec

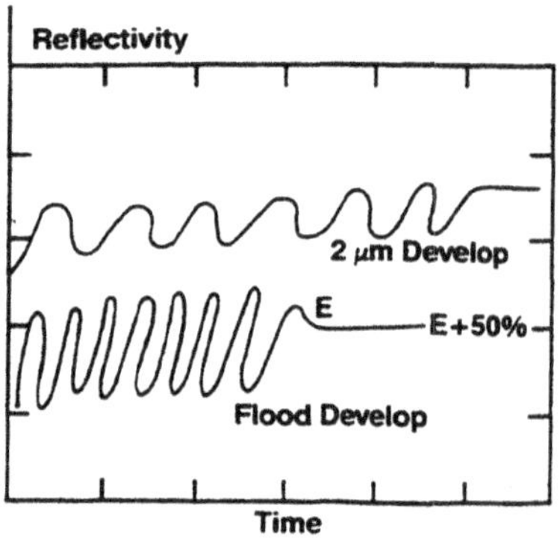

FIGURE 15-4-16. Reflectivity signal of 2-μm line and time to develop versus a flood exposed region endpoint. The endpoint to clean out small geometries is longer than the flood exposed region due to intraproximity effects.

If a flood exposed area is monitored by endpoint, at least a 50% longer develop time is required (Fig. 15-4-16). If a 1-cm^2 area is flood UV exposed at 48 mJ/cm^2, it will develop out at the same time (139 sec) as the 1-μm line exposed at 75 mJ/cm^2. Thus, a step wedge or decreased dose can be used to an exact endpoint to correct for the longer develop time required for < 2-μm features. In a production line, the automated developer endpoint control will boost yield.[88] Improved process control by a computer feedback loop will reduce the risk of:

- Incorrect dose
- Wrong developer
- Incorrect prebake conditions
- Resist thickness variations
- Developer temperature excursions
- Developer exhaustion

Since some optical exposure tools do not monitor the dosimetry of the source with a feedback loop, the resist itself serves as its own process monitor by measuring the rate of dissolution versus dose. The bleaching of DQN monitored by UV spectrophotometry can also be used as an internal dosimeter.

To sum up, Hryhorenko[1] assessed the contributions of all process steps with major emphasis on developer controls, for a DQN resist image developed in alkaline solution (Table 15-4-11) and subsequently anisotropically etched. Precise control of exposure dose, and developer time, temperature, and concentration is necessary for reducing the largest linewidth deviations.

For production setup and control, the statistical analysis and design of experiments to minimize resist linewidth biases can save considerable experimentation for lithographic process optimization. An example of a three-dimensional (variable) regression analysis of prebake and dose effects on linewidths is shown in Fig. 15-4-17.[71] A 70°C prebake and a dose of 90 mJ/cm^2 achieves zero image bias. Critical to the adjustment of the process steps are the accurate and precise

TABLE 15-4-11. Assessment of Process Variations to Linewidth Control[a]

Step	Percent variation	Δ linewidth (μm)
Thickness	10	0.05
Prebake time	10	0.01
Prebake temperature	10	0.02
Expose dose	10	0.25
Develop time	10	0.15
Develop temperature	10	0.02
Postbake temperature	10	0.11
Anisotropic etch time	10	0.15/1.0 μm
Anisotropic etch temperature	10	0.05
Developer concentration	5	0.21
Developer concentration	2	0.08
Develop temperature	±0.1 °C	0.15
Develop time	±1 sec/60 sec	0.02
		S.D. 0.13

[a] From Ref. 1.

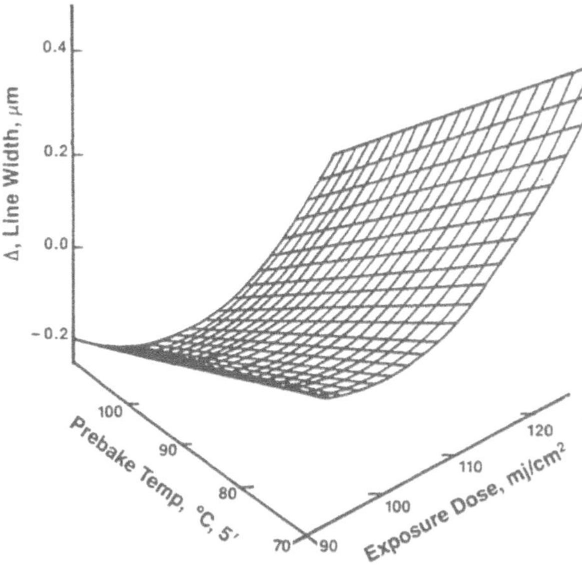

FIGURE 15-4-17. Three-dimensional plot of the result of regression analysis of prebake versus exposure dose versus linewidth. From Ref. 77, reprinted with permission of *Solid State Technology*, published by Technical Publishing, a company of Dun and Bradstreet.

measurements of linewidths on masks,[81] in resists,[82] and in etched wafers. Manual optical[81–83] or automated laser photometric[85] methods are accurate to ±50 nm. Calibration by electron microscopic standards is necessary.[84] On uncoated (gold) wafers, automated SEM measurements at low energy will be the dominant metrology tools for submicrometer geometries.[82] In fact, a 1.0-μm resist image line has been accurately measured by SEM[86] to 1.002 μm with a precision of ±0.003 μm.

15-4-3. Etching Process Controls

Wet etching suffers from a large isotropic bias, proportional to the film thickness. The tolerance is on the same order as the linewidth (1 ± 1 μm) due to:

- Variation in substrate thickness
- Loss of adhesion during etching
- Batch processing

Etch endpoint detect has reduced the tolerance to less than the bias.[36,39] The biases and tolerances of aluminum wet etching were reduced from ±2 μm to ±1 μm with an IR endpoint monitor.[40] An electrochemical cell was also disclosed to precisely measure the endpoint but was never commercialized.[41]

Wet etching bias variations are similar in size to those encountered in positive resist DQN development (previously discussed). For substrate films < 0.3 μm thick, wet etching is a low-cost process for etching patterns to within a ±0.3-μm variance. Etch monitors by reflected light improve process control. For opaque films, a reference film is used as an interference etch monitor.[60]

Dry Etching. Dry etching occurs as either an isotropic or an anisotropic process or as a mixture of both directionalities.[13,89] Selectivity is important in ceasing the etching, before the underlying films are eroded. Since several parameters[34] such as pressure, power, electrode configuration, and flow rate influence

TABLE 15-4-12. Comparison of Bias and Tolerance of Etching (μm)

Film thickness	Wet etch	Dry isotropic	Dry anisotropic
1.0	1.0 ± 0.8	0.4 ± 0.6	0.2 ± 0.4
0.3	0.3 ± 0.7	0.2 ± 0.4	0.1 ± 0.1

the rate of etching, etch monitors are essential for commencing and ending the etching. Dry etching has successfully reduced the bias (true value) and tolerance of etching (Table 15-4-12) to a fraction of the substrate film thickness. Straight wall profiles can be produced in substrate films as thick as 2–3 μm.

The etching of polysilicon on top of thin gate oxide (Fig. 15-4-18) is a good example of the control of a dry etch process.[89] The specific requirements for polysilicon etching are as follows:

- Linewidths to 2.0 μm
- Thickness of 0.4–0.5 μm
- Maximum etch linewidth loss 0.1 μm
- Maximum gate loss 200 Å
- Maximum oxide loss 100 Å

Armstrong and Maleham[42] used a Freon 115 and CCl_4 etchant in a two-stage etching monitored by a laser endpoint detect (Fig. 15-4-19) and achieved less than a 100-Å SiO_2 film loss and $< \pm 0.1$-μm edge variation of the etched polysilicon. In order to achieve run-to-run reproducibility in a two-stage process, they found it necessary to clean the plasma chamber of polymer deposits with a short O_2 ash

FIGURE 15-4-18. Example of anisotropic reactive ion beam etching of a sandwich of tantalum silicide/polysilicon on thermal SiO_2. From Ref. 89, reprinted with permission of *Solid State Technology*, published by Technical Publishing, a company of Dun and Bradstreet.

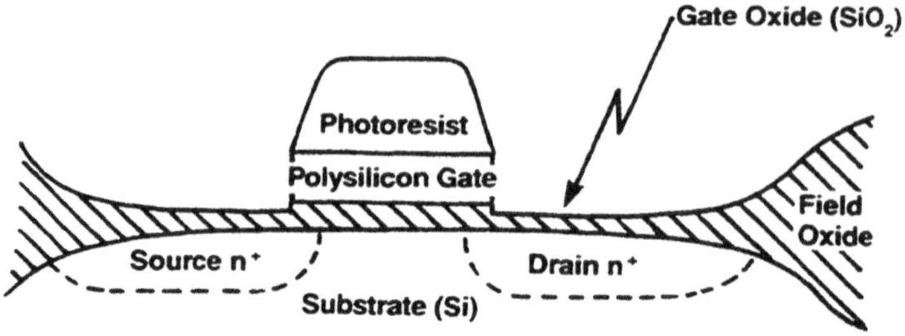

FIGURE 15-4-19. MOS structure after polysilicon dry etching. From Ref. 42, copyright 1983 and courtesy of Pergamon Press.

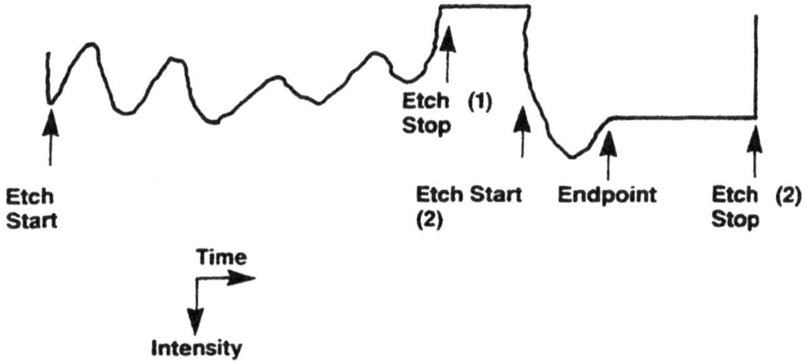

FIGURE 15-4-20. Laser signal output for etching polysilicon with two etchants of different selectivity. From Ref. 42, copyright 1983 and courtesy of Pergamon Press.

cycle. Omission of this step led to a rapid reduction in the etch selectivity relative to the underlying oxide during the CCl_4 etch due to the liberation of adsorbed fluorine species from the walls of the reactor.

In another example of polysilicon etching, Korman[44] monitored the emission spectra of the etching gas, Cl_2, in a batch planar reactor to sense the SiO_2 etching. The wafer plate was rotated to improve wafer-to-wafer etch uniformity (Table 15-4-13). The decrease in the SiCl* emission spectra gives an indication of

TABLE 15-4-13. Oxide Loss and Polysilicon Linewidth Variation[a]

Wafer number	Average oxide (Å)	Oxide loss (Å)	Polysilicon linewidth (μm)
1	836 ± 28	94	6.31 ± 0.25
2	738 ± 11	172	6.36 ± 0.27
3	767 ± 40	141	6.59 ± 0.19
4	791 ± 30	139	6.59 ± 0.11
5	874 ± 40	56	6.14 ± 0.12
6	756 ± 22	169	5.18 ± 0.27
		129 ± 46	

[a] From Ref. 44.

TABLE 15-4-14. Uniformity and Reproducibility of Etch Results[a]

		Poly-Si	SiO_2	Al
Uniformity of etched depth	Single wafer	Within ±2.7%	Within ±1.8%	Within ±4.2%
	Among wafers	Within ±4.7%	Within 2.3%	Within ±8.0%
Reproducibility of etched depth	Single chamber	Within ±3.7%	Within ±1.7%	—
Uniformity of pattern width	Among wafers	Within ±0.11 μm	Within ±0.11 μm	Within ±0.1 μm

[a] From Ref. 45.

when to drop the RF power to prevent excess oxide loss. Winkler[43] concluded that the batch mode of etching in a planar reactor has difficulties with good uniformity: ±6% across wafers and a 0.2-μm linewidth loss, in addition to the gate oxide erosion. None of the batch systems met the linewidth and uniformity criteria. Because of variation in polysilicon thickness and the averaging of the product concentration in the emission monitor spectrum, overetching is a common problem. Better linewidth control is obtained with single-wafer processing.

Itakura and co-workers[45] described a multiturret single-wafer etcher with good uniformity and reproducibility (Table 15-4-14). As wafer size increased to 250-nm diameter, single-wafer processing becomes more competitive in cost with batch processing (Fig. 15-4-21).[19,47,87] Hayes and Pandhumsoporn[46] found that a single-wafer etcher with laser endpoint detect will have a better control over batch etching; which suffers from load effects and lateral etch loss > 0.1 μm.

A more statistical approach was employed by Bergeron and Duncan[48] in defining process controls for isotropic and anisotropic polysilicon etching. Response surface methodology[71] (RSM), which treats dependent variables as polynomials, was used to find the optimum conditions and tolerable limits for the following polysilicon etching variables:

- Electrode spacing
- Pressure
- Power
- Flow rate
- Temperature
- Gas composition

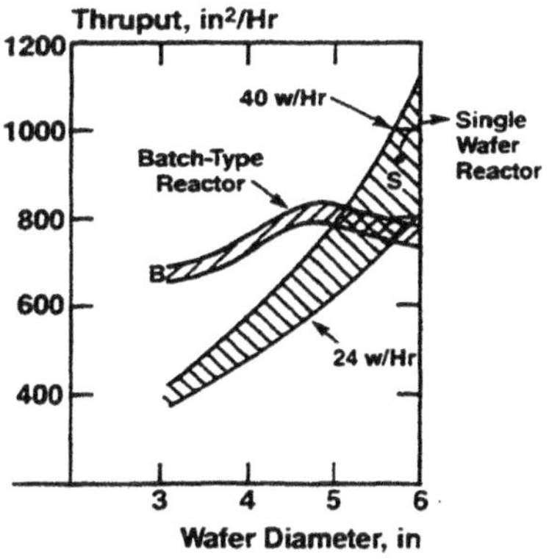

FIGURE 15-4-21. Single (S)-wafer processing becomes more competitive with batch (B) processing as wafer size increases. From Ref. 47, reprinted with permission of *Solid State Technology*, published by Technical Publishing, a company of Dun and Bradstreet.

Caution must be exercised in interpreting the true anisotropy, since even isotropic systems can produce nearly straight walls in a parallel plate etcher. To prove anisotropy,[49] the SEM analysis must show both edges of the etched feature. The width of the etched feature must be compared with the width of the resist pattern before etching.[49]

Inherent to plasma etching is the reproducibility of the erosion of resist. Excess resist loss over steps must be avoided to prevent pinhole formation. Even though the activation energy for resist etching is low (4–8 kcal/mole), wafer surface temperatures[50,51] can easily reach 100°C. Cooled reactor plates and conductive bonding of the wafers to the plates are necessary to moderate the rise in temperature. In addition, the successive use of a reactor can lead to even higher temperatures[52] due to retained heat. Fluorooptic[53] thermometers were used to accurately evaluate different methods of attaching wafers to a water-cooled electrode. Wafers glued with conductive paint only lost 500 Å of resist thickness, while wafers sitting on the electrodes lost 5000 Å.[51]

In plasma etching, the integrity, uniformity, and etch reproducibility require accurate endpoint detection. For example, the degree of undercutting by O_2 plasma in a trilayer lift-off is monitored by the O/O_2^* emission intensity ratio.[76] Is endpoint detection sufficient, or should the monitor supply data to a

TABLE 15-4-15. Plasma Etching System and Wafer Parameters[a]

	Intrinsic	Interactive
System parameters	Pressure	Pressure
	Molar gas flow	Surface contamination
	Temperature	Transducer shift
	Electrode material	Window opacity
	Electrode spacing	Outgassing films
	Manifolds	Dielectric films
	RF power	Pump speed
	RF frequency	Air leak
	DC bias	Trap efficiency
	Wafer position	Exhaust conductance
	Orientation	Pump concentration
Wafer parameters	Film thickness	Surface temperature
	Film uniformity	Resist decomposition
	Film composition (dopants, alloys)	Profile modification
	Underlying films	Outgassing
	Resist	Removal
	Type	Penetration
	Thickness	Pinholes
	Bake/harden	Substrate damage
	Profile	Etching
	Linewidth uniformity	Contamination
	Backside coating	Radiation
	Wafer thickness	Film etch rate
	Wafer loading	Anisotropy
	Resist pattern factor	Isotropy
	Planarization	Etched linewidths
		Wafer uniformity

[a] From Ref. 54.

TABLE 15-4-16. Endpoint Techniques[a]

	Signal sensitivity				Changeable for film/ etchant	Detailed to diagnostics
	Batch	Surface area	Film	Etchant		
Laser interferometry	No	No	No	No	—	No
Emission spectroscopy	Yes	Yes	Yes	Yes	Yes	Yes
Mass spectroscopy	Yes	Yes	Yes	Yes	Yes	Yes
Pressure	Yes	Yes	No	No	—	No
Mass flowmeter	Yes	Yes	No	No	—	No
Discharge impedance	Yes	Yes	No	No	—	No

[a] From Ref. 57.

microprocessor to correct a plasma condition? Bersin[(54)] has arranged the plasma system and wafer parameters into categories of intrinsic (over which the process engineer has tool or microprocessor control) and interactive parameters, which occur between the wafer and the system (Table 15-4-15).[(54)]

Interactive system parameters change conditions slightly, as the next wafers are processed. Process monitors have to be developed to control the process in real time.[(55,56)] Endpoint monitors are now being developed to provide on-line kinetic and diagnostic information (Table 15-4-16).[(57)] Pressure monitors[(58)] are not consistent and do not correlate with the stoichiometry of reaction. They are ineffective at ceasing reactions at the interface of an etchable sublayer.[(59)] The monitoring of the actual etch rate of the film has proved more useful as a direct method of etch stop. The monitoring of opaque films can be accomplished by interference techniques against reference materials.[(60)]

Nonuniformity in etching is consistent with a nonuniform substrate film thickness. For films other than oxides, depositions are made in a batch mode, where variations of $\pm 5\%$ in film thickness are common. Aluminized wafers can vary $\pm 10\%$ in thickness. In a typical batch etcher, with a uniformity of $\pm 10\%$ etch rate, the thinnest films are overetched by 40%.[(54)] This results in a 4% linewidth loss of $\pm 0.04\ \mu m$. For a 4:1 selectivity ratio of resist to substrate-isotropic etch rate, this results in a ± 0.2-μm etched linewidth bias and at 1:1 selectivity, a loss of $\pm 0.4\ \mu m$ in linewidth. Thus, with ever-increasing wafer size, single-wafer processing with precise endpoint, plasma diagnostics, and the control of interactive parameters, are required for better process control and uniformity.

15-5. PROCESS TROUBLESHOOTING

The daily operation of a lithographic process is the equivalent of the operation of a chemical factory producing ultrapure radioactive materials. The resist materials and associated process chemicals (liquids, solids, and gases) have passed through strict quality control and have not aged or become contaminated with foreign particulates. In the meantime, the complex interaction and physical conversion of the

TABLE 15-5-1. Troubleshooting Resist and Processing

Liquid resist									
Storage temp.	Gas evolution upon opening	Opaque particles	Gel particles	Different UV absorption	Different viscosity	Shelf life	Filtration–dilution		
Wafer surface									
Topography	Surface particles	Water-rinse stains	Adhesion promotor	Storage condition	Storage container	Humidity	Thickness		
Apply film									
Backside coating	Comets	Striations	Film thickness	Film uniformity	Cobwebs	Swirls	Lip	Dewets	Pinholes
Prebake									
Time	Atmosphere	Underlayer	Scratches	Shrinkage	Boat	Oven recovery	Batch size	Temperature	
Expose									
Substrate temp.	Mask	Dose	Intensity uniformity	Resist thickness	Reflectivity	Wavelength	Atmosphere		
Develop									
No image	Image missing	Large images only open	Scum in images	Hazy resist surface	Ragged image	Oversize	Undersize	Swollen–reticulation	Pinholes
Developer concern	Developer temp.	Developer filtration	Developer discoloration	Rinse	Batch size	Agitation	Spray nozzle	Resist remaining	

				Contamination					
Pinholes	Solid particulates	Discoloration	Dewets	Electrical shorts					
				Postbake					
Time	Temperature	Atmosphere	Batch size	Shrunken images	Cracked resist	Image rounding	Underlayer		
				Wet etch					
Does not etch	Large patterns only etch	Missing images	Notches in pattern	Unetched areas	Bubbles on surface	Etch filtration	Halo around images		
Etch temp.	Etch concn	Batch size	Endpoint	Etch volume	Agitation	Discoloration of surface	Shelf life	Wafer backside etching	
				Dry etch					
Endpoint	Polymer redeposit	Resist ablation	Resist flow	Pinholes	Nonuniform	Unetched areas	Underlayer etching	Sidewall deposits	
				Lift-off					
Metal flake redeposits	Unlifted areas	Notches	Nicks	Sidewall burrs	Conical studs	Missing pages	Solvent concern	Solvent temp.	Rinse
				Wet stripping					
Substrate attack	Redeposits	Stains	Gelatinous products	Filtration	Temperature	Stripper concn			
				Dry stripping					
Radiation damage	Metallic residues	Substrate oxidation	Batch uniformity	Resist type	Resist heating	Batch size			

radiation, plasma, and physical forces into chemical products are interrupted by entropic factors that give rise to opaque and adsorbed particles. Localized random defects lower the final wafer or chip electrical yield. In the clean room, the lithographic process equipment and personnel are themselves the largest generators of particulates. Other random and process factors give rise to radial or lateral linewidth variations.

The successful operation of a lithographic line depends on the ability of sensors, metrology tools, electrical tests, and personnel to recognize the incipient symptoms or problems before they become permanent defects. As a reminder of defects or symptoms, a comprehensive troubleshooting outline of the total lithography processes is given in Table 15-5-1. The most difficult problem will be the nondestructive determination of pinholes (see Chapter 6) on a routine basis. Pinholes $< 1.0\,\mu m$ in diameter are difficult to detect and usually require some permanent destructive method. Defects the size of 100 nm will be as fatal as the inability to overlay to $\pm$100-nm accuracy. Better clean rooms, process monitors, and robotics will be ever-increasing in cost and complexity.

The separation of process defects falls into four major categories:

1. Process related
2. Problems in process materials
3. Equipment malfunctions
4. Contamination

Process related and contamination are the major contributors to the daily yield losses. High productivity and rapid electrical testing can speed up the correcting of the processing defects. Attacking and solving problems in lithography can be done in three basic ways:

1. Equipment related
2. Process dependent
3. Material variations

For example, in order to obtain vertical profiles in the resist, one can use either a beam with a high MTF, a multilayer resist process, or a resist–developer combination with high contrast. The interaction of tool and materials can also lead to new or unexpected findings, such as the oxygen effect of resists exposed in air versus inside a vacuum chamber. The first order of business is to try and identify the source of the problem. Are there independent or dependent variables operating at the same time? Can the problem be traced solely to a tool or a materials problem? Process problems are more difficult to track and crack, since they are more random in nature and do not occur at the same time, even on a single wafer. On-line controls are necessary to prevent random process fluctuations.

15-6. SUMMARY

The basic object of process controls is to achieve steady-state conditions in prebaking, exposing, and developing and to achieve a dynamic balance of conditions in plasma etching. Loading effects and radial disparities in batch processing are corrected in a first approximation by a monitor or send-ahead wafer. Automation may be regarded as a direct means of increasing productivity and yields. On-line computer control will improve production control, eliminate operator errors, and reduce contamination from excessive handling operations. The concept of total resist process automation, including rapid thickness measurements, will be a costly investment but will pay back in higher yields.

Both on-line feedback or detection controls and off-line quality controls of resist and process chemicals serve to eliminate systematic process errors or defects. The net of all the resist process steps and controls is the process reproducibility. For a typical batch process of producing 2-μm geometries, the etched linewidth bias is $\pm 0.5\ \mu$m for a batch monitored process with a single-wafer align and expose,

$$\sqrt{(\Delta\ \text{align}, 0.3)^2 + (\Delta\ \text{image}, 0.3)^2 + (\Delta\ \text{etch}, 0.3)^2}$$

With improved endpoints, single-wafer processing, and diagnostic controls, the innovations should reduce the deviations to $\pm 0.15\ \mu$m to satisfy the device linewidth control requirements for the 1990s,

$$\sqrt{(0.2)^2 + (0.2)^2 + (0.2)^2}$$

Precise exposure and development will be necessary to control the major factors of linewidth deviations. Etching endpoint and etchstop monitors safeguard against the secondary problems of sublayer erosion in multilayer insulators. Single-wafer plasma etching will dominate the direct process mode and will rely heavily on feedback controls.

In support of process controls, advanced metrology equipment to measure linewidths, overlay, and adsorbed particulates will be necessary to improve the yields of submicrometer device processing. In the complementary photomask production, more automated defect inspection and better linewidth metrology, will be required to support processing.[61–70]

The three "T"s of processing—time, temperature, and thickness—are essential to control and monitor the exposure, develop, and etch steps. Prebake and postbake time should reach steady-state temperatures. With high-activation-energy (10 kcal/mole) processes such as resist developing or wet etching, temperature controls to ± 0.2°C will hold linewidths under $\pm 0.2\ \mu$m. Resist erosion during plasma etching should be controlled by proper electrode cooling and good wafer bonding on the backside of the wafer. More automated resist thickness measurements are necessary for satisfying the alignment requirements of precise steppers. Finally, the dynamic measurements of the developing resist (thickness) or the etching of the substrate correct for the prior fluctuations in prebake-exposure and the internal concentration depletion of the developer or plasma etchant. Single-wafer processing will be necessary for overcoming the lateral nonuniformity of current batch process equipment (baking, developing, and etching).

REFERENCES

1. E. Hryhorenko, *Kodak Microelectron. Semin. Proc. 1979* **G-102**, 18 (1980).
2. G. Solvendy, Editor, *Handbook of Industrial Engineering*, Wiley, New York, 1982, Section 8.3.5.
3. R. Reichelderfer, *Solid State Technol.* April 1982, p. 160.
4. T. Chang, D. Kyser, and C. Ting, *Solid State Technol.* May 1982, p. 60.

5. A. Wat and K. Chen, *SPIE Proc.* **394**, 58 (1983).
6. M. Nakase and T. Shinozaki, *IEEE Trans. Electron Devices* **ED-28**, 1416 (1981).
7. C. Ausschnett, T. Bruner, and D. Cronin, *SPIE Proc.* **394**, 64 (1983).
8. D. Yen, L. Linholm, and M. Buehler, *J. Electrochem. Soc.* **129**, 2313 (1982).
9. I. Stemp, K. Nicholas, and H. Brockman, *IEEE Trans. Electron Devices* **ED-26**, 729 (1979).
10. D. Angel, *Semicond. Int.* April 1983, p. 72.
11. N. Armstrong and J. Maleham, *Vacuum* **33**, 291 (1983).
12. W. Oldham, A. Neureuther, and D. Kim, *IEEE Trans. Electron Devices* **ED-26**, 717 (1979), **ED-27**, 1455 (1980), **ED-28**, 1410 (1981).
13. N. Viswanathan, *J. Vac. Sci. Technol.* **16**, 388 (1979).
14. T. Chang, C. Codella, and R. Lange, *IEEE Trans. Electron Devices* **ED-28**, 1427 (1981).
15. M. Eib and F. Jones, *J. Vac. Sci. Tech.* **B1**, 1327 (1983).
16. W. Loveless, *Solid State Technol.* Sept. 1977, p. 47.
17. K. Harris, P. Sanland, and R. Singleton, *Solid State Technol.* Feb. 1984, p. 159.
18. D. Awamura, *Jpn. Electron. Eng.* Sept. 1983, p. 64.
19. M. Furuchi and N. Kimura, *Jpn. Semicond. Technol. News* **2**(5), 25 (1983).
20. W. Bayless, *Solid State Technol.* Feb. 1981, p. 132.
21. C. Ting and K. Liauw, *J. Vac. Sci. Technol.* **B1**, 1225 (1983).
22. C. Mogab, *J. Electrochem. Soc.* **124**, 1262 (1977).
23. V. Marriott, *SPIE Proc.* **394**, 144 (1983).
24. W. Arden, H. Keller, and L. Mader, *Solid State Technol.* July 1983, p. 143.
25. T. Shankoff, J. Bruning, and R. Johnston, *Polym. Eng. Sci.* **20**, 1102 (1980).
26. P. Frasch and K. Saremski, *IBM J. Res. Dev.* **26**, 561 (1982).
27. H. Rottmann, *J. Appl. Photogr. Eng.* **8**, 101 (1982).
28. D. Widmann and H. Binder, *IEEE Trans. Electron. Devices* **ED-22**, 467 (1975).
29. M. Nakase and Y. Matsumoto, *Photogr. Sci. Eng.* **23**, 215 (1979).
30. M. Cowan, *Solid State Technol.* May 1982, p. 57.
31. M. Borovicka, SPE RETEC Photopolymers, Ellenville, N.Y., 1979, p. 20.
32. A. Broers, *IEEE Trans. Electron Devices* **ED-28**, 1268 (1981).
33. J. Levine and H. Schick, *SPIE Proc.* **221**, 142 (1980).
34. H. Naguib, *Vacuum* **33**, 285 (1983).
35. D. Elliott, *Integrated Circuit Fabrication Technology*, McGraw–Hill, New York, 1982, p. 340.
36. H. Kleinknecht and H. Meir, *J. Electrochem. Soc.* **125**, 798 (1978).
37. P. Heiman and R. Schutz, *Electrochem. Soc. Ext. Abstr.* **83-1**, 257 (1983).
38. B. Lin, *IEEE Trans. Electron Devices* **25**, 419 (1978), U. S. Patent 4,142,107 (1979), IBM.
39. M. Sternheim, W. van Gelder, and A. Hartman, *J. Electrochem. Soc.* **130**, 655 (1983).
40. H. Moritz, *Solid State Technol.* Aug. 1975, p. 54.
41. B. James, Proc. Semicond. Processes Prod., June 15, 1971, p. 111.
42. N. Armstrong and J. Maleham, *Vacuum* **33**, 291 (1983).
43. U. Winkler, *Solid State Technol.* April 1983, p. 169.
44. C. Korman, *Solid State Technol.* April 1982, p. 115.
45. H. Itakura, H. Komiya, and K. Ukai, *Solid State Technol.* April 1982, p. 209.
46. J. Hayes and T. Pandhumsoporn, *Solid State Technol.* Nov. 1980, p. 71.
47. T. O'Neill, *Semicond. Int.* April 1981, p. 67.
48. S. Bergeron and B. Duncan, *Solid State Technol.* Aug. 1982, p. 98.
49. P. Parry and A. Rodde, *Solid State Technol.* April 1979, p. 125.
50. E. Egerton, A. Nef, W. Milkinsun, W. Cook, and D. Bard, *Solid State Technol.* Aug. 1982, p. 84.
51. R. Bond, S. Dzioba, and H. Naguib, *J. Vac. Sci. Technol.* **18**, 335 (1981).
52. S. Chung, *Solid State Technol.* April 1978, p. 114.
53. P. Kolodner and J. Tyson, *Appl. Phys. Lett.* **40**, 782 (1982).
54. R. Bersin, Kodak Microelectronics Seminar, 1981, p. 127.
55. R. Fink and W. Giles, *Solid State Technol.* Nov. 1982, p. 107.
56. A. Weiss, *Semicond. Int.* Dec. 1982, p. 61.
57. A. Weiss, *Semicond. Int.* Sept. 1983, p. 98.
58. U. S. Patent 4,352,596 (1982), IBM.
59. E. Bulat, G. Doyle, and B. Gerlent, *Semicond. Int.* Nov. 1982, p. 113.
60. U. S. Patent 4,367,044 (1983), IBM.
61. P. Singer, *Semicond. Int.* March 1983, p. 48.

62. D. Angel, P. Johnson, and M. Yvic, *Semicond. Int.* Dec. 1978, p. 101.
63. K. Harris, P. Sanland, and R. Singleton, *Solid State Technol.* Feb. 1984, p. 159.
64. C. Mallory, D. Perloff, T. Hasan, and R. Stanley, *Solid State Technol.* Nov. 1983, p. 121.
65. K. Murray, *Semicond. Int.* Dec. 1982, p. 70.
66. Y. Uchiyama, D. Awamura, and K. Nakashima, *SPIE Proc.* **394**, 232 (1983).
67. P. Burggraaf, *Semicond. Int.* Aug. 1981, p. 57.
68. G. Carver, T. Russell, and L. Linholm, *Solid State Technol.* Sept. 1980, p. 85.
69. S. Lim and D. Ridley, *Solid State Technol.* Feb. 1983, p. 99.
70. U. S. Patent 4,303,341 (1981), RCA.
71. L. Goldman, *Solid State Technol.* June 1979, p. 74.
72. D. Elliott and M. Hockey, *Solid State Technol.* June 1979, p. 53.
73. J. Shaw and M. Hatzakis, *J. Electrochem. Soc.* **126**, 2026 (1979).
74. J. Greeneich, *J. Electrochem. Soc.* **122**, 970 (1975).
75. A. Neureuther, Y. Sakai, and J. Reynolds, *J. Electrochem. Soc.* **131**, 623 (1984).
76. B. Soller, P. Shuman, and R. Ross, *J. Electrochem. Soc.* **131**, 1353 (1984).
77. M. Johnson and K. Lee, *Solid State Technol.* Sept. 1984, p. 281.
78. I. Hawakaya, *Jpn. Semicond. Technol. News* **3**(4), 15 (1984).
79. U. S. Patent 4,384,139 (1982), IBM.
80. T. Pampolone, M. Hannifan, S. Jain, and C. Krieger, *J. Electrochem. Soc.* **131**, 2670 (1984).
81. H. Rottmann, *IBM J. Res. Dev.* **26**, 553 (1982).
82. P. Singer, *Semicond. Int.* March 1980, p. 39.
83. D. Nyyssonen, *Appl. Opt.* **16**, 2223 (1977).
84. D. Nyyssonen, *Semicond. Int.* March 1980, p. 39.
85. K. Nakasawa, S. Yoshida, A. Tanimoto, S. Murakami, and T. Yamaguchi, *Proc. Kodak Microelectron. Semin.* **G-130**, 69 (1980).
86. G. Tora-Lira and R. Melen, *Microelectron. Manuf.* Feb. 1985, p. 19.
87. G. Mathad, *Solid State Technol.* April 1985, p. 221.
88. L. Lauchlan, K. Sautter, and T. Batchhelder, *Solid State Technol.* April 1985, p. 33.
89. A. Baudrant, A. Passerat, and D. Bollinger, *Solid State Technol.* Sept. 1983, p. 183.

16

Nonresist Processes

16-1. INTRODUCTION

Since organic-based resists and etching processes were introduced into semiconductor lithography, there has always been a motivation to replace them with "resistless" processes that directly form circuit components. The direct deposition of metal lines and the direct doping of silicon eliminates all of the resist process steps, some process equipment and the associated organic chemical operations. If we expend 100 mJ/cm^2 to expose an organic resist, can the same energy be used to decompose metal or silicon precursors into useful circuit patterns? For specialized applications, we would like to deposit conformal coatings at different thicknesses on the same layer and we desire the capability of repairing defects in a circuit or photomask such as a metal short or a missing image. Another cost motivation would be the elimination of wet chemical processes and the ensuing waste disposal. The successful elimination of wet development can be analogous to the success of the instant developing "Polaroid" process (over aqueous processing in photographic chemistry).

Anytime we print an image in a resist, we distort the image by the optical or electron modulation. The transfer of the resist into the substrate also involves isotropic distortion in some cases. Thus, a bias of 0.3–1.0 μm in an image 1.5 μm wide by 1 μm is probable with conventional resist lithography.

The direct formation of an image in or on a semiconductor film on a wafer involves an energy-assisted (additive) deposition or (subtractive) etching–sublimation process. The beam of energy as shown in Fig. 16-1-1 can be focused to a spot or it can be collimated and masked to flood expose a chip on a wafer. Generally, metals are easier to deposit from gaseous metal alkyls (lower activation energy) than insulators deposited from organic precursors. Conversely, metals are easier to remove than semiconductors and insulators. Secondly, gas-phase precursors are more abundant than a layer of a liquid or solid material (Fig. 16-1-2) and more energy is evolved in the phase changes to form a final solid pattern. On the silicon wafer, four major types of reactions (Fig. 16-1-3) result in material deposition below or on the silicon surface and the direct removal of selective film areas. Depending on the dose, the reactions on the solid surface (Table 16-1-1) can involve a variety of chemical or physical reactions. Although gaseous reagents are involved, the preliminary step occurs through physical adsorption of the reactants and the reaction proceeds in a monomolecular layer. At the high doses (> 1 J/cm^2),

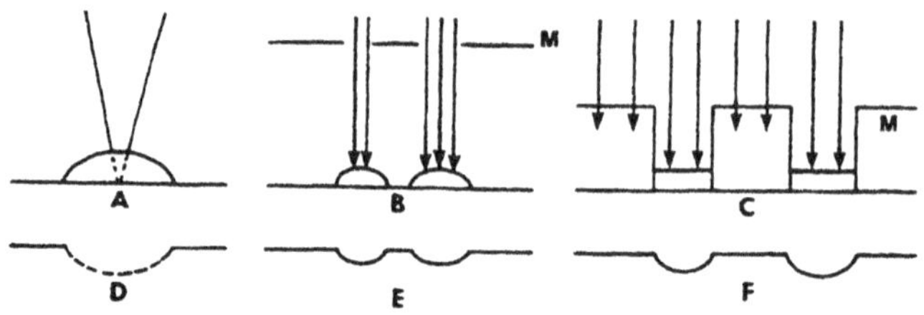

FIGURE 16-1-1. Use of directed energy beams to deposit materials (A–C) or to selectively remove portions of a film (D–F). The focused beam (A, F) uses no mask but requires a long writing time. Exposure with mask (M) in cases (B, C, E, F) requires a durable mask and a powerful uniform beam over an area of 50-200 mm.

localized damage to the silicon may occur. In this case, the beam intensity must be controlled and the dose reproducible within each pulse. As the beam interacts with the substrate, the side effects, such as beam heating beyond the intended size, must be determined and controlled. The application of directed beams should consider the following factors to assess their efficiency, cost, throughput, resolution, and processing area:

- Efficiency of reaction
- Reproducibility of dose
- Durability of mask
- Overlay accuracy
- Beam spot placement accuracy
- Cost of exposure
- Permanent damage to device

While many processes can be envisioned with a focused spot at the circuit level, they become more difficult to operate at the flood chip level especially if perforated stencil masks are required. For example, two complementary masks are required to handle the "donut hole" problem of an isolated image in a foil mask for ion-beam or *e*-beam proximity print.[1] The wafer and process considerations are as follows:

1. *Energy absorbed.* The reactants (Figs. 16-1-2 and 16-1-3) can be a gas, liquid, or solid in contact with a silicon wafer. At elevated temperatures, simple gases such as F_2, Cl_2, XeF_2, and H_2O will react directly with Si or SiO_2. An order-

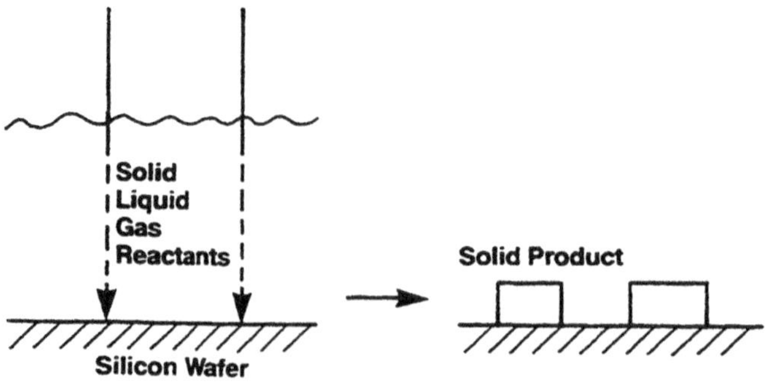

FIGURE 16-1-2. Schematic of material phase changes occurring with directed beam imaging on solid silicon surface.

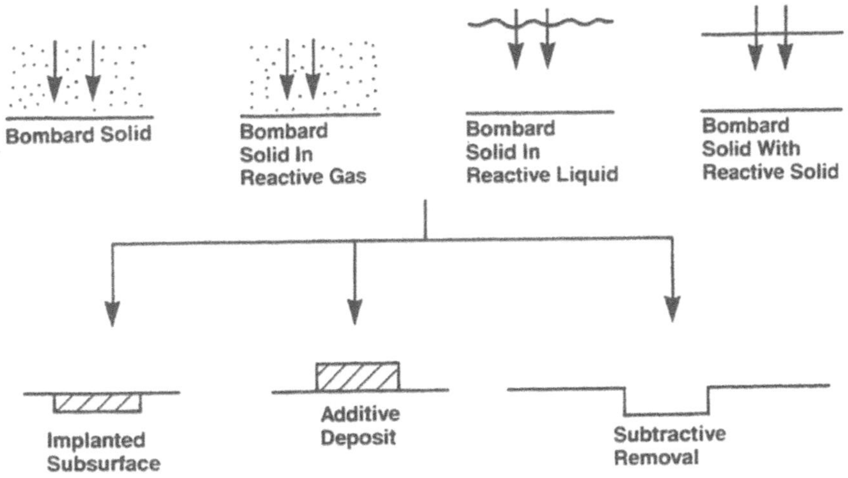

FIGURE 16-1-3. Direct patterning of silicon wafer using photons, electrons, or ion beams.

of-magnitude increase in the rate of etching can be obtained ($> 1\ \mu m/min$) if the reactant is dissociated into radicals or "hot" molecules. Since the gases generally absorb in the 200- to 300-nm region with low molar extinction coefficients ($\varepsilon < 100$ liters/mole-cm), path lengths of 1–10 cm are needed. In the liquid-phase reactions, much shorter path lengths are required since the molarity and density are an order of magnitude larger. Solid films have even higher density (3–10 ×) and are the most effective absorbers (disregarding reflection) but if bimolecular reactions are involved the immobility of the matrix may require absorption of extra heat to melt or vaporize the products from the reactive site. In general, the absorption (A) of photons,

$$A_{\text{photons}} > A_{\text{ions}} \ggg A_{\text{electrons}}$$

is higher than ions or electrons but in the end similar chemical reactions ensue once

TABLE 16-1-1. Radiation Effects on Solids

Dose (J/cm^2)	Effects
10^{-2}–10^{-1}	Physicochemical
	Photochemistry
	Electron excitation
	Ionization
	Energy transfer
	Desorption of molelcules
	Polymerization
	Depolymerization
10^{-1}–10^{2}	Crystal damage
	Annealing
	Diffusion enhancement
	Surface alloying
10^{2}–10^{4}	Sublimation
	Evaporation

the precise energy is absorbed and cascaded to a low-lying and dissociative excited state. Photons may suffer reflectivity loss ($> 30\%$ on Si) depending on the refractive index of the layers and the exposing wavelength. In some cases, the reactions suffer from the absorption of photons of insufficient energy (< 5 eV) or the products may act as inner filters and absorb the incoming radiation from the surface-absorbed photons. Likewise, large-mass ions (> 40 in atomic number) do not penetrate very deeply into a film ($< 0.1\ \mu m$) while electrons (10–200 keV) are lightly ($< 10\%$) absorbed and penetrate too deeply into the substrate to cause electrical device damage. Infrared (10.6-μm CO_2 lasers) is effectively absorbed by gases with > 3-cm path length and with at least a 100-mTorr vapor pressure. However, the IR lasers cannot be focused to a spot smaller than their wavelength (10 μm).

2. *Energy conversion.* In general, most of the gaseous reactants such as CF_3X, Cl_2, O_2, and metal organic vapors dissociate with high quantum[2] efficiency ($\Phi > 0.5$) at low bond dissociation energies (2–5 eV). For example, F_2 requires only 1.5 eV.[2] The activation of a solid such as silicon requires the creation of an electron hole by excitation into the conductor band (at 4–5 eV, < 350 nm) or by exceeding the work function of the metal. For every photon or particle absorbed, one molecule of reactant will react with at least one or not less than 50% of a molecule of the solid to form the desired product. In the photoinduced etching of silicon, the quantum yield (Φ) of the reaction should be at least 0.5 in moles of SiF_4 formed,

$$2F_2 + Si \xrightarrow{h\nu} SiF_4$$

If any amplification by a bimolecular reaction or catalyst can occur to raise the quantum yield (> 1), then the beam can be spread out over a larger area to expose more than one chip at a time. By forming a volatile product, the etching reaction is not inhibited by product absorption. Because of the extensive work going on in gas lasers[3] and plasma chemistry,[4] the quantum efficiency of many gases has been determined and etchants can be selected for maximum energy conversion into hot reactants. For a reasonable process time, the rates of removal or deposition must be 10^3 higher than normal semiconductor processes ($> 0.1\ \mu$m/min).

On an energy scale, metals require less excitation (2–3 eV) to deposit or remove than inorganic materials. Crystalline materials require more energy to overcome the lattice forces. Transition metals such as Au, Ni, Ag, and Sn have low reduction potentials and are the easiest material to be electrolessly or electrolytically deposited.

On the silicon wafer, the thermal conductivity of the solid phase determines the "image distortion" due to the radial transport of heat beyond the beam diameter (especially if a spot is $< \mu$m in diameter). A considerable number of thermal calculations have been performed on laser[5]- or *e*-beam[6]-induced heat transfer on silicon wafers. In order to achieve a spatial confinement of deposits of metal to a micrometer size, the metal precursors' heat of reaction should be thermally neutral or endothermic.[4, 211]

3. *Dose/area and masking.* Although personalized circuit repair or addressed chip level exposures are sufficient for customized processing, the direct processing of silicon wafers (of 5–7.5 cm in diameter) requires throughputs of > 15 wafers/hr at

less than \$10/wafer exposure cost. Broad-beam or full-field lasers, ion- or *e*-beams are difficult to enlarge beyond chip (7.5 mm × 7.5 mm) size especially if other than chemical (> 0.1 J/cm^2, Table 16-1-1) reactions are required. With pulsed beam operations, there are difficulties in monitoring the input dose and time delays may be necessary to allow heat or bimolecular transport reactions to occur. Focusing of a Gaussian spot and poor overlapping in writing a line, especially with lasers, lead to image distortion especially at the 1-μm level. In the case of an beam writer, a spot expose time of nanoseconds for a 1-μm spot is necessary for a throughput of at least 15 (100-mm) wafers/hr.[7]

Although full chip or wafer exposure through a proximity[1] or projection mask[8] has been used for resist exposure and at higher power[9] for packaging applications at 10-μm dimensions, the traditional difficulties of alignment, heat distortion of the mask, and fabrication are more of a problem at the high powers required for any thermally induced etching or deposition process. Stencil foil[1, 8] masks are the most difficult to fabricate and maintain without a transparent backing or if thicknesses of greater than 1 μm in stopping power are required.

4. *Implementation.* The major problems of the expose source are the stability of the gun, focal length, vacuum conditions (handling and contamination), precision of the stepping table and the settling time. All of these technical tool factors drive the cost of a high-power exposure source to \$2–5000 K[10] and the cost of exposure per wafer into the \$10–100 range. At least a minimum exposure of one chip/sec is required for any form of resistless lithography. At present, we can expect this form of lithography to be implemented in photomask repair where it will have time to mature into applications in circuit repair or in circuit personalization. Since delivery of a high dose (> 1 J/cm^2) is a major technical limitation, the technical blending of various disciplines of physics, electronics, and chemistry is required to perfect high-speed chemical reactions (μm/sec rates of deposition of removal) in an anisotropic cell (manner) in order for direct patterning (nonresist process) to be scaled from a 1- to 10-μm spot (10^{-4} cm) to the 10-cm diameter of a wafer. The first challenge is to find or induce high-speed reactions.

In this chapter, I will concentrate on photoinduced reactions especially involving lasers and then discuss *e*- and ion-beam reactions. Both additive and subtractive processes (as shown in Fig. 16-1-3) occurring in the solid, liquid, or gas will be examined.

16-1-1. *Elementary Considerations*

The development of high-power lasers has provided a unique source for photochemical and thermally induced reactions. By depositing the excitation energy in nanosecond pulses, the rates of reactions can be enhanced by 10^6. Some of the basic characteristics of laser light are as follows.[11]

The laser is an interesting optical source because it is spatially and temporally coherent radiation. Temporal coherence means that the light is perfectly monochromatic and spatial coherence means that the laser emits a narrow beam with an angular divergence θ, limited by diffraction:

$$\theta = cg/d \qquad (16\text{-}1\text{-}1)$$

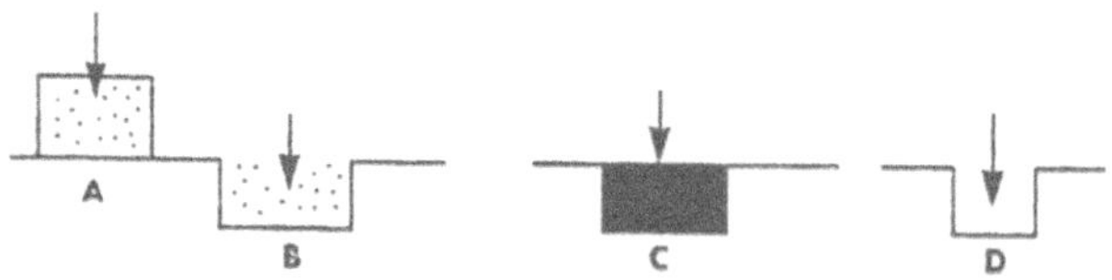

FIGURE 16-1-4. Physical and chemical transformations on a silicon surface. (1) Chemical reactions above (A) or below (B) the surface of silicon; (2) phase changes in subsurface such as melting (C); (3) sublimation of surface regions (D).

where g is the wavelength, d is the output diameter, and c is a constant of 1.22 for a uniform beam and $2/f$ for a Gaussian beam. Because of the high degree of collimation, all of the emitted energy can be focused into a small spot of size ϕ, which is limited only by the diffraction of the focusing lens:

$$\phi = cfg/D \tag{16-1-2}$$

where f is the focal length of the lens and D is its aperture. For an F/1 condensor, ϕ is about one wavelength (0.2–1.0 μm). In contrast, Hg arc sources emit light in 4π steradians. An argon laser emitting a 1-W single line has a brilliance of 10^9 W/cm^2/steradian compared to 10^3 W/cm^2/steradian for a 200-W Hg lamp.

16-1-2. *Chemical and Thermal Energy Requirements*

There are three basic modes of transforming reactions on or in a solid silicon wafer. Depending on the energy absorbed, the deposited energy can transform the patterned area shown in Fig. 16-1-4 by (A, B) chemical reaction, (C) phase transformation (including diffusion), and (D) sublimation. If we assume an area of a spot to be patterned on silicon 2 μm wide $\times$ 1 μm $\times$ 1 μm deep, we can make an estimate of the dose required for the three cases A, B, and C. In Table 16-1-2 are listed the thermal properties of semiconductor materials used in case C of phase melting (ΔH_f) and for case D of vaporization (ΔH_{vap}). For the physical melting of silicon, an absorption of 11 J/cm^2 is required. The following examples of resistless lithography illustrate the enormous energetics of the localized chemical reactions or the phase transformations.

TABLE 16-1-2. Thermal Properties of Silicon Process Materials[a,b]

Material	ΔH_{vap}	ΔH_f	D	Mass	Moles	Molecules	J/cm^2
Polymer	~5	~2	1.2	2.4	2.4	1.4	0.2
Si	105	11	2.3	4.6	16	9.9	3
SiO_2	30	2.1	2.3	4.6	7.5	4.7	6
Al	68	2.6	2.7	5.4	10	6.2	3

[a] From *Lange Handbook of Chemistry*, McGraw-Hill, 1967.
[b] ΔH_{vap}, heat of vaporization in kcal/mole; ΔH_f, heat of fusion in kcal/mole; D, density in g/cm^3; Mass, in g $\times 10^{-12}$ for volume of 2 μm^3; Moles, $\times 10^{-10}$; Molecules, $\times 10^{14}$; J/cm^2, for vaporization assuming no heat loss.

Cases A, B—Chemical Transformation

For a photochemical reaction in a volume of 2 μm^3, 5×10^{14} quanta or more are required for the 5×10^{14} molecules in the substrate. At 300 nm, which is sufficient to produce the photovoltaic effect in Si,[12]

$$Si \rightarrow Si^+ + e$$

An absorbed dose of 1 J/cm^2 supplies 1.5×10^{18} quanta. Assuming a quantum efficiency of 0.1, a minimum dose of 100–500 mJ/cm^2 will be required for the photochemical activation of silicon compounds. Secondly, reactions emanating from the gas phase will require a transport time and an activation energy to deposit on the substrate. The reaction kinetics will be a function of the vapor pressure for a first-order reaction.

Case C—Phase Changes

The energy needed for melting (ΔH_f) silicon and inorganic films (Table 16-1-2) is about tenfold lower than the energy required for vaporization (ΔH_{vap}). In the calculation, I have assumed 100% absorption of energy (no reflectivity loss) and 100% utilization (no heat transfer to the unexposed regions). In order to compensate for thermal losses, realistically, we increase the incident dose to the range of 1–10 J/cm^2 for inorganics and 0.1–1 J/cm^2 for organic phase changes (melting, diffusion, permeation crystallization, precipitation).

Case D—Sublimation

If the processes are performed in vacuum or at low pressures, then we derive the benefit of enhanced evaporation. As calculated in Table 16-1-2 for a ΔH_{vap} of about 50 kcal/mole, an absorbed dose of 10 J/cm^2 will vaporize a volume 1 μm deep of 2 μm^3 of inorganic materials. For example, at a dose of 5 J/cm^2, Al_2O_3 is vaporized from the surface of aluminum to allow the etching of Al in Cl_2 gas to commence.[12, 186] Because of the lower ΔH_{vap}, organic polymers readily sublime (< 1 J/cm^2).[12, 187] Unfortunately, they may decompose if we try to redeposit a film but in the cases of dry developing a positive resist, this facilitates their removal. Polymers of low molecular weight can be laser sublimed and deposited as resists[13] and organic dyes can be readily sublimed from PMMA films to create an opaque conformable mask (a built-on mask) for deep-UV lithography.[12, 187] At doses > 5 J/cm^2, the radial heat loss is a major problem for micrometer-size spots since the energy deposited spreads in an isotropic fashion and may distort the image and produce Gaussian-shaped holes.

TABLE 16-1-3. Laser Sources

Laser	Wavelength (μm)	Power
Excimer (KrF, Ar)	0.2–0.4	5 mJ/pulse
Ar or Kr	0.28, 0.4–0.6	3 W
Ruby or Nd	0.7–1.2	1 J/pulse
CO_2	10.6	10^5 W

16-1-3. *Laser Sources*

High-energy pulsed or finely focused gas lasers can supply energy from 0.1 to 10^5 J/cm^2 in pulses as short as 10^{-8} sec. The most popular laser sources and their properties are given in Table 16-1-3. Both pulsed and continuous wave (CW) are available. In a pulsed laser, a power of 3 mW or 3×10^{-3} J in 10^{-4} sec or 10^{-7} J in a 1-μm (1×10^{-8} cm^2) spot is equivalent to 1 J/cm^2. More or less energy can be supplied by varying the pulse width. For an *e*-beam, an incident dose of 10^{-4} C/cm^2 is about 2 J/cm^2 at 25 keV. At doses of 10^{-3} to 1 C/cm^2, silicon is sufficiently lattice damaged and readily undergoes wet chemical etching. Doses of 10–100 J/cm^2 are required for (alloy) welding and evaporation since the *e*-beam penetrates deeply (> 10 μm) into the substrate, while the absorption of photons is more favorably confined to a 1-μm depth.

In the following sections, I will discuss photon-induced reactions of gas, liquid, and solid precursors for depositing or removing semiconductor device materials.

16-2. GAS–SOLID PHOTON-INDUCED REACTIONS

16-2-1. *Deposition*

While many gaseous reactants can be decomposed to generate "plasma"-type etchants, a select few can also be decomposed such as in chemical vapor deposition (CVD) to deposit metals[153] and silicon insulators such as polycrystalline silicon, SiO_2, and Si_3N_4.[159–161] The repair of pinholes in chrome photomasks is an immediate application of metal deposition.[177, 182, 185] In the additive process, metals, silicon and derivatives, and polymers can be deposited by exciting the reactant (Fig. 16-2-1) in the gas phase,[14] a few (1–10) centimeters above the silicon surface (Fig. 16-2-1) or in a path,[15] 1–10 cm, along the exposure axis (Fig. 16-2-2). The gas may also condense[16] on the surface to a depth of 100 Å (Fig. 16-2-3) and be exposed. Table 16-2-1 lists metals and insulators which have been deposited. Gaseous metal alkyls[16] or carbonyls[14] decompose with unity quantum yields (Φ) by dissociation at 200–300 nm,

$$Al_2(CH_3)_6 \rightarrow 2\,Al + 3\,CH_3CH_3$$

$$Zn(CH_3)_2 \rightarrow Zn + CH_3CH_3$$

$$Cr(CO)_6 \rightarrow Cr + 6\,CO$$

The aluminum formed by decomposition of the metal alkyl is free of carbon since the intermediate radicals do not absorb the KrF (< 254 nm) radiation.[201]

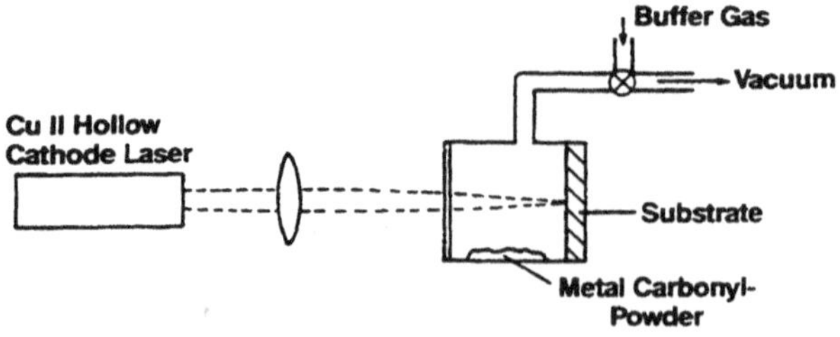

FIGURE 16-2-1. Laser-assisted metal deposition from vapor of metal carbonyl. From Ref. 14, copyright 1981 and courtesy of AIP.

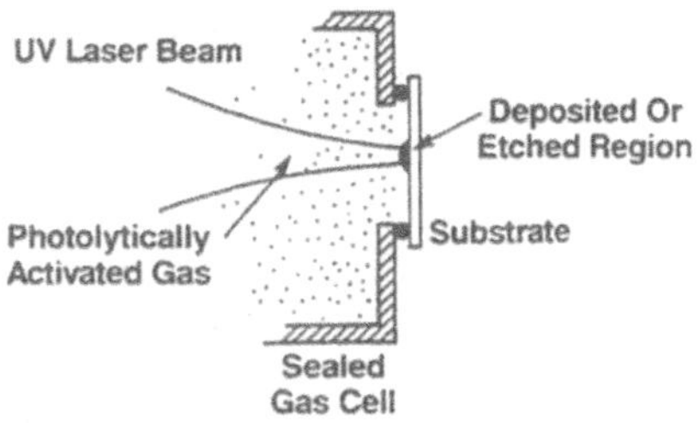

FIGURE 16-2-2. Experimental configuration in UV-assisted laser etching of semiconductor films. From Ref. 15, copyright 1980 and courtesy of IEEE.

Since the molar extinction coefficients of the gases are low, the gas should have a vapor pressure of at least 0.3 Torr and an optical path length of 3–10 cm.[16] While the deposits formed from precursors of metal organics may contain[16] carbon ($< 1\%$), a pure metal such as zinc can be deposited from vapors of the metal by nonphotochemical reactions. Large-area ($> 1\ cm^2$) deposition may require incoherent sources (Hg lamp).[21, 33]

In the mechanistic steps of gas-phase reactions and depositions, the condensation of the gaseous reactants (Fig. 16-2) serves as a prenucleation center[16] for depositing the first 100 Å of metal.[26] The rate of formation of the solid deposit can be calculated by using the Langmuir[27] adsorption mechanism and gas-phase kinetics.[28] If we assume N adsorbed (metal alkyl molecules) per unit area with a mean time of stay of T, a photon striking the gaseous reactant produces a solid metal (M) per unit area. Then for a first-order growth rate,

$$dM/dt = \Phi f N \tag{16-2-1}$$

where Φ is the quantum yield, and f is the number of photons per unit area per unit time. Since every photon absorbed removes a molecule N on the surface,

$$dN/dt = F - N/T - dM/dt \tag{16-2-2}$$

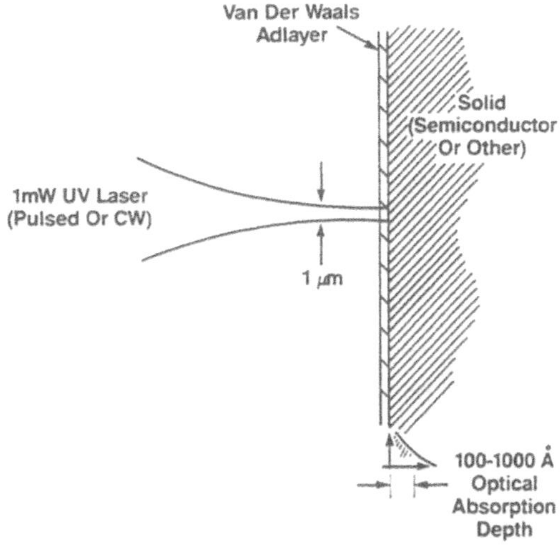

FIGURE 16-2-3. UV laser-induced reactions at the gas–solid interface. From D. Ehrlich, R. Osgood, and T. Deutsch, *J. Vac. Sci. Technol.* **21**, 23 (1982), courtesy of AIP.

TABLE 16-2-1. Photodeposited Semiconductor Materials

Material	Source	Substrate	Laser	Reference
Cr	$Cr(CO)_6$	Glass	Ar, 546 nm	17 (1970)
Cd, Al	$Cd(CH_3)_2$	Glass	Ar, 257 nm	18 (1979)
Cd	$Cd(CH_3)_2$	InP	Ar, 257 nm	19 (1980)
Zn	Zn vapor	SiO_2	KrF, 249 nm	20 (1980)
Cr, Mo, W	$Cr(CO)_6$	Si	CuI, 260 nm	14 (1981)
Bi, Ga, Ge, Si, Sn, Zn	$M(CH_3)_2$	Si	Ar, 257 nm	16 (1982)
Zn, Se, InSe	$Zn(CH_3)_2$	Quartz	Hg[a], 1000 W	21 (1982)
Si	SiH_4	Si	CO_2, 10.6 mm	22 (1978)
Si	$SiCl_4$	SiO_2	CO_2, 10.6 mm	23 (1980)
Si	SiH_4	SiO_2	Ar, 257 nm	24 (1981)
Ge, Si	SiH_4	Si	Ar, 193 nm	25 (1982)
Al	Tri(isoBu)Al	SiO_2	Ar, 257 nm	169 (1984)
Cu	$Cu(acac)_2$	SiO_2	Ar, 193 nm	172 (1985)
Au	AuSR	Si	Ar, 546 nm	193 (1985)

[a] Incoherent source.

where F is the number of gas molecules which impinge on the surface per unit area per unit time and is a function of the pressure given by the universal gas law,

$$PV/RT = F(t) \tag{16-2-3}$$

Substitution of Eq. (16-2-2) into (16-2-1) and integrating, we obtain

$$M = \{F/[1 + (\Phi Tf)]\}\, t \tag{16-2-4}$$

The rate of deposition (R),

$$R = VM/t \tag{16-2-5}$$

where V is the volume of the gas alkyl molecule. The rate of deposition is a function of the vapor pressure P [Eq. (16-2-3) and the quantities related to M in Eq. (16-2-4) A plot of the deposition rate[16] as a linear function of pressure of $Cd(CH_3)_2$ is shown in Fig. 16-2-4. The rate of deposition (0.1 μm/sec at 10^4 W/cm^2) of Cd is proportional to $1/\Phi$, since the quantum yield, Φ,

$$\Phi = \text{Metal moles formed/number of } Cd(CH_3)_2 \text{ decomposed} \tag{16-2-6}$$

and the rate of deposition (0.1 μm/min to 1.0 μm/min) is also proportional[16] to the incident light intensity of the beam. The absorption of light by $Cd(CH_3)_2$ results in decomposition [Eq. (16-2-6), Fig. 16-2-5]. If a powerful high-intensity laser is used in a parallel arrangement, rates as high as 10 μm/min could be obtained.[16] The above equations are also applicable to photon, *e*-, or ion-beam deposition of polymers derived from vinyl or silyl monomers.

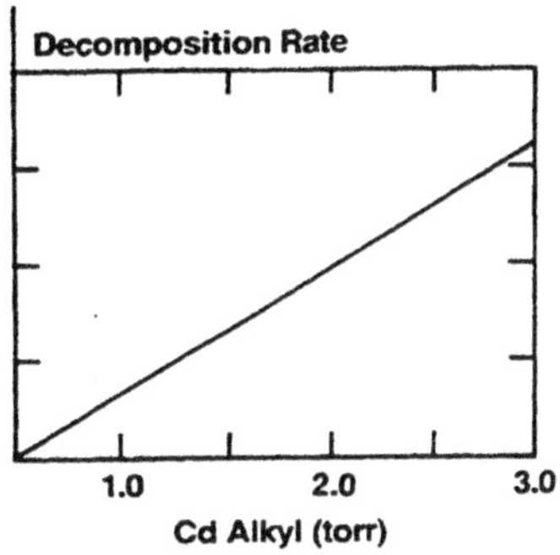

FIGURE 16-2-4. Laser decomposition of Cd vapor at different vapor pressures. From Ref. 16, courtesy of AIP.

Once an initial layer has been decomposed, the outward anisotropic growth of the pattern continues on the fresh metal surface. The surface nucleation step contributes to the high resolution of the process. Since the reaction depends on a surface layer being activated, a step height six times the linewidth (2-μm line on 10-μm step; Fig. 16-2-6) could be written as well as films thicker in width than height (2 to 1 aspect ratio). The rate of deposition is also influenced by the (adhesion) mean time to stay (T) [Eqs. (16-2-2) and (16-2-4)] of the gas reactant. However, once a nucleating layer has been formed, Deutsch and co-workers[(16)] showed that it was not necessary for further irradiation of the surface.

Instead of decomposing metal organic vapors[(201)] and risking carbon contamination in the deposits, Coombe and Wodarczk[(20)] found that pure Zn metal could be deposited from pure metal vapors on the laser-irradiated spot. The metal vapors condense along the laser (249 nm) beam with a dose of only 20 mJ/cm^2. A plausible explanation for the phenomenon is the laser desorption of foreign contaminants from the silicon surface. A clean surface will increase the mean time of stay, T [Eq. (16-2-2)]. In 1971, Genequand[(29)] showed that gases adsorbed on surfaces readily desorb with doses less than 10 mJ/cm^2. Laser desorption of hydrocarbons on glass is also used in chemical analysis.[(30)] Thus, the laser removal of normal laboratory contaminants induces a nucleating site for metal deposition.

The quality of the photodeposited metals ranges from discontinuous crystallites from Cr, Mo, and W carbonyls[(14)] to continuous films from $Cd(CH_3)_2$.[(16)] Deposition of an alloy of Zn on Al was also achieved.[(50)] The codeposition of materials such as Al–Cu will be a new step for the photodeposition process. However, the electrical conductivity of the photodeposited metal is poor

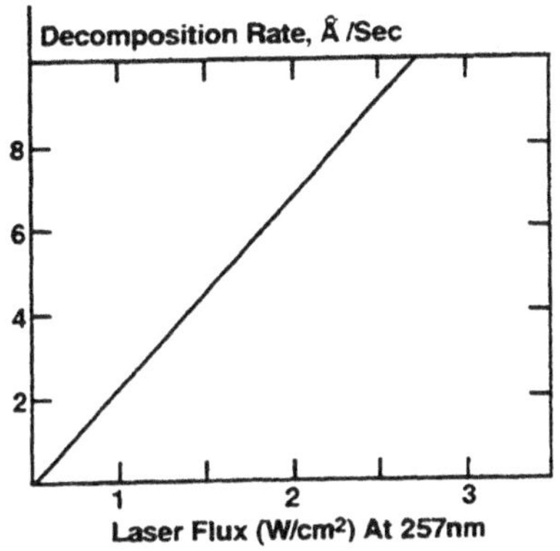

FIGURE 16-2-5. Cadmium deposition rate versus flux of laser beam. From Ref. 16, courtesy of AIP.

FIGURE 16-2-6. Laser deposition of Zn metal line of 2-μm width over 10-μm step. From Ref. 16, courtesy of AIP. Scale bar: 10μm.

possibly due to the nucleation of the metal deposited at room temperature. Since the temperature is kept low, the metal fims consist of smaller droplets which increases the resistance.

A major limitation on the use of laser direct write techniques has been the lack of metal-organic precursors with sufficiently high vapor pressures. Recently, spin-on solid precursors such as gold mercaptides have been suggested as suitable precursors for localized metal deposition.[108, 198] However, some difficulties in controlling linewidths were noted and they were traced to the highly exothermic nature of the chemical reduction reaction (37 kcal/mole) that produces the metal product. For a typical exothermic reaction, at high laser intensities and rapid heating, the rate of heat release from the reaction can be competitive with the rate of energy deposition by the laser. The resulting feature size under these conditions is considerably larger than the irradiated zone. Alternatively, the feature size may be reduced by using long exposures at low power so that the rate of reaction heat generation is negligible.[198, 201]

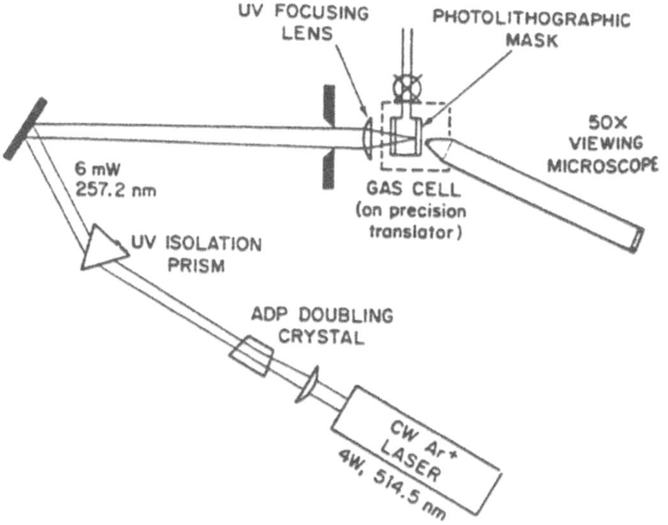

FIGURE 16-2-7. Apparatus for UV laser repair of chrome photomasks by deposition from the gas phase. From Ref. 32, copyright 1980 and courtesy of IEEE.

The potential applications of the time-dependent photodeposition process are in the area of circuit repairs (opens), personalization, deposition over steps,[16] and mask repair (opens).[31] A typical setup for mask repair of opens is shown in Fig. 16-2-7.

Although photodeposits of 2–3 μm in width were readily made with the single-element focusing lens, the limited resolution of the viewing microscope did not allow location of defects smaller than 10 μm.[32] The scaleup of the photodeposition process may be applicable to first-level packaging whose circuit size will gradually shrink to < 25 μm with "sub-mil" lithography and whose electrical requirements of the deposits are not as severe as in semiconductor lithography.

Besides metals, the deposition of other materials has been primarily concerned with mimicking chemical vapor-deposited films. The low-temperature vapor deposition of SiO_2 and Si_3N_4 has been commercialized,[33]

$$\begin{aligned} SiH_4 + 2N_2O &\xrightarrow[Hg]{h\nu} SiO_2 + 2N_2 + 2H_2 \\ 3SiH_4 + 4NH_3 &\xrightarrow[Hg]{h\nu} Si_3N_4 + 12H_2 \end{aligned} \qquad (16\text{-}2\text{-}7)$$

The reaction involves the Hg-sensitized step,[34]

$$Hg(^1S_0) + h\nu(2537\ \text{Å}) \rightarrow Hg(^3P_1)$$

$$Hg(^3P_1) + M \rightarrow Hg(^3P_0) + M^*$$

The excited Hg triplet donor transfers 4.5 eV of energy to the eventual silicon precursors. The photochemical deposition is used in a process (to replace plasma CVD) where the radiation (plasma) may damage the device. In most incoherent light (Hg) deposition, the rates are slow (100 Å/min) because of the low effective output of the sources. Boyer and co-workers[35] increased the rate of deposition and eliminated Hg sensitization by using a 193-nm Ar laser to (photosensitize by N_2O) decompose SiH_4 and form SiO_2,

$$N_2O + h\nu \rightarrow N_2 + O^1D$$

$$O^1D + SiH_4 \rightarrow SiO_2 + H_2O$$

The rate of SiO_2 deposition was independent of the substrate temperature and films of better quality than CVD-deposited oxide were formed. Other materials, such as

TABLE 16-2-2. Laser-Induced Oxidation

Substrate	Laser	Reference
Si, ⟨111⟩	Nd, 1.1 μm, 4.7 J/cm^2	37 (1980)
Si, ⟨100⟩	Ar, 546 nm, 5 W	38 (1981)
Si, ⟨100⟩	Nd, 266 nm, 4 J/cm^2	39 (1981)
Si, ⟨100⟩	Nd, 1.1 μm, 3 J/cm^2	40 (1981)
GaAs	Ruby, 0.7 μm, 1 J/cm^2	41 (1981)
Si, ⟨100⟩	CO_2, 10.6 μm, 5 W	42 (1982)

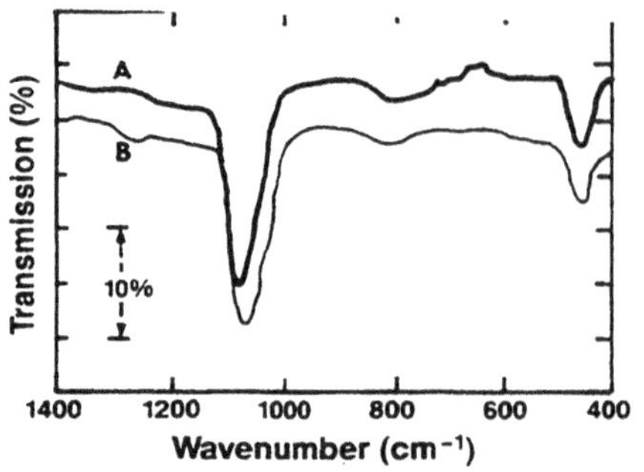

FIGURE 16-2-8. Comparison in IR bands of (A) laser-grown and (B) thermally grown oxidized silicon. From Ref. 42, courtesy of AIP.

silicon (Table 16-2-2), have been deposited by the CO_2 or the Ar[25] laser decompositions of SiH_4 or $SiCl_4$. Carbon has been deposited from acetylene and silicone monomers.[36] Because of the availability of short UV lasers (193 nm), the deposition of these layers will be a distinct possibility for applications as oxygen-resistant polymers in multilayer resists or as adhesion promoters.

Finally, the localized growth of SiO_2 can be achieved by oxidation or diffusion of oxygen into a Si surface (Table 16-2-2). The depth and rate of growth depend upon the incident dose. Depths of 10–1000 Å are grown[38] at a slow rate (1 Å/sec) at room temperature. Preheating the wafers to 300°C speeds up the rate of growth and greater depths (1–2 μm) of oxidation can be achieved at doses[39] of 5 J/cm^2. The depth was monitored by using an ^{18}O reactant. The laser-grown films are quite comparable to the furnace-grown layer (Fig. 16-2-8).

After absorption of the intense laser beam, localized melting of silicon produces a spatial growth of oxide in the spot (Fig. 16-2-9). Since surface nucleation is not involved such as in photodeposition[188] of metals, the resolution is limited by the heat transfer beyond the absorbed profile.

16-2-2. *Subtractive Etching and Removal*

The removal of a metal or insulator by a photochemically excited gas is very similar to plasma-induced etching[4] since the plasma generates a complex mixture of neutrals, radicals, and ions while a photon produces excited singlets, triplets, and vibrationally (especially with CO_2 lasers) excited atoms or fragments.[3] The photon coupling is more effective than plasma and, in fact, the laser exposure inside a plasma reactor has been suggested as a means of enhancing the rates of etching (> 0.1 μm/min) in a low-pressure reactor.[43, 44, 179, 181] The first "plasma etcher" invented in 1960 used photodecomposed F_2O gas to etch SiO_2.[45, 46] Since no ionic

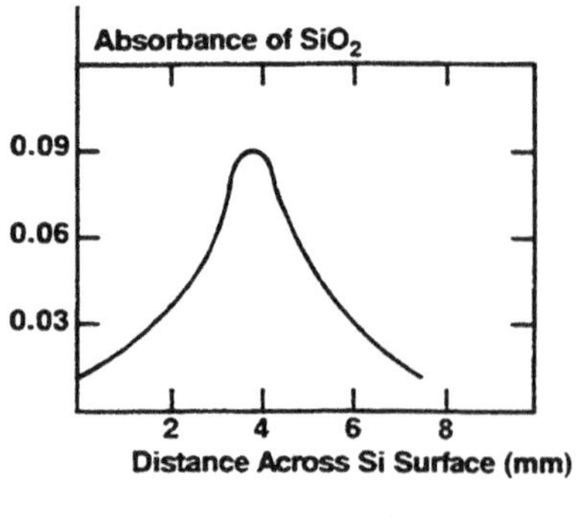

FIGURE 16-2-9. Spatial profile of CO_2 laser-induced oxide growth monitored by IR absorbance of SiO_2. A 1.5-mm aperture was used. From Ref. 42, courtesy of AIP.

TABLE 16-2-3. Laser-Assisted Gas Etching

Substrate	Etchant	Laser	Reference
SiO_2	HF	CO_2	42 (1977)
SiO_2	SF_6, XeF_2	CO_2	51 (1981)
Si	Cl_2	Ar, 546 nm	50 (1981)
SiO_2	SF_6	CO_2	53 (1982)
Si	SF_6	CO_2	53 (1982)
SiO_2	SF_6	Ar, 457 nm	53 (1982)
Te, Ta	SF_6	CO_2	54 (1982)
Mylar	O_2	KrF, 193 nm	55 (1982)
Ga As	Cl_2	KrF, 183 nm	155 (1984)
SiO_2	NF_3	ArF, 193 nm	187, 189 (1985)
Al	Cl_2	KrF, 193 nm	186 (1985)
Si	NF_3	KrF, 193 nm	173 (1985)

species are present, as in plasma etching, photon etching should not induce radiation damage.

High-power CO_2 lasers efficiently decompose a host of gases.[3] Steinfeld and co-workers[47] at MIT decomposed CF_3Br and produced rates as high as 1.2×10^3 μm/min [10^3 higher than the fastest magnetron plasma etchers (1 μm/min)].[48, 49] Other workers at MIT[50] used a focused Ar laser to excite Cl_2 and etch grooves (4 μm $\times$ 10 μm) in silicon wafer at a rate of 25×10^3 μm/min. Instead of using a beam perpendicular to the surface (and causing localized heating), Ehrlich and co-workers[50] and later Chung[51] used parallel beams close to the surface to excite HCl and Cl_2, respectively, to etch silicon. However, the rates are only 0.1–1.0 μm/min. Table 16-2-3 is a compilation of laser-assisted gas etching.[163–166, 185, 190]

The ablative removal[55] of Mylar is a high-resolution process since the eroding beam does not suffer forward scattering as in normal resist exposure. As the ablative material volatizes, the path length diminishes.[191, 196] Besides the localized heating, the pressure and mean free path of the photoexcited etchant can introduce isotropic sidewall erosion. Once radicals are produced in the gas phase, they can

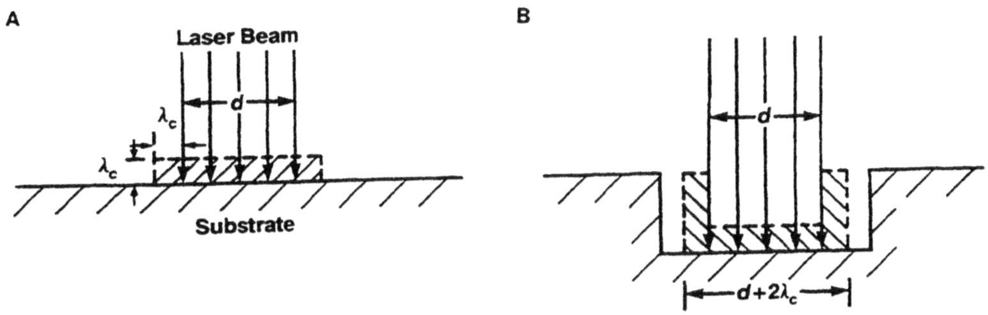

FIGURE 16-2-10. Model of adsorbed etchant molecules on a silicon surface excited by an IR laser (A) before and (B) after etching to some depth. Collisional deactivation can confine the chemically active molecules that are available to react with the substrate in a small region above the surface as indicated in the shaded region. λ_c is the mean free path for collisional deactivation as determined by the gas pressure. From Ref. 54, courtesy of IBM.

diffuse randomly toward the surface to cause chemical etching beyond the focal point. Chuang[54] calculated, using the lifetime of excited SF_6, the mean free path at $P = 266\ P_a$ (Torr). The mean free path Γ_c for molecular collision is < 0.1 mm (10^{-2} cm) and the lifetime before collision is $< 0.25\ \mu$sec. At this rate, the etchant can be deactivated within a radius of 10^{-2} cm. At 13,000 P_a, $\Gamma_c = 2 \times 10^{-4}$ cm. Consequently, as shown in Fig. 16-2-10, with a CO_2 laser beam of diameter d, the diameter of the etched area can be $d + 2\Gamma_c$. Since the CO_2 laser beam is limited by diffraction to a 10-μm spot, deep-UV lasers are required to reduce d and obtain better anisotropic etching.

16-2-3. *Summary of Gas–Solid Reactions*

The immediate application of these processes is in the area of circuit or mask repair at the 10-μm level. Doses as low as a normal resist dose ($\geqslant 100$ mJ/cm^2) can produce etchants or metal deposits. The anisotropic growth of metals by nucleation or desorption of contaminants are fundamental findings for future ultrahigh nanometer (100 Å) resolution.

The kinetics of heterogeneous gas-phase–solid reactions determine the rates of removal and deposition of materials. The removal of metals such as Al–Cu or Cr has not been investigated. Since photon-assisted plasma etching can induce orders-of-magnitude rate increases, the process may be useful in normal plasma etching to reduce resist heating and erosion.

The scaleup of these processes from a 1-μm spot to a whole wafer requires high-intensity sources as powerful as the CO_2 laser but operating at much shorter deep-UV (200 nm) wavelength. As the first step, a perforated foil mask may be useful for direct photon deep-UV patterning for semiconductor device fabrication. Further advances in organic material precursors (reactants) will prompt more investiations of gas–solid reactions.

16-3. LIQUID–SOLID REACTIONS

16-3-1. *Deposition Including Oxidation*

The deposition of a metal from a liquid is based on the processes of masked electroplating (Fig. 16-3-1).[171] The resist acts as a mask in (a) for electroplating conductor. In case (b), the surface is presensitized with $SnCl_2$ which adheres to only the hydrophilic substrate and not to the hydrophobic organic resist.[56, 57] In the late 1960s, the genesis of this technology was initiated for printed circuit boards. In 1970, the demand for finer lines of conductors, beyond the 5-mil limits of silk screening, fostered research into photo-assisted deposition.

The generation of metal patterns on dielectric substrates used an additive electroless process termed photoselective metal deposition (PSMD).[58] As shown in Fig. 16-3-2, the PSMD process used the photooxidation of Sn^{2+} to SnO_2 instead of the normal heating step,[56, 57]

$$Sn + O_2 \xrightarrow{\Delta} SnO_2$$

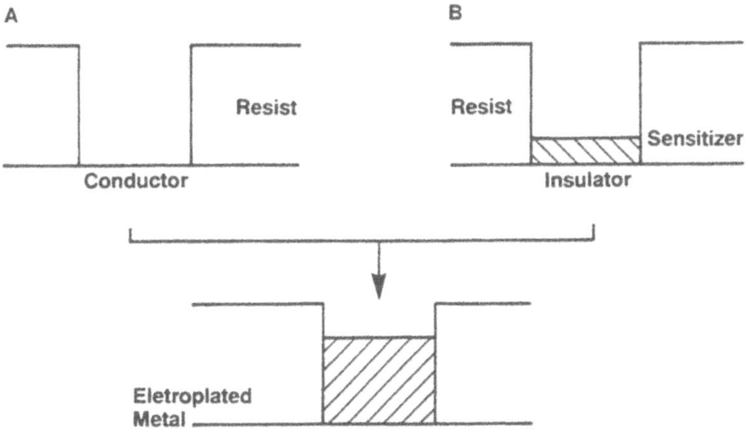

FIGURE 16-3-1. Principles of electrolytic and electroless deposition of metals with a resist mask to define pattern.

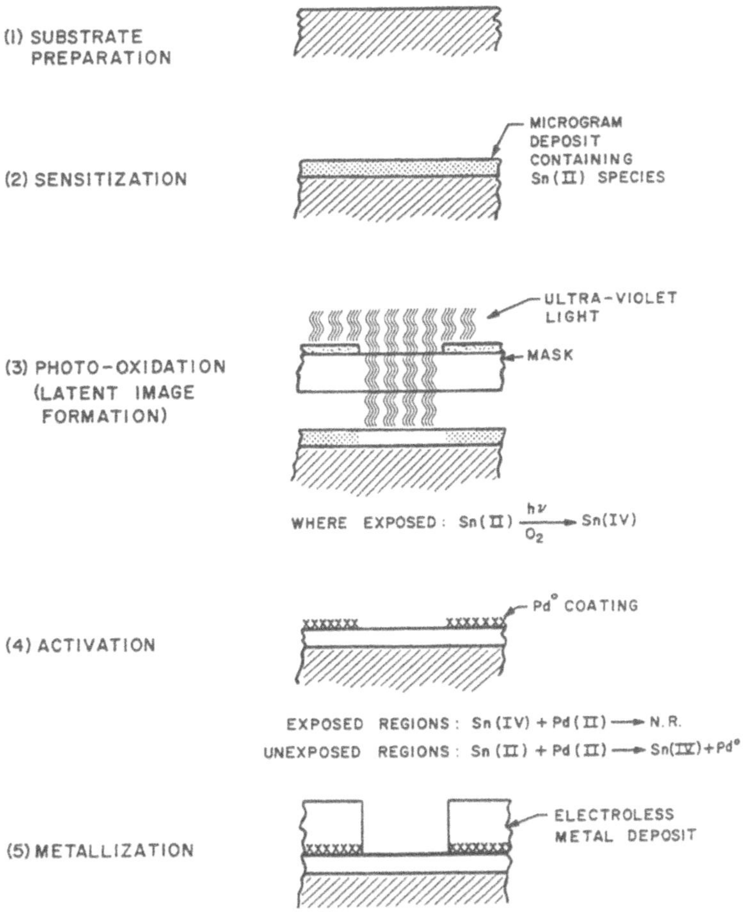

FIGURE 16-3-2. Schematic of PSMD UV sensitized electroless deposition. From Ref. 58, reproduced by permission of the Electrochemical Society Inc.

The destruction or oxidation of other sensitizing salts such as Ag, Hg, or Cu have formed the bases for larger dimensional photodeposition, 0.5–10 mil (10–250 μm) (Table 16-3-1).

Processes which do not require sensitizer activation are based on the excitation of the surface of a semiconductor. Upon illumination, silicon (p-type) or GaAs (300 nm) produces surface electrons (e^-_{surf}) by excitation into the conduction band.[192, 195, 197, 210] The reduction of a metal ion (M^+) on the illuminated silicon surface can be written as

$$M^{n+} + ne^-_{surf} \rightarrow M_{surf}$$

The rectifying nature of the semiconductor electrolytic junction acts to separate the photogenerated electron–hole pairs at the surface. The redox potential of the plating solution of Ni is -0.42 V (versus SCE), while depending on the pH, the band gap for p-Si in an aqueous electrolyte is -0.6 to $+0.5 \pm 0.4$ V.[63] The metal redox potentials are well within the band gap of Si or GaAs and illumination at 254 nm is sufficient to exceed the work function (photoelectric effect) of Si or most metals. However, sometimes the metal plating solution absorbs light and slows down the rate.

Excitation of a p-Si in a nickel plating solution with a 200-W Hg lamp (20 mJ/cm^2) produced a plating rate of 450 Å/min with a resolution of about 2 μm. The deposits, however, were microscopic metal clusters and had poor adhesion. Annealing in H_2 improved the adhesion and conductivity.[63] Instead of using the photovoltaic effect, workers at IBM focused an Ar laser (Fig. 16-3-3) on an Ni-plated glass cathode and induced localized heating and boiling with a dose of 1–100 J/cm^2.[64, 200] To explain the disruption of the electrolyte layer and the transport of fresh metal ions to the oxidizing cathode, the thermal dependency was calculated from the Peltier Effect [on the Gibbs–Helmholtz (F) free energy equation],

$$i = \exp(-nF\varepsilon_0/kT) \tag{16-3-1}$$

When the localized temperature rises, the plating current (i) increases since the disruption of the localized electrolyte boundary layer drives the cathode potential in favor of high current flow.[180] Because of the high temperature sensitivity, the control of image size is critically dependent on the laser rise time, the beam spot size,

TABLE 16-3-1. Metal-Photoassisted Liquid Deposition

Deposited	Substrate	Light	Reference
Cu, Ni	Epoxy ($SnCl_2$)	254 nm	58 (1971)
Cu	Epoxy (Hg^+, Ag^+)	365 nm	59 (1971)
Cu, Ni	Epoxy (Cu^{2+})	254 nm	60 (1975)
Cu	Cu on PVC[a]	250 nm	61 (1980)
Cu, Ni, Au	Sn, polymers[b]	254 nm	62 (1981)
Cu, Ni, Au	GaAs, p-Si	350 nm	63 (1981)
Cu, Ni, Au	Ni cathode	546 nm	64 (1982)

[a] Cu ions impregnate polyvinyl chloride.
[b] Photooxidized polymers as parylene (Union Carbide).

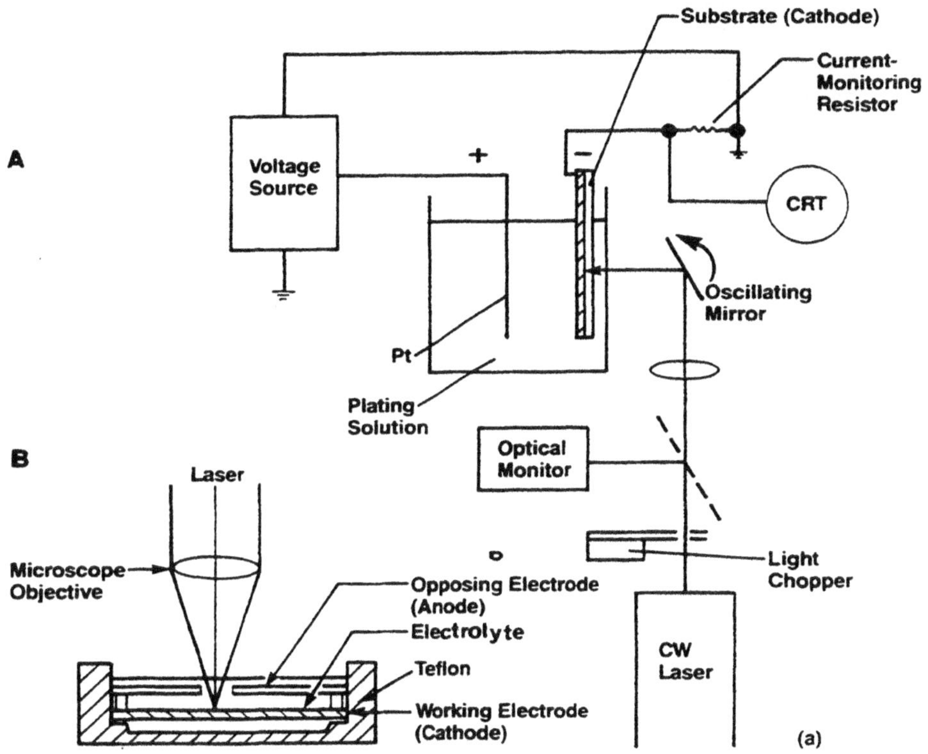

FIGURE 16-3-3. (A) Experimental arrangement for laser-assisted electroplating. (B) Cell for use with microscope objective to selectively focus laser beam for fine electroplating. From Ref. 64, courtesy of IBM.

and the thermal diffusivity (k) of the metal substrate. For a laser beam of 10- to 100-μm diameter, which is large in diameter when compared to the thermal diffusion spread ($\sqrt{kt_0}$, where t_0 is the laser pulse width), the local temperature rise was found to increase by $T^{1/2}$, consistent with a one-dimensional heat flow. As the plating thickness builds up, however, the heat flow in a radial direction lowers the temperature (Fig. 16-3-4) and causes overplating. For smaller spots ($< 10\ \mu$m), the heat flow becomes a three-dimensional problem.

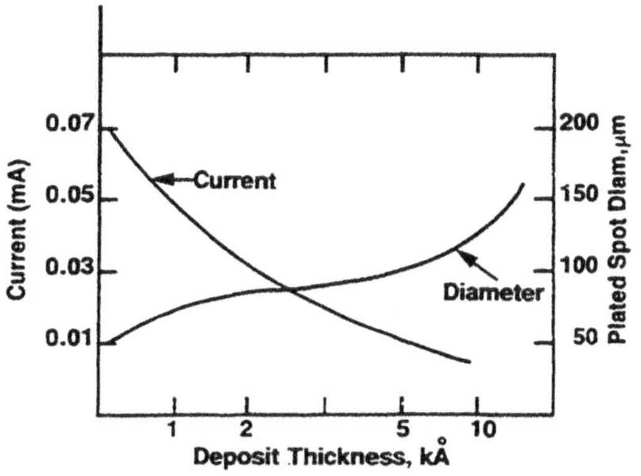

FIGURE 16-3-4. Influence of metal thickness buildup in laser-assisted electroplating. From Ref. 65, reproduced by permission of the Electrochemical Society Inc.

For additive applications, the electrolytic photoanodic oxidation of silicon was one of the first proofs of the photovoltaic effect.[12] Irradiation of silicon in the presence of H_2O_2 or H_3PO_4 produces SiO_2[67] and phosphosilicate glass,[68] respectively. At the same time, diffusion masks could be formed by projection of light patterns on silicon to induce localized anodic oxidation.[69]

16-3-2. *Etching at Liquid–Solid Interface*

The electrochemical etching of silicon requires the presence of holes according to the reaction[70]

$$Si + 2HF + 2e^{+} \rightarrow SiF_2 + 2H^{+}$$

In the 1960s, early efforts in laser optical memories were directed at storing optical bits on *n*-type silicon held at an anode. The exposure of silicon was also used to etch silicon.[12] In later work in the 1970s, laser holograms were recorded (*p*-type silicon was more difficult to etch). The systems were popular for instant image development.[71, 72]

The rate of etching of low-resistivity *n*-type silicon is proportional to the $I^{1/2}$ power of the irradiation.[72] Although some anisotropic etching can be achieved, the engraved surface of ⟨111⟩ material is pitted from preferential crystalline axis etching. The ⟨100⟩ surface is more suitable for fine engravings. Aberrational distortion by the liquid–air interface is a general problem for liquid–solid systems which limits resolution to about 2 μm. The diffusion length of electron–hole pairs (as well as radial heat transfer for spots < 10 μm) are other limiting factors on an ideal surface activated process. In oxide films, pinholes can also be detected by etching through the oxide and by monitoring the electromotive current flow.[73] A summary of the various photoactivated etchants is given in Table 16-3-2. Although the rates of etching (0.1–1.0 μm/min) are only equivalent to normal wet etching, the low-intensity[12, 73, 81] (20–100 mJ/cm^2) etching has the best chance for maintaining anisotropic resolution. Instead of exciting the band gap, workers at IBM again used a focus laser at 1–10 J/cm^2 to cavitate electrothinning (Fig. 16-3-3). An increase in

TABLE 16-3-2. Photoactivated Liquid Etchants

Substrate	Etchant	Light	Reference
n-Si, *p*-Si	HF, HNO_3	W	12 (1961)
SiO_2/Si	HF	Xe	74 (1969)
Au	CHI_3 in ϕ^a	Xe	75 (1969)
SiO_2	Hexafluoroisoproponal	Hg	76 (1970)
n-Si	HF	Ar, 151 nm	77 (1970)
Cu	CHI_3 in CH_3OH	Ar, 515 nm	71 (1971)
GaAs	H_2SO_4–H_2O_2	Hg	78 (1972)
Si_3N_4	ϕ_2NF CH_3OH	Ar, 515 nm	79 (1976)
Si	HF	Ar, 515 nm	73 (1979)
GaAs	KOH	Ar, 515 nm	80 (1981)
GaAs	H_2SO_4	He, 632 nm	81 (1981)
Fe	H_2SO_4	Ar, 515 nm	82 (1981)
Cu	$K_2Cr_2O_7$	Ar, 488 nm	152 (1984)

^{a} ϕ is benzene.

the etch rate of stainless steel by a factor of 10^4 was achieved in etching 160-mm holes.[82]

In summary, liquid–solid reactions have relied upon the principles of electroplating and electrochemical oxidation. The application of laser heating to locally enhance the etch rates by magnitudes is a significant finding and may find applicability in repair of shorts on masks and circuits. Most of the findings are applicable to circuit board and packaging applications at circuit features > 1 mil in size. If no heating is involved, anisotropic etching on a receptive surface is possible providing the depth of etching is less than the focus depth of the optical system. Optical image distortion by the air–liquid interface is a limiting factor.

16-4. SOLID–SOLID REACTIONS

In a subtractive process, the solid reactant may be an etching precursor such as a polymer containing an acid. For additive processes, the solid precursor may be a metal film cast on a transparent backing, a solid metal salt, or a diffusant contained as a spun-on polymer.

16-4-1. *Sublimography*

The deposition of a solid from a supported film in contact or in close proximity to a wafer has been called sublimography.[85] Resists, such as plating masks, can be deposited onto a surface instead of using the silk screen process which is limited to about 5 mil. In the 1970s, sublimable organic dyes were used for optical information storage (sublimed out of film and not deposited) and the same principle was applied in 1985 to form a portable conformable mask in PMMA resist.[187] One of the first examples of sublimography was reported by Moreau and Schmidt who deposited 2-μm images of chrome or dyed polymers by a pulsed laser

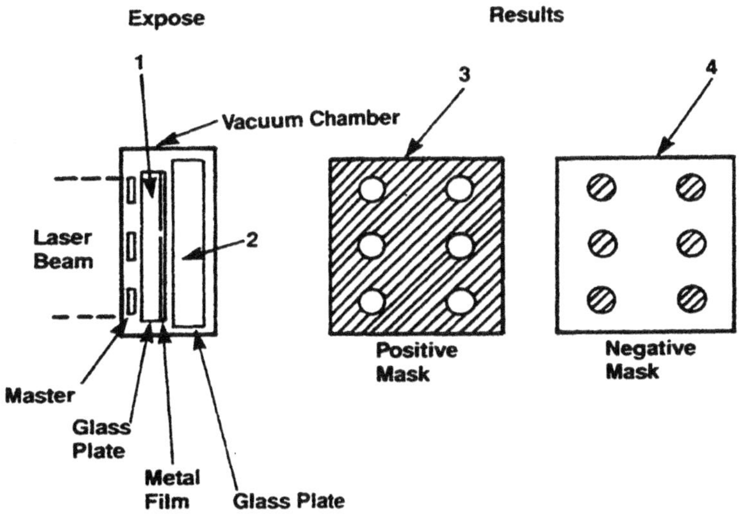

FIGURE 16-4-1. Simultaneous laser-assisted deposition (contact printing) of a chrome photomask by selective sublimation of chrome from one inverted film and redeposition on a glass substrate in contract with the chrome film. From Ref. 83.

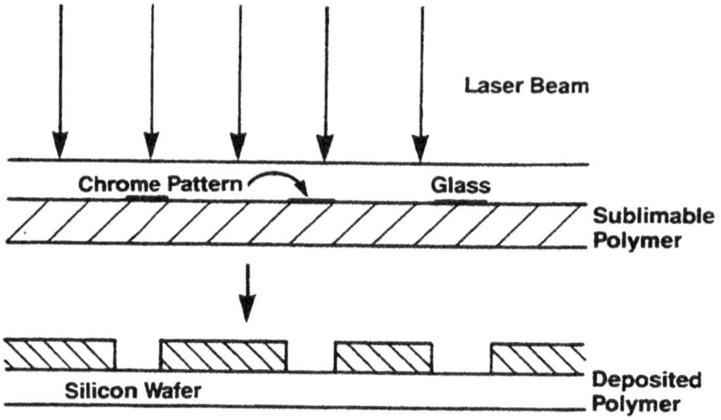

FIGURE 16-4-2. Sublimography in the additive mode. A polymer or resist precursor is deposited on a transparent substrate. A powerful laser beam heats the resist and deposits the polymer on a silicon wafer. The source layer is configured in a proximity printing mode.

irradiation.[83] A negative and positive chrome mask image were simultaneously fabricated (Fig. 16-4-1).

Since the heat of vaporization of polymers (organics in general) is only 0.1–1.0 J/cm^2 (Table 16-1-2) compared to metals (> 5 J/cm^2), organic materials can be transferred using a conventional chrome mask in a contact mode (Fig. 16-4-2). In another patent[84, 85] for depositing polymer patterns on printed circuits, dyed nitrocellulose was evaporated by masked Xe flash lamps. These processes may find applicability in extending the resolution of silk-screened compositions from the present 5-mil limit to 1 mil or below. In a negative tone, siloxane polymers are directly removed by an excimer laser, and fired to form SiO_2 via holes or diffusion masks.[170]

16-4-2. *Fill-and-Fire Resists*

At the printed circuit level, filled resists for the deposition of metals or conductors are a hybrid process which tries to circumvent silk screening plating or lift-off using a resist. The resist filler serves the purpose of defining an image on a surface followed by firing to burn off the resist and leave a conductive pattern or resistor behind (Fig. 16-4-3). Representative fill-and-fire formulations are given in Table 16-4-1. Since the formulations have been intended for patterns of 10–15 mil in size,

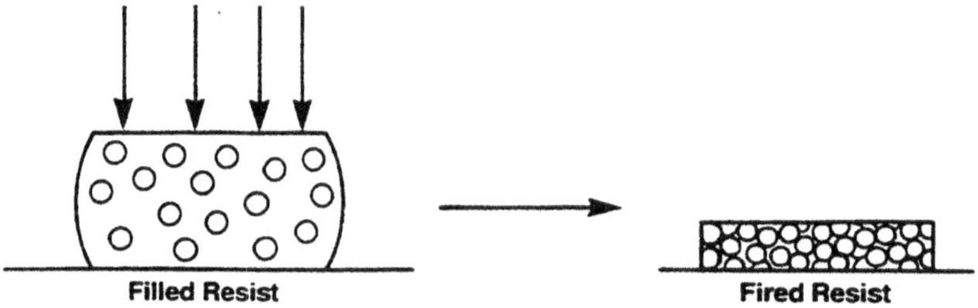

FIGURE 16-4-3. Fill-and-fire resist. The resist is loaded with appropriate conductor and insulator, patterned, and then fired in air to burn off resist and leave pattern of insulator or conductor.

TABLE 16-4-1. Fill-and-Fire Resists

Final deposit	Vehicle	Reference
Sn, Ni	Metal resinate in resist	98 (1971)
Ni electroless	Pd in resist	90 (1973)
SiO_2, fire	SiO_2 in KPR	91 (1970)
SiO_2, fire	SiO_2 in KMER	92 (1967)
SiO_2, fire	SiO_2 in KTFR	93 (1969)
Phosphors	Filled AZ-1350	94 (1968)
SiO_2, fire	Silylacrylate	95 (1980)

the resolution is limited by the particle size of the additives, the opacity of the cast film, the loading of the filler, and the final fired thickness. Specialized additives common to paint formulations are necessary to prevent agglomeration of inorganic or metallic fillers.(86)

Once an image is formed in a filled resist, the pattern is posttreated with hot air, ozone, or an oxygen plasma to remove the organic components.(88) By heating in air, metal powders are oxidized and serve as sensitization sites for electroless plating.(87)

Although most patterns are formed as negative resist images by direct irradiation, patterns can also be formed by lift-off of metals, polymers, or glass particles dusted on top of a positive resist image. In the "mid-film" process (Fig. 16-4-4), sprayed copper powder or glass powder becomes preferentially embedded on the surface of a resist. A final firing burns off the resist and forms a conductor of metal oxide.(87, 96) In other processes, an organosilicon polymer is coated on top of a plasma-erodable resist. The composite is burned and the excess organosilicon removed with a CF_4 plasma etchback.(174) A dry variation of this process is to dust glass powder on top of an imaged KTFR rubber resist. The resist is heated above 200°C to "pop-off" the undesired land and leave glass patterns.(97)

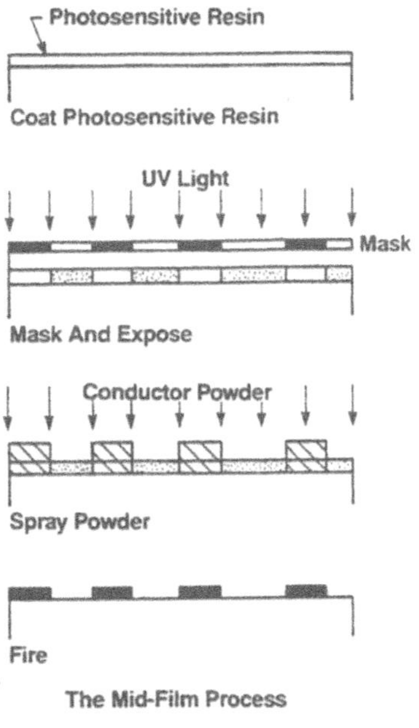

FIGURE 16-4-4. Mid-film process of depositing metal conductors on top of resist pattern followed by resist firing. From H. Naguib, K. Kavanaugh, and L. Hobbs, *Solid State Technology*, Oct. 1980, p. 109, reprinted by permission of Technical Publishing, a company of Dun and Bradstreet.

16-4-3. *Metal Silicides*

Metals (M) will alloy with silicon and form silicides (MSi) under the action of localized heat from a laser beam. Silicides have been formed by the laser exposure of Mo,(99) Pd,(100) Co,(101) and Ni.(102) Pulsed laser annealing leads to the formation of several silicide phases and substantial silicon precipitation.(98) On the other hand, CW laser annealing induces the formation of single-phase homogeneous layers free of precipitated silicon and well-defined interfaces. Therefore, the basic mechanisms of silicide formation are strongly dependent on the annealing temperatures and the effective annealing times.(98)

Besides silicides, alloys can be formed such as Zn–Al, by the laser decomposition of $Zn(CH_3)_2$ gas, followed by laser drive-in.(103) The alloyed area can also act as a wet etch resistant mask for etching Al films.

16-4-4. *Metal Salt Decomposition*

In his book on light-sensitive systems published in 1964, Kosar(212) listed about 50 derivatives of Hg, Pb, Cd, Ni, Cu, Zn, and Ag salts which undergo self-oxidation–reduction to form the free metal. The must famous of these is AgBr, but silver acetate or glutamate will also photodeposit metal,(104)

$$AgOOCH_3 \rightarrow Ag + CO_2 + CH_4$$

The precursors can exist as liquids, solids, or condensed gases, either as pure materials or dispersed in a matrix. For example, $SnCl_4$ vapor is photolyzed at 254 nm to deposit tin on GaAs devices. A side product, Cl_2 gas, etches the surface of GaAs and should be avoided. Tetramethyltin as a precursor does not produce an etching product.(203)

As long as continuous amorphous precursor films can be formed, most metal halides or metal organic salts decompose under the action of light or high-energy radiation to form metal deposits.(193) Doses as high as 1 J/cm^2 for photodecomposition and 10 C/cm^2 for the *e*-beam-induced decomposition of Ca(167) and Al(168) fluorides are required.

16-4-5. *Generation of Etchants from Solid Precursors*

The photogeneration of etchants by photolysis of solid film precursors was developed at General Electric(105) for a Navy project to form dry-developed "Polaroid" images in metals. The etchants (Fig. 16-4-5) were derived from cellulose acetate cast films containing diazonium fluoroborates, trityldifluoroamine (ϕ_3CFN_2), *N*-halosuccinimides, or haloforms (CHX_3). For example, HF can be generated from irradiation of benzene diazonium fluoride,

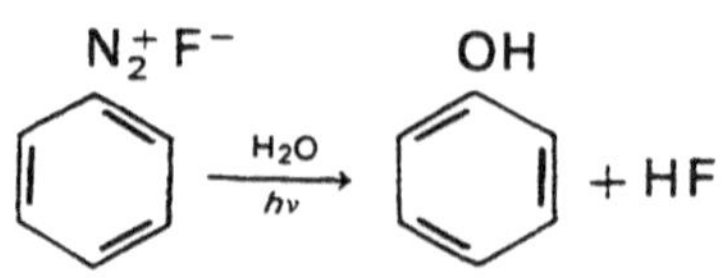

The generation of the etchant at the surface of the metal is similar to the HF per-

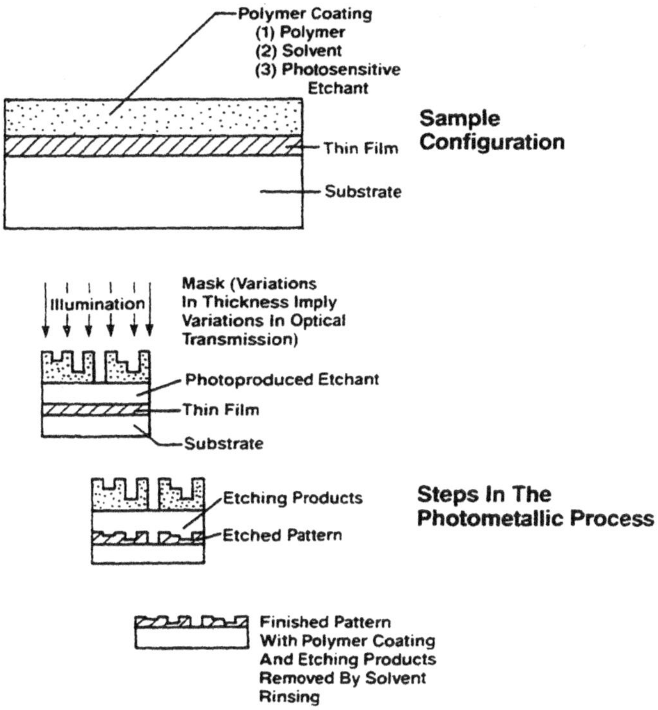

FIGURE 16-4-5. Photoassisted etching of metals by generation of etchant in resist layer. From Ref. 106, courtesy of AIP.

meation etching through a latent photoresist image discussed in Chapter 13. The advantages of the process are:

1. A continuous variation in depth (halftones) can be produced in metal, or SiO_2 films proportional to the dose.
2. Elimination of the necessity for good resist adhesion or etch resistance.
3. Ease of removal since the polymers are not cross-linked.

The disadvantages of the process are the slow speed (10–60 min) to etch 5000 Å SiO_2. Laser exposure should be able to speed up the rate of reaction, since thick polymer films (> 1 mil) have to be used to trap sufficient moisture. The resolution is limited to about 10 μm. An improved system which simultaneously generates HF and water consists of the following[(107)]:

- Etchant—trifluorocresol
- Binder—polystyrene
- Photosensitizer—benzoin
- Water precursors—nitrosodimethylaniline or benzyl cyanide

The decomposition product etches 0.5-μm-thick SiO_2 in 60 sec at an exposure dose of 100 mJ/cm^2. As long as sufficient moisture is present, the etching products can diffuse away from the etch zone.

16-5. DIRECT LASER TREATMENTS

Over the last decade, CO_2, ruby, Nd–Yag, and focused Ar lasers have been used to treat metal films and silicon wafers for a variety of phase change processes. Currently, deep-UV excimer gas lasers are the popular sources to etch metals,[202] oxides,[194] and polymer films.[196] The evaporation of metal fims (hole burning) to record optical bits was undertaken by many electronic companies interested in optical memory systems.[108] A dose of about 2–3 J/cm^2 (Table 16-1-2) will evaporate most metallic films from a transparent substrate.

In order to absorb the photon energy, the metal film must be at least 150 Å thick (Fig. 16-5-1) or the film will lose too much of the incident energy by reflection.[66] For a CO_2 laser, the absorption coefficient (A) for a highly reflective metal can be approximated by

$$A_{max} = 1/(1+n) \tag{16-5-1}$$

where n is the refractive index of the metal.[109] Because of the metallic surface, much of the energy will be reflected. Since the pulse times are on the order of 10^{-7} to 10^{-3} sec and the thermal diffusivity (k) on the order of 1 μm/sec, the absorbed photon will be thermalized and propagated to a depth greater than 1 μm at the end of the irradiation. In the case of large-area irradiation ($> 5\ \mu$m in diameter), the radial heat flow can be neglected. For the specific case of a Pd film on top of a silicon wafer, Liau and co-workers[110] treated the Pd film as a one-dimensional heat source. For spots smaller than 5 μm, the radial heat flow beyond the beam cannot be neglected. The need for a three-dimensional model is apparent from the elementary consideration of the aluminum thermal diffusion length, $\sqrt{kt}$. For a 70-nsec laser pulse, $\sqrt{kt}$ is approximately 2 μm. Aboelfotoh and von Gutfeld[66] performed computer simulations to the basic heat balance transport equation,

$$K_i \nabla^2 T - pc\frac{\partial T}{\partial t} + A = 0 \tag{16-5-2}$$

where K_i, p, and c are the thermal conductivity, density, and heat capacity, respectively, and A is the heat input in terms of power/volume. In Fig. 16-5-2, the temperature rise for laser heating of a 150-Å-thick aluminum film on plastic is shown. The curves indicate that the presence of the Mylar (DuPont trademark) substrate

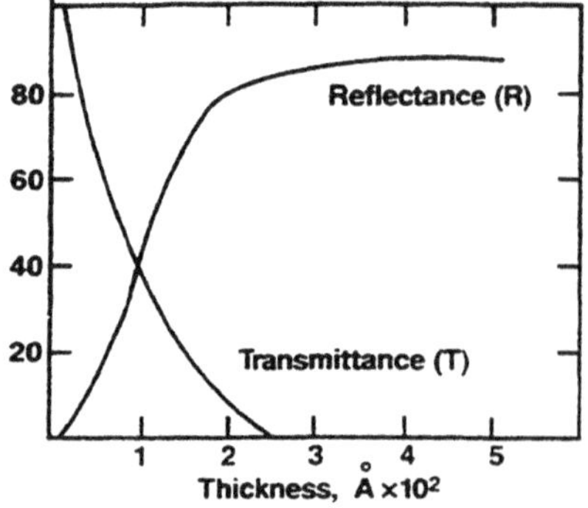

FIGURE 16-5-1. Variations of reflectance and transmission of Al films at 514 nm. From Ref. 66, courtesy of AIP.

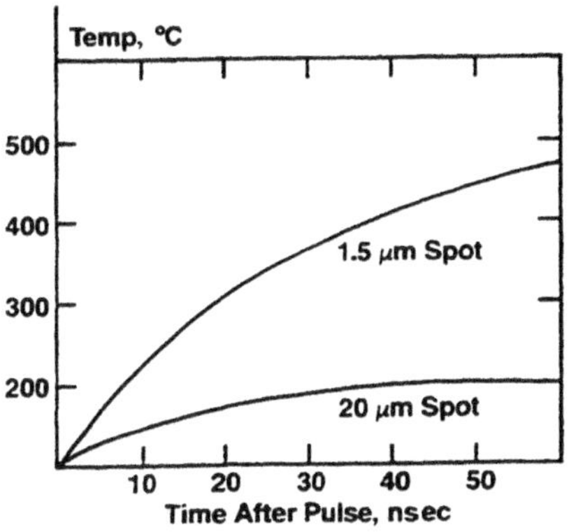

FIGURE 16-5-2. Temperature rise in 150-Å-thick Al film after laser impact. From Ref. 66, courtesy of AIP.

results in a lower film temperature (2.4×) after a 70-nsec pulse. The analytical separation of the deposited energy into the heat losses to the substrate and by radial transfer are shown in Fig. 16-5-3. The radial heat loss is approximately equal to the heat loss to the underlying substrate. Thus, a one-dimensional treatment will greatly overestimate the maximum temperature rise in the film. The heat (E) required for vaporizing a metal is

$$E = m(C_P \Delta T + \Delta H_f + \Delta H_{vap}) \qquad (16\text{-}5\text{-}3)$$

where C_P is heat capacity.

For chromium films (0.1 μm) on glass, a dose of 1 J/cm^2 is required.[111] For a spot of 5-μm radius, the removal of an extra chrome defect with an Nd–Yag laser takes 50 nsec. Since radiation after this time serves only to heat the substrate, it is desirable that the pulse be no longer than this time. At lower powers, metal silicides were formed by the surface melting, mixing, and rapid resolidification (see Section 16-4-3).[98–102]

In silicon annealing, lasers have been more effective than thermal annealing in reducing damages of ion implantation, dislocation loops, precipitates and in the transformation of amorphous Si to single crystals.[112] In Fig. 16-5-4, the melt front as calculated by Baeri and co-workers[113] is shown. The thermal conductivity spreads the energy into the depth[187] even during the duration of the pulse. If sufficient energy is introduced, a surface layer will melt.[114] After the heat flows away, the molten surface refreezes. Pulse beams with pulse lengths between 10 and 100 nsec introduce localized energy into a thin silicon layer. The precise control of

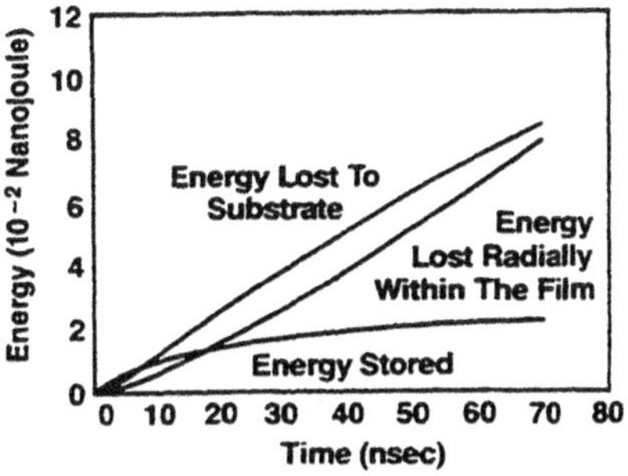

FIGURE 16-5-3. Thermal transport components of energy deposited in 15-nm-thick Al film on Mylar after receiving a 70-nsec 100-mW laser pulse. From Ref. 66, courtesy of AIP.

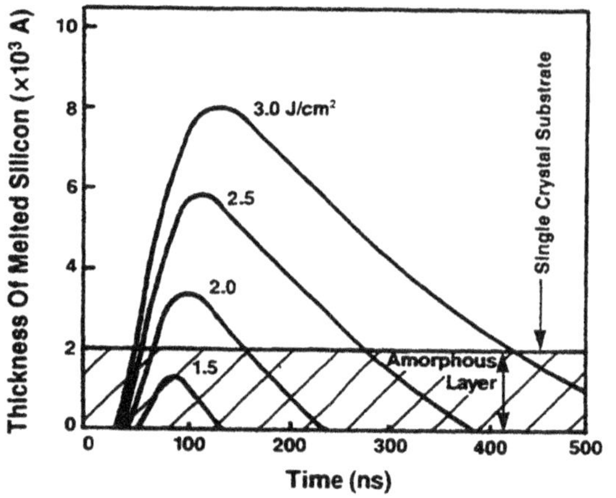

FIGURE 16-5-4. Laser heating and melting of silicon. The melt front penetration versus time and laser pulse energy[113] density is shown. The crystallization of the melted and refrozen material depends on whether the melt has penetrated through the amorphous layer to the crystalline substrate. From Ref. 114, courtesy of AIP.

junction depth, dopant diffusion, as well as lowering the defect concentration, will become a routine process for VLSI devices.[114]

In the thick-film circuit area, the laser has been widely used in electronics production.[115] Popular operations are as follows:

- Resistor trim
- Ceramic scribe and cut
- Microsolder
- Wafer scribe
- Wafer dice
- Insulation strip
- Desolder
- Photomask fabrication

The development of large-diameter beams (> 100 mm) and durable masks is needed to augment the localized laser processes.

16-6. SUMMARY OF PHOTOASSISTED PROCESSES

Although most of the direct image processes were developed for printed circuits, the basic principles, materials, and processes can be applied to semiconductor and submil dimensions in packaging. The availability and application of powerful laser sources will be the significant limiting factor. A summary of laser-assisted processes for pattern definition (additive and subtractive methods) is given in Table 16-6-1. For nonthermal exposure, the activation of surface monolayers is an important reaction for high-resolution lithography. Currently, the most immediate application of direct patterning is in the deposition or removal of fine metal lines for photomask repair. Deposition can be produced by gas-phase decomposition or electroless plating photosensitization. The removal of metals by chemical reaction needs improvement since etchants which produce volatile products have not been found.

Since the radiation is confined to surface layers, the contrast of the systems are very high (Table 16-6-2). The resolution of photon processes is about 1–2 μm. Indeed, submicrometer lithography could achieve wavelength-limited images

TABLE 16-6-1. Demonstrated Rates and Linewidths for Laser Patterning

Process	Temperature rise (K)	Rate (μm/sec)	Smallest line (μm)
Photodeposition	0–10	1–5	0.7
Photopolymerization	5	1	0.8
Photoelectrochemical etching of GaAs	1	0.01–0.1	1
Photochemical dry etching	0–10	0.01–0.1	2
Thermal dry etch	1000–5000	4000	0.3
Thermal deposition	600–1500	200–300	0.2
Thermal doping (Si)	1200–1400	1	0.2
Thermal wet etch	100–200 (Al)	10	1.5

[a] From Ref. 158.

(0.2 μm) if the optical diffraction and lens aberrations are minimized. Surface-activated reactions do not suffer from "thick" 1-μm film optical distortion encountered with resist exposure.

For high-power lasers, localized memory repair, mask repair, and customizing metallurgy are probable uses. Lower-power lasers will find use in the deep UV as light sources for normal resist exposures. Direct patterning will find immediate application at the circuit or chip level involving maskless exposure. Masked exposure through durable and/or perforated masks will require new innovations and sophisticated laser optics.

The thermal energy of the laser has been extensively used in silicon wafer processing. The control of the three-dimensional heat transport is the limiting resolution factor.

Now for the future implementation of direct imaging processes, new materials and processes requiring perhaps image amplification to reduce the dose to 1 mJ/cm^2 are necessary to spread the beam over large surface areas. The present low throughput remains as the prominent technical challenge for this new area of unconventional semiconductor lithography.

TABLE 16-6-2. Effective Contrast for Laser Direct-Write Systems[a]

Process	Contrast
Etching	
Si/Cl_2	22
Ge/Cl_2	18
Si/HCl	5.8
Thermal deposition	
SiH_4	11
$SiCl_4/H_2$	30
Doping	
Si	48
Metal nucleation	18–23

[a] From Ref. 158.

16-7. ELECTRON-BEAM-ASSISTED PATTERNING

16-7-1. *Deposition by Electron Beams*

Prior to 1960, *e*-beams were primarily used for scanning electron microscopy and for the thermal machining of metals. These processes included evaporation, soldering, welding, and *e*-beam cutting. In the early 1960s, researchers at Cambridge University[117] in England and at the Stanford Research Institute[116] began to consider the scanning electron microscope for direct microcircuit fabrication. These workers concentrated upon utilizing surface chemical reactions and focused evaporation for inducing polymer formation as etching. Glass precursors of organosilicone monomers derived from silicone pump oil, triphenyl silanol, and tetraorthosilicate were adsorbed and polymerized by the *e*-beam.[118-121] Inorganic monomers were such precursors as stannous chloride, tetrabutyl tin, tetrabutyl lead, and tungsten hexacarbonyl.[120] To deposit Ni, monel metal masked the irradiated patterns of adsorbed $Ni(CO)_6$.[122]

The deposition from adsorbed gases depends on the vapor pressure and substrate temperature.[118] Most monomers (small molecules) can be decomposed by *e*-beams since excess energy (20–150 keV) is available. The absorption cross section is weak (1%) for the low-density and low-atomic-number carbon precursors. Since the *e*-beam operates at low pressures ($< 10^{-5}$ mm), lower-vapor-pressure monomers can be used than in laser exposure. However, *e*-beam guns are more easily poisoned by organic vapors (Chapter 9). The backscattering of electrons from the substrate induces lateral image spread unless the silicon wafer is deliberately thinned to $< 1.0\ \mu m$ in thickness.[123]

From the solid phase, the deposition relies on the well-known decomposition of metal alkyls,[201] metal acetylonates,[199] and metal halides,[184]

$$2\,AgCl \rightarrow 2\,Ag + Cl_2$$

Since a high dose ($10^{-2}\ C/cm^2$) is required to deposit 0.8 μm of Ag,[124] full wafer writing is not practical. Electron-beam irradiation of silanes deposits SiO_2 and SiN_4.[162] Electron-beam annealing instead of CW laser annealing offers some additional advantages[134]:

1. Depth of penetration is more easily controlled.
2. Surface oxide films do not cause interference.
3. There may be higher cost and radiation damage.

16-7-2. *Electron-Beam-Induced Etching*

Although these processes require higher doses ($> 1\ J/cm^2$, $10^{-2}\ C/cm^2$) than photon-activated processes, they were examined in the early 1970s for their potential in silicon-induced etching. The bombardment-enhanced etching rate (BEER) of insulators has been investigated for the various insulators (Table 16-7-1) wherein the damaged silicon or insulator etches more rapidly ($> 3\times$) in a plasma[183] or liquid etchant than the unirradiated region. Since Si_3N_4 films required hot phosphoric acid etchant at 180°C, no organic resists were able to withstand the

TABLE 16-7-1. BEER Process

Insulator	Etchant	Reference
SiO_2, Si_3N_4	Zr, HF	125 (1971)
SiO_2	HF	126 (1971)
Al/SiO_2	HF	127 (1971)
Al_2O_3	NaOH	128 (1971)
TiO/SiO_2	NaOH	129 (1974)
Fe_2O_3	HCl	130 (1974), 137
Si_3N_4	HF	131 (1975)
Si	N_2 [a]	132 (1981)
Ni plate	Polymers[b]	133 (1969)

[a] Grow Si_3N_4.
[b] Irradiate polymer.

high temperature. To solve this problem, Lepselter[(123)] deposited Zr on top of Si_3N_4 or SiO_2 and *e*-beam induced the following displacement reactions:

$$Zr + SiO_2 \rightarrow ZrO_2 + Si$$

$$3Zr + Si_3N_4 \rightarrow Zr_3N_4 + 3Si$$

Since there is no significant beam spreading in the thin (0.05 μm) Zr layer, anisotropic images were formed by removing the reaction products in HF. Subsequently, transistor fabrication by the BEER technique was described by O'Keefe and Handy.[(135)] Radiation damage was not apparent since the electron energies (1–30 keV) were too low to cause damage to the silicon, whose damage threshold is 45 keV. Since a dose of 1 C/cm^2 produced a 3 to 1 increase in HF in etch rate of exposed to unexposed SiO_2 or Si_3N_4, a thicker than normal (0.5 μm) thermal oxide film was required to compensate for the background loss. Tapered images (70°) were formed due to unexposed oxide removal. If an aperture mask of an aluminum (photolith patterned on the oxide) is used, profiles close to 90° can be obtained. Another modification of the process to etch Si is the growth of an oxide mask induced by *e*-beam bombardment.[(151)]

Since the BEER process requires very high doses (1 C/cm^2 or 2 J/cm^2), the effective *e*-beam intensity tends to decrease as semiconductor resolution requirements (< 2 μm) are increased. A full-field photocathode *e*-beam exposure tool was developed at Westinghouse in the 1970s.[(136)] The photocathode pattern of Ti on glass emits electrons when UV irradiated. The projected pattern of a chip was focused on a silicon wafer. Although the exposure time per chip was reduced from hours to minutes, the photocathode system still suffers from burnout (poor cathode life) and overlay limitations.

In the BEER process, the most critical parameter is the depth and spread of the penetrating beam. The depth of penetration is related to the accelerating *e*-beam voltage (V_a) by the relation[(2)]

$$Z = (0.0456/p)\ V_a 1.75 \qquad (16\text{-}7\text{-}1)$$

where p is 5 for the density of SiO_2. Based on optical absorption studies, Kammlott and Sinclair[(137)] concluded that the electron irradiation induces localized crystallization. The accelerating voltage must be chosen to allow penetration to ground but of sufficient absorption to obtain localized heating above 500°C. In the majority of cases, the *e*-beam serves as a localized heat source to induce physical and thermal chemical reactions in thin films. Above a dose of 1 C/cm^2, the direct *e*-beam machining of a silicon wafer can be achieved. The process is monitored by an *in situ* X-ray probe[138] of the surface undergoing large area ($> 10\ \mu m$) removal. At 25 keV, the smallest regions which can be melted are on the order of 10 μm, while at 50 to 150 keV, the smallest regions which can be melted are two to three times greater. The fast transient melting[114] and refreezing of a thin silicon surface layer results in drastic changes in the distribution of impurities in the silicon (a miniature form of zone refining). A scanning CW beam can be used to form amorphous CVD silicon deposits.[114]

To sum up, the use of an *e*-beam to fabricate parts without resists has the best potential for submicrometer lithography since the beam can be focused, aligned, and moved with 0.05-μm accuracy. However, the high dose required ($> 10^{-4}$ C/cm^2) and the backscatter effects are major impediments to their practical utilization. Low throughput and radiation damage remain as economic and reliability concerns.

16-8. ION-BEAM PROCESSES

Since the absorbed ions do not suffer from the severe scattering problems of electrons or photons, projected masked or focused ion beams have been cited[139] as the ideal source and images with 90° anisotropic profiles have been formed in polymeric resists. In an ideal silicon fabrication process, the doping ions from focused beams of As, B, and P are directly implanted by a scanning ion beam. Unfortunately, the focusing of ion beams of other than light atoms has not been able to achieve the sufficient beam intensity[140] for sufficient throughput and chip size beam deflection. Although ion-beam irradiation has not received as much attention as *e*-beam lithography, there have been some attempts at resistless lithography:

1. Proton bombardment of SiO_2 to induce etching[141] (10 to 1 etch rate ratio).
2. Bombard Si in presence of N_2 or O_2 and grow Si_3N_4 or SiO_2.[142, 199]
3. Ion implant with B, P, or As to a micrometer-size image to produce *n*- or *p*-type conductivity.[143, 176]
4. Bombard with atomized metal ions along a path of a grounding *e*-beam to induce localized metal deposition (Fig. 16-8-1).[144, 154]
5. Etch GaAs with a Cl_2^+ beam.[145]
6. Focus UV on metal, generate ejected electrons, ionize the ambient gas, and accelerate the ionized gas back onto the metal to be etched.[146]
7. Step and repeat foil mask ion-beam-induced etching.[148, 178, 204, 205]
8. Damage silicon and anodize to grow SiO_2.[148]
9. Deposit metal patterns[157, 203] or silicone[153, 156, 175] from irradiation of metal organics.
10. Reversal etching of Cr in a plasma after implant with As^+.[174]

An interesting combination of ion and *e*-beams was applied to the localized

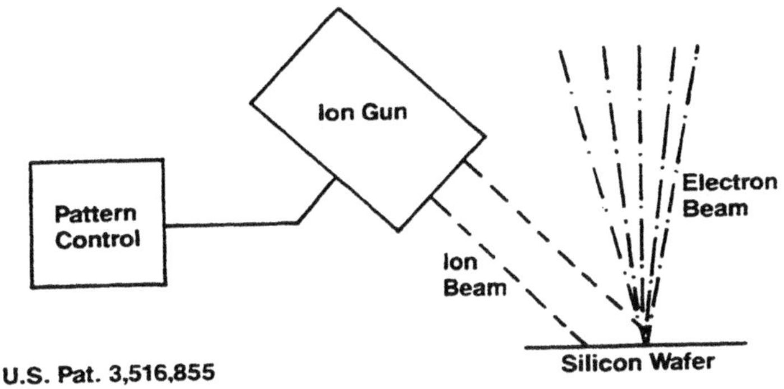

FIGURE 16-8-1. Ion-beam neutralization of the charge buildup during *e*-beam-induced etching of thick insulators.

deposition of metal for packaging applications. The system as shown in Fig. 16-8-1 uses a metal vapor which is ionized to create metal vapor ions. By impinging an *e*-beam on the substrate, a high negative charge concentration is built up which attracts and neutralizes the metal ion vapor inducing localized deposit.

The most successful attempt to employ ion-beam-induced etching has been the work of Stengl[147, 206] and co-workers at Sacker who developed an (H^+ or Ar^+, 60–100 keV) ion projection step-and-repeat system using a 10× nickel foil mask. For resist exposure, a dose of 10^{12} ions/cm^2(10^{-6} J/cm^2) is required, while enhanced SiO_2 etching needs 10^{16-18} ions/cm^2(0.1–1.0 J/cm^2). The alignment is accomplished by signals derived from backscattered ions to a layer-to-layer accuracy of ± 0.1 μm. Radiation damage to silicon has to be carefully controlled by the choice of energy level (keV). Since an etch rate ratio of 3 was achieved, the profiles in oxide are about 70° which is similar to the results obtained with *e*-beam-enhanced etching (BEER).[135] Ions,[141] like electrons, can induce selective[150] etching in oxide. However, ion damage can be confined to < 0.1 μm in depth. At doses of 10^{13} ions/cm^2, 50-keV Ar ions were used to damage a top layer of oxide. Afterwards the top surface (Fig. 16-8-2) was overcoated with a resist, patterned, and etched. The HF etching induced a severe lateral undercut ($D/L \ll 1$, Chapter 13) to produce 30° steps for further phosphosilicate glass growth.[149]

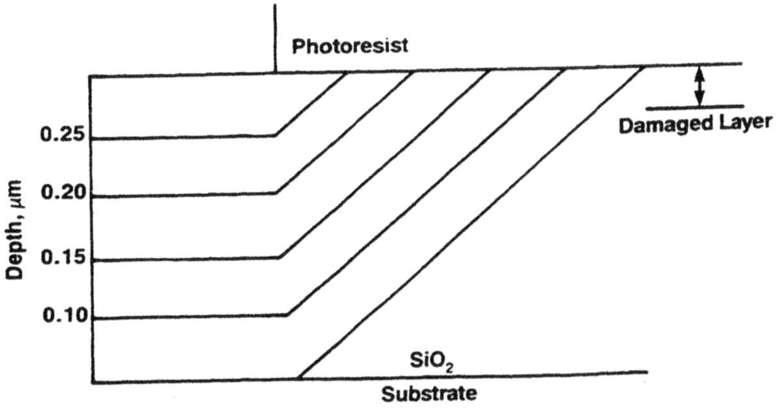

FIGURE 16-8-2. Calculated edge profile in etching of radiation-damaged SiO_2 at an etch rate ratio of 3 to 1. From Ref. 149, reproduced by permission of the Electrochemical Society Inc.

The most successful commercial application of ion-beam resistless lithography has been in the repair of photomask defects. After the scanning ion beam locates a clear or opaque defect, the beam sputters away the extraneous material or clear defects are filled with condensed deposits.(177, 207–209)

Ion-beam lithography is a technology which is in its infancy. The wide variety of sources and the nonscattering features make this technology a strong contender for resistless lithography. Advances in sources, projectors, and durable masks are required before future implementation.

16-9. SUMMARY AND OUTLOOK OF RESISTLESS DIRECT PATTERNING LITHOGRAPHY

Since high doses ($> 1\ J/cm^2$) are required for present processes and throughput and/or resolution is severely limited, direct resistless imaging will be primarily used for off-line applications such as in the repair of circuits or masks and in low-density "submil" packaging applications. An assessment of the present state of the art is given in Table 16-9-1 for processing silicon wafers without a resist using reactions of a gas etchant on a solid surface, a liquid with a solid surface, and reactions involving solids.

The most advanced system is the ion foil mask projection system.(147) The critical factors in direct write are threefold:

1. Optics—Complexity of mask, flatness, durability, overlay, and beam stability

TABLE 16-9-1. Direct Patterning Process Capabilities

a. Resolution

	Gas–solid		Liquid–solid		Solid state	
	Deposit	Remove	Deposit	Remove	Deposit	Remove
Photon	2 μm	λ	5 μm	λ	3 μm	λ
Electron	<1 μm	1 μm	—	—	2 μm	2 μm
Ion beam	<1 μm	1 μm	—	—	2 μm	1 μm

b. Processing area[a]/throughput[a]

	Gas–solid	Liquid–solid	Solid state
Photon	No	No	Yes
Electron	Yes	—	No
Ion beam	No	—	No

c. Projection mask exposure capability

	Gas–solid	Liquid–solid	Solid state
Photon	Yes	No	Yes
Electron	Yes	—	Yes
Ion beam	Yes	—	Yes

[a] < 10 sec per 7.5-mm chip.

2. Process—Throughput, radiation damage, and competition from alternative resist processes
3. Material—Dose (chemical and thermal effects), surface limiting rates, heats of fusion, vaporization, or activation energy

Since the dose requirements for resistless exposure are orders of magnitude higher than resist exposure doses (0.1 J/cm^2), new chemical reactions or process amplifications are required before direct patterning will be a viable contender.

REFERENCES

1. P. Nehmiz, H. Bohlen, and J. Greschner, *J. Vac. Sci. Technol.* **19**, 971, 1291 (1981).
2. J. Calvert and J. Pitts, *Photochemistry*, Wiley, New York, 1966, p. 226.
3. J. Steinfeld, *Laser-Induced Chemical Processes*, Plenum Press, New York, 1981.
4. D. Flamm and V. Donnelly, *Plasma Chem. Plasma Proc.* **1**, 317 (1981).
5. Z. Liau, B. Tsaur, and J. Mayer, *Appl. Phys. Lett.* **34**, 221 (1979).
6. G. Basile and A. Moisan, Electron Beam and Ion Beam Science and Technology, Eighth International Conferences, Electrochemical Society, 1978, p. 618.
7. R. Moore, G. Caccoma, H. Pfeiffer, E. Weber, and O. Woodard, *J. Vac. Sci. Technol.* **19**, 950 (1981).
8. G. Stengl, R. Kaitna, H. Loschner, R. Rieder, P. Wolf, and R. Sacher, *J. Vac. Sci. Technol.* **19**, 1164 (1981).
9. U. S. Patent 3,378,401 (1968), 3M.
10. J. Edison, *IEEE Spectrum* July 1981, p. 27.
11. M. Lacombat, J. Massin, G. Dubroeucq, and M. Brevignon, *Solid State Technol.* Aug. 1980, p. 115.
12. J. Braun, *J. Electrochem. Soc.* **108**, 588 (1961).
13. W. Moreau and P. Schmidt, *Electrochem. Soc. Ext. Abstr.* Fall 1970, p. 455.
14. R. Solanki, P. Boyer, J. Mahan, and G. Collins, *Appl. Phys. Lett.* **38**, 572 (1981).
15. D. Ehrlich, R. Osgood, and T. Deutsch, *IEEE J. Quantum Electron.* **QE-16**, 1233 (1980).
16. D. Ehrlich, R. Osgood, and T. Deutsch, *J. Vac. Sci. Technol.* **21**, 23 (1982).
17. J. Ameen and A. Tencza, *IBM Tech. Discl. Bull.* **12**, 1286 (1970).
18. T. Deutsch, D. Ehrlich, and R. Osgood, *Appl. Phys. Lett.* **35**, 175 (1979).
19. D. Ehrlich, R. Osgood, and T. Deutsch, *Appl. Phys. Lett.* **36**, 916 (1980).
20. R. Coombe and F. Wodarczk, *Appl. Phys. Lett.* **37**, 846 (1980).
21. W. Johnson and L. Schlie, *Appl. Phys. Lett.* **41**, 798 (1982).
22. C. Christensen and K. Laben, *Appl. Phys. Lett.* **32**, 254 (1978).
23. V. Baranauskas, C. Mammana, R. Klinger, and J. Greene, *Appl. Phys. Lett.* **36**, 930 (1980).
24. D. Ehrlich, R. Osgood, and T. Deutsch, *Appl. Phys. Lett.* **39**, 957 (1981).
25. R. Andreatta, C. Abele, J. Osmunden, D. Lubben, and J. Greene, *Appl. Phys. Lett.* **40**, 183 (1982).
26. V. Daneu, R. Osgood, and T. Deutsch, *Opt. Lett.* **6**, 563 (1981).
27. G. Pannetier and P. Souchay, *Chemical Kinetics*, Elsevier, Amsterdam, 1967, p. 304.
28. R. Christy, *J. Appl. Phys.* **31**, 1680 (1980).
29. P. Genequand, *Surf. Sci.* **25**, 643 (1971).
30. S. Allen, J. Porteus, and W. Faith, *Appl. Phys. Lett.* **41**, 416 (1982).
31. L. Maissel, in *Handbook of Thin Film Technology*, edited by L. Maissel and R. Glang, McGraw-Hill, New York, 1970, pp. 13–28.
32. D. Ehrlich, R. Osgood, D. Silversmith, and T. Deutsch, *IEEE Electron Device Lett.* **EDL-1**, 101 (1980).
33. Tylan Co., *Solid State Technol.* Dec. 1982, p. 29.
34. Ref. 2, p. 31.
35. P. Boyer, G. Roehe, W. Ritchie, and G. Collins, *Appl. Phys. Lett.* **40**, 716 (1982).
36. G. Leyendecker, D. Bauerle, P. Geittner, and H. Lydtes, *Appl. Phys. Lett.* **39**, 921 (1981).
37. K. Hoh, H. Koyama, K. Uda, and Y. Muira, *Jpn. J. Appl. Phys.* **19**, L375 (1980).
38. I. Boyd, J. Wilson, and J. West, *Thin Solid Films* **83**, L173 (1981).
39. Y. Liu, S. Chiang, and F. Bacon, *Appl. Phys. Lett.* **38**, 1005 (1981).
40. S. Chiang, Y. Liu, and R. Riehl, *Appl. Phys. Lett.* **39**, 752 (1981).

41. M. Matsuura, M. Oshida, A. Suzuki, and K. Hara, *Jpn. J. Appl. Phys.* **20**, L726 (1981).
42. I. Boyd and J. Wilson, *Appl. Phys. Lett.* **41**, 162 (1982).
43. U. S. Patent 4,183,780 (1980), IBM.
44. U. S. Patent 3,551,213 (1970), Bell.
45. J. Ligenza, *J. Appl. Phys.* **31**, 5940 (1960).
46. U. S. Patent 3,122,463 (1961), Bell.
47. J. Steinfeld, T. Anderson, C. Reiser, D. Denison, L. Hartsough, and J. Hollahan, *J. Electrochem. Soc.* **127**, 514 (1980).
48. Y. Horike, *Jpn. J. Appl. Phys.* **20**, L817 (1981).
49. J. Coburn, *Plasma Chem. Plasma Process* **2**, 1 (1982).
50. D. Ehrlich, R. Osgood, and T. Deutsch, *Appl. Phys. Lett.* **38**, 1018 (1981).
51. T. Chuang, *J. Vac. Sci Technol.* **19**, 638 (1981).
52. K. Daree and W. Kaiser, *Glass Technol.* **18**, 19 (1977).
53. F. Houle and T. Chuang, *J. Vac. Sci. Technol.* **20**, 734 (1982).
54. T. Chuang, *IBM J. Res. Dev.* **26**, 144 (1982).
55. R. Srinivasan, *Appl. Phys. Lett.* **40**, 582 (1982).
56. N. Feldstein and T. Lancsek, *RCA Rev.* **32**, 306 (1971).
57. U. S. Patent 3,672,975 (1972), RCA.
58. J. D'Amico, M. DeAngelo, J. Henrickson, J. Kenney, and D. Sharp, *J. Electrochem. Soc.* **118**, 1695 (1971), **120**, 1469 (1973).
59. C. Janssen, H. Jonker, and A. Molenaar, *Plating* (East Orange, N.J.) Jan. 1971, p. 42.
60. U. S. Patent 3,907,621 (1975), Photocircuits.
61. S. Ohno and K. Maryujama, *Am. Chem. Soc. Org. Coat. Abstr.* **43**, 268 (1980).
62. U. S. Patent 4,259,435 (1981).
63. R. Micheels, A. Darrow, and R. Rauk, *Appl. Phys. Lett.* **39**, 418 (1981).
64. R. von Gutfeld, R. Acosta, and L. Romankiw, *IBM J. Res. Dev.* **26**, 136 (1982), U. S. Patent 4,217,183 (1980).
65. J. Puippe, R. Acosta, and R. von Gutfeld, *J. Electrochem. Soc.* **128**, 2540 (1982).
66. M. Aboelfotoh and R. von Gutfeld, *J. Appl. Phys.* **43**, 3792 (1972).
67. U. S. Patent 3,345,275 (1967), Westinghouse.
68. U. S. Patent 3,345,274 (1967), Westinghouse.
69. P. Schmidt, J. Oroshnik, and C. Hardman, *Solid State Electron.* 7, 631 (1964).
70. D. Turner, *J. Electrochem. Soc.* **105**, 402 (1958).
71. P. Magill and C. Speicher, *J. Electrochem. Soc.* **118**, 1000 (1971).
72. R. Wales, *J. Electrochem. Soc.* **483** (1969).
73. J. Deines, J. Philbrick, M. Poponiak, and D. Dove, *Appl. Phys. Lett.* **34**, 746 (1979).
74. U. S. Patent 3,482,974 (1969), GE.
75. U. S. Patent 3,482,975 (1969), GE.
76. U. S. Patent 3,494,768 (1970), GE.
77. A. Dalisa, W. Zwicker, D. Debitteto, and P. Harneck, *Appl. Phys. Lett.* **17**, 208 (1970).
78. F. Kuhn-Kuhnenfeld, *J. Electrochem. Soc.* **119**, 1063 (1972).
79. U. S. Patent 3,935,117 (1976), Fuji.
80. H. Hoffman, J. Woodall, and T. Chappell, *Appl. Phys. Lett.* **38**, 564 (1981).
81. F. Ostenmayer and P. Kohl, *Appl. Phys. Lett.* **39**, 76 (1981).
82. U. S. Patent 4,283,259 (1981) and Ref. 64.
83. W. Moreau and P. Schmidt, *Electrochem. Soc. Ext. Abstr.* Fall 1970, p. 455.
84. U. S. Patent 3,547,629 (1970), American Screen.
85. U. S. Patent 3,502,871 (1978), Print Arts Res. Lab.
86. T. C. Patton, *Paint Flow and Pigment Dispersion*, Wiley, New York, 1979.
87. H. Naguib, K. Kavanagh, and L. Hobbs, *Solid State Technol.* Oct. 1980, p. 109.
88. U. S. Patent 2,914,404.
89. U. S. Patent 3,622,322 (1971), RCA.
90. U. S. Patent 3,779,758 (1973), Photocircuits.
91. U. S. Patent 3,542,550 (1970), IBM.
92. U. S. Patent 3,355,291 (1967), TI.
93. U. S. Patent 3,474,718 (1969), Sperry Rand.
94. U. S. Patent 3,387,975 (1968), Sony.
95. A. Pastor, G. Tanogooniami, R. Pastor, S. Wong, and R. Chew, *Thin Solid Films* **67**, 9 (1980).

96. U. S. Patent 3,390,992 (1968), Rockwell.
97. U. S. Patent 3,551,176 (1970), Hughes.
98. G. Bomchil, O. Benashel, A. Golanski, F. Ferrieu, G. Auvert, A. Perio, and J. Pfister, *Appl. Phys. Lett.* **41**, 46 (1982).
99. Z. Liau, B. Tsau, and J. Mayer, *Appl. Phys. Lett.* **34**, 221 (1979).
100. M. Bosch, A. Dayem, T. Harrison, and A. Lemons, *Appl. Phys. Lett.* **41**, 363 (1982).
101. S. Sastoh, H. Iswara, and S. Furikawa, *Appl. Phys. Lett.* **37**, 203 (1980).
102. T. Shibata, A. Wakita, T. Sigmon, and J. Gibbons, *Appl. Phys. Lett.* **38**, 399 (1981).
103. D. Ehrlich, R. Osgood, and T. Deutsch, *Appl. Phys. Lett.* **38**, 399 (1981).
104. H. Taber, S. Nara, and K. Matsuyama, *J. Electrochem. Soc.* **121**, 69 (1974).
105. U. S. Patents 3,346,384 (1967), 3,482,975 (1969), 3,485,630 (1969), 3,489,563 (1970), 3.494,768 (1970), 3,520,685 (1970), 3,520,687 (1970), GE.
106. J. Burgess and D. Schaefer, *J. Vac. Sci. Technol.* **6**, 143 (1969).
107. U. S. Patent 3,992,208 (1976), Fuji.
108. D. Maydan, *Bell Syst. Tech. J.* **30**, 1761 (1971).
109. V. Zaleckas and J. Koo, *Appl. Phys. Lett.* **31**, 616 (1977).
110. Z. Liau, B. Tsaur, and J. Mayer, *Appl. Phys. Lett.* **34**, 24 (1979).
111. R. Kaplan, *SPIE Advances in Laser Engineering and Applications* **247**, 2 (1980).
112. J. Narayan and F. Young, *Appl. Phys. Lett.* **35**, 330 (1979).
113. P. Baeri, S. Campiasano, G. Fati, and E. Rimini, *J. Appl. Phys.* **50**, 788 (1979).
114. W. Brown, *J. Vac. Sci. Technol.* **20**, 734 (1982).
115. C. Goodridge, *Circuits Manuf.* July 1981, p. 59.
116. K. Shoulders, Publication PB171027, U. S. Department of Commerce Office of Technical Service, Sept. 1960.
117. W. Nixon, *Microelectron. Reliab.* **3**, 153 (1964).
118. R. Christy, *J. Appl. Phys.* **33**, 1884 (1962).
119. G. Hill, *Microelectron. Reliab.* **4**, 109 (1965).
120. U. S. Patent 3,113,896 (1962).
121. T. Woodman, *Br. J. Appl. Phys.* **16**, 359 (1965).
122. U. S. Patent 3,378,401 (1968), 3M.
123. M. Lepselter, U. S. Patent 3,442,701 (1969), Bell.
124. J. Ballantyne and W. Nixon, *J. Vac. Sci. Technol.* **10**, 1094 (1973).
125. U. S. Patent 3,585,091 (1971), Bell.
126. U. S. Patent 3,615,935 (1971).
127. U. S. Patent 3,580,749 (1971), Westinghouse.
128. U. S. Patent 3,615,953 (1971).
129. U. S. Patent 3,799,77 (1974), Westinghouse.
130. U. S. Patent 3,833,396 (1974), Bell.
131. U. S. Patent 3,867,148 (1975), Westinghouse.
132. B. Chin and G. Ehrlich, *Appl. Phys. Lett.* **38**, 253 (1981).
133. U. S. Patent 3,436,468 (1969), TI.
134. H. Smith, E. Ligeon, and A. Bontemp, *Appl. Phys. Lett.* **37**, 1036 (1980).
135. T. O'Keefe and R. Handy, *Solid State Electron.* **11**, 261 (1968).
136. U. S. Patents 3,585,433 (1971), 3,588,570 (1972), 3,686,020 (1973), 3,672,987 (1974), Westinghouse.
137. G. Kammlott and W. Sinclair, *J. Electrochem. Soc.* **121**, 929 (1974).
138. U. S. Patent 3,355,568 (1967), Hitachi.
139. J. Bartelt, C. Slayman, J. Wood, J. Chen, C. McKenna, C. Minning, J. Coakley, R. Holman, and C. Perrigo, *J. Vac. Sci. Technol.* **19**, 1166 (1981).
140. V. Wang, J. Ward, and R. Seliger, *J. Vac. Sci. Technol.* **19**, 916, 1158 (1981).
141. T. O'Keefe, *J. Electrochem. Soc.* **112**, 149C (1965).
142. U. S. Patent 3,563,809 (1970), Hughes.
143. R. Kubens, C. Anderson, R. Seliger, R. Jullens, E. Stevens, and I. Lagando, *J. Vac. Sci. Technol.* **19**, 916 (1981).
144. U. S. Patent 3,516,855 (1970), IBM.
145. R. Barker, T. Mayer, and R. Burton, *Appl. Phys. Lett.* **40**, 583 (1982).
146. U. S. Patent 3,551,213 (1970).
147. G. Stengl, R. Kartna, H. Loschner, R. Rieder, P. Wolf, and R. Sacher, *J. Vac. Sci. Technol.* **19**, 1164 (1981).

148. U. S. Patent 3,801,390 (1974), Bell.
149. G. Bell and J. Hoepfner, Proc. Electrochem. Soc., 1976, p. 47.
150. J. Gotzlich and H. Ryssel, *J. Electrochem. Soc.* **128**, 617 (1981).
151. D. Day, *IEEE Electron Device Lett.* **EDL-5**, 32 (1984).
152. European Patent Application EP 97,819 (1984), *Chem. Abstr.* **100**, 201856 (1984).
153. Japanese Patent 58,209,123 (1983), *Chem. Abstr.* **101**, 31145 (1984).
154. K. Gamo, *Jpn. J. Appl. Phys.* **23**, L293 (1984).
155. P. Brewer, S. Halle, and R. Osgood, *Appl. Phys. Lett.* **45**, 475 (1984).
156. H. Hiroaka, *SPIE Proc.* **469**, 127 (1984).
157. T. Kanayama, Ext. Abstr. 1984 Int. Conf. Solid State Dev., p. 27.
158. D. Ehrlich and J. Tsao, *J. Vac. Sci. Technol.* **B1**, 979 (1983).
159. S. Allen, *SPIE Proc.* **459**, 42 (1984).
160. J. Tsao and D. Ehrlich, *SPIE Proc.* **459**, 2 (1984).
161. R. Jan and S. Allen, *SPIE Proc.* **459**, 75 (1984).
162. K. Emery, L. Thompson, J. Rocca, and G. Collins, *SPIE Proc.* **459**, 75 (1984).
163. F. Houle, *J. Chem. Phys.* **79**, 4237 (1984), *SPIE Proc.* **459**, 110 (1984).
164. G. Loper and M. Tabat, *SPIE Proc.* **459**, 121 (1984).
165. J. Gee and P. Hargis, *SPIE Proc.* **459**, 132 (1984).
166. F. Houle, *SPIE Proc.* **385**, 127 (1984).
167. P. Mankewich, H. Craighead, T. Harrison, and A. Daymen, *Appl. Phys. Lett.* **44**, 468 (1984).
168. A. Muray, M. Issacson, and I. Aesida, *Appl. Phys. Lett.* **45**, 589 (1984).
169. J. Tsao and D. Ehrlich, *Appl. Phys. Lett.* **45**, 617 (1984).
170. R. Krchnavek, H. Giligen, and R. Osgood, *J. Vac. Sci. Technol.* **B2**, 641 (1984).
171. J. Maes, A. Van Nu, and G. Hut, *Reliab.* **17**, 325 (1978).
172. F. Houle, C. Jones, T. Baum, C. Pico, and G. Kovac, *Appl. Phys. Lett.* **46**, 204 (1985).
173. W. Holber, G. Retsken, and R. Osgood, *Appl. Phys. Lett.* **46**, 201 (1985).
174. T. Yamazaki, Y. Suzuki, and H. Nakata, *J. Vac. Sci. Technol.* **17**, 1384 (1980).
175. Japanese Patent 59,145,528, *Chem. Abstr.* **102**, 70256 (1985).
176. D. Tennant, *J. Vac. Sci. Technol.* **B3**, 458 (1985).
177. P. Singer, *Semicond. Int.* April 1985, p. 66.
178. H. Yamaguchi, A. Shisame, S. Haraichi, and T. Miyauchi, *J. Vac. Sci. Technol.* **B3**, 71 (1985).
179. M. Takai, J. Tsuchimoto, N. Junuchi, N. Hiroyuki, K. Gamo, and S. Namba, *Jpn. J. Appl., Phys.* Part 2, **23**, L852 (1984).
180. P. Heard, J. Cleaver, and H. Ahmed, *J. Vac. Sci. Technol.* **B3**, 87 (1985).
181. Japanese Patent 59,155,935, *Chem. Abstr.* **102**, 88669 (1984).
182. J. Randall, D. Ehrlich, and J. Tsao, *J. Vac. Sci. Technol.* **B3**, 87 (1985).
183. U. S. Patent 4,496,449 (1985), Colromm.
184. M. Green, C. Aidinis, and O. Fakulujo, *J. Appl. Phys.* **57**, 631 (1985).
185. R. Osgood and T. Deutsch, *Science* **227**, 709 (1985).
186. G. Koren, F. Ho, and J. Ritsko, *Appl. Phys. Lett.* **46**, 1006 (1985).
187. K. Li and M. Oprysko, *Appl. Phys. Lett.* **46**, 997 (1985).
188. G. Fisnanick, M. Gross, J. Hopkins, M. Fennell, K. Schnoes, and A. Katzer, *J. Appl. Phys.* **57**, 1139 (1985).
189. S. Yokoyama, Y. Yamakazi, and M. Hirose, *Appl. Phys. Lett.* **47**, 389 (1985).
190. D. Ehrlich, *Solid State Technol.* Dec. 1985, p. 81.
191. J. Bjorkholm, L. Eichner, J. White, R. Howard, and H. Craighead, *J. Appl. Phys.* **58**, 2098 (1985).
192. R. Tenne, V. Marcu, and Y. Prior, *Appl. Phys.* **A37**, 205 (1985).
193. T. Baum and C. Jones, *Appl. Phys. Lett.* **47**, 538 (1985).
194. C. Fiori and R. Devine, *Appl. Phys. Lett.* **47**, 361 (1985).
195. F. Ostermayer, P. Kohl, and R. Lum, *J. Appl. Phys.* **58**, 4390 (1985).
196. J. Brannon, J. Lankard, A. Baise, F. Burns, and J. Kaufman, *J. Appl. Phys.* **58**, 2036 (1985).
197. C. Ashby and R. Biefeld, *Appl. Phys. Lett.* **47**, 62 (1985).
198. M. Gross, G. Fissnick, P. Gallager, K. Schnols, and M. Fennell, *Appl. Phys. Lett.* **47**, 923 (1985).
199. D. Day, G. Middleton, T. James, J. White, and V. Mifsud, *J. Electrochem. Soc.* **131**, 407 (1985).
200. R. Gutfeld and D. Vigliotti, *Appl. Phys. Lett.* **46**, 1003 (1985).
201. T. Mottoka, S. Gorbathsen, D. Lubben, and J. Greene, *J. Appl. Phys.* **58**, 4397 (1985).
202. D. Ehrlich, J. Tsao, D. Silversmith, J. Sedlackik, R. Mountain, and U. Grabber, *IEEE Electron Device Lett.* **EDL-5**, 32 (1984).

203. S. Stensey, S. Eskeldsen, and G. Sorensen, *Appl. Phys. Lett.* **46**, 1101 (1985).
204. U. S. Patent 4,523,971 (1985), IBM.
205. U. S. Patent 4,530,734 (1985).
206. G. Stengl, H. Loeschner, W. Maurer, and P. Wold, *SPIE Opt. Eng.* **537**, 138 (1985).
207. B. Ward, D. Shaver, and M. Ward, *SPIE Opt. Eng.* **537**, 110 (1985).
208. K. Orvek and C. Huffman, *Nucl. Instrum. Methods Phys. Res.* **B7-8**, 501 (1985).
209. German Patent 3,420,353 (1984), *Chem. Abstr.* **102**, 158083 (1985).
210. R. Spiedel and Y. Benzhang, *Optik* (Stuttgart) **68**, 363 (1984).
211. C. Fisanck, J. Hopkins, M. Gross, M. Fennel, and K. Schnoes, *Appl. Phys. Lett.* **46**, 1184 (1985).
212. J. Kosar, *Light Sensitive Systems*, Wiley, New York, 1965, pp. 1–49.

17

Summary and Future Processes

17-1. CRITICAL LITHOGRAPHY

In this chapter I would like to restate the significant materials and processes which have aided the progress of semiconductor lithography. I shall also project into the future with processes for the next decade of submicrometer technology.

The interaction of materials, processes, and equipment forms the basic dynamics (Fig. 17-1-1) of semiconductor lithography. In the past 20 years, the semiconductor industry was driven by circuit design and the basic device physics. Now that the fundamentals are well understood, the mass production of smaller and denser circuits will not only be dependent on the performance of the exposure tools and etching equipment but also on the choice of process and the durability of the resist materials. In this regard, the performance of a resist material can be perceived from three distinct viewpoints:

1. The equipment engineer desires a high-speed resist that produces high throughput.
2. The materials engineer sums up the performance of a resist in one word —resistance—plasma, thermal, and adhesive.
3. The process engineer requires good linewidth control, minimum pattern biases and tolerances, good alignment, and wide process latitude (Fig. 17-1-2).

To meet the challenges of shrinking linewidths, significant new lithographic equipment (Table 17-1-1), processes (Table 17-1-2), and resist materials (Table 17-1-3) have been developed. In the equipment arena, dry etching tools, optical scanners, and *e*-beam exposure have been the key advances. All of the tools are required to be operated in ultraclean rooms and themselves not be generators of particulates or invisible contaminants. Optical lithographic tools face the biggest challenge of delivering submicrometer images with 0.1-μm overlay requirements at decreasing depths of focus.

In the imaging and process area, anisotropic plasma etching deserves all of the accolades as the most significant lithographic process innovation. So many processes that have spun off require plasma etching to succeed. For example, multilayer resists are the key to extending lithography to submicrometer dimensions. Improved linewidth control has been achieved with development and etching endpoint monitors. With the ever-increasing wafer size, the days of batch processing

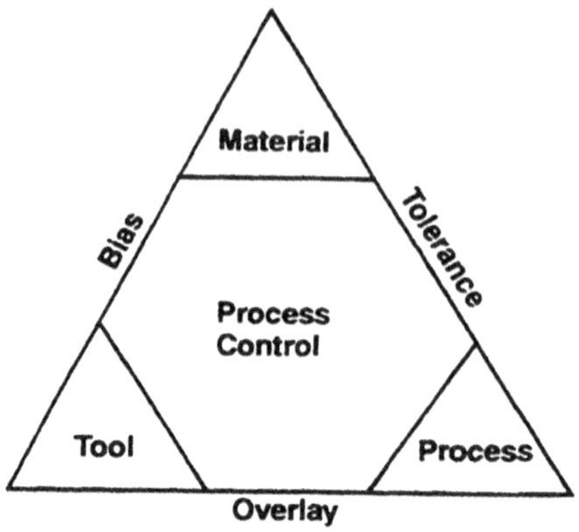

FIGURE 17-1-1. The interaction of tool, resist, and process in lithography.

are limited. Nevertheless, single-wafer processes will require very reproducible conditions.

Although many different types of polymers have been examined, only a handful stand out. The positive DQN photoresists are the current workhorse materials but if multilayer resists are required, silicone-containing materials will likely be used for all different types of photon and high-energy resists.

The major performance objectives of a resist are to limit the lateral spread of the images and to produce devices economically using:

1. *Speed* (sensitivity)—the need is most apparent for *e*-beam, followed by X-ray, and optical exposure. A speed of 0.1 $\mu C/cm^2$ for *e*-beam and 1 mJ/cm^2 for X-ray or deep UV is desirable. High speed will always be the first or second priority in a resist material.

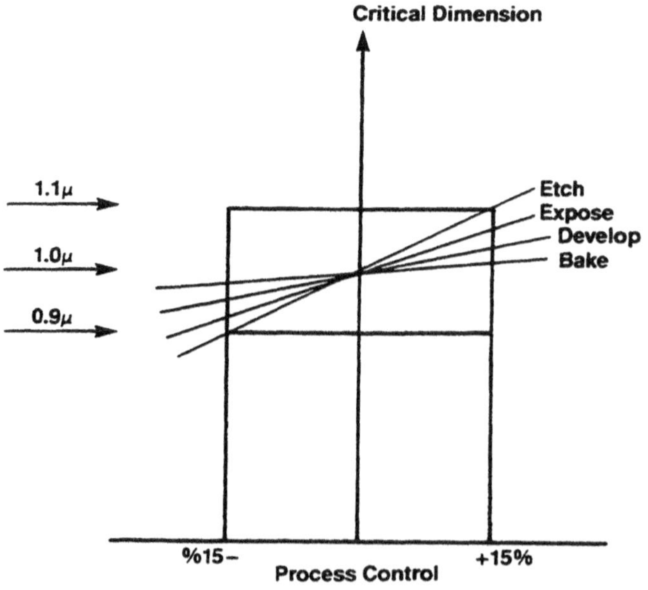

FIGURE 17-1-2. Lithography with a wide process latitude.

TABLE 17-1-1. Significant Lithographic Equipment

Process step	Innovation
Apply	Spin coating, vapor HMDS
Expose	Optical reflective scanner
	Refractive step and repeat
	Autoalignment
Etch	Reactive sputter etcher
Strip	Plasma strip
Process control	Autoendpoint
	Develop and etch
	Autolinewidth
	Class 1 clean rooms
Harden resist	Flood deep UV

2. *Process resistance*—the image transfer step continues to put more stress on the resist (Table 17-1-4). Bombarding ions, radicals, heat, and strong developers test the adhesive, plasma, and thermal strength of the resist on a variety of surfaces. Additional material resistance can be embodied in a multilayer configuration.
3. *Resolution*—the pitch of images approaches the practical film thickness limit of 1 μm. Diffraction, interference, backscattered electrons, and topographical distortions all require a resist to succeed with a lower MTF. High contrast, multilayer resists, and high doses augment high resolution. The demands on the resist performance increase as the device size decreases. The image biases and overlay tolerances of today are the dimensions of tomorrow (Table 17-1-4).

TABLE 17-1-2. Key Lithographic Processes

Process	Innovation
Multilayer resist	Lift-off
	RIE mask
	Thermal stability
	Proximity *e*-beam
	Interference reduction
Dry image processing	Multilayer resists
	Dry develop
	Dry conformal deposit
	Dry strip
Film thinning monitors	Linewidth control
	Dosimetry
	Endpoint
	Smaller process spreads
	Process controls
	Reproducibility
Lift-off	Deposit multiple metals
	Self-aligned vias

TABLE 17-1-3. Key Resist Materials

Resist	Type and performance
Positive photoresists	Diazoquinones (aqueous process, good resolution, spectral sensitivity, and dry process resistance)
Positive radiation resists	PMMA (high resolution, many spin-offs, stepping stone to improved materials)
	Polysulfones (high speed)
Negative photoresists	Azide resists (superior adhesion, innovations for dry develop, and thermally stable resists)
	PMMA dry film (bridge-plate holes, process quality control, thermally stable derivatives)
Negative radiation resists	Silicone (dry develop, multilayer resists, high speed)

One, ideal, single film resist does not exist today with outstanding performance in speed, resolution, stability, and solubility (after 250°C lift-off of metals). Today, resist materials perform with help from process innovations (e.g., multilayers) or tool conditions (low RIE power, high dose) to limit resist erosion or enhance resolution (usually at the expense of throughput or extra process steps).

The lithographic process section of fabricating a semiconductor device has been the scene of a concentrated interdisciplinary effort of electrical engineers, physicists, and chemists. For example, from a chemist's viewpoint, all the major disciplines (Fig. 17-1-3) of physical, polymer, organic, inorganic, and analytical chemistry are involved in understanding, controlling, and improving resist materials and lithographic processes. In the next two sections I will examine the resist materials and the accompanying processes that have been critical to the progress of semiconductor lithography and the outlook for continued growth in better materials and processing.

TABLE 17-1-4. Increasing Demands on Resist Performance

	1960	1970	1980	1990
Thickness (μm)	1–5	1–2	1–2	0.5–1
Resolution (μm)	$5–10 \pm 2$	$3–5 \pm 0.7$	$2–4 \pm 0.5$	$0.75–1.5 \pm 0.3$
Thermal stability (°C)	75	75	150	250
Pinholes/cm^2 (1 μm)	100	10	1–5	0.2
Dose				
UV (mJ/cm^2)	100–200	75	50	10–25
EB (μC/cm^2)	50	25	1–10	0.1–1.0
Etch resistance	Acids	Acids	CF_4	Halogens
Developers	Organic	Organic	Alkali	Alkali/plasma
Commercial suppliers	2	4	20	35

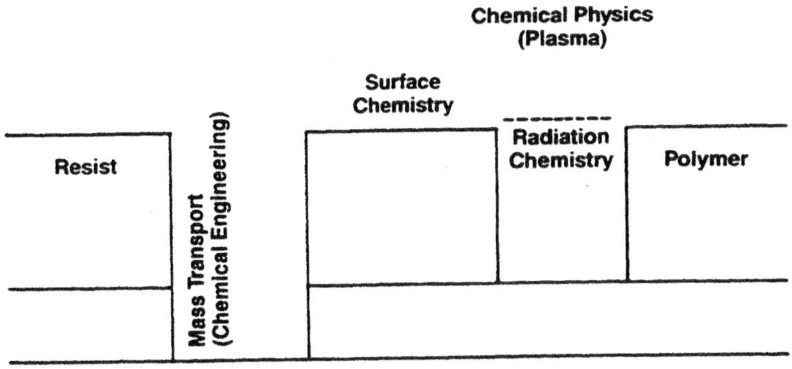

FIGURE 17-1-3. The disciplines of chemistry which interact in resist processing.

17-2. RESIST MATERIALS

Resist materials span the spectrum from single component to formulations containing five to six additives:

- Single polymers
- Bipolymers
- Terpolymers
- Mixtures of polymers
- Mixtures of molecules and polymers
- Mixtures of monomers and polymers
- Pre-cross-linked polymers
- Monomers
- Inorganic compounds
- Multiple layers
- Posttreated resists
- Cationic, anionic, or radical sourced
- Photosensitized polymers
- Multiple films
- Organometallic
- Charge transfer complexes
- Plasma deposited

The resists have evolved from negative-working low-resolution materials (Table 17-2-1) into high-resolution multilayer or top surface imaging materials by combinations of materials or technical advances in new polymeric substances. The improved performance of resists has been a series of material and process innovations which are optimized in five critical areas: speed, resolution, plasma resistance, thermal resistance, and adhesion (Table 17-2-2). Improved materials usually result from a combination of ideas, patents, and developments in other technologies such as high-temperature materials or other polymeric fields. For any of the desired resist properties, enhancement by copolymerization or by posttreatments appears to be the primary means of enhancement. For high-sensitivity resists, free radical mechanisms have been exhausted and cationic catalysis offers the

TABLE 17-2-1. Evolution and Development of Resist Materials and Processes

UV → Electron beam → X-ray → Ion beam

Negative resists → Positive resists → Multilayers
Organic based → Organometallic → Inorganic

Standard formulations (azide–rubber, DQN–novolak) → Posttreatment of standard → New mixtures of old components
↓
New components

Liquid apply → Dry deposit
Organic developers → Aqueous developers → Dry developers
Plasma ash → Plasma etch → Reactive ion etch
Wet dip etch → Spray etch

1970	1980	1990
Negative photoresist	Positive photoresist	Positive Photoresist
Single layer	Single layer	Multiple layer
Near-UV contact print	Near-UV projection	Mid-UV projection

TABLE 17-2-2. Best State-of-the-Art Performance in Resists

Speed	Image (monomer amplification) PMMA
	Cationic amplification
	Polysulfones
	Silicones
	Epoxy
	Multilayer resist
	Optimization of molecular weight and dispersity
	Developers
Resolution	Multilayer resists
	Top surface imaging
	Ultrathin resist layer
	Proximity contrast correction
	High-γ resist-developer
Thermal	High T_g
	Arylcyclics
	Pre-cross-linked
	Silicones
Dry etch resistance	Metal inorganic
	Silicone (O_2)
	Aryl
	Polymerizing gas
Adhesion	Low-T_g polymers
	Polymerizing adhesion promoters
	Etch-graft promoters
	Adhesive underlayers

possibility of image amplification. In many cases, trade-offs of speed versus stability are likely as materials are bounded by the limits of atomic mass, bond energies, and kinetic mass transport factors. Entropy, in the form of contaminants, limits the lithographic yield and forces trade-offs in resist thickness for resolution. Although inorganic resists possess the excellent qualities of etch and thermal resistance, basic improvements in spectral response and speed are required for their consideration as submicrometer lithographic resists.

I will now highlight the four chapters on positive and negative resists. Ideally, one would desire a bipolar resist capable of rapid exposure by UV and *e*-beam and imaged with the same developer. The introduction of one family of resist material into a process line would decrease the learning curve, enhance automation, decrease process control, and eventually improve yield. Positive resists are preferred because of the nature of exposure while negative cross-linking resists tend to be overexposed to enhance the adhesion to the substrate. In addition, the unexposed space in a positive resist is not subject to the same scattering angles as the exposed region of negative resists. For *e*-beam writing of low-density patterns, a positive tone would be preferred. Image reversal, of course, could be an added bonus. For a later comparison, the desired properties of an "ideal resist" are given in Table 17-2-3.

I shall now examine the performance of the four major resist classes.

17-2-1. Positive Photoresists

The diazoquinone (DQN) formulation has been the most significant resist system ever developed. Its outstanding attributes are: (1) wide spectral sensitivity, (2) insensitive to oxygen during exposure, (3) curable for high heat resistance, (4) aqueous processing, (5) reversible to a negative resist, and (6) extendable resolution by portable conformable masking. As long as steppers are able to resolve fine lines, exposure in the 300- to 450-nm region will probably continue to use DQN as the primary choice of resists.

Areas which are constantly being improved by additives or by modification of the sensitizer or the novolak resin include:

1. *Photospeed*—controlled molecular weight resins, and better sensitizers. For enhanced mid-UV absorption, alkyl sulfone esters were used to blueshift the π–π transition of DQ. The quantum yield of 0.21 could be extended by

TABLE 17-2-3. Performance of an "Ideal Resist"

Process	Performance
1. Dry deposit	Conformal, rapid deposition, pinhole-free film (0.1–3.0 μm) adhesive, uniform ($\pm$50 Å)
2. Prebake	None
3. Expose	0.1 $\mu C/cm^2$ *e*-beam 1 mJ/cm^2 X-ray, UV
4. Develop	Dry or self (no film loss)
5. Resistance	Heat 250°C Plasma 1/5 (SiO_2/Al/Si) etch rate
6. Resolution	Pitch < thickness
7. Strippability	Plasma, aqueous for lift-off

5× to achieve 1–10 mJ/cm^2. Complementary developers, which inhibit unexposed loss, increase resolution and sensitivity.

2. *Resolution*—developers and resin which provide inhibited (induction non-linear) dissolution increase the chance for top-surface imaging. Silicone-containing DQN offers bilayer image transfer by anisotropic oxygen ion etching. A bilayer resist, with a (PCM) deep-UV transfer, is a lower cost alternative if a stable, planarizing layer can be found. Dyed multilayer resists will counteract interference and allow steppers and scanners to achieve submicrometer resolution.
3. *Thermal resistance*—transparent resins, alkyl diazoquinones, or single-component DQ polymers may achieve deep-UV absorption if sufficient bleaching occurs. Posttreatments (Table 17-2-4) may be necessary to enhance plasma and thermal resistance.

In the deep-UV region, we can expect aromatic polymers with photolabile groups to comprise deep-UV resists. With image amplification, speeds of 1 mJ/cm^2 would be a throughput bonus for future step-and-repeat optical scanners.

TABLE 17-2-4. Postprocessing Resists

Treatment	Examples
Solvent soak	Chlorobenzene soak of DQN prior to develop for lift-off profile Dry resist without prebake
Heat resist	Pre-cross-link polyacrylates Depropagate polyolefin sulfones Image amplify onium salts in carbonate resists Harden DQN resists Dry develop (sublime silicone monomers) Bake DQN for MLR underlayer or adhesion Bake DQN to reduce standing waves Bake polyolefin sulfone/novolak prior to develop Melt resist for taper via holes by plasma etching Bake amine/DQN for decarboxylation image reversal
Irradiate resist (UV, *e*-beam, ion beam, X-ray)	Harden DQN for thermal stabilization Darken DQN or PMMA to make photomasks Flood expose DQN/amines for image reversal
Plasma	Dry develop Top surface treat to improve DQN thermal stability Insolubilize resist in MLR to prevent interfacial mixing Etch-back resist to planarize substrate
Chemical	Incorporate Si or Sn, B, Ti or metals in DQN for bilayers Graft acrylic acid into PMMA for image amplification NH_3 treat DQN for image reversal Soak DQN in HCHO to stabilize thermally Cross-link DQN or PMMA acid with dichlorocompounds Incorporate O_2 plasma-resistant monomers for top surface imaging

17-2-2. Negative Photoresists

Negative photoresists have been the pioneering resists of lithography for the manufacture of printing plates, printed circuits, and the first transistors. They were based on photosensitized polyvinylcinnamate and azide crosslinked rubbers. From a process viewpoint, the resists were insensitive to overdevelopment, exhibited excellent adhesion, and could resolve 3- to 4-μm geometries. Although azide-based rubber resists are sensitive to oxygen during exposure, the PVC resist is insensitive to oxygen and can withstand 250°C, if deep-UV hardened. If the azide sensitizer is incorporated in an aqueous developable phenolic polymer, improved resolution can be obtained since little swelling is observed. Azides were also successfully used to cross-link high-temperature polymers and form the basis of dry develop resists.

For printed circuits and packaging lithography, the most outstanding negative photoresist is the monomer–polymer combination of PMMA dry film resists. The precast resist avoids the toxic solvents of liquid-cast films and allows the manufacturer to produce a highly controlled product.

The basic swelling of organic negative photoresists has been addressed by:

1. Charge transfer ionic polymers
2. Inorganic resist
3. Aqueous based developers
4. Image reversed DQN
5. Cationic catalyzed decarbonylation

Cationic decarbonylation changes the polarity of a resist by molecular elimination and the formation of a hydrophilic polymer which is not cross-linked. In addition, high sensitivity is obtained in the onium salt-catalyzed image amplification. These new insolubilizing systems are needed to promote negative resists to the performance of their positive counterparts.

Photopolymerization of monomers or image amplification by posttreatment of images with silicone monomers offers top-surface imaging with dry O_2^+ image transfer capability in a bilayer configuration.

17-2-3. Positive Radiation Resists

PMMA resist paved the way for *e*-beam, X-ray, deep-UV, lift-off, and ion-beam exposure. Subsequently, many new resists have been cloned from PMMA and the synthetically accessible analogues. For enhancing speed and resolution, the effect of molecular weight and dispersity was first successfully demonstrated with PMMA. Some of the unique lithographic processes or materials which originated from PMMA are as follows:

1. Introduction of electron-withdrawing groups for secondary electron capture. Introduction of halogens or metals to favorably absorb X rays.
2. Pre-cross-linked resists to enhance thermal stability and offer the wide process latitude of a negative cross-linked resist to overdevelopment.
3. Total all-dry processing from plasma deposition to dry develop, etch, and strip (Fig. 17-2-1).
4. Image amplification to 10^{-9} C/cm^2 by postreaction of irradiated PMMA with vinyl monomers followed by solvent development.
5. Top-surface imaging by ion implant and O^+ develop.

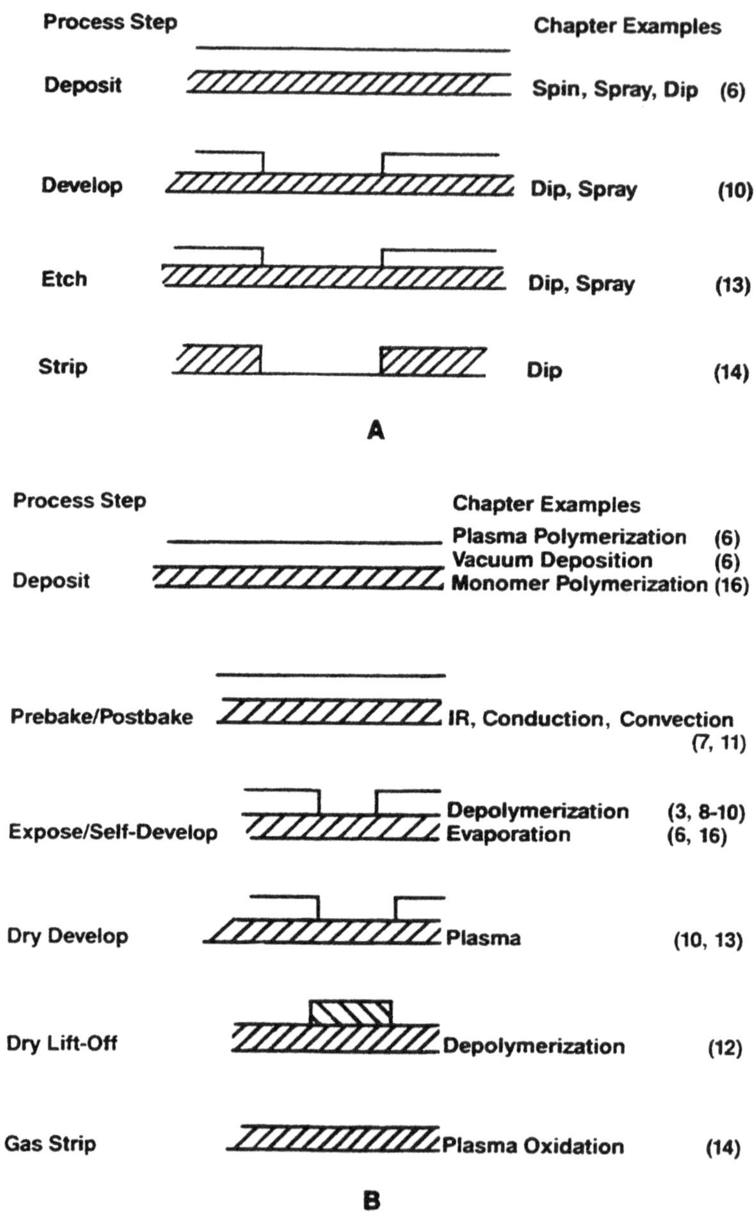

FIGURE 17-2-1. (A) The present state of wet processes in lithography and (B) the future all-dry process lithography.

6. The basic concepts of multilayer resists for deep-UV PCM or O_2^+ dry develop were founded on the basis of PMMA resists.
7. Laser or deep-UV dry develop.

While PMMA and derivatives remain as fair sensitivity resists with high resolution, a basic need exists for better dry process stability and for enhanced sensitivity. These improvements can be realized in several basic ways:

1. Posttreatments of copolymers with metal-organic cross-linking agents to enhance RIE and thermal stability.
2. Additives to enhance speed, resolution, and stability.
3. Improved developers for speed and improved resolution.
4. Copolymerization or new side chains to amplify speed.
5. Image amplification processes.

Other faster resists (Table 17-2-5) have been based on low-ceiling-temperature unzippable polyaldehydes and polysulfones. Although polyolefin sulfones lack thermal and plasma resistance, they are one of the fastest resists. Image amplification by the generation of Lewis acids in polyaldehydes and polycarbonates shows promise for high speeds.

Two-component resists, which combine the speed of polyolefin sulfones with the etch resistance of novolaks, are a useful resist for chrome mask making or direct silicon write. The formulations are a classic example of material improvement by combination of old systems.

In order for X-ray exposure to succeed as a lower cost alternative to flood-optical or *e*-beam exposure, single film "processable" resists of 1 mJ/cm^2 are a necessity. Mask (*e*-beam)-making positive resists lack dry etch resistance and have poor endpoint control.

17-2-4. Negative Radiation Resists

While negative radiation resists have struggled with the low-resolution aspects of swelling, ion beams and X rays have helped as nonscatter (backscatter) sources, but no fast, single film resist has been successful at the 1- to 2-μm thickness level. Negative resists have served to fabricate 10× master masks by *e*-beam exposure.

In order to enhance resolution, new insolubilization mechanisms are required to control the swelling. Charge transfer resists are an initial example of a moderate-speed system without swelling, and onium salt-decarbonylation of polyvinylphenol esters shows promise as high-speed and high-resolution resists that can also be postsilylated for bilayer processing. For bilayer processing and reduced proximity effects, metalloorganic silicone polymers offer high speed and oxygen resistance, but more thermally (silicone) stable materials are necessary for high-temperature lift-off masking of soluble high-temperature underlayers (e.g., soluble polyimides). Negative-working inorganic resists require speed enhancement for multilevel masking. We can expect more organometallic negative resists as monomolecular Langmuir films for nanolithography.

17-2-5. Complementary Lithographic Materials

Complementary lift-off underlayers and dyed layers including bleachable contrast-enhancing layers (Table 17-2-6) are indispensible aids to improve resist perfor-

TABLE 17-2-5. High-Sensitivity Resists

Dose	UV	X-ray	Ion beam	Electron beam
10^{-8} C/cm^2 (EB) 0.1 mJ/cm^2 (UV, X-R) 10^9 ions/cm^2 (IB)	AgBr (N)[a] PMMA-acrylic acid (N)			AgBr (N) PMMA-acrylic acid (N)
10^{-7} C/cm^2 (EB) 1 mJ/cm^2 (UV, X-R) 10^{10} ions/cm^2 (IB)		AgBr (N) PMMA-acrylic acid (N) Acetylenic monolayers (N)	AgBr (N) PMMA-acrylic acid (N)	Polyepoxides (N) Polysilicones (N) Polyhaloacrylates (P) Polyallyls (N)
10^{-6} C/cm^2 (EB) 10 mJ/cm^2 (UV, X-R) 10^{11} ions/cm^2 (IB)	Azide + polymers (N) Polyhalostyrenes (N) Poly(t-butylcarbonato) styrene + onium salts (N, P)	Polyhaloacrylates (P) Vinyl monomers + polyhaloacrylates (N)	Polyepoxides (N) Polyhalostyrenes (N)	Polysulfones (P) Polyaldehydes (P)

[a] Grafting posttreatment:

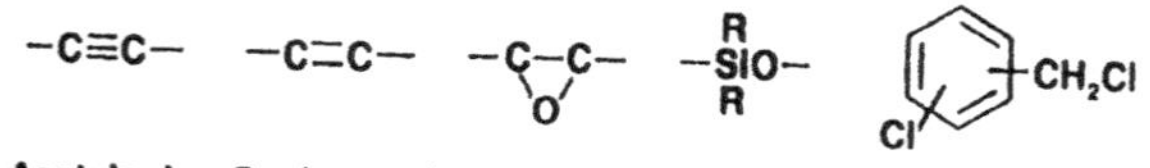

Acetylenic Conjugated Epoxy Siloxy Chloroaryl

Negative Resist Functional Groups

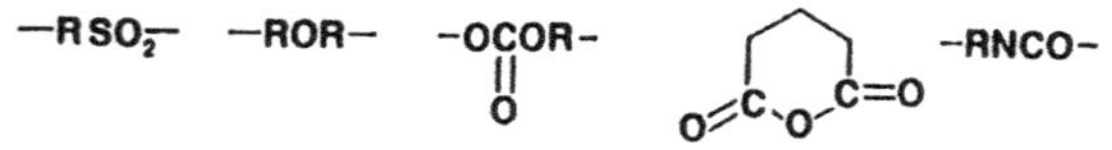

Sulfone Ether Carbonate Anhydride Isocyanate

Positive Resist Functional Groups

TABLE 17-2-6. Complementary Use of Polymers in Lithography

			Topcoatings			
Uses	Oxygen effect suppression	Thermal mold	Contrast enhancement	Electron-beam charge release	Photomask release	Photomask protection
Examples	Polyvinylbutyral on KTFR Mylar on Riston	PMMA/DQN	Nitrone/PVP	Polyvinylcarbazole	Silicone or fluoropolymer	Nylon
			Middle polymer layers			
Uses	Oxygen barrier layer in MLR	Interface mixing prevention	Adhesive layer			
Examples	Polysilicones, polystannyl methacrylates	Polyvinyl alcohol	Polybutadiene			
			Bottom layers			
Uses	Antireflective planarizing layer	Planarizing layer MLR	Adhesion promoter	Release layer in lift-off		
Examples	Dyed polyimide	Polyimide or baked DQN	Baked KTFR	Polyarylsulfone		
			Miscellaneous single layers			
Uses	Paint on diffusion	Insulating–passivation	Photomask repair	Sacrificial etch-back RIE layer	Alpha protection	
Examples	Polysilica	Polyimide	Dyed polymers	Polyimide	Polyimide	

TABLE 17-2-7. Application of Dyes in Lithography

Dye	Application
Organic soluble absorbing 300–450 nm	Antireflective underlayers in MLR
	DQN resist additive to reduce standing waves
Bleaching dye (200–450 nm)	Contrast enhancement topcoat
	Resist dosimeter
Fluorescent dye	Plasma etch rate monitor of eroding resist
	Absorbed on developed resist to detect residues
Conductive dye	Eliminate *e*-beam charging as topcoat additive
Absorbed dyes	Organic photomasks

mance. For example, high-temperature underlayers in multilayer resist stacks permit fast but unstable resists to complete the successful etching or lift-off of a pattern. Organic dyes (Table 17-2-7) suppress standing waves, bleach out diffraction effects as components of contrast-enhancing layers, and extend the lifetime of present optical tools. From a resist and process requirement, high-temperature, stable, nonsublimable dyes that blend with resists are needed for withstanding dry etching processing or process conditions in excess of 150°C.

17-2-6. Overall Rating of Resists

In Table 17-2-8, a rating of all resist materials and accompanying process configuration is given. If we accept the dominant factors of resolution as the primary requisite for submicrometer lithography, then multilayer resists are the primary choice. The object in this case such as top-surface imaging is to make the image layer (disappear) and be at least 1/10 in thickness as the desired resolution. Overall, positive near- and mid-UV photoresists will continue for the near future as the best choice for an economical resist process. In the future, the exposure tool choice will in turn dictate the resist to fill the gap of speed (throughput) and resolution. Other thermal and plasma deficiencies will have to be made up by multilayer structures.

17-3. PROCESS LITHOGRAPHY

In the process area, the daily resist process serves to produce an image and transfer the resist pattern into a permanent semiconductor device. Some of the basic conditions to consider in choosing a resist to fit the process and process equipment are given in Table 17-3-1. In fact, one should reverse engineer the considerations and start with the stripping conditions and assess if they will permanently damage the substrate or device that will be processed. Next, the image transfer conditions, especially the selectivity of plasma erosion, will narrow the choice of resist and/or will require a multilayer resist configuration. Past experiences also come to bear with choosing different families of resist and in determining the degree of process controls or latitude that will be necessary to achieve the device image biases, tolerances, and overlay requirements. Once the resist or resist candidates have been selected, we can turn our attention to more detailed process considerations to make our final commitments to a resist, a process, a particular lithographic equipment, clean rooms, and metrology tools.

TABLE 17-2-8. Rating of Different Resist Systems[a]

	Pos. near-UV	Pos. mid-UV	Pos. deep-UV	Neg. near-UV	Neg. mid-UV	Neg. deep-UV	CEL	BIL	TRL	SDR	PDR	EB pos.	EB neg.	X-R	IB	InR	IR	TSI	IA
Exposure-throughput	5	4	2	5	4	3	3	2	5	1	1	2	3	1	1	1	3	3	5
Processing costs	5	5	5	4	4	4	3	2	1	4	3	5	4	4	4	3	4	2	2
Resolution	3	4	4	2	3	3	5	5	5	4	4	4	2	5	5	4	4	5	4
Defects	5	5	4	3	3	3	4	5	5	4	3	4	2	5	5	3	5	4	3
Thermal stability	4	4	3	2	2	3	4	5	5	2	3	3	3	4	4	5	4	4	3
Plasma stability	4	4	3	3	3	3	4	5	5	2	3	3	3	3	3	5	4	4	3
Soluble for lift-off	5	5	5	1	1	1	5	5	5	4	3	5	1	3	3	1	4	4	4
Resist costs	4	3	3	4	4	4	3	3	3	3	3	3	3	2	4	2	3	3	3
Total	35	34	29	24	24	24	31	32	34	24	23	29	21	27	29	24	31	29	27
Ranking	1	2	5	8	7	7	4	3	2	8	9	5	10	6	5	8	4	5	6

[a] CEL, contrast enhancement layer; BIL, bilevel resist; TRL, trilevel resist; SDR, self-developing resist; PDR, plasma developable resist; EB pos., *e*-beam positive resist; EB neg., *e*-beam negative resist; X-R, X-ray resist; UB, ion-beam resist; InR, inorganic resists; IR, image reversal (DQN, MIRP); TSI, top surface imaging; IA, image amplification.

TABLE 17-3-1. Factors to Consider before Selection of Resists

Stability of substrate to stripper
Stripping conditions (radiation damage)
Etchant conditions, etch rate of resist, depth and time of etching, concentration of etchant, step coverage of resist, etch bias
Lift-off thickness, metal deposition temperature, lift-off solvent, pitch, step coverage
Exposure tool throughput–resist dose, overlay required, radiation damage to derive, resist contrast, substrate reflectivity, depth of focus, keV, mask or maskless exposure, resist polarity, wafer size
Resist polarity, single or multilayer, layer mixing, profile, image bias, thermal and plasma resistance, wavelength of exposure, developer
Substrate adhesive properties, resist primer and adhesion, primer conditions, coating uniformity, step height, planarization, registration mark, prebake conditions
Postbake, hardening conditions, postchemical treatment
Resist cost, toxicity, shelf life, storage, filtration, Material Safety Data Sheet, clean room conditions, relative humidity, dispenser, rinse, viscosity grade
Masks, part number, pellicle, wafer batch size, polarity, turnaround time, defect level
Tool compatibility of resist, spin-apply, prebake oven, exhaust, clean room conditions, compatibility of resist liquids with tool components

17-3-1. Coating Techniques

The semiconductor industry relies heavily upon spin coating since the planarizing action of the radial forces produces ultrasmooth films. No other technique produces films as uniform as spin coating. However, the planarization does hamper single film resists cast over topography. On the other hand, multilayer resists take advantage of the planarizing action to define a planar imaging layer. While alternate coating techniques have not achieved the uniformity of spin coating they do offer cost savings in the volume of resist transferred.

After the physics of spin coating on flat surfaces has been defined, the next order requiring a model and basic improvements are the planarization of resists over steps. Toward this end, conformal coatings deposited by gas-phase techniques will someday be perfected. For maintaining the current levels of yields for future submicrometer devices, the importance of clean surfaces and low particulate levels in a clean room cannot be overemphasized. Better pinhole detectors and contamination monitors are required.

Once the resist is deposited, its thickness, in general, controls or influences:

1. Resolution equal to image layer thickness
2. Resist filtration, 1/5 of thickness
3. Resist thickness, 2× step height
4. Pinhole size, 1/5 of thickness
5. Lift-off, 2× thickness
6. Exposure time $\propto$ thickness
7. Prebake time $\propto$ thickness
8. Develop time and etch time $\propto$ thickness
9. Uniformity of thickness influences alignment
10. Step coverage, planarization, influences registration (*e*-beam) and linewidth control

Adhesion, as measured by lack of failure, is an Edisonian area of resist lithography. No improved promoters have appeared in the last decade and more effort on better resist primers for polysilicon and silicon nitride surfaces should be expended.

Due to toxicological considerations, the choice of casting solvents continues to be depleting. Solventless coatings and dry film resists will become more important for submil (5–25 μm) packaging lithography of semiconductor chips.

17-3-2. Prebake and Postbake

As wafer diameters increase to 200 mm, the removal of the casting solvent to a consistent level in a single wafer and within wafers will probably require single-wafer processing. Hot plate baking has been the most consistent method. Only physical changes should occur in prebake. In postbake, or predevelop bake, deliberate physical and chemical changes are induced to strengthen resist performance.

As in any chemical or physical manufacturing processes, analytical monitors are necessary to control and quantify the extent of prebake or postbake. The control of prebake or postbake temperature is more important than time as long as a steady-state temperature is achieved. For establishing and maintaining an isothermal bake process, measurement of the temperature of the surface of the resist is required. Rapid curing should not be traded for the poor performance of a resist in the loss of adhesion, poor resolution, development thinning, or solvent outgassing in metal lift-off or ion implantation. Because of potential linewidth distortion, the upper temperature limit of postbaking should be avoided. Posttreatment of diazoquinones with UV, plasma, chemicals, or heat furthers the plasma and the thermal stability of the novolak resists.

In the future, we can expect an all-dry lithography with no need for heating or drying cycles. Film stresses still persist which can only be relieved by exceeding the T_g of the resist polymer. Inorganic resists, if perfected, would be concurrent with the present state of the art of vacuum deposition techniques.

17-3-3. Expose

The rating of all exposure tools and associated masks is given in Table 17-3-2. UV exposure is the dominant form followed by *e*-beam. X-ray and ion beam require another era of basic improvements. The usable exposed linewidth is about 4× the overlay accuracy.

For images to 1.0 μm and high volume–low part numbers, reflective 1× scanners have high throughput and suitable overlay. Below the 1-μm barrier, step-and-repeat refractive (5× operating at 365 nm) compete with mix-and-match tools to about 0.6 μm. Then major problems arise in 1× masks such as defects and the inability to hold dimensions over 250 × 250-mm glass plates. From 0.6 to 0.3 μm, reflective step-and-repeat (3–5×) operating with deep-UV resists will try to extend optical lithography to its limits. Below 0.3 μm, X-ray and *e*-beam proximity printing will compete for the mass production of memory circuits. Electron beams will handle the traditional custom circuits and make complementary masks for all other forms of exposure.

The key to high-volume lithography relies heavily upon high-quality durable

TABLE 17-3-2. Rating of Exposure Tools and Associated Masks and Resists

	Contract–proximity	1/1 UV scan	4/1 UV scan/step	5/1 UV step	10/1 UV step	EB	XR	IB	EB proximity
Minimum Feature size	1	2	3	4	4	5	4	3	3
Registration	1	2	3	3	3	4	3	2	4
Throughput	4	5	3	3	2	1	1	1	3
Mask cost and simplicity	2	2	3	4	4	3	1	1	1
Topography sensitivity	2	3	3	3	2	4	4	4	3
Resist simplicity and cost	4	2	2	3	3	1	1	2	3
Tool cost	5	3	2	3	3	1	2	2	1
Ease of operation	5	4	3	3	3	4	3	2	3
Defect susceptibility	1	3	4	4	5	4	4	4	3
Outlook for future sub-micrometer development	1	2	4	3	3	5	3	2	2
Total	26	28	30	33	32	32	26	23	26
Rating (1 = best)	4	3	2	1	1	1	4	5	4

masks which can survive the daily thermal and mechanical stresses. Which form of mask exposure (flood UV, X-ray, *e*-beam, and ion beam) will dominate will depend on three major factors: (1) Can the mask be made with an edge resolution better than 1/10 the printed linewidth? (2) Will the mask be flat enough and will it maintain its flatness and images during exposure? (3) Can an alignment scheme be developed which does not use the source of exposure as the alignment wavelength? Mask fabrication will be the proving ground for new resists and processes.

Optical exposure requires flatter wafers, autofocus, and autoalign to compensate for the decreasing focal lengths. Lower-temperature processes to prevent wafer warpage and more planar devices are required to minimize the focus and alignment errors. For submicrometer lithography, field-by-field or die-by-die alignment will be necessary. The mix and match between scanners and steppers will be driven by overlay requirements.

The interproximity effects are major problems for all forms of optical and *e*-beam exposure on Si wafers. Multilayer resists appear to be the practical solution.

Resist systems for high resolution can be classified into three categories by the relation that resolution (R) is a function of wavelength (Γ), the numerical aperture (NA), the K factor (Fig. 17-3-1), and the thickness factor, t,

$$R = K(\Gamma/\mathrm{NA})\, t$$

where $K = 0.3$ for top surface imaging, 0.5 for mid-surface imaging (multilayer resists), and 0.8 for bottom surface imaging in single film resists. The driving force for resolution will be to make the thickness of the image layer at least one-fourth the desired minimum linewidth (resolution).

For all exposure tools, the throughput is limited by the intensity of the source and the resist speed. In optical exposure, the speeds of resist and the intensity of the sources are adequate, except in the deep-UV region. Electron-beam and ion-beam exposure can use more speed, especially in novolak-based resists. X-ray exposure requires a good single film resist to take advantage of its high resolution and offset

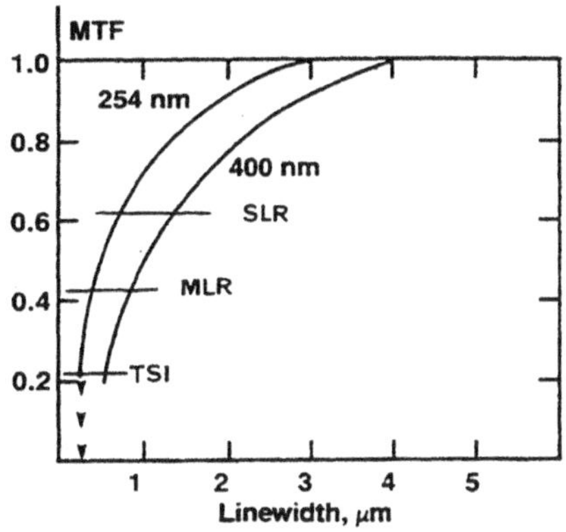

FIGURE 17-3-1. Resolution modulation transfer function for resist imaging in UV at 254 and 400 nm for top surface imaging (TSI), multilayer resist middepth imaging (MLR), and single film resist (SLR). Resolution is proportional to the thickness of the resist image layer.

its lower throughput and X-ray exposure can also offset the added costs of multilayer resists required for future optical and *e*-beam processes.

More attention will have to be paid to alignment marks and detection schemes for submicrometer device requirements (± 0.1 μm, 3σ). From a circuit design, self-aligning features in a device, e.g., via hole etching followed by lift-off of metal contacts, will reduce some of the critical overlay requirements.

17-3-4. Develop

The develop process in the case of a positive resist is a facsimile of etching. The developer responds to the absorbed and scattered radiation. New developers, especially for diazoquinone resists, are always needed to improve contrast and photosensitivity. For reproducible processing, the time, temperature, concentration, and rate of development are essential monitors. With shrinking linewidths, batch processes will be replaced by single-wafer processing since biases of a nanometer are required. Positive resists require tight process controls and the added dimension of profile control for dry processing, while negative resists have a wide process latitude but are limited by swelling and backscattered radiation.

Dry development is still in the test tube stages. Will dry development ever be needed or when will the liquid phase be inhibited by mass transport? Although nanosize devices have been made by liquid development, gross adhesion failure in liquid processing may force the rapid development of dry systems.

From an automation viewpoint, an endpoint system for spray development is needed. Automatic linewidth measurements will be required in the future to determine if the printed image meets specifications or the wafer will have to be reworked. The thinning of most resists will not be allowed since pinholes will be a dominant problem of submicrometer lithography.

17-3-5. Subtractive Processing

Wet etching in a batch process was sufficient for 3- to 10-μm dimensions with little process controls. With the introduction of thickness, time, and etch rate monitors, process biases could be reduced to less than 1 μm. Wet etching has served the needs of batch processing and will continue to be used for cleaning or touch-up of dry processes.

Dry etching has been the most important tool development in semiconductor lithography. The anisotropic dry etching has enabled etch biases to be smaller than the printed and developed image. Multilayer resists would not have been successful without anisotropic dry transfer. Dry etching grew out of plasma stripping and sputter etching. The selection of molecular ion beams for dry etching single wafers will evolve out of the current batch-to-batch irreproducibility.

Dry etching of silicon relies on polymerizing gases derived from the resist or electrode to protect the walls from erosion during etching. About 15 variables have to be controlled for precise dry etching. Although many papers on plasmas have been published, a unified rate model of plasma etching is needed. At present, each reactor configuration and etchant is modeled by experimental kinetics. A critical understanding of the chemical physics of radical and ion transport from a mass transport and surface chemistry viewpoint is needed for integrating the choice of an etchant, reactor configuration, and process controls for a uniform and reproducible manufacturing process. For the future, dry etching has to overcome the nonuniform supply of etchant in the concentration-depleted conditions and approach steady-state kinetics in a uniform electric field.

Dry etching of metal and alloys requires high-temperature and reactive ion-resistant masks. New etchants are required to handle the transition element metallurgies. With the increasing number of multilayer films on the substrate, high selectivity and etch stops are continuing problems in dry etching. The control of the resist profile (before and after etching) adds a new dimension to lithographic processing. Lastly, radiation damage from dry etching is always under study as a potential reliability problem. Downfield reactors may be necessary to prevent radiation damage.

17-3-6. Additive Processes

Although multilayer resists were originally developed for lift-off, they were quickly copied and integrated as dry etch masks. A single-film high-temperature soluble resist has not been available. Multilayer resists and soluble underlayers (e.g., polyimides) are required for successful lift-off. In addition, precise evaporation and stable layers are also a basic requirement. The added costs of multilayer resists has to be compensated by the extendability of present optical tools. Due to the resist undercut profile and evaporation angle limitations, additive lift-off of metals is reaching a pitch limit of about 2 μm.

17-3-7. Stripping of Resists

This process should be the simplest step but can be troublesome since remaining monomolecular contaminants can be randomly distributed across the wafer. Initially, wet stripping used toxic materials until forced to respond with safer disposable formulations. Although hot acids are still used for stripping resists and cleaning wafers, they suffer from a poor batch life.

Dry plasma stripping was the forerunner of all plasma processing. The study of the attack of oxygen on polymers paved the way for multilayer resists and dry development. In a similar fashion, plasma ashing gave birth to plasma etching. Although dry stripping is an efficient means of disposal, radiation damage is still a problem for some devices. Downfield stripping, in which the excited gas is transported to the wafer surface, shows promise as a radiationless means of removal. Since irreproducibility and nonuniformity are evident in barrel reactors, we can expect individual resist stripping of large wafers, as is done in Al etching, to be introduced into manufacturing processes using high strip rates for sufficient throughput.

17-3-8. Resistless Processes

Future processing may circumvent the resist process altogether. Selective ion implantation, metal deposition, and material removal appear as a perpetual, computerized process converting raw inorganics and metals into a semiconductor device. However, the high doses and small exposure fields are major impediments to submicrometer resolution at a respectable throughput, if any. Thus, resistless lithography will find its primary application at the chip level in the repair of masks and in the repair of circuits at the packaging level.

17-3-9. Process Controls

For on-line processing, the three "T"s of processing—time, thickness, and temperature—are essential to control and monitor in the exposure–develop–etch cycles (Fig. 17-3-2). What process latitude is necessary is dictated by the allowable process bias (Table 17-3-3). The most critical areas of process control are the exposure time, the develop time and temperature, and the etch time. Hopefully, all three are monitored by on-line endpoint process control units. In the future, the drive toward submicrometer geometries will require tenfold better process control and the reduction of image biases, especially in the expose, develop, and etch process sectors. For off-line materials control, both the manufacturer and the user have to meet functional and analytical specifications to ensure a reproducible process, even in spite of all the process controls. The adoption of one family of resists for all process levels and the commonality of process tools and conditions will reduce the learning cycles and improve the final yields.

Improved endpoints for micrometer geometries, single-wafer processing, and diagnostic computer controls will be necessary to reduce the final device biases to the range ± 0.1–$0.2\ \mu m$ for submicrometer circuit fabrication. More automated defect inspection, linewidth, overlay, and mask checks will be used to support

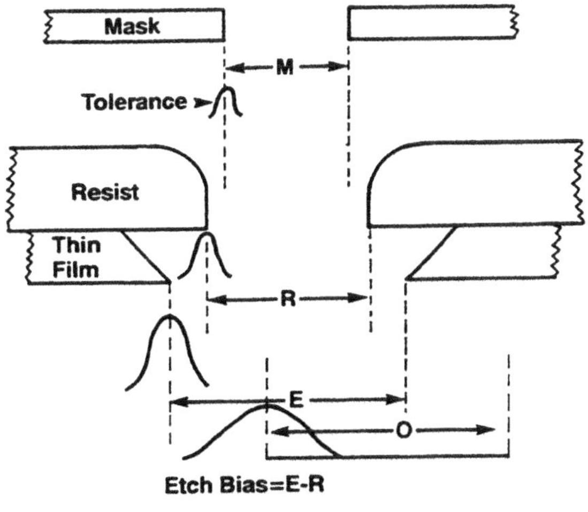

FIGURE 17-3-2. Minimum process biases of resist (R), etching (E), and overlay (O) which are controlled by the three "T"s of processing—time, temperature, and thickness.

TABLE 17-3-3. Critical Areas of Resist-Process Control

Process step	Typical image bias (μm)
Thickness	0.05
Prebake time	0.01
Prebake temp.	0.02
Expose dose	0.20
Develop time	0.15
Develop temp.	0.10
Postbake temp.	0.10
Anisotropic etch time	0.15/1.0
Anisotropic etch temp.	0.05
Developer concn	0.21 (DQN)
Developer temp.	0.12
Develop time	0.05
Overlay	0.20–1.0
	±0.5 (1985)
	±0.1 (1995)

processing. Process controls can benefit from a basic understanding of the chemistry and physics of the process. In this regard, computer modeling and process simulation will statistically determine the critical process controls and process conditions.

The inability to achieve acceptable yields in submicrometer lithography will be dominated by entropic forces which give rise to the irreversible adsorption of foreign solids and invisible films of contaminants. Their detection on a nondestructive basis of defects (pinholes and particulates), their origin and control or elimination will perhaps be a greater technical challenge than shrinking lithography to make submicrometer dimensions.

The economic driving force for lithography is the lower cost per bit (Table 17-3-4) brought about by increasing circuit density. By extending present equipment and improving yield, lithography answers the challenge of the semiconductor industry to provide submicrometer circuitry without escalating cost and complexity.

TABLE 17-3-4. Lithography Trends and Technology Economics[a]

	1970	1980	1990
Complexity (number of functions)	1K	10K	100K
Image size (μm)	10	3	1
Overlay (μm)	2	1	0.4
Defect density per cm^2	10	1	0.2
Expose equipment cost ($)	100K	800K	1–3M
Exposure/hr per 150-mm wafers	100	50	20
Cost/bit (microcent)	100	20	1

[a] Adapted in part from M. Long, "Toward Submicron Lithography," SPIE Vol. 275 Semiconductor Microlithography VI (1981), p. 2.

17-4. POSTLUDE

The interaction of materials, processes, and equipment is the dynamics of semiconductor lithography (Fig. 17-1-1). Each one of these process sectors drives each other and will equally share in defining circuitry for the next decades of semiconductor lithography. Toward this objective, I hope this book not only serves as a guide to solving lithography problems but that it also stimulates the reader to ask better questions as to their origin. Hopefully, semiconductor lithography will continue to be a learning experience for new and experienced technologists eager to apply chemical and physical principles to improve a multidisciplinary process.

I gratefully acknowledge the cooperation and courtesy of the many publishers of textbooks and technical journals in semiconductor technology for permission to use copyrighted material.

Index

GPSR Compliance

The European Union's (EU) General Product Safety Regulation (GPSR) is a set of rules that requires consumer products to be safe and our obligations to ensure this.

If you have any concerns about our products, you can contact us on ProductSafety@springernature.com

In case Publisher is established outside the EU, the EU authorized representative is:

Springer Nature Customer Service Center GmbH
Europaplatz 3
69115 Heidelberg, Germany

Batch number: 10761693

Printed by Printforce, the Netherlands